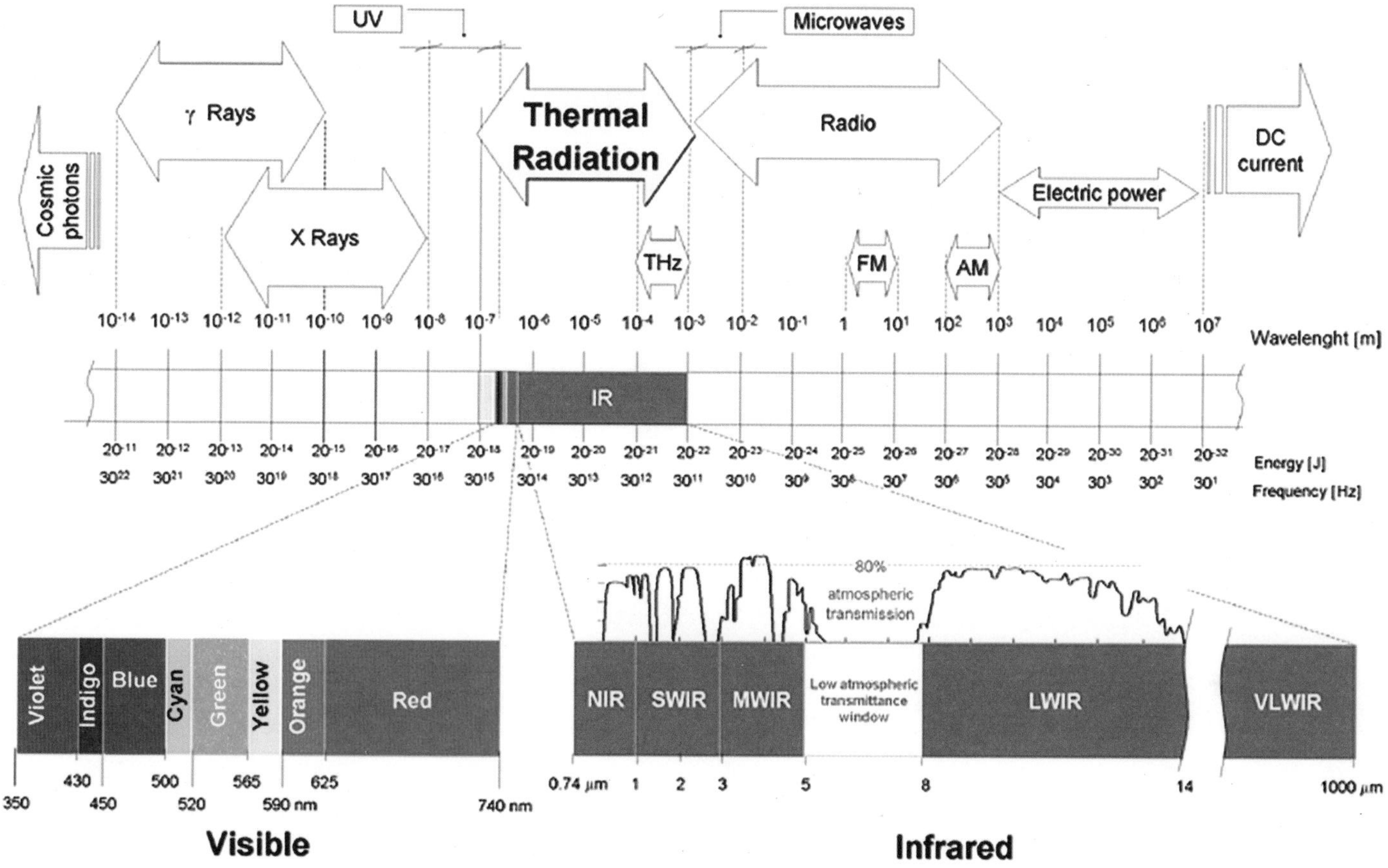
UV
Microwaves
Cosmic photons
γ Rays
X Rays
Thermal Radiation
Radio
DC current
Electric power
THz
FM
AM
Wavelenght [m]
IR
Energy [J]
Frequency [Hz]
80%
atmospheric transmission
Violet
Indigo
Blue
Cyan
Green
Yellow
Orange
Red
350
430
450
500
520
565
590 nm
625
740 nm
NIR
SWIR
MWIR
Low atmospheric transmittance window
LWIR
VLWIR
0.74 μm
1
2
3
5
8
14
1000 μm
Visible
Infrared

Third Edition

# Fundamentals of MICROFABRICATION AND NANOTECHNOLOGY

## VOLUME II

## Manufacturing Techniques for Microfabrication and Nanotechnology

Third Edition

# Fundamentals of MICROFABRICATION AND NANOTECHNOLOGY

## VOLUME II

## Manufacturing Techniques for Microfabrication and Nanotechnology

Marc J. Madou

CRC Press
Taylor & Francis Group
Boca Raton London New York

CRC Press is an imprint of the
Taylor & Francis Group, an **informa** business

CRC Press
Taylor & Francis Group
6000 Broken Sound Parkway NW, Suite 300
Boca Raton, FL 33487-2742



Printed and bound in India by Replika Press Pvt. Ltd.
10 9 8 7 6 5 4 3 2 1

International Standard Book Number: 978-1-4200-5519-1 (Hardback)



**Visit the Taylor & Francis Web site at**
**http://www.taylorandfrancis.com**

**and the CRC Press Web site at**
**http://www.crcpress.com**

*I dedicate this third edition of Fundamentals of Microfabrication to my family in the US and in Belgium and to all MEMS and NEMS colleagues in labs in the US, Canada, India, Korea, Mexico, Malaysia, Switzerland, Sweden and Denmark that I have the pleasure to work with. The opportunity to carry out international research in MEMS and NEMS and writing a textbook about it has been rewarding in terms of research productivity but perhaps even more in cultural enrichment. Scientists have always been at the frontier of globalization because science is the biggest gift one country can give to another and perhaps the best road to a more peaceful world.*

# Contents

# Roadmap

*Manufacturing Techniques for Microfabrication and Nanotechnology* consists of three parts and ten chapters.

In Part I, we review different forms of lithography, detailing those that differ most from the miniaturization processes used to fashion integrated circuits (ICs). Chapter 1 starts with a short historical note about the origins of lithography, followed by a description of photolithography, including developments that have allowed the printing of the ever-shrinking features of modern ICs. After reviewing the limits of photolithography, we detail alternative next-generation lithographies (NGEs) in Chapter 2, including extreme ultraviolet (EUV), x-ray, and charged-particle (electron and ion) lithographies, followed by lithography techniques in the early research and development (R&D) stage. Lithography approaches in the R&D stage include very thin resist layers and block copolymers, zone plate array lithography (ZPAL), quantum lithography (two-photon lithography), and proximal probe-based techniques such as atomic force microscopy (AFM), scanning tunneling microscopy (STM), dip-pen lithography (DPL), near-field scanning optical microscopy (NSOM), and apertureless near-field scanning optical microscopy (ANSOM). Also considered are holographic lithography, plasmonic lithography, and lithography with superlenses ("Pendry's dream"). As an example of a lithography method that is capable of patterning resist on nonplanar substrates (e.g., patterning a resist on the surface of a cylinder) we detail soft lithography.

In Part II we deal with material removal processes, and additive technologies are the focus of Part III. We review gas and vapor phase dry etching processes, including chemical, physical (purely mechanical), and chemical-physical (mostly chemical with some mechanical assist) in Chapter 3. Chapter 4 consists of two parts: one covers wet *chemical etching* and the other *wet bulk micromachining*, the latter being chemical etching as well but optimized for the manufacture of ICs, nano-, and micromachines. Chapter 5 discusses thermal removing techniques where thermal energy, provided by a heat source, melts and/or vaporizes material from the work piece. Examples of thermal techniques covered in this chapter are electric discharge machining (EDM) (including μ-EDM), laser- and electron-beam machining (LBM and EBM), and plasma arc cutting. In Chapter 6, we cover traditional and nontraditional mechanical removing techniques. Traditional mechanical material removing technologies covered are mechanical precision machining and abrasive wheel machining. In nontraditional mechanical machining, we deal with ultrasonic drilling, electrolytic in-process dressing (ELID) grinding, water jet, abrasive water jet, abrasive jet machining, and focused ion-beam milling (FIB). Also in this chapter, we highlight recent developments toward miniaturized mechanical manufacturing equipment, the so-called desktop factories.

In Part III we cover additive (forming) processes, where materials are added, usually in a selective manner, to a work piece or a device under construction. Physical vapor deposition processes (PVD) and chemical vapor deposition (CVD) are topics of Chapter 7. In the same chapter we survey thin film properties and detail surface micromachining, an important application of thin film deposition techniques. In surface micromachining, features are built up, layer by layer, on the surface of a substrate (e.g., a single-crystal silicon wafer). Dry etching defines the surface features in the $x,y$-plane, and wet etching releases them from the plane by undercutting. In surface micromachining, shapes in the $x,y$-plane are unrestricted by the crystallography of the substrate. In Chapter 8 we discuss additive chemical, photochemical, and electrochemical processes. The chemical forming methods covered in Chapter 8 are geared toward BioMEMS applications. In photochemical forming, photoenergy solidifies a material into a 3D shape, as illustrated in rapid prototyping and microphotoforming (stereolithography). In the case of electrochemical forming, the energy at the work piece is electrochemical. Chapter 9 covers thermal forming techniques where thermal energy provided by a heat source "transforms" a material as in crystallization, sintering, alloying, annealing, decomposition, and pyrolizing. Replication methods based on plastic molding and LIGA—a sophisticated type of replication technique combining x-ray lithography with electroplating and plastic molding—are discussed in Chapter 10. Plastic molding techniques include RIM, thermoplastic injection molding (IM), and compression molding or hot embossing.

**Note to the Reader:** *Manufacturing Techniques for Microfabrication and Nanotechnology* was originally composed as part of a larger book that has since been broken up into three separate volumes. *Manufacturing Techniques for Microfabrication and Nanotechnology* represents the second volume in this set. The other two volumes include *Solid-State Physics, Fluidics, and Analytical Techniques in Micro- and Nanotechnology* and *From MEMS to Bio-NEMS: Manufacturing Techniques and Applications.* Cross-references to these books appear throughout the text and will be referred to as Volume I and Volume III, respectively. The interested reader is encouraged to consult these volumes as necessary.

# Author

Dr. Madou is the Chancellor's Professor in Mechanical and Aerospace Engineering (MEA) at the University of California, Irvine. He is also associated with UC Irvine's Department of Biomedical Engineering and the Department of Chemical Engineering and Materials Science. He is a Distinguished Honorary Professor at the Indian Institute of Technology Kanpur, India, and a World Class University Scholar (WCU) at UNIST in South Korea.

Dr. Madou was Vice President of Advanced Technology at Nanogen in San Diego, California. He specializes in the application of miniaturization technology to chemical and biological problems (bio-MEMS). He is the author of several books in this burgeoning field he helped pioneer both in academia and in industry. He founded several micromachining companies.

Many of his students became well known in their own right in academia and through successful MEMS start-ups. Dr. Madou was the founder of the SRI International's Microsensor Department, founder and president of Teknekron Sensor Development Corporation (TSDC), Visiting Miller Professor at UC Berkeley, and Endowed Chair at the Ohio State University (Professor in Chemistry and Materials Science and Engineering).

Some of Dr. Madou's recent research work involves artificial muscle for responsive drug delivery, carbon-MEMS (C-MEMS), a CD-based fluidic platform, solid-state pH electrodes, and integrating fluidics with DNA arrays, as well as label-free assays for the molecular diagnostics platform of the future.

To find out more about those recent research projects, visit http://www.biomems.net.

# Acknowledgments

I thank all of the readers of the first and second editions of Fundamentals of Microfabrication as they made it worthwhile for me to finish this completely revised and very much expanded third edition. As in previous editions I had plenty of eager reviewers in my students and colleagues from all around the world. Students were especially helpful with the question and answer books that come with the three volumes that make up this third edition. I have acknowledged reviewers at the end of each chapter and students that worked on questions and answers are listed in the questions sections. The idea of treating MEMS and NEMS processes as some of a myriad of advanced manufacturing approaches came about while working on a WTEC report on International Assessment Of Research And Development In Micromanufacturing (http://www.wtec.org/micromfg/report/Micro-report.pdf). For that report we travelled around the US and abroad to visit the leading manufacturers of advanced technology products and quickly learned that innovation and advanced manufacturing are very much interlinked because new product demands stimulate the invention of new materials and processes. The loss of manufacturing in a country goes well beyond the loss of only one class of products. If a technical community is dissociated from manufacturing experience , such as making larger flat-panel displays or the latest mobile phones, such communities cannot invent and eventually can no longer teach engineering effectively. An equally sobering realization is that a country might still invent new technologies paid for by government grants, say in nanofabrication, but not be able to manufacture the products that incorporate them. It is naïve to believe that one can still design new products when disconnected from advanced manufacturing: for a good design one needs to know the latest manufacturing processes and newest materials. It is my sincerest hope that this third edition motivates some of the brightest students to start designing and making things again rather than joining financial institutions that produce nothing for society at large but rather break things.

# Part I

# Lithography

(a) (b)

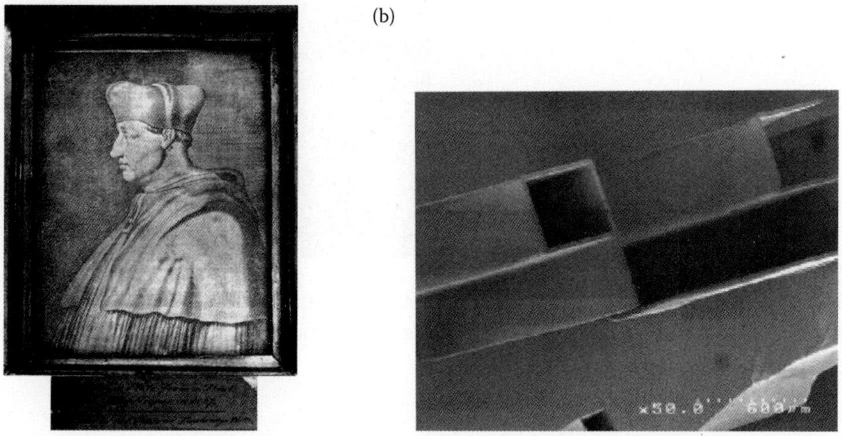

(a) Engraving of Cardinal d'Amboise. The earliest example of photolithography followed by wet etching. (Photograph from the Science Museum. Courtesy of the Royal Photographic Society.) (b) Ultrahigh-aspect-ratio SU-8 photoresist structures. Microfeatures with designed thickness of 40 μm and height of 2000 μm. (Courtesy of Dr. Wanjun Wan, Louisiana State University.)

> There are those who still believe that America can move from the strongest industrial core economy in the world to essentially a service economy without losing its greatness, its dynamism, and industrial health. I am absolutely certain we cannot.
>
> **David Halberstam**
> *The Next Century* (1991)

# Introduction to Part I

**Chapter 1 Photolithography**
**Chapter 2 Next-Generation Lithographies (NGLs) and Lithography Research**

---

## Introduction to Part I

For many miniaturization tasks, lithography, the technique used to transfer copies of a master pattern onto the surface of a solid material such as a silicon wafer, is of central importance. In Part I of this book, we review different forms of lithography, detailing those that differ most from the miniaturization processes used to fashion integrated circuits (ICs). Chapter 1 starts with a short historical note about the origins of lithography, followed by a description of photolithography, including developments that have allowed the printing of the ever-shrinking features of modern ICs. After reviewing the limits of photolithography, we detail alternative next-generation lithographies (NGLs) in Chapter 2, including extreme ultraviolet (EUV), x-ray, and charged-particle (electron and ion) lithographies, followed by techniques in the early research and development stage.

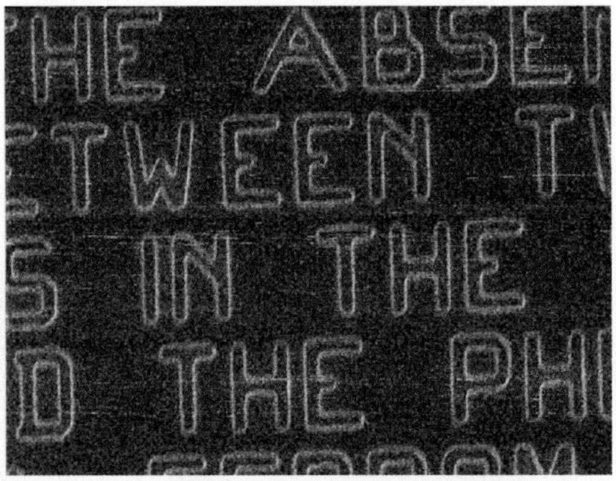


Text by William Gilbert (from De Magnete), written using a 100-keV electron microscope in a common resist (PMMA). The height of each letter is 250 nm. The pattern was then imaged using the same electron microscope. (From http://www.lancs.ac.uk/staff/koltsov/nanoimage.html.)

# 1

# Photolithography



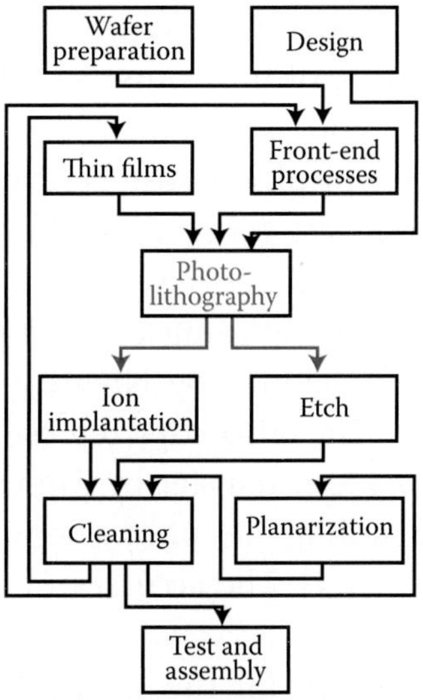


The image of an IC, micro-, or nanomachine pattern is transferred to a substrate in a process known as photolithography. Substrates are first coated with a thin emulsion of photoactive organic material called a *photoresist*. A pattern is then aligned and projected onto the substrate using a light source (e.g., deep UV). Subsequently, the resulting image is developed and inspected.

## Introduction

We start this chapter with a short historical note about the origins of photolithography and follow with details on all the steps involved, including resolution enhancement techniques (RET), that have allowed the printing of the ever-shrinking features of modern integrated circuits (ICs). In this photolithography treatise we expand especially on those aspects that are of most importance to microelectromechanical systems (MEMS) and nanoelectromechanical systems (NEMS) and that differ the most from typical lithography processes for IC fabrication.

## Historical Note: Lithography's Origins

Within the space of 40 years two developments in photography laid the foundations for the photoresists we use today. In 1782, the Swiss pastor Jean Senebier (1742–1808) of Geneva investigated the property of certain resins that become insoluble in turpentine after exposure to sunlight. Inspired by Senebier, Nicéphore Niépce, after experimenting with various resins in sunlight, managed to copy an etched print on oiled paper by placing it over a glass plate coated with bitumen (asphalt) dissolved in lavender oil (France, 1822). In so doing, he resurrected an ancient Egyptian embalming technique that involved the use of what is now known as Syrian asphalt. After two or three hours of sunlight, the unshaded areas in the bitumen became hard compared with the shaded areas, which remained more soluble and could be washed away with a mixture of turpentine and lavender oil. Niépce's concoction, we will learn below, corresponds to a negative-type photoresist. Five years later, in 1827 (talk about fast turnaround time!), using strong acid, the Parisian engraver Lemaître made an etched copy of an engraving of Cardinal d'Amboise from a plate developed by Niépce (see (a) on title page of Part I). The latter copy represents the earliest example of pattern transfer by photolithography and chemical milling [i.e., photochemical milling (PCM); see Chapter 4]. The accuracy of the technique was 0.5–1 mm.[1]

The word *lithography* [Greek for the words *stone* (*lithos*) and to *write* (*gráphein*)] refers to the process invented in 1796 by Aloys Senefelder. Senefelder found that stone (he used Bavarian limestone), when properly inked and treated with chemicals, could transfer a carved image onto paper. As a result of the chemical treatment of the stone, image and nonimage areas became oil receptive (water repellent) and oil repellent (water receptive), respectively, attracting ink onto the image area and attracting water on nonimage areas.

Photomasking, followed by chemical processing, led to the photolithography now used in fabricating integrated circuits (ICs) and in miniaturization science. Not until World War II, more than 100 years after Niépce and Lemaître, did the first applications of the printed circuit board (PCB) [also printed wiring board (PWB)], invented in 1943 by the Austrian Paul Eisler, come about. Interconnections were made by soldering separate electronic components to a pattern of "wires" produced by photoetching a layer of copper foil that was laminated to a plastic board. Jack Kilby of Texas Instruments and Robert Noyce of Fairchild Semiconductor developed the first IC in the late 1950s. By 1961, Fairchild Semiconductor had introduced the first commercial IC in which photoetching processes produced large numbers of transistors on a thin slice of Si. Now, ICs are ubiquitous, and we find them in microprocessors, audio and video equipment, dishwashers, garage openers, security systems, automobiles, and so on. With the first ICs, patterns had a resolution not better than 5 μm. Today photolithography, x-ray lithography, and charged particle lithography all achieve submicrometer-printing accuracy. Intel introduced the Prescott Pentium IV chip, an IC chip incorporating 90-nm-sized features using UV photolithography, in the fourth quarter of 2003. Chips based on a 65-nm technology became available in 2006, and the 32-nm node was achieved in 2009.

For details on the projected feature size of future ICs, see the International Technology Roadmap for Semiconductors (ITRS) in Appendix 1A of Chapter 1 in Volume I or on the web at http://public.itrs.net. In the ITRS, technology modes have been defined, i.e., the feature sizes that have to be in volume manufacturing at a fixed date (year of production).

## Photolithography Overview

### Introduction

The most widely used form of lithography is photolithography. In the IC industry, pattern transfer from masks onto thin films is accomplished almost exclusively via photolithography. The combination of accurate registration and exposing a series of successive patterns leads to complex multilayered ICs. This essentially two-dimensional (2D) process has a limited tolerance for nonplanar topography, creating a major constraint for building non-IC miniaturized systems, which often exhibit extreme topographies. Photolithography has matured rapidly, and its ability to resolve ever-smaller features is constantly improving. For the IC industry, this continued improvement in photolithography resolution has impeded the adaptation of alternative, higher-resolution lithography techniques, such as x-ray lithography. Research during the past 10 years in high-aspect-ratio resist features to satisfy needs of both IC and non-IC miniaturization is also finally improving dramatically photolithography's capacity to cover wide ranges of topography. Performance of a photolithographic process is determined by its resolution, the minimum feature size that can be transferred with high fidelity; the registration, how accurately patterns on successive masks can be aligned; and throughput, the number of wafers that can be transferred per hour (a measure of the efficiency of the lithographic process).

Photolithography and pattern transfer involve a set of process steps summarized in Figure 1.1. As an example, we use an oxidized Si wafer and a negative photoresist system. For simplicity, not all the steps are detailed in this figure, as they will be covered in the subsequent text. A short preview follows.

An oxidized wafer (Figure 1.1a) is coated with a 1-μm-thick negative photoresist layer (Figure 1.1b). After exposure (Figure 1.1c), the wafer is rinsed in a developing solution or sprayed with a spray developer, which removes the unexposed areas of photoresist and leaves a pattern of bare and photoresist-coated oxide on the wafer surface (Figure 1.1d). The photoresist pattern is the negative image of the pattern on the photomask. In a typical next step after development, the wafer is placed in a solution of HF or HF + $NH_4F$, meant to attack the oxide but not the photoresist or the underlying silicon (Figure 1.1e). The photoresist protects the oxide areas it covers. Once the exposed oxide has been etched away, the remaining photoresist can be stripped off with a strong acid such as $H_2SO_4$ or an acid-oxidant combination such as $H_2SO_4$-$Cr_2O_3$, attacking the photoresist but not the oxide or the silicon (Figure 1.1f). Other liquid strippers include organic solvent strippers and alkaline strippers (with or without oxidants). The oxidized Si wafer with the etched windows in the oxide (Figure 1.1f) is ready now for further processing, which might entail a wet anisotropic etch of the Si in the etched windows with $SiO_2$ as the etch mask.

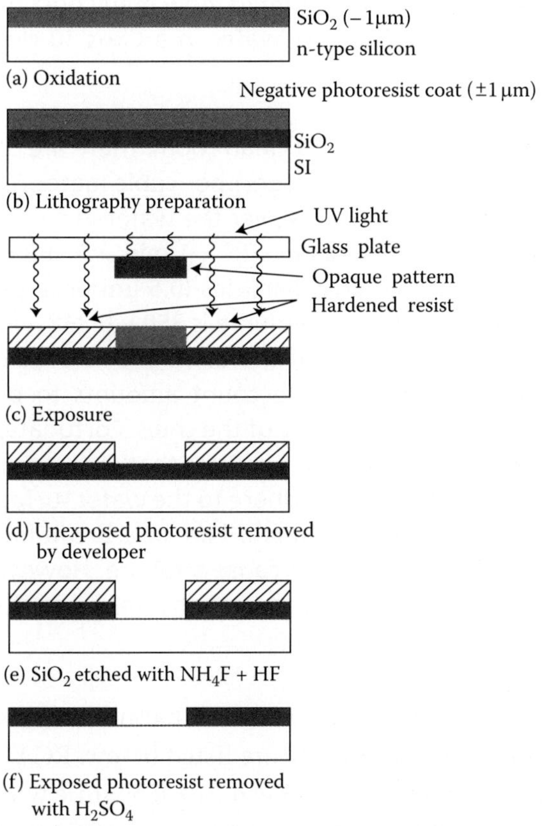


**FIGURE 1.1** Basic photolithography and pattern transfer. Example uses an oxidized silicon wafer and a negative photoresist system. Process steps include exposure, development, oxide etching, and resist stripping. Steps (a) through (f) are explained in the text.

## Wafer Cleaning and Contaminants: The Clean Room

An important step, even before lithography proper, is wafer cleaning. Contaminants include solvent stains (e.g., methyl alcohol, acetone, trichloroethylene, isopropyl alcohol, and xylene), dust from operators and equipment, and smoke particles. Stains or films may lead to adverse effects during oxidation and evaporation processes. Particulates, chunks of granular matter, may cause undesirable masking effects and scratches on the photomask during contact printing (see Figure 1.6). Solvent stains and other contaminants on a Si wafer can be easily observed in dark field microscopy (see Volume III, Chapter 6 on metrology and MEMS, NEMS modeling). All lithography processes take place inside a semiconductor clean room, which is a specially constructed enclosed area environmentally controlled with respect to airborne particulates, temperature (±0.1°F), air pressure, humidity (from 0.5–5% relative humidity), vibration, and lighting. In Table 1.1, some common sources of clean room contaminants are listed, and in Figure 1.2, the clean room classification system is elucidated. In a Class 1 clean room, the particle count does not exceed 1 particle

**Table 1.1** Some Common Clean Room Contaminant Sources

| |
|---|
| • Location: a clean room near a refinery, smoke stack, sewage plant, cement plant spells big trouble |
| • Construction: the floor is an important source of contamination. Also, items such as light fixtures must be sealed, and room construction tolerances must be held very tight |
| • Wafer handling: transfer box |
| • Process equipment: never use fiberglass duct liner; always use 100% polyester filters, eliminate all nonessential equipment |
| • Chemicals: residual photoresist or organic coatings, metal corrosion |
| • Attire: only proper attire and dressing in the anteroom |
| • Electrostatic charge: clean room must have a conductive floor |
| • Furniture: only clean room furniture |
| • Stationary: use ballpoint pen instead of lead pencil, only approved clean room paper |
| • Operator: no eating, drinking, smoking, chewing gum, or makeup of any kind |

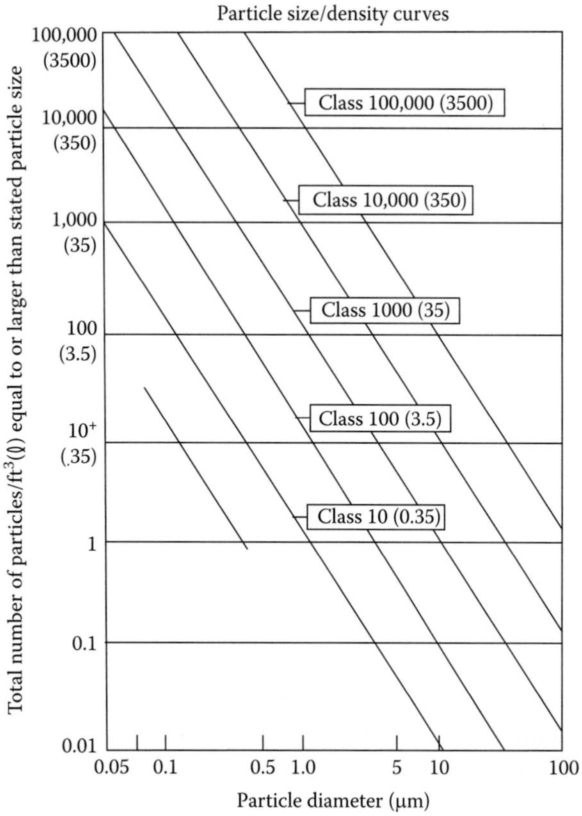


**FIGURE 1.2** U.S. Federal Standard 209b for clean room classification. The bottom solid line shows the definition of Class 1. (From Cunningham, J.A. 1992. The remarkable trend in defect densities and chip yield. *Semicond Int* 15:86–90. With permission.[2])

per cubic foot with particles 0.5 μm and larger, and in a Class 100 clean room, the particle count does not exceed 100 particles per cubic foot with particles 0.5 μm and larger. The allowable contamination particle size in IC manufacture has been decreasing hand in hand with the ever-decreasing minimum feature size. With a 64-kB dynamic memory chip [dynamic random access memory (DRAM)], for example, one can tolerate 0.25-μm particles, but for a 4-MB DRAM, one can only tolerate 0.05-μm particles. The smallest feature sizes in these two cases are 2.5 and 0.5 μm, respectively. As a reference point, a human hair has a diameter of 75–100 μm (depending on age and race); tobacco smoke contains particles ranging from 0.01–1 μm; and a red blood cell ranges from 4–9 μm.

#### Example 1.1

**Problem:** If a 125-mm diameter wafer is exposed for 1 minute to an air stream under a laminar-flow condition at 30 m/min, how many dust particles will land on the wafer in a Class 10 clean room?

**Solution:** In a Class 10 clean room, there are 350 particles (0.5 μm or larger) per cubic meter. The air volume that passes over the wafer in 1 min is (30 m/min) × $\pi(0.125\ \text{m}/2)^2 \times 1$ min = 0.368 m³. The number of dust particles (0.5 μm or larger) contained in that air volume is 350 × 0.368 = 128 particles. Therefore, if there are 200 IC chips on the wafer, the particle count amounts to one particle on each of 64% of the chips. Fortunately, only a fraction of the particles that land on the surface of the wafer adhere to the wafer surface, and of those only a fraction are at a circuit location critical enough to cause a failure. However, the calculation demonstrates the importance of the clean room.

Many different dry and wet methods for wafer cleaning currently in use are listed below. RCA1 and RCA2, developed by W. Kern in 1965 while working for the Radio Corporation of America (RCA), which are well known, use mixtures of hydrogen peroxide and various acids or base followed by deionized (DI) water rinses. Others include vapor cleaning; thermal treatment, for example, baking at 1000°C in a

**Table 1.2** Wet versus Dry Cleaning Attributes

| Attribute | Wet | Dry |
|---|---|---|
| Particle removal | + | – |
| Metal removal | + | – |
| Heavy organics, i.e., photoresist | + | – |
| Light organics, i.e., outgassed hydrocarbon residues | + | + |
| Throughput | + | – |
| Process repeatability | + | + |
| Water usage | – | + |
| Process chemical cleanliness | – | + |
| Environmental impact, purchase, and disposal cost | – | + |
| Single wafer use applicability | – | + |

Dry cleaning usually requires wet followup. UV ozone can effectively remove light organic contamination.[3]

*Source:* Iscoff, R. 1991. Wafer cleaning: can dry systems compete? *Semicond Int* 14:48–54. With permission.[3]

vacuum or in oxygen; and plasma or glow discharge techniques, for example, in Freons with or without oxygen. Mechanical methods include ultrasonic agitation, polishing with abrasive compounds, and supercritical cleaning. Ultrasonic cleaning, which is excellent for removing particulate matter from the substrate, is unfortunately prone to contamination and mechanical failure of deposited films. Attributes of wet versus dry cleaning techniques are compared in Table 1.2. Except for environmental concerns, wet etching still outranks other cleaning procedures.

The prevalent RCA1 and RCA2 wet cleaning procedures are as follows:

- RCA1: Add 1 part of $NH_3$ (25% aqueous solution) to 5 parts of DI water; heat to boiling, and add 1 part of $H_2O_2$. Immerse the wafer for 10 minutes. This procedure removes organic dirt (resist).
- RCA2: Add 1 part of HCl to 6 parts of DI water; heat to boiling, and add 1 part of $H_2O_2$. Immerse the wafer for 10 minutes. This procedure removes metal ions.

The second RCA cleaning process is required to keep oxidation and diffusion furnaces free of metal contamination. Both cleaning processes leave a thin oxide on the wafers. Before a further etch of the underlying silicon is attempted, oxide must be stripped off by dipping the wafer in a 1% aqueous HF solution for a very short time. Water spreads on an oxide surface (hydrophilic) and beads up on a bare Si surface (hydrophobic). This behavior can be used to establish whether any oxide remains.

In most IC labs, processing a wafer previously exposed to KOH is not allowed as it is feared that the potassium will spoil the IC fabrication process. In more lenient environments, carefully cleaned wafers using RCA1 and RCA2 are allowed.

Supercritical cleaning (Figure 1.3) with $CO_2$ is especially suited for microstructure cleaning.[4] These fluids possess liquid-like solvative properties and gas-like diffusion and viscosity that enable rapid penetration into crevices with complete removal of organic and inorganic contaminants contained therein. During wet cleaning of surface micromachined structures, thin, mechanical microstructures tend to stick to one another or to the substrate through surface tension (stiction) (Chapter 7). Consequently, dry vapor phase and supercritical cleaning with low or no surface tension are preferred.

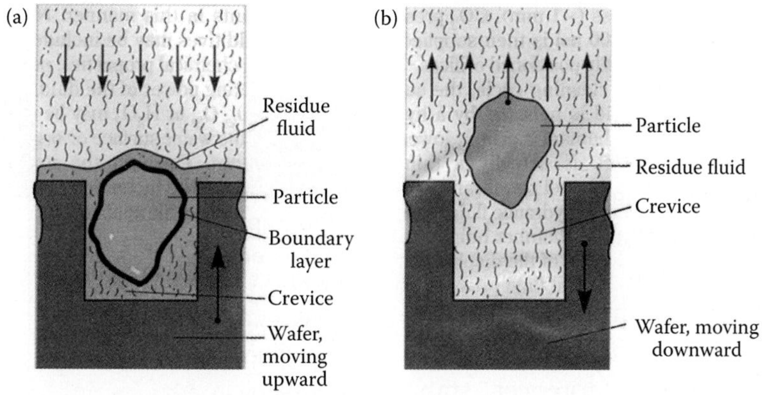


**FIGURE 1.3** (a) Wet cleaning. (b) Supercritical cleaning.

Vapor phase cleaning also uses significantly fewer chemicals than wet immersion cleaning.[5]

We will learn more about the emerging importance of supercritical fluids as developers for completely dry resist processes further below.

Wafer cleaning has become a scientific discipline in its own right with journals such as *Microcontamination* and *The Magazine for Ultraclean Manufacturing Technology*, books such as *Handbook of Contamination Control in Microelectronics: Principles, Applications and Technology*[6] and the *Handbook of Semiconductor Wafer Cleaning Technology*,[7] and dedicated conferences (e.g., the Microcontamination Conference). Visit the Semiconductor Subway (http://www-mtl.mit.edu/semisubway.html) for frequent updates in this important area.

## Silicon-Based Device Manufacturing Costs

The manufacturing of a Si-based device involves many processing steps, and each step adds to the cost of the finished wafer/device. In the most general terms, the finished wafer costs depend on the number of masks used, the complexity of the devices, and the clean room requirements. The costs increase with the number of layers in a device in a nonlinear fashion as defects are introduced in each new layer, but with each new layer more defects accumulate in the underlying layers as well. Cost per device/chip further increases the smaller the feature size because of more stringent requirements for the lithography steps and overall process control. In first world countries, an added concern is that semiconductor fabrication processes are highly toxic; therefore, environmental and worker protection laws need to be considered. In the case of an IC fab, the current cost is in the $2–3 billion range. A typical fab line occupies one city block and employs a few hundred workers. The most profitable period is typically in the first 18 months to 2 years. It is important to note that for large-volume ICs, packaging and testing constitute the largest costs, and that for low-volume ICs, design costs may swamp out the manufacturing costs. In the case of non-IC micro- and nanomachines, testing and packaging become yet more significant contributors to costs. Moreover, standardization is much more difficult, and multiphysics design software packages required for these devices are not yet sophisticated enough.

In Table 1.3 we list the relative costs of IC production processes (does not include design, test, and packaging); lithography is the most expensive step (at 35%), and cleaning adds another 20%.

Variability in the IC manufacturing processes listed in Table 1.3 can lead to deformations or nonconformities in finished products. The product yield is defined as the percentage of devices or circuits that meet a nominal performance specification. One distinguishes between a functional (hard) yield (e.g., opens, shorts, particles) and a parametric (soft) yield (e.g., speed, noise, immunity, power consumption). Functional yield is affected by the presence of defects. Defects result from a number of random sources such as contamination from equipment or processes, handling, mask imperfections, and airborne particles. Functional yield $Y$ is a function of $A_c$, a critical area in which, if a defect occurs, it has a high probability of inducing a fault, and $D_0$, the density of defects per unit area, or:

$$Y = f(A_c, D_0) \tag{1.1}$$

This relation depends on circuit geometry, density, feature sizes, number of processing steps, and so on, and is sometimes represented as:

$$Y = \frac{1 - e^{-D_0 A_c}}{D_0 A_c} \tag{1.2}$$

Manufacturing issues that influence hard yield $Y$ include photoresist shrinking/tearing, variations in material deposition, variations in temperature, variations in oxide thickness, impurities, variations between lots, variations across the wafer, and so on.

**Table 1.3** Relative Costs of IC Production Processes

| Manufacturing Process Step | Percentage of Wafer Processing Cost per Square Centimeter* |
|---|---|
| Lithography | 35% |
| Multilevel materials and etching | 25% |
| Furnaces and implants | 15% |
| Cleaning/stripping | 20% |
| Metrology | 5% |

*Excludes packaging, test, and design costs.

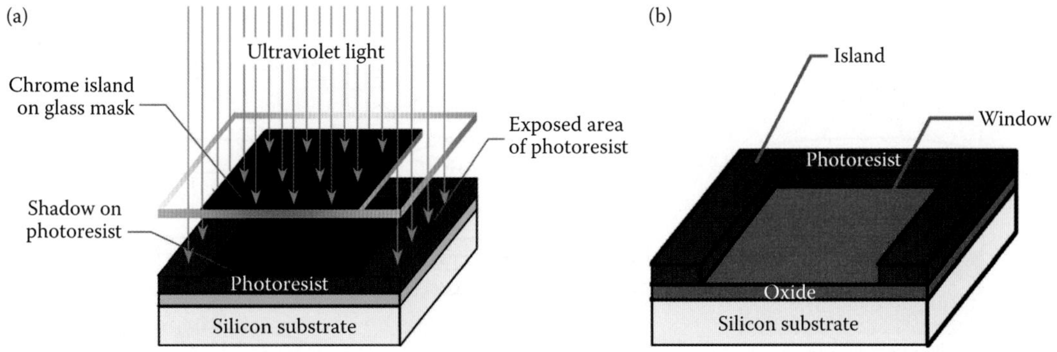


**FIGURE 1.4** (a) Photomask in an exposure setup. (b) Resulting pattern in case a negative photoresist is used.

The parametric yield measures the "quality" of functioning systems. Even in a defect-free manufacturing environment, random process variations can lead to varying levels of system performance. Parametric yield may include variations in threshold voltage (because of oxide thickness, ion implantation, poly variations), variations in R, C (because of changes in doping or poly-Si, metal variations in height and width), and shorts and opens (because vias are not cut all the way through, or undersized vias that have too much resistance or oversized vias that short).

## Masks

### Standard Photolithography Masks

The stencil used to repeatedly generate a desired pattern on resist-coated wafers is called a *mask*. In typical use, a photomask—a nearly optically flat glass [transparent to near ultraviolet (UV)] or quartz plate (transparent to deep UV) with an absorber pattern metal (e.g., an 800-Å-thick chromium layer)—is placed above the photoresist-coated surface, and the mask/wafer system is exposed to UV radiation (Figure 1.4). The absorber pattern on the photomask is opaque to UV light, whereas the glass or quartz is transparent. The absorber pattern on the mask is generated by e-beam lithography, a technique that yields higher resolution than photolithography. In e-beam lithography, a pattern drawn on a computer-aided design (CAD) system is exposed onto the mask. Like resists, masks can be positive or negative. A *positive* or *dark field* mask is a mask on which the pattern is clear with the background dark. A *negative* or *clear field* mask is a mask on which the pattern is dark with the background clear. A light field or dark field image, known as *mask polarity* (Figure 1.5), is then transferred to the semiconductor surface. This procedure results in a 1:1 image of the entire mask onto the silicon wafer.

Masks making direct physical contact (also referred to as *hard contact*) with the substrate are called *contact masks*. Unfortunately, these masks degrade faster because of wear than noncontact, proximity masks (also referred to as *soft contact masks*), which are slightly raised, e.g., 10–20 μm, above the wafer. The defects resulting from hard contact masks on both the wafer and the mask make this method of optical pattern transfer unsuitable for *very large scale integration* (VLSI) manufacturing. In VLSI, ICs have between 100,000 and 1 million components, and in *ultralarge scale integration* (ULSI), there are more than 1 million circuit elements on a single chip. In 2005, Intel began manufacturing computer chips with an average feature size of 65 nm (the so-called *65-nm technology node*, see Volume I, Appendix 1A). This new generation of chips has more than half a billion transistors and constitutes a ULSI chip.

We review hard contact masks because they are still used in research and development, in mask making itself, and for prototyping. Contact mask

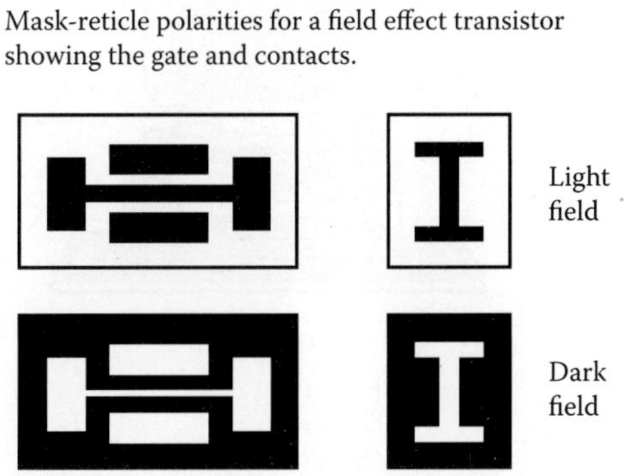


**FIGURE 1.5** Light field and dark field.

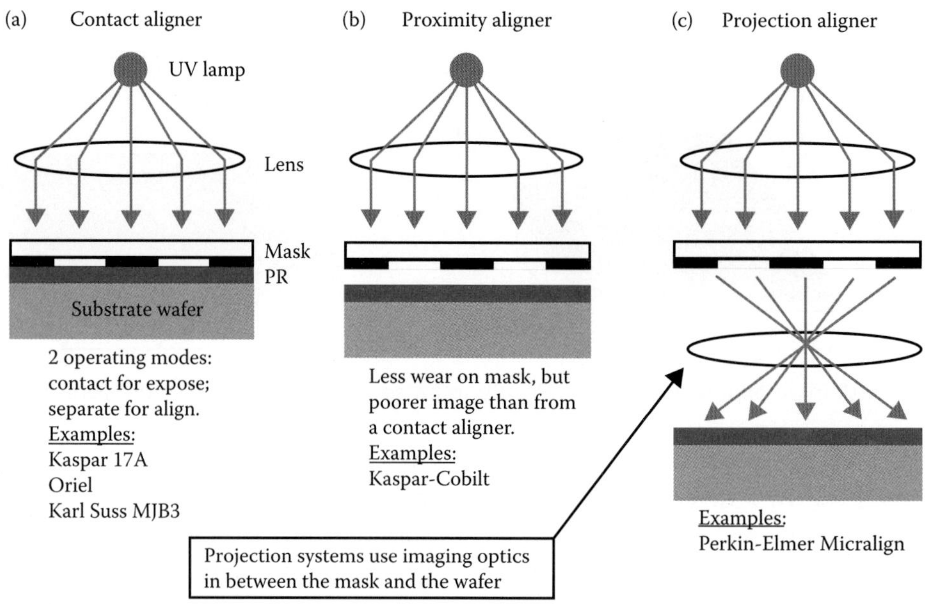


**FIGURE 1.6** (a) Contact printing, (b) proximity printing, and (c) projection printing.

and proximity mask printing are collectively known as *shadow printing*. A more reliable method of masking is *projection printing*, where, rather than placing a mask in direct contact with (or in proximity of) a wafer, the photomask is imaged by a high-resolution lens system onto the resist-coated wafer. In the latter case, the only limit to the mask lifetime results from operator handling. The imaging lens can reduce the mask pattern by 1:5 or 1:10, making mask fabrication less challenging. In Figure 1.6, we compare contact, proximity, and projection lithography, and example commercial systems are listed also.

The design of electron beam-generated masks for ICs and miniaturized machines is generally fairly straightforward and requires some suitable CAD software and a platform on which to run it (Figure 1.7). Electron-beam lithography is maskless and is discussed later in this chapter; mask design and suitable CAD software will be addressed in Volume III, Chapter 6.

### Gray-Tone Lithography Masks

As described above, photolithography constitutes a binary image transfer process: the developed pattern consists of regions with resist (1) and regions without resist (0). In contrast, in grayscale lithography, the partial exposure of a photoresist renders it soluble to a developer in proportion to the local exposure dose, and as a consequence, after development, the resist exhibits a surface relief.

Gray-tone masks (GTMs), letting varying amounts of light pass through, are of great potential use in

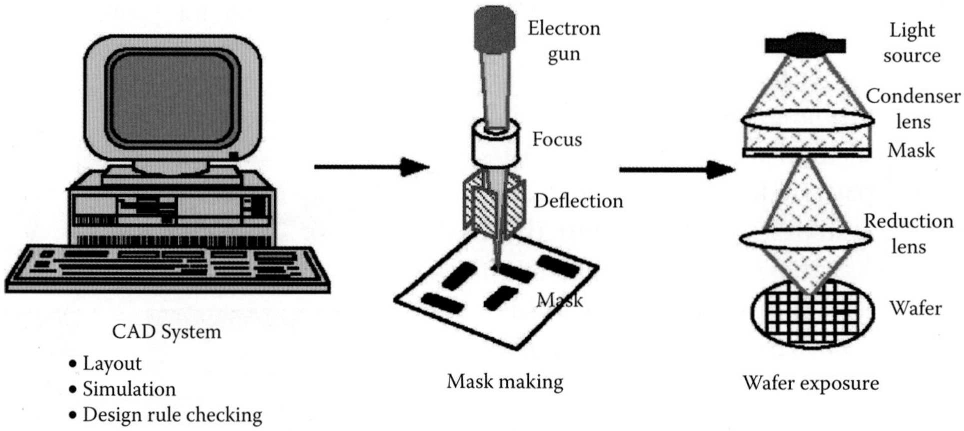


**FIGURE 1.7** E-beam writing of a photomask and subsequent projection printing.

miniaturization science as they allow for the mass production of micromachines with varying topography. The possibility of creating profiled micro-3D structures offers tremendous additional flexibility in the design of microelectronic, optoelectronic, and micromechanical components. In optical integrated circuits (OICs), for example, GTMs can be used for the fabrication of tapered waveguides, gratings with saw tooth profiles (blazed gratings), microlenses, and so on.[8] As an example of GTM use, in Figure 1.8, we show experiments with a Rohm and Haas dual-tone resist.[9,10] The right-hand path in Figure 1.8a

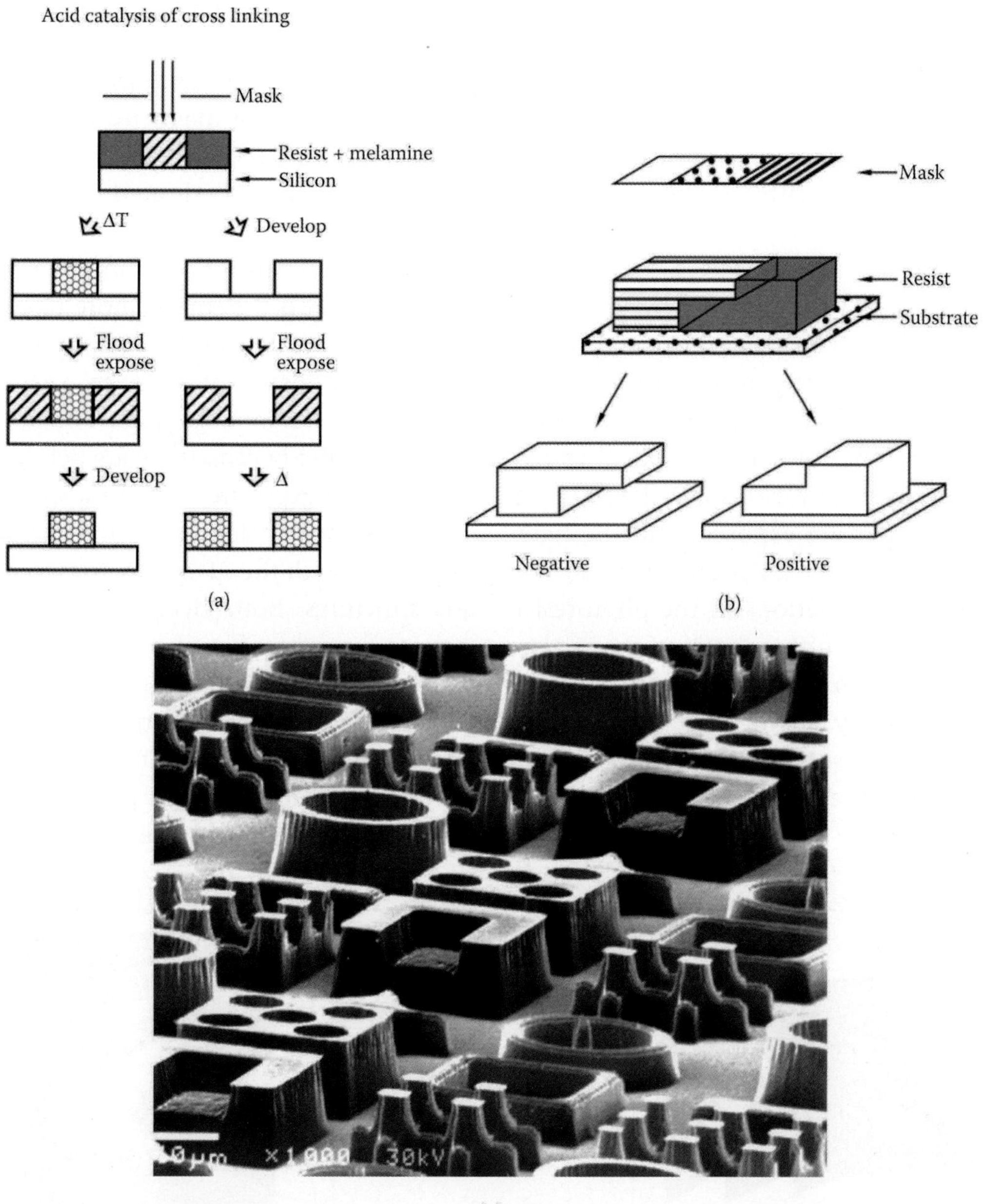


**FIGURE 1.8** Rohm and Haas dual-tone resist. (a) Process options for Rohm and Haas dual-tone resist. The right-hand path produces positive-tone images in which the remaining resist is cross-linked and does not flow when exposed to temperatures above the $T_g$ of the novolak-base resin. The left-hand path provides thermally stable negative-tone images. (b) Exposure of the Rohm and Haas resist with a mask exhibiting nominally 0% transmission, 50% transmission, and 100% transmission allows generations of patterns with controlled variations in thickness and/or overhangs and cantilevered structures. (c) Example of the types of features that can be printed using the Rohm and Haas resist. Note the structures with full thickness and 50% thickness. (Courtesy of Drs. Feely, Rohm, and Haas. Feely, W. E. 1988. *Technical digest: solid state sensor and actuator workshop.* Hilton Head Island, SC: IEEE.[9])

produces positive-tone images in which the remaining resist is cross-linked and does not flow when exposed to temperatures above the $T_g$ of the novolak-base resin. The left-hand path provides thermally stable negative-tone images. The mask used here has only 0% or 100% transmission fields. Exposure of the same Rohm and Haas resist with a GTM, exhibiting nominally 0% transmission, 50% transmission, and 100% transmission, allows generations of patterns with controlled variations in thickness and/or overhangs and cantilevered structures. An example of the types of features that can be printed using the Rohm and Haas resist with a GTM is shown in Figure 1.8c. Note the structures with full thickness and 50% thickness.

The fabrication of 3D components by gray-tone technology is performed in three principal steps. The first and most critical step deals with the realization of the GTM itself. The challenge in this step is the definition of zones of variable optical transmission that represent the various gray levels. In a second step, the wafers are exposed, and the light intensity modulation by the gray-tone areas on the mask generates depth variations in the photoresist (Figure 1.9). The profile in the photoresist can also be transferred proportionally into the substrate during a third step, i.e., a dry etching step.[11]

For each substrate/resist combination, the dry etching process has to be established to obtain resist/substrate selectivity close to 1:1.

One approach to make multilevel photoresist patterns directly, without a physical mask, is by variable-dose e-beam writing, in which the electron dosage (current × dwell time) is varied across the resist surface.[13] Variable-dose e-beam writing is slow and costly, and GTMs are a desirable alternative, especially if high throughput production is required. A laser writer can produce the same result but at a lower resolution.

Possible methods for making variable transmission masks include magnetron sputtering of amorphous carbon onto a quartz substrate. Essentially any transmittance ($T$) desired in the $0\% < T < 100\%$ can be achieved by controlling the film thickness ($t$) in the $200 \text{ nm} > t > 0 \text{ nm}$ range with subnanometer precision.[14] Perhaps more elegantly, gray levels may be created by the density of dots that will appear as transparent holes in a chromium mask. These dots are small enough not to be transferred onto the wafer because they are below the resolution limit of the exposure tool. In a CAD program, the dot matrices associated with the gray levels are then combined with the 2D geometries to define the final 3D structures. Both electron-beam and laser-beam pattern generators have been used for manufacturing GTMs. The electron-beam pattern generator is better adapted to this application because of its flexibility and higher resolution.[11] Optimization of the exposure and the development parameters improves the resist depth and profile. The resist profile may then be

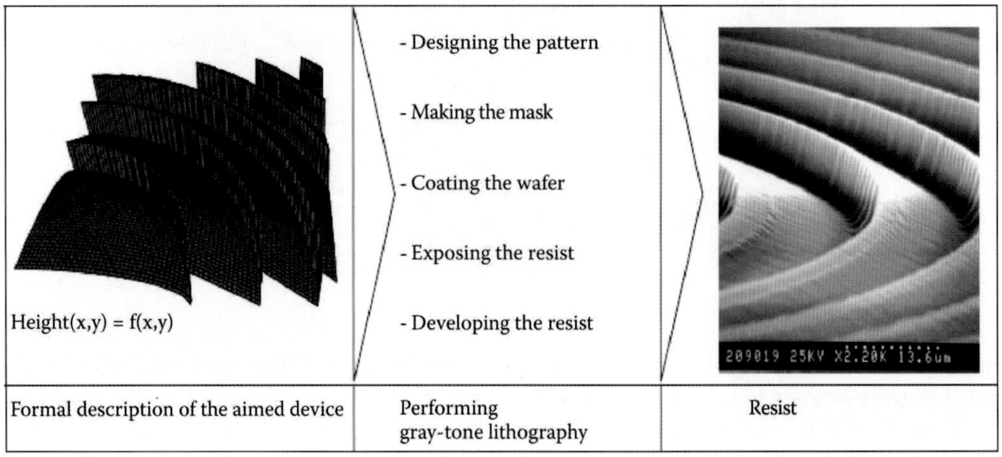


**FIGURE 1.9** Process flow chart for transferring the image on a gray-tone mask into a substrate. Gray-tone lithography is an inexpensive one-step lithographic method to fabricate 3D microstructures using a raster-screened (gray-tone) photomask. (Reimer, K., R. Engelke, U. Hofmann, P. Merz, K.T.K.v. Platen, and B. Wagner. 1999. *MF-Conf., micromachine technology for diffractive and holographic optics.* Santa Clara, CA: SPIE.[12])

used directly; it may function as a mold; or it is conformally transferred into the substrate using plasma etching with controlled selectivity. Figure 1.10 shows a blazed grating proportionally transferred from a patterned resist to the underlying fused silica. Blazed gratings with a period of 16 μm and 8 μm covering a surface of 3 × 3 mm$^2$ were fabricated with an eight-level GTM. A reflection grating in which the grooves are asymmetric with respect to the surface normal is said to be blazed. A blazed grating is used to direct the light only into the diffracted order of interest with high efficiency. From the atomic force measurements of the grating profile shown here, the distribution of light in the various diffraction orders was computed and found to be in agreement with the experimental optical measurements. Compared with multiple mask techniques, the GTM approach allows the realization of different depths from the same GTM.

Another attractive way to implement grayscale lithography is with high-energy beam-sensitive (HEBS) glass. HEBS glass turns dark on exposure to an electron beam; the higher the electron dosage, the darker the glass turns. In HEBS glass, a top layer, a couple of micrometers thick, contains silver ions in the form of silver-alkali-halide $(AgX)_m$ $(MX)_n$ complex nanocrystallites that are about 10 nm or smaller in size and are dispersed within cavities of the glass

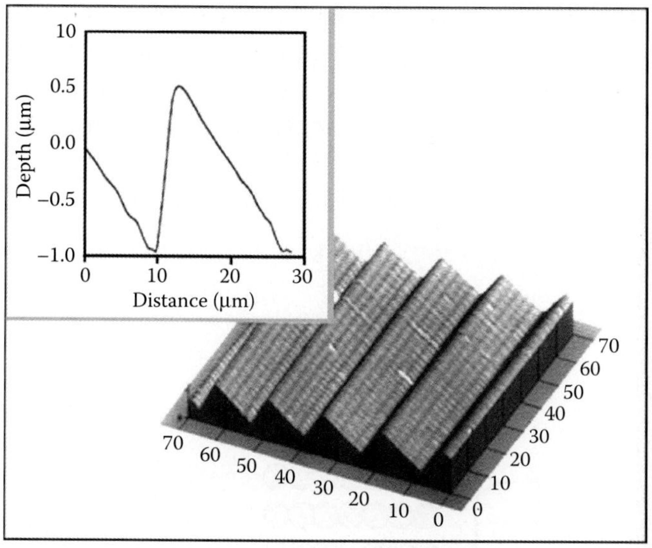


**FIGURE 1.10** Atomic force microscope profile of a 16-μm pitch blazed grating in fused silica: the grating was proportionally transferred from resist to fused silica. A sawtooth (blazed) profile notably increases the efficiency of an optical grating.

$SiO_4$ tetrahedron network. Chemical reduction of the silver ions produces opaque specks of silver atoms on exposure to a high-energy electron beam (>10 kV). A key part in the development of a grayscale process is the characterization of the resist thickness as a function of the optical density in the mask for a given lithographic process. It is desirable to use photoresists that exhibit a low contrast to achieve a wide process window (see "Contrast and Experimental Determination of Lithographic Sensitivity," this chapter). Ideally, the resist response can be linearized to the optical density within the mask. Using an HEBS glass mask, Sure et al.[8] fabricated 3D silicon tapers to couple light from an optical fiber into an optical waveguide for an OIC. Fabrication involves writing a single grayscale mask in HEBS glass with a high-energy electron beam, UV grayscale lithography, and inductively coupled plasma etching (see Figure 1.11a).[8] In Figure 1.11b we show an example HEBS mask for the fabrication of an array of lenses (http://www.canyonmaterials.com).

### *Inexpensive Masks and Maskless Optical Projection Lithography for Research and Development*

In miniaturization science, one often is looking for low-cost and fast-turnaround methods to fabricate masks. This may involve manually drawn patterns on cut-and-peel masking films and photoreduction, affording fast turnaround without relying on outside photomask services. Alternatively, it may involve direct writing on a photoresist-coated plate with a laser plotter (~2-μm resolution).[15] Simpler yet, using a drawing program such as Canvas® (Deneba Systems, Inc.), Freehand® (Macromedia, Inc.), Illustrator® (Adobe Systems, Inc.), or L-Edit® (Tanner Research, Inc.), a mask design can be created on a computer and saved as a Postscript® file to be printed with a high-resolution printer on a transparency.[16] A common laser printer has a resolution of 72 lines per inch (lpi) and between 300 and 1200 dots per inch (dpi) (see Figure 1.12). The average dot size for a 1200-dpi printer is 21 μm, and if we print a vertical line approximately 100 μm wide with this printer, approximately 5 dots make up the 100-μm width, but the space between successive lines of 5 dots is 332 μm. Thus, this vertical

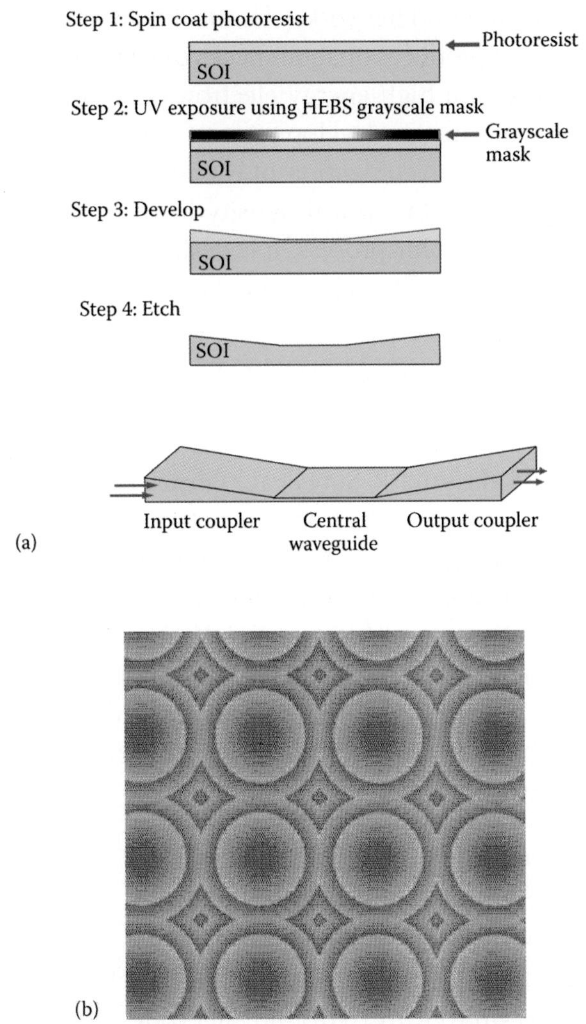


**FIGURE 1.11** (a) Fabrication steps for a silicon taper. (Sure, A., T. Dillon, J. Murakowski, C. Lin, D. Pustai, and D.W. Prather. 2003. Fabrication and characterization of three-dimensional Si tapers. *Optics Express* 11:555–61. Drawing by Mary Amasia.[8]) (b) HBS mask for the fabrication of a diffractive microlens array (http://www.canyonmaterials.com).

line will look like a dotted line. Using a typesetting type of printing can avoid these resolution problems (16,000 dpi).

The transparency with the printed image may then be clamped between a presensitized chrome-covered mask plate (i.e., a vendor such as Nanofilm has preapplied the resist) and a blank plate to make a traditional mask from it. After exposure and development, the exposed plate is put in a chrome etch for a few minutes to generate the desired metal pattern, and the remaining resist is stripped off. Simpler yet, the printed transparency may be attached to a quartz plate to be used as a mask directly.

Lithography is still mostly carried out using masks, but because of problems caused by masks, such as expense and time in fabricating them, contamination introduced by them, their disposal, and the difficulties in their alignment, research into maskless optical projection lithography (MOPL) is growing rapidly and broadly. One MOPL approach already on the market (http://www.intelligentmp.com) is based on the digital micromirror device (DMD) chip from Texas Instruments Inc. (TI) and relies on the same spatial light modulation technology used in digital light processing projectors and high-definition television (see Chapter 7, Example 7.10). Enormous simplification of lithography hardware is feasible by using the movable mirror arrays in a DMD chip to project images on the photoresist. This technique is capable of fabricating micromachined elements with any surface topography and can, just like e-beam lithography or laser writing, be used for

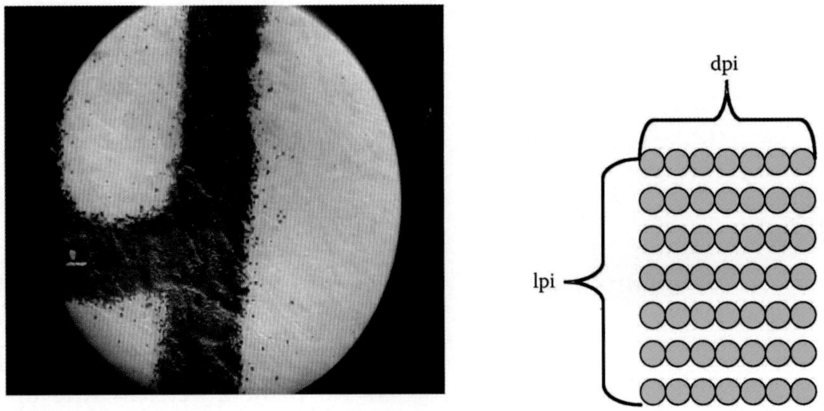


**FIGURE 1.12** Laser printer resolution. Image of a laser-printed letter. Note the scatter of small dots. Technology is not so precise as to place dots in a straight line, and the effective dpi might be partially the result of overspray. (From Dr. James Lee, Ohio State University.)

implementing maskless binary and grayscale lithography. The resolution of DMD-based maskless photolithography (about 5 μm) is significantly less than with e-beam lithography (0.25 μm) or laser writers (<1.0 μm), but it is a parallel technique, and for many applications, i.e., in microfluidics, the lower resolution might not be an obstacle. Using DMD technology, on-chip electronics, in conjunction with a control computer, can generate a pattern by turning individual mirrors "on" or "off." DMD chip's micromirrors are mounted on tiny hinges that enable them to tilt either toward the light source in the projection lens (ON) or away from it (OFF) (see Example 7.10), creating a light or dark pixel on the projection surface. The bit-streamed image code entering the semiconductor directs each mirror to switch on and off at a rate of up to 10,000 times/s. This pattern is then demagnified and focused on the wafer to expose the photoresist while light from the remaining off mirrors is deflected away from the optics. The use of refractive optics prevents the extension of MOPL to wavelengths much shorter than 157 nm. The unique capability of representing a grayscale is probably the most essential merit of this type of maskless lithography. When a mirror is switched on more frequently than off, it reflects a light gray pixel; a mirror that is switched off more frequently reflects a darker gray pixel. In this way, the mirrors in a DMD system can reflect pixels in up to 1024 shades of gray to convert the video or graphic signal entering the DMD chip into a highly detailed grayscale image. The method to fabricate the micromirrors used in DMDs is based on MEMS and is detailed in Chapter 7 on surface micromachining. In Figure 1.13 we show a photo of Intelligent Micro Patterning's lithography system, the SF-100, based on TI's DMD with a line-width resolution of 5 μm (http://www.intelligentmp.com/SF100Photos.htm).

Maskless zone-plate-array lithography (ZPAL)[17] is another optical maskless technology. This technology operates on the principle of diffraction rather than refraction. Instead of a single large numerical aperture (NA) lens, an array of thousands of nanofabricated Fresnel zone plate lenses is used, each focusing a beam of light onto the substrate. A computer-controlled array of micromechanical spatial-light modulators turns the light to each lens on or off as the substrate is scanned under the array, thereby printing the desired pattern in a "dot-matrix" fashion.

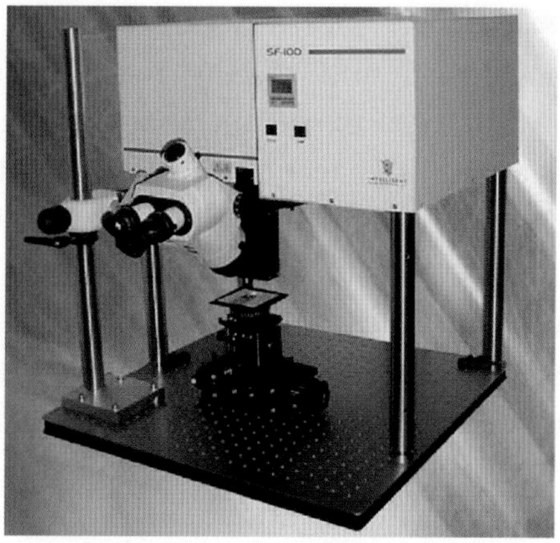

**FIGURE 1.13** Intelligent MicroPatterning's SF-100 lithography system based on DMDs (http://www.intelligentmp.com/SF100Photos.htm).

ZPAL is reviewed in Chapter 2 as a nanolithography technique that can be adapted to work with an extreme UV light source.

## Photoresist Deposition

### *Introduction*

Lithography is the most expensive step in microelectronics technology, representing up to 35% of the wafer manufacturing cost (see Table 1.3). Within lithography, photoresist coating is one of the more expensive steps. Photoresists may cost $100/L (and it is about $40/gallon for a typical developer), and several of the resist deposition processes used waste significant amounts of material. For silicon ICs, the resist thickness after a prebake (see below for what a prebake does) typically ranges between 0.5 and 2 μm. For miniaturized 3D structures, much greater resist thicknesses are often required, and complex topography might also call for a conformal resist coat over very high-aspect-ratio features. For thick resist coats, techniques such as casting and the use of thick sheets of dry photoresists replace the ineffective resist spinners. For conformal coating, resist spraying or, better yet, electrodeposition (ED) of photoresist might be preferable. In this section we emphasize methods used to deposit liquid photosensitive materials in ICs, MEMS, and NEMS, including spin coating, spray coating, dip coating, meniscus coating, electrodeposited (electrophoretic) photoresist, and casting methods such as roller, curtain, and extrusion coating. Photoresist coating is an additive technology, and in Chapter 8 we present a broader review of available processes for the deposition of organic films in general, including photoresists and nonphotosensitive organics for applications such as biosensors, membranes, packaging, protein arrays, DNA arrays, and so on. Very thin photoresist coatings (<<1 μm) are reviewed further below under the section on Thin Photoresists, and in Chapter 10 we explore very thick photoresist coatings (>>20 μm) for molding applications. Lamination of dry photoresist films to a substrate and a comparison of wet versus dry photoresists are covered below in the section on Permanent and Dry Resists.

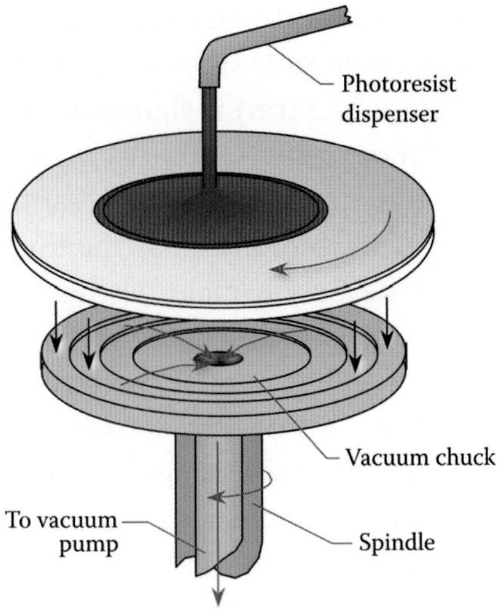

**FIGURE 1.14** Resist spinner.

### *Spin Coating*

A common step before spinning on a resist with Si as the substrate is the growth of a thin layer of oxide on the wafer surface by heating it to between 900 and 1150°C in steam or in a humidified oxygen stream (see Figure 1.1a). Dry oxygen also works, but wet oxygen and steam produce faster results. The oxide can serve as a mask for a subsequent wet etch or boron implantation. As the first step in the lithography process itself, a thin layer of an organic polymer, a photoresist sensitive to UV radiation, is deposited on the oxide surface (see Figure 1.1b). The photoresist is dispensed onto the wafer lying on a wafer platen in a resist spinner (see Figure 1.14).[18] A vacuum chuck holds the wafer in place. A speed of about 500 rpm is commonly used during the dispensing step, enabling the spread of the fluid over the substrate. After the dispense step it is common to accelerate to a relatively high speed to thin the fluid to near its final desired thickness. Typical spin speeds for this step range from 1500–6000 rpm, depending on the properties of the fluid (mostly its viscosity), as well as the substrate. This step can take from 10 seconds to several minutes. The combination of spin speed and time selected for this step will generally define the final film thickness. At these speeds, centrifugal force causes the solution to flow to the edges, where it builds up until expelled when surface tension is exceeded. The resulting polymer thickness, $T$, is a function of spin speed, solution concentration, and molecular weight (measured by intrinsic viscosity). The empirical expression for $T$ is given by:

$$T = \frac{KC^{\beta}\eta^{\gamma}}{\omega^{\alpha}} \tag{1.3}$$

where $K$ = the overall calibration constant
$C$ = the polymer concentration in g/100-mL solution
$\eta$ = the intrinsic viscosity
$\omega$ = the number of rotations per minute

Once the various exponential factors ($\alpha$, $\beta$, and $\gamma$) have been determined, Equation 1.3 can be used to predict the thickness of the film that can be spun for various molecular weights and solution concentrations of a given polymer and solvent system.[19]

The resist spinning process is of primary importance to the effectiveness of pattern transfer. The quality of the resist coating determines the density of defects transferred to the device under construction. The photoresist film, after application to the substrate, must have a uniform thickness and must be chemically isotropic so that its response to exposure and development is uniform. The coating thickness of the thin, glassy resist film depends on the chemical resistance required for image transfer and the fineness of the lines and spaces to be resolved. The coating process should take place in Class 100 laminar flow air conditions, and a minimum spin time of 30 seconds is necessary for coating uniformity. The photoresist should also be filtered just before application. The application of too much resist results in edge covering or run-out, hillocks, and ridges, reducing manufacturing yield. Application of too little resist may leave uncovered areas. Optimization of the "regular" photoresist coating process in terms of resist dispense rate, dispense volume, spin speed, ambient temperature, venting of the resist spin station, and humidity presents a growing challenge.

To ensure reproducible line widths and development times in subsequent steps, a resist thickness of 1 μm, with a repeatability of ±10 Å (±0.1%) is becoming the norm. In such cases, even small changes in air draft over the resist-coated substrate could influence the results. In Table 1.4 we present a handy spin coating-troubleshooting chart based on data presented at http://www.brewerscience.com/products/cee/technical/spintheory. In Appendix 1A a more detailed spin coating process troubleshooting list is presented.

Since 2005, computer chips with an average feature size of 65 nm have been on the market, and for these, as well as smaller feature sizes, yet better control over thinner and thinner resist layers is sought. An on-line film thickness monitor, possibly a technique based on reflection spectroscopy, will become essential for statistical process control of such demanding photoresist coatings.[18] The need for an alternative photoresist deposition technique might arise as the amount of waste material generated by spin coating is high, with most of the resist solution (>95%) thrown off the substrate during the spin casting process (the wasted resist must be disposed of as a toxic material). Also inherent to this process is the formation of edge beads, which require an additional removal process (see Figure 1.15a) before subsequent process steps. The edge of a wafer might exhibit resist ridges that are about 10 times the mean thickness on the rest of the substrate. A typical bead removal solution is AZ edge bead remover solvent (http://www.glue.umd.edu/~nima/doc/AZEBRSolventMSDS.pdf). With the current trend toward larger substrate formats, these limitations will be exacerbated. The main obstacle to using spin coating in MEMS is caused by varying topography: deeply etched features cause a physical obstruction to the solution flow, preventing complete coverage and often causing striation or resist thickness variation. For example, resist thickness variations can occur on the near and far sides of a cavity or in cavities at different locations on the substrate. Sizes and shapes of the cavities also have influence on the resist uniformity and coating defects. For substrates with moderate topography, alternative coating techniques, like spray coating (see the next section below), offer better prospects. Wafers with extreme topography, e.g., as encountered in radio frequency-MEMS, require a process with yet higher shape conformality. For the latter application we will see below that ED photoresists are a possible solution.

A good review on spin coating can be found at http://www.brewerscience.com/cee/technical/spintheory.html.

**Table 1.4** Spin Coating Troubleshooting Chart

| | | |
|---|---|---|
| **Film too thin** | | |
| Spin speed too high | → | Select lower speed |
| Spin time too long | → | Decrease time during high-speed step |
| **Film too thick** | | |
| Spin speed too low | → | Select higher speed |
| Spin time too short | → | Increase time during high speed step |
| Exhaust volume too high | → | Adjust exhaust lid or house exhaust damper |

*Source:* Based on http://www.brewerscience.com/cee/technical/spintheory.html.

### *Alternative Photoresist Deposition Methods*

In this section, we compare alternatives to spin coating (Figure 1.15a). We follow closely James Webster's treatise of the subject.[20]

*Spray Coating* In spray coating, the substrates to be coated pass under a spray of photoresist solution. A diagram depicting this process is shown in Figure 1.15b. The spray system includes an ultrasonic spray nozzle that generates a distribution of droplets in the micrometer range. The resist is pushed out of a pressurized tank via a supply pipe to the spray head. The spray head has a defined aperture where the spray mist is formed. The shape of the spray pattern can be adjusted by a secondary air cushion. Compared with electrostatic spraying (see

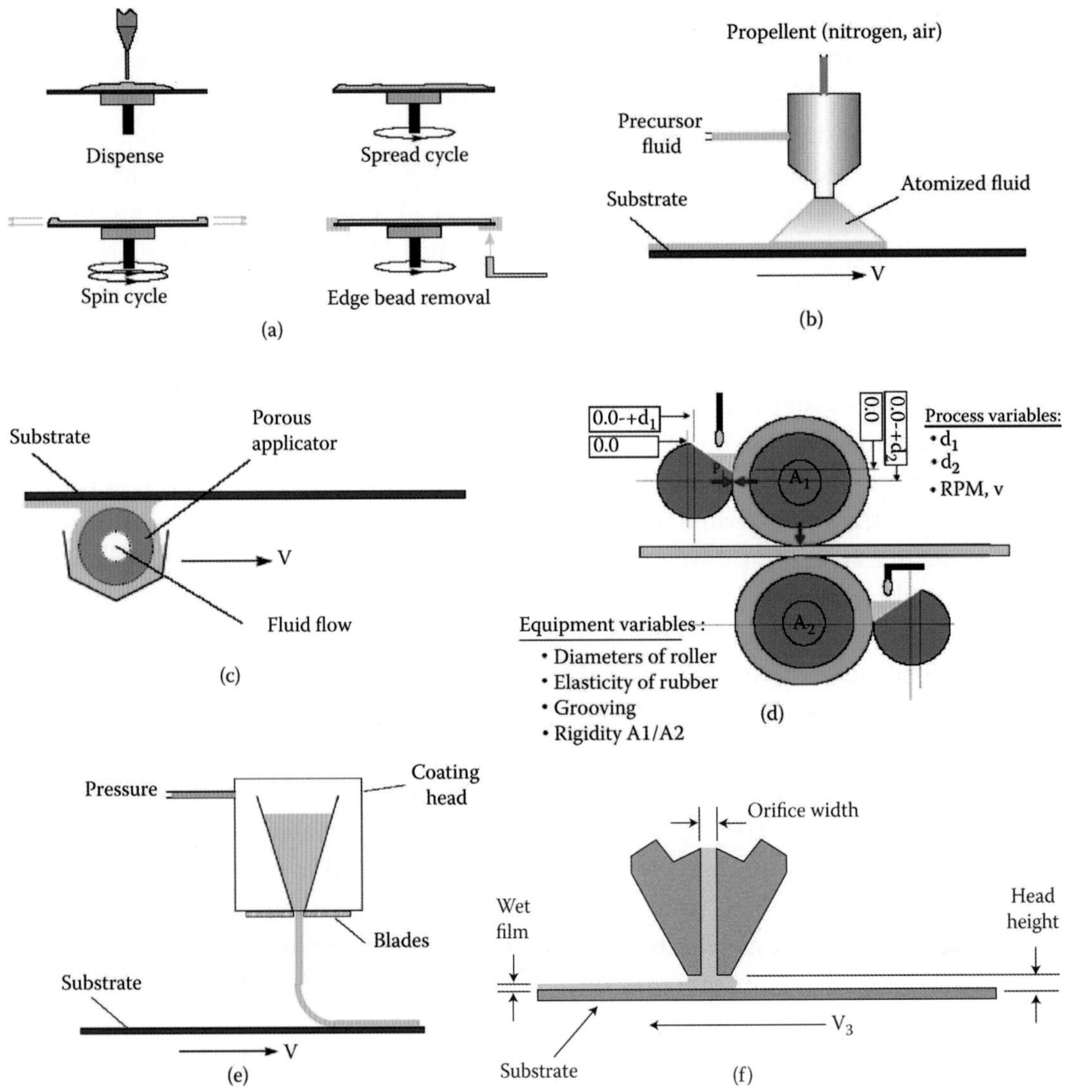


**FIGURE 1.15** Photoresist deposition methods: spin coating (a), spray coating (b), meniscus coating (c), roller coating (d), curtain coating (e), and extrusion coating (f) V represents a movement in the direction of the arrow. (Webster, R.J. 1998. *Thin film polymer dielectrics for high-voltage applications under severe environments.* Master's thesis. Virginia Polytechnic Institute and State University.[20])

the coating method described immediately below), the droplets formed in the spray mist are not as small. This can lead to problems in the coating. To avoid the undesired overspray the spray gun must be kept close to the substrate (e.g., 5 cm compared with 30 cm in electrostatic spray). Compared with spin coating, spray coating does not suffer from the resist thickness variation caused by the centrifugal force because the resist droplets are supposed to stay where they are deposited. Spray coating is often used where thicker photoresist films are required, and the uniformity of spray coatings can be held within 5% tolerance when mechanized spray arms are used. Another major advantage is its ability to coat uniformly over nonuniform surfaces, making the technique appropriate for MEMS. During spray coating, the wafer may be rotated slowly while the swivel arm of the spray-coating unit is moved across the wafer. A low spin speed (30–60 rpm) is used to minimize the centrifugal force effects. The rotation allows for resist coverage in all the angles of MEMS cavities. Importantly, sprayed coatings do not have the internal stress forces that are common to spin-coated films. However, control of the deposited film thickness is not as precise as with spin- and extrusion-coated substrates (see last coating method reviewed in this section), and some waste of photoresist solution must be accepted as part of this processing technique. It is difficult to deposit layers of resist thicker than 20 μm with spray coating. The process can be automated and may coat substrates double-sided.

*Electrostatic Spraying* Electrostatic spraying or electrodeposition is a variant of spray coating. During the atomization of the resist by air or nitrogen pressure, the droplets formed are given a static charge by applying a large voltage (e.g., 20 kV). The charge causes the droplets to repel each other, maintaining the integrity of the mist of resist formed. Because the droplets follow the electric flux lines, the amount of overspray can be minimized by a set of charged wires keeping the cone of spray to within the desired coating area, thus minimizing overspray. The substrate to be coated can be passed horizontally or vertically through the spray. The latter allows for simultaneous coating on both sides of substrates.

*Meniscus Coating* Meniscus coating is another viable coating method, especially attractive for large area substrates such as flat panel displays. The method is illustrated in Figure 1.15c. Photoresist solution is pumped through a porous tube (10-μm pores), and a gravity-assisted laminar flow of the liquid is established around the perimeter of the tube. The inverted substrate is made to touch and pass the solution flowing around the tube, and the meniscus of the resist solution adheres to the surface of the moving substrate. Excess fluid that does not adhere to the substrate is collected and recirculated to the coating head. Important process parameters include the separation of the coating head to the substrate, coating velocity, solids content in the resist solution, and evaporation rate of the carrier solvent. These variables control the coating uniformity and thickness. Two important advantages of meniscus coating are the reduction in coating materials loss (only 5–10% loss) and the avoidance of edge buildup (unlike spin coating). Coating techniques involving intimate contact between the substrate and the photoresist supply source, such as meniscus coating or silkscreen printing, lead to good coverage if the adhesion of the resist to the substrate surface is adequate.

*Silkscreen Printing* Silkscreening can also be used to apply thin, even films of resist for small production runs. Silkscreening of organics is covered in Chapter 8 on chemical, photochemical, and electrochemical-based forming (see Figure 8.10).

*Plasma-Deposited Resist* Another interesting method is the deposition of resist from the gas phase, i.e., plasma-deposited photoresist. In this method, a monomeric coating precursor is required, which may be evaporated at ambient temperatures and forms a polymer after deposition on the substrate. The latter technique is covered in Chapters Seven and Eight (see for example Figure 8.9).

*Electrodeposition (Electrophoresis)*[21] Electrophoretic photoresist deposition (ED) is an appropriate technique for the coating of substrates with extreme topography. The ED process requires electrically conductive substrates and electrical biasing during the resist-coating process. The coating solution contains charged micelles comprising resist, solvent, and dye and photoinitiator molecules. The sizes of the micelles are in the 50–200 nm ranges, and coulombic repulsion keeps the particles sufficiently separated to avoid flocculation and settlement. The electrophoretic process was explained in detail in Volume I, Chapter 6; it suffices to mention here that the solution must be rather resistive for a strong enough electrophoretic field to be established. The ED of the photoresist occurs onto a cathodic (negative) or anodic (positive) polarized conductive substrate.[22] In the case of a cataphoretic resist emulsion, the ionized polymer forms positively charged micelles. When the electric field is applied, the positively charged micelles migrate by electrophoresis toward the cathode. When the micelles reach the cathode, their positive surface charges are neutralized by hydroxide ions produced by the electrolysis of water at that electrode. The micelles then become destabilized and coalesce on the surface of the cathode to form a self-limiting, insulating photoresist film. The growth of the film is self-limiting because the resist layer is nonconductive, and once sufficient resist has been deposited to insulate the substrate surface, no further "plating" can occur. Both positive and negative photoresist chemistries for ED are available (e.g., 2400 ED and Eagle 2100 ED from Shipley Ltd.). Typical coating thicknesses, highly dependent on the voltage and the temperature, are in the range of 5–10 μm, but specific resist systems can be deposited up to about 35 μm on both sides of the substrate simultaneously. The main advantage is that it yields a pore-free deposit of the resist, even at very low thicknesses. Although the most extreme surface topographies can be coated with this technique, this process needs metal plating of substrates and has some associated process complexities, such as the requirement for wafer electrical biasing during the resist-coating process.

*Dip Coating* In dip coating, one dips the substrate into a solution of liquid resist to apply a coating. The method is simple in that the substrate after cleaning and drying is dipped into a tank containing the photoresist and withdrawn slowly. A thin coating of the resist adheres to the substrate and is fixed by mild baking. The rate of withdrawal and the viscosity and solid contents of the resist are adjusted to obtain the desired thickness. However, it is not possible to get thicknesses greater than about 8–10 μm through this method. The technique does not enable much thickness or uniformity control either, although this method is commonly used in small-volume shops or for odd-sized substrates. The benefits of dip coating are obvious. As a double-sided coating method, all types of formats can be coated. It allows coating thick plates, as well as thin sheets and even wires. Simultaneous coating of multiple sheets increases the productivity significantly. A high and consistent coating quality can be ensured because the few and basic coating parameters can easily be maintained. Dip coating, a process often used to make needle-type biosensors, is covered in more detail in Chapter 8. In the dip-coating process, multiple dip coatings and drying cycles follow each other.

*Roller Coating* Roller, curtain, and extrusion coating are all variations of directly casting the coating solution on the substrate. Roller coating allows for single- or double-sided coating, and the equipment is less expensive than that used in curtain coating (see next section). Double-sided roller coating is illustrated in Figure 1.15d. Grooved rubber rolls are used to transfer the liquid resist, which is constantly fed into a nip between a doctor bar or roller and the coating roll, to the substrate surface. The pressure between the doctor and coating roller affects the final resist film thickness, as does the pitch of the grooves on the roller, the nonvolatile content of the resist, and the coating roller-to-substrate pressure (see Figure 1.15d). Roller coating can produce very even coatings of low thickness. Excess resist flows back to a sump for recycling via an automatic viscosity controller and filter unit, thus limiting wastage. Roller coaters are often used in-line with a photoresist dryer. The roller-coating technique can also be useful for the application of thick photoresist films to thin substrates. In general, however, it is incapable of producing uniform coatings less than 5 μm in thickness. Other advantages of roller coating include low photoresist waste and adaptability to automation.

*Curtain Coating* Curtain coating is illustrated in Figure 1.15e; the substrate is moved on a conveyor through a sprayed "curtain" of the resist. The liquid resist is pumped into a head from which the only exit is a thin nip on the head's underside. The resist forced through this nip forms a curtain of resist through which the substrates to be coated are passed. This method can only coat single-sided. The distance from the resist delivery point (the slit) to the substrate is large compared with other coating methods. This method gives high transfer efficiency; however, the thin layer of resist falling through the air results in a significant loss of solvent. Typically, thickness in the 25–60 μm range can be obtained with curtain coating. Undeposited material is recirculated back to the coating head. By careful control of material viscosity, belt speed, and pump speed, reproducible thicknesses can be achieved and maintained over the substrate surface with less than 10% variation in the overall thickness.

*Extrusion Coating* A method closely related to curtain coating is extrusion coating, as illustrated in Figure 1.15f. In extrusion coating, the extrusion head is positioned at a short, predetermined height above the substrate. Film thicknesses from less than 1 μm to greater than 150 μm in a single coating pass have been demonstrated with uniformities better than ±3% on substrates up to 350 mm × 400 mm.[20] An important commercial advantage of extrusion coating is that it can be easily integrated into high throughput microelectronic fabrication lines. However, there are some disadvantages associated with extrusion coating, such as the variation in the substrate surface uniformity, which is transferred to the extrusion head during the deposition process. Furthermore, edge beads form along the leading edge of extrusion-coated substrates, although not to the extent encountered in spin-coated films. A thin curtain of the resist is made to fall on the substrate, which is moving horizontally at a controlled rate. As it falls on the substrate, the liquid curtain gently covers it just like a cloth and gets fixed as a film during conveyor movement and drying. This method obviously requires a lot of controls to achieve good results and, if performed well, can yield very satisfactory coatings. It should be noted, however, that coating the two sides of the laminate requires two operations. Control schemes exist in which the fluid flow into the extrusion head and the internal head pressure are measured and passed to a computer that determines the proper motion profile for the extrusion head, thereby improving the uniformity of the coating along the leading edge. The resist extrusion process has been proposed as a lower cost alternative to spin coating for producing highly uniform thin films on large area substrates.

There is no forced drying during roller, curtain, and extrusion coating other than evaporation. Therefore, the coating material has time to flow and planarize over surface features. The degree of coverage into deep features is highly dependent on the surface wettability and the solution viscosity.

A number of the photoresist deposition processes reviewed above are depicted in Figure 1.15.

In Table 1.5, three coating methods of photoresist—spin, spray, and electrodisposition—on large topography MEMS surfaces are compared.[23] Characteristics of each method, as well as advantages and disadvantages, are outlined.

### *Soft Baking or Prebaking*

After resist coating, the resist still contains up to 15% solvent and may contain built-in stresses. Therefore,

**Table 1.5** Comparison of Spin Coating, Spray Coating, and Electrodeposition (ED) of Photoresists

| | Spin Coating | Spray Coating | ED |
|---|---|---|---|
| Process | Simple<br>Difficult to automate process | Simple<br>Possible for batch production | More complicated (equipment, solution)<br>Batch production |
| Surface materials | Insulating or conductive | Insulating or conductive | Only on conductive layer<br>Electrical connection to wafer |
| Parameters | Viscosity<br>Spin speed | Solid content of solution<br>Resist dispensed volume<br>Scanning speed<br>Spray pressure | Voltage<br>Temperature |
| Photoresist | Several commercially available types | Resist solution with viscosity <20cSt<br>Use much less resist | Special ED resist<br>Resist bath needs frequent refreshing |
| Resist uniformity | Difficult to control<br>Poor reproducibility<br>Dependent on position of cavities in wafer | Controllable<br>Reproducible<br>Independent on position of cavities | Reproducible and good uniformity (best among three methods) |
| Suitable application | Transfer patterns to the bottom of etched cavities<br>One level etched and large cavities preferable | Transfer patterns to bottom of etched cavities<br>Cavities with comparable size preferable | Transfer patterns that run in and across cavities<br>Metal patterning preferable |

*Source:* Boellaard, E., P.N. Pham, L.D.M.v.d. Brekel, and G.J. Bertens. 2003. Electrodepositable photoresist processing for RF-devices in MEMS technology. *MEPTEC Rep* Quarter Two: 29–31; and Pham, N.P., E. Boellaard, P.M. Sarro, and J.N. Burghartz. 2002. *SAFE 2002.* Veldhoven, the Netherlands: STW.[22,23]

the wafers are soft baked (also pre-exposure baked or prebaked) at 90–100°C for about 20 minutes in a convection oven or at 75–85°C for 1–3 minutes with a vacuum hot plate to remove solvents and stress and to promote adhesion of the resist layer to the substrate. This is a critical step in that failure to sufficiently remove the solvent will affect the resist profile. Excessive baking destroys the photoactive compound and reduces sensitivity. Thick resists may benefit from a longer bake time. The resist thickness, for both negative and positive resists, is typically reduced by 10–25% during soft baking. Hot plating the resist is faster, more controllable, and does not trap solvent like convection oven baking. In convection ovens, the solvent at the surface of the resist is evaporated first, and this can cause an impermeable resist skin, trapping the remaining solvent inside. Therefore, heating in a convection oven must proceed slowly to avoid solvent bursts. On a hot plate, a smooth substrate surface is needed for good thermal contact and heating uniformity, and the hot plate should be perfectly horizontal. In this case, the temperature increase starts at the bottom of the wafer and works upward, more thoroughly evaporating the coating solvent and leading to a generally much faster and more suitable approach for automation. Commercially, microwave heating or IR lamps are also used in production lines. The optimization of the prebaking step may substantially increase device yield.

## Exposure and Postexposure Treatment

After soft baking, the resist-coated wafers are transferred to an illumination or exposure system where they are aligned with the features on the mask (see Figure 1.16). For any lithographic technique to be of value, it must provide an alignment technique capable of a superposition precision of mask and wafer that is a small fraction of the minimum feature size of the devices under construction. In the simplest case, an exposure system consists of a UV lamp illuminating the resist-coated wafer through a mask without any lenses between the two. The purpose of the illumination is to deliver light with the proper intensity, directionality, spectral characteristics, and uniformity across the wafer, allowing a nearly perfect transfer or printing of the mask image onto the resist in the form of a latent image.

In photolithography, wavelengths of the light source used for exposure of the resist-coated wafer range from the very short wavelengths of extreme UV (10–14 nm) to deep UV (150–300 nm) to near UV (350–500 nm). In near UV, one typically uses the g-line (435 nm) or i-line (365 nm) of a mercury lamp. The brightness of shorter-wavelength sources is severely reduced compared with that of longer-wavelength sources, and the addition of lenses further reduces the efficiency of the exposure system. For example, the total collected deep UV power for a 1-kW mercury-xenon lamp in the 200–250 nm range is only 30–40 mW; the additional optics absorb more

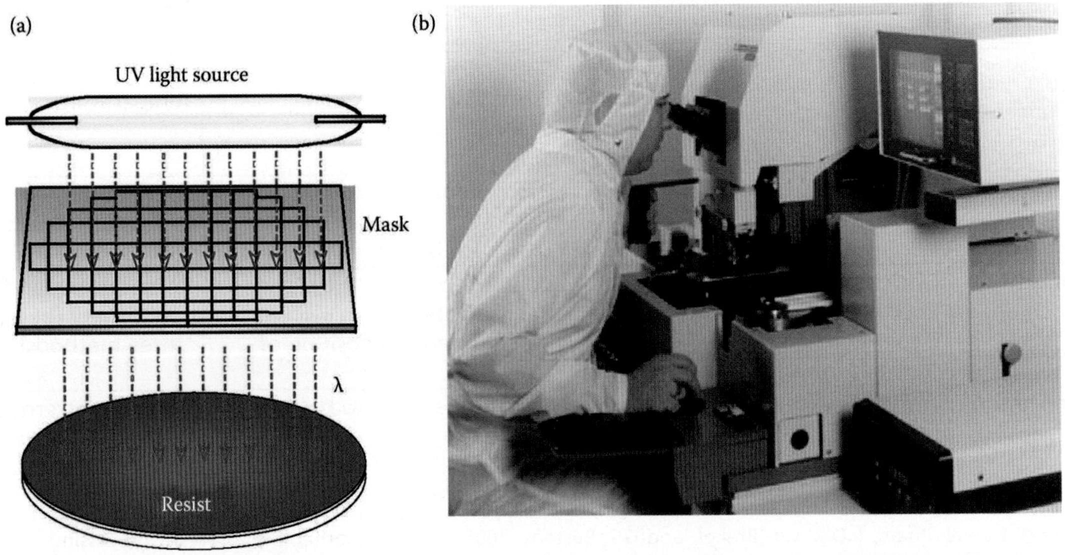


**FIGURE 1.16** (a) Exposure of resist-coated wafer to UV light source. (b) An exposure station.

energy of the short wavelengths passing through them. As a consequence, with shorter wavelengths, higher resist sensitivity is required, and newer deep UV sources that produce a higher flux of deep UV radiation must be used. For example, a KrF excimer laser with a short wavelength of 248 nm and a power of 10–20 W at that wavelength is used. In general, the smallest feature that can be printed using projection lithography is roughly equal to the wavelength of the exposure source; in this example, 248 nm would be expected. The same laser, in combination with sophisticated resolution-enhancing techniques (RETs), may be used to produce more advanced circuits with transistor gate features of 160 nm and less. RET methods (see below) enable one to go a bit beyond the conventional Rayleigh diffraction limit. Other exposure systems now available include two deep UV excimer lasers, the ArF at 193 nm and the $F_2$ at 157 nm, and extreme UV lithography at 13.4 nm. The current generation of lithography uses 193-nm light[24] from ArF lasers. In the case of extreme UV, a plasma or synchrotron source and all-reflective reduction optics (4×) (at this wavelength all materials absorb!) are used. Refractive optical elements are too absorbing at those wavelengths.[25] Extreme UV as a commercial tool is in the exploratory phase only and is discussed in detail in Chapter 2 on next generation lithographies (NGLs). Step and scan printing extreme UV systems are expected to come on-line by the end of this decade. In Figure 1.17 we show the various technologies under investigation for the development of pilot and production lines of large scale integrated (LSI) circuitry.

The incident light intensity (in W/cm$^2$) multiplied by the exposure time (in seconds) gives the incident energy (J/cm$^2$) or dose, D, across the surface of the resist film. Radiation induces a chemical reaction in the exposed areas of the photoresist, altering the solubility of the resist in a solvent either directly or indirectly via a sensitizer. During the latent image-forming reaction, the sensitizer in the resist usually *bleaches*; in other words, exposed resist is rendered transparent to the incoming wavelength. This bleaching allows the use of thick films with high absorbency because light will reach the substrate through the bleached resist. The absorbency of the unexposed resist should not reach 40% to avoid degradation of the image profile through the resist depth because

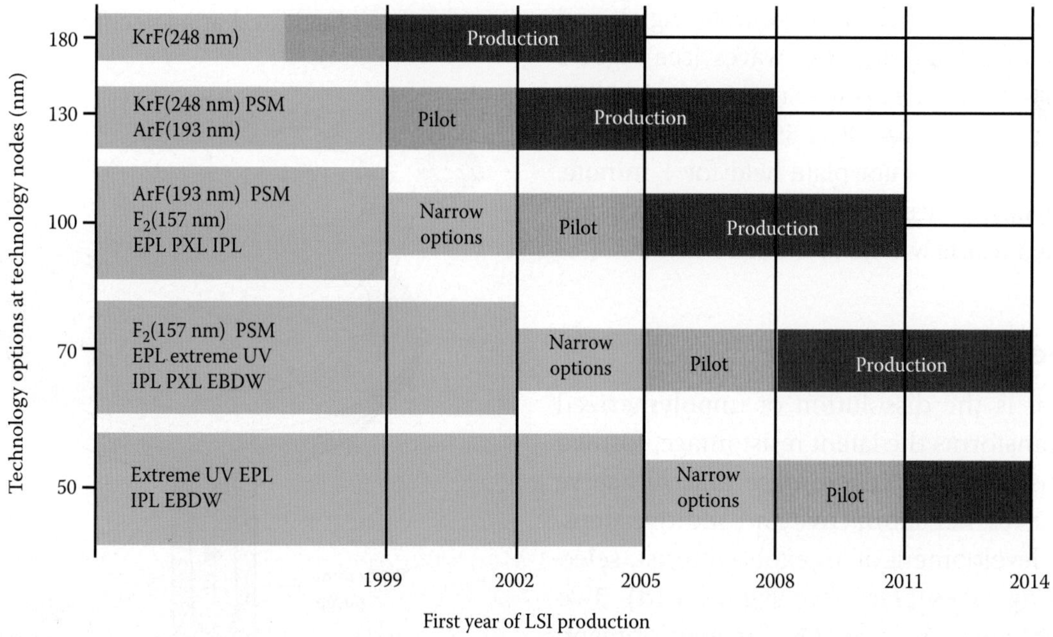


**FIGURE 1.17** Various technologies under investigation for the development of pilot and production lines of LSI circuitry. These include KrF (248 nm), KrF excimer laser lithography with wavelength 248 nm; ArF (193 nm), ArF excimer laser lithography with wavelength 193 nm; $F_2$ (157 nm), $F_2$ excimer laser lithography with a wavelength of 157 nm; PSM, phase-shifting mask applied to KrF, ArF, and $F_2$; EPL, electron-projection lithography; PXL, proximity X-ray lithography; IPL, ion-projection lithography; EBDW, electron-beam direct writing. (Ito, T., and S. Okazaki. 2000. Pushing the limits of lithography. *Nature* 406:1027–31.[26])

too large a percentage of the light is absorbed in the top layer. On the other hand, with the absorbency far less than 40%, exposure times required to form the image become too long. The smaller the dose needed to "write" or "print" the mask features onto the resist layer with good resolution, the better the lithographic sensitivity of the resist.

A postexposure treatment of the exposed photoresist is often desired because the reactions initiated during exposure might not have run to completion. To halt the reactions or to induce new ones, several postexposure treatments are in use: postexposure baking (PEB), flood exposure with other types of radiation, treatment with a reactive gas, and vacuum treatment. PEB (sometimes in a vacuum) and treatment with reactive gas are used in image reversal and dry resist development. In the case of a chemically amplified resist, the PEB is most critical. Although reactions induced by the catalyst that forms during exposure take place at room temperature, their rate is greatly increased by baking at 100°C. The precise control of this type of PEB critically determines the subsequent development itself. A PEB improves line-width control by improving adhesion, reduction of scumming (resist left behind after development), increasing contrast and resist profile (higher edge-wall angle), and reducing the effects of standing waves (see Figures 1.69c and 1.70) in regular positive resist. A typical PEB used for the positive resist OCG 895i involves a temperature of 115°C on a hot plate held for 1 minute. Image reversal, dry resist development, and chemically amplified resists will be discussed below.

## Development

Development is the dissolution of unpolymerized resist that transforms the latent resist image, formed during exposure, into a relief image that will serve as a mask for further subtractive and additive steps. During the development of an exposed resist, selective dissolving takes place (see Figure 1.1d). Two main technologies are available for development: wet development, illustrated in Figure 1.18, is widely used in circuit and miniaturization manufacture in general, and dry development, which is starting to replace wet development for some of the ultimate line-width resolution applications.

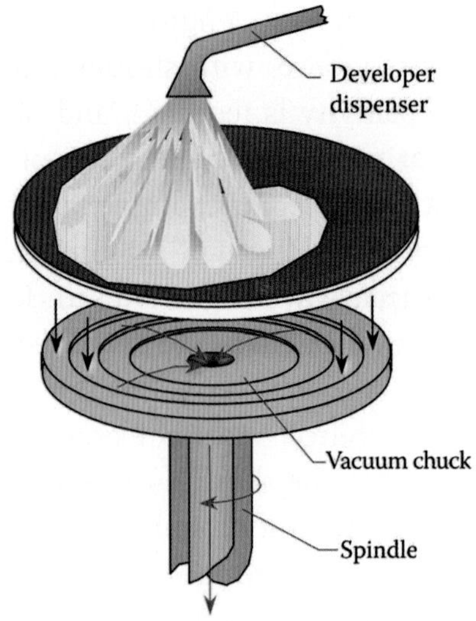


**FIGURE 1.18** Wet development.

Wet development by solvents can be based on at least three different types of exposure-induced changes: variation in molecular weight of the polymers (by cross-linking or by chain scission), reactivity change, and polarity change.[27] Two main types of wet development setups are used: immersion (Figure 1.18) and spray developers (Figure 1.19). During batch immersion developing, cassette-loaded wafers are batch-immersed for a timed period in a developer bath and agitated at a specific temperature.

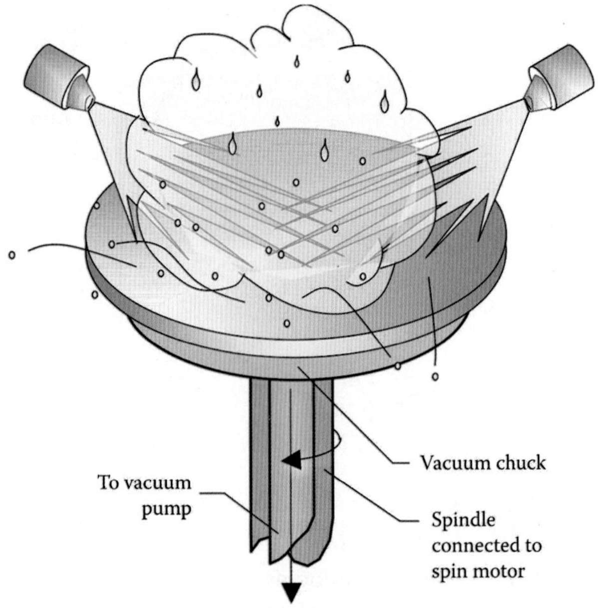


**FIGURE 1.19** Spray developer: fresh developing solution is directed across wafer surfaces by a fan-type spraying nozzle. The renewal of developer allows uniform developer strength to be maintained.

During batch spray development, fan-type sprayers direct fresh developing solution across wafer surfaces. Positive resists are typically developed in aqueous alkaline solutions, and negative resists are developed in organic ones. For alkaline developers, simple solutions of NaOH (Shipley 351) or KOH (AZ/Clariant 400K) may be used, but because of the possibility of mobile ion contamination in metal oxide semiconductor (MOS) devices, developers free of metal ions are preferred. The latter are usually tetramethyl ammonium hydroxide (TMAH) based (Shipley CD-26, MF-321, or AZ/Clariant 300-MIF). Each developer has a different dilution, and some require longer development times than others. Developers are generally matched to a type of photoresist. Although they may be interchangeable to some extent, changing the type of developer used in a process will usually change the exposure time necessary to resolve the pattern. It should be noted that all of the above-mentioned alkaline developers etch aluminum. If the micromachine or microchip under construction contains aluminum features, Shipley Microposit Developer Concentrate, a mixture of proprietary alkaline salts (mostly phosphates) with the slowest aluminum etch rate, is a better developer choice.

Aqueous development is highly favored for health reasons. The aqueous development rate depends on the pH of the developer and the temperature, which needs to be controlled to within ±0.5°C.[19] Surfactants and other wetting agents added to the developer ensure uniform wetting, and buffers provide a more stable operating window and a longer lifetime. The more newly developed negative resists also may be developed in aqueous solutions.

The use of organic solvents leads to some swelling of the resist (especially for negative resists) and loss of adhesion of the resist to the substrate. Dry development overcomes these problems, as it is based either on a vapor phase process or a plasma.[28] In the latter, oxygen-reactive ion etching is used to develop the latent image. The image formed during exposure exhibits a differential etch rate to oxygen-reactive ion etching rather than differential solubility to a solvent.[19] Dry developed resists should not be confused with dry film resists, which are resists that come in film form and are laminated onto a substrate rather than spin coated.

With continued pressure by the U.S. Environmental Protection Agency for a cleaner environment, dry development and dry etching (see Chapter 3) are becoming the predominant technologies to use. Dry developed resists, such as the DESIRE process, where the surface of the exposed resist is treated with a silicon-containing reagent, will be discussed below.

## Descumming and Postbaking

A mild oxygen plasma treatment, so-called *descumming*, removes unwanted resist left behind after development. Negative and, to a lesser degree, positive resists leave a thin polymer film at the resist/substrate interface. The problem is most severe in small (<1 μm) high-aspect-ratio structures where the mass transfer of a wet developer is poor. Patterned resist areas are also thinned in the descumming process, but this is usually of little consequence.

Before etching the substrate or adding a material, the wafer must be postbaked. Postbaking or hard baking removes residual coating solvent and developer and anneals the film to promote interfacial adhesion of the resist that has been weakened either by developer penetration along the resist/substrate interface or by swelling of the resist (mainly for negative resists). Hard baking also improves the hardness of the film and avoids solvent bursts during vacuum processing. Improved hardness increases the resistance of the resist to subsequent etching steps. Postbaking frequently occurs at higher temperatures (120°C) and for longer times (e.g., 20 min) than soft or prebaking. The major limitation for heat application is excessive flow or melt, which degrades wall profile angles and makes it more difficult to remove the resist. Postbake induces some stress and resist shrinkage. Special care needs to be taken when the baking temperature is above the glass transition temperature, $T_g$, when impurities are easily incorporated into the resist because of the plastic flow of the resist. Positive resists withstand higher heating temperatures than negative resists, but their stripping proves more difficult. Descumming and postbaking both follow step d in Figure 1.1.

Resist does not withstand long exposure to etchants well. As a consequence, with 1:7 buffered

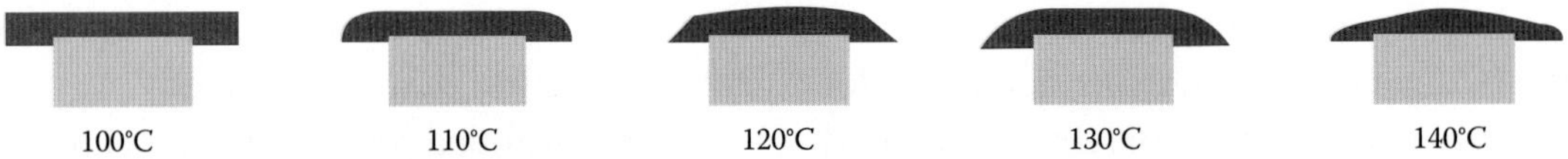


**FIGURE 1.20** Tailoring resist sidewalls through resist reflow.

HF (a mixture of 1 part 49% aqueous HF solution and 7 parts $NH_4F$ that is used to strip $SiO_2$), the postbake sometimes is repeated after 5 minutes of etching to prolong the lifetime of the resist layer (see Figure 1.1e). Also, postbaking should be prolonged before electroplating but is not needed for processes in which a soft resist is desired, e.g., in metal liftoff patterning (see Figure 1.37). As mentioned before, photoresist will undergo plastic flow with sufficient time and/or temperature, and this resist reflow may be used for tailoring resist sidewalls (Figure 1.20).

## Resists

The principal components of photoresists are a polymer (base resin), a sensitizer, and a casting solvent. The polymer changes structure when exposed to radiation; the solvent allows spin application and formation of thin layers on the wafer surface; sensitizers control the chemical reactions in the polymeric phase. Resists without sensitizers are single-component or one-component systems, whereas sensitizer-based resists are two-component systems. Solvent and other potential additives do not directly relate to the photoactivity of the resist.

### *Resist Tone*

If the photoresist is of the type called *positive* (also *positive tone*), the photochemical reaction during exposure of a resist weakens the polymer by rupture or scission of the main and side polymer chains, and the exposed resist becomes more soluble in developing solutions (e.g., 10 times more soluble). In other words, the development rate, $R$, for the exposed resist is about 10 times faster than the development rate, $R_0$, for the unexposed resist. If the photoresist is of the type called *negative* (also *negative tone*), the reaction strengthens the polymer by random cross-linkage of main chains or pendant side chains, becoming less soluble (slower dissolving). Exposure, development, and pattern-transfer sequences for negative and positive resists are shown in Figure 1.21. Figure 1.22 illustrates chain scission (positive resists) and cross-linking (negative resists).

*Positive Resists* Two well-known families of positive photoresists are the single component poly (methylmethacrylate) (PMMA) (Figure 1.23) resists and the two-component diazonaphtoquinone (DNQ)

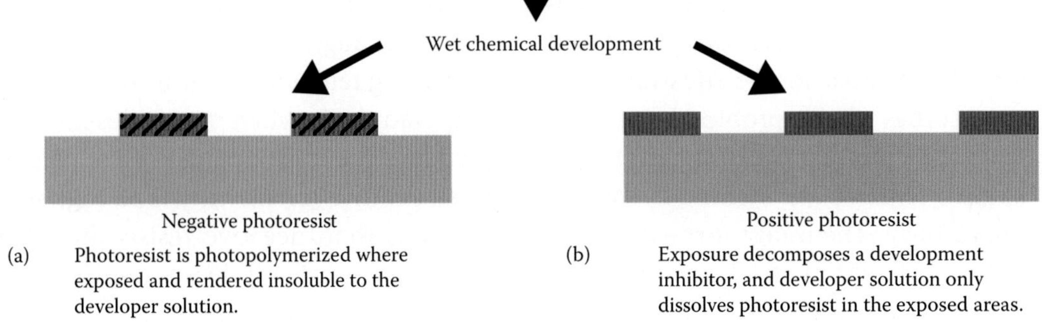

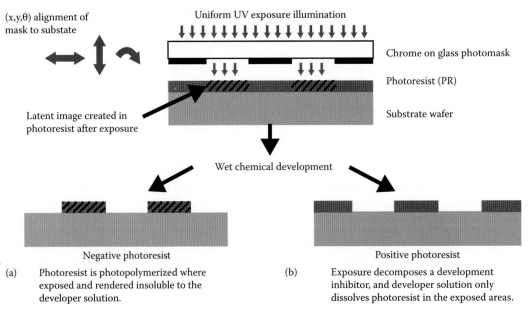


**FIGURE 1.21** Positive and negative resist: exposure, development, and pattern transfer. (a) Negative resists remain in the exposed region. (b) Positive resists develop in the exposed region.

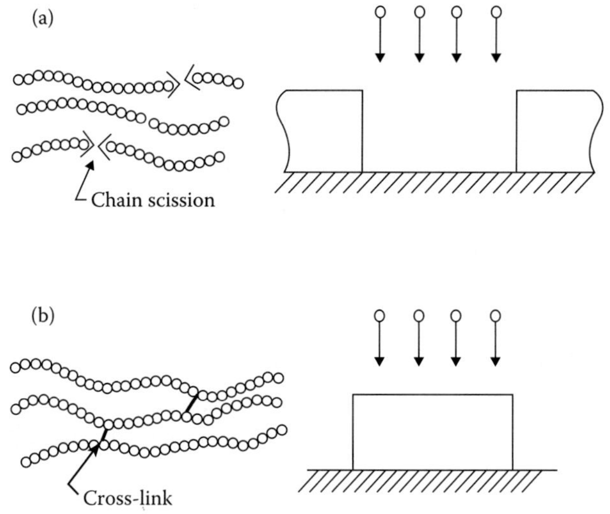


**FIGURE 1.22** (a) Polymer chain scission (positive resists) and (b) cross-linking (negative resists).

(Figure 1.24) resists composed of a photoactive component such as DNQ ester (20–50 wt%) and a phenolic novolak resin.

PMMA becomes soluble through chain scission under deep UV illumination. The maximum sensitivity of PMMA lies at 220 nm; at greater than 240 nm the resist becomes insensitive. PMMA resin by itself constitutes a rather insensitive or slow deep UV photoresist requiring doses >250 mJ/cm$^2$. Exposure times of tens of minutes were required with the earliest PMMAs available.[28] By adding a photosensitizer such as t-butyl benzoic acid, the UV spectral absorbency of PMMA is increased, and a 150-mJ/cm$^2$ lithographic sensitivity can be obtained. PMMA is also used in electron beam, ion beam, and x-ray lithography. Although the resolution of PMMA is very good, its plasma etch tolerance is very low.

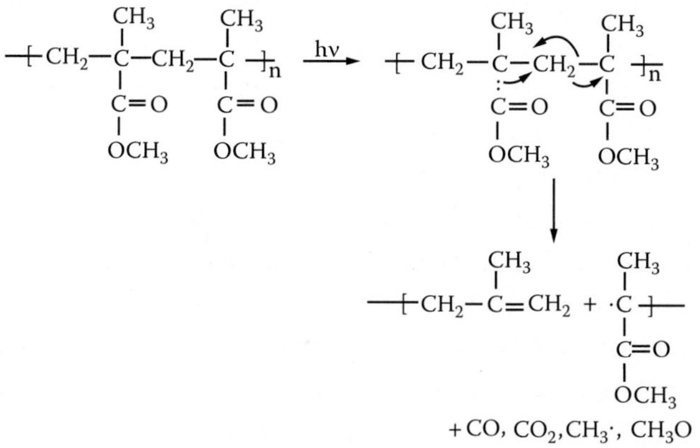

**FIGURE 1.23** Poly(methylmethacrylate) or PMMA. Photo-induced chain scission of PMMA resist.

Also, dissociation of PMMA changes the chemistry of a plasma etch and often leads to polymeric deposits on the surface of the substrate.

The DNQ resist system is a "workhorse," near-UV, two-component positive resists, which photochemically transform into a polar, base-soluble product.[24] The hydrophilic novolak resin (N) is in itself alkali soluble because of the OH groups. Diazonaphthaquinone (DQ) is a hydrophobic and nonionizable compound, and when phenolic resins are impregnated with DQ, they become hydrophobic and are rendered insoluble. The addition of 20–50 wt% DQ forms a complex with the phenol groups of the novolak resin and reduces the solubility rate of the unexposed resist to less than 1–2 nm·s$^{-1}$. During exposure, DNQ undergoes photolysis that destroys the inhibitory effect of DQ on film dissolution (Figure 1.24). The photolysis causes the DNQ to undergo a reaction forming a base-soluble carboxylic acid that can be rapidly developed in aqueous solution of hydroxide ions (e.g., 1% NaOH). In contrast with cross-linked resists, the film solubility is controlled by chemical and polarity difference rather than molecular size. The matrix novolak resin itself is a condensation product of a cresol

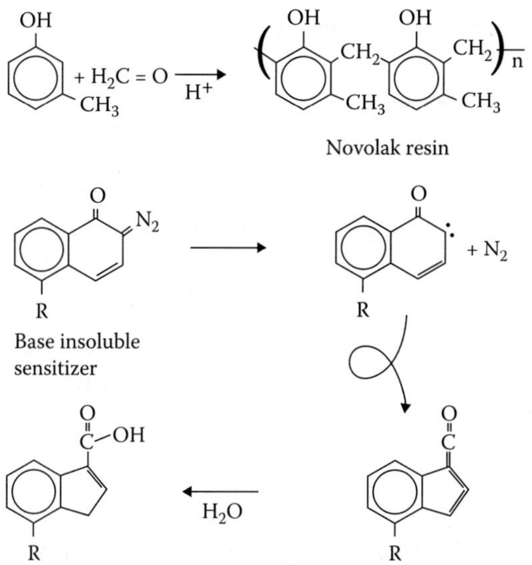


**FIGURE 1.24** Phenolic novolak resin and the influence of diazonaphthoquinone ester (DNQ) on it. The novolak matrix resin is prepared by acid copolymerization of cresol and formaldehyde. The base-insoluble sensitizer, a DNQ, undergoes photolysis to produce a carbine, which then undergoes a rearrangement to form a ketene. The ketene reacts with water, present in the film, to form a base-soluble, indenecarboxylic acid photoproduct.

isomer (*para-cresol*) and formaldehyde consisting of hydrocarbon rings with two methyl groups and 1 OH group attached. Phenolic resins are readily cross-linked by thermal activation into rigid forms (bakelite was the first thermosetting plastic). A novolak resin absorbs light less than 300 nm, and the DNQ addition adds an absorption region around 400 nm. The 365-, 405-, and 435-nm mercury lines can all be used for exposure of DNQ. The intense absorption of aromatic molecules prevents the use of this resist at exposing wavelengths less than about 300 nm; at those shorter wavelengths, linear acrylate and methacrylate copolymers have the advantage. Positive attributes of novolak-based resist are that the unexposed areas are essentially unchanged by the presence of the developer. Thus, line width and shape of a pattern are precisely retained. Moreover, novolak is a long-chain aromatic ring polymer that is fairly resistant to chemical attack. Therefore, the photoresist is a good mask for the subsequent plasma etching. Moreover, unlike radical-based systems, it shows no oxygen inhibition.

Most positive resists are soluble in strongly alkaline solutions (a fact which is taken advantage of for stripping of the resist; see Figure 1.1f and text below) and develop in mildly alkaline ones (as shown in Figure 1.1d). Some typical industrially used developers for positive resists are KOH (e.g., a 0.05–0.5 N aqueous solution and a surfactant), TMAH, ketones, and acetates. As mentioned above, besides changing the molecular weight of the resist, radiation-induced reactions may also change the resist's solubility by altering its hydrophilicity (polarity) or its reactivity. Typical casting solvents for positive resists are Cellosolve* acetate, methyl Cellosolve, and aromatic hydrocarbons.

*Negative Resists* The first negative photoresists were based on free radical-initiated photo cross-linking processes of main or pendant polymer side chains, rendering the exposed parts insoluble. They were the very first types of resists used to pattern semiconductor devices and still comprise the largest segment of the overall photoresist industry, being widely used to define circuitry in printed wiring boards (PWBs).[29] A negative photoresist becomes insoluble in organic (more traditional negative resists) or water-based developers (newer negative resist systems) on exposure to UV radiation. The insoluble layer forms a "negative" pattern that is used as a stencil (usually temporarily) to delineate many levels of circuitry in semiconductors, MEMS, and PWBs. The insolubilization of radiated negative resists can be achieved in one of two ways: the negative resist material increases in molecular weight through UV exposure (traditional negative resists), or it is photochemically transformed to form new insoluble products (newer negative resist products). The increase in molecular weight is generally accomplished through photoinitiators that generate free radicals or strong acids, facilitating polymeric cross-linking or the photopolymerization of monomeric or oligomeric species. Photochemical transformation of negative photoresists may also lead to hydrophobic or hydrophilic groups, which provide another means of inducing preferential solubility between the exposed and unexposed resist film.

Commonly used negative-acting, two-component resists are bis(aryl)azide rubber resists (Figure 1.25), whose matrix resin is cyclized poly(*cis*-isoprene), a synthetic rubber. Cyclized rubber is obtained by acid treatment of poly(*cis*-isoprene), a process that leads to some ring formation in the polymer and stiffens it, thereby raising its glass transition temperature ($T_g$). For use in a photoresist, a resin should have a molecular weight of around 100,000–200,000 to ensure proper viscosity, melting point, softening point, and stiffness. To get a molecular weight of 100,000 using isoprene (M = 68.12 g/mol), one needs chains of an average length of n = 100,000/68.12 or 1,468 monomer units. This represents a molecule that is too long for proper photolithographic resolution, so the chains need to coil up to make them shorter and to increase the mechanical stiffness of the resist. Therefore, *cis*-isoprene is used to curl up the chains into rings; *trans*-isoprene would not work because of steric hindrance of the $CH_3$ groups.

The earliest negative photoresists, introduced in the 1960s, were based on the photo cross-linking of

* Cellosolve is a trade name for solvents based on esters of ethylene glycol; these solvents have been identified as possible carcinogens.

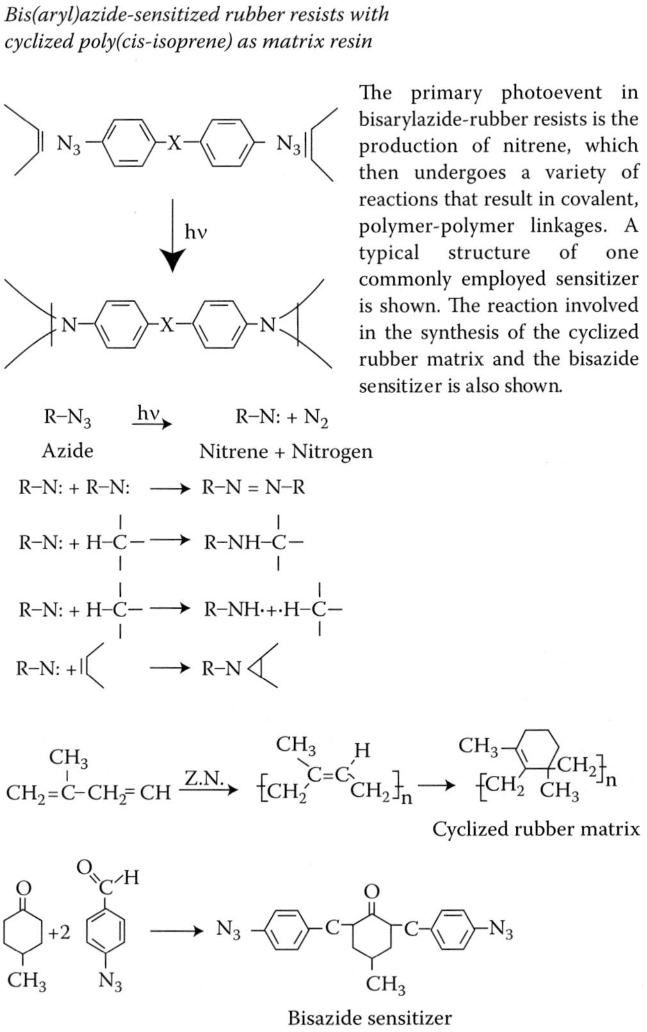
*Bis(aryl)azide-sensitized rubber resists with cyclized poly(cis-isoprene) as matrix resin*
The primary photoevent in bisarylazide-rubber resists is the production of nitrene, which then undergoes a variety of reactions that result in covalent, polymer-polymer linkages. A typical structure of one commonly employed sensitizer is shown. The reaction involved in the synthesis of the cyclized rubber matrix and the bisazide sensitizer is also shown.
hv
Azide
Nitrene + Nitrogen
Z.N.
Cyclized rubber matrix
Bisazide sensitizer

**FIGURE 1.25** Bis(aryl)azide-sensitized rubber resists with cyclized poly(*cis*-isoprene) as matrix resin. The primary photoevent in bis(aryl)azide rubber resists is the production of nitrene, which then undergoes a variety of reactions that result in covalent, polymer-polymer linkages. A typical structure of one commonly used sensitizer is shown.

"cyclized rubber" using an additive that contained two azide groups such as bis(aryl)azide. When this bis-azide is photolyzed, it loses nitrogen and produces two highly reactive nitrene moieties.

The nitrene intermediates undergo a series of reactions that results in the cross-linking of the resin. The nitrenes can form bonds with the cyclized rubber in a variety of ways, the most common of which is aziridine ring formation. Oxidation of the nitrene intermediate, with oxygen from the ambient or dissolved in the polymer to form azoxy and nitrose products, often competes with the addition and insertion reactions, reducing the extent of the polymer-polymer cross-linking. This phenomenon is called oxygen inhibition. In other words, polymerization can be inhibited by the quenching of the cross-linking reactions through scavenging of the nitrene photoproduct by oxygen. This competing reaction represents a disadvantage, as exposure must be carried out under a nitrogen blanket or in a vacuum. This type of resist was successfully used in the semiconductor industry when printed in "hard contact," minimizing the effect of the oxygen inhibition. With the introduction of soft contact (proximity mode) and off-contact (projection mode), the resist showed severe film thinning (scalping) as a direct consequence of the reduced cross-linking density near the air/polymer interface. Today, cyclic poly(*cis*-isoprene)-based resists play only a small roll in the semiconductor industry.

Another disadvantage of negative resists is that the resolution is limited by film thickness. The cross-linking process starts topside, where the light hits the resist first. Consequently, overexposure is needed to render the resist insoluble at the substrate interface. The greater the desired resist thickness, the greater the overdose needed for complete polymerization and the larger the scattered radiation. Scattered radiation at the resist/substrate interface in turn reduces the obtainable resolution. Moreover, the organic solvent developer swells the cross-linked negative image, which further degrades the resolution. In a practical situation, this leads to a 2–3-μm maximum resolution in a 1-μm-thick resist layer. To improve the resolution of a negative resist, thinner resist layers can be used; however, when using thin layers of negative resist, pinholes become problematic. Xylene is the most commonly used aromatic solvent for negative resists, although almost any organic solvent will do. Aromatic solvent developers may pose environmental, health, and safety concerns. Newer negative resists are water developable.

An example of a commercial, two-component negative photoresist is the Kodak KTFR [an azide-sensitized poly(isoprene) rubber] with a lithographic sensitivity (also photospeed) of 75–125 $mJ/cm^2$. Negative photoresists, in general, adhere very well to the substrate, and a vast amount of compositions are available (stemming from research and development work in paints, UV curing inks, and adhesives all based on polymerization hardening). Negative resists are highly resistant to acid and alkaline aqueous solutions, as well as to oxidizing agents. As a consequence, a given thickness of negative resist is more chemically resistant than a corresponding thickness of positive resist. This chemical resistance ensures better retention of resist features even during a long, aggressive wet or dry etch. Negative resists also are more sensitive than positive resists but exhibit a lower contrast ($\gamma$ smaller; see "Contrast and Experimental Determination of Lithographic Sensitivity," this chapter).

A comparison of negative and positive photoresist features is presented in Table 1.6. This table is not exhaustive and is meant only as a practical guide for selection of a resist tone. The choice of whether to use a negative or a positive resist system depends on the needs of the specific application, such as resolution, ease of processing, speed, and cost. The choice of resist tone will even depend on the specific intended pattern geometry, which is known as the *optical proximity effect*. For example, an isolated single line most easily resolves in a negative resist (higher resolution line), whereas an isolated hole or trench is most easily defined in a positive resist. Because traditional negative resists used to have a line-width limit of only about 2–3 μm and because the industry moved away from organic solvent-based systems in favor of less toxic, water-based developers, positive resists gained in popularity. However, traditional negative resists continue to be used in the production of PWBs and low-cost, high-volume chips

Table 1.6 Comparison of Traditional* Negative and Positive Photoresists

| Characteristic | Resist Type | |
|---|---|---|
| | Positive Resist | Negative Resist |
| Adhesion to silicon | Fair (priming required) | Excellent (priming not required) |
| Available compositions | Many | Vast |
| Baking | In air (+) | In nitrogen (–) |
| Contrast $\gamma$ | Higher, e.g., 2.2 | Lower, e.g., 1.5 |
| Cost | More expensive | Less expensive |
| Developer | Temperature sensitive (–) and aqueous based (Ecologically sound) | Temperature insensitive (+) and organic solvent (–) |
| Developer process window | Small | Very wide, insensitive to overdeveloping |
| Influence of oxygen | No (+) | Yes (–) |
| Liftoff | Yes [usually with multiple layer resist (MLR)] | Yes [even in single layer resist (SLR)] |
| Mask type | Dark-field: lower defects | Clear-field: higher defects |
| Opaque dirt on clear portion of mask | Not very sensitive to it | Causes printing of pinholes |
| Photospeed | Slower | Faster |
| Pinhole count | Higher | Lower |
| Pinholes in mask | Prints mask pinholes | Not so sensitive to mask pinholes |
| Plasma etch resistance | Not very good | Very good |
| Proximity effect | Prints isolated holes or trenches better | Prints isolated lines better |
| Residue after development | Mostly at <1 μm and high aspect ratio | Often a problem |
| Resolution | High | Low (>1 μm) |
| Sensitizer quantum yield $\Phi$ | 0.2–0.3 | 0.5–1 |
| Step coverage | Better | Lower |
| Strippers of resist over<br>Oxide steps<br>Metal steps | <br>Acid<br>Simple solvents | <br>Acid<br>Chlorinated solvent compounds |
| Swelling in developer | No | Yes |
| Thermal stability | Good | Fair |
| Wet chemical resistance | Fair | Excellent |

*Newer resist systems are discussed under "Photolithography Resolution Enhancement Technology," this chapter.

because they require only small amounts of sensitizers and therefore are substantially less expensive than positive resists. Moreover, great progress has been made in improving the resolution of new types of water-soluble negative resists. These are used in new generations of ICs and in high-aspect-ratio miniaturized systems.[29,30] In working with different resists, it is also important to be aware of such properties as shelf life, flash point, and threshold limit value (TLV) rating. The flash point is the temperature at which the resist vapors ignite in the presence of an open flame. The TLV is the toxicity rating that specifies the maximum ambient concentration (in parts per million) to which a worker can be safely exposed during a normal workday.

Table 1.7 lists some common positive and negative resists used in various lithography strategies along with their lithographic sensitivities. For charged particles (e-beam lithography and ion-beam lithography), sensitivity is expressed in coulombs per centimeter square ($C/cm^2$); for photons (optical and x-ray), joules per centimeter square ($J/cm^2$) is used. Ideally, in charged-particle lithography, one should select a resist with sensitivity in the range of $10^{-5}$–$10^{-7}$ $C/cm^2$, and in photon lithography, 10–100 $mJ/cm^2$, to minimize the exposure duration.

Resist research and development are producing better resolution in both positive and negative resists. Results have been especially impressive in the latter. Negative resists used to be relegated to low-resolution chips and PWBs, but newer systems offer wider processing latitude and higher resolution. Progress has been so swift that new negative resists are now used in advanced complementary metal oxide silicon (CMOS) logic device manufacture. Several of these new negative resists also enable very high-aspect-ratio microfabrication.

Finally, in Table 1.8, we list the development of photoresists as they keep up with the scale of integration in ICs. In the inset of Table 1.8, we show the photoresist evolution with the type of lithography. In this inset, SCALPEL, ion-projection lithography, and x-ray lithographies are next generation lithographies (NGLs) that will be reviewed in Chapter 2; PSM stands for phase-shifting masks and OAI for off-axis illumination, two resolution-enhancing techniques (RETs) that are reviewed further below.

### *Permanent Resists and Dry Resists*

*Permanent Resists* Resists typically are removed (stripped) once they have served their function as temporary stencils. Some negative resists, hardened through UV exposure, are used as permanent components of miniature devices. Two prominent examples in this category are polyimide and SU-8-based resists.

Polyimides are a class of polymers synthesized from two monomers: a dianhydride and a diamine.

**Table 1.7** Typical Negative and Positive Photoresists and Their Lithographic Sensitivity

| Class of Resist | Resist Name | Tone (Polarity) | Lithographic Sensitivity |
|---|---|---|---|
| Optical | CAMP-6 (OCG) | Positive | 100 $mJ/cm^2$ |
| | APEX-E (IBM and Shipley) | Positive | 75 $mJ/cm^2$ |
| | Kodak 747 | Negative | 9 $mJ/cm^2$ |
| | AZ-1350J | Positive | 90 $mJ/cm^2$ |
| | PR102 | Positive | 140 $mJ/cm^2$ |
| | XP-2198 (Shipley) | Positive | 30 $mJ/cm^2$ |
| | KRF (from UCB-JSR) | Negative | 20–30 $mJ/cm^2$ |
| e-Beam | COP [Copolymer-(α-cyano ethyl acrylate-α-amido ethyl acrylate)] | Negative | 0.5 $\mu C/cm^2$ |
| | GeSe (germanium selenide) | Negative | 80 $\mu C/cm^2$ |
| | PBS [poly-(butene-1-sulfone)] | Positive | 1 $\mu C/cm^2$ |
| | PMMA | Positive | 100 $\mu C/cm^2$ |
| X-ray | COP | Negative | 100 $mJ/cm^2$ |
| | DCOPA | Negative | 14 $mJ/cm^2$ |
| | PBS | Positive | 170 $mJ/cm^2$ |
| | PMMA | Positive | 6500 $mJ/cm^2$ |

TABLE 1.8 Development of Photoresists during the Past 38 Years

| Year | Scale of Integration | Minimum Line Width | Photoresist |
|---|---|---|---|
| 1970 | 1 Kb | 10 μm | Rubber-based negative |
| 1973 | 4 Kb | 8 μm | Rubber-based negative |
| 1976 | 16 Kb | 5 μm | Rubber-based negative |
| 1979 | 64 Kb | 2 μm | Rubber-based negative |
| 1982 | 256 Kb | 2 μm | Novolak-type positive |
| 1985 | 1 Mb | 1.2 μm | High-resolution positive photoresist of DQ-novolak type |
| 1988 | 4 Mb | 0.8 μm | High-resolution positive photoresist and multilayer process |
| 1992 | 16 Mb | 0.5 μm | Deep UV resist, multilayer process |
| 1995 | 256 Mb | 0.25 μm | I-line (356 nm) resist |
| 1999 | 1 Gb | 0.18 μm | KrF (248 nm) resist |
| 2002 | 4 Gb | 0.13 μm | ArF (248/193 nm) resist |
| 2005 | 16 Gb | 0.1 μm | F2 (193/157 nm) resist |
| 2008 | 64 Gb | 0.07 μm | Extreme UV (13 nm) resist |

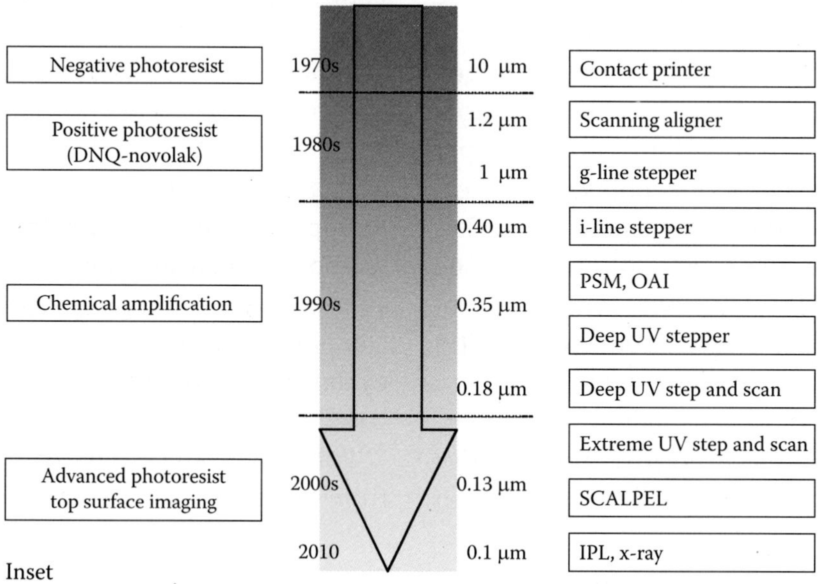


Inset

Commercial products are supplied as soluble polyamic acid (PAA) intermediates, which undergo a thermal imidization with the evolution of water to form an insoluble polyimide (Figure 1.26). Polyimide is used as an interlayer dielectric because of its desirable characteristics such as a high thermal stability with a glass transition temperature, $T_g >$ 300°C, a low dielectric constant ε, good chemical stability, and ease of processing. However, its tendency to absorb moisture causes the polyimide layers to swell and their dielectric constants to increase significantly. Various polyimide analogs have been developed to avoid the latter undesirable property.[31] A common approach to reduce the dielectric constant of polyimides is the inclusion of organofluorine components, in the form of pendant perfluoroalkyl groups. The polyamic acid precursor for fluorinated polyimides is commonly based on hexafluorodianhydrideoxydianiline (HFDA-ODA). The inclusion of fluorinated monomers in the polyimide backbone reduces moisture absorption and the dielectric constant. Unfortunately, these modified polyimides generally show an increased susceptibility to chemical attack, making their use in multilayer fabrication questionable. However, newer formulations have improved the chemical resistance of polyimides and have demonstrated a unique wet etch capability for making vias with aspect ratios approaching 1.2:1. Polyimide can be "wet etched" using a strong alkaline solution

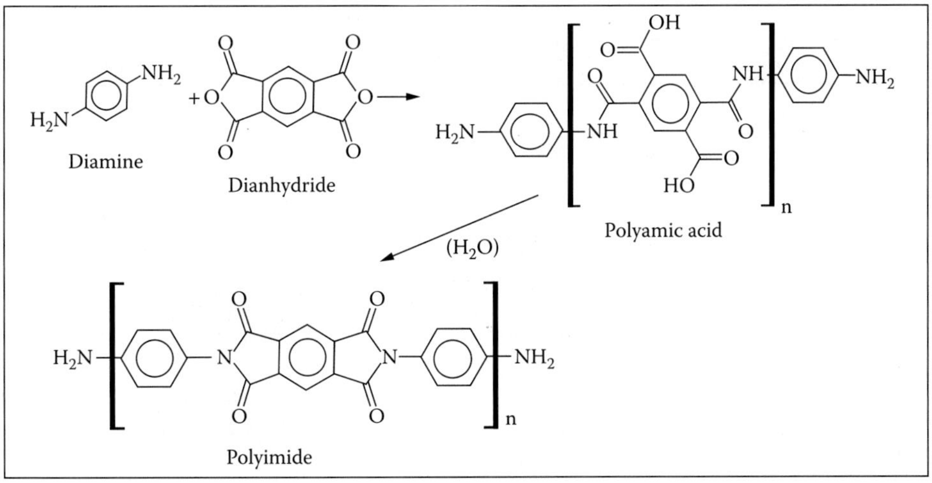


**FIGURE 1.26** Typical imidization process used to form polyimides from polyamic acid.

such as TMAH. For dry etching of polyimide, one can use plasma-enhanced chemical vapor deposition (PECVD) silicon nitride as a mask with an $O_2$ plasma etch (possibly adding some $CF_4$). Tougher polyimide formulations are commercially available in polyamic acid form from Amoco under the trade name Ultradel. They have a slightly modified chemical structure based on hexafluorodianhydride-aminophenoxy-biphenyl (HFDA-APBP). The newest polyimides are extremely versatile materials and are used, for example, as stress-release layers in multichip modules, low dielectric insulation layers,[32] and even as optical waveguides based on fluorinated polyimides[33] (Figure 1.27; see also Chapter 3 on dry etching, Example 3.4). Ultradel 9000D is a soluble, preimidized, fluorinated polymer with properties optimized for integrated optical applications.

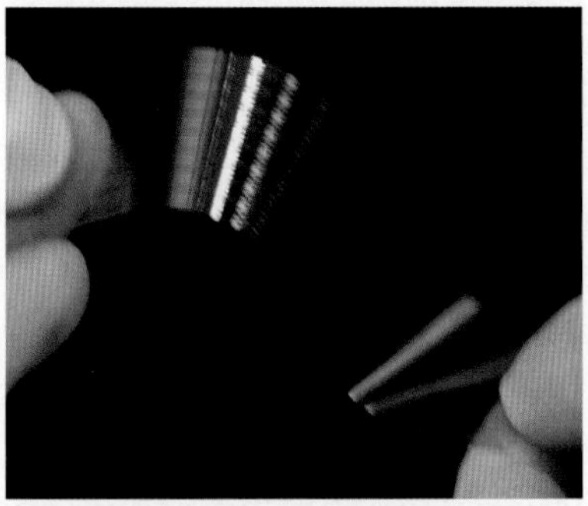

**FIGURE 1.27** Flexible polyimide waveguide from Mitsui Chemicals. (Shioda, T. 2007. Next generation packaging technology which supports the ubiquitous devices. Application of the flexible optical waveguide to the consumer products. *Electr Parts Mater* 46:38–42.[35])

Typical absorption losses in the near-IR range from 0.2–0.4 dB/cm. In MEMS, polyimides have been used, for example, as flexible hinges in mechanical miniaturized structures (see Figure 7.86).[34]

Both nonphotosensitive polyimides and photosensitive polyimides (negative and positive) are available. Photosensitive polyimides have been processed in mass production lines using g-line (436 nm) and i-line (365 nm) exposure tools. Using photosensitive polyimides in manufacturing reduces the number of process steps dramatically compared with the use of nonphotosensitive polyimides. This is illustrated in Figure 1.28 for making polyimide vias in packaging. Positive tone polyimides, as used in Figure 1.28, are relatively new and were developed by HD Microsystems, a joint venture between Hitachi Chemical and Dupont (http://www.hdmicrosystems.com). The positive tone polyimides are more environmentally friendly than negative tone polyimides because of reduced organic solvents and associated volatile organic compounds. These positive tone resists also enable the process to use industry-standard TMAH developers. Polyimide films very strongly absorb UV light less than 350 nm. This absorbency is a result of the polymer's high aromaticity that is also responsible for the exceptional thermal properties of polyimides, which allows for processing at greater than 400°C. Polyimides in packaging and MEMS are further reviewed in Volume III, Chapter 4 on packaging.

Whereas in IC fabrication, sacrificial resist layers of 1 μm or less are common, in miniaturization

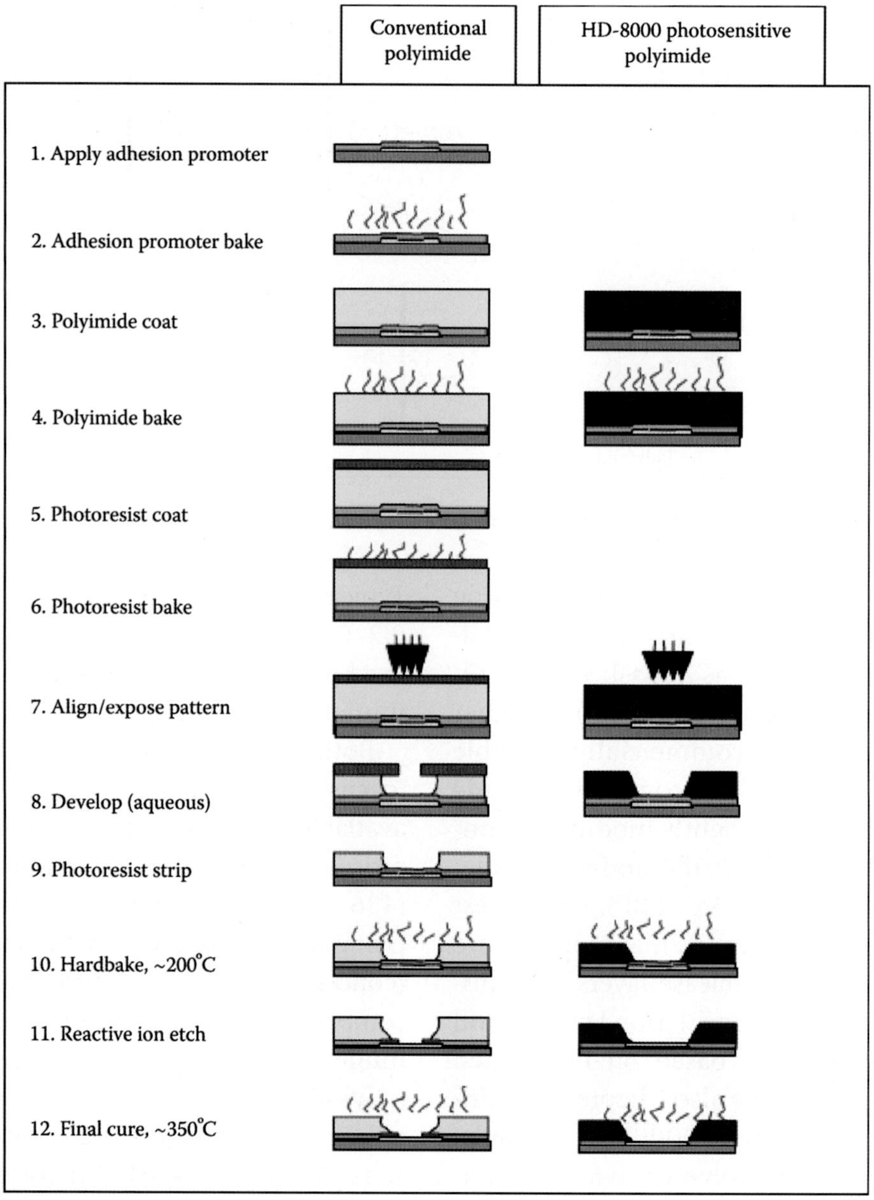


**FIGURE 1.28** Comparison of conventional polyimide process and positive acting photosensitive polyimide process in packaging. The photosensitive polyimide process eliminates multiple steps and decreases cycle time. (Flack, W., S. Kulas, and C. Franklin. 2000. Process characterization of an aqueous developable photosensitive polyimide as a broadband stepper. SPIE. 3999–45.[36])

science, 3D structures frequently require thick permanent resist layers, capable of high resolution and high aspect ratios. Relatively new, chemically amplified deep UV resist systems, such as SU-8, can be applied in very thick layers (>100 μm) and have excellent sensitivity, high resolution, low optical absorption, high aspect ratio, and good thermal and chemical stability. These properties make SU-8 a good candidate for molds and other permanent MEMS structures. SU-8 is an acid-catalyzed negative photoresist, made by dissolving EPON®-SU-8 resin (registered trademark of Shell Chemical Company, http://www.shell.com) in an organic solvent such as cyclopentanone solvent or GBL (γ-butyrolactone) and adding a photoinitiator. The viscosity, and hence the range of thicknesses accessible, is determined by the ratio of solvent to resin. The EPON resist is a multifunctional, highly branched epoxy derivative that consists of bisphenol-A novolak glycidyl ether (Figure 1.29). On average, a single molecule contains eight epoxy groups, which explains the "8" in the name SU-8. The material has become a major

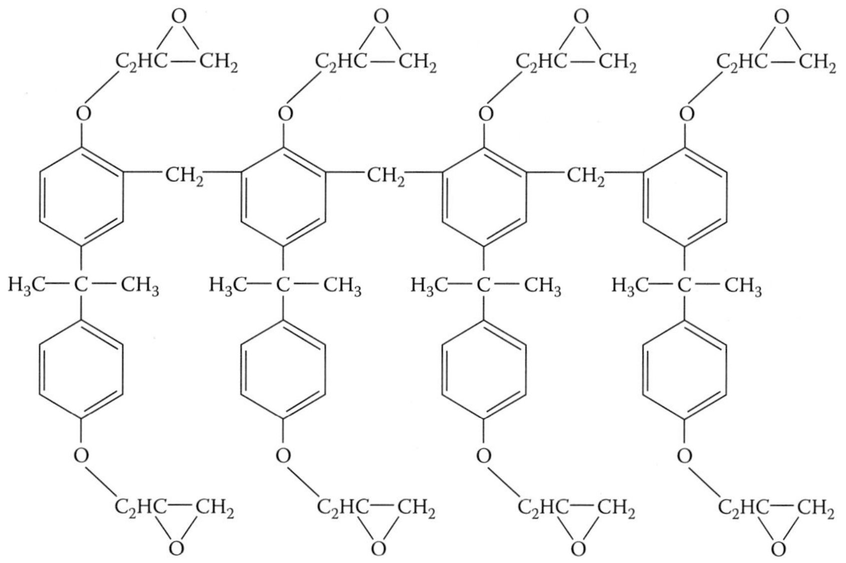

**FIGURE 1.29** Glycidyl ether of bisphenol A: SU-8: cross-linking reaction of carboxy functional polymer with diepoxide.

workhorse in miniaturization science. In a chemically amplified resist like SU-8, one photon produces a photoproduct that in turn causes hundreds of reactions to change the solubility of the film. Because each photolytic reaction results in an "amplification" via catalysis, this concept is dubbed "chemical amplification".[37] We touch briefly on resist amplification here, in the context of SU-8 photoresist only, but will discuss the underlying principles in more detail in "Strategies for Improved Resolution through Improved Resist Performance" (this chapter).

Scientists at IBM discovered that certain photoinitiators, such as onium salts, polymerize low-cost epoxy resins such as EPON®-SU-8. An onium salt, on UV exposure, generates a strong Lewis acid and catalyzes the cationic polymerization of the resin. These salts are called photochemical acid generators (PAGs), and their action constitutes excellent dissolution inhibition of a phenolic resin. In contrast with conventional free-radical initiators, an onium salt cationic photoinitiator, for example, triphenylsulfonium hexafluoroantimonate, is oxygen insensitive and stable over a wide temperature range. On UV exposure of the resist, Lewis acids are released within the resist matrix, forming a latent image—a 3D distribution of the catalytic photoproduct—and image formation is realized after an activating postexposure baking (PEB), as shown in Figure 1.30. Not only UV light in the 365–436-nm range but also electrons and x-rays initiate a high level of cross-linking density, converting the SU-8 photoepoxy into a strong polymer with a $T_g$ of more than 200°C (the $T_g$ of the unexposed resist is 55°C).[38] During the PEB, the generated photoacid initiates the ring opening of the epoxy groups, and extensive cross-linking makes the SU-8 insoluble. An SU-8 soft bake is typically carried out at 95°C, and a hard bake at 200°C. Hot plating, on a level hot plate, is preferred over an oven heat treatment. For developing a 130-μm-thick SU-8 film, a 5-minute dip in undiluted propylene glycol methyl ether acetate, a rinse with isopropyl alcohol, and repeat of this procedure until all the unpolymerized material is dissolved works best.

SU-8 resist, patented by IBM in 1992, was originally developed for e-beam lithography[39] and became commercially available in 1996. Because of its aromatic functionality and highly cross-linked matrix, the SU-8 resist is thermally stable and chemically very inert. After a hard bake, it withstands nitric acid, acetone, and even NaOH at 90°C, and it is more resistant to prolonged plasma etching and better suited as a mold for electroplating than PMMA.* The low molecular weight [~7000 ± (1000)] and multifunctional nature of the epoxy gives it the high cross-linking propensity, which

* PMMA or poly(methyl 2-methylpropenoate) is the synthetic polymer of methyl methacrylate. This thermoplastic and transparent plastic is sold by the trade names Plexiglas, Limacryl, R-Cast, Perspex, Plazcryl, Acrylex, Acrylite, Acrylplast, Altuglas, Polycast, and Lucite and is commonly called acrylic glass or simply acrylic

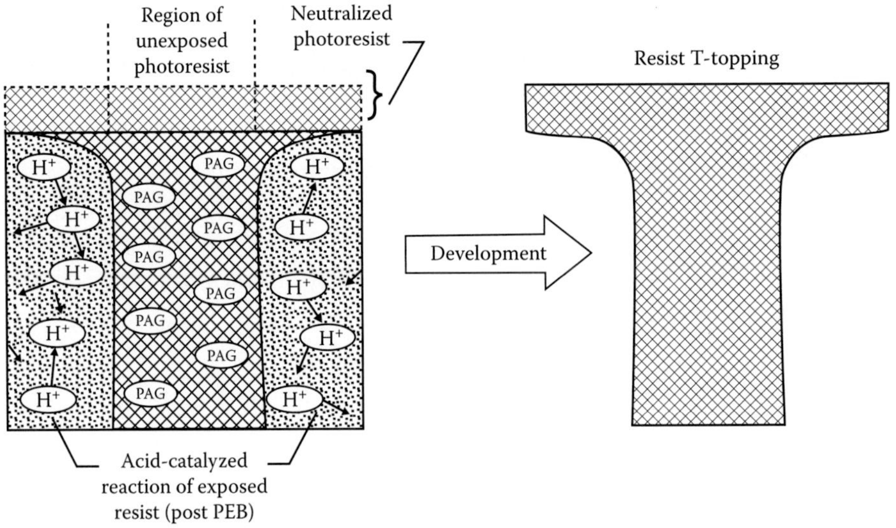


**FIGURE 1.30** SU-8 is a negative-type amplified photoresist. Where light strikes the photochemical acid generators (PAGs), a Lewis acid is released within the resist matrix, forming a latent image—a 3D distribution of the catalytic photoproduct—and image formation is realized after an activating postexposure bake (PEB). After development a T-topped structure often results (see text).

also reduces the solvent-induced swelling typically associated with negative resists. As a result, very fine feature resolution, unprecedented for negative resists, has been obtained, and epoxy-based formulations are now used in high-resolution semiconductor devices, such as 0.35 μm CMOS logic (e.g., with CGR* chemically amplified negative resist). Low molecular weight characteristics also translate into high contrast and high solubility. Because of their high solubility (up to 85% solids by weight in solvents such as methyl-iso-butyl ketone), very concentrated resist casting formulations can be prepared. The increased concentration benefits thick film deposition (up to 500 μm in one coat) and planarization of extreme topographies. The high epoxy content promotes strong SU-8 adhesion to many types of substrates and makes the material highly sensitive to UV exposure. On the negative side, strong adhesion makes stripping of the exposed SU-8 material currently one of the most problematic aspects. Stripping of SU-8 may be carried out with hot 1-methyl-2-pyrrolidon, plasma, or laser ablation. There are other issues to be resolved with this resist; for example, thermal mismatch of SU-8 on a Si substrate (the thermal expansion coefficient α. for silicon is 2.361 ppm/K vs. 21–52 ppm/K for SU-8) produces stress and may cause film cracking. Moreover, the absorption spectrum of SU-8 shows much higher absorption coefficients at shorter wavelengths. As a result, lithography using a broadband light source tends to result in overexposure at the surface of the resist layer and underexposure at the bottom. The resulting developed photoresist tends to have a negative slope, which is not good for mold applications: the mold sidewall should have a positive or at least a vertical slope for easy release of the molded part from the mold. The exaggerated negative slope at the top of the resist structure surface is often called *T-topping* (see Figure 1.30). UV light shorter than 350 nm is strongly absorbed near the surface, creating locally more acid that diffuses sideways along the top surface. Selective filtration of the light source is often used to eliminate these undesirable shorter wavelengths (<350 nm) and thus obtain better lithography results. For example, Reznikova et al. used a 100-μm-thick SU-8 resist layer to filter exposure radiation at 334 nm,[40] and Lee et al.[41] reported using a Hoya UV-34 filter to eliminate the T-top (overexposed top part). Nearly vertical sidewalls can be achieved using a Hoya UV-34 filter. However, there might still be a beaked feature created at the very top of the resist features. This results from the diffraction at the interface between the

* CGR, chemically amplified resist based on poly(p-hydroxystyrene) or its copolymers with styrene or vinylcyclohexane with powderlink crosslinker and a photochemical acid generator.

mask and the photoresist, and it can be eliminated by filling this gap with index matching oil (see results by Wanjun Wan in Figure 1.31). Besides its sensitivity and relatively low cost (<$100/L), aspect ratios up to ~25 for lines and trenches have been demonstrated in SU-8-based contact lithography.[42] When patterned at 365 nm, the wavelength at which the photoresist is the most sensitive, total absorption of the incident light in SU-8 is reached at a depth of 2 mm. In principle, resist layers up to 2 mm thick can be structured.[43] Wan et al.[44] recently confirmed this astounding potential experimentally. Wan's group at Louisiana State University, using both wavelength optimization by patterning using a filtered i-line (365 nm) and air gap compensation (with glycerin or a Cargille refractive index matching fluid), demonstrated aspect ratios greater than 190 (for a feature with a 6-μm thickness and a height of 1150 μm) and structures as high as 2 mm. In the inset in the header of this chapter we reproduce one result of Wan et al., and more are displayed in Figure 1.31.

New resists like SU-8 have enabled miniaturization scientists to make LIGA (German acronym for *Lithographie, Galvano-formung, Abformung*)-like structures less expensively. LIGA is the ultimate lithography technique for making high-aspect-ratio features. In LIGA, a synchrotron x-ray source is used for exposure, and resist walls with the ultimate in wall straightness and wall smoothness are obtained (see Chapter 10). For many applications, photolithography with a much less expensive UV exposure system and a thick photoepoxy, such as SU-8, suffices. This alternative method to LIGA is sometimes referred to as *poor man's LIGA*. The example SU-8 application from Mimotec (http://www.mimotec.ch) shown in Figure 1.32 is a typical LIGA-like application, in

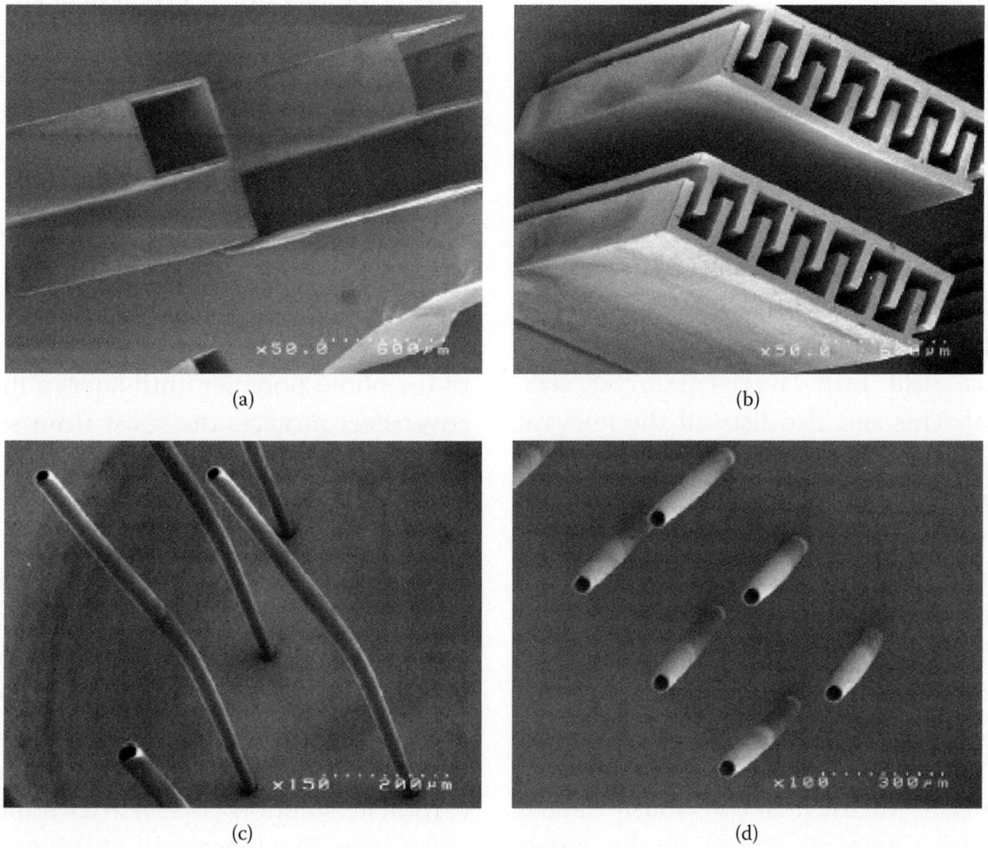


**FIGURE 1.31** Ultrahigh-aspect-ratio structures obtained using a filtered light source (using a thick PMMA plate as filter) and with gap compensation (using an index matching solution). (a) Micro features with designed thickness of 7 μm and height of 2000 μm. (b) Micro comb structures with designed thickness of 40 μm and height of 2000 μm. (c) Micro cylinders 1150 μm tall and designed wall thickness of 6 μm and an internal diameter of 20 μm. As the thickness is reduced, the strength of the cylinders becomes too low to maintain their upright position. (d) Micro cylinders 1150 μm tall and designed wall thickness of 9 μm and internal diameter 27 μm.

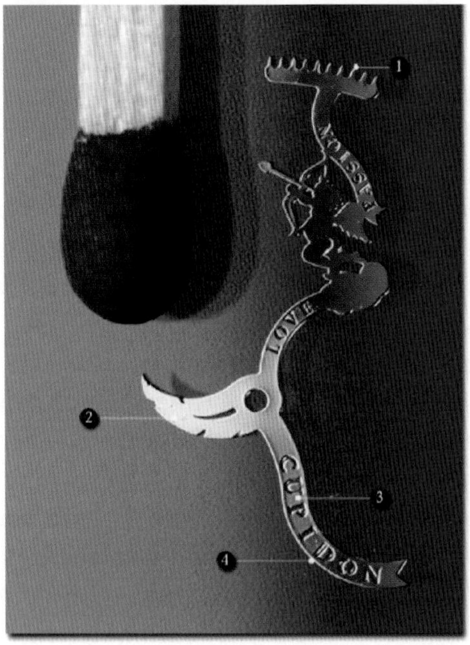


**FIGURE 1.32** Typical SU-8 application. The metal structure shown here has been made by plating in an SU-8 mold. (http://www.mimotec.ch). 1) Rake: cogwheel actuation; 2) cam reader; 3) microinscriptions; and 4) rest position.

which a sacrificial cavity in SU-8 resist has been electroplated to make a metal positive replicate of the cavity. The advantages of the poor man's LIGA method are that it is inexpensive and can be quickly brought into play, all of which makes the process very competitive compared with LIGA.

To find the latest updates on SU-8 applications and processing tips, visit http://aveclafaux.freeservers.com/SU-8.html. This site also lists all the relevant physical properties of SU-8 accompanied with literature references. Another resource with plenty of SU-8 information is the MEMScyclopedia at http://memscyclopedia.org/su8.html.

The first two companies that bought an SU-8 sales license from IBM were MicroChem Corporation[45] and Sotec Microsystems.[46] The latter company, now defunct, developed conductive and magnetic SU-8 and used dry SU-8 films to laminate over large cavities and channels[42]. MicroChem (previously named Microlithography Chemical Corp.) is the oldest manufacturer of the SU-8 photoresist, licensed by IBM. SU-8 Gerstel Engineering Solutions, Ltd. (http://www.gersteltec.ch) not only sells SU-8 under the name SU-8 GM-10## with different viscosities (e.g., GM-1040, GM-1060, and GM-1070) but also offers prototyping and mass production of SU-8 microsystems.

*Dry Resist* Most resists in IC and MEMS fabrication are deposited as liquids, whereas resists used in printed wiring board (PWB) manufacture are usually dry film resists that come in rolls (ranging from 2–60 in. wide and 125–1000 ft. long) and are laminated onto the substrate instead of being spin coated onto it (Figure 1.33). Dry film resist formulations are sandwiched between a polyolefin release sheet and a polyester base, rolled up on a support core (Figure 1.33A and B). These protective covers shield the film from environmental oxygen and facilitate handling. The dry film resist layer comes in thicknesses starting from 25 μm and goes up to 100 μm. Resist thicknesses of 1–1.5 mils are common for imaging purposes, and thicker resists (1.5–2.0 mils) are used as plating, rather than etch, resists. Dry film resists offer advantages such as excellent adhesion on most substrates, no liquid handling because there is no solvent, high process speed, excellent thickness uniformity over a whole substrate (even substrates with holes), simple handling, no formation of edge beads, low exposure energy, low cost, short processing time, and near vertical sidewalls. The photoresist film is applied by removing the polyolefin sheet before lamination of the resist onto the substrate. The polyester top cover sheet remains on top of the photo polymer until later in the process. This coversheet protects the resist from scratches, keeps contaminants from the surface, and prevents the exposure tool from touching and perhaps sticking to the resist. Conformation of the resist to the substrate is achieved by heating under pressure in a hot-roll or cut-sheet laminator (Figure 1.33C). The heat and pressure of the laminating rollers cause the dry film to soften and adapt to surface topologies. The resist is then exposed to a UV light source. Modern dry film resists are developed in a simple sodium carbonate solution (1–2%) after removal of the top cover layer. The biggest disadvantage of dry resist is its relatively low resolution compared with liquid resists as illustrated in Figure 1.34. Two major reasons for this poorer resolution are the many-times thicker resist coating and the fact that the mask is positioned on top of a thick protective Mylar cover

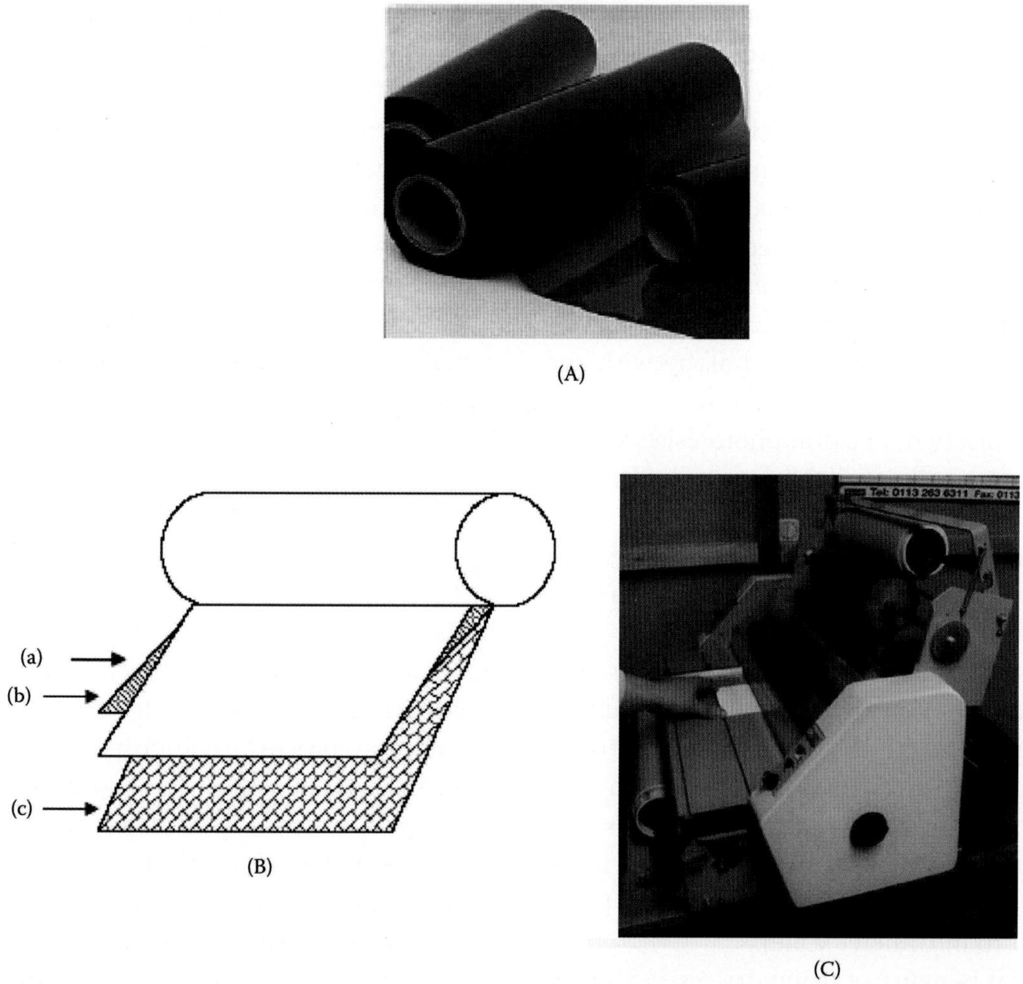


**FIGURE 1.33** (A) Dry film resist roll. (B) Dry film photoresist composition; three-layer structure, (a) polyethylene separation sheet, (b) photoresist, and (c) polyester support. (C) Dry lamination equipment.

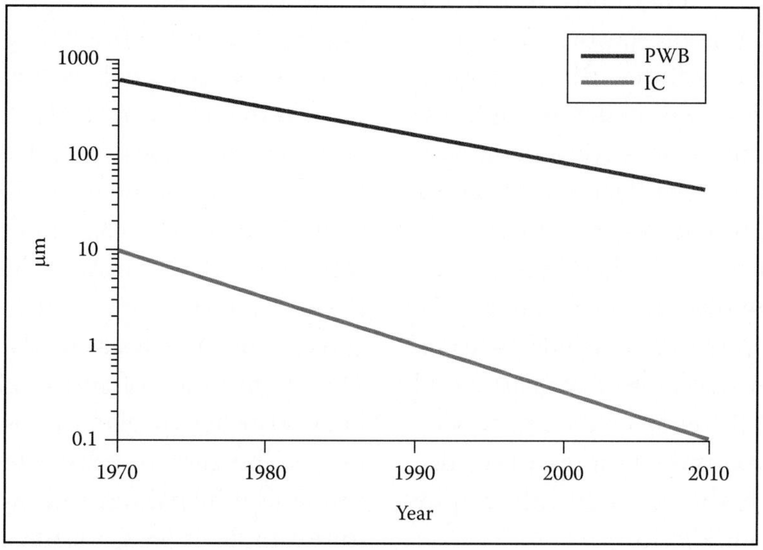


**FIGURE 1.34** Comparison of resolution obtained with liquid resist compared with dry resists as a function of time. (Drawn by Mary Amasia from UC Irvine.)

film. Under highly controlled conditions, one can achieve an opening (space) of 25 μm with a 25-μm-thick dry film, but under production conditions, it is rather difficult to achieve such resolutions. By removing the top cover sheet from the photoresist before exposure, higher resolutions are possible. The removal of 25 μm of separation between the photoresist and the exposure tool decreases the chances of light diffusion. Also, foreign particles that collect on the coversheet are removed along with the coversheet.

There are a variety of dry film photoresists widely used and commercially available; examples include Riston®, Ordyl BF 410, Etertec® 5600, DF 4615, and DFR-15. All are used in the manufacture of circuit boards and can be made thick; additionally, all are candidates for broad use in MEMS as well.

Today it is difficult to obtain high-aspect-ratio devices with these resists because of their resolution limitation, high optical absorbency, and the difficulty in obtaining layers thicker than 50 μm in a single coat.[29] However, for many applications, the poorer resolution of dry resists (25–75 μm) is no obstacle for their usage in mesoscale devices (100s of micrometer to millimeter scale).

The potential benefits of using dry resist films as permanent components in the mass production of biosensors and microfluidics were recognized and described, for example, by Madou et al.[47] These authors suggest that continuous, web-based manufacturing may finally make ubiquitous, disposable miniaturized devices such as biosensors and microfluidics possible (see also Volume III, Chapter 1, Example 1.1). Dry resist film materials are less expensive than silicon and form a convenient substrate; they come in rolls so that large sheets can be processed. In the future, this author envisions a continuous lithographic process, including exposure and development, taking place between a dry resist supply roll and a pick-up reel. This would be especially attractive if higher resolution dry resist could be formulated. Dry film SU-8, as Louis Guerin suggested many years ago, holds the promise of continuous high-aspect-ratio, high-resolution lithography (http://www.geocities.com/guerinlj).

At MicroChem (http://www.microchem.com/about/index.htm), progress is being made making SU-8 dry resist films, and Dupont already offers the MX series dry film for MEMS applications. The new MX products feature an aspect ratio of 3:1. At a thickness of 50 μm, resolution is 16–17 μm, which is better than any other dry film product. Additionally, these films can be multilaminated to build up the thickness (http://tyvek.com.mx/APL/en_US/products/mx/index.html).

Nonspin liquid resist-coating methods such as roller coating or spraying (see above) also will play an important role in MEMS manufacturing of the future. Large substrate sheets can be accommodated; the throughput is higher than with dry resists; and the resolution is considerably better.

### *Glass Transition Temperature of a Resist ($T_g$)*

Resists must meet several rigorous requirements: good adhesion, high sensitivity, high contrast, good etching resistance (wet or dry etching), good resolution, easy processing, high purity, long shelf life, minimal solvent use, low cost, and a high glass transition temperature, $T_g$. Most resists are amorphous polymers that exhibit viscous flow with considerable molecular motion of the polymer chain segments at temperatures above the glass transition. At temperatures below $T_g$, the motion of the segments is halted, and the polymer behaves as a glass rather than a rubber. If the $T_g$ of a polymer is at or below room temperature, the polymer is considered a rubber; if it is above room temperature, it is considered to be a glass. Because above $T_g$ the polymer flows easily, heating the resist film above its $T_g$ for a reasonable amount of time enables the film to anneal into its most stable energetic state. In the rubber state, it is easy to remove the solvent from the polymer matrix; that is, soft bake the resist. Extreme attention needs to be given to the cleanliness of the working environment with the resist in this state. When softening the resist at or above $T_g$, it may be easier to remove solvent, but the resist tends to pick up impurities. The importance of resist reflow, as we will learn later, also lies in planarizing of undesirable topography. The glass transition temperature of a resist is influenced by baking and exposure; the $T_g$ for SU-8, for example, is 55°C for the unexposed material but 200°C after exposure and hard baking. In the fabrication of carbon-MEMS structures from photoresist by pyrolysis (Volume III, Chapter 5), we make sure that the temperature always stays below the $T_g$ to avoid collapse of the resist features.

In general, polymers that crystallize are not useful as resists because the formation of crystalline segments prevents the formation of uniform high-resolution isotropic films.[19]

## Wafer Priming

Resists, especially positive resists, do not adhere well to a Si wafer. The native silicon dioxide on a Si surface, typically 20–50 Å thick, forms long-range hydrogen bonds with water adsorbed from the air. When resist is spun onto such a surface, it adheres to the water molecules rather than to the surface, resulting in poor adhesion. This effect is more pronounced when the humidity is high or if the wafer has been previously immersed in water. Good humidity control (at ~40% relative humidity) and annealing are required to remove surface water and prepare a silicon wafer for resist coating. In the early 1960s, photoresist was spun directly onto the wafer surface. This worked for the large geometry circuits at the time, but when geometries reached the 20-μm range, further feature decrease became intimately linked to the quality of the resist adhesion. To faithfully reproduce the smallest features of a circuit design, a better photoresist/Si wafer contact had to be devised. To this end, surface priming with reactive silicone primers was introduced. A typical resist adhesion promoting primer is 1,1,1,3,3,3-hexamethyldisilazane (abbreviated HMDS) [formula: $(CH_3)_3SiNHSi(CH_3)_3$], developed and patented by IBM. Other primers include trichlorophenylsilane (TCPS) [formula: $C_6H_5SiCl_3$] and bistrimethylsilylacetamide (BSA) [formula: $(CH_3)_3SiNCH_3COSi(CH_3)_3$]. Reactive Si–NH–Si functional groups in HMDS react with the oxide surface in a process known as *silylation*, and a strong siloxane bond to the surface (Si-O-Si) is created (Figure 1.35). At room temperature, unhydrolyzed silane does not react with surface silanols [$Si(s)–OH + (CH_3)_3SiOCH_3 \rightarrow$ no reaction], but with some surface-adsorbed water present, a small amount of reaction occurs as a result of methoxysilane hydrolysis, followed by reaction with surface silanols [$Si(s)–OH + \text{Surface } H_2O + (CH_3)_3SiOCH_3 \rightarrow (CH_3)_3SiOH + Si(s)–OH$ and $(CH_3)_3SiOH + Si(s)–OH \rightarrow Si(s)–O–Si(CH_3)_3 + H_2O$]. When silylation is carried out with HMDS, a strong catalytic effect is observed. The amine adsorbs onto the relatively acidic surface silanols, rendering the oxygen attached to the surface silicon atom highly

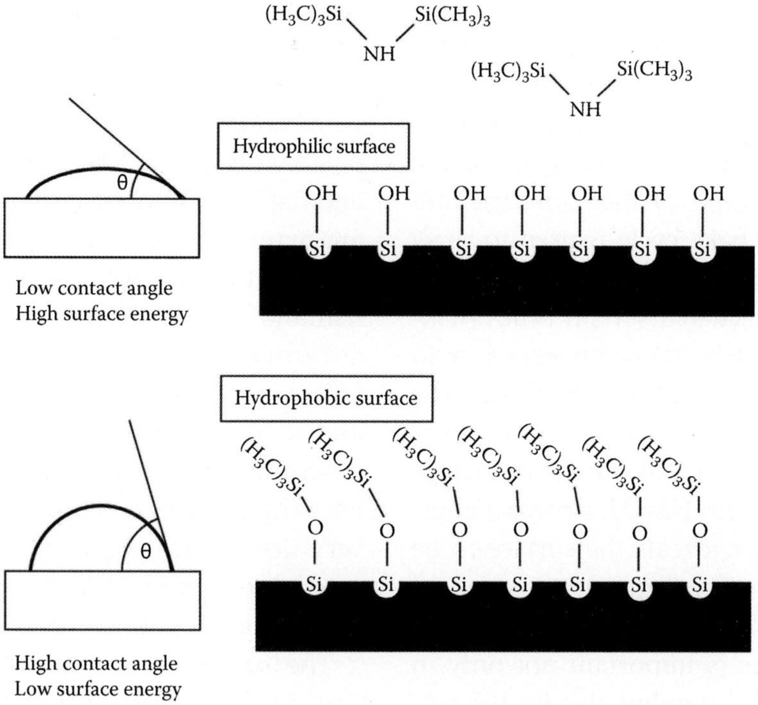


**FIGURE 1.35** HMDS reaction mechanism monitored by the contact angle θ of a drop of water on the treated surface.

nucleophilic. This oxygen then attacks the relatively electrophilic silane silicon, resulting in surface silylation. The more hydrophobic the wafer surface, the better the resist adhesion, and the methyls, it is assumed, bond/adhere to the photoresist, enhancing the photoresist adhesion. The process works not only on silicon dioxide but also on other oxides (e.g., $Al_2O_3$). It should be noted that HMDS is extremely flammable and a suspected carcinogen; it should be handled with care. A dehydration bake of the $SiO_2$ surface at 200–250°C for 30 minutes (with optional vacuum) removes adsorbed water from the silanol groups at the silicon surface, which then can react with the amino groups of the HMDS vapor. The primer may then be applied by dipping the wafer in a 1–10% HMDS solution in xylene or by spin coating it onto the wafer. Baking the wafers to dehydrate and then removing them from the oven and transporting them to a dip solution or a spinner poses a serious problem, however, because on removal from the bake oven, the surface may rehydrolyze. Further shortcomings of this approach include a tendency to generate and deposit particles and HMDS wastage. Today priming of wafers is performed in a process that combines a bake and prime in the same vacuum. The wafer is exposed to HMDS vapor in a stream of dry nitrogen in a vacuum chamber while heating the wafer for approximately 40–60 seconds. This method provides the best adhesion and constitutes the best overall process as it remedies several problems: 1) no rehydrolyzation can occur because the unprimed wafers are not exposed again to the atmosphere before priming; 2) the same vacuum that is used during the bake cycle is used to draw monolayer amounts of HMDS; and 3) the interior of the chamber may be evacuated several times (with intermediary nitrogen backfills) before exposure to HMDS. Thus, at the time of HMDS exposure, the amount of oxygen present in the chamber (and hence the chance of a flammable mix forming) is very low. After treating a Si wafer with HMDS, a contact angle test with a drop of water reveals the surface to be much more hydrophobic (see Figure 1.35).

The chemical modification of oxide surfaces using organofunctional silanes is important not only in affecting better resist adhesion but also for the fabrication of water-repellent and antifouling coatings, DNA arrays, protein arrays, enzyme-based sensors, and so on.

Sputtering of the surface (Chapter 3), for some applications, presents an attractive alternative to vapor priming; the microroughness at the surface induced by sputtering provides for mechanical adhesion of the resist to the substrate. Primers for GaAs include monazoline C, trichlorobenzene, and xylene.

## Resist Stripping

We now turn to the last step of the photolithographic process: photoresist stripping, as illustrated in Figure 1.1f.

### *Wet Stripping*

Photoresist stripping, in slightly oversimplified terms, is organic polymer etching. The primary consideration is complete removal of the photoresist without damaging the device under construction. Turning back to Figure 1.1, once the exposed oxide has been etched away in (e), the remaining photoresist can be stripped off with a strong acid such as $H_2SO_4$ or an acid-oxidant combination such as $H_2SO_4$–$Cr_2O_3$ attacking the photoresist but not the oxide or the silicon (Figure 1.1f). Other liquid strippers include organic solvent strippers and alkaline strippers (with or without oxidants). Acetone can be used if the postbake is not too long or occurs at a low enough temperature. With a postbake of 20 minutes at 120°C, acetone is still fine. But with a postbake at 140°C, the resist develops a tough "skin" and has to be burned away in oxygen plasma. There are many commercial strippers available; some are specific for positive resists (e.g., ACT-690C from Ashland), others for negative resists, and still others are universal strippers (e.g., ACT-140 from Ashland). Other popular commercial strippers are Piranha and RCA clean. The RCA1 clean for organics was described above. To make up a Piranha solution, measure 5 parts of $H_2SO_4$ in a Pyrex beaker, and very slowly add 1 part of $H_2O_2$. Note that this mixture is exothermic. When cool, it may be refreshed by very slowly adding more $H_2O_2$.

The ozone-water method is a Piranha alternative. In this process, water of 25–95°C, depending on the application, is sprayed onto the substrate as it is rotated. At the same time, dry ozone gas is injected into the reaction chamber. The ozone diffuses through the thin boundary layer of water to react with the organics on the substrate. Strip rates exceed 1.2 μm/min.[48]

The oxidized Si wafer with the etched windows in the oxide (Figure 1.1f) now awaits further processing. This might entail a wet anisotropic etch of the Si in the oxide windows with $SiO_2$ as the etch mask.

### *Dry Stripping*

Dry stripping or oxygen plasma stripping, also known as ashing, has become more popular as it poses fewer disposal problems with toxic, flammable, and dangerous chemicals. Wet stripping solutions lose potency in use, causing stripping rates to change with time. Accumulated contamination in solutions can be a source of particles, and liquid phase surface tension and mass transport tend to make photoresist removal difficult and uneven. Dry stripping is more controllable than liquid stripping, less corrosive with respect to metal features on the wafer, and, more importantly, it leaves a cleaner surface under the right conditions. Finally, it does not cause the undercutting and broadening of photoresist features that can be caused by wet strippers.

In solid-gas resist stripping, a volatile product forms either through reactive plasma stripping (e.g., with oxygen), gaseous chemical reactants (e.g., ozone), and radiation (UV), or a combination thereof (e.g., UV/ozone-assisted). Plasma stripping uses a low-pressure electrical discharge to split molecular oxygen ($O_2$) into its more reactive atomic form (O). This atomic oxygen converts an organic photoresist into a gaseous product that may be pumped away. This type of plasma stripping belongs in the category of chemical dry stripping and is isotropic in nature (see Chapter 3). In ozone strippers, ozone, at atmospheric pressure, attacks the resist. In UV/ozone stripping, UV helps to break bonds in the resist, paving the way for a more efficient attack by the ozone. Ozone strippers have the advantage that no plasma damage can occur on the devices in the process. Reactive plasma stripping currently is the predominant commercial technology because of its high removal rate and throughput. Some different stripper configurations are barrel reactors, downstream strippers, and parallel plate systems. These prevalent stripping systems are reproduced in Figure 1.36.[49] More details will be provided in Chapter 3.

Posing new challenges to modern stripping processes are shrinking feature sizes, increasing aspect ratios, and stripping of resist over some of the newest dielectrics ($\sigma_r < 3$) and copper interconnects. Especially novel dielectric materials, such as fluorinated silicate glass, carbon-doped oxides, and fluorinated amorphous carbon, are easily attacked or degraded by stripping.[48] In practice, combinations of dry and wet etching often form the most successful strategy.

Dry stripping of resist has been so successful that it has accelerated the use of plasmas in other lithography steps such as development and deposition of resist. For example, it was the study of the attack of oxygen on polymers that led to the development of dry developed resists as used in the DESIRE process (see page 75). Such dry developed resists are vital for future submicrometer lithography where

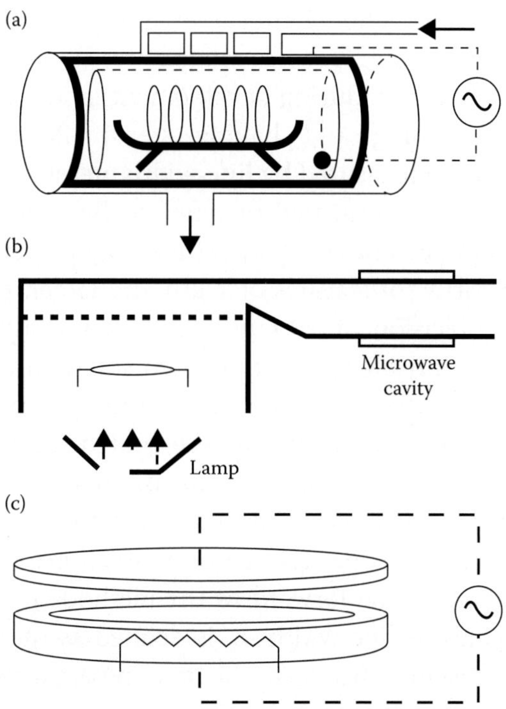


**FIGURE 1.36** Various dry stripping reactors: barrel reactor (a), downstream etchers (b), parallel plate systems (c). (From Flamm, D.L. 1992. Dry plasma resist stripping. Part I: overview of equipment. *Solid State Technol* 35: 37–39. With permission.[49])

underetching and broadening of features are most critical. Although not detailed in the current short treatise, inspection and metrology techniques (see Volume III, Chapter 6) play a crucial role at various points in the lithography process.

## Critical Dimension, Overall Resolution, Line Width Metrology

The absolute size of a minimum feature in an IC or a miniature device, whether it involves a line width, spacing, or contact dimension, is called the *critical dimension* (CD). The overall resolution of a process describes the consistent ability to print a minimum size image, a CD, under conditions of reasonable manufacturing variation.[50] Many aspects of the process, including hardware, materials, and processing considerations, can limit the resolution of lithography. Hardware limitations include diffraction of light or scattering of charged particles (in the case of charged-particle lithography or hard x-rays), lens aberrations, mechanical stability of the system, and so on. The resist material properties that impact resolution are contrast, swelling behavior, thermal flow, and chemical etch resistance. The most important process-related resist variables include swelling (during development) and stability (during etching and baking steps). Resolution frequently is measured by line-width measurements using either transmitted or reflected light or other metrology techniques. Optical techniques perform satisfactorily for features of 1 μm and larger, providing a precision of ±0.1 μm (at 2 σ, i.e., all data points within plus and minus two standard deviations). By 1998, devices with features as small as 0.25 μm were launched and necessitated equipment requirements for line-width measurement with a precision of at least ± 0.02 μm (at 3 σ). Scanning electron microscopes (SEMs) or atomic force microscopes have come forward as the methods to reach these goals. A line width, L, is defined as the horizontal distance between the two resist-air boundaries in a given cross-section of the line at a specified height above the resist/substrate interface. Because different measurements may measure the line width of the same line at different heights of the cross-section, the measuring technology used always needs to be identified. The successful performance of devices depends on the control of the size of critical structures across the entire wafer and from one wafer to another, referred to as line-width control. A rule of thumb is that the dimensions must be controlled to tolerances of at least ± 1/5 of the minimum feature size. Typically, a series of features with known sizes across a substrate is measured and then plotted as a function of position on the wafer. The standard deviation at the 1 or 2 σ level is adopted as the line-width control capability of the particular exposure/resist technology. Plotting these data as a function of time enables line managers to maintain optimum performance on a manufacturing line.[50]

In Volume III, Chapter 6, the metrology tools used in IC, MEMS, and NEMs are discussed: laser interferometers for accurate alignment of mask to wafer (with a 5-nm "resolution"), film thickness measurement with profilometers, SEM, scanning transmission electron microscope (STEM), scanning probe microscope (SPM) with vertical resolution of 1 Å, and lateral resolution depending on tip sharpness.

## Lithographic Sensitivity and Intrinsic Resist Sensitivity (Photochemical Quantum Efficiency)

### Lithographic Sensitivity

A distinction must be made between the intrinsic sensitivity of a resist (that is, the resist's response to radiation) and the lithographic sensitivity when defining the measurement of the efficiency that translates resist exposure into a sharp image. In the literature, the values given for the lithographic sensitivity of a resist show a tremendous spread as a result of the complex relationship between the intrinsic resist sensitivity and the dose required to successfully process that resist. This relationship involves the intrinsic resist sensitivity and the bandwidth of the optical exposure system, baking conditions, resist thickness, developer composition, and development conditions. To reproduce a reported lithographic sensitivity, all these parameters need to be duplicated exactly. The best way to determine lithographic sensitivity is experimentally, as we will explain next.

## Intrinsic Sensitivity of a Resist (Photochemical Quantum Efficiency)

A first indication of the intrinsic sensitivity of a resist to a certain wavelength can be deduced from the spectral-response curve of the resist. If the resist absorbs strongly in ranges where the radiation source shows strong emission lines, relatively short exposure times can be expected. Practical limits confine resist sensitivity: too sensitive a resist might mean an unacceptably short shelf life, and yet clearly the resist should be insensitive to the yellow and green light of the clean room.

High intrinsic resist sensitivity is a sought-after characteristic. To increase resolution of photolithography, the shortest possible wavelengths must be used. At those wavelengths exposure sources become less bright and optics absorb more. Because the total energy incident on a resist is a function of light source intensity, time, and absorption efficiency of the exposure optics, a decrease in intensity and an increase in light absorption require compensation through longer exposure times. This results in a smaller throughput of wafers per hour; conversely, a more sensitive resist decreases the exposure time, resulting in a higher throughput.

The intrinsic sensitivity or photochemical quantum efficiency, Φ, of a resist is defined as the number of photoinduced events divided by the number of photons required to accomplish that number of events:

$$\Phi = \frac{\text{Number of photon-induced events}}{\text{Number of photons absorbed}} \tag{1.4}$$

For polymer resins where the polymer undergoes scission or cross-linking without the need for light-absorbing sensitizers (one-component resists), a *G*-value is introduced. The *G*-value corresponds to the number of scissions or cross-links produced per 100 eV of absorbed energy. For scission reactions, the symbol $G(s)$ is used; for cross-linking, the symbol $G(x)$ is used. In contrast to lithographic sensitivity, the measurement of intrinsic radiation sensitivity as expressed through Φ, $G(s)$, or $G(x)$ is reliable, and values from different sources agree relatively well.

The experimental determination of the quantum efficiency of a one-component resist is a complex undertaking. Samples of the polymer must be exposed to a known dose of gamma radiation, and the molecular weight of the irradiated samples must be measured either by membrane osmometry or gel permeation chromatography. Quantitative analysis of the molecular weight versus dose in polymers that undergo scission or cross-linking leads to an important relationship for a better understanding of resist exposure. We will use such a relationship when exploring x-ray lithography for the creation of "high-rise" PMMA resist structures (»10 µm high; see Chapter 10 on LIGA). Let us consider a positive resist here. For a positive resist sample of weight $w$ (in grams) containing $N_0$ molecules before exposure, the definition of average molecular weight $M_n{}^0$ is given by[28,51]:

$$M_n^0 = \frac{wN_A}{N_0} \tag{1.5}$$

where $N_A$ is Avogadro's number. Expressing the dose, $D$, in eV/g, the total number of scissions produced in the sample, $N^*$, is proportional to the absorbed dose, or:

$$N^* = KDw \tag{1.6}$$

where $K$ is a constant dependent on the polymer structure, generally expressed in terms of a *G*-value. $G(s)$ (for positive resists) and $G(x)$ (for negative resists), like Φ, are figures of merit used to compare one resist material with another. With $K$ expressed in terms of $G(s)$, Equation 1.6 can be rewritten as:

$$N^* = \left[\frac{G(s)}{100}\right]Dw \tag{1.7}$$

in which we divide by 100 to express the number of events per 100 eV. As scission occurs, the number of molecules increases, and the new average molecular weight after exposure to dose $D$ is then given by:

$$M_n' = \frac{wN_A}{N_0 + N^*} \tag{1.8}$$

where the total mass of the polymer is assumed to remain constant during exposure.

By substituting Equations 1.5 and 1.7 into Equation 1.8, we obtain:

$$M_n' = \frac{N_A}{\frac{N_A}{M_n^0} + \left[\frac{G(s)}{100}\right] D} \quad (1.9)$$

which is independent of the sample mass $w$. Rearranging Equation 1.9 we obtain:

$$\frac{1}{M_n'} = \frac{1}{M_n^0} + \left[\frac{G(s)}{100}\right] D \quad (1.10)$$

From Equation 1.10, we conclude that a linear relationship exists between the inverse of the average molecular weight and the exposure dose $D$. The intercept on the $y$-axis yields $1/M_n^0$, and the slope allows one to calculate $G(s)$. There is a very high correlation between $G(s)$ values for gamma radiation [the radiation commonly used to determine $G(s)$] and sensitivity for electrons, ions, and x-rays.

The $G(s)$ of polymers commonly used as one-component, positive resist systems ranges from 1.3 for some PMMAs to approximately 10 for certain poly(olefin sulfones). A PMMA with a $G(s)$ value of 1 has a corresponding photochemical quantum yield for scission, Φ (Equation 1.4), of 0.02.[28] For some polymers, both scissioning and cross-linking events occur simultaneously on exposure. It is possible, even in the latter case, to uniquely determine both scission efficiency $G(s)$ and cross-linking efficiency $G(x)$.[51]

For one-component negative resists the figure of merit for intrinsic sensitivity, $G(x)$, expressed as number of cross-links per 100 eV absorbed dose, ranges from 0.1 for poly(ethylene) to approximately 10 for polymers containing oxirane groups (epoxy groups) in their side chains.

For a two-component positive system such as DNQ, Φ in Equation 1.4 corresponds to the quantum efficiency, that is, the number of sensitizer molecules converted to photoproduct, divided by the number of absorbed photons required to accomplish that conversion. Quantum efficiency in this case can easily be measured by using a narrow-bandwidth radiation source and a UV-Vis spectrophotometer. The quantum efficiency, Φ, of typical DNQ sensitizers ranges from 0.2–0.3 (compared with 0.02 for PMMA). Because of the high opacity (i.e., high nonbleachable absorption) of novolak resins in the deep UV region (200–300 nm) region, other resists like PMMA are used for shorter wavelength exposures (e.g., for deep UV, e-beam, and x-ray lithographies). The quantum efficiency of the bis(aryl) azide sensitizers in negative resist systems ranges from 0.5–1, making negative resists more sensitive than positive resists.

## Resist Profiles

### Overview of Profile Types

Manipulation of resist profiles is one of the most important concerns of a lithography engineer. Depending on the final objective, one of the three resist profiles shown in Figure 1.37 is attempted. A re-entrant, undercut, or a reverse resist profile (resist sidewall >90°) is required for metal liftoff (see below for more details). Some authors confusingly call slopes >90° *overcut* [28]; most, including this author, refer to this type of resist profile as an *undercut* (Figure 1.37a). Shallow resist angles (<90°) enable continuous deposition of thin

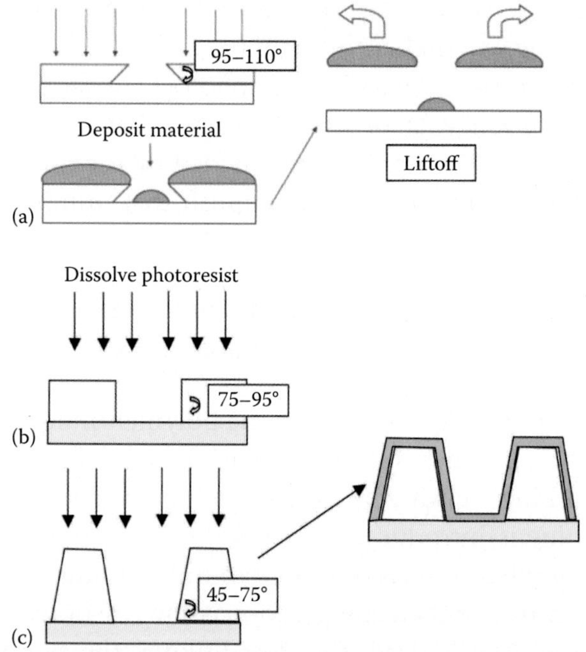


**FIGURE 1.37** The three important resist profiles. (a) Re-entrant, undercut, or a reverse resist profile (resist sidewall >90°) is required for metal liftoff. (b) A vertical (75–90° resist sidewall angle) slope is desirable for a perfect fidelity transfer of the image on the mask to the resist. (c) Shallow resist angles (45–75° < 90°) enable continuous deposition of thin films over the resist sidewalls.

films over the resist sidewalls (Figure 1.37c). A vertical (90° resist sidewall angle) slope is desirable for a perfect fidelity transfer of the image on the mask to the resist (Figure 1.37b). If resist sidewalls are perpendicular, or if the resist layer is undercut, a deposited metal is likely to be noncontinuous: there is a separation between the desired and the undesired metal for a liftoff process. In the case of a shallow slope, the deposit is continuous.

In general, after development of negative and positive resists, three different photoresist wall profiles may be obtained as summarized in Figure 1.37 and tabulated in Table 1.9. In the table, $R$ is the development rate of the exposed region, $R_0$ is the development rate of the unexposed region, and $\gamma$ is the resist contrast. The resist contrast will be explained in greater detail later in this chapter. This table also lists typical applications for each resist profile. First, we will concentrate on the dependence of the resist profile of a positive resist on exposure dose and development mode. In a developer-dominated process, "force" developed with $R/R^0 < 5$, a shallow outward sloping resist profile results, and thinning of the entire resist layer occurs. For positive resists, the shallow angle is the most typical profile with a 45–75° resist wall angle. With a quenched developer, $R/R_0 = 5–10$, and a moderate dose, a straight resist wall profile results (~90°, e.g., 75–95°). In the vertical wall case, the removal of the laterally exposed region has been inhibited, and a perfect pattern transfer of the mask features onto the resist is obtained. An undercut profile is difficult to achieve in positive resists because the optical exposure dose (and hence the development rate of the system) is greater at the surface than at the resist/substrate interface, resulting in a normal profile with shallow resist angles. An undercut profile is desirable for liftoff processes in which deposited layers are lifted from the substrate

**TABLE 1.9 Photoresist Profiles Overview**

| Profile | Dose | Developer Influence | $R/R_0$ | $\gamma$ | Uses |
|---|---|---|---|---|---|
| (A) Positive resists undercut (a) 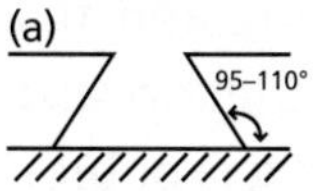  | High (often with backscatter radiation) | Low | >10 | >6 | Ion implant; liftoff; not good for plasma etching; often only obtained through image reversal |
| Vertical (b) 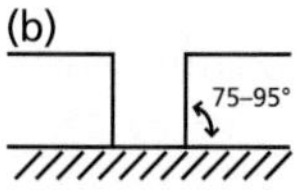  | Normal dose | Moderate | 5–10 | <4 | Liftoff; reactive ion etch; wet etch; ion beam etch; perfect fidelity |
| Normal or overcut (c) 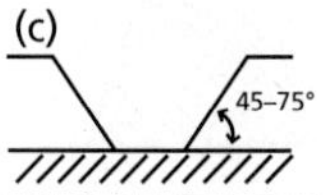  | Low | Dominant | <5 | <3 | Typical for positive resists; wet etch; metallization; <20% resist loss |
| (B) Negative resists undercut 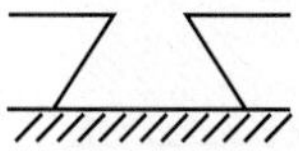 | Dominant | Little influence | <0.1 | <3 | Permanent resists; larger devices; MEMS |

(A) Positive resist. (a) Desired resist profile for liftoff, that is, exposure-controlled profile also called *undercut*. (b) Perfect fidelity image transfer by applying a normal exposure dose and relying moderately on the developer. (c) Receding photoresist structure with thinning of the resist layer, that is, developer control also called *overcut*, the normal profile for positive resists. (B) Negative resist. Profile is mainly determined by the exposure. Development swells the resist slightly but otherwise has no influence on the wall profile for the older types of resists. The undercut profile is the normal profile for the newer types (aqueous developable) of negative photoresists.

*Source*: Based on Moreau, M.W. 1988. *Semiconductor lithography*. New York: Plenum Press. With permission.[28]

by dissolving the underlying resist structure (see next section). In the extreme case of overexposure and a fast developer, the result is a lip, retrograde or inward wall-angle undercut (>90, e.g., 95–110°). This profile might come about if there is a lot of light back scatter from shiny surfaces at the interface of the substrate and the bottom of the resist layer. A ratio of $R/R_0 > 10$ qualifies as a fast developer. However, liftoff profiles with positive resists are more readily formed with multilayer resist (MLR) systems or with a postexposure soaking procedure (see below). With negative resists, forming more insoluble products at the resist surface than at the resist/substrate interface, one more easily obtains an undercut profile; a single-layer resist (SLR) will do, and no complicated MLR systems are required in this case.

A more rigorous, mathematical treatment of resist profiles is presented under Mathematical Expression for Resist Profiles.

We now detail the more unusual profiles for positive and negative resists a bit further. Not all photons strike the resist/substrate interface at the same angle; especially with a high overdose, scattering at a reflective interface may cause broadening of the radiation profile at the substrate/resist interface. The scattered radiation profiles for overexposed positive and negative resists are shown in Figure 1.38.

Time-independent organic solvent development (with a strong developer) of overexposed negative resists from outside the scattered region reveals the scattered radiation zone because cross-linking and further swelling inhibit its removal. On the other hand, development of overexposed positive resists using aqueous alkaline is time dependent and rapidly removes the exposed region, edge-scattered radiation zones, and part of the photoresist top unless the developer is quenched. The time-dependent aspect of this process enables the operator to tailor positive resist profiles. With negative resists, the exposed regions remain, as they are rendered insoluble; with positive resists, the exposed region develops, and the unexposed regions usually remain soluble. Swelling in traditional negative resists is one of the reasons these resists used to be limited to the manufacture of devices with minimum feature size of about 3 μm. Scattered radiation and swelling result in a broadening of the remaining resist features. Positive resists do not exhibit this swelling because of a different dissolution mechanism. The oxygen effect, quenching the cross-linking of negative resists as discussed earlier, usually means a disadvantage for negative resists but can be turned into an advantage to improve resolution. We already know that oxygen can scavenge the photogenerated reactive nitrene species and that this reaction eliminates the precursors for cross-linking and insolubilization. Excluding oxygen from the top surface of the resist by flushing with an inert gas or blocking with a polymer topcoat causes the oxygen dissolved in the polymer film to move laterally from the unexposed dark areas into the light zone. The light zone

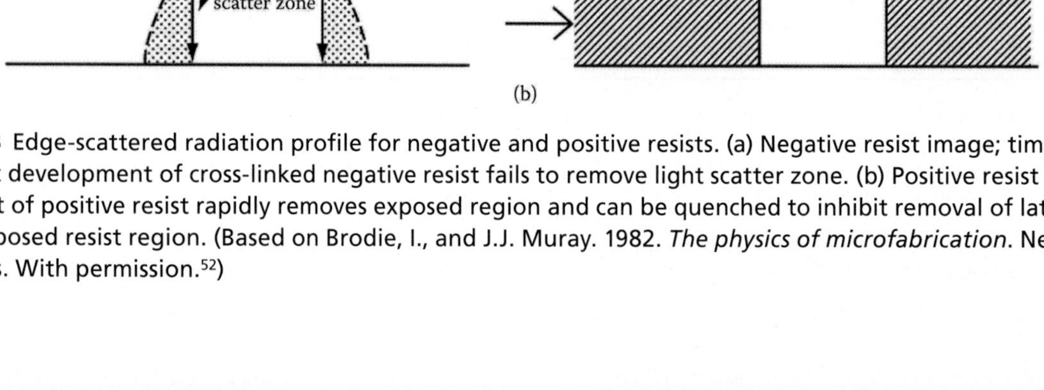

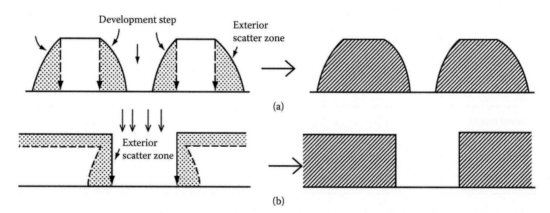


**FIGURE 1.38** Edge-scattered radiation profile for negative and positive resists. (a) Negative resist image; time-independent development of cross-linked negative resist fails to remove light scatter zone. (b) Positive resist image; development of positive resist rapidly removes exposed region and can be quenched to inhibit removal of lateral scattered exposed resist region. (Based on Brodie, I., and J.J. Muray. 1982. *The physics of microfabrication*. New York: Plenum Press. With permission.[52])

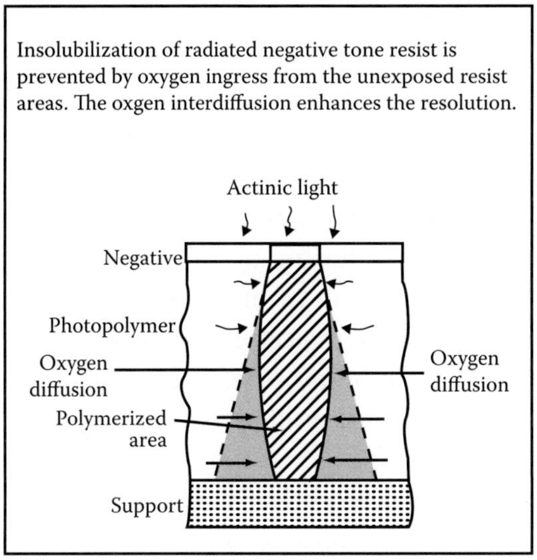


**FIGURE 1.39** Lightly scattered zones do not insolubilize. (From Moreau, W.M. 1988. *Semiconductor lithography.* New York: Plenum Press. With permission.[28])

becomes an oxygen sink, and the lightly scattered zones (Figure 1.39) do not insolubilize, leading to better resolution.[28]

## Liftoff Profile

The creation of a straight photoresist wall or, better yet, that of a lip or undercut can be taken advantage of in a so-called lift-off process. Lift-off is important, for example, for patterning catalytic metals such as Pt, frequently used in chemical sensors but not easily patterned directly. In the process sequence, shown in Figure 1.40, a solvent dissolves the positive photoresist underneath the deposited metal, starting at the edge of the unexposed photoresist, and lifts off the metal. It is important that there be a discontinuity or gap in the metal deposit so that solvent can get at the uncoated resist wall. This is accomplished by depositing the metal with a line-of-sight–type technique such as thermal evaporation, which is described in Chapter 7. In a line-of-sight deposition technique, a vertical or inward sloping wall will receive little or no metal deposit, leaving a gap for the resist solvent to dissolve the unexposed resist and lift off the metal on top of it. One disadvantage of this technique is the rounded profile, a result of shadowing, associated with deposited features and temperature limitations (see Figure 1.40). A more desirable profile for a conductor line has a rectangular cross-section minimizing electrical resistance. The latter is one reason why liftoff in IC fabrication, where contact resistance is of prime concern, is used with discretion. Also with liftoff,

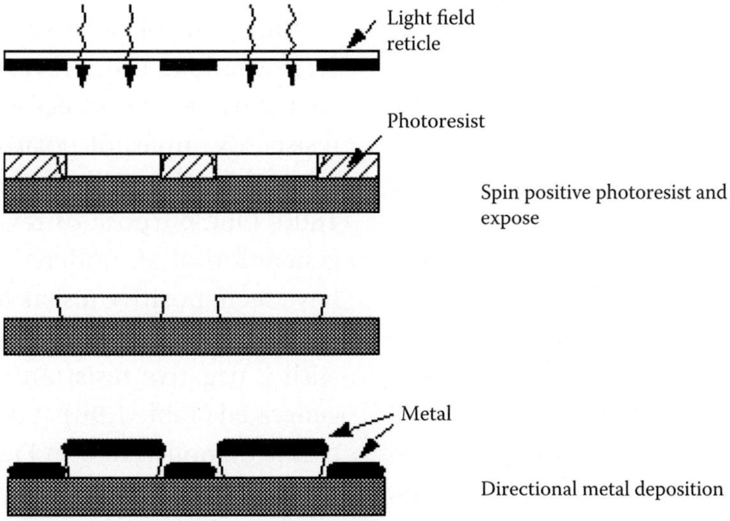

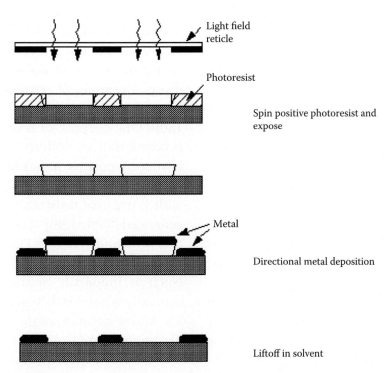

**FIGURE 1.40** Example of liftoff sequence with positive resist for the construction of a Pt-based electrochemical sensor electrode. Rounding of deposited features through shadowing is observed (see text).

the metal deposition technique is limited to temperatures less than 200–300°C, where resist begins to degrade.[53]

In a liftoff process, one can either use a negative resist, such as the alkaline-developing ZPN1100, or modify a positive resist to exhibit the desired undercut. We know already that an undercut does not readily form with a positive resist: therefore, some "tricks" are in order. The first trick is to "presoak" the positive resist surface with an aromatic solvent (e.g., chlorobenzene) to convert a surface layer. This layer will develop at a much slower rate than the bulk of the resist film, thereby providing an undercut during development in alkaline solution. The second method, image reversal, follows further below.

A presoak process to develop an undercut on a positive resist (Shipley 1827) involves the following steps—with typical example materials, equipment, and process parameters:

1. Dehydration bake of Si wafer to remove all moisture. Typically this is done in a natural convection oven at 200°C for 30 minutes.
2. Wafer priming using HMDS.
3. Spin coat resist using a Solitec 5100 spinner. Spin to 1000 rpm for 5 seconds to spread the resist and then at 7000 rpm for 35 seconds to complete the spin cycle.
4. Soft bake at 90°C for 20 minutes in a natural convection oven.
5. Expose the wafer for 15 seconds in a Kasper Contact Mask Aligner.
6. A 5-minute chlorobenzene soak. Chlorobenzene diffuses into the photoresist top layer, causing it to swell. A gel is formed to the depth of the diffusion, which develops much slower than the bulk of the resist. This causes the developer to undercut the photoresist structures and produces the desired profile. After the soak, blow-dry the wafers with a nitrogen gun.
7. Develop the wafers using Microposit MF319 from Shipley. Development rates are increased with exposure time and decreased by soaking time. The exposed regions develop faster; the undercut of the structure is formed when the development front passes the gel layer, and fast lateral development of the unexposed regions begins. Use a development time of 5 minutes with mild agitation. Then place the wafers in DI water for approximately 2 minutes, and blow-dry with nitrogen.
8. Deposit desired material.
9. Liftoff. Dissolution of the photoresist is the final step in the liftoff process. Positive resist is very soluble in acetone, which has been used traditionally. Soak the wafers in acetone for 5 minutes with mild agitation.

There are many different techniques to make undercut resist profiles with a positive tone resist, including:

1. Chlorobenzene induced lip in single-layer photoresist (see above)
2. Multilayer resists (e.g., a bilayer) where top layer develops more slowly
3. Special undercoatings that develop faster than resist in developer
4. Trilayer methods
5. Image reversal resists (see next section)

### Image Reversal

Because swelling does not take place during the development of positive photoresists, several process variations aim at reversing the tone of the image so that the resists can act as high-resolution negative resists. Examples of positive photoresists that may be used for this purpose are OCG 895i and Shipley 1800. One purpose of reversal processes is for the generation of an undercut profile for liftoff. As we have seen, positive resists more readily lead to overcut resist profiles as shown in Table 1.9Ac, whereas with a negative resist an undercut is more readily generated (Table 1.9B). A typical image-reversal process is demonstrated in Figure 1.41a. After a diazo-positive resist has been patterned, an amine vapor (such as imidazole or triethanolamine, or, more generally, a base) is diffused into the exposed areas. The amine neutralizes the byproduct of the photodecomposition (a carboxylic acid in this case) and makes these exposed areas highly resistant to further change by exposure to light and highly insensitive to further development. The base inducing the

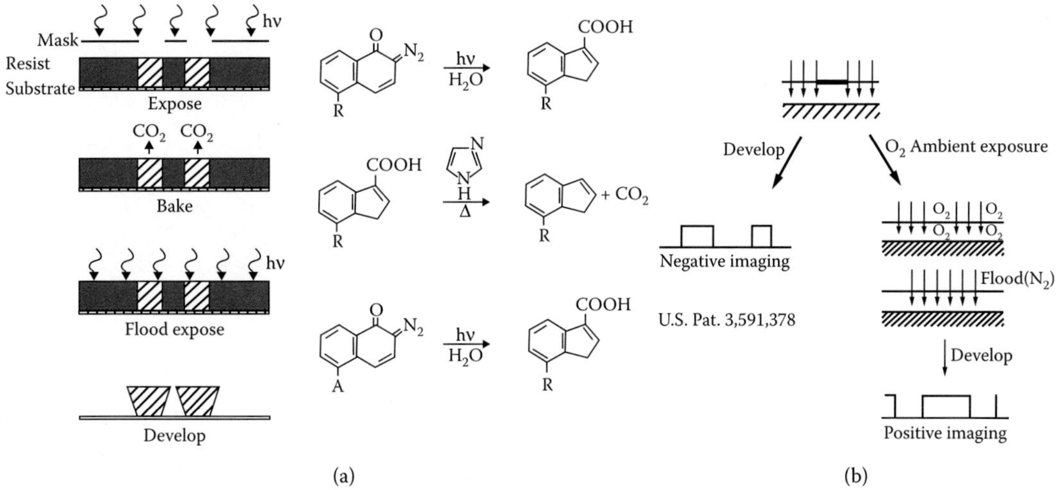


**FIGURE 1.41** Image reversal process sequence. (a) Positive photoresist. (b) Negative photoresist. [(a) Based on Alling, E., and C. Stauffer. 1988. Image reversal photoresist. *Solid State Technol* 31:37–43;[54] (b) based on Moreau, W.M. 1988. *Semiconductor lithography.* New York: Plenum Press. With permission.[28]]

image reversal may also be added to the resist formulation before coating. A subsequent flood exposure makes the areas adjacent to the neutralized image soluble in conventional positive photoresist developers. The net result is a negative image of the mask with an undercut profile and improved resolution because the resist does not swell as negative resists do.

Image reversal with a negative resist, such as KTFR, is possible as well (see Figure 1.41b). We have already discussed that remaining oxygen in a resist can improve resolution by preventing insolubilization in light-scattered zones outside the exposed area (Figure 1.39). Image reversal is an extension of this insolubilization. By working at low intensity, oxygen flooding can scavenge all photogenerated azide polymerization precursors, making the exposed areas soluble in a developer as they are prevented from polymerizing. A subsequent flood exposure under a nitrogen blanket, at higher intensity, initiates the polymerization in the previously nonexposed areas.[28]

A dual-tone resist chemistry, based on a phenolic resin with diazonaphthoquinone as the acid generator and hydroxymethylmelamine as the hardening agent, was developed in 1986 by Feely and coworkers at Rohm and Haas (Figure 1.42).[10] The increase in molecular weight for the exposed dual-tone Rohm and Haas photoresist is the result of a condensation reaction between phenol formaldehyde resins and amino-based cross-linkers. Initially, the Rohm and Haas process drew a lot of attention because it promised to have many miniaturization science applications. Earlier, in Figure 1.8a, we illustrated the process flow options for this resist.[9]

Exposure produces the latent image, which results in the "normal" positive resist relief image when developed at this stage (right-hand path in Figure 1.8a). If, on the other hand, after exposure, the film is heated to induce the acid-catalyzed reaction of the melamine with the phenolic resin, the exposed areas become insoluble, and image reversal is initiated. The reversal is completed by a subsequent flood exposure, solubilizing the nonexposed areas and rendering a negative-tone image of the mask (left-hand path in Figure 1.8a). The cross-linked nature of the image imparts resistance to swelling and dissolution by aqueous and organic solvents, as well as dimensional and thermal stability to temperatures >300°C. By using three-tone masks that have opaque, transparent, and partial transmission (50%), stepped (positive-tone process) and cantilevered (negative-tone process) structures are produced (Figure 1.8b). By appropriate processing of image reversal systems, positive, negative, and vertical walls become possible as well. Figure 1.8c showcases an example of the types of microstructures that can be printed with the Rohm and Haas system. In principle, this resist could be the basis of a complete microstructural universe.

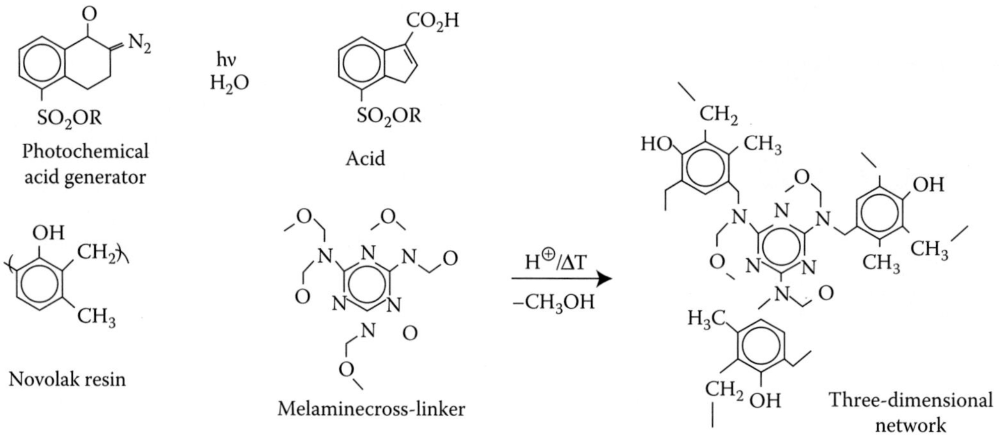


**FIGURE 1.42** Cross-linking of a novolak resin through acid-catalyzed condensation with a melamine for negative aqueous-base development.

## Contrast and Experimental Determination of Lithographic Sensitivity

The resolution capability of a resist, defined as the smallest line width to be consistently patterned, is directly related to resist contrast $\gamma$. For positive resists, the contrast is related to the rate of chain scission and the rate of change of solubility with molecular weight. The latter is very solvent dependent. After development, the thickness of the exposed resist layer decreases until, at a critical dose $D_P$, the film is completely removed. Lithographic sensitivity, $D_P$, and contrast can be obtained from the response curve—a plot of normalized film thickness versus log $D$ (dose) (Figure 1.43a). To construct a curve as shown in Figure 1.43a, a series of positive resist pads of known area are subjected to varying doses and developed in a solvent that does not attack the unexposed film. The thickness of the remaining film in the exposed area is then measured and normalized to the original thickness and plotted as a function of cumulative dosage. Contrast $\gamma_p$ is determined from the slope of this sensitivity or exposure response curve as:

$$\gamma_p = \frac{1}{(\log D_p - \log D_p^0)} = \left[\log \frac{D_p}{D_p^0}\right]^{-1} \quad (1.11)$$

And the lithographic sensitivity $D_p$ is the $x$-axis intersection. For a given developer, $D_p$ corresponds

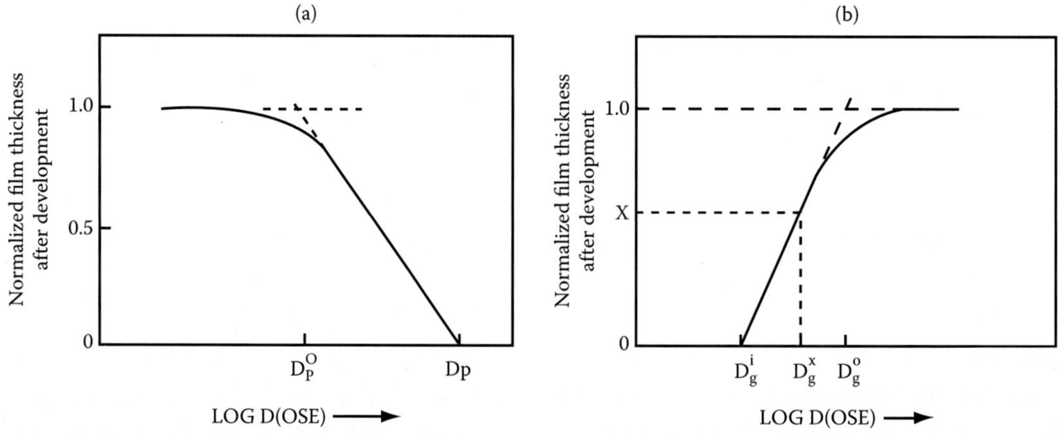


**FIGURE 1.43** Typical response curves or sensitivity curves. (a) For a positive resist. Contrast $\gamma_p$ is determined from the slope. The contrast for a positive resist is markedly solvent dependent. A typical contrast value for a positive optical resist is 2.2. (b) For a negative resist. The value of $D\gamma^x$ usually occurs at 0.5–0.7 normalized thickness as shown in Figure 1.44. The slope determines the contrast, $\gamma_n$. A typical value for the contrast of a negative optical resist is 1.5. (From Thompson, L.F., C.G. Willson, and M.J. Bowden, eds. 1994. *Introduction to microlithography.* Washington, DC: American Chemical Society. With permission.[55])

to the dose required to produce complete solubility in the exposed region while not affecting the unexposed resist. $D_p^0$ is the dose at which the developer first begins to attack the irradiated film. For a dose less than $D_p$ but higher than $D_p^0$, another developer could "force develop" the resist. In force developing, the developer attacks or thins the original, unexposed resist (Table 1.9Ac). The above describes how the profile of a positive resist can be manipulated by the operator.

For a negative resist, contrast relates to the rate of cross-linked network formation at a constant input dose. This is simpler than in the case of a positive resist, where contrast is also very solvent dependent. Consequently, if one negative resist has a higher cross-linking rate compared with another, it also possesses the higher contrast of the two. With negative resists, the onset of cross-linking, as evidenced by gel formation, is not observed until a critical dose $D_g^i$ (also called the *interface gel dose*) has been reached (see Figure 1.43b). In Figure 1.43b, we show the response or sensitivity curve: the normalized developed film thickness versus log dose. Below the interface gel dose, no image can form, as the film thickness is insufficient to serve as an etching mask. At higher doses, the image thickness increases until the thickness of the image equals that of the resist before exposure (in reality, it remains thinner as the film shrinks as a result of cross-linking). The latter dose is shown in Figure 1.43b as $D_g^0$, the dose required to reach 100% polymerization of initial film thickness (before exposure). Contrast, $\gamma_n$, is obtained from the slope of this curve as:

$$\gamma_n = \frac{1}{(\log D_g^0 - \log D_g^i)} = \left[\log \frac{D_g^0}{D_g^i}\right]^{-1} \qquad (1.12)$$

Lithography sensitivity ($D_g^x$) defines the dose cross-linking the film to the required thickness for optimal resolution. The required dose is sometimes defined as the dose resulting in dimensional equality of clear and opaque features (corresponding to nominally equal structures on the mask) imaged in the resist. The so-defined lithographic sensitivity can be determined separately from a plot of feature size versus dose for an opaque and a clear feature of equal size (Figure 1.44). This dose, $D_g^x$, corresponding to the lithographic sensitivity transposed on the *x*-axis of Figure 1.43b, fixes the required cross-linked film thickness after development on the *y*-axis (usually 0.5–0.7 times the normalized thickness). The lithographic sensitivity also may be taken as the dose $D_{0.7}^x$, at which 70% of the original film is retained after development.[27]

Resists with higher contrast result in better resolution than those with lower contrast, and this can be explained as follows. In an exposure, energy is delivered in a diffused manner because of diffraction and scattering effects. Some areas outside the mask-defined pattern will receive an unintended dose higher than $D_g^i$ but lower than $D_g^x$. The resulting resist profile will exhibit some slope after development. The higher the contrast of the resist, the more vertical the resist profile. Because line width is measured at a specified height above the resist/substrate surface, as the resist profile becomes less vertical, the resist line width represents the original mask dimension less accurately.

Values of $D_p$ and $D_g^x$ are figures of merit used only to compare different resists. For lithographic sensitivity numbers to have any value they must be accompanied by a detailed description of the conditions under which they were measured.

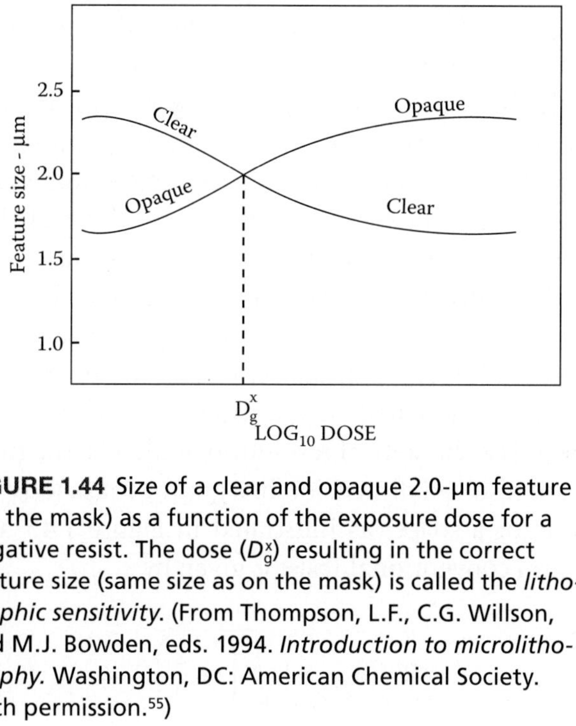


**FIGURE 1.44** Size of a clear and opaque 2.0-µm feature (on the mask) as a function of the exposure dose for a negative resist. The dose ($D_g^x$) resulting in the correct feature size (same size as on the mask) is called the *lithographic sensitivity*. (From Thompson, L.F., C.G. Willson, and M.J. Bowden, eds. 1994. *Introduction to microlithography*. Washington, DC: American Chemical Society. With permission.[55])

## Resolution in Photolithography

### Introduction

A line-width measurement made with a SEM or another metrology tool from the many options reviewed in Volume III, Chapter 6 is used to determine the resolution of a lithographic system. Correct feature size must be maintained within a wafer and from wafer to wafer, as device performance depends on the absolute size of the patterned structures. The term *critical dimension* (CD) refers to a specific minimal feature size and is a measure of the resolution of a lithographic process. We first consider the theoretical resolution limits of different photolithography printing techniques; in the subsequent section, we review how one can go beyond those conventional limits by using resolution-enhancing techniques (RETs). With reference to Figure 1.6 where we compare contact, proximity, and projection printing, we now delve into a mathematical formulation of the expected resolution for each of these printing techniques.

## Resolution in Shadow Printing and Self-Aligned Masks

### *Introduction*

In the shadow printing mode, including contact and proximity arrangements of mask and wafer, optical lithography has a resolution with limits set by a variety of factors. These include diffraction of light at the edge of an opaque feature in the mask as the light passes through an adjacent clear area, alignment of wafer to mask, nonuniformities in wafer flatness, and debris between mask and wafer. Figure 1.45 illustrates a typical intensity distribution of light incident on a photoresist surface after passing through a mask containing a periodic grating consisting of opaque and transparent spaces of equal width, *b*.[51] Diffraction causes the image of a perfectly delineated edge to become blurred or diffused. The theoretical resolution, *R*, that is, the minimum resolved dimension in a grating mask ($b_{min}$ for a line or a space) as illustrated in Figure 1.45 and using a conventional resist is given by:

$$R = b_{min} = k\sqrt{\lambda\left(s+\frac{z}{2}\right)} \tag{1.13}$$

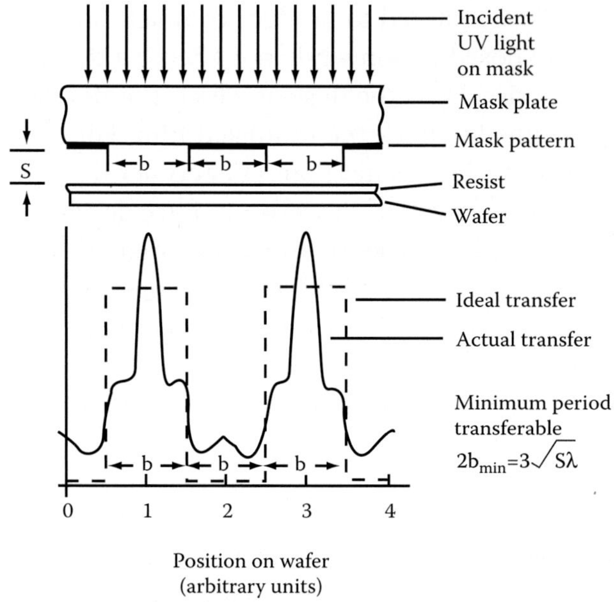


**FIGURE 1.45** Light distribution profiles on a photoresist surface after light passes through a mask containing an equal line and space grating. (From Thompson, L.F., C.G. Willson, and M.J. Bowden, eds. 1994. *Introduction to microlithography*. Washington, DC: American Chemical Society. With permission.[55])

where

$b_{min}$ = half the grating period and the minimum feature size transferable

s = the gap between the mask and the photoresist surface

λ = the wavelength of the exposing radiation

z = the photoresist thickness

k = a constant that theoretically is ~1.5

Depending on the relative magnitude of *s* and *z*, one distinguishes between contact and proximity printing. The square root relation in contact and proximity printing is a consequence of the near-field or Fresnel diffraction theory valid in the near-field region just below the mask openings; this contrasts with the far-field behavior (Fraunhofer diffraction) of projection lithography (see below). These three types of lithography are compared in Figure 1.46 (see also Volume I, Chapter 5 on far-field and near-field imaging).

Self-aligned printing can be regarded as an extreme form of contact printing where there is no gap between the mask and the substrate.

### *Contact Printing*

In contact printing, a photomask is pressed against the resist-covered wafer with pressures in the range

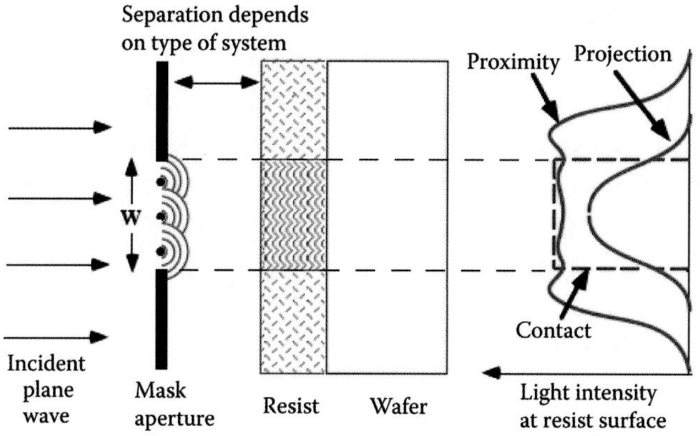


**FIGURE 1.46** Proximity and contact printing are near-field, whereas projection is far-field or Fraunhofer diffraction (see Volume I, Chapter 5).

of 0.05–0.3 atm, and the *s* in Equation 1.13 is zero. In this case, Equation 1.13 reduces to:

$$R = b_{min} = k\sqrt{\lambda \frac{z}{2}} \tag{1.14}$$

With $\lambda = 400$ nm and a 1-μm-thick resist, we conclude that a resolution higher than 1 μm is possible. For thinner resist layers, that is, *z* very small and with shorter wavelengths (e.g., 248 nm), the resolution capability of contact printing further improves. Equation 1.14 clarifies the need to use shorter wavelength and thinner resist layers to achieve higher resolution. However, the theoretical maximum resolution is seldom achieved because only diffraction effects were taken into account to derive Equation 1.13. The other factors mentioned above (e.g., wafer flatness, mask alignment) and shadowing effect (penumbral blur) usually conspire to make the resolution worse. Typical contact printers are the Kasper and the Cobilt 800. The required contact between mask and wafer also causes mask damage and contamination, rendering the method unsuitable for most modern microcircuit fabrication.

### *Proximity Printing*

In proximity printing, spacing of the mask removed from the substrate (by at least 10 μm) minimizes defects that result from contact. On the other hand, diffraction of the transmitted light reduces the resolution. The degree of reduction in resolution and image distortion depends on the wafer-to-substrate distance, which may vary across the wafer. For proximity printing, Equation 1.13, with $s \gg z$, can be rewritten as:

$$R = b_{min} = k\sqrt{\lambda s} \tag{1.15}$$

The technology parameter *k* theoretically is 1.5. Based on Equation 1.15, for a gap of 10 μm using 400-nm exposing radiation and a *k* value of 1.5, the resolution limit is about 3 μm.[56] More typical mask and wafer separations are in the range of 20–50 μm. The smallest features resolvable in a practical UV proximity exposure measure about 2–3 μm for most processes. A typical instrument used for proximity printing is the Canon PLA-600FA (the same setup can also be used for contact printing). Resolution is not as good as in contact printing (see above) or projection printing (see below), and for dimensions less than 2 μm optical projection methods are used.

The main disadvantage of proximity printing is a severe reduction in resolution as a result of diffraction spreading. Resolution can be enhanced by either decreasing the gap at the risk of contact and defect generation or by reducing the wavelength. Using wavelengths in the deep UV or even extreme UV range will not suffice for optical proximity printing to compete with projection printing. However, using x-rays with a wavelength of about 1 nm, feature sizes less than 0.2 μm can be produced with proximity methods. This makes 1× proximity x-ray a promising candidate for NGL.

### *Resolution with Self-Aligned Masks*

The most desirable fabrication processes involve in situ deposited masks, also called *self-aligned* or *conformable* masks. These masks, forming molecular contact with the machine under construction, offer superior resolution as light has no chance to diffract between mask and substrate. They may be regarded as an extreme case of contact printing—that is, *s* is zero with the mask in atomic contact over the whole device under construction. Conformable masks comprise either permanent or sacrificial layers produced in intermediate process steps on the device itself rather than on separate quartz plates.

Figure 1.47 presents an important example of self-alignment with the so-called SALICIDE process, i.e., the **s**elf-**al**igned sil**icide** process, a process in which

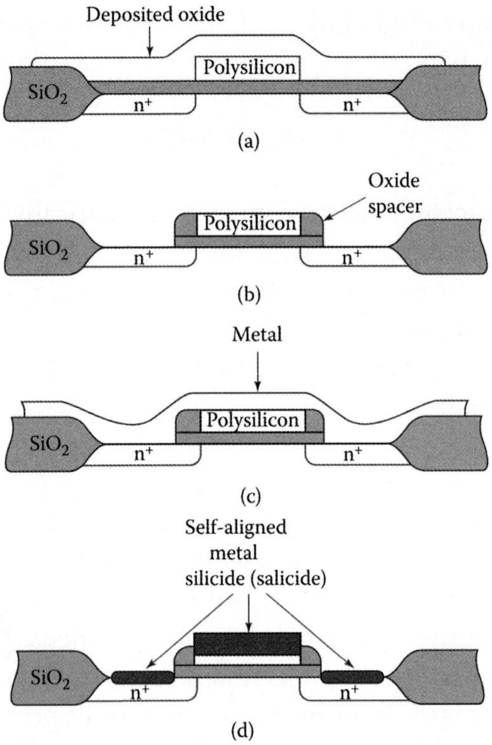


**FIGURE 1.47** Example of a self-aligned silicide (SALICIDE) process in the formation of a MOS transistor. (a) An oxide is deposited over the polysilicon gate. (b) The deposited oxide is RIE patterned, leaving the sidewalls of the poly-Si covered with oxide (oxide spacer). (c) Metal is deposited over the structure and heated to form silicides where naked Si is exposed. (d) Unreacted metal is etched away, leaving silicide automatically aligned to gate, source, and drain region.

silicide contacts are formed only in those areas in which deposited metal is in direct contact with silicon. This self-aligning process is commonly implemented in MOS/CMOS processes in which ohmic contacts to the source, drain, and poly-Si gate are formed by the "salicide" process. The salicide process begins with deposition of a thin transition metal layer over fully formed and patterned semiconductor devices (e.g., transistors). The wafer is heated, allowing the transition metal to react with exposed silicon in the active regions of the semiconductor device (e.g., source, drain, gate) forming a low-resistance transition metal silicide. The transition metal does not react with the silicon oxide or nitride insulators present on the wafer. Following the reaction, any remaining unreacted transition metal is removed by chemical etching, leaving silicide contact metal automatically aligned to gate, source, and drain region.

We explained in Volume I, Chapter 4 that an important advantage of the SALICIDE process in new generation CMOS circuitry is that the entire junction area is used for contact formation, which translates into a lower overall contact resistance. The silicide contact materials on the transistors are then further contacted through chemical vapor deposition (CVD)-deposited plugs of pure W filling the contact vias (see Volume I, Chapter 4).

## Projection Printing

### *Resolution*

In projection printing, wafer contact is completely avoided; a high-resolution lens projects an image of the photomask onto the photoresist-covered wafer. Imaging in this case is far field or Fraunhofer diffraction dominated, as shown in Figure 1.46. A schematic of a projection lithography system is presented in Figure 1.48.

In Volume I, Chapter 5, we calculated (based on the Maxwell equations) the Rayleigh criterion for far-field resolution as:

$$R = 0.61\lambda/NA$$

with $\lambda$ the wavelength of the light used for the pattern transfer, and NA the numerical aperture of the imaging lens system.

Diffraction in a small opening transforms normally incident light into an angular spread as

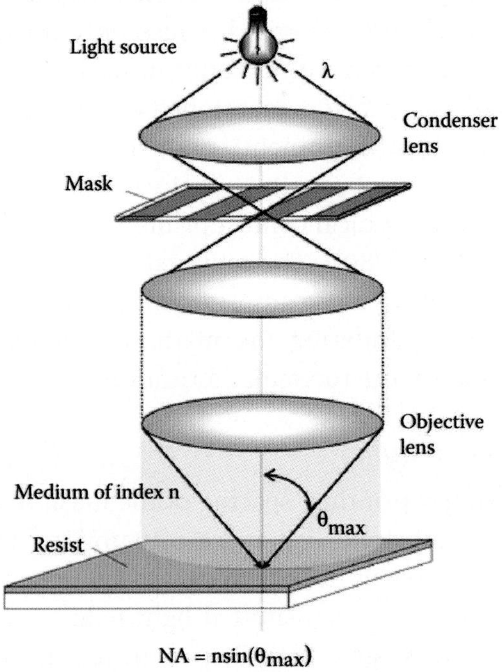


**FIGURE 1.48** Optical projection printing. The basic components of a generic optical projection system.

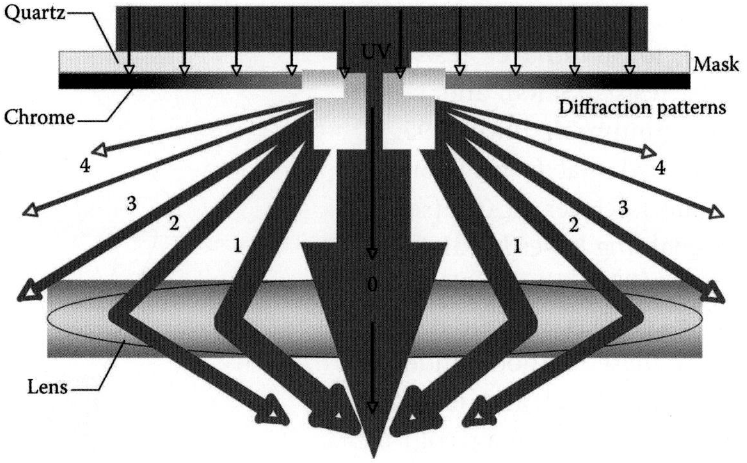


**FIGURE 1.49** The numerical aperture (NA) of a lens tells us how much of this diffracted light the lens can capture and image (see Volume I, Equation 5.3: NA = nsinθ).

illustrated in Figure 1.49. The numerical aperture of a lens determines how much of this diffracted light the lens can capture and image (see Volume I, Chapter 5: NA = nsinθ). Because of this spreading effect, fewer diffracted orders are captured by the lens and form the image. This means that information about the smallest pattern is lost because the smaller the pattern the wider the spread. In Figure 1.50 we show the effect of including increasing numbers of diffracted orders on the image of a slit of width *w*. The more diffracted orders we can pick up, the higher the fidelity of the resulting image. As explained in Volume I, Chapter 5, for absolute image fidelity, we need an NA = 1. Pendry's near-field amplifying superlens (NFSL) offers that prospective. A near-field superlens illustrated in Volume I, Chapter 5 enables subdiffraction limit imaging so that even nanometer objects can be imaged with light of say 365-nm wavelength.

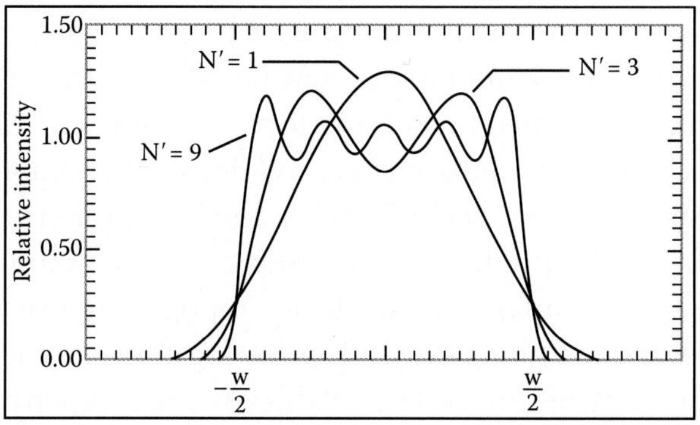


**FIGURE 1.50** The higher the number of diffracted orders a lens can pick, i.e., the higher the NA of a lens, the better the resolution of the image.

The NA of lenses in projection aligners ranges from ~0.16–0.80 (see Figure 1.80). With a good lens, possessing an NA of 0.5, the Rayleigh expression suggests that the best resolution one can obtain roughly equals the size of the wavelength of the exposure system. Although the Rayleigh criterion is too simplistic—it only takes into account characteristics of the optical system—it correctly predicts that a larger NA or a shorter wavelength leads to better resolution.

The derivation of R = 0.61λ/NA (Volume I, Chapter 5) was based on the concept of a "point source" something that does not really exist. Here we want to introduce the practical resolution of a photolithography system starting from the grating equation. From the grating formula $d = \frac{N\lambda}{2n\sin\theta} = \frac{N\lambda}{2NA}$, we calculated that the resolution with incoherent illumination is better than with coherent light [R = 0.5 λ/NA for coherent light (Volume I, Chapter 5) vs. R = 0.25 λ/NA for incoherent light (Volume I, Chapter 5)]. We generalize these results using a dimensionless constant $k_1$ to define the practical limiting resolution *R* in projection printing as:

$$R = k_1 \frac{\lambda}{NA} \tag{1.16}$$

where $k_1$ is the experimentally determined dimensionless parameter that depends on resist parameters, process conditions, mask aligner optics, and so on.

To achieve higher resolutions, we want to work at shorter wavelengths λ, higher NA, and smaller $k_1$.

The $k_1$ parameter fluctuates between 0.3 for surface imaging resists to 0.5 for multilayer resists (MLRs), 0.75 for single layer resists (SLRs), and 1.1 for reflective surfaces like aluminum. Up until 1995, production engineers demanded that $k_1$ be about 0.7 for SLRs; today, its value is closer to 0.5 (see Figure 1.81). Technologies enabling lower $k_1$ values, besides surface imaging and MLRs, include phase-shift masking, antireflective coatings, and off-axis illumination. We will review these methods under resolution-enhancing technique (RET) and learn, for example, that with RET one can obtain a resolution about one-half of the exposure wavelength or even better. The better resolution afforded by a lower $k_1$ makes for a narrower process window and thus a more difficult process; it is expected that one cannot go much lower than a $k_1$ of 0.25 (see Figure 1.81).

### Depth of Focus

To obtain good line-width control, the latent image must remain in focus throughout the depth of the resist layer. A certain amount of defocus tolerance is allowed. The defocus tolerance or depth of focus (DOF or δ) of an optical system is given by:

$$\mathrm{DOF} = \pm\delta = \pm k_2 \frac{\lambda}{\mathrm{NA}^2} \tag{1.17}$$

And by using Equation 1.16:

$$\mathrm{DOF} = \pm\delta = \pm k_2 \frac{R^2}{k_1^2\lambda} \tag{1.18}$$

In Equations 1.17 and 1.18, $k_2$ is a process-dependent constant hovering around 0.5. For a detailed mathematical derivation of the DOF equations we refer to Bowden.[56] From Equation 1.16, we know that to achieve better resolution we must reduce λ and increase NA. As deduced from the DOF equations, the penalty is a reduction in DOF, perhaps becoming so small that a focused image through the total depth of a typical 1.0–1.5-μm-thick photoresist cannot be achieved anymore. A good NA of a lens for a g-line (436 nm) lithography system is 0.54. With a $k_1$ factor of 0.8, this leads to a resolution of 0.65 μm (based on Equation 1.16). With i-line (365 nm) lithography, a resolution of 0.65 μm can be achieved with a 0.45-NA lens while exhibiting a superior DOF of 0.9 μm compared with 0.7 μm for the g-line. Wide-field i-line steppers with variable numeric apertures (Figure 1.51) are available with a 22 × 22-mm field, a 5× reduction ratio, and a resolution of better than 0.35 μm (e.g., the Nikon NSR-2205i14E, http://www.nikon.co.jp/main/eng/news/dec14e_97.htm). This type of equipment allows one to balance resolution, DOF, and wafer throughput for different applications. Because the patterns on wafers have their own topology and wafers may not be perfectly flat, a large DOF is needed to ensure that the image stays in focus over the entire field. If even a small part is out of focus, the final product will be ruined.

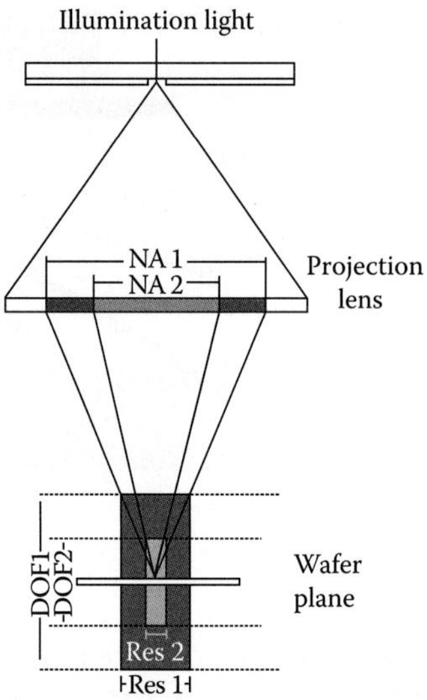


**FIGURE 1.51** Variable numeric aperture stepper.

Photography enthusiasts easily relate to this discussion (see Figure 1.52): a high aperture setting on a camera leads to sharp images for objects that are close to the camera (see rose), whereas a small aperture keeps objects in focus all the way to infinity (see stones).

Microlithographic (or practical) DOF is defined as the total defocus allowable for a desired tolerance on a minimum feature size. It may be different from the values estimated from Equations 1.17 and 1.18. The practical DOF must encompass device topography, resist thickness, wafer flatness, and focus tilt errors. For sub-0.5-μm lithography, only a small amount of residual nonplanarity in device topography can be tolerated without negatively affecting critical dimension control. Small DOF values require expensive planarization processes to bring all IC features in focus within the DOF of the optical system. Some of these planarization processes will be reviewed

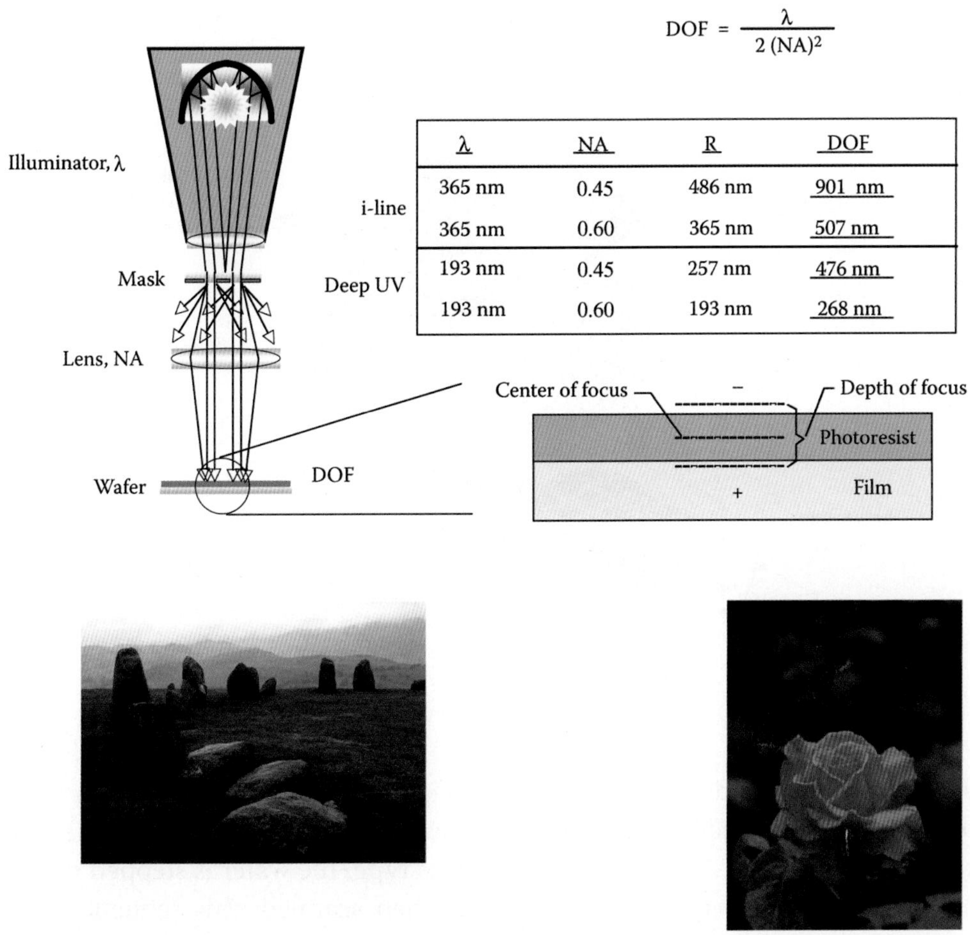


| λ | NA | R | DOF |
|---|---|---|---|
| 365 nm | 0.45 | 486 nm | 901 nm |
| 365 nm | 0.60 | 365 nm | 507 nm |
| 193 nm | 0.45 | 257 nm | 476 nm |
| 193 nm | 0.60 | 193 nm | 268 nm |

**FIGURE 1.52** The concept of depth of focus (DOF) in lithography and photography.

further below. Thin film imaging (TFI) methods, reviewed below, also allow one to work around DOF problems by imaging in thin film resist layers and then transferring that image to underlying thicker resist layers. Miniaturized structures often possess more extreme topologies than ICs, making planarization an even bigger challenge. During the past 20 years, progress in expanding the DOF of semiconductor equipment has not kept pace with the decreasing CDs. For miniaturization science, the former is more important. We will learn how topographical masks and using x-ray and e-beam lithography have enabled the higher DOFs needed for high-aspect-ratio miniaturized 3D machines.

### *Types of Projection Methods*

Since 1973, when Perkin-Elmer (http://instruments.perkinelmer.com) first introduced its scanning projection system for lithography, optical projection of mask patterns has become the standard lithography method. Some of the many projection systems in use include projection scanners (e.g., Perkin-Elmer's Micralign 700 series), 1:1 and reduction (e.g., 5:1 and 10:1 times, often denoted as 5× and 10×), step-and-repeat projection systems (e.g., 10× Electro-Mask), step and scan systems [e.g., ASML's (http://www.asml.com) PAS 5500/900], and double-sided mask aligners [e.g., the SUSS MicroTec MA/BA6 Mask Aligner (http://www.suss.com/products/mask-aligner/ma-ba6-gen2.html)].

A scanning projection system exposes the wafer in a single scan without the benefit of midscan realignment to local alignment marks. The wafer and the mask move simultaneously and continually on an air-bearing carriage through a light arc covering the whole mask and wafer at once. With a deep UV light source, resolution of 1 μm can be obtained, DOF of ±6 μm is possible, and an overlay accuracy of ±0.25 μm (1 σ) has been reached. In practice, these scanners are mainly used for alignment of patterns with CDs in the 3-μm range and for high throughput applications (e.g., 100 wafers/hour).

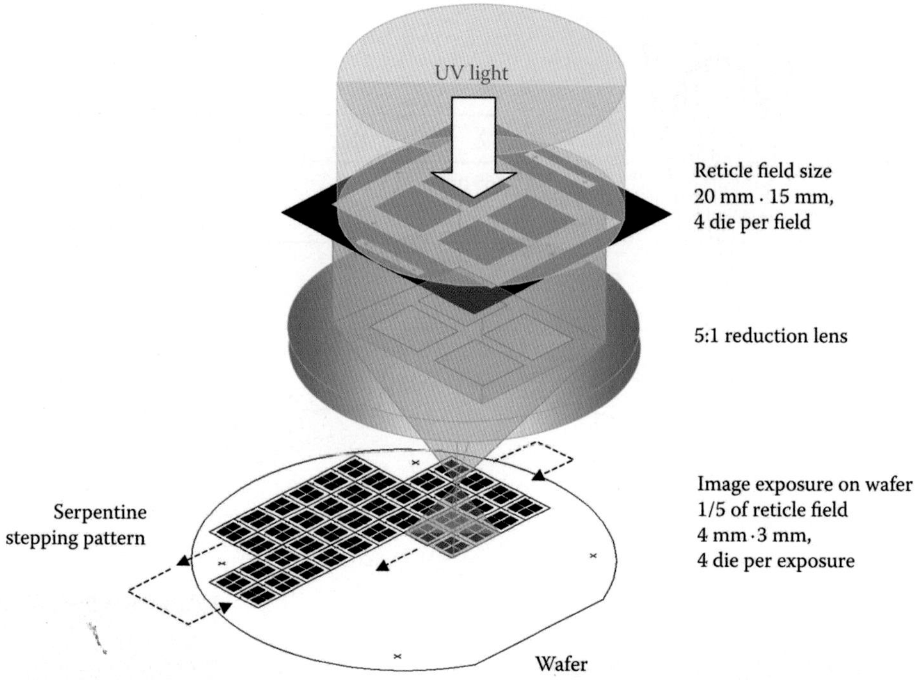


**FIGURE 1.53** Stepper exposure field.

A stepper system, also step-and-repeat, exposes one small part of the wafer followed by a new exposure at the next position (Figure 1.53). By stepping and repeating, the entire wafer is covered with the reticle pattern. Steppers have the ability to align to each field and make adjustments to *x*, *y*, rotation and focus, and tilt. They offer higher alignment accuracies but are slower than scanners. Most steppers use reduction lenses (e.g., with a 10:1 or a 5:1 reduction) rather than one-to-one projection printing. The reduction printer exposes part of the wafer to a pattern from a mask 5–10 times larger than the projected image. The reduction process makes reticle inaccuracies less significant, consequently improving the resolution and resulting in easier mask making. The only drawback pertains to the size of the image field: the higher the reduction ratio, the smaller the image field. Because of lens imperfections and diffraction considerations, projection techniques have a lower resolution for pattern transfer than that provided by a contact or self-aligned mask; however, CDs of 65 nm have already been achieved today with projection lithography.

With steppers, the full exposure field (pretty much the whole lens) is illuminated, and the reticle image is projected in one flash at each location across the wafer. In newer step and scan systems (Figure 1.54), used in 193-nm lithography, only a narrow slit of the wafer is illuminated. In tools of this type, the wafer is stepped to a new field, which is then scanned; this sequence continues until all the fields have been scanned. This method projects a part of the image from the reticle onto the wafer and uses only a small fraction of the lens area. By synchronously moving the reticle stage (up to 1 m/s) and the wafer stage (up to 25 cm/s) in opposite directions, the whole reticle pattern is imaged onto the wafer. With step and scan systems, larger image fields can be exposed (see Figure 1.54), resulting in higher productivity; because a smaller area of the lens is used (only a slit over the center of the lens), each lens can be better optimized for higher resolution and lower distortion.

A mask aligner of particular interest to the non-IC miniaturization engineer is a double-sided mask aligner. With this equipment, features can be aligned on opposite sides of the wafer. An example is the SUSS MA-150 RH with a top-to-bottom precision alignment accuracy of 1 μm (1σ) in production. Another double-sided top and bottom aligner is the 600 series EV640 by Electronic Visions Co. (http://www.elvisions.com). Double-sided mask alignment is discussed in more detail under "Mask Alignment in Projection Printing," this chapter.

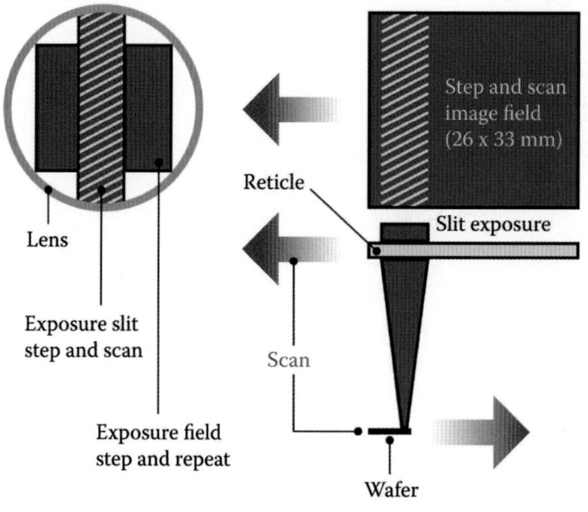


**FIGURE 1.54** Step and scan systems. With wafer steppers, the full image is exposed in one flash, whereas with the step and scan systems, the wafer stage and reticle stage move simultaneously in opposite directions during the exposure to scan the image onto the wafer.

### *Modulation Transfer Function*

In the projection printing of a grating, a series of light intensity maxima and minima are produced, as illustrated in Figure 1.55. Because of mutual interference, dark regions in the image never reach complete darkness, and the maximum brightness never corresponds to 100% transmission. The quality of the image transfer can be expressed by the modulation index (also simply modulation) *M* defined as:

$$M = \frac{I_{max} - I_{min}}{I_{max} + I_{min}} \tag{1.19}$$

In this equation, $I_{max}$ and $I_{min}$ are the peak and trough intensities, respectively. *M* reveals the degree to which diffraction effects cause incident radiation to fall on the resist between the images of two slits in a mask. Ideal optics would give a modulation *M* equal to 1. In practice, exposure systems behave less ideally with an $M < 1$.

The optical imaging quality (that is, the capability of reproducing a mask feature on a wafer surface) for a given projection system can be characterized in terms of the modulation transfer function (MTF) curve. The MTF of an exposure system is defined as the ratio of the modulation in the image plane to that in the object or mask plane (that is, $M_{im}/M_{mask}$) as a function of spatial frequency (ν) or number of line pairs per millimeter on the mask. Because the intensity in the mask plane at the center of an opaque feature is essentially zero, we can equate $M_{mask} = 1$. Consequently, we can also equate MTF and $M_{im}$ as shown in Figure 1.55 where the curve is normalized to 100%. An MTF curve is obtained by imaging a series of gratings with different spatial frequencies placed in the object plane, and the corresponding modulation $M_{im}(\nu)$ in the image plane is measured.[52]

The MTF of an exposure system depends on NA, λ, mask feature size, and the degree of spatial coherency of the illuminating system. Because NA and λ are fixed by the system hardware design, a plot of MTF versus feature size, parametrically changing the spatial coherency, often is used to compare a system's imaging capability. Coherence (σ) varies from $\sigma = 0$ for coherent radiation to $\sigma = \infty$ for fully incoherent illumination. Only point sources are completely coherent with all light waves impinging perpendicular on the mask. In reality, we deal with light sources of finite size as illustrated in Figure 1.56, resulting in partially coherent light. The degree of coherence of the light on the resist plane is dependent on that size, and the resulting light is usually only partially coherent. The definition of coherence is:

$$S = NA_c/NA_o \text{ or also } S = s/d \tag{1.20}$$

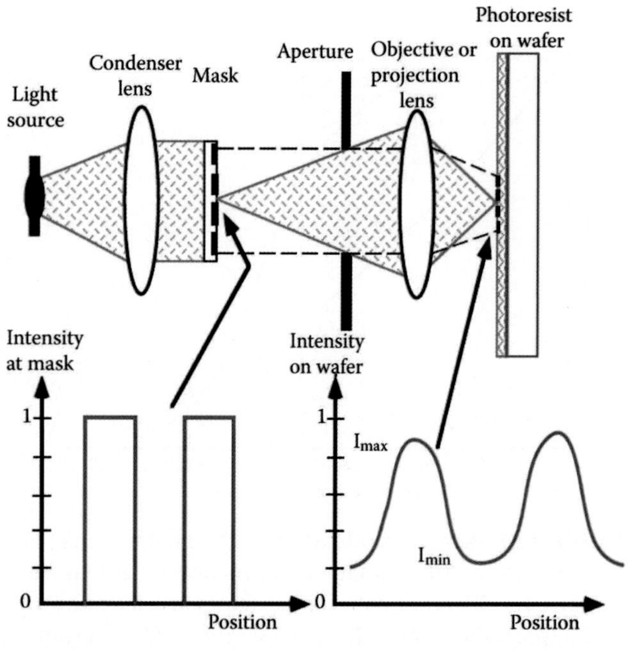


**FIGURE 1.55** Modulation and modulation transfer function (MTF).

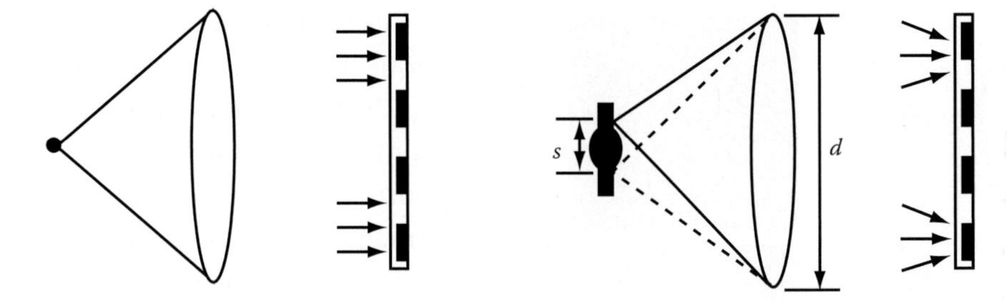


**FIGURE 1.56** Coherent and incoherent light.

where $NA_c$ and $NA_o$ are the NA of condenser and objective lens, respectively, $s$ is the size of the light source, and $d$ is the diameter of the lens (see Figure 1.56).

MTF curves (also the modulation $M_{im}$ in the image plane) for coherent, partially coherent, and incoherent illumination are shown in Figure 1.57. Contrary to the coherent case in which MTF remains constant up to the cutoff frequency $\nu_{max} = NA/\lambda$, the MTF curve for the incoherent case decreases monotonically as the frequency increases up to a maximum value $\nu_{max}$ of $2NA/\lambda$. Because $\nu_{max} = 1/2b_{min}$ and $R = b_{min}$, the maximum resolution for coherent light is predicted to be $0.5\lambda/NA$; for incoherent light it is $0.25\lambda/NA$ (see also the derivation of Equations 5.6 and 5.9 in Volume I). Completely incoherent light produces an MTF curve that has a greater resolution limit, but the MTF value only slowly increases as the feature size increases. For these reasons, partially coherent light often is preferred.[56] Based on various tradeoffs, a coherence value of ~0.7 typically is selected for a practical exposure. The cutoff frequency shown in Figure 1.57 defines the resolution limit of the exposure system. It is proportional to the NA of the lens and inversely proportional to the wavelength. Good sources for a more detailed treatise on the MTF are Bowden,[56] Sheaths,[57] and Wolf.[50]

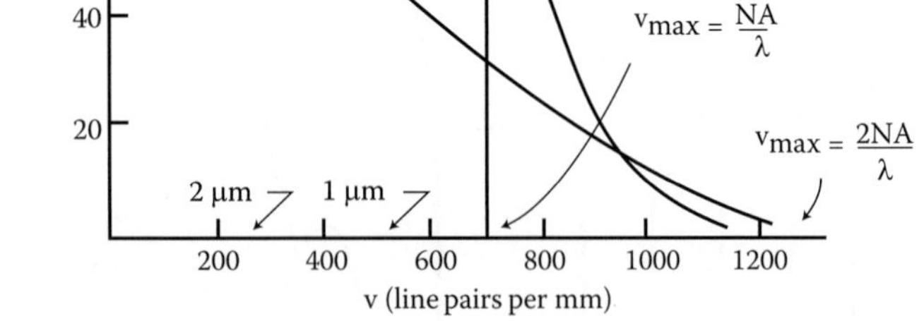

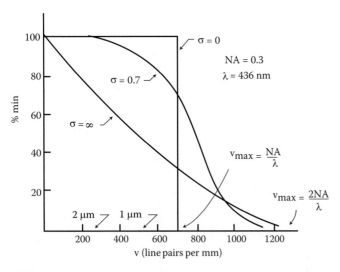


**FIGURE 1.57** Percent modulation in the image plane, % $M_{im}$, modulation of an image as a function of spatial frequency ν for different coherency factors. σ = 0 (coherent), σ = 0.7 (partially coherent), and σ = infinite (incoherent). (From Brodie, I., and J.J. Muray. 1982. *The physics of microfabrication.* New York: Plenum Press. With permission.[52])

### *Critical MTF Values*

For the practical resolution of a resist to approach the Rayleigh limit of the optical system, the resist should have an infinite contrast. Because a resist always has a finite contrast value, a greater degree of modulation is needed before an adequate image can be formed. Thus, the minimum MTF of an optical system to adequately define an image in a resist depends on the resist contrast value γ and is defined as the critical MTF, or $CMTF_{resist}$. The relationship between the resist contrast and CMTF is given by:

$$CMTF_{Resist} = \frac{10^{\frac{1}{\gamma}} - 1}{10^{\frac{1}{\gamma}} + 1} \tag{1.21}$$

A resist material with γ = 2 results in a CMTF value of 0.52. For an exposure system to adequately print a given feature size, the MTF of the feature size must be larger than or equal to the $CMTF_{resist}$ of the resist used. If the MTF of an exposure system is known for various feature sizes, knowledge of the resist contrast and Equation 1.21 will allow prediction of the smallest features printable when applying that system.

### *Mask Alignment in Projection Printing*

*Fiducial Marks* Thus far we have not dwelled much on the alignment or registration of the wafer to the mask in so-called aligners or printers. In projection lithography, the mask, or reticle, is held in place above the projection lens by a reticle stage. The projection lens focuses the high-resolution images of the reticle patterns precisely onto a wafer that is positioned and held in place by the wafer stage. During this critical alignment process, the wafer stepper aligns the mask image to alignment or fiducial marks on the wafer. This ensures that the pattern pictured by each mask layer is precisely aligned and "overlays" the previous layer. In the process of manufacturing a miniaturized device, several thin films are stacked on top of one another; each is patterned differently; and each wafer goes through the wafer stepper 20–30 times. For a detailed description of mask alignment equipment, we refer the reader to the specialized literature.[58] We discuss the types of errors that can be expected from mask alignment below.

Mask alignment errors can have catastrophic effects on the performance of integrated circuits and miniature machines, often rendering them incapable of operating. For example, a device that has a minimum line width between 1 and 1.5 μm can only tolerate a variation of ±0.25 μm with respect to the alignment of masks without encountering greatly increased device failure rates. Product failure is caused mainly by poor alignment between the image being projected and the pre-existing patterns on the wafer. To align one layer with a previously fabricated layer when performing photolithography, appropriate fiducial marks are created on the mask. Different laboratories and foundries use different marks, and these may need to be placed in a specific position on the design. Additional requirements include a scribe lane around each chip to indicate where the wafer is to be cut when it is diced. The inclusion of a unique mask number and an indication of layer names on the mask make it possible to tell by looking how far through the fabrication process a wafer has progressed (http://www.dbanks.demon.co.uk/ueng). In Figure 1.58, we illustrate a mask set with many of the common features: fiducial marks, arrows to point to the marks, test structures, and mask and layer numbers. Optical vernier (Figure 1.59) patterns created on the different levels to be aligned are another common and useful feature. Vernier scales allow us to resolve alignment errors more accurately than the minimum feature size for a given process. To gain this fine resolution, vernier scales depend on accurately spaced lines. A pictorial tutorial on how to use vernier scales in lithography can be found on the Internet at http://www.schlenkent.com/vernier.htm.

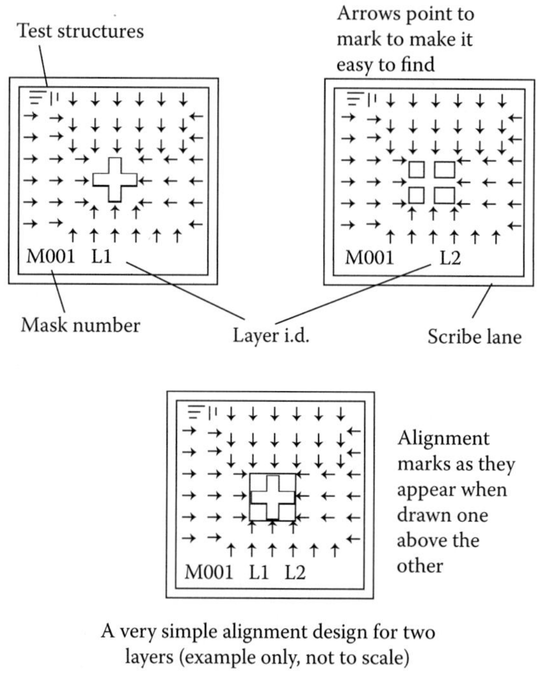


**FIGURE 1.58** A simple set of fiducial marks and other important guiding features often found on masks. (Based on Banks at http://www.dbanks.demon.co.uk/ueng.)

Types of errors encountered in projection printing are illustrated in Figure 1.60. They range from misalignment to mask error, optical distortion, wafer or

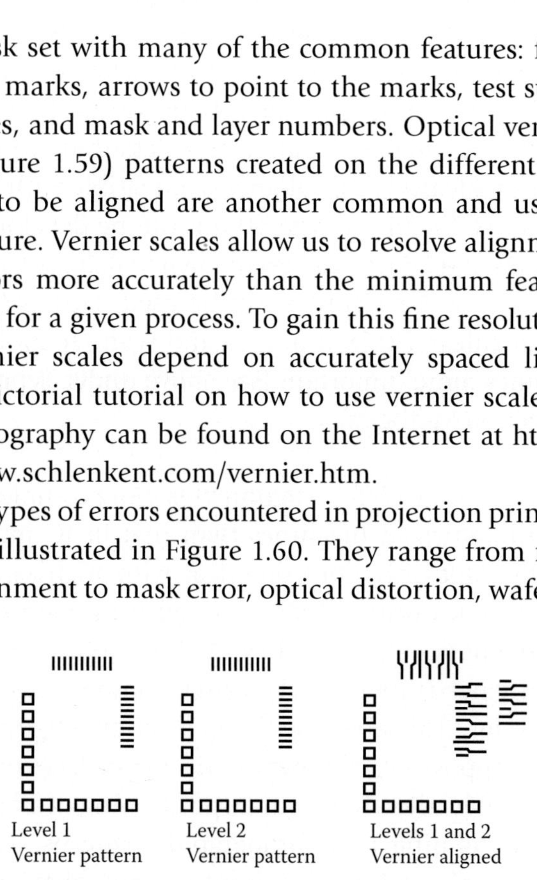


**FIGURE 1.59** Standard optical vernier test pattern for overlay evaluation. (From Moreau, W.H. 1988. *Semiconductor lithography*. New York: Plenum Press. With permission.[28])

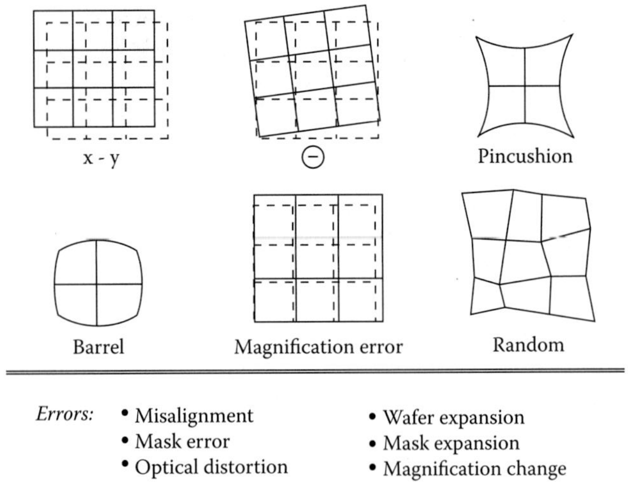


**FIGURE 1.60** Sources of errors in projection printing.

mask expansion, and magnification change. Pattern registration capability is the degree to which the pattern being printed can "fit" mask alignment marks and previously printed patterns. With an optical stepper, the step-and-repeat operation is performed by laser interferometer-controlled stages to a positioning accuracy of <40 nm, thereby allowing the registration of successive layers of a semiconductor device with similar precision. Because interferometric schemes enable sensitivities at less than 10 nm, it appears feasible that aligners compatible with line widths of 0.1 μm or even 50 nm can be developed. For CDs of 0.25 μm and even down to 0.20 μm, conventional wafer steppers can be used. However, at design rules of 0.18 μm, step and scan technology becomes more important (see above under Types of Projection Methods).

*Alignment in MEMS* Alignment is more complex in 3D miniaturized machines than that in IC manufacture. Not only does one deal with high-aspect-ratio 3D features, which can cause problems for alignment systems with low DOF, but one also frequently needs to align 3D features on both sides of the wafer. The objective is to position alignment marks opposite each other on the two surfaces of the same wafer, allowing accurate positioning of all later feature-defining photolithographic patterns. There are several options for front-to-back alignment. One not too elegant or desirable way is by etching holes through the silicon wafer. Another better way is to use infrared to see through the silicon or, with visible light, to use a double-sided mask alignment system defining marks on opposite sides of the wafer with a set of mirrors, two light sources, and two masks. In the latter double-sided mask aligner, mask 2 is first aligned to mask 1; then the wafer is inserted and aligned, and UV exposure of both sides can begin. Wafers polished on both sides are used to minimize light scattering. Two popular commercial double-sided alignment and exposure systems are by SUSS, Germany, and Electronic Visions, Austria. The operation of the SUSS MA-150 is illustrated in Figure 1.61. Cross-hair marks on the mask are aligned to cross-hair marks on the back of the wafer. In the alignment process the alignment marks on the mask are first viewed by two microscope objectives, and the image is electronically stored. The wafer is then loaded with the back-side alignment marks facing the microscope objectives, and the wafer stage is moved and adjusted until the marks are aligned to

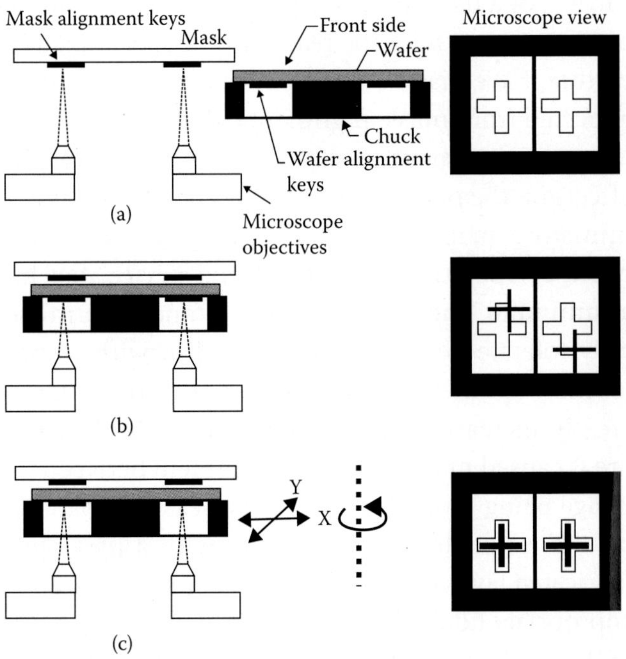


**FIGURE 1.61** Double-sided alignment scheme for the SUSS MA-150 production mode system: the image of mask alignment marks is electronically stored (a); the alignment marks on the back side of the wafer are brought in focus (b); the position of the wafer is adjusted by translation and rotation to align the marks to the stored image (c). The right side illustrates the view on the computer screen as the targets are brought into alignment. [Adapted from product technical sheet (SUSS, Munich, Germany) and Maluf, N. 2000. *An introduction to microelectrochemical systems engineering*. Boston: Artech House.[59]]

the electronically stored image. With the alignment finished, exposure of the mask on the front side of the wafer is completed either in contact or proximity mode. On a SUSS MA-150 RH, with robotic bottom-side alignment (BSA), a standard deviation less than 0.33 μm on front-to-back alignment has been demonstrated; of course, the alignment accuracy depends on a multitude of process parameters, but in a typical production mode this mask aligner can achieve an accuracy of better than 1 μm (3σ).

Because two-sided mask aligners and infrared microscopes are costly, researchers have been looking into less expensive but still accurate alternatives. In the simplest, least-accurate approach, wafer flats can be used as reference for the double-sided alignment with, at best, 5-μm accuracy. In a slightly more sophisticated procedure, White and Wenzel[60] use a simple laboratory jig as shown in Figure 1.62. In this approach, mask 1, containing only alignment marks, is contact printed onto photoresist-coated mask 2 while both are positioned snugly against the three pins on the jig. After developing and etching the alignment marks on mask 2, the individual alignment patterns from the two masks are transferred onto the opposite faces of a semiconductor wafer coated on both sides with photoresist. This is accomplished by sandwiching the resist-coated Si wafer between the two alignment masks (again set snugly against the three pins of the jig), exposing each wafer surface (directly for one side and through the large hole in the jig for the other side). The alignment patterns then are etched into the wafer and used in a conventional one-sided mask aligner. The authors estimate the predictable alignment errors to be less than 1 μm across a 250-μm-thick, 2-in. wafer. Kim et al.[61] describe a different front-to-back alignment technique involving the visual alignment of stepper crossbars (alignment keys) to the center of a transparent thin diaphragm etched from the back of the wafer. These authors claim a <1-μm alignment error, but no experimental evidence is presented. Tatic-Lucic et al.[62] elaborated on the latter type of double-sided alignment method and demonstrated a <2-μm error across a 4-in. wafer. In this upgrade, only alignment marks on the front of the wafer formed in a low-stress insulating layer are required. This insulating layer with the alignment marks is later etched from the back so that a freestanding diaphragm with a cavity underneath it results. This cavity makes the front-side alignment marks visible from the back. This allows visual alignment from both sides of the wafer with a traditional GCA 4800 stepper. Yet another alternative to facilitate double-sided alignment is the pickup of capacitive signals between conductive metal fingers on the mask and ridges on a small area of the Si wafer. This technique, requiring close proximity between mask and wafer, is only applicable for 1:1 proximity lithography as used, for example, in LIGA (see Chapter 10).

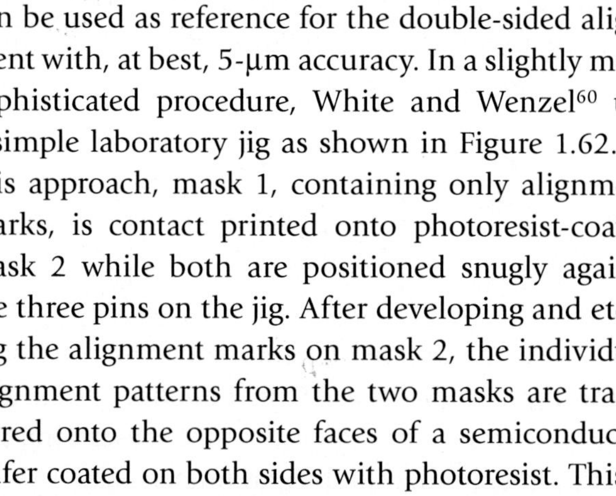

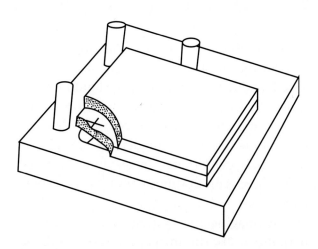

**FIGURE 1.62** Sketch of two-sided alignment jig with the two alignment masks in place for contact printing marks onto mask 2 (top mask). The cross represents an alignment mark on mask 1. (Courtesy of Dr. Richard White, University of California, Berkeley.)

There are numerous problems associated with using classical exposure tools for fabricating highly nonplanar miniature machines. Therefore, maskless exposures to photons (in air) or electrons (in vacuum) with high-precision linear and rotary positioning stages and numerically controlled beam direction and stage position are gaining popularity with non-IC machinists. Nonplanar lithography methods are discussed in Chapter 2 on next generation lithographies.

### *Mathematical Expression for Resist Profiles*

The slope of the resist edge (*dz*/*dx*) of an image of a mask feature of width *W* (see Figure 1.63), for a case where the dose *D* is low and the developer influence dominant, can be written as:

$$\frac{dz}{dx} = \left(\frac{dz}{dD}\right)\left(\frac{dD}{dx}\right) \tag{1.22}$$

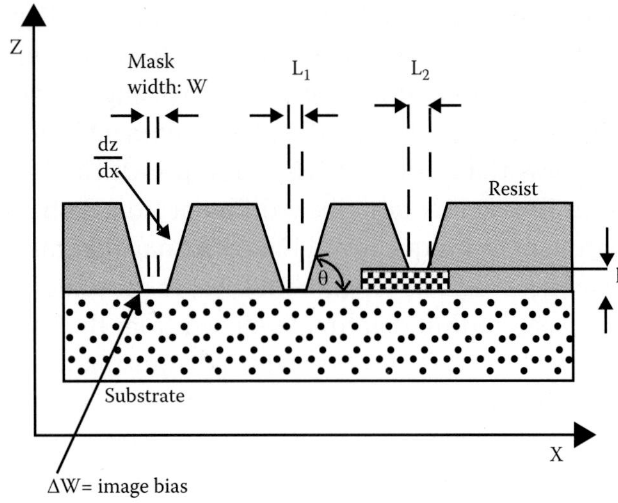


**FIGURE 1.63** Edge slope of a positive resist, *dz/dx* primarily determines the image bias, Δ*W*. The change of line width of a feature developed in resist with a topographic feature of height *h* is given by $L_2 - L_1 = \Delta CD$.

where *z* is the thickness dimension and *x* is the lateral dimension in the resist.

The term *dz/dD* (developer elution term) is only resist- and resist-processing dependent. Additionally, the term *dD/dx*, the derivative of the energy or dose absorbed, also called the *intensity profile*, is a term dependent only on the exposure system and the object that is being imaged. The response of a positive resist to dose *D* is given by $\gamma_p$, the contrast of the resist, and the developer-process term in Equation 1.22 can be approximated by[63]:

$$\frac{dz}{dD} = -\frac{\gamma_p}{D_p} \tag{1.23}$$

where $\gamma_p$ is the contrast of a positive resist as given by Equation 1.11, that is, the slope of the curve of thickness remaining versus log of exposure dose. In Equation 1.11, $D_p^0$ represents the exposure energy below which no resist removal by the solvent takes place, and $D_p$ embodies the sensitivity of the resist or the exposure energy at which no resist remains after development. Exposure energy $D_p^0$ is independent of the thickness *t* of the resist. Resist sensitivity $D_p$, on the other hand, will vary strongly with the thickness, predominantly through the absorbance of the film:

$$D_p \approx D_T.10^{\alpha t} \tag{1.24}$$

with α the resist absorbance per unit thickness and $D_T$ a threshold exposure that depends on resist type and is equal to or smaller than $D_p^0$. With these further simplifications we can rewrite Equation 1.23 as:

$$\frac{dz}{dD} \approx \frac{1}{(a+\alpha t)D_p} \tag{1.25}$$

where *a* is a constant for the resist and α is the absorption coefficient. The *dD/dx* term in Equation 1.22, or the intensity profile, depends on the object and the imaging system. The intensity profile of the image is affected by the wavelength of exposure λ, the numerical aperture NA, the depth of focus of the exposure tool DOF, and the uniformity of illumination and can be written as:

$$\frac{dD}{dx} = \frac{2NA}{\lambda\left[1-k\left(\frac{DOF \times NA^2}{\lambda}\right)\right]^2} \tag{1.26}$$

where *k* is a parameter depending on the coherence of the light source. Based on Equations 1.25 and 1.26, we can rewrite Equation 1.22, describing the resist profile, in identifiable parameters, as:

$$\frac{dz}{dx} = \frac{2NA}{\lambda(a+\alpha t)\left[1-k\left(\frac{DOF \times NA^2}{\lambda}\right)\right]^2} \tag{1.27}$$

It follows that, for a steeper edge slope, *dz/dx*, the thickness and the absorption of the resist should be reduced, and the resist contrast $\gamma_p$ should be as high as possible; finally the shorter the exposure wavelength λ and the higher the NA, the sharper the image profile. Because a higher NA means a lower DOF, which deteriorates image profiles at larger depths, there is an optimum NA for each exposure/resist system.

When imaging the grating of Figure 1.55 with a period p = 2b, the image intensity will be sinusoidal [D ≈ 1/2 (1 + M cos 2πx/p)], where *x* is the lateral dimension, and *M* is the normalized modulation. The maximum image edge slope is then simply dD/dx = πM/p, and Equation 1.27 may then be written with the maximum of the $dz/dx$ term expressed as a function of the normalized modulation $M$, or:

$$\frac{dz}{dx} = \frac{\pi M}{2(a+\alpha t)D_p b} \quad (1.28)$$

From this expression, we would want to work with the largest possible image modulation $M$.[63] Equation 1.27 also implies that a high-contrast resist with a low absorbency will have wider exposure latitude and can tolerate a larger intensity variation in exposure system output ($D_p$).

The image bias or error, $\Delta W$, shown in Figure 1.63, is primarily a function of the resist edge slope and grows smaller as $dz/dx$ becomes steeper. At a high radiation dose (>100 mJ/cm$^2$), the profile of a positive resist is dominated by the absorption of radiation, the reflected photons, and the quantum yield [$\Phi$ or $G(s)$] of the photochemical reaction (see Table 1.9Aa). The absorbed energy per depth $z$ dominates the rate of dissolution. In a developer-dominated (also force-developed) resist profile (also overcut), the profile recedes, and thinning of the whole resist layer occurs (see Table 1.9Ac). This thinning does not occur in the case of insolubilized traditional negative resists, rendering those resists less susceptible to overdevelopment (see Table 1.9B). For traditional negative resists, $\gamma_n$ is not influenced by the solvent, and apart from some uncontrollable swelling of negative resists during development, the resist profile cannot be manipulated in the development step. By using a normal exposure dose with moderate developer influence, a "perfect" image transfer can be accomplished with a positive resist as shown in Table 1.9 Ab. At those moderate doses, both the developer elution term and the energy absorption term contribute to the formation of the resist profile.

From Figure 1.63, we can see how the exposure gradient results in the developed positive resist profile having a larger opening on top, forming a so-called normal or overcut profile. This profile results in a difference of dimension in subsequent patterning steps over features with different topography. The change in CDs, $\Delta CD$, relates to profile angle $\theta$ and topography height $h$, and can be expressed as:

$$\Delta CD = L_1 - L_2 = 2h\,(\tan\theta)^{-1} \quad (1.29)$$

A lithography simulator such as PROLITH/2 (http://www.finle.com/index.asp) or Sample (http://cuervo.eecs.berkeley.edu/Volcano/docs/sample3D.html) may be used to perform resist profile simulations.[64] Experimentally determined parameters are fed into the simulated models to make the predicted resist profiles more accurate. An example is the measurement of deprotection reaction parameters of chemically amplified (CA) resists.[65] The concentration of the protection groups determines the solubility of the resist, and consequently the simulation parameters related to the deprotection reaction will have a significant influence on the accuracy of profile simulations of CA resists. The measurement is carried out by incorporating a PEB system inside a Fourier transform infrared (FTIR) spectrometer.

## Planarization

### *Introduction*

Microlithography process latitude can become severely limited by nonsmooth topography. Resist films crossing over steps have their local thickness altered; thinning occurs over high features, and pileup happens in low-lying regions. During exposure, the thin resist may get overexposed and the thick regions underexposed. Moreover, resist pileup regions may exhibit/standing wave effects (see below), leading to resolution losses. Reflective notching on buried topographical features could decrease resolution even further (see Figure 1.64). This rough topography creates problems for uniformity of subsequent layers, creating a need to planarize surfaces after each new layer. As seen above, practical DOF should encompass device topography, resist thickness, wafer flatness, focus, and tilt errors for the projected image to remain sharp over the whole wafer. Shallow DOF of high NA-optical-exposure tools, standing waves, and reflective notching can be solved by using very thin resist layers on planar substrates or by planarizing resist layers on nonplanar substrates. Chip makers stack metal lines four layers high, and five- and six-level devices are possible. These "high-rise" chips need rigorous interlevel planarization. The SEM image in Figure 1.65a is a cross-section of a modern IC device showing the various metallization lines (yellow) embedded

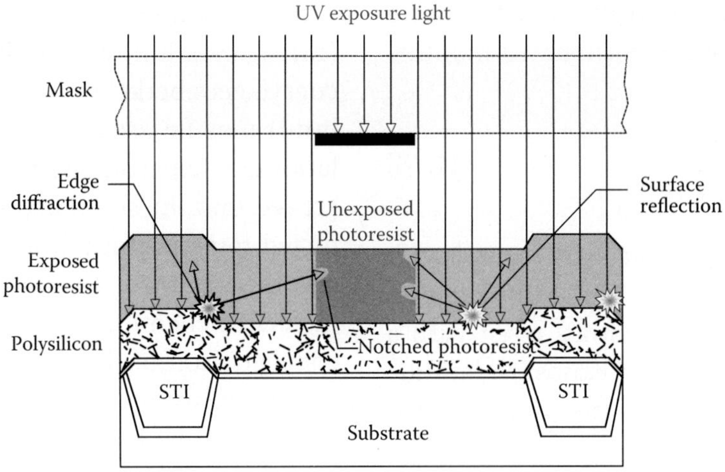


**FIGURE 1.64** Reflective notching degrades resist profile. STI, shallow trench isolation.

in passivation material (e.g., silicon nitride). In Figure 1.65b we show an SEM section through 4-MB dynamic random access memory (DRAM) cells. The section was made by using a focused ion beam microscope a technique capable of performing corrective microsurgery on IC chips. The extremely high features often encountered in micromachining pose further challenging planarization problems. Strategies often rely on an interplay of deposition and removal of sacrificial layers (e.g., resist); some of the terminology used in this section will become better understood after reading Chapter 3 on dry etching and Chapter 7 that covers additive processes.

### *Planarization Strategies*

Planarization provides a means of having very flat wafers even after previous patterning steps, and allows for very high-resolution (high *R*) lithography with limited depth of field (low DOF). Planarization methods provide degrees of smoothing rather

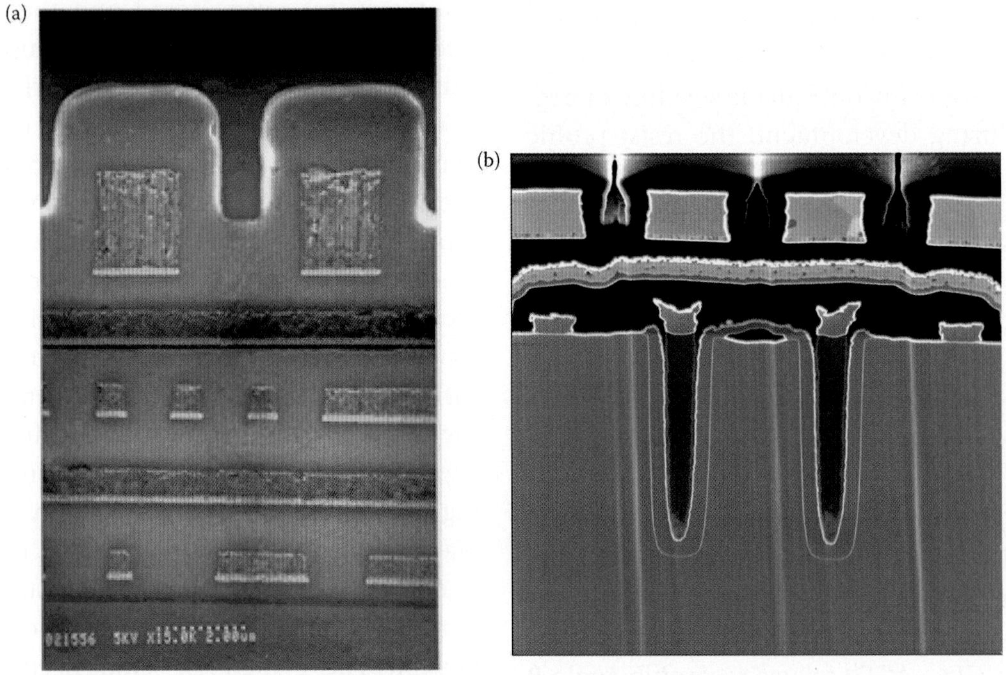


**FIGURE 1.65** (a) SEM image of a cross-section of a modern IC device showing the various metallization lines (yellow) embedded in passivation material (silicon nitride, for example. (Courtesy of IBM.) (b) SEM section through 4-MB DRAM cells. The section was made using a focused ion beam microscope. (Courtesy of FIB International.)

than true absolute flatness. In IC fabrication, the chemical-mechanical polishing (CMP) process is used for interlevel dielectric layers (IDL), shallow trench isolation (STI—see Figure 1.64), and copper metallization (the "damascene" process).

A major cause for nonsmooth topography is chemical vapor deposition (CVD) of dielectric layers. CVD methods experience difficulty depositing thin layers conformably. More film tends to build on flat than vertical surfaces, and the pronounced tendency for narrow spaces to be covered (shallow trench isolation) before being completely filled in creates voids. Smoothing in this case should entail two components: filling in gaps and planarizing a film's top surface.

In the planarizing etch-back process, hilly contours left behind by, for example, a CVD oxide deposition technique, are planarized by spin-on glasses (SOGs) or resist sacrificial layers, after which both sacrificial layer and oxide are etched back simultaneously. As its name implies, a SOG dielectric is spun onto a wafer as a liquid and then cured at elevated temperatures. The etch-back process is adjusted, usually by adding oxygen to the etch gases to cause the sacrificial layer and the underlying film, principally oxide or other dielectrics such as PECVD silicon nitride, to etch at equal rates (see Chapter 3). In the etch-back planarization, photoresists are more popular than SOG because they tend to provide a more planar result.[66] SOG often is used in the dielectric sandwich approach for narrow gap filling. Using many thin dielectric layers piled on top of one another until the narrow spaces between metal lines are completely filled averts voids. In this sandwich, layers of CVD oxide alternate with coats of a SOG dielectric. Etching back with photoresist may then planarize the top surface of the film.

There are two types of SOGs: silicates (inorganic) and siloxanes (organic). The film properties of both resemble those of low-temperature CVD glasses. Silicate films tend to absorb water, resulting in a higher relative dielectric constant. In both cases, films tend to crack because of tensile strength as they are densified at 300–500°C. Film stress may be as high as 200 MPa in tension, but annealing lowers tension considerably. Silicate glasses are more prone to cracking, but doping with phosphorus makes them more crack resistant. Even phosphosilicate glasses can only be applied in layers of 1000 Å; thus, often two or three coats are required. It is easier to apply siloxane in a thicker film, for example, greater than 3000 Å. Hitachi Chemical Company America, Ltd. (http://www.hitachi.com) has developed an organosiloxane SOG able to deposit films thicker than 1 mm. As gaps in shallow trench isolation become increasingly narrow, the gap-filling to planarize a surface becomes more difficult. At gap spacings less than 0.3 μm, current SOGs fail to fill the gap completely and are prone to leave a small circular or teardrop-shaped void. Such voids appear with increasing frequency as the gap size decreases and as the gap aspect ratio increases. The effect is most pronounced with re-entrant angles.[67]

Replacing the oxides with a type of dielectric that fills and planarizes much the same as a liquid film may help to eliminate the severe topologies that CVD oxides create. Polyimides fit that dielectric by forming crack-free films that have virtually no pinholes, absorb stress, and generally exhibit a much lower dielectric constant than oxides. Currently, polyimides that will planarize a large pitch geometry to a 90% level with a single coat are available. Reluctance to expand the use of polyimide at this point is linked to its poor barrier properties toward moisture and ions. Also, the prevailing tendency is to trust inorganic CVD oxides more than organic polyimides.[66]

Because the electrical properties of SOGs are poor compared with CVD silicon oxides, an ideal solution would be if the CVD process itself filled the spaces between metal lines and planarized the top surface without breaking the vacuum within a single processing chamber. One example of this type of processing is electron cyclotron resonance CVD (ECRCVD) (see Figure 7.45 illustrating planarization using ECRCVD). ECRCVD uses ECR to generate high-density plasma that can deposit CVD $SiO_2$ at high rates while temperatures and pressures remain low. As deposition proceeds, the wafer is radio frequency-biased so that argon atoms from the plasma can simultaneously sputter-etch the substrate. The sputtering keeps submicrometer spaces open until they are filled. By carefully balancing the two processes, one can deposit a film and sputter it back in such a way that high-aspect-ratio spaces are completely filled while the oxide surface planarizes. If planarization needs improvement, ECRCVD can be combined with an etch back.

CMP combines mechanical action with chemical etching; it typifies a planarization technique where a wafer with an uneven surface is polished in an abrasive suspended in an alkaline solution on a polishing pad (see Figure 1.66). Material removal rate is controlled by the slurry flow and pH, applied pressure on the polishing head, rotational speed, and operating temperature. Although slow (with removal rates less than 100 nm/min), CMP results in large surfaces with a roughness less than 1 nm.

The CMP process has been around for a long time but became popular only in the 1990s when it became one of the crucial steps in the manufacturing of IC chips. The reason CMP became crucial is because of its ability to achieve defect-free global planarization of surfaces. The chemicals in the slurry have a "softening" effect on the surface to be polished, thus reducing damage when the abrasives abrade the surface. Oxides, polysilicon, and metal topography can be planarized this way. For oxide or silicon polishing, an alkaline slurry of colloidal silica ($SiO_2$ particles in a KOH solution or $NH_4OH$) is continuously fed to the pad/wafer interface. CMP is also popular with micromachinists. For example, Sniegowski[68] is using CMP to enhance the manufacturability of polysilicon surface micromachinery.

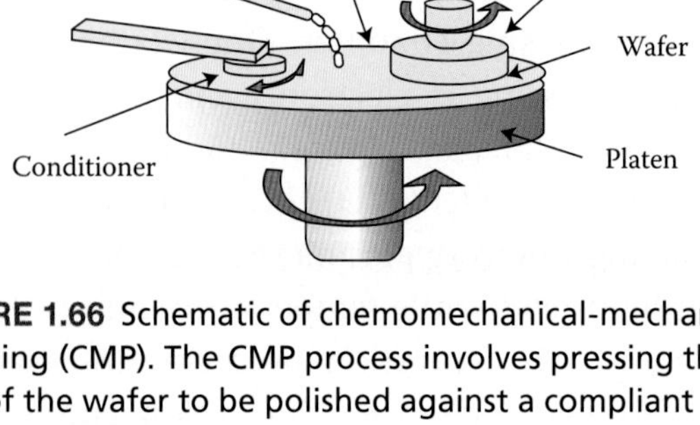

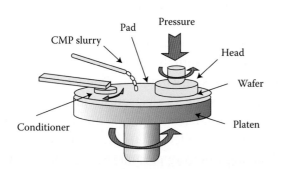

**FIGURE 1.66** Schematic of chemomechanical-mechanical polishing (CMP). The CMP process involves pressing the face of the wafer to be polished against a compliant polymeric polishing pad and generating relative motion between the two surfaces. A slurry consisting of abrasives and chemicals is fed in between the interface between the wafer and the pad.

CMP planarization alleviates processing problems associated with the fabrication of multilevel structures, eliminates design constraints linked with nonplanar topography, and provides an avenue for integrating different process technologies. A typical CMP setup looks very similar to a lapping machine, but the precision is much higher.

The damascene process (named after an art form from the Damascus, Syria area) for inlaid metal objects (see Chapter 8, Figures 8.41 and 8.42) used for copper metallization is also dependent on CMP as one of its key steps.

A thin film of Au, Al, or Cu can effectively planarize when melted briefly with a laser. Planarization occurs rapidly because of the high surface tension and low viscosity of clean liquid metals.[69]

## Photolithography Resolution Enhancement Technology

### Introduction

Since the mid-1980s, the demise of optical lithography has been predicted as being only a few years away, but each time a new resolution limit approaches, some new method extends the useful life of the technology. The three main technologies involved in printing ICs and other miniaturized devices are resist technology, mask technology, and the exposure tool, all of which need to be addressed to optimize lithography resolution.

From the previous section, we learned that optimized production lithography uses projection printing operated close to the conventional Rayleigh diffraction limit with features about the size of the wavelength of the exposure source printed. In this section, we learn about resolution enhancement technology (RET) enabling subdiffraction printing through improvements in resist strategies, mask engineering, and exposure tools. However, the cost of RET is often prohibitive, and in the past, printing with shorter wavelengths has proven more economical.[25] But with CDs of 65 nm, one is forced to rely on RET. Miniaturization science is well suited to contribute to the development of new RET techniques, especially in the area of new mask engineering.

## Strategies for Improved Resolution through Improved Resist Performance

### *Chemically Amplified Resists*

Quantum yields for typical positive-tone resists are 0.2–0.3; for negative-tone resists they are 0.5–1 (see Table 1.6). As several photons are required for one useful scission (positive resist) or polymerization (negative) event, this places a fundamental limit on the photosensitivity of these resists. The situation is worse at shorter exposing wavelengths. Typical near-UV positive resists (e.g., DNQ-sensitized novolak resins) are not useful for deep UV lithography because of the strong unbleachable absorption at less than 300 nm (reminder: bleaching is the decrease in optical density during exposure enabling further penetration of the resist as the top layers are being cleared). The short wavelength absorption coefficients of the base resin and the added sensitizers are too high to allow uniform imaging through practical resist thicknesses (0.5–1 μm for ICs).

In the late 1970s, commercially available resists performed poorly at short wavelengths; those with a good plasma etching resistance, frequently based on novolak resins, had excessive unbleachable absorption, whereas those with an acceptable transparency, such as methacrylate-based resists, had insufficient etching resistance. This is of even greater concern in miniaturization science where often much thicker resist layers are used. Short wavelengths do not penetrate thick resist films completely, and very long exposure to dry or wet etchants, needed to create the desired features, degrades the resist. An additional problem, touched on before, is that deep UV optics absorb more light, thus making less light available at the resist plane.

To circumvent the intrinsic sensitivity limitation of low quantum yield resists, chemical amplification was proposed in 1980 and developed at IBM (http://www.ibm.com) in the early 1980s.[37,70–72] In this approach, a single photon initiates a cascade of chemical reactions of the sort that characterizes a silver halide photographic emulsion system. This amplification or gain is based on the photogeneration of a catalytic photoproduct, often an acidic species that catalyzes the scission of the base resin for a positive-tone resist or the cross-linking of the resin for a negative-tone resist. The overall quantum efficiency of the catalyzed reaction is higher than the efficiency for the initial acid generation.[73] Within a resist, which has been sensitized with an onium salt such as diphenyliodonium hexafluoroarsenate, a Lewis acid is released on photolysis. Compared with conventional free-radical initiators, the onium salts have excellent thermal stability and are not sensitive to oxygen.[29] In the case of a positive resist, the released acid, on baking at 100°C, may catalyze the cleavage or scission of the resist, making it more soluble. This type of imaging involves the usual formation of a latent image during exposure. However, the latent image, a 3D distribution of the catalytic photoproduct, does not immediately generate a concomitant change in dissolution rate, as do regular resists. Image formation only takes place after an activating thermal step—to diffuse the photogenerated acid (PGA) and complete the reaction—in a postexposure bake (PEB) at 100°C. A typical turnover rate for one acid catalyst molecule is in the range of 800–1200 cleavages. Resists amplified this way may attain a photosensitivity of 5 mJ/cm$^2$ or better. Tenfold sensitivity increases are common. Both the acid-catalyzed scission and cross-linking reactions are highly dependent on the PEB temperature, time, and method. The control of these parameters represents a significant difference with conventional resists. In Figure 1.67, a schematic representation of a generalized chemically amplified resist process is shown, as well as typical chemicals. Depending on the developer being polar or nonpolar, a positive or a negative image can be generated. Their dual-tone nature makes these resist systems more flexible. Chemical amplification resists are continually improving and today constitute an important foundation for the design of advanced resist systems for use in short-wavelength (<300 nm) lithographic IC technologies and miniaturization science applications. It was the advent of chemical amplification that made 248-nm lithography finally possible.

The pioneering chemical amplification work involved a catalytic deprotection scheme based on poly(tert-butoxycarbonyloxystyrene) and a triphenylsulfonium hexafluoroantimonate onium salt (Figure 1.68), in which a thermally stable, acid-labile

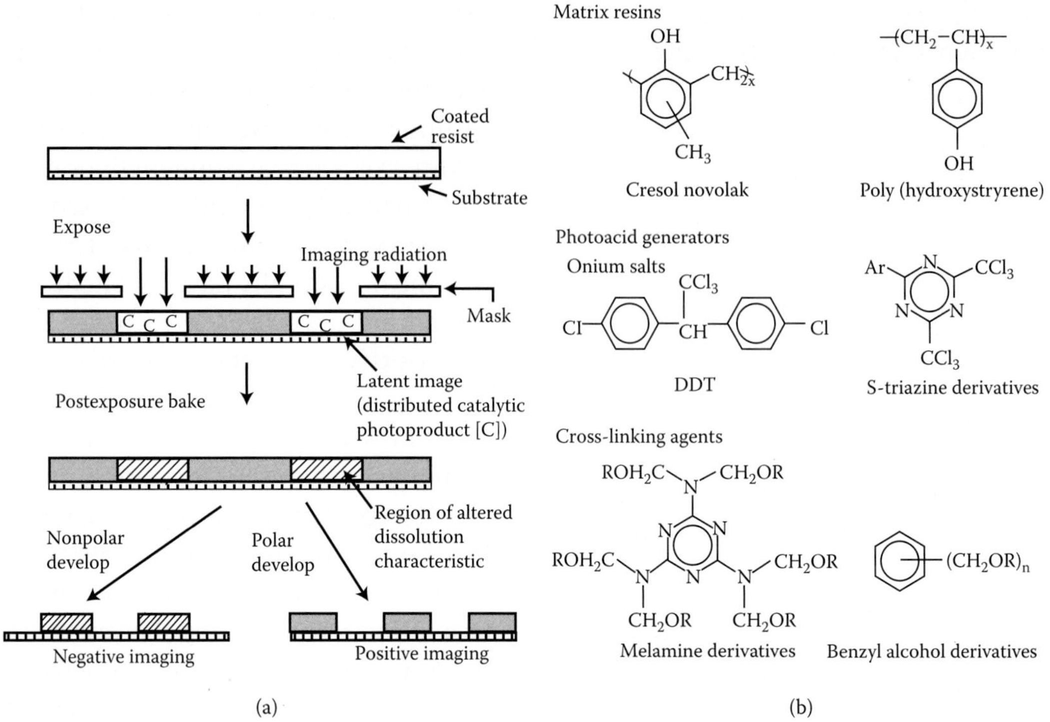


**FIGURE 1.67** (a) Schematic representation of a generalized chemically amplified resist process. Depending on the developer being polar or nonpolar a positive or negative image can be created. (b) Typical chemicals used in amplification type resists. [Part (a) based on Reichmanis, E., L.F. Thompson, O. Nalamasu, A. Blakeney, and S. Slater. 1992. Chemically amplified resists for deep-UV lithography: a new processing paradigm. *Microlithogr World* November/December:7–14;[74] Part (b) based on Lamola, A.A., C.R. Szmanda, and J.W. Thackeray. 1991. Chemically amplified resists. *Solid State Technol* 53:53–60. With permission.[75]]

tert-butoxycarbonyl group (tBOC) is used to mask the hydroxyl functionality of poly(vinylphenol).[24]

In this amplification scheme, development is based on polarity changes rather than molecular weight changes through scission and polymerization. On exposure to deep UV and subsequent baking to diffuse the photogenerated acid (PGA) and complete the reaction, the acid cleaves the labile tBOC protecting group to form a polar polyvinyl phenolic polymer. Deprotection (cleavage) of pendant groups

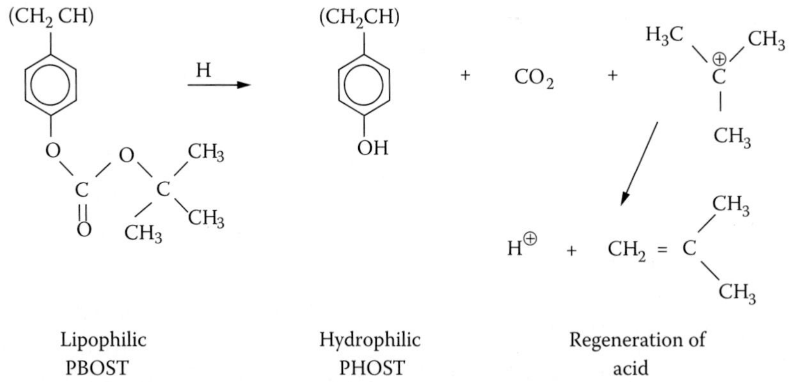


Acid-catalyzed deprotection for polarity change (tBOC resist); cleavage of a pendant group to convert a lipophilic polymer (PBOCST) to a hydrophilic polymer (PHOST)

**FIGURE 1.68** Chemical amplification with a catalytic deprotection scheme based on poly(tert-butoxycarbonyloxystyrene) (PBOCST) and a triphenylsulfonium hexafluoroantimonate onium salt, in which a thermally stable, acid-labile tert-butoxycarbonyl group (tBOC) is used to mask the hydroxyl functionality of poly(vinylphenol). (Reichmanis, E., O. Nalamasu, F.M. Houlihan, and A.E. Novembre. 1999. Radiation chemistry of polymeric materials. *Polym Int* 48:1053–59.[24])

in tBOC induces a polarity change and allows for dual-tone (positive/negative) imaging. By using a nonpolar solvent, the unexposed resist is removed and a negative image formed. A polar solvent is used to remove the exposed area, resulting in a positive image. By the mid-1980s, tBOC was used in the production of millions of 1-MB DRAM devices at IBM.[37]

We encountered dual-tone resists earlier when reviewing the novolak/DNQ system by Rohm and Haas (see Figure 1.8).

The chemical amplification resist family today has become prolific and can be grouped into categories according to imaging mechanisms: deprotection (see tBOC), depolymerization, rearrangement, intramolecular dehydration, condensation, and cationic polymerization (SU-8). Cationically polymerizing monomers such as epoxies and vinyl compounds and condensation reactions between phenol formaldehyde resins and amino-based crosslinkers increase the molecular weight after exposure. Changes in polarity are achieved through the acid-catalyzed deprotection of a variety of esters.

Although chemically amplified resists provide reasonable throughput for e-beam and deep UV lithographies (short wavelengths), they often suffer from poor environmental stability and too high a sensitivity to processing conditions. Environmental effects (oxygen) cause surface inhibition by neutralization of photogenerated acid in the top resist layer. A resist topcoat can be used to minimize this effect. Unfortunately, such a multilayer resist further complicates the photoresist process.

### *Antireflection Coatings: Thin Film Interference Effects*

When using single-layer resists, the deleterious effects of thin film interference effects need to be overcome for good CD control, especially of submicrometer features. The most important thin film interference effects are schematically illustrated in Figure 1.66. The effects manifest themselves as nonvertical resist profiles, line-width variations, reflective notching, scumming (underexposed resist leaving organics behind after development), and alignment inaccuracies. Solutions to thin film interference effects involve the use of antireflective coatings and the use of MLRs (see below). Antireflective coatings overcome interference effects caused by reflections from either the top or bottom of the resist (Figure 1.69a and b). Example antireflective coatings are the ARC coatings (trademark from Brewer; http://www.brewerscience.com), liquid resins that are cast into a film before coating the photosensitive resist layer. A key design element is for the cast film to absorb light very efficiently at specific wavelengths. In designing the light-absorbing properties of ARC coats, two strategies were pursued. One is the mixing of dyes with a polymer resin. The second is to design the polymer with molecular chains that intrinsically absorb light.

When a resist is exposed to monochromatic radiation, standing waves are formed in the resist as a consequence of coherent interference from reflecting substrates creating a periodic intensity distribution in the direction perpendicular to the surface (see Figures 1.69c and 1.70). The standing wave effect leads to light intensity variations, perpendicular to the resist film, of as much as a factor of three from one position to another. The standing wave effect is a strong function of resist thickness, and exposure variations resulting from variation in resist thickness in the vicinity of steps result in changes in

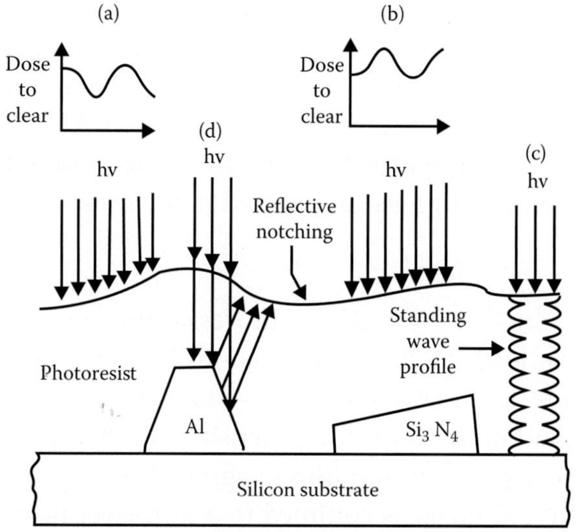


**FIGURE 1.69** Thin film interference effects. Light reflected from resist/Si (a) or $Si_3N_4$/Si (b) interfaces through various resist or nitride thicknesses changes the dose necessary to clear the resist. Also, standing-wave profile (c) and reflective notching effects (d) are shown. (From Horn, M.W. 1991. Antireflection layers and planarization for microlithography. *Solid State Technol* 31:57–62. With permission.[76])

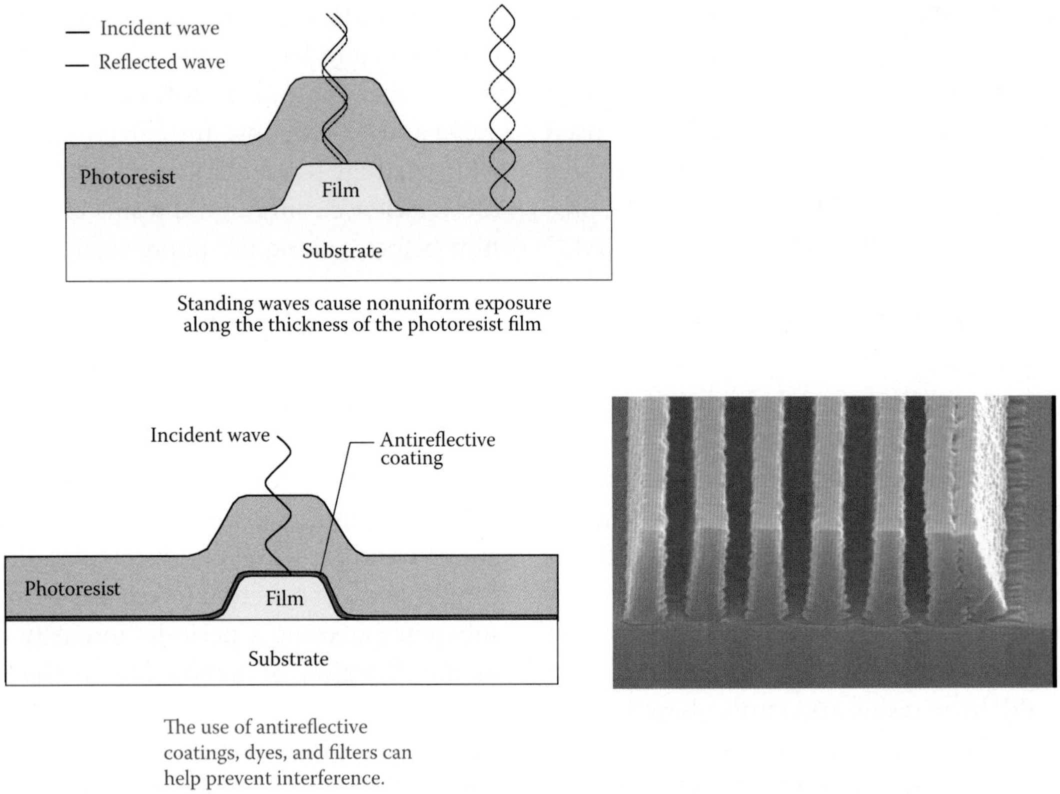
Incident wave
Reflected wave
Photoresist
Film
Substrate
Standing waves cause nonuniform exposure along the thickness of the photoresist film
Incident wave
Antireflective coating
Photoresist
Film
Substrate
The use of antireflective coatings, dyes, and filters can help prevent interference.

**FIGURE 1.70** Standing waves and antireflective coatings.

line width. Because standing waves are the result of coherent interference of monochromatic light, using broadband illumination can decrease the effect. To minimize standing wave effects, one also can use thinner resists (<0.3 μm), but these cause a degradation in line-width control over varying topography. Reflective notching (Figures 1.69d and 1.70) comes about when light is reflected from the more reflective topological features buried in the resist.

### *Thin Film Imaging*

*Introduction* Variations in resist thickness over steps lead to line-width control problems (see Figure 1.71). This lack of line-width control and the limitations of low source intensity and shallow DOF are overcome in thin film imaging (TFI). In TFI, the projected image is confined to a thin layer near the surface of the resist. Because the thin exposed layer is optically isolated from the substrate, the previously mentioned deleterious effects such as standing waves and scattering from underlying topography are avoided. There is often a need for a thick planarizing resist underneath because of topography and because a very thin film by itself usually does not have the required etch resistance. As a consequence, TFI usually involves a surface modified thick resist in the case of top surface imaging (TSI) or a multilevel resist stack (bilayer, trilayer, or multilayer). After the image has been transferred onto the thin top surface layer, it is transferred to the underlying resist layer(s). Pattern transfer often involves dry etching in oxygen plasma, that is, "dry development." We will discuss single-layer resist TSI first, and then discuss multilayer resist methods.

*Thin Film Imaging in Single-Layer Resist—Top Surface Imaging* Top surface imaging (TSI) is one approach to thin film imaging (TFI). In TSI, only the top surface of an SLR is modified under light exposure. A single layer of a thick, opaque resist is used, and exposure leads to differential diffusion/reaction rates of a silicon-containing compound into either exposed or unexposed areas (depending on the resist tone). It was Taylor et al.[77] who discovered that small amounts of Si (about 10%) drastically lower the etching rates of organic polymers. Areas containing Si convert to $SiO_x$ in oxygen plasma and thus become plasma resistant.

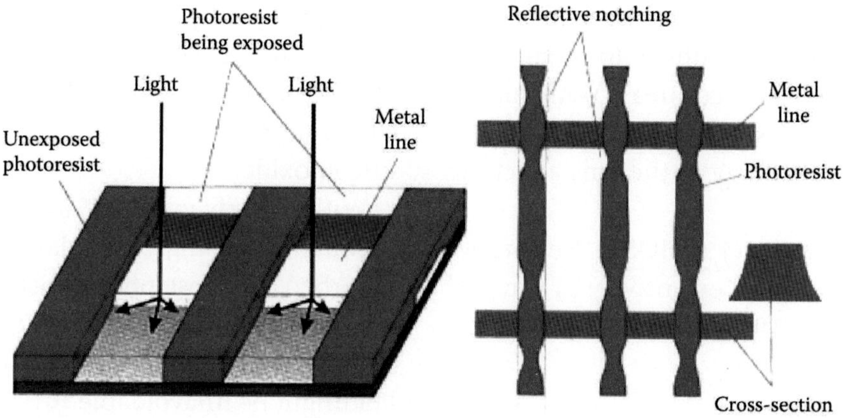


**FIGURE 1.71** The dimensional variation of resist lines going over steps is a well-known phenomenon: reflection of light by the nonvertical sidewalls of the step causes increased incoupling of light into the resist adjacent to the step, leading to a reduction in line width. Similar effects may be observed on reflective materials with a particularly coarse-grained structure.

The DESIRE (dry etching of silylated image resist) process by UCB (Belgium, http://www.ucb.be), illustrated in Figure 1.72, exemplifies a single-layer resist TSI scheme. In this case, a short exposure of a thick layer of planarizing positive DQN-novolak resist produces latent images of indenecarboxylic acid. In a postexposure bake, in a vacuum, the layer is converted into a film enabling selective silylating. In the silylating process, Si atoms only diffuse/react on/into the surface of the exposed resist, locally forming a thin Si-containing resist layer. The polymer film is typically exposed and silylated to a depth of at least 1000 Å to provide sufficient differential plasma etch resistance for pattern transfer. An oxygen plasma etch subsequently removes the unexposed, unreacted resist, and the silylated mask remains because of the formation of a protecting silicon oxide layer; a negative-tone image of the mask results. The DESIRE process in Figure 1.72 illustrates not only dry resist development but also image reversal.

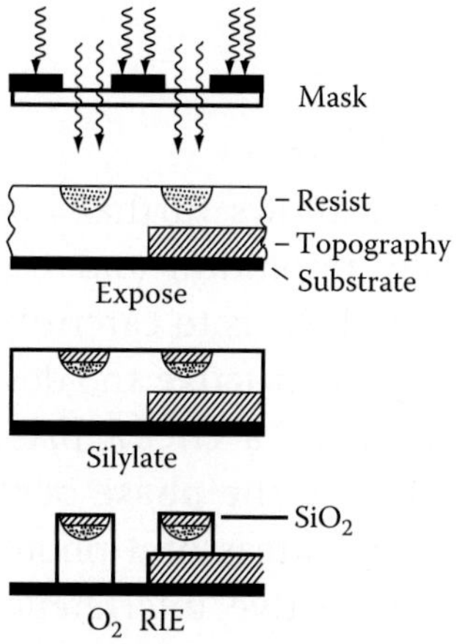


**FIGURE 1.72** The DESIRE (dry etching of silylated image resist) process developed by UCB Chemical Company. Exposure and silylation are confined to the upper layers of the resist coating.

A major advantage of TFI techniques is that only the surface of the resist requires exposure, which, coupled with the anisotropic nature of plasma etching, allows for the use of thick resist layers that planarize the underlying topography to a higher degree than resists with more typical thicknesses could. The lithography process also remains unaffected by substrate reflection, and resist thickness variation as light does not have to penetrate the thickness of the resist. Image quality with DESIRE is excellent. An advantage of TSI over bilayer or multilayer TFI techniques is that only one single layer is spun on before silylation. DESIRE made it from research and development into manufacturing in the late 1980s at TI (http://www.ti.com).[78] It was used for the top metal level patterning in DRAM manufacture. This level had the most reflectivity and most topography. The resist that TI applied, using a standard coater, was the Plasmask® 200-g from Japan Synthetic Rubber (JSR, http://necsv01.keidanren.or.jp/A2J/data/06025.html). Two additional machines, a silylation tool and a dry develop etcher, were developed for the TSI process. Both liquid and vapor silylation are feasible and have been compared at IBM.[78] The use of a dry etch in development leads occasionally to so-called *grass residue* in open areas (see Chapter 3, Figure 3.23). The grass is assumed to stem from sputtered silicon from silylated areas. With higher plasma density systems, less grass forms. This can be explained by the lower energy of the ions attacking the resist surface.

Pure poly(hydroxystyrene) (PHOST) (sensitivity 22 mJ/cm$^2$) has been studied extensively as a model TSI resist for use with 193-nm optical projection lithography.[79]

*Multilayer Resist Thin Film Imaging* In the bilayer TFI approach shown in Figure 1.73, the resist stack consists of a thick, planarizing bottom layer and a thin, top imaging overlayer. These layers function synergistically to achieve higher resolution otherwise impossible to obtain with thick, planarizing single-layer resists. After exposure and development of the thin imaging overlayer, the pattern is transferred to the planarizing underlayer. An intermediate isolation layer sometimes separates the planarizing and imaging layers, preventing their mixing. Depending on the exposure dose and development condition (wet etching) of the planarizing resist layer, an undercut or overcut may be generated. Alternatively, if good fidelity of the mask pattern is desired, dry directional etching (RIE) may be used (see Figure 1.73).

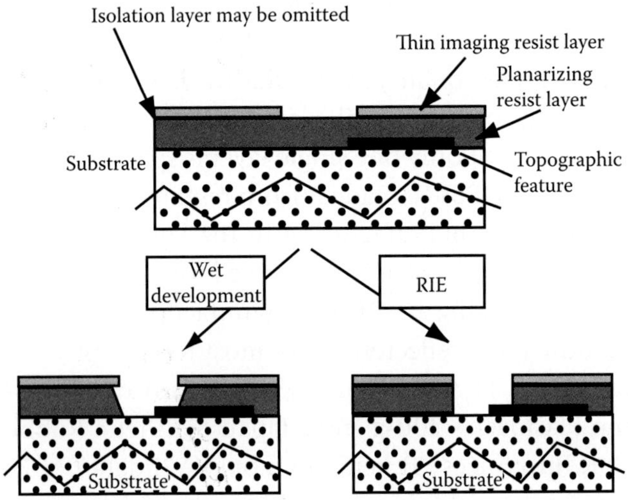


**FIGURE 1.73** Multilayer resist scheme. Thick planarizing underlayer for planarizing of wafer topography, optional intermediate isolation layer, and thin imaging overlayer for optimum resolution. Wet etching (development) and dry etching (RIE) produce different resist profiles.

In a trilayer system, a thin imaging layer is spun onto a thin oxygen dry etch-resistant hard mask coated on top of a thick planarizing layer. The hard mask may be a plasma-deposited or spin-on silicon oxide.

With the advent of 193-nm lithography, TFI will become increasingly popular as radiation is readily absorbed by most photoresists at this short wavelength. Once extreme UV is used as the exposure radiation of choice (see below), widespread use of TFI might be unavoidable.[78]

## Strategies for Improved Resolution through Improved Mask Technology

### *Background*

Besides improved resists, mask engineering also provides important contributions to RET. Interestingly, Richard Feynman in the second of his now famous lectures on micromachining imagined a micromachined mask with miniature levers and valves that could be used over and over to generate different patterns.[80] The mask techniques described above are timid attempts to use micromachining technology for mask engineering. In one scenario, by controlling the phase and the amplitude of the light at the image plane in a phase-shifting mask, resolution and depth of focus are greatly improved.

In miniaturization science, where we often print high-aspect-ratio features, masks with an expanded DOF are favored—but masks with increased resolution remain of greater importance in IC manufacture.

### *Phase-Shifting Masks*

One method that allows further improvement of photolithography resolution and DOF at a given wavelength ($\lambda$) and NA is to carefully control light diffraction using constructive and destructive interference to help create a circuit pattern. The idea of selectively altering the phase of the light passing through certain areas of a photomask to take advantage of destructive interference to improve resolution and DOF in optical lithography was first proposed in 1982.[81] The method is based on building PSMs. Using such a mask one can control both the amplitude and the phase of the light and, in particular, arrange the mask so that light with the opposite phase emerges from adjoining mask features. Destructive interference can be used to cancel some of the image-spreading effects of diffraction.

In Figure 1.74, we illustrate the effect for three nearby mask apertures; a classical transmission mask is added for comparison.[82] The amplitude profile at the plane of the classical transmission mask consists of three square-cornered features with positive amplitude. The amplitude at the wafer level is broadened and rounded but does not change sign. With a PSM, a shifter layer results in a reversal of the phase of the light at the mask. As the photoresist only shows sensitivity toward the intensity of the light and not to the sign, the three bright features develop identically, and the areas where the light of opposite phases causes destructive interference form very dark contrast lines.

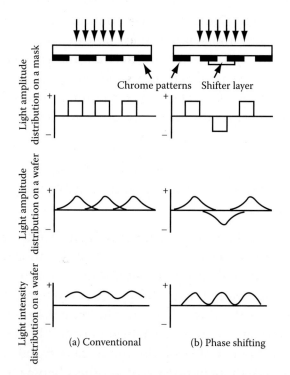


**FIGURE 1.74** Comparison of a conventional transmission mask (a) and a "Levenson-type" phase-shifting mask (b). Both contain opaque chrome (Cr) regions, but the light passing through some of the apertures of the phase-shifting mask also passes through the transparent material of the phase shifter. This reverses the phase of the light at the mask, giving rise to destructive interference at the photoresist plane. (From Levenson, M.D. 1992. Phase-shifting mask strategies: isolated dark lines. *Microlithogr World* 6:6–12. With permission.[82])

One type of PSM is an alternating aperture PSM (AAPSM), also called a *hard* shifter. Going beyond the traditional chrome-on-quartz approach, AAPSMs accomplish the task of shifting the phase of the light by etching regions of the quartz substrate to the precise depth (which depends on the wavelength of the light to be used to expose the wafer) where shifting is desired and have become known as Levenson-type PSMs. The etched areas cause the light to become 180° out of phase with the light passing through the unetched regions (Figure 1.75).

An important alternative PSM is the embedded attenuated PSM (EAPSM). EAPSMs, or *soft* shifter masks, are similar to ordinary (binary) masks in that a quartz substrate is coated with a material in which the design is etched. The most common material used today is molybdenum silicide. Unlike chrome, molybdenum silicide allows a small percentage of the light to pass through; however, the amount passing through is "soft" and does not expose the resist on the wafer. Where it does pass through, the light is 180° out of phase compared with light passing through the quartz alone. Where the coating and quartz meet, light interferes in such a way as to sharpen the edges of the design. The most difficult task in the case of EAPSMs is to create a thin film structure with the desired transmission (4–8%) and phase (180°) at a specific wavelength (i-line or deep UV) that can be patterned, inspected, and repaired using currently available tools.[83] It has been

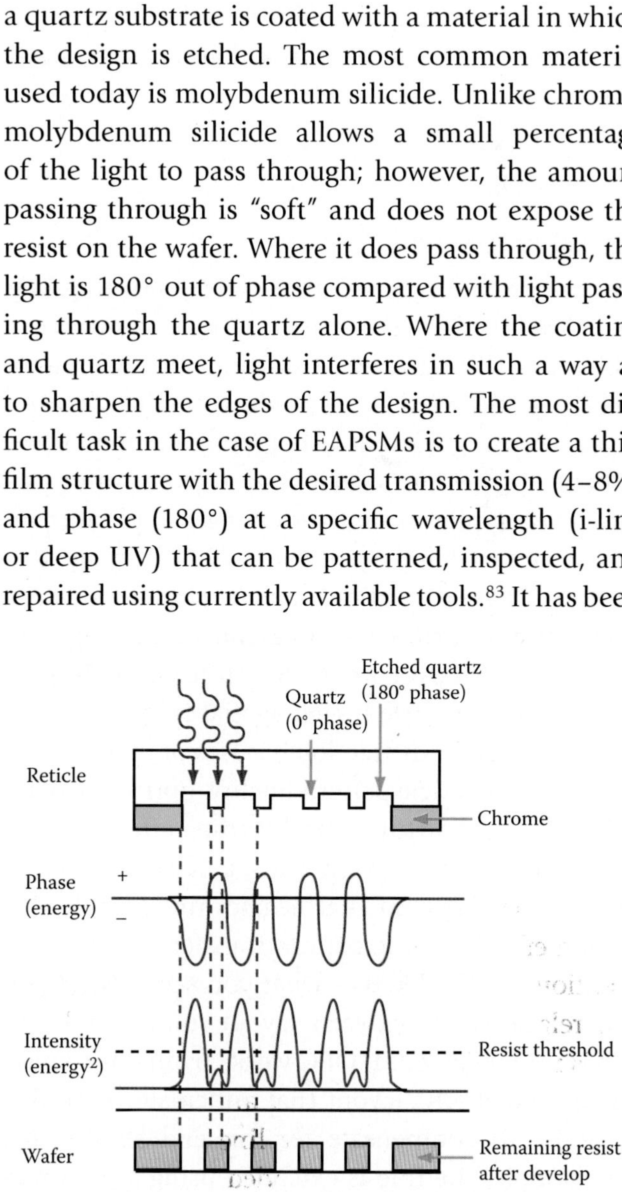


**FIGURE 1.75** A hard shifter alternating aperture PSM or AAPSM.

estimated that a PSM mask for 100-nm work could cost up to $50,000. This makes RET often synonymous for Really Expensive Technology!

How narrow a feature can one make using PSMs? Oki Electric Industries (http://www.oki.de) sells a metal semiconductor field effect transistor (MESFET) device with a 0.18-μm gate patterned by i-line lithography, ordinarily considered capable of only 0.5 μm resolution![82] Thus, PSMs can easily improve resolution by 50–100%. Using 248-nm deep UV lithography researchers at MIT's Lincoln Laboratory have reported sub-100-nm features using phase-shifting masks. Because the same method is applicable to 193-nm radiation, the approach appears promising for further extending photolithography.[84]

The use of TSI and PSM techniques can reduce $k_1$ in Equation 1.16 to 0.4, well below the 0.61 expected from Rayleigh's criterion.

### *Optical Proximity Correction*

Optical proximity correction (OPC) printing was first developed in the early 1970s as a means of addressing lithography distortions in IC manufacturing. The goal is to produce smaller features in an IC using current lithography equipment by enhancing the "printability" of a given wafer pattern. OPC applies systematic changes to photomask geometries to compensate for nonlinear image distortions. These distortions include line-width variations (nonuniform CDs), line-end shortening, and rounding of sharp corners. Sharp features, in general, are lost because higher spatial frequencies are lost as a result of diffraction. Causes of these distortions include reticle pattern fidelity, optical proximity effects, and diffusion and loading effects during resist and etch processing. Thus, a mask incorporating OPC is a system that negates undesirable distortion effects during pattern transfer. In OPC, the traditional (WYSIWYG—what you see is what you get) relationship between layout, mask, and silicon is no longer valid. OPC works by making small changes to the IC layout that anticipate these distortions. To compensate for line-end shortening, for example, the line is extended using a hammerhead shape that results in a line in the resist that is much closer to the original intended layout. To compensate for corner rounding, serif shapes are added to (or subtracted from) corners to produce corners on the silicon that are closer to the ideal layout (see Figure 1.76). Determining the optimal type, size, and symmetry (or lack thereof) is very complex and depends on neighboring geometries and process parameters. A sophisticated computer program is necessary to properly implement OPC. OPC improves the resolution by decreasing $k_1$.

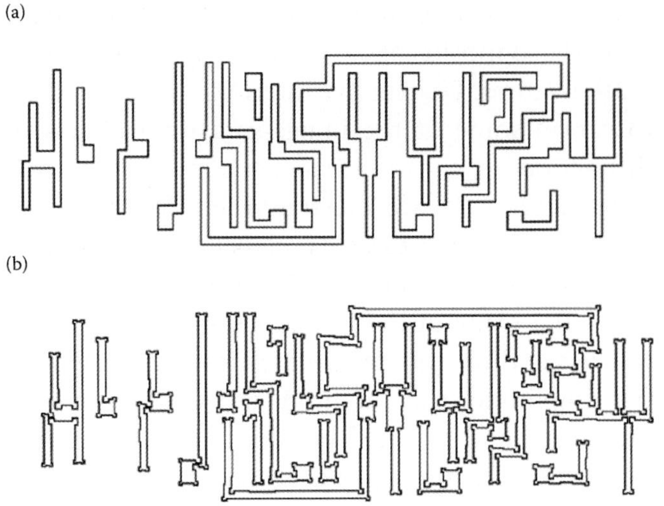


**FIGURE 1.76** Optical proximity correction (OPC). (a) Mask of a standard cell before OPC. (b) Mask after OPC.

## Strategies for Improved Resolution through Improved Exposure Technology

### *Background*

Under "Improved Exposure Equipment," we describe progress in decreasing $k_1$ and NA in Equation 1.16 to further enhance resolution $R$ by modifying the exposure equipment. Off-axis illumination (OAI) is one of the three major resolution enhancement technologies that have enabled optical lithography to push practical resolution limits far beyond what was once thought possible (the others being phase-shifting masks and optical proximity corrections; see above). We will briefly review OAI and Köhler illumination and finish this chapter by introducing immersion lithography.

### *Off-Axis Illumination and Köhler Illumination*

With OAI, diffraction information is distributed selectively over the lens pupil, influencing the impact

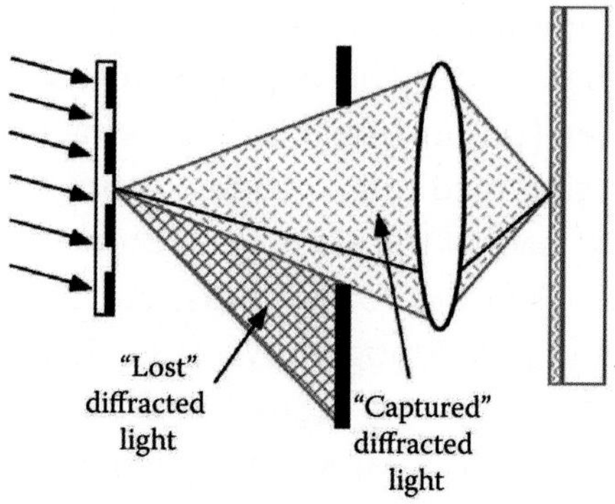


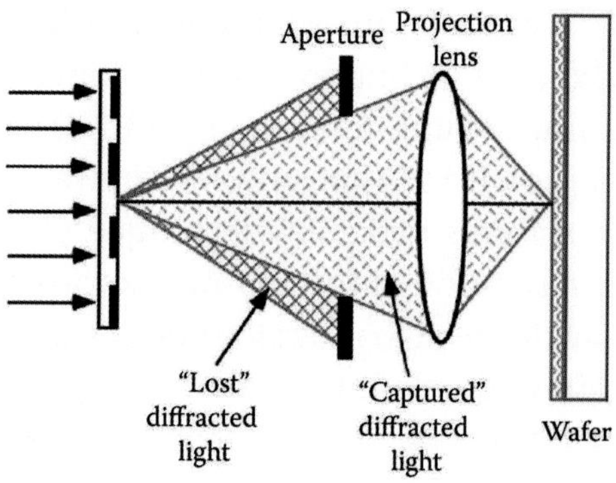


**FIGURE 1.77** Off-axis-illumination (OAI).

of aberrations on imaging. Off-axis illumination allows some of the higher order diffracted light to be captured and hence can improve resolution (by decreasing $k_1$). OAI refers to any illumination shape that significantly reduces or eliminates the "on-axis" component of the illumination, that is, the light striking the mask at near-normal incidence. By tilting the illumination away from normal incidence, the diffraction pattern of the mask is shifted within the objective lens. This enables higher diffracted orders to be included in the lens aperture, which increases the resolution (see Figure 1.77). The increased focal depth for OAI can be understood in terms of the path difference, which is now shorter for all the waves arriving at the resist surface. OAI can improve imaging resolution by 20% and DOF by 40%.

Köhler illumination, named after August Köhler, the man who invented it in 1983, is also known as double diaphragm illumination because it uses both a field and an aperture iris diaphragm to set up the illumination in microscopy. If the light path is set up properly, one has the advantages of an evenly illuminated field, essential to avoid shadows, glare, and incorrect contrast. Köhler illumination in lithography achieves the same goals by using a collector lens in front of the light source (see Figure 1.78). This "captures" diffracted light equally well from all positions on the mask. This improves the resolution by bringing $k_1$ down.

### *Immersion Lithography*

From the discussion up until this point, the resolution limit for 193-nm exposure systems may be calculated using the Rayleigh equation with $\lambda = 193$ nm, NA = 0.93, and $k_1 = 0.25$ (see Equation 1.16) as:

$$R = 0.25 \times 193/0.93 = 52 \text{ nm} \qquad (1.30)$$

From this calculation, a highly optimized ArF exposure system has an absolute optimal

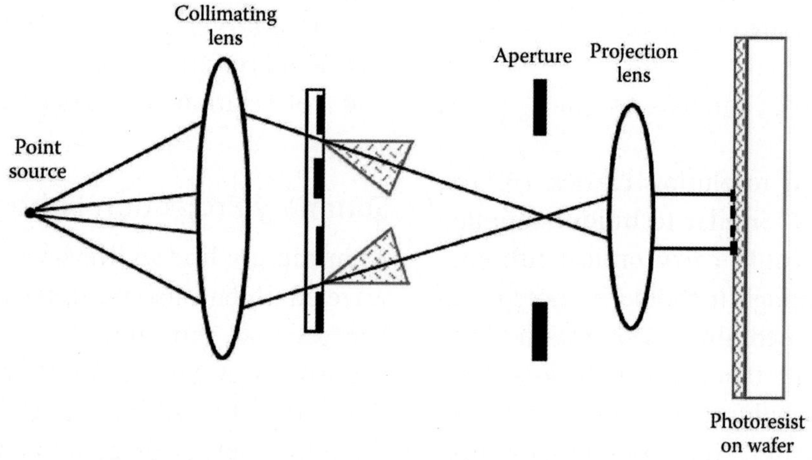


**FIGURE 1.78** Köhler illumination in lithography.

resolution of 52 nm, sufficient for the 65-nm line widths introduced in 2005 but not capable of meeting the 45-nm line widths forecast in 2007. The technical challenges with 157-nm and shorter wavelength exposure systems make any technique that can improve the resolution of the 193-nm exposure systems and delay the need to move to shorter wavelengths an important development.

The NA is determined by the acceptance angle of the lens and the index of refraction of the medium surrounding the lens: NA = $n\sin\theta$ (Volume I, Equation 5.3), where $n$ is the index of refraction and $\theta$ is the acceptance angle of the lens. Because the sine of any angle is always <1 and $n = 1$ for air, the physical limit for an air-based system is clear, but what if a medium with a higher index of refraction is substituted for air? Microscopy has for years used oil as the medium surrounding the lens and the sample for resolution enhancement, and it is somewhat surprising that the semiconductor industry has taken this long to seriously consider the merits of replacing air with an alternative.

The medium between the lens and the wafer being exposed needs to have an index of refraction >1, have low optical absorption at 193 nm, be compatible with photoresist and the lens material, and be uniform and noncontaminating. Surprisingly, ultrapure water may meet all these requirements. Water has an index of refraction $n = 1.47$, absorption of <5% at working distances of up to 6 mm, is compatible with photoresist and lens, and in its ultrapure form is noncontaminating. Plugging $n = 1.47$ into Equation 5.3 (Volume I) and assuming $\sin\theta$ can reach 0.93, then the resolution limits for 193-nm immersion lithography are:

$$R = 0.25 \times 193/1.470.93 = 35 \text{ nm} \quad (1.31)$$

A 35-nm theoretical resolution carries 193-nm exposure beyond 2007. Similar techniques applied to 157-nm exposure could carry optical lithography even further, although it should be noted that water is not a usable medium at 157 nm and suitable mediums are still being researched (a high index fluid such as polyfluoropolyether is a candidate medium in this case). The resolution enhancement from immersion lithography is about 30–40% (depending on materials used), or about one technology node. The depth of focus is also 40–70% better (proportional to the refractive index of the imaging medium considered) than a corresponding "dry" tool at the same resolution.[85]

In Figure 1.79 we illustrate a schematic of an immersion lithography setup. It is estimated that such a setup may cost $30 million.

There are a large number of practical issues yet to resolve to implement immersion lithography. The stage on a 193-nm exposure tool steps from location to location across the wafer scanning the reticle image for each field. To achieve high throughput, the stage must accelerate rapidly, move accurately to the next field location, settle, scan the image, and then step to the next location all in a short period. Maintaining a consistent bubble-free liquid between the lens and the wafer is very difficult. One other issue that is likely to be significant for immersion lithography is temperature control. Variations in temperature cause variations in $n$ and therefore image distortion. Maintaining temperature uniformity with a rapidly moving stage and a pulsed laser passing through the fluid will likely be a significant challenge. A more complete listing of issues includes:

- Very big lenses (hence expensive)
- Field size reduction
- Mechanical issues and hydrodynamics
- Stage vibrations transferred to lens
- Heating of immersion liquid on exposure
- New defect mechanisms at wafer level
- Interaction of photoresist with immersion liquid
- Fluid contamination
- Polarization effects degrading contrast

## Summary Projection Lithography

As minimum line widths have shrunk, the exposing wavelength has also periodically shrunk. Table 1.10 lists the year, minimum line-width generation, and exposure wavelength for state-of-the-art ICs since the mid-1980s. At 1.2-µm and larger line widths, the g-line output of mercury lamps ($\lambda$ = 436 nm) was used; at the 0.8-µm (800 nm) generation, the

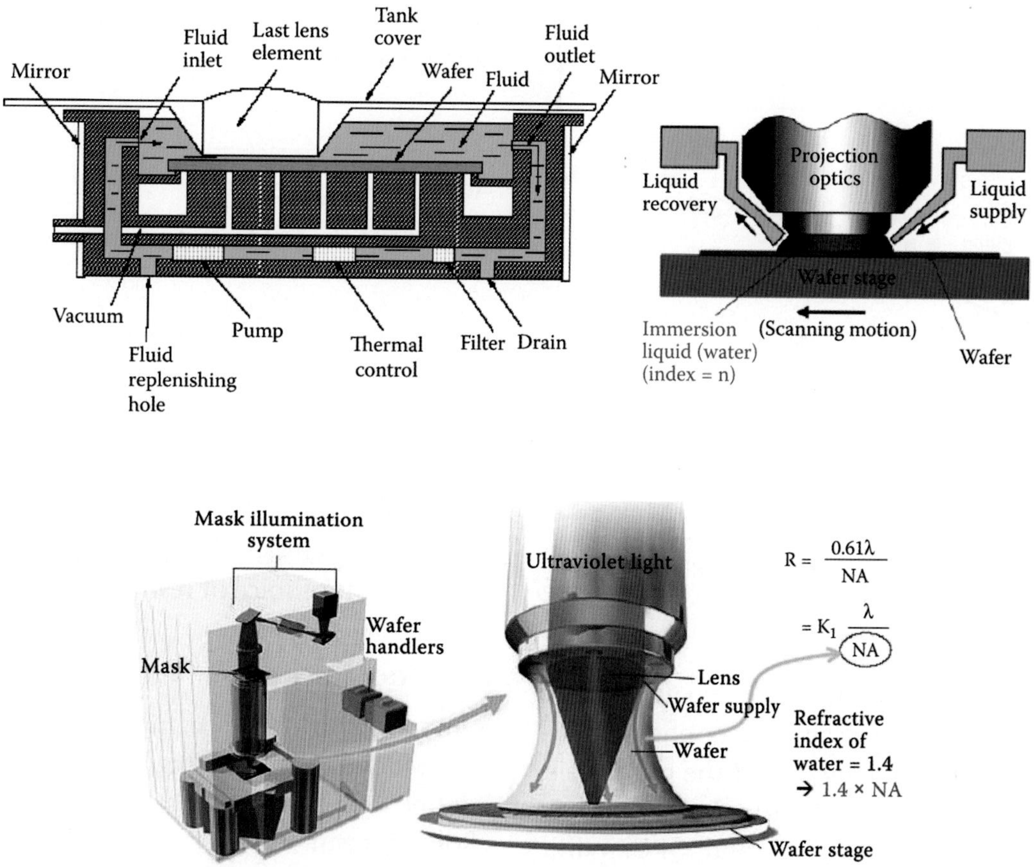


**FIGURE 1.79** Schematics of an immersion lithography setup.

i-line output of mercury lamps (λ = 365 nm) was introduced for critical layers, and i-line use continued to the 350-nm line width, where early adopters began to use KrF excimer lasers (λ = 248 nm) as the exposure source. KrF use has surprised many observers by persisting through the 130-nm line-width generation. With 90-nm line widths now entering production, KrF is finally running out of stream, and ArF excimer lasers are being introduced (λ = 193 nm). Beyond ArF, there are fluorine excimer lasers ($F_2$) with λ = 157 nm, but there are a number of technical challenges to overcome. At less than the 157-nm wavelength, the optical exposure systems must change to all reflecting optics because of high levels of absorption in refractive lens at shorter wavelengths. The introduction of an all-reflective lens exposure system introduces a number of technical challenges.

**Table 1.10** Minimum Line Width and Exposure Wavelength versus Year

| Year | Line Width (nm) | Wavelength (nm) |
|---|---|---|
| 1986 | 1200 | 436 |
| 1988 | 800 | 436/365 |
| 1991 | 500 | 365 |
| 1994 | 350 | 365/248 |
| 1997 | 250 | 248 |
| 1999 | 180 | 248 |
| 2001 | 130 | 248 |
| 2003 | 90 | 248/193 |
| 2005 (fcst) | 65 | 193 |
| 2007 (fcst) | 45 | 193 |

At the same time that exposure wavelengths have been reduced, improvements in lens design have led to improvements in the NA of the exposure systems lens, as shown in Figure 1.80. In the mid-1980s an NA value of approximately 0.4 was typical; today 248-nm exposure systems are available with an NA greater than 0.8. The physical limit to NA for exposure systems using air as a medium between the lens and the wafer is 1; the practical limit is somewhere around 0.9, with recent reports suggesting that an NA as high as 0.93 may be possible for ArF systems in the future.

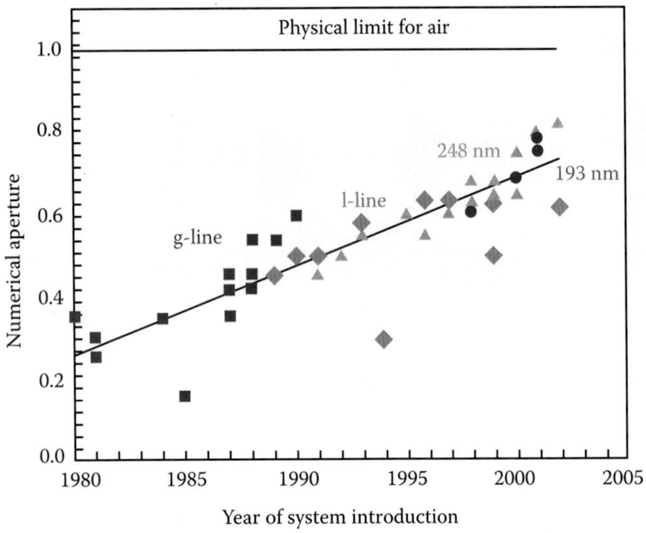


**FIGURE 1.80** The evolution of the numerical aperture in lithography systems as a function of time.

The third element in the Rayleigh equation is $k_1$, a complex factor of several variables in the photolithography process such as the quality of the photoresist and the use of resolution enhancement techniques such as phase shift masks, OAI, and OPC. While exposure wavelengths have been decreasing and NA has been increasing, $k_1$ has been decreasing as well (see Figure 1.81). The practical lower limit for $k_1$ is thought to be about 0.25. In Figure 1.81, we plot the evolution of $k_1$ during the past 18 years.

From this chapter it is clear that photolithography remains the only viable IC, MEMS, and NEMS production technology. As we realize now it is very urgent that we look beyond photolithography for the next generations of ICs and NEMS. This is the problem we tackle in Chapter 2 on NGLs and lithography research.

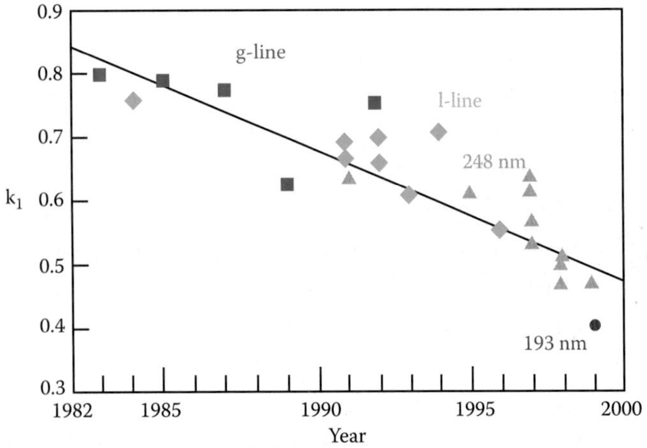


**FIGURE 1.81** The evolution of $k_1$ versus time.

## Acknowledgments

Special thanks to Maria Amasia and Genis Turon.

## Appendix 1A: Spin Coating Process Troubleshooting

### Film Too Thin

Spin speed too high: select lower speed. Spin time too long: decrease time during high-speed step. Inappropriate choice of resin.

### Film Too Thick

Spin speed too low: select higher speed. Spin time too short: increase time during high-speed step. Exhaust volume too high: adjust exhaust. Inappropriate choice of resin.

### Air Bubbles on Wafer Surface

Air bubbles in dispensed fluid. Dispense tip is cut unevenly or has burrs or other defects.

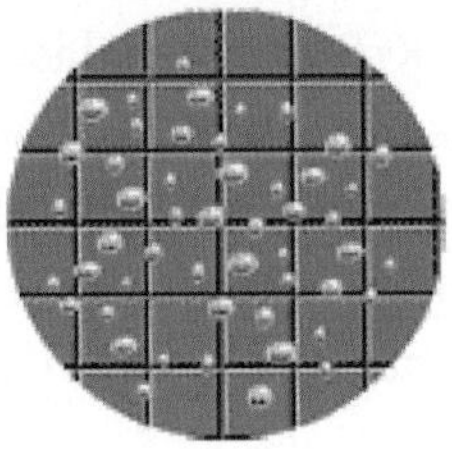

### Swirl Patter

Spin bowl exhaust rate is too high. Fluid strikes substrate surface off center. Spin speed and acceleration setting is too high. Spin time is too short.

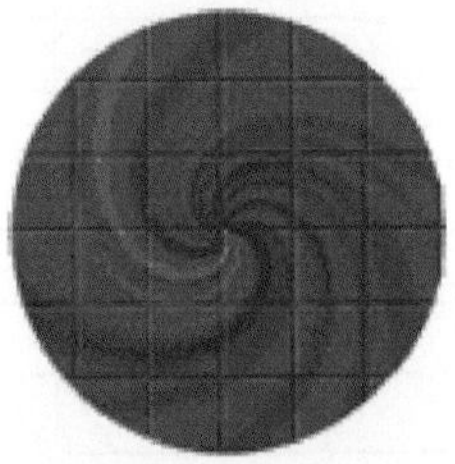

### Center Circle

If the circle is the same size as the spin chuck: proper chuck selection should be based on substrate size and rigidity. The proper chuck diameter is 1/4–1-in. smaller than the substrate diameter.

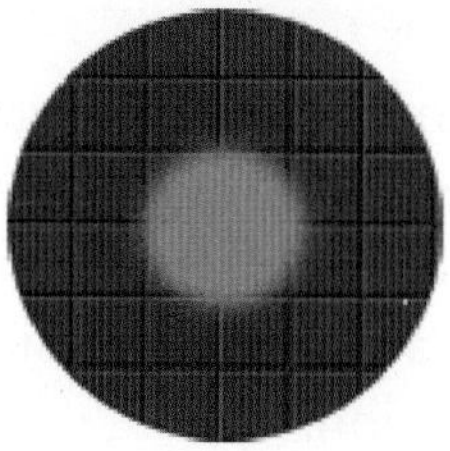

### Uncoated Areas

Insufficient dispense volume.

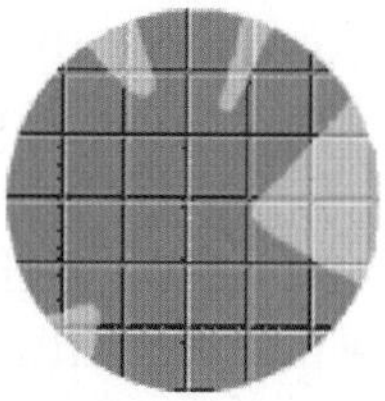

### Pinholes

Air bubbles. Particles in fluid. Particles exist on substrate surface before dispense.

### Comets, Streaks, or Flares

Fluid velocity (dispense rate) is too high. Spin bowl exhaust rate is too high. Resist sits on wafer too long before spin. Spin speed and acceleration setting are too high. Particles exist on substrate surface before dispense. Fluid is not dispensed at the center of the substrate surface.

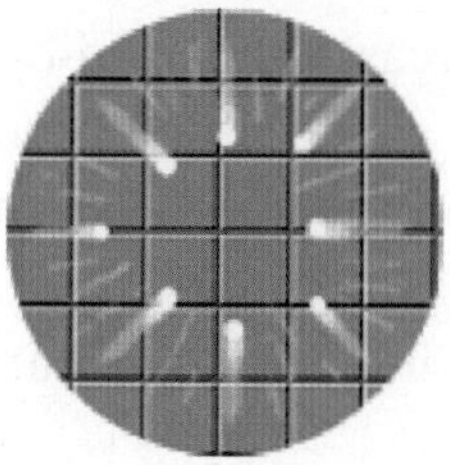

### Poor Reproducibility

Variable exhaust or ambient conditions: adjust exhaust lid to fully closed. Substrate not centered properly: center substrate before operation. Insufficient dispense volume: increase dispense volume. Inappropriate resin material. Unstable balance in speed/time parameters: increase speed/decrease time or vice versa.

### Poor Film Quality

Exhaust volume too high: adjust exhaust lid or house exhaust damper. Acceleration too high: select lower acceleration. Unstable balance in speed/time parameters: increase speed/decrease time or visa versa. Insufficient dispense volume: increase dispense volume. Inappropriate resin material: contact resin manufacturer.

---

## Questions

1.1: Analyze each of the expressions for photolithography resolution and explain how to improve resolution in each case. What are the advantages and disadvantages of using e-beam lithography compared to typical photolithography using UV radiation? What is the most likely next-generation lithography?

1.2: An exposure is performed with coherent light using a step and repeat projection printing system. The light source has a wavelength of 365 nm (i-line of a Hg arc lamp).
The pattern is a grating with a line-to-line spacing of 1 μm.
- (a) Calculate the minimum value of the numerical aperture (NA) that will provide contrast at the image plane (the plane of the resist).
- (b) What is the maximum value of the numerical aperture, above which there will be no improvement in image quality?
- (c) Calculate the depth of field of the image for cases (a) and (b).

1.3: Lift-off refers to the process of exposing a pattern into a photoresist, depositing a thin film such as a metal or dielectric over the entire area, and then washing away the photoresist to leave behind the film only in the patterned area. Why is it easier to obtain a lift-off profile with a negative resist than with a positive resist?

1.4: Which of the following statements are NOT correct?
   (a) Short exposure wavelengths can create standing waves in a layer of photoresist.
   (b) Regions of constructive interference create increased exposure.
   (c) Standing waves can impair the structure of the resist, but they can be eliminated by use of multiple wavelength sources or postbaking.
   (d) Standing waves effects are most noticeable at the center of the resist.
   (e) The primary components of a positive photoresist are
       1. Nonphotosensitive base phenolic resin
       2. Photosensitive dissolution inhibitor
       3. Coating solvent
   (f) Projection lithography resolution is limited by exposure wavelength, resist thickness, and diffraction and dispersion of light.
   (g) Proximity lithography resolution is limited only by exposure wavelength and resist thickness.

1.5: Indicate which line in this graph corresponds to **negative** photoresist and which corresponds to **positive** photoresist.

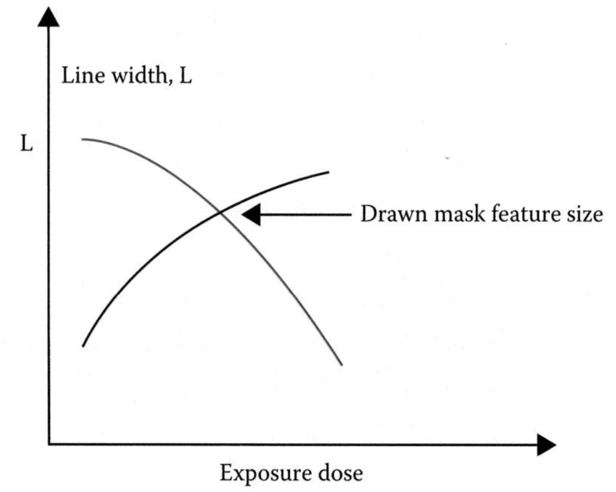


1.6: Why is it easier to find good data from the literature on resist sensitivity than on lithographic sensitivity?

1.7: Sketch the lithography steps involved in generating a staircase resist pattern that is oriented in the same direction on both sides of a Si wafer. The steps of the staircase need to be aligned within 2 μm and your laboratory cannot afford a >$250,000 double-sided mask aligner.

1.8: Present a comparison of negative and positive photo resists. Also explain what a permanent resist is. Describe what happens chemically to positive and negative resists when exposed to UV radiation.

1.9: A polyimide photoresist requires 100 mJ/cm$^2$ per micron of thickness to be developed properly. If a lamp provides 1000 W/m$^2$, how long does it take to expose a 20-μm-thick film?

1.10: List four major factors in which IC technology differs from MEMS or miniaturization science.

1.11: Why does one prefer a high $T_g$ in a resist and why are there no crystalline materials used in lithography?

1.12: Provide a simple rule of thumb for dimensional tolerances in lithography. Why is the resolution with incoherent light larger than that for coherent illumination? How is the depth of focus (DOF) of an imaging system influenced by the numerical aperture NA of the imaging lens, the resolution of the system, and the wavelength of the exposing light?

1.13: Demonstrate with some simple sketches how you would pattern a small Pt electrode for an electrochemical sensor using wet etching. The substrate is an oxidized Si wafer. (Remember that Pt is very difficult to wet etch.)

1.14: Mark the INCORRECT statement(s).
   (a) If we want to do the lift-off process, we do not need a postbake.
   (b) Before we spin dry a wafer, we need to make sure the vacuum is off.
   (c) To increase the pump-down speed for the metal evaporation process, we need to turn on the roughing and foreline pump at the same time.
   (d) Ultrasonic vibration can help the lift-off process.

(e) Four-point resistivity measurements can be used to determine the doping concentration of the water.

(f) $O_2$ plasma etching can be used to remove the photoresist.

(g) After the fabrication is completed, we can deposit a thin layer of $Si_3N_4$ to protect the Al wires.

1.15: How is the diffraction limit expression derived? Hint: Use the mathematical expression for the intensity variation of the Fraunhofer diffraction pattern of a circular aperture (Airy disk) with radius a.

1.16: How does immersion lithography influence the value of the DOF?

1.17: What are the pros and cons of proximity, contact, and projection lithography?

1.18: Explain what happens chemically to positive and negative photoresist when exposed to UV radiation.

1.19: List the most important methods that can be employed to improve resolution in photolithography.

## References


1. Harris, T. W. 1976. *Chemical milling.* Oxford, UK: Clarendon Press.
2. Cunningham, J. A. 1992. The remarkable trend in defect densities and chip yield. *Semicond Int* 15:86–90.
3. Iscoff, R. 1991. Wafer cleaning: can dry systems compete? *Semicond Int* 14:48–54.
4. Bok, E., D. Kelch, and K. S. Schumacher. 1992. Supercritical fluids for single wafer cleaning. *Solid State Technol* 35:117–19.
5. Singer, P. H. 1992. Trends in wafer cleaning. *Semicond Int* 15:34–39.
6. Tolliver, D. L. 1988. *Handbook of contamination control in microelectronics: principles, applications, and technology.* Park Ridge, NJ: Noyes Publications.
7. Deal, B. E. 1993. *Handbook of semiconductor wafer cleaning technology.* Park Ridge, NJ: Noyes Publications.
8. Sure, A., T. Dillon, J. Murakowski, C. Lin, D. Pustai, and D. W. Prather. 2003. Fabrication and characterization of three-dimensional silicon tapers. *Optics Express* 11:555–61.
9. Feely, W. E. 1988. *Technical digest: solid state sensor and actuator workshop.* Hilton Head Island, SC: IEEE.
10. Feely, W. E., J. C. Imhof, and C. M. Stein. 1986. The role of the latent image in a new dual image, aqueous developable, thermally stable photoresist. *Polym Eng Sci* 26:1101–04.
11. Six, P. 1995. Phase masks and grey-tone masks. *Semiconductor Fabtech 5.*
12. Reimer, K., R. Engelke, U. Hofmann, P. Merz, K. T. K. v. Platen, and B. Wagner. 1999. *MF-Conf., micromachine technology for diffractive and holographic optics.* Santa Clara, CA: SPIE.
13. Stauffer, J. M., Y. Oppliger, P. Regnault, L. Baraldi, and M. T. Gale. 1992. Electron beam writing of continuous resist profiles for optical applications. *J Vac Sci Technol* B10:2526.
14. Windt, D. L., and R. A. Cirelli. 1999. Amorphous carbon films for use as both variable-transmission apertures and attenuated phase shift masks for deep ultraviolet lithography. *J Vac Sci Technol* B17:930–32.
15. Arnone, C. 1992. The laser-plotter: a versatile lithographic tool for integrated optics and microelectronics. *Microelectron Eng* 17:483–86.
16. Duffy, D. C., J. C. McDonald, O. J. A. Schueller, and G. M. Whitesides. 1998. Rapid prototyping of microfluidic systems in poly(dimethylsiloxane). *Anal Chem* 70:4974–84.
17. Smith, H. I., R. Menon, A. Patel, D. Chao, M. Walsh, and G. Barbastathis. 2006. Zone-plate-array lithography: a low-cost complement or competitor to scanning-electron-beam lithography. *Microelectron Eng* 83:956–961.
18. Metz, T. E., R. N. Savage, and H. O. Simmons. 1992. In situ control of photoresist coating processes. *Semicond Int* 15:68–69.
19. Thompson, L. F. 1994. *Introduction to microlithography.* Eds. L. F. Thompson, C. G. Willson, and M. J. Bowden. Washington, DC: American Chemical Society.
20. Webster, R. J. 1998. *Thin film polymer dielectrics for high-voltage applications under severe environments.* Master's thesis, Virginia Polytechnic Institute and State University.
21. Merricks, D. 1995. *Special polymers for electronics and optoelectronics.* Ed. J. A. Chilton and M. T. Goosey. London: Chapmann & Hall.
22. Boellaard, E., P. N. Pham, L. D. M. v. d. Brekel, and G. J. Bertens. 2003. Electrodepositable photoresist processing for RF-devices in MEMS technology. *MEPTEC Rep* Quarter Two:29–31.
23. Pham, N. P., E. Boellaard, P. M. Sarro, and J. N. Burghartz. 2002. *SAFE 2002.* Veldhoven, the Netherlands: STW.
24. Reichmanis, E., O. Nalamasu, F. M. Houlihan, and A. E. Novembre. 1999. Radiation chemistry of polymeric materials. *Polym Int* 48:1053–59.
25. Harriott, L. R. 1999. Next generation lithography. *Mater Today* 2:9.
26. Ito, T., and S. Okazaki. 2000. Pushing the limits of lithography. *Nature* 406:1027–31.
27. Le Barny, P. 1987. *Molecular engineering of ultrathin polymeric films.* Eds. P. Stroeve and E. Franses. New York: Elsevier.
28. Moreau, W. M. 1988. *Semiconductor lithography.* New York: Plenum Press.
29. Shaw, J. M., J. D. Gelorme, N. C. LaBianca, W. E. Conley, and S. J. Holmes. 1996. Negative photoresists for optical lithography. *IBM J Res Dev* 41:81–94.
30. O'Brien, M. J., and D. S. Soane. 1989. *Microelectronics processing.* Eds. D. W. Hess, and K. F. Jensen. Washington, DC: American Chemical Society.
31. Ando, S., T. Matsuura, and S. Sasaki. 1994. *Polymers for microelectronics: resists and dielectrics.* Eds. L. F. Thompson, C. G. Willson, and S. Tagawa. Washington, DC: American Chemical Society.
32. Studt, T. 1992. Polyimides: hot stuff for the '90s. *Res Dev* August:30–31.
33. Furuya, A., F. Shimokawa, T. Matsuura, and R. Sawada. 1994. Fluorinated polyimide fabrication by magnetically controlled reactive ion etching (MC RIE). *J Micromechan Microeng* 4:67.

34. Suzuki, K., I. Shimoyama, and H. Miura. 1994. Insect-model based microrobot with elastic hinges. *J Microelectromech Syst* 3:4–9.
35. Shioda, T. 2007. Next generation packaging technology which supports the ubiquitous devices. Application of the flexible optical waveguide to the consumer products. *Electr Parts Mater* 46:38–42.
36. Flack, W., S. Kulas, and C. Franklin. 2000. Process characterization of an aqueous developable photosensitive polyimide on a broadband stepper. SPIE 3999-45.
37. Ito, H. 1996. Chemical amplification resists: history and development within IBM. *IBM J Res Dev* 41:69–80.
38. LaBianca, N. C., J. D. Gelorme, E. Cooper, E. O'Sullivan, and J. Shaw. 1995. *JECS 188th Meeting*. Chicago: JECS.
39. Angelo, R., J. Gelorme, J. Kuczynski, W. Lawrence, S. Pappas, and L. Simpson. 1992. U.S. Patent 5,102,772, IBM.
40. Reznikova, E. F., J. Mohr, and H. Hein. 2005. Deep photolithography characterization of SU-8 resist layers. *Microsyst Technol* 11:282–91.
41. Lee, S. J., W. Shi, P. Maciel, and S. W. Cha. 2003. *University/government/industry microelectronics symposium*. Boise, ID: IEEE.
42. Guerin, L. J. 1999. *Industrial exhibition*. Montreux, Switzerland.
43. Bertsch, A., H. Lorenz, and P. Renaud. 1999. 3D Microfabrication by combining microstereolithography and thick resist UV lithography. *Sensors Actuators* 73:14–23.
44. Yang, R., and W. Wang. 2005. A numerical and experimental study on gap compensation and wavelength selection in UV-lithography of ultra-high aspect ratio SU-8 microstructures. *Sensors Actuators Chem* 110:279–88.
45. MicroChem Corp. 2000. http://www.microchem.com.
46. Sotec Microsystems. 2000. http://mems.isi.edu/mems/yp/LOU.html.
47. Madou, M., and J. Florkey. 2000. From batch to continuous manufacturing of microbiomedical devices. *Chem Rev* 100:2679–91.
48. Braun, A. E. 1999. Photoresist stripping faces low-k challenges. *Semicond Int* October:64–74.
49. Flamm, D. L. 1992. Dry plasma resist stripping. Part I: overview of equipment. *Solid State Technol* 35:37–39.
50. Wolf, S., and R. N. Tauber. 2000. *Silicon processing for the VLSI era*. Sunset Beach, CA: Lattice Press.
51. Willson, C. G. 1994. *Introduction to microlithography*. Eds. L. F. Thompson, C. G. Willson, and M. J. Bowden. Washington, DC: American Chemical Society.
52. Brodie, I., and J. J. Muray. 1982. The physics of microfabrication. New York: Plenum Press.
53. Sze, S. M. 1988. *VLSI technology*. New York: McGraw-Hill.
54. Alling, E., and C. Stauffer. 1988. Image reversal photoresist. *Solid State Technol* 31:37–43.
55. Thompson, L. F., C. G. Willson, and M. J. Bowden, eds. 1994. *Introduction to microlithography*. Washington, DC: American Chemical Society.
56. Bowden, M. J. 1994. *Introduction to microlithography*. Eds. L. F. Thompson, C. G. Willson, and M. J. Bowden. Washington, DC: American Chemical Society.
57. Sheats, J. R., and B. W. Smith, eds. 1998. *Microlithography: science and technology*. New York: Marcel Dekker.
58. Rothschild, M., A. Forte, R. Kunz, C. Palmateer, and J. Sedlacek. 1997. Lithography at a wavelength of 193 nm. *IBM J Res Dev* 41:49–55.
59. Maluf, N. 2000. *An introduction to microelectrochemical systems engineering*. Boston: Artech House.
60. White, R. M., and S. W. Wenzel. 1988. Inexpensive and accurate two-sided semiconductor wafer alignment. *Sensors Actuators* 13:391–95.
61. Kim, E. S., R. S. Muller, and R. S. Hijab. 1992. Front-to-backside alignment using resist-patterned etch control and one etching step. *J Microelectromech Syst* 1:95–99.
62. Tatic-Lucic, S., and Y.-C. Tai. 1994. Novel extra-accurate method for two-sided alignment on silicon wafers. *Sensors Actuators* A41–42:573–77.
63. Bruning, J. H. 1980. Optical imaging for microfabrication. *J Vac Sci Technol* 17:1148–55.
64. Oldham, W. G., S. N. Nandgaonkor, A. N. Neureuther, and M. O'Toole. 1979. A general simulator for VLSI lithography and etching processes: Part I. Application to projection lithography. *IEEE Trans Electron Devices* 26:717–22.
65. Sekiguchi, A., M. Isono, and T. Matsuzawa. 1999. Measurement of parameters for simulation of 193 nm lithography using Fourier transform infrared baking system. *Jpn J Appl Phys* 38:4936–41.
66. Comello, V. 1990. Planarizing leading edge devices. *Semicond Int* November:60–66.
67. Wiesner, J. R. 1993. Gap filling of multilevel metal interconnects with 0.25-μm geometries. *Solid State Technol* October:63–64.
68. Sniegowski, J. S. 1996. *Micromachining and microfabrication process technology II*. Austin, TX: SPIE.
69. Ong, E., H. Chu, and S. Chen. 1991. Metal planarization with an excimer laser. *Solid State Technol* August:63–68.
70. Ito, H., and C. G. Willson. 1982. Chemical amplification in the design of dry developing resist materials. *Technical Papers, Regional Technical Conference, Society of Plastic Engineers* 331–53.
71. Ito, H., C. G. Willson, and J. M. J. Fréchet. 1982. New UV resists with negative or positive tone. Digest of technical papers of 1982 symposium on VLSI technology.
72. Ito, H., C. G. Willson, and J. M. J. Fréchet. 1985. Positive- and negative-working resist compositions with acid generating photoinitiator and polymer with acid labile groups pendant from polymer backbone. U.S. Patent 4,491,628, IBM.
73. Reichmanis, E., F. M. Houlihan, O. Nalamasu, and T. X. Neena. 1994. *Polymers for microelectronics: resists and dielectrics*. Eds. L. F. Thompson, C. G. Willson, and S. Tagawa. Washington, DC: American Chemical Society.
74. Reichmanis, E., L. F. Thompson, O. Nalamasu, A. Blakeney, and S. Slater. 1992. Chemically amplified resists for deep-UV lithography: a new processing paradigm. *Microlithogr World* November/December:7–14.
75. Lamola, A. A., C. R. Szmanda, and J. W. Thackeray. 1991. Chemically amplified resists. *Solid State Technol* 53:53–60.
76. Horn, M. W. 1991. Antireflection layers and planarization for microlithography. *Solid State Technol* November:57–62.
77. Taylor, G. N., T. M. Wolf, and L. E. Stillwagon. 1984. Role of inorganic materials in dry-processed resist technology. *Solid State Technol* 27:145.
78. Seeger, D. E., D. C. La Tulipe Jr., R. R. Kunz, C. M. Garza, and M. A. Hanratty. 1997. Thin-film imaging: past, present, prognosis. *IBM J Res Dev* 41:105–18.
79. (a) Rothschild, M., R. B. Goodman, M. A. Hartney, M. W. Horn, R. R. Kunz, J. H. C. Sedlacek, and D. C. Shaver. 1992. Photolithography at 193 nm. *J. Vac. Sci. Technol. B* 10:2989; (b) Hibbs, M., R. R. Kunz, and M. Rothschild. 1995. 193-nm Lithography at MIT Lincoln Lab. *Solid State Technol.* 38:69.
80. Feynman, R. P. 1993. Infinitesimal machinery. *J Microelectrochem Syst* 2:4–14.

81. Levenson, M. D., N. S. Visnawathan, and R. A. Simpson. 1982. Improving resolution in photolithography with a phase-shifting mask. *IEEE Trans Electron Devices* 29: 1828–36.
82. Levenson, M. D. 1992. Phase-shifting mask strategies: isolated dark lines. *Microlithogr World* 6:6–12.
83. Van Den Broeke, D. 1996. Transferring phase-shifting mask technology into mainstream manufacturing. *Semiconductor Fabtech* 6.
84. Fritze, M., D. Astolfi, H. Liu, C. K. Chen, V. Suntharalingam, D. Preble, and P. W. Wyatt. 1999. Sub-100 nm KrF lithography for complementary metal-oxide-semiconductor circuits. *J Vac Sci Technol* B17:345–49.
85. Switkes, M., M. Rothschild, R. R. Kunz, S-Y. Baek, D. Coles, and M. Yeung. 2003. Immersion lithography: beyond the 65 nm mode with optics. *Microlithogr World* May:4.

# 2

# Next-Generation Lithographies and Lithography Research

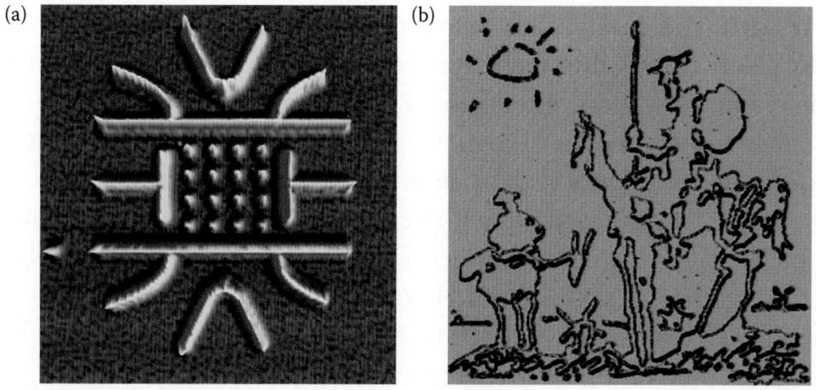


(a) AFM image of quantum dot structure created on a GaAs wafer using oxidation lithography, 2.5-μm scan. (Courtesy of D. Graf and R. Shleser, Ensslin Group, ETH Zurich, http://www.asylumresearch.com.) (b) AFM image of the nanolithographically patterned polycarbonate with scratch lithography, 5-μm scan. Copy of Pablo Picasso's "Don Quixote" (http://www.asylumresearch.com).

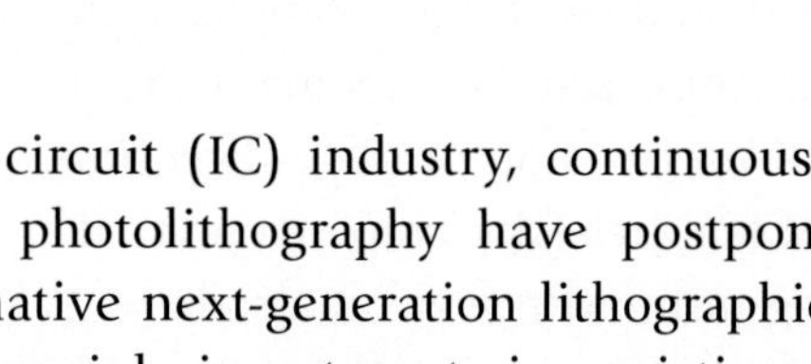

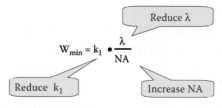


Resolution of photolithography system.

## Outline



## Introduction

In the integrated circuit (IC) industry, continuous improvements to deep UV (DUV) photolithography have postponed the industrial adoption of alternative next-generation lithographies (NGLs) because of the huge financial investment in existing photolithography equipment. DUV photolithography is expected to continue until the 45-nm node through resolution-enhancing techniques (RET) such as off-axis illumination (OAI), phase shift masks (PSM), and optical proximity correction (OPC) as reviewed in Chapter 1. The 32-nm node is viewed as beyond the scope of DUV lithography. There are no clear lithography solutions listed in the International Technology Roadmap for Semiconductors (ITRS roadmap) for 2010 and beyond (Volume I, Appendix 1A).

In this chapter we introduce as possible NGLs: extreme ultraviolet lithography (EUVL), x-ray lithography, charged-particle beam lithography based on electrons and ions (such as electron and ion projection techniques), and imprint lithography [nanoimprint lithography (NIL) and step-and-flash imprint lithography (SFIL)]. These techniques are those that are regarded today as important technologies for the beyond the DUV lithography era, either for mask making or for IC production. We also review lithography approaches in the research and development stage, including lithography based on very thin resist layers and block copolymers, zone plate array lithography (ZPAL), quantum lithography (two-photon lithography), and proximal probe-based techniques such as atomic force microscopy (AFM), scanning tunneling microscopy (STM), dip-pen lithography (DPL), near-field scanning optical microscopy (NSOM), and apertureless near-field scanning optical microscopy (ANSOM). Also considered are holographic lithography and plasmonic lithography. Lithography with superlenses ("Pendry's dream") was reviewed in Volume I, Chapter 5 and will not be repeated here. As an example of a lithography method that is capable of patterning resist on nonplanar substrates (e.g., patterning a resist on the surface of a cylinder) we detail soft lithography.

It is possible that some of the alternative lithography tools we treat in the research and development category will emerge as serious NGLs in the coming years.

IC and miniaturization science are taking increasingly separate paths in adopting preferred lithography strategies. For ICs, throughput and finer geometries are needed, and batch processing is a prerequisite. In miniaturization science, modularity, good depth of focus (DOF), extension of the $z$-direction, that is, the height of features (skyscraper-type structures), incorporating nontraditional materials (e.g., gas-sensitive ceramic layers, polymers), and replication methods catch the spotlight, and batch fabrication is not always a prerequisite. While reading the current chapter, these important differences in characteristics of IC manufacture and miniaturization science should be kept in mind.

## Moore's Law

### Introduction

The dramatic increase in performance and cost reduction in the electronics industry derives from progress in IC fabrication and packaging. Speed and performance improvement of IC chips and their associated packages, and hence higher-level hardware such as computers, all are based on the minimum printable feature size [critical dimension (CD)].

Moore's Law states that the number of transistors per square centimeter roughly doubles every 18–24 months without increase in cost. Recent trends show that feature size decreases about 30% every 2 years, and one needs more and more upfront investment money to fabricate. Moore's Law—an observation, really—might hit an even bigger snag with CDs in the 32-nm mode, where DUV lithography runs out of steam and quantum effects start wreaking havoc.

After introducing Moore's Law, we analyze Ray Kurzweil's optimistic speculations about the future of Moore's Law.

### Moore's Law Detailed

Control of the thickness ($z$) of films deposited on a substrate and control of the depth of a cavity etched in a substrate can be achieved with remarkable precision, down to 20 Å. The width and length ($x$, $y$) dimensions of a deposit or etch are more difficult to control and depend on the type of lithography used. The minimum $x$, $y$ feature is determined primarily by the precise focus of the energy source that allows discrimination between areas of exposed resist and nonexposed resist. Other key factors are the tolerances on the mask, the ability to align the mask to the wafer and align subsequent layers of masks to create the proper vertical geometry, and the ability to control the rate and direction of etching and deposition.

Gordon Moore, inventor of the IC and former chairman of Intel, first described the rate of progress in reducing feature sizes for ICs when, in 1965, he observed that the surface area of a single transistor reduced by approximately 50% every 12 months. In 1975, he revised this to every 2 years, not every

**Table 2.1** CDs, DRAM Bits, MPU Transistor Density, and Die (Chip) Sizes Based on the SIA Road Map*

| | 1997 | 1999 | 2002 | 2005 | 2008 | 2011 |
|---|---|---|---|---|---|---|
| Minimum feature size: CD in nm | 250 | 180 | 130 | 100 | 70 | 50 |
| DRAM (bits) | 256M | 1G | 4G | 16G | 64G | 256G |
| MPU transistors/cm$^2$ | 3.7M | 6.2M | 18M | 39M | 84M | 180M |
| DRAM chip size (mm$^2$) | 280 | 400 | 560 | 790 | 1120 | 1580 |
| MPU chip size (mm$^2$) | 300 | 340 | 430 | 520 | 620 | 750 |

*Source:* Semiconductor Industry Association. 1997. *The National technology roadmap for semiconductors: technology needs.* http://www.rennes.supelec.fr/ren/perso/gtourneu/enseignement/roadmap97.pdf. For updates visit http://www.sia-online.org.[2]

*See Figure 2.1 and Volume I, Chapter 4.

18 months as is often quoted.[1] Although only an engineer's rule of thumb, this is now widely known as Moore's law. In Table 2.1, based on the Semiconductor Industry Association (SIA) road map[2] (see also Volume I, Appendix 1A), we summarize the minimum feature size (CD), dynamic random access memory (DRAM)* bits, microprocessor transistors per square centimeter, and DRAM and microprocessor die sizes as a function of the year of their introduction.[3] With decreasing CDs comes more rigid CD control: a 180-nm line width requires ±14-nm CD control, and a 50-nm line width requires ±4-nm CD control.

Historically, DRAMs have driven the progress of microlithography more than microprocessors, which exhibit more relaxed design rules (see Table 2.1). The state of the art in ICs today is a 2-GB DRAM with 60-nm features (Samsung). Intel (Integrated Electronics) introduced the Core 2 Quad "Kentsfield" chip in January 2007, a chip featuring a 65-nm technology mode.

In Figure 2.1, we illustrate Moore's law by plotting DRAM bit capacity and minimum feature size on a logarithmic $y$-axis with time on the $x$-axis (see also Volume I, Chapter 4).

Looking ahead, IC developers relying on Si complementary metal oxide semiconductor (CMOS) architecture today will switch more to alternative CMOS materials and architectures. At least five fundamental technical roadblocks need to be addressed to scale down CMOS transistors further (beyond 45 nm):

1. Carrier mobility limitations/clustering of dopant atoms: As CMOS transistors shrink, mobility of carriers decreases, and to compensate, smaller devices need a higher amount of dopant atoms; however, above a certain limit the dopant atoms clump together, forming clusters that are not electrically active. The concentration cannot be increased further, and today's chips are close to the maximum.
2. Tunneling through the gate dielectric: The transistor gates that control the flow of electrons have become so small that they are prey to undesirable quantum effects such as tunneling through extremely small barriers (e.g., a gate oxide of 1.2 nm). High-k dielectrics are used to minimize this tunneling current.
3. Doping nonuniformity: Doping nonuniformity over a wafer surface is becoming statistically significant and appears in transistors, showing a varying degree of doping (see also point 1).
4. Interconnects: Interconnects are becoming the dominant portion of circuit speed and process complexity. By volume ICs have become all wires.
5. Power dissipation: The latest Pentium chips dissipate more heat than a stove-top cooking surface (100 W/cm$^2$).

A few comments and additional details on the above technical challenges follow; for yet more details, consult Volume I, Chapter 4.

Transistors with feature sizes between 0.5 and 0.25 μm are operated at low supply voltages to attain adequate gain and low leakage currents. A small supply voltage keeps the dissipated power within reasonable limits, makes for faster transistor switching, and allows the implementation of smaller transistors. The latter is an example of favorable downscaling because, as transistors get smaller, the insulation

* Dynamic random access memory. A type of memory component used to store information in a computer system. "Dynamic" means the DRAMs need a constant "refresh" (pulse of current through all of the memory cells) to keep the stored information.

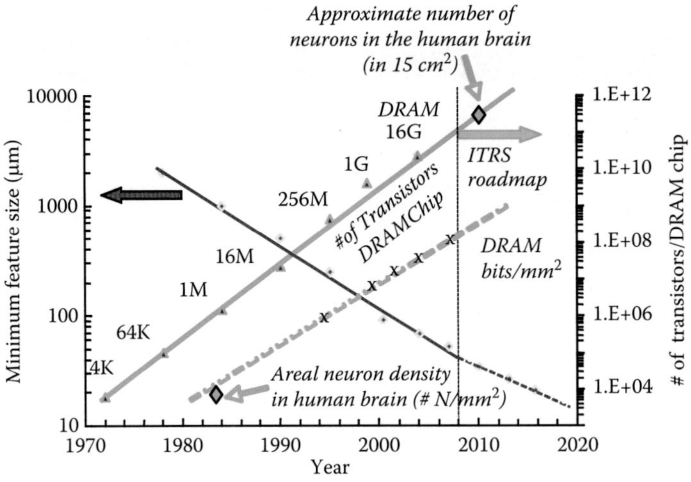


**FIGURE 2.1** Semilog plot of DRAM bits and minimum feature size as a function of time. (Courtesy of Dr. Rashid Bashir, Purdue University.) See also Volume I, Chapter 4.

layer between the gate and the source/drain becomes thinner. With too high a supply voltage, electrons would "leak" through the thin insulation and compromise system stability. Eventually, the thin insulation layer will be only a couple of molecules thick, at which point transistor scaling will reach an abrupt end. In 1.8-V logic, threshold voltage is only 0.45 V. As this threshold becomes lower and lower, it will be easier for a minimal flow of electrons to be mistaken as a logical 1 instead of 0, thus causing operational problems.

Below 0.25 μm, cooling becomes necessary to maintain the same adequate transistor characteristics. Cooling of chips might be implemented using methods derived from miniaturization science, such as the use of miniature thermoelectric coolers or Joule-Thomson refrigerators (see Volume III, Chapter 8 on actuators). In the 30–50-nm range, quantum effects begin playing an important role. Quantum effects, as we saw in Volume I, Chapter 3 on quantum mechanics, can be exploited, and major efforts are underway to manufacture quantum boxes, lines, and dots for the next generation of ICs (gigascale integration or 1 billion devices per chip).

Interconnections pose another major challenge for further miniaturization of ICs. Like transistors, metal connection lines are scaled down with every new chip generation, but unlike transistors, their performance decreases as they become thinner because of increased resistance and capacitance. In this case, downscaling is disadvantageous. Connection lines were once aluminum, but copper is now replacing aluminum (e.g., in the Coppermine Pentium III) as interconnect material in building devices that not only run faster—because of copper's higher conductivity—but are also resistant to electromigration. Copper also reduces the number of steps required in manufacture. A more drastic advancement would be the use of superconducting materials for implementing interconnections, but currently no superconductor demonstrates the desired behavior under the temperature and current density conditions that exist in modern ICs. Optical interconnects are another possibility, although these have size limitations (they cannot be thinner than the wavelength of light they transmit), and the conversion of light pulses to electrical ones and vice versa could seriously compromise their performance. If optical interconnections were implemented, a switch from Si to GaAs would also be in order because GaAs offers far superior light source-making capabilities than Si. In Volume I, Chapter 5, we saw how the use of photonic crystals and metal plasmonics can circumvent size limitations of traditional optical waveguides. Photonic crystals and plasmonics offer perhaps the first real possibility for the integration of electronics and optics on the same chip.

One further comment is in order here. Ultimately, all the technological barriers listed above might be overcome, but the real question is, at what cost? Moore's second law states that capital costs increase faster than revenue and that the rate of technological progress is controlled by financial realities. Specifically, Moore gloomily predicted that the cost of building a chip manufacturing plant doubles with every other chip generation, or roughly every 36 months. This is often referred to as Moore's "second law." The projected cost of an IC lab in 2015 is ~$100 billion.

## Kurzweil: A View from an Optimist

To explore what the future of Moore's law holds, we consider Ray Kurzweil's optimistic predictions.[4] Kurzweil (Figure 2.2) and others observed that computing power had been growing exponentially long before the invention of the transistor. From Kurzweil we learn that from 1900 onward the speed of computation has doubled every three years, and that by 2020, computers will equal the human brain in terms of memory capacity and computing speed (see Figure 2.3).

Kurzweil believes that Moore's law is an illustration of a broader underlying law, *The Law of Accelerating Returns*, which is based on *The Law of Time and Chaos*. The latter, as quoted from Kurzweil, reads: "In a process, the time interval between salient events (that is, events that change the nature of the process, or significantly affect the future of the process) expands or contracts along with the amount of chaos." As a consequence, the *Law of Accelerating Returns* (a sublaw of the *Law of Time and Chaos*) states that as order exponentially increases, time exponentially speeds up, or the interval between salient events grows shorter as time passes.* It is assumed that another technology will take over when Moore's law, based on photolithography, is no longer valid. New technology may involve quantum devices, three-dimensional (3D) chips, superconductors, photonic crystals, plasmonic devices, and metamaterials. All that is needed for this evolution to take place is the growing order of the evolving technology itself and enough surrounding chaos—that is, diversity of options or alternatives—from which to draw new ideas. This is similar to natural evolution, which needed the many options provided by a rich and varied environment to develop sophisticated and intelligent life forms. It follows that Moore's law, describing the speeding up of technology developments in the IC world, and Taniguchi's laws (see Chapter 6), describing the speeding up of technology developments in mechanical machining (tool development), are both examples of the *Law of Accelerating Returns*. Similarly, in data recording, the rate of progress has exponentially increased. It started slowly, perhaps with crude cave drawings, then pictorial script followed by symbolic script, then pen and paper followed by a colossal milestone, which speeded up things considerably, i.e., the Gutenberg printing press. Then magnetic recording in computers started with recording using a copper-wound coil; when this became too crude, the invention of the thin film read/write head by IBM's L. Romankiw in 1973 took over.[5] In magnetic data storage with thin film read/write, the ultimate limit at room temperature is projected to be 40 Gb/in.[2,6] A next leap forward in recording density is expected when proximal probes prove their mettle as read/write heads; 45 $Gb/in.^2$ (Terris et al.[7]) and 400 $Gb/in.^2$ (Kraus and Chou[8]) have been demonstrated already. Using a 32 × 32 (1024) array of AFM cantilevers, King et al.[6] thermomechanically wrote data bits 40 nm in diameter and pitch (this corresponds to a data density of 0.4 $Tb/in.^2$), and they read those bits thermally at a speed of 480 Mb/s. Figure 2.4 shows a picture of a thin film read/write head (a) and an STM-based read/write head (b) (IBM's millipede).

Yet another example of the *Law of Accelerating Returns* is exemplified in the Human Genome Project started in the mid-1980s (see Volume III,

**FIGURE 2.2** Ray Kurzweil is on the left, but does Bill believe him?

* The other sublaw of the *Law of Time and Chaos* is the *Law of Increasing Chaos*. With chaos exponentially increasing, time exponentially slows down. The events developing after the Big Bang are an illustration of this law; most salient events occurred immediately after the Big Bang and became less frequent thereafter.

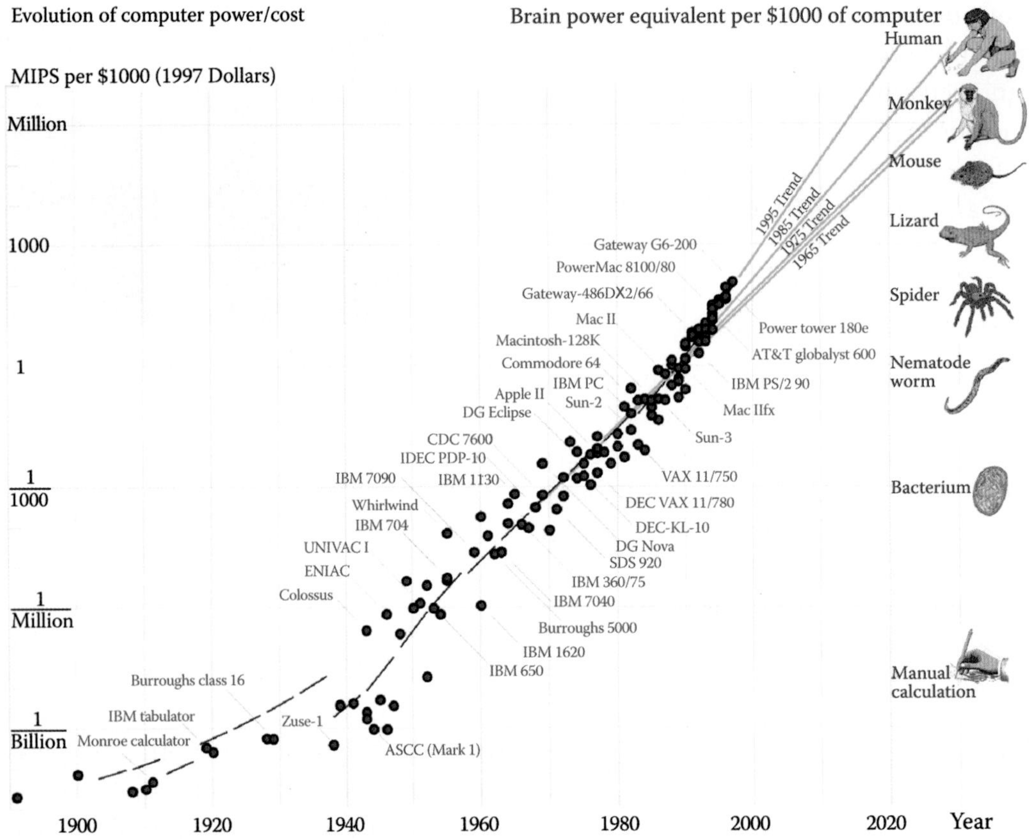


**FIGURE 2.3** What $1000 of computing buys. The number of million instructions per second (MIPS) is a general measure of computing performance. (Based on Ray Kurzweil.)

Chapter 2). The speed of the process of writing down all 3.5 billion bits of the human genetic code exponentially increased since the start of the project, causing the project to finish ahead of schedule. For some additional reading from Kurzweil's work, see *The Singularity Is Near: When Humans Transcend Biology*.[9]

## Next-Generation Lithographies

### Introduction

Having set a rather optimistic forecast for the future of miniaturization sciences, we now introduce a set of possible next-generation lithographies

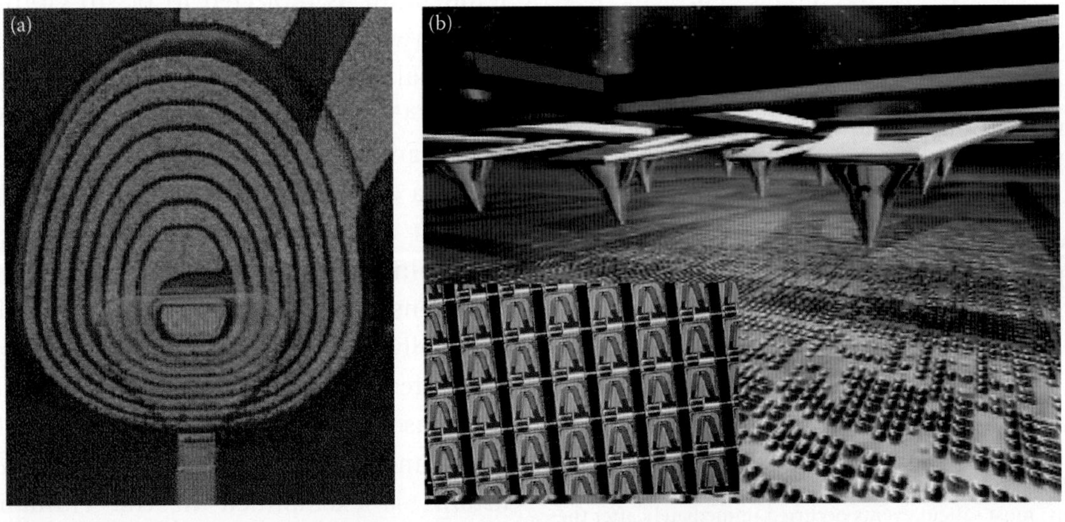


**FIGURE 2.4** (a) Thin film read/write head. (b) STM-based read/write head (IBM's millipede).

(NGLs). This includes extreme ultraviolet lithography (EUVL), x-ray lithography, charged-particle-beam lithography, including scattering with angular limitation projection electron-beam lithography (SCALPEL) and ion projection lithography (IPL), as well as imprint lithography [nanoimprint lithography (NIL) and step-and-flash imprint lithography (SFIL)]. In the context of electron-beam lithography (EBL), we also look at progress made with microelectromechanical system (MEMS)-based arrays of field emitters.

## Extreme Ultraviolet Lithography

As mentioned before, the 32-nm node is viewed as beyond the scope of DUV lithography [even when applying all the resolution-enhancing technologies (RETs) or taking advantage of immersion lithography, which we learned about in Chapter 1]. EUVL is one of the most likely NGLs to achieve the 32-nm node.

Extreme ultraviolet lithography (EUVL), using wavelengths in the 10–14-nm range to carry out projection imaging, is the most natural extension of optical projection lithography as, in principle, it only differs in terms of the wavelength. This type of radiation is also referred to as *soft x-ray radiation* and *vacuum UV*. Sources for this type of radiation are laser-produced plasmas and synchrotrons.

To appreciate the merits of EUVL, we take as a starting point the expressions for resolution (*R*):

$$R = k_1 \frac{\lambda}{NA} \tag{1.16}$$

and depth of focus (DOF):

$$DOF = \pm\delta = \pm k_2 \frac{\lambda}{NA^2} \tag{1.17}$$

In Chapter 1 we learned that in practice, both resolution *R* and DOF are experimentally determined and are influenced by both resist and exposure systems. In large-scale IC manufacture with g- and i-line light, $k_1$ values used to be around 0.6 (see Figure 1.81) and DOFs were larger than 0.5 μm. At that time, these parameters led to the desired CD control within a tolerable process window. The push for higher resolution with 248-nm and 193-nm radiation made it necessary to achieve smaller $k_1$ values, and this led to processing problems; it became more difficult to control the CD, and it created an intolerably low DOF. The RETs discussed in Chapter 1 enhance resolution and the effective DOF, but, as we discussed, they are often not very manufacturer-friendly (people only half-jokingly refer to RET as "real expensive technology"). With EUVL, on the other hand, one surmounts these problems because of its shorter wavelength. The technique is capable of printing sub-100-nm features while maintaining a DOF of 0.5 μm or larger, and has *k*-values that make the process control less demanding. Unfortunately, things are not that simple—EUV is strongly absorbed in virtually all materials, and, consequently, imaging must be carried out in a vacuum; also, all camera optics and masks used must be reflective rather than refractive. New resists and processing techniques must be developed as well.[10–12]

In Figure 2.5 a schematic of an EUVL setup is shown. A laser is directed at a xenon gas and heats up the gas to create a plasma, radiating light at 13.6 nm—too short for the human eye to see. That EUV light is directed onto a condenser, gathering the light and directing it onto the mask. The mask is a representation of one level of the IC under construction and is made by patterning an absorber on some parts of a mirror but not on others (Figure 2.6). The pattern on the mask is reflected onto a series of four to six curved mirrors, reducing the size of the image and focusing the image onto the resist-coated silicon wafer. The entire process takes place in a vacuum because at these short wavelengths even air absorbs the light.

With respect to EUVL reflectivity, it is well known that most materials have very low reflectivity for near-normal incidences at those short wavelengths; therefore, multilayer Bragg reflectors with multiple reflecting coatings are used, as illustrated in Figure 2.6. The concave and convex mirrors in EUVL, including the mask, are coated with multiple layers (e.g., 80 of them) of molybdenum and silicon (another pair of material used is Mo-Be) with dissimilar EUVL optical constants, providing a resonant reflectivity when the period of the layers is close to λ/2. With magnetron-sputtered Mo-Si-based reflectors, peak reflectivities of up to 70% at 13.4 nm

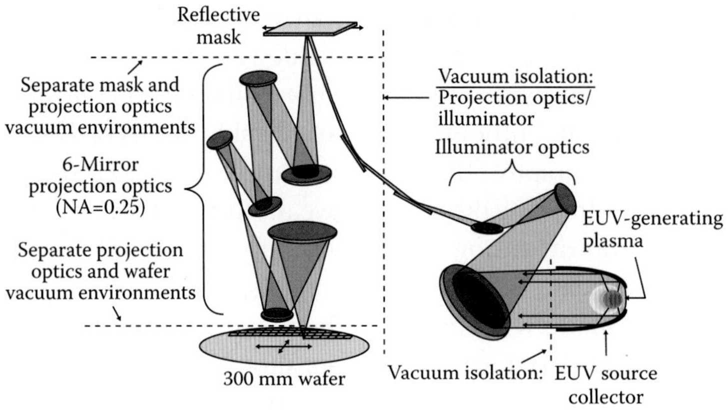


**FIGURE 2.5** Schematic of an EUVL lithography setup. (From SEMATECH's Next Generation Lithography Workshop brochure.)

have been achieved (the other 30% is absorbed by the mirror). Without the coating, the light would be almost totally absorbed before reaching the wafer. The mirror surfaces have to be nearly perfect; even small defects in coatings can destroy the shape of the optics and distort the printed circuit pattern.[10]

The multilayer stacks in EUVL masks today have a demonstrated defect level of ~$10^{14}$ defects/cm$^2$, still two orders of magnitude above target. Prototype EUVL cameras designed for 13.4-nm radiation have an numerical aperture (NA) = 0.1 and a four-mirror set with multilayer Mo-Si-reflecting coatings.[10] These step-and-scan 4× cameras have a resolution better than 100 nm with a ring field of 26 × 1.5 mm. In the step-and-scan system, the mask and wafer are simultaneously scanned in opposite directions, with the mask moving four times faster than the wafer. The camera wave-front error at 13.4 nm needs to be less than 1 nm so that the surface figure (that is, the basic shape of each of the mirrors) will be accurate to 0.25 nm rms or better. Achieving these accuracies will also require improved metrology techniques because current ones cannot confirm the accuracies required. The fact that EUVL masks are reflective has advantages and disadvantages. Advantages are that no thin and brittle membranes are needed as required in the case of x-ray lithography—just a solid substrate (e.g., a Si wafer) coated with the required reflective coatings and a patterned EUVL absorber; because of the 4:1 reduction, mask making is also easier than with x-rays. A major disadvantage is that the defect density in the reflective coating needs to be made so small that there are few techniques available yet to accomplish such defect-free films. Current resists absorb all the EUVL photons within less than 100 nm. Approaches for EUVL resists include silylated single-layer resist, refractory bilayer resists, and trilayer resists.

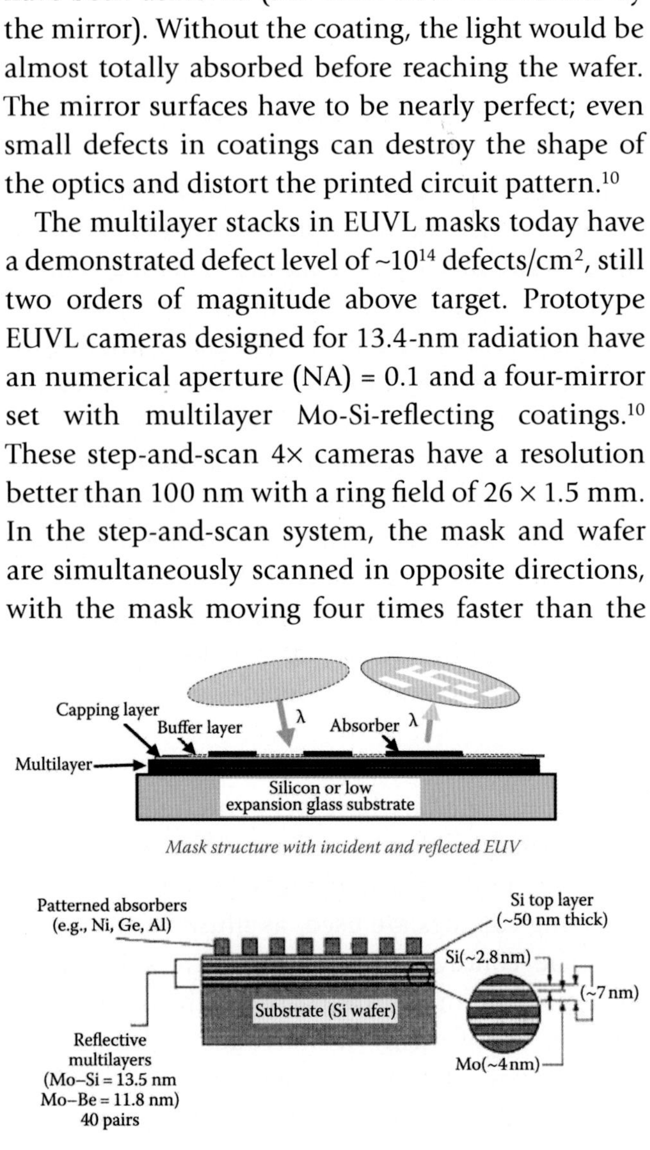


**FIGURE 2.6** Two schematic representations of EUVL reflective masks. In EUVL, only reflective optics will work (see text).

The Extreme Ultraviolet LLC (EUV LLC) was created to spearhead the research involved in bringing EUVL to the marketplace. The consortium includes Intel Corporation, Motorola Corporation, Advanced Micro Devices Corporation, IBM, Infineon, Micron Technology, and the Sandia and Lawrence Livermore National Laboratories (http://www.sandia.gov/media/NewsRel/NR2001/euvlight.htm). Intel is leading this consortium, and an introduction of EUVL is planned for 2009. For the past couple of years there has been a growing realization that the resolution capabilities of the EUVL wavelength are being countered by the effects of electrons released after absorption, and it is feared that a 13.4-nm resolution might be impossible. On the other hand, on May 30, 2007, NEDO announced, that using EUVL, it had drawn circuit patterns with just 26-nm line widths (isolated and dense lines) using EUVL (http://www.nedo.go.jp/english/archives/190820/190820.html).

## X-Ray Lithography

### *Introduction*

X-ray lithography uses a shadow printing method similar to optical proximity printing. The x-ray wavelength (4–50 Å) is much shorter than that of UV light (2,000–4,000 Å) (Table 2.2). Hence diffraction effects are reduced and higher resolution can be attained. For example, for an x-ray wavelength of 5 Å and a gap of 40 μm, $R$ is equal to 0.2 μm. In contrast with electron lithography and ion-beam lithography, no charged particles are directly involved in x-ray exposures, which makes the need for vacuum less stringent (see Figure 2.7).[13] Like with all high-energy radiation (see EUVL), a vacuum is preferred to avoid absorption of photons by gas molecules, formation of ozone, and corrosive reactions with the exposure station's hardware. Another advantage of x-rays is that one can use flood exposure of resist-coated wafers, ensuring higher throughput than when writing with a thin electron or ion beam. The method is also referred to as deep x-ray lithography (DXRL).

**Table 2.2** Light Sources for Various Types of Lithography

| Wavelength (nm) | Source | Range |
|---|---|---|
| 436 | Hg arc lamp | g-line |
| 405 | Hg arc lamp | h-line |
| 365 | Hg arc lamp | i-line |
| 248 | Hg/Xe arc lamp, KrF excimer laser | Deep UV (DUV) |
| 193 | ArF excimer laser | DUV |
| 157 | F2 laser | Vacuum UV (VUV) |
| ~10 | Laser produces plasma sources | Extreme UV (EUV) |
| ~1 | x-ray tube, synchrotron | x-ray |

The three main classes of sources for x-ray lithography are electron impact tubes, laser-based plasmas, and synchrotrons (Figure 2.8). In miniaturization science, we are mainly concerned with synchrotron radiation. For reasons that are explained below, in the case of ICs, less-expensive alternative x-ray sources might suffice.

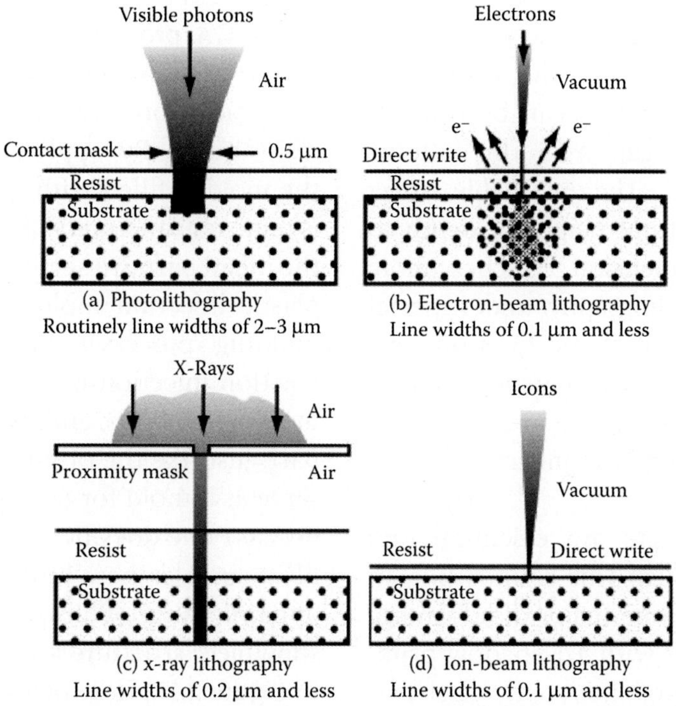


**FIGURE 2.7** A comparison of (a) photolithography, (b) electron-beam, (c) x-ray lithography, and (d) ion-beam. (Based on Brodie, I., and J.J. Muray. 1982. *The Physics of Microfabrication*. New York: Plenum Press. With permission.[13])

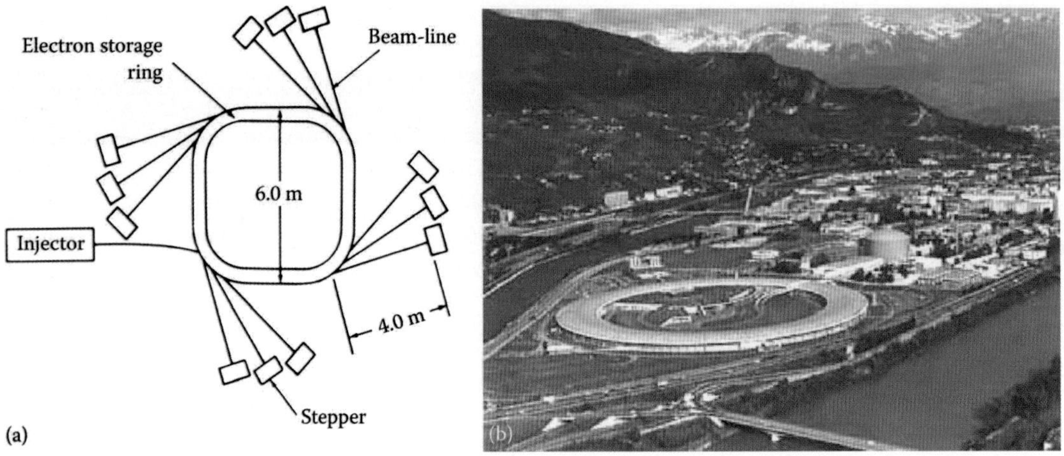


**FIGURE 2.8** (a) Schematic of a synchrotron. (b) Grenoble synchrotron (http://www.esrf.eu).

### *Major Features of X-Ray Lithography*

X-ray lithography is superior to optical lithography because of its use of shorter wavelengths and its very large DOF, as well as because exposure time and development conditions are not as stringent. Reproducibility is high because results are independent of substrate type, surface reflections, and wafer topography. Another important benefit is that x-ray lithography is immune to low atomic number ($Z$) particle contamination (dust). With an x-ray wavelength on the order of 10 Å or less, diffraction effects generally are negligible, and proximity masking can be used, increasing the lifetime of the mask. With a standard 50-μm proximity gap and using synchrotron x-rays, 0.25-μm patterns can be printed; by decreasing the proximity gap to 25 μm, patterns of 0.15 μm can be resolved.[14] The obtainable aspect ratio, defined as the structural height or depth to the minimum lateral dimension, reaches more than 100. With UV photolithography, under special conditions, an aspect ratio of about 25 is possible at most (see SU-8 in Chapter 1). An aspect ratio of 100 corresponds to the aspect ratio attainable by wet chemical anisotropic etching of monocrystalline Si (see Chapter 4).

In x-ray lithography, there are essentially no optics involved, and although this sounds like an advantage, it also presents one major disadvantage because one can only work with 1:1 shadow printing. No image reduction is possible, so the mask fabrication process is very complicated. In the United States, IBM remains the only major champion of x-ray lithography for NGL.[15] At the end of 1999, IBM fabricated several PowerPC 604e microprocessor batches to demonstrate the viability of the method.[16] In Japan, there are still many players involved, such as NTT, Toshiba, Mitsubishi, and NEC.[17]

### *LIGA*

The LIGA technique (a German acronym for *Lithographie, Galvanoformung, Abformung*) was invented about 25 years ago.[18] LIGA exploits all of the advantages of x-ray lithography listed above, and the process is schematically illustrated in Figure 2.9. The LIGA process involves a thick layer of resist (from micrometers to centimeters), high-energy x-ray radiation, and resist development to make a resist mold. By applying galvanizing techniques, the mold is filled with a metal. The resist structure is removed, and metal products result. Alternatively, the metal part can serve as a mold itself for precision plastic injection molding. Several types of plastic molding processes have been tested, including reaction injection molding, thermoplastic injection molding, and hot embossing. The so-formed plastic part, just like the original resist structure, may also serve as a mold for fast and cheap mass production because one does not rely on a new x-ray exposure. LIGA enables new building materials and a wider dynamic range of dimensions and possible shapes. A showpiece structure for LIGA technology is pictured in Figure 2.10, in which an ant holds a Ni gear.

Of particular interest to miniaturization science is the possibility of creating 3D shapes with slanted

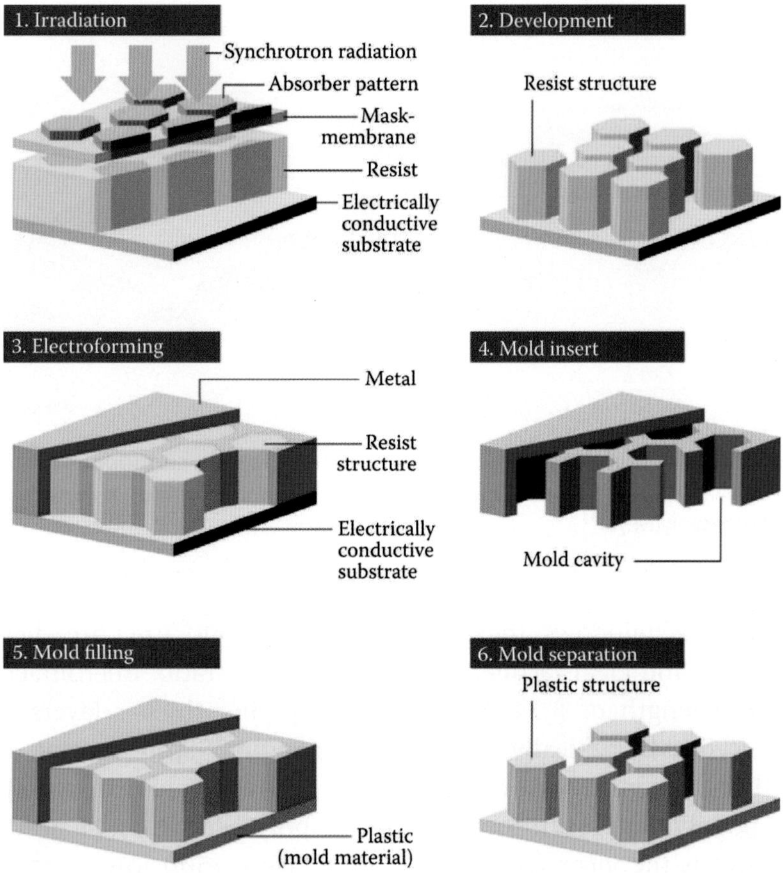


**FIGURE 2.9** The LIGA process.

sidewalls (see Figure 2.11)[19] and step-like structures. Two German companies, IMM (http://www.imm-dueck.de) and Jenoptik Mikrotechnik GmbH (http://www.jo-mikrotechnik.com), have developed an x-ray scanner that allows for continuous tilt angles of the mask/substrate assembly with respect to the x-ray beam, rotation of the mask/substrate, and that contains an internal mask alignment system for different masks for multiple exposures. The unprecedented precision attainable with LIGA makes this technique stand out against other 3D lithography methods such as DUV with SU-8 (see Chapter 1). Using the IMM/Jenoptik scanner, a vertical resist wall with a precision better than 0.05 μm was demonstrated, and even resist walls inclined at a 45° angle were made with an accuracy of 1 μm over a height of up to 500 μm.[18] LIGA is further discussed in Chapter 10.

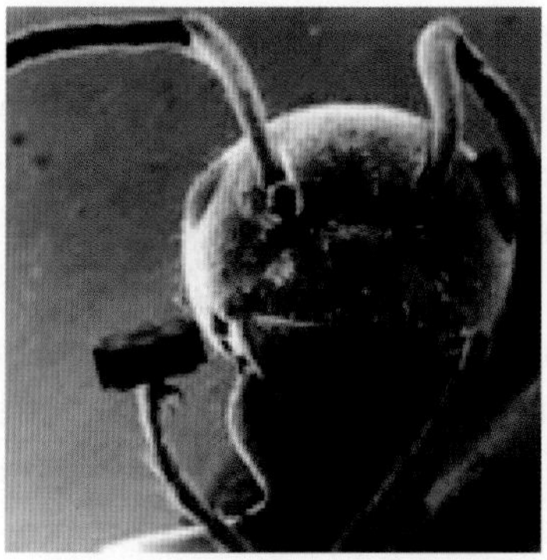

**FIGURE 2.10** Ant with gear. (From Forschungszentrum Karlsruhe, Program Microsystem Technologies. With permission.)

### *X-Ray Resists*

An x-ray resist should have high sensitivity to x-rays, high resolution and resistance to chemical, ion, and/or plasma etching, thermal stability of >140°C, and a matrix or resin absorption of less than 0.35 $\mu m^{-1}$ at the wavelength of interest. No presently available resist meets all those requirements. One material predominantly used in x-ray lithography is positive poly(methylmethacrylate) (PMMA), a material better known by its trade names Plexiglas™ and Lucite™. Another name commonly used for polymers based on PMMA is *acrylics*. Clear sheets of the material are

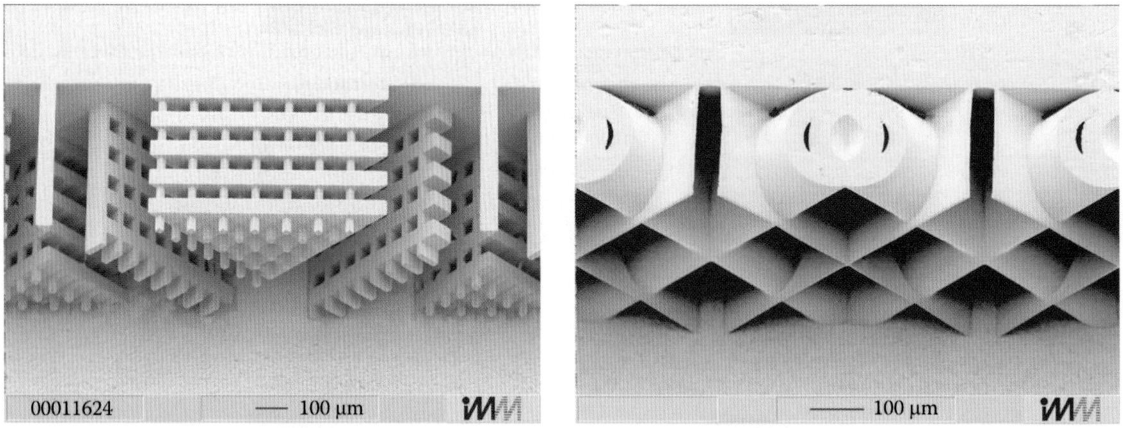


**FIGURE 2.11** LIGA structures obtained at IMM using an x-ray scanner enabling continuous tilt angles of the mask/substrate assembly. (Ehrfeld, W., and A. Schmidt. 1998. Recent developments in deep x-ray lithography. *J Vac Sci Technol* B16:3526–34.[19]) (Courtesy IMM.)

used to fabricate "unbreakable" windows, inexpensive lenses, machine guards, and clear lacquers on decorative parts. At a wavelength of 8.34 Å, the lithographic sensitivity of PMMA typically hovers about 2–6.5 J/cm$^2$, a rather low sensitivity implying a small throughput. A possible approach to make PMMA more x-ray sensitive is the incorporation of x-ray-absorbing high atomic number atoms. Another approach, discussed in Chapter 1, involves the use of chemically amplified photoresists. Other x-ray resists explored for LIGA applications are poly(lactides)[20] and some SU-8 epoxy resin modifications.[21] These resists show a considerably enhanced sensitivity and reduced stress corrosion compared with PMMA.[19]

Negative x-ray resists exhibit inherently higher sensitivities compared with positive x-ray resists, although their resolution capability is limited by swelling. Poly(glycidyl methacrylate-co-ethyl acrylate) (PGMA), a negative e-beam resist, has been used in x-ray lithography. In general, resist materials sensitive to e-beam exposure are also sensitive to x-rays and function in the same way; that is, materials that are positive in tone for electron-beam radiation typically also are positive in tone for x-ray radiation. A strong correlation exists between the resist sensitivities observed with these two radiation sources, suggesting that the reaction mechanisms might be similar for both types of irradiation.[22]

The IC industry requires a typical resist layer not more than 1–2 μm thick. Thicker layers, between 10 and 1,000 μm, are dictated by the need for high-aspect-ratio micromachines. The technology of applying thicker layers of photoresist remains challenging. Spin coating of multiple resist layers (for relatively thin coats), resist casting with in situ polymerization of mildly cross-linked PMMA (for layers >500 μm), and plasma polymerization of PMMA with a possibility of subsequent diamond grinding of the resulting layers are some of the techniques currently in use (see also Chapters 1 and 10).

In the case of exposures with very high energy x-rays (hard x-rays), the associated wavelength is measured in angstroms, not in nanometers. At those energies, almost every type of polymer becomes a "resist," and even "resistless" lithography becomes possible as thin films can be etched, vaporized, or ion-implanted directly.

### *X-Ray Masks*

Another challenge in x-ray lithography, besides the low sensitivity of the resists and the high cost of sufficiently bright x-ray sources, is the mask making—already complex for producing DRAMs, but even more complex for 3D structures with high aspect ratios. In Table 2.3, the procedure for making an optical mask is compared with the procedure involved in making an x-ray mask. Technologies developed for manufacturing masks for submicrometer circuitry with x-ray lithography do not directly transfer to the fabrication of x-ray

TABLE 2.3 Optical versus X-Ray Mask

| Optical Mask | X-Ray Mask |
|---|---|
| Mask design: CAD | Mask design: CAD |
| Substrate preparation: | Substrate preparation: |
| Quartz | Thin membrane substrate (Si, Be, Ti) |
| Thin metal film deposition | Deposit plating base (50 Å Cr then 300 Å Au) |
| Pattern delineation: | Pattern delineation: |
| Coat substrate with resist | Coat with resist |
| Expose pattern (optical, e-beam) | Expose pattern (optical, e-beam) |
| Develop pattern etch Cr layer | Develop pattern |
| Strip resist | Absorber definition: Electroplate Au (~15 μm for hard x-rays) Strip resist |
| Cost: $1–3K | Cost: $4–12K |
| Duration: 3 days | Duration: 10 days |

masks for building micromachines. An x-ray mask basically consists of a pattern of x-ray-absorbing material (a material with a high atomic number, *Z*, such as gold, tungsten, or tantalum silicide) on a membrane substrate transparent to x-rays (a low *Z* material, e.g., Ti, Si, SiC, $Si_3N_4$, BN, Be) (see Figure 2.12). The membrane must have windows or other features in certain areas for alignment purposes. Additionally, the material also needs to be thermomechanically stable to a few parts in 10.[23] Mechanical stress in the absorber pattern can cause in-plane distortion of the supporting thin membrane, requiring a high Young's modulus material. Also, humidity or the x-ray exposure itself might distort the membrane.

A fabrication sequence for an x-ray mask is shown in Figure 2.13. In this case, the x-ray absorber pattern is ion beam-etched in a Au-film after the pattern first has been written in a 300 Å PMMA layer with an e-beam. The absorber film typically consists of two metal layers: a thin layer of chromium for adhesion to the substrate, topped by a thicker layer of gold. The higher the required aspect ratio of the exposed resist, the thicker the gold layer of the mask absorber pattern must be to maintain a good contrast. The x-ray mask shown in Figure 2.13 only has a 400-Å-thick gold layer, which adequately covers DRAM manufacture but not high-aspect-ratio 3D structures, where 5–15 μm of gold absorber might be required for a 500-μm-thick resist and up to 50 μm for 10-cm-thick resists. A proximity x-ray mask, as shown in Figure 2.13, is placed close to the substrate, and its pattern is reproduced by exposure to x-rays. Various other schemes to make LIGA x-ray masks are explored in Chapter 10.

FIGURE 2.12 X-ray mask structure: Ta patterns on SiC or SiN membrane. The absorber Ta is 300 nm thick (stress ~ 0 MPa). Membrane SiC or SiN is 2 m thick (stress = 100 MPa ~ 200 MPa), 20 ~ 30 mm square. The minimum pattern width is 50 nm. Pattern position accuracy: 3σ < 30 nm (http://www.ntt-at.com/products_e/x-ray_masks/index.html).

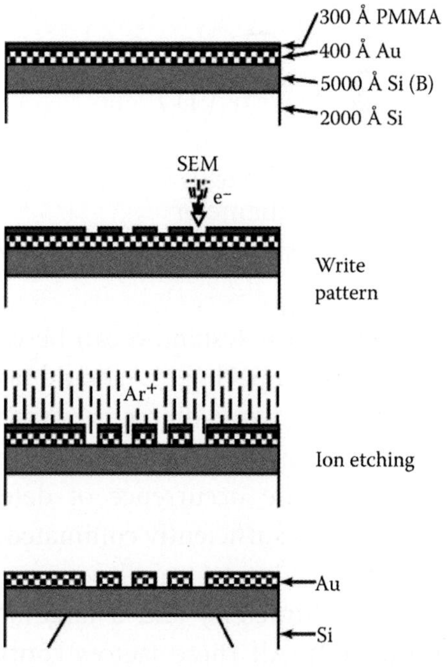


FIGURE 2.13 Fabrication of a silicon membrane-based x-ray mask with a gold absorber pattern. For use in high-aspect-ratio micromachining, the gold absorber layer must be between 5 and 15 μm. (Based on Brodie, I., and J.J. Muray. 1982. *The physics of microfabrication.* New York: Plenum Press. With permission.[13])

*Why Use a Synchrotron to Generate X-Rays?*

The full power of x-ray lithography for miniaturization science only materializes when using hard, collimated synchrotron radiation. To appreciate this, we will explore the procedure using a less intense, less collimated beam from an electron-beam bombardment source. A schematic diagram of an x-ray exposure system using an electron-beam bombardment-based x-ray source is shown in Figure 2.14.

The mask is typically offset above the wafer by about 10 μm after alignment. A proximity scheme, rather than a contact mask, is a useful feature, given the fact that an x-ray mask can cost up to $13,000. Because the x-ray source is finite in size and separated by a distance, *D*, the edge of the mask does not cast a sharp shadow but rather has a blurry region associated with it, known as penumbral blur, *d*. Image blurring limits the ultimate resolution power of an x-ray exposure system, as shown in Figure 2.14. As diffraction effects can be ignored, simple geometric considerations can be used for relating the image to the pattern on the mask. From Figure 2.14, we estimate that the blurring, δ, at the resist plane is given by:

$$\delta = s\left(\frac{d}{D}\right) \tag{2.1}$$

where s = mask-to-wafer gap
d = the source diameter
D = source-to-substrate distance

In a high-resolution system, δ can be controlled to within 0.1 μm. Spacing, *s*, should allow the accommodation of large-diameter masks while avoiding the high risk of contacting the resist and greatly increasing the occurrence of defects. The x-ray source must be sufficiently collimated. In practice, this translates into a small source diameter d (e.g., a few millimeters) and a large source-to-mask distance (*D*). All these factors contribute to minimizing penumbral blurring. Conventional e-beam-generated x-ray sources have sizes of a few millimeters and are about 40 cm away from the mask. Unfortunately, a large distance required for adequate collimation results in prohibitively

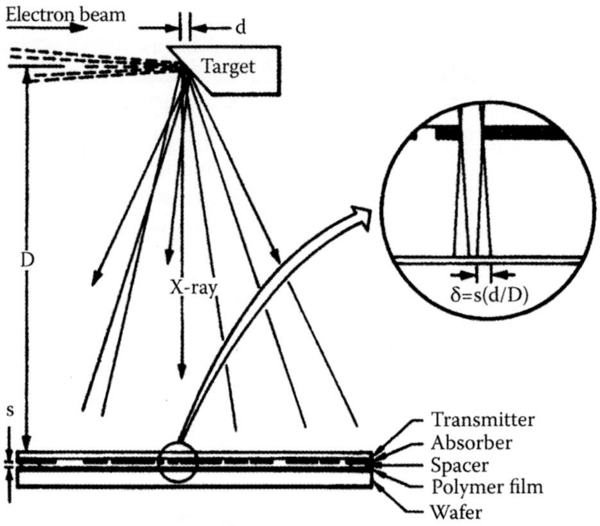


**FIGURE 2.14** X-ray lithography with an electron-beam x-ray source. Inset, extent of penumbral effect calculated from geometric considerations. (Based on Brodie, I., and J.J. Muray. 1982. *The physics of microfabrication.* New York: Plenum Press. With permission.[13])

long exposure times (e.g., hours) as a result of the weak intensity of these sources. With synchrotron radiation, on the other hand, penumbral blurring does not limit the spatial resolution. Because of the high collimation of synchrotron radiation, rather large distances between the mask and the wafer can be tolerated (about 1 mm for 1-μm line width patterns). In the electron storage ring or synchrotron, a magnetic field constrains electrons to follow a circular orbit, and the radial acceleration of the electrons causes electromagnetic radiation to be emitted forward. Thus, the radiation is strongly collimated in the forward direction and can be assumed to be parallel for lithographic applications. Because of the much higher flux of usable collimated x-rays, shorter exposure times become possible. The Advanced Light Source (ALS) synchrotron in Berkeley, for example, can deliver a flux of 0.4 W/cm$^2$ at 30 m for 3–9-keV radiation (http://www.als.lbl.gov). Especially if one wants to generate a highly collimated photon flux in the spectral range required for precise deep-etch x-ray lithography in thick resist layers, synchrotron radiation comes close to being the ideal source because of its intensity, tunability, small source size, and small divergence.[24]

As mentioned in Chapter 1, the IC industry today favors improved photolithography over x-rays. Despite huge efforts, until recently, x-ray lithography seemed abandoned as a candidate for NGL. The prohibitive cost of introducing a new type of industrial lithography remains a strong deterrent, pushing existing photolithography to its absolute resolution limit. For example, by using UV phase-shifting mask lithography, planar IC features approaching x-ray lithography resolution have been made. Even in the micromachining field, there are continued attempts to squeeze more out of classical photolithography. Using techniques such as DUV lithography and deep dry etching with dense plasmas, LIGA-like, high-aspect-ratio features have been produced, hemming in the potential for broader use of x-ray lithography even for building 3D miniaturized machines. But x-ray lithography might well make a comeback.[25] At recent lithography conferences, four in five NGL presentations were on x-rays. Highlights are there is no need for a synchrotron anymore as new types of light sources have been developed (e.g., from a laser); x-ray masks are cheaper than EUV masks; and 15-nm resolution has been established.[17]

## Charged-Particle Beam Lithography

### *Introduction*

In this section, we introduce lithographies based on charged-particle beams. Charged-particle versus non–charged-particle approaches were compared earlier in Figure 2.7. Both narrow-beam direct writing and flood exposure projection systems with electrons and ions will be considered. The mask fabrication process is significantly simpler in the case of narrow-beam lithography compared with flood exposure-based lithography.

In direct write systems, the computer-stored pattern is directly converted to address the writing charged particle beam, enabling the pattern to be exposed sequentially, point by point, over the whole wafer. In other words, the mask is a *software mask.* Electron-beam (e-beam) and ion-beam (i-beam) lithographies involve high current density in narrow electron or ion beams. The smaller the beam sizes, the better the resolution, but more time is spent writing the pattern. This sequential (scanning)-type system exposes one pattern element or pixel at a time. Within that area, the charged-particle beam delivers maximum current ($i$), which is limited primarily by the source brightness and column design. The experimental setup imposes a limit on the speed at which the writing beam can be moved and modulated, resulting in a "flash" time in seconds ($t$). The maximum dose (in coulombs/cm$^2$) deliverable by a particular beam is given by:

$$D_{max} = it/A \tag{2.2}$$

with $A$ the pixel area in square centimeters. It will then be necessary to work with resists that react sufficiently fast at $D_{max}$ to produce a lithographically useful, 3D image (latent or direct image). The e-beam method displays a large DOF, as active focusing over various topographies is possible. The continued development of better charged-particle beam sources keeps widening the possibilities for nanoscale engineering through direct write lithography, etching, depositing, analyzing, and modifying a wide range of materials, well beyond the capability of classical photolithography. Table 2.4 lists some i-beam and e-beam applications.

Flood exposure of a mask in a projection system (that is, parallel exposure of all pattern elements at the same time, as we saw in DUV) is possible with ions and electrons as well. In principle, these methods can take advantage of the excellent resolution of charged-particle beams while providing the throughput levels necessary for IC manufacture. Exposure masks are fabricated from heavy metals on semitransparent organic or inorganic membranes. We will learn how SCALPEL and IPL have the potential of making e-beam and i-beam high throughput. The high cost

**Table 2.4** Electron- and Ion-Beam Applications

| Electron-Beam Applications | Ion-Beam Applications |
|---|---|
| Nanoscale lithography | Micromachining and ion milling |
| Low-voltage scanning electron microscopy | Microdeposition of metals |
| Critical dimension measurements | Maskless ion implantation |
| Electron beam-induced metal deposition | Microstructure failure analysis |
| Reflection high-energy electron diffraction (RHEED) | Secondary ion mass spectroscopy |
| Scanning auger microscopy | |

of mask fabrication and the instability of the mask as a result of heating have postponed commercial acceptance of these high-energy exposure systems. Moreover, with i- and e-beams, flood exposure is limited to chip-size fields because of difficulties in obtaining broad, collimated, charged-particle beams. The most prevalent use of charged-particle beams remains the narrow-beam scanning mode.

### Electron-Beam Lithography

*Overview* Electron beam lithography (EBL) refers to a lithographic process that uses a focused beam of electrons to form the circuit patterns needed for material deposition on (or removal from) a substrate. In contrast with optical lithography, which uses light for the same purpose, electron lithography offers higher patterning resolution than optical lithography because of the shorter wavelength associated with the 10–50-keV electrons that it uses. Electrons are easily generated and can be used to either directly write the desired structure to the resist in electron-beam direct write or in electron projection lithography (EPL) flood exposure where a much wider beam is used (such as in SCALPEL). We also analyze the status of micromachined electron emission sources and electron emission source arrays.

*Direct Write e-Beam Lithography* Direct write EBL is a high-resolution patterning technique in which high-energy electrons (10–100-keV) are focused into a narrow beam and used to expose electron-sensitive resists. There are two basic ways to scan an electron beam. In raster scanning, the patterns are written by an electron beam that moves through a regular pattern. The beam scans sequentially over the entire area and is blanked off where no exposure is required. On the contrary, in vector scanning, the electron beam is directed only to the requested pattern features and hops from features to features. Therefore, time is saved in a vector scan system. The direct write technique was first developed in the 1960s using existing scanning electron microscope (SEM) technology. A typical setup involves 30-keV electrons from a Hitachi HL-700F EBL. As a research solution, several groups are still "Rube Goldberging" their standard SEM to create customized e-beam writing systems. For example, Rosolen[25] modified a Hitachi S2500 with a purpose-built pattern generator and alignment system (see also Figure 6.14 in Volume III for a schematic of an SEM and a photo of a commercial SEM machine). The instrument did not require alignment marks on the sample and was able to compensate for positional errors caused by the sample stage and mask tolerances. The EBL method, like x-ray lithography, does not limit the obtainable feature resolution by diffraction because the quantum mechanical wavelengths of high-energy electrons are exceedingly small. EBL exhibits some other attractive attributes compared with photolithography. These include:

1. Precise control of the energy and dose delivered to a resist-coated wafer
2. Deflection and modulation of electron beams with speed and precision by electrostatic or magnetic fields
3. Imaging of electrons to form a small point of <100 Å, as opposed to a spot of 5000 Å for light
4. No need for a physical mask; only a *software mask* is required
5. The ability to register accurately over small areas of a wafer
6. Lower defect densities
7. Large DOF because of continuous focusing over varying topography; at 30 keV, electrons will travel on average >14 μm deep into a PMMA resist layer

Some of the disadvantages of EBL include:

1. Electrons scatter quickly in solids, limiting practical resolution to dimensions >10 nm
2. Electrons, being charged particles, need to be held in a vacuum, making the apparatus more complex than for photolithography
3. The slow exposure speed: an electron beam must be scanned across the entire wafer (for a 4-in. wafer with a high feature density this requires ~1 hour)
4. High system cost

The resolution of EBL tools is not simply the spot size of the focused beam; it also is affected by forward scattering of the e-beam inside the resist and substrate (at relatively smaller angles) and by backscattering from the substrate exposing the resist over a greater area than the beam spot size (Figure 2.15). This phenomenon is called the "proximity effect," and hence the developed pattern is wider than the scanned pattern due to the interactions of the primary beam electrons with the resist and the substrate, as illustrated in Figure 2.16. Line width variations caused by local feature density are an immediate result. Proximity correction algorithms are used to achieve more uniform resist exposure with EBL. Such corrections are computer intensive and time consuming, however, and make a slow technique even slower.

The biggest disadvantage of direct write electron lithography is its low throughput (approximately 5 wafers/hour at less than 0.1-μm resolution). As a result, the use of EBL has been limited to mask making and direct writing on wafers for specialized applications, for example, small batches of custom ICs.

Writing with an e-beam can be additive or subtractive. In lithography the technique is subtractive; as an example of additive e-beam writing, e-beam-induced metal deposition from a metal organic gas [e.g., W deposition from $W(CO)_6$] has been used for the formation of microstructures of various geometries (see Figure 9.29). These devices are made one by one rather than in a large batch. Usually, this type of slow, expensive fabrication technique prohibits commercial acceptance. Some microstructures, especially intricate microsystems, might be worth the bigger price tag. In that case, serial microfabrication techniques may not be as prohibitive as they would be in the case of ICs.

*Electron Emission Sources* Three types of electron emission (field, thermionic, and photoemission) (Figure 2.17) underwrite the principle for electron emission source construction.[26] Schottky emission (SE) and cold field emission (CFE) have been in common use, especially for nanometer-sized beams for electron focusing systems. Emission of electrons from a metal under the influence of a field occurs in both SE and CFE. During SE, a blunt tungsten emitter tip coated with a low work function material (ZrO) is heated to 1800 K, and thermionic emission takes place; that is, heat thermally excites the electrons enough to bring them out of the material. In other words, the field emission is helped along by thermal excitation of the electrons (the current in this case is given by the Richardson-Dushman equation).[27] During CFE, a much smaller tungsten wire (radius of <0.1 μm) is used, and a very high field causes electrons to tunnel out of the material. In CFE sources, electrons tunnel from various energies below the Fermi level. With SE cathodes, thermally excited electrons (nontunneling electrons) escape over a field-lowered potential energy barrier (the current in this case is given by the Fowler-Nordheim equation).[27] Both SE and CFE sources display similar energy spreads, but their energy distributions are mirror images (Figure 2.18).[26]

With photo emitters, energy is transferred to electrons by incident photons, and photoelectrons generated close to the surface are able to escape.

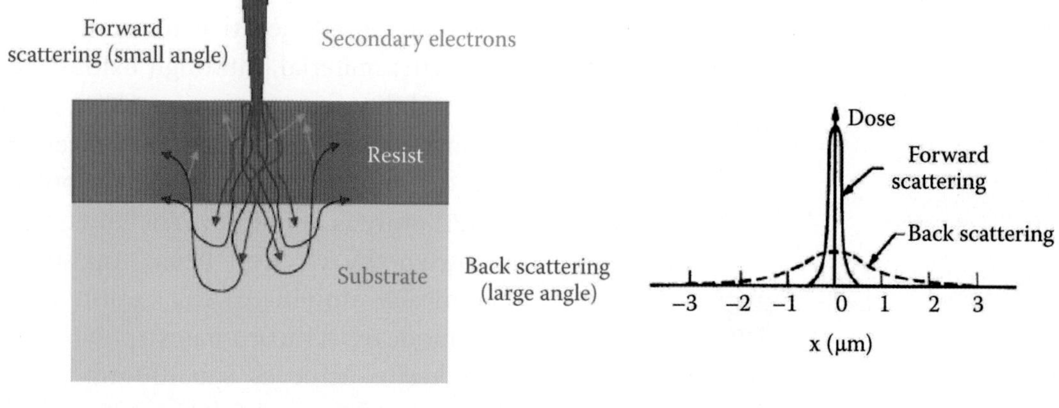


**FIGURE 2.15** Scattering of the e-beam inside the resist and substrate and backscattering from the substrate exposing the resist over a greater area than the beam spot size.

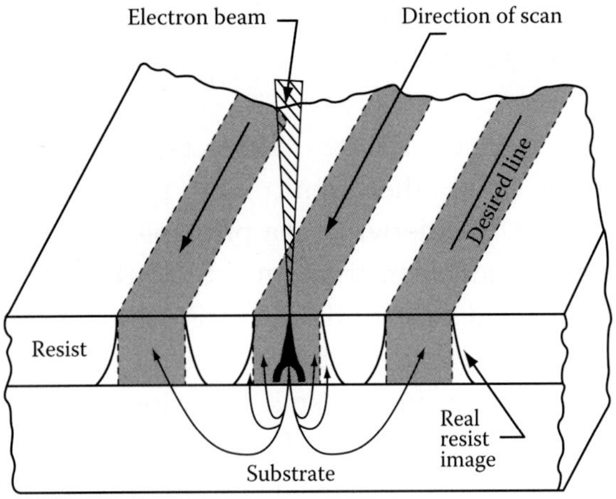


**FIGURE 2.16** Proximity effect in direct write e-beam lithography.

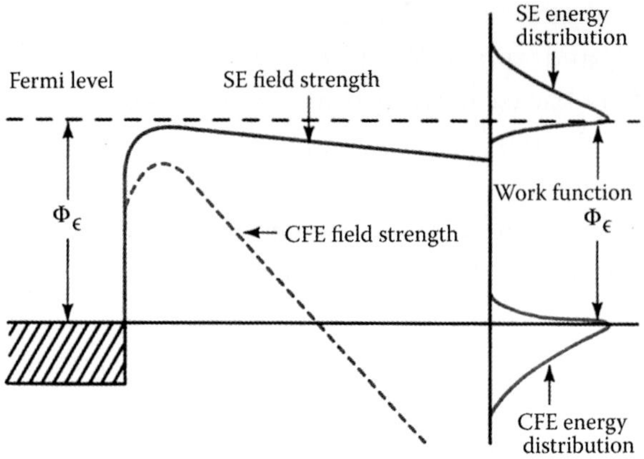


**FIGURE 2.18** SE and CFE energy distributions.

Attaining high current levels in a submicrometer electron beam at low voltages (500 eV–1 keV) is of interest for EBL and SEM (see also Chapter 6 in Volume III). When sensitive biological samples or electron beam-sensitive resists are involved, SEM pictures must be made at voltages less than 1 keV. At these low voltages, high currents are required to attain the needed detail and to minimize edge effects. Traditional SEM cathodes using tungsten hairpin filaments are very limited at low voltages; they cannot supply enough current in submicrometer beams at the low-voltage end. The high brightness SE and CFE sources, including $LaB_6$ and $CeB_6$, can do the job (Table 2.5). SEM instruments with beam diameters of 1 nm have been made using such cathodes. For further reading about EBL, refer to Brewer.[28]

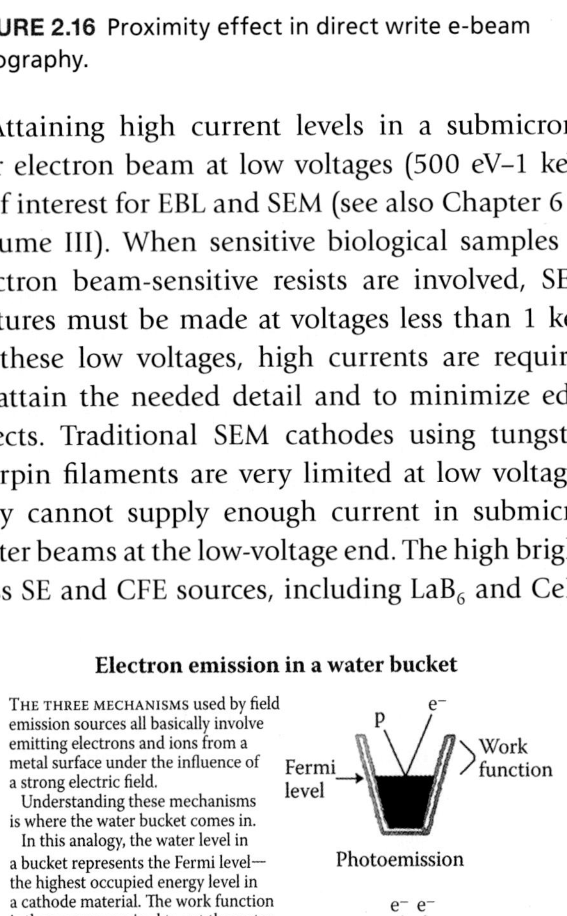
Electron emission in a water bucket
THE THREE MECHANISMS used by field emission sources all basically involve emitting electrons and ions from a metal surface under the influence of a strong electric field.
Understanding these mechanisms is where the water bucket comes in.
In this analogy, the water level in a bucket represents the Fermi level—the highest occupied energy level in a cathode material. The work function is the energy required to get the water droplets (electrons) from the top of the liquid out of the bucket. This is the distance equivalent to the potential energy barrier.
In photoemission, photon energy excites electrons at the Fermi level of the cathode material and can impart enough kinetic energy to allow the electrons to escape from the bucket.
In thermionic emission, heat thermally excites the electrons, providing enough energy to boil the electrons off and out of the bucket.
In field emission a high electric field can thin the side of the bucket enough so that the electrons can tunnel through it.
p
e⁻
Fermi
level
Work
function
Photoemission
e⁻ e⁻
Fermi
level
Work
function
Thermionic emission
Fermi
level
Work
function
e⁻
Field emission

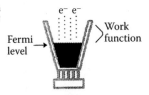

**FIGURE 2.17** Photoemission, thermionic emission, and field emission.

*Electron-Beam Resists* Numerous commercial e-beam resists are produced for mask making and direct write applications. Bombardment of polymers by electrons causes bond breakage; thus, in principle, any polymer material can function as a resist. However, the important considerations include sensitivity, tone, resolution, and etching resistance. PMMA exemplifies an inexpensive positive e-beam resist with a high-resolution capability and a moderate glass transition temperature $T_g$ (114°C). Microposit SAL601 (Shipley) is an often-used negative e-beam resist. SAL601, being novolak based, has much better dry etch resistance than PMMA resists. The same materials act as x-ray resists as well. This is not coincidental, as there is a strong relation between x-ray and e-beam sensitivity. A copolymer of glycidyl methacrylate and ethyl acrylate (COP) is another frequently used negative resist in mask manufacture. This material, although exhibiting good thermal stability, has (as is typical for acrylates) poor plasma-etching resistance. The measured G value (see Chapter 1) of representative polymers in the COP family is about 10. Whelan et al.[29] developed a low-energy electron beam top surface imaging chemically amplified (AXT) resist. This AXT positive-mode resist incorporates a poly(hydroxystyrene) base resin, achieves sub-100-nm resolution with 2-keV electrons, and features a sensitivity less than 1 μC/cm$^2$.

**Table 2.5** Comparisons of Electron Sources

| Working Principle | Gun Material | Brightness (B) ($A/cm^2$ Sr) | Energy Width (eV) | Filament Temperature | Gun Vacuum (torr) |
|---|---|---|---|---|---|
| **Electron Source** | | | | | |
| **Thermionic Emission** | | | | | |
| Electron emission at high temperature | W | ~$10^5$ | 2–3 | ~3000 K | $10^{-5}$–$10^{-6}$ |
| | $LaB_6$ | ~$10^6$ | 2–3 | 2000–3000 K | $10^{-7}$–$10^{-8}$ |
| **Field Emission** | | | | | |
| Electron tunneling in high field | W | $10^5$–$10^{10}$ | 0.2–0.5 | Room | <$10^{-9}$ |

Streblenchenko et al.[30] patterned $FeF_2$ and $CoF_2$ with 100-keV electrons from a scanning transmission electron microscope using a 0.5-nm diameter electron probe. The fluoride resist films are prepared by thermal evaporation on thin carbon films. During electron bombardment, fluorine escapes and the transition metal coalesces. The resolution in a 20-nm-thick, very small-grained $CoF_2$ film is about 5 nm. An electron dose of 1000 $C/cm^2$ at 100 keV removes 90% of all the fluorine in the 20-nm-thick $CoF_2$ film. Arbitrarily shaped nanometer-scale magnetic structures have been written in such $CoF_2$ films.

For further reading, see *Handbook of Microlithography Micromachining and Microfabrication*, edited by P. Rai-Choudhury (published by SPIE Press), and Joe Nabity's web site (http://www.jcnabity.com) describing the Nanometer Pattern Generation System. This system is the top-selling SEM lithography system at research institutions in North America, and its use is becoming widespread worldwide.

*Electron Projection Lithography: Scattering with Angular Limitation Projection Electron Beam Lithography* Electron projection lithography (EPL) is electron beam lithography (EBL) with a wide electron beam, using a mask for high throughput. Scattering with angular limitation projection lithography or SCALPEL, a projection electron beam technique, using a 4× reduction and a step-and-scan writing strategy, is the most prominent EPL technique. The aspect of SCALPEL (1995, Bell Laboratories and Lucent Technologies) that differentiates it from previous attempts at projection EBL is the use of a mask as illustrated in Figure 2.19. As in the case of x-ray lithography, the mask consists of a low atomic number membrane covered with a layer of a high atomic number material: the pattern is delineated in the latter. Although the mask is almost completely electron-transparent at the energies used (100 keV), using the difference in electron-scattering characteristics between the membrane and the patterned material generates contrast. The membrane scatters electrons weakly and to small angles, whereas the patterned layer scatters them strongly and to high angles. An aperture in the back-focal (pupil) plane of the projection optics blocks the strongly scattered electrons, forming a high-contrast aerial image at the wafer plane.

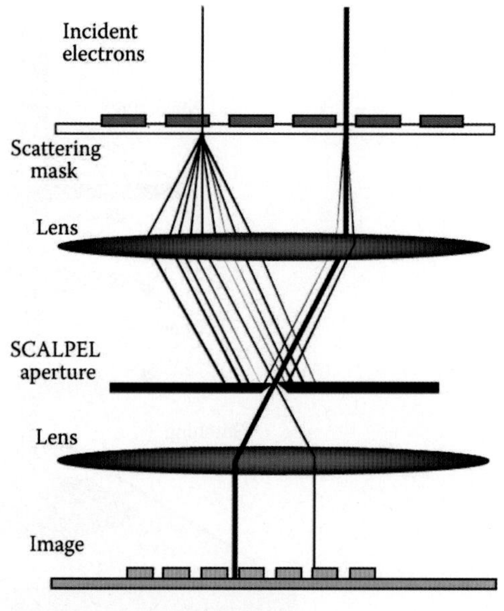


**FIGURE 2.19** Scattering with angular limitation projection lithography or SCALPEL, a projection electron beam technique.

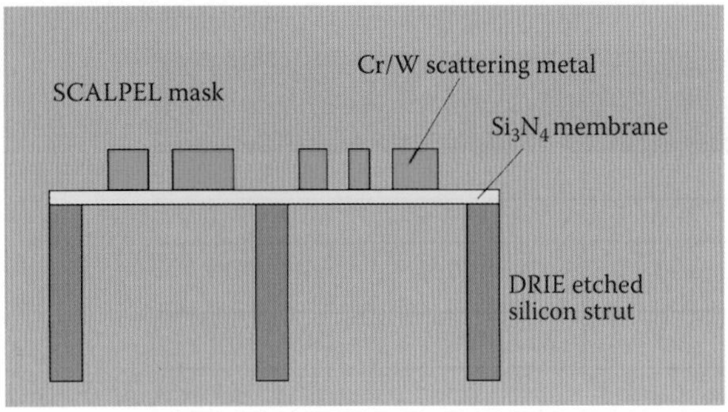


**FIGURE 2.20** SCALPEL mask.

The mask is a continuous thin membrane (~100 nm) made of a low atomic number material like silicon nitride, patterned with a 25–50-nm-thick layer of a high atomic material such as tungsten (Figure 2.20). To minimize image distortion, thick silicon struts support the thin silicon nitride membrane. A broad beam (2–3 mm in diameter) of high-energy electrons (~100 keV) passes through both membrane and W pattern with minimal energy loss, leading to minimal heating effects. In SCALPEL, the aperture, rather than the mask, absorbs unwanted energy. Thus, the functions of contrast generation and energy absorption are divided between the mask and the aperture. This means that very little of the incident energy is absorbed by the mask, minimizing thermal instabilities in the mask. It should be noted that, although the membrane scatters electrons weakly compared with the scatterer, a significant fraction of the electrons passing through the membrane are scattered sufficiently to be stopped by the SCALPEL aperture. Image field is still limited today to 1 × 1 mm at the mask level. The small images at the wafer are assembled (also stitched together) from the individual portions on the mask through a combination of interferometrically monitored mechanical motions of the mask and wafer stages and electronic deflections of the beam (see Figure 2.21). The relatively low cost masks in SCALPEL [$27,000 vs. $40,000 for a phase shift mask (PMS), $40,000–59,000 for an EUV mask, and $5,000–10,000 for standard binary masks] may make this technique a cost-effective solution for lithography at 130 nm and less.[31,32]

In 1999, IBM introduced Projection Reduction Exposure with Variable Axis Immersion Lenses (PREVAIL).[33] In this system, the optical axis of the electron system is shifted, so aberrations are reduced, enabling larger scan fields. The lens system shifts the electron optical axis along a preset curvature and precisely deflects the electron beam to follow the variable axis curvature. The SCALPEL

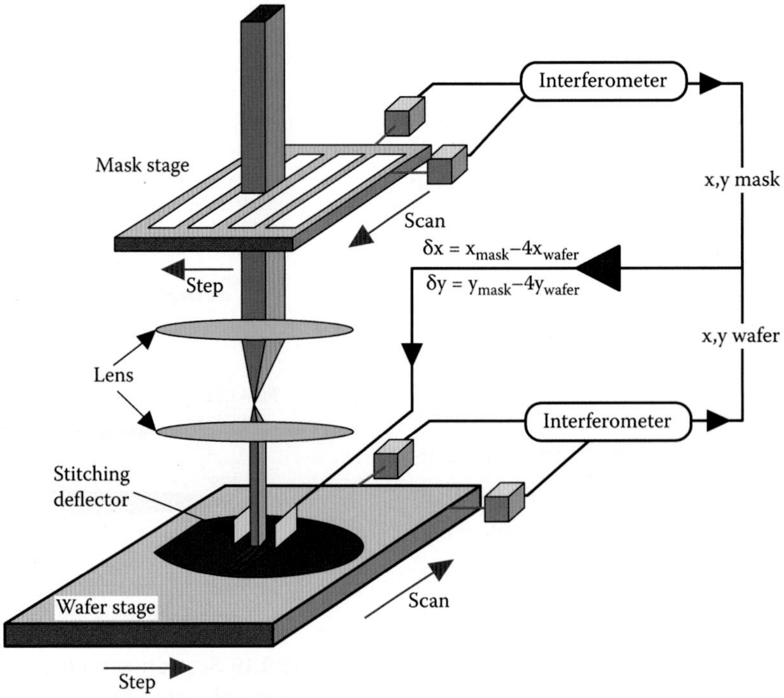


**FIGURE 2.21** SCALPEL writing strategy involves step-and-scan writing.

proof-of-concept system was the first system to implement sequential illumination of the mask in an e-beam reduction projection system by mechanical scanning of reticle and wafer at a 4:1 speed ratio underneath a stationary beam. PREVAIL carries this concept further by combining electronic beam scanning with continuous stage motions.[34] This approach provides a significantly larger effective field size needed to achieve commercially viable throughput levels.

*Micromachined Electron Emission Sources* There have been many attempts at decreasing the accelerating voltage in direct write EBL because lower energy electrons have a more confined lateral backscattering range. Moreover, it was found that low-energy EBL leads to higher sensitivity—that is, for a given resist, a lower exposure dose is required than with higher energy electrons.[35] Low-energy e-beams require thinner resist layers, as the penetration depth falls off very quickly (e.g., in a typical resist, 2 keV will penetrate more than 120 nm). Micromachining is ideally suited to deliver these low-energy electron writing tools. We only present some examples here of micromachined e-beam tools. A first example involves the work by Zlatkin et al.,[36] who developed an elegant array of focused electron writing beams operating at 300 eV or less. The emitters used are cold field emission (CFE) sharpened tungsten tips, although thermionic or Schottky emitters would be feasible as well (for a description of different types of emission sources see above). The emitters are positioned several millimeters above the micromachined extraction holes of a lens array fashioned in a single crystal Si substrate. Because of the relatively large distance between the emitter and the extraction anode (a 1-μm aperture), no precise alignment is necessary. The setup and a detail of the lens system are shown in Figure 2.22.

Each lens has a total area of 1 × 1 cm on a 500-μm-thick silicon wafer and is generally built up from two 0.5-μm thin Si layers. The two thin Si layers are separated from each other and the Si substrate by insulating silicon oxide. An anisotropic etch into the single crystal Si side opposite the thin layered sandwich structure exposes the oxide/Si/oxide/Si/oxide thin layered structure. A short isotropic etch of the exposed oxide is followed by a 1-μm dry-etch bore in the first thin Si layer (first aperture, 1-μm diameter). The small aperture is used to extract electrons from the emitter and to spatially confine the initial electron beam. Another isotropic etch through the second oxide is followed by the larger dry etched aperture (50 μm) in the second thin Si layer. This

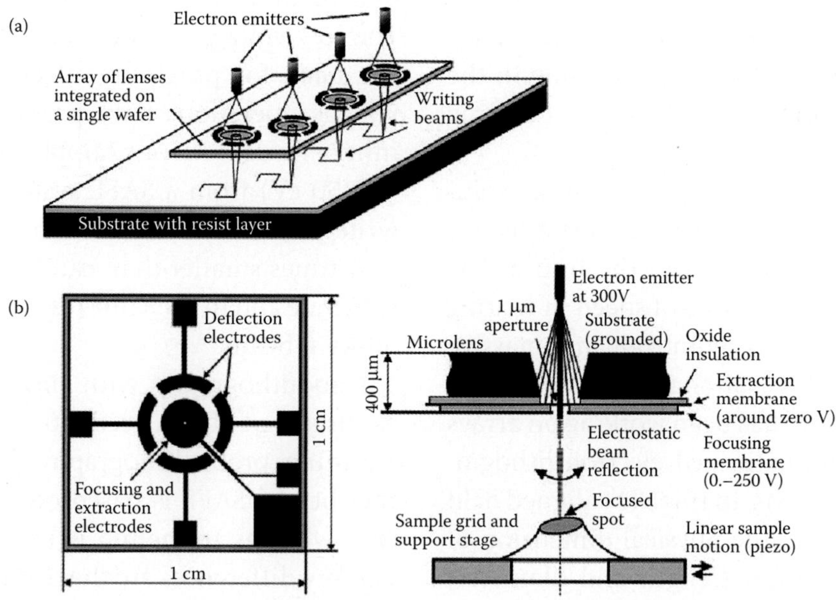


**FIGURE 2.22** Array of focused electron writing beams operating at 300 eV or less. Setup (a) and detail of the lens system (b). (Redrawn from Zlatkin, A., and N. Garcia. 1999. Functional scanning electron microscope of low beam energy with integrated electron optical system for nanolithography. *Microelectron Eng* 46:213–17.[36])

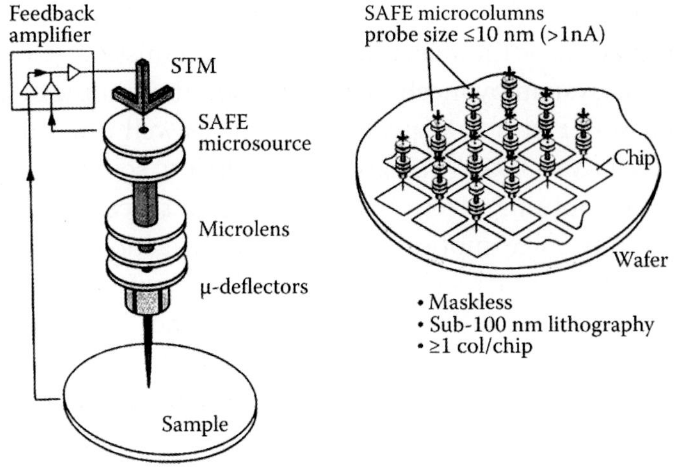


**FIGURE 2.23** A microcolumn based on STM aligned field emission (SAFE) and arrayed microcolumn lithography. (From Editorial. 1993. Novel electron-beam lithography system being explored at Cornell's NNF. *Solid State Technol* 36:25–26. With permission.[37]) Acronyms are explained in the text.

second aperture is used to focus the electron trajectories and is well aligned with the first aperture. Changing the voltage on the extraction and focusing electrodes on the thin Si membranes can change the focal length. To deflect the electron beam for scanning, a set of four electrodes symmetrically placed around the extraction and focus apertures is energized. The four electrodes are placed on top of another insulator layer on the focusing membrane. Applying potentials to opposite pairs of deflection quadrants deflects the electron beam laterally, and the focused spot can be swept over an area either to create an SEM image or to pattern a design in EBL mode. The aim is to eventually be able to image and write in the sub-100-nm range. Imaging in the 30-nm range was demonstrated already, but writing is still inadequate (limited to >200 nm).

Micromachining presents other potential solutions to make "nanolithography" a cost-effective and adequately fast proposition in the future. The next example addresses the issue of speed in writing with small e-beam systems by introducing massive parallelism. At Cornell's National Nanofabrication Facility, a research group has been working on arrays of microfabricated, miniaturized electron lithography systems based on STMs. In this STM aligned field emission (SAFE) system, the physical dimension of the electron beam column (length and diameter) are on the orders of millimeters. A field emission tip is mounted onto an STM; the STM feedback principle is used for precision *x*, *y*, and *z* piezoelectric alignment of the tip to a miniaturized electron lens to form a focused probe of electrons as shown in Figure 2.23.[37] Because many electron-optic aberrations scale with size, microfabrication techniques enable lenses with negligible aberrations, resulting in exceptionally high brightness and resolution. STM controls also allow for stability of the emission by automatically adjusting the *z*-position via a piezo element.

An array of these microcolumns, each with a field emission tip as the source with individual STM sensors and controls, can generate patterns in parallel, one or more columns per chip (see Figure 2.23). The low voltage of operation of these tips obviates the need for proximity-effect corrections, as low-voltage operations have proven to eliminate proximity effects.[37,38] For example, low-energy electrons (15–50 eV) from a SAFE system have been used to write patterns with 23-nm feature sizes, more than four times smaller than can be written on the same substrate and in the same resist with a tightly focused 50-kV e-beam.[38]

Nanolithography with ultrasharp field emitters as discussed above needs to be distinguished from scanning probe lithography (SPL), discussed further below. SAFE is based on an emission technology, whereas tunneling is the mechanism in SPL. The key differences are the distance between tip and substrate (1 nm with SPL vs. 100 nm or more with sharp field emitter tips) and the electron generation mechanism (tunneling vs. emission).

Besides the application described above, micromachined emission sources are also used or are intended to be used for flat panel displays (see Spindt cathodes in Figure 2.42), high-temperature and high-radiation amplifiers, vacuum gauges, and radio frequency oscillators and amplifiers.[39]

### *Ion-Beam Lithography*

*Introduction* In ion-beam lithography, resists are exposed to energetic ion bombardment in a vacuum. As in the case of e-beam systems, ion-beam lithography offers direct write and flood exposure fabrication opportunities. Direct write ion-beam lithography consists of point-by-point exposures with a narrow beam scanning source (e.g., 20 nm in diameter) of liquid gallium metal. In ion projection lithography (IPL), flood exposure, with a broad ion beam of $H^+$, $He^{2+}$, or $Ar^+$ (e.g., 1–2 $cm^2$ in size), of a stencil mask is used to pattern a substrate. Ion-beam lithography uses ions of a kinetic energy from a few keV up to several MeV. Ion lithography achieves higher resolution than optical, x-ray, or electron beam techniques because ions come with a smaller wavelength and therefore undergo almost no diffraction and scatter much less than electrons because the secondary electrons produced by an ion beam are of lower energy and have a short diffusion range. The total spread including forward and backward scattering of the "stiffer" ion beams is typically less than 10 nm, and they only require about 1–10% of the electron dose to expose a resist. The resist can be an ordinary PMMA resist that absorbs most of the ions during exposure. Therefore, radiation damage to sensitive, underlying structures is minimized and smaller compared with EBL and x-ray lithography. There is also the possibility of a resistless wafer process. However, the most important application of focused ion beams remains the repair of masks for optical or x-ray lithography, a task for which commercial systems are available.

*Focused Ion-Beam and Deep Ion-Beam Lithography* For ion-beam construction, liquid metal ion (LMI) sources are becoming the choice for producing high-current-density submicrometer ion beams. With an LMI source, liquid metal (typically gallium) migrates along a needle substrate. By applying an electrical field, a jet-like protrusion of liquid metal forms at the source tip. The gallium-gallium bonds are broken under the influence of the extraction field and are uniformly ionized without droplet or cluster formation. LMI sources hold extremely high brightness levels ($10^6$ A/$cm^2$ sr) and a very small energy spread, making them ideal for producing high-current-density submicrometer ion beams. Beam diameters of less than 50 nm and current densities up to 8 A/$cm^2$ are the norm. In addition to Ga, other pure element sources are available, such as indium and gold. By adopting alloy sources, the list expands to dopant materials such as boron, arsenic, phosphorus, silicon, and beryllium.

Compared with photons (x-rays and DUV light) or electrons, ions chemically react with the substrate, allowing a greater variety of surface modifications such as patterned doping. The ion-beam spot size is the smallest possible—smaller than UV, x-ray, or electron-beam spots. The smallest focused ion-beam (FIB) spots reached are about 4–8 nm, accomplished by using a two-lens microprobe system and a single-isotope gallium ion source.[40, 41] Ion-beam lithography experiences the same drawbacks as an electron-beam system in that it requires a serially scanned beam and a vacuum.

FIBs can be used to perform maskless implantation and metal patterning with submicrometer dimensions. FIB also has been applied to milling in IC repair, maskless implantation, circuit fault isolation, and failure analysis (see Table 2.4). In Figure 2.24, deposition, milling, and imaging are illustrated. Micromachining applications of ion-beam technology are reviewed in more detail in Chapter 6 on mechanical energy-based removing (see Figure 6.32). As a machining tool, FIB is very slow. Except for research, it may take a long time to become an accepted "micromachining tool." For additional reading on ion-beam lithography in general refer to Selinger et al.[42]; for more specific reading on FIB-induced deposition, see Brodie.[13]

Using high-energy (2 MeV) protons, deep ion-beam lithography (DIBL) in PMMA produces submicrometer (300 nm) walls with an aspect ratio approaching 100. Three-dimensional complex microstructures with smooth walls and corners have

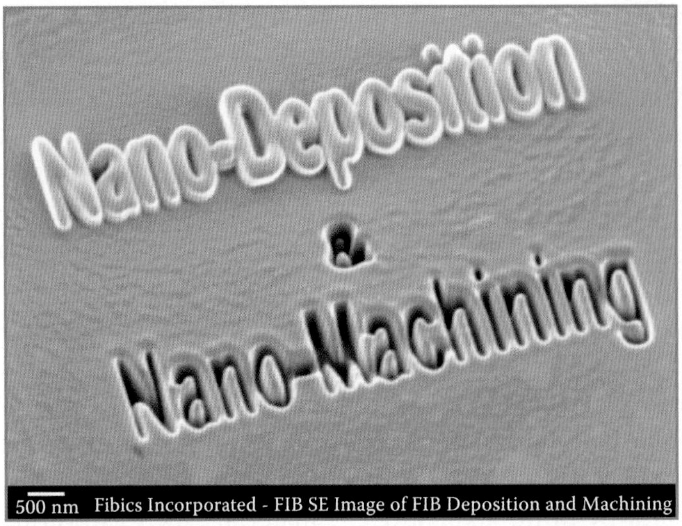


**FIGURE 2.24** FIB deposition can produce features 200 nm or less in thickness. FIB milling can produce even finer results. Approximately 30 minutes of FIB time was required to produce this structure (http://fibics.com/Micromachining.html).

been produced this way. The range of 2-MeV protons in PMMA is 63 μm.[43] Multiple exposures at different ion energies (e.g., 0.6 MeV and 2 MeV) allow production of multilayer structures in single-layer resists such as SU-8.[44]

FIB systems have been produced commercially for approximately 20 years, primarily for large semiconductor manufacturers. FIB systems operate in a similar fashion to an SEM. When operated at low-beam currents, the system can be used for imaging, and when operated at high-beam currents, it can be used for site-specific sputtering or milling (see Figure 2.24).

*Ion-Projection Lithography* Ion projection lithography (IPL), illustrated in Figure 2.25, is another of the candidates for high-throughput lithography dedicated to future 32-nm and sub-32-nm IC generations. In ion projection lithography, a stencil mask, consisting of an Si membrane with a fine pattern of holes, is illuminated with a uniform beam of hydrogen or helium ions at about 10 kV. The image of this mask is projected through an ion optical column and demagnified by 4× on the wafer. Ions are very efficient at exposing resists so that the exposure time per die is well under a second even for comparatively insensitive (1 μC/cm$^2$) resists. Protons or helium ions are generated by a radio frequency-driven filament (ion source). An extraction system consists of

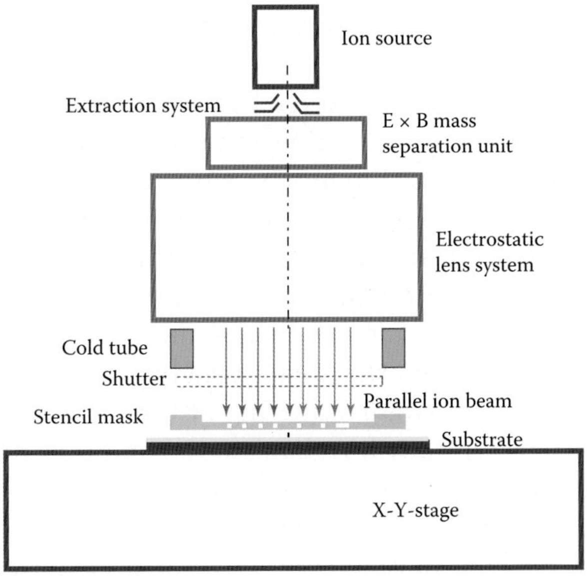


**FIGURE 2.25** Ion projection lithography (IPL).

an **E** × **B** mass separation unit and an electrostatic lens system.

The ion flood lithography mask typically consists of a silicon membrane a few micrometers thin (3 μm) with pattern openings that allow protons to pass through.[45] The mask fabrication process involves silicon-on-insulator (SOI) and dry etching.[46] The holes are defined by EBL and reactive ion etching, whereby an oxide is used as a mask. Current IPL approaches, like SCALPEL, are 4× technologies. DOF is large and may reach up to 500 μm. The optics in IPL are all electromagnetic, and the

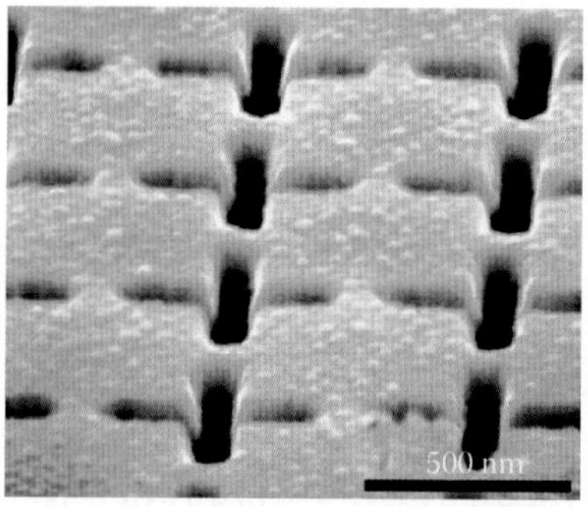


**FIGURE 2.26** IPL mask 80-nm feature sizes etched into a Si membrane mask. The patterns were defined using a Leica VB6 EBL. (From http://www.cnf.cornell.edu/image.)

potential for a 50 × 50-mm exposure field exists. The membranes for this work typically consist of low-stress silicon nitride or Si. An example of a Si stencil mask created at the Cornell Nanoscale Science and Technology Facility for masked ion-beam lithography is shown in the electron micrograph in Figure 2.26.

Ion-beam lithography has at least two advantages over EBL: 1) it has almost two orders of magnitude higher resist sensitivity, and 2) it has negligible ion scattering in the resist and very low backscattering from the substrate. A major problem is the potential for damage to sensitive electronic functions from the high energy ions.

Efforts in IPL are especially strong in Europe (e.g., at IMS, http://www.ims.co.at).

## Imprint Lithography

Nanoimprint lithography (NIL) and step-and-flash imprint lithography (SFIL) are techniques that use hard molds instead of the soft molds used in soft lithography (see below). Stephen Chou at Princeton University invented NIL in 1994, with the aim of overcoming the diffraction-limited minimal feature sizes obtained in semiconductor manufacturing based on DUV lithography (http://www.princeton.edu/~chouweb). It is possible to break through the diffraction-controlled size limit of photolithography, for example, by using electron beam lithography (EBL). But the infrastructure of this approach, we know by now, costs billions of dollars. NIL patterns a resist by deforming the resist shape through embossing (with a mold/stamp/template) rather than by altering resist chemical structures through radiation (with particle beams). After imprinting the resist, a dry anisotropic etch is used to remove the residual resist layer in the compressed area to expose the substrate underneath. In NIL, a template (the mold/stamp/template) is made of a hard material (usually Ni or Si) and is pressed against a layer of polymer. High temperature and pressure conditions mold and harden the polymer layer. As illustrated in Figure 2.27, the NIL technique is based on the craft of hot embossing, with an adaptation to modern semiconductor needs. In this example the technique is used to pattern a metal film via liftoff. The method relies on the excellent replication fidelity obtained with polymers and combines thermoplastic molding with common pattern transfer methods. Once a solid stamp with a nanorelief on the surface is fabricated, it can be used for the replication of many identical surface patterns. The resolution of the NIL process is a direct function of the resolution of the original template/stamp fabrication process.

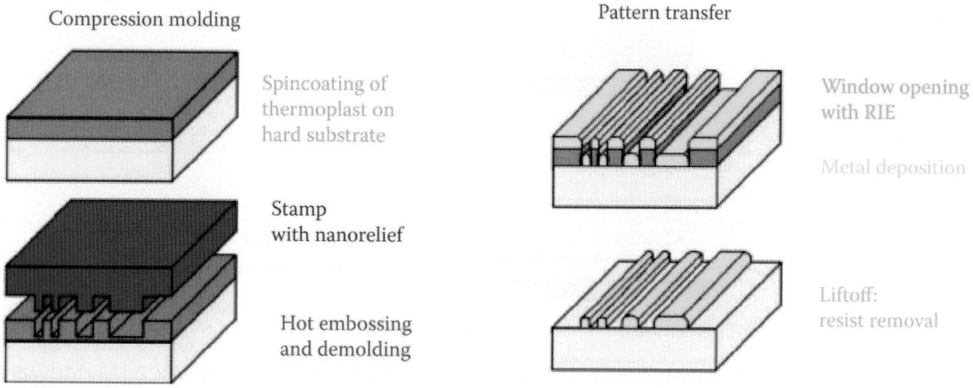


**FIGURE 2.27** Schematic illustration of the nano imprint lithography (NIL). The technique is used here to pattern a metal film via liftoff.

Electron beam writers that provide high resolution but lack the throughput required for mass production are used to make them.

The University of Texas–Austin developed its own version of nanoimprint lithography, i.e., step-and-flash imprint lithography (SFIL), in 1998.[47] In 2001, the SFIL concept was licensed to Molecular Imprints, Inc., a company that develops SFIL-capable tools (S-FIL® systems) and related processes (http://www.molecularimprints.com). The SFIL method is distinct from the original NIL in its use of UV-assisted nanoimprinting that molds photocurable liquids in a step-and-repeat, die-by-die fashion rather than by heat-assisted molding of full, polymer-coated wafers. As shown in Figure 2.28, in SFIL, a hard but transparent template/stamp/mold (fused silica) is used to mold a polymer photoresist layer. The fused silica surface of the stamp, coated with a release layer, is gently pressed into the thin layer of low viscosity photoresist. The photoresist is subsequently exposed to UV light through the transparent template to harden it. On separation of the fused silica template, one layer of the circuit pattern is left on the wafer surface. Molecular Imprints uses a family of photocurable, low viscosity materials called MonoMat™ as its imprint resists. MonoMat™ is composed of an organic monomer that polymerizes in seconds using low-cost, broadband UV light sources. Whereas other imprint tools use spin-on imprint resist, S-FIL® systems dispense low viscosity imprint resist in very small droplets, one field at a time, immediately before the imprint step. Consequently, both the monomer dispenses and imprint functions happen in a step-and-repeat fashion. A residual layer of polymer between features is eliminated by a dry etch process, and a perfect

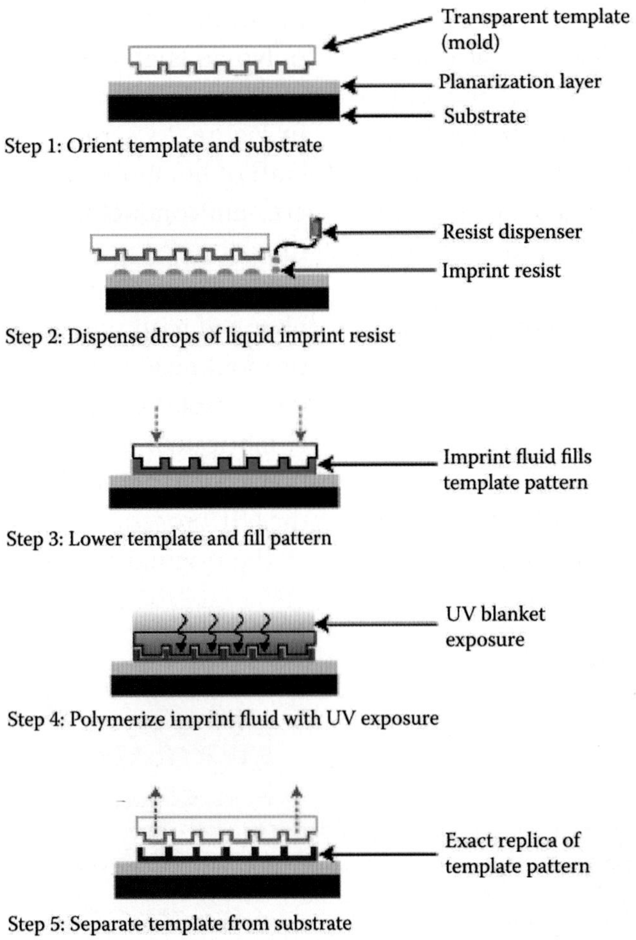


**FIGURE 2.28** Schematic of step-and-flash imprint lithography (SFIL). (Colburn, M., S. Johnson, M. Stewart, S. Damie, T.C. Bailey, B. Choi, M. Wedlake, T. Michaelson, S.V. Sreenivasan, J. Ekerdt, and C.G. Wilson. 1999. Step and flash imprint lithography: a new approach to high-resolution patterning. *Proc SPIE* 3676:379–89.[47])

replica of the pattern is ready to be used in semiconductor processing for etch or deposition. Only the template fabrication process, typically accomplished with an e-beam writer, limits the resolution of the features. It has been demonstrated that the mold templates do not deteriorate even after 1500 imprints with sub-100-nm feature sizes. Although more progress is needed for widespread adoption of NIL and SFIL, sub-100-nm feature size devices can be reliably fabricated at a reasonable throughput. Implementing SFIL for high-volume manufacturing of the gate or contact level of transistors is not feasible in the near term, but it is one of the few methods currently available for low-volume prototyping at the 32-nm node. Sub-20-nm features have been made already that exceed the present requirements of the ITRS. For example, using Molecular Imprints' Imprio 250 system, Toshiba was able to print 18-nm isolated features and 24-nm dense features with <1-nm critical dimension uniformity and <2-nm line edge roughness.[48] The Toshiba demonstrations also showed lower defectivity and improved results for imprint lithography. Defectivity levels of as low as <0.3 defects/cm$^2$ are approaching those of immersion lithography, and device overlay results were also within Toshiba's required specifications. However, the approach faces alignment challenges, and critics say the cost of making the reticles will be high. Unlike conventional lithography, with mask patterns that are four times larger than the printed image, nanoimprint is a 1:1 template technology.

Others argue that imprinting processes are economical because of the simplicity of the equipment involved and the potential for high throughput. As the rate of improvements in optical lithography decelerates and as the costs of manufacturing continue to escalate, there is an increased interest in printing and molding as alternative processes for microfabrication. As imprint lithography depends on another lithography technique (perhaps an NGL) to make the master, it does not really qualify as an NGL itself. Whereas it may initially have targeted semiconductor applications, the SFIL process is now being explored for manufacturing of several emerging technologies, such as photonic crystals, micro/nano-optical components, and nanopatterned magnetic media for future hard disk drives. Progress in these new areas has been such that SFIL is likely to find its first commercial manufacturing application in one of these emerging technologies well before it would be required for high-volume, sub-50-nm semiconductor lithography. A notable feature of nanoimprint technologies is their relatively low cost, which allows researchers to explore applications of nanopatterning that would never be economically feasible given the extraordinary cost associated with EUV lithography or even current-generation 193-nm steppers. This is very similar to the current applications of soft lithography (see below) with this difference that using a hard stamp renders NIL and SFIL more industrial manufacturing capable.

Thermal NIL using thermoplastic polymer films is focused on applications such as biochips, life sciences, storage media, and optical devices, whereas SFIL is focused on LSI and magnetic storage applications, as summarized in Figure 2.29. Hitachi, with NEDO support, has also developed a sheet or roll-to-roll prototype NIL machine and demonstrated the capability to process 15-m sheets.[49] The goal is to create a continuous imprinting process by using belt molding (nickel-plated molds) and polystyrene sheets for large geometry applications, such as membranes for fuel cells, batteries, and possibly for displays. Currently the prototype tool does not

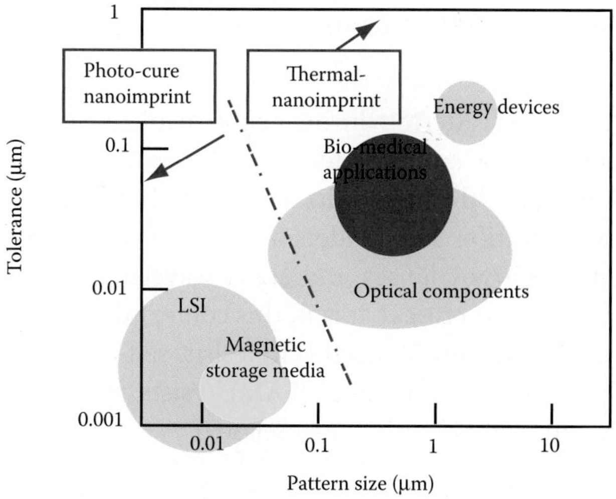


**FIGURE 2.29** Usage domains of photocure (SFIL) and thermal nanoimprinting (NIL). (Based on a figure by Hitachi.)

offer the desired throughput, but the concept has been demonstrated, with future efforts focused on throughput, reliability, and repeatability.

## Lithography Techniques in the Research and Development Stage

### Introduction

In this section we cover the more futuristic lithography methods, some of which could cause a paradigm shift in the improvement of CD printing capability by perhaps a factor of 10. These might eventually enable devices that are a few nanometers in size and have switching speeds in the terahertz range. We review lithography with ultrathin resist layers and block copolymers, zone plate array lithography (ZPAL), proximal probe-based lithographies such as atomic force microscopy (AFM), scanning tunneling microscopy (STM), dip-pen lithography (DPL), quantum lithography (two-photon 3D lithography), near-field scanning optical microscopy (NSOM), apertureless near-field scanning optical microscopy (ANSOM), holographic lithography, plasmonic lithography, and lithography with superlenses. In terms of an important lithography method that is capable of patterning resist on nonplanar substrates (e.g., patterning a resist on the surface of a cylinder), we detail soft lithography, in which a soft mask can be bent around nonplanar substrates.

## Very Thin Resist Layers

### Introduction

Radiation scattering during exposure of a resist limits its resolution to no better than the resist thickness. Using ever thinner resist layers further improves the resolution and renders a low DOF—associated with high NA lenses and short wavelength lithography—less critical. It is in this context that physisorbed Langmuir-Blodgett (LB) resists, self-assembled chemisorbed monolayer (SAM) resists, and ultrathin film (UTF) resists are explored.

### Langmuir-Blodgett Resists

LB resist films were considered for some time for the finest lithography applications such as ultraresolution nanolithography. In this method, LB resist films are prepared by transferring organic monolayers floating on a water surface onto solid substrates. The procedure is explained in Chapter 8 on chemical, photochemical, and electrochemical-based forming (see Figure 8.13). Multilayer films of PMMA deposited by the LB technique, for example, have been investigated as a potential e-beam resist for nanolithography. The range of backscattered electrons limits the resolution of e-beam resists through the proximity effect, and nanometer-thick layers minimize this effect. Despite extensive work in the early 1980s to automate LB resist deposition systems, the technology is not envisioned as a viable manufacturing option because of its low mechanical and chemical stability.[50,51] Other major problems concerning LB films need to be addressed before the method can become viable, and research in this area focuses on speeding up the coating process, simplifying surface cleaning procedures, and improving etching resistance. Because the bonding of an LB film with the substrate is based on physisorption, the adhesion is weak, and chemisorbed SAMs have become much more popular (see next section).

### Self-Assembled Monolayers

*Molecular self-assembly* is a chemical process in which molecules spontaneously organize to form larger ordered structures. Self-assembly is one aspect of supramolecular chemistry, i.e., chemistry beyond the molecule or the chemistry of the intermolecular bond.[52] Self-assembly is studied extensively with an eye toward mimicking nature's spectacular use of this bottom-up, automated manufacturing approach (see Volume III, Chapter 4 on packaging, assembly, and self-assembly). In one special type of self-assembly, self-assembling precursor molecules, from a solution or the vapor phase, react at interfaces to produce layers of monomolecular thickness that are chemically bonded to solid surfaces. Such layers belong to a class of materials known as *self-assembled monolayers*. The thermodynamically favorable bond formation involves chemisorption, which results in monolayers that are more stable than those possible with physisorbed LB films (see above). Therefore, SAMs potentially make for better resist candidates.

Homogeneous and densely packed molecular layers incorporate reactive groups that form bonds with the substrate on one side, whereas an organic group (e.g., an alkyl group, R) on the other side imparts the desired chemical functionality to the surface modified with the thin film. Changing even one atom at the end group of self-assembling molecules is sufficient to dramatically alter macroscopic properties such as wettability, biocompatibility, and adhesion. For example, fluorinated SAM precursors form monolayers ~1-nm thick with the same low wettability and resistance to adhesion typical of thick samples of Teflon™.[53] Many SAM films can be molecularly engineered to be patterned by various types of energetic radiation, including DUV, soft x-rays, ion beam, and low-energy electrons.[54] Scratching and microcontact printing (see below) can also be used to pattern SAMs. Structures less than 20 nm have been obtained by exposure with STMs and conventional EBL systems.[55] SAMs have successfully been deposited on metals like gold, aluminum, titanium, zirconium, silver, copper, and platinum, as well as on $SiO_2$, GaAs, and other surfaces.[53] The quality of the films depends strongly on surface pretreatment. Pinhole density on a Au surface of less than $5/mm^2$ has been reported (see Müller and references therein[55]). Alkane-thiols and dialkyldisulfides are typical precursor materials for SAMs on gold, the most extensively studied substrate. The formation of long chain ω-substituted dialkyldisulfides on gold was first demonstrated in 1983.[56] Films of better quality are formed by the adsorption of alkylthiols.[57] Alkanethiols and dialkyldisulfides are lipid-like organic molecules having the general formula HS-$(CH_2)_n$-X and X-$(CH_2)$-SS-$(CH_2)_m$-Y, respectively, where n and m indicate alkyl chain length and X,Y the end groups (e.g., -$CH_3$, -azobenzene, -OH).[58] Organosilanes are often used to form SAMs on Si. These molecules form a Si-O-substrate siloxane bond on the Si surface, and the alkyl group R is responsible for the ordered nature of the film. For $n > 6$ ($n$ is number of carbon groups in the R chain), SAMs have substantial crystalline order at the air/monolayer interface.[53, 58]

Microcontact printing, discussed below, also takes advantage of SAMs, and so does DNA and protein patterning, discussed in Chapter 8. Microcontact printing can be considered a merging of top-down manufacturing techniques (i.e., traditional lithography used to make the elastomeric stamp) with bottom-up manufacturing (i.e., proteins used to ink the stamp as building blocks). Protein and DNA patterning has evolved into an important research application for lithography. It is easy to imagine arrays with several different enzymes, antibodies, or DNA probes immobilized precisely onto a small transducer surface as a diagnostic panel for clinical applications. Different approaches for patterning organic materials are reviewed in Chapter 8.

### *Ultrathin Film Resist Layers*

Polarity changes are at the heart of the newest ultrathin film (UTF) resist strategies. UTF resists represent the next evolutionary step in surface imaging technologies (see Chapter 1). In UTF, only the top monolayer of the resist changes, opening up the potential for further resolution improvement. Calvert et al.[59] use a few monolayers (<10 Å) of organosilanes, which are chemisorbed onto a substrate. These chemisorbed films are easier to prepare and more robust than the physisorbed LB films (see above). DUV irradiation of these films cleaves organofunctional groups from the film and produces an extremely hydrophilic surface (water contact angle < 10°) to which colloidal Pd/Sn catalyst does not adhere. Subsequent electroless deposition occurs only on those surface regions that were not irradiated. Thus, patterns of electroless Ni, Co, and Cu are produced on a variety of substrates, including silicon, quartz, alumina, metals, and polymers. These metal patterns serve as an efficient plasma-etching mask for pattern transfer. Fabrication of 0.3-μm line widths has been demonstrated in this way.[50, 59]

## Block Copolymer Lithography

### *Block Copolymers Background*

Self-assembly of block copolymers is widely investigated as a simple, inexpensive means for patterning high-density arrays of nanoscale features.[60] Block copolymers comprise two or more different monomer units, strung together in long sequences, which can self-assemble into highly ordered lattices with unit cell dimensions of 10–100 nm. This length scale reaches well below the limits of conventional optical lithography. The simplest block copolymer architecture is the diblock. In diblock copolymers, two different types of polymer chains are connected at one end with a covalent bond (in a triblock copolymer, three different polymer chains are involved, etc.). The immiscibility between the unlike segments drives them to self-assemble into periodic domains, a process termed "microphase separation." This self-assembly process is driven by an unfavorable mixing enthalpy and a small mixing entropy while the covalent bond between the two blocks prevents macrophase separation. The size and shape of the periodic domains are dependent on the molecular weight and composition of the copolymer and typically assume morphologies of spheres, cylinders, and lamellae. These morphologies that represent different phases are dictated by the Flory-Huggins interaction parameter $\chi$, the degree of polymerization $N$ (i.e., number of monomers per chain, proportional to the polymer molecular weight), and the volume fraction $f$ of the blocks.[61] The Flory-Huggins interaction parameter $\chi$ is a dimensionless parameter, which expresses, in units of kT, the energy change when one takes a molecule of pure $A$ from an environment of pure $A$ and puts it in an environment of pure $B$; it is inversely proportional to temperature and reflects the interaction energy between the different segments of the polymer chains. As $\chi$ increases, phase separation becomes more likely, and as $N$ increases, phase separation becomes less so. From Figure 2.30, we glean that if the volume fractions of the blocks are close to equal, a layered morphology is observed. When moving toward less equal block ratios, the observed morphologies go through a bicontinuous gyroid structure, hexagonally packed cylinders, and finally body-centered cubic-packed spherical domains. Thus, microphase separation leads to a wonderful variety of exotic ordered phases, increasing rapidly with the number

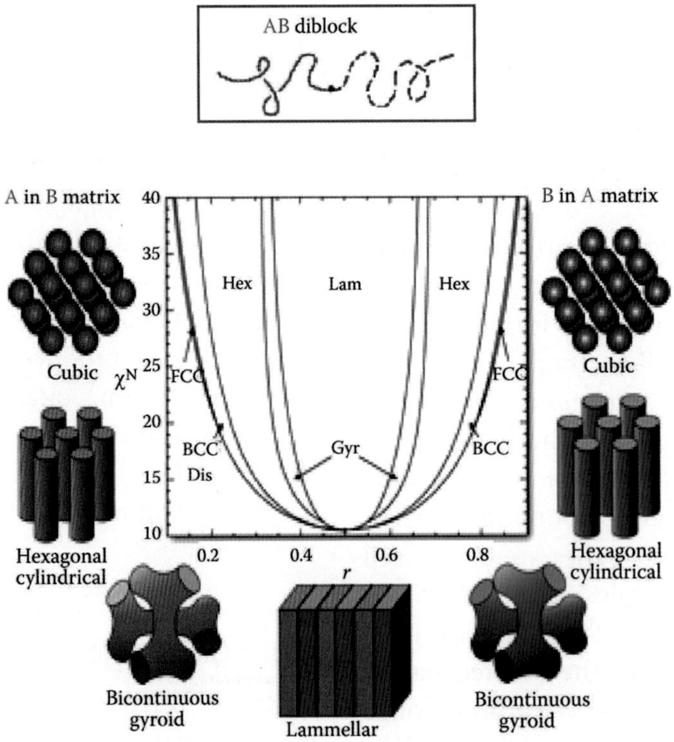


**FIGURE 2.30** Schematic diblock copolymer phase diagram: f = volume fraction of $A$ in $B$, $\chi$ = Flory-Huggins interaction parameter, N = diblock degree of polymerization. Known equilibrium mesophases are spheres, cylinders, gyroid, and lamellae, as well as the disordered (homogeneous) state at small interblock segregation strength ($\chi N$). (Diagram adapted from Matsen, M.W., and F.S. Bates. 1996. Unifying weak- and strong-segregation block copolymer theories. *Macromolecules* 29:1091–98.[62])

of blocks (with a triblock copolymer, a veritable "zoo" of structures results).

### Block Copolymers Lithography

The previous section applied to bulk block copolymers. What about applying thin layers of block copolymers to surfaces? On surfaces, the microphase separation also results in nanoscale structures (domains) with sub-100-nm length scales, unless the surface has too strong an interaction with the polymer blocks that it prevents them from assembling. Block copolymer lithography refers to the use of block copolymers in the form of thin films, in which the domain structure provides a template for additive or subtractive pattern transfer operations, as illustrated in Figure 2.31. The idea of using block copolymer thin films as lithographic masks, as shown here, was first proposed in 1995 by Mansky et al.[63] In Figure 2.31, we compare traditional lithography with block copolymer lithography. The material chosen for the block copolymer lithography demo is from IBM work, where a diblock copolymer, in which polystyrene and PMMA are tied together in one chain (referred to as PS-b-PMMA, with total molecular weight in the range of 60,000 g/mol), is used.[64] When spun onto the surface of a silicon wafer, the two polymers separate, and although the molecules stretch out, the chemical bonds keep them attached. Subsequent heat treatment above the polymer glass transition temperature completes the microphase separation. Typical annealing conditions are 200°C (in a vacuum or $N_2$) for times ranging between 30 minutes and 24 hours, depending on the application. In the end, PMMA ends up concentrated in small cylinders surrounded on all sides by polystyrene. Thus, the diblock copolymer forms on its own into a nearly

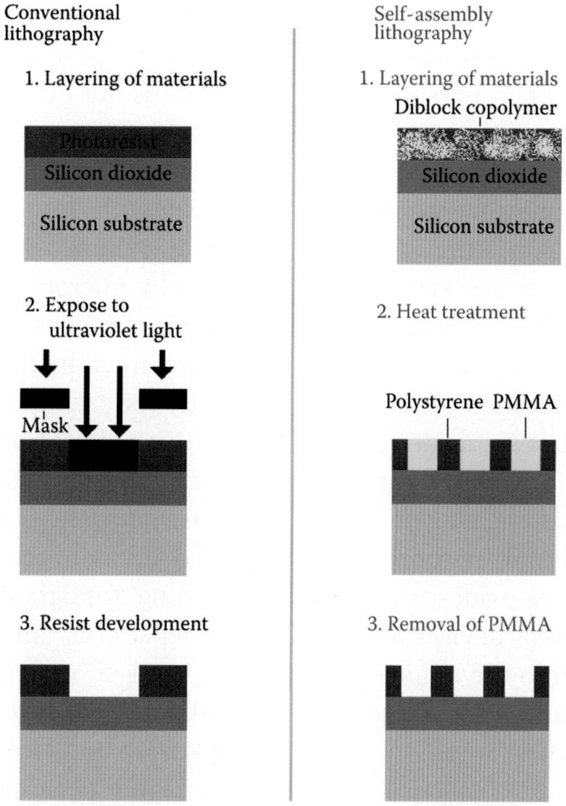


**FIGURE 2.31** Traditional lithography compared with block copolymer lithography. Conventional lithography exposes a photoresist to ultraviolet light. A solvent etchant then removes the exposed part of the photoresist (for a positive type photoresist). Self-assembly patterning occurs when a diblock copolymer is heated, thereby separating the two polymers in the material into PMMA and polystyrene domains. The PMMA can then be selectively etched away, and the template of cylindrical holes can be transferred into the silicon dioxide. (Adapted from Stix, G. 2004. *Nano patterning: IBM brings closer to reality chips that put themselves together.* http://www.fractal.org/Fractal-Research-and-Products/Nano-patterning.htm.[65])

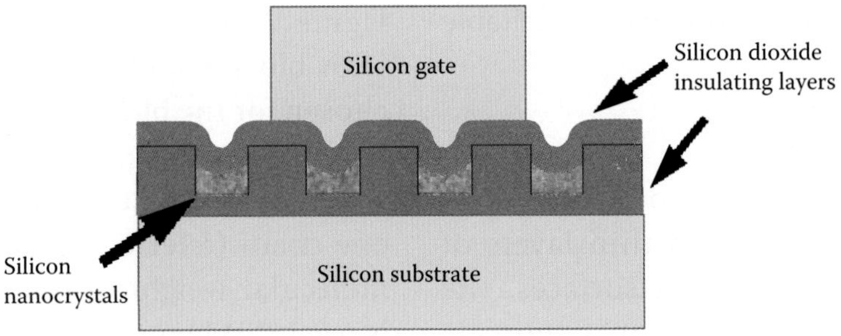


**FIGURE 2.32** Flash memory: a layer of self-assembled silicon nanocrystals is inserted into an otherwise standard device as part of a novel IBM manufacturing process. (Based on Stix, G. 2004. *Nano patterning: IBM brings closer to reality chips that put themselves together.* http://www.fractal.org/Fractal-Research-and-Products/Nano-patterning.htm; and Black, C.T., R. Ruiz, G. Breyta, J.Y. Cheng, M.E. Colburn, K.W. Guarini, H.-C. Kim, and Y. Zhang. 2007. Polymer self assembly in semiconductor microelectronics. *IBM J Res Dev* 51: 605–33.[65,64])

perfect honeycomb-like template without the need of UV exposure or a mask.

Finally, the IBM team created an array of 20-nm-wide pores in the polystyrene matrix by removing the PMMA using an organic etching solvent (acetic acid). The same team used the thus-obtained perforated polystyrene structure to create the flash memory cells illustrated in Figure 2.32.[64] After the PMMA etch, a subsequent etching step was used to transfer the same honeycomb pattern into an underlying layer of silicon dioxide. Next the polystyrene template is removed, and subsequently, nanocrystal formation is accomplished by conformal a-Si (amorphous or nanocrystalline Si)* deposition over the whole wafer, and a dry anisotropic a-Si etch is used to remove all the deposited silicon except for that deposited in the holes in the oxide matrix. All that is left in the end are nanocrystalline cylinders surrounded by silicon dioxide. The final steps place an insulating layer of silicon dioxide (7-nm-thick control or program oxide) over the top of the Si nanocrystal islands and a polysilicon control gate is defined on top of the control oxide using conventional lithography. These flash devices are programmed by injecting charge into the nanocrystals (through the program oxide) and erased by expelling charge from the nanocrystals.

Creating closely spaced holes for a flash memory as demonstrated here is exceedingly difficult to accomplish with ordinary lithographic and deposition methods. Forming nanocrystalline silicon using these conventional techniques creates elements of different sizes. In contrast, the self-assembled nanocrystals are evenly spaced and of uniform size, improving their durability and their capacity to retain a charge while allowing the cylinders to shrink to dimensions even smaller than 20 nm.

For most lithography applications, we must also be able to control the orientation of the self-assembled pattern with respect to the underlying substrate surface. Whereas the self-assembly process itself is driven by microphase separation of the constituent copolymer blocks, the ultimate pattern orientation is determined by the relative strength of the surface affinity for each block. Control of block copolymer film thickness plays a critical role in determining vertical and lateral ordering and alignment of the nanostructures. A variety of approaches for controlled domain orientations have been attempted, including topographic prepatterns and/or chemical prepatterns, an external electric field, thermal field, and several other methods as reviewed by Segalman,[66] Hawker and Rusell,[67] Black et al.,[64] Stoykovich and Nealy,[68] and Hamley.[60]

In Figure 2.33, the patterned wettability (polar vs. nonpolar stripes) of the surface is used to align the nanodomains. Substrates are often preferentially wetted by one of the blocks of the copolymer, and the domains, e.g., lamellae or cylinders, tend to orient parallel to the substrate (see Figure 2.33a and b, where all the lamellae align parallel to

* Nanocrystalline silicon (nc-Si) is an allotropic form of silicon that is similar to amorphous silicon (a-Si), in that it has an amorphous phase. However, nc-Si has small grains of crystalline silicon embedded within the amorphous phase. This is in contrast to polycrystalline silicon (poly-Si), which consists solely of crystalline silicon grains, separated by grain boundaries. The term nanocrystalline silicon refers to a range of materials around the transition region from the amorphous to the microcrystalline phase in a silicon thin film.

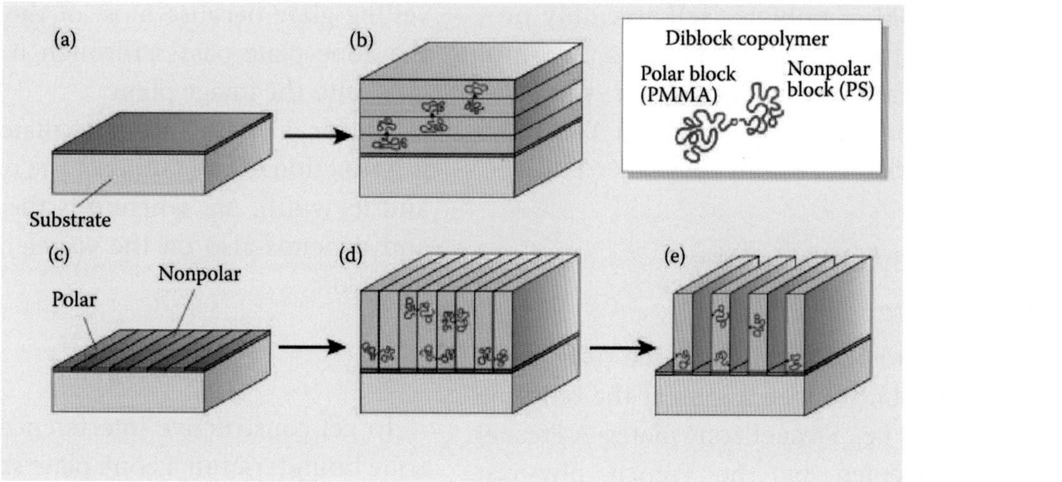


**FIGURE 2.33** Patterned wettability of the surface causes the alignment of the nanodomains of a block copolymer. PMMA, polymethylmetacrylate; PS, polystyrene.

the substrate). Chemical modification can be used to neutralize or overcome surface and interfacial forces that tend to drive the domains to form parallel to the plane of the film. Controlling the interactions at the interfaces and scaling of the prepatterns can achieve parallel and perpendicular alignments of the lamellae achieved as shown in Figure 2.33c and d: the polar PMMA block aligns over the polar stripes and the nonpolar polystyrene blocks over the nonpolar stripes. The block copolymers on the surface arranged themselves into the underlying pattern without imperfections. In Figure 2.33e we show a finished structure after etching away the PMMA.

In a significant accomplishment, Kramer's group[69] (University of California, Santa Barbara) has patterned single crystals of 10-nm block copolymer spheres across an entire 2-in.-diameter silicon wafer. The silicon wafer was first patterned with photolithography to produce 12-μm-wide hexagonal wells. A single layer of block copolymer spheres was then self-assembled inside those wells. Assessing the quality of the self-assembly of the block copolymer in perfect crystal patterns is very challenging because of the small feature size. Whereas high-resolution microscopy can resolve individual block copolymer domains, this technique is limited to extremely small sampling areas (~$10^{-6}$ mm$^2$). The Santa Barbara team used grazing-incidence small-angle x-ray scattering (GISAXS) to check for the crystal quality over macroscopic areas (~2 mm$^2$), and they established it as nearly perfect, demonstrating that single crystals resulted in each well. These results clarify why copolymers constitute a viable avenue toward building photonic crystals.

### *Summary/Applications*

The obvious interest in using block copolymers for patterning is derived from the fact that they can self-assemble to form dense arrays of nanostructures with dimensions and spacings that are difficult or even impossible to create by other means or are prohibitively expensive to fabricate using conventional lithographic materials and processes. Polymer self-assembly can define only a limited set of pattern geometries. However, within these constraints the materials provide a straightforward means for achieving feature sizes (<20 nm), pitches (<40 nm), and densities (~$10^{11}$/cm$^2$) that are currently not achievable by optical lithography.

The use of block copolymer films to pattern dense arrays of features is more than a decade old, and other reviews have covered this topic in more detail (for example, see Segalman,[66] Hawker and Rusell,[67] Stoykovich and Nealy,[68] and Hamley[60]). These review articles reveal that patterns of dense periodic arrays fabricated with block copolymer templates have been demonstrated, and in some cases are being commercialized for the fabrication of quantum dots, magnetic storage media, flash memory devices (see Figure 2.32), semiconductor capacitors, nanowires, photonic crystals, and nanopores. At IBM, a polymer self-assembly process was successfully integrated into a 200-mm semiconductor fabrication facility at the IBM Thomas J. Watson Research Center in Yorktown Heights, New York.[64]

## Zone Plate Array Lithography

### *Fresnel Zone Plate*

Before getting into the details of zone plate array lithography (ZPAL), we take a look at the building blocks of a ZPAL, i.e., Fresnel zone plates. A Fresnel zone plate, invented by the French physicist Augustin-Jean Fresnel, is a single plate consisting of alternating opaque and transparent circular zones as illustrated in Figure 2.34. Zone plates use constructive interference of light rays from adjacent zones (Fresnel zones) to form a focus. The zones are spaced so that diffracted light constructively interferes at the desired focus. The Fresnel zone plate is a relative of the pinhole camera in that it does not use mirrors or lenses for its imaging properties. Because a diffraction lens can be manufactured using planar technologies, it is easier to make than a traditional refractive lens. The zone plate is especially useful in the UV and x-ray regions of the spectrum, for which other imaging devices are hard to find. Self-supporting gold zone plates have been manufactured for these spectral regions. Moreover, it comes with a higher refractive index, allowing for better contrast at the image plane. Unfortunately, the zone plate has low efficiency and suffers from veiling glare because most of the light incident on the zone plate passes through it undiffracted and falls onto the image plane.

The focal length *f* of a zone plate with many zones is a function of the radius of the outermost zone, $r_n$, and its width, $\Delta r_n$, where n is the number of zones and depends also on the wavelength λ (see Figure 2.34), or:

$$f = \frac{2r_n \Delta r_n}{\lambda} \tag{2.3}$$

To get constructive interference at the focus, the zone boundaries in a zone plate should switch from opaque to transparent at radii where:

$$r_n^2 = n\lambda f + \frac{n^2\lambda^2}{4} \tag{2.4}$$

$$r_n = \sqrt{n\lambda f + \frac{n^2\lambda^2}{4}} \tag{2.5}$$

and simplifying when the zone plate is small compared to the focal length:

$$r_n = \sqrt{n\lambda f} \tag{2.6}$$

where *n* is a positive integer, $r_n$ is the radius of the nth zone, *f* is the primary focal length, and λ is the wavelength of the illumination source. A Fresnel zone plate functions very much like an optical lens. Its optical properties do not depend on the material used, though, but are determined at the design time. For example, the diffraction limited resolution for

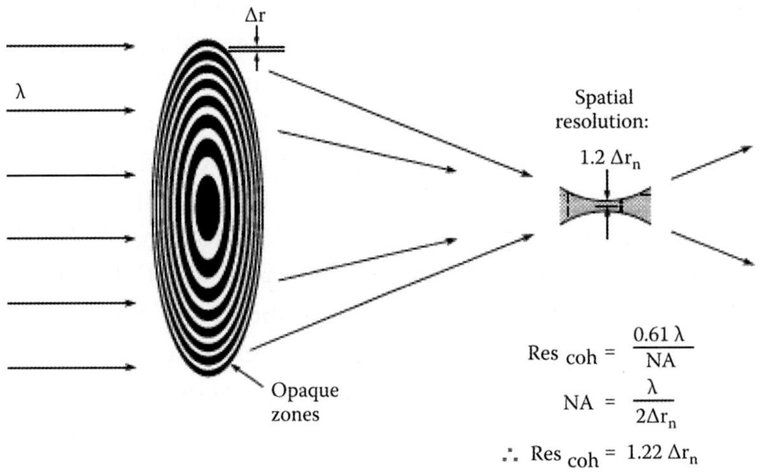


**FIGURE 2.34** A Fresnel lens. (From http://xradia.com.)

coherent light $Res_{coh}$ of a zone plate is governed by the Rayleigh principle as:

$$Res_{coh} = 0.61\frac{\lambda}{NA} \text{ and with}$$

$$NA = \frac{\lambda}{2\Delta r_n} \text{ we obtain:} \tag{2.7}$$

$$Res_{coh} = 1.22\Delta r_n$$

The resolution depends only on $\Delta r_n$, the outermost zone width. NIL and EBL can be used for the fabrication of large arrays of high NA diffractive Fresnel lens arrays.

In optical lithography, further increasing the NA of a lens is difficult and costly (see Figure 1.80 where a maximum NA of around 0.9 is projected). In diffractive optics, making a high NA lens is not significantly different than making a low NA lens. At MIT, Fresnel lenses with an NA of 0.7–0.9 have been achieved.[70]

### *Zone Plate Array Lithography Detailed*

Zone plate array lithography (ZPAL) requires no masks but rather uses arrays of individually targetable optical beams as shown in Figure 2.35. Beamlets of photons, which can be rapidly turned on and off,

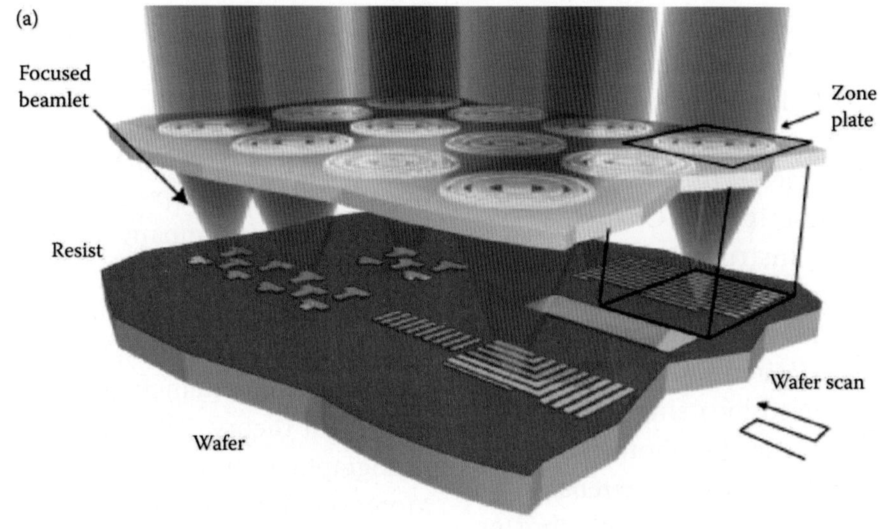


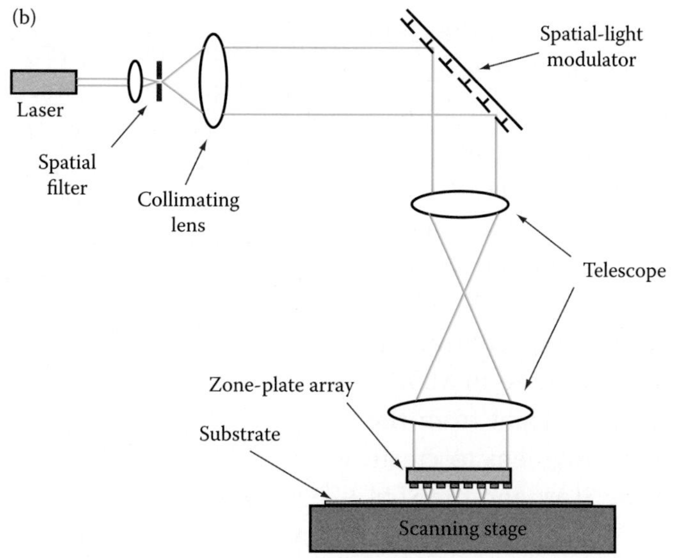


**FIGURE 2.35** (a) Lumarray (http://www.lumarray.com) has demonstrated fully multiplexed ZPAL writing. (b) A laser beam is passed through a spatial filter and a collimating lens onto a digitally controlled programmable spatial light modulator (SLM), which replaces the photomask.[70,71]

are projected through diffractive Fresnel zone plate lenses (see above), allowing myriad complex shapes to be fabricated. The zone plates are made using lithography techniques, and the shutters for each beamlet are micromechanical movable mirrors. ZPAL was originally developed at MIT's nanostructures laboratory in the mid to late 1990s.[70] As shown in Figure 2.35a, each zone plate is responsible for one unit cell of photoresist exposure corresponding to the diameter of one individual zone plate. The writing with the shutters on and off results in dot-matrix-type lines, and moving the wafer in a serpentine fashion under the focused beamlets results in a full pattern.

ZPAL is an example of maskless optical projection lithography (MOPL, see Chapter 1); a laser beam is passed through a spatial filter and a collimating lens onto a digitally controlled programmable spatial light modulator (SLM), which replaces the photomask (see Figure 2.35b). In one of the MOPL systems described in Chapter 1, the SLM used is the Texas Instrument micromirror array (see DMD™, Example 7.10). In operation, a desired pattern is fed into the SLM, and the movable mirrors create the desired image. This image is demagnified and focused by a single reduction lens on the wafer to expose the photoresist. The use of refractive optics prevents the extension of this type of MOPL to wavelengths much shorter than 157 nm.

In the ZPAL system, the MIT team uses the Grating Light Valve™ (GLV)* from Silicon Light Machines (http://www.siliconlight.com) as the light modulator. In contrast to MOPL, not a single refractive reduction lens is used. Instead, an array of (high NA) diffractive Fresnel lenses ("the zone plates") is used to focus the incident light onto an array of spots on the photoresist-coated substrate. The light source in the ZPAL system preferably is a continuous wave laser (a pulsed laser source, as used in MOPL, severely limits system throughput). Light from the laser is passed through a collimating lens to create a uniform beam incident on the SLM, and the SLM breaks that beam up into individually controllable beamlets. A telescope is placed such that each beamlet is normally incident on one zone plate in the array. A Fourier filter within the telescope ensures that there is sufficient contrast between the on and off states. Each zone plate behaves like a microscopic lens focusing the light into a tight on-axis spot in its focal plane. The diffractive Fresnel lens optics in ZPAL allow the technique to scale to much shorter wavelengths than what is possible with MOPL (even x-rays are feasible).

In Chapter 1 we used a dimensionless constant $k_1$ to define the practical limiting resolution $R$ in projection printing as:

$$R = k_1 \frac{\lambda}{NA} \tag{1.16}$$

where $k_1$ depends on resist parameters, process conditions, mask aligner optics, and so on. The MIT team and a startup company called Lumarray (http://www.lumarray.com) have demonstrated fully multiplexed ZPAL writing, multilevel alignment, and resolution corresponding to $k_1 < 0.3$ [this is without any resolution-enhancing technology (RET) applied to it].[71] In theory, ZPAL could use 193-nm optical lithography to inexpensively produce 45-nm chips. Moreover, this technology should theoretically work with 157-nm lithography and EUVL to create 20-nm ICs. ZPAL's main competition at this point is regular 193-nm lithography enhanced with immersion lithography, a combination that could etch 45-nm features (see Chapter 1). ZPAL, however, might be a less-expensive and an easier solution. As in optical projection lithography, to increase NA, immersion lithography can also be applied to ZPAL; in experiments at MIT, a resolution improvement of 1.36 was achieved this way.[71]

## Quantum Lithography

### *Background*

In optical projection lithography, a reduced-size image of complicated patterns is reproduced onto a photoresist. If light always followed the law of geometrical optics, there would be a point-to-point

* The Grating Light Valve™ (GLV) technology comprises a series of microscopic ribbons on the surface of a silicon chip; the GLV device is a diffractive MEMS that acts a dynamic, tunable grating that can switch, attenuate, and modulate laser light.

relationship between the object (mask) and the image planes (resist). But, as we saw in Volume I, Chapter 5, light is a wave, and resolution is limited to a finite spot size, defined by the "point-spread function". This point-spread function determines the spatial resolution of the imaging setup and limits the ability to produce demagnified images (see Rayleigh limit of 0.6 $\lambda$/NA, Volume I, Chapter 5). It has been demonstrated that quantum lithography with entangled $N$-photon states beats this Rayleigh diffraction limit by a factor of $N$; thus, in two-photon lithography the resolution is improved by a factor of two (as if one used a classical source with wavelength $\lambda/2$).[72,73] Einstein, Poldosky, and Rosen (EPR) described the entangled two-particle state according to the principle of quantum superposition in 1935, and they pointed out a surprising consequence: the momentum $\Delta P_x$ (position $\Delta x$) for neither photon is known (see Heisenberg uncertainty principle, Volume I, Chapter 3); however, if one particle is measured to have a certain momentum (position), the momentum (position) of its "twin" is known with certainty, despite the distance between them (this is known as the EPR paradox).[74] The entangled photon pairs come out from a point of the object plane, undergo two-photon diffraction, and result in twice-narrower point-spread function on the image plane as shown in Figure 2.36.

## Two-Photon 3D Lithography

Two-photon lithography is an intrinsic 3D lithography technique that enables the fabrication of structures difficult to produce by conventional single-photon processes and with far greater spatial resolution than most other 3D microfabrication techniques. As we discuss in Chapter 8 in the section on stereolithography, the use of laser rapid prototyping by polymerizing a resin point by point enables the computer-aided manufacture (CAM) of 3D polymer MEMS structures. Linear absorption of laser light by a monomer for cross-linking typically limits the resolution to the wavelength of the laser that is used. Moreover, deep penetration of the photons into the resin bulk is impossible because of the strong absorption of the light, so that the process is carried out necessarily thin layer by thin layer (see Figure 8.30). Using two-photon excitation changes all this as it enhances the resolution and can penetrate deep into the bulk of the resin. When high-intensity light shines on a material, the probability for two-photon absorption is proportional to the square of the field intensity (see Volume III, Chapter 6 for an illustration of this effect), and thus is greatest at the center of a Gaussian laser spot. Using a high NA lens, the laser spot can be made less than 1 μm in diameter, and significantly narrower structures can be obtained because polymerization starts as soon as a certain threshold of intensity has been reached.

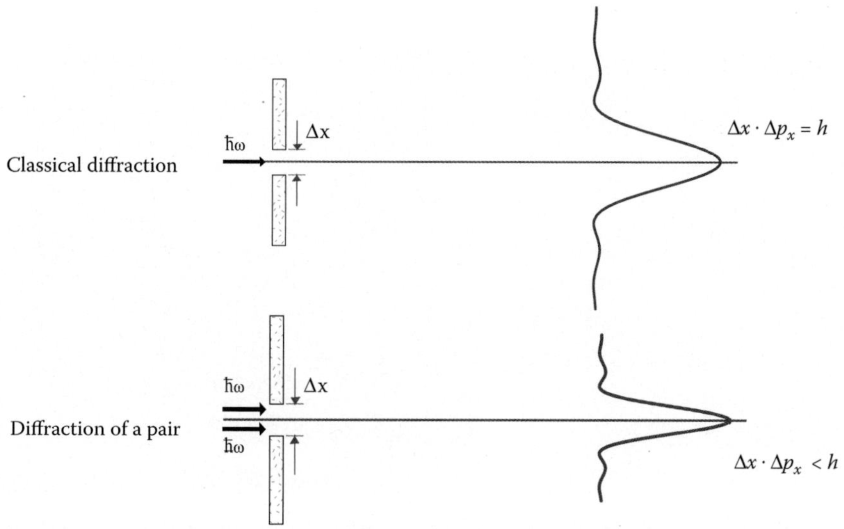


**FIGURE 2.36** The entangled photon pair comes out from a point of the object plane, undergoes two-photon diffraction, and results in twice narrower point spread function on the image plane.

Therefore, by tightly focusing a femtosecond laser beam into a resin, subsequent photo-induced reactions such as polymerization occur only in the close vicinity of the focal point, allowing the fabrication of subwavelength lithography and 3D polymer structures. Importantly, because the energy required to initiate polymerization is twice that of the incoming radiation, it is possible to excite deep into the monomer as the material is transparent to one-photon excitation, whereas at the focus, where the intensity is high, two-photon excitation occurs.

Parthenopoulos and Rentzepis[75] first demonstrated the 3D microfabrication capability possible through two-photon polymerization in optical memory in 1989, and Wu et al.[76] pioneered the technique's utility for lithography in 1992. In two-photon lithography one often works with epoxy resins and urethane acrylates. Several two-photon photoinitiators with very large two-photon absorption cross-sections to be added to base polymers have been developed.[77] In these photoinitiator molecules the photon energy that is absorbed is emitted as photons of a higher energy (up-converted fluorescence). The up-converted fluorescence is then used to induce various photochemical reactions such as photopolymerization and photodissociation of the base resin. Two-photon photolithography experiments using this idea recently have achieved 120-nm spatial resolution using femtosecond laser pulses with a 780-nm central wavelength.[78] Below we will see that using two-photon lithography combined with apertureless near-field scanning optical microscopy (ANSOM), a resolution nearly a factor of 2 better than the resolution achieved in far-field two-photon lithography experiments can be achieved. We will also learn that combining a metallic ANSOM tip and femtosecond laser pulses, lithographic patterning with resolution of 70 nm in a SU-8 photoresist can be obtained.

As we clarified above, energy transferred to the photosensitive resin material by two photons of wavelength $\lambda$ corresponds to that transferred by a single photon of wavelength $\lambda/2$. At the same time one takes advantage of long-wavelength exciting light with larger penetration depth, which enables 3D patterning of bulk material. Scanning the focus within the material allows one to produce microscale and submicroscale structures. Subwavelength structures can be produced because the absorption profile for the two-photon process is narrower than the beam profile. Some examples of 3D two-photon lithography are shown in Figure 2.37. The two-photon 3D lithography has also been used to make photonic crystals. In Chapter 8 on photochemical forming, we will come back to this two-photon 3D lithography process and its use in stereolithography or microphotoforming.

Quantum lithography is in its infancy, and many experimental challenges remain. The main problem is how to generate $N$-state entangled photons

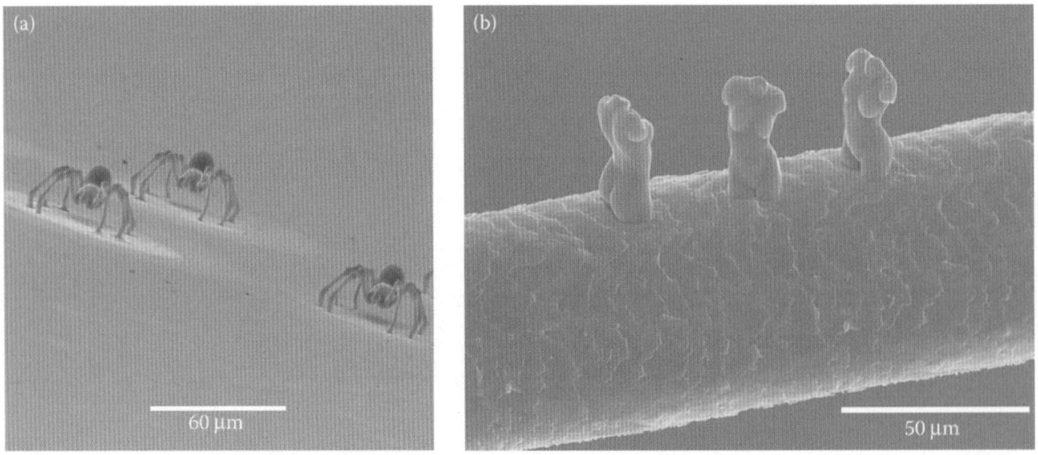


**FIGURE 2.37** Two-photon 3D lithography. (a) Microspider array, the body of the spider is above the substrate; it is supported by eight 1-μm thick legs (http://www.laser-zentrum-hannover.de). (b) Venus micromodels on a human hair (http://www.laser-zentrum-hannover.de).

in a reasonable way. The next problem is how to develop an *N*-photon-sensitive photoresist. Third, the exposed field is very small. However, if we could overcome all three problems, we can beat the Rayleigh limit and approach the Heisenberg limit instead, thus approaching a resolution on the order of the atomic size.

## Plasmon Lithography

Atwater and coworkers from Caltech introduced plasmon lithography in 2002. Plasmon lithography is based on plasmon resonance occurring in nano-sized metallic structures, allowing for the replication of patterns with a resolution limit considerably below the diffraction limit.[77] As we discovered in Volume I, Chapter 5 on photonics, when metallic nanoparticles are excited in an electromagnetic field, they can exhibit collective electron oscillations known as surface plasmon oscillations. With the diameter of the particle much smaller than the wavelength of the incoming light, the oscillating charges produce an oscillating dipole field around the metal particles. At resonance, i.e., when the excitation frequency coincides with the natural relaxation frequency of the nanoparticles ($\omega_p$), this results in a strongly enhanced local electrical field. One can choose the nanoparticles such that their resonance frequency falls within the sensitivity range of an underlying photoresist. The field enhancement at the interface of the particle and the resist is then used to locally expose the resist, as schematically illustrated in Figure 2.38. A transparent mask with a configuration of metal nanoparticles deposited on it is brought into intimate contact with a thin layer of photoresist. Kik et al.[79] used spray deposition to obtain ~41-nm-diameter silver nanoparticles. To ensure that the local field intensity is enhanced directly below the metal particles, the illumination beam should be at a glancing angle to the resist (p-polarized light). Although the whole resist film is illuminated, development of the resist only affects the areas that received locally enhanced exposure. In this way an image of the mask pattern is transferred onto the resist layer. The choice of nanoparticle material and size depends on the required wavelength; as stated earlier,[79] the resonance wavelength should fall within the resist sensitivity window. For example, to be able to use standard g-line photoresist, the resonance wavelength should lie between 300 nm and 460 nm. Additionally, to obtain high-field enhancement, long electron scattering times ($\tau$) are desirable. Silver is a good candidate, with long reported electron scattering times up to 10 fs and a resonance occurring at a wavelength of ~360 nm in air. The particles should also be small enough (<40 m) to attain confined dipolar enhanced fields at the resist interface. Particles that are too large develop multipolar oscillations as we detailed in Volume I, Chapter 5.

From Volume I, Chapter 5 we know that the resonance frequency of the metal nanoparticles is not a fixed quantity. For a spherical particle, for example, the surface plasmon resonance occurs at the wavelength $\lambda$ for which $\varepsilon'$ (dielectric constant of the particle) = $-2\varepsilon_d$ (dielectric constant of the medium). Consequently, the wavelength at which resonance occurs can also be tuned by changing the refractive index of the surrounding medium. Using this method Kik et al.[79] have shown that the local field enhancement occurring around metal nanoparticles when they are excited at the surface plasmon resonance frequency can be used to print nanoscale features in thin resist layers. Feature sizes less than $\lambda/10$ were generated in a parallel fashion using visible illumination and standard g-line photoresist. Further below we analyze ANSOM

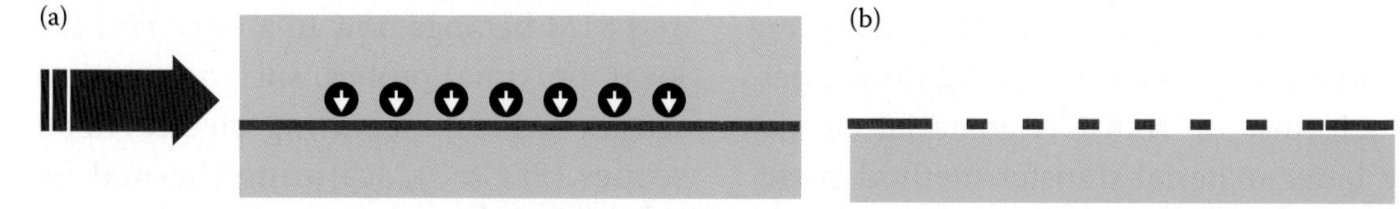


**FIGURE 2.38** Schematic representation of plasmon printing showing glancing angle illumination using polarized visible light, producing enhanced resist exposure directly below the metal nanostructures in the mask layer (a), and the resulting pattern in the resist layer after development (b).

lithography, where a scanning metal tip is used to generate plasmons for exposing the underlying resist.

In Volume I, Chapter 5, we discussed how Pendry had the amazing insight that a superlens could be made from a flat slab of a negative refractive index material, which not only would bring incident rays to a focus but also has the capacity to amplify the near-field radiation so that it could contribute to image formation, thus removing the wavelength limitation completely. Thus, a superlens would prevent image degradation and beat the diffraction limit established by Abbe! We also discussed the development of a poor man's near-field superlens ($\varepsilon < 0$ and $\mu = 1$) in 2003 by Zhang's group at the University of California, Berkeley, who showed that optical evanescent waves can be enhanced as they pass through a silver superlens.[80] This team imaged objects as small as 40 nm across with their superlens, which itself is just 35-nm thick.[81] In contrast, current optical microscopes can only resolve objects down to around 400 nm. With the superlens, using 365-nm illumination, features of a few tens of 10 nm were imaged, clearly breaking Abbe's diffraction limit.

## Scanning Proximal Probe Lithography

### *Introduction*

Proximal probe techniques rely on the use of nanoscale probes, positioned and scanned in the immediate vicinity of a material surface. Their development is often viewed as a first step toward nanotechnology because they demonstrate the feasibility of building purposeful structures one atom or one (macro) molecule at a time. Proximal probes used in lithography might involve electrical methods where a scanning tunneling microscope (STM) tip generates a local field/current that modifies the region directly under the tip (e.g., SiH → Si). A second approach involves mechanical methods where a scanning force microscope (SFM/AFM) tip scrapes, thermally deforms, or transfers material at the surface; the latter material transfer method corresponds to dip-pen lithography (DPL). Third, it may involve a near-field scanning optical microscope (NSOM) tip or apertureless near-field scanning optical microscopy (ANSOM) that exposes photoresist under the tip only.

The use of single proximal probe tips for lithography poses a serious drawback in terms of processing speed. To use these techniques in the manufacture of ICs and data storage devices, it is necessary to devise a scheme for parallel processing by making arrays of these proximal probes.[82]

### *STM/AFM Background*

The scanning tunneling microscope (STM) images the surface of conducting materials with atomic-scale detail (see Volume III, Chapter Six). STM was invented by Gerd Binnig and Heinrich Rohrer of IBM's Zurich Laboratory in 1981 (they received the 1986 Nobel Prize for their invention).[83] In general, STM works by bringing a small conducting probe tip up to a conducting surface. When the probe is very close to the surface (<10 Å) and operating voltages in the ±10-V range are applied, very small currents are produced because the electrons in the probe and the surface have wave functions extending beyond the physical surface boundaries. To the extent that these spillover wave functions overlap, a measurable current results. The interesting part about this current is that it depends exponentially on the spacing between the two conductors (as well as the voltage) (see Volume I, Chapter 3 in the section on tunneling). A piezoelectric transducer accomplishes the *z*-axis distance variation between tip and sample. By changing the distance over 1 Å, the current changes by a factor of 10. In practice, the current is kept constant through a feedback mechanism, and the probe moves up and down over the surface following the atomic contours it "sees." The images produced by the STM come from the electronic structure and from the geometry of the sample. Up to 100 times more powerful than SEMs, STMs measure objects in the angstrom range.

Over time, many proximal probe were developed, and STM belongs now to a large new family of very local, proximal probes, such as atomic force microscopes (AFMs), scanning electrochemical microscopes (SECMS), scanning thermal microscopes, scanning capacitance microscopes, magnetic force microscopes, and scanning pH probes, enabling microscopy of almost any type of material and property (for more details see Volume III, Chapter 6 on metrology). The common feature of these instruments is that their resolution is not determined by visible light used for the interaction with the probed object, as in conventional microscopy.[84]

Besides their use for imaging and characterization of surfaces, proximal probes can also be used to modify solid surfaces. Below we introduce some examples of this type of proximal lithography.

### *Scanning Probe Lithography*

*Introduction* In lithography, STM can produce nanostructures at higher resolution but at lower speed, whereas AFM can operate at higher scanning speeds but offers lower resolution. AFM can be used to image and pattern any kind of material, making AFM the more widely used technique. AFM/STM lithography can be classified as force assisted (including thermomechanical), bias assisted, and material transfer assisted.

*Scratch Lithography—Mechanical Effect* The simplest (and perhaps the crudest method) to pattern a surface with a proximal probe is by scratching in scratch lithography (see inset in the heading of this chapter: copy of Pablo Picasso's "Don Quixote" in polycarbonate patterned with scratch lithography). In this mechanical process, simply drawing an AFM tip across a surface with enough force may cut a trench, or the piezomechanism driving the AFM cantilever may be modulated in the *z*-direction to tap the tip on the surface as it is drawn over it. Cuts as narrow as 20 nm and 2 nm deep have been made in III–V semiconductor surfaces.[85] One difficulty with the scratching approach has been that probe tips often break following direct collision with a solid surface. As a possible solution, carbon nanotubes have been mounted on the tip of silicon cantilevers, enabling lines as narrow as 10 nm (Figure 2.39). Carbon nanotubes constitute the strongest material known along the axial direction and yet are highly elastic and flexible along the radial direction (see also Volume III, Chapter 3).[86]

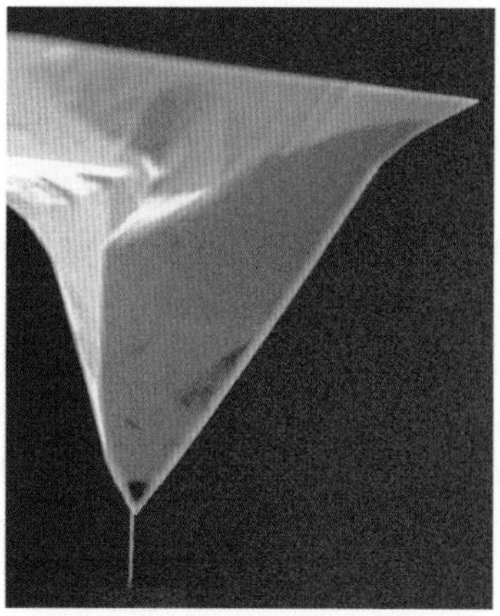

**FIGURE 2.39** Nanotube mounted on the micromachined tip of a Si cantilever is used as a nanopencil for lithography of 20–10-nm lines (courtesy of Dr. M. Meyyappan, NASA Ames). SEM picture of nanotube mounted on tip of Si cantilever (http://cnst.rice.edu/pics.html).

The Naval Research Laboratory (NRL) worked with proximal probes to pattern thin films of chemically amplified negative e-beam resist (SAL601 from Shipley). Resist films of 30–70-nm thick were patterned with typical tip-sample voltages from −15 to −35 V, resulting in minimum feature sizes of 23 nm.[54] Using SAMs as resists (see above), it was shown that the lower the exposure threshold energy of the film, the better the lithographic resolution. There are indications that, with these low voltages (~4 V), SAM resists will eventually yield sub-10-nm CDs.

In a variation on scratch lithography, a resistively heated AFM probe writes a data bit by scanning over a polymer surface. The combined heat and mechanical force of the tip causes the polymer to soften and flow, thus facilitating the writing of data bits in a storage medium (see the millipede chip developed by IBM for thermal writing in a polymer for memory; Figure 2.4b).

*Electrical Field-Induced Chemical Modification—Electrochemical Effect* In an alternative approach, surface oxides were induced on Si and GaAs by slightly increased tip voltages on samples held in a wet nitrogen atmosphere.[38] In this method, the water meniscus formed at the tip-sample gap is dissociated by the negative tip bias, and the $O^-$ and $OH^-$ oxidative ions react with the substrate to form localized oxide nanostructures. Because the molecular volume of the oxides is usually larger than that of the substrate materials, raised nanopatterns are formed after the oxidation reaction. These thin oxides, although only a few monolayers thick, are sufficiently robust to act as a mask for subsequent reactive ion etching of the substrate.[54] The oxidation process is fairly general and may also be applied to Ti and Cr. At NRL, oxide features with lateral dimensions as small as 10 nm have been achieved.[54] Multilayer resist films for nanopatterning with SPL have also been developed. Sugimura et al.[87] work with a three-layer resist to pattern insulating substrates such as thermally grown $SiO_2$ with current injection from a scanning probe tip. The process sequence is sketched in Figure 2.40.

The bottom layer of the three-layer resist consists of 20 nm of amorphous Si (a-Si) and is prepared by ion-beam sputtering; a second layer consists of an intermediate 2-nm-thick Si oxide and is prepared by photo-oxidation of the top layer of the a-Si. The top resist layer is 2-nm thick and consists of an octadecylsilyl self-assembled monolayer (ODS-SAM). To pattern the resist on top of the insulating $SiO_2$, a bias voltage is applied between the AFM probe and the conductive a-Si layer, which is made positive. When a scanning probe is operated in the presence of atmospheric water vapor, the probe and sample are automatically connected via a tiny water column created by capillary forces of the adsorbed water. This assembly serves as a minute electrochemical cell. As a result of the electrochemical anodization reaction, the monolayer resist becomes degraded in the region where the probe has passed, and the underlying photo oxide grows thicker. A subsequent etch in 0.5 wt% HF removes the exposed oxide, and that pattern is then transferred to the underlying a-Si by an etch of the a-Si in an aqueous solution of 25 wt% tetramethylammonium hydroxide (TMAH). The latter solution etches Si but not $SiO_2$. The a-Si is etched, and the etch stops when the underlying insulator is reached. The underlying thermal oxide can now be etched in HF and the resist stripped. Minimum feature sizes reached so far are about 50 nm.

*Electron Scanning Probe Lithography-Bias Effect* Lithography using electrons from a scanning microprobe offers several potential advantages over writing with a traditional e-beam source.[84,88] One important benefit is that the low energy of electrons in SPL (<50 eV) compared with those in EBL (10–100 keV) avoids the detrimental effects of electron backscattering, thereby virtually eliminating proximity effects and thus producing enhanced resolution and superior pattern fidelity.[54] Moreover, SPL using an STM can be operated in air, whereas EBL relies on a vacuum. Because of the small tip-to-sample distance,

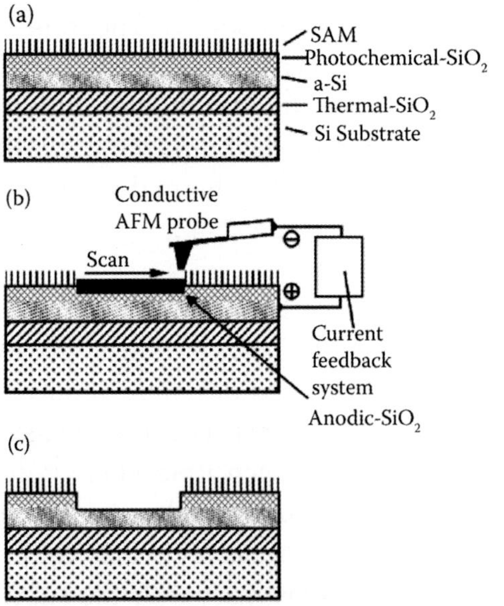

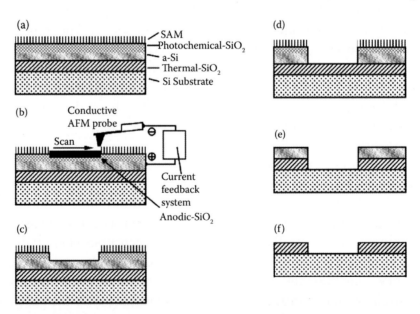


**FIGURE 2.40** Schematic of AFM lithography on $SiO_2$ using a multilayered resist system. (a) Cross-section. (b) Exposure by drawing patterns into the ODS-SAM layer by current injection from an AFM probe. (c) First development step by HF etching to remove the $SiO_2$ just formed. (d) Second development step by TMAH etching to remove the a-Si in the exposed area. (e) Pattern transfer with HF to remove the thermal oxide in the exposed area. (f) Resist removal, i.e., removal of all the remaining a-Si in TMAH etch. (Based on Sugimura, H., O. Takai, and N. Nakagiri. 1999. Multilayer resist films applicable to nanopatterning of insulating substrates based on current-injecting scanning probe lithography. *J Vac Sci Technol* B17:1605–08.[87])

extremely small spot sizes are achievable so that the exposure dose may be confined to a beam diameter of less than 10 nm. The method enables wider exposure latitude than EBL, a fact demonstrated in Figure 2.41.[89] In this figure, line width is plotted versus dose for SAL601 resist using both EBL and SPL lithography systems. From the lower slope of the SPL curve, one deduces that SPL has higher dose latitude than EBL—in other words, SPL is less sensitive to dose variations. On the other hand, SPL is less sensitive and requires a higher dose to write the same feature size. The mechanism of electron bombardment in SPL is very different from that in EBL. In EBL, the mean free path of the bombarding electrons is long compared with the resist thickness. In contrast, low-energy electrons in SPL have a mean free path less than 2 nm and travel through the resist under the influence of an electrical field, undergoing a number of scattering events before reaching the resist/substrate interface.[89] The latter makes it clear why a thinner resist will result in a better resolution. The influence of secondary electrons (i.e., all electrons emanating from the resist/substrate) in proximal probe lithography was investigated, for example, by Völkel et al.[90]

Writing speeds of an SPL tool of 0.1–100 μm/s are typical, and several thousands of micrometers per second have been demonstrated.[91] However, very

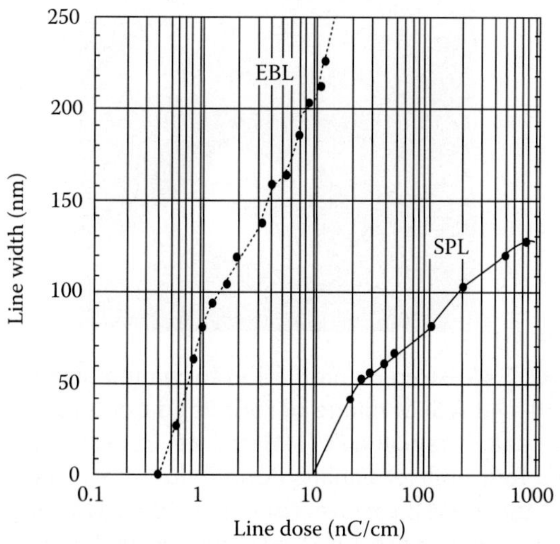


**FIGURE 2.41** Line width versus dose for SAL601 resist using both EBL and SPL lithography systems. (Wilder, K., C.F. Quate, B. Singh, and D.F. Kyser. 1998. Electron beam and scanning probe lithography: a comparison. *J Vac Sci Technol* B16:3864–73.[89])

fast scanning will remain limited because the close tip-to-substrate distance (typically 1 nm with STM) is prone to cause tip crashes.

With an ultrasharp tip (radius < 10 nm) in the emission mode rather than the tunneling mode, one may stay as far as 100 nm away from the surface. This approach not only enables faster writing (limited by the response of the piezoelectric material only) without fear of crashing the tip as frequently, but the high beam current also contributes to the possibility for exposure of large area patterns. Even at a distance of ~100 nm, the beam diameter remains small at ~30 nm.[55] Structures of lateral dimension of ~20 nm have been created in SAMs using ultrasharp field emitters.[92] The use of arrays of micromachined electron emission sources to improve throughput was covered earlier (see Figures 2.22 and 2.23) as a type of lower energy EBL with less lateral backscattering. In this context a useful array of field emitters for EBL is the so-called Spindt cathode array[93] shown in Figure 2.42. The use of such an array in a field emission display is shown in Chapter 7 (Figure 7.90a). The cathode tips in Figure 2.42 are molybdenum, and electrons are extracted from the tip at gate voltages as low as 20 V (70–100 V is more typical). In the newest embodiments of Spindt arrays, the molybdenum tip is replaced with carbon nanotubes. The use of nanotubes as field emitters was first proposed by Walt de Heer, André Chatelain, and Daniel Urgate in 1995.[94] Carbon nanotubes are excellent field emitters because of their chemical, structural, and electronic properties, which afford important aspects of robustness that have been lacking in the conventional metal and silicon field emitter arrays.

### *Material Transfer Methods: Atom Lithography or Mechanosynthesis and Dip Pen Nanolithography*

*Atom Lithography or Mechanosynthesis* As we just learned, proximal probe equipment locally modifies surfaces[95] and is used to expose resists and to oxidize and mechanically scribe surfaces. Also, a variety of direct atomic manipulations have been demonstrated with proximal probes.[95] The electric field strength in the vicinity of a probe tip is very strong and inhomogeneous (e.g., a field of 2 V Å$^{-1}$ concentrated around the probe tip). This field can manipulate atoms, including sliding of atoms over

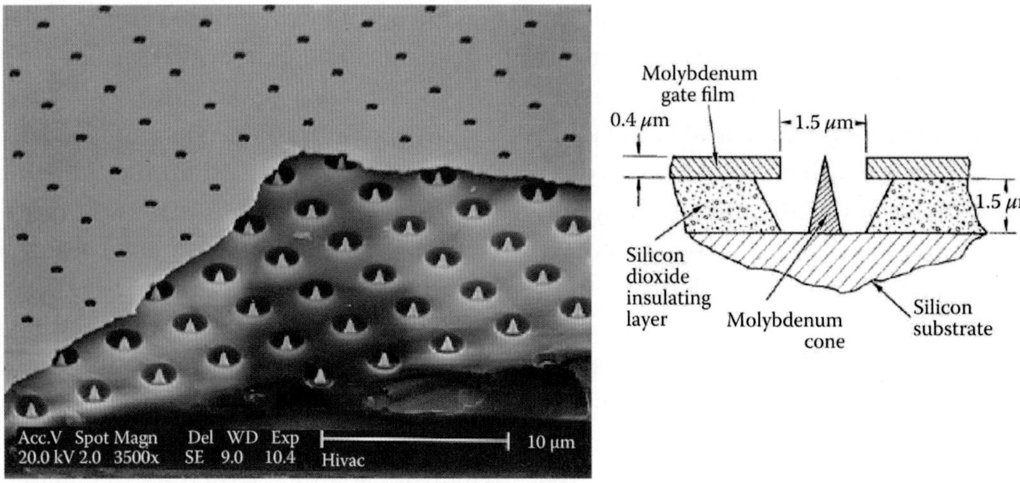


**FIGURE 2.42** Spindt field emitter array. For its fabrication see Chapter 7, and for its application in a field emission display see Volume III, Chapter 10.

surfaces and transferring atoms by pick (erase) and place (write). Drexel calls this type of machining mechanosynthesis.[96] Although both positive and negative voltages have been used for the deposition, a tip with a negative polarity is usually more stable with higher transfer probability. These atomic manipulation processes may be classified in parallel processes and perpendicular processes. In parallel processes, an adsorbed atom or molecule is induced to move along the surface (sliding); in perpendicular processes, the atom or molecule is transferred from the surface to the tip of a proximal probe or vice versa. Patterning surfaces at the atomic level is demonstrated in Figure 2.43a where a 60 × 48-Å image of four Pt adatoms is shown. The four atoms were herded into a linear array.[95] Figure 2.43b shows seven Pt atoms compacted together. The theoretical resolution of a lithography technique based on these atomic probes is a single atom. Even with "only" a 100-Å resolution, a single memory bit could be stored in an area that measures 100 Å on a side. This will enable bit storage of $10^{12}$ bits/cm$^2$ as compared with $10^9$ bits/cm$^2$ with conventional technology.

In Figure 2.44 we show linear Au chains 2–20 atoms long, written by Ho et al.[97] from the University of California at Irvine on a NiAl (110) surface via the manipulation of single atoms with an STM. Differential conductance (dI/dV) images of these gold chains reveal 1D electronic density oscillations at energies 1.0–2.5 eV above the Fermi energy. Thus, Ho et al. were able to make a direct correlation between the geometric structure and the electronic properties of the Au chains, where the former is precisely known and the latter is directly measured. The same team, again using STM, also connected Au chains to a copper phthalocyanine (CuPc) molecule.[98] A typical artificial molecule assembly is CuPc@2$Au_3$, consisting of a CuPc molecule and two $Au_3$ (three Au atoms in each chain) chains. When the separation between the Au chains is six lattice constants [of the underlying NiAl (100) substrate]

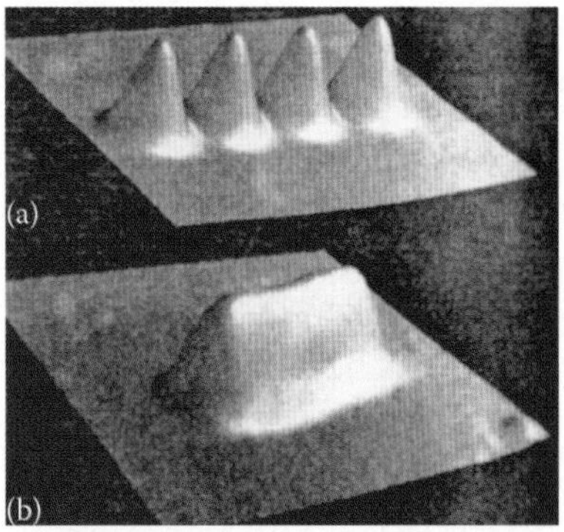


**FIGURE 2.43** Atom manipulation for nanomatching. (a) A 60 × 48 Å STM image of four Pt adatoms assembled into a linear array on a Pt (111) surface. Pt atoms were herded four unit cells apart along a close-packed direction of the Pt (111) surface. (b) A 40 × 40 Å STM image of a compact array of seven Pt adatoms. (From Stroscio, J.A., and D.M. Eigler. 1991. Atomic and molecular manipulation with the scanning tunneling microscope. *Science* 254:1319–26. With permission.[95]) For more pictures of STM atom manipulation, visit http://www. almaden.ibm. com:80/vis/stm/gallery.html.

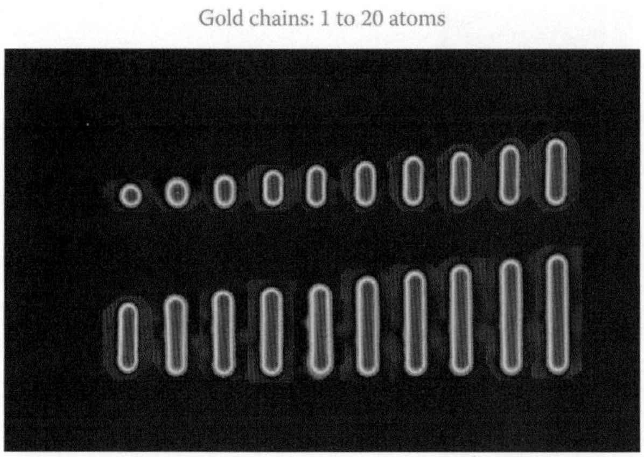


**FIGURE 2.44** STM topographic images from a single Au atom to a linear chain of 20 atoms, arranged in single-atom increments from left to right. The images are cut from 20 separate scans, each taken with a sample bias voltage between 2.0 and 2.5 V and a tunneling current between 1.0 and 1.5 nA. Each chain has an apparent height between 2.4 and 2.7 Å. The chains between 12 and 20 atoms long were constructed and imaged with a tip different than the tip used to construct and image the other chains. (Wallis, T.M., N. Nilius, and W. Ho. Electronic density oscillations in gold atomic chains assembled atom by atom. *Phys Rev Lett* 89:236802-1–2-4.[97])

or more, a Pc molecule is unable to bridge the gap between the chains. With five lattice constants, the gap is small enough for a Pc molecule to fill it, and the molecule adsorbs symmetrically with two opposite benzene rings attached to the Au chains (see the schematic in Figure 2.45j).

This work by Ho's team is contributing to the detailed understanding of the electron transport through a molecule that bridges the gap between two metal electrodes. Electron transport through single-molecule junctions is determined by the local electronic structure of the nanoscale region that includes the molecule and a number of metal atoms in proximity to the molecule-metal contacts. As we will detail in Volume III, Chapter 3, a detailed understanding of the contact between metal-contacting electrodes and molecules is essential for the understanding of molecular electronics.

One major drawback to bear in mind with atom placing or removing techniques for micromachining is the time involved in generating even the simplest of features. If one wanted to deposit a metal line 10 μm long, 1 μm wide, and 0.5 μm high, $10^{16}$ atoms would need to be manipulated. Even at a deposition rate of $10^9$ atoms/s, this would take more than 100 days. In the late 1980s, it took an IBM team 1 week to spell out the IBM logo with an STM, putting Xe atoms in place on a Ni surface. Only use of massive parallelism can help this cause.

*Dip Pen Lithography* AFMs suffer from an annoying problem: water from the air tends to attach itself to the tip, and the greater the relative humidity, the more water that collects. This water meniscus interferes with high-resolution AFM imaging. In 1999, Chad Mirkin, a Northwestern University researcher, realized that this meniscus could be turned into an asset. Because water is always moving from the tip to the substrate surface or vice versa, he decided to use it to float ink molecules onto the substrate. The principle of dip pen nanolithography (DPN) is shown in Figure 2.46.[99] DPN is named after the 4000-year-old dip pen, with an AFM tip for a sharp point.

A reservoir of "ink" is stored on the cantilever holding the scanning probe tip, which is manipulated across the surface, leaving lines and patterns behind. Lines as thin as 15 nm have been drawn. The attainable resolution depends strongly on the substrate roughness, the writing speed, and the relative humidity. Smooth surfaces, fast scan rates (and thus shorter pen/paper contact), and lower relative humidity (narrower water meniscus) all make for higher-resolution DPN. Thus far, most of the work has been carried out with a sulfur-containing compound called 1-octadecanethiol on gold surfaces. However, many inks and many different types of surfaces can be written on, making this an interesting nanolithography option. A review of different ink/substrate combinations in DPN was listed by Ginger et al.[100] An additional attractive feature of DPN lies in the fact that the same AFM can be used to write and image the results. The downside is that DPN is slow, but to offset this negative aspect, arrays of dip pens, similar to the millipede chip developed by IBM (see Figure 2.4b), can parallelize and speed up the technique: the newest array from Chad Mirkin's group has 55,000 pens.[101] Much needed now is the development of an automated ink-coating system for inking individual tips with its own ink; perhaps such a system

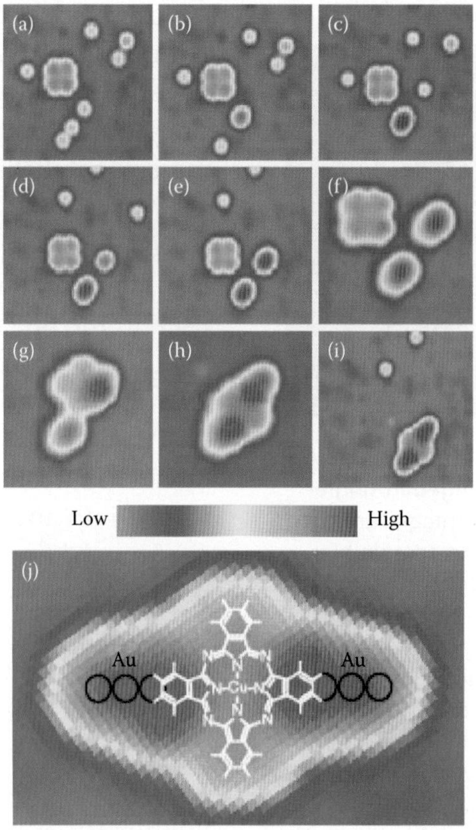


**FIGURE 2.45** Assembly sequence and schematic for a structure consisting of a CuPc molecule and two $Au_3$ chains ($2Au_3$). All images were taken with $V_{bias} = 1$ V, I = 1 nA. (a) CuPc molecule (appears as a protruding four-lobe structure) on the NiAl surface together with seven Au atoms, which appear as round protrusions. Four of the atoms have been aligned along the same Ni trough. Image size is 74 Å by 74 Å. (b) One of the atoms has been manipulated along the Ni trough to create a $Au_2$ chain. (c) First $Au_3$ is created; another atom is positioned five Ni-Ni lattice sites (5 × 2.89 Å) away from $Au_3$, marking the starting position for another $Au_3$ chain. (d) A second $Au_2$ chain is created. (e) Assembly of the second $Au_3$ chain and the $2Au_3$ junction is completed. (f) Zoom-in image of the $2Au_3$ and CuPc molecule; the area of image is 47 Å × 47 Å. The image was taken with a tip modified by a CuPc molecule adsorbed on it. (g) The CuPc molecule has been moved into the junction between the two $Au_3$ chains. As is clear from the image, the molecule is adsorbed only on one of the chains. (h) The position of the molecule was adjusted by manipulating the leftmost lobe of the molecule in (g) toward the lower $Au_3$ chain. The molecule is adsorbed symmetrically between the two chains, forming CuPc@$2Au_3$. (The tip still has a CuPc molecule adsorbed on the apex.) (i) Image of same structure taken after removing the molecule adsorbed on the tip. The area corresponds to that of (e). (j) Schematic showing the relation between the internal molecular structure, adsorption geometry, and STM image. Black circles represent the Au atoms comprising the two $Au_3$ chains. (From Nazin, G.V., X.H. Qiu, and W. Ho. 2003. Visualization and spectroscopy of a metal-molecule-metal bridge. *Science* 302:77–81.[98])

can be achieved through the integration of microfluidic technology. Besides its use in nanolithography one can envision the use of DPN for placing of molecules on a surface precisely where one wants them as in protein and DNA arrays and to build 3D nanostructures. Mirkin et al.[102] have shown that through the use of DPN, protein nanoarrays can be constructed with high fidelity of specific protein binding to proper sites on a Si substrate. NanoInk (http://www.nanoink.net) is a start-up company commercializing the DPN technology.

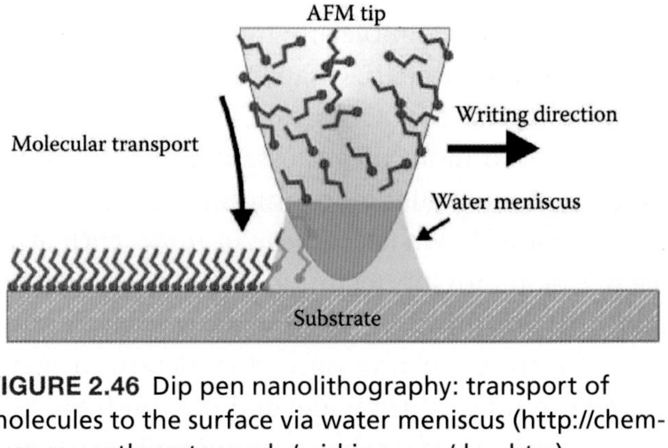


**FIGURE 2.46** Dip pen nanolithography: transport of molecules to the surface via water meniscus (http://chem-groups.northwestern.edu/mirkingroup/dpn.htm).

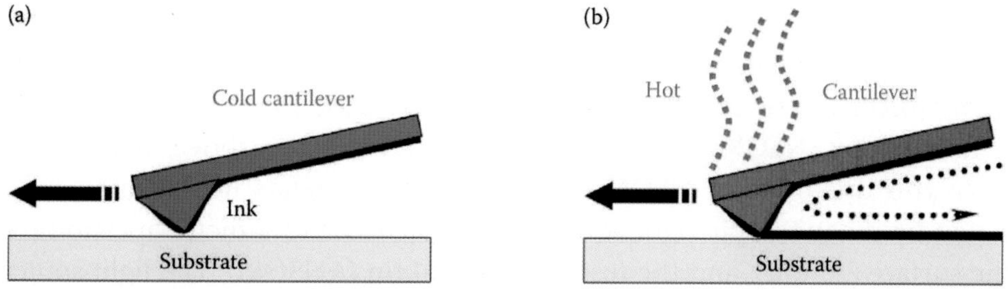


**FIGURE 2.47** Diagram illustrating thermal dip pen nanolithography. When the cantilever is cold (a), no ink is deposited. When the cantilever is heated (b), the ink melts and is deposited onto the surface. (Image courtesy of Naval Research Laboratory, http://gtresearchnews.gatech.edu/newsrelease/tdpn.htm.)

In Figure 2.47, we illustrate an extension of DPN, a writing method dubbed "thermal dip pen nanolithography." In conventional DPN, the probe tip is coated with a liquid ink, which then flows onto the surface to make patterns wherever the tip makes contact. This approach has been hampered by an inability to turn the ink flow on and off. A thermal-DPN (tDPN) method developed by Georgia Tech's William King and NRL's Lloyd Whitman solves this specific problem by using easily melted solid inks and special AFM probes with built-in heaters that allow writing to be turned on and off at will.[103,104] Moreover, conventional DPN cannot be used in a vacuum because liquid inks evaporate, whereas the solid materials used in the thermal process bond to surfaces, allowing them to be used in vacuum environments that are part of conventional semiconductor manufacturing. The thermal materials also provide sharper features because they do not spread out like liquid inks.

In Figure 2.49 a topographic image of a surface scanned with a heated AFM cantilever tip for 256 sec in each of four 500-nm squares is shown. The cantilever temperature is shown for each of the four scans. No writing is observed from the two low-temperature scans. The scan at 98°C resulted in light deposition. Robust deposition occurred during the final scan when the cantilever temperature was increased to 122°C.

Combining thousands of individually controlled AFM pens into arrays could allow writing of faster and more complex semiconductor patterns. One of the first inks used in thermal writing was octadecylphosphonic acid, which melts at about 100°C. But the tDPN technique has been used also to apply other materials, including solders and polymers.

*Near-Field Optical Scanning Microscope and Apertureless Near-Field Scanning Optical Microscopy*

*Introduction: Breaking the Diffraction Limit* To obtain a resolution better than Abbe's optical microscopy limit ($\Delta x \sim \frac{\lambda}{2NA} \sim 200$ nm, based on Volume I, Equation 5.1), complicated and costly electron microscopy (with resolution of a few nanometers) or STMs (with atomic resolution) are required. Unfortunately, these techniques sacrifice many of the advantages associated with traditional optical microscopes (including nondestructiveness, low cost, high speed, reliability, versatility, accessibility, ease of use, informative contrast, spectroscopy, and

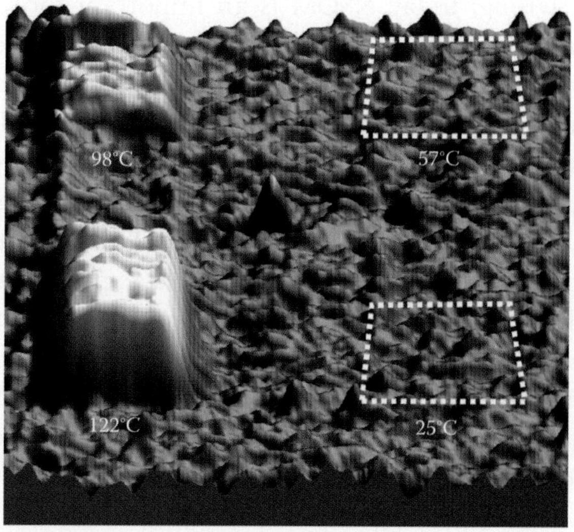


**FIGURE 2.48** A topographic image of a surface scanned with a heated AFM cantilever tip for 256 s in each of four 500-nm squares. The cantilever temperature is shown for each of the four scans. No deposited material is observed from the two low-temperature scans. The scan at 98°C resulted in light deposition. Robust deposition occurred during the final scan when the cantilever temperature was 122°C. (Courtesy of Naval Research Laboratory.)

real time). As seen in Volume I, Chapter 5 on photonics, combining a scanning proximal probe technique with optical microscopy in so-called scanning near-field optical microscopy, NSOM provides an attractive solution to this dilemma. In NSOM, the sample is illuminated by a nanoscopic light source located close to the surface (10 nm), and the resolution is dictated by the source diameter *a*, or $\Delta x \sim a \leq$ 100 nm. This is achieved by using nanoscale apertures in NSOM or by using apertureless techniques (e.g., a scattering metal tip) in ANSOM or scattering scanning near-field optical microscopy. A typical NSOM probe is a commercial single-mode optical fiber, tapered and coated by a thin CrAl film with an aperture of ~100 nm. In ANSOM, light interaction over nanoscale dimensions is enhanced with the use of scattering nanoscale tips, nanospheres, and so on. The conceptual idea of near-field optics and near-field microscopy came from the seminal proposal of E. H. Synge. It was Synge (in 1928) who proposed that if source or detector is scanned close to a surface, resolution is dependent on probe size and probe-sample distance instead of the Abbe limit.[105] Synge corresponded about his idea with Einstein, who accepted the concept but was doubtful that it could be reduced to practice. But Einstein was wrong; today NSOM is an important tool for nanotechnologists. NSOM constitutes a form of lensless optics with subwavelength resolution, independent of the wavelength of the light being used! The light coming from an imaged object in the near field contains a considerable amount of nonpropagating, evanescent waves containing high lateral spatial resolution, information that decays exponentially within a distance comparable to a wavelength. It is this evanescent wave information, the information-rich light, we capture in NSOM and ANSOM, thus enhancing the resolution. Several NSOM implementations using apertures and SNSOM were illustrated in Volume I, Chapter 5.

Aperture-based probes are the best developed and come with a typical resolution of 50–100 nm. Tip fabrication involves heating and pulling and chemical etching of glass optical fibers. Apertureless probes enhance local scattering and feature a 1–20-nm resolution. NSOM and SNSOM microscopy are reviewed in Volume III, Chapter 6.

*NSOM and ANSOM Lithography* Besides their utility for imaging applications, NSOM and ANSOM may also be used for direct write optical nanolithography.[106] From the above, near-field scanning optical microscopy refers to the interaction of light with a sample close to a metal aperture (NSOM) or a solid metal tip (ANSOM). The light source is held in place in a manner similar to those used for other scanning proximal probe microscopes. To write patterns into a photoresist, one uses aperture NSOM in illumination mode or ANSOM in plasmon mode. The first NSOM lithography results obtained in several labs dealt with nanometric tips on standard positive photoresist and on self-assembled monolayers of alkanethiols and using laser sources visible up to the UV range. The NSOM approach to improved spatial resolution in lithography unfortunately is limited because of the very small apertures, through which only a very small fraction of the light can be transmitted. ANSOM overcomes this problem by using sharp metallic tips instead of fiber apertures to achieve nanometer-scale resolution. In ANSOM, light is polarized along the sharp axis of the metallic tip, and this induces a high concentration of surface plasmons that results in strong enhancement of the electromagnetic field in the local vicinity of the tip—typically more than a few tens of nanometers (see plasmonic printing with metal nanoparticles discussed above)! Yin et al.[107] demonstrated near-field two-photon optical lithography by using 120-fs laser pulses at 790 nm in an apertureless near-field optical microscope and produced lithographic features, even without optimization, of 70 nm in a commercial SU-8 resist. Thus, combining ANSOM with two-photon absorption can provide a spatial resolution as high as $1/10\ \lambda$, nearly a factor of two better than the resolution achieved in previous far-field two-photon lithography experiments.

Limitations of NSOM and ANSOM lithography include very low working distances, very shallow depth of field, and very long scan times for patterning large surfaces with a high-resolution image.

## Holographic Lithography

### Holography Background

As this section concerns holographic lithography, a short reminder about what holography is, is in order. Holography comes from the Greek words for *whole* (holo) and *graphy* (or graphic, image), standing for "the entire image." The Hungarian physicist Dennis Gabor invented holography in 1948 and received the 1971 Nobel Prize for Physics for it. However, the field of holography only took off after 1962 with the invention of the laser, providing the coherent light needed to make a hologram. In holography two laser beams are used, a reference and an object beam, to snap pictures of an object. If these two beams are coherent, optical interference between the reference beam and the object beam, as a result of the superposition of the light waves, produces a series of fringes on standard photographic film. These fringes form a type of diffraction grating on the film, which is called a hologram. To reconstruct the image in the imaging medium one can go about it two different ways. First, on using the same reference light that was used to record the hologram, one obtains a virtual image of the object. Second, if one uses the phase conjugate light of the reference light to shine through the recording medium, the image of the object is reproduced.

### *Holographic Lithography or Interference Lithography*

As mentioned above, a hologram is a complex interference pattern that can recreate a 3D image. In Figure 2.49 we show how a generic hologram is recorded. The "partially silvered" mirror in this figure reflects and transmits fractions of the incident light. This is useful because the reflected and transmitted beams are coherent (i.e., have a definite common phase) even when the incident light has a time-varying random phase. From Volume I, Chapter 5 we indeed recall that reflection from a surface produces a "phase-flip" of 180° in the reflected light. This is true for a 180° reflection. However, when light is reflected at an angle other than 180°, the phase change in the reflected light is the same as the reflection angle. Thus, light reflected at 90° has a 90° change in phase. In holographic lithography the specific object is the mask.

In holographic or interference lithography (Figure 2.50), a holographically constructed photomask replaces the standard photomask.[108,109] In the holographic recording or construction phase of the hologram (Figure 2.49), one uses the interference of two mutually coherent beams. A well-collimated flood laser (object beam) passes through the photomask and is diffracted by the mask features. This signal beam contains the amplitude and phase information of the photomask. When this diffracted object beam passes through the holographic recording layer, it interacts with the reference beam to create the interference pattern. The reference beam converts the amplitude and phase information into intensity information, which is stored in the photosensitive holographic medium (Figure 2.50a). The light-sensitive recording layer stores the holographic image data as variations in the refractive index of the photopolymer. Pattern printing or image reconstruction is accomplished by scanning a collimated laser illumination beam—the phase conjugate of the reference beam—to create the hologram. By interaction with the recorded hologram, the latter beam generates an image of the original photomask at precisely its original position in space (Figure 2.50b). When a

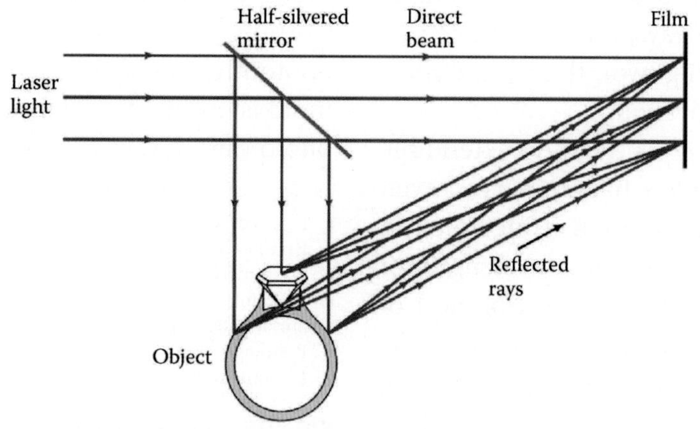


**FIGURE 2.49** Recording a hologram. (From John G. Cramer.)

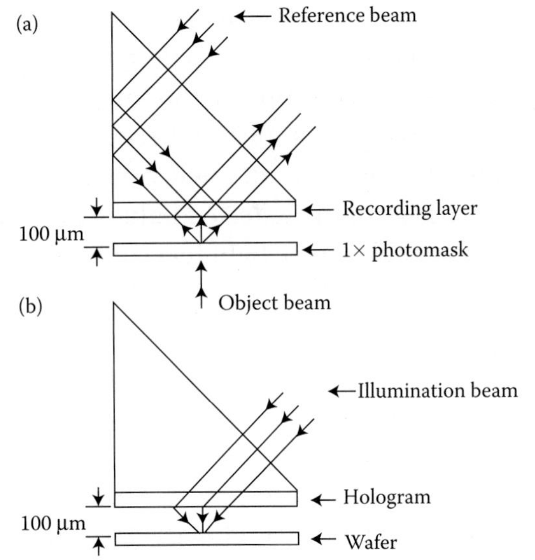


**FIGURE 2.50** Basic arrangement for holographic lithography. (a) The photomask pattern forms a hologram in the polymer recording layer. (b) Using the illumination beam, the high-resolution holographic mask image is reconstructed into a printable masking layer in the resist-coated substrate. (From Brook, J., and R. Dandliker. 1989. Submicrometer holographic photolithography. *Solid State Technol* 32:91–94; and Omar, B., S. Clube, F. Hamidi, M.D. Struchen, and S. Gray. 1991. Advances in holographic lithography. *Solid State Technol* September:89–94.[108,109])

photoresist-coated substrate resides in this plane, a copy of the original mask can be printed. The image is diffraction limited, but because the holographic mask and the wafer can lie very close, a high NA and thus a very good resolution are possible (0.3 μm has been reported). Only the size of the photomask itself restricts the image field size. The holographic image of the original mask exactly overlays the wafer surface with a full field (large area recording).

Holographic lithography may also use the interference pattern of multiple coherent lasers. Simple grating patterns, 2D dot arrays, and 3D lattice patterns can be fabricated using two-, three-, and four-beam interference, respectively.

Holographic lithography is readily extendible to electrons. Spacings of a few nanometers to even less than 1 nm have been reported using electron holograms.[110] This is because the wavelength of an electron is always shorter than for a photon of the same energy. Atom lithography is also possible: the momentum of an atom is even larger than for electrons or photons, allowing even smaller wavelengths, per the de Broglie relation.

A benefit of using interference lithography is the quick generation of dense features over a wide area without loss of focus. A major drawback of interference lithography is that it is, in general, limited to patterning arrayed features. Recently, holographic lithography has emerged as a very promising technique for the inexpensive fabrication of high-quality 3D photonic crystals. Together with direct laser writing approaches, holographic lithography offers the attractive possibility of controlled incorporation of functional elements using direct laser writing in a second step.

## Soft Lithography

### *Introduction*

Soft lithography refers to a series of methods that use a patterned elastomer as a stamp, mold, or mask to generate micropatterns and microstructures instead of using a rigid photomask. Soft lithography has become an important tool in the nanofabrication arsenal.[111] Today the name is a collective term for a set of new techniques: microcontact printing (μ-CP), microtransfer molding (μ-TM), micromolding in capillaries (MIMIC), and microreplica molding.[112] The method is a very exciting research tool and may offer advantages over conventional methods for patterning of nonplanar substrates, unusual materials, and large areas.

In soft lithography, a master mold is first made by lithographic techniques, and an elastomeric stamp [e.g., polydimethylsiloxane (PDMS); see Figure 2.51] is cast from this master mold. A simple example procedure for making a PDMS (Sylgard 184, Dow Corning, http://www.dowcorning.com) stamp from a photolithographically patterned resist layer as master mold is outlined in Figure 2.52. A thin layer of SU-8 photoresist [SU-8 (50) from MicroChem, Newton,

- Low cost
- Easy fabrication
- Elastomeric/soft
- Moldable
- Surface chemistry
- Bonding
- Biocompatible

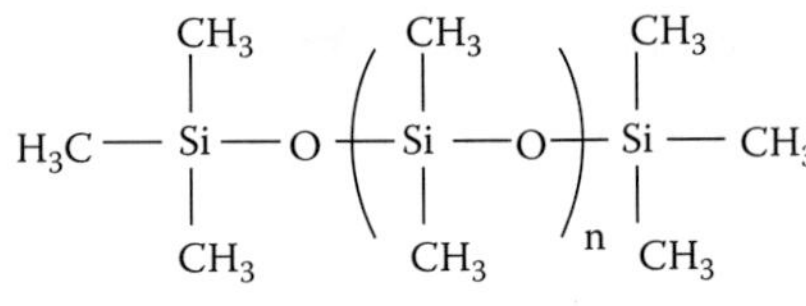

**FIGURE 2.51** Polydimethylsiloxane (PDMS). (Sylgard 184, Dow Corning, http://www.dowcorning.com.)

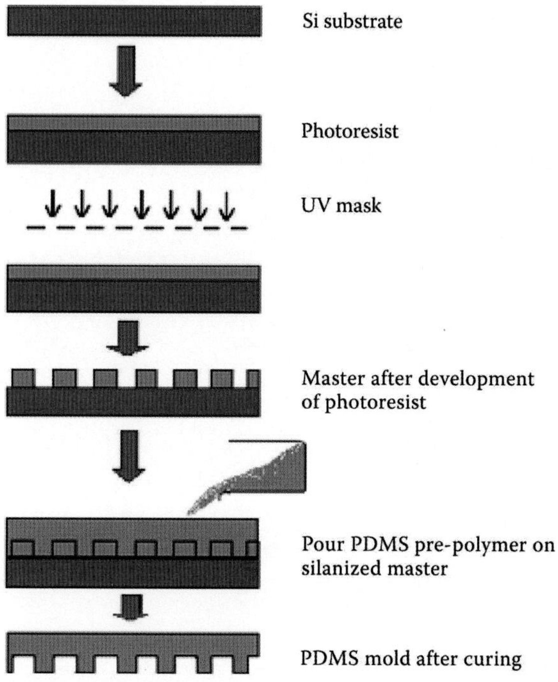


**FIGURE 2.52** Making a PDMS mold. (Based on Xia, Y., and G.M. Whitesides. 1998. Soft lithography. *Angew Chem Int Ed* 37:550–75; and Xia, Y., and G.M. Whitesides. 1998. Soft lithography. *Ann Rev Mater Sci* 28:153–84.[113,114])

MA] is coated on a Si wafer. The resist is patterned by UV lithography. To make things really fast (<24 h) and inexpensive (<\$20), the contact mask for the UV lithography may be a transparency on which—using drawing software such as Freehand—a design is printed with a high-resolution printer (>3300 dpi). After development, the photoresist is treated with (tridecafluoro-1,1,2,2-tetrahydrooctyl)-1-trichlorosilane (Hüls Chemicals, http://www.degussa.com) vapor to facilitate PDMS removal once cured. A 10:1 ratio of a PDMS mix is cast on the photoresist film and cured for 1 hour at 60°C in an oven.[111]

PDMS has many properties that make it suitable as a stamp material. It 1) provides a surface that has a low interfacial free energy (~21.6 dyn/cm), 2) is chemically inert, 3) is nonhygroscopic (does not swell with humidity), 4) passes gas easily (don't start laughing now), 5) has good thermal stability (~186°C in air), 6) is optically transparent down to ~300 nm, 7) is isotropic and homogeneous, 8) is durable (stamps may be used > 50 times over several months without degradation), and 9) has interfacial properties that are easy to modify.[114]

Many PDMS stamps can be generated, and each one may be used many times. The rubber stamp can be used in a number of different ways. Some different soft lithography processes are outlined below. For further reading on soft lithography, consult Xia et al.[113,114] and visit the soft lithography homepage at University of Washington/Washington Technology Center (http://www.ee.washington.edu/research/microtech/cam/CAMsoftlithhome.html). Another excellent site to follow the experimental steps involved in soft lithography is on JoVE (http://www.jove.com), the "YouTube" for experimental procedures.

### *Microcontact Printing*

In microcontact printing (μ-CP), the PDMS rubber stamp is coated with an ink of the molecules (e.g., alkylthiols or a protein) that one wants to print in selected patterns on a solid substrate. During stamping, only the raised parts of the stamp collect the "ink." The inking of the substrate consists of self-assembled monolayers (SAMs) on the solid surface formed by covalent chemical reaction. The inked areas are self-passivating and exhibit very low interfacial tension that repels additional molecular layers; therefore, SAM forms only in areas of conformal contact between polymer and substrate. STM studies of the transferred SAMs reveal an achievable monolayer order indistinguishable from that found for SAMs prepared from solution.[53] The SAM pattern acts as a highly localized and efficient barrier to some wet etches. This lithographic technique—once the master is made—is not subject to diffraction or DOF limitations. The deformability of the elastomeric stamp allows it to accommodate rough surfaces. The method even works on rounded substrates (such as optical fibers or lenses) with radii of curvature of less than 10 μm. PDMS-based elastomers do not adhere to novolak or PMMA-based polymers, allowing convenient replication of masters formed by EBL.[53] The technique has been used, for example, to build an antibody grating on a Si wafer by inking the rubber stamp with an antibody solution.[115] The antibody grating alone produces insignificant diffraction, but on immunocapture of cells the optical phase change produces diffraction.

In Figure 2.53 we illustrate a generalized scheme for microcontact printing of proteins.

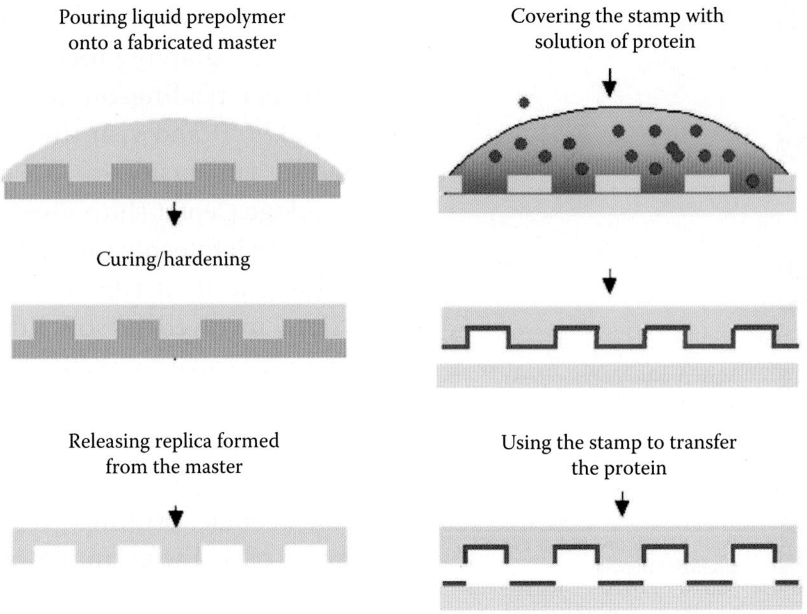


**FIGURE 2.53** Microcontact printing of protein.

### *Microtransfer Molding*

In microtransfer molding (μ-TM), illustrated in Figure 2.54, the rubber mold is filled with a polymer precursor, and the rubber stamp is pushed up against a substrate (e.g., a Si wafer or another flat sheet of PDMS).[116] The polymer in the rubber stamp relief is cured and transferred to the substrate, and the stamp is peeled off. Schueller et al.[111] used this technique to make curved glassy carbon microgrids. In this case, a photoresist was used as polymer precursor and cured while sandwiched between a patterned PDMS rubber stamp and a planar PDMS sheet. To obtain the curved carbon grid, the PDMS sandwich was deformed against a curved cylinder during curing. After removing the two thin strips of PDMS, the free-standing and curved photoresist grid was pyrolized (carbonized, see Volume III, Chapter 5) at high temperature in an argon atmosphere (Figure 2.55).

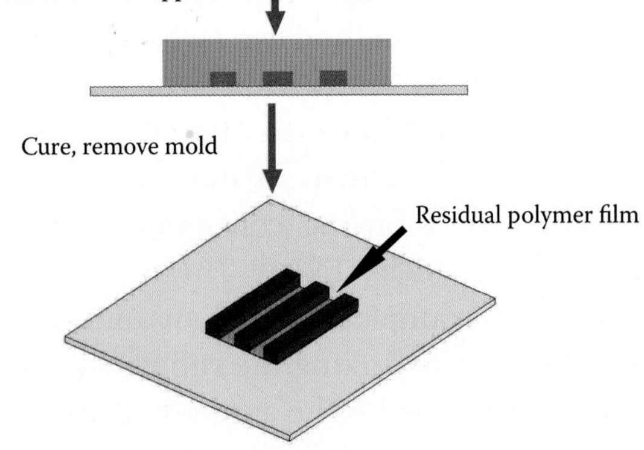

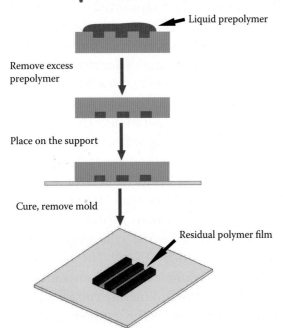


**FIGURE 2.54** Schematic description of microtransfer molding (μ-TM). (Based on Xia, Y., and G.M. Whitesides. 1998. Soft lithography. *Angew Chem Int Ed* 37:550–75; and Xia, Y., and G.M. Whitesides. 1998. Soft lithography. *Ann Rev Mater Sci* 28:153–84.[113,114]) PU is prepolymer (polyurethane).

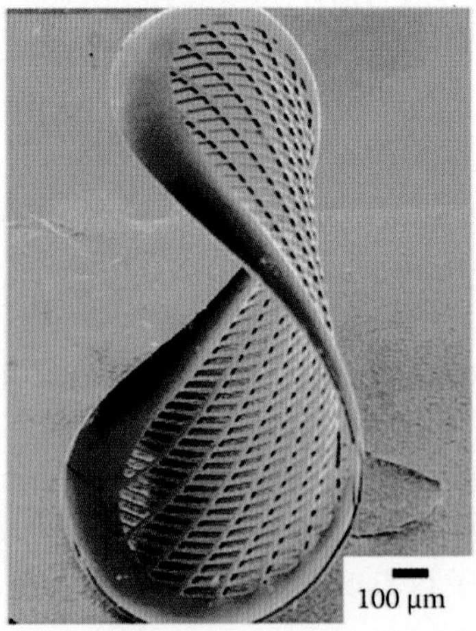


**FIGURE 2.55** Curved glassy carbon structure. (Schueller, O.J.A., S.T. Brittain, and G.M. Whitesides. 1999. Fabrication of glassy carbon microstructures by soft lithography. *Sensors Actuators A* A72:125–39.[111] Courtesy of Dr. G.M. Whitesides, Harvard University.)

### Micromolding in Capillaries

In micromolding in capillaries (MIMIC), illustrated in Figure 2.56, the rubber stamp is pushed up against a substrate, and liquid is applied to access holes in the mold. The liquid wicks into the cavities formed by the rubber mold against the substrate. Sometimes this process is pressure, electro-osmosis, or vacuum assisted.[117] The polymer is cured, and the stamp is removed. MIMIC has been used to fabricate polymeric transistors.[118]

### Microreplica Molding

In microreplica molding, the master mold is replicated in PDMS by casting and curing the PDMS prepolymer (polyurethane, PU); this negative replica is then oxidized in an oxygen plasma for 1 minute and exposed to fluorinated silane for 2 hours to provide a surface with low adhesion to PDMS. PDMS is then cast against this negative replica, cured, and peeled away to reveal a positive replica of the original master.[109]

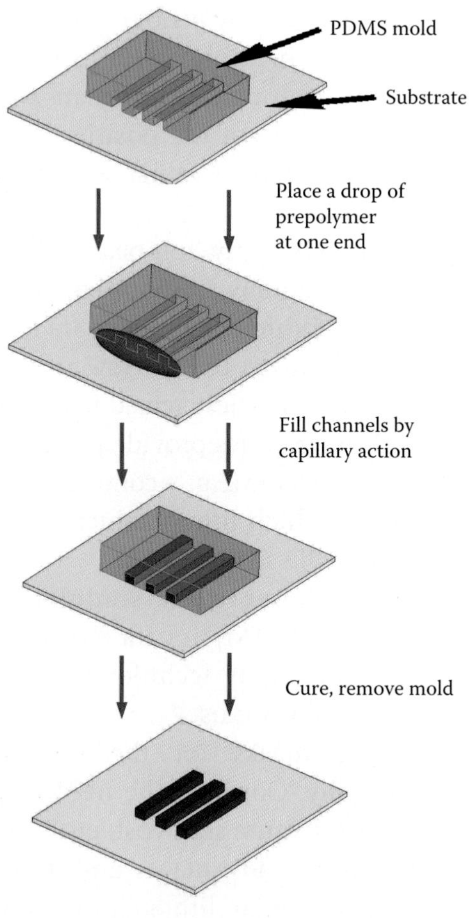


**FIGURE 2.56** Schematic description of micromolding in capillaries (MIMIC). (Based on Xia, Y., and G.M. Whitesides. 1998. Soft lithography. *Angew Chem Int Ed* 37:550–75; and Xia, Y., and G.M. Whitesides. 1998. Soft lithography. *Ann Rev Mater Sci* 28:153–84.[113,114])

### Soft Lithography Summary

Soft lithography is capable of generating structures as small as 30 nm and is especially attractive as a potential method for fabricating devices on nonplanar substrates. The major advantage of soft lithography is its short turnaround time. It is possible to go from design to production of replicated structures in less than 24 hours. The method is low cost, and, unlike photolithography, soft lithography is applicable to almost all polymers and thus to many materials (e.g., carbon, glasses) that can be prepared from polymeric precursors.

Some of the disadvantages of PDMS-based soft lithography include 1) shrinkage during curing (~1%), 2) swelling by nonpolar solvents such as toluene and hexane, 3) thermal expansion, 4) level of defects in printing SAMs, 5) softness of the material limits the achievable aspect ratio through sagging, and 6) deformation of the soft elastomeric stamps. These factors limit the accuracy in registration across a large area and may limit the practical utility in nanofabrication and multilayer devices.[114] Because soft materials are used, deformation of the stamp/mold, low reproducibility (because of distortion), and defects (yield) are problems that prevent this technology from being a viable manufacturing technique. The technique constitutes a very handy research tool but most likely will not reach the commercial manufacturing stage.

Soft lithography technology inspired new approaches to fabricate nanodevices that do not have commercial potential, i.e., NIL and SFIL (see above). For example, Krauss et al.[115] used NIL to fabricate a 400-Gb/in.$^2$ storage device containing sub-10-nm minimum features. As we saw earlier, in NIL the master mold may be an electron beam and RIE-patterned $SiO_2$ layer (e.g., 10-nm-wide features with 40-nm period and 75-nm tall). This mold is then imprinted into a 90-nm-thick PMMA film on a Si disk. During the hot embossing, the mold and resist-coated disk are heated to 175°C and pressed together at 4.4 MPa for 10 minutes. After cooling to room temperature, the mold is separated from the disk, resulting in replication of the nano-CD pattern in the PMMA film. Surface Logix, Inc. is the first commercial entity based entirely on the exploitation of soft lithography.

## Questions

*Most of the questions by Mr. Chuan Zhang, UC Irvine.*

2.1: Extreme ultraviolet lithography (EUVL) is one of the most promising future lithography techniques. Explain why EUVL increases the resolution dramatically. The high cost is the major obstacle for the commercialization of this technology; list the major reasons for the high cost of EUVL.

2.2: Compare the differences between traditional projection photolithography and extreme ultraviolet lithography. Why do reflective optics have to be used in an EUVL system?

2.3: Why does x-ray lithography have a high depth of focus? LIGA is less promising as a future MEMS technique due to the requirement of a synchrotron source. List the competing technologies and explain why they might replace LIGA.

2.4: Is the resolution of e-beam lithography given by the spot size of the focused beam? Explain why or why not. What are proximity effects, and how can one minimize their impact on the resolution?

2.5: What are the differences between thermionic emission (also called Schottky emission or SE) and cold field emission (CFE)? What are the advantages of low-energy electron beam lithography?

2.6: What is the difference between direct write e-beam lithography and electron projection lithography? Describe the working principle of SCALPEL (scattering with angular limitation projection electron beam lithography).

2.7: Describe the major procedures in nanoimprint lithography.

2.8: How can a diblock copolymer be used in lithography? List some of the merits of block copolymer lithography.

2.9: How does a Fresnel zone plate work? What are the major merits of Fresnel zone plates over diffractive optics lenses? Can they be applied in lithography?

2.10: What is the definition of scanning proximal probe lithography? Which technologies can be counted as scanning proximal probe lithography?

2.11: How does near-field optics break Abbe's diffraction limit? What is the difference between NSOM and ANSOM lithography?

2.12: What are the major steps in soft lithography? Explain the reasons why they have not been widely used in industry.

2.13: How would you make a conical shaped PMMA structure that is 150 μm tall and features a top angle of 45°?

2.14: Design a miniaturized device incorporating both a lift-off process and a self-aligned mask step in its manufacture.

2.15: Why can only proximity masking be used in the case of x-ray lithography? What about projection printing with x-rays? Sketch the process for fabricating an x-ray mask. What are some of the positive attributes of x-ray lithography? What are the negative attributes?

2.16: Compare UV, x-ray, ion-beam, and electron-beam lithography. Summarize in a comparison Table. Which techniques are used mostly in the IC industry today? How are the photons or charged particles created in each case?

2.17: Most lithographies provide projected shapes. A truly 3D shape (e.g., a contact lens) is harder to make. Which lithographies are capable of providing 3D shapes?

2.18: How could one make scanning tunneling microscopy (STM) into a mass production ready lithography technique? What type of resists would you use?

2.19: Compare masks for the next-generation lithography. Consider the need for vacuum, materials, feature size, cost, etc.

2.20: What are the advantages and disadvantages of using e-beam lithography compared to typical photolithography?

2.21: What difficulties do you expect in trying to implement immersion lithography commercially?

## References


1. Gibbs, W. W. 1997. Gordon E. Moore: Parts 1–3: The birth of the microprocessor. *Scientific American* (September 22, 1997); see also Moore, G. E. 1999. A pioneer looks back at semiconductors. *IEEE Design & Test of Computers* 16:8–14.
2. Semiconductor Industry Association. 1997. *The National technology roadmap for semiconductors: technology needs.* http://www.rennes.supelec.fr/ren/perso/gtourneu/enseignement/roadmap97.pdf
3. Singh, R. N., A. E. Rosenbluth, G. L.-T. Chiu, and J. S. Wilczynski. 1997. High-numerical aperture optical designs. *IBM J Res Dev* 41:39–48.
4. Kurzweil, R. 1999. *The age of the spiritual machine.* New York: Penguin Group.
5. Romankiw, L. T. 1993. Think small: one day it may be worth a billion. *Interface* Summer:17–57.
6. King, W. P., T. W. Kenny, K. E. Goodson, G. L. W. Cross, M. Despont, U. Düürig, H. Rothuizen, G. K. Binnig, and P. Vettiger. 2002. Design of atomic force microscope cantilevers for combined thermomechanical writing and ready in array operation. *Journal of Microelectromechanical Systems* 11: 765–74.
7. Terris, B. D., H. J. Mamin, M. E. Best, J. A. Logan, D. Rugar, and S. A. Rishton. 1996. Nanoscale replication for scanning probe data storage. *Appl Phys Lett* 69:4262–64.
8. Kraus, P. R., and S. Y. Chou. 1997. Nano-compact disks with 400 Gbit/in$^2$ storage density fabricated using nanoimprint lithography and read with proximal probes. *Appl Phys Lett* 71:3174–76.
9. Kurzweil, R. 2005. *The singularity is near.* New York: Viking.
10. Bjorkholm, J. E. 1998. EUV lithography: the successor to optical lithography. *Intel Technol J* Q3:1–8.
11. Wu, B. and A. Kumar. 2009. *Extreme Ultraviolet Lithography*. Columbus, OH: McGraw-Hill.
12. Bakshi, V., ed. 2009. *EUV Sources for Lithography*. SPIE press monograph. Bellingham, WA: SPIE Publications.
13. Brodie, I., and J. J. Muray. 1982. *The physics of microfabrication.* New York: Plenum Press.
14. Arden, W., and K. H. Muller. 1989. Light vs. x-rays: how fine can we get? *Semicond Int* 12:128–31; see also Vora, K. D., B. Y. Shew, E. C. Harvey, J. P. Hayes, and A. G. Peele. 2008. Sidewall slopes of SU-8 HARMST using deep x-ray lithography. *J Micromech Microeng* 18:1–7.
15. Cataldo, A. 1999. IBM uses x-ray lithography to build prototypes. *EE Times.* http://www.eet.com/story/OEG19990414S0031
16. Dejule, R. 1999. Next-generation lithography tools: The choices narrow. *Semiconductor International.* http://www.semiconductor.net/article/197958-Next_Generation_Lithography_Tools_The_Choices_Narrow.php
17. Bourdillon, A., and Y. Vladimirsky. 2006. *X-ray lithography—On the sweet spot.* San Jose, CA: UHRL.
18. Ehrfield, W., P. Hagmann, A. Maner, and D. Münchmeyer. 1986. Fabrication of microstructures with high aspect ratios and great structural heights by synchrotron radiation lithography galvanoforming and plastic moulding (LIGA process). *Microelectronic Engineering* 4:35–56.
19. Ehrfeld, W. and A. Schmidt. 1998. Recent developments in deep x-ray lithography. *J Vac Sci Technol* B16:3526–34; see also Vora, K. D., B. Y. Shew, E. C. Harvey, J. P. Hayes, and A. G. Peele. 2008. Sidewall slopes of SU-8 HARMST using deep x-ray lithography. *J Micromech Microeng* 18:1–7.
20. Wollersheim, O., H. Zumaque, J. Hormes, J. Langen, P. Hoessel, L. Haussling, and G. Hoffman. 1994. Radiation chemistry of poly(lactides) as new polymer resists for the LIGA process. *J Micromech Microeng* 4:84–93.
21. Kouba, J., R. Engelke, M. Bednarzik, G. Ahrens, H. U. Scheunemann, G. Gruetzner, B. Loechel, H. Miller, and D. Haase. 2007. SU-8: Promising resist for advanced direct LIGA applications for high aspect ratio mechanical microparts. *Microsystem Technologies* 13:311–17.
22. O'Brien, M. J., and D. S. Soane. 1989. *Microelectronics processing.* Eds. D. W. Hess, and K. F. Jensen. Washington, DC: American Chemical Society.
23. Bley, P., W. Menz, W. Bacher, K. Feit, M. Harmening, H. Hein, J. Mohr, W. K. Schomburg, and W. Stark. 1991. Three-dimensional mechanical microstructures. *The 4th International Symposium on Microprocess Conference*, pp. 384–391, Kanazawa, Japan.
24. Levinson, H. J. 2005. *Principles of Lithography* (2nd edition). Bellingham, WA: SPIE Publications.
25. Rosolen, G. C. 1999. Automatically aligned electron beam lithography on the nanometer scale. *Appl Surface Sci* 144–45:467–71.
26. Lindquist, J., D. Ratkey, and P. Fischer. 1990. Do you have the right beam for the job? *Res Dev* June:91–98.
27. Wolf, B. 1995. *Handbook of Ion Sources*. New York: CRC Press.
28. Brewer, G., ed. 1980. *Electron-beam technology in microelectronic fabrication.* New York: Academic Press.
29. Whelan, C. S., D. M. Tanenbaum, D. C. La Tulipe, M. Isaacson, and H. G. Craighead. 1997. Low energy electron beam top surface image processing using chemically amplified AXT resist. *J Vac Sci Technol* B15: 2555–60.
30. Streblechenko, D., and M. R. Scheinfein. 1998. Magnetic nanostructures produced by electron beam patterning of direct write transition metal fluoride resists. *J Vac Sci Technol* A16:1374–79.
31. Farrow, R. C., J. A. Liddle, S. D. Berger, H. A. Huggins, J. S. Kraus, R. M. Camarda, R. G. Tarascon, C. W. Jurgensen, R. R. Kola, and L. Fetter. 1992. Alignment and registration schemes for projection electron lithography. *J Vac Sci Technol* B10:2780–83.
32. Harriott, L. R. 2001. Limits of lithography. *Proceedings of IEEE* 89:366–74; see also Harriott, L. R. 1997. Scattering with angular limitation projection electron beam lithography for sub-optical lithography. *J Vac Sci Technol* B15:2130.
33. Pfeiffer, H. C. PREVAIL: IBM's e-beam technology for next generation lithography. *IBM Micronews* 6:41–44.
34. Bindra, A. E-beam projection prevails in nanometer lithography. *Electronic Design.* http://electronicdesign.com/Articles/Index.cfm?AD=1&ArticleID=1345
35. Lee, Y.-H., R. Browning, and R. F. W. Pease. 1992. E-beam lithography at low voltages. *Proc SPIE* 1671:155–165.
36. Zlatkin, A., and N. Garcia. 1999. Functional scanning electron microscope of low beam energy with integrated electron optical system for nanolithography. *Microelectron Eng* 46:213–17.
37. Editorial. 1993. Novel electron-beam lithography system being explored at Cornell's NNF. *Solid State Technol* 36:25–26.
38. Marrian, C. R. K., and E. A. Dobisz. 1992. Scanning tunneling microscope lithography: a viable lithographic technology? *SPIE* 1671:166–75.

39. Temple, D., W. D. Palmer, L. N. Yadon, J. E. Mancusi, D. Vellenga, and G. E. McGuire. 1998. Si field emitter cathodes: fabrication, performance, and applications. *J Vac Sci Technol* A16:1980–90.
40. Editorial. 1991. Ion beam focused to 8-nm width. *Res Dev* September: 23.
41. Langford, R. M., P. M. Nellen, J. Gierak, and Y. Fu. 2007. Focused ion beam micro- and nanoengineering, pp. 417–23, in *Focused Ion Beam Microscopy and Micromachining*, Volkert, C. A., and A. M. Minor, eds. *MRS Bulletin*. Special issue.
42. Seliger, R. L., J. W. Ward, V. Wang, and R. L. Kubena. 1979. A high-intensity scanning ion probe with submicrometer spot size. *Appl Phys Lett* 34:310–12.
43. Sanchez, J. L., J. A. van Kan, T. Osipowicz, S. V. Springham, and F. Watt. 1998. A high resolution beam scanning system for deep ion beam lithography. *Nucl Instruments Methods Phys Res* B136–38:385–89.
44. van Kan, J. A., J. L. Sanchez, B. Xu, T. Osipowicz, and F. Watt. 1999. Micromachining using focused high energy ion beams: deep ion beam lithography. *Nucl Instruments Methods Phys Res* B148:1085–98.
45. Seidel, P., J. Canning, S. MacKay, and W. Trybula. 1998. Next generation lithography. *Semiconductor Fabtech* 7:147.
46. Tejeda, R., G. Frisque, R. Engelstad, E. Lovell, E. Haugeneder, and H. Löschner. 1999. Finite element simulation of ion-beam lithography mask fabrication. *Microelectron Eng* 46:485–88.
47. Colburn, M., S. Johnson, M. Stewart, S. Damie, T. C. Bailey, B. Choi, M. Wedlake, T. Michaelson, S. V. Sreenivasan, J. Ekerdt, and C. G. Wilson. 1999. Step and flash imprint lithography: a new approach to high-resolution patterning. *Proc SPIE* 3676:379–89.
48. Mutschler, A. S. 2007. Toshiba validates 22-nm CMOS imprint litho. *Electronic News*. http://www.edn.com/article/CA6491602.html
49. http://www.hitachi.com/ICSFiles/afieldfile/2007/08/07/r2007_technology_tp.pdf
50. Calvert, J. M., C. S. Dulcey, M. C. Peckerar, J. M. Schnur, J. H. J. Georger, G. S. Calabrese, and P. Sricharo-Enchaikit. 1991. New surface imaging techniques for sub-0.5 micrometer optical lithography. *Solid State Technol* 34:77–82.
51. Stroeve, P., and E. Franses, eds. 1987. *Molecular engineering of ultrathin polymeric films*. London: Elsevier.
52. Lehn, J.-M. 1990. Perspectives in supramolecular chemistry: from molecular recognition towards molecular information processing and self-organization. *Angew Chem Int Ed Engl* 29:1304–19.
53. Biebuyck, H. A., N. B. Larsen, E. Delamarche, and B. Michel. 1997. Lithography beyond light: microcontact printing with monolayer resists. *IBM J Res Dev* 41:159–70.
54. Marrian, C. R. K., and E. S. Snow. 1996. Proximal probe lithography and surface modification. *Microelectron Eng* 32:173–89.
55. Müller, H. U., C. David, B. Völkel, and M. Grunze. 1995. Nanostructuring of alkanethiols with ultrasharp field emitters. *J Vac Sci Technol* B13:2846–49.
56. Nuzzo, R. G., and D. L. Allara. 1983. Adsorption of bifunctional organic disulfides on gold surfaces. *J Am Chem Soc* 105:4481–83.
57. Bain, C. D., E. B. Troughton, Y. T. Tao, J. Evall, G. M. Whitesides, and R. G. Nuzzo. 1989. *J Am Chem Soc* 111:321.
58. Delamarche, E., B. Michel, H. A. Biebuyck, and C. Gerber. 1996. Golden interfaces: the surface of self-assembled monolayers. *Adv Mater* 8:719–29.
59. Calvert, J. M., W. J. Dressick, C. S. Dulcey, M. S. Chen, J. H. Georger, D. A. Stenger, T. S. Koloski, and G. S. Calabrese. 1994. *Polymers for microelectronics: resists and diaelectrics*. Eds. L. F. Thompson, C. G. Willson, and S. Tagawa. Washington, DC: American Chemical Society.
60. Hamley, I. W., ed. 2004. *Developments in block copolymer science and technology*. Hoboken, NJ: John Wiley and Sons.
61. Flory, P. J. 1941. Thermodynamics of high polymer solutions. *J Chem Phys* 10:51–61.
62. Matsen, M. W., and F. S. Bates. 1996. Unifying weak- and strong-segregation block copolymer theories. *Macromolecules* 29:1091–98.
63. Mansky, P., P. Chaikin, and E. L. Thomas. 1995. Monolayer films of diblock copolymer microdomains for nanolithography applications. *J Mater Sci* 30:1987–92.
64. Black, C. T., R. Ruiz, G. Breyta, J. Y. Cheng, M. E. Colburn, K. W. Guarini, H.-C. Kim, and Y. Zhang. 2007. Polymer self assembly in semiconductor microelectronics. *IBM J Res Dev* 51:605–33.
65. Stix, G. 2004. *Nano patterning: IBM brings closer to reality chips that put themselves together.* http://www.fractal.org/Fractal-Research-and-Products/Nano-patterning.htm.
66. Segalman, R. A. 2005. Patterning with block copolymer thin films. *Mater Sci Eng* 48:191–226.
67. Hawker, C. J., and T. P. Russell. 2005. Block copolymer lithography: merging "bottom-up" with "top-down" processes. *MRS Bull* 30:952–66.
68. Stoykovich, M. P., and P. F. Nealey. 2006. Block copolymers and conventional lithography. *Materials Today* 9:20–29.
69. Stein, G. E., E. J. Kramer, X. Li, and J. Wang. 2007. Single-cystal diffraction from two-dimensional block copolymer arrays. *Phys Rev Lett* 98:086101.
70. Smith, H. I. 1996. A proposal for maskless, zone-plate-array nanolithography. *J Vac Sci Technol* B14:4318–22.
71. Menon, R., A. Patel, D. Gil, and H. I. Smith. 2005. Maskless lithography. *Mater Today* 8:26–33.
72. Angelo, M. D., M. V. Chekhova, and Y. H. Shih. 2001. Two-photon diffraction and quantum lithography. *Phys Rev Lett* 87:013602.
73. Boto, A. N., P. Kok, D. S. Abrams, S. L. Braunstein, C. P. Williams, and J. P. Dowling. 2000. Quantum interferometric optical lithography: exploiting entanglement to beat the diffraction limit. *Phys Rev Lett* 85:2733–36.
74. Einstein, A., B. Podolsky, and N. Rosen. 1935. Can quantum-mechanical description of physical reality be considered complete? *Phys Rev Lett* 47:777–80.
75. Parthenopoulos, D. A., and P. M. Rentzepis. 1989. Three-dimensional optical storage memory. *Science* 245:843–45.
76. Wu, E.-S., J. Strickler, R. Harrell, and W. W. Webb. 1992. Two-photon lithography for microelectronic application. *SPIE Proc* 1674:776–82.
77. Prasad, P. N. 2004. *Nanophotonics*. Hoboken, NJ: Wiley-Interscience.
78. Tanaka, T., H. B. Sun, and S. Kawata. 2002. Rapid sub-diffraction limit laser micronanoprocessing in a threshold material system. *Appl Phys Lett* 80:312–14.
79. Kik, P. G., S. A. Maier, and H. A. Atwater. 2002. Plasmon printing—a new approach to near-field lithography. *Mat Res Soc Symp Proc* 705:66–71.
80. Fang, N., and X. Zhang. 2003. Imaging properties of a metamaterial superlens. *Appl Phys Lett* 82:161.

81. Fang, N., H. Lee, C. Sun, and X. Zhang. 2005. Sub-diffraction-limited optical imaging with a silver superlens. *Science* 308:524–37.
82. Watanabe, F., M. Arita, T. Motooka, K. Okano, and T. Yamada. 1998. Diamond tip arrays for parallel lithography and data storage. *Jpn J Appl Phys* 37:L562–64.
83. Behm, R. J., N. Garcia, and H. Rohrer. 1989. *Proceedings: NATO ASI*. Dordrecht, the Netherlands: Kluwer.
84. Stroscio, J., and W. Kaiser, eds. 1993. *Scanning tunneling microscopy*. Boston: Academic Press.
85. Magno, R., and B. R. Bennett. 1997. Nanostructure patterns written in III-V semiconductors by an atomic force microscope. *Appl Phys Lett* 70:1855–57.
86. Wong, E. W., P. E. Sheehan, and C. M. Lieber. 1997. Nanobeam mechanics: elasticity, strength, and toughness of nanorods and nanotubes. *Science* 277:1971–75.
87. Sugimura, H., O. Takai, and N. Nakagiri. 1999. Multilayer resist films applicable to nanopatterning of insulating substrates based on current-injecting scanning probe lithography. *J Vac Sci Technol* B17:1605–08.
88. Soh, H. T., K. W. Guarinin, and C. F. Quate. 2001. *Scanning probe lithography*. Dordrecht, the Netherlands: Kluwer.
89. Wilder, K., C. F. Quate, B. Singh, and D. F. Kyser. 1998. Electron beam and scanning probe lithography: a comparison. *J Vac Sci Technol* B16:3864–73.
90. Völkel, B., A. Gölzhäuser, H. U. Müller, C. David, and M. Grunze. 1997. Influence of secondary electrons in proximal probe lithography. *J Vac Sci Technol* B15:2877–81.
91. Sugimura H., and N. Nakagiri. 1997. AFM lithography in constant current mode. *Nanotechnology* 8:A15–A18.
92. Gölzäuser, A. 1999. Nanolithography and electron holography with ultrasharp field emitters. *Appl Surface Sci* 141:264–73.
93. Spindt, C. A., I. Brodie, L. Humphrey, and E. R. Westerberg. 1976. Physical properties of thin-film emission cathodes with molybdenum cones. *J Appl Phys* 5284.
94. de Heer, W. A., A. Chatelain, and D. Ugarte. 1995. A carbon nanotube field-emission electron source. *Science* 270:1179–80.
95. Stroscio, J. A., and D. M. Eigler. 1991. Atomic and molecular manipulation with the scanning tunneling microscope. *Science* 254:1319–26.
96. Drexler, K. E. 1992. *Nanosystems: molecular machinery, manufacturing, and computation*. New York: Wiley.
97. Wallis, T. M., N. Nilius, and W. Ho. 2002. Electronic density oscillations in gold atomic chains assembled atom by atom. *Phys Rev Lett* 89:236802-1-2-4.
98. Nazin, G. V., X. H. Qiu, and W. Ho. 2003. Visualization and spectroscopy of a metal-molecule-metal bridge. *Science* 302:77–81.
99. Piner, R. D., J. Zhu, F. Xu, S. Hong, and C. A. Mirkin. 1999. Dip pen nanolithography. *Science* 283:661–63.
100. Ginger, D. S., H. Zhang, and C. A. Mirkin. 2004. The evolution of dip-pen lithography. *Angew Chem Int Ed* 43:30–45.
101. Salaita, K., Y. Wang, J. Fragala, R. A. Vega, C. Liu, and C. A. Mirkin. 2006. Massively parallel dip-pen nanolithography with 55000-pen two-dimensional arrays. *Angew Chem Int Ed* 45:7220–23.
102. Lim, J. H., D. S. Ginger, K. B. Lee, J. Heo, J. M. Nam, and C. A. Mirkin. 2003. Direct-write dip-pen nanolithography of proteins on modified silicon oxide surfaces. *Angew Chem Int Ed* 42:2309–12.
103. King, W. P., T. W. Kenny, K. E. Goodson, G. L. W. Cross, M. Despont, U. Durig, H. Rothuizen, G. Binnig, and P. Vettiger. 2001. Atomic force microscope cantilevers for combined thermomechanical data writing and reading. *Appl Phys Lett* 78:1300–02.
104. King, W. P., and K. E. Goodson. 2002. Thermal writing and nanoimaging with a heated atomic force microscope cantilever. *J Heat Transfer* 124:597.
105. Synge, E. H. 1928. A suggested method for extending microscopic resolution into the ultra-microscopic region. *Phil Mag* 6:356–62.
106. Likodimos, V., M. Labardi, L. Pardi, M. Allegrini, and M. Giordano. 2003. Optical nanowriting on azobenzene side-chain polymethacrylate thin films by near-field scanning optical microscopy. *Appl Phys Lett* 18:3313–15.
107. Yin, X., N. Fang, and X. Zhang. 2002. Near-field two-photon nanolithography using an apertureless optical probe. *Appl Phys Lett* 81:3663–65.
108. Brook, J., and R. Dandliker. 1989. Submicrometer holographic photolithography. *Solid State Technol* 32:91–94.
109. Omar, B., S. Clube, F. Hamidi, M. D. Struchen, and S. Gray. 1991. Advances in holographic lithography. *Solid State Technol* September:89–94.
110. Dunin-Borkowski, R. E., and T. Kasama. 2004. The prospect of three-dimensional induction mapping inside magnetic nanostructures by combining electron holography with electron tomography. *Microsc Microanal* 10:1010–11.
111. Schueller, O. J. A., S. T. Brittain, and G. M. Whitesides. 1999. Fabrication of glassy carbon microstructures by soft lithography. *Sensors Actuators A* A72:125–39.
112. Xia, Y., J. A. Rogers, K. E. Paul, and G. M. Whitesides. 1999. Unconventional methods for fabricating and patterning nanostructures. *Chem Rev* 99:1823–48.
113. Xia, Y., and G. M. Whitesides. 1998. Soft lithography. *Angew Chem Int Ed* 37:550–75.
114. Xia, Y., and G. M. Whitesides. 1998. Soft lithography. *Ann Rev Mater Sci* 28:153–84.
115. St. John, P. M., R. Davis, N. Cady, J. Czajka, C. A. Batt, and H. G. Craighead. 1998. Diffraction-based cell detection using a microcontact printed antibody grating. *Anal Chem* 70:1108–11.
116. Zhao, X.-M., Y. Xia, and G. M. Whitesides. 1996. Fabrication of three-dimensional micro-structures: microtransfer molding. *Adv Mater* 8:837–40.
117. Jeon, N. L., I. S. Choi, B. Xu, and G. M. Whitesides. 1999. Large-area patterning by vacuum-assisted micromolding. *Adv Mater* 11:946–50.
118. Rogers, J. A., Z. Bao, and V. R. Raju. 1998. Nonphotolithographic fabrication of organic transistors with micron feature sizes. *Appl Phys Lett* 72:2716–18.
119. Duffy, D. C., J. C. McDonald, O. J. A. Schueller, and G. M. Whitesides. 1998. Rapid prototyping of microfluidic systems in poly(dimethylsiloxane). *Anal Chem* 70: 4974–84.

# Part II

# Pattern Transfer with Subtractive Techniques

Stretching across the Nazca plains—like a giant map or blueprint left by ancient astronauts lie the famous Nazca Lines of Peru.

What subtractive, removing techniques were involved in drawing the monkey above?

**Marc J. Madou**

## Introduction to Part II



## Introduction to Part II

In manufacturing, removing and forming are the two primary processes. Removing processes destroy cohesion among particles as exemplified in wet chemical etching, electrical discharge machining (EDM), traditional mechanical turning, and laser drilling. Forming creates an original shape from a molten mass, a gaseous state, a solution, a plasma, or solid particles. During forming processes, cohesion is created among individual particles; examples include plastic molding, thin film deposition by evaporation or sputtering, and electrochemical metal deposition. In this book, we are interested in forming and removing processes enabling the manufacture of precision miniature machines. Materials are either removed from (subtracted) or added to (formed) a work piece or device under construction (usually in a selective manner).

Manufacturing techniques are often categorized according to the energy used at the work piece as mechanical, chemical, or thermal. Whereas mechanical machining mostly involves material removal,*

* An example of a mechanical additive technique is physical vapor deposition (see Chapter 7) where material is deposited by direct line-of-sight impingement deposition. The depositing material might be generated through the application of thermal energy (e.g., thermal evaporation) or mechanical energy to the source material (e.g., sputter deposition).

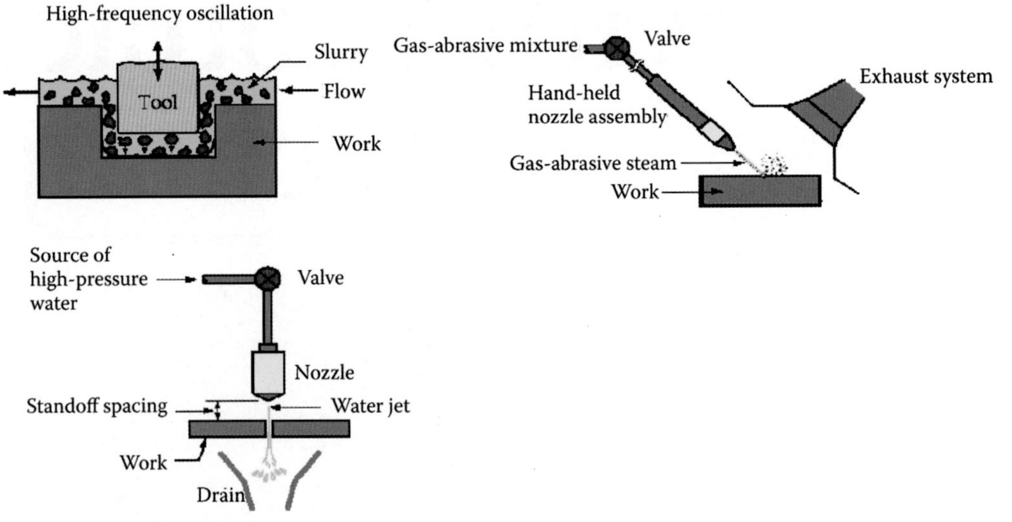


**Mechanical machining:** ultrasonic machining (S), abrasive jet-machining (S), and water-jet machining (S).

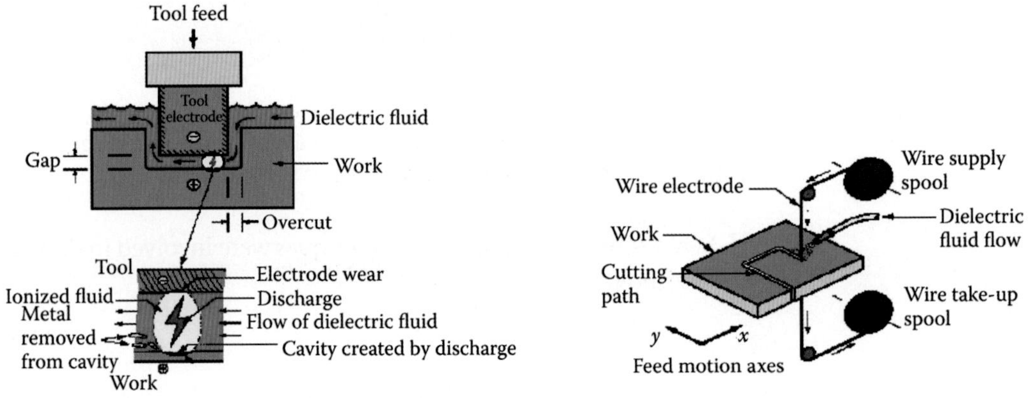


**Thermal machining:** electrical discharge machining (EDM) (S) and electrical discharge wire cutting (ED-WC) (S).

**FIGURE II.1** Examples of manufacturing processes grouped according to the energy used at the work piece as mechanical, chemical, and thermal. S = subtractive, A = additive.

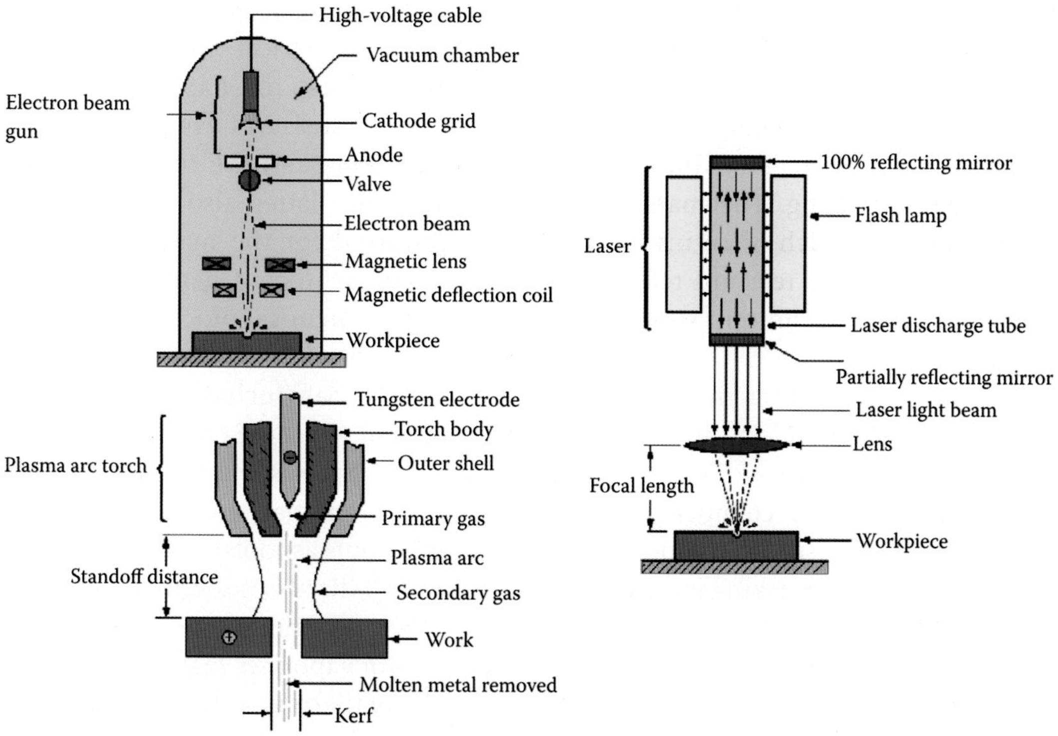


**Thermal machining:** electron-beam machining (EBM) (S shown here but can also be A), laser-beam machining (LBM) (S shown here but can also be A); also called high-energy beam machining.

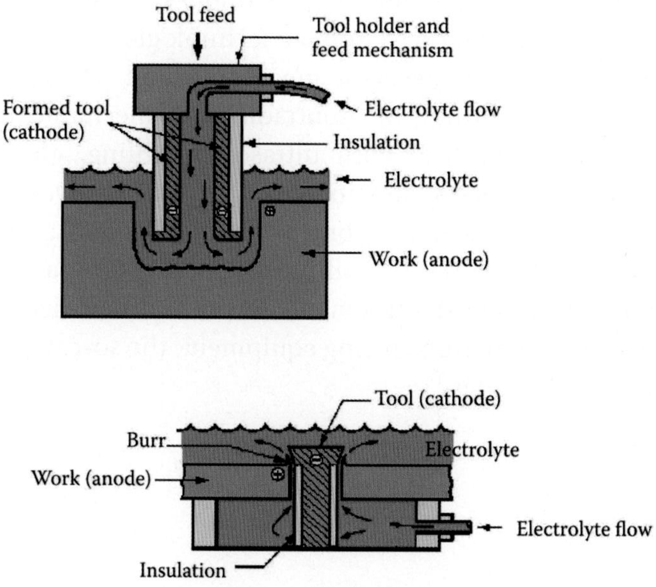


**Chemical machining:** electrochemical machining (ECM) (S shown here but can also be A).

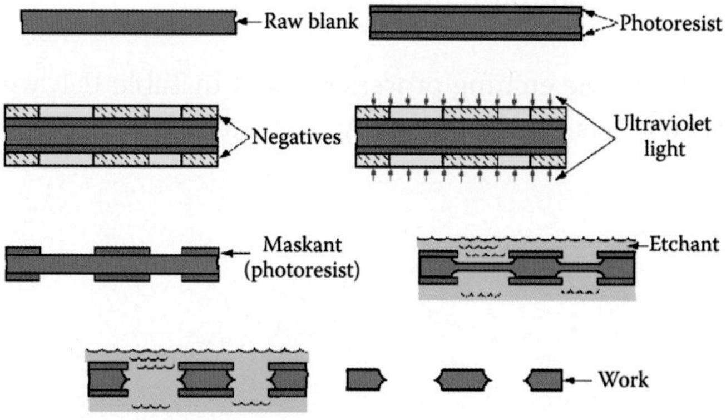


**Chemical machining:** chemical milling (CM) (S).

**FIGURE II.1** (*Continued*)

chemical and thermal can be additive (A) or subtractive processes (S). In Part II of this book we deal with material removal processes only; additive technologies are the focus of Part III.

Some examples of subtractive techniques are illustrated in Figure II.1. In categorizing machining techniques, some authors group e-beam, laser, and ion-beam machining under high-energy beam machining.

Table II.1 lists the most important subtractive processes encountered in miniaturization science and includes wet and dry etching, laser machining, ultrasonic drilling, electrical discharge machining (EDM), traditional precision machining, and focused ion-beam (FIB) milling. The manufacturing category, applications (which often involve Si), typical material removal rates, relevant references, and a few remarks on the technique supplement the list.

One of the most widely used subtractive techniques for integrated circuits (ICs) and micro- and nanomachining listed in this table is etching. Etching can be described as pattern transfer by chemical/physical removal of a material from a substrate—often in a pattern defined by a protective masking layer such as a resist or an oxide. Etch parameters—dry or wet—include (for more details see Table 3.5):

1. Etch rate: Rate at which material is removed from the substrate
2. Uniformity: Etch rate constancy across the substrate
3. Throughput: Amount of substrates/wafers etched during one process cycle
4. Directional control: Control of the horizontal and vertical etch rate—isotropic or anisotropic
5. Selectivity of etch:
   a. Controlled by etchant composition
   b. Controlled by etch rate

From the etching processes listed in Table II.1, we review gas and vapor phase dry etching processes, including chemical, physical (purely mechanical), and chemical-physical (mostly chemical with some mechanical assist) in Chapter 3.

Chapter 4 consists of two parts: one deals with wet *chemical etching* and the other with *wet bulk micromachining*, the latter also being chemical etching but optimized for the manufacture of ICs, nano-, and micromachines. Chemical etching techniques cover purely chemical processes as in *chemical milling* (CM), *photochemical machining* (PCM), *electrochemical machining* (ECM) (including μ-ECM), and combinations thereof. Of these chemical etching techniques, chemical milling is the simplest: it only requires a liquid solution that selectively dissolves a specific material (pools as tools).

Chapter 5 discusses thermal removing techniques where thermal energy, provided by a heat source, melts and/or vaporizes material from the work piece. Examples of thermal techniques covered in this chapter are EDM (including μ-EDM), laser- and e-beam machining (LBM and EBM), and plasma arc cutting.

In Chapter 6, we cover traditional and nontraditional mechanical removing techniques. Traditional mechanical material removing technologies covered are mechanical precision machining and abrasive wheel machining. In nontraditional mechanical machining we deal with ultrasonic drilling, electrolytic in-process dressing (ELID) grinding, water jet, abrasive water jet, abrasive jet machining, and focused ion beam (FIB) milling. In Chapter 6, we also highlight recent developments toward miniaturized mechanical manufacturing equipment: the so-called desktop factories.

**Table II.1** Subtractive Processes Important in ICs, Micro-, and Nanomachining

| Category | Subtractive Technique | Applications | Typical Etch Rate | Remark | References |
|---|---|---|---|---|---|
| Chemical | Dry chemical etching | Resist stripping, isotropic features | Si etch rate: 0.1 μm/min (but with more recent high-density plasmas up to 6 μm/min) | Resolution better than 0.1 μm, loading effects | 1, 2<br>Chapter 3 |
| | Physical/<br>Chemical etching | Very precise pattern transfer | Si etch rate: 0.1–1 μm/min (but with more recent high-density plasmas up to 6 μm/min) | Most important dry etching technique | 2, 3<br>Chapter 3 |
| | Vapor phase etching with $XeF_2$ (and other interhalogens) | Isotropic Si etching in presence of Al and $SiO_2$ | Si etch rate: 1–3 μm/min | Very selective, fast, and very simple | 4<br>Chapter 3 |
| | Chemical machining (CM) | Shallow removal (up to 12 mm) on large flat or curved surfaces; blanking of thin sheets; in Si spheres, domes, grooves, etc. | General 0.025–0.1 mm/min; in case of Si polishing 50 μm/min with stirring (RT, acid) | Little control, simple; low tooling and equipment cost; suitable for low production cost | 5<br>Chapter 4 |
| | Anisotropic wet chemical etching | Si angled mesas, nozzles, diaphragms, cantilevers, bridges | 1 μm/min on (100) Si (90°C, alkaline) | With etch-stop more control, simple | 5<br>Chapter 4 |
| | Wet photoetching | Etches p-type layers in p-n junctions | Etches p-Si up to 5 μm/min (acid) | No electrodes required | 6<br>Chapter 4 |
| Electrochemical | Electrochemical etching (ECM) | Complex shapes with deep cavities; etches p-Si and stops at n-Si in an p-n junction); etches n-Si of highest doping (in n/n+) | Highest rate of material removal. V: 5–25 dc; A: 1.5–8 $A/mm^2$; 2.5–12 mm/min, depending on current density; etches p-Si at 1.25–1.75 μm/min [(100) plane] | Complex (requires contacts) expensive tooling and equipment; high-power requirement; medium to high production quantity | 7, 8<br>Chapter 4 |
| | Electrochemical grinding (ECG) | Cutting off and sharpening of hard materials, such as tungsten-carbide tools. Also used for honing | Higher removal rates than grinding; A: 1–3 $A/mm^2$ | | 9<br>Chapter 4 |
| | Photoelectrochemical etching | Etches n-Si in p-n junctions, production of porous Si | Si etch rate:<br>5 μm/min (acid) | Complex, requires electrodes and light | 9<br>Chapter 4 |
| Thermal | Laser machining (with and without reactive gases) | Cutting and hole making in thin materials; circuit repair, resistor trimming, hole drilling, labeling of Si wafers | Drilling a hole in Si with a Nd:YAG (400-W laser): 1 mm/s (3.5 mm deep and 0.25-mm diameter) | Expensive equipment, does not require vacuum, heat-affected zone; laser beams can focus to a 1-μm spot, etch with a resolution of 1 $μm^3$ | 10<br>Chapter 5 |
| | Electron-beam machining | Cutting and small hole making in thin films; etching of circuits in microprocessors, large number of simple holes in hard to machine materials | Material removal rate is ~0.1 mg/s | Vacuum is needed, very expensive equipment, fast, drilling rates of 2000 holes/s | 11<br>Chapter 5 |

*(Continued)*

**Table II.1** Subtractive Processes Important in ICs, Micro-, and Nanomachining (continued)

| Category | Subtractive Technique | Applications | Typical Etch Rate | Remark | References |
|---|---|---|---|---|---|
| | Electrical discharge machining (EDM) | Shaping and cutting complex parts made of hard materials; some surface damage may result; drilling holes and channels in hard brittle metals | V: 50–380 dc; A: 0.1–500; typical removal rate for metals: 0.3 $cm^3$/min | Poor resolution (>50 μm), only conductors, expensive tooling and equipment | 12 Chapter 5 |
| Mechanical | Mechanical turning, drilling, and milling | Almost all machined objects surrounding us | Removal rates of turning and milling of most metals: 1–50 $cm^3$/min, for drilling: 0.001–0.01 $cm^3$/min | Prevalent machining technique | 13 Chapter 6 |
| | Physical dry etching, sputter etching, and ion milling | Si surface cleaning, unselective thin film removal | Si etch rate: 300 Å/min | Unselective and slow, plasma damage | 3 Chapter 3 |
| | Water-jet machining (WJM) | Cutting all types of nonmetallic materials 25 mm and thicker | Varies considerably with work piece material | | 14 Chapter 6 |
| | Abrasive water-jet machining (AWJM) and abrasive jet machining (AJM) | Single or multilayer cutting of metallic and nonmetallic materials; cutting, slotting, deburring, flash removal, and cleaning of metallic and nonmetallic materials | Up to 7.5 m/min | Tends to round off sharp edges; some hazards because of airborne particles | 14 Chapter 6 |
| | Abrasive grinding | | | | 15 Chapter 6 |
| | Electrolytic in-process dressing (ELID) grinding | | | | 16 Chapter 6 |
| | Ultrasonic drilling | Holes in quartz, silicon nitride bearing rings | Typical removal rate of Si: 1.77 mm/min | Especially useful for hard, brittle materials | 17 Chapter 6 |
| | Focused ion-beam (FIB) milling | Microholes, circuit repair, microstructures in arbitrary materials | Typical Si etch rate: 1 μm/min | Long fabrication time: >2 h including setup | 18 Chapter 6 |

Different-colored entries correspond to different chapters where more details can be found. RT, room temperature; chem. etch., chemical etching; iso., isotropic; anis., anistropic.
*Source:* Kalpakjian, S., and S. R. Schmidt. 2003. *Manufacturing processes for engineering materials*, 4th ed. Upper Saddle River, NJ: Prentice Hall. See also http://www.engineershandbook.com.

## References


1. Flamm, D. L. 1993. Feed gas purity and environmental concerns in plasma etching: Part I. *Solid State Technol* October:49–54.
2. Pandhumsoporn, T., M. Feldbaum, P. Gadgil, M. Puech, and P. Maquin. 1996. *Micromachining and microfabrication process technology II* Austin, TX: SPIE.
3. Manos, D. M., and D. L. Flamm, eds. 1989. *Plasma etching: an introduction.* Boston: Academic Press.
4. Chang, F. I., R. Yeh, G. Lin, P. B. Chu, E. Hoffman, E. J. Kruglick, K. S. J. Pister, and M. H. Hecht. 1995. Gas-phase silicon micromachining with xenon difluoride. *Proceedings: SPIE microelectronic structures and microelectromechanical devices for optical processing and multimedia applications* 2641:117–28.
5. Kern, W., and C. A. Deckert. 1978. *Thin film processes.* Eds. J. L. Vossen, and W. Kern. Orlando, FL: Academic Press.
6. Yoshida, T., T. Kudo, and K. Ikeda. 1992. *Proceedings: IEEE Micro Electro Mechanical Systems (MEMS '92).* Travemunde, Germany: IEEE.
7. Jackson, T. N., M. A. Tischler, and K. D. Wise. 1981. An electrochemical P-N junction etch-stop for the formation of silicon microstructures. *IEEE Electron Device Lett* EDL-2:44–45.
8. Meek, R. L. 1971. Electrochemically thinned N/N+ epitaxial silicon-method and applications. *J Electrochem Soc* 118:437.
9. Engineer's Handbook.
10. Watanabe, Y., Y. Arita, T. Yokoyama, and Y. Igarashi. 1975. Formation and properties of porous silicon and its applications. *J Electrochem Soc* 122:1351–55.
11. Chryssolouris, G. 1991. *Laser machining.* New York: Springer Verlag.
12. Kalpajian, S. 1984. *Manufacturing processes for engineering materials.* Reading, MA: Addison-Wesley.
13. DeVries, W. R. 1992. *Analysis of material removal processes.* New York: Springer Verlag.
14. Lorincz, J. November 2009. Waterjets: evolving from macro to micro. *Manufacturing Engineering.* Society of Manufacturing Engineers and http://www.drolsenslab.com/
15. Salmon, S. February 2010. What is abrasive machining? *Manufacturing Engineering.* Society of manfacturing Engineers.
16. Ohmori, H., and I. D. Marinescu, eds. 2011. *Handbook of ELID technologies.* Boca Raton, FL: CRC Press, *The 3rd International conference on Ultraprecision and ELID-Grinding,* 8–9 September, 2010, Toledo, OH and www.nexsys.co.jp/contents_e/technology.htm
17. Moreland, M. A. 1992. *Engineered materials handbook.* Ed. S. J. Schneider. Metals Park, OH: ASM International.
18. Vasile, M. J., C. Biddick, and S. Schwalm. 1994. Microfabriction by ion milling: the lathe technique. *J Vac Sci Technol* B12:2388–93.

# 3

# Dry Etching

Tek-Vac Model DRIE-1200-LL-ICP: High Density Plasma Etching System for 200 mm substrates

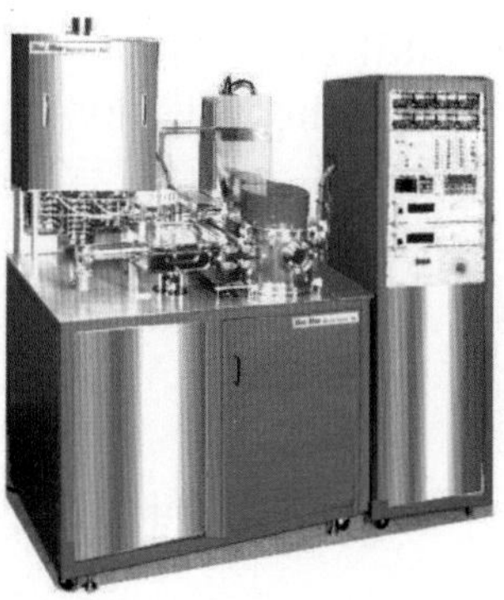

Etching in IC, micro- and nano-fabrication

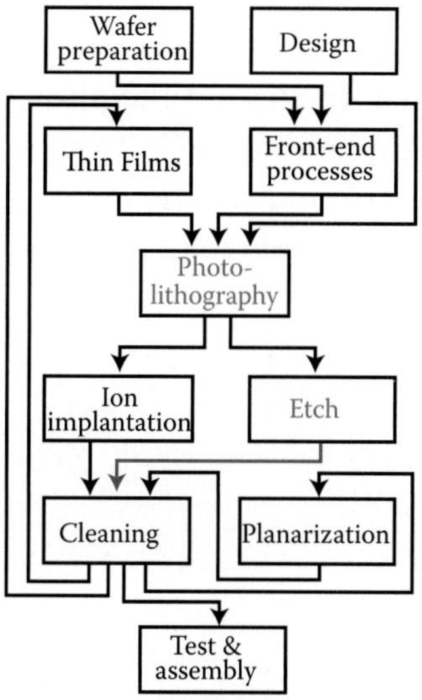


## Outline



I think there is a world market for maybe five computers.

**Thomas Watson**
*Chairman, IBM, 1943*

Science has a way of getting us to the future without consulting the futurists and visionaries.

**Robert Park**
*Voodoo Science*

## Introduction

Lithography steps precede a number of subtractive and additive processes. Materials are either removed or added to a device, usually in a selective manner, transferring the lithography patterns onto integrated circuits (ICs) micro- or nanomachines. This chapter deals with material removal by dry etching processes. In dry etching, the solid substrate is etched by gaseous species. While introducing resist stripping in Chapter 1, we learned that the popularity of dry stripping of resists (ashing) and the need for better control of critical dimension (CD) precipitated major research and development efforts in all types of plasma-based dry etching processes. The preference of dry etching over wet etching methods is based on a variety of advantages: fewer disposal problems, less corrosion of metal features in the structure, and less undercutting and broadening of photoresist features; that is, better CD control (nanometer dimensions compared with 3 μm or larger for wet etching) and, under the right circumstances, a cleaner resulting surface. With the new generations of ICs, all with sub-100-nm geometries, surface tension precludes a wet etchant from reaching down between photoresist features, whereas dry etching avoids this problem. In Figure 3.1, dry anisotropic and wet isotropic etchings for patterning a thin film of $SiO_2$ on a Si substrate are compared. In step (a) the photoresist protects parts of the underlying $SiO_2$ from dry or wet etching. In steps (b) and (c) the etching proceeds until the underlying Si substrate is reached, undercutting the photoresist in wet etching but not in dry etching. In step (d) the resist is stripped. As in IC manufacture, dry etching has evolved into an indispensable technique in miniaturization science. After the commercial successes with wet-etched Si micromachined products in the 1980s and early 1990s, the maturation of flexible, generic, and fast anisotropic deep reactive ion etching (DRIE) techniques in the late 1990s spawned a second wave of Si miniaturized commercial products.

## Dry Etching: Definitions and Jargon

In dry etching, substrates are etched without wet chemicals or rinsing and enter and leave the etching system in a dry state. The term *dry etching* covers a family of methods by which a solid surface is etched in the gas or vapor phase physically by ion bombardment,

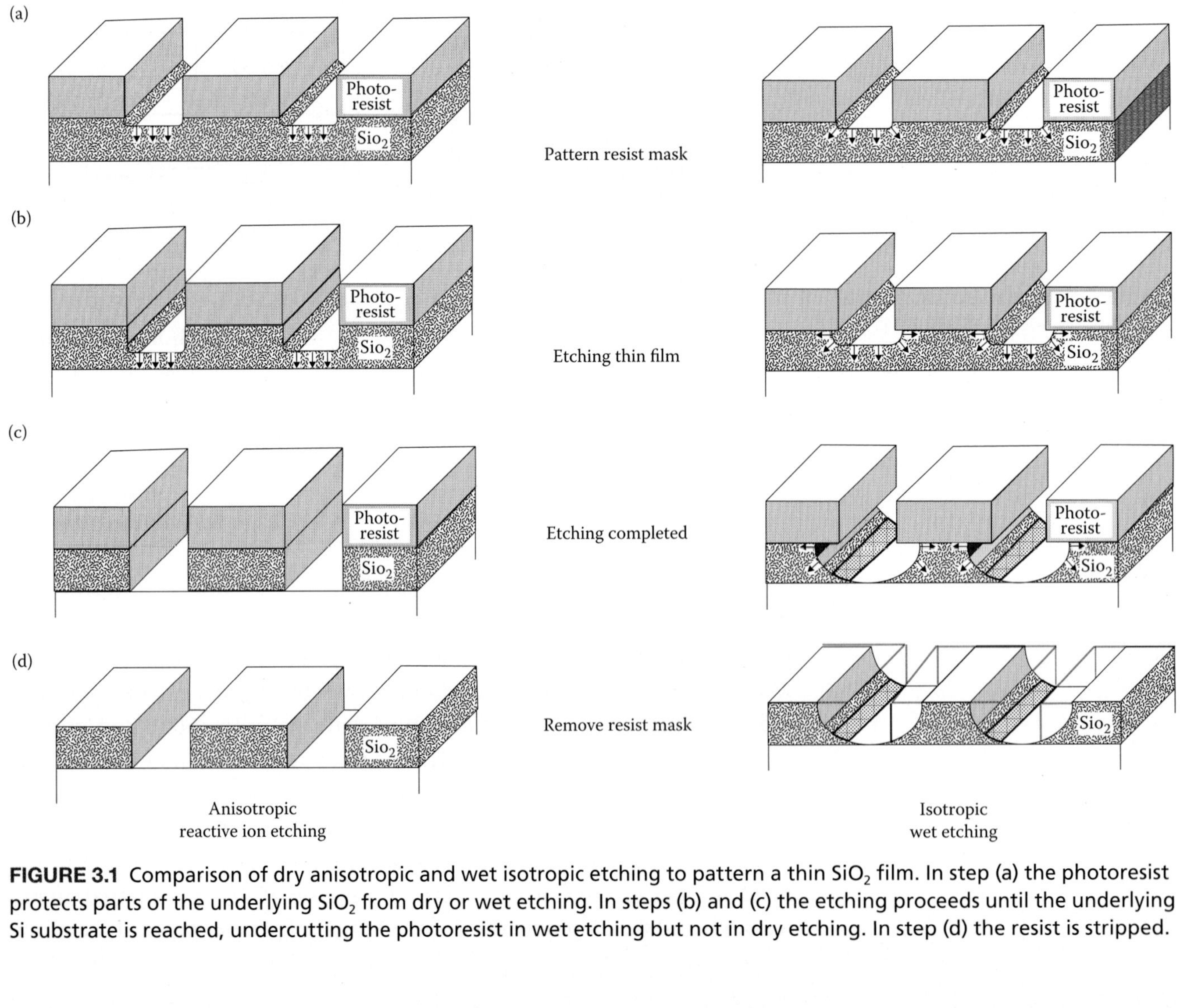


**FIGURE 3.1** Comparison of dry anisotropic and wet isotropic etching to pattern a thin $SiO_2$ film. In step (a) the photoresist protects parts of the underlying $SiO_2$ from dry or wet etching. In steps (b) and (c) the etching proceeds until the underlying Si substrate is reached, undercutting the photoresist in wet etching but not in dry etching. In step (d) the resist is stripped.

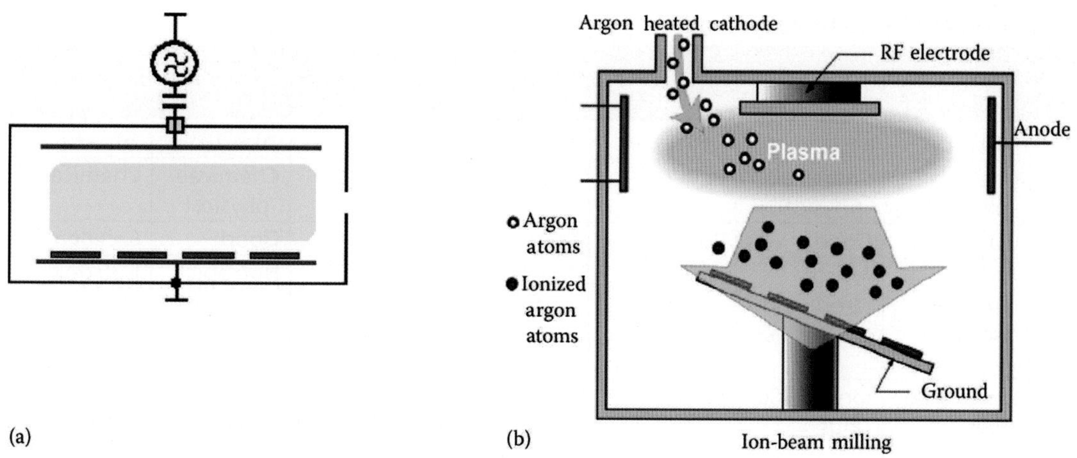


**FIGURE 3.2** Plasma-assisted dry etching is categorized according to the specific setup as either *glow discharge* etching (diode setup) (a) or *ion-beam* etching (triode setup) (b).

chemically by a chemical reaction through a reactive species at the surface, or by combined physical and chemical mechanisms or reactive ion etching (RIE). A key ingredient in dry etching involves the creation of plasma. Plasma-assisted dry etching is categorized according to the specific setup as either *glow discharge* etching (diode setup) or *ion-beam* etching (triode setup). Using glow discharge techniques, a broad plasma is generated in the same part of the vacuum chamber where the substrate is located; when using ion-beam techniques, plasma is generated in a separate chamber or in another part of the same vacuum chamber from which ions are extracted and directed in a beam toward the substrate by a number of electrode grids. In Figure 3.2 we compare a dry etching diode and dry etching triode setup.[1] It is also common to differentiate between 1) physical sputter/ion etching and ion-beam etching (IBE)[2] or also ion-beam milling (IBM), 2) chemical plasma etching (PE), and 3) synergetic reactive ion etching (RIE). Each of these techniques will be discussed in detail further below, but here we present a short dry etching synopsis:

1. In physical sputter/ion etching and ion-beam milling (IBE), etching occurs as a consequence of a purely physical effect, namely, momentum transfer between energetic $Ar^+$ ions and the substrate surface. Some type of chemical reaction takes place in all the other dry etching methods, but not in this one.
2. In chemical plasma etching (PE), neutral chemical species such as chlorine or fluorine atoms generated in the plasma diffuse to the substrate, where they react to form volatile products with the layer to be removed. The only role of the plasma is to supply gaseous, reactive etchant species.
3. In the case of physical/chemical etching (RIE), line-of-sight-impacting ions damage the surface, inducing highly anisotropic chemical reactions of the surface with plasma neutrals, or, alternatively, a passivating layer is cleared by the ion bombardment, clearing horizontal surfaces only and again resulting in highly anisotropic etching.

We can also distinguish plasmas produced in a DC field (0 Hz), an AC field (50 kHz and up), a radiofrequency (RF) field (13.6–27 MHz), and in a microwave field (300 MHz–10 GHz). Finally, dry plasma etching systems can also be divided into single-wafer and batch reactors.

In Table 3.1, we review the most common dry etching methods and clarify the associated jargon in the legend. Of the techniques listed, PE and RIE are the most widely used in IC manufacture and micro- and nanomachining.

In selecting a dry etching process, the desired shape of the etch profile and the selectivity of the etching process require careful consideration. Figure 3.3 shows different possible dry etch profiles, where, depending on the etching mechanism used, nondirectional, i.e., isotropic, or directional, i.e., vertical, or anisotropic etch profiles are obtained. In dry etching, anisotropic etch profiles—directional or vertical—can be generated in single crystalline, polycrystalline, and amorphous materials. The anisotropy of dry etching is not a result of aniso-

TABLE 3.1 Some Popular Dry Etching Systems

| | CAIBE | RIBE | IBE | MIE | MERIE | RIE | Barrel Etching | PE |
|---|---|---|---|---|---|---|---|---|
| Pressure (Torr) | ~$10^{-4}$ | ~$10^{-4}$ | ~$10^{-4}$ | $10^{-3}$–$10^{-2}$ | $10^{-3}$–$10^{-2}$ | $10^{-3}$–$10^{-1}$ | $10^{-1}$–$10^{0}$ | $10^{-1}$–$10^{1}$ |
| Etch Mechanism | Chemical/ physical | Chemical/ physical | Physical | Physical | Chemical/ physical | Chemical/ physical | Chemical | Chemical |
| Selectivity | Good | Good | Poor | Poor | Good | Good | Excellent | Good |
| Profile | Aniso-tropic or isotropic | Aniso-tropic | Aniso-tropic | Aniso-tropic | Aniso-tropic | Isotropic or aniso-tropic | Isotropic | Isotropic or aniso-tropic |
| Commercial equipment | Ionfab 300 Plus | Ionfab 300 Plus | Ionfab 300 Plus | MIE-710, Materials Research | Tegal 6000 MERIE | Unaxis Plasma-Therm 790 | Polaron K1050X from Quorum Techno-logies | Technics PE-IIA |

CAIBE, chemically assisted ion-beam etching; RIBE, reactive ion-beam etching; MIE, magnetically enhanced ion etching; MERIE, magnetically enhanced reactive ion etching; RIE, reactive ion etching; PE, plasma etching. See text.

tropy in the etching rate in different crystallographic directions, as in the case of anisotropic wet chemical etching in single crystal Si (Chapter 4); rather, the degree of anisotropy is controlled by the plasma conditions. Selectivity of a dry etch refers to the difference in etch rate between the mask and the film to be etched, again controllable by plasma conditions.

At low pressures, in the $10^{-3}$–$10^{-4}$ Torr range, anisotropic etching is easy, but accomplishing selectivity is difficult. The reason is that etching in this regime is mechanical in nature: the substrate is hit directionally, and atoms are ejected like a billiard ball. Physical ion-beam etching (IBE) at those pressures results in tapered profiles as shown in Figure 3.3a, low selectivity, and low etch rates because the mechanical energy transfer is inefficient. At higher pressures (1 Torr), in plasma etching (PE), chemical effects dominate, leading to better selectivity and higher etch rates than in the case of ion beams, but etched features are isotropic (Figure 3.3b). Chemical etching reactions are fast and efficient compared with mechanical etching, where atoms are ejected by mechanical force. Reactive ion etching (RIE) enables profile control as a result of the synergistic combination of physical ion sputtering with the chemical activity of neutral reactive species with a high etch rate and high selectivity. In the extreme case of anisotropic etching, perfect vertical sidewalls result (no lateral undercut) with perfect retention of

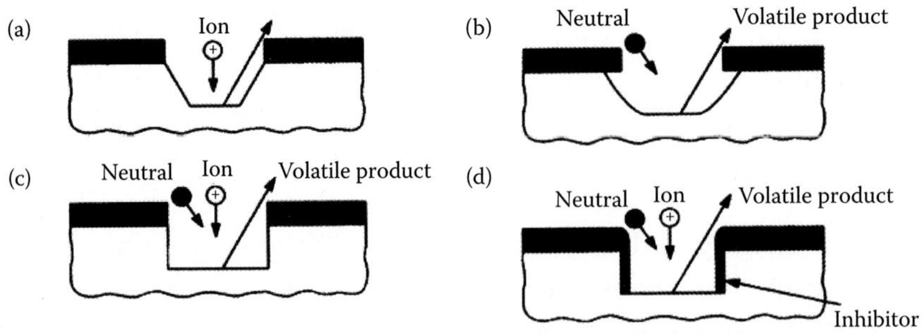


**FIGURE 3.3** Important dry etching profiles associated with different dry etching techniques. (a) Sputtering. (b) Chemical etching in plasma at low voltage and relative high pressure leads to isotropic features and lateral undercuts. (c) Ion-enhanced etching: physical-chemical is the most perfect image transfer as the undercutting is limited by the combined action of physical and chemical etching. The low pressure and high voltage lead to directional anisotropy. (d) Ion-enhanced inhibitor. Sidewalls are protected from undercutting by a surface species (e.g., a polymer), which starts etching when hit by a particle such as an ion, a photon, or an electron. Because very few particles hit the sidewalls (high voltage and high pressure), undercutting is suppressed. (From Flamm, D. L. 1993. Feed gas purity and environmental concerns in plasma etching: Part I. *Solid State Technol* October:49–54. With permission.[75])

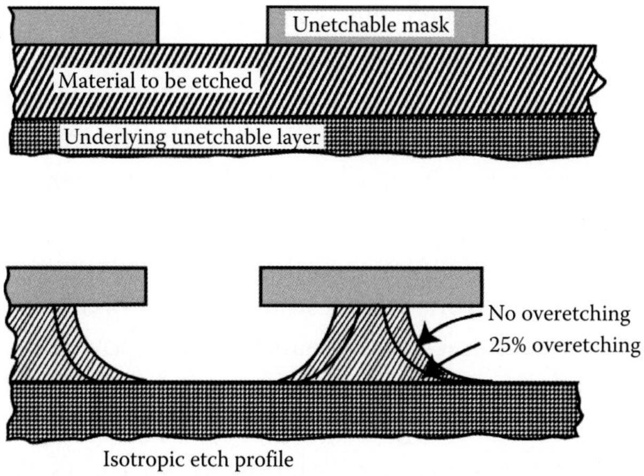


**FIGURE 3.4** Example illustrating the loss of dimensionality incurred by an isotropic etch when the depth is of the order of the lateral dimensions involved. Making the initial lithography smaller compensates for the loss of dimensionality by overetching.

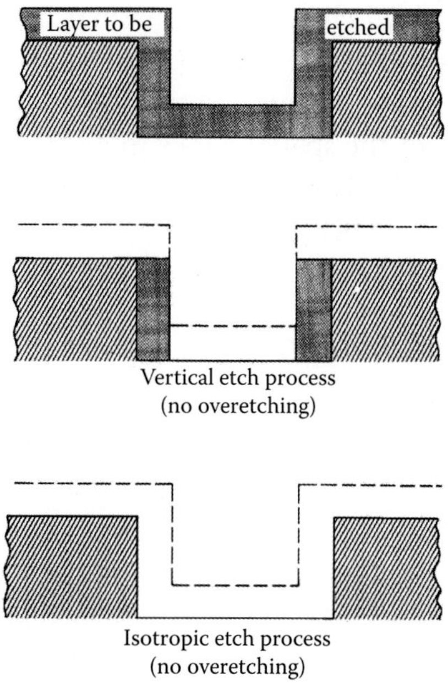


**FIGURE 3.5** An example of dry etching illustrating a case where an isotropic etch is preferred to a vertical etch process. To clear the layer on the vertical walls, an isotropic etch only requires a minimal overetch, whereas an anisotropic etch requires a substantial amount of overetching.

the CDs (see Figure 3.3c and d). In the case of isotropic etching, dimensionality losses are incurred with an etch depth the order of the lateral dimensions as illustrated in Figure 3.4. To ensure uniform etch rates over the whole wafer and between wafers, isotropic overetching may be required, further compromising the dimensionality. Figure 3.4 compares the etch profile of no overetching with a 25% overetch. Making the initial lithography smaller compensates for the loss of dimensionality incurred by overetching.

Even though isotropic etching leads to nonvertical sidewalls and loss of dimensionality, there remain some advantages to their use. Often a vertical sidewall is not the desired edge profile. For example, with line-of-sight deposition methods, such as resistive evaporation (see Chapter 7), a tapered sidewall is easier to cover than a vertical wall. Another example in which a nonvertical process has an advantage is illustrated in Figure 3.5. To take away the layer shown on the vertical walls, an anisotropic etchant would require extensive overetching, whereas an isotropic etchant will remove the material quickly.

## Plasmas or Discharges

### Introduction

Plasmas, defined as ionized gases that are neutral with equal numbers of ions and electrons, are commonly referred to as the fourth state of matter. A more precise definition is that plasma is a quasineutral gas of charged and neutral particles exhibiting collective behavior. The different states of matter generally found on earth are solid, liquid, and gas, and we have learned to work, play, and rest using these familiar states of matter. But in 1879, Sir William Crookes, an English physicist, identified a fourth state of matter, now called plasma. Plasmas in the stars and in the tenuous space between them make up more than 99% of the visible universe and perhaps most of which is not visible. Here on Earth, plasmas are important in film etching, sputter deposition, fluorescent lamps, welding arcs, and gas lasers (to name only a few).

Most dry etching systems also find their common base in plasmas or discharges, areas of high energy electric and magnetic fields that rapidly dissociate a suitable feed gas (e.g., sulfur hexafluoride, $SF_6$ for Si etching) to form neutrals, energetic ions ($SF_5^+$ and $F^-$), photons, electrons, and highly reactive radicals ($F^\bullet$). A surface in contact with plasma is exposed to fluxes of atoms, molecules, ions, electrons, and photons. This bombardment stimulates the production of outgoing fluxes of neutrals, ions, electrons, and photons, leading to a modified or etched surface layer.[3] We will start our foray into the physics and chemistry of plasmas by looking at the simplest plasma setup—a DC-diode glow discharge in a low-pressure argon atmosphere. After a discussion of the spatial zones in a glow discharge, we explain the Paschen curve and then introduce radiofrequency (RF) plasmas.

## Physics of DC Plasmas

The simplest plasma reactor consists of opposed parallel-plate electrodes in a chamber maintainable at low pressure, typically in the order of 1 mbar, as illustrated in the reactor shown in Figure 3.6a. The electrical potentials established in the reaction chamber, filled with an inert gas such as argon at a reduced pressure, determine the energy of ions and electrons striking the surfaces immersed in the discharge. To explain the generation of a DC plasma consider the setup shown in Figure 3.6a where we have simply put two electrodes (anode and cathode) at opposite ends of an evacuated glass tube and applied 1.5 kV between them. With the electrodes separated by 15 cm this results in a 100 V/cm field. Electrical breakdown (rapid ionization of a medium following the application of an overvoltage) of the argon gas in this reactor will occur when electrons, accelerated in the existing field, transfer an amount of kinetic energy greater than the argon ionization potential (i.e., 15.7 eV) to the argon neutrals. Elastic collisions deplete very little of the electron's energy and do not significantly influence the molecules because of the great mass difference between electrons and gas molecules. Inelastic collisions, on the other hand, excite the molecules of the gas or ionize them by completely removing an electron. Thus, such energetic inelastic collisions may generate a second free electron and a positive ion for each successful strike. Both free electrons re-energize, creating an avalanche of ions and electrons that results in a gas breakdown emitting a characteristic beautiful blue glow (in the case of argon; for air or nitrogen a pink color results from excited nitrogen molecules). The excitation/relaxation processes are responsible for the plasma glow. Avalanching requires the ionization of 10–20 gas molecules by one secondary electron. At the beginning of a sustained gas breakdown, a current, typically several hundred milliamps, starts to flow, and the voltage between the two electrodes decreases from 1.5 kV to about 150 V. Newly produced electrons are accelerated toward the anode, and with sufficient voltage, the gas rapidly becomes filled with positive and negative particles throughout its volume, i.e., it becomes ionized. The discharge current builds up to the point where the voltage decrease across a current-limiting resistor is equal to the difference between the supply voltage (1.5 kV) and the electrode potential difference (150 V). To sustain plasma, a mechanism to generate additional free electrons must exist after the plasma-generating electrons have been captured at the positively charged anode. Plasma-sustaining electrons are generated through the emission of secondary electrons (Auger electrons) when the

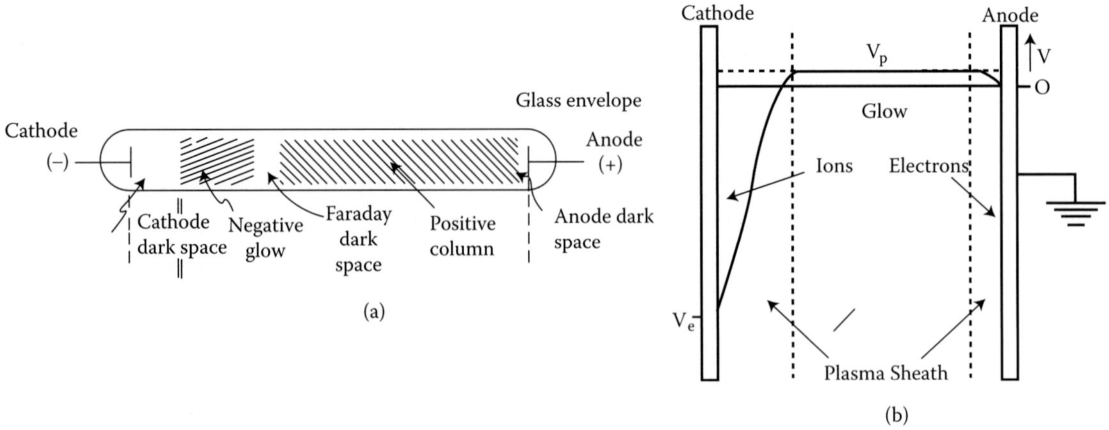


**FIGURE 3.6** Voltage distribution in a DC diode discharge in equilibrium. (a) Structure of the glow discharge in a long tube-based DC diode system. (b) Voltage distribution in a real DC diode system where the anode is brought much closer to the cathode than in (a) and is positioned in the negative glow region (so-called obstructed glow).

cathode is struck by ions. The continuous generation of those "new" electrons prompts a sustained current and a stable plasma glow. The discharge reaches a steady state when electron generation and loss processes balance each other. Electrons are lost by 1) drift and diffusion to the chamber walls, 2) recombination with positive ions, and 3) attachment to neutral molecules to form negative ions. Once this equilibrium is reached, the glow region of the plasma, being a good electrical conductor, sustains a very low field only, and its potential is almost constant. Most of the space between the electrodes is now filled with a soft glow as a result of the relaxation of excited species; however, certain dark spaces are observed near the electrodes. Most of the potential drops reside at the electrode surfaces where electrical double layers are formed in "sheath fields," counteracting the loss of electrons from the plasma (Figure 3.6b). The most significant dark space is seen at the cathode. This dark space, called the "ion sheath," is a positive space charge region, which is key to most dry etching processes. The cathode sheath, the first few millimeters in front of the cathode, corresponds to the cathode dark space (also called *Crookes* or *Hittorf* region). Positive ions entering this sheath move slower toward the cathode than electrons move away from it, leaving this sheath with an excess of positive charges. Thus, if a conductor is placed on the cathode, it is bombarded with ions and may be etched. If an insulating material, such as a dielectric or an oxide-coated silicon wafer, is placed on the cathode (or the anode for that matter), the process stops. The dielectric rapidly charges up, and the field in the gas drops, causing the generation of electrons to cease. For this reason, DC plasmas are rarely useful for etching silicon wafers. The solution to this problem lies in the use of an AC power source of sufficiently high frequency to reverse polarity before the dielectric becomes fully charged. Typically, radiofrequency (RF) is used, although the minimum effective frequency is in the range of >50 kHz (see below under RF plasmas).

In Table 3.2 we summarize the cathode dark space characteristics.

The light distribution shown in Figure 3.6a is only observed when the interelectrode distance is large compared with the size of the electrodes. The "Faraday dark space" and the "positive column" are characterized by a low electric field to conduct the electrons toward the anode. These two regions are often present

TABLE 3.2 Cathode Dark Space Characteristics

| Characteristics | Remark |
|---|---|
| Fewer electrons | Highly mobile electrons are accelerated quickly away from the negative cathode. |
| More positive ions | As electrons escape, they leave a high concentration of positive ions. These are attracted to the negative cathode. |
| Low conductivity | Positive ions move slowly. The result is that the field is higher across the cathode dark space. |
| Ions are accelerated | When the ions drift into the field of the dark space, they are accelerated toward the cathode. |
| Ions collide with the cathode and with neutral species on their way to the cathode | Therefore, the cathode is bombarded both with ions and with neutral species such as radicals and gas molecules. The ionic bombardment releases secondary electrons, which sustain the glow discharge. At low pressures and high voltages, the bombardment can cause sputtering. |
| Secondary electrons are created here | These electrons quickly accelerate to the ionization energy. As they leave the high field region and encounter ionizing collisions, they lose energy and then only have enough energy to excite. This is the edge of the negative glow region. |
| The dark space increases as the pressure is reduced (longer mean free path, higher energies, fewer collisions) | Eventually the dark space extends nearly across the chamber; the mean free path becomes too large to maintain a sufficient number of collisions and the glow is extinguished. |
| A conductor placed on the cathode | Will be subjected to ion bombardment. |
| An insulator placed on the cathode | The material charges up, and all the applied voltage exists across the material, and the plasma is extinguished. |

*Source:* Based on a class by Dr. Lynn Fuller and Dr. Richard Lane, Microelectronic Engineering, Rochester Institute of Technology.

in glow discharges used as lasers ("positive column lasers") and as fluorescence lamps. However, for most of the other applications of glow discharges (e.g., sputtering, deposition, chemical etching, analytical chemistry), the distance between cathode and anode is very short, so that normally (i.e., at typical discharge conditions) only a short anode zone is present beside the cathode dark space and the negative glow region, where the slightly positive plasma potential returns back to zero at the anode (see Figure 3.6b). This situation is termed an *obstructed glow*. Virtually all processing plasmas used in industry operate with a short interelectrode distance (e.g., twice that of the cathode dark space width) using the negative glow as principal plasma.[4] For electrodes of 12 cm in diameter, the interelectrode distance will typically be less than 1 cm. In the negative glow region, the potential is nearly constant and slightly positive, and hence the electric field is very small. It is this type of discharge that we depict in Figure 3.6b, with only a cathode dark space, negative glow, and anode dark space. If the interelectrode distance becomes significantly more than twice the dark space thickness, the discharge is extinguished. The latter characteristic can be applied to prevent discharge from contacting areas in a reactor where it is not desired.[5]

A striking characteristic of plasma is its permanent positive charge with respect to the electrodes—a result of the random motion of electrons and ions. The positive charge of plasma can be understood from kinetic theory, which predicts that for a random velocity distribution, the flux of ions, $j_i$, and electrons, $j_e$, on a surface is given by:

$$j_{i,e} = \frac{n_{i,e} \; <v_{i,e}>}{4} \tag{3.1}$$

where $n$ and $<v>$ are the densities and average velocities, respectively. Because ions are heavier than electrons (typically 4,000–100,000 times as heavy), the average velocity of electrons is larger. Consequently, the electron flux is larger than the ion flux, and the plasma loses electrons to the walls, thereby acquiring a positive charge. The bombarding energy of ions is proportional to the potential difference between the plasma and the surface being struck by the ions. The reason behind the asymmetric voltage distribution at the anode and cathode as seen in Figure 3.6b is as follows: electrons near the cathode rapidly accelerate away from it because of their relatively light mass; ions, being more massive, accelerate toward the cathode more slowly. Thus, on average, ions spend more time in the Crookes dark space, and, at any instant, their concentration is greater than that of electrons. The net effect is a very large field in front of the cathode compared with that in front of the anode and in the glow region itself. Consequently, the greatest part of the voltage between the anode and the cathode, $V_e$, is dropped across the Crookes dark space where charged particles (ions and electrons) experience their largest acceleration. This region is also called the *cathode-fall* region.

Glow discharges represent nonequilibrium conditions as the average electron energy, $<v_e>$, also called *electron temperature*:

$$<v_e> \; = kT_e \text{ (e.g., 1–10 eV)} \tag{3.2}$$

is considerably higher, in the range of 1–10 eV, than the average ion energy, $<v_i>$, also called *ion temperature*:

$$<v_i> \; = kT_i \text{ (e.g., 0.04 eV)} \tag{3.3}$$

which is much colder, typically having energies of 0.02–0.1 eV. Thus, a discharge or plasma cannot be described adequately by one single temperature but rather by the ratio of electron and ion temperature [$T_e$ (electrons)/$T_i$ (ions) = 10–100]. High-temperature electrons in a low-temperature gas are possible because of the small mass of electrons ($m_e$) compared with ions and neutrals. In collisions between electrons and argon atoms ($M_A$), as a result of an $m_e/M_A$ ratio of $1.3 \times 10^{-5}$, electrons have a very poor energy transfer and stay warmer longer than the heavier ions and neutrals. In plasma with an ionization fraction of $10^{-4}$, there are many more cold gas atoms than ions, and the ions quickly reach the background gas temperature as a result of the many collisions between ions and neutrals.[4] Electrons can attain a high average energy, often many electronvolts (equivalent to tens of thousands of degrees above the gas temperature), permitting electron-molecule collisions to excite high-temperature-type reactions, forming free radicals in the low-temperature neutral gas. Generating the same reactive species without plasma would require temperatures in the ~$10^3$–$10^4$ K range, destroying resists and damaging most inorganic films.

Plasma is typically weakly ionized: the number of ions is small compared with the number of reactive neutrals such as radicals. The ratio between ionized and neutral gas species in a glow discharge plasma is in the order of $10^{-6}$–$10^{-4}$. This fact, as we will see below, is crucial in understanding which entities are responsible for the etching of a substrate placed in a glow discharge. There are 1 million times more radicals than ions or electrons because radicals form more easily and their lifetime is much longer. We will see that ions do not etch, but radicals do. Ions affect the process by energetic (physical) bombarding of the surface, influencing the chemical processes of etching. The degree of ionization in plasma depends on a balance between the rate of ionization and the rate at which particles are lost by volume recombination and by losses to the walls of the apparatus. Wall losses generally dominate volume recombination. Accordingly, the occurrence of a breakdown in a given apparatus depends on the gas pressure (particle density), the type of gas, electric field strength (electron velocity), surface-to-volume ratio of the plasma, and distance between the anode and the cathode (see also Paschen's law in the section below). In air at atmospheric pressure (760 Torr) the mean free path is very short, and it is difficult for electrons to acquire enough energy to ionize gases. In such case, only extremely high electric fields may create a plasma in the form of arcing (as in lightning; see Figure 3.7) instead of a steady glow discharge.

**FIGURE 3.7** Discharge in lightning.

The choice of a dry etching technique depends on the efficiency or "strength" of the particular plasma. Plasma strength is evaluated by parameters such as average electron energy, $<v_e>$ (Equation 3.2), average ion energy (Equation 3.3), electron density (e.g., between $10^9$ and $10^{12}$ $cm^{-3}$), plasma ion density (e.g., $10^8$–$10^{12}$ $cm^{-3}$), neutral species density (e.g., $10^{15}$–$10^{16}$ $cm^{-3}$), and ion current density (e.g., 1–10 $mA/cm^2$). A quantity of particular use in characterizing plasma's average electron or ion energy is the ratio of the electrical field to the pressure:

$$kT_{i,e} \sim \frac{E}{p} \tag{3.4}$$

With increasing field strength, the velocity of free electrons or ions increases because of acceleration by the field (~E), but velocity is lost by inelastic collisions. Because an increase in pressure decreases the electron or ion mean free path, resulting in more collisions, the electron or ion energy decreases with increasing pressure (~1/P).

To use the described DC plasmas for 1) dry etching and 2) sputter deposition, one uses one of the following setups:

1. In the etching arrangement, the substrate to be etched is placed on the cathode of the DC diode system. An argon plasma in this setup produces sufficiently energetic ions (between 200 and 1000 eV) to induce physical etching, that is, ion etching or sputtering of the material on the cathode.
2. The same setup is used for sputter deposition where atoms from the target (cathode) are ejected like a billiard ball and deposited on the substrate laying on the anode (see Chapter 7).

The low pressure in sputter etching (10–200 mTorr) gives rise to a long mean free path, and the externally applied voltage concentrates across the cathode plasma sheath where ions are accelerated before hitting the cathode. Each ion will collide numerous times with other gas species before transversing the plasma sheath, as the sheath thickness is larger than the mean free ion path. As a result of these collisions, ions lose much of their energy and move across the sheath with a drift velocity that is less than "free fall" velocity. But these vertical velocities are still very large compared with the random thermal velocities in the glow discharge region. The resulting bombarding ion flux, $j_i$, is given by:

$$j_i = qn_i\mu_i E \quad (3.5)$$

where E = electric field
$n_i$ = ion density
$\mu_i$ = ion mobility
q = charge

Reducing the pressure in the reactor increases the mean free path because particles have more of a chance to accelerate before colliding with another particle or wall. Consequently, ions accelerating toward the cathode at lower pressures can gain more energy before a collision takes place. A typical sputtering profile was illustrated earlier in the schematic in Figure 3.3a. In the opposite high pressure mode, reactive species such as radicals generated by the DC plasma may combine with the substrate to form volatile products, chemically etching the substrate in so-called *plasma etching*. In this case, ions striking the surface have lower energy (e.g., less than 10–20 eV) as a result of the lower sheath voltages and short mean free path because of the relatively high pressures (typically >0.1 Torr). Aside from generating the radicals through energetic collisions in the plasma, ions contribute very little or nothing to the etching process itself. The fast etching radicals reach the etch site via diffusion from the plasma rather than by electromigration. A typical plasma etching profile was shown in Figure 3.3b.

## Spatial Zones in a DC Glow Discharge

Introducing Figure 3.6a we noticed that the structure of plasma discharges in tubes that are long compared to the electrode sizes is complex. When a DC glow discharge is sustained in a long, narrow cylindrical tube filled with a rare gas at low pressure, the visible light emitted from the glow is distributed as shown in Figure 3.6a and, in more detail, in Figures 3.8 and 3.9. In Figure 3.9 we also show the potential distribution and the ion and electron density along the length of such a discharge tube (a DC "gas diode"). Regions described as "glows" emit

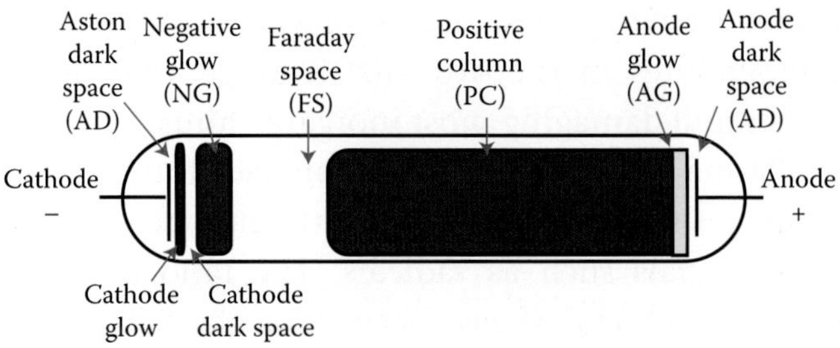


FIGURE 3.8 Detailed structure of a glow discharge in a DC gas diode.

significant light; regions labeled as "dark spaces" do not. Photons are created by the excitation/relaxation of the gas molecules, and when excitation/relaxation is the dominant process the plasma glows. The glow discharge can be subdivided into several regions between the anode and the cathode. Besides the Crookes or Hittorf dark space (cathode DS), the Aston DS, the Faraday DS, and the anode DS, a cathode glow can be observed, a negative glow (NG) region, the positive column (PC), and the anode glow (AG). The "dark" regions are not really dark, just less luminous than the bright regions. As the discharge becomes more extended (i.e., stretched horizontally in the geometry of the figure), the positive column may become striated, and for monatomic gases, the cathode glow may become striated as well. That is,

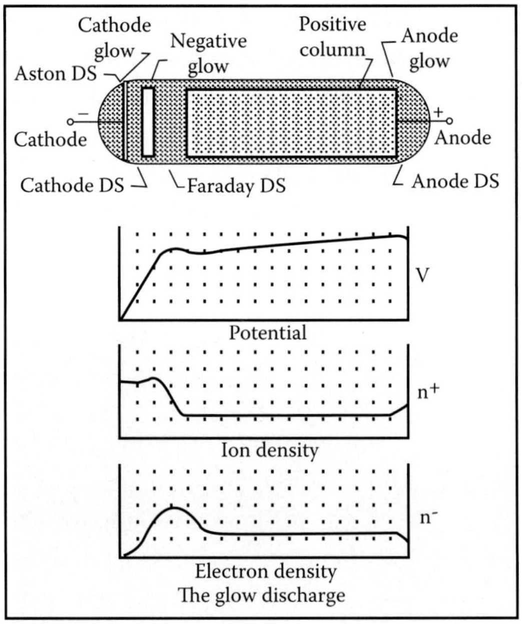


FIGURE 3.9 Detailed structure of a glow discharge in a DC diode with potential distribution and the ion and electron density along the length of the discharge tube (see also Figure 3.8).

additional alternating dark and bright regions may form. Alternatively, compressing the discharge horizontally results in fewer regions. The positive column will be compressed, whereas the negative glow will remain the same size, and, with small enough gaps, the positive column will disappear altogether.

The cathode dark space has the strongest electric field in the discharge. This region is filled with a positive space charge of ions moving in the opposite direction of the electrons. Electrons leaving the cathode with a few electronvolts of energy do not have the capacity of excitation/relaxation, resulting in a very thin first dark layer at the cathode (Aston dark space-AD). However, electrons emitted from the cathode rapidly attain energy suitable for excitation/relaxation, giving rise to a small bright layer; that is, the first cathode layer or cathode glow. After this thin excitation layer, the electrons are exhausted, and a second dark layer arises. Electrons then re-energize in the field and may form a second thin cathode layer, vaguer than the first, and so on. In Figures 3.8 and 3.9 we show one Aston dark space and one cathode glow layer only. If there is only one cathode glow, the second Aston dark space corresponds to the Crookes, Hittorf, or cathode dark space. A plurality of Aston dark spaces and thin cathode glow layers are only clearly observable in monoatomic gases; in polyatomic gases other energy dissipation processes take over.[6] In the main part of the cathode DS (Crookes or Hittorf dark space), electrons have gained too much energy from the electric field for efficient excitation, and luminescence becomes very weak.

The negative glow is the bright segment adjacent to the cathode dark space. In this brightest region of the glow discharge, there are many more electrons (a result of charge multiplication), and they do not gain any further energy, as this region is equipotential or field free.

The electron energy in the negative glow region is much more suitable for excitation/relaxation, and a bright light with a color characteristic for the discharge gas (blue in the case of argon) results. The electrons are exhausted by this effort and drift out into the Faraday dark space and the positive column at low velocity.

If the distance between anode and cathode is sufficiently large, a long positive column is present. There are now much fewer ions around, so the electrons can be accelerated in the gentle field. As they speed up, their number density goes down (the main losses are through diffusion to the walls). When they have sufficient energy to excite the molecules of the gas, the emitted light becomes evident. This light is produced with minimal excitation and consists only of light emitted by neutral atoms that have been excited above their resonance energy, which explains its different character from the light emitted by the negative glow. The positive column is less bright than the negative glow, and, although of another color, it is also characteristic of the discharge gas (purple in the case of argon). This region is of no importance in thin film processing technology but, as mentioned before, is very important for plasma discharges used as light sources (e.g., a fluorescent Hg-Ar discharge lamp) and in lasers.[4]

Close to the anode, electrons are attracted and accelerated, and positive ions are repelled. The accelerating electrons excite gas molecules, and an anode glow results.

The described discharge regions vary in radiation intensity, potential and electrical field distribution, space charge, and current density, and their occurrence and position depend on the discharge parameters such as pressure, voltage, current, kind of gas, and, as we have seen, the distance between cathode and anode. We refer to work by Coe et al.,[7,8] Raizer et al.,[9] and Bogaerts et al.[10–12] for a more in-depth study of the various phenomena involved in low-pressure noble gas discharges.

## Paschen's Law

The preceding section on DC plasmas introduces the concept of avalanche and gas breakdown in a DC diode system. To enhance our understanding of plasmas in general, and specifically to understand better what happens when the electrode distance or pressure in the discharge chamber is changed, we now introduce Paschen's law.

When the electric field between a pair of metal electrodes in air (at room temperature and atmospheric pressure) increases to more than, roughly, 3 MV $m^{-1}$, *electrical breakdown* occurs. Analogous to the process described above for the argon-filled chamber, free electrons accelerate in the electric field, and if they pick up enough energy before hitting air molecules, ionization results and more ions and electrons are created. This can lead to yet further ionization or an avalanche destroying the insulating properties of the air between the electrodes and leading to the passage of a high current, accompanied by the emission of heat, light, and sound—that is, a spark discharge. A spark discharge in a localized region (e.g., near a sharp point) is called a *corona* discharge.

The voltage for electrical breakdown depends on the geometry of the electrode gap, the gas, and the pressure. In 1889, Paschen published a paper describing a generalized form for that relationship, and this has become known as *Paschen's law*.[13] The law essentially states that the breakdown characteristics of an electrode gap are a function of the product of the gas pressure and the gap distance, usually written as $V = f(P \times d)$, where the breakdown voltage $V$ is a function of the pressure $P$ and the gap distance $d$. As the pressure or gap is reduced, the breakdown voltage decreases slowly to a minimum and then rises steeply, as illustrated in Figure 3.10, representing the Paschen's law curve in air and other gases. The sharp increase in the breakdown voltage on the left side in Figure 3.10 (low $P \times d$ side) occurs because the electrode spacing is too small for ionization to occur. The number of gas molecules in the space between the electrodes is proportional to $P \times d$. When the pressure is too low or the distance too small, most electrons reach the anode without

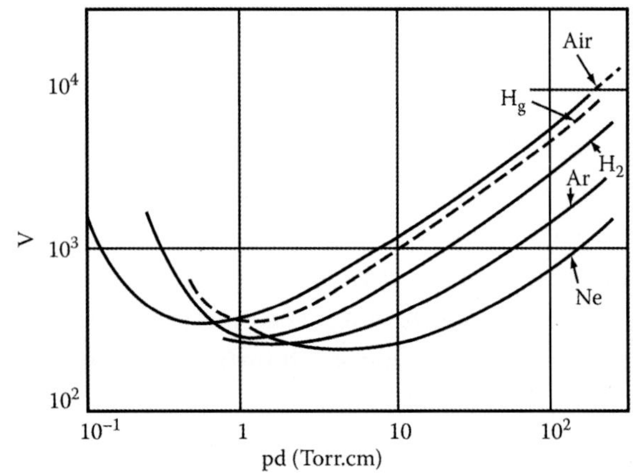


**FIGURE 3.10** A DC breakdown voltage as a function of gas pressure *P* and electrode spacing *d* for plane parallel electrodes in air and some other gases. Such curves are determined experimentally and are known as *Paschen curves.*

**TABLE 3.3** Minimum Breakdown Voltage ($V_{min}$) and Pressure/Distance Product ($P \times d$) at That Minimum Voltage for Various Gases as Deduced from Their Paschen Law Curves

| Gas | $V_{min}$ in Volts (V) | $P \times d$ at $V_{min}$ (Torr cm) |
|---|---|---|
| Air | 327 | 0.567* |
| Ar | 137 | 0.9 |
| $H_2$ | 273 | 1.15 |
| He | 156 | 4.0 |
| $CO_2$ | 420 | 0.51 |
| $N_2$ | 251 | 0.67 |
| $N_2O$ | 418 | 0.5 |
| $O_2$ | 450 | 0.7 |
| $SO_2$ | 457 | 0.33 |
| $H_2S$ | 414 | 0.6 |

*At 1 atm = 760 mmHg or 760 Torr, d = 7.460 $10^{-4}$ cm or ~8 μm.
*Source:* Naidu, M. S., and V. Kamaraju. 1995. *High voltage engineering*. New York: McGraw-Hill.[14]

colliding with gas molecules. As a consequence, the lower the pressure or the shorter the *d*, the higher the value of *V* required to give electrons enough energy to cause breakdown of the gas. For air, the minimum breakdown voltage ($V_{min}$) is about 300 V at ~8-μm gap distance (see also Table 3.3). On the right side of this plot (high $P \times d$ side), the slow increase in breakdown voltage with increasing electrode distances and higher pressures arises because electron collisions with gas atoms become more and more frequent so that for the electrons to gather sufficient energy to overcome the ionization potential of the inert gas, more potential needs to be applied.

Some microdevices such as electrostatic micromotors, microionizers for ion mobility spectrometers, and microswitches operate in air without sparking, as they operate on the left side of the Paschen curve (see Volume III, Chapter 8). Our intuition would tend to predict that the smaller the gap distance, the smaller the breakdown voltage. This typifies how linear scaling can be misleading when predicting the behavior of microdevices. Microstructures, with dimensions of a few micrometers, operate in air as if surrounded by a reduced pressure environment.

Different gases exhibit similar "Paschen" behavior, with the curves more or less shifted from the air curve, depending on the mean free path of the gas molecules involved (see Figure 3.10). Paschen's law—$V = f(P \times d)$—should be more correctly stated as V = f(δ × d), where δ is the density of gas molecules, which is, of course, affected by the temperature and the pressure of the gas. Much research has been done to provide a theoretical basis for Paschen's law and to develop a greater understanding of the breakdown mechanism. This has proven to be a formidable task, as many factors have an effect on the breakdown in a gap, such as radiation, dust, and surface irregularities. The electrode material also affects the sparking voltage, with cathodes of barium and magnesium having higher voltages than aluminum, for example.

In Table 3.3 we list the minimum breakdown voltage ($V_{min}$) and the product of the pressure and electrode spacing ($P \times d$) at that such voltage for various common gases as deduced from Paschen's law curves.[14]

A good text and reference book on electrical breakdown of gases, although old, is written by Meek and Craggs.[15]

## Physics of RF Plasma Reactors

AC voltages overcome the problem of charge accumulation on a dielectric in a DC plasma system. The positive charge that accumulates as a result of ion bombardment during one-half of the AC cycle can be neutralized by electron bombardment during the next half cycle. The frequency of AC must be high enough so the half period will be shorter than the charge-up time of the dielectric. Although this time will vary because of conditions and dielectric materials, for most applications the frequency must be greater than 50 kHz. At low frequencies, both electrons and ions can follow the changing field, so that a glow discharge is the same as DC, except that the polarity reverses twice each cycle. At high frequency, the massive ions cannot respond to the frequency changes, whereas electrons can. By far the most common RF used is 13.56 MHz, allowed by the Federal Communications Commission (because of its noninterference with radio-transmitted signals). The RF power supply is rated between 1 and 2 kW. In RF-generated plasma, a radiofrequency voltage applied between the two electrodes causes free electrons to oscillate and collide with gas molecules, leading to a sustainable plasma. The RF-excited discharges can be sustained without relying on the emission of secondary electrons from

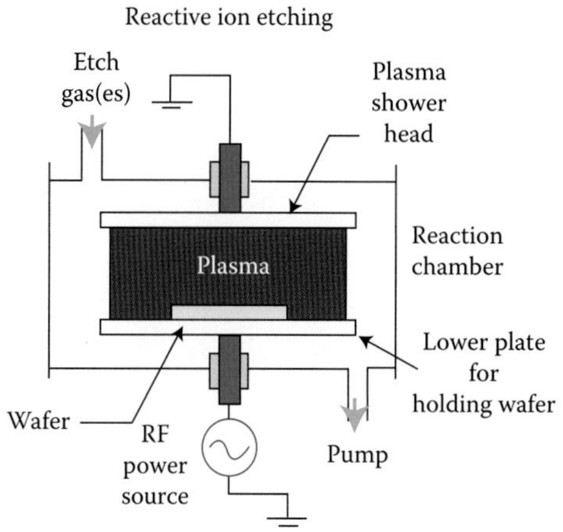


**FIGURE 3.11** Two-electrode setup (diode) for RF ion sputtering or sputter deposition. For ion sputtering, the substrates are located on the cathode (target); for sputter deposition (see Chapter 7), the substrates to be coated are located on the anode.

the cathode. Electrons pick up enough energy during oscillation in an RF field to cause ionization, thus sustaining the plasma at lower pressures than in DC plasma (e.g., 10 vs. 40 mTorr). The RF breakdown voltage of an AC plasma exhibits the Paschen behavior of a DC plasma; that is, a minimum in the required voltage as a function of the pressure with the mean free path of electrons, $\lambda_e$, substituting the spacing, *d*, between electrodes.

In the simplest case of RF ion sputtering, the substrates to be etched are laid on the cathode (target) of a discharge reactor, for example, a planar parallel-plate reactor as shown in Figure 3.11. The reactor consists of a grounded anode and powered cathode or target enclosed in a low-pressure gas atmosphere (e.g., $10^{-1}$–$10^{-2}$ Torr of argon). The RF plasma, formed at low gas pressures, consists of positive cations, negative anions, radicals, and photons (the UV photons create the familiar plasma glow). As with DC plasma, neutral species greatly outnumber electrons and ions—the degree of ionization only being on the order of $10^{-4}$–$10^{-6}$ for parallel-plate gas discharges. With one of the two electrodes capacitively coupled to the RF generator, this electrode automatically develops a negative DC bias and becomes the cathode with respect to the other electrode (see Figure 3.12a). This DC bias (also called *self-bias* $V_{DC}$) is induced by the plasma itself and is established as follows. When initiating an AC plasma arc, electrons, being more mobile than ions, charge up the

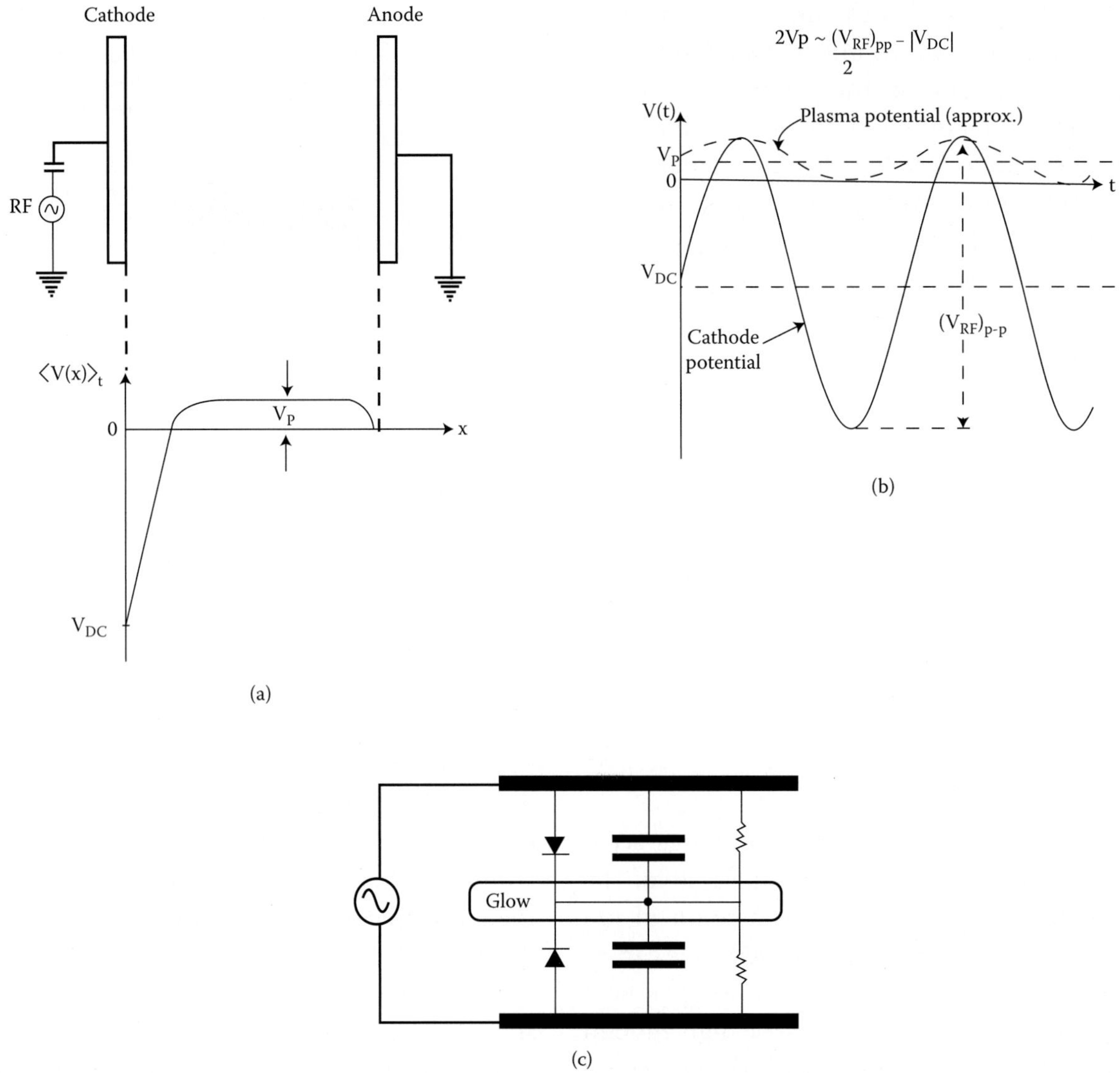


**FIGURE 3.12** An RF plasma. (a) Approximate time-averaged potential distribution for a capacitively coupled planar RF discharge system. (b) Potential distribution in glow discharge reactors—$V_p$: plasma potential; $V_{DC}$: self-bias of cathode electrode; $(V_{RF})_{pp}$: peak-to-peak RF voltage applied to the cathode. (c) Equivalent electrical circuit of an RF plasma.

capacitively coupled electrode; because no charge can be transferred over the capacitor, the electrode surface retains a negative DC bias.

The energy of charged particles bombarding the surface in a glow discharge is determined by three different potentials established in the reaction chamber: the plasma potential $V_p$, that is, the potential of the glow region, the self-bias $V_{DC}$, and the bias on the capacitively coupled electrode $(V_{RF})_{pp}$ (see also Figure 3.12b). The following analysis clarifies how these potentials relate to one another and how they contribute to dry etching.

The self-bias buildup, $V_{DC}$, on an insulating electrode by the electron flux (Equation 3.1) is given by:

$$V_{DC} = \frac{kT_e}{2e} \ln \frac{T_e m_i}{T_i m_e} \qquad (3.6)$$

where $T_e$ and $T_i$ are the electron and ion temperatures defined by Equations 3.2 and 3.3, and $m_e$ and $m_i$ are the electron and ion masses, respectively.[16] In the plasma, there are equal amounts of hot electrons ($T_e = 2 \times 10^4$ K) and cold positive ions ($T_i = T_{gas} = 500$ K). For a common RF power density of 4 $kWm^{-2}$ the self-bias is about 300 V, so that the powered electrode (the cathode) is bombarded with 300 eV of positive ions.[17]

The electron loss creates an electric sheath field in front of any surface immersed in the plasma, counteracting further electron losses. This sheath or dark space forms a narrow region between the conductive glow region and the cathode (the Crookes dark space) where most of the voltage ($V_{DC}$ in Figure 3.12a) is dropped as in the DC plasma case. The cathode, being capacitively coupled, effectively acts as an insulator for DC currents. The thickness of the sheath is typically 0.01–1 cm, depending on pressure, power, and frequency. The other electrode is grounded and conductive (no charge buildup and no voltage buildup) and automatically becomes the anode with respect to the capacitively coupled electrode.

The time-average of plasma potential $V_p$, DC cathode potential (self-bias potential) $V_{DC}$, and peak-to-peak RF voltage $(V_{RF})_{pp}$ applied to the cathode are approximately related as (see Figure 3.12b):

$$2V_p \sim \frac{(V_{RF})_{PP}}{2} - |V_{DC}| \quad (3.7)$$

Clearly, the magnitude of the self-bias depends on the amplitude of the RF signal applied to the electrodes. In the RF discharge, the time-averaged RF potential of the glow region, referred to as the *plasma potential* $V_p$, is more significantly positive with respect to the grounded electrode than in the case of a DC plasma.

Positive argon ions from the plasma are extracted by the large field at the cathode and sputter that electrode at near-normal incidence, with energies ranging from a few to several hundred electronvolts, depending on plasma conditions and chamber construction. One of the most important parameters determining plasma condition is total reactor pressure. As pressure is decreased to less than 0.05–0.1 Torr, the total ion energy, $E_{max}$, increases as both the self-bias voltage and the mean free path of the bombarding ions increase. Subsequently, ion-substrate bombardment energy increases sharply with decreasing pressure. The maximum energy of positive ions striking a substrate placed on the cathode is proportional to:

$$E_{max} = e(|V_{DC}| + V_p) = eV_T \quad (3.8)$$

with $V_T = |V_{DC}| + V_p$, whereas the maximum energy for a substrate on the grounded electrode (i.e., the anode) is proportional to:

$$E_{max} = eV_p \quad (3.9)$$

Typical values are 300 eV for the cathode and less than 20 eV for the anode.[17] Equations 3.8 and 3.9 can be deduced from an inspection of Figure 3.12a. The situation where the wafers are placed on the cathode is referred to as *reactive ion etching* or *reactive sputter etching*. Chamber construction, especially the ratio of anode to cathode area, influences the rates of $V_T/V_p$ and, consequently, as calculated from Equations 3.8 and 3.9, the energy of the sputtering ions on these respective electrodes. Etching of the anode should be avoided, that is, $V_p$ should be kept small (e.g., <20 eV). To deduce the influence of the geometry of the anode and cathode on dry etching, we compare the sheath voltages in front of the anode and cathode using the Child-Langmuir equation, expressing the relationship between the ion-current flux, $j_i$, the voltage drop, $V$, over the sheath thickness of the dark space, $d$, and the mass of the current-carrying ions, $m_i$. The relation can be deduced from Equation 3.5, assuming the presence of a space-charge limited current:

$$j_i = \frac{KV^{\frac{3}{2}}}{\sqrt{m_i}d^2} \quad (3.10)$$

in which $K$ is a constant.[18] The current density of the positive ions must be equal on both the *anode* $j_i(P)$ and cathode $j_i(T)$; that is, $j_i = j_i(P) = j_i(T)$, resulting in the following relation for the sheath voltages:

$$(\text{cathode})\frac{V_T^{\frac{3}{2}}}{d_T^2} = (\text{anode})\frac{V_P^{\frac{3}{2}}}{d_P^2} \quad (3.11)$$

The plasma behaves electrically as a diode (large blocking voltage drop toward the capacitively coupled cathode and small voltage drop on the anode/plasma interface), in parallel with the sheath capacitance. As soon as any electrode tends to become positive relative to the plasma, the current increases dramatically, causing the plasma to behave as if a diode were present in the equivalent electrical circuit of the plasma. Hence a planar, parallel setup also is called a *diode* setup. The equivalent electrical circuit representing an RF plasma is represented in Figure 3.12c. The dark spaces in plasma are areas of limited conductivity and can be modeled as capacitors:

$$C \sim A/d \tag{3.12}$$

The plasma potential is determined by the relative magnitudes of the sheath capacitances, which, in turn, depend on the relative areas of anode and cathode. An RF voltage will split between two capacitances in series according to:

$$\frac{V_T}{V_P} = \frac{C_P}{C_T} \tag{3.13}$$

Using Equations 3.12 and 3.13, we can write:

$$\frac{V_T}{V_P} = \left(\frac{A_P}{d_P}\right)\left(\frac{d_T}{A_T}\right) \tag{3.14}$$

and substituting into Equation 3.11, we obtain:

$$\frac{V_T}{V_P} = \left(\frac{A_P}{A_T}\right)^4 = R^4 \tag{3.15}$$

where $A_p$ is the anode area and $A_T$ the cathode area. If there were two symmetric electrodes, both blocked capacitively, sputtering would occur on both surfaces. If the area of the cathode were significantly smaller than the other areas in contact with the discharge, the plasma potential would be small, and little sputtering would occur on the anode, whereas the cathode would sputter very effectively. Because the cathode in a setup such as that represented in Figure 3.11 usually is large (>1 m$^2$), allowing many silicon wafers or other substrates to be etched simultaneously, the grounded area needs to be larger yet. In sputtering systems, the anode and the entire sputtering chamber are grounded, creating a very small anode dark space where hardly any sputtering takes place. Other researchers have assumed conservation of the positive ion current; that is, $j_i(P) \times A_p = j_i(T) \times A_T$ instead of current density, and they find an exponent 2 in deriving Equation 3.15. The exponential in Equation 3.15 in practical systems is usually closer to 2.[5]

In a setup where the electrode dimensions are large compared with the electrode distance, the anode and cathode areas are considered equal, and little difference is made between $V_T$ and $V_p$.[5]

Higher ion energies ($V_T$ large) translate into lower etch selectivity and can be a cause of device damage (Table 3.4). Consequently, a key feature for good etch performance (Table 3.5) is effective ionization to produce very high quantities of low-energy ions and radicals at low pressures. Selectivity as high as 40:1 might be required in the future for ICs and even more for micromachines. It is generally believed that "radiation" damage of plasma can be minimized by keeping ion energies low. It is also generally believed that a lot of the radiation damage can be annealed out. In reality, very little is understood of the damage plasma can do. The higher the etch rate, the better the wafer throughput. Good selectivity, uniformity, and profile control are more easily achieved at lower etch rates. Trenches of various depths are made in Si in the manufacture of metal oxide semiconductor (MOS) devices. Shallow

**Table 3.4** Device Damage

| |
|---|
| Some "flavors" of device damage: |
| • Alkali (sodium) and heavy metal contamination |
| • Catastrophic dielectric breakdown |
| • Current-induced oxide aging |
| • Particulate contamination |
| • UV damage |
| • Temperature excursions that can activate metallurgical reactions |
| • "Rogue" stripping processes that simply do not remove all the residue |
| • Plasma-induced charges, surface damage, ion implantation |

**Table 3.5** Etch Performance*

| |
|---|
| Etch performance is judged in terms of the following: |
| 1. Etch rate: Rate at which material is removed from the substrate |
| 2. Uniformity: Etch rate constancy across the substrate |
| 3. Throughput: Amount of substrates/wafers etched during one process cycle |
| 4. Directional control: Control of the horizontal (CD) and vertical etch rate (profile)—isotropic or anisotropic |
| 5. Surface quality (residue, microloading effects) |
| 6. Device damage |
| 7. Particle control |
| 8. Reproducibility |
| 9. Selectivity of etch: |
| a. Controlled by etchant composition |
| b. Controlled by etch rate |

*See also Introduction to Part II.

trenches (0.5–1.5 μm) are used to aid in reducing the effects of lateral diffusion during processing and to make flat structures (planarization). Deeper trenches (1.5–10 μm) are used to create structures that become capacitors and isolation regions in integrated circuits. For the shallowest trenches, photoresist is adequate as a mask. For deeper etching, a silicon dioxide mask may be needed. In micromachining, one would like to obtain aspect ratios of 100 and beyond; thus, the difficulty in masking keeps mounting. With aspect ratios above 2:1 or 2.5:1, *the etching action* at the bottom of a trench tends to slow down or stop altogether.

In answer to these identified needs, equipment builders have come up with low-energy, high-density plasmas, for example, magnetrons, inductively coupled plasmas (ICPs),[19] and electron cyclotron resonances (ECRs)[20] (see below Figure 3.36).

## Physical Etching: Ion Etching or Sputtering and Ion-Beam Milling

### Introduction

Bombarding a surface with inert ions (e.g., argon ions), in a setup as shown in Figure 3.11, translates into ion etching or sputter etching. With ions of sufficient energy impinging vertically on a surface, momentum transfer (sputtering) causes bond breakage and ballistic material ejection, throwing the bombarded material across the reactor to deposit on an opposing collecting surface, provided the surrounding pressure is low enough. The kinetic energy of the incoming particles largely dictates which events are most likely to take place at the bombarded surface, that is, physisorption, surface damage, substrate heating, reflection, sputtering, or ion implantation. At energies less than 3–5 eV, incoming particles are either reflected or physisorbed. At energies between 4 and 10 eV, surface migration and surface damage results. At energies >10 eV (e.g., from 5–5000 eV), substrate heating, surface damage, and material ejection, that is, sputtering or ion etching, take place. At yet higher energies, >10,000 eV, ion implantation, that is, doping, occurs. The energy requirements for these various processes are summarized in Table 3.6 (see also Figure 3.15).

**Table 3.6** Energy Requirements Associated with Various Physical Processes

| Ion Energy (eV) | Reaction |
|---|---|
| <3 | Physical adsorption |
| 4–10 | Some surface sputtering |
| 10–5000 | Sputtering |
| 10–20 K | Implantation |

The deposition phenomenon mentioned above can be used to deposit materials in a process called *sputter deposition* (Chapter 7). A low pressure resulting in a long mean free path is required for material to leave the vicinity of the sputtered surface without being backscattered and redeposited.

In what follows, we shall consider two purely physical dry etching techniques at energies >10 eV, that is, sputtering or ion etching and ion-beam milling.

### Sputtering or Ion Etching

The gradient in the potential distribution around the target lying on the cathode (see Figure 3.11) accelerates the ions, prompting them to impinge on the substrate in a direction normal to the surface. The etch rate in the direction of the impinging ions ($V_z$) becomes a strong function of $E_{max}$ (Equations 3.8 and 3.20). The impinging ions erode or sputter etch the surface by momentum transfer. This offers some advantages: volatility of the etch products is not as critical as for dry chemical etching, and only a billiard ball effect plays a role in physical etching. As a consequence, differences in etch rates for various materials are not large (sputter yields for most are within a factor of three), and the method entails directional anisotropy. Physical etching is inherently nonselective because the ion energy required to eject material is large compared with differences in chemical bond energy and chemical reactivity. When no reactive etching processes are available for a given material, physical etching is always an option. Directional anisotropy remains as long as the dimensions of the surface topographical structures are small compared with the thickness of the sheath between the bulk plasma and the etched surface. However, etch rates are slow, typically a hundred to a few hundred angstroms per minute. The use of a magnetron can help improve the etch rate and minimize heating of substrates. In both DC and RF diode sputtering, most electrons do not cause ionization events with Ar atoms. They end up being collected by the anode, substrates, and so on, where they cause unwanted heating. A magnetron adjusts this situation by confining the electrons with magnetic fields near the target surface; consequently, current densities at the target may increase from 1 mA/cm$^2$ to 100 mA/cm$^2$.

As sputter etching is nonselective, it introduces a masking problem. Another hurdle is the need for rather high gas pressure to obtain sufficiently large ion currents, resulting in a short mean free path, $\lambda_i$, of the ions. With a mean free path smaller than the interelectrode spacing, considerable redeposition of sputtered atoms on the etching substrate lying on the cathode can occur. Electrical damage from ion bombardment also must be taken into account when critical electronic components reside on the substrate.

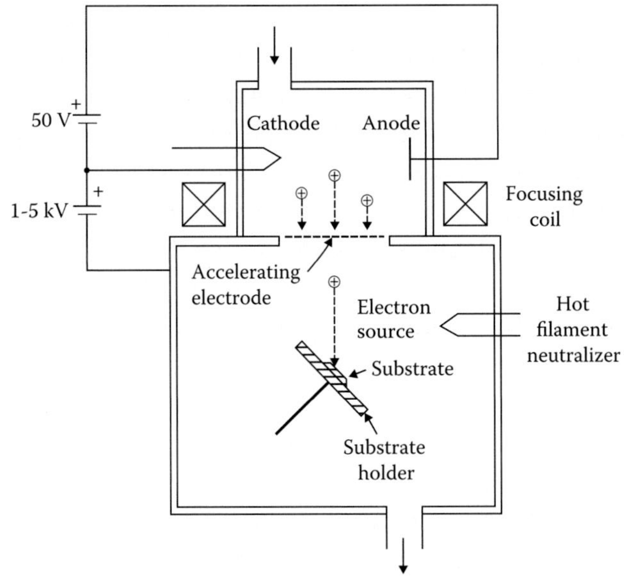


**FIGURE 3.13** Ion-beam etching (IBE) apparatus (triode). In an ion-beam apparatus, the beam diameter is approximately 8 cm. Substrates are mounted on a moveable holder allowing etching of large substrates. Coils focus the ion beam and densify the ion flux in magnetically enhanced ion etching (MIE).

## Ion-Beam Etching or Ion-Beam Milling

Ion-beam milling (IBM) is a process wherein the plasma source for ion etching is decoupled from the substrate, which is placed on a third electrode. Decoupled plasma is also called *remote* plasma, and the equipment needed is called a *triode* setup. The plasma source can be an RF discharge or a DC source (Penning source). In a typical DC triode setup, as shown in Figure 3.13, control of the energy and flux of ions to the substrate in the substrate chamber happens independently. The sample being etched can be rendered neutral by extracting low-energy electrons from an auxiliary thermionic cathode (i.e., a hot filament neutralizer), thus making this DC equipment usable for sputtering insulators, as well as conductors. In ion-beam milling, as in ion etching, noble gases are generally used as they exhibit higher sputtering yields as a result of heavy ions and avoid chemical reactions. The argon pressure in the upper portion of the chamber can be low ($10^{-4}$ Torr), resulting in a large mean free ion path. Adding a hot wire (Kaufman source) or a hollow cathode can enhance ionization at those low pressures. In a Kaufman source, electrons are emitted by a hot filament (typically tantalum or tungsten) and accelerated by a potential difference between the cathode filament and an anode. A hollow cathode is a metal cylinder with a hole several millimeters in diameter. When the pressure is decreased, the hollow cathode starts to emit extremely strong light (negative glow) when the cathode sheath thickness reaches about the same dimensions as the diameter of the hole. The hollow cathode bounces electrons back and forth within the negative glow, thereby enhancing the plasma density considerably beyond that of a normal discharge.[5]

The discharge voltage in IBE must be larger than the gas ionization potential (15.7 eV for argon) and typically is operated at several times this value, about 40–50 eV, to establish a glow discharge. Ions are extracted from the upper chamber by sieve-like electrodes (extraction grids), formed into a beam, accelerated, and fired into the lower chamber where they strike the substrate. Achievable ion energies range from about 30 eV to several keVs. Beam divergence through collisions with residual gas molecules in the substrate chamber usually limits the lowest practical ion energy to a few hundred electronvolts. Typical etch rates with argon ions of 1 keV of energy and an ion current density of 1.0 mA/cm$^2$ are in the range of 100–3000 Å/min for most materials such as silicon, polysilicon, oxides, nitrides, photoresists, and metals. Inert ion-beam etching (IBE) in an ion miller is, in principle, capable of very high resolution (<100 Å), but aspect ratios are usually less than or equal to unity, and selectivity is poor. The use of the extraction grid also increases the potential for contaminating the ion beam with sputtered grid material.

Higher plasma ion densities may be created by using magnetic coils in a magnetron sputtering machine to increase the electron path length. Just as the ionospheric van Allen belts are confined by the Earth's dipole field, plasma remains confined within the magnetic envelope generated by the coils, thus preventing electron loss to plasma-exposed surfaces. This principle is exploited, for example, in magnetically enhanced ion etching (MIE) (see coils in Figure 3.13). A magnetic field is applied so that electrons cannot pass directly from the anode to the cathode, but rather follow helical paths between collisions, greatly increasing their path length and ionization efficiency. To understand this enhancement in ionization efficiency, consider an electron moving with a velocity $v_e$. This velocity will be affected by both the electric and the magnetic fields, and the force on the electron is given by the Lorentz force (see Volume I, Equation 3.64):

$$F = e\,(V_e \times B) \qquad \text{(Volume I, Equation 3.64)}$$

in which all quantities, except for the electron charge, are vectors. The direction of the force vector on the electron is perpendicular to both the magnetic field and the direction of the velocity; consequently, the electron will spiral around the magnetic field lines (see Figure 3.14).

With an angle $\phi$ between the momentary electron velocity vector and the magnetic field vector, the radius $r$ of the electron orbit is:

$$r = \frac{m_e v_e}{eB\sin\phi} \tag{3.16}$$

where $m_e$ is the mass of the electron. Equation 3.16 is obtained by equating the magnetic and centripetal forces. When the electron is moving at right angles to the magnetic field, this simplifies to:

$$r = \frac{m_e v_e}{eB} \tag{3.17}$$

The circular motion of electrons in the presence of a magnetic field can also be described in terms of an angular frequency $\omega_c$ (also the cyclotron frequency), which can be calculated from the electron velocity $v_e$ and the radius $r$ of the electron helix as described in Equation 3.17:

$$\omega_c = \frac{v_e}{r} \tag{3.18}$$

At an electron energy of 2 eV, the mean electron velocity is $9 \times 10^5$ m s$^{-1}$ (see Equation 3.2). According to Equation 3.17, with a magnetic field $B$ of $5 \times 10^{-3}$ Tesla (T) applied, electrons will be forced into a circular path with a radius of about 1 mm.[21] At a higher electron temperature of 100 eV, the cyclotron radius of an electron in a field of 0.01 T (or 100 Gauss) is 3.2 mm. These cyclotron radii are of the order of the mean free path and smaller than the smallest dimension of the average dry etch reactor. By lengthening the electron path considerably, electrons are prevented from reaching the reactor walls where

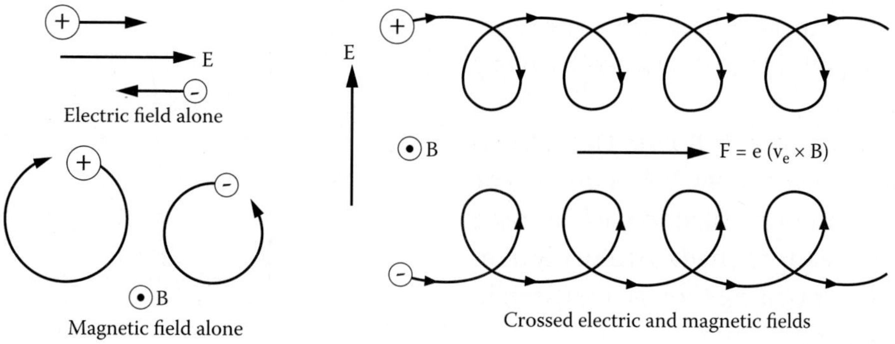


**FIGURE 3.14** Electron path with electric field alone and with magnetic field alone and finally with crossed electric and magnetic fields.

they would be lost. Any velocity component of an electron parallel to the magnetic field will change the circular path into a helix. In principle, ions also will move in helical paths; however, the radius $r$ (see Equation 3.17), because of the heavy mass of ions, is so large that the effect can be neglected. Clearly, from the above, a magnetic field can lower wall recombination by confining the electrons, hence making relatively low field strengths adequate for obtaining high plasma densities. When plasma density is increased with magnetic confinement, the degree of ionization is between $10^{-2}$ and $10^{-4}$, compared with $10^{-4}$–$10^{-6}$ for simple plate discharges, resulting in an increased etch rate.

The cyclotron resonance frequency of an electron estimated from Equation 3.18 is about $10^8$ $s^{-1}$, which is in the microwave range. If the RF excitation frequency is in the same range, resonance takes effect, causing an additional increase in the RF energy transfer efficiency. Microwave energy may be injected into plasma to energize the ECR ion source as reviewed further below (Figure 3.36b).

When inert argon ions in a triode setup as shown in Figure 3.13 are replaced with reactive ions, reactive ion-beam etching (RIBE) occurs. The ions not only transfer momentum to the surface but they also react directly with the surface, which means a chemical/physical mechanism is involved. Direct reactive ion etching of a substrate is the exception rather than the rule, and it is the radicals generated in the plasma that usually dominate the chemical reactions at the surface.

Systems using a hot filament to produce a plasma have a relatively short lifetime in the presence of reactive gases, as conventional filament materials tend to etch away (e.g., tungsten in the presence of halides forms volatile tungsten halides).[21] The latter type of plasma source problem is avoided by using microwave excitation.

IBE may cover an area of 3–8 cm. This also is referred to as *showered ion etching*. Yet another type of IBE, a maskless technique called *focused ion beam* (FIB), in which the beam is made extremely narrow and used as a direct writing tool, is discussed in more detail in Chapter 6.

Because of its low etch rate and low selectivity, IBE is used more for fundamental mechanistic studies rather than for patterning. For example, in IBE, the substrate can be tilted relative to the direction of ion bombardment, and, as the ion energy at the substrate and the angle of incidence are known, fundamental information, such as sputter yields, can be obtained.[21] The sputter yield $S$ is defined as the mean number of atoms removed from the surface per incident ion. The sputtering yield depends on the incident ion energy, the mass of the incident ion, the mass of the substrate atom to be etched away, the angle of the incident ion with respect to the substrate, the crystallinity and the crystal orientation of the substrate, the temperature of the substrate during etching, and the partial pressure of oxygen in the residual gas. The sputter yield as a function of the incident ion energy is shown in Figure 3.15. The sputtering yield increases with the incident ion energy and reaches a broad maximum at energies between 5 and 50 keV. Beyond 50 keV, the sputtering yield decreases because of the deeper penetration of the ions in the substrate—that is, ion implantation.[22] The angular dependence of the sputtering yield $S$ will be considered below when we discuss etching profiles in physical etching.

In Chapter 7 we will learn more about sputter deposition. In sputter deposition the substrate on the cathode is replaced by a sputter target cathode, and materials sputtered from that target are deposited on the substrate now placed on the anode.

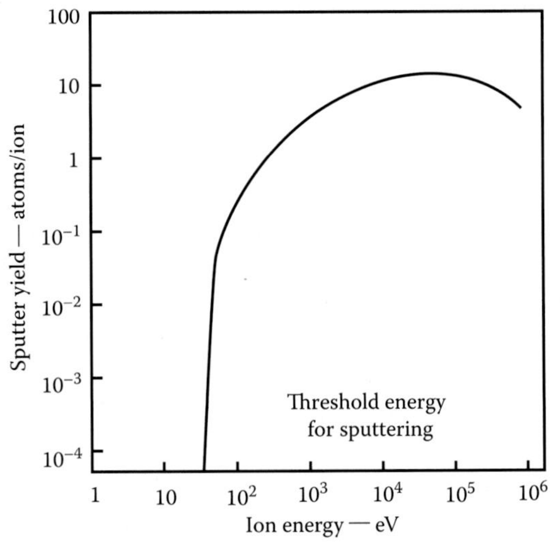


**FIGURE 3.15** Sputter yield as a function of the energy of the bombarding ion.

## Etching Profiles in Physical Etching

### *Introduction*

The ideal result in dry or wet etching is usually the high-fidelity transfer of the mask pattern onto the substrate, with no distortion of critical dimensions (CDs). Isotropic etching (dry or wet) always enlarges features and thus distorts CDs. Chemical anisotropic wet etching is crystallographic. As a consequence, CDs can be maintained as long as features are strategically aligned along certain lattice planes (see Chapter 4 on wet chemical etching and wet bulk micromachining: pools as tools). With sputtering, the anisotropy is controllable by the plasma conditions.

As can be seen from Figure 3.3a, ion etching and ion milling do not lead to undercutting of the mask, but the walls of an etched cut are not necessarily vertical. A variety of factors contribute to this loss of fidelity in pattern transfer, and they are either caused by involatile sputtering reaction products or by special ion-surface interactions. We shall briefly review these dry physical ion etching problem areas.

### *Faceting Due to Angle-Dependent Sputter Rate*

In Figure 3.15 sputter yield *S* is shown as a function of the energy *E* of the bombarding ion. The sputter yield is a complex function of the binding energy *U* of the material and its atomic number $Z_{material}$ as well as the nature of the sputtering gas (with its own atomic number $Z_{gas}$). *S* increases for heavier gases and also depends on the angle of incidence of the ions (θ). The dependence of *S* on different factors is summarized in Table 3.7. Many different phenomena may occur when an ion hits a substrate/target.

**Table 3.7** Sputtering Yield S Dependence on Various Factors

| Factor | | Effect on Sputter Yield S |
|---|---|---|
| Target material | Binding energy U | Higher U means lower S |
| | Atomic number $Z_{material}$ | Higher Z means lower S |
| Sputtering gas | Atomic number $Z_{gas}$ | S increases for heavier gases |
| | Incident energy E | S increases for higher energy |
| Geometry | | S reaches a maximum 20–30° of glancing |

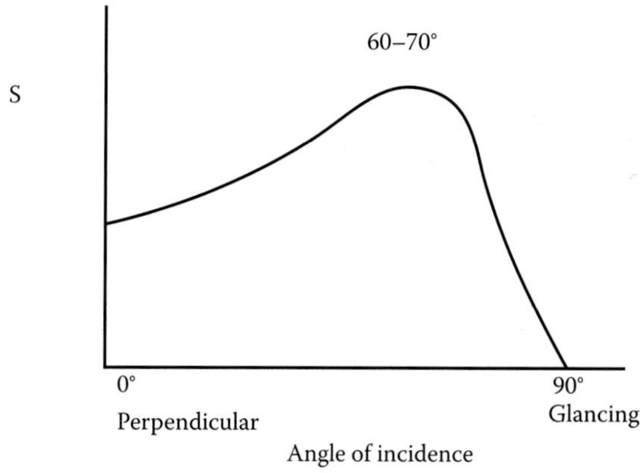


**FIGURE 3.16** Sputter yield as a function of angle of incidence of bombarding ions.

The ions may bounce back, implant, absorb on the surface, or reflect by a grazing collision sequence.

In Figure 3.16 we plot the sputter yield *S* as a function of θ, the angle at which ions are directed at the surface (0° is perpendicular and 90° is glancing). Sputtering is often most efficient at about 20–30° off glancing. In sputtering (at energies <50 keV), the incoming ions may be modeled as hard spheres colliding with the substrate, with 95% of the incident energy going into the substrate (thus making cooling of the target absolutely required) and 5% of the incident energy carried off by substrate/target atoms. The atoms come off with a cosine distribution. It intuitively can be understood that 1) at a glancing angle the incoming ions do not eject atoms from the substrate/target (S = 0), 2) hitting the substrate perpendicularly with ions leads to fewer atoms being ejected from the substrate/target, and 3) that in between there are a maximum number of atoms being kicked out.

Even when starting out with a vertical mask sidewall, ion sputtering exhibits a tendency to develop a facet on the mask edge at the angle of maximum etch rate. This corner faceting is detailed in Figure 3.17a. The corner of the mask, always a little rounded even when the mask walls are very vertical, etches faster than the rest and is worn off. Faceting at the mask corner arises because the sputter yield *S* for materials is a function of the angle at which ions are directed at the surface [$S = f(\theta)$, see Figure 3.16]. The sputter-etch rate of resist, for example, reaches a maximum at an incidence angle of about 60°, more than twice the rate at normal incidence.[22] Sloped mask sidewalls may eventually be followed by sloped etch steps in the

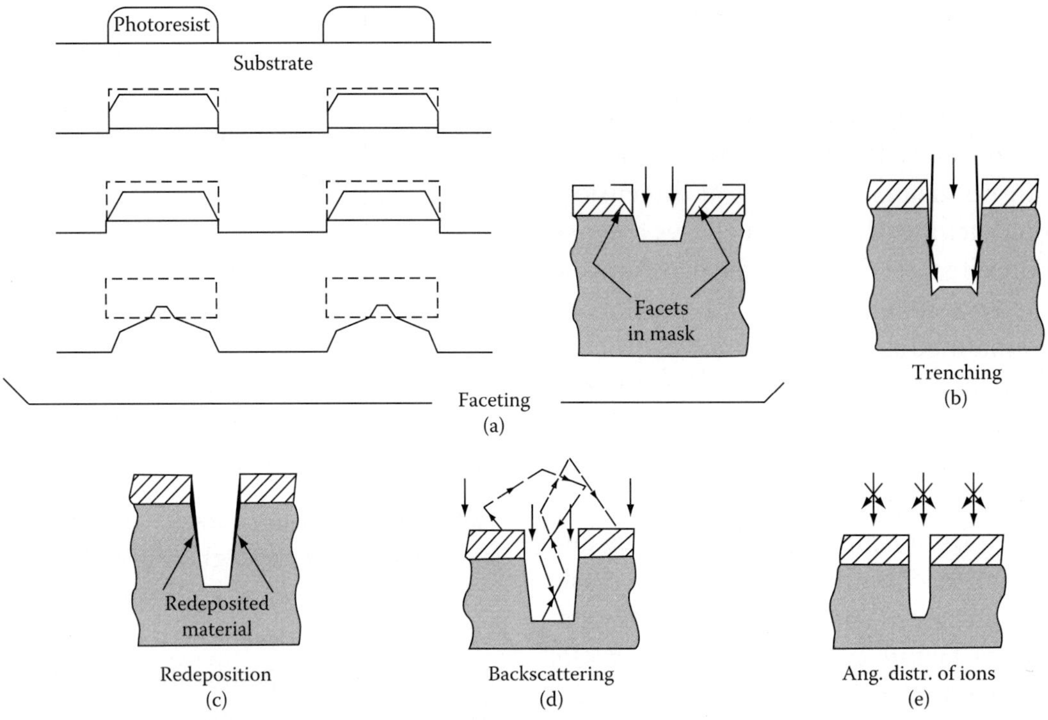


**FIGURE 3.17** Limitations of dry physical etching. (a) Faceting. Sputtering creates angled features. An angled facet (~60°) in the resist propagates as the mask is eroded away. Sloped walls may also be created in the underlying substrate. (b) Ditching or trenching as a result of glancing incidence of ions. (c) Redeposition of material sputtered from the bottom of a trench. (d) Backscattering. (e) Angular distribution of incident ions. (Based on Lehmann, H. W. 1991. *Thin film processes II*. Eds. J. L. Vossen, and W. Kern. Boston: Academic Press.[1])

substrate. Faceting of the substrate itself will proceed along its own preferred sputtering direction angle. The faceting is more pronounced with an applied bias as a result of the increased electric field at corners. Usually, faceting affects only the masking pattern, and making the mask sufficiently thick minimizes its influence on the fidelity of the pattern transfer.

It should be noted that some of the disadvantages of physical etching, such as resist corner faceting, can sometimes be exploited. For example, a gently sloping edge is advantageous to facilitate metal coverage or planarization because a tapered sidewall is easier to cover than a vertical wall, especially when using line-of-site deposition methods. The method most often used to obtain such a taper is a controlled resist failure, that is, erodible or sacrificial masks (Figure 3.18). In other words, the "negative effect" of faceting described in connection with Figure 3.17a is put to good use.

### *Ditching or Trenching*

When the slope of the side of the mask is no longer completely vertical, some ions will collide at a glancing angle with the sloping edges before they arrive at the etch surface. This gives a local increase in etch rate, leading to ditches (Figure 3.17b). For this mechanism to be active, there must be a sizable fraction of ions with at least slightly off-vertical trajectories, or the sidewall must have a slight taper as shown in Figure 3.17b.[23] For example, the taper of the sidewall could result from redeposition (see next section) or faceting (see above). Because ditching is a

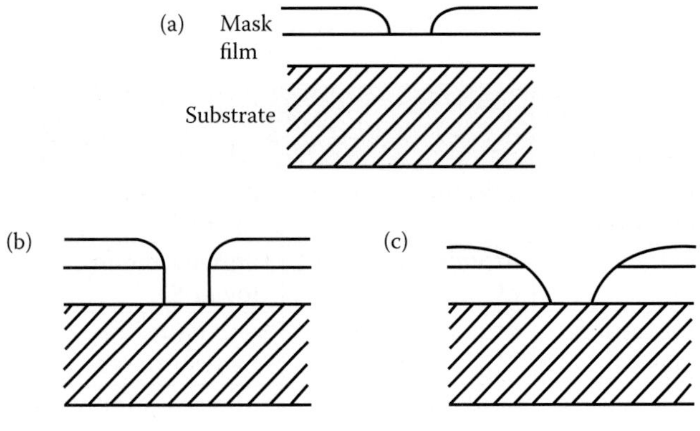


**FIGURE 3.18** Erodible or sacrificial masks. (a) Patterned masking structure prior to etching. (b) Nonerodible mask. (c) Partially erodible mask.

small effect (e.g., 5%), it often goes unnoticed unless thick layers are etched.

### *Redeposition*

Another sputtering limitation, already alluded to, is the redeposition of involatile products on step edges (Figure 3.17c). Redeposition involves sputtered involatile species from the bottom of the trench settling on the sidewalls of the mask and etched trench. The phenomenon manifests itself mainly on sloped sidewalls.

By tilting and rotating the substrate during etching (Figure 3.19), etch profiles can be improved. The reasons for tilting and rotating improvements are a combination of shadowing the bottom of the step (to reduce trenching), partially etching the sidewalls of the mask (to reduce redeposition), and gaining more nearly vertical edges on the etched profiles in the substrate. Especially with aspect ratios exceeding unity, redeposition becomes problematic, and reactive gas additives are necessary to generate volatile etch products. With aspect ratios greater than 2:1 or 2.5:1, the etching action at the bottom of too-fine features tends to slow down or stop altogether. Reactive additives bring us into the realm of chemical-physical etching (discussed further below). For micromachining, tilting and rotating of substrates are recurring topics. They are some of the desirable modifications of standard equipment used for micromachining applications (for example, see the fabrication of Spindt field emitters shown in Figure 2.42).

Redeposition may also be used to control sidewall profiles. The material redeposited on the sidewall protects the sidewalls, whereas the material redeposited at the bottom gets removed during the etching process. With the proper control of the redeposition, one can achieve a highly anisotropic profile.

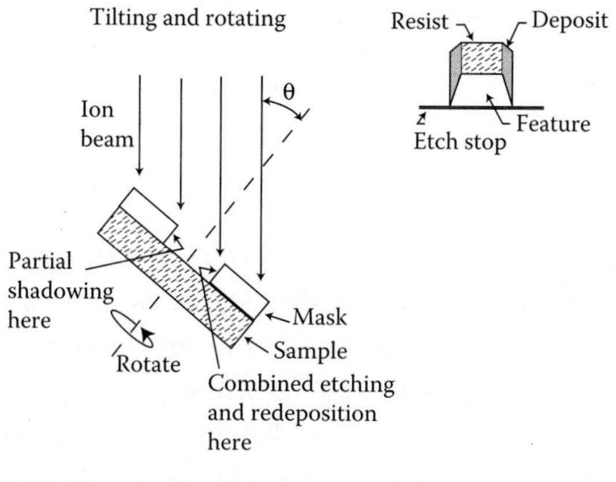


**FIGURE 3.19** Left: By tilting and rotating the substrate during etching etch profiles can be improved. Right: Redeposition may be used to control sidewall profiles.

### *Backscattering*

Backscattering, illustrated in Figure 3.17d, is a form of redeposition associated with involatile etch products. A fraction of the sputtered and involatile species from the surface is backscattered onto the substrate after several collisions with gas phase species. This indirect redeposition may involve contaminants from the walls and fixtures in the vacuum chamber. Backscattering fixes the upper pressure limit for ion-enhanced etching. Significant redeposition can take place at pressures as low as 10 mTorr.[1]

### *Angular Distribution of Incident Ions*

Sheath scattering and/or field nonuniformities cause off-vertical ion trajectories. Most ions impinge perpendicularly on the substrate. Scattering of a small fraction of ions in the sheath causes a distribution of impingement angles.[1] This scattering in the sheath is responsible for hourglass-shaped etch profiles that have been observed in trench etching of silicon (see Figure 3.17e). Besides sheath scattering, there is a second mechanism active that may lead to skewing of ion directionality; namely, inhomogeneities in the electrical field at the substrate surface. With etching conductors, the bending of electric field lines caused by surface topography has the effect of enhancing ion flux at feature edges and leads to ditching. On the other hand, with etching insulators, charging effects may cause appreciable ion fluxes to the sidewalls of a trench and contribute to lateral etching. The latter will again lead to hourglass-shaped trenches. The importance of the effect is very much related to the electrical conductivity of the masking and etching surfaces, with the greatest significance for strongly insulating materials.[24]

## Temperature Control during Dry Etching

During dry etching, heavy ion bombardment of the substrate generates a lot of heat. To protect photoresist masks, temperature control is required. This could in principle be achieved by water cooling the wafer-chuck (pedestal or cathode) although the low pressure makes for poor heat transfer between the substrate and the wafer-chuck, so helium backside cooling is required. Alternatively a clamp ring or electrostatic chuck (E-chuck) is used to hold the wafer.

### Physical Etching Summary

In physical etching, ion etching or sputtering, and ion-beam milling, argon or other inert ions extracted from the glow discharge region are accelerated in an electrical field toward the substrate, where etching is purely impact controlled. Sputtering is inherently nonselective because large ion energies compared with the differences in surface bond energies and chemical reactivities are involved in ejecting substrate material. The method is slow compared with other dry etching means, with etch rates limited to several hundreds of angstroms per minute compared with thousands of angstroms per minute, and higher for chemical and ion-assisted etching (as high as 6 μm/min with the newest deep RIE equipment). Sputter etching tends to form facets, ditches, and hourglass-shaped trenches and frequently redeposits material in high-aspect-ratio (>2:1) features. Electrical damage to the substrate from ion bombardment and implantation can be problematic. However, some of the reversible plasma damage can be removed by a thermal anneal. With the continuing increase in device complexity, which includes layers of different chemical composition, inert ion etching and ion-beam sputtering will continue to find new applications.

## Plasma Etching (Radical Etching)

### Introduction

In reactive plasma etching (PE), which corresponds to a high-pressure RIE extreme, reactive neutral chemical species such as chlorine or fluorine atoms and molecular species generated in the plasma diffuse to the substrate where they form volatile products with the layer to be removed. The only role of the plasma is to supply gaseous, reactive etchant species. Consequently, if the feed gas were reactive enough, no plasma would be needed. At pressures of $>10^{-3}$ Torr, the neutrals strike the surface at random angles, leading to isotropic, rounded features. A dry chemical etching regime can be established by operating at low voltages, eliminating impingement of high-energy ions on the sample, and facilitating surface etching almost exclusively by chemically active, neutral species formed in the plasma. The reaction products—volatile gases—are removed by the vacuum system. The volatility of the formed reaction product introduces a major difference with sputtering, where involatile fragments are ejected like a billiard ball and may be redeposited close by.

### Reactor Configurations

Reactive plasma etching is one extreme of RIE and basically follows the same process we encountered when discussing dry stripping of resists (Chapter 1, Dry Stripping). Three different popular configurations for plasma etching (barrel reactor, downstream etcher, and parallel-plate system) were shown in Figure 1.36. In resist etching, the process is called *ashing*. Depending on the configuration, high-energy ion bombardment of the substrate can be prevented, more or less, and plasma-induced device damage avoided. For example, in a barrel reactor, the substrates are shielded by a perforated metal shield to reduce substrate exposure to charged high energetic species in the plasma. In a downstream stripper, the geometry of the reactor allows reactant generation and stripping to take place in two physically separated zones (triode-type configuration with a remote plasma). In parallel-plate strippers, the substrates are placed inside the plasma source, which leads to higher ion damage compared with the two previous methods.

### Reaction Mechanism

The plasma etching process can be broken down into as many as five primary steps, as illustrated in Figure 3.20. The first step is the production of the reactive species in the gas phase (1). In a glow discharge, a gas such as $CF_4$ dissociates to some degree by impact with energetic particles in the plasma with an average energy distribution between 1 and 10 eV. In the dissociation, reactive species such as

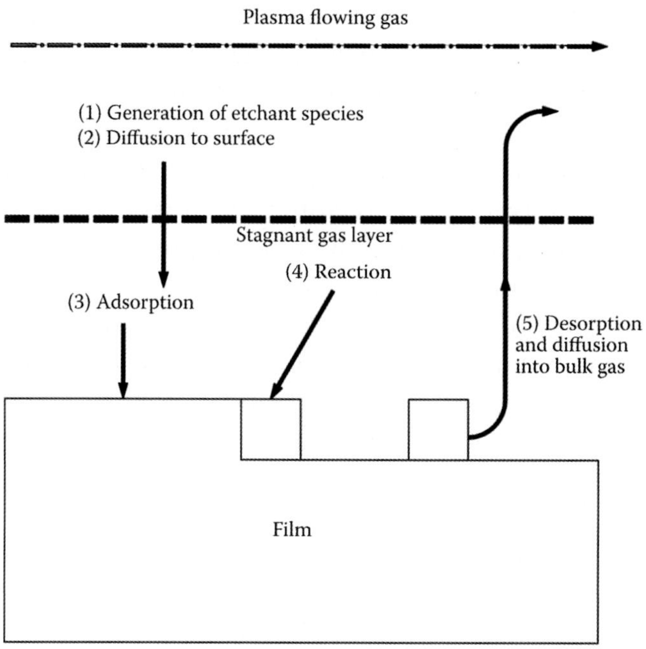


**FIGURE 3.20** Primary process steps occurring in a plasma etch process. See text for details.

$CF_3^+$, $CF_3$, and F are formed. This step is vital because most of the gases used to etch thin films do not react spontaneously with the film; for example, $CF_4$ does not etch silicon. In a second step, the reactive species diffuse to the solid (2) where they become adsorbed (3), diffuse over the surface, and react with the surface (4). The etching species must get to the surface to react with a thin film or substrate molecules. The mechanics of getting to the surface can limit aspect ratio and determines undercutting and uniformity. Adsorption also can affect aspect ratio. The reaction is strongly temperature-dependent. Finally, the reaction products leave the surface by desorption and diffusion to the bulk (5). Desorption can stop the etching process if the reacted species are not volatile, and diffusion to the bulk gas phase can lead to nonuniform etching as a result of dilution of unreacted etching species. As in a parallel resistor combination, total resistance is determined by the smallest resistance—the reaction with the smallest rate constant determines the overall reaction rate.

Some of the reaction steps listed above occur in the gas phase and are termed *homogeneous* reactions, whereas others occur at the surface and are called *heterogeneous* reactions. For homogeneous reactions, the plasma only plays the role of creating highly reactive species from the plasma gas. Radicals are more abundant in a glow discharge than ions because they are generated at a lower threshold energy (e.g., <8 eV), which leads to a higher generation rate. Moreover, the uncharged radicals have a longer lifetime. Low-energy ions rarely act as the reactant themselves; instead, neutrals are responsible for most reactive etching (chemical etching) at pressures greater than 0.001–0.005 Torr. Radicals and molecules formed in the plasma are not inherently more chemically reactive than ions, but they are present in significantly higher concentrations. Heterogeneous reactions display even more complexity than homogeneous reactions. In principle, all the species generated in the plasma may influence the reaction rate at the surface; nonreactive species may decrease the reaction rate by blocking surface sites, and adsorption of radicals may enhance the reaction rate. Radicals and other neutrals reach the surface by diffusion, whereas ions are accelerated toward the surface by the negative potential on the substrate electrode. Besides chemically active species and ions, the effect of electron bombardment and irradiation by visible and UV radiation emanating from the plasma requires consideration. Chemical etching occurs at low bias, and because, in principle, no highly energetic ions bombard the surface, sputtering itself is not an important surface-removal mechanism, and radiation damage to the substrate is reduced. Unfortunately, the term *reactive ion etching* is used indiscriminately for all chemical dry etching, even though ions themselves are not the major reactive species. Radicals and molecules also serve as the primary depositing species for all types of films in plasma-enhanced chemical vapor deposition (PECVD) (see Chapter 7). Ions directly participate in chemical etching only in RIBE, where ion reactions at very low pressures can etch at modest rates of less than 400 Å/min. RIBE is an example of physical/chemical dry etching where the same ion has both a physical (ion impact) and a chemical (reactive etching) component. Working with remote plasma in a triode-type system (e.g., a downstream stripper) further reduces ion bombardment of and current flux to the wafer. The steps of reactant generation and etching are separated efficiently because charged species (mainly electrons and positive

$O_2^+$ ions in the case of pure oxygen plasma used for stripping resist) suffer a much higher loss than reactive neutrals such as atomic oxygen as a result of the presence of plasma excitation.

In the absence of crystallographic effects (typically seen with III–V compounds but not with Si), chemical dry etching leads to isotropic profiles only when ions do not assist the reaction. In the case where ions assist, anisotropy is induced (see below under Energy-Driven Anisotropy or Damage Mechanism). Isotropic etching of the reactive species leads to mask undercutting. In some cases mask undercutting is required—for example, in device fabrication involving lift-off or when layers on sidewalls must be cleared.

Because of the chemical nature of the etching process, a high degree of control over the relative etch rates of different materials (i.e., selectivity) can be obtained by choosing suitable reactive gases. During stripping, oxygen plasma removes photoresists by oxidizing the hydrocarbon material to volatile products. Fluorine compounds are used for silicon etching, and many materials are susceptible to chlorine etching. For example, aluminum is etched in chlorine but not in fluorine because aluminum chlorides are volatile, whereas aluminum fluorides are involatile. Mixtures of gaseous compounds such as $CF_4$ (fluorocarbon)-$O_2$ assist in patterning silicon, silicon dioxide, silicon nitride, and so on.

## Macro- and Microscopic Loading Effects

### *Macroscopic Loading Effects*

In dry etching, the number of radicals in the plasma is in the same range as the number of atoms to be removed; in wet etching, on the other hand, the number of etchant molecules might be $10^5$ times higher than the number of atoms to be removed. A "loading effect" occurs in dry etching as the etchant is depleted by reaction with the substrate material. As a result, the etch rate is inversely proportional to the Si area that is exposed to the plasma.[2] The more purely chemical the etching process, the bigger the loading effect. With lower pressures, the loading effect becomes smaller. Conventional plasmas can sustain enough radicals to etch at rates of 1000 Å/min. With more effective power sources, such as a cyclotron or magnetron, $10^4$ Å/min can be achieved.

The loading effect brings an important limitation of dry etching to light: as the etch rate becomes dependent on wafer loading, uniformity is jeopardized. If the supply of reactant limits the etch rate, small variations in flow rate or gas distribution uniformity may lead to nonuniformities in the etch rate. The gas flow $F$ is the most important parameter to control in this regard. Gas flow rates are on the order of 5–200 sccm (standard cubic centimeter per minute), with a maximum flow rate set by the maximum pumping speed and the desired working pressure.[1] The symmetry of the gas flow (i.e., the relative position of gas inlet and pumping port) has to be optimized for a given reactor configuration. Hence a minimum flow rate of the reactant gas prevents the process from being limited by reactant supply. A utilization factor, $U$ (the ratio of rate of formation of etch product to the rate of etch gas flow), may be defined, and $U \geq 0.1$ is suggested for uniform etching. To illustrate: a 500-Å/min etch rate of a 3-in. Si wafer in a $CF_4$ plasma corresponds to a removal rate of $5 \times 10^{19}$ Si atoms/min, or a $SiF_4$ evolution rate of approximately 2 sccm. This means that the $CF_4$ flow rate should be at least 20 sccm.[1,25]

Closely linked to the utilization factor $U$ is the residence time $\tau$ in seconds of the feed gas in the dry etch reactor. The residence time is given by:

$$\tau = \frac{PV}{F} \qquad (3.19)$$

where V = reactor volume in liters
P = steady-state pressure in Torr
F = flow rate of the feed gas in Torr-liters per second

(Most gas flow meters are calibrated in standard cubic centimeters per minute and 1.0 Torr-L/s = 79.0 sccm.)

The residence time represents the average length of time a molecule of gas spends in the chamber irrespective of any chemical reaction that might occur.

Etching uniformity is also impacted by the relative reactivity of the wafer surface with respect to the cathode material used. Resulting nonuniformities in this case are referred to as *bull's eyes* because circular interference patterns appear on an etched wafer.

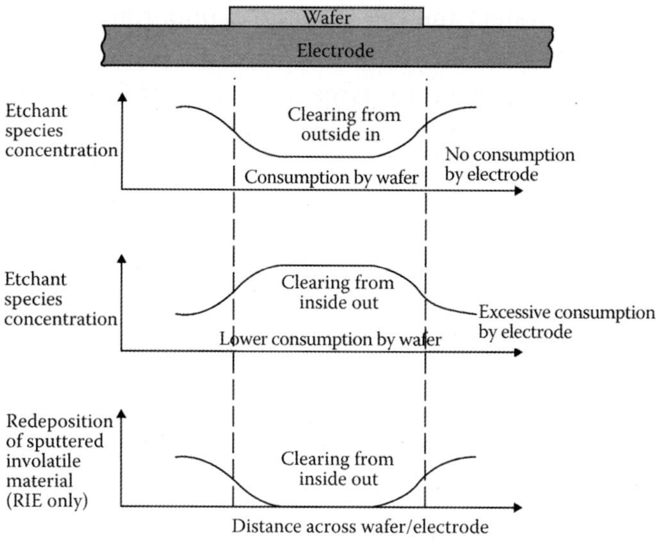


**FIGURE 3.21** Chemical and physical effects leading to nonuniform etching across a wafer (bull's eye effect). (From Elwenspoek, M., H. Gardeniers, M. de Boer, and A. Prak. 1994. *Micromechanics*. Report No. 122830, Twente, the Netherlands: University of Twente.[25])

If an aluminum electrode is used for Si or poly-Si etching in $SF_6$ plasma, a very pronounced bull's eye effect will be evident. The striking nonuniformity in this etching pattern results from a lower consumption of reactant species above the aluminum electrode because aluminum only mildly reacts with reactive species formed by $SF_6$. By applying an electrode material that consumes the fluorine reactive species as fast as Si itself (e.g., a Si cathode), concentration gradients of reactant species at the edge of the wafer are avoided, resulting in uniform etching (see Figure 3.21).

### *Microloading Effects during Plasma Etching*

The loading effect may also be observed while etching different features on the same wafer. In this case one refers to local loading or microloading effect. For example:

1. Density of the unmasked areas may vary over small distances, resulting in differences in etch rate analogous to macroscopic loading.
2. Etch rates in DRIE show a dependence on the aspect ratio of the feature being etched; the etch rate is diffusion limited and decreases with increasing aspect ratio. In cases where a high-aspect-ratio trench is etched in the vicinity of a wide trench, the latter locally decreases the availability of etching species at the expense of the narrower trench.[26] The effect is also known as "RIE lag" or aspect-ratio-dependent etching (ARDE). Lower etch rate for higher AR trenches (smaller widths) is illustrated in Figure 3.22.

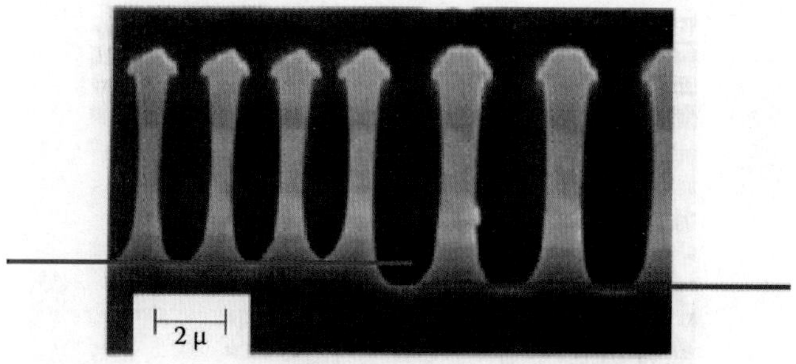


**FIGURE 3.22** Illustration of RIE lag. The probability of reactant species reaching the bottom is reduced. (See red lines; wider trenches etch deeper.)

Possible mechanisms for explaining the RIE lag are the depletion or trapping of the reactant species (conductance limitations) as the species travel to the bottom of the trench, thus causing distortion of the ion paths as a result of charging and shadowing effects involving neutrals or ions. In all cases the result is the reduction of the probability of reactant species reaching the bottom.

## Silicon Grass or Black Silicon

A common issue of nonuniformity with RIE, especially in the case of deep dry chemical etching (DRIE; see below), is the micrograss or black silicon structures observed, for example, at the bottom of a deep Si trench (Figure 3.23). Jansen et al.[2,27] associated the effect, with all types of micro- and nanomasking from materials deposited or grown on the Si, for example, native oxide or dust (that is, contamination already present on the wafer before etching). Si spikes formed as a result of dirty wafers before etching are easily prevented by a precleaning step, and because the etch is very selective, the native oxide needs to be removed completely. More grass may form during etching because of redeposition of mask material.[2]

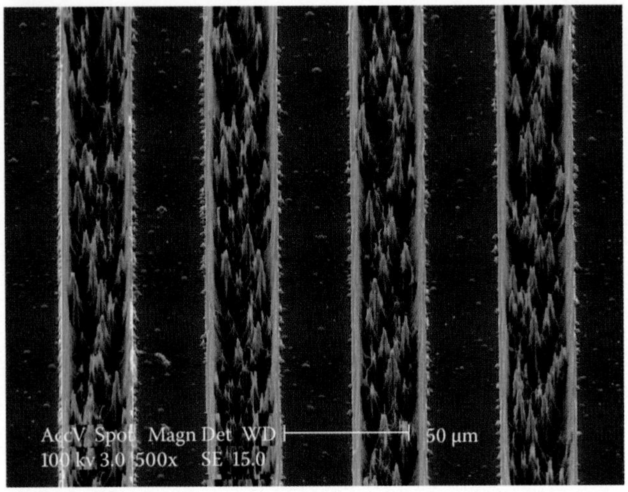


**FIGURE 3.23** Silicon "micrograss" structures, which form by plasma etching of silicon in an atmosphere that causes small pieces of silicon dioxide to sputter and then block (mask) etching where they land.

This source of black Si is more pronounced with higher ion energies; therefore, working with a lower self-bias is in general advantageous. Grass formation during RIE was also explained as a consequence of the sharpening of the ion angular distribution with the increasing aspect ratio of a trench during etching.[23] To further minimize the latter types of nonuniformities, wafers should be positioned away from the edges of electrodes to eliminate edge effects caused by changing sheath thicknesses and varying the angles of incidence. A good thermal contact between wafers and cathode is also important. Local temperature variations caused by nonuniform heat-sinking can lead to large etch nonuniformities, particularly in chemically dominated processes.[25]

## Atmospheric Plasma Etching

### *Introduction*

Until recently, most plasma science experiments were carried out under vacuum conditions. However, atmospheric-pressure plasmas enjoyed a growth spurt in the 1980s driven by new applications such as high power lasers, MEMS electrostatic switches, novel plasma processing applications and sputtering, EM absorbers and reflectors, remediation of gaseous and other pollutants and waste streams, sterilization, decontamination, and other biomedical applications. Recently, new hardware designs have enabled the stable operation of nonthermal plasmas at atmospheric pressure, and publications on atmospheric-pressure glow discharge plasmas have grown rapidly. We introduce two examples of atmospheric plasmas: one operates in an argon plasma arc, whereas the other is arc free.

The main characteristics of a plasma, such as the breakdown voltage and the voltage current characteristic and structure of the discharge, depend on the geometry of the electrodes and the vessel, the gas used, and the electrode material. In the highly

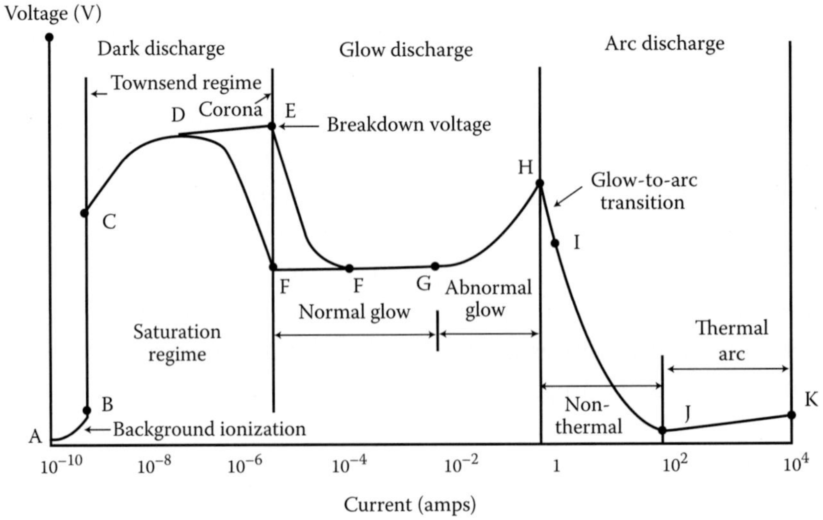


**FIGURE 3.24** Voltage current plot of a glow discharge.

nonlinear voltage current plot of plasma, as shown in Figure 3.24, we discern several regimes:

**A to B**: The applied electrical field E accelerates residual ions and electrons as a result of background radiation in the reactor (e.g., from cosmic rays, radioactive materials in the building). These ions and electrons move toward the electrodes.

**B to C**: As the applied V increases, eventually all the available residual charges are swept up. This causes the current through the plasma to saturate.

**C to E**: As the applied V increases beyond C, the E field is now large enough that the remaining electrons can ionize gas. As V increases further, secondary electrons also cause ionization.

**D to E**: In the so-called Townsend regime, coronal discharges occur at various edges and sharp points of the electrodes in the gas. This is the result of the E field concentration at such sites.

**A to E**: Electrical breakdown occurs. The voltage is now high enough to cause ionization to occur because of secondary electrons in the gas. Additional secondary electrons are generated at the cathode as a result of ion impact on it. There is a sudden increase of the current by several orders of magnitude, causing an avalanche.

**F to G**: After breakdown from E to F, the gas enters the normal glow where the voltage V is almost independent of the current over several orders of magnitude. As the current increases, the fraction of the cathode occupied by plasma increases until the plasma covers the entire cathode surface at G.

**G to H**: In the abnormal glow regime, further increases in the applied voltage cause increases in the current. This is the region of operation for most sputtering and other plasma systems.

**H to K**: At H, the electrodes have become sufficiently hot that the cathode can now emit electrons thermionically, and a second avalanche can happen in the arc region (this is the regime used in spectrochemistry). If the DC power supply is capable (low internal R), a transition from glow to arc happens in the region from H to I. In this arc regime from I to K, the discharge V decreases as I increases. The point J marks the transition from a nonthermal to a thermal arc (electrons have the same temperature as the gas).

At atmospheric pressures, plasmas are weakly ionized, and the mean-free path for electron-molecule collisions is very short (<1 μm). These plasmas are typically thermal with electrons and gas at the same temperature. At atmospheric pressure a small perturbation can easily cause a glow-to-arc transition. The transition time is very short, and this makes it hard to stabilize an atmospheric glow discharge. Some tricks to prevent an arc include the use of inert gas to lower breakdown voltage, high-frequency power sources and operation in the dielectric barrier discharge mode, and a structured cathode (e.g., a pin cathode). Stabilization of high-pressure (atmospheric-pressure) plasmas is most easily achieved if the plasma is confined to "small" spaces. Interesting pattern formation and self-organization has been observed in these microplasmas.

### *Atmospheric Downstream Plasma*

In atmospheric plasma jet etching, reactive chemical species are generated in a DC arc between two electrodes in a noble gas (such as argon) at atmospheric pressure. A stream of reactant gas (the gas jet) such as $CF_4$ then flows into the arc and onto the wafer, which is located up to 12 in. away to avoid damage, as shown in Figure 3.25.[28] Although atmospheric plasma sources have been used previously for industrial powder spray and plasma cutting (see Chapters 5 and 9), they have not been used in microelectronic processing for fear of damaging and contaminating the silicon wafer surface. Among the concerns were possible overheating as a result of high power density, contamination from chemical erosion of electrodes, and nonuniform etching.

Atmospheric downstream plasma (ADP) processing overcomes these problems by using a magnetically controlled, inert gas, DC arc-plasma discharge.

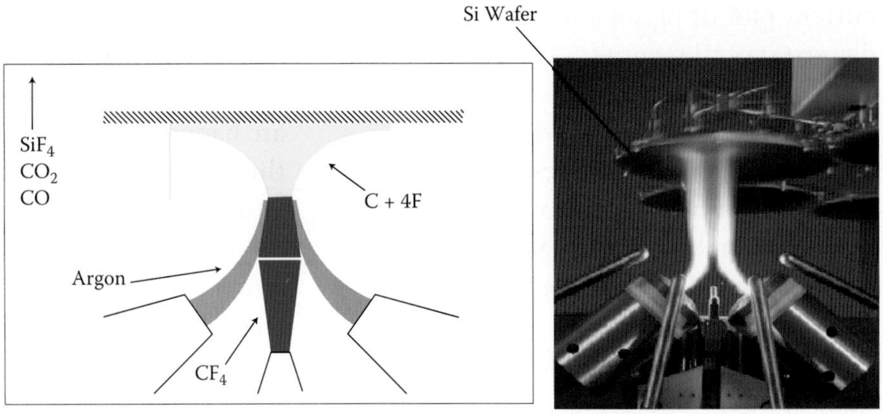


**FIGURE 3.25** Atmospheric downstream plasma processing of a Si wafer. (From Tru-Silicon Technologies, http://www.trusi.com/frames.asp?.)

Plasma jet etching is an atmospheric variant of plasma etching capable of producing very high etch rates without substrate damage. By deploying the plasma in a downstream configuration, ADP prevents ion impact damage while keeping wafer temperatures low. This technology may improve the performance and reduce the cost of bulk material removal operations, such as mechanical grinding, lapping, and wet chemical etching [especially chemomechanical polishing (CMP); see Chapter 1, Figure 1.66]. ADP etching for back-side damage removal after grinding and wafer thinning is a potential major application. Other applications include damage-free dicing and via etching (for the latter two applications, an aluminum mask is used). The trend to thinner wafers and thinner chips will not stop with the current 200–250-μm thickness, but is forecast to continue reducing to 100–150 μm and even thinner (see Figure 3.26), driven by the demand for portable electronic devices, smart cards, multichip modules, PC Card (originally PCMCIA Card for People Can't Memorize Computer Industry Acronyms or more seriously for Personal Computer Memory Card Interface Association), and so on. Wafer strength requirements of large-diameter wafers limit back-grinding to thickness of 200–250 μm; therefore, new technology is essential. Tru-Silicon Technologies is one of the companies providing this type of equipment (for a tutorial on applications of ADP, see http://www.trusi.com/frames.asp?).

### *Atmospheric Arc-Free Plasma Etching*

A picture of an atmospheric, arc-free (!), plasma jet invented at the University of California, Los Angeles is shown in Figure 3.27.[29] The "cold flame" (T ~ 75°C) shown here is impinging on a finger (http://www.seas.ucla.edu/prosurf). This jet was developed for

**FIGURE 3.26** A 50-μm thin silicon wafer produced using atmospheric downstream plasma (ADP). (From Tru-Silicon Technologies.)

**FIGURE 3.27** A picture of an atmospheric, arc-free (!), plasma jet invented at University of California, Los Angeles. The "cold flame" (T ~ 75°C) shown here is impinging on a finger (http://www.seas.ucla.edu/prosurf).

etching materials and depositing materials at atmospheric pressure and temperatures between 100 and 275°C. Gas mixtures containing helium, oxygen, and carbon tetrafluoride are passed between an outer, grounded electrode and a center electrode, driven by 13.56-MHz RF power at 50–500 W. At a flow rate of 51 L/min, a stable, arc-free discharge was produced. The discharge extends out through a nozzle at the end of the electrodes, forming a plasma jet. Materials placed 0.5 cm downstream from the nozzle are etched at high rates: 8.0μm/min for Kapton ($O_2$ and He only), 1.5 μm/min for silicon dioxide, 2.0 μm/min for tantalum, and 1.0 μm/min for tungsten.[29] The plasma jet produces a large flux of atoms and/or radicals, depending on the gas fed to the device. The setup may also be used for deposition. For example, tetraethoxysilane was used to deposit high-quality silicate glass films. It should be noted that like ADP this is a downstream plasma source. The ions and electrons are rapidly consumed by collisions before impinging on the substrate.

## Physical/Chemical Etching

### Introduction

The most useful plasma etching is neither entirely chemical nor physical. By adding a physical component to a purely chemical etching mechanism, the shortcomings of both sputter-based and purely chemical dry etching processes can be surmounted.

In physical/chemical techniques, ion-surface interactions promote dry etching. A first type is found in reactive ion beam etching (RIBE). RIBE constitutes a rather exceptional case of energy-driven anisotropy where ions are reactive and etch the surface directly. More common in energy-driven etching is *chemically assisted ion-beam etching* (CAIBE). In this case, ion bombardment induces a reaction by making the surface more reactive for the neutral plasma species, for example, by creating surface damage. In inhibitor-driven anisotropy, ions clear the surface of film-forming reaction products, allowing etching with reactive neutrals to proceed on the cleared areas. We will also learn that silicon dry etch anisotropy may be influenced by local dopant concentrations and that a wide variety of existing gas compositions crucially influences the etch profile. To help the reader summarize matters, a list of simplifying rules is provided at the end of this section.

### Energy-Driven Anisotropy or Damage Mechanism

During energy-driven anisotropy, bombardment by ions (<1000 eV) disrupts an unreactive substrate and causes damage, such as dangling bonds and dislocations, resulting in exposed surface atoms, which are more reactive toward etchant species (electrons or photons also can induce surface activation). Ion bombardment is mainly in the vertical direction so that the etch rate in the vertical direction is much higher than in the lateral direction (anisotropy!). Besides "energy-driven anisotropy," the process is also referred to as the *damage mechanism*. In Figure 3.3c, a typical resulting etch profile was shown. Vacuum pumping removes the volatile reaction products. This type of etching is referred to as *reactive ion etching* (RIE) when it involves reactive chemicals in a diode-type reactor, and as *chemically assisted ion-beam etching* (CAIBE) when it involves chemical reactants (e.g., $Cl_2$) introduced over the substrate surface in a triode-type setup. Figure 3.11 shows a parallel-plate setup for RIE, and a hexode reactor is shown in Figure 3.28. The hexode reactor is designed for batch processing. In this design, the cathode has the shape of a hexagon surrounded by the cylindrical chamber walls forming the anode. As many as 24 100-mm-diameter or 18 150-mm-diameter wafers

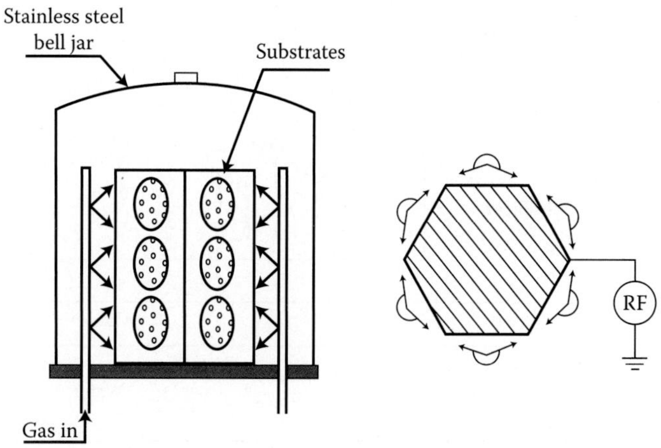


**FIGURE 3.28** Schematic diagram of a multiwafer RIE etcher, the so-called hexode reactor.

can be mounted on the slightly tilted sidewalls of the hexagonal cathode. The latter is slowly rotated during operation to increase uniformity.[1]

The etch rate reached by RIE is substantially higher than in ion etching. For example, the etch rate of Si in Ar sputtering hovers around 100 Å/min compared with 2000 Å/min for a reactive gas such as $CCl_2F_2$. Chemically assisted ion etching can lead to accurate transfers of the mask pattern to the substrate and to a fair selectivity in etching different materials. The directional anisotropy ensues from operating at low pressures and high voltages. Under these conditions, the mean free path of reacting molecules typically grows larger than the depth to be etched, resulting in the horizontal surfaces being hit and etched by reactive neutrals more frequently than the sidewalls. The need for energetic impinging ions, as in the case of physical etching, is reduced as the complex on the surface activates more easily. In Figure 3.29 we illustrate the energy-driven anisotropy process or damage mechanism.

Reactive ion-beam etching (RIBE), like CAIBE, involves chemical reactants in a triode setup. There is an important difference, however. In the case of RIBE, the reactive ions are introduced through the ion source itself; in CAIBE, the reactive gas is fed over

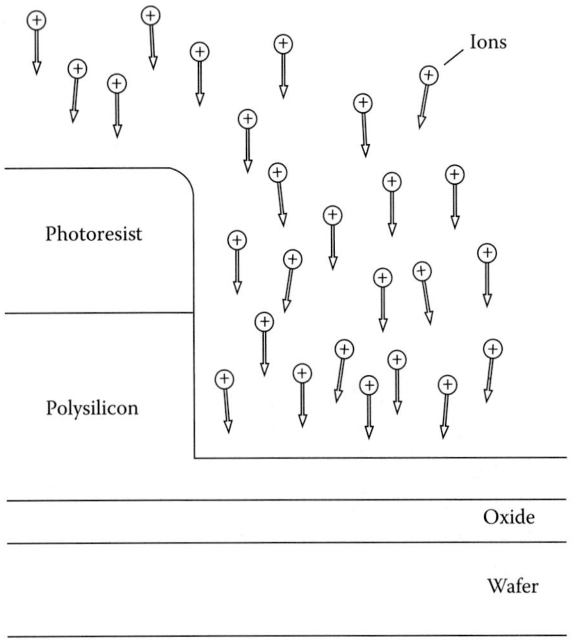


**FIGURE 3.29** Energy-driven anisotropy or damage mechanism. (Drawing by Genis Turon.)

the substrate to be etched, and unreactive ions are generated in the ion source. At very low pressures in RIBE systems, reactive ions, substituting Ar ions, can sustain a modest etch rate of less than 400 Å/min. In CAIBE, ion bombardment of a substrate in the presence of a reactive etchant species leads to a synergism where fast directional material removal rates greatly exceed the separate sum of chemical attack and sputtering rates. At one point, this type of etching was thought to be caused by direct chemical reactions between the ions and the surface material (as with RIBE). However, in most practical situations for etch rates ranging from 1,000–10,000 Å/min that is impossible because the ion flux is much lower than the surface removal rates. The synergistic effect of chemical attack and sputtering rates is illustrated in Figure 3.30; the etch rate of $Ar^+$ ion bombardment alone or the chemical etch rate of $XeF_2$ alone does not add up to the combined effect.

CAIBE is particularly effective in achieving highly anisotropic etch profiles because of the independent control of the physical and chemical etching components. To get an idea of how smooth a sidewall can be achieved by CAIBE, consider the work on vertical mirror facets in an InGaN/AlGaN laser diode structure. Atomic force microscope (AFM) measurements show that the laser diode vertical sidewalls etched in a CAIBE system (using $Cl_2$ and $BCl_3$ as the reactive species) exhibit a root mean squared roughness of only 40–60 Å, and the inclination angle is within ±2° of vertical.[30] Using $N_2/H_2/CH_4$ chemistry, extremely smooth sidewalls were also observed in CAIBE etching of InP.[31] To reduce surface damage, the aim is to increase the chemical component, decrease the ion acceleration energy, and work with a high ion flux. In CAIBE, as in other ion-beam techniques, the ion acceleration energy and the ion density can be independently controlled. With an ICP or ECR, a highly dense and uniform ion beam (e.g., $Ar^+$) is generated. ICP and ECR plasmas are discussed below.

The distinction between RIBE and CAIBE is not absolute because the presence of a reactive gas in CAIBE will produce some beam ions from species that back-diffuse into the broad-beam ion source. In a magnetically enhanced RIE (MERIE) system, a higher ion flux is obtained at lower energy through

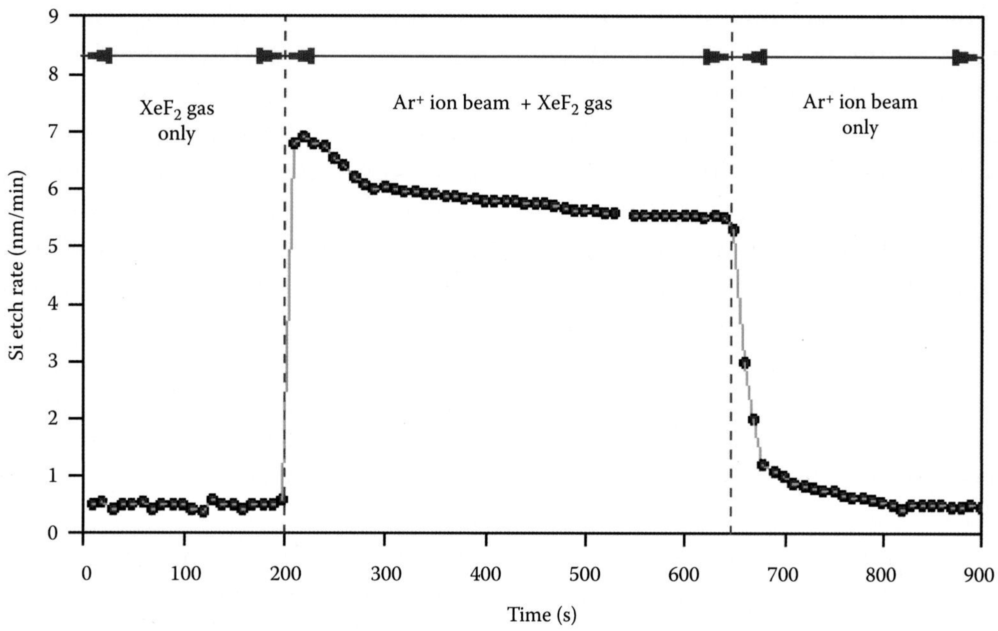


**FIGURE 3.30** Etch rate experiment in RIE. Etch rate of $Ar^+$ bombardment alone or chemical etch of $XeF_2$ alone does not add up to their combined synergistic effect.

magnetic confinement of the plasma. For typical RIBE and CAIBE performance characteristics also visit http://www.oxfordplasma.de/technols/ibe.htm.

A hypothetical etch profile for Si and $SiO_2$ etching with a reactive gas and positive ion bombardment is shown in Figure 3.31a. In the absence of ionic bombardment, the etch rate is assumed to be zero for $SiO_2$; consequently, the $SiO_2$ film etches with perfect anisotropy (vertical sidewalls). The silicon etch rate in the absence of ionic bombardment, on the other hand, is assumed to be finite but small; consequently, the etched feature ends up having a profile with slanted sidewalls. Mathematically, the above can be expressed in terms of etch depth $Z$, undercut $X$, the etch rate under bias $V_Z$, and the etch rate without bias $V_X$, as:

$$\frac{V_x}{V_z} = \frac{X}{Z} \tag{3.20}$$

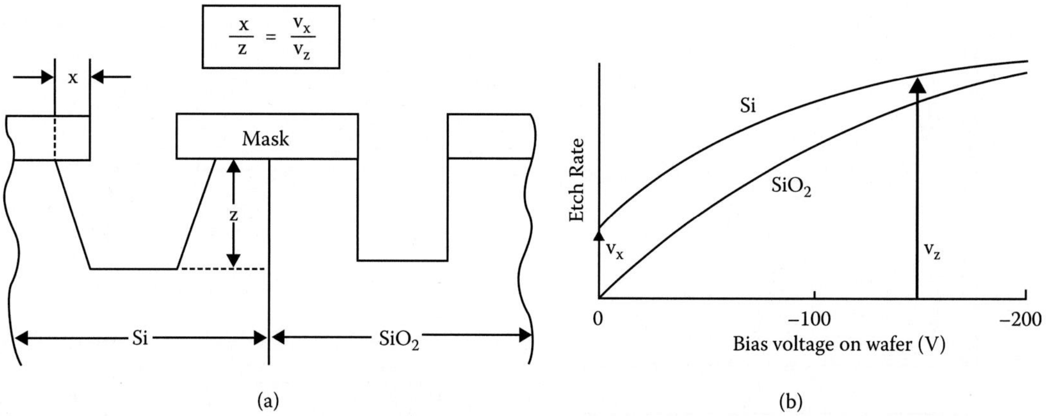


**FIGURE 3.31** Relationship between the shape of the etched wall profile and the dependence of the etch rate on the wafer potential. (a) Etch profiles. (b) Etch rate bias dependence. The etch rate of $SiO_2$ in the *z*-direction, $V_z$, increases with increasing wafer bias. Assuming no chemical etching component is present in the $SiO_2$ reaction, that is, $V_X = 0$ (the etch rate equals 0 at bias = 0), an ideal vertical profile results. The etch rate of Si also is bias-dependent, but some reaction occurs at $V_X = 0$. In other words, a chemical component to the reaction does not need ion bombardment resulting in nonvertical walls (see text). (From Manos, D. M., and D. L. Flamm, eds. 1989. *Plasma etching: An introduction.* Boston: Academic Press.[76])

(See Figure 3.31b.) The anisotropy of RIE can further be enhanced by the use of helium back-side cooling (as low as −120°C might be required). The low-temperature suppression of chemical attack of silicon by fluorine atoms exemplifies how lower temperatures further improve anisotropy and critical dimension control. Although suppressing the isotropic fluorine reaction, the same low temperature influences ion-assisted reactions only slightly and hence improves the anisotropy of profiles; the etch rate at $V_X = 0$ decreases or becomes zero.

## Inhibitor-Driven Anisotropy or Blocking Mechanism

Inhibitor-driven anisotropy embodies another example of a physical/chemical etching technique (Figure 3.3d). In this case, etching leads to the production of a surface-covering agent in a so-called blocking mechanism. Ion bombardment clears the "passivation" from horizontal surfaces; reaction with neutrals proceeds on these cleared surfaces only, and the etch proceeds again mainly in the vertical direction as shown in Figure 3.32. The protective film may originate from involatile etching products or from film-forming precursors that adsorb during the etching process. Passivating gases, such as $BCl_3$ and some halocarbons (Freons™ such as $CCl_4$ and $CF_2Cl_2$), are sources of inhibitor-forming species. In the latter case, the reactive gas component appears to be adsorbed on the surface (for example, a polymer is formed), where it is subsequently dissociated by electron, ion, or photon bombardment, clearing the surface for reaction with the reactive neutrals.

An example of etch profile manipulation with inhibitor-driven chemistry using an idealized Si sample is illustrated in Figure 3.33. At the top of this figure, we show the silicon etch rate as a function of percentage of $H_2$ in $CF_4$ for a biased and an unbiased wafer. We remember from earlier in this chapter that the silicon etch rate increases as the wafer bias is increased; thus, the etch rate curve for the biased wafer lies above the one for the unbiased wafer. If $H_2$ is added to a $CF_4$ feed gas, the Si etch rate decreases, and at some value of $H_2$ concentration, the nonbombarded surface etch rate decreases to zero ($V_X = 0$ at 10% $H_2$ in Figure 3.33), whereas the bombarded surface continues to etch. The decrease in etch rate stems from an increase in the amount of passivating polymerization. Aggressive fluorine reacts with the hydrogen so that carbon compounds polymerize more readily. As before, Equation 3.20 can be used to calculate the ratio of underetch $X$ to etch depth $Z$. At $V_X = 0$, that ratio, of course, is zero.

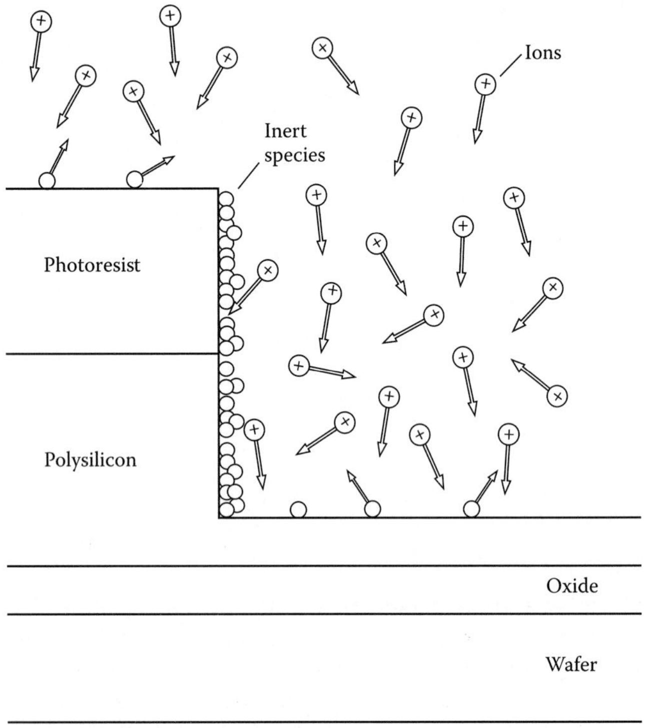


**FIGURE 3.32** Inhibitor-driven anisotropy or blocking mechanism. (Drawing by Genis Turon.)

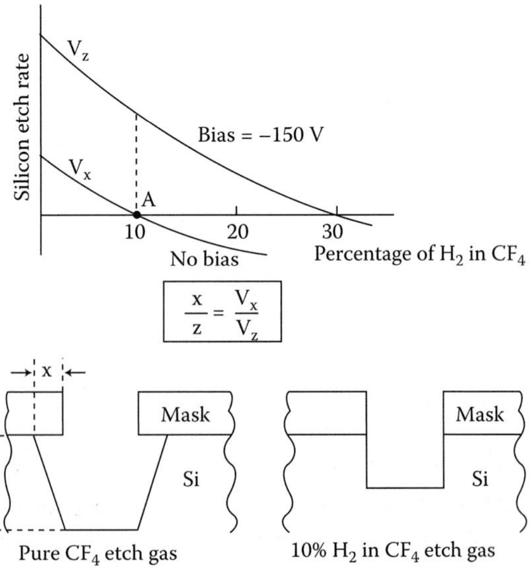


**FIGURE 3.33** Trench profile manipulation by decreasing the fluorine-to-carbon ratio (through hydrogen introduction). (From Manos, D. M., and D. L. Flamm, eds. 1989. *Plasma etching: An introduction.* Boston: Academic Press.[76])

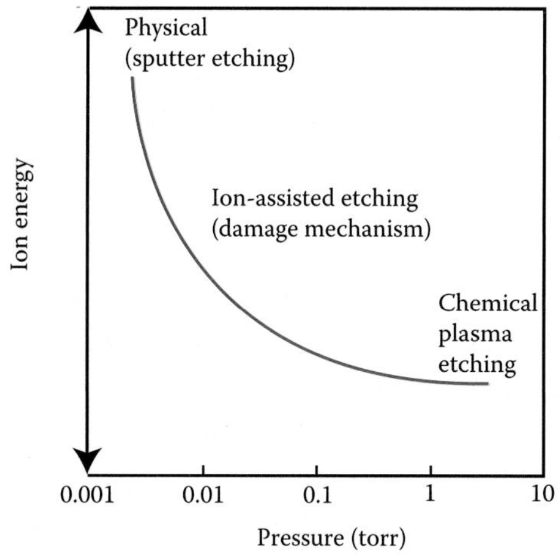


**FIGURE 3.34** Ion energy versus pressure for a plasma. (Redrawn by Genis Turon from Flamm, D. L. 1992. Dry plasma resist stripping part I: overview of equipment. *Solid State Technol* 35:37–39. With permission.[36])

Similarly, aluminum etching in $CCl_4 + Cl_2$ or $CHCl_3 + Cl_2$ plasmas presents a good example of an inhibitor system. Even though $Cl_2$ gas and chlorine atoms formed in these plasmas are rapid chemical etchants for clean aluminum, aluminum in these plasma mixtures can afford near-vertical profiles with excellent line-width control.

Another typical example of a sidewall mechanism is the etching of phosphorus-doped Si, which etches isotropically in $Cl_2$ plasma but etches anisotropically when $C_2F_6$ is added to the source gas. Qualitatively the effect is accounted for by assuming that the two gases dissociated in the plasma are:

$$C_2F_3 + e \rightarrow 2CF_3 + e \qquad \text{Reaction 3.1}$$

$$Cl_2 + e \rightarrow 2Cl + e \qquad \text{Reaction 3.2}$$

and the possible surface reactions are:

$$xCl + Si \rightarrow SiCl_x \text{ (etching)} \qquad \text{Reaction 3.3}$$

$$CF_3 + Cl \rightarrow CF_3Cl \text{ (recombination)} \qquad \text{Reaction 3.4}$$

Reaction 3.3 results in etching of the silicon surface, and Reaction 3.4 results in recombination without material removal. Ion bombardment enhances Reaction 3.3, and Reaction 3.4 is assumed to be dominant on the sidewalls. The recombination reaction acts as a sidewall passivant.

In practice, the anisotropy brought about by the directed ion flux in plasma allows the final etched feature to be within 10% of its dimensions on the mask, and submicron resolutions become feasible. A desire to move away from sidewall passivating chemistries to avoid CD loss is prominent, especially in the IC industry. Thick sidewall coatings not only reduce CD control, but they also prove difficult to strip. Processes involving little or no sidewall thickness increase are sought for tight CD control in submicron devices. In this respect, wafer cooling (to as low as −120°C) presents an attractive way of obtaining sidewall protection without the need for additional chemistries. At low substrate temperatures, reaction products are involatile and can serve as very thin sidewall inhibitors. Generally, at lower temperatures, lateral etch rate can be suppressed while using simpler chemistries. In micromachining applications where feature size is in general larger than in ICs and aspect ratios are more extreme, applying passivating chemistries is almost a necessity, but wafer cooling can help the process.

## Dopant-Driven Anisotropy

Although not commonly acknowledged, silicon plasma etch anisotropy may be influenced by local dopant concentrations. For example, Li et al.[32] have demonstrated that Cl-based plasma etch can be used to etch lightly doped p- or n-type silicon anisotropically and heavily doped n-type silicon isotropically. These authors, by forming buried $n^+$ layers beneath a lightly doped epitaxial layer, were able to selectively undercut structures above the $n^+$ regions.[32] Schwartz and Schaible[33] also described the dopant dependence of dry plasma etching.

It is more commonly known that the Si etch rate depends on the electronic properties of the Si substrate.[34] The n-type silicon (e.g., doped with P, As, and so on) etches faster than intrinsic silicon, which etches faster than p-silicon (e.g., doped with B, Ga, and so on). Dopant concentration effects on etch rates only become apparent at concentrations greater than $10^{19}$ $cm^{-3}$. The doping effect is not chemical in nature because it is absent if the dopants are not electrically activated. The effect depends on the electronic structure of the surface, and band-bending effects at the semiconductor surface have explained it. Using ECR etching, Juan et al.[35] demonstrated that the $n^{++}$ Si etch increases, but the $p^{++}$ Si etch rate decreases with respect to lightly doped silicon. Thus, it was established that, with a high microwave power or high RF power, as well as an increased temperature, a large difference in etch rates of $p^{++}$ and $n^{++}$ Si results.

## Ion Energy versus Pressure Relationship in a Plasma: Comparison of Different Dry Etch Technologies

In the preceding sections, we described two extreme cases of dry etching—ion etching (purely physical) and plasma etching (purely chemical)—and the intermediate case of chemical/physical etching. In general, three factors control the etch rate in a plasma reactor: 1) neutral atom and free radical concentration, 2) ion concentration, and 3) ion energy. Ion and radical concentrations control the reaction rate, whereas ion energy provides the necessary activation and controls the degree of anisotropy. The respective contribution to chemical and physical action of plasma can be manipulated by varying voltage and gas pressure. For good CD control, anisotropy is required, and positive ions need to be formed at low pressures ($<10^{-2}$ Torr) to strike the surface at normal incidence. Etching at low pressures, with a long mean free path length of the ions, $\lambda_i$, is inherently more anisotropic (i.e., directional) and less contaminating because etch reaction byproducts are more volatile at lower pressures and are easier to remove. At these lower pressures, ion density drops off quickly, causing a lower etch rate and lower wafer throughput. Increasing the power or wafer bias increases the etch rate, as the remaining ions become more energetic. Higher ion energies can cause additional problems in terms of device damage. What is needed is a plasma source operating at low pressure with very high ion density. Operating at low pressures also reduces the effects of chemically induced macro- and microloading (see above). With high pressure, short $\lambda_i$, and low voltage, one gets isotropic chemical etching. The pressure/voltage relationship for a plasma is schematically represented in Figure 3.34.[36] Thus, sputtering and dry chemical etching represent the two extremes of a continuous dry etching spectrum (Table 3.8) with physical etching by sputtering with inert argon ions at one end and chemical etching with reactive neutral species at the other.

In Tables 3.9 and 3.10 we contrast the features of different etching techniques, highlight the underlying mechanism, and give a typical application. Entries concerning endpoint detection will be clarified further below.

## Gas Compositions in Dry Etching

Most dry bulk etching of silicon is accomplished using fluorine free radicals, generated from fluorine-containing gases and by forming volatile $SiF_4$ during the etch (the byproducts in dry etching should always have the highest possible vapor pressure). As opposed to chlorine- and bromine-based processes, fluorine plasma reactions with silicon usually proceed spontaneously, not requiring ion bombardment. This method results in high etch rates, and by themselves fluorine radicals produce profiles that are nearly isotropic. Often, chlorofluorocarbons (CFCs) are used to produce polymer deposition in parallel with the etching process. Oxidizing additives increase etchant concentration and suppress excessive polymerization. The addition of oxygen to $CF_4$ in Si etching typifies the procedure.[37] Oxygen, at concentrations less than 16%, reacts with $CF_X$ radicals to enhance F atom formation [increase of fluorine-to-carbon ratio (F/C)] and eliminate polymerization. At yet higher concentrations, adsorbed oxygen on

**Table 3.8** Dry Etching Spectrum

| Types of Etching | Methods | Geometry | Selectivity ↓ | Excitation Energy ↓ | Pressure ↑ |
|---|---|---|---|---|---|
| Gas/vapor etching | Chemical | Isotropic | Very high | None | High (760–1 Torr) |
| Plasma etching | Chemical | Isotropic | High | 10–100s of watts | Medium (>100 mTorr) |
| Reactive ion etching | Chemical and physical | Directional | Fair | 100s of watts | Low (10–100 mTorr) |
| Sputtering etching | Physical | Directional | Low | 100–1000s of watts | Low (~10 mTorr) |

1 Torr = 1 mmHg.

TABLE 3.9 Comparison of Different Dry Etch Technologies

| | Chemical Etch | RIE | Physical Etch |
|---|---|---|---|
| Examples | Stripping | Plasma patterned etches | Argon sputtering |
| Etch rate | High to low | High, controllable | Low |
| Selectivity | Very good | Reasonable, controllable | Very poor |
| Etch profile | Isotropic | Anisotropic, controllable | Anisotropic |
| Endpoint | By time or visual | Optical | By time |

the surface depresses the etch rate.[37] Radical scavengers, such as hydrogen, increase the concentration of inhibitor former and reduce etchant concentration of fluorine in the etchant (decrease of F/C) (see also example in Figure 3.33). Heinecke[38] realized that the etch ratio of $SiO_2/Si$ could be increased either by adding $H_2$ to the $CF_4$ feed gas or by using $CHF_3$ or, in general, fluorocarbons displaying a smaller F/C ratio than $CF_4$. Adding hydrogen to fluorocarbon gases helps promote $CF_X$ film growth (such as $CF_2$, $C_2F_4$) for selective $SiO_2$ etching. Carbon accumulates less on oxide surfaces than on Si surfaces, as $SiO_2$ surfaces directly react with hydrogen. By adding hydrogen, carbon blocking increases (especially on the Si) because hydrogen scavenges fluorine, forming HF and preventing reaction with carbonaceous species on the surface.

The same approach helps to optimize the selectivity on other oxide/nonoxide systems (e.g., $SiO_2$/resist, $SiO_2/Si_3N_4$, and $Ta_2O_5$/Ta). The F/C ratio should not be made smaller than 2, or contamination of the whole system with $C_XF_y$ polymer sets in. Factors that tend to control polymerization and selectivity are temperature, hydrogen concentration, pressure, and ion bombardment. Some etchants, such as Cl atoms, do not readily etch through thin native oxide films on Si, Nb, and Al, for example. These native oxides prevent the onset of etching unless small amounts of native oxide etchants, such as $BCl_3$, are added. Inert gases such as argon or helium help stabilize the plasma, enhance anisotropy, improve uniformity, or reduce the etching rate by dilution. Because helium has a high thermal conductivity, it also improves heat transfer between wafers and the supporting electrodes.[25]

From the above, the F/C ratio in dry etching is a key parameter to control. With an F/C ratio >3, etching is dominant, and with an F/C <2, polymerization takes over. In Figure 3.35 we show the F/C ratio for a number of different etchant gases (for $CF_4$, the F/C ratio is obviously 4 and for $C_2F_4$ it is 2) and different applied voltages. A negative bias on a surface exposed to the plasma increases, at a constant F/C ratio, the etching behavior over

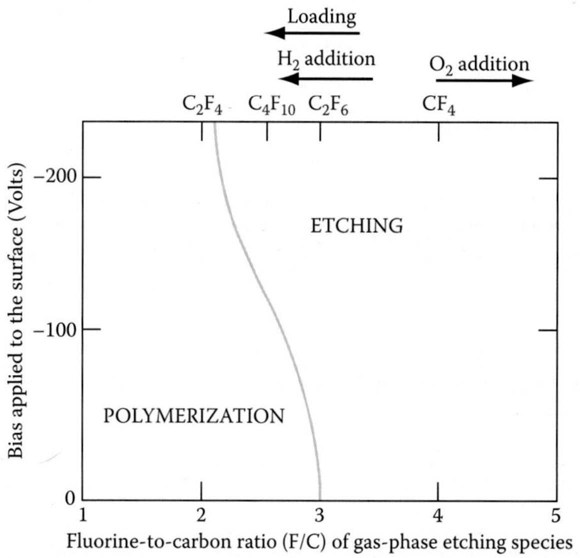


**FIGURE 3.35** F/C ratio, DC bias, and polymerization. (Drawing by Genis Turon.)

TABLE 3.10 Dry Etch Mechanism and Their Application

| | Reactive Ion Etch (RIE) | | |
|---|---|---|---|
| Pure Chemical Etch | Blocking mechanism | Damaging mechanism | Pure Physical Etch |
| No ion bombardment | Light ion bombardment | Heavy ion bombardment | Only ion bombardment |
| Photoresist strip | Single crystal Si etch | | |
| Ti strip | Polysilicon etch | Oxide etch | Sputtering etch |
| Nitride strip | Metal etch | Nitride etch | |

polymerization tendency. This effect is caused by the enhanced energy of the ions striking the surface, resulting in polymer sputtering. In an etch of an oxide film, the oxygen byproducts react with C to release more F, and the F/C ratio increases shifting to a more aggressive etching regime. When etching Si or a silicide, no oxygen is released, and F is consumed, lowering the F/C ratio and thus favoring polymerization. For selectivity, etching in the etch-dominant region but near the polymerization region is preferred.

Table 3.11 lists frequently used reactive gases and typical applications.

The Environmental Protection Agency mandate to eliminate the use of CFCs changed the type of gases used for dry etching purposes. This evolution in gas mixtures is captured in Table 3.12.[39]

The primary new etchant gases are $SF_6$, $NF_3$, and $SiF_4$ (fluorine-based etch); $Cl_2$, $BCl_3$, and $SiCl_4$ (chlorine-based etch); and $Br_2$ (bromine-based etch). For very aggressive etch chemistry at low pressures (e.g., to etch deep trenches in Si), HBr has also gained popularity. Other new etchant gases include $SiBr_4$, HI, $I_2$, and even nonhalogenated gases such as $CH_4$ and $H_2$. Table 3.13 lists etchants appropriate for etching some common electronic materials.

Table 3.14 presents an overview of the etch rates or etch ratios for a variety of important microfabrication materials in some popular reactive gases and gas mixtures.[40] Examples include etching of photoresist on Si, $SiO_2$, aluminum, and so on in $O_2$ plasma and etching of silicon with a metal mask (e.g., Al) in $CF_4$ plasma. Metal masks influence the etch rate of silicon or even $SiO_2$ by catalytic action of fluorinated metal surfaces, leading to excess production of free radicals.

Finally, Table 3.15 reviews mask materials listing their suitability for a number of gases.[25] A plus sign in this table corresponds with a low etch rate and high selectivity; a minus sign stands for high etch rate and low selectivity.

For further reading on gas composition for dry etching and stripping, see Flamm,[42–44] Jansen et al.,[2] Peters,[39] and Moreau.[40]

## Simplifying Rules

Some simple rules help interpret Tables 3.11–3.15 when specifying a choice of dry etchant and mask for a specific application. These rules are a set of semiempirical observations; they should not be looked on independently, and some state the same phenomenon in slightly different ways.[41]

**Table 3.11** Frequently Used Reactive Plasma Gases

| Etchant Purpose | Composition (Additive-Etchant): Application |
|---|---|
| Oxide etchant: etches through oxide to initiate etching | $C_2F_6$–$Cl_2$:$SiO_2$<br>$BCl_3$–$Cl_2$:$Al_2O_3$<br>$CCl_4$–$Cl_2$:$Al_2O_3$ |
| Oxidant: increases etchant concentration and suppresses polymerization | $O_2$–CF4:Si<br>$N_2O$–$CHF_3$:$SiO_2$<br>$O_2$–$CCl_4$:GaAs, InP |
| Inert gas: stabilizes plasma, dilutes etchant, improves heat transfer | Ar–$O_2$: organic material removal<br>He–$CF_3Br$:Ti, Nb |
| Inhibitor former: improves selectivity, induces anisotropy | $C_2F_6$–$Cl_2$:Si<br>$BCl_3$–$Cl_2$:GaAs, Al<br>$H_2$–$CF_4$:$SiO_2$<br>$CHF_3$–$SF_6$:Si<br>$O_2$ (50%)–$SF_6$:Si |
| Water and oxygen scavenger: prevents inhibition, improves selectivity | $BCl_3$–$Cl_2$:Al<br>$H_2$–$CF_4$:$SiO_2$ |
| Radical scavenger: increases film formation and improves selectivity | $H_2$–$CF_4$:$SiO_2$<br>$CHF_3$–$SF_6$:Si<br>$CF_3Br$–$SF_6$:Si<br>$CF_2Cl_2$–$SF_6$:Si |

*Source:* Most data from Flamm, D. L. 1993. Feed gas purity and environmental concerns in plasma etching: Part I. *Solid State Technol* October:49–54. With permission.[75]

TABLE 3.12 Evolution in Gas Mixtures for Dry Etching

| Material Being Etched | Conventional Chemistry | New Chemistry | Benefits |
|---|---|---|---|
| Poly-Si | $Cl_2$ or $BCl_3/CCl_4$ + sidewall passivating gases<br>$Cl_2$ or $BCl_3/CF_4$ + sidewall passivating gases<br>$Cl_2$ or $BCl_3/CHCl_3$ + sidewall passivating gases<br>$Cl_2$ or $BCl_3/CHF_3$ + sidewall passivating gases | $SiCl_4/Cl_2$<br>$BCl_3/Cl_2$<br>$HBr/Cl_2/O_2$<br>$HBr/O_2$<br>$Br_2/SF_6$<br>$SF_6$<br>$CF_4$ | No carbon contamination<br>Increased selectivity to $SiO_2$ and resist<br>No carbon contamination<br>Higher etch rate |
| Al | $Cl_2$<br>$BCl_3$ + sidewall passivating gases<br>$SiCl_4$ | $SiCl_4/Cl_2$<br>$BCl_3/Cl_2$<br>$HBr/Cl_2$ | Better profile control<br>No carbon contamination |
| Al-Si (1%)-Cu (0.5%) | Same as Al | $BCl_3/Cl_2 + N_2$ | $N_2$ accelerates Cu etch rate |
| Al-Cu (2%) | $BCl_3/Cl_2/CHF_3$ | $BCl_3/Cl_2 + N_2 + Al$ | Additional aluminum helps etch copper |
| W | $SF_6/Cl_2/CCl_4$ | $SF_6$ only<br>$NF_3/Cl_2$ | No carbon contamination<br>Etch stop over TiW and TiN<br>No carbon contamination |
| TiW | $SF_6/Cl_2/O_2$ | $SF_6$ only | |
| $WSi_2$, $TiSi_2$, $CoSi_2$ | $CCl_2F_2$ | $CCl_2F_2/NF_3$<br>$CF_4/Cl_2$ | Controlled etch profile<br>No carbon contamination |
| Single crystal Si | $Cl_2$ or $BCl_3$ + sidewall passivating gases | $CF_3Br$<br>$HBr/NF_3$ | Higher selectivity trench etch |
| $SiO_2$ (BPSG) | $CCl_2F_2$<br>$CF_4$<br>$C_2F_6$<br>$C_3F_8$ | $CCl_2F_2$<br>$CHF_3/CF_4$<br>$CHF_3/O_2$<br>$CH_3CHF_2$ | CFC alternatives |
| $Si_3N_4$ | $CCl_2F_2$<br>$CHF_3$ | $CF_4/O_2$<br>$CF_4/H_2$<br>$CHF_3$<br>$CH_3CHF_2$ | CFC alternatives |
| GaAs | $CCl_2F_2$ | $SiCl_4/SF_6$<br>$SiCl_4/NF_3$<br>$SiCl_4/CF_4$ | Fluorine provides etch stop on AlGaAs |
| InP | None | $CH_4/H_2$<br>HI | Clean etch<br>Higher etch rate than with $CH_4/H_2$ |

*Source:* Peters, L. 1992. Plasma etch chemistry: the untold story. *Semicond Int* May:660–70. With permission.[39]

1. **F/C ratio**: During etching, polymerization occurs simultaneously. Etching stems from the fluorine, and polymerization from the hydrocarbons. The dominant process will depend on the gas stoichiometry, reactive-gas additions, the amount of material to be etched, and the electrode potential. Adding hydrogen causes HF to form and the F/C ratio to decrease, leading to more polymerization and less etching (see example in Figure 3.33). A decrease in the fluorine concentration can also occur by overloading the reactor, leading to overconsumption of fluorine and favoring of polymerization. Adding oxygen to a fluorine carbon mixture leads to formation of CO and $CO_2$ reaction products, increasing F/C, and thus the etch rate, while decreasing the polymerization tendency. In other words, the addition of oxygen to a gas mixture reduces the tendency of Freons™ to polymerize and increases the concentration

TABLE 3.13 Plasma Etchants for Microelectronic Materials

| Material | Common Etch Gases* | Dominant Reactive Species | Product | Comment | Vapor Pressure in Torr at 25°C |
|---|---|---|---|---|---|
| Aluminum | Chlorine-based | Cl, $Cl_2$ | $AlCl_3$ | Toxic gas and corrosive gases | $7 \times 10^{-5}$ |
| Copper | Chlorine forms low pressure compounds | Cl, $Cl_2$ | $CuCl_2$ | Toxic gas and corrosive gases | $5 \times 10^{-2}$ |
| Molybdenum | Fluorine-based | F | $MoF_6$ | | 530 |
| Polymers of carbon, photoresists (PMMA and polystyrene) | Oxygen | O | $H_2O$, CO, $CO_2$ | Explosive hazard | $H_2O$ = 26 CO, $CO_2$ > 1 atm |
| III-V and II-VI compounds | Alkanes | | | Flammable gas | |
| Silicon | Fluorine- or chlorine-based | F, Cl, $Cl_2$ | $SiF_4$, $SiCl_4$ | Toxic gas | $SiF_4$ > 1 atm $SiCl_4$ = 240 |
| $SiO_2$ | $CF_4$, $CHF_3$, $C_2F_6$, $C_3F_6$ | $CF_X$ | $SiF_4$, CO, $CO_2$ | | $SiF_4$ > 1 atm CO, $CO_2$ > 1 atm |
| Tantalum | Fluorine-based | F | $TaF_3$ | | 3 |
| Titanium | Fluorine- or chlorine-based | F, Cl, $Cl_2$ | $TiF_4$ $TiF_3$ $TiCl_4$ | | $TiF_4$ = 2.10[4] $TiCl_4$ = 16 |
| Tungsten | Fluorine-containing | F | $WF_6$ | | 1000 |

*Common chlorine containing gases: $BCl_3$, $CCl_4$, $Cl_2$, and $SiCl_4$. Common fluorine containing gases: $CF_4$, $SF_4$, and $SF_6$.

*Source:* From Flamm, D. L. 1993. Feed gas purity and environmental concerns in plasma etching: Part I. *Solid State Technol* October:49–54; Elwenspoek, M., H. Gardeniers, M. de Boer, and A. Prak. 1994. Report No. 122830, University of Twente, Twente, the Netherlands.; Moreau, W. M. 1988. *Semiconductor lithography.* New York: Plenum Press; and Hess, D. W., and D. B. Graves. 1989. *Microelectronics processing: chemical engineering aspects.* Eds. D. W. Hess, and K. F. Jensen. Washington, DC: American Chemical Society.[75,25,40,41]

of halogen etchants ensuing from these gases. Adding oxygen to improve the F/C ratio leads to aggressive resist etching. Gases such as $NF_3$ and $ClF_3$ allow high fluorine concentrations (for aggressive Si etching) without the addition of oxygen, thus avoiding resist attack.

2. **Selective versus unselective dry etching**: The polymerizing point of the gas (i.e., the composition of the gas where polymerization takes over) primarily determines selectivity. The closer one works to the polymerization point, the better the selectivity. Factors that

TABLE 3.14 Frequently Used Reactive Plasma Gases, Reported Etch Rates, and Etch Ratios

| Film/Underlayer (F/U) | Etch Ratios |
|---|---|
| $Si_3N_4$ over AZ-2400 resist | $CF_4$-10 |
| $SiO_2$ over AZ-1350 resist | $CF_4/H_2$-10 |
| Poly-Si over PBS resist | $CF_4$-15 |
| PSG over AZ-1350 | $SF_6$-10 |
| $SiO_2$ over Si underlayer | $CF_3Cl$-30 |
| $Si_3N_4$ over $SiO_2$ underlayer | $NF_3$-50 |
| Poly-Si over $SiO_2$ underlayer | $CCl_4$-10 |
| Si over $SiO_2$ underlayer | $SF_6$-30 |
| Etchant | Material and Etch Rate (Å/min) |
| Ar | Si-124, Al-166, Resist-185, Quartz-159 |
| $CCl_2F_2$ | Si-2200, Al-1624, Resist-410, Quartz-533 |
| $CF_4$ | Si-900, PSG-200, Thermal $SiO_2$-50, CVD $SiO_2$-75 |
| $C_2F6$-$Cl_2$ | Undoped Si-600, Thermal $SiO_2$-100 |
| $CCl_3F$ | Si-1670 |

*Source:* Moreau, W. M. 1988. *Semiconductor lithography.* New York: Plenum Press. With permission.[40]

TABLE 3.15 Mask Materials in Dry Etching

| Mask Material | $SF_6$ | $CHF_3$ | $CF_4$ | $O_2$ | $N_2$ |
|---|---|---|---|---|---|
| Si | – | –\+ | – | + | + |
| $SiO_2$ | +\– | – | +\– | + | + |
| $Si_3N_4$ | +\– | – | +\– | + | + |
| $Al/Al_2O_3$ | + | + | + | + | + |
| W | – | – | – | + | + |
| Au | + | + | + | + | + |
| Ti | – | – | – | + | + |
| Resist | +\– | +\– | +\– | – | + |
| CFs | + | + | + | + | – |

*Source:* Elwenspoek, M., H. Gardeniers, M. de Boer, and A. Prak. 1994. *Micromechanics*. Report No. 122830, Twente, the Netherlands: University of Twente.[25]

have a tendency to increase the polymerization rate increase selectivity. Decreased temperature, high hydrogen concentration, low power, high pressure, and high monomer concentration all increase polymerization and thus selectivity. Typical unselective etchants for Si and polysilicon, used for noncritical etching, are $CF_4$ and $CF_4$-$O_2$. Less toxic gases such as $CF_4$ and $SF_6$ have preference over fluorine and are more selective, leading to polymerization. The most commonly used gases for selective Si etching are $Cl_2$, $CCl_4$, $CF_2Cl_2$, $CF_3Cl$, $Br_2$, and $CF_3Br$, along with mixtures such as $Cl_2$-$C_2F_6$. Small additions of halogens significantly increase the selectivity of fluorine-based etchants, especially the selectivity of silicon over silicon dioxide. For pure chlorine, the etch rate of $SiO_2$ is so low that even a native oxide can prevent the Si from etching. The high selectivity of $SiO_2$/Si is a major objective of many plasma processes trying to mimic the wet HF etch efficiency in this regard.

3. **Substrate bias**: A negative bias on a surface exposed to the plasma increases, at a constant F/C ratio, the etching behavior over polymerization tendency. This effect is caused by the enhanced energy of the ions striking the surface, resulting in polymer sputtering.
4. and 5. **Dry etching of III–V compounds**: Group III halides, particularly fluorides, tend to be involatile. As a result, F plasmas are usually substituted for chlorine-containing plasmas, and elevated substrate temperatures are used to further help volatilize the chlorides (fourth rule). The chemical composition (Ga/As ratio) of the different atomic planes in GaAs varies, and crystallographic etch patterns are observed under etch conditions in which chemical processes dominate (fifth rule). The latter establishes a major difference with dry Si etching, whereas crystallography does not play a role at all.
6. **Metal etching**: Chlorocarbons and fluorocarbons are typically used to etch metal films because they can reduce native metal oxides chemically. Oxygen and water vapor must be rigorously excluded during etching because of the high stability of the metal-oxide bond. Also, because of the stability of such bond, ion bombardment is essential. Because $AlF_3$ is not volatile, chlorine-containing gases are preferred for Al etching, whereas tungsten can be etched in fluorine ($SF_6$ or $NF_3$/chlorine).
7. **Organic films**: To retain an organic mask while etching away materials such as $SiO_2$ and $Si_3N_4$, one must work in etching conditions close to the polymerization point so that some loss of resist is compensated by condensation of reaction product, for example, $CF_4$-$C_2H_4$ and $CF_4$-$H_2$. We have already discussed that $CF_4$-$O_2$ plasmas severely degrade resist materials.
8. **Carbon-containing additives**: Carbon-containing additives generally degrade the selectivity of Cl- and Br-based inorganic chemistries for polysilicon etching over $SiO_2$. The Br atom attack on $SiO_2$ is thermodynamically unfavorable, whereas reactions between carbon halogen-bonded species and $SiO_2$ are exothermic. With HBr, a selectivity of 300:1 of polysilicon

over silicon dioxide can be achieved when completely avoiding carbon-containing gases or photoresists; the presence of resists or carbon traces in gases severely diminishes this ratio.

Leading suppliers of plasma etching equipment are Applied Materials, Inc. (Santa Clara, CA; http://www.appliedmaterials.com) and Lam Research Corporation (Fremont, CA; http://www.lamrc.com).

## Deep Reactive Ion Etching

### Introduction—Need for High-Density Plasmas

For building high-aspect-ratio micromachines and nanomachines with dry etching, until the mid-1990s, one had to battle three major problems. The first was the low etch rates with which to contend; typical Si dry etch rates of 1 μm/min required more than 5 hours to etch 300 μm deep! A second impediment was the inability to maintain high-aspect ratios over etch depths of a few tens of microns. Third, masking layers performed poorly and could not protect during a long etch.[45] New, high-density plasma sources provide an answer to each of these problems. As a result, for many applications, deep reactive ion etching (DRIE) presents an attractive MEMS solution today. An important example is the fabrication of high-aspect-ratio micromolds, which may be made much less expensive with deep DRIE than fabricating them with LIGA—an alternate high-aspect-ratio technique (see Chapter 10).

Ideally, dry etching stations generate high plasma densities (>$10^{11}$/cm$^3$) to achieve a high etch rate while operating at low pressure (1–20 mTorr). Low pressure increases ion directionality and discourages microloading. Normally low pressure also means less ionization and low plasma density. However, as we saw earlier, the use of magnetic fields forces electrons to spin with a very small gyroradius. This makes for a longer travel distance and thus more chance for electrons to collide with atoms, making a higher plasma density at low pressure possible. The source also needs to produce ions uniformly in energy and distribution while keeping the energy of the ions low to prevent damaging active electronics. The magnetic field reduces the charge difference in the sheath region, lowering the DC bias and therefore reducing the ion energy.

The two high-density plasma sources discussed here are inductively coupled plasma (ICP) stations [also called transformer coupled plasma (TCP)] and electron cyclotron resonance (ECR) machines. Both systems work at low pressures (a few mTorr) and can independently control ion flux and ion energy. In Figure 3.36 we show a schematic of an ICP and an ECR station.

### Inductively Coupled Plasma

In inductively coupled plasmas, a coil, operating at the standard 13.56 MHz, drives the

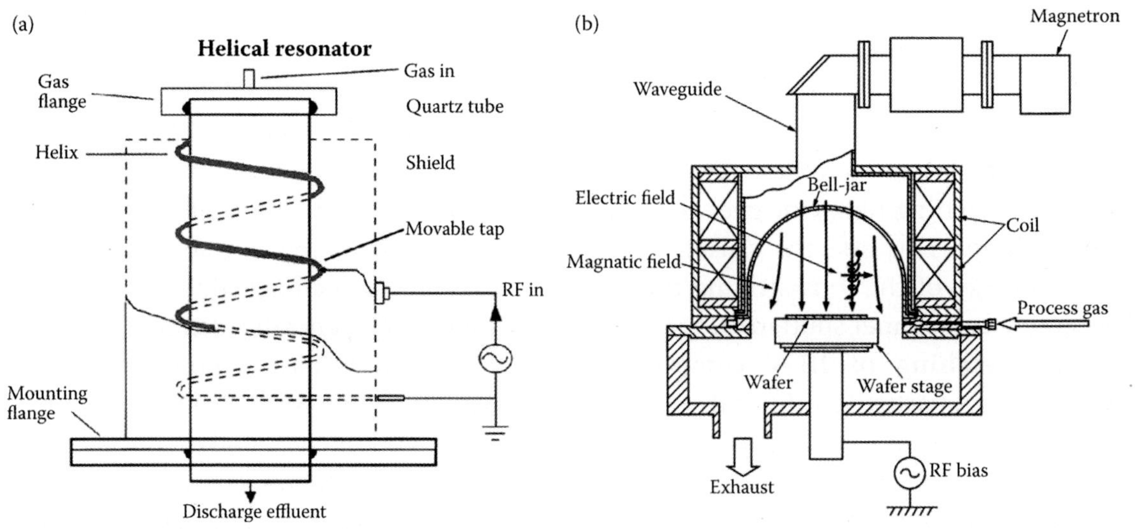


**FIGURE 3.36** (a) ICP station. (b) ECR station.

plasma inductively (Figure 3.36a). Such an ICP source creates high-density, low-pressure, low-energy plasmas by coupling ion-producing electrons to the magnetic field arising from the rf voltage. Electrons travel a long way without collisions with the chamber walls or electrodes. The upper part of the ICP chamber, where the plasma is generated, is either ceramic or quartz. The RF source in the upper chamber generates the plasma and controls the ion density, whereas the RF bias in the lower chamber controls the ion bombardment energy. Thus, the ion energy and ion density are effectively controlled independently. By cooling the wafer chuck to liquid nitrogen temperatures (77 K) and using a helium gas flow under the wafer for efficient heat transfer, wafer temperature can be maintained at cryogenic temperatures during etching.[46] The cryogenic cooling of the chuck results in condensation of reactant gases and protects the sidewalls from etching, which renders a more anisotropic process.

The Alcatel RF ICP source is one of the systems advanced as an ideal source for deep anisotropic Si etching. The electron density in this plasma source reaches $>10^{12}/cm^3$, and using fluorine-based gases, silicon etches at rates of up to 6 μm/min with etch uniformity better than ±5% while maintaining a $Si/SiO_2$ selectivity of more than 150:1. Etch depths greater than 250 μm and profile angles of ±1° were demonstrated while maintaining aspect ratios of 9:1.

Using an ICP etching station from Surface Technology Systems (STS), Klaassen et al.[45] were able to etch 300 μm deep in Si at 5 μm/min using a photoresist layer as mask of a 6-μm thickness only. Polymerization of the photoresist mask on the sidewalls of the etched trenches slows the lateral etch rate and results in high anisotropy. The Si to photoresist selectivity was determined at about 50:1. However, cryogenic cooling (e.g., −100°C) may introduce undesirable thermal stresses, and an important, much-touted alternative is the "Bosch process" patented by Robert Bosch GmbH (Stuttgart, Germany).[47]

The Bosch advanced silicon etch (ASE) typically achieves high-aspect ratios of 20–30:1, and etch rates of ~3 μm/min are standard.[47] In this process carried out, for example, in an STS or PlasmaTherm high-density plasma system, one switches (pulses) between sequential passivation (85 sccm $C_4F_8$, 8 s, 0 W platen, 600 W coil, 16 mTorr) and etching (80 sccm $SF_6$, 6 s, 8 W platen, 600 W coil, 33 mTorr) (see Table 3.16). In the passivation step, a ~10-nm-thick fluorocarbon deposits and energetic $SF_x^+$ ions remove this layer at the bottom of a trench, as sketched in Figure 3.37.

**Table 3.16** Characteristics of the DRIE Process Used in the STS and PlasmaTherm Systems

| | |
|---|---|
| $SF_6$ flow | 30–150 sccm |
| $C_4F_8$ flow | 20–100 sccm |
| Etch cycle | 5–15 s |
| Deposition cycle | 5–12 s |
| Pressure | 0.25–10 Pa |
| Temperature | 20–80°C |
| Etch rate | 1.5–4 μm/min |
| Sidewall angle | 90° ± 2° |
| Selectivity to photoresist | ~100 to 1 |
| Selectivity to $SiO_2$ | ~200 to 1 |

*Source:* Maluf, N. 2000. *An introduction to microelectromechanical systems engineering.* Boston: Artech House.[48]

The repetitive alternation of etch and passivation steps results in a very directional etch. Some scalloping near the top of the trench is observed, but in general the sidewalls exhibit good surface planarity with roughness less than 50 nm.[48] The scalloping effect is illustrated in Figure 3.38. The experiment depicted in this figure also demonstrates that changing the length of the sequential passivation and etching cycles can control scalloping.

Using the Bosch process, Chow et al.[49] fabricated through-wafer electrical interconnects by dry etching 20-μm holes through a 400-μm wafer (see Example 3.1). By etching from both sides, the required aspect ratio is reduced by two, from 40:1 to 20:1. Each etch takes 180 minutes.

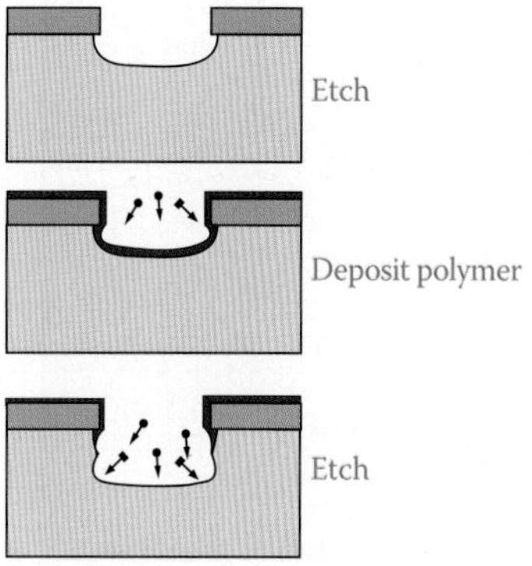


**FIGURE 3.37** Illustration of the Bosch process.

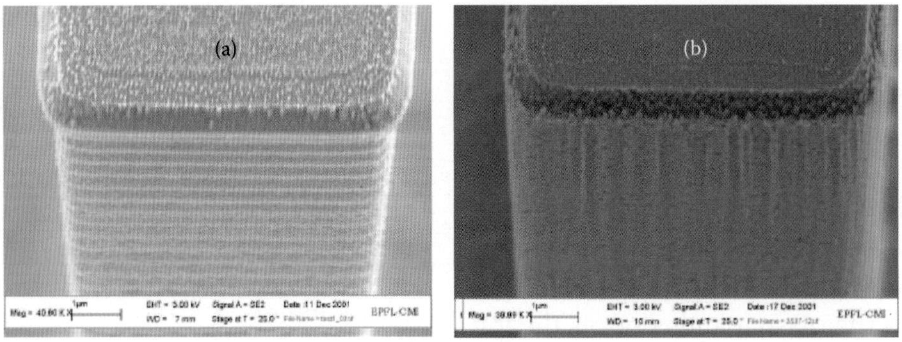


**FIGURE 3.38** Sidewall roughness at the top of a deep anisotropic etching of Si (Bosch process) as a function of pulse duration: $SF_6/C_4F_8$ = 7s/2s (a) and $SF_6/C_4F_8$ = 3s/1s (b).

Ayon et al.[26] tackled the issue of microloading in the Bosch process by increasing the ion flux through control of the applied source coil power. Ions reaching the wafer surface during the etch cycle promote more efficient removal of the protective polymer coating at the bottom of the trench, obviating the microloading effect. To etch ultradeep anisotropic silicon trenches in the range of 300–500 μm, Ayon et al.[26] compensate the reduction of neutral species flux reaching the feature bottom by increasing the $SF_6$ flow rates during the etching cycle (thus replenishing the etching species while the higher pressure also facilitates the removal of etching byproducts). This method results in a net improvement in passivation and etch rates and is known as the MIT 59 method (Figure 3.39). The method has already been applied in processes like the fabrication of accelerometers, gyros, microdisk-drive armatures, SCALPEL mask membranes, microgas turbines, tethered motors, power MEMS, and, in general, for etching through wafers.[26,50] A deep silicon back-side Bosch etch also allows fabrication of high-aspect-ratio and MEMS devices integrated with complementary metal oxide semiconductor (CMOS) circuitry (see Example 3.3).[51]

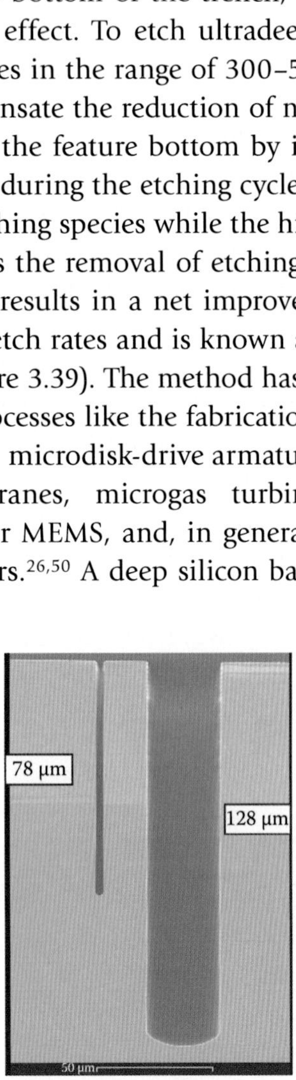


**FIGURE 3.39** MIT 59 method. See text.

An extensive database for the Bosch process conditions required to achieve prescribed profiles, as well as targeted silicon etch rates and uniformity, was presented in the *Journal of the Electrochemical Society* by Ayon et al.[52,53]

## Electron Cyclotron Resonance

New, high-density plasma sources may also be based on microwave discharges (Figure 3.36b). An RF source controls the ion energy, and a microwave power source controls the ion flux. Microwaves are electromagnetic waves in the 1–100-GHz (corresponding to wavelengths from 30 to 0.3 cm) range. Waves in this frequency range are transmitted by the use of waveguides, which are typically made of metal tubes of rectangular or circular cross-section. Cavities with metallic conducting walls resonate at specific frequencies and harmonics that are easy to calculate. The energy dissipated within the cavity dictates the frequency range above and below the resonant frequency that the cavity will accept and is characterized by the quality factor Q:

$$Q = \frac{f_0}{2\Delta f} \tag{3.21}$$

where $\pm\Delta f$ are the frequencies at which the amplitude is one-half that at the resonant frequency, $f_0$, for the same input power. In the case of an empty waveguide, the dissipation of energy inside the cavity is determined by the skin depth, δ, of the electrical field into the metal walls, which is given by (see also Volume I, Chapter 5, Equation 5.215):

$$\delta = \sqrt{\frac{\rho}{\pi \mu f}} \quad (3.22)$$

where ρ is the resistivity in Ωm, μ the permittivity of free space (4π × 10$^{-7}$ Henry/m), and *f* the frequency in hertz. For the generation of microwave plasmas, a cavity in the form of a cylinder with radius *a* and length *l* is generally used operating in $TM_{0,1}$ mode, that is, with the electric field parallel to the axis of the cylinder. In this case:

$$\lambda_0 = 2.61a \quad (3.23)$$

where $\lambda_0$ is the microwave wavelength. The resonant frequency of the cavity changes as a result of the change in the dielectric constant of the ionized gas, and Q decreases because of the increased power loss in generating the plasma.

In a quartz chamber filled with a gas at low pressure (0.1–10 Torr), the microwave power will cause breakdown when free electrons within the quartz chamber gain sufficient energy to ionize the gas. A static magnetic field generated by solenoid coils is also applied to the quartz chamber, and electrons move in a circular motion at the ECR frequency for that magnetic field (see Equation 3.18). In an ECR etching station, the ECR discharge creates a region, or surface layer, within the quartz chamber where the electron cyclotron frequency and the microwave frequency are equal and the electric field component is perpendicular to the static magnetic field. Under these conditions, the electrons within the ECR region are accelerated until they lose energy by an ionization or excitation collision with a neutral gas molecule. Varying the input microwave power and the gas pressure controls the plasma density. The maximum plasma density is reached if the microwave input power is such that the plasma frequency is equal to the microwave frequency, which is itself equal to the ECR frequency. The interaction of the magnetic field and the microwaves results in an intense, high-density plasma maintainable at low pressure. In operation, a 2.45-GHz microwave frequency is generated by a magnetron and injected into an etching chamber, enclosed by a quartz bell jar, through a waveguide. The maximum number of electrons (and ions) that can be formed within the quartz chamber at 2.45 GHz is about 7.5 × 10$^{10}$ electrons/cm$^3$.

In many ECR systems, the wafers are placed downstream of the plasma to further limit their exposure to this intense discharge. Wafers can also be biased to control ion bombardment energy. Like MERIE, ECR and ICP produce higher plasma densities with low-energy density, reducing charge-up damage.

Juan et al.[54] used ECR to etch polyimide molds for the fabrication of electroplated microstructures in one of several reported "pseudo-LIGA" processes (see also Chapter 10). A fast polyimide etch rate of 0.91 μm/min and a high selectivity over a Ti etch mask of 3150:1 were reached in an oxygen plasma. Polyimide (Dupont Pyralin® PI-2611) was spun on in multiple coatings and cured at 380°C for 1 hour. According to the authors, etching with an ECR source is 10 times faster and six times more selective than conventional RIE. The combination of magnetic confinement and microwave power, leading to an order of magnitude greater dissociation efficiency at much lower pressures than RIE, reduces scattering of reactive species and promotes vertical profiles and smooth surface morphology. Cooling down to −130°C helps maintain a vertical etch profile by suppressing spontaneous chemical reactions. This way, lateral resonator elements 32-μm thick with a 1-μm gap have been fabricated in polyimide, and electroplated Ni accelerometers with an aspect ratio of 11:1 were formed from these polyimide molds. Compared with LIGA, where the average wall roughness is <50 nm, the walls produced by this process still appear rough, but for many applications they might be adequate.

For a review on etching high-aspect-ratio trenches in silicon by DRIE, we refer to Jansen et al.[2] This article features an excellent review of possible trench shapes. Equipment for DRIE is available from STS (Abercarn, Wales, UK; http://www.stsystems.com), Plasma Therm Inc. (St. Petersburg, FL; http://www.plasmatherm.com/index.html), and Alcatel, S.A. (Paris, France; http://www.alcatel.com).

## Vapor-Phase Etching without Plasma ($XeF_2$)

As far back as 1983, it was demonstrated that etching of Si with $XeF_2$ in a CAIBE reactor enabled sidewall tailoring by changing the partial pressure of $XeF_2$, the ion energy, and the current density.[55] At the higher partial pressures of $XeF_2$, isotropic etching and undercutting became predominant. Isotropic etching of Si with xenon difluoride ($XeF_2$) actually does not require a plasma to generate the etching species. $XeF_2$ is a white solid with a room temperature vapor pressure of about 4 Torr, which reacts readily with Si. The Si etch occurs in the vapor phase at room temperature and at pressures between 1 and 4 Torr, established by a vacuum pump throttled to the right pressure. The etch rate is very dependent on the feature size in Si to be etched (see microloading). Hoffman et al.[56] observed silicon etch rates as high as 10 μm/min but worked at more typical rates of 1–3 μm. The extreme selectivity of $XeF_2$ to silicon over silicon dioxide and silicon nitride are well documented, and Hoffman et al.[56] have shown that $XeF_2$ also displays extreme etching selectivity over aluminum and photoresist. One important problem is that $XeF_2$ reacts with water to form water and HF. The latter may unintentionally react with Si to form $SiO_2$. The exothermic etch reaction, according to Chang et al., is given by:

$$2XeF_2 + Si \rightarrow 2Xe + SiF_4 \qquad \text{Reaction 3.5}$$

with only Si in the solid phase.

A 50-nm Al mask or a single layer of hard-baked photoresist suffices as deep etch masks. The simplicity of the process, the fast etch rate, as well as the resistance of even very thin layers of oxide, nitride, and Al metal make this etching process a possible choice for etching micromachined structures in the presence of CMOS electronics (in a "post-CMOS procedure"). The etched surfaces have a granular structure (10 μm and smaller feature size), and this etchant is not suited when polished mirror surfaces are required. In Figure 3.40a, an Al hinge 5 μm wide and 1.1 μm thick, holding an oxide

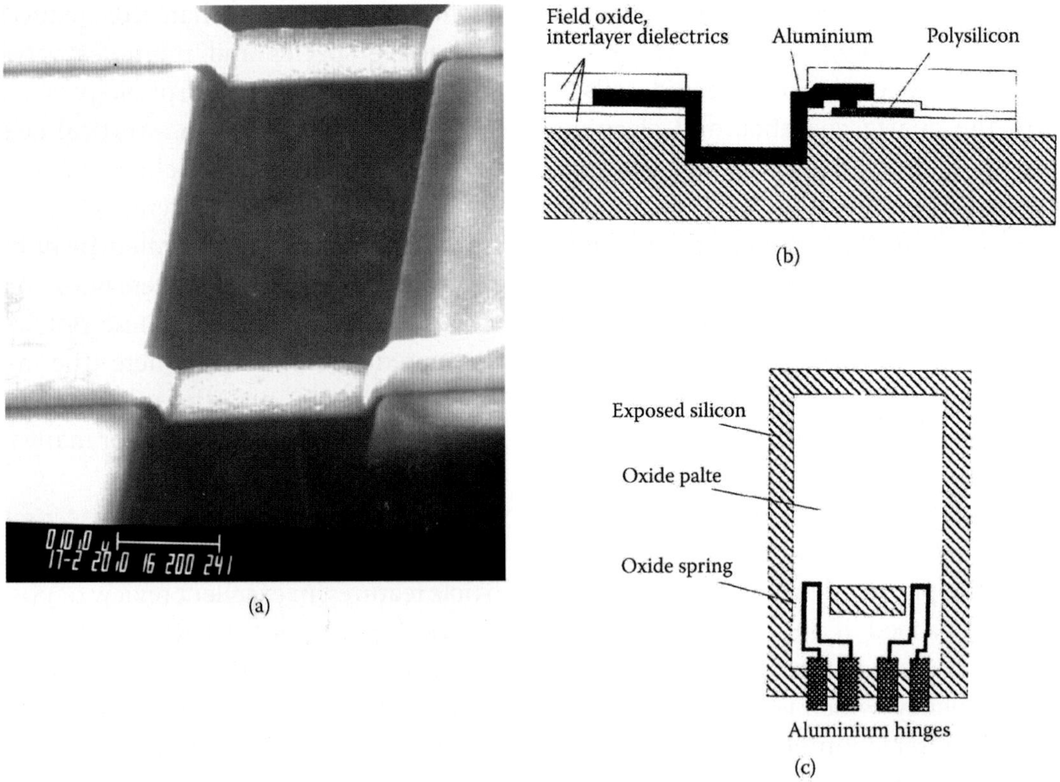


**FIGURE 3.40** (a) Al hinge (5 μm wide and 1.1 μm thick) holding a suspended oxide plate (plate is on the right side of the photograph). The Si is etched from underneath the Al hinge and oxide plate by $XeF_2$. The suspended oxide plate is a 200-μm square. (Courtesy of Dr. Kris Pister.) (b) Al hinge contacts a poly-Si piezoresistor embedded in the oxide plate to protect it from the $XeF_2$ etchant. (c) The piezoresistive accelerometer design shown enables rotation of the oxide plate out of the plane to detect orthogonal acceleration. (From Hoffman, E., B. Warneke, E. Kruglick, J. Weigold, and K. S. J. Pister. 1995. *Proceedings: IEEE micro electro mechanical systems (MEMS '95).* Amsterdam, the Netherlands: IEEE. With permission.[56])

plate 200-μm square suspended over a Si etch pit, is shown. The etch pit underneath the suspended plate is nearly isotropic and exhibits a surface roughness of several microns. The conductivity of the Al metal was unaffected by the $XeF_2$ etch. The hinge contacts a polysilicon strain gauge, sandwiched between oxide layers, to protect it from the $XeF_2$ etchant (see Figure 3.40b). The flexible hinge enables rotation of the oxide plate out of the plane of the wafer. Hoffman et al.[56] suggest that this contraption might function as a piezoresistive accelerometer, with hinges providing mechanical support and electrical connectivity between strain gauges embedded in the oxide plate proof mass and the wafer (see Figure 3.40c).

The rough Si surfaces of a $XeF_2$ etch in principle may be avoided by using interhalogen gases such as $BrF_3$ and $ClF_3$ with xenon gas as diluent.[57] The interhalogens are formed from single-element feed gases (e.g., bromine and fluorine) and result in perfectly isotropic etches and smooth silicon surfaces. Wang et al.[58] measured etch rate ratios for thin films with respect to Si in $BrF_3$ for several materials:

1. Si:LPCVD silicon dioxide = 3000:1
2. Si:silicon nitride = 400–800:1 (depending on the Si concentration)
3. Si:hard-baked AZ4400 and AZ1518 photoresist = 1000:1
4. Si:Al, Cu, Au, and Ni ≥ 1000:1

A first commercial instrument developed for $XeF_2$ etching is the Xetch™ from Xactix (http://www.xac tix.com).

## Dry Etching Models—In Situ Monitoring

A serious impediment to more effective use of dry etching is the large number of parameters that affect the process.[59] Just consider pressure and temperature. Pressure influences the ion-to-neutral ratio and fluxes of these particles, the sheath potentials and energy of the ions bombarding surfaces, electron energy, and the relative rate of chemical kinetics. Temperatures, both of gas and surface, have a profound influence on discharge chemistry as well. Only the surface temperature is controllable and influences selectivity, etch rates, degradation of resist masks, and surface roughness (higher temperature leads to a rougher surface).

Also, electrode area ratio, RF frequency, and power will alter plasma and electrode potentials, thereby changing ion energies. At lower frequencies, where ions can respond directly to the oscillating field, ions can attain the maximum energy corresponding to the maximum field across the sheath. As a result, for a constant sheath potential, ion bombardment energies are higher at frequencies less than 50 kHz. Nonuniformity can be caused by gradients in etchant concentration. Such gradients are responsible for the dreaded bull's eye etching pattern, where the etch rate decreases monotonically from the wafer periphery to its center (see above). Pressure, power, and gas flow must be adapted until the specified uniformity is obtained.

Predicting the exact relation between the flux and the energy of particles striking a wafer surface and the etch rate still is not state of the art, largely because no equilibrium is attained in these dry etching reactors, and for most reactions the kinetics are not known. Consequently, in situ monitoring of the plasma to improve etching uniformity is an important area for further research.[60,61] Besides measuring these controlling parameters, techniques such as optical emission spectroscopy (OES) to map in situ etchant concentration are coming to the foreground as sensitive and effective endpoint detectors. It is well known that the color of a plasma changes when etching different materials. Emissions can emanate from etchants, etch products, or their fragments. An optical spectrometer is set to the line of a reaction product of interest, and one follows its intensity during an etch cycle. Some etch endpoint wavelengths are summarized in Table 3.17.

Mass spectrometry allows one to monitor the concentration of a wide variety of species; it is limited, however, in that the results are obtained downstream from the real process in the reactor.[62] Laser-induced fluorescence (LIF) has been used to measure ground-state ion and radical concentrations. LIF is more powerful than OES, but it is expensive and requires a lot of experimental expertise; consequently, it is used more in the research lab than on a production line.[63] In imaging

**Table 3.17** Etch Endpoint Wavelength

| Film | Etchant | Wavelength (Å) | Emitter |
|---|---|---|---|
| Al | $Cl_2$, $BCl_3$ | 2614<br>3962 | AlCl<br>Al |
| Poly-Si | $Cl_2$ | 2882<br>6156 | Si<br>O |
| $Si_3N_4$ | $CF_4/O_2$ | 3370<br>3862<br>7037<br>6740 | $N_2$<br>CN<br>F<br>N |
| $SiO_2$ | $CF_4$ and $CHF_3$ | 7037<br>4835<br>6156 | F<br>CO<br>O |
| PSG, BPSG | $CF_4$ and $CHF_3$ | 2535 | P |
| W | $SF_6$ | 7037 | F |

of radicals interacting with a surface (IRIS), light from a dye laser optically excites specific radicals in the incident and reflected molecular beam striking a surface. The light emission of radicals in incident and reflected beam is measured and used to extract information on their density. By taking the difference of the emission data, it is possible to determine the intensity of scattered radicals and the reactivity of specific radicals with particular surfaces.[3] Laser interferometry (for thin transparent films such as polysilicon, silicon dioxide, and silicon nitride) and laser spot reflectance (for opaque, highly reflecting metal films) are used to monitor film thickness and film disappearance, respectively. Also, the temperature of the wafer during reactive ion etching affects process quality in terms of photoresist integrity, selectivity, etch rate, and etch residues. An in situ temperature monitor enables one to improve the performance during the plasma etch.[64] With Langmuir probes, one can, in principle, determine the complete electron energy distribution of a plasma. In such an experiment, a small wire probe is inserted into the plasma and biased with respect to some other electrode. By recording the current-voltage plot of the probe, charged particle densities and energies can be deduced. In practice, Langmuir and other electrostatic probes are difficult to apply, invasive, and perturb the plasma.

Bosch-Charpenay et al.[50] integrated an FTIR sensor in an STS Multiplex ICP deep reactive ion etcher. This enabled in situ, real-time measurements of DRIE etch depth and eroding photoresist thickness measurements of MEMS devices using infrared reflectance analysis. Typical capabilities of the system are summarized in Table 3.18.

## Examples

Several examples of dry-etched micromachines are presented below. They include deep dry etching to fabricate interconnects through single crystal Si, an example of single crystal reactive etching and metallization (SCREAM), which relies on a combination of anisotropic and isotropic dry etching, a variant on the SCREAM process from the Carnegie Mellon University, dry etching of polymeric materials, and a cantilever micromachining process combining wet and dry etching.

**Table 3.18** Summary of FTIR Sensor Capabilities

| | |
|---|---|
| MEMS structures that can be measured | Pillars, holes, trenches, trench grids, honeycombs of etched width less than ~100 μm |
| Etch depth precision (reproducibility) | 0.05–0.5 μm depending on the structure |
| Etch depth accuracy (as compared with SEM) | ~1–2 μm (estimated) |
| Maximum measurable etch depth | >200 μm |
| Current spot size* | ~800 μm diameter |
| Measurement time | ~1 s |
| Measurement spot range† | ±5 mm in both *x*- and *y*-direction |
| Probing IR wavelength | 2–10 μm (1000–5000 $cm^{-1}$) |

*Optical diffraction limits preclude spot sizes smaller than ~400 μm. MEMS devices smaller than this minimum spot size can, however, still be measured, provided they are geometrically isolated from other DRIE etched features.

†Translation range of infrared spot over sample is strictly related to equipment geometry at the etch chamber—in this case a small overhead view port into the STS Multiplex ICP system.

*Source:* Bosch-Charpenay, S., J. Xu, J. Haigis, P. A. Rosenthal, P. R. Solomon, and J. M. Bustillo. 2000. *Technical Digest Solid-state sensor and actuator workshop.* Hilton Head Island, SC: Transducers Research Foundation, Inc.[50]

**Example 3.1: Via Etching in Si**

Fabrication of high-density electrical, through-wafer interconnects on silicon substrates remains a challenging proposition. In MEMS, through-wafer interconnects greatly facilitate their integration with active electronics (see Making Vias through Silicon in Volume III, Chapter 4). Different approaches to fabricating those 3D contact lines are compared in Volume III, Chapter 4 on packaging. Chow et al.[49] use the Bosch process to etch a high-aspect-ratio through-wafer via and subsequently coat an insulator and then fill in a signal-conducting layer.[49] Etching the via is the first and most critical step. A 10-μm-thick photoresist (Shipley SPR220-7) is coated on both sides of the Si wafer, and a double-sided aligner is used to expose 20-μm circles on both sides (±1 μm accuracy). The authors let the wafers then sit for 8 hours to absorb water. The moisture, they claim, helps ensure a shorter development time (90 s in Shipley LDD26W), resulting in straighter resist walls. Using a high-density plasma etcher from STS, a Bosch etch procedure (such as shown in Table 3.16) is performed for 120 minutes from the front side of the wafer. Resist is then coated and baked again, and another 180-minute etch from the back is used to punch through the vias (Figure 3.41). The diameter of the resulting vias varies from 24 μm (top and bottom) to 18 μm in the middle. By etching from both sides of the wafer, the required aspect ratio is reduced from 40:1 to 20:1, and it is possible to etch a 20-μm circle through a 400-μm-thick wafer. For electrical isolation of the via, thermal silicon dioxide (2 μm) is grown at 1100°C. For signal conduction, low-pressure chemical vapor deposition (LPCVD) polysilicon is deposited into the via and doped by phosphorous diffusion. The resistance of the resulting via is 14 Ω, and its low capacitance (<1 pF) makes it attractive for high-speed applications.

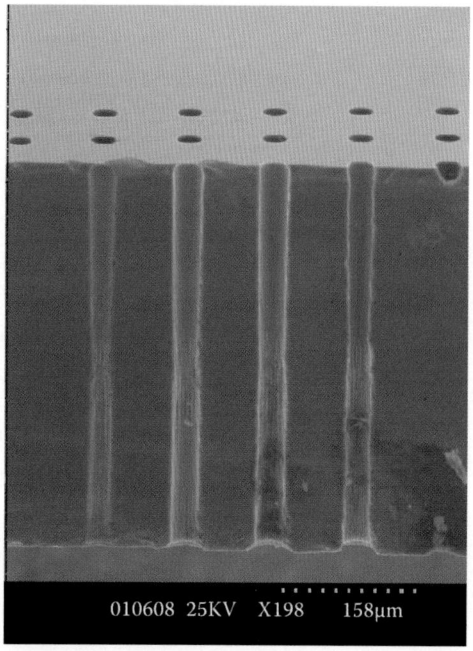


**FIGURE 3.41** SEM cross-section after plasma etching vias through the wafer. (From Arnold, J. C., and H. H. Sawin. 1991. Charging of pattern features during plasma etching. *J Appl Phys* 70:5314.[24])

**Example 3.2: Single Crystal Reactive Etching and Metallization**

As a second example of dry etching in micromachining, we follow some of the work by the Cornell Nanofabrication Laboratory making laterally driven micromachines.[65,66] In laterally driven micromachines, as illustrated in Figure 3.42a, beams and structures of various shapes should preferably be of a narrow and thick construction and able to move freely and in parallel to the supporting substrate while being held rigidly above it. Increased thickness of beams permits greater driving power for a given voltage and increases stiffness of the device perpendicular to the plane of motion. Thin, suspended beams (e.g., 2 μm wide) are fabricated with a combination of dry and wet etching in polysilicon (surface micromachining, Chapter 7). However, thick single crystal beams greater than 20 μm in materials such as quartz, Si, and GaAs can best be made with dry etching, especially because thick lateral moving microstructures of arbitrary shape, given crystallographic limitations, are very difficult to realize using wet bulk micromachining (Chapter 4). The monitoring of the resonant frequency variation of small resonator devices made this way (see Figure 3.42a) can be used to measure monolayer mass changes, as well as pressure, temperature, acceleration, and so on. In the measurement of mass, for example, the sensitivity of the device is a function of resonant frequency and mass of the resonating structure:

$$\text{Sensitivity} \sim \frac{\text{Resonant frequency}}{\text{Mass}} \quad (3.24)$$

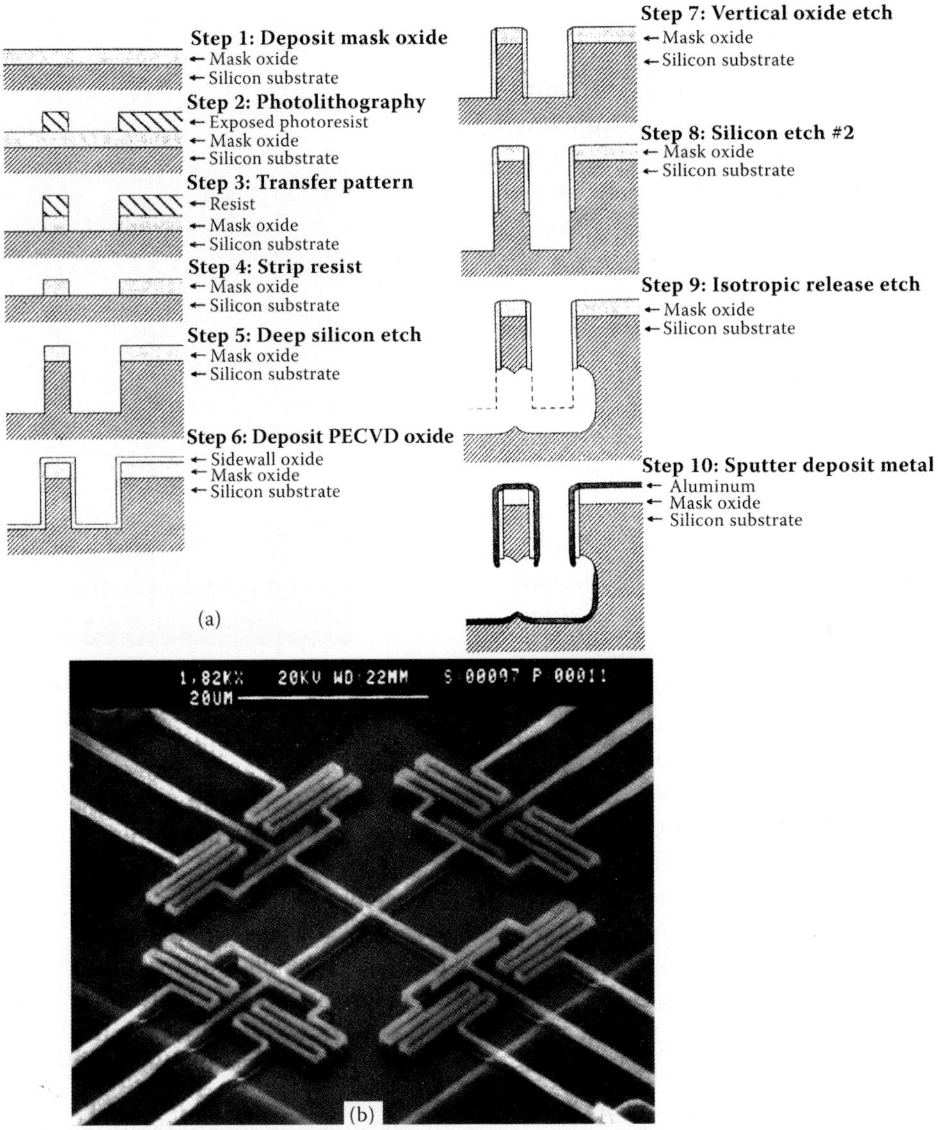


**FIGURE 3.42** (a) Si-SCREAM process (see text). The cross-section of a typical beam made using the SCREAM process is illustrated in step 10. This drawing shows a suspended beam and its associated parallel-plate capacitor. The released beam is free to move, whereas the plate on the right is static and can be used to measure the motion of the beam. (From K. A. Shaw et al., *Sensors Actuators A*, 40:63–70 (1994). With permission.) (b) A representative SCREAM structure. SEM micrograph of a Si *x,y* stage with parallel-plate capacitive drives and springs. (From Yao, J. J., S. C. Arney, and N. C. MacDonald. 1992. Fabrication of high-frequency two-dimensional nanoactuators for scanned probe devices. *J Microelectromech Syst* 1:14–22. With permission.[67])

To build the most sensitive detector, the mass must be minimized and the resonant frequency maximized. The mechanical resonant frequency of an ideal resonator is given by:

$$f_0 = \left(\frac{k}{m}\right)^{\frac{1}{2}} \quad (3.25)$$

with $k$ the spring constant and $m$ its mass; in other words, the resonant frequency is proportional to (stiffness/mass)$^{1/2}$. A large spring stiffness must be designed and $m$ minimized to make $f_0$ as high as possible. For some of the very stiff, low mass structures made at Cornell, $f_0$ was 5 MHz, with k = 55 N/m and m = 2.0 × $10^{-13}$ kg. The above simple considerations illustrate the power of micromachining in enabling more sensitive detectors. For more details on building MEMS resonant beams, see Volume III, Chapter 8 on actuators.

Lateral resonant structures as shown in Figure 3.42a are used to make sets of comb electrodes, for example, to drive precision *x-y* microstages for accurate positioning or as comb drives in accelerometers. Scanning tunneling microscopes (STMs), 40 × 40 μm in size, have been manufactured using SCREAM technology.[67] In the latter case, field emitter tips were integrated on an *x-y* stage with moving beams with nominal cross-sectional dimensions of 250 nm in width and 1000 nm in height. Displacements in the *x-y* directions through parallel-plate capacitive drives measured ±200 nm for an applied voltage in the 50-V range.

An elegant Cornell dry etching process that obtains such laterally driven, single crystal micromachines is called *SCREAM*. This technology uses reactive ion etching to both define and release structures. SCREAM portrays an important technique from several points of view. It is a self-aligned, single-mask process, run at temperatures lower than 300°C, completed in less than 8 hours, and can be carried out in the presence of active circuitry integrated on the same chip. The example given here is the fabrication of single crystal silicon-released beam structures (see Figure 3.42).[67]

The steps of the SCREAM process are illustrated in Figure 3.42b. As a starting substrate, an arsenic-doped, 0.005-Ωcm, n-type (100) silicon wafer is used. A masking layer of PECVD silicon dioxide, 1–2 μm thick, is coated with resist (step 1). Photolithography on the photoresist layer creates the desired pattern (step 2) to be transferred onto the silicon dioxide using $CHF_3$-based, magneton reactive ion etching (MERIE) (step 3). The photoresist is then stripped in an oxygen barrel etcher (step 4). The silicon dioxide mask pattern is subsequently transferred to the silicon substrate using a $Cl_2/BCl_3$ anisotropic etch. The etch depth, depending on the intended micromachine, ranges between 4 and 20 μm (step 5). Following the silicon etch, silicon dioxide is conformably deposited using PECVD. This oxide, at a thickness of 0.3 μm, will serve as a protection for the sidewalls during release (step 6). The next step is interesting and crucial to the process. The objective is to remove the oxide from the bottom and the top of the structure without removing the oxide from the sidewalls. This selective clearing of oxide is carried out in an anisotropic $CF_4/O_2$ etch at 10 mTorr. It removes 0.3 μm of oxide from the mesas and the trench bottoms while leaving the sidewall oxide intact (step 7). In the next step, a second deep Si etch ($BCl_3/Cl_2$) deepens the trench further (3–5 μm deeper) to expose some Si sidewall underneath the oxide-covered part (step 8). The release is accomplished with an $SF_6$ isotropic etch at 90 mTorr. The RIE step etches the Si out from underneath the beams, releasing them in the process (step 9). An etchant such as $SF_6$ hardly etches the oxide at all and gives good selectivity to the process. A layer of aluminum is then sputter-deposited using DC magnetron sputter deposition (see Chapter 9) (step 10). The Al can be contacted and used as a drive electrode in micromechanical resonator structures. Resonator elements with beam elements 0.5–5 μm wide and aspect ratios >10 have been achieved using this Si SCREAM process.

A variation of the CMOS-compatible Si SCREAM process is used at EG&G's IC Sensors (Milpitas, CA; http://www.meas-spec.com/myMeas/sensors) for the manufacture of accelerometers.

The SCREAM process can also be used to fabricate GaAs micro- and nanomachines. Figure 3.43a outlines the SCREAM process steps in the case of a GaAs-based device.[66] Specifically, the process sequence for the fabrication of a cantilever beam structure identical to the Si one in Figure 3.42a is shown.[68] Considering two identical structures enables us to better compare the degree of complexity of micromachining in those two materials. Single crystal GaAs, in some respects, is an attractive material for micromechanics because of its optoelectronic properties. With GaAs, one can envision the possibility of vertically etched mirrors, lenses, detectors, waveguides, and semiconductor lasers, all adding up to the possibility of a totally integrated optical bench. On the other hand, compared with silicon, GaAs possesses a low fracture strength (its yield strength is considerably lower: 2 vs. 7 GPa) and is more difficult to passivate than Si, as its own oxides are either leaky or chemically unstable. The process example illustrates the difficulty in passivating GaAs versus passivating Si (a dual-layer dielectric is required to passivate GaAs). A more complete comparison of silicon versus GaAs in micromachining is presented in Volume III, Chapter 5 on selected materials for MEMS and NEMS.

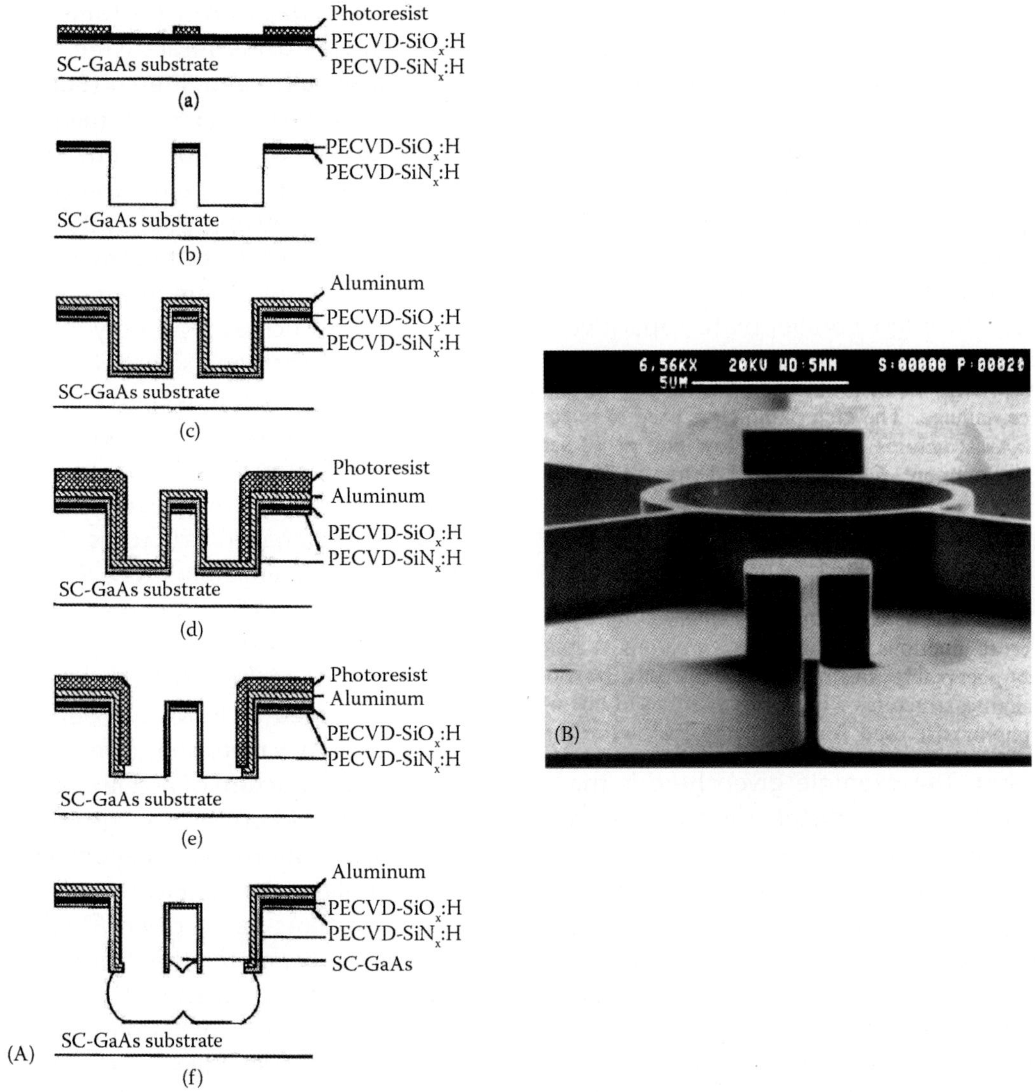


**FIGURE 3.43** (A) Formation of an SC-GaAs cantilever beam with aluminum electrodes adjacent to each side of the beam using PECVD-SIN$_x$:H for insulation and for the top and sidewall etch mask. (From Zhang, Z. L., G. A. Porkolab, and N. C. MacDonald. 1992. *Proceedings: IEEE micro electro mechanical systems (MEMS '92).* Travemunde, Germany: IEEE. With permission.[68]) (B) SEM micrograph showing SC-GaAs circular and angled straight-line features after the CAIBE step. (From Zhang, Z. L., and N. C. MacDonald. 1993. Fabrication of submicron high-aspect-ratio GaAs actuators. *J Microelectromech Syst* 2:66–73. With permission.[66])

For creating a simple movable beam, CAIBE produces deep vertical trenches, and RIE laterally undercuts and releases the vertically etched structures from the GaAs substrate.[68] This process produces aspect ratios of 25:1 with trenches of 20 μm vertical depth to 400 nm lateral width. PECVD-SiN$_X$:H (200 nm) and a second layer of SiO$_X$:H (150 nm) are used to mask the GaAs substrate (Figure 3.43Aa). PECVD-SiN$_X$:H on GaAs easily results in structural buckling and layer cracking, but nevertheless is preferred as a conformable layer because the surface of PECVD-SiO$_X$:H on GaAs is too rough as a result of the diffusion of gallium into PECVD-SiO$_X$:H. The PECVD-SiN$_X$:H deposition temperature of 100°C is selected to obtain simultaneously good dielectric properties and low residual stress in the film. To further protect the GaAs during etching, an additional sacrificial PECVD-SiO$_X$:H is deposited. The thermal expansion mismatch between the two dielectrics is less than between either of the dielectrics and the GaAs. The pattern-producing freestanding beams in GaAs are then printed in a photoresist layer on top of the bilayer dielectric. Plasma etching with $CHF_3/O_2$ transfers this image into the dielectric layer. The pattern in the dielectric stack subsequently is transferred to the single crystal GaAs by CAIBE (Figure 3.43Ab) using a 500-eV, 0.1-mA/cm$^2$ Ar ion beam and a $Cl_2$ flow of 10 ml/min. After the CAIBE process, the remaining photoresist is stripped in oxygen plasma, and a 300-nm PECVD layer of $SiN_X$:H is conformably deposited. Subsequently, using DC magnetron sputtering, a 400-nm layer of aluminum is deposited (Figure 3.43Ac). Photoresist is spun on the Al layer, and the aluminum side-electrode pattern is printed in the resist, developed (Figure 3.43Ad), and then transferred to the Al by a $Cl_2/BCl_3$ etch. The latter step removes the aluminum from the areas that have been cleared of resist. Before the Al deposition, contact windows are formed to allow electrical contact with both the GaAs substrate and the movable GaAs beams. To clear the PECVD-$SiN_X$:H (from the second deposition of this compound) from the bottom of the trenches and to retain the dielectric elsewhere, a highly anisotropic $CHF_3/O_2$ is applied (Figure 3.43Ae). Finally, the GaAs structures are released from the substrate using a chlorine-based RIE (Figure 3.43Af; the top and sidewalls of the GaAs structure remain protected by the PECVD-$SiN_X$:H and PECVD-$SiO_X$:H mask. At the end of the beam-releasing etch, the PECVD-$SiO_X$:H is consumed, that is, it acts as a sacrificial mask. Finally, the photoresist for the Al patterning is stripped in oxygen. In a simplification of the process described in Figure 3.43A, Zhang et al.[66] later discovered an improved PECVD-$SiN_xH_y$ conformal coating enabling a single dielectric mask for the GaAs SCREAM process. Figure 3.43B demonstrates circular and angled straight-line features in single crystal-GaAs produced with the simplified process.

### Example 3.3: Post-CMOS Processing for High-Aspect-Ratio Integrated Silicon Microstructures

Fedder et al.[69] demonstrated the use of a maskless SCREAM-like approach of combined anisotropic and isotropic dry etching to fabricate thin laminated aluminum/silicon dioxide (or dielectrics in general) mechanical structures using the top-level aluminum layer in standard CMOS as an etch mask for the underlying Si.[69] In this process, an anisotropic silicon dioxide etch ($CHF_3/O_2$) is followed by an anisotropic Si etch ($SF_6/O_2$), and finally an undercutting isotropic Si etch is used to release the movable mechanical elements. This post-CMOS process is illustrated in Figure 3.44. The etching masks are the interconnect metals from preceding CMOS processing, so no post-CMOS lithography is required, hence the name "maskless post-CMOS micromachining." An aspect ratio of 4:1 can be achieved. The microstructures consist of up to three metal layers that are electrically isolated by dielectrics. Both lateral and vertical CMOS accelerometers with fully differential full-bridge capacitive interface circuits have been fabricated using this post-CMOS micromachining process, and a triaxial microstage has also been realized.[51] A major drawback with this type of CMOS-MEMS is that the residual stress and thermal expansion coefficient in the thin metal/dielectric layers cause curling or bending.

More recently, with the improvements in deep RIE, the same Carnegie Mellon University-based research group has revisited their SCREAM-like

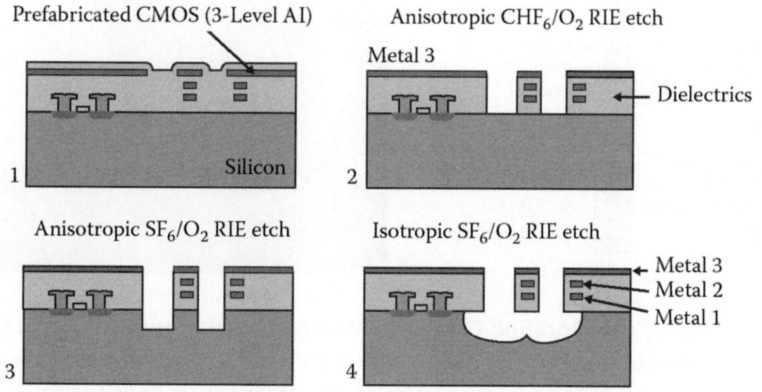


**FIGURE 3.44** Cross-sectioned view of thin-film post-CMOS micromachining. Illustration of the process flow showing use of the uppermost metallization level as mask for silicon etching, first with an anisotropic etch and then an isotropic etch to free movable structures.

approach by incorporating a front and a back-side anisotropic Bosch etch. The process flow is illustrated in Figure 3.45A. In step 1, a deep anisotropic etch leaves a 10–100-μm thick single-crystal Si membrane underneath the CMOS microstructural layer. The central idea is to leave a thick layer of single crystal backing the thin CMOS multilayer such that the mechanical properties are dominated by the underlying single-crystal Si, and electrical connections continue to be provided by the CMOS microstructure layer. In step 2, a front-side anisotropic etch is carried out to etch through the thin CMOS dielectric layers. In step 3, the microstructure is released by an anisotropic Si etch instead of the isotropic etch typically used in the SCREAM process. The resulting thick Si, which remains underneath the CMOS layer, results in a totally flat released microstructure (no curling). Moreover, the proof mass and comb-finger capacitance for CMOS-based inertial sensors are significantly increased because of the higher *z*-dimensions. A close-up of one corner of a comb-drive actuator fabricated using the new process is shown in Figure 3.45B.

### Example 3.4: Dry Etching of Polymeric Materials

As we learned so far, to obtain the optimum anisotropy in high-aspect-ratio devices, one cools the substrate, uses high RF power density, works at the lowest possible pressure, uses high gas flow rate, relies on sidewall-passivating additives, or, as in the Bosch process, one switches back and forth between etching and passivation. We give examples here of the anisotropic power of MERIE by Furuya et al.[70]

Furuya et al.,[70] at NTT, made a fluorinated polyimide microgrid (Figure 3.46) by using MERIE in oxygen plasma. The grid elements consist of 18 pole-shaped, gold-covered electrodes connected at the base by a common electrode. Individual poles measure 15 μm in diameter and are 100 μm high. The microgrid is formed by pairs of grid elements facing each other. A voltage can be applied between opposing electrodes of the grid-element pairs. A microgrid like this, the authors suggest, can be used in an electrostatic ion drag micropump and in microelectrophoresis devices. The fluorinated polyimide used in this study etches rapidly (3–5 μm/min) and with high selectivity (2600 over a Ti mask) because MERIE produces a high-density, low-energy plasma and because the fluorine in the polyimide produces a high concentration of oxygen radicals in the plasma. From the F/C rule presented above, we predict that a fluorocarbon/oxygen atmosphere leads to a more aggressive and less selective etch. In the current case, by adding the fluorine to the polyimide instead of to the etchant gas, a faster etch rate occurs without losing the high selectivity.

The morphology of the fluorinated polyimide surface, as revealed by atomic force microscopy

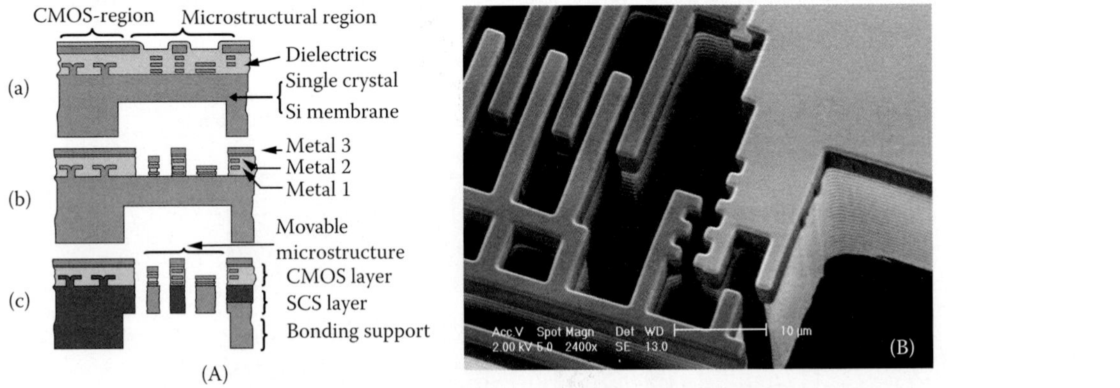


**FIGURE 3.45** (A) The process flow for the modified CMOS-MEMS micromachining process; in step (a) the CMOS chip with back-side etch is shown; in step (b) the anisotropic dielectric etch is illustrated; and in step (c) the anisotropic silicon etch for the release is demonstrated. (B) SEM of a corner of a comb-drive actuator fabricated in Hewlett-Packard 0.5-μm CMOS followed by post-CMOS Si DRIE micromachining. The upper 5-μm-thick composite layer is the CMOS three-metal interconnect stack. The underlying Si is 25-μm thick and illustrates the sidewall scalloping caused by the Si DRIE. (Courtesy of Dr. G. Fedder, CMU.)

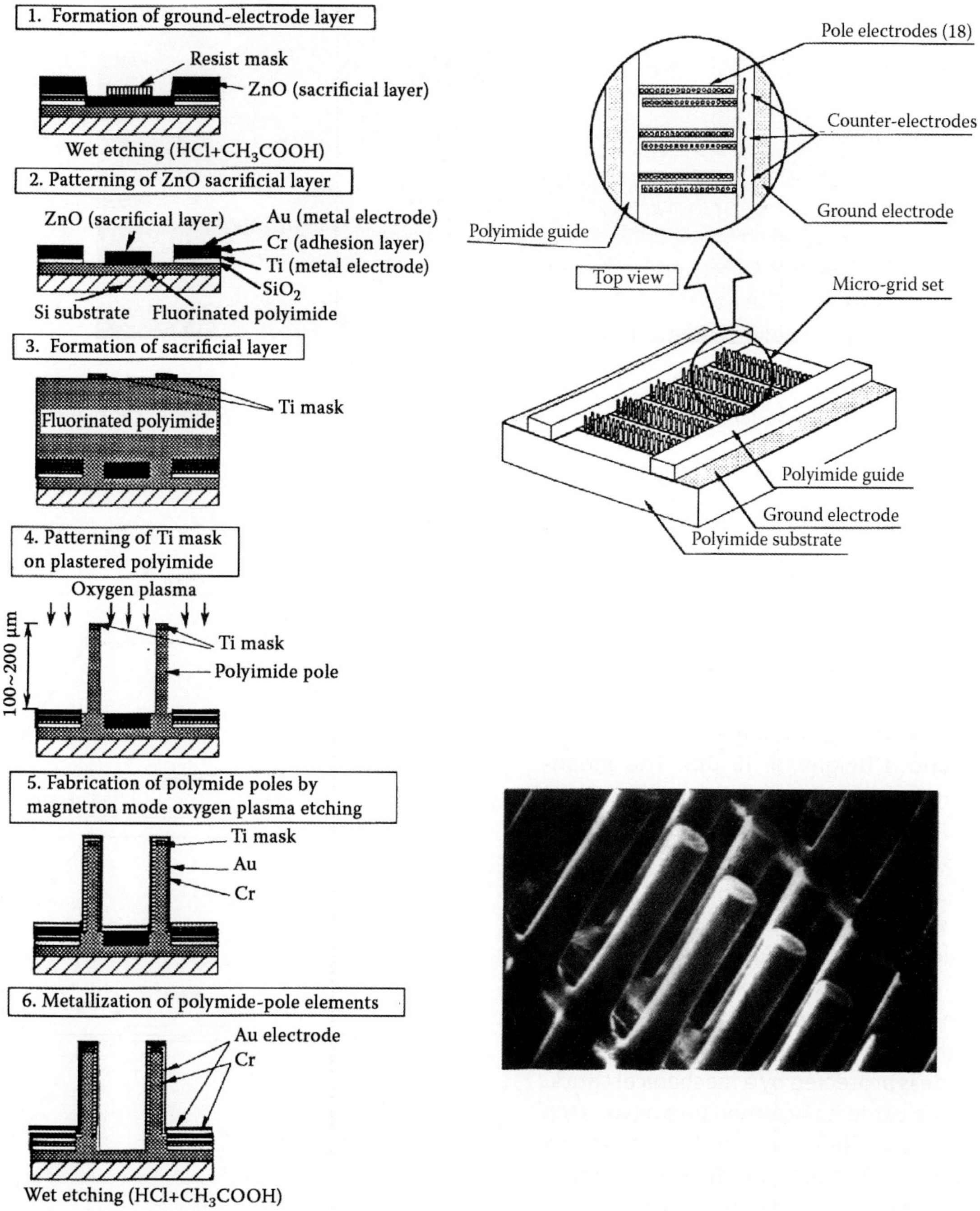


**FIGURE 3.46** An array of polyimide posts made by cryogenic MERIE. (Furuya, A., F. Shimokawa, T. Matsuura, and R. Sawada. 1993. *Proceedings: IEEE micro electro mechanical systems (MEMS '93).* Fort Lauderdale, FL: IEEE. With permission.[70])

is smooth at a high etching rate, that is, at high magnetic fields. The NTT fluorinated polyimide is chemically stable, and the fluorine content makes the material extremely transparent, so it can be used to make optical components such as optical waveguides and lenses. The optical quality (and the high-aspect ratio of the produced polyimide structures) could make this fabrication technology competitive for many x-ray lithography-based LIGA structures (see Chapter 10).

### Example 3.5: Combination Wet and Dry Etching

Brugger et al.[71] developed a micromachining process, combining wet and dry etching, to fabricate a meander-type cantilever with an

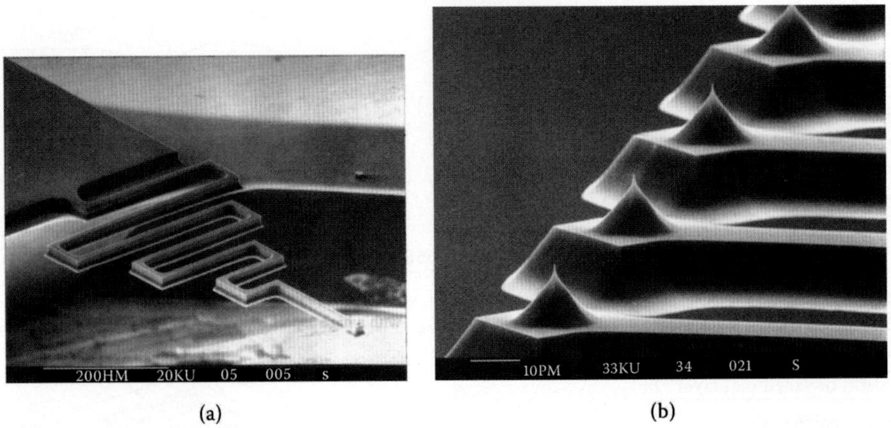


**FIGURE 3.47** Si cantilever with square cross-section, designed as a meander shape. (a) Meander cantilever with tip. (b) Close-up view of tips. (From Brugger, J., R. A. Buser, and N. F. de Rooij. 1992. Silicon cantilevers and tips for scanning force microscopy. *Sensors Actuators A* A34:193–200. With permission.[71])

integrated high-aspect-ratio tip for microprobe applications. The crucial part of an atomic force microscope (AFM) or scanning tunneling microscope (STM) microprobe is the tip, which for some applications (e.g., IC trench probing) should have a curvature radius ideally in the range of 50 Å or less, an angle of aperture of about 5°, and a height of 10 μm. The meander structure and integrated tip are presented in Figure 3.47, and the process sequence is in Figure 3.48.

First, a 30-μm thin Si membrane is anisotropically etched in a double-sided polished 280-μm-thick Si wafer (Figure 3.48a). The etching occurs at the back with 40% KOH at 60°C through an opening etched with buffered HF in a 1.5-μm thermal oxide. During the KOH etch, the front side is protected by a mechanical chuck. The front-side oxide is patterned into a two-step profile using two lithographic and two buffered HF etching steps. The 0.75 μm-thick oxide in the two-step profile acts as the mask for the cantilever, and a 1.5-μm-thick oxide serves as the mask for the tip (Figure 3.48a). The topside is etched with 15-μm- deep RIE using chlorine/fluorine gas mixtures ($C_2ClF_5/SF_6$) to preshape the cantilevers (Figure 3.48b). By adjusting the RIE parameters, vertical sidewalls of the cantilever can be obtained. The 0.75-μm-thick oxide that covered the cantilever is completely removed afterward, and the remaining oxide cap, formerly 1.5 μm and now 0.75 μm thick, serves as a mask for the tip etching. Then a $C_2ClF_5/SF_6$ gas mixture is applied again to RIE the tip (Figure 3.48c). In the final step (Figure 3.48d), an isotropic

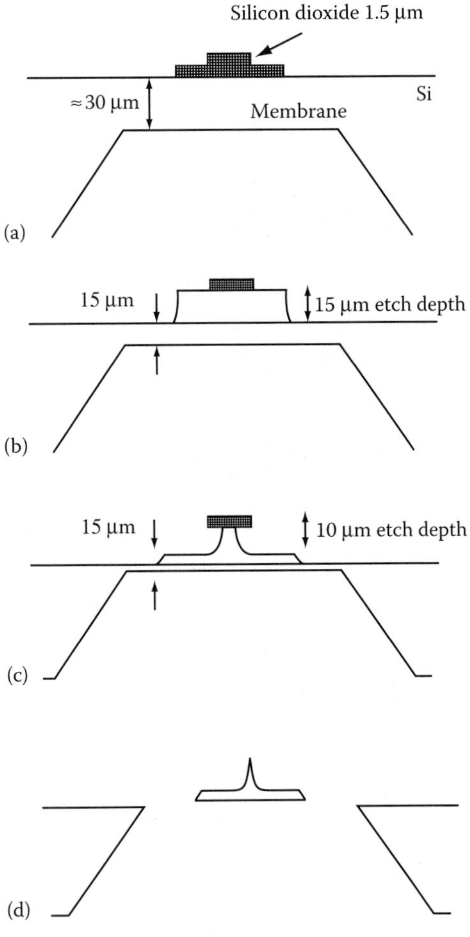


**FIGURE 3.48** Fabrication process for a silicon meander beam with integrated tip. (a) Patterning of oxide mask onto preprocessed 30-μm thin membrane. (b) Dry etching of rough cantilever shape. (c) Dry etching of the tip. (d) Final wet etching step releasing the cantilever, forming the tip and smoothing the surface. (From Brugger, J., R. A. Buser, and N. F. de Rooij. 1992. Silicon cantilevers and tips for scanning force microscopy. *Sensors Actuators A* A34:193–200. With permission.[71])

$HNO_3$:HF:$CH_3COOH$ etch lasting 2 minutes forms the tips (using the oxide cap as mask), releases the cantilever by etching through the remaining Si membrane, smoothes the dry-etched rough surface, and cleans the remaining organic photoresist. Mixtures of $HNO_3$, HF, with $CH_3COOH$ and/or water etch the silicon isotropically and exhibit high etch rates (10–300 μm/min). The $SiO_2$ mask withstands the three successive etch steps: cantilever RIE etching, tip RIE etching, and 1–2-min wet etching with HF:$HNO_3$:$CH_3COOH$ sharpening the tip. Tip heights up to 20 μm with opening angles of approximately 5–10° and tip radii estimated to be 40 nm can be obtained this way.

A postoxidation process yielding Si tips with curvature radii of less than 10 nm by several consecutive oxidation and HF etching steps, exploiting an anomaly of the oxidation behavior at regions with high geometric curvature, has been demonstrated by Marcus et al.[72,73] and by our own group[74] (see also Volume I, Chapter 4, on the dependence of oxidation rate of Si on geometry). By applying this oxidation-sharpening technique to the tips produced in the current example, further sharpening would result. A rigid cantilever structure rather than a meander structure as shown in Figure 3.48a would also be of more use in a practical STM.[74]

## Acknowledgments

Special thanks to Genis Turon, Robert Gorkin, and Xavier Casadevall.

## Questions

3.1: How is a DC plasma created, and how does an RF plasma differ? Why is a plasma always positive with respect to the reactor vessel walls? In which etching setup would you prefer to etch an insulator?

3.2: Detail the different dry etching profiles available and explain how they are obtained.

3.3: Consider a single electron in an electric field between two parallel plates located 10 centimeters apart (d = 10 cm). Assume the potential varies sinusoidally between 1000 V and –1000 V at a frequency of 13.5 MHz. Calculate the maximum kinetic energy of the electrons.

3.4: Using the figure below as a model, assume that the silicon wafer has a bias of –150 V. A reactive ion etch is used to etch 10 μm wide grooves.

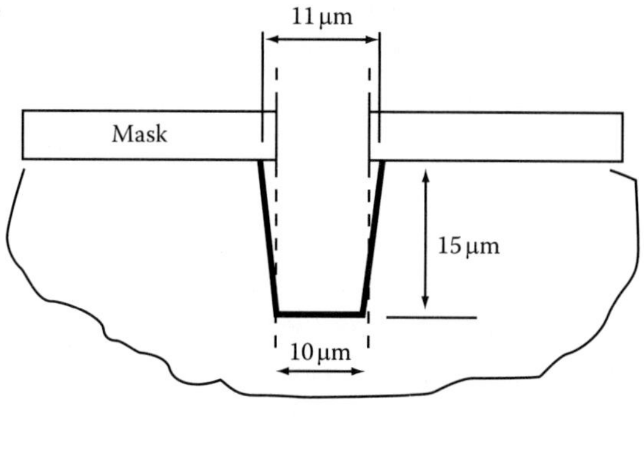


After 1 hour, the shape of the grooves is measured and found to have the following topology:

(a) If the bias were increased to –200 V and the depth of the groove was maintained at 15 μm, would the width at the top of the groove be less than or greater than 11 μm?
Explain.

(b) If the same etch procedure was used to etch $SiO_2$, qualitatively describe the shape of the grooves in the $SiO_2$ if a –150 V bias is applied.

3.5: Explain the DC breakdown voltage vs. electrode distance curve (Paschen's law) and how it is relevant to dry etching. How is miniaturization of an electrode set equivalent to creating a local vacuum?

3.6: Use SCREAM to fabricate a capacitive accelerometer. You may rely on the newest deep dry etching technology.

3.7: Why is reactive ion etching (RIE) somewhat of a misleading name? In physical-chemical etching, there are at least three types of ion-surface interactions which may promote dry etching. Detail each of them.

3.8: Develop a process sequence to fabricate a capacitive pressure sensor using both wet and dry etching techniques.

3.9: Explain loading effects in wet and dry etching. How do you avoid the loading effect? What is the bull's-eye effect? What is the difference between loading and micro-loading?

3.10: Discuss the etch profiles in physical etching (ion sputtering). Also draw profiles exhibiting faceting, ditching, and redepositioning.

3.11: Explain the F/C ratio effect in dry etching and present a couple of means to influence it. How can cryogenic cooling lead to similar etching behavior as inhibitor-driven chemistry?

3.12: Explain the ion energy versus pressure curve for a plasma and how different etching mechanisms are enabled by it. Describe the differences between high-pressure plasmas and reactive ion etching systems. Explain for what applications each is the preferred process.

3.13: You have only a high-pressure plasma etch and an ion mill at your disposal. Select which one to use in the following applications. Justify your answer.

(a) Etching a 5000 Å polysilicon layer that serves as the upper electrode of a large square capacitor. The capacitor dielectric is 50 Å of $SiO_2$.

(b) Anisotropic patterning of a thin layer of $YBa_2Cu_3O_7$ on a thick insulating film.

(c) Recessing the channel of a GaAs FET. For this application the residual etch damage must be minimized.

3.15: Sketch the voltage distribution for an RF plasma discharge in equilibrium and contrast it with the voltage distribution in a DC plasma.

3.16: How would you monitor a dry etching process? Why does endpoint detection of a reactant species require a loading effect?

3.17: Why is the grounded electrode (anode) in an AC sputtering station made as large as possible? Use equations to illustrate your answer.

3.18: Explain how magnetic coils enable higher ion density plasmas. Use equations to explain.

3.19: Compare dry versus wet etching techniques.

3.20: How would you make the structure below with the least expensive equipment, the highest selectivity, and the fastest process?

## Further Reading

Boxman, R. L., D. M. Sanders, and P. J. Martin. 1995. *Handbook of vacuum arc science and technology*. Park Ridge, NJ: Noyes Publications, Chapter 1.

Brown, S. D. 1967. *Basic data of plasma physics, 1966*. Cambridge, MA: MIT Press.

Cobine, J. D. 1958. *Gaseous conductors*. Mineola, NY: Dover Publications.

Nasser, E. 1971. *Fundamentals of gaseous ionization and plasma electronics*. Hoboken, NJ: Wiley-Interscience.

Roth, J. R. 1995. *Industrial plasma engineering*. Bristol, UK: Institute of Physics Publication, Chapters 8–13.

## References

1. Lehmann, H. W. 1991. *Thin film processes II*. Eds. J. L. Vossen, and W. Kern. Boston: Academic Press.
2. Jansen, H., H. Gardeniers, M. de Boer, M. Elwenspoek, and J. Fluitman. 1996. A survey on the reactive ion etching of silicon in microtechnology. *J Micromech Microeng* 6:14–28 and Jansen, H. V., M. de Boer, R. Legtenberg, and M. Elwenspoek. 1994. The black silicon method: a universal method for determining the parameter setting of a fluorine-based RIE in deep Si trench etching with profile control. *Proceedings of micromechanics Europe (MME'94)*, 60–64.
3. Oehrlein, G. S., M. F. Doemling, B. E. E. Kastenmeier, P. J. Matsuo, N. R. Rueger, M. Schaepkens, and T. E. F. M. Standaert. 1999. Surface science issues in plasma etching. *IBM J Res Dev* 43:181.
4. Rossnagel, S. M. 1991. *Thin film processes II*. Eds. J. L. Vossen, and W. Kern. Boston: Academic Press.
5. Konuma, M. 1991. *Film deposition by plasma techniques*. Berlin: Springer-Verlag.
6. Sugawara, M., B. L. Stansfield, S. Handa, K. Fujita, S. Watanabe, and T. Tsukamoto. 1998. *Plasma etching fundamentals and applications*. Oxford, UK: Oxford University Press.
7. Coe, S. E., J. A. Stocks, and A. J. Tambini. 1993. An investigation of the cathode region of a fluorescent lamp. *J Phys D Appl Phys* 26:1203–10.
8. Coe, S. E., and G. G. Lister. 1992. Modeling of the negative glow and Faraday dark space of a low-pressure noble gas discharge. *J Appl Phys* 71:4781–87.
9. Raizer, Y. P., and M. N. Shneider. 1997. The nonmonotonicity of transition from Faraday dark space to positive column and the emergence of standing strata behind the cathode region of glow discharge. *High Temperature* 35:15–21.
10. Bogaerts, A., and R. Gijbels. 1996. Mathematical description of a direct current glow discharge in argon. *Fresenius J Anal Chem* 355:853–57.
11. Bogaerts, A., and R. Gijbels. 1996. Two-dimensional model of a direct current glow discharge: description of the electrons, argon ions, and fast argon atoms. *Anal Chem* 68:2296–303.
12. Bogaerts, A., and R. Gijbels. 1998. Fundamental aspects and applications of glow discharge spectrometric techniques. *Spectrochim Acta (Part B)* 53:1–42.
13. Paschen, F. 1889. Ueber die zum Funkenübergang in Luft, Wasserstoff und Kohlensáure bei verschiedenen Drucken erforderliche Potentialdifferenz. *Wied Ann* 37:69–96.

14. Naidu, M. S., and V. Kamaraju. 1995. *High voltage engineering*. New York: McGraw-Hill.
15. Meek, J. M., and J. D. Craggs. 1953. *Electrical breakdown of gases*. Oxford, UK: Clarendon Press.
16. Brodie, I., and J. J. Muray. 1982. *The physics of microfabrication*. New York: Plenum Press.
17. van Roosmalen, A. J. 1984. Review: dry etching of silicon oxide. *Vacuum* 34:429–36.
18. Wolf, S., and R. N. Tauber. 1987. *Silicon processing for the VLSI era*. Sunset Beach, CA: Lattice Press.
19. Comello, V. 1993. ICP etchers start to challenge ECR. *R & D* April:79–80.
20. Singer, P. 1991. ECR: is the magic gone? *Semicond Int* July:46–48.
21. van Roosmalen, A. J., J. A. G. Baggerman, and S. J. H. Brader. 1991. *Dry etching for VLSI*. New York: Plenum Press.
22. Taniguchi, N. 1989. *Energy-beam processing of materials: advanced manufacturing using various energy sources*. Oxford, UK: Clarendon Press.
23. Schutz, R. J. 1988. *VLSI technology*. Ed. S. M. Sze. New York: McGraw-Hill.
24. Arnold, J. C., and H. H. Sawin. 1991. Charging of pattern features during plasma etching. *J Appl Phys* 70:5314.
25. Elwenspoek, M., H. Gardeniers, M. de Boer, and A. Prak. 1994. *Micromechanics*. Report No. 122830. Twente, the Netherlands: University of Twente.
26. Ayon, A. A., X. Zhang, and R. Khanna. 2000. *Solid-state sensor and actuator workshop*. Hilton Head Island, SC: Transducers Research Foundation, Inc.
27. Jansen, H., H. Gardeniers, M. de Boer, M. Elwenspoek, and J. Fluitman. 1996. A survey on the reactive ion etching of silicon in microtechnology. *J. Micromech Microeng* 6:14–28.
28. Editorial. 1996. *Introducing plasma jet etching*. San Jose, CA: IPEC Precision, Inc.
29. Jeong, J. Y., S. E. Babayan, V. J. Tu, J. Park, I. Henins, R. F. Hicks, and G. S. Selwyn. 1998. Etching materials with an atmospheric-pressure plasma jet. *Plasma Sources Sci Technol* 7:282–85.
30. Kneissl, M., D. Hostetter, D. P. Bour, R. Donaldson, J. Walker, and N. M. Johnson. 1998. Dry-etching and characterization of mirrors on III-nitride laser diodes from chemically assisted ion beam etching. *J Crystal Growth* 189/190:846–49.
31. Carlstrum, C. F., S. Anand, and G. Landgren. 1999. Extremely smooth surface morphologies in $N_2/H_2/CH_4$ based low energy chemically assisted ion beam etching of InP/GaInASP. *Thin Solid Films* 343–344:374–77.
32. Li, Y. X., P. J. French, P. M. Sarro, and R. F. Wolffenbuttel. 1995. *Proceedings: IEEE electro mechanical systems conference*. Amsterdam, the Netherlands: IEEE.
33. Schwartz, G. C., and P. M. Schaible. 1979. Reactive ion etching of silicon. *J Vac Sci Technol* 16:410–13.
34. Oehrlein, G. S. 1990. *Reactive ion etching: handbook of plasma processing technology*. Ed. S. M. Rossnagel. Park Ridge, NJ: Noyes.
35. Juan, W. H., J. W. Weigold, and S. W. Pang. 1996. Dry etching and boron diffusion of heavily doped, high aspect ration SI trenches. *Proceedings: SPIE*. Austin, TX: SPIE.
36. Flamm, D. L. 1992. Dry plasma resist stripping part I: overview of equipment. *Solid State Technol* 35:37–39.
37. Mogab, C. J., A. C. Adams, and D. L. Flamm. 1979. Plasma etching of Si and $SiO_2$: the effect of oxygen additions to $CF_4$ plasmas. *J Appl Phys* 49:3796–803.
38. Heinecke, R. H. 1975. Control of relative etch rates of $SiO_2$ and Si in plasma etching. *Solid State Electron* 18: 1146–47.
39. Peters, L. 1992. Plasma etch chemistry: the untold story. *Semicond Int* May:660–70.
40. Moreau, W. M. 1988. *Semiconductor lithography*. New York: Plenum Press.
41. Hess, D. W., and D. B. Graves. 1989. *Microelectronics processing: chemical engineering aspects*. Eds. D. W. Hess, and K. F. Jensen. Washington, DC: American Chemical Society.
42. Flamm, D. L. 1992. Dry plasma resist stripping Part II: physical processes. *Solid State Technol* 35:40–42.
43. Flamm, D. L. 1993. Feed gas purity and environmental concerns in plasma etching: part II. *Solid State Technol* November:43–50.
44. Flamm, D. L. 1992. Dry plasma resist stripping part III: production economics. *Solid State Technol* 35:43–48.
45. Klaassen, E. H., K. Petersen, J. M. Noworolski, J. Logan, N. I. Maluf, J. Brown, C. Storment, W. McCulley, and G. T. A. Kovacs. 1995. Proceedings: Eighth international conference on solid-state sensors and actuators, and Eurosensors IX. Stockholm, Sweden: IEEE.
46. Kovacs, G. T. A., N. I. Maluf, and K. E. Petersen. 1998. Bulk micromachining of silicon. *Proc IEEE* 8:1536–51.
47. Laermer, F., and A. Schilp. 1996. Method of anisotropically etching silicon. US Patent 5,501,893.
48. Maluf, N. 2000. *An introduction to microelectromechanical systems engineering*. Boston: Artech House.
49. Chow, E. M., A. Partridge, C. F. Quate, and T. W. Kenny. 2000. *Solid-state sensor and actuator workshop*. Hilton Head Island, SC: Transducers Research Foundation, Inc.
50. Bosch-Charpenay, S., J. Xu, J. Haigis, P. A. Rosenthal, P. R. Solomon, and J. M. Bustillo. 2000. *Technical digest: solid-state sensor and actuator workshop*. Hilton Head Island, SC: Transducers Research Foundation, Inc.
51. Xie, H., L. Erdmann, X. Zhu, K. J. Gabriel, and G. K. Fedder. 2000. *Technical digest: solid-state sensor and actuator workshop*. Hilton Head Island, SC: Transducers Research Foundation, Inc.
52. Ayon, A. A., R. Braff, C. C. Lin, H. Sawin, and M. A. Schmidt. 1999. Characterization of a time multiplexed inductively coupled plasma etcher. *J Electrochem Soc* 146: 339–49.
53. Ayon, A. A., R. Braff, R. Bayt, H. H. Sawin, and M. A. Schmidt. 1999. Influence of coil power on the etching characteristics in a high density plasma etcher. *J Electrochem Soc* 146:2730–36.
54. Juan, W. H., S. W. Pang, A. Selvakumar, M. W. Putty, and K. Najafi. 1994. *Technical digest: 1994 solid-state sensor and actuator workshop*. Hilton Head Island, SC: Transducers Research Foundation, Inc.
55. Chinn, J. D., I. Adesida, and E. D. Wolf. 1983. Chemically assisted ion beam etching for submicron structures. *J Vac Sci Technol* B1:1028–32.
56. Hoffman, E., B. Warneke, E. Kruglick, J. Weigold, and K. S. J. Pister. 1995. *Proceedings: IEEE micro electro mechanical systems (MEMS '95)*. Amsterdam, the Netherlands.
57. Köhler, U., A. E. Guber, W. Bier, M. Heckele, and T. Schaller. 1996. *Proceedings: SPIE miniaturized systems with micro-optics and micromechanics*. San Jose, CA: SPIE, 18–22.
58. Wang, X.-Q., X. Xang, K. Walsh, and Y.-C. Tai. 1997. *Proceedings: Transducers '97*. Chicago.
59. Mucha, J. A., and D. W. Hess. 1983. *Introduction to microlithography*. Eds. L. F. Thompson, C. G. Willson, and M. J. Bowden. Washington, DC: ACS Symposium Series.
60. Economou, D., E. S. Aydil, and G. Barna. 1991. In situ monitoring of etching uniformity in plasma reactors. *Solid State Technol* 34:107–11.

61. Elta, M. 1993. Developing "smart" controllers for semiconductor processes. *Res Dev* February:69–70.
62. Oehrlein, G. S. 1986. Reactive-ion etching. *Phys Today* 39:26–33.
63. Deshmukh, V. G. I., and T. I. Cox. 1988. Physical characterization of dry etching plasmas used in semiconductor fabrication. *Plasma Physics Controlled Fusion* 30: 21–33.
64. Duek, R., N. Vofsi, M. Haemek, S. Mangan, and M. Adel. 1993. Improving plasma etch with wafer temperature readings. *Semicond Int* July:208–10.
65. Zhang, Z. L., and N. C. MacDonald. 1991. *Sixth international conference on solid-state sensors and actuators (Transducers '91)*. San Francisco, CA.
66. Zhang, Z. L., and N. C. MacDonald. 1993. Fabrication of submicron high-aspect-ratio GaAs actuators. *J Microelectromech Syst* 2:66–73.
67. Yao, J. J., S. C. Arney, and N. C. MacDonald. 1992. Fabrication of high frequency two-dimensional nanoactuators for scanned probe devices. *J Microelectromech Syst* 1:14–22.
68. Zhang, Z. L., G. A. Porkolab, and N. C. MacDonald. 1992. *Proceedings: IEEE micro electro mechanical systems (MEMS '92)*. Travemunde, Germany.
69. Fedder, G. K., S. Santhanam, M. L. Reed, S. C. Eagle, D. F. Guillou, M. S.-C. Lu, and L. R. Carley. 1997. Laminated high-aspect-ratio microstructures in a conventional CMOS process. *Sensors Actuators* A57:103–10.
70. Furuya, A., F. Shimokawa, T. Matsuura, and R. Sawada. 1993. *Proceedings: IEEE micro electro mechanical systems (MEMS '93)*. Fort Lauderdale, FL: IEEE.
71. Brugger, J., R. A. Buser, and N. F. de Rooij. 1992. Silicon cantilevers and tips for scanning force microscopy. *Sensors Actuators A* A34:193–200.
72. Marcus, R. B., and T. S. Ravi. 1993. Method for making tapered microminiature silicon structures, Bell Communications Research. US Patent 5,201,992.
73. Marcus, R. B., T. S. Ravi, T. Gmitter, K. Chin, D. Liu, W. J. Orvis, D. R. Ciarlo, C. E. Hunt, and J. Trujillo. 1990. Formation of silicon tips with <1 nm radius. *Appl Phys Lett* 56:236–38.
74. Editorial. 1991. Tips for STM. TSDC (Teknekron). Menlo Park, CA.
75. Flamm, D. L. 1993. Feed gas purity and environmental concerns in plasma etching: Part I. *Solid State Technol* October:49–54.
76. Manos, D. M., and D. L. Flamm, eds. 1989. *Plasma etching: an introduction.* Boston: Academic Press.

# 4

# Wet Chemical Etching and Wet Bulk Micromachining—Pools as Tools

Fear of things invisible is the natural seed of that which every one in himself calleth religion.

**Thomas Hobbes**
*Leviathan (1651)*

*Pools as Tools, Nankoweap,* by Jerome Grimmer

Outline



Ignorance more frequently begets confidence than does knowledge: it is those who know little, and not those who know much, who so positively assert that this or that problem will never be solved by science.

**Charles Darwin**
*The Descent of Man (1871)*

The human mind treats a new idea the way the body treats a strange protein; it rejects it.

**P.B. Medawar**

## Introduction

Chapter 4 is split in two parts: one part deals with *wet chemical etching* and the other with *wet bulk micromachining*. Both these subtractive manufacturing techniques cover purely chemical etching, photochemical etching, electrochemical etching, and combinations thereof. Wet bulk micromachining is derived from chemical etching, but wet etching and bulk micromachining are typically used in very different industrial settings. In chemical etching, one carves isotropic features, large (milling) and small (blanking), out of metals, glasses, polymers, and ceramics. The technique is mostly used in more traditional industrial manufacturing settings such as aerospace, electronics, automotive, medical, computer, instrument manufacturing, and decorating. Wet bulk micromachining, on the other hand, is used to sculpt isotropic and anisotropic microstructures principally in semiconductors substrates. It is a technique championed in the integrated circuit (IC) industry and in micro- and nanomachining.

We start with a historical note covering both chemical etching and wet bulk micromachining.

## Historical Note

The ancient Egyptians etched copper jewelry with citric acid as long ago as 2500 BC. The Hohokam people, of what is now Arizona, etched snail-shell jewelry with fermented cactus juice around 1000 BC. The earliest use of wet etching of a substrate, using a mask (wax) and etchants (acid-base), appears in the late fifteenth or early sixteenth century for decorating armor (Figure 4.1).[1,2] Engraving hand tools were not hard enough to work the armor, and more powerful acid-based processes took over. By the early seventeenth century, chemical etching to decorate arms and armor was a well-established process. Some pieces stemming from that period have been found with designs as small as 0.5 mm. This is remarkable as masking involved cutting the maskant (linseed oil paint) with a scribing tool and peeling the maskant off where etching (with vinegar) was desired. This purely chemical-driven etching process is today also called chemical milling (larger pieces) or chemical blanking (smaller pieces).

**FIGURE 4.1** Armor pieces have been found where the chemical milling led to designs as small as 0.5 mm. A maskant (wax) is cut with a scribing tool and peeled off where etching is desired. Elements of a Light-Cavalry Armor, ca. 1510 Italian (Milan). Steel, etched, and gilded; weight 19 lb., 13 oz. (9 kg). Gift of William H. Riggs, 1913. The Metropolitan Museum of Art. http://www.metmuseum.org/TOAH/hd/rarm/ho_14.25.716.htm.

Harris[1] describes in detail the improvements that, by the mid-1960s, made chemical milling a valuable and reliable method of manufacturing, especially popular in the aerospace industry (M. Sanz of North American Aviation patented chemical milling in 1953 for use on aerospace parts, and in 1956, Turco Inc. purchased the patent rights). The method enabled many parts to be produced more easily and cheaply than by other means and, in many cases, provided design and production configurations not previously possible.

Through the introduction of photosensitive masks by Niépce in 1822 (see Chapter 1), chemical milling in combination with lithography became a reality, and yet smaller feature sizes were within reach. This type of chemical etching is called photochemical milling (PCM). In PCM, photomasks are used to mask parts typically smaller and thinner than those used in chemical milling. The inventor of the printed circuit board (PCB) is generally assumed to be the Austrian engineer Paul Eisler (1907–1995), who, working in England, made a printed circuit board in 1936 as part of a radio set (Eisler is sometimes referred to as the "father of the PCB"). In about 1943 the United States began to use the technology on a large scale to make rugged radios for use in World War II. After the war, in 1948, the United States released the invention for commercial use. Printed circuits did not become commonplace in consumer electronics until the mid-1950s. A PCB basically consists of a planar (flat) substrate that has electronic components mounted on it that are interconnected by conductive tracks. The high volumes required for this product provided strong impetus for the development of more sophisticated PCM methods, particularly in the design of better etching equipment. Today the two most important applications of lithography-based chemical milling remain the manufacture of PCBs and, by 1961, the fabrication of Si-integrated circuitry. The latter application of chemical milling became better known as *wet bulk micromachining*. In this case, the substrates are typically smaller yet (Si wafer size) than in PCM, and the etch profile can be isotropic or anisotropic. The method used to produce a "cut" when etching silicon integrated circuits (ICs) is the same as that used for photochemical milling of metals in the aerospace industry, but the IC and microelectromechanical system (MEMS)/nanoelectromechanical system (NEMS) industries have refined the technique to reach minimum feature sizes orders of magnitudes smaller and to obtain much higher aspect ratios (height-to-width ratio of the features as high as 400 in anisotropic etching of Si). Whereas the device sizes in microfabrication are often much larger than in IC processing, the aspect ratios in MEMS/NEMS are generally much greater than in IC fabrication (see Figure 4.2).

Isotropic etching in silicon semiconductor processing has been used since its introduction in the early 1950s. One finds representative work from that period in an impressive series of papers by Robbins and Schwartz[3–6] on chemical isotropic etching and in Uhlir's early work on electrochemical isotropic etching.[7] The usual chemical isotropic etchant used for silicon was HF in combination with $HNO_3$, with or without acetic acid or water as diluent. The early work on isotropic etching in an electrochemical cell (i.e., *electropolishing*) was carried out mostly in nonaqueous solutions to avoid black or red deposits that formed on the silicon surface in aqueous solutions.[8] Turner[9] showed that, if a critical current density is exceeded, silicon can be electropolished in aqueous HF solutions without the formation of those deposits.

In the mid-1960s, Bell Telephone Laboratories started work on anisotropic Si etching in mixtures of, at first, KOH, water, and alcohol and later in KOH and water. The need for high-aspect-ratio cuts in silicon arose in the fabrication of dielectrically isolated structures in integrated circuits, such as those for beam leads. Both chemical and electrochemical anisotropic etching methods were pursued.[10–20] In the mid-1970s, a new surge of activity in anisotropic etching was associated with the work on V-groove and U-groove transistors.[21–23]

In Volume I, Chapter 1, we mentioned that the first use of Si as a micromechanical element can be traced back to a discovery and an idea from the mid-1950s and early 1960s, respectively. The discovery was the large piezoresistance in Si and Ge by Smith in 1954.[24] The idea stemmed from Pfann et al.,[25] who in 1961 proposed a diffusion technique for the fabrication of Si piezoresistive sensors for stress, strain, and pressure. This led eventually to

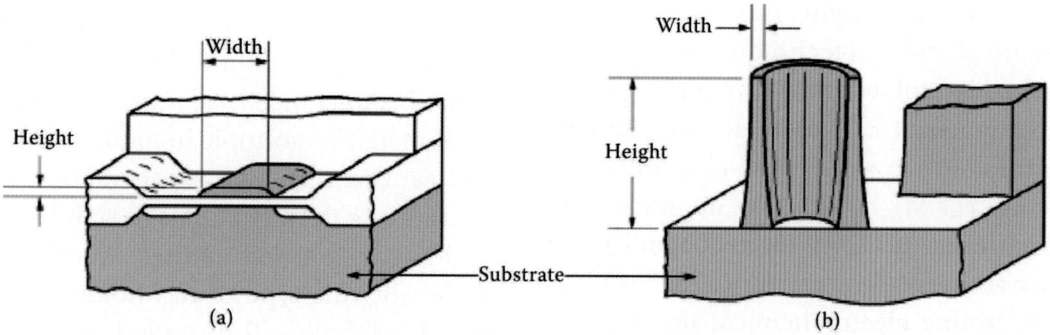


**FIGURE 4.2** Aspect ratio (height-to-width ratio) typical in fabrication of integrated circuits (a) and microfabricated components (b).

the development of commercial Si micromechanical pressure sensors, accelerometers, and gyros made by wet bulk micromachining.

## Chemical Etching

### Introduction

Under chemical etching we review a number of different processes where material removal through etching of preferentially exposed (masked) surfaces produces stress-free parts in many types of substrates. The energy source at the work piece in this subtractive process is purely chemical (chemical milling or CM), photochemical (photochemical milling or PCM and photofabrication), or electrochemical (all types of electrochemical removal processes).

In the simplest embodiment of chemical etching, i.e., chemical milling or chemical machining, the method relies on nonlithographic masking, in which a masking layer is physically carved out to open up a pattern. A scribing tool, of the surgical knife type, is used to cut through the maskant along the line defined by the edge of a template.

In PCM, photomasks are used to mask parts typically much smaller and thinner than those used in chemical milling. The process uses photolithography just as in the etching of Si and other semiconductors, in which case one refers to wet bulk micromachining (see below). Photofabrication is a type of photochemical milling where ultraviolet exposure through a mask results in a strong modification of the solubility in an optically clear solid layer. This permits the direct photochemical production of thick three-dimensional (3D) glass, plastic, or ceramic structures without the need for a separate pattern transfer to an underlying substrate.

Electrochemical removal techniques use an electrical field in an electrolyte that destroys the atomic bonds of the material.[26] Under electrochemical removal techniques, we will review electrochemical machining (ECM), including microelectrochemical machining (μ-ECM), electrochemical jet etching, laser-assisted electrochemical jet micromachining, scanning electrochemical microscope (SECM), and through-mask electrochemical micromachining and electrochemical grinding (ECG).

### Chemical Milling

Chemical milling or chemical machining involves material removal through chemical etching of preferentially exposed (masked) surfaces, producing stress-free parts in many metals. It is the oldest among nontraditional machining processes and has been used to engrave metals and hard stones. The method is also referred to as immersion chemical blanking/milling, and it is a purely chemical process that dissolves material from the unmasked (unprotected) areas of metallic parts immersed in a tank filled with heated and agitated chemical reagents. The term "blanking" denotes small, thin work pieces, and "milling" indicates relatively large work pieces. The method produces clean, sharp, well-defined edges with no burrs (often a source of fatigue failure in highly stressed parts). The work piece does not require finishing of the surface, and no electric current to remove material is involved. Although most common with aluminum and titanium, several other metals have been successfully chemically milled, with different metals etched in different etch solutions. The method relies on nonlithographic masking, in which a masking layer is physically carved out to open up a pattern as illustrated in Figure 4.3.[1] The mask, an organic layer—usually a tape or chemically resistant paint—covers a large area substrate that may be up to 2 cm thick. The etch factor, $A/R$, or eat-back ratio, is identified in Figure 4.3a, with $R$ the etch depth and $A$ the overhang of the maskant. A scribing tool, of the surgical knife type, is used to cut through the maskant along the line defined by the edge of a template (Figure 4.3b). Lateral tolerances of ±0.25–0.5 mm are typical for chemical milling, an accuracy mainly determined by the manual scribing process. For the IC industry and micromachining, this lateral tolerance would be very poor performance, but for many industrial applications, this tolerance is adequate. Chemical machining of metals is isotropic in nature, and aspect ratios are low. Creating high-aspect-ratio structures (>1), e.g., to make deep channels in stainless steel sheets, would necessitate laminating several of these sheets. Obviously, this type of technology is better suited for etching shallow features in large, single-substrate devices. Chemical milling has at least two key

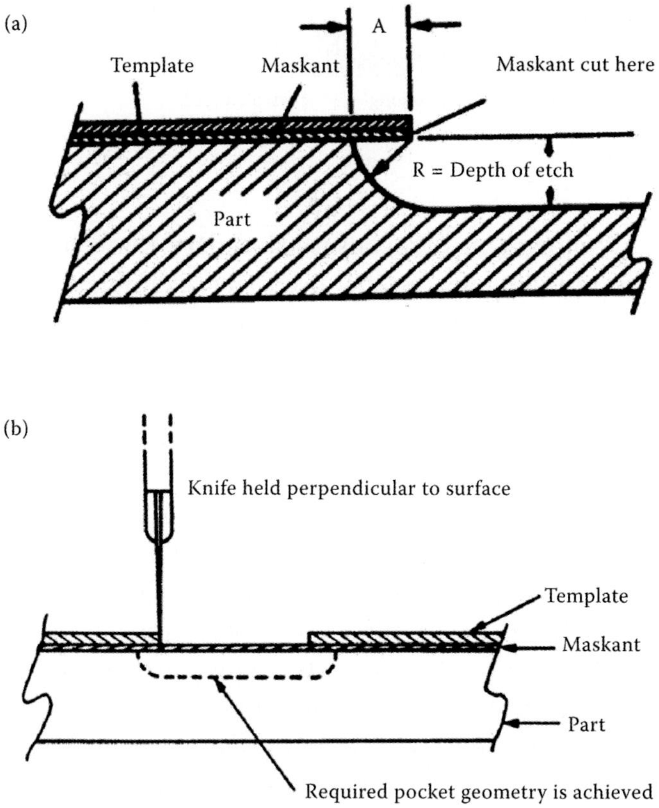


**FIGURE 4.3** Chemical milling showing the effect on the work piece. (a) Undercut caused by chemical milling. (b) Scribing of the maskant. (From Harris, T. W. 1976. *Chemical milling.* Oxford: Clarendon Press. With permission.[1])

advantages over traditional mechanical milling. First, mechanical milling is difficult on thin metal sheets because the tool tends to pull the thin metal foil, severely damaging the work piece. Second, chemical milling can produce complexly shaped parts, such as the inside leading edge of an airplane wing, which a milling machine could not reach.

Chemical milling of metals is widely applicable, simple, and requires little operator skill. Practical cutting (etching) depths are generally limited to less than 2 cm because the etchant process is relatively

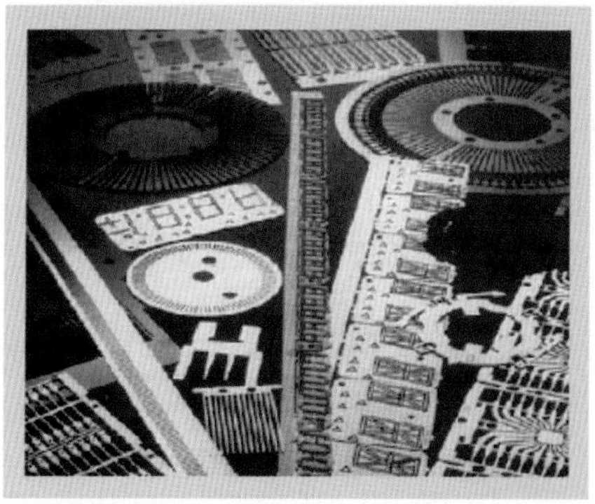

**FIGURE 4.5** Various parts made by chemical blanking. Note the fine detail. (Courtesy of Buckabee-Mears St. Paul.)

slow, and undercutting occurs behind the edges of the protective mask. Industrial applications are mainly in the shaping of thin metal foils and large shallow areas. Most modern aerospace designs use chemical milling to remove about half the weight of their products. It is used extensively on commercial and military airplanes, satellites, and the space shuttle. Large parts can be accommodated; aluminum airplane wing parts, for example, are chemically machined in 50-ft.-long tanks. In Figure 4.4 we show a chemical milling tank and a piece that has been thinned using three consecutive masking steps. Example products made by chemical blanking are shown in Figure 4.5.

A good further review with many relevant details on wet etching as it relates to the IC and electronics industry is presented by Romankiw.[27]

## Photochemical Milling

Photochemical milling (PCM) is an engineering production technique for the manufacture of burr-free and stress-free flat metal components by selective chemical etching through a photographically

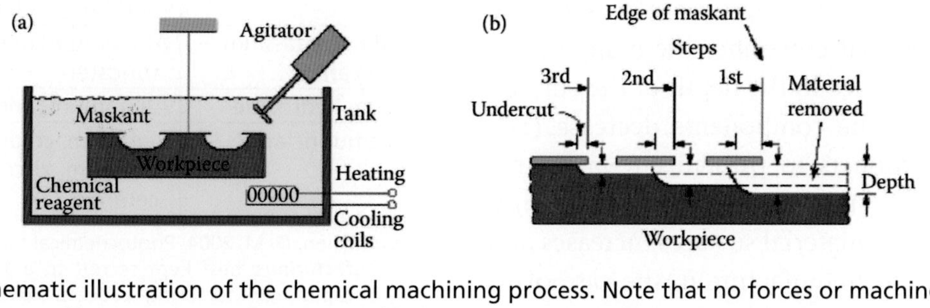


**FIGURE 4.4** (a) Schematic illustration of the chemical machining process. Note that no forces or machine tools are involved in this process. (b) Stages in producing a profiled cavity by machining; note the undercut.

**FIGURE 4.6** Photochemical milling (PCM) of metal parts.

produced mask. In PCM of metals, illustrated in Figure 4.6, photomasks are used to mask parts typically much smaller (a maximum of about 60 × 60 cm) and thinner (<0.5 mm) than those used in chemical etching (see above). The process uses photolithography just as in the etching of Si and other semiconductors as discussed in the section below on wet bulk micromachining. To ensure high accuracy, oversized drawings (100×) of a design are reduced photographically to actual size. The photomask is then used to contact print the component image onto a metal sheet covered with a photosensitive coating. The photosensitive layer may be a dry resist layer such as Etertec® 5600 from Eternal (http://www.eternal-group.com) or Riston CM206 from DuPont (http://www2.dupont.com/Imaging_Materials/en_US/tech_info/product_selector.html) (see also Chapter 1). Careful masking is crucial to achieve precise parts in both chemical and photochemical cutting. However, because of the absence of scribing errors in photochemical milling, the photosensitive maskant is extremely accurate and leads to far superior tolerances for small intricate parts. Photochemical milling was designed specifically to alleviate the lateral-accuracy problem of chemical milling, achieving it extraordinarily well; lateral accuracies (e.g., for printed circuit board production) are ±0.013 mm. Another reason why the tolerances in PCM are so much better than in chemical milling is related to the attempted etch depths. For example, the geometry of the cut produced with photoetching printed circuit boards or integrated circuits might be similar to the chemical-milling cut of the aerospace industry, but one must remember the many orders in magnitude difference in the depth of the cut.[1] As the dimensions of the components decrease (i.e., from precision products to printed circuit boards and to integrated circuits), the necessity for chemical cleanliness of the material surfaces increases and requires clean-room manufacture in the case of IC fabrication.

Photoetching of metals is less generic than chemical milling but can be used, for example, with the following metals (in order of increasing difficulty): copper, nickel, carbon steel, stainless steel, aluminum, titanium, and molybdenum.[28,29] Etching rates—with sodium hydroxide for aluminum, hydrochloric and nitric acids for steel, and iron chloride with nitric acid for copper—vary from $1.3–7.6 \times 10^{-3}$ cm/min. Some PCM etchants are listed in Table 4.1.

Photochemical etching of metals, like chemical etching, is isotropic, and aspect ratios are consequently low. Normally, the diameter of a hole or width of a slot needs to be equal to or exceed the metal thickness. A rule of thumb for PCM tolerance is ±10% of the metal substrate thickness. This guideline, however, does not express the ultimate capabilities of the PCM process because variable etch factors and reduced sheet size can enhance dimensional control.

Photoetching of metals requires moderate investments and competence in two fields: metal

**Table 4.1** Example PCM Etchants and Materials That Are Etched by Them

| Aqueous Etchants | Material Etched in the PCM Process |
|---|---|
| Acidified ferric chloride | Aluminum, Alloy 42, copper and copper alloys, HyMu, Inconels, Invar, Kovar, Permalloy, Monel, Mumetal, nickel, Nimonics, phosphor bronze, stainless steels and other steels, tin |
| Acidified cupric chloride | Beryllium copper, copper and copper alloys including brass and bronze, lead |
| Alkaline potassium ferricyanide | Aluminium, molybdenum, and tungsten |
| Sodium hydroxide | Aluminium, anodized aluminium |
| Hydrofluoric acid | Beryllium, columbium (niobium), titanium, zirconium, glasses, and ceramics |

*Source:* Allen, D. M. 2004. Photochemical machining: from 'manufacturing's best kept secret' to a $6 billion per annum, rapid manufacturing process. *Annals of the CIRP*, 53: 559–572.

TABLE 4.2 PCM and Chemical Milling Compared

| Fabrication Method or End Product | Typical Material Etched | Thickness of Material Etched | Underlying Material or Support |
|---|---|---|---|
| Photochemical machining (PCM) | Metals, glasses, or ceramics | 2 mm (typical maximum) | Not applicable |
| Printed circuit boards (PCBs) | Electroformed foil (a conductor) | 17.5 μm (typical) | Rigid epoxy fiber glass or other insulator such as flexible polyimide |
| Integrated circuits (ICs) | Silicon dioxide or silicon nitride | 0.01–0.1 μm (typical) | Silicon |
| Chemical milling | Aluminum, titanium, and aerospace alloys | 0.1–10 mm (typical) | Not applicable |

etching and lithography. PCM is a fast, efficient, cost-competitive, burr- and stress-free method of producing metal components. Applying photochemical etching to large surface areas is a very expensive proposition,[30] and, in that case, chemical milling may be a better approach. Common industries served include aerospace, electronics, automotive, computer, and instrument manufacturing. Major applications are lead frames, PCBs, shadow masks for color televisions, heat sinks, shields, covers, lids, gaskets, springs, encoders, disks, grids, screens connectors, panels, shims, bus bars, and combs.

In Table 4.2 we compare PCM on metals, glasses, and ceramics with PCM on printed circuit boards and ICs and chemical milling in Al, Ti, and aerospace alloys.

A typical foundry for PCM in the United States is PhotoMachining, Inc. (http://www.photomachining.com/apps.html). Good information on Photochemical Milling is available at http://www.pcmi.org. Standard tool production time in these foundries is 2–3 days but can be several hours if necessary. Tooling is produced and stored in an environmentally controlled facility to ensure accuracy and quality. Modern digital photography uses customer-supplied IBM/Mac-based programs, as well as all AutoCAD versions and DXF and GbR files. In Figure 4.7 some typical PCM products are shown.

## Photofabrication

### *Introduction*

Photofabrication is a subtractive process in which ultraviolet exposure through a mask results in a strong modification of the solubility in an optically clear solid layer. This permits the direct photochemical production of thick 3D structures without the need for a separate pattern transfer to an underlying substrate. In photofabrication, photosensitive glasses, plastics, and ceramics are patterned into intricate 3D shapes. As opposed to most photolithography processes, where photoresist is a sacrificial layer only, in photofabrication the photosensitive material itself becomes a permanent part of the microdevice under construction. The photosensitive materials in photofabrication are permanent, and in the case of polymers this typically involves negative photoresists (see also Chapter 1). The method offers simple, inexpensive, high-aspect-ratio 3D microstructures for small to medium batch sizes without the need to resort to more expensive methods such as Si wet bulk micromachining or LIGA (see Chapter 10). The industrial application of photofabrication, creating sharp images in ceramics, plastics, and glasses, preceded the Si microfabrication era. Given the current interest in high-aspect-ratio 3D structures, this technique is worth revisiting.

**FIGURE 4.7** Mechanical metal parts produced by photochemical milling: leaf springs, connector arrays, masks, free standing grids, laser and IR target apertures, precision pin holes, and encoders.

For most photoplastics, aspect ratios do not exceed 3:1, but, with photosensitive glasses and ceramics, it is possible to make a 0.13-mm-diameter hole through a 2.7-mm-thick plate, with an aspect ratio of ~21:1; this is no small achievement. We will mostly detail photosensitive glasses here. A few words on photosensitive plastic sheets suffice because the topic of dry permanent resists was covered already in Chapter 1.

Note stereolithography, as discussed in Chapter 8, is an additive technique in which plastic shapes are generated by laser writing in a molten plastic mass; the technology is sometimes referred to as *microphotoforming* and should not be confused with the subtractive photofabrication discussed here.

### *Photosensitive Glasses/Ceramics—Photography in Glass*

Electrochromic glasses temporarily change color when exposed to strong light. Photosensitive glasses, on the other hand, develop an invisible, permanent "latent image" after UV illumination through a mask. Subsequent heat treatment can develop the latent image into a permanent structural and/or color change with accompanying differences in solubility. In the mid-1940s, Donald Stookey invented both photosensitive and electrochromic glasses at Corning Glass Works (now Corning, Inc.) To read up on Stookey's career, consult *Profiles in Ceramics* (http://www.ceramicbulletin.org/months/Mar00/StookeyProfile.pdf)[31] and *Explorations in Glass: An Autobiography S. Donald Stookey.*[32] A typical photosensitive glass composition, FotoForm® glass, developed by Corning Glass Works, was $SiO_2$, 81.5; $Li_2O$, 12.0; $K_2O$, 3.5; $Al_2O_3$, 3.0; $CeO_2$, 0.03; and Ag, 0.02 (all in weight percent). On exposure to UV light, photoelectrons released by cerium, the photosensitizing ingredient, are trapped at specific sites in the silicate-glass network to form a latent image. By heating the exposed glass to around 500°C, the electrons are released from the traps, and silver ions are reduced to silver atoms, which aggregate into minute colloidal islands. The minute silver clumps serve as nuclei on which crystals, such as lithium metasilicate, sodium bromide, or sodium fluoride, precipitate. If sufficient numbers of fairly large crystals form (~4 μm), an opalescent image results, whereas the masked regions remain glassy and transparent. In HF, opal regions etch 15–30 times faster than the unexposed glass. No masking is required during the etching process. In this way, by exposing a photosensitive glass plate to UV and etching in HF after heat treatment, a wide variety of microstructures can be made. Through-holes in photosensitive glass were experimentally found to have a sidewall taper of only 2–3°; this means that substantially parallel holes may be cut, having a very high depth-to-diameter ratio. The sloping effect can be reduced even further if the HF etching occurs from both sides of the glass simultaneously. Etching from both sides produces a profile with two truncated cones and gives a minimum overall increase of hole diameter. Although etching of holes with diameters appreciably less than one-tenth of the thickness of the glass is not recommended, aspect ratios of 20 are possible. In contrast, in photoetching of metal, as we have seen above, holes cannot be made with a diameter less than the thickness of the sheet. If sidewall taper in photosensitive glass is desired, it can be increased to 20–30°.

Further heat treatment of the finished photosensitive glass pieces, Stookey discovered by accident in 1957, converts the glass to a stronger, partially crystalline material, e.g., Fotoceram®. Stookey left a sample of lithium silicate glass containing silver in an oven at too high a temperature. He expected to find a puddle of glass once he realized his mistake, but instead found a piece of white ceramic because the glass had crystallized with a fine grain size.

The maximum sheet thickness of FotoForm glass available was 0.6 cm, but several pieces can be thermally bonded, forming excellent seals and converting the glass to Fotoceram glass. One disadvantage of the bonding process is that it causes the photoglasses to become dark, which makes the approach less attractive for optical applications.

Corning's photosensitive FotoForm glass found an important application in fluidic logic in the mid-1960s (see for example Figure 4.8), but as that application started fading the production of FotoForm glass and Fotoceram unfortunately was discontinued.[31]

Fluidic components, providing functions such as flip-flops, logic elements such as OR-NOR and AND gates, proportional amplifiers, and fluid resistors, exemplified the capabilities of photofabrication with FotoForm glass in the mid-1960s. The fluidic-control device in Figure 4.8 shows the individual layers of an AND device after through-etching.[1]

Fusion of stacks of photosensitive glass plates provided a permanently sealed and rugged unit, complete with the necessary input and output vents. Through-etching provides the connection between the different planes. The geometrical accuracy of these fluidic devices was maintained within ±0.025 mm, and the tolerance that may generally be expected on this type of etched glass product was ±0.05 mm on etched edges up to 12.7 mm in length. With hole spacings up to 25 mm apart, accuracy to within 0.025 mm center to center was possible.

Few of the Si and plastic microfluidic structures produced today have reached the degree of complexity achieved with photofabrication methods of almost four decades ago when they were applied to fluidic logic (see Figure 4.8). "Fluidic" elements made in the 1960s could achieve many of the functions of an electronic circuit, and they were robust, simple to fabricate, and suited to operating in a hostile environment. Of course, they were much slower (milliseconds instead of nanoseconds) and much larger than their electronic counterparts. The latter spelled the eventual demise of microfluidic "computing" in the late 1960s. In the 1990s, several micromachined fluidic devices were developed

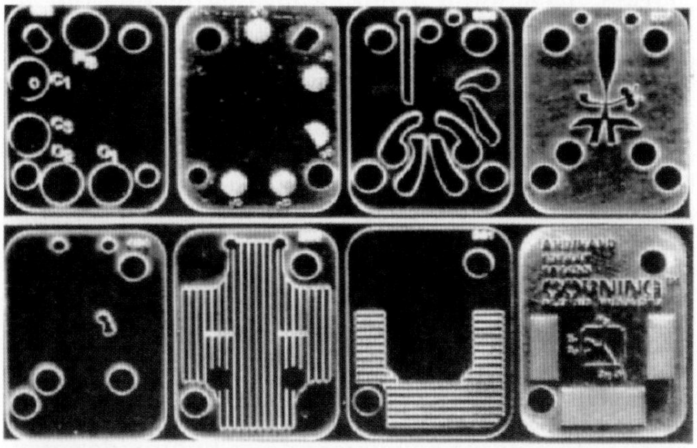

FIGURE 4.8 Fluidic-control device using chemically milled photosensitive glass. The individual layers of an AND device after through-etching with HF. (From Harris, T. W. 1976. *Chemical milling.* Oxford: Clarendon Press. With permission.[1]) See also Volume III, Figure 1.2.

for analytical/biomedical applications, without reference to this earlier sophisticated work in photosensitive glasses. Early in the development of these analytical/biomedical applications, many fluidic devices were attempted in Si. However, the relatively large size of fluidic elements, the need for a robust environment, the relatively low numbers of required devices, and the fact that electronics usually do not need to be integrated render the use of silicon for this application hardly justifiable. The photofabrication machining approach, which is inexpensive and simple, coupled with the chemical inertness of glass, represents a more attractive approach. Fluidic oscillators, fabricated using the more expensive and hardly accessible LIGA technique, are even more difficult to justify when compared with photoformed devices.[33,34] The latter misapplication of miniaturization tools was eventually recognized, and photoformed glass was reintroduced as the best solution for this application by one of the inventors of LIGA technology[35] (see Volume III, Figure 1.2). In July 1996 the commercial entity "Mikroglas" was founded out of the Institute of Microtechnology in Mainz, Germany. Mikroglas obtained the exclusive rights for the use of Foturan™ glass from Glaswerk-Schott (http://www.us.schott.com/english). In 1999, Mikroglas started with the development of microfluidic components for chemical and biotechnological applications. Products include chemical microreactors and whole equipment to use these microreactors (http://www.mikroglas.com/index_e.html).

Unlike ceramics, Foturan is pore-free, has higher temperature stability and chemical resistance than plastics, and has a higher breaking strength than silicon. Processing of Foturan consists of exposure to ultraviolet light, crystallization through heat treatment, and etching as illustrated in Figure 4.9a. In Figure 4.9b we show example hole plates made this way. The thermal expansion coefficient of the processed Foturan can be precisely varied between 8 and 16 ppm/°C. Variously shaped recesses can readily be formed on both sides of the "wafer," as well as 40:1 aspect ratio through holes. In addition, the surfaces can be metallized and patterned by conventional techniques. The material can be processed in sheets of up to 250 mm × 200 mm. Besides for making photosensitive glass ornaments,

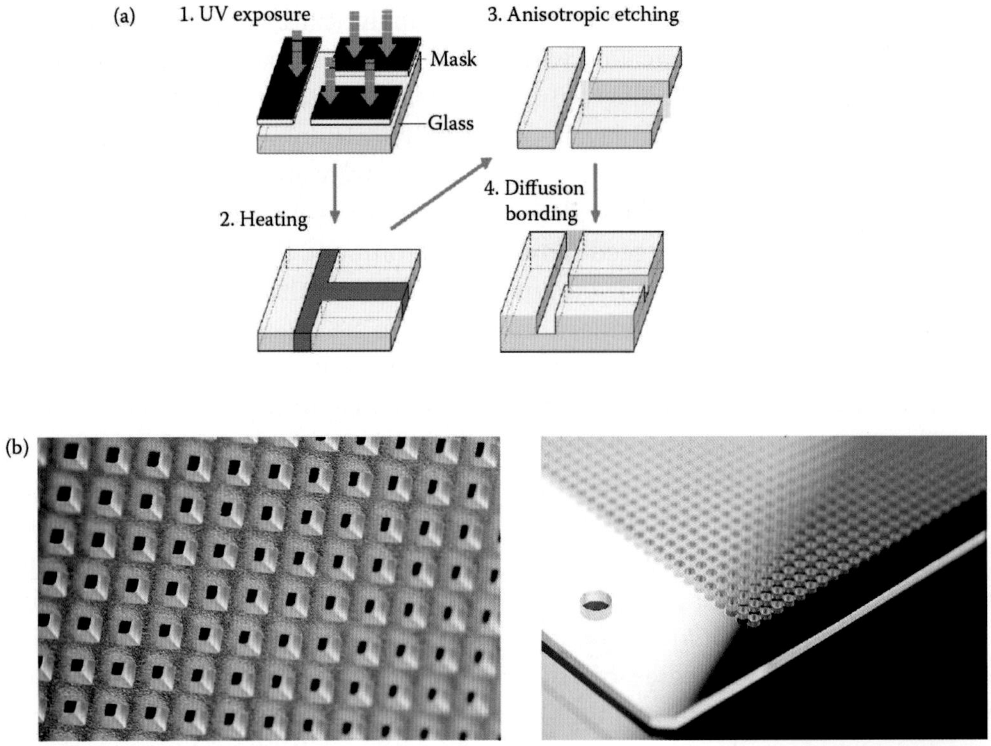


**FIGURE 4.9** (a) The photofabrication process in Foturan. (b) Hole plates shapes, round or rectangular; diameter of holes, down to 50 μm; thickness of plates, 0.1 up to 2.5 mm and up to 100 holes/mm$^2$ (http://www.mikroglas.com).

photosensitive glass has been used to make microlenses (±100 μm in diameter), fluidic elements (see Figure 4.8), spacers in photomultiplier tubes, color filters, integrated optics cell sheets in gas-discharge displays, and charge plates and nozzles in ink-jet printers.

In a very intriguing further development of photosensitive glasses, it was found that the different properties of opal and glass areas on photosensitive glass enable the direct formation of electrical circuitry on glass. Immersing a sample with an opal pattern in a molten salt bath and heating it in a hydrogen atmosphere forms a conducting silver film on the opal regions. The film that forms on the glassy regions remains nonconducting.[31] This discovery could be used in the construction of DC and AC electrokinetic devices with complex patterns of metal electrodes on a glass substrate (see Volume I, Chapter 7).

*Suggestion*: A most useful new MEMS material would be a Foturan-type photosensitive glass with a thermal expansion coefficient matching that of Si (like Corning 7740 glass used in anodic bonding). Such a product would enable the packaging of all types of Si sensors (see Volume III, Figure 4.9) combining a Si micromachined sensing element with a photofabricated glass package.

### Photosensitive Plastics

Although dry photoresists and photosensitive plastics are covered in Chapter 1 and Volume III, Chapter 4, some additional photosensitive plastics are briefly covered here. In the early 1960s, DuPont started developing a photosensitive nylon sold in sheet form, either bonded to a substrate such as carbon steel, aluminum, or a polyester Cronar film, or freestanding. The alcohol-soluble nylon is made photosensitive by adding a cross-linking agent such as methylene bis-acrylamide and an activator such as benzophenone. On exposure to UV, the benzophenone forms free radicals, which react with both of the other compounds, hence insolubilizing the nylon; in other words, the material acts as a negative tone resist. The unexposed areas are then washed away with a dilute aqueous sodium hydroxide spray, forming channels and cavities. By using the benzophenone in a low concentration, about 5% by weight, a layer with a relatively low absorption coefficient results, and thick layers can be insolubilized.

Stencils with 1-mm-high vertical sidewalls with a taper of 2° can be produced this way, and 25-μm-deep, 25-μm-wide channels do not show any undercutting. The aspect ratio did not exceed 3:1, and the operating temperature range is limited to −100 and +200°F. To seal cavities or channels (e.g., for the fabrication of fluidic devices), the first nylon sheet can be laminated with another Cronar film (http://www2.dupont.com/Products/en_RU/Cronar_en.html).[28]

Riston®, Fodel®, and VACREL® photopolyer films are other DuPont dry photoresists (see Figure 1.33). These complex photosensitive systems are designed to polymerize and harden when exposed to ultraviolet light. Fodel can pattern 75-μm vias on a 150-μm pitch, and conductors can be patterned with 50-μm lines on a 100-μm pitch. Fodel extends the density capability of the thick film process to allow densities typically achievable using more costly thin film processes.

Negative-tone photosensitive resins such as SU-8 are particularly suited for high-aspect ratio, high-thickness photofabrication. As we saw in Chapter 1, SU-8 is an epoxy-type, near-UV photoresist based on EPON SU-8 epoxy resin (from Shell Chemical) that was originally developed and patented by IBM. The main advantage of photosensitive resins over glasses and plastics is the fact that they are deposited in liquid form, thus conforming better to existing 2D and 3D structures. Other advantages include exquisite control over the thickness of the layer, higher attainable aspect ratios, higher feature density, and easier alignment to existing features.

### *Conclusions*

Photofabrication is an attractive alternative for microfabrication of all types of high-aspect-ratio, passive elements, especially for intricate shapes that are >1 $cm^2$ in size and are only produced in modest volumes. The current interest in micromachined fluidic elements for analytical purposes is reviving interest in photosensitive glasses, plastics, and resins. Photosensitive glasses are also being considered for the fabrication of photonic crystals (see Volume I, Chapter 5). It certainly makes more sense to use photofabrication of insulators than to fabricate fluidic elements from Si because the latter, after etching, must be passivated. This passivation may undergo electrical breakdown, especially in an aqueous environment. We expect dry photoresist formulations to continue to improve, with tolerances approaching those of traditional spin-on photoresists (see Chapter 1). Applications such as RF-MEMS and fluidics might also lead to the introduction of novel photoformable ceramics and glasses.

## Electrochemical Removal Techniques

### *Introduction*

The term *electrochemical machining* is used to describe electrochemical removing and forming processes with or without masking the substrate. Both processes involve electrochemical energy. Despite performance, speed, and cost advantages, electrochemical milling (ECM) is used little in the IC field when other techniques, especially dry processes, are available. This seems to be connected in part with the overall trend toward dry processing and the disproportionately large number of electrical engineers versus chemical and electrochemical engineers in the IC world. LIGA and other replication techniques described in Chapter 10 have reinvigorated interest in electrodeposition, however. Both electrochemical removing and forming techniques are of extraordinary importance in miniaturization science—without electrochemical removal there would have been, for example, no porous Si (PS) work, and without electroplating of Ni master molds, there would be no plastic micromolding. Here we will review subtractive electrochemical processes only. Additive electrochemical processes are covered in Chapter 8. In electrochemical material removal, an electrical field in an electrolyte destroys the atomic bonds of the material.[26] Under electrochemical removal techniques, we will review electrochemical grinding (ECG), ECM including μ-ECM, electrochemical jet etching, laser-assisted electrochemical jet micromachining, scanning electrochemical microscope (SECM), and through-mask electrochemical micromachining.

### *Electrochemical Grinding*

*Introduction* Grinding usually constitutes a mechanical machining process that removes small

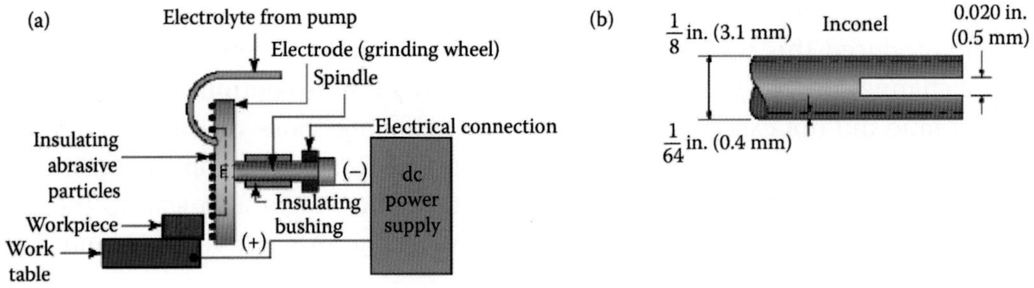


**FIGURE 4.10** (a) Schematic illustration of the electrochemical grinding process. (b) Thin slot produced on a round nickel alloy tube by this process.

amounts of material from a metallic work piece in the form of tiny chips through the contact of small, hard, sharp, nonmetallic particles often embedded in a grinding wheel (see Chapter 6). In electrochemical grinding (ECG), the abrasive action of an electrically conductive wheel, the cathode, accounts only for 10% of the metal removal; the remainder is electrochemical. This technique should not be confused with electrolytic in-process dressing or ELID, where the material removal process (any type of material in this case) is purely mechanical but an electrochemical dissolution process of the metal matrix dresses the abrasive wheel continuously (see Chapter 6).

*Electrochemical Grinding Operating Principles* In ECG, an electrically conductive abrasive wheel, the cathode, and an anode work piece are connected to a low-voltage DC power source (Figure 4.10).[36]

An electrolyte is flushed onto the abrasive wheel, and the abrasive particles in the wheel's surface remove electrochemically dissolved material, always keeping the surface fresh for further electrochemical attack. Very low DC voltages (usually between 3 and 15 V) are used in ECG; therefore, spark discharge is minimized and little or no heat is developed in the cut. All hard conductive materials are candidates for ECG (e.g., sharpening of tungsten-carbide tool bits). The method is also good for fragile parts that would be damaged by conventional mechanical grinding. The method produces parts free of residual stresses and heat damage and surface finishes ranging from 0.1–0.5 μm. The surfaces are also free of scratches and burrs. This process provides not only burr-free machining but also considerably less cutting wheel wear. ECG is similar to electrochemical machining (see below) except that the cathode is a specially constructed grinding wheel instead of a tool shaped like the contour to be machined.

Compositron Corporation is a service company providing an electrochemical grinding homepage (http://www.evercut.com), and Everite Machine Products Co. sells ECG equipment (e.g., Everite Model EG618; http://www.everite.net).

### *Electrochemical Machining*

Electrochemical machining (ECM) historically followed electrochemical grinding (ECG). In ECM one uses a cathode electrode shaped to provide the complementary structure in an anode work piece (Figure 4.11). A highly conductive electrolyte stream separates the cutting tool from the work piece, and metal removal is accomplished by passing a DC current of up to 100 A/cm$^2$ through the salt solution cell. As the cathode tool approaches the anode work piece it erodes its complementary shape in it. Thus, complex shapes may be made from a material such as

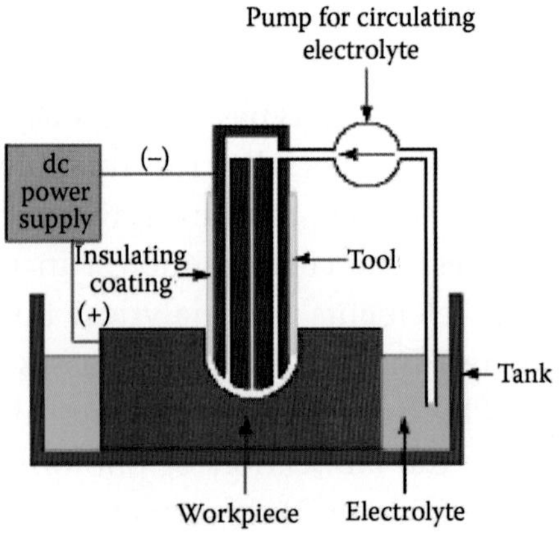


**FIGURE 4.11** Schematic illustration of the electrochemical machining process. This process is the reverse of electroplating, described in Chapter 8.

soft copper and used to produce negative duplicates of it. The process is also called electrochemical sinking. Machines having current capacities as high as 40,000 A and as low as 5 A are available. The pressurized electrolyte (concentrated solutions of inorganic salts such as sodium chloride, potassium chloride, and sodium nitrate) passes at high speed (10–60 m/s) through the gap (about 0.1–0.6 mm) between the work piece and the tool to prevent metal ions from plating onto the cathode tool and to remove the heat that is generated as a result of the high current flow. The cathode is advanced into the anode work piece at a rate matching the dissolution rate, which is between 0.5 and 10 mm/min when applying current densities of 10–100 A/cm$^2$. The supply voltage commonly used in ECM ranges from 5–20 V; the lower values are used for finish machining (creating a final smooth surface), and the higher voltages are used for rough machining. The rate of material removal is the same for hard or soft materials, and surface finishes are between 0.3 and 1 μm. These cutting speeds and surface finishes are comparable with those of electrical discharge machining (EDM; a thermal subtractive technique described in Chapter 5).

The cathode tool must have these four characteristics: be machinable, be rigid (high Young's modulus), be a good conductor, and have good corrosion resistance. The three most common cathode materials used are copper, brass, and stainless steel. The outside of the electrode is insulated to prevent machining actions on the sides. This makes the fabrication of these tools challenging. Because there is no contact between the tool and the work, the tool does not have to be harder than the work, as in traditional machining methods. Hence, this is one of the few ways to machine very hard material; another is spark-discharge machining (see Chapter 5).

The advantages of ECM are that it has relatively low running costs, as there is little scrap, no tool wear, and unskilled labor or automatic machine operation can be used. Because shear is not involved in this process, the surface of the work is, as in chemical milling, free from imperfections and burrs. It is also used in deburring already machined components. A very good surface finish is usually obtained. The disadvantages of ECM are that the machine is initially very expensive—several times the cost of an EDM machine; large systems may cost up to $600,000 (about the same as multiaxis milling machines). The tool is also difficult to make because its sides have to be insulated. The electrolyte has to be well contained, and the most common electrolyte, sodium chloride, corrodes the equipment, tool, and work. Another issue is that the electrochemical erosion reduces sharp surfaces and profiles so the method is not suited for making products with sharp edges. Also, irregular cavities may not be produced to the desired shape with acceptable dimensional accuracy, and designs should make provisions for small taper for holes and cavities to be machined. Overall, the technique is best suited for mass production of high-precision, small, and complex shapes in hard-to-machine materials (accuracy better than 10 μm).[37] To date, ECM has found its widest acceptance in the aircraft, automotive, and business machine industries. In the aircraft industry, ECM is used extensively for shaping turbine blades. In the automotive industry, the technique is used to make connecting rods, pistons, and fuel-injection nozzles. Some example products are shown in Figure 4.12.

The fact that metal removal in ECM is not achieved by mechanical shearing (as in mechanical machining) or by melting and vaporization of the metal (as in EDM or laser machining; see Chapter 5) means that, as with chemical machining, no thermal damage occurs and no residual stresses are produced on the worked surface. For this reason, ECM is often selected for highly stressed or fatigue-sensitive applications in the aerospace industries.

Pulsed electrochemical machining (PECM) is a refinement of ECM, where current pulses are used instead of DC currents. This improves fatigue life and eliminates the recast layer left on die and mold surfaces by EDM (see Chapter 5).

Sometimes, chemical and electrochemical removal machining are in competition. Chemical machining might be preferred when very large work pieces have to be machined because of the large currents involved in ECM, and also for ultrasmall parts, where the electric connection causes problems for ECM.

Typical ECM service providers are Köppern (http://www.koeppern.com/en_ecm_set.htm) and Raycon/AMCHEM (Ann Arbor, MI). A handbook

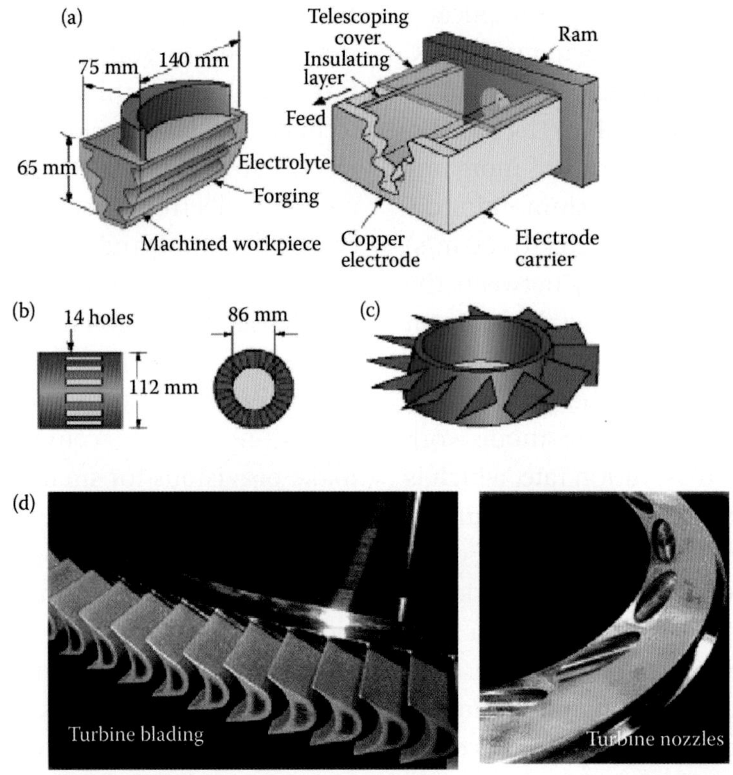


**FIGURE 4.12** Typical parts made by electrochemical machining. (a) Turbine blade made of a nickel alloy, 360 HB. (From *Metal handbook,* 9th ed., vol. 3. Materials Park, OH: ASM International, 1980, p. 849). (b) Thin slots on a 4340-steel roller-bearing cage. (c) Integral airfoils on a compressor disk. (d) Because of the low forces on the work piece, ECM can be used to make holes at very large angle to a surface; in the example, we show turbine nozzle holes (http://www.barber-nichols.com).

covering ECM is the *Non-Traditional Machining Handbook* by Carl Sommer.[38]

### *Electrochemical Micromachining or Microelectrochemical Machining*

*Introduction* The application of ECM in thin film processing and in the fabrication of microstructures is referred to as *electrochemical micromachining* (EMM) or *microelectrochemical machining* (μ-ECM).[39] Different from ECM, the cathode does not necessarily have the shape of the contour desired in the anode work piece. The 3D shaping in EMM may involve maskless or through-mask material removal, and we will examine examples of both approaches. The tool may also be connected to a CNC machine to produce even more complex shapes with a single tool as illustrated in Figure 4.13.

In conventional ECM, the gap between cathode tool and anode work piece is typically about 150 μm; in μ-ECM, the gap is closer to 15–20 μm, and feature sizes change from 150–200 μm to 15–20 μm as we move from the ECM to the μ-ECM domain. The major challenge in moving from the conventional ECM to the μ-ECM domain is to control the size of the reaction region. Methods to accomplish this include:

1. Reduce the size of electrodes—μ-EDM is used.
2. Shield the electrode—for stray currents.
3. Gap control strategies.
4. Use ultrashort-pulsed voltages having time duration in the ranges of nanoseconds.

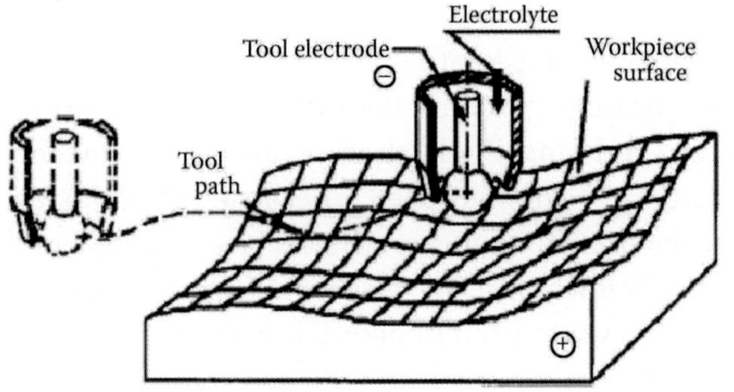


**FIGURE 4.13** The ECM microcathode is connected to a CNC machine to produce more complex shapes with a single tool.

With EMM, most metals, alloys, and conducting ceramics of interest in the microelectronics and MEMS/NEMS industry can be anodically dissolved in a variety of neutral salt electrolytes, such as sodium nitrate, sulfate, or chloride. The dissolved metal ions form hydroxide precipitates in these electrolytes and remain suspended in solution so they may easily be filtered out. The latter minimizes safety problems and waste disposal. With hydrogen evolution generally as the main cathodic reaction, metal plating on the cathode is avoided. EMM is currently receiving considerable attention in the electronics and other high-technology industries, particularly as an alternative, "greener" method of processing metallic parts.[39] Compared with chemical etching, the EMM process offers better control, less aggressive chemicals, and thus minimizes safety and environmental concerns.[39]

*Electrochemical Jet-Etching and Laser-Assisted Electrochemical Jet-Etching* Thin film patterning by maskless EMM may be accomplished by highly localized material removal induced by the impingement of a fine electrolytic jet emanating from a small nozzle.[39] Investigation of jet EMM at IBM demonstrated that neutral salt solutions can be effectively used for high-speed micromachining of many metals and alloys. An example of such an investigation, using a nozzle diameter of 200 μm and a 5 M sodium nitrate solution, is shown in Figure 4.14. Through-slots were micromachined in a 50-μm-thick stainless steel foil to create the IBM logo, which demonstrates the feasibility of using an

**FIGURE 4.14** IBM logo made by jet EMM. Through-slots machined in a 50-μm-thick stainless sheet using a 5 M sodium nitrate electrolyte jet. The jet nozzle diameter is 200 μm. (Courtesy of Datta, M. J., 1995. Fabrication of an array of precision nozzles by through-mask electrochemical micromachining. *Electrochem Soc* 142:3801–05.[44])

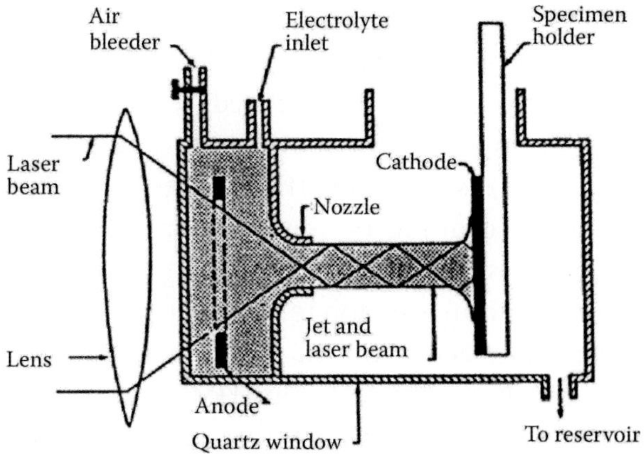


**FIGURE 4.15** Schematic representation of laser-enhanced electrochemical jet etching/plating showing the coincident laser beam and solution jet with the jet acting as the waveguide. (From Romankiw, L. T., and T. A. Palumbo. 1987. *Proceedings: symposium on electrodeposition technology, theory and practice*. Pennington, NJ: The Electrochemical Society, Inc. With permission.[27])

electrolytic jet for generating complicated patterns in metallic foils and substrates.

An interesting variation on electrochemical jet etching is a combination of a fluid-impinging jet and laser illumination (see Figure 4.15). In laser-enhanced electrochemical jet etching, properly chosen lasers, whose energy is not absorbed by the etching solution but is absorbed by the solid, cause local heating of the substrate (up to 150°C), resulting in highly increased etching. The jet is used as a light pipe for the laser and at the same time as a means for the local high rate of supply of ions. For stainless steel, etch rates of 10 μm/s have been demonstrated using laser-enhanced electrochemical jet machining.

For more background on electrochemical jet etching and laser-enhanced jet etching, refer to Romankiw et al.[27] and references therein and Datta et al.[40] Several of these techniques offer tremendous opportunities for the development of alternative micromachining methods.

Water jet etching is a mechanical process, and as such it is covered in Chapter 6 on mechanical energy-based removing. Water jet-guided laser etching without the electrochemical component is a purely thermal technique and is covered in Chapter 5. In this important method we will see that a fine water-jet again guides the laser beam, provides cooling for the work piece, and expels the molten material.

*Scanning Electrochemical Microscope* The scanning electrochemical microscope (SECM) is a scanned probe microscope (SPM) related to the familiar scanning tunneling (STM) and atomic force microscopes (AFM) (see Volume III, Chapter 6). All SPMs operate by scanning or "rastering" a small probe tip over the surface to be imaged. In SECM, imaging occurs in an electrolyte solution with an electrochemically active tip. In most cases, the SECM tip is an ultramicroelectrode (UME), and the tip signal is a Faradaic current from an electrochemical reaction at the surface. In some SECM experiments, the tip is an ion-selective electrode (ISE). In Chapter 1, we saw how a scanning tunneling microscope may induce electrochemical anodization reactions (an oxidation) on surfaces such as Si and GaAs. An SECM can also be used for local etching and deposition with high resolution in the *x*, *y*, and *z* dimensions, basically forming a high-resolution ECM setup (Figure 4.16).

Metal and semiconductor surfaces can be etched by the generation of a suitable etchant at the tip, for example $Br_2$, which will etch holes in GaAs.[41] In many cases, dissolution of semiconductors can be further enhanced by irradiation of the surface, creating holes in the valence band, which react with the lattice of the semiconductor. Besides electrochemical micromachining, SCEM is also used for chemical microscopic imaging and the measurement of physicochemical constants and coefficients.

In 2008, a review on the technology appeared in the *Annual Review of Analytical Chemistry*.[42] For an update on SECM on the Internet, visit http://www.msstate.edu/Dept/Chemistry/dow1/secm/secm.html and http://www2.warwick.ac.uk/fac/sci/chemistry/cim/research/electrochemistry/research/secm.

*Through-Mask Electrochemical Micromachining* Through-mask electrochemical micromachining (EMM) involves selective metal dissolution from unprotected areas of a one- or two-sided photoresist-patterned work piece that is made the anode in an electrolytic cell.[39,43] The technique is faster than chemical milling and provides better control on a micro- and macroscale of shape and surface texture of anodically dissolved materials. The performance of through-mask EMM depends largely on the successful control of voltage distribution in the cell, hydrodynamics, and surface kinetics.[43] In wet isotropic chemical etching, the material is removed both vertically and laterally at the same rate, and the etch boundary usually recedes at a 45° angle relative to the surface. In EMM, however, the metal-removal rate in the lateral direction may be significantly reduced through proper manipulation of mass transport and current distribution.[39]

Datta[44] described a cost-effective, high-speed process for the fabrication of precision nozzles for ink-jet printer heads using through-mask EMM. The process involves fabrication of a series of rombezoidal nozzles in a metal foil and is applicable to a variety of materials, including high-strength, corrosion-resistant materials such as conducting ceramics. For the

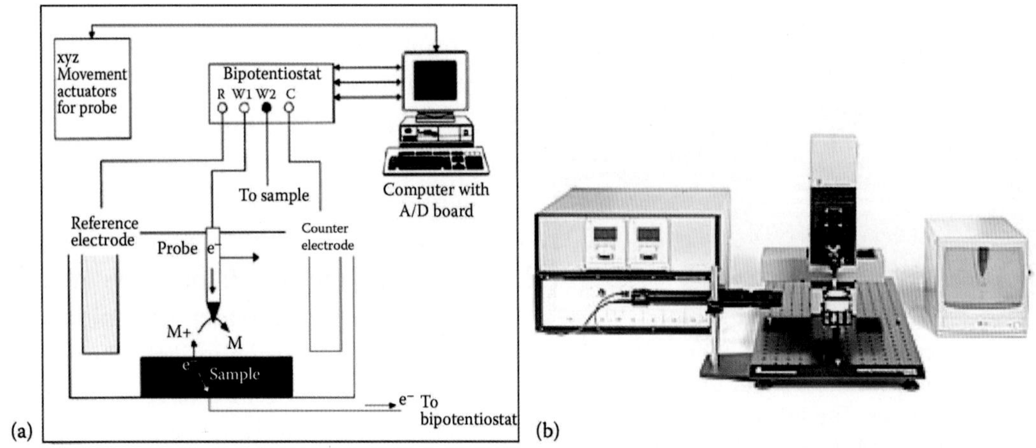


**FIGURE 4.16** (a) A schematic setup for SECM (http://www.msstate.edu/Dept/Chemistry/dow1/secm/secm.html) and (b) a typical commercial product (http://www.princetonappliedresearch.com/products/SurfaceScanning.cfm).

fabrication of the example nozzles in Figure 4.17, a cleaned 25-μm-thick stainless steel foil was laminated on both sides with 25-μm-thick dry photoresist. On one side, the photoresist was then exposed and developed, and the photoresist on the back remained as a protective insulating layer. The photoresist pattern was an array of circular openings 55 μm in diameter. Direct- and pulsed-voltage experiments were then performed with a neutral salt solution of sodium chloride and glycerol mixture as the electrolyte. Pulsating-voltage EMM was found to be more effective in providing dimensional etching uniformity over the whole nozzle array. This result is based on the application of extremely high peak currents (voltages), which, in addition to giving directionality, enables breakdown and elimination of inhibiting layers, thus facilitating activation of all the openings at the same time. A desired nozzle angle of 27° was produced with an etch factor (ratio of straight-through etch to undercut) of 2 (see Figure 4.17). This way, Datta demonstrated the feasibility of fabricating arrays of hundreds to thousands of precision nozzles with microsmooth surfaces in copper and stainless steel foils. Alternate ink-jet nozzle fabrication techniques in silicon are shown in Figure 4.45.

Through-mask EMM has also been used to fabricate controlled topographies on titanium to improve its biocompatibility.[45] Arrays of very smooth, high-precision cavities 30 μm in diameter were fabricated. Such structures affect attachment and growth of both soft and hard tissues. Biocompatibility has been linked to surface structures in the micrometer and nanometer range (see also Volume III, Chapter 4).[46]

*Summary* The driving force for metal dissolution reactions in EMM is derived from an external current; as a consequence, the electrolyte does not have to be as aggressive as in chemical etching. In fact, even neutral salt solutions are applicable. Metal-removal rates in EMM are determined by the machining current, which can be very high if proper hydrodynamic conditions prevail within the electrochemical cell; hence EMM etch rates are generally orders of magnitude higher than those in chemical etching. More importantly, through manipulation of mass transport and current distribution, metal-removal rate in the lateral direction can be significantly influenced, thus providing better wall slope control in EMM. Other EMM benefits include minimized safety and environmental concerns. All the listed features make the EMM process a cost-effective, environmentally friendly processing technology. Newly emerging technologies such as advanced microelectronic packaging, microsensors, and MEMS/NEMS provide

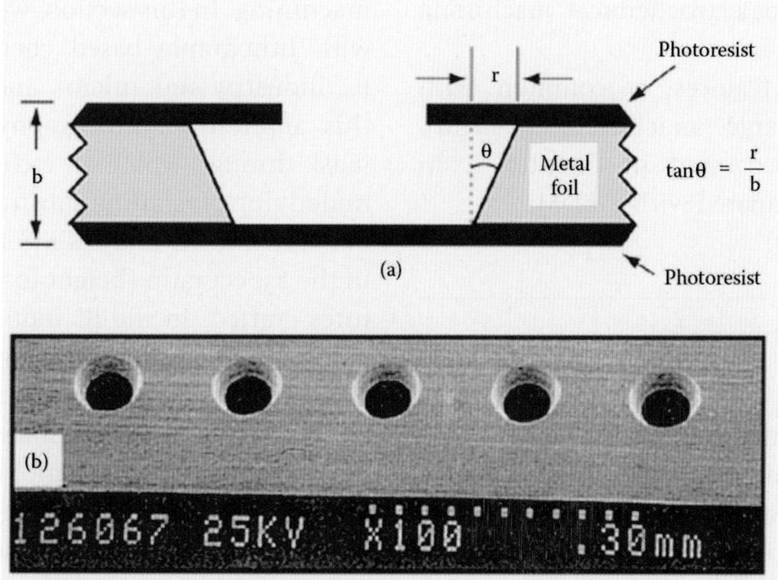


**FIGURE 4.17** (a) One-side through-mask EMM with angular sidewall. (b) SEM of nozzles fabricated in 25-μm-thick stainless steel. (Courtesy of Datta, M. J., 1995. Fabrication of an array of precision nozzles by through-mask electrochemical micromachining. *Electrochem Soc* 142:3801–05.[44])

**Table 4.3** Comparison of Electrochemical Machining (ECM) with Microelectrochemical Machining (μ-ECM)

| Major Machining Characteristics | ECM | μ-ECM |
|---|---|---|
| Voltage | 10–30 V | <10 V |
| Current | 150–10,000 A | <1 A |
| Current density | 20–200 A/cm² | 75–100 A/cm² |
| Power supply, DC | Continuous/pulsed | Pulsed |
| Frequency | Hz–KHz range | KHz–MHz range |
| Electrolyte flow | 10–60 m/s | <3 m/s |
| Electrolyte type | Salt solution | Natural salt or dilute acid/ alkaline solution |
| Electrolyte temperature | 24–65 °C | 37–50 °C |
| Electrolyte concentration | >20 g/L | <20 g/L |
| Size of the tool | Large to medium | Micro |
| Interelectrode gap | 100–600 μm | 5–50 μm |
| Operation | Maskless | Mask/maskless |
| Machining rate | 0.2–10 mm/min | 5 μm/min |
| Side gap | >20 μm | <10 μm |
| Accuracy | ±0.1 mm | ±0.02–0.1 mm |
| Surface finish | Good: 0.1–1.5 μm | Excellent: 0.05–0.4 μm |
| Problems caused by waste disposal/ toxicity | Low | Low to moderate |

ample opportunities for further investigation of EMM. With a better understanding of the principles involved in manipulating high-rate anodic dissolution of metals from narrow and deep cavities, EMM will become an increasingly important processing technology for micro- and nanofabrication.[39]

In Table 4.3 we compare electrochemical machining (ECM) with microelectrochemical machining (μ-ECM).

μ-ECM has some features in common with microelectrical discharge machining (μ-EDM), a thermal technique covered in Chapter 5. In Table 5.2 μ-ECM is compared with μ-EDM.

## Wet Bulk Micromachining

### Introduction

As mentioned in the introduction to this chapter, wet bulk micromachining is chemical etching used to sculpt isotropic and anisotropic 3D microfeatures in single crystal semiconductor substrates. Removal of material from the surface of a crystalline material (such as Si), immersed in an alkaline or acidic solution, occurs as a result of complicated chemical reactions between the surface atoms and the etchant molecules (see Figure 4.18a). The isotropy or anisotropy of the etching refers to the orientation dependency of the etch rate: for the isotropic case, there is no orientation dependence; for the anisotropic case, there is orientation dependence (see Figure 4.18b).

Wet bulk micromachining is a technique championed in the IC industry and in micro- and nanomachining. In this section, we are mostly concerned with lithography-based chemical etching for the IC industry and micro- and nanofabrication. In this application, lithography is used to create a mask through which an etchant can etch away the underlying crystalline substrate. A major difference between the IC and MEMS fields in this regard is in the aspect ratio (height-to-width ratio) of the features crafted. In the IC industry, one deals mostly with very small, flat structures with aspect ratios of 1–2 (see Figure 4.2). In the microfabrication field, structures typically are somewhat larger, and aspect ratios (height-to-width ratio) may be as high as 400. In wet bulk micromachining, features are sculpted in the bulk of materials such as silicon, quartz, SiC, GaAs, InP, and Ge by orientation-independent (isotropic) or orientation-dependent (anisotropic) wet etchants. The technology uses pools of liquid as tools,[1] instead of the (dry) plasmas studied in

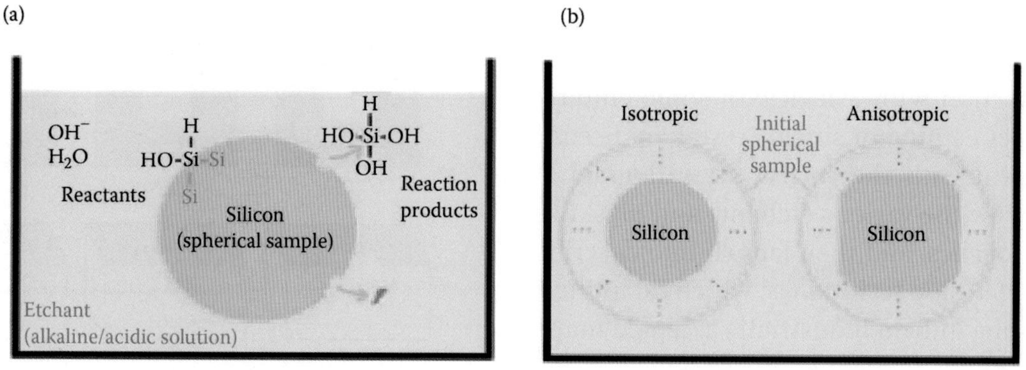


**FIGURE 4.18** (a) Schematic illustration of the dissolution process during wet etching. (b) Wet etching can be isotropic or anisotropic. (Gonsalvez, M. 2003. *Atomistic modelling of anisotropic etching of crystalline silicon.* PhD diss., Helsinki University of Technology.[47])

Chapter 3. In Figure 4.19 we show some examples of high-aspect-ratio MEMS structures created using wet anisotropic etching of properly masked <100>- and <110>-oriented silicon wafers.

A vast majority of wet bulk micromachining work is based on single-crystal silicon and glass as a companion material [7740 Jena glass (Pyrex) is often used for packaging; see Volume III, Chapter 4], mostly for mechanical-type sensors such as pressure, acceleration, and gyros. There has been some work on quartz, crystalline Ge, SiC, and GaAs, and a minor amount on GaP and InP.

Wet bulk micromachining, together with surface micromachining (Chapter 7), form the two principal commercial Si micromachining tool sets used today. A third technique, micromolding—often from a lithography-defined master—also is becoming more commercially viable (Chapter 10).

The emphasis in this chapter is on the wet etching process itself. Other machining steps typically used in conjunction with wet bulk micromachining include many of the alternative subtractive techniques described elsewhere in Part II, the additive processes described in Part III, and the bonding processes that are part of the packaging processes dealt with in Volume III, Chapter 4.

A historical note on Si wet bulk micromachining, clarifying the importance of Si as a sensor material,

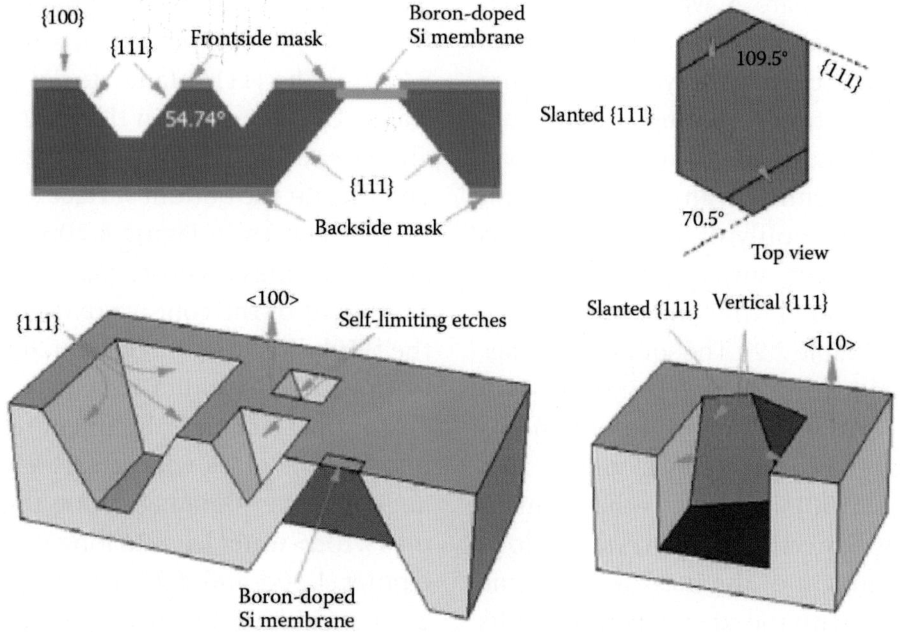


**FIGURE 4.19** Examples of structures created using wet anisotropic etching in a Si <100> wafer and in a Si <110> wafer. The significance of boron doping will be clarified further below.

heads this chapter (an expanded treatise on this subject can be found in Volume I, Chapter 1). We start the current section with a review of some empirical data on wet etching of single-crystal Si, discuss mask alignment with the correct crystallographic direction, review the various etchants and masking material approaches, and then launch into describing different etch stop techniques, which catapulted micromachining into an industrially accepted manufacturing method. Subsequently, we review models explaining wet isotropic and anisotropic etching and pick up the discussion anew of anodic polishing, photoetching, and formation of porous Si in HF solutions. Next comes a discussion of the technical issues encountered in bulk micromachining, such as IC incompatibility and extensive real-estate usage. We conclude with some example applications of wet bulk micromachining.

The first MEMS modeling software was developed around wet bulk micromachining processes and mechanical MEMS structures. Such programs are reviewed in Volume III, Chapter 6.

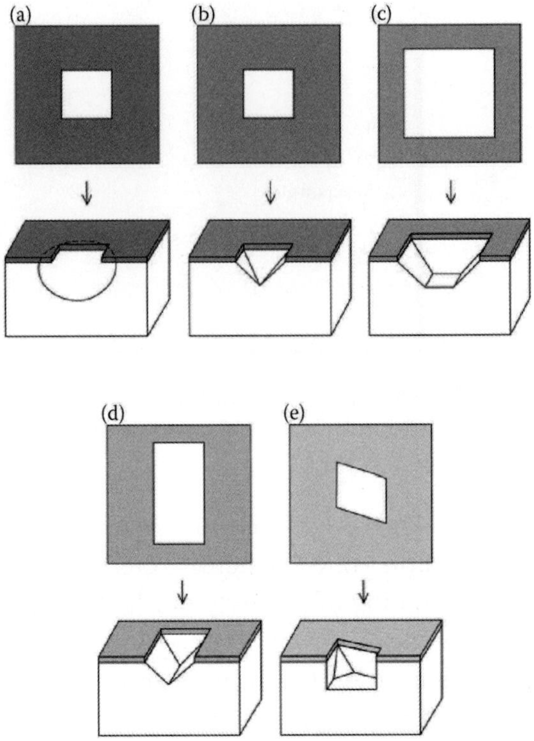


**FIGURE 4.20** Isotropic and anisotropic etched features in [100] and [110] wafers. (a) Isotropic etch. (b–e) Anisotropic etch. (a–d) [100]-oriented wafers. (e) [110]-oriented wafer.

## Empirical Data on Wet Etching of Single-Crystal Si

### *Isotropic and Anisotropic Etched Shapes in Single-Crystal Si*

In Figure 4.20, we compare a wet isotropic etch (Figure 4.20a) with examples of anisotropic etches on masked flat pieces of single-crystal Si (Figure 4.20b–e). In the anisotropic etching examples, a square (Figure 4.20b and c) and a rectangular pattern (Figure 4.20d) are defined in an oxide mask with sides aligned along the family of <110> directions on a [100]-oriented silicon surface. For a refresher on the use of < >, [], and () (the Miller indices), see Volume I, Chapter 2 (Table 2.2). The square openings are precisely aligned (within 1° or 2°) with the [110] directions on the (100) wafer surface to obtain pits that conform exactly to the oxide mask rather than undercutting it. Most [100] silicon wafers have a main flat parallel to a [110] direction in the crystal, allowing for a rough alignment of the mask (see Volume I, Figure 4.3). Etching with the square pattern results in a pit with well-defined {111} sidewalls (at angles of 54.74° to the surface) and a (100) bottom. Volume I, Equation 4.2, $W_0 = W_m - 2z\cot 54.74° = W_m - \sqrt{2}z$ allows one to calculate the dimensions of the hole at the bottom of the pit. For a larger square opening in the mask, the point where the {111} sidewalls of the pit intersect lays deeper inside the crystal. If the oxide opening is wide enough, that is, $W_0 > \sqrt{2}\,l$ (with $l = 600$ μm for a typical 6-in. wafer, this means $W_0 >$ 849 μm), the {111} planes do not intersect within the wafer. The etched pit in this particular case extends all the way through the wafer, creating a small square opening on the bottom surface (see Volume I, Figure 4.8). As shown in Figure 4.20b–d, no underetching of the etch mask is observed as a result of the perfect alignment of the concave oxide mask opening with the [110] direction. In Figure 4.20a, an undercutting isotropic etch (acidic) is shown. In the case of an anisotropic etch, misalignment results in pyramidal pits, but the mask will be severely undercut. A rectangular pattern aligned along the <110> directions on a <100> wafer leads to long V-shaped grooves (see Figures 4.20d and 4.21) or an open slit, depending on the width of the opening in the oxide mask.

Using a properly aligned mask (see Volume I, Chapter 4) on a [110] wafer, grooves with four

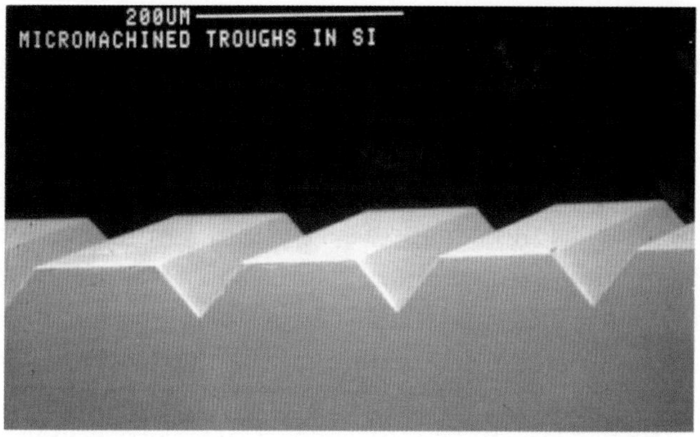


FIGURE 4.21 Long V-shaped grooves in a (100) Si wafer.

vertical and two slanted {111} planes result (see Figures 4.20e and 4.22a). Even a slight mask misorientation leads to all skewed sidewalls instead (Figure 4.22b). Before the emergence of slanted {111} bottom planes, the groove is defined by four vertical (111) planes and a horizontal (110) bottom (Figure 4.19, bottom right), and between there is a competition between {110} and {100} bottom planes. Self-stopping occurs when the tilted end {111} planes intersect at the bottom of the groove (Figure 4.22d). It is easy to etch very long, narrow grooves deeply into a [110] silicon wafer (Figure 4.23). However, it is impossible to etch a short, narrow groove deeply into it[48] because the narrow dimension of the groove

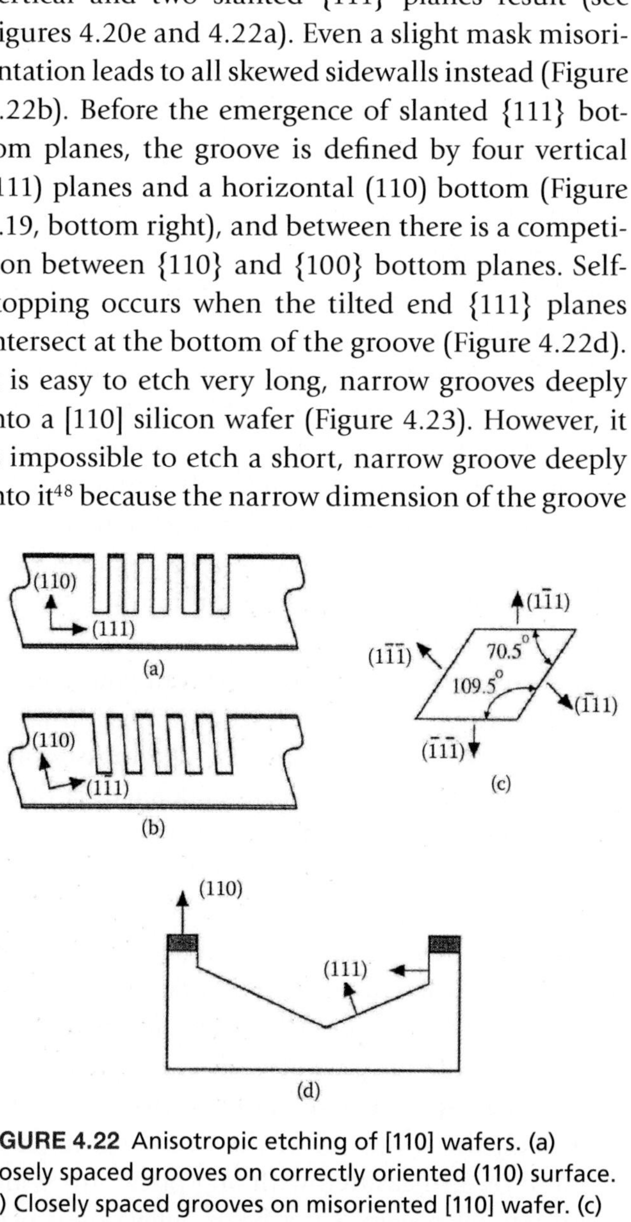


FIGURE 4.22 Anisotropic etching of [110] wafers. (a) Closely spaced grooves on correctly oriented (110) surface. (b) Closely spaced grooves on misoriented [110] wafer. (c) Orientations of the {111} planes looking down on a (110) wafer. (d) Shallow slanted (111) planes eventually form the bottom of the etched cavity.

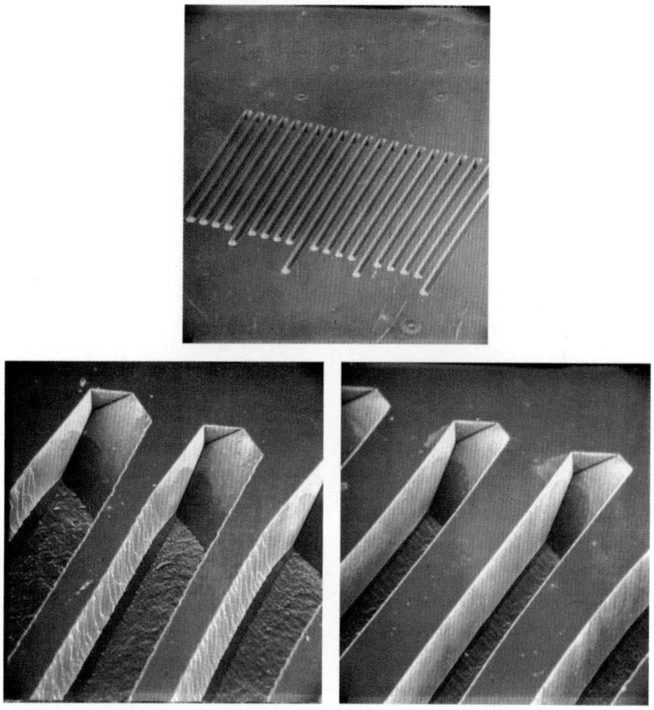

FIGURE 4.23 Test pattern of U-grooves in a <110> wafer to help in the alignment of the mask; final alignment is done with the groove that exhibits the most perfect long perpendicular walls.

is quickly limited by slow-etching {111} planes that subtend an angle of 35.26° to the surface and cause etch termination. At a groove of length L = 1 mm on the top surface, etching will stop when it reaches a depth of 0.289 mm, that is, $D_{max} = L/2\sqrt{3}$. For very long grooves, the tilted end planes are too far apart to intersect in practical cases, making the end effects negligible compared with the remaining trench-shaped part of the groove. Tuckerman and Pease[49] demonstrated the use of such trenches as liquid cooling fins for integrated circuits. The set of long, narrow grooves in a <110> wafer shown in Figure 4.23 can also be used as a test pattern to help in the alignment of the mask; final alignment is done with the groove that exhibits the most perfect long perpendicular walls (see also Figure 4.25).

The orientation of the wafer is of extreme importance, especially when machining surface structures by undercutting. Consider, for example, the formation of a bridge in Figure 4.24a.[50] When using a (100) surface, a suspension bridge cannot form across the etched V-groove; two independent truncated V-grooves flanking a mesa structure result instead.

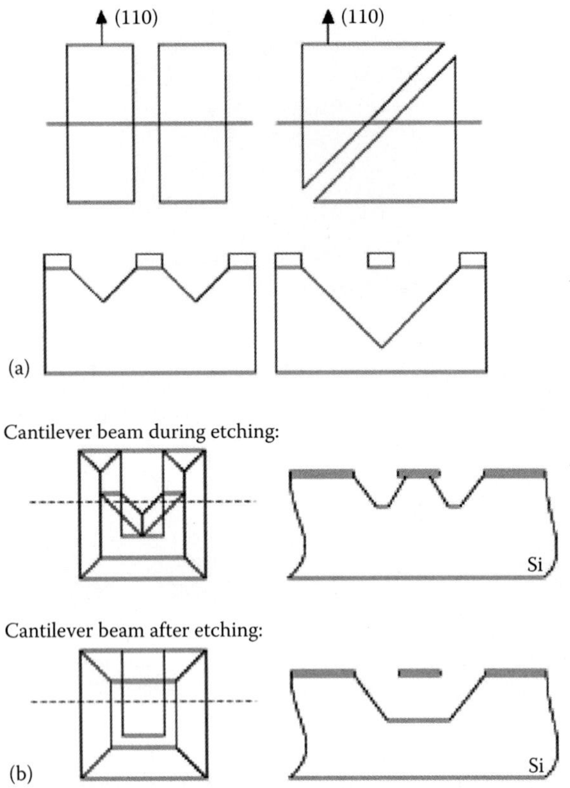


**FIGURE 4.24** How to make a suspension bridge from a [100] Si wafer (a) and a diving board from a [110] Si wafer (b). (Barth, P. W., P. J. Shlichta, and J. B. Angel. 1985. *Third international conference on solid-state sensors and actuators*. Philadelphia.[50])

To form a suspended bridge, the V-groove must be oriented away from the [110] direction. Contrast this with a (110) wafer where a microbridge crossing a V-groove with a 90° angle will be undercut. Convex corners will be undercut by etchant, allowing formation of cantilevers as shown in Figure 4.24b. The diving board shown in Figure 4.20b forms by undercutting, starting at the convex corners. For more details, Table 4.4 is in the section on selection of [100]- or [110]-oriented silicon wafers, where we compare the main characteristics of etched features in [100]- and [110]-oriented wafers.

### *Alignment Patterns—Wagon-Wheel Masking Patterns*

We just learned how a mask pattern may be aligned with the wafer flat to obtain a specific desired anisotropic etched feature. When alignment of a pattern along a certain crystallography direction is more critical, pre-etch alignment targets become useful to delineate the planes of interest because the wafer flats often are aligned to ±1° only. To find the proper alignment for the mask, a test pattern of closely spaced lines can be etched (see Figure 4.23). As discussed above, the groove with the best vertical walls determines the proper final mask orientation. Along this line, Ciarlo[51] made a set of lines 3 mm long and 8 μm wide, fanned out like spokes in a wagon wheel at angles 0.1° apart. This target was printed near the perimeter of the wafer and then etched 100 μm into the surface. Again, by evaluating the undercut in this target, the correct crystal direction could be determined. Alignment with better than 0.05° accuracy was accomplished this way.

The wagon-wheel pattern technique has also been used to obtain detailed experimental data on crystal orientation dependence of etch rates. For example, Seidel et al.[52] used a wagon-wheel or star-shaped mask (e.g., made from CVD-$Si_3N_4$, $SiH_4$, and $NH_3$ at 900°C), consisting of radially divergent segments with an angular separation of 1° (Figure 4.25a). By selectively masking the wafer surface material with material that is inert to the etchant (in this case, $Si_3N_4$) the removal of material is forced to occur vertically at the nonmasked areas and laterally under the masking spokes (Figure 4.25b). Beautiful patterns are formed as a result of the sidewalls being etched at different rates (Figure 4.25c). The etch rate for each sidewall in the wheel (distance $r$) is obtained from the lateral underetch rate (distance $w$) from $r = w \sin\theta$ (Figure 4.25b).

Seidel et al.[52] found an etch pattern emerging on a <100>-oriented wafer covered with wagon-wheel mask as shown in Figure 4.26a. The blossom-like figure is the result of the total underetching of the passivation layer in the vicinity of the center of the wagon wheel, leaving an area of bare, exposed Si. The radial extension of the bare Si area depends on the crystal orientation of the individual segments, leading to a different amount of total underetching. The observation of these blossom-like patterns is used for qualitative guidance of etching rates only. To establish quantitative numbers for the lateral etch rates, the width, $w$, of the overhanging passivation layer is measured with an optical line-width measurement system. Laser beam reflection is used to identify the crystal planes, and ellipsometry is used to monitor the etching rate of the mask itself.

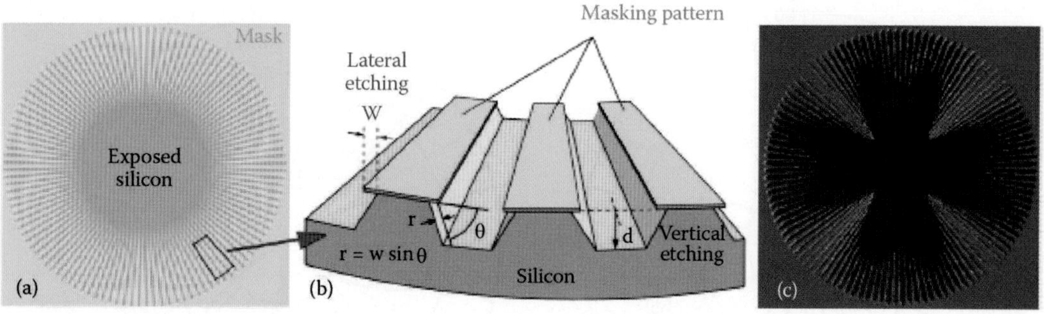


**FIGURE 4.25** (a) Wagon-wheel masking pattern on silicon before etching. (b) Vertical etching (distance d) and lateral underetching (distance w). (c) Flower pattern generated during anisotropic etching. (Gonsalvez, M. 2003. *Atomistic modelling of anisotropic etching of crystalline silicon.* PhD diss., Helsinki University of Technology.[47])

Lateral etch rates determined in this way on <100>- and <110>-oriented wafers at 95°C in EDP (470 mL water, 1 L ethylene-diamine, 176 g pyrocatechol) and at 78°C in a 50% KOH solution are shown in Figure 4.27 (these two etchants are detailed further below). Etch rates shown are normal to the crystal surface and are conveniently described in a polar plot in which the distance from the origin to the polar plot surface (or curve in two dimensions) indicates the etch rate for that particular direction. Note the deep minima at the {111} planes. It can also be seen that, in KOH, the peak etch rates are more pronounced. A further difference is that with EDP the minimum at {111} planes is steeper than with KOH. For both EDP and KOH, the etch rate depends linearly on misalignment. All the above observations have important consequences for the interpretation of anisotropy of a given etch (see below). The difference between KOH and EDP etching behavior around the {111} minima has the direct practical consequence that it is more important for etching in EDP to align the crystallographic direction more precisely than in KOH.[53]

When determining etch rates without using underetching masks but using vertical etching of beveled silicon samples, results are different from working with masked silicon.[54] The etch rates on open areas of beveled structures are much larger than in underetching experiments with masked silicon, and different crystal planes develop. Herr et al.[54] conclude that crevice effects may play an important role in anisotropic etching. The model of Elwenspoek et al.,[55,56] analyzed below, is the only model that predicts such a crevice effect. The authors explain why etching rates slow down and increase anisotropy when etchants are in a small restricted crevice area and are not refreshed fast enough.

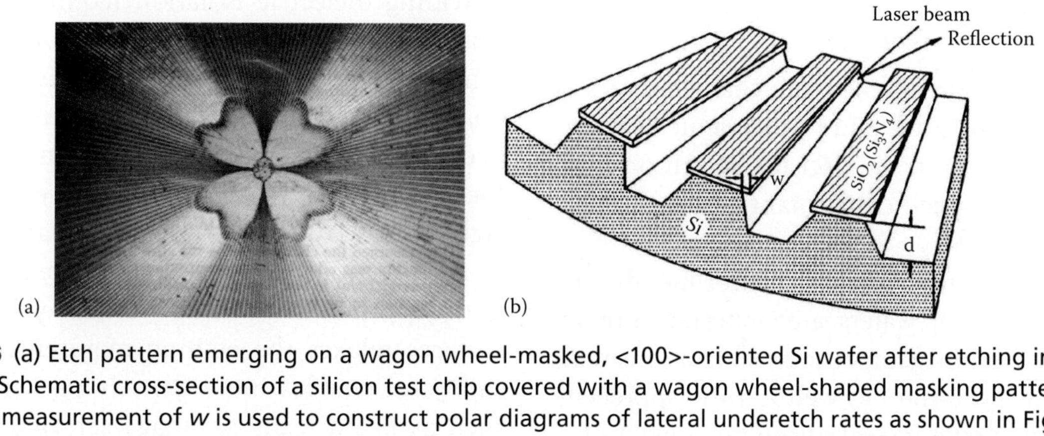


**FIGURE 4.26** (a) Etch pattern emerging on a wagon wheel-masked, <100>-oriented Si wafer after etching in an EDP solution. (b) Schematic cross-section of a silicon test chip covered with a wagon wheel-shaped masking pattern after etching. The measurement of *w* is used to construct polar diagrams of lateral underetch rates as shown in Figure 4.27. (From Seidel, H., L. Csepregi, A. Heuberger, and H. Baumgartel. 1990. Anisotropic etching of crystalline silicon in alkaline solutions–part I: orientation dependence and behavior of passivation layers. *J Electrochem Soc* 137:3612–26. With permission.[52])

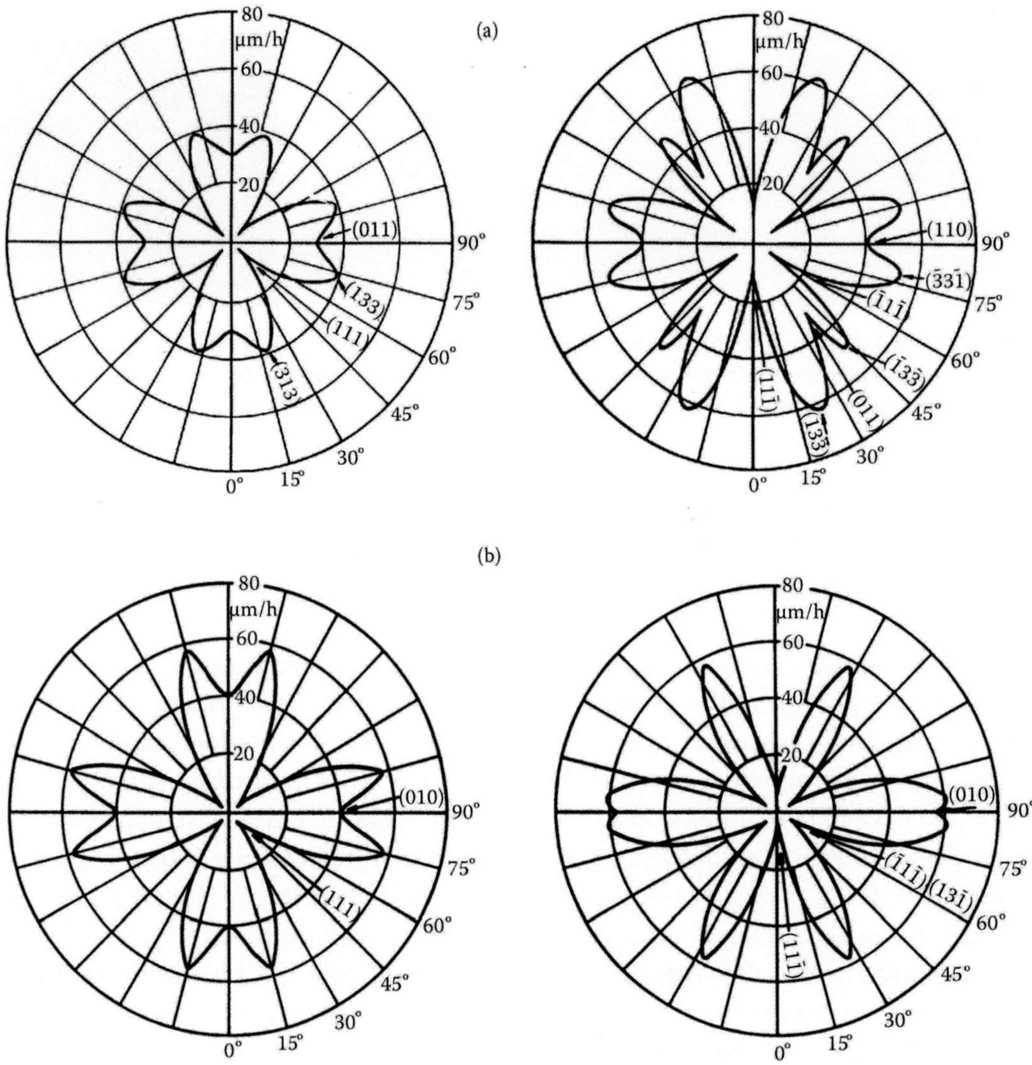


**FIGURE 4.27** Lateral underetch rates as a function of orientation for (a) EDP (470 ml water, 1 l ED, 176 g pyrocatechol) at 95°C and (b) KOH (50% solution) at 78°C. Left, <100>- and right, <110>-oriented Si wafers. (From Seidel, H., L. Csepregi, A. Heuberger, and H. Baumgartel. 1990. Anisotropic etching of crystalline silicon in alkaline solutions–part I: orientation dependence and behavior of passivation layers. *J Electrochem Soc* 137:3612–26. With permission.[52])

### *Selection of [100]- or [110]-Oriented Silicon Wafers*

In Table 4.4, we compare the main characteristics of etched features in [100]- and [110]-oriented wafers. This table may help in the decision which orientation to use for a specific microfabrication application at hand.

From this table, it is obvious that for membrane-based sensors, [100] wafers are preferred, but an understanding of the geometric considerations with [110] wafers is important as well to fully appreciate all the possible single-crystal silicon micromachined shapes and to better understand corner compensation schemes (see below). Moreover, processes for providing dielectric isolation require that the silicon be separated into discrete regions. To achieve a high component density with anisotropic etches on [100] wafers, the silicon must be made very thin because of the aspect ratio limitations as a result of the sloping walls (see above). With vertical sidewall etching in a [100] wafer, the etch mask is undercut in all directions to a distance approximately equal to the depth of the etching. Vertical etching in [110] surfaces relaxes the etching requirement dramatically and enables more densely packed structures such as beam leads or image sensors. Kendall[48,57] describes and predicts a wide variety of applications for [110] wafers, such as fabrication of trench

**Table 4.4** Matrix for Optimal Selection of Wafer Type

| [100] Orientation | [110] Orientation |
|---|---|
| Inward sloping walls (54.74°) | Vertical {111} walls |
| The sloping walls cause a lot of real estate loss | Narrow trenches with high-aspect ratio are possible |
| Flat bottom parallel to surface is ideal for membrane fabrication | Multifaceted cavity bottom ({110} and {100} planes) makes for a poor diaphragm |
| Bridges perpendicular to a V-groove bound by (111) planes cannot be underetched | Bridges perpendicular to a V-groove bound by (111) planes can be undercut |
| Shape and orientation of diaphragms convenient and simple to design | Shape and orientation of diaphragms awkward and more difficult to design |
| Diaphragm size, bounded by nonetching {111} planes, is relatively easy to control | Diaphragm size is difficult to control (the <100> edges are not defined by nonetching planes) |

capacitors, vertical multijunction solar cells, diffraction gratings, infrared interference filters, large area cathodes, and filters for bacteria.

### *Laser Spoiling*

A laser can be used to melt or "spoil" the shallow {111} surfaces, making it possible to etch deep vertical-walled holes through a [110]-oriented wafer as shown in Figure 4.28a.[50,58,59] The absorbed energy of a neodymium-doped yttrium aluminium garnet (Nd:YAG) laser beam causes a local melting or evaporation zone, enabling etchants to etch the shallow {111} planes in the line of sight of the laser. Etching proceeds until "unspoiled" (111) planes are encountered. Some interesting resulting possibilities, including partially closed microchannels, are shown in Figure 4.28c.[58,60] Note that with this method it is possible to use [111] wafers for micromachining. The light of the Nd:YAG laser is very well suited for this micromachining technique as a result of the 1.17 eV photon energy, just exceeding the bandgap energy of Si. Details on this laser machining process can be found, for example, in Alavi et al.[60,61]

## Wet Isotropic and Anisotropic Etching

### *Introduction*

Besides for the shaping of 3D microstructures, wet etching is used for cleaning, removing surface damage, polishing, and characterizing structural and compositional features.[7] The most important parameters in wet chemical etching are bias (undercut), tolerance, etch rate, anisotropy, selectivity, overetch, feature size control, and loading effects. These controlling parameters are very similar to the parameters controlling dry chemical etching (see Chapter 3). The emphasis in this chapter is on wet etching of single-crystal Si. Wet chemical etching of Si provides a higher degree of selectivity than dry etching techniques. Wet etching is usually also faster: a few microns to tens of microns per minute for isotropic

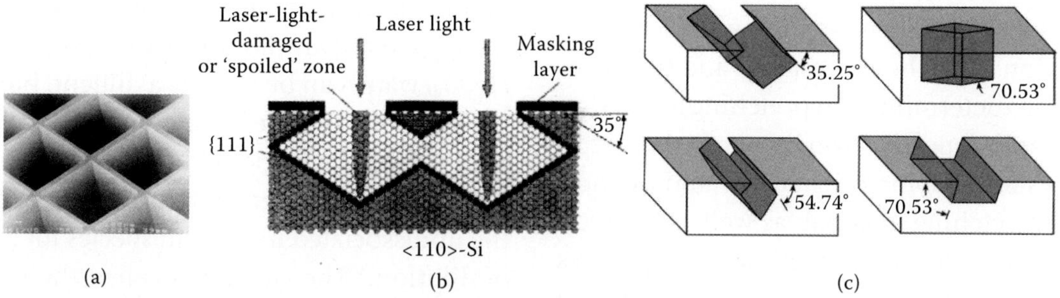


**FIGURE 4.28** Laser/KOH machining. (a) Holes through a (110) Si wafer created by laser spoiling and subsequent KOH etching. The two sets of (111) planes making an angle of 70° are the vertical walls of the hole. The (111) planes making a 35.26° angle with the surface tend to limit the depth of the hole, but laser spoiling enables one to etch all the way through the wafer. (b) After laser spoiling, the line-of-sight (111) planes are spoiled, and etching proceeds until unspoiled (111) planes are reached. (c) Some of the possible features rendered by laser spoiling. (From Schumacher, A., H.-J. Wagner, and M. Alavi. 1994. Mit Laser und Kalilauge. *Technische Rundschau* 86:20–23. With permission.[58])

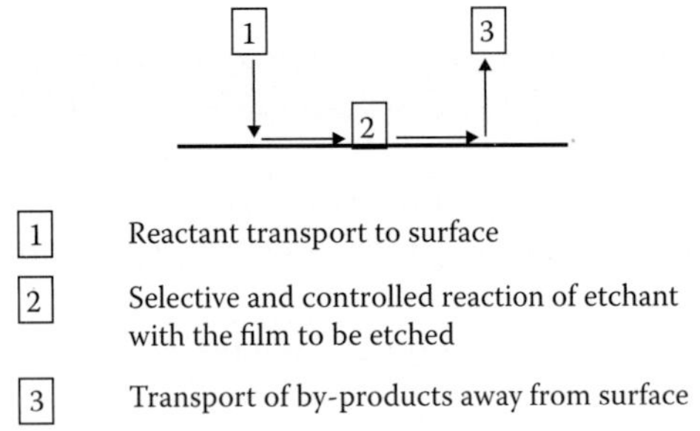
1
2
3
1 Reactant transport to surface
2 Selective and controlled reaction of etchant with the film to be etched
3 Transport of by-products away from surface

**FIGURE 4.29** Wet etching processes.

etchants and about 1 μm/min for anisotropic wet etchants versus 0.1 μm/min in typical dry etching. More recently, however, rates of up to 6 μm/min were achieved using high-density plasmas for dry etching of Si (see Chapter 3).

Modification of the nature of the wet etchant and/or operating temperature can also alter the selectivity to the silicon dopant concentration and type and, especially when using alkaline etchants, to crystallographic orientation. Etching proceeds by reactant transport to the surface (1), surface reaction (2), and reaction product transport away from the surface (3) (see Figure 4.29). If (1) or (3) is rate determining, etching is diffusion limited, and etching is controlled by the hydrodynamics of the system; thus, the process may be enhanced by stirring. Mass transport-determined etching, being isotropic, gives rise to rounded features. This also means that the surface reactions (2) are fast and that the etch rate is not sensitive to the nature of the surface, e.g., its crystallographic orientation or morphology. Diffusion-limited processes have lower activation energies (in the order of a few kilocalories per mole) than reaction rate-controlled processes and therefore are relatively insensitive to temperature variations. If a surface reaction (2) is the slowest step and thus constitutes the rate-determining step, etching is reaction rate limited and depends strongly on the temperature, the crystallographic orientation, and/or morphology of the etching material, as well as solution composition. In other words, with surface kinetics important the etch rate is very likely to be sensitive to surface orientation. In general, reaction rate limitation is preferred because it is easier to reproduce a temperature setting than a stirring rate. The best generic etching apparatus has both a good temperature controller and a reliable stirring facility.[62,63]

Preferential or selective etching (also *structural etching*) is a type of etching that is usually isotropic but exhibits some anisotropy.[64] These etchants are used to produce a difference in etch rate between different materials or between compositional or structural variations of the same material on the same crystal plane. These types of etches are often the fastest and simplest techniques to delineate electrical junctions and to evaluate the structural perfection of a single crystal in terms of, for example, slip and stacking faults. The artifacts introduced by the defects etch into small pits of characteristic shape. Most of the etchants used for this purpose are acids with an oxidizing additive.[65–69]

### *Isotropic Etching*

*Usage of Isotropic Etchants* When etching silicon with aggressive acidic etchants, rounded isotropic patterns form. The method is widely used for:

1. Removal of work-damaged surfaces
2. Rounding of sharp anisotropically etched corners (to avoid stress concentration)
3. Removing of roughness after dry or anisotropic etching
4. Creating structures or planar surfaces in single-crystal slices (thinning)
5. Patterning single-crystal, polycrystalline, or amorphous films
6. Delineation of electrical junctions and defect evaluation (with preferential isotropic etchants)

For isotropic etching of silicon, the most commonly used etchants are mixtures of nitric acid ($HNO_3$) and hydrofluoric acids* (HF). In the case of $HNO_3$, water can be used as a diluent, but acetic acid ($CH_3COOH$) is preferred because it prevents the dissociation of the nitric acid better and thus preserves the oxidizing power of $HNO_3$, which depends on the undissociated nitric acid species for a wide range of dilution.[4] The etchant is called the HNA system; we will return to this etch system below.

* Be careful when working with HF; it is deceptive because it looks like water, and even when not a strong acid, it penetrates the skin (adsorption) and slowly attacks bones.

*Simplified Reaction Scheme* In acidic media, the Si etching process involves hole injection into the Si valence band by an oxidant, an electrical field, or photons. Nitric acid in the HNA system acts as an oxidant; other oxidants such as $H_2O_2$ and $Br_2$ also work.[70] The holes attack the covalently bonded Si, oxidizing the material, followed by a reaction of the oxidized Si fragments with $OH^-$ and subsequent dissolution of the silicon oxidation products in HF. It was Kooij et al.[71] who, from electrochemical measurements on the Si/HF/$HNO_3$ system, concluded that the holes, in the absence of photons and an applied field, are introduced by $HNO_3$:

$$HNO_3 + H_2O + HNO_2 \rightarrow 2HNO_2 + 2OH^- + 2h^+ \quad \text{Reaction 4.1}$$

The holes ($h^+$) in Reaction 4.1 are generated in an autocatalytic process; $HNO_2$ generated in the above reaction re-enters into the further reaction with $HNO_3$ to produce more holes. With a reaction of this type, there is an induction period before the oxidation reaction takes off until a steady-state concentration of $HNO_2$ has been reached. This induction period has been observed at low $HNO_3$ concentrations.[70] After hole injection, $OH^-$ groups attach to the oxidized Si species to form $SiO_2$, liberating hydrogen in the process:

$$Si^{4+} + 4OH^- \rightarrow SiO_2 + H_2 \quad \text{Reaction 4.2}$$

HF dissolves the $SiO_2$ by forming the water-soluble $H_2SiF_6$. The overall reaction of HNA with Si is then:

$$Si + HNO_3 + 6HF \rightarrow H_2SiF_6 \text{ (water-soluble)} + HNO_2 + H_2O + H_2 \text{ (bubbles)} \quad \text{Reaction 4.3}$$

The simplification in the above electrochemical component of the reaction scheme is that only holes are taken into account. In the actual Si acidic corrosion reaction, both holes and electrons are involved. The question of hole and/or electron participation in Si corrosion will be considered after the introduction below of the model for the Si/electrolyte interfacial energetics. We will learn from that model that the rate-determining step in acidic etching involves hole injection in the valence band, whereas in alkaline anisotropic etching, it involves electron injection in the conduction band by surface states. The reactivity of a hole injected in the valence band is significantly greater than that of an electron injected in the conduction band. The observation of isotropy in acidic etchants and anisotropy in alkaline etchants centers on this difference in reactivity.

Isotropic etching can be performed chemically, in very aggressive solutions containing HF and an oxidizing agent such as $HNO_3$, or electrochemically, in much less aggressive fluoride-containing solutions. Both methods offer the possibility of forming porous Si. Porous Si formation is easier to control using the electrochemical method. For electrochemical etching, an electrical contact must be made to the back side of the sample, which limits the possibilities of the process (see below).

*Iso-Etch Curves* By the early 1960s, the isotropic HNA silicon etch was well characterized. Schwartz and Robbins published a series of four very detailed papers on the topic between 1959 and 1976.[3–6] Most of the material presented below is based on their work.

HNA etching results, represented in the form of iso-etch curves, for various weight percentages of the constituents are shown in Figure 4.30. For this work,

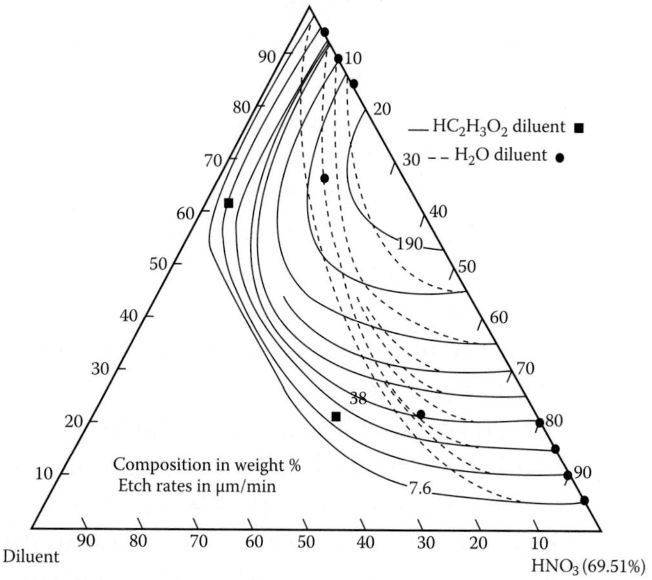


**FIGURE 4.30** Iso-etch curves. (Robbins, H., and B. Schwartz. 1960. Chemical etching of silicon-II: the system HF, $HNO_3$, $H_2O$, and $HC_2C_3O_2$. *J Electrochem Soc* 107:108–11,[4] recalculated for one-sided Si etching and expressed in microns per minute.)

normally available concentrated acids of 49.2 wt% HF and 69.5 wt% $HNO_3$ are used. Water as diluent is indicated by dashed-line curves, whereas acetic acid is noted by solid-line curves. Acetic acid is less polar than water and helps prevent the dissociation of $HNO_3$, thereby allowing the formation of more of the species directly responsible for the oxidation of Si. A typical formulation for HNA is 250 mL of HF, 500 mL of $HNO_3$, and 800 mL of $CH_3COOH$. When used at room temperature, this formulation results in an etch rate of about 4–20 μm/min, increasing with agitation.[72] In Figure 4.30 we have recalculated the curves from Robbins and Schwartz[4] to express the etch rate in microns per minute and divided the authors' numbers by 2 as we are considering one-sided etching only. The highest etch rate is observed around a weight ratio HF-$HNO_3$ of 2:1 and is nearly 100 times faster than anisotropic etch rates. Adding a diluent slows down the etching. From these curves, the following characteristics of the HNA system can be summarized:

1. At high HF and low $HNO_3$ concentrations, the iso-etch curves describe lines of constant $HNO_3$ concentrations (parallel to the HF-diluent axis); consequently, the $HNO_3$ concentration controls the etch rate. Etching at those concentrations tends to be difficult to initiate and exhibits an uncertain induction period (see above). In addition, it results in relatively unstable silicon surfaces proceeding to slowly grow a layer of $SiO_2$ over time. The etch is limited by the rate of oxidation; therefore, it tends to be orientation-dependent and affected by dopant concentration, defects, and catalysts (sodium nitrate is often used). In this regime, the temperature influence is more pronounced, and activation energies for the etching reaction of 10–20 kcal/mol have been measured.
2. At low HF and high $HNO_3$ concentrations, iso-etch curves are lines parallel to the nitric-diluent axis; that is, they are at constant HF composition. In this case, the etch rate is controlled by the ability of HF to remove the $SiO_2$ as it is formed. Etches in this regime are isotropic and truly polishing, producing a bright surface with anisotropies of 1% or less (favoring the <110> direction) when used on <100> wafers.[73] An activation energy of 4 kcal/mol is indicative of the diffusion-limited character of the process; consequently, in this regime, temperature changes are less important.
3. In the region of maximal etch rate, both reagents play an important role. The addition of acetic acid, as opposed to the addition of water, does not reduce the oxidizing power of the nitric acid until a fairly large amount of diluent has been added. Therefore, the rate contours remain parallel with lines of constant nitric acid over a considerable range of added diluent.
4. In the region around the HF vertex, the surface reaction rate-controlled etch leads to rough, pitted Si surfaces and sharply peaked corners and edges. In moving toward the $HNO_3$ vertex, the diffusion-controlled reaction results in the development of rounded corners and edges, and the rate of attack on (111) planes and (110) planes becomes identical in the polishing regime (anisotropy less than 1%; see point 2).

In Figure 4.31, we summarize how the topology of the Si surfaces depends strongly on the composition of the etch solution. Around the maximum etch rates the surfaces appear flat with rounded edges, and very slow etching solutions lead to rough surfaces.[6]

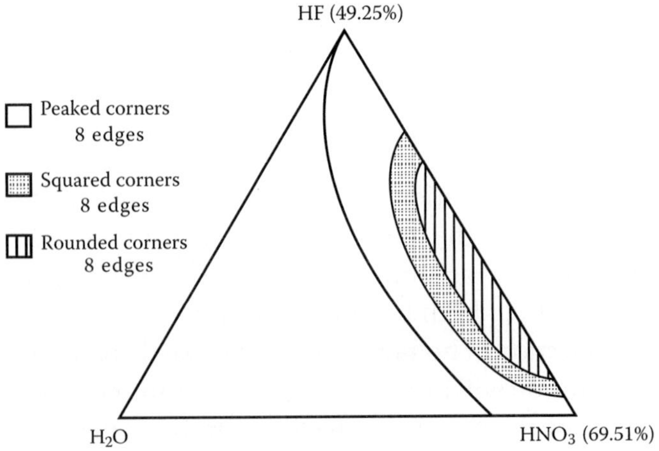


**FIGURE 4.31** Topology of etched Si surfaces. (From Schwartz, B., and H. Robbins. 1976. Chemical etching of silicon-IV: etching technology. *J Electrochem Soc* 123:1903–9. With permission.[6])

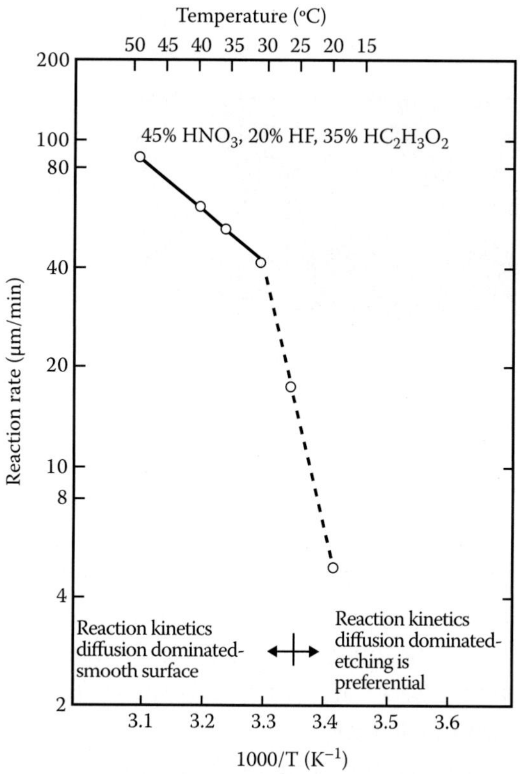


**FIGURE 4.32** Etching an Arrhenius plot. Temperature dependence of the etch rate of Si in HF:$HNO_3$:$CH_3$:COOH (1:4:3). (From Schwartz, B., and H. Robbins. 1961. Chemical etching of silicon-III: a temperature study in the acid system. *J Electrochem Soc* 108:365–72. With permission.[5])

*Arrhenius Plot for Isotropic Etching* Schwartz and Robbins[5] studied the effect of temperature on the reaction rate in the HNA system in detail. An Arrhenius plot for etching silicon in 45% $HNO_3$, 20% HF, and 35% $HC_2H_3O_2$, culled from their work, is shown in Figure 4.32. Increasing the temperature increases the reaction rate. The graph shows two straight-line segments, indicating higher activation energy at less than 30°C and a lower one above this temperature. In the low temperature range, etching is preferential, and the activation energy is associated with the oxidation reaction. At higher temperatures, the etching leads to smooth surfaces, and the activation energy is lower and associated with diffusion-limited dissolution of the oxide.[5]

*Loading Effects in Isotropic Etching* Two HNA solutions that etch silicon isotropically at a reasonable rate (around 2 μm/min) and leave smooth walls [roughness in the order of 5 nm, as determined from scanning electron microscope (SEM) pictures] are shown in Table 4.5.[74] The etch rates given in this table are averages because the etch rate depends on the mask opening size and structure density on the wafer, resulting from the diffusion-controlled nature of the etching process.

**TABLE 4.5** Compositions of Two Typical Isotropic HNA Solutions

| | Vol % HF (50%) | Vol % $HNO_3$ (69%) | Vol % Water | Etch Rate (μm/min) |
|---|---|---|---|---|
| Sol 1 | 11 | 22 | 67 | 3 |
| Sol 2 | 5 | 15 | 80 | 2 |

*Source:* Tjerkstra, R. W. 1999. *Isotropic etching of silicon in fluoride containing solutions as a tool for micromachining*. PhD diss., University of Twente, Twente, the Netherlands.[74]

With isotropic etchants, the etchant moves downward and outward from an opening in the mask, undercuts the mask, and enlarges the etched pit while deepening it (Figure 4.33). The resulting isotropically etched features show more symmetry and rounding when agitation accompanies the etching (the process is diffusion limited). This agitation effect is illustrated in Figure 4.33. With agitation, the etched feature approaches a more ideal round cup; without agitation, the etched feature resembles a rounded box.[75] However, the flatness of the bottom of the rounded box generally is poor because the flatness is defined by agitation. Therefore, sometimes solutions are on purpose not stirred during etching because stirring enhances the etch rate in the direction of fluid flow, yielding badly shaped channels.

As we saw above, the etching of silicon in HF/$HNO_3$ solutions involves two processes: the oxidation of the silicon surface and the subsequent removal of the oxidized silicon by the fluoride in the solution. In the case of isotropic etching, the rate of the latter process is determined by the transport of fluoride ions to the surface (diffusion limited). This can lead to loading effects as illustrated during the isotropic etch of a spiral in Figure 4.34a.

We observe that the channels on the rim of the spiral have a larger diameter than the channels in the middle of the spiral. This is because on the rim of the spiral there is no silicon that etches and consumes fluoride ions so the ions above the mask material next to the outermost channels can diffuse to these channels. This leads to nonuniformities in

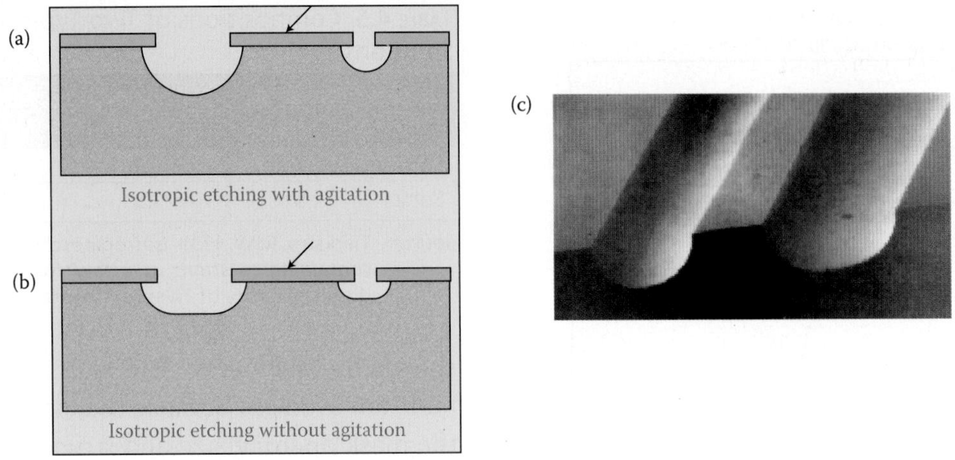


**FIGURE 4.33** Isotropic etching of silicon with (a) and without (b) etchant solution agitation; (c) Examples of isotropic etching in Si. Left groove with agitation, right groove no agitation.

the widths of the channels on the rim of the spiral. Stirring the solution during etching does not necessarily help because, as we just saw, that enhances the etching in the direction of fluid flows, yielding badly shaped channels. In an attempt to solve the loading effect problem, the spiral may be surrounded with sacrificial channels with the same width and the same distance from each other as the channels in the spiral ("compensation structures") as illustrated in Figure 4.34b. The etching of these extra structures takes away the extra fluoride that would otherwise cause the broadening of the outer channels. From work by Kuiken et al.,[76] Tjerkstra[74] concluded that for short etch times (etch times in the order of seconds) the bulk concentration is reached at a distance of around $3(Dt)^{1/2}$, in which $D$ is the diffusion coefficient and $t$ the time.

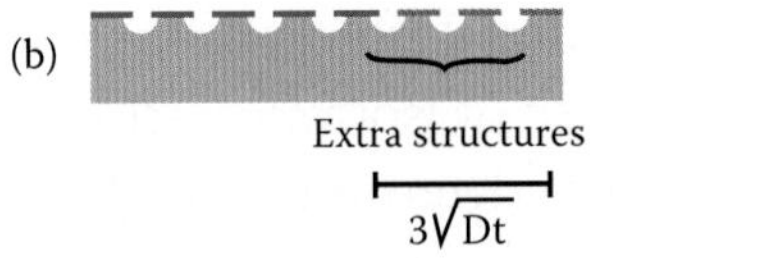


**FIGURE 4.34** Side view of anisotropically etched spiral, illustrating the loading effect. (a) The fluoride ions in the area next to the etching spiral can diffuse to the outer channels. These channels etch faster than the other channels because of the extra supply of fluoride. (b) To reduce this effect, extra structures are placed around the spiral that reduce the extra supply of fluoride. This results in a uniform channel width (Tjerkstra, R. W. 1999. *Isotropic etching of silicon in fluoride containing solutions as a tool for micromachining.* PhD diss., University of Twente, Twente, the Netherlands.[74])

The enhanced etching of structures at the rim of the spiral in Figure 4.34 may also be a temperature effect. Etching is the result of an exothermic reaction; thus, the temperature of the solution increases during etching. This causes the appearance of fluid flow, as shown in Figure 4.35. The channels at the rim of the spiral may get an extra supply of fresh solution and thus etch faster.

In Chapter 3 we encountered loading effects also when we discussed dry etching, where they are more prevalent because the etchant concentration is much lower in the gas phase.

*Masking for Isotropic Silicon Etchants* Acidic etchants are very fast; for example, an etch rate for Si of

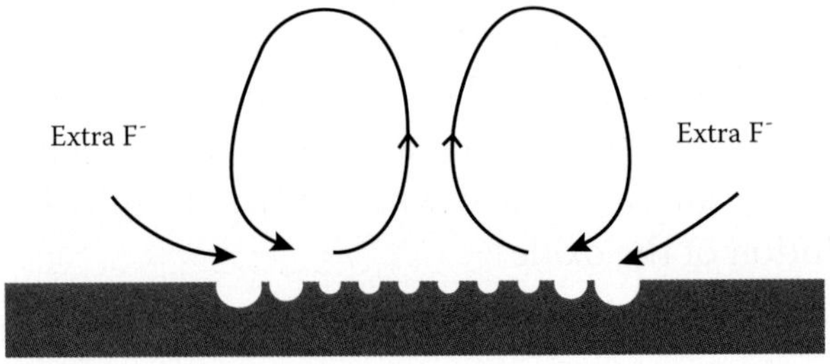


**FIGURE 4.35** Enhanced etching caused by fluid flow resulting from the increased temperature of the etching solution. (Tjerkstra, R. W. 1999. *Isotropic etching of silicon in fluoride containing solutions as a tool for micromachining.* PhD diss., University of Twente, Twente, the Netherlands.[74])

TABLE 4.6 Masking Materials for Acidic Etchants

| | Etchants | | |
|---|---|---|---|
| Masking | Piranha (4:1, $H_2O_2$: $H_2SO_4$) | Buffered HF (5:1$NH_4F$: conc.HF) | HNA |
| Thermal $SiO_2$ | | 0.1 μm/min | 300–800 Å/min. Limited etch time, thick layers often are used because of ease of patterning |
| CVD (450°C) $SiO_2$ | | 0.48 μm/min | 0.44 μm/min |
| Corning 7740 glass | | 0.063 μ/min | 1.9 μ/min |
| Photoresist | Attacks most organic films | OK for a short while | Resists do not stand up to strong oxidizing agents like $HNO_3$ and are not used |
| Undoped Si polysilicon | Forms 30 Å of $SiO_2$ | 0.23–0.45 Å/min | 0.7–40 μm/min at RT [at a dopant concentration $<10^{17}$ $cm^{-3}$ (*n* or *p*)] |
| Black wax | | | Usable at room temperature |
| Au/Cr | OK | OK | OK |
| LPCVD $Si_3N_4$ | | 1 Å/min | Etch rate is 10–100 Å/min; preferred masking material |

The many variables involved necessarily means that the given numbers are approximate only.

up to 50 $\mu min^{-1}$ can be obtained with 66% $HNO_3$ and 34% HF (volumes of reagents in the normal concentrated form).[64,77] Isotropic etchants are so aggressive that the activation barriers associated with etching the different Si planes are not differentiated; all planes etch equally fast, making masking a real challenge.

Although $SiO_2$ has an appreciable etch rate of 300–800 Å/min in the HF:$HNO_3$ system, thick layers of $SiO_2$ are often used as a mask, especially for shallow etching, as the oxide is so easy to form and pattern. A mask of nonetching Au or $Si_3N_4$ is needed for deeper etching. Photoresists do not stand up to strong oxidizing agents such as $HNO_3$, and neither does Al.

Silicon itself is soluble to a small extent in pure HF solutions; for a 48% HF, at 25°C a rate of 0.3 Å/min was observed for n-type, 2-Ω-cm (111)-Si. It was established that Si dissolution in HF is not the result of oxidation by dissolved oxygen. Diluted HF etches silicon at a higher rate because the reaction in aqueous solutions proceeds by oxidation of Si by $OH^-$ groups.[78] A typically buffered HF (BHF) solution, also called buffered oxide etch, has been reported to etch Si at radiochemically measured rates of 0.23–0.45 Å/min, depending on doping type and dopant concentration.[79] In BHF, HF is buffered with $NH_4F$ to control the pH value (−log[$H^+$]) at constant level and to replenish the depletion of fluoride ions to maintain a stable etch performance. BHF is the preferred $SiO_2$ etch:

$$SiO_2 + 4HF + 2NH_4F \rightarrow (NH_4)_2\,SiF_6 + 2H_2O$$

$$SiO_2 + 3HF_2^- + H^+ \rightarrow SiF_6^{2+} + 2H_2O \qquad \text{Reaction 4.4}$$

By reducing the dopant concentration (*n* or *p*) to less than $10^{17}$ atoms/$cm^3$, the etch rate of Si in HNA is reduced by ~150.[80] The doping dependence of the etch rate provides yet another means of patterning a Si surface (see next section). A summary of masks that can be used in acidic etching is presented in Table 4.6.

A simple way to detect the end-point for BHF* etching of a layer of $SiO_2$ on Si is to monitor the wetting properties of the surface. Whereas $SiO_2$ wets, Si is hydrophobic.

*Dopant Dependence of Silicon Isotropic Etchants* The isotropic etching process is fundamentally a charge-transfer mechanism. With isotropic etching Si at open circuit, we will learn further below, the chemical etch may have both chemical and electrochemical components. In the latter case, mobile charge carriers are injected or extracted from the solid, making the reaction sensitive to an applied potential. This also explains the etch rate dependence on dopant type and dopant concentration. Typical etch rates with an HNA system (1:3:8) for n- or p-type dopant concentrations more than $10^{18}$ $cm^3$ are 1–3 μm/min. As presented in the preceding section, a reduction of the etch rate by 150 times

* HF is inert in contact with skin but attacks bone.

is obtained in n- or p-type regions with a dopant concentration of $10^{17}$ $cm^{-3}$ or smaller.[81] This presumably is because of the lower mobile carrier concentration available to contribute to the charge transfer mechanisms. In any event, heavily doped silicon substrates with high conductivity can be etched more readily than lightly doped materials. Dopant-dependent isotropic etching can also be exploited in an electrochemical setup as described in the next section. Although doping changes the chemical etch rate, attempts to exploit these differences for industrial production have failed.[81] This situation is different in the case of electrochemical isotropic etching (see next section).

*Preferential Etching* A variety of additives to the HNA system, mainly oxidants, can be included to modify the etch rate, surface finish, or isotropy, rendering the etching baths preferential. It is clear that the effect of these additives will only show up in the reaction-controlled regime. Only additives that change the viscosity of the solution can modify the etch rate in the diffusion-limited regime, thereby changing the diffusion coefficient of the reactants.[70,82] We will not review the effect of these additives any further; refer to Table 4.7 and the cited literature for more information.[65–69]

*Problems Associated with Isotropic Etchants* There are several problems associated with isotropic etching of Si. First, it is difficult to mask with high precision using a desirable and simple mask such as $SiO_2$ (etch rate is 2–3% of the silicon etch rate). Second, the etch rate is very agitation sensitive in addition to being temperature sensitive. This makes it difficult to control lateral and vertical geometries. Electrochemical isotropic etching (see below) and the development of anisotropic etchants in the late 1960s (see below) were able to overcome many of these problems.

A comprehensive review of isotropic etchants solutions can be found in Kern et al.,[64] including a review of different techniques practiced in chemical etching such as immersion etching, spray etching, electrolytic etching, gas-phase etching, and molten salt etching (fusion techniques). Table 4.7 lists some isotropic and preferential etchants and their specific applications.

*Electrochemical Isotropic Silicon Etch—First Example of an Etch Stop* For the electrochemical etching of Si, holes are needed at the substrate surface. In p-type Si, these holes are automatically present in the material. In n-type Si or with intrinsic Si, holes can be generated by, for example, illuminating the substrate with photons having energies larger than the bandgap (1.1 eV). The thus-created holes can be forced to the surface by applying a potential to an ohmic contact to the Si. In this case, a high-temperature or extremely aggressive chemical etching process is replaced by an electrochemical procedure using a much milder solution, thus allowing even a simple photoresist mask to be used.[64] In electrochemical acidic etching, with or without illumination of the corroding Si electrode, an electrical power supply is used to drive the chemical reaction by supplying holes to the silicon surface [working electrode (W-EL), see Figure 4.36a]. A voltage is applied across the silicon wafer, and a counter electrode (C-EL, usually platinum) is arranged in the same etching solution. Oxidation is promoted by a positive bias applied to the silicon, causing an accumulation of holes in the silicon at the silicon/electrolyte interface. Under this condition, oxidation at the surface proceeds rapidly while the oxide is readily dissolved by the HF solution. No oxidant such as $HNO_3$ is needed to supply the holes; excess electron-hole pairs are created by the electrical field at the surface and/or by optical excitation, thereby increasing the etch rate. This technique proved successful in removing heavily doped layers, leaving behind the more lightly doped membranes in all possible dopant configurations: p on $p^+$, p on $n^+$, n on $p^+$, and n on $n^+$.[87,88] This electrochemical etch-stop technique is demonstrated in Figure 4.36b.[89] A 5% HF solution is used; the electrolyte cell is kept in the dark at room temperature; and the distance between the Si anode and the Pt cathode in the electrochemical cell is 1–5 cm. Instead of using HF, one can substitute $NH_4F$ (5 wt%) for the electrochemical etching as described by Shengliang.[90] Shengliang reports a selectivity of n-type silicon to $n^+$-type silicon (0.001 Ω cm) of 300 with the latter etchant. In Figure 4.36b, the current density versus applied voltage across the anode and cathode during

TABLE 4.7 Isotropic and Preferential Defect Etchants and Their Specific Applications

| Etchant | Application/Material | Remark/Reference |
|---|---|---|
| HF 8 vol%, $HNO_3$ 75 vol%, and $CH_3COOH$ 17 vol% | n- and p-type Si, all planes, general etching | Planar etch; e.g., 5 µm/min at 25°C |
| 1 part 49% HF, 1 part (1.5M-$CrO_3$) (by volume) | Delineation of defects on (111), (100), and (110) Si without agitation | Yang etch[65] |
| 5 vol parts nitric acid (65%), 3 vol parts HF (48%), 3 vol parts acetic acid (96%), 0.06 parts bromine | Polishing etchant used to remove damage introduced during lapping | So-called CP4 etchant; Heidenreich, US Patent 2619414 |
| HF 48% | $SiO_2$ | Etch rate is 20–2000 nm/min; etch rate for Si is 0.3 Å/min for n-type 2 Ω cm (111); etch rate for Al is 5 nm/min |
| HF:$NH_4F$ (buffered HF 28 ml HF, 170 ml $H_2O$, 2113 g $NH_4F$) | $SiO_2$ | Etch rate is 100–500 nm/min at 25°C |
| 1HF, 3$HNO_3$, 10 $CH_3COOH$ | Delineates defects in (111) silicon; etches $p^+$ or $n^+$ and stops at $p^-$ or $n^-$ | Dash etch[83]; p- and n-type Si at 1300 Å/min in the [100] direction and 46 Å/min in the [111] direction at 25°C |
| 1HF, 1(5M-$CrO_3$) | Delineates defects in (111); needs agitation; does not reveal etch pits on (100) well | Sirtl etch[84] |
| HF:$H_2O_2$ | Titanium | 880 nm/min |
| 2HF,1(0.15M-$K_2Cr_2O_7$) | Yields circular (100) Si dislocation etch pits; agitation reduces etch time | Secco etch[69] |
| 60 ml HF, 30 ml $HNO_3$, 30 ml (5M-$CrO_3$), 2 g $Cu(NO_3)_2$, 60 ml $CH_3COOH$, 60 ml $H_2O$ | Delineates defects in (100) and (111) Si; requires agitation | Jenkins etch[85] |
| $H_2O_2$ | Tungsten | 20–100 nm/min |
| 34 g $KH_2PO_4$, 13.4 g KOH, 33 g $K_3FE(CN)_6$ and $H_2O$ to make up 1 L | Tungsten | 160 nm/min |
| 1 ml HCl, 9 ml saturated $CeSO_4$ solution | Chromium | 80 nm/min |
| 1 ml HCl, 1 ml glycerine | Chromium | 80 nm/min |
| 2HF, 1(1M-$CrO_3$) | Delineates defects in (100) Si without agitation; works well on resistivities 0.6–15.0 Ω cm n- and p-types) | Schimmel etch[86] |
| 2HF, 1(1M-$CrO_3$), 1.5 $H_2O$ | Works well on heavily doped (100) Si | Modified Schimmel |
| HF/$KMnO_4$/$CH_3COOH$ | Epitaxial Si | |
| 3 ml HCl, 1 ml $HNO_3$ | Gold | 25–50 µm/min Aqua regia |
| 4 g KI, 1 g $I_2$, and 40 ml $H_2O$ | Gold | 0.5–1 µm/min |
| $H_3PO_4$ Etchants | $Si_3N_4$ | Etch rate is 5–10 nm/min 160–180°C |
| KOH + alcohols | Polysilicon | 85°C |
| $H_2SO_4$/$H_2O_2$ | Organic layers | >1000 nm/min |
| Acetone | Organic layers | >4000 nm/min |
| $H_3PO_3$/ $HNO_3$/$HC_2H_3O_2$ | Aluminum | Etch rate is 660 nm/min 40–50°C |
| $HNO_3$/BHF/water | Si and polysilicon | 0.1 µm/min for single-crystal Si |

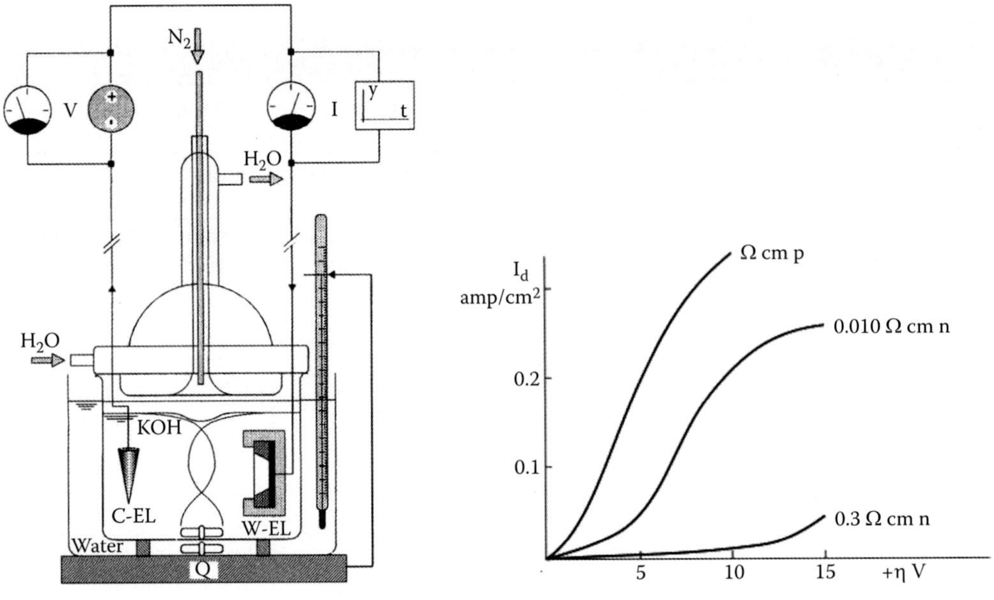


(a) (b)

**FIGURE 4.36** (a) Electrochemical etching apparatus. W-EL, working electrode (Si); C-EL, counter electrode (e.g., Pt); Q, heat supplied. (b) Current-voltage (*I-V*) curves in electrochemical etching of silicon of various doping. Etch rate dependence on dopant concentration and dopant type for HF-anodic etching of Si. (From van Dijk, H. J. A., and J. de Jonge. 1970. Preparation of thin silicon crystals by electrochemical thinning of epitaxially grown structures. *J Electrochem Soc* 117:553–54. With permission.[89])

dissolution is plotted (*I-V*). The current density is related to the dissolution rate of silicon. It can be seen that p-type and heavily doped n-type materials can be dissolved at relatively low voltages, whereas n-type silicon with a lower doping level does not dissolve at the same low voltages. Experiments in this same setup with homogeneously doped silicon wafers show that n-type silicon of about $3 \times 10^{18}$ $cm^{-3}$ (<0.01 Ω cm) completely dissolves in these etching conditions, whereas n-type silicon of donor concentrations lower than $2 \times 10^{16}$ $cm^{-3}$ (>0.3 Ω cm) barely dissolves. For p-type silicon, dissolution is initiated when the acceptor concentration is higher than 5 × 1015 cm−3 (<3 Ω cm), and the dissolution rate further increases with increasing acceptor concentration.

For more precise electrochemical isotropic etching work, the two-electrode system from Figure 4.36a is replaced with a three-electrode system and a potentiostat. The latter is an instrument used to control the potential of the semiconductor working electrode precisely with respect to a third electrode, the so-called reference electrode (REF) (for example, see Figure 4.46 for a three-electrode and potentiostat setup, and for more theoretical electrochemistry background see Volume I, Chapter 7). The current I passing between the W-EL and the C-EL is recorded as a function of the potential. No current passes through the reference electrode. The potential voltage of the W-EL can be scanned over a given potential range at a certain rate (mV/s). A typical *I-V* curve, also called a cyclic voltammogram, for a p-type silicon sample in an HF solution is shown in Figure 4.37, where the reference electrode used is an Ag/AgCl electrode.

The shape of the curve shown here is determined by the speed at which reactions occur at the silicon surface, and the rate of transport of reacting particles. In general the *I-V* curve for a Si electrode in an HF solution can be divided into three ranges: the cathodic range, the porous range, and the electropolishing range. In the cathodic range, i.e., the potential range negative to the open-circuit potential, the surface hole concentration in the Si electrode is low; therefore, no etching can take place. For p-type Si electrodes in the dark there is also no hydrogen evolution for lack of electrons. If an n-type Si electrode were used, copious hydrogen evolution would be observed in this same voltage range because n-type electrodes have plenty of electrons in the conduction band. The porous range is the potential range

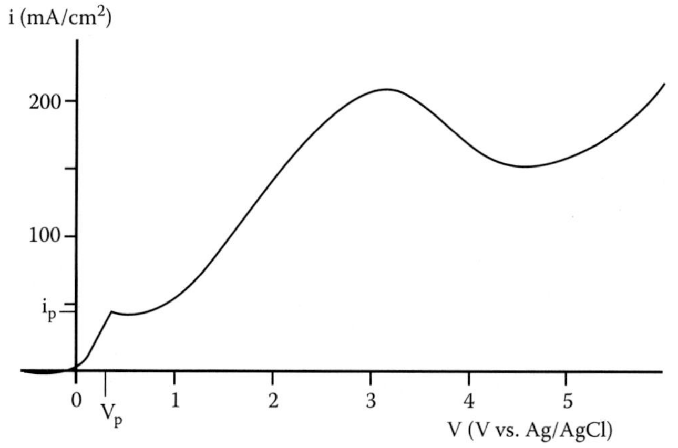


**FIGURE 4.37** *I-V* curve of a p-type silicon wafer in a 5% HF solution. Resistivity of the wafer 0.01–0.018 Ωcm, scan rate 500 mV/s. (From Tjerkstra, R. W. 1999. *Isotropic etching of silicon in fluoride containing solutions as a tool for micromachining.* PhD diss., University of Twente, Twente, the Netherlands.[74])

before the potential of the first current peak ($V_p$ in Figure 4.37) where porous Si forms. During the formation of porous Si, hydrogen gas is evolved, and the morphology of the porous Si depends, among other factors, on the doping density of the wafer, the HF concentration, and the current density. The potential range after the first current peak ($V_p$) is called the electropolishing range. Here, the current density exceeds the so-called critical current density, $I_p$, marked by the first peak in the *I-V* curve. The silicon etches, leaving behind a smooth (electropolished) surface. Porous Si does not form in this range. The critical current density depends on the fluoride concentration in the electrolyte solution, the dopant concentration in the semiconductor, the hydrodynamics of the system, and the temperature.[75] For a given Si electrode, the critical current density is given as[91]:

$$I_p = Ac^{1.5}e^{\frac{-E_a}{kT}} \qquad (4.1)$$

where the constant, *A*, is 3300 A cm$^{-2}$ (wt% HF)$^{-1.5}$, *c* is HF concentration, the activation energy $E_a$ is 0.345 eV, *k* is Boltzmann constant, and *T* is the temperature. When the current density is lower than $I_p$, a porous layer is formed on the Si substrate.

Many researchers have investigated the electrochemical etching of silicon in HF solutions, and initially the following reaction mechanisms were suggested as[9]:

$$Si + 2HF + 2h^+ \rightarrow SiF_2 + 2H^+ \qquad \text{Reaction 4.5}$$

In this scenario, in the porous Si range, two holes are necessary to dissolve each silicon atom. Thus, it was assumed that the silicon dissolves as an Si(II) species, after which it is further oxidized by water or disproportionates to an Si(IV) and Si(0) species. One ends up with the soluble $SiF_6^{2-}$ complex [Si(IV)] and porous Si [Si(0)]. The porous Si deposited on the surface was assumed to be amorphous because it stemmed from the disproportionation reaction of $SiF_2$. The hydrogen evolution in the porous Si range was believed to be caused by a reaction of the silicon subfluoride:

$$SiF_2 + 2H_2O \rightarrow SiO_2 + 2HF + H_2 \qquad \text{Reaction 4.6}$$

However, porous Si was shown to be monocrystalline, and this theory had to be discarded.[92,93] It was also discovered that the freshly etched Si surface is H-terminated, making the Si surface hydrophobic.[94] Current reaction mechanisms are generally based on the assumption that the presence of a fluoride ion close to the substrate surface causes the Si-Si backbonds to be polarized, thus facilitating the breaking of these bonds.[95–97] The backbonds are broken one at a time, leaving an H-terminated surface and an $SiH_xF_y$ compound that reacts with water and HF to form $SiF_6^{2-}$ and $H_2$. The overall reaction is then:

$$Si + 2H^+ + 2h^+ \rightarrow Si(IV) + H_2 \qquad \text{Reaction 4.7}$$

At pH < 3, fluoride is mainly present in the form of HF or $HF_2^-$, so these components are also assumed to be involved in the reactions. It is generally assumed that the electropolishing of silicon takes place in two steps. In the first step an oxide layer is formed, and this oxide layer is subsequently dissolved by HF. Dissolution rates and thickness of the oxide layer have been studied using infrared[98] and current-time measurements.[96,99–102] The second peak in the *I-V* curve in Figure 4.37 is thought to mark a change from one type of oxide to another.[103] The overall reaction of the dissolution of Si in the polishing range is:

$$Si + 2H_2O + 4h^+ \rightarrow SiO_2 + 4H^+ \qquad \text{Reaction 4.8}$$

$SiO_2 + 6HF \rightarrow SiF_6^{2-} + 2H_2O + 2H^+$ Reaction 4.9

For the dissolution of n-type silicon, holes have to be generated by illuminating the sample with photons having energy larger than the bandgap. If the photon flux is high enough, an *I-V* curve similar to that shown in Figure 4.37 can be obtained. If the photon flux is lower than the critical current density, porous Si forms and a potential-independent photocurrent is observed (see further below).

The acidic electrochemical technique has not been used much in micromachining and is primarily used to polish surfaces. Because the etching rate increases with current density, high spots on the surface are more rapidly etched, and very smooth surfaces result. This method of isotropic electrochemical etching has some major advantages that could make it a more important micromachining tool in the future. Such advantages include very smooth etched surfaces (e.g., with an average roughness $R_a$ of 7 nm), room temperature and IC compatible process, usage of simpler resist schemes because the process is much milder than etching in HNA, and etching can be controlled simply by switching a voltage on or off. We will pick up the discussion of anodic polishing, photoetching, and formation of porous Si in HF solutions after gathering more insight into various etching models.

### *Anisotropic Etching*

*Introduction* Anisotropic etchants shape, also "machine," desired structures in crystalline materials by etching much faster in one direction than another. When carried out properly, anisotropic etching results in geometric shapes bounded by the slowest etching and perfectly defined crystallographic planes. Anisotropic wet etching techniques, dating back to the 1960s at the Bell Laboratories, were developed mainly by trial and error. Figure 4.38 shows the cross-section of a variety of shapes formed in single-crystal Si using anisotropic etching. The thinned membrane shown in Figure 4.38a with diffused resistors may be used for a piezoresistive pressure sensor. In this application, the wafer is selectively thinned from a starting thickness of 300–500 μm to form a diaphragm having a final thickness of 10–20 μm with precisely controlled lateral dimensions and a thickness control of the order of 1 μm or better. A typical procedure involves the steps summarized in Table 4.8.[53]

The development of anisotropic etchants solved the lateral dimension control lacking in isotropic etchants. Lateral mask geometries on planar photoengraved substrates can be controlled with an accuracy and reproducibility of 0.5 μm or better, and the anisotropic nature of the etchant allows this accuracy to be translated into control of the vertical

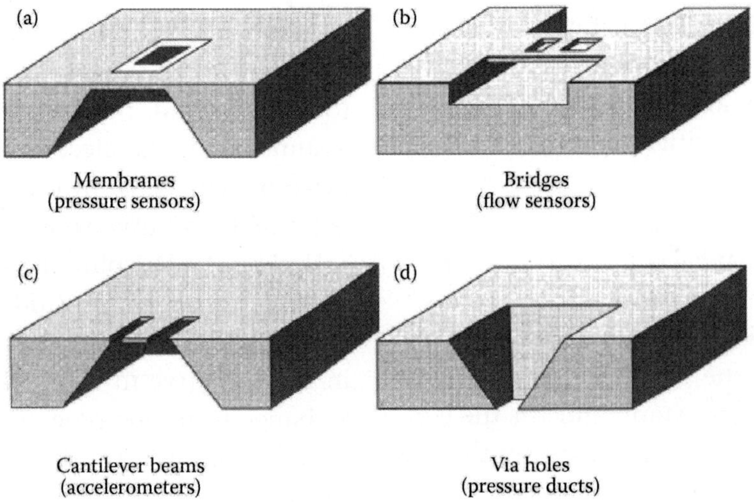


**FIGURE 4.38** Anisotropic wet etching in bulk Si. Membranes for (a) pressure sensors, (b) bridges for flow sensors, (c) cantilevers for accelerometers, and (d) via holes for pressure duct are shown.

**Table 4.8** Summary of the Process Steps Required for Anisotropic Etching of a Membrane in Si

| Process | Duration | Process Temperature (°C) |
|---|---|---|
| Oxidation | Variable (h) | 900–1200 |
| Spinning resist at 5000 rpm | 20–30 s | Room temperature |
| Prebake | 10 min | 90 |
| Exposure | 20 s | Room temperature |
| Develop | 1 min | Room temperature |
| Postbake | 20 min | 120 |
| Strippping of oxide (BHF:1:7) | ±10 min | Room temperature |
| Stripping resist (acetone) | 10–30 s | Room temperature |
| RCA1 ($NH_3$ (25%)+ $H_2O$+$H_2O_2$:1:5:1) | 10 min | Boiling |
| RCA2 (HCl+$H_2O$+ $H_2O_2$:1:6:1) | 10 min | Boiling |
| HF-dip (2% HF) | 10 s | Room temperature |
| Anisotropic etch | From minutes up to 1 day | 70–100 |

*Source:* Elwenspoek, M., H. Gardeniers, M. de Boer, and A. Prak. 1994. *Micromechanics*. Report No. 122830. Twente, the Netherlands: University of Twente.[53]

etch profile. Different etch stop techniques, needed to control the thickness of an active sensor component (such as the membrane or cantilever structures in Figure 4.38), are available. The invention of etch stop techniques truly made the application of the bulk micromachined structures shown in Figure 4.38 commercially manufacturable and viable.

Although anisotropic etchants solve problems of lateral control associated with isotropic etching, they are not problem-free. These etchants are slow—even in the fast etching <100> direction—with etch rates of about 1 µm/min or less. That means that etching through a wafer is a time-consuming process: to etch through a 300-µm-thick wafer, one needs 5 hours. They also must be run hot to achieve these etch rates (80–115°C), precluding many simple masking options. Like the isotropic etchants, their etch rates are temperature sensitive; however, they are not particularly agitation sensitive, which is considered to be a major advantage.

A wide variety of etchants have been used for anisotropic etching of silicon, including alkaline aqueous solutions of KOH, NaOH, LiOH, CsOH, RbOH, $NH_4OH$, and quaternary ammonium hydroxides, with the possible addition of alcohol. Alkaline organics such as ethylene-diamine, choline (trimethyl-2-hydroxyethyl ammonium hydroxide), hydrazine, and sodium silicates with additives such as pyrocatechol and pyrazine are used as well. Etching of silicon is possible without the application of an external voltage and is dopant insensitive over several orders of magnitude; however, in a curious contradiction to its suggested chemical nature, it has been shown to be bias-dependent.[104,105] This contradiction will be explained with the help of the etching models presented below.

Alcohols such as propanol and isopropanol butanol typically slow the attack on Si.[106,107] The role of pyrocatechol[108] is to speed up the etch rate through complexation or chelation of the reaction products. Additives such as pyrazine and quinone have been described as catalysts by some,[109] but this is contested by others.[110] The etch rate in anisotropic etching is reaction rate controlled and thus temperature-dependent. The etch rate for all planes increases with temperature, and the surface roughness decreases with increasing temperature; thus, etching at higher temperatures gives the best results. In practice, etch temperatures of 80–85°C are used to avoid solvent evaporation and temperature gradients in the solution.

### *Selected Anisotropic Etchant Systems*

*Overview* In choosing an anisotropic wet etchant, a variety of issues must be considered:

- Ease of handling
- Toxicity
- Etch rate
- Desired topology of the etched bottom surface
- IC compatibility
- Etch stop availability
- Etch selectivity over other materials
- Mask material and thickness of the mask

The principal characteristics of four different anisotropic etchants are listed in Table 4.9. The most commonly used are KOH[10–23,52,110–118] and ethylenediamine pyrocatechol + water (EDP)[108,109,119];

**Table 4.9** Principal Characteristics of Four Different Anisotropic Etchants

| Etchant/ Diluent/ Additives/ Temperature | Etch Stop | Etch Rate (100) μm/min | Etch Rate Ratio | Remarks | Mask (Etch Rate) |
|---|---|---|---|---|---|
| KOH (water) 85°C 44 g/100 ml | B > $10^{20}$ $cm^{-3}$ reduces etch rate by 20 | 1.4 | 400 for (100)/(111) and 600 for (110)/(111) | IC incompatible, avoid in eyes, etches oxide fast, lots of $H_2$ bubbles | Photoresist (shallow etch at room temperature); $Si_3N_4$ (<1 nm/min) $SiO_2$ (28 Å/min) |
| Ethylenediamine pyrocatechol (water) 115°C 750 ml/120 g/240 ml | ≥7 × $10^{19}$ $cm^{-3}$ reduces the etch rate by 50 | 1.25 | 35 for (100)/(111) | Toxic, ages fast, $O_2$ must be excluded, few $H_2$ bubbles, silicates may precipitate | $SiO_2$ (2–5 Å/min), $Si_3N_4$ (1 Å/min), Ta, Au, Cr, Ag, Cu are not attacked, Al at a 0.33 μm/min |
| Tetramethyl ammonium hydroxide (TMAH) (water) 90°C | >4 × $10^{20}$ $cm^{-3}$ reduces etch rate by 40 | 1 | From 12.5–50 (100)/(111) | IC compatible, easy to handle, smooth surface finish, few studies | $SiO_2$ etch rate is four orders of magnitude lower than (100) LPCVD $Si_3N_4$ |
| $N_2H_4$ (water, isopropyl alcohol) 100°C 100 ml/100 ml | >1.5 × $10^{20}$ $cm^{-3}$ practically stops the etch | 2.0 | 10 (100)/(111) | Toxic and explosive, OK at 50% water | $SiO_2$ (<2 Å/min) and most metallic films; does not attack aluminum |

Given the many possible variables, the data in the table are only typical examples.

hydrazine-water is rarely used.[120,121] Quaternary ammonium hydroxide solutions such as tetramethyl ammonium hydroxide-water (TMAHW) and tetraethyl ammonium hydroxide-water (TEAHW) have become popular.[122,123] Each has its advantages and problems. NaOH is not used much anymore.[124]

*Potassium Hydroxide* The simple KOH water system is the most popular Si anisotropic etchant. A KOH etch, in near-saturated solutions (1:1 in water by weight) at 80°C, produces a uniform and bright surface. Williams and Muller[125] used 50 wt% KOH at 80°C for a reported (100) etch rate of 1.4 μm/min. Nonuniformity of etch rate gets considerably worse at more than 80°C. An abundance of bubbles are seen emerging from the Si wafer while etching in KOH. The etching selectivity between Si and $SiO_2$ is not very good in KOH because it etches $SiO_2$ too quickly. KOH is also incompatible with the IC fabrication process (e.g., aluminum bond pads are quickly attacked and damaged) and can cause blindness when it gets in contact with the eyes. The etch rate for low index planes is maximal at around 4 M (see Figure 4.39a and c from Peeters[126] and Lambrechts et al.[127])

Herr[54] found that the high-index crystal planes exhibit the highest etch rates for 6 M KOH and that for lower concentrations the etch bottoms disintegrate into microfacets. In 6 M KOH, the etch-rate order is (311) > (144) > (411) > (133) > (211) > (122). These authors could not correlate the particular etch rate sequence with the measured activation energies. This is in contrast to lower activation energies corresponding to higher etching rates for low index planes as shown further below in Figure 4.40. Their results obtained on large open-area structures differ significantly from previous ones obtained by underetching special mask patterns. The vertical etching rates obtained here are substantially higher than the underetching rates described elsewhere, and the etch rate sequence for different planes is also significantly different. These results suggest that crevice effects may play an important role in anisotropic etching.

To create vertical (100) faces, as shown in Volume I, Chapter 4 (Figure 4.10), in general only KOH works (not EDP or TMAHW), and it has to be carried out in high-selectivity conditions (low temperature, low concentration: 25 wt% KOH, 60°C). Interestingly, high-concentration KOH (45 wt%) at higher temperatures (80°C) produces a smooth sidewall, controllable and repeatable at an angle of 80°. EDP produces 45° angled planes, and TMAHW usually makes a 30° angle.[129]

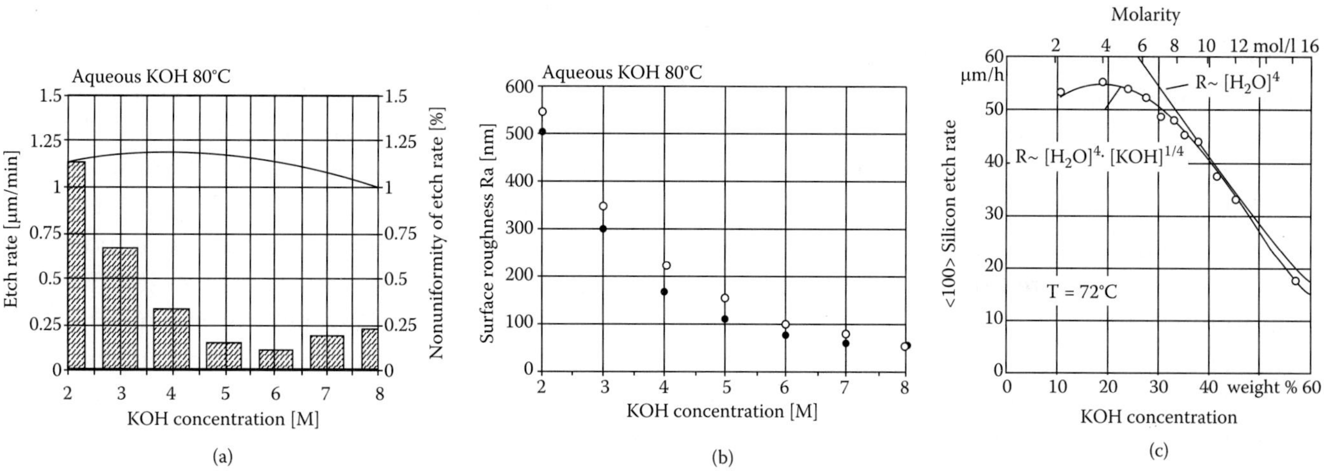


**FIGURE 4.39** Anisotropic etching of silicon. (a) Silicon (100) etch rate (line) and nonuniformity of etch rate (column) in KOH at 80°C as a function of KOH concentration. The etch rate for all low index planes is maximal at around 4 M. (b) Silicon (100) surface roughness (Ra) in aqueous KOH at 80°C as a function of concentration for a 1-hour etch time (thin line) and for an etch depth of 60 µm (thick line). (c) Silicon (100) etch rate as a function of KOH concentration at a temperature of 72°C. (a and b from Peeters, E. 1994. Process development for 3D silicon microstructures, with application to mechanical sensor design. Ph.D. diss., Catholic University of Louvain, Belgium. With permission. c from Seidel, H., L. Csepregi, A. Heuberger, and H. Baumgartel. 1990. Anisotropic etching of crystalline silicon in alkaline solutions–part I: orientation dependence and behavior of passivation layers. *J Electrochem Soc* 137:3612–26. With permission.[126,52])

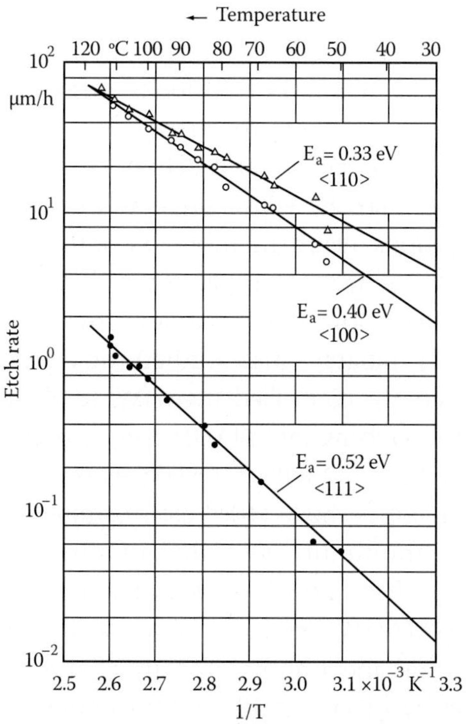


**FIGURE 4.40** Vertical etch rates as a function of temperature for different crystal orientations: (100), (110), and (111). Etch solution is EDP (133 ml $H_2O$, 160 g pyrocatechol, 6 g pyrazine, and 1 l ED). (From Seidel, H., L. Csepregi, A. Heuberger, and H. Baumgartel. 1990. Anisotropic etching of crystalline silicon in alkaline solutions–part I: orientation dependence and behavior of passivation layers. *J Electrochem Soc* 137:3612–26. With permission.[52])

Besides KOH,[129] other hydroxides have been used, including NaOH,[104,124] CsOH,[130] and $NH_4OH$.[130] A major disadvantage of KOH is the presence of alkali ions, which are detrimental to the fabrication of sensitive electronic parts.

*Ethylenediamine Pyrocatechol* With ethylenediamine pyrocatechol (EDP; sometimes referred to as EPW, for ethylenediamine, pyrocatechol, and water), a variety of masking materials (e.g., $SiO_2$, $Si_3N_4$, Au, Cr, Cu, Ag, and Ta) can be used. EDP is less toxic than hydrazine (see below), no sodium or potassium contamination occurs, and the etch rate of $SiO_2$ is much slower than with KOH. The ratio of etch rates of Si and $SiO_2$ using EDP can be as large as 5000:1, corresponding to about 2 Å/min of $SiO_2$ compared with 1 µm/min of Si, which is much larger than the ratio for Si to $SiO_2$ in KOH, which is at the highest 400:1.[131] Importantly, the etch rate slows down at a lower boron concentration than with KOH. A typical fastest-to-slowest hierarchy of Si etch rates with EDP at 85°C according to Barth[132] is (110) > (411) > (311) > (511) > (211) > (100) > (331) > (221) > (111). Petersen uses 750 ml of ethylenediamine, 120 g of pyrocatechol, and 100 ml of water. This cocktail, at

115°C, results in an etch rate of 0.75 μm/min, with an (100)/(111) etch-rate ratio of 35:1.[75]

Ethylenediamine in EDP reportedly causes allergic respiratory sensitization, and pyrocatechol is described as a toxic corrosive. The material is also optically dense, making end-point detection harder, and it ages quickly; if the etchant reacts with oxygen, the liquid turns a red-brown color, and it loses its useful properties. If cooled down after etching, precipitation of silicates in the solution will occur. Sometimes, precipitation during etching can happen, spoiling the results. When preparing the solution, the last ingredient added should be water because water addition causes the oxygen sensitivity. All the above make the etchant difficult to handle.

In terms of etching and masking layers, amine gallates are similar to EDP but perhaps safer.[133] Amine gallate etchants are not used much but appear promising. They are composed of a mixture of ethanolamine, gallic acid, water, pyrazine, hydrogen peroxide, and a surfactant. Etch rates as high as 2.3 μm/min have been measured on a (100) Si plane, and etch stops at lower boron concentration than it takes to stop EDP have been observed ($>3 \times 10^{19}$ cm$^{-3}$). Pyrazine and peroxide can be added to increase the etch rate, but they affect the surface roughness negatively.

*Ammonium Hydroxide-Water and Tetramethyl Ammonium Hydroxide-Water* Efforts continue to find anisotropic etchants that are more compatible with complementary metal oxide semiconductor (CMOS) processing than alkali hydroxides and that are neither toxic nor harmful. Two examples are ammonium hydroxide-water (AHW) mixtures[130] and tetramethyl ammonium hydroxide-water (TMAHW) mixtures (Tabata et al.[123] and Schnakenberg et al.[130]). Kern[77] used AHW (9.7% in water) and achieved 0.11 μm/min etch rates on (100) Si at temperatures of 85–92°C. Schnakenberg et al.[130] reported their best AHW results with a 3.7 wt% solution at 75°C for well-stirred etching baths. For the same solution, these authors demonstrated a boron-dependent etch stop at $1.3 \times 10^{20}$ cm$^{-3}$ with a selectivity of 1:8000 (see also under Etch-Stop Techniques, this chapter). Ammonia-based etchants have not become widely used for several reasons, including their slow etch rate, tendency to lead to rough surfaces (hillocks), and rapid evaporative losses.[72] A TMAHW [$(CH_3)_4NOH$] solution, on the other hand, is one of the more useful wet etchants for silicon. TMAHW does not decompose at temperatures less than 130°C, a very important feature from the viewpoint of production; it is nontoxic, not expensive, and can be handled easily. TMAHW solutions also exhibit excellent selectivity to silicon oxide and silicon nitride masks. The etchant is so selective for Si over $SiO_2$ that it is advisable to remove the thin native oxide of Si in HF before attempting a TMAHW etch. The solution is often already present in the clean room because it is used in many positive photoresist developers. At a solution temperature of 90°C and 22 wt% TMAH, a maximum (100) silicon etch rate of 1.0 μm/min is observed, 1.4 μm/min for (110) planes [this is higher than those observed with EDP, AHW, hydrazine water, and tetraethyl ammonium hydroxide (TEA), but slower than those observed for KOH] and an anisotropy ratio, AR(100)/(111), of between 12.5 and 50.[134] A disadvantage is that TMAHW occasionally results in the formation of pyramidical hillocks at the bottom of the etched cavity. From the viewpoint of fabricating various silicon sensors and actuators, a concentration greater than 22 wt% is preferable because lower concentrations result in more pronounced roughness on the etched surface. However, higher concentrations give a lower etch rate and lower etch ratio (100)/(111). Tabata[135] also studied the etching characteristics of pH-controlled TMAHW. To obtain a low aluminum etching rate of 0.01 μm/min, pH values less than 12 for 22 wt% TMAHW were required. At those pH values the Si (100) etching rate is 0.7 μm/min. The aluminum etch rate can also be reduced by adding silicon powder to the etchant.[136] The etch rate for TMAHW begins to slow down for boron doping levels above approximately $1 \times 10^{19}$ cm$^{-3}$ and decreases by a factor of 40 for $2 \times 10^{20}$ cm$^{-3}$.[137]

*Hydrazine ($H_2N_4$)* Etch rates with hydrazine-water mixtures are on the order of 2 μm/min, and similar masking materials can be used as with EDP.[75] The (100)/(111) etch ratios are lower than those for KOH or EDP. Hydrazine-water is very reducing and explosive at high hydrazine concentrations (rocket fuel), is a suspected carcinogen, and is difficult to dispose so its use should be avoided for safety reasons. However, a 50 wt% hydrazine-water solution is stable and is commonly used, and, according to Mehregany,[121] excellent surface quality and sharply defined corners are obtained in Si. Also on the positive side, the etchant has a very low $SiO_2$ etch rate and will not attack most metal masks except for Al, Cu, and Zn. According to Wise, on the other hand, Al does not etch in hydrazine either, but the etch produces rough Si surfaces.[138]

*Arrhenius Plots for Anisotropic Etching* A typical set of Arrhenius plots for <100>, <110>, and <111> silicon etching in an anisotropic etchant (e.g., EDP) is shown in Figure 4.34.[52] It is seen that the temperature dependence of the etch rate is large and that the slope differs for the different planes, that is, (111) > (100) > (110). Lower activation energies in Arrhenius plots correspond to higher etch rates. The anisotropy ratio (AR) derived from this figure is:

$$AR = \frac{(hkl)_1 \text{ etch rate}}{(hkl)_2 \text{ etch rate}} \tag{4.2}$$

The AR is approximately 1 for isotropic etchants and can be as high as 400:200:1 for (110)/(100)/(111) in 50 wt% $KOH/H_2O$ at 85°C. Generally, the activation energies of the etch rates of EDP are smaller than those of KOH. Price found that (111) planes always etch slowest, and the selectivity with respect to (100) in KOH etching can be greatly increased by adding isopropyl alcohol (IPA), a less polar diluent, used to saturate the solution.[107] The sequence for (100) and (110) etching rate can be reversed, for example, to 50:200:8 in 55 vol% ethylene/diamine/$H_2O$ (also at 85°C). The (110) Si plane etches eight times slower and the (111) eight times faster in $KOH/H_2O$ than in $ED/H_2O$, whereas the (100) etches at the same rate.[111] Working with alcohols and other organic additives often changes the relative etching rate of the different Si planes. Along this line, Seidel et al.[52,110] found that the decrease in etch rate by adding isopropylalcohol to a KOH solution was 20% for the <100> planes but almost 90% for <110> planes. As a result of the much stronger decrease of the etch rate on a (110) surface, the etch ratio of (100)/(110) is reversed.

Misalignment will change the etch rate greatly; a 1° misalignment on the [111] direction may increase the etch rate on (near) the (111) surface by 300%.[139]

*Si Surface Roughness in Anisotropic Etching* Aniso-tropic etchants frequently leave too rough a Si surface behind, and a slight isotropic etch is used to "touch up" the surface. The average surface roughness, $R_a$, of Si continuously decreases with increasing KOH concentration as can be gleaned from Figure 4.39b. The silicon etch rate, $R$, as a function of KOH concentration is shown in Figure 4.39a and c.[52] Because the difference in etch rates for different KOH concentrations is small, a highly concentrated KOH (e.g., 7 M) is preferred in obtaining a smooth surface on low index planes. Except at very high concentrations of KOH, the etched (100) plane becomes rougher the longer one etches. This is thought to be because of the development of hydrogen bubbles, which hinder the transport of fresh solution to the silicon surface, causing "micromasking," which results in hillock formation.[72,140] Average roughness, $R_a$, is influenced strongly by fluid agitation. Stirring can reduce the $R_a$ values more than an order of magnitude, probably caused by the more efficient removal of hydrogen bubbles from the etching surface when stirring.[141] Ohwada et al.[142] noted that the use of ultrasonic agitation essentially eliminated surface roughness in KOH etching. Baum et al.[143] investigated the surface finish of (100) Si in KOH etching with an atomic force microscope and confirmed that mild ultrasonic agitation improved the surface finish considerably. Hillock formation may also be suppressed by the addition of a suitable oxidizing agent that competes with hydrogen evolution, such as by adding ferricyanide or peroxydisulfate ions. Bressers et al.[144] report a drastic reduction in hillock formation when adding 18 mM $K_3Fe(CN)_6$ to a 4 M KOH solution at 70°C, and Klaassen et al.[145] accomplish the same by adding 5 g/L ammonium peroxydisulfate to a 5-wt% TMAHW solution. Baum et al.[143] found that the inclusion in the KOH bath of oxygen and/or isopropanol results in root mean square roughness values smaller than 20 nm. The effectiveness of these additives has been related to

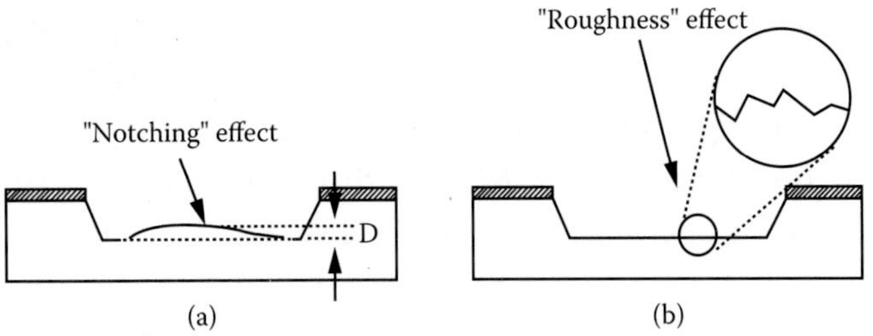


**FIGURE 4.41** (a) Macroscopic roughness (notching effect) and (b) microscopic roughness on a Si membrane.

changes of the contact angle between the liquid/gas/etching interface.

A distinction must be made between macroscopic and microscopic roughness (Figure 4.41). Macroscopic roughness, also referred to as notching or pillowing, results when centers of exposed areas etch with a seemingly lower average speed compared with the borders of the areas; therefore, the corners between sidewalls and (100) ground planes are accentuated. Thus, membranes or double-sided clamped beams (microbridges) tend to be thinner close to the clamped edges than in the center of the structure. This difference can be as large as 1–2 μm, which is considerable if one is etching 10–20-μm-thick structures. Notching increases linearly with etch depth but decreases with higher concentrations of KOH. The microscopic smoothness of originally mirror-like polished wafers can also be degraded into microscopic roughness. It is this type of short-range roughness we referred to in discussing Figure 4.39b above. For more background on metrology techniques to measure surface roughness, see Volume III, Chapter 6.

*Masking for Anisotropic Etchants* Etching through a whole wafer (400–600 μm) takes several hours (a typical wet anisotropic etch rate is 1.1 μm/min), definitely not a fast process. When using KOH, $SiO_2$ cannot be used as a masking material for features requiring that long an exposure to the etchants. The $SiO_2$ etch rate as a function of KOH concentration at 60°C is shown in Figure 4.42. There is a distinct maximum at 35 wt% KOH of nearly 80 nm/h. The shape of this curve will be explained further below based on Seidel et al.'s etching model. Experiments have shown that even a 1.5-μm-thick oxide is not sufficient for the complete etching of a 380-μm-thick wafer (6 hours) because of pinholes in the oxide.[128]

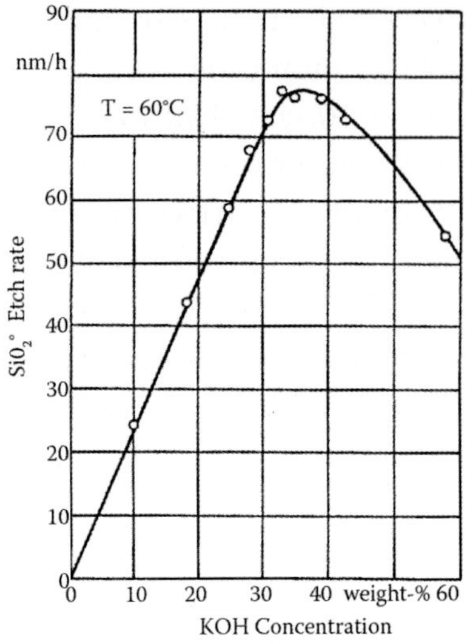


**FIGURE 4.42** The $SiO_2$ etch rate in nanometers per hour as a function of KOH concentration at 60°C. (From Seidel, H., L. Csepregi, A. Heuberger, and H. Baumgartel. 1990. Anisotropic etching of crystalline silicon in alkaline solutions–part I: orientation dependence and behavior of passivation layers. *J Electrochem Soc* 137:3612–26. With permission.[52])

The etch rate of thermally grown $SiO_2$ in KOH-$H_2O$ somewhat varies and apparently depends not only on the quality of the oxide but also on the etching container and the age of the etching solution, as well as other factors.[48] The Si/$SiO_2$ selectivity ratio at 80°C in 7 M KOH is 30 ± 5. This ratio increases with decreasing temperature; reducing the temperature from 80 to 60°C increases the selectivity ratio from 30 to 95 in 7 M KOH.[57] Thermal oxides are under strong compressive stress because in the oxide layer one silicon atom takes nearly twice as much space as in single-crystalline Si (see also Volume I, Chapter 4). This might have severe consequences; for example, if the oxide mask is stripped on one side of the wafer, the wafer will bend. Atmospheric pressure chemical vapor-deposited (APCVD) $SiO_2$ tends to exhibit pinholes and etches much faster than thermal oxide. Annealing of APCVD oxide removes the pinholes, but the etch rate in KOH remains greater by a factor of 2–3 than that of thermal oxide. Low-pressure chemical vapor-deposited (LPCVD) oxide is a mask material of comparable quality as thermal oxide. The etch rate of $SiO_2$ in EDP is smaller by two orders of magnitude than in KOH.

For prolonged KOH etching, a high-density silicon nitride mask has to be deposited. An LPCVD nitride generally serves better for this purpose than less-dense plasma-deposited nitride.[146] With an etch rate of less than 0.1 nm/min, a 400-Å layer of LPCVD nitride suffices to mask against KOH etchant. The etch selectivity Si/$Si_3N_4$ was found to be better than $10^4$ in 7 M KOH at 80°C. The nitride also acts as a good ion-diffusion barrier, protecting sensitive electronic parts. Nitride can easily be patterned with photoresist and etched in $CF_4/O_2$-based plasma or, in a more severe process, in $H_3PO_4$ at 180°C (10 nm/min).[147] Nitride films are typically under a tensile stress of about $1 \times 10^9$ Pa. If, in the overall processing of the devices, nitride deposition does not pose a problem, KOH emerges as the preferential anisotropic wet etchant. For dopant-dependent etching, EDP is the better etchant and generally better suited for deep etching because its oxide etch rate is negligible (<5 Å/min).

Summarizing, oxide and nitride films mask for anisotropic etchants to varying degrees with both mask types being used. When these layers are used to terminate a Si etch in the [100] direction, a low etch rate of the mask layer allows overetching of silicon to compensate for wafer thickness variations. A KOH solution etches $SiO_2$ at a relatively fast rate of 1.4–3 nm/min; therefore, $Si_3N_4$ or Au/Cr must be used as a mask against KOH for deep and long etching.

*Back-Side Protection of Silicon Wafer during Etching* In many cases it is necessary to protect the back of a wafer from an isotropic or anisotropic etchant. The back is either mechanically or chemically protected. In the mechanical method, the wafer is held in a holder, often made from Teflon®. The wafer is fixed between Teflon-coated O-rings that are carefully aligned to avoid mechanical stress in the wafer. In the chemical method, waxes or other organic coatings are spun onto the back of the wafer. Two wafers may also be glued back to back for faster processing.

## Etch-Stop Techniques

### *Introduction*

In many cases, it is desirable to stop etching in silicon when a certain cavity depth or a certain membrane thickness is reached. Nonuniformity of etched devices resulting from nonuniformity of the silicon wafer thickness can be high. Taper of double-polished wafers, for example, can be as high as 40 μm![127] Even with the best wafer quality, the wafer taper is still around 2 μm. The taper and variation in etch depth lead to intolerable thickness variations for many applications. High-resolution silicon micromachining relies on the availability of effective etch-stop layers. It is the existence of impurity-based etch stops in silicon that has allowed micromachining to become a high-yield commercial production process. Etch-rate control typically requires monitoring and stabilization of the following parameters:

- Etchant composition
- Etchant aging
- Stabilization with $N_2$ sparging (especially with EDP and hydrazine)
- Taking account of the total amount of material etched (loading effects)
- Etchant temperature
- Diffusion effects (constant stirring is required, especially for EDP)
- Stirring also leads to a smoother surface through bubble removal
- Trenching (also pillowing) and roughness decrease with increased stirring rate
- Light may affect the etch rate (especially with n-type Si)
- Surface preparation of the sample can have a big effect on etch rate (the native oxide retards etch start; a dip in dilute HF is recommended)

With a good temperature, etchant concentration, and stirring control, the variation in etch depth typically is 1% (see Figure 4.43).[127] A good pretreatment of the surface to be etched is a standard RCA clean (see Volume II, Chapter 1, Wafer Cleaning and Contaminants) combined with a 5% HF dip to remove the native oxide immediately before etching in KOH.

In the early days of micromachining, one of the following techniques was used to etch a Si structure anisotropically to a predetermined thickness. In the simplest mode, the etch time was monitored (in Table 4.9 we listed some etch rates for different etchants) or a bit more sophisticated; the infrared

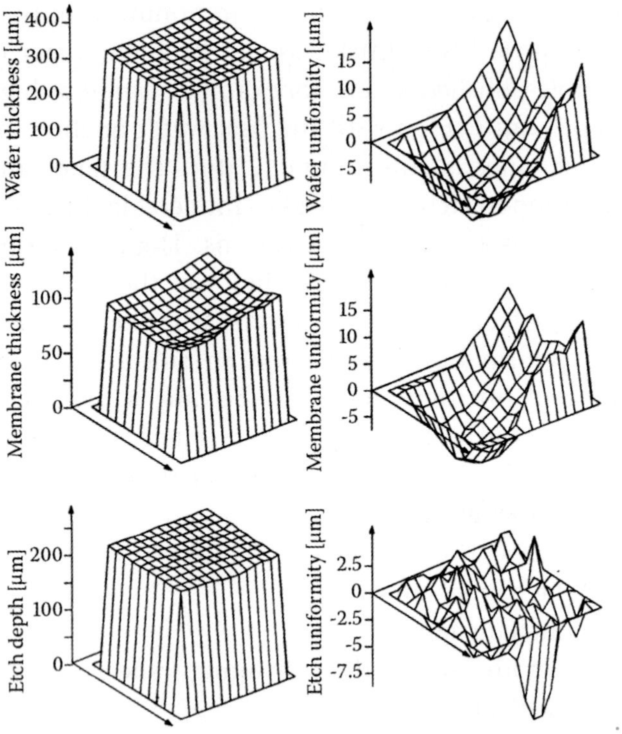


**FIGURE 4.43** Map of wafer thickness, membrane thickness, and etch depth. (From Lambrechts, M., and W. Sansen. 1992. *Biosensors: microelectrochemical devices.* Philadelphia: Institute of Physics Publishing. With permission.[127])

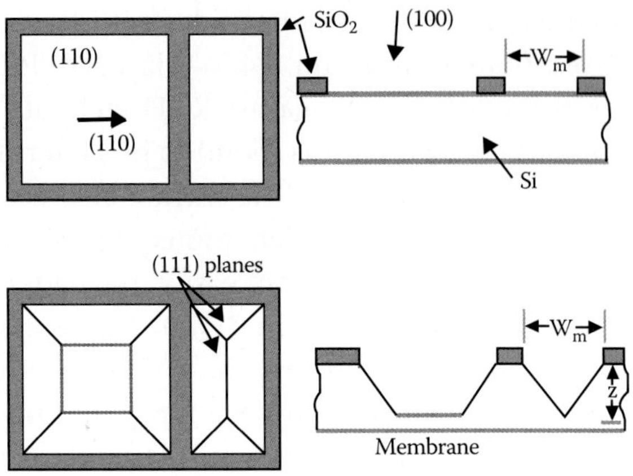


**FIGURE 4.44** V-groove technique to monitor the thickness of a membrane. At the precise moment the V-groove is developed, the membrane has reached the desired thickness.

transmittance through the etching membrane was followed. For thin membranes, the etch stop cannot be determined by a constant etch time method with sufficient precision. The spread in etch rates becomes critical if one etches membranes down to thicknesses of less than 20 μm; it is almost impossible to etch structures down to less than 10 μm with a timing technique. In the V-groove technique, V-grooves with precise openings (see Volume I, Equation 4.2, $W_0 = W_m - 2z\cot 54.74^\circ = W_m - \sqrt{2}z$) were used such that the V-groove stopped etching at the exact moment a desired membrane thickness was reached (see Figure 4.44).[148]

One can also design wider mask openings on the wafer's edge so that the wafer is etched through at those sites at the moment the membrane has reached the appropriate thickness. Although Nunn and Angell[149] claimed that an accuracy of about 1 μm could be obtained using the V-groove method, none of the mentioned techniques are found to be production worthy. Nowadays the above methods are almost completely replaced by etch-stop techniques based on a change in etch rate-dependent on doping level or the chemical nature of a stopping layer.

### *Boron Etch Stop*

The most widely used etch-stop technique is based on the fact that anisotropic etchants, especially EDP, do not attack heavily boron-doped ($p^{++}$) silicon layers. Selective $p^{++}$ doping is typically implemented using gaseous or solid boron diffusion source with a mask (such as silicon dioxide). The maximum practical depth achievable is 15 μm. The boron etch-stop effect was first noticed by Greenwood in 1969.[150] He assumed that the presence of a p-n junction was responsible. In 1971 Bogh[151] found that an impurity concentration of about $7 \times 10^{19}/cm^{-3}$ resulted in the etch rate of Si in EDP decreasing sharply (see also Table 4.9) but without any requirement for a p-n junction. For KOH-based solutions, Price[107] found a significant reduction in etch rate for boron concentrations greater than $5 \times 10^{18}$ cm$^{-3}$. The model by Seidel et al. (see below) provides an elegant explanation for the etch stop at high boron concentrations.

From the above, it follows that a simple boron diffusion or implantation, introduced from the front of the wafer, can be used to create beams and diaphragms by etching from the back. A boron etch-stop technique is illustrated in Figure 4.45A for the fabrication of a micromembrane nozzle.[152] The $SiO_2$ mesa in Figure 4.45Ab leads to the desired boron $p^{++}$ profile. The anisotropic etch from the back clears the

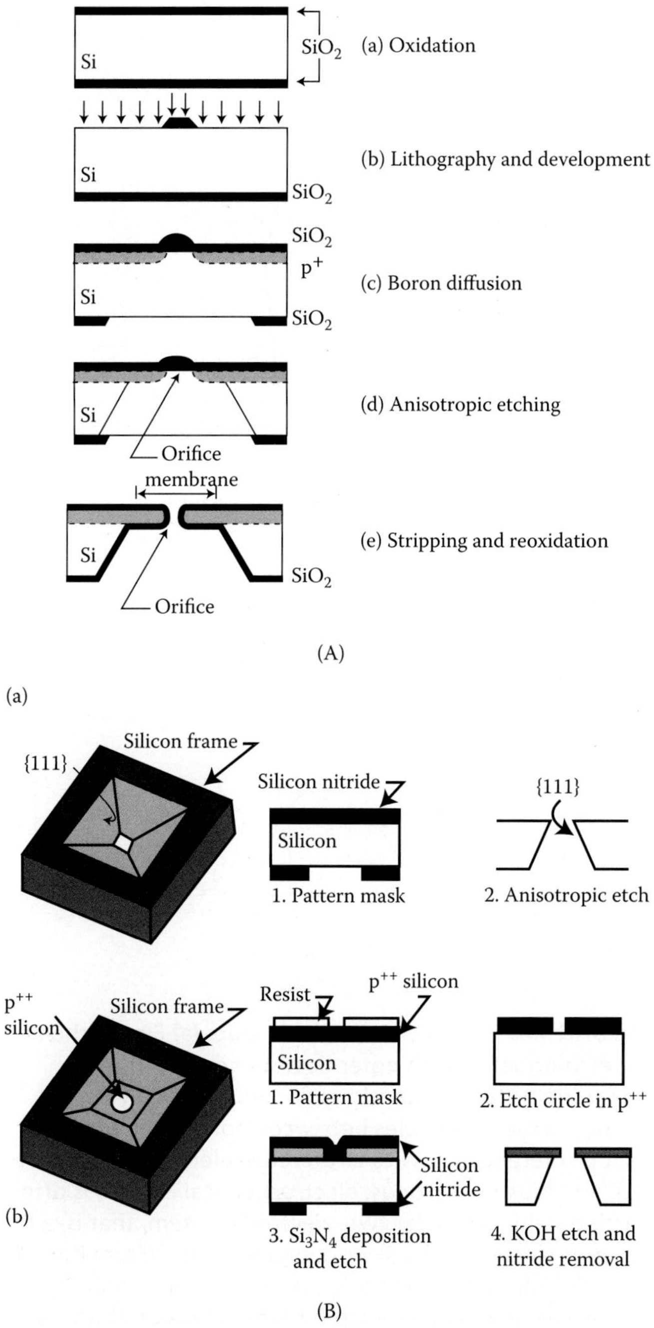


**FIGURE 4.45** (A) Illustration of the boron etch stop in the fabrication of a membrane nozzle. (From Brodie, I., and J. J. Muray. 1982. *The physics of microfabrication.* New York: Plenum Press. With permission.[152]) (B) Two alternate methods to fabricate nozzles. (a) Etch stop is based on silicon nitride. (b) Etch stop is based on boron etch stop.

lightly doped p-type Si (Figure 4.45Ad). Stripping and reoxidation produce an orifice in the suspended membrane (Figure 4.45Ae). Two alternative ways of fabricating nozzles—one square and one circular—are illustrated in Figure 4.45B. The method shown in Figure 4.45Ba uses silicon nitride as an etch-stop layer to etch a square nozzle. The side of the backside opening in the silicon nitride must be larger than 71% of the wafer thickness to etch all the way through the wafer (Volume I, Equation 4.2, $W_0 = W_m - 2z\cot 54.74^\circ = W_m - \sqrt{2}z$). The approach shown in Figure 4.45Bb is again based on a $p^{++}$ etch-stop layer. The difference with the approach in Figure 4.45A is that a uniform $p^{++}$ doping profile is first established here, and that layer is subsequently etched into a circular pattern. Layers of $p^{++}$ silicon with a thickness of 1–20 μm can easily be fabricated, and the boron etch stop is very effective; it is not critical when the operator takes the wafer out of the etchant. One important practical note is that the boron etch stop may become badly degraded in EDP solutions that were allowed to react with atmospheric oxygen. Because boron atoms are smaller than silicon, a highly doped, freely suspended membrane or diaphragm will be stretched; the boron-doped silicon is typically in tensile stress, and the microstructures are flat and do not buckle. Whereas doping with boron decreases the lattice constant, doping with germanium increases the lattice constant. A membrane doped with B and e still etches much more slowly than undoped silicon, and the stress in the layer is reduced. A stress-free, dislocation-free, and slow etching layer (±10 nm/min) is obtained at doping levels of $10^{20}$ $cm^{-3}$ boron and $10^{21}$ $cm^{-3}$ germanium.[110,153]

One major disadvantage of the boron etch-stop technique is that the extremely high boron concentrations are not compatible with standard CMOS or bipolar techniques, so they can only be used for microstructures without integrated electronics. Another limitation of this process is the fixed number and angles of (111) planes one can accommodate. The etch stop is less effective in KOH compared with EDP. Besides boron, other impurities have been tried for use in an etch stop in anisotropic etchants. Doping Si with germanium has hardly any influence on the etch rate of either the KOH or EDP solutions. At a doping level as high as $5 \times 10^{21}$ $cm^{-3}$, the etch rate is barely reduced by a factor of two.[110]

By burying the highly doped boron layer under an epitaxial layer of lighter doped Si, the problem of incompatibility with active circuitry can be avoided; ±1% thickness uniformity is possible with modern epilayer deposition equipment. A widely used method of automatically measuring the epi-thickness is with IR instruments, especially Fourier transform infrared (see also epitaxy in Chapter 7).[154]

### Electrochemical Etch Stop

For the fabrication of piezoresistive pressure sensors, the doping concentration of the piezoresistor must be kept smaller than $1 \times 10^{19}$ cm$^{-3}$ because the piezoresistive coefficients decrease considerably above this value, and reverse breakdown becomes an issue. Moreover, high boron levels compromise the quality of the crystal by introducing slip planes and tensile stress and prevent the incorporation of integrated electronics. As a result, a boron stop often cannot be used to produce well-controlled thin membranes unless, as suggested above, the highly doped boron layer is buried underneath a lighter-doped Si epilayer. Alternatively, a second etch-stop method, an electrochemical technique, can be used. In this case, a lightly doped p-n junction is used as an etch stop by applying a bias between the wafer and a counter electrode in the etchant, a technique first proposed by Waggener in 1970.[16] Other early work on electrochemical etch stops with anisotropic etchants such as KOH and EDP was performed by Palik et al.,[155] Jackson et al.,[156] Faust and Palik,[157] and Kim and Wise.[158] In electrochemical anisotropic etching, a p-n junction is made, for example, by the epitaxial growth of an n-type layer (phosphorus-doped, $10^{15}$ cm$^{-3}$) on a p-type substrate (boron-doped, 30 Ω-cm). This p-n junction forms a large diode over the whole wafer. The wafer is usually mounted on an inert substrate, such as a sapphire plate, with an acid-resistant wax and is partly or wholly immersed in the solution. An ohmic contact to the n-type epilayer is connected to one pole of a voltage source, and the other pole of the voltage source is connected via a current meter to a counter electrode in the etching solution (see Figure 4.46a). In this arrangement, the p-type substrate can be selectively etched away, and the etching stops at the p-n junction, leaving a membrane with a thickness solely defined by the thickness of the epilayer. The incorporation of a third electrode (a reference electrode) in the three-terminal method, together with the use of a potentiostat depicted in Figure 4.46a, allows for a more precise determination of the silicon potential with respect to the solution than a two-terminal setup as we illustrated in Figure 4.36a.

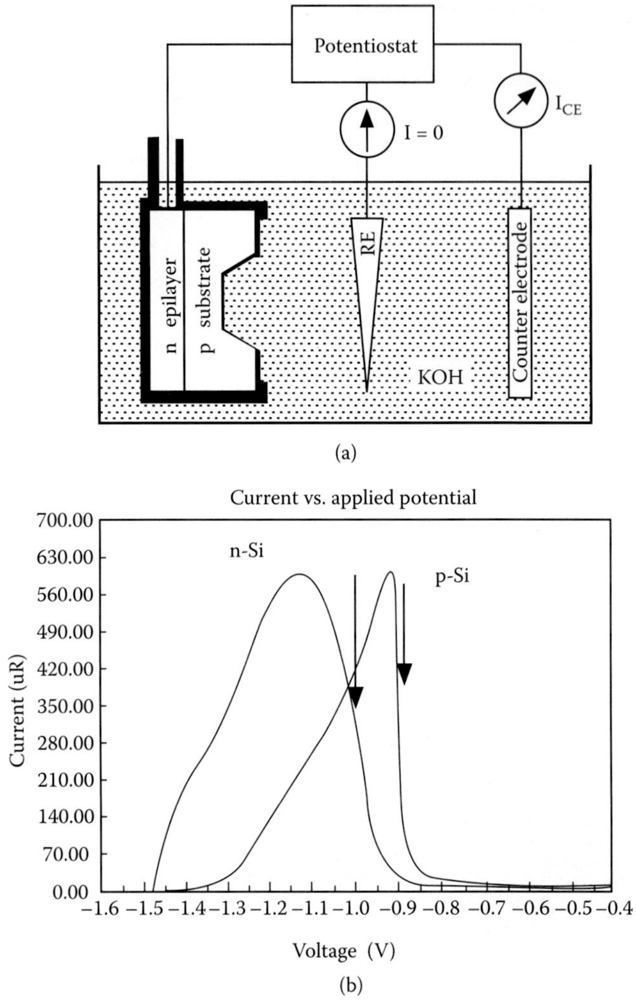


**FIGURE 4.46** Electrochemical etch stop. (a) Electrochemical etching setup with potentiostatic control (three-electrode system). Potentiostatic control, mainly used in research studies, enables better control of the potential as it is referenced now to a reference electrode such as an SCE. In industrial settings, electrochemical etching is often carried out in a simpler two-electrode system, that is, a Pt counter electrode and Si working electrode. (From Kloeck, B., S. D. Collins, N. F. de Rooij, and R. L. Smith. 1989. Study of electrochemical etch-stop for high-precision thickness control of silicon membranes. *IEEE Trans Electron Devices* 36:663–69.[159]) (b) Cyclic voltammograms of n- and p-type silicon in an alkaline solution at 60°C. Flade potentials are indicated with an arrow.

At the Flade potential, marked with an arrow in Figure 4.46b, the oxide growth rate equals the oxide etch rate; a further increase of the potential results in a steep fall of the current as a result of complete passivation of the silicon surface. At potentials positive of the Flade potential, all etching stops. At potentials negative of the Flade potential, the current increases as the potential becomes more positive. This can be explained by the formation of an oxide etching faster than it forms; that is, the silicon is etched away. Whereas electrochemists like to talk about the Flade potential, physicists like to discuss matters in terms of the flat-band potential. The flat-band potential is the applied potential at which there are no more fields within the semiconductor; that is, the energy band diagram is flat throughout the semiconductor (see Volume I, Chapter 4). The passivating $SiO_2$ layer is assumed to start growing as soon as the negative surface charge on the silicon electrode is cancelled by the externally applied positive bias, a bias corresponding to the flat-band potential. At these potentials, the formation of $Si(OH)_x$ complexes does not lead to further dissolution of silicon because two neighboring Si-OH HO-Si groups will react by splitting off water, leading to the formation of Si-O-Si bonds. As can be learned from Figure 4.46b, the value of the Flade potential depends on the dopant type. Consequently, if a wafer with both n- and p-type regions exposed to the electrolyte is held at a certain potential in the passive range for n-type and the active range for p-type Si, the p-type regions are etched away, whereas the n-type regions are retained. In the case of the diode shown in Figure 4.46a, where, at the start of the experiment, only p-type Si is exposed to the electrolyte, a positive bias is applied to the n-type epilayer ($V_n$). This reverse biases the diode, and only a reverse bias current can flow. The potential of the p-type layer in this regime is negative to the flat-band potential, and active dissolution takes place. At the moment that the p-n junction is reached, a large anodic current can flow, and the applied positive potential passivates the n-type epilayer. Etching continues on the areas where the wafer is thicker until the membrane is reached there, too. The thickness of the Si membrane is thereby solely defined by the thickness of the epilayer; neither the etch uniformity nor the wafer taper will influence the result. A uniformity of better than 1% can be obtained on a 10-μm-thick membrane. The current versus time curve can be used to monitor the etching process; at first, the current is relatively low, limited by the reverse bias current of the diode. Then, as the p-n junction is reached, a larger anodic current can flow until all the p-type material is consumed and the current decreases again to a low plateau value. The plateau indicates that a current associated with a passivated n-type Si electrode has been reached. The etch procedure can be stopped at the moment that the current plateau has been reached, and because the etch stop is basically one of anodic passivation, it is sometimes called an *anodic oxidation* etch stop. Registering an $I$ versus $V$ curve as in Figure 4.46b will establish an upper limit on the applied voltage, $V$,[160] and such curves are used for in situ monitoring and controlling of the etch stop. A crude endpoint monitoring can be accomplished by the visual observation of cessation of hydrogen bubble formation accompanying silicon etching. Palik et al.[161] presented a detailed characterization of Si membranes made by electrochemical etch stop.

Using the same anodic oxidation etch stop in a hydrazine-water solution at 90 ± 5°C and a simple two-terminal electrochemical cell, Hirata et al.[162] obtained a pressure sensitivity variation of less than 20% from wafer to wafer (pressure sensitivity is inversely proportional to the square of membrane thickness; see Volume III, Chapter 8). A great advantage of this etch-stop technique is that it works with Si at low doping levels of the order of $10^{16}$ $cm^{-3}$. Because of the low doping levels, it is possible to fabricate structures with a very low, or controllable, intrinsic stress. Moreover, active electronics and piezomembranes can be built into the Si without problems. A disadvantage is that the back of the wafer with the aluminum ohmic contact has to be sealed hermetically from the etchant solution, which requires complex fixturing and manual wafer handling. The fabrication of a suitable etch holder is no trivial matter. The holder 1) must protect the epicontact from the etchant, 2) must provide a low-resistance ohmic contact to the epilayer, and 3) must not introduce stress into wafers during etching.[53,126] Stress introduced by etch holders easily leads to diaphragm or wafer fracture and etchant seepage through to the episide.

Using a four-electrode electrochemical cell, controlling the potentials of the epitaxial layer and the silicon wafer separately, as shown in Figure 4.47, can further improve the thickness control of the resulting membrane by directly controlling the p/n bias

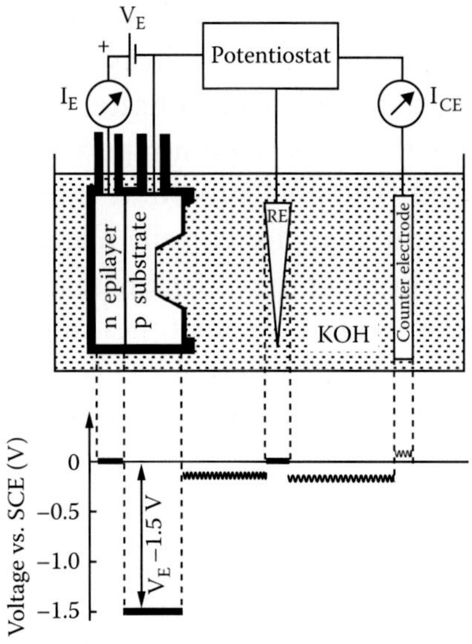


**FIGURE 4.47** Four-electrode electrochemical etch-stop configuration. Voltage distribution with respect to the SCE reference electrode (RE) for the four-electrode case. The fourth electrode enables an external potential to be applied between the epitaxial layer and the substrate, thus maintaining the substrate at etching potentials. (From Kloeck, B., S. D. Collins, N. F. de Rooij, and R. L. Smith. 1989. Study of electrochemical etch-stop for high-precision thickness control of silicon membranes. *IEEE Trans Electron Devices* 36:663–69. With permission.[159])

voltage. The potential required to passivate n-type Si can be measured using the three-electrode system in Figure 4.46a, but this system does not take into account the diode leakage. If the reverse leakage is too large, the potential of the n-type Si, $V_n$, will approach the potential of the p-type Si, $V_p$. If there is a large amount of reverse diode leakage, the p-type region may passivate before reaching the n-type region, and etching will cease. In the four-electrode configuration, the reverse leakage current is measured separately via a p-type region contact, and the counter electrode current ($I_{CE}$) may be monitored for endpoint detection. The four-electrode approach allows etch stopping on lower quality epilayers (larger leakage current) and should also enable etch stopping of p-type epi on n-type substrates. Kloeck et al.[159] demonstrated that, using such an electrochemical etch-stop technique and with current monitoring, the sensitivity of pressure sensors fabricated on the same wafer could be controlled to within 4% standard deviation. These authors used a 40% KOH solution at 60 ± 0.5°C. Without the etch stop, the sensitivity from sensor to sensor on one wafer varied by a factor of two.

The etching solution used in electrochemical etching can be either isotropic or anisotropic. An electrochemical etch stop used in isotropic media was discussed above. In this case, $HF/H_2O$ mixtures are used to etch the highly doped regions of $p^+p$, $n^+n$, $n^+p$, and $p^+n$ systems.[87,88,163–165] The rate-determining step in etching with isotropic etchants does not involve reducing water with electrons from the conduction band, as it does in anisotropic etchants, and the etch-stop mechanism, as we will learn below, is very different. In isotropic media, the etch stop is simply a consequence of the fact that higher conductivity leads to higher corrosion currents, and the etching slows down on lower conductivity layers. A major advantage of the KOH electrochemical etch is that it retains the anisotropic characteristics of KOH without needing a heavily doped $p^+$ layer to stop the etching.[166]

In Figure 4.48 we illustrate the etch-stop techniques discussed so far. Figure 4.48a and b, represents two types of boron-dependent etch stops: diffused boron etch stop and buried layer boron

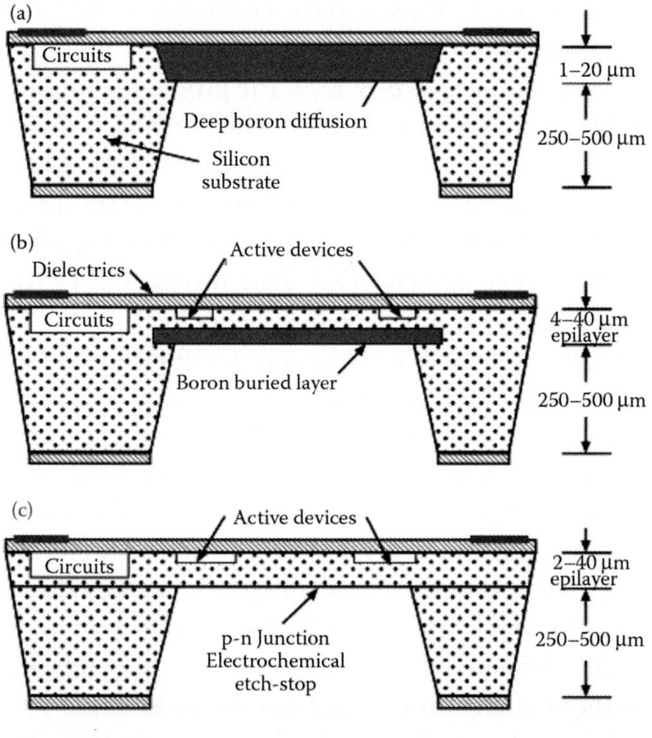


**FIGURE 4.48** Typical cross-sections of bulk micromachined wafers with various methods for etch-stop formation shown. (a) Diffused boron etch stop. (b) Boron etch stop as a buried layer. (c) Electrochemical etch stop. (From Wise, K. D. 1985. *Micromachining and micropackaging of transducers.* Eds. C. D. Fung, P. W. Cheung, W. H. Ko, and D. G. Fleming. New York: Elsevier. With permission.[138])

etch stop. In Figure 4.48c we illustrate the electrochemical etch stop.[138] A comparison of the boron etch stop and the electrochemical etch stop reveals that the IC compatibility and the absence of built-in stresses, both results of the low dopant concentration, are the main assets of the electrochemical etch stop.

### *Photoassisted Electrochemical Etch Stop (for n-Type Silicon)*

A variation on the electrochemical diode etch-stop technique is the photoassisted electrochemical etch-stop method illustrated in Figure 4.49.[168] An n-type silicon region on a wafer may be selectively etched in an HF solution by illuminating and applying a reverse bias across a p-n junction, driving the p-type layer cathodic and the n-type layer anodic. Etch rates up to 10 μm/min for the n-type material and a high-resolution etch stop render this an attractive potential micromachining process. Advantages also include the use of lightly doped n-type Si, bias- and illumination intensity-controlled etch rates, in situ process monitoring using the cell current, and the ability to spatially control etching with optical masking or laser writing. Using this method, Mlcak et al.[167] prepared stress-free cantilever beam test samples. They diffused boron into a $10^{15}$ $cm^{-3}$ (100) n-type Si substrate through a patterned oxide mask, leaving exposed a small n-type region that defines two p-type cantilever beams (see Figure 4.49).

The boron diffusion resulted in a junction 3.3 μm underneath the surface. An ohmic contact on the back of the wafer was used to apply a variable voltage across the p-n junction, and both p and n areas were exposed to the HF electrolyte. The exposed n-type region was etched to a depth of 150 μm by shining light on the whole sample. The resulting n-type Si surface was at first found to be rough because porous Si up to 5 μm in height forms readily in HF solutions. The Si surface could be made smoother by etching at higher bias [4.3 V vs. saturated calomel electrode (SCE)] and higher light intensity (2 $W/cm^2$) to a finish with features of the order of 0.4 μm in height. Smoothing could also be accomplished by a 5-second dip in $HNO_3$:HF:$H_3COOH$ or a 30-second dip in 25 wt%, 25°C, KOH. Yet another way of removing unwanted porous Si is a 1000°C wet oxidation to make oxidized porous Si (OPS) followed by an HF etch.[168] The formation of porous Si and OPS is detailed further below.

The photoassisted electrochemical etching of n-type Si is commercially exploited at Boston Microsystems Inc. (http://www.bostonmicrosystems.com).

### *Photoinduced Preferential Anodization (for p-Type Si)*

Electrochemical etching requires the application of a metal electrode to apply the bias. The application of such a metal electrode often induces contamination

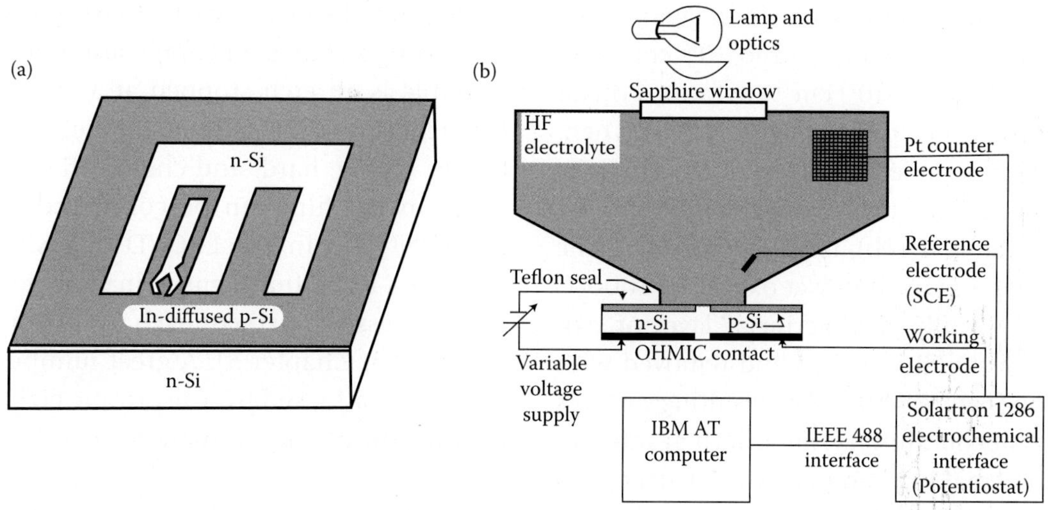


**FIGURE 4.49** Photoelectrochemical etching. (a) Schematic of the photoelectrochemical etching experiment apparatus. (b) Schematic of the spatial geometry of the diffused p-type silicon layer into n-type Si used to form cantilever beam structures. (From Mlcak, R., H. L. Tuller, P. Greiff, and J. Sohn. *Proceedings: IEEE micro electro mechanical systems (MEMS '93).* Fort Lauderdale, FL: IEEE. With permission.[167])

and constitutes at least one extra process step, and extra fixturing is needed. With photoinduced preferential anodization (PIPA), it is not necessary to deposit metal electrodes. Here, one relies on the illumination of a p-n junction to bias the p-type Si anodically, and the p-type Si converts automatically into porous Si, whereas the n-type Si acts as a cathode for the reaction. The principle of photobiasing for etching purposes was known and patented by Shockley as far back as 1963.[169] In US patent 3,096,262, he writes that "light can be used in place of electrical connections ... for biasing of the sample ... This means a small isolated area of p-type material on an n-type body may be preferentially biased for removal of material beyond the junction by etching." The method was "reinvented" by Yoshida et al.[168] in 1992 and in 1993 by Peeters et al.[170,171] The latter group called the method PHET, for photovoltaic electrochemical etch-stop technique, and the former group coined the PIPA acronym.

In PIPA, etch rates of up to 5 μm/min result in the formation of porous layers that are readily removed with Si etching solutions. An important advantage of the technique is that very small and isolated p-type islands can be anodized at the same time. The method also lends itself to fabrication of 3D structures using p-type Si as sacrificial layers.[168] A disadvantage of the technology is that one cannot control the process very well because the current cannot be measured for endpoint detection. Application of PIPA to form a microbridge is shown in Figure 4.50. First, a buried p-type layer doped to $10^{18}$ $cm^{-3}$ and an n-type layer doped to $10^{15}$ $cm^{-3}$ are formed on an n-type substrate using epitaxy (Figure 4.50a). Then, the p-type layer is preferentially anodized in 10% HF solution under 30 $mW/cm^2$ light intensity for 180 minutes (Figure 4.50b), forming porous Si. The porous Si is then oxidized in wet oxygen at 1000°C (Figure 4.50c). Finally, the sacrificial layer of oxidized porous Si (OPS) is etched and removed with an HF solution (Figure 4.50d). The resulting surfaces of the n-type silicon are very smooth. It is interesting to consider making complicated 3D structures by going immediately to the electropolishing regime instead. Yoshida et al.[168] believe that porous Si as a sacrificial intermediate is more suitable for fabricating complicated structures.

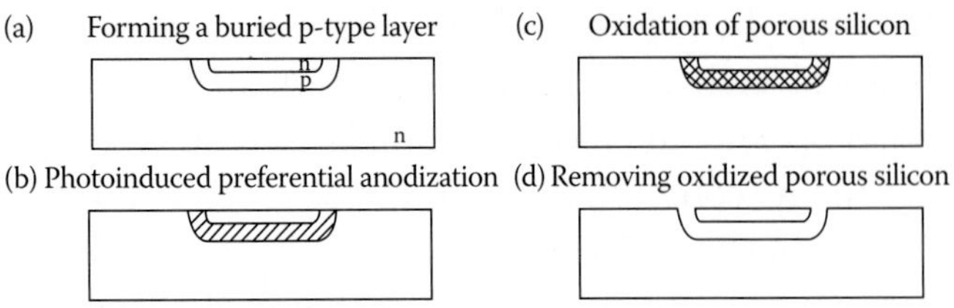


**FIGURE 4.50** Photoassisted electrochemical etch stop (for n-type Si). Fabrication process for a microbridge and SEM picture of Si structure before and after PIPA. (From Yoshida, T., T. Kudo, and K. Ikeda. 1992. *Proceedings: IEEE micro electro mechanical systems (MEMS '92).* Travemunde, Germany: IEEE. With permission.[168])

The authors are probably referring to the fact that electropolishing is much more aggressive and could not be expected to lead to the same retention of the shape of the buried, sacrificial p-type layers. Peeters et al.[160] carry out their photovoltaic etching in KOH, thus skipping the porous Si stage of Yoshida et al. They, like Yosida et al., stress the fact that, in one single etch step, this technique can make a variety of complex shapes that would be impossible with electrochemical etching techniques. These authors found it necessary to coat the n-type Si part of the wafer with Pt to get enough photovoltaic drive for the anodic dissolution; this metallization step makes the process more akin to photoelectrochemical etching (PEC), and some advantages of the photobiasing process are lost.

### *Etch Stop at Thin Insoluble Films*

Yet another distinct way (the fourth) to stop etching is by using a change in composition of material. An example is an etch stopped at a $Si_3N_4$ diaphragm (see Figure 4.45Ba for an example). Silicon nitride is very strong, hard, and chemically inert, and the stress in the film can be controlled by changing the Si/N ratio in the LPCVD deposition process. The stress turns from tensile in stoichiometric films to compressive in silicon-rich films (for details, see Volume III, Chapter 5). A great number of materials are not attacked by anisotropic etchants. Hence a thin film of such a material can be used as an etch stop.

Another example is the $SiO_2$ layer in a silicon-on-insulator (SOI) structure. A buried layer of $SiO_2$, sandwiched between two layers of crystalline silicon, forms an excellent etch stop because of the good selectivity of many etchants of Si over $SiO_2$. The oxide does not exhibit the good mechanical properties of silicon nitride and consequently is rarely used as a mechanical member in a microdevice. As with PIPA, no metal contacts are needed with an SOI etch stop, greatly simplifying the process over an electrochemical etch-stop technique.

We classified SOI micromachining under surface micromachining in Chapter 7, where more details about this promising micromachining alternative are presented.

## Chemical Etching Models for Single-Crystal Si

### *Introduction*

Conflicting data abound in the literature for the anisotropic etch rates of the different Si planes, especially for the higher index planes. This is not too surprising, given the multiple parameters influencing individual results: temperature, stirring, size of etching feature (i.e., crevice effects and loading effects), etchant concentration, addition of alcohols and other organics, surface defects, complexing agents, surfactants, pH, cation influence, and so on. Even the reason(s) why acidic media lead to isotropic etching whereas alkaline media lead to anisotropic etching was, until recently, not explicitly addressed in any of the models surveyed. More rigorous experimentation and standardization will be needed, as well as improved etching models, to better understand the influence of all these parameters on etch rates. In the following section we will present recent atomistic and fluctuating energy level models (see Volume I, Chapter 4) to explain anisotropic and isotropic etching of silicon. It is our hope that the reading of this section will inspire more detailed semiconductor electrochemistry work on Si electrodes. The refining of an etching model will be of invaluable help in writing more predictive Si etching software code.

### *Atomistic Model—Gosalvez-Kelly*

The origin of the macroscopic anisotropy in the etching of Si lies in the site-specificity of the etch rates at the atomistic scale. Surface sites on different crystallographic planes react with an etchant in a different manner because their surface orientations are different. As a consequence, as pointed out by Gosalvez, the macroscopic etch rate is different for different crystal orientations.[47] The removal of Si surface atoms is a complex process that involves chemical and electrochemical reactions. As shown in Figure 4.51, the etching process takes place through sequential oxidation and etching reactions. During etching of Si, the surface, as we show here, is predominantly H-terminated, and this is true for both isotropic[94] and anisotropic etching.[172] The H-termination of the surface was established from both ex situ (after etching the surface) and in situ (during etching) infrared spectroscopy by Kelly et al.[172] In other words, removal of the native oxide from Si leaves the surface hydrogen-terminated.

Once the substitution of H (hydrophobic) by OH (hydrophilic) has taken place, a fast sequence of chemical steps (represented as "etching" in Figure 4.51) leads to the removal of the silicon atom as a $Si(OH)_4$ product. The triggering effect of the reaction by the OH ligand is attributed to the difference in electronegativities between the Si and O atoms, resulting in the polarization and weakening of the backbonds of the Si atom, which as a result, becomes more vulnerable to further attack. Although the hydroxyl group plays an important role in catalyzing the removal of the surface atom by favoring backbond attack, the real active species attacking the backbonds are polar molecules: water in alkaline etchants and HF in acidic etchants.[172]

In an alkaline medium, the first step involving the nucleophilic attack by an $OH^-$ ion on the Si-H bond may be represented as[172]:

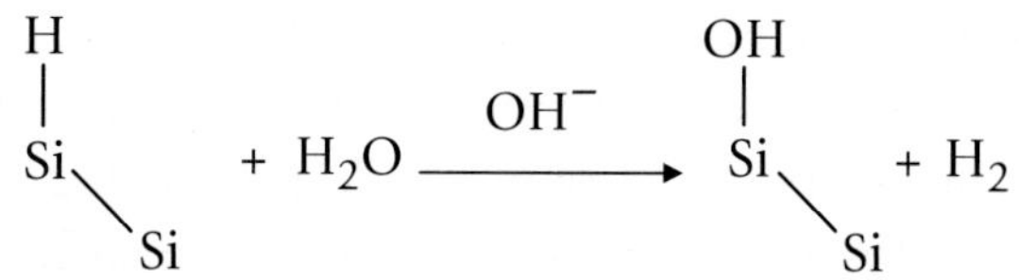


Reaction 4.10

The departing hydride reacts with water to give hydrogen gas, and the Si surface atom is hydroxylated. It is the presence of the OH group that polarizes the Si-Si backbond. With the OH adding to the positively polarized surface atom, the backbond is

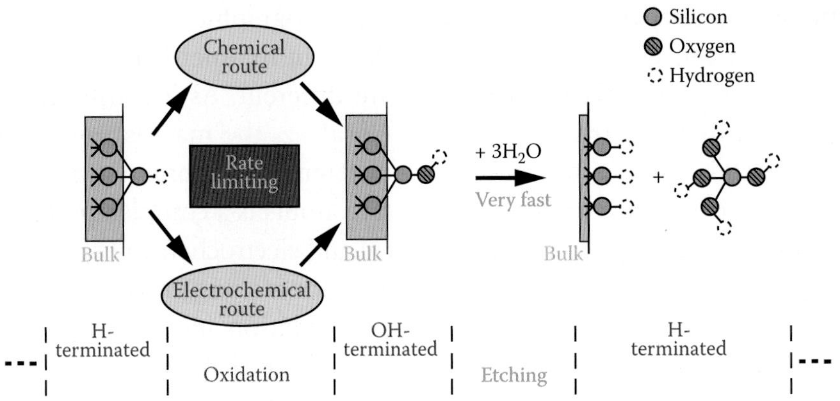


**FIGURE 4.51** Schematic representation of the oxidation and etching reactions that sequentially occur for the removal of Si surface atoms. The rate limiting process (oxidation) can take place through chemical and electrochemical routes. Only the nearest underlying bulk atoms in the neighborhood of the surface site are depicted. (Gonsalvez, M. 2003. *Atomistic modelling of anisotropic etching of crystalline silicon.* PhD diss., Helsinki University of Technology.[47])

attacked by water, and an H atom attaches itself to the subsurface atom, restoring the hydrogen termination of the Si surface as shown in Reaction 4.11[172]:

Reaction 4.11

Whether the surface is hydrogenated (hydrophobic) or hydroxylated (hydrophilic) during etching depends on the relative rates of Reactions 4.10 and 4.11. The fact that in practice a predominantly H-termination is found is determined by kinetics rather than thermodynamics, and this implies that the site specificity of the whole process is a result of the rate-limiting oxidation reaction alone.[47]

The number of backbonds associated with the attacked Si atom is universally accepted as a major source for surface specificity. In the case of a (100) surface, each atom has a dihydride termination and is bonded to two subsurface Si atoms. In the (111) case, the surface is monohydride terminated with three backbonds. A dihydride is chemically less stable than a monohydride, so one expects Reaction 4.10 to be faster for a (100) surface.[173] Because it should be easier to dislodge a (100) Si atom with only two backbonds, the rate of Reaction 4.11 also should be faster than for a (111) surface. This qualitatively explains already why one expects a (111) plane to etch much more slowly than a (100) plane in alkaline media.[172] In the model of Seidel et al.[52] (see below) the energy levels of the electrons in the backbonds are associated with surface states within the bandgap of Si. The energy level of these surface states is assumed to vary for different surface orientations, being lowest for {111} planes.

We return now for a moment to Figure 4.51 and draw the readers' attention to the fact that chemical and electrochemical reaction routes provide two alternative mechanisms for the oxidation of the hydrogen-terminated sites before the removal of the resulting hydroxyl-terminated silicon. For open-circuit etching of Si, i.e., with no external circuit connecting the Si to a counter electrode, there are two possible mechanisms: chemical and electroless. The two are distinguished by the fact that the latter involves the participation of free charge carriers, giving rise to measurable currents and allowing for the possibility of controlling the etching process with a biasing potential.[172] In the electrochemical route of Si oxidation, an oxidizing agent in solution extracts/injects mobile charge carriers from/into the solid. When, for example, an oxidizing agent extracts bonding electrons from the valence band, i.e., injects holes, and the holes stay localized at the surface, they will cause bond rupture. It is important to realize that these holes are mobile so that the two reactions, the reduction of the oxidant and the oxidation of the solid, can be considered as two spatially and temporally independent electrochemical reactions. This form of etching is generally referred to as electroless because it does not involve two separate electrodes (the same process is involved in metal corrosion). A reaction like this, involving the exchange of charge carriers, can lead to external currents when an external circuit is fashioned and can be influenced by a bias applied between the etching electrode (Si) and a counter electrode (e.g., Pt). In chemical etching, bonds in the etching agent and the solid are broken, and new bonds are formed in a localized and synchronous reaction, not involving free carriers. In this case, the reaction rate cannot be influenced by an applied bias. In the oxidation subprocess of Figure 4.51, the main difference between the chemical and the electrochemical routes lies in the size of the activated complex. In chemical oxidation, we saw (Reaction 4.10) that the hydroxyl ion acts as a catalyst. It is postulated that the hydroxyl ion reduces the energy barrier of a sizable pentavalent transition state as shown schematically in Figure 4.52.[174]

The fact that both water (reactant) and $OH^-$ (catalyst) are involved in the reaction depicted in Figure 4.52 is made clear from the fact that etching of single-crystal silicon is possible in pure boiling water, where there is a small concentration of $OH^-$ present, but not in concentrated HF, where the concentration of $OH^-$ is too low to enable the oxidation subprocess in Figure 4.52. In this case, the reaction stops after the H-termination.[47] Besides the number of backbonds of the site under attack, Gosalvez concludes that the site specificity of anisotropic etching is also determined by the geometry of the transition complex; in other words, the steric hindrance of the OH-termination controls the reaction rate as well.[47] The larger the size and the more rigid the transition complex, the larger one expects the anisotropy to be. As a result of these constraints, the chemical oxidation route is very anisotropic.

The Si etching process has been found to depend on the applied potential; thus, the chemical reaction in Figure 4.52 cannot be the only oxidation subprocess. Under anodic potential, the dissolution rate becomes enhanced, and the ratio of $H_2/Si$ is reduced because of the contribution of the electrochemical reaction path.[105] Therefore, there also must be an electrochemical route involving free charge carriers. We will detail our understanding of the electrochemical route when analyzing the Seidel et al. model (see below); here we just take advantage of the experimental evidence that the anodic current is found to be only weakly dependent on the $OH^-$ concentration but strongly so on the water concentration (see Equation 4.3, page 275),[52,110] suggesting that water is the only active species. This is illustrated in Figure 4.53, where the process is depicted as occuring through the dissociation of the surface hydride group and subsequent reaction with a water molecule.

In contrast with the sizable pentavalent transition state involved in the chemical oxidation route (see Figure 4.52), in the electrochemical oxidation route the small surface hydride leaves plenty of space for the reaction with water molecules, and the dissociation might occur at any surface site, making the electrochemical route rather isotropic compared with the chemical route.[47]

The surface morphology of the etched surfaces also turns out to be orientation-dependent when examined at the micrometer and smaller length scales. Pyramidal hillocks and round shallow pits,

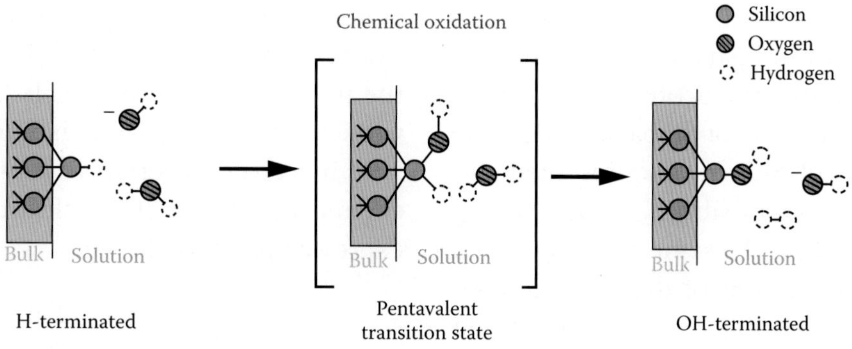


**FIGURE 4.52** Schematic representation of the chemical oxidation reaction. Only the nearest underlying bulk atoms in the neighborhood of the surface site are shown. (Gonsalvez, M. 2003. *Atomistic modelling of anisotropic etching of crystalline silicon.* PhD diss., Helsinki University of Technology.[47])

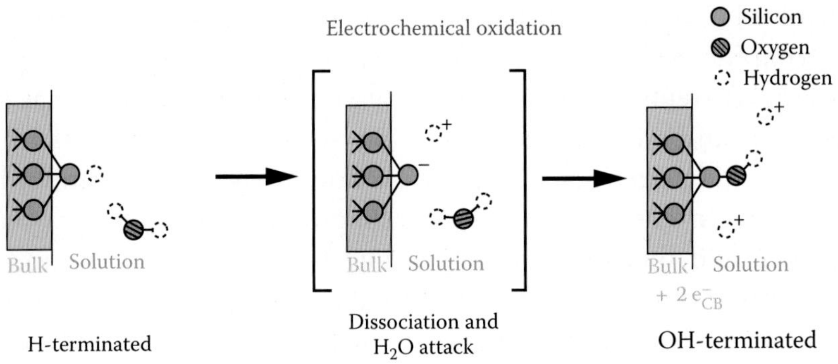


**FIGURE 4.53** Schematic representation of the electrochemical oxidation reaction. Only the nearest underlying bulk atoms in the neighborhood of the surface site are shown. (Gonsalvez, M. 2003. *Atomistic modelling of anisotropic etching of crystalline silicon.* PhD diss., Helsinki University of Technology.[47])

for example, are typical for {100} planes, whereas triangular pits are found on {111} surfaces. Gosalvez writes that the macroscopic anisotropy and the surface morphology are two manifestations of the same microscopic site specificity observed at different length scales.[47] The surface microstructure during wet etching of Si then depends on the relative importance of the chemical and electrochemical oxidation routes. The etching process is most anisotropic under cathodic bias, where the etching process is completely chemical, and it becomes increasingly more isotropic as the anodic bias is increased. At sufficiently negative bias one can even eliminate the surface pits on {111} planes, or a totally chemical oxidation process results in perfectly flat {111} surfaces. In Gosalvez's view, under the usual alkaline etching conditions, without biasing potentials, the etching process is very anisotropic, and he claims that the electrochemical oxidation route has only minor effects (mainly the nucleation of pits on {111} planes). Within Gosalvez's elegant model, the etch stop at passivating anodic potentials is a result of an increased OH-termination. The high number of hydroxyl ligands on the Si surface produces the formation of silicon oxide through oxygen insertion into the Si-Si backbonds. The etching mechanism of $SiO_2$ in alkaline media is slow compared with that of silicon, so the etching process is stopped. This is the same effect observed when increasing the amount of dissolved oxygen in the solution. Very revealing, the effect of oxygen and other oxidizing agents, added to the etching solution, is to make the etching process more isotropic, improving the surface finish of {100} surfaces but negatively affecting that of (111), just in the same manner as the biasing potentials. In both cases, mobile charge carriers make the process more dependent on water molecules only rather than requiring the water/$OH^-$ complex that makes a chemical process more anisotropic.

### *Seidel et al.'s Model*

*Anisotropy of Etch Rates* Seidel et al.'s model is based on the fluctuating energy level model of the silicon/electrolyte interface and assumes the injection of electrons in the conduction band during the etching process. The construction of energy diagrams of the type discussed here was introduced in Volume I, Chapter 7. Consider the situation of a piece of Si immersed in an alkaline solution without applied bias (open circuit conditions). After immersion of the silicon crystal into the alkaline electrolyte, a negative excess charge builds up on the surface as a result of the higher original Fermi level of the $H_2O/OH^-$ redox couple in the etchant as compared with the Fermi level of the solid. In other words, as explained in Volume I, Chapter 7, the electrochemical potential difference is equalized across the interface; the Fermi levels line up; or also, the work function difference is equalized. The Fermi level in a given material is equivalent to the electrochemical potential of the electrons in that material expressed per particle, or:

$$E_F = \frac{\bar{\mu}_e}{N_A} \qquad (7.79)$$

(see Volume I, Chapter 7). This leads to a downward bending of the energy bands on the solid surface for both p- and n-type silicon (Figure 4.54a and b). The downward bending is more pronounced for p-type than for n-type because of the initially larger difference of the Fermi levels between the solid and the electrolyte.

Next, hydroxyl ions cause the Si surface to oxidize, consuming water and liberating hydrogen in the process. From the measurement of the amount of Si dissolution and $H_2$ generation, a stoichiometric ratio of $2H_2/1Si$ was found,[108] and thus the following detailed steps, based on Palik,[128] were suggested:

$$Si + 2OH^- \rightarrow Si(OH)_2^{2+} + 2e^-$$

$$Si(OH)_2^{2+} + 2OH^- \rightarrow Si(OH)_4 + 2e^-$$

$$Si(OH)_4 + 4e^- + 4H_2O \rightarrow Si(OH)_6^{2-} + 2H_2$$

Reaction 4.12

Silicate species were observed by Raman spectroscopy.[175] Thus, the overall silicon oxidation reaction consumes four electrons first injected into the conduction band, where they stay near the surface as a result of the downward bending of the energy bands (see Figure 4.54). Raley et al.[176] first presented evidence for injection of four electrons rather than a mixed hole and electron mechanism. These authors could explain the measured etch-rate dependence on hole concentration only by assuming that the proton or water reduction reaction is rate determining and that a four-electron injection mechanism with the conduction band was involved. The injected electrons are highly "reducing" and react with water to form hydroxide ions and hydrogen:

$$4H_2O + 4e^- \rightarrow 4H_2O^- \qquad \text{Reaction 4.13}$$

$$4H_2O^- \rightarrow 4OH^- + 4H^+ + 4e^- + 2H_2 \qquad \text{Reaction 4.14}$$

It is thought that the hydroxide ions in Reaction 4.14, generated directly at the silicon surface, react in the oxidation step. The hydroxide ions from the bulk of the solution may not play a major role because they will be repelled by the negatively charged Si surface, whereas the hydroxide ions formed in situ do not need to overcome this repelling force. This would explain why the etch rates for an EDP solution with an $OH^-$ concentration of 0.034 M are nearly as large as those for KOH solutions with a hundred-fold higher $OH^-$ concentration of 5–10 M.[52] The hydrogen formed in Reaction 4.14 can inhibit the reaction, and surfactants may be added to displace the hydrogen (IBM, US Patent 4,113,551, 1978). Additional support for the involvement of four water molecules (Reaction 4.13) comes from the experimentally observed correlation between the fourth power of the water concentration and the silicon etch rate for concentrated KOH solutions. The Si etch rate, *R*, as a function of KOH concentration at 72°C, was shown in Figure 4.39a and c.[52] The etch rate has a maximum at about 20 wt% KOH of about 55 μm/h. The best fit for this experimentally determined etch rate, for most KOH concentrations, is[52,110]:

$$R = k[H_2O]^4\,[KOH]^{\frac{1}{4}} \qquad (4.2)$$

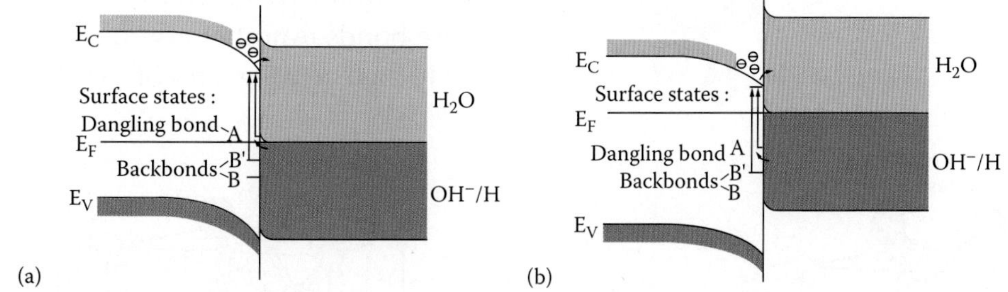


**FIGURE 4.54** Band model of the silicon/electrolyte interface for moderately doped silicon (electrolyte at pH > 12): p-type silicon (a) and n-type silicon (b). We assume no applied bias and no illumination. The energy scale functions in respect to the saturated calomel electrode (SCE), an often-used reference in electrochemistry. Notice that p-type Si exhibits more band bending because its Fermi level is lower in the bandgap. For simplicity, we show only one energetic position for surface states associated with dangling bonds and backbonds; in reality, there will be new surface states arising during reactions as the individual dangling bonds and backbonds are taking on different energies as new Si-OH bonds are introduced.

The weak dependence of the etching curve on the KOH concentration (~1/4 power) supports the assumption that the hydroxide ions involved in the oxidation reactions are mostly generated from water. A strong influence of water on the silicon etch rate was also observed for EDP solutions. In molar water concentrations of up to 60%, a large increase of the etch rate occurs.[108] The driving force for the overall Reaction 4.12 is given by the larger Si-O binding energy of 193 kcal/mol as compared with a Si-Si binding energy of only 78 kcal/mol. The role of cations, $K^+$, $Na^+$, and $Li^+$, and even complicated cations such as $NH_2(CH_2)_2NH_3^+$ can probably be neglected.[52]

The four electrons in Reaction 4.13 are injected into the conduction band in two steps. In the case of {100} planes, there are two dangling bonds per surface atom for the first two of the four hydroxide ions to react with, injecting two electrons into the conduction band in the process. As a consequence of the strong electronegativity of the oxygen atoms, the two bonded hydroxide groups on the silicon atom reduce the strength of the two silicon backbonds (Figure 4.55). With two new hydroxide ions approaching, two more electrons (now stemming from the Si-Si backbonds) are injected into the conduction band, and the silicon-hydroxide complex reacts with the two additional hydroxide ions. Seidel et al.[52] claim that the step of activating the second two electrons from the backbonds into the conduction band is the rate-limiting step, with an associated thermal activation energy of about 0.6 eV for {100} planes.

The electrons in the backbonds are associated with surface states within the bandgap (see Figure 4.54). The energy level of these surface states is assumed to vary for different surface orientations, the lowest being for {111} planes. The thermal activation of the backbonds corresponds to an excitation of the electrons out of these surface states into the conduction band. Because the energy for the backbond surface state level is the lowest within the bandgap for {111} planes, these planes will be hardest to etch. The {111} planes have only one dangling bond for a first hydroxide ion to react with. The second step, which is the rate-limiting step, involves breaking three lower energy backbonds (Figure 4.56).

The lower energy of the backbond surface states within the Si bandgap for {111} Si atoms can be understood from the simple argument that their energy level is raised less by the electronegativity of a single binding hydroxide ion compared with two in the case of the silicon atoms in {100} planes. The high etch rate generally observed on {110} surfaces is similarly explained by a high energy level of the backbond-associated surface states for these planes. Elwenspoek et al.[55,56] do not accept this "two versus three backbonds" argument (see below). They point out that the silicon atoms in the {110} planes also have three backbonds, and activation energy in these crystallographic directions should be comparable with that of {111} planes in contrast to experimental evidence. Seidel et al. would probably counterargue here that the backbonds and the energy levels of the associated surface states are not necessarily the same for {111} and {110} planes because that energy will also be influenced by the effect of the orientation of these bonds. Another argument in favor of the relatively high etching rates of {110} planes is the easier

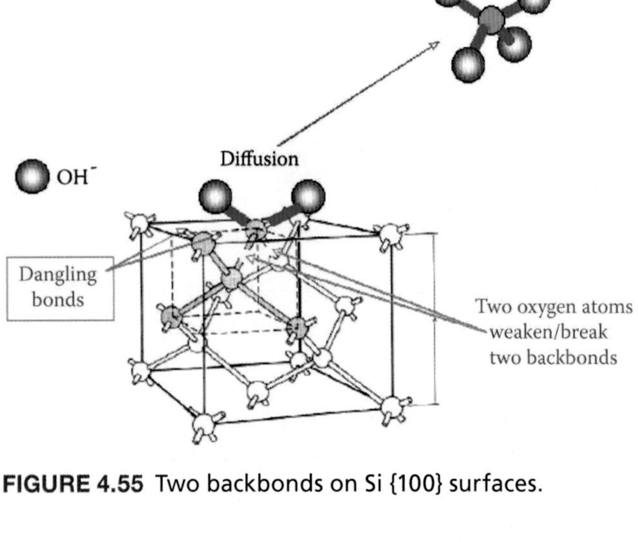


**FIGURE 4.55** Two backbonds on Si {100} surfaces.

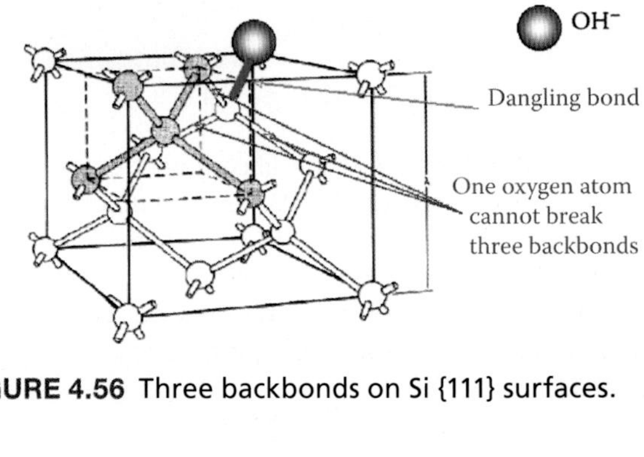


**FIGURE 4.56** Three backbonds on Si {111} surfaces.

penetrability of {110} surfaces for water molecules along channels in that plane.

The final step in the anisotropic etching is the removal of the reaction product $Si(OH)_4$ by diffusion. If the production of $Si(OH)_4$ is too fast for solutions with a high water concentration, the $Si(OH)_4$ leads to the formation of a $SiO_2$-like complex before $Si(OH)_4$ can diffuse away. This might be observed experimentally as a white residue on the wafer surface.[177] The high pH values in anisotropic etching are required to obtain adequate solubility of the $Si(OH)_4$ reaction product and to remove the native oxide from the silicon surface. From silicate chemistry it is known that for pH values greater than 12 the $Si(OH)_4$ complex will undergo the following reaction by the detachment of two protons:

$$Si(OH)_4 \rightarrow SiO_2(OH)_2^{2-} + 2H^+ \quad \text{Reaction 4.15}$$

$$2H^+ + 2OH^- \rightarrow 2H_2O \quad \text{Reaction 4.16}$$

Pyrocatechol in an ethylenediamine etchant acts as a complex-forming agent for reaction products such as $Si(OH)_4$, converting these products into more complex anions:

$$Si(OH)_4 + 2OH^- + 3C_6H_4(OH)_2 \rightarrow$$

$$Si(C_6H_4O_2)_3^{2-} + 6H_2O \quad \text{Reaction 4.17}$$

There is evidence by Abu-Zeid et al.[178] of diffusion control contribution to the etch rate in EDP, probably because the hydroxide ion must diffuse through the layer of complex silicon reaction products (see Reaction 4.17). The same authors also found that the etch rate depends on the effective silicon area being exposed and on its geometry (crevice and loading effects). That is why the silicon wafer is placed in a holder, and the solution is vigorously agitated to minimize the diffusion layer thickness. For KOH solutions, no effect of stirring on etching rate was noticed. Stirring here is mainly used to decrease the surface roughness, probably through removal of hydrogen bubbles.

The influence of alcohol on the KOH etching rate in Equation 4.3 is the result of a change in the relative water concentration and its concomitant pH change; it does not participate in the reaction (this was confirmed by Raman studies by Palik et al.[175]). The reversal of etch rates for {110} and {100} planes through alcohol addition to KOH/water etchants can be understood by assuming that the alcohol covers the silicon surface,[175] thus canceling the "channeling advantage" of the {110} planes. In the case of EDP, alcohol has no effect because the water concentration can be freely adjusted without significantly influencing the pH value as a result of the incomplete dissociation of EDP.

For the etching of $SiO_2$ shown in Figure 4.42, Seidel et al. propose the following reaction:

$$SiO_2 + 2OH^- \rightarrow SiO_2(OH)_2^{2-} \quad \text{Reaction 4.18}$$

At KOH concentrations up to 35%, a linear correlation occurs between $SiO_2$ etch rate and KOH concentration. The $SiO_2$ etch rate in KOH solutions exceeds those in EDP by close to three orders of magnitude. For higher concentrations, the etch rate decreases with the square of the water concentration, indicating that water plays a role in this reaction. Seidel et al. speculate that at high pH values the silicon electrode is highly negatively charged (the point of zero charge of $SiO_2$ is 2.8), repelling the hydroxide ions while water takes over as reaction partner. An additional reason for the decrease is that the hydroxide concentration does not continue to increase with increasing KOH concentration for very concentrated solutions. The decrease of the $Si/SiO_2$ etch rate ratio with increasing temperature and pH of the solution (up to 35%) follows out of the larger activation energy associated with the $SiO_2$ etch rate (0.85 eV) and its linear correlation with the hydroxide concentration, whereas the silicon etch rate mainly depends on the water concentration.

The effect of water concentration and pH value on the etching process in the Seidel et al. model is summarized in Table 4.10.[179]

*Doping Dependence of Si Anisotropic Etching* A good model for anisotropic etching of Si should also be able to explain the fact that the anisotropic etchant systems of Table 4.9 exhibit drastically reduced etch rates for high boron concentrations in silicon ($\geq 5 \times 10^{19}$ $cm^{-3}$ solid solubility limit). Other

**Table 4.10** Effect of Water Concentration and pH Value on the Characteristics of Si Etching

| | $- H_2O +$ | $-$ pH $+$ |
|---|---|---|
| $SiO_2$ etch rate | No effect | $- \Leftrightarrow +$ |
| Si etch rate | $- \Leftrightarrow +$ | Little effect |
| Solubility | No effect | $- \Leftrightarrow +$ |
| Si/$SiO_2$ ratio | $- \Leftrightarrow +$ | $+ \Leftrightarrow -$ |
| Diffusion effects | $- \Leftrightarrow +$ | $+ \Leftrightarrow -$ |
| Residue formation | $- \Leftrightarrow +$ | $+ \Leftrightarrow -$ |
| $p^+$ etch stop | $- \Leftrightarrow +$ | $+ \Leftrightarrow -$ |
| p, n etch stop | $- \Leftrightarrow +$ | $+ \Leftrightarrow -$ |

*Source:* Seidel, H. 1990. *Technical digest: IEEE solid-state sensor and actuator workshop.* Hilton Head Island, SC.[179]

impurities (P, Ge) also reduce the etch rate but at much higher concentrations (see Figure 4.57[52]). Boron typically is incorporated using ion implantation (thin layers) or liquid/solid source deposition (thick layers >1 μm). These doped layers are used as very effective etch stop layers (see above). Hydrazine or EDP, which display a smaller (100) to (111) etch rate ratio (~35) than KOH, exhibits a stronger boron concentration dependency. The etch rate in KOH is reduced by a factor of 5–100 for a boron concentration larger than $10^{20}$ $cm^{-3}$. When etching in EDP, the factor climbs to 250.[151] With TMAHW solutions, the Si etch rate decreases to 0.01 μm/min for boron concentrations of about $4 \times 10^{20}$ $cm^{-3}$.[180]

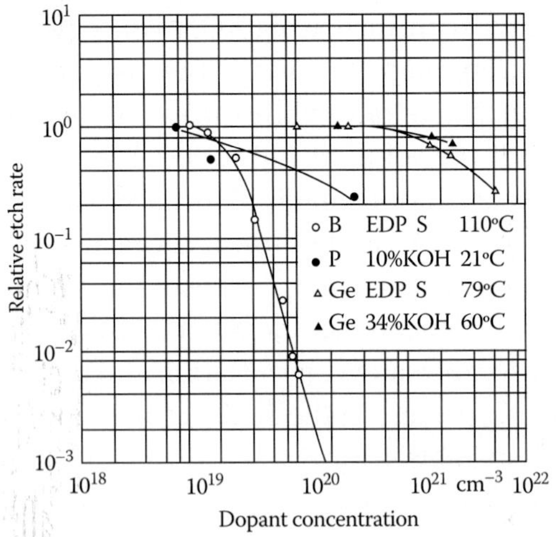


**FIGURE 4.57** Relative etch rate for (100) Si in EDP and KOH solutions as a function of concentration of boron, phosphorus, and germanium. (From Seidel, H., L. Csepregi, A. Heuberger, and H. Baumgartel. 1990. Anisotropic etching of crystalline silicon in alkaline solutions–part I: orientation dependence and behavior of passivation layers. *J Electrochem Soc* 137:3612–26. With permission.)

Seidel et al.'s model, we shall see, gives the most plausible explanation for this effect.

Some of the different mechanisms that have been invoked to explain the doping dependence of Si anisotropic etching follow.

1. Several observations suggest that doping leads to a more readily oxidized Si surface. Highly boron- or phosphorus-doped silicon in aqueous KOH spontaneously can form a thin passivating oxide layer.[128,155] The boron oxides and hydroxides initially generated on the silicon surface are not soluble in KOH or EDP etchants.[75] The substitutional boron creates local tensile stress in the silicon, increasing the bond strength so that a passivating oxide might be more readily formed at higher boron concentrations. Boron-doped silicon has a high defect density (slip planes), encouraging oxide growth.
2. Electrons produced during the oxidation of silicon are needed in a subsequent reduction step (hydrogen evolution in Reaction 4.14). When the hole density passes $10^{19}$ $cm^{-3}$, these electrons combine with holes instead, thus stopping the reduction process.[155] Seidel et al.'s model follows this explanation (see below).
3. Silicon doped with boron is under tension as the smaller boron atoms enter the lattice substitutionally. The large local tensile stress at high boron concentration makes it energetically more favorable for the excess boron (>$5 \times 10^{19}$ $cm^{-3}$) to enter interstitial sites. The strong B-Si bonds bind the lattice rigidly. With high enough doping, the high binding energy can stop etching.[75] This hypothesis is similar to item 1, except that no oxide formation is invoked.

The model by Seidel et al. provides an elegant explanation for the etch stop at high boron concentrations. At moderate dopant concentration, we saw that the electrons injected into the conduction band stay localized near the semiconductor surface as a result of the downward bending of the bands (Figure 4.58). The electrons there have a small probability of recombining with holes deeper into the crystal even for p-type Si. This situation changes when the doping level in the silicon increases

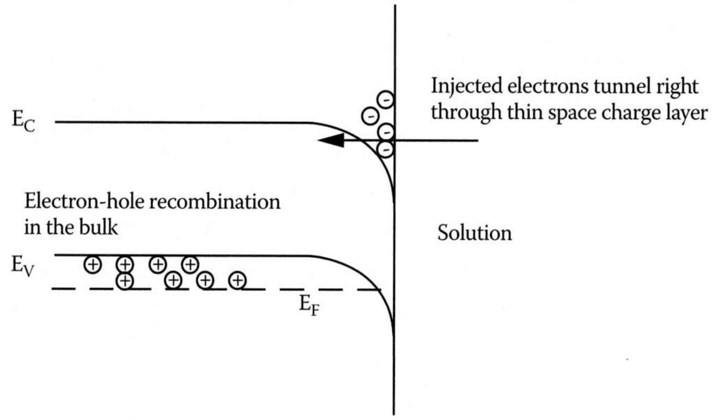


**FIGURE 4.58** Si/electrolyte interface energetics at high doping level explaining etch-stop behavior.

further. At a high dopant concentration, silicon degenerates and starts to behave like a metal. For a degenerate p-type semiconductor, the space charge thickness shrinks, and the Fermi level drops into the valence band as indicated in Figure 4.58.

The injected electrons shoot (tunnel) right through the thin surface charge layer into deeper regions of the crystal, where they recombine with holes from the valence band. Consequently, these electrons are not available for the subsequent reaction with water molecules (Reaction 4.13), the reduction of which is necessary for providing new hydroxide ions in close proximity to the negatively charged silicon surface. These hydroxide ions are required for the dissolution of the silicon as $Si(OH)_4$. The remaining etch rate observed within the etch stop region is then determined by the number of electrons still available in the conduction band at the silicon surface. This number is assumed to be inversely proportional to the number of holes and thus the boron concentration. Experiments show that the decrease in etch rate is nearly independent of the crystallographic orientation, and the etch rate is proportional to the inverse fourth power of the boron concentration in all alkaline etchants.

*Model of Seidel et al. Summary* The key points of the Seidel et al. model can now be summarized as follows (see also Table 4.10)[179]:

1. The rate-limiting step in anisotropic etching of Si is the water reduction.
2. Hydroxide ions required for oxidation of the silicon are generated through reduction of water at the silicon surface. The hydroxide ions in the bulk do not contribute to the etching because they are repelled from the negatively charged surface. This implies that the silicon etch rate will depend on the molar concentration of water and that cations will have little effect on the silicon etch rate.
3. The dissolution of silicon dioxide is assumed to be purely chemical with hydroxide ions. The $SiO_2$ etch rate depends on the pH of the bulk electrolyte.
4. For boron concentrations in excess of $3 \times 10^{19}$ $cm^{-3}$, the silicon becomes degenerate, and the electrons are no longer confined to the surface. This prevents the formation of the hydroxide ions at the surface and thus causes the etching to stop.
5. Anodic biases will prevent the confinement of electrons near the surface as well and lead to etch stop as in the case of a $p^+$ material.

This model applies well for lower index planes (i.e., {nnn} with $n < 2$), where high etch rates always correspond to low activation energies. However, for higher index planes (i.e., {n11} and {1nn} with n = 2, 3, 4), Herr et al.[54] found no correlation between activation energies and etch rates. For higher index planes, we must rely mainly on empirical data.

Several chemical models explaining the anisotropy in etching rates for the different Si orientations have been proposed. In the different etching models, different Si crystal properties have been correlated with the anisotropy in silicon etching:

1. According to Gandhi[181] and Kern,[77] the silicon oxidation reaction steps involve injection of holes into the Si (raising the oxidation state of Si), hydroxylation of the oxidized Si species, complexation of the silicon reaction products, and dissolution of the reaction products in the etchant solution. In this reaction scheme, etching solutions must provide a source of holes and hydroxide ions, and they must contain a complexing agent that solubilizes reacted Si species in the etchant solution, for example, pyrocatechol forming the soluble $Si(C_6H_4O_2)_3^{2-}$ species. This older model still seems to guide the thinking

of many micromachinists, although Gosalvez's atomistic model and Seidel et al.'s energy level-based model of the silicon/electrolyte interface are more powerful and predictive.

2. Because the {111} Si planes present the highest density of atoms/cm$^2$ to the etchant and because the atoms are oriented such that three bonds are below the plane (Figure 4.56), it is possible that these bonds become chemically shielded by surface-bonded OH or oxygen, thereby slowing the etch rate for {111} planes. In this model, it is assumed that the anisotropy is caused by differences in activation energies and backbond geometries on different Si surfaces.[182] Seidel et al.[52,110] support this explanation. They detail a process to explain anisotropy based on the difference in energy levels of backbond-associated surface states for different crystal orientations.
3. It also has been suggested that etch rate correlates with available bond density; thus, the surfaces with the highest bond density etch faster.[107] The available bond densities in Si and other diamond structures follow the sequence 1:0.71:0.58 for the {100}/{110}/{111} surfaces. However, Kendall[48] commented that bond density alone is an unlikely explanation because of the magnitude of etching anisotropy (e.g., a factor of 400) compared with the bond density variations of at most a factor of two.
4. Kendall[48] explains the slow etching of {111} planes based on their faster oxidation during etching; this does not happen on the other faces, he speculates, because of greater distance of the atoms on planes other than (111). Because the {111} planes oxidize faster, these planes may be better protected against etching. The oxidation rate in particular in oxygen follows the sequence {111} > {110} > {100}, and the etch rate often follows the reverse sequence (see also Volume I, Chapter 4 on Si oxidation in an oxygen gas stream). In the most used KOH-$H_2O$ wet etching system, however, the sequence is {110} > {100} > {111}.
5. Besides the number of backbonds of the silicon atom under attack, Gosalvez concludes that the site specificity of anisotropic etching is also determined by the geometry of the transition complex; in other words, the steric hindrance of the OH-termination controls the reaction rate as well. The larger the size and the more rigid the transition complex, the larger one expects the anisotropy to be.
6. Madou (see below) explains why Si etches anisotropically in alkaline etchants and isotropically in acidic etchants based on the pH dependency of the energy levels at the silicon/electrolyte interface.
7. Finally, Elwenspoek et al.[55,56] (see below) propose that it is the degree of atomic smoothness of the various surfaces that is responsible for the anisotropy of the etch rates. Basically, this research group argues for the kinetics of smooth faces [the (111) plane is atomically flat] being controlled by a nucleation barrier absent on rough surfaces. The latter, therefore, would etch faster by orders of magnitude. This model also contributes to the understanding why certain etchants etch isotropically and others etch anisotropically.

#### *Elwenspoek et al.'s Model*

Elwenspoek et al.[55,56] introduced an alternative model for anisotropic etching of Si, a model built on theories derived from crystal growth. According to these authors, the Seidel et al. model does not clearly explain the fast etching of {110} planes. Those planes, having three backbonds like the {111} planes, should etch equally slowly. The activation energy of the anisotropic etch rate depends on the etching system used; for example, etching in KOH is faster than in EDP, even when the pH of the solution is the same. Seidel et al. attribute this dependence to diffusion that plays a greater role in EDP than in KOH solutions. However, Elwenspoek et al. point out that, at least for slow etching, the etch rate should not be diffusion controlled but governed by surface reactions. With surface reactions, diffusion should have a minor effect, analogous to growth at low pressure in an LPCVD reactor (see Chapter 7). Another comment focuses on the lack of understanding why certain etchants etch isotropically and others etch anisotropically.

Elwenspoek et al. note that crystal surfaces are not ideal and highlight the parallels in the process of etching and growing of crystals; slowly growing crystal planes also etch slowly! A key to understanding both processes, growing or dissolution (etching), pertains to the concept of the energy associated with the creation of a critical nucleus on a single-crystalline smooth surface, that is, the free energy associated with the creation of an island (growth) or a cavity (etching). Etching or growing of a material starts at active kink sites on steps. Kink sites are atoms with as many bonds to the crystal as to the liquid. Kinetics depends critically on the number of such kink sites (see Figure 4.59). Remember that in Volume I, Chapter 2 we calculated that at room temperature a certain number of vacancies ($n_{eq}$) must be present in an ideal crystal at any temperature above 0 K (Volume I, Equation 2.59 and Figure 2.35).

A group of vacancies and adatoms can lead to a cavity (etching) or an island (growth), respectively. This aspect has remained neglected in the discussion of etch rates of single crystals until now. The free energy change, $\Delta G$, involved in creating an island or digging a cavity (of circular shape in an isotropic material) of radius $r$ on or in an atomically smooth surface, is given by:

$$\Delta G = -n\Delta\mu + 2\pi r\gamma \tag{4.3}$$

where $n$ is the number of atoms forming the island or the number of atoms removed from the cavity, $\Delta\mu$ is the chemical potential difference between silicon atoms in the solid state and in the solution, and $\gamma$ is the step free energy. The step free energy in Equation 4.3 will be different at different crystallographic surfaces. This can be easily understood from the following example. A perfectly flat {111} surface in the Si diamond lattice has no kink positions (three backbonds, one dangling bond per atom), whereas on the {001} face every atom has two backbonds and two dangling bonds; that is, every position is a kink position. Consequently, creating an adatom-cavity pair on {111} surfaces costs energy: three bonds must be broken and only one is reformed. In the case of {001} faces, the picture is different. Creating an adatom-cavity pair now costs no energy because two bonds must be broken to remove an atom from the {001} face, but the energy is returned by placing the atom back on the surface. In Equation 4.3 n can be further written out as:

$$n = \pi r^2 h\rho \tag{4.4}$$

where $h$ is the height of the step (down for a cavity and up for an island), $r$ is the diameter of the hole or island, and $\rho$ is the density (atoms/cm$^3$) of the solid material. The result is:

$$\Delta G = -\pi r^2 h\rho\Delta\mu + 2\pi r\gamma \tag{4.5}$$

where $\Delta\mu$ is counted positive, and $\gamma$ is positive in any case. In Figure 4.60 we show a plot of $\Delta G$ versus $r$. Equation 4.6 exhibits a maximum at:

$$r^* = \frac{\gamma}{h\rho\Delta\mu} \tag{4.6}$$

At $r^*$, the free energy is:

$$\Delta G^* = \Delta G(r^*) = \frac{\pi\gamma^2}{h\rho\Delta\mu} \tag{4.7}$$

Consequently, an island or an etched cavity of critical size exists on a smooth face. If by chance a cavity

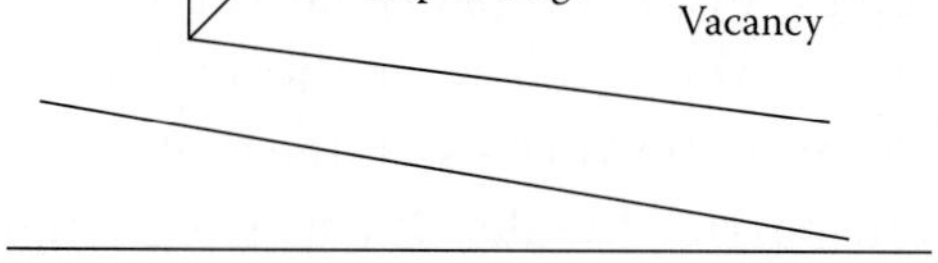

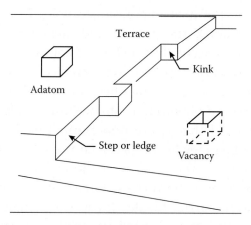


**FIGURE 4.59** Perspective drawing of a crystal showing terraces, step or ledge, kinks, adatoms, and vacancies.

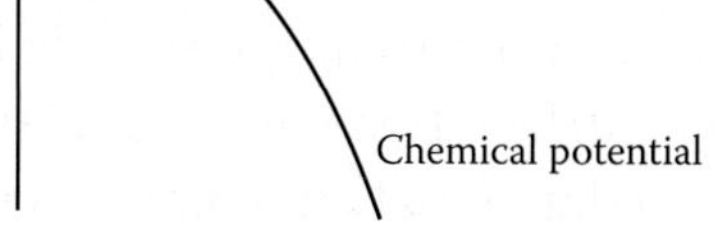

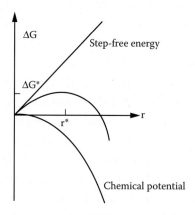


**FIGURE 4.60** A plot of $\Delta G$ versus $r$ based on Equation 4.5 exhibits a maximum.

is dug into a crystal plane smaller than $r^*$, it will be filled rather than allowed to grow, and an island that is too small will dissolve rather than continue to grow because that is the easy way to decrease the free energy. With $r = r^*$, islands or cavities do not have any course of action, but in case of $r > r^*$, the islands or cavities can grow until the whole layer is filled or removed. Importantly, if $r^*$ is equal or smaller than the diameter of a single atom, every atom that leaves the surface will lead to a whole layer filled (a terrace) or removed. In the case of crystal growth, the reverse of etching, this phenomenon is called kinetic roughening. For isotropic etching $\Delta\mu$ is very large; therefore, for all crystal orientations $r^*$ is very small. On the other hand, in light of this nucleation barrier theory, to remove atoms directly from flat crystal faces, such as the {111} Si faces, seems very difficult because the created cavities increase the free energy of the system, and filling of adjacent atoms is more probable than removal; in other words, a nucleation barrier has to be overcome. That is, in anisotropic etching, $\Delta\mu$ is small, or, based on Equation 4.7, a sizable nucleation barrier must be overcome. The growth and etch rates, $R$, of flat faces are proportional to:

$$R \sim e^{-\frac{\Delta G^*}{kT}} \tag{4.8}$$

Because $\Delta G^*$ is proportional to $\gamma^2$ (see Equation 4.7), the activation energy is different for different crystallographic faces, and both the etch rate and the activation energy are anisotropic. If $\Delta G^*/kT$ is large, the etch rate will be very slow, as is the case for large step free energies. Both $\Delta\mu$ and $\gamma$ depend on the temperature and type of etchant, and these parameters might provide clues to understanding the variation of etch rate, degree of anisotropy, temperature dependence, and so on, giving this model perhaps more bandwidth than Seidel et al.'s model. According to Elwenspoek et al., the chemical reaction energy barrier and the transport in the liquid are isotropic, and the most prominent anisotropy effect is caused by the step free energy (absent on rough surfaces) rather than the surface free energy. However, the surface free energy and the step free energies are related: when comparing flat faces, those having a large surface free energy have a small step free energy, and vice versa. The most important difference in these two parameters is that the step free energy is zero for a rough surface, whereas the surface energy remains finite.

Flat faces grow and etch with a rate proportional to $\Delta G^*$, which predicts that faces with a large free energy to form a step will grow and etch much slower than faces with a smaller free energy. Elementary analysis indicates that the only smooth face of the diamond lattice is the (111) plane. There may be other flat faces, but with lower activation energies, as a result of reconstruction and/or adsorption, prominent candidates in this category are {100} and {110} planes. On the other hand, a rough crystal face grows and etches with a rate directly proportional to $\Delta\mu$. The temperature at which $\gamma$ vanishes and a face transitions from smooth to rough is called the *roughening transition temperature* $T_R$.[183,184] Above $T_R$, the crystal is rough on the microscopic scale. Rough crystal faces grow and dissolve with a rate proportional to $\Delta\mu$ and therefore proceed faster than flat surfaces. Imperfect crystals, for example, surfaces with screw dislocations, etch even faster with $R$ proportional to $\Delta\mu^2$.

For the state of a surface slightly above or below the roughening temperature, $T_R$, thermal equilibrium conditions apply. Etching, in most practical cases, is far from equilibrium, and kinetic roughening might occur. Kinetic roughening occurs if the super- or undersaturation of the solution is so large that the thermally created islands or cavities are the size of the critical nucleus.[184] One can show that if the super- or undersaturation is larger than $\Delta\mu_c$, given by:

$$\Delta\mu_c = \frac{\pi f_0 \gamma^2}{kT} \tag{4.9}$$

($f_0$ being the area one atom occupies in a given crystal plane), the growth and etch mechanism changes from a nucleation barrier-controlled mechanism to an isotropic growth/etch mechanism. The growth rate and etch rate become proportional to the chemical potential difference ($\Delta\mu$). Thus, it can be expected that if the undersaturation becomes high enough, even the {111} faces could etch isotropically, as they do in acidic etchants. If the undersaturation becomes so large that $\Delta\mu \ll kT$, the nucleation barrier breaks down. Each single-atom cavity acts as a nucleation site made in vast numbers by thermal fluctuations. The face in question etches with a rate comparable to the etch rate of a rough surface. This situation is called *kinetic roughening*. If all faces are kinetically rough, the etch rate becomes isotropic.

The binding energy $\Delta E$ of an atom in a crystal slice with orientation (*hkl*) divided by *kT* (Boltzmann constant × absolute temperature) is known as the $\alpha$ factor of Jackson of that crystal face,[185] or:

$$\alpha = \frac{\Delta E}{kT} \tag{4.10}$$

At sufficiently low temperature, where entropy effects can be ignored, $kT\alpha$ is proportional to the step free energy $\gamma$, and the number of adatom-cavity pairs is then proportional to $\exp(-\alpha)$. This number is very small on the {111} silicon faces at low temperature, but 1 on the {001} silicon faces at any temperature. The consequence for {111} and {001} planes is that, at sufficiently low temperatures, the first are atomically smooth and the latter are atomically rough.

Isotropic etching requires conditions of kinetic roughening because the etch rate is no longer dominated by a nucleation barrier but by transport processes in solution and the chemical reaction. To test this aspect of the model, Elwenspoek et al. show that there is a transition from isotropic to anisotropic etching if the undersaturation becomes too small. This can occur if one etches with an acidic etchant very long or if one etches through very small holes in a mask (crevice effect). In both cases, anisotropic behavior becomes evident as aging or limited transport of the solution causes the undersaturation to become very small. No proof is available to indicate that acidic etchants are much more undersaturated than the alkaline etchants. Still, the above explains some phenomena that Seidel et al.'s model fails to address. Another nice confirmation of Elwenspoek et al.'s model is in the effect of misalignment on etch rate. A misalignment of the mask close to smooth faces implies steps; there is no need for nucleation to etch. Because the density of steps is proportional to the angle of misalignment, the etch rate should be proportional to the misalignment angle, provided the distance between steps is not too large. Nucleation of new cavities becomes very probable. This has been observed for the etch rate close to the <111> directions.[52]

Where Elwenspoek et al.'s model becomes a bit murky is in the classification of which surfaces are smooth and which ones are rough. As mentioned above for single-crystal Si, only the {111} planes are smooth at low temperatures. At this stage, the model does not explain anything more than other models; every model has an explanation for the slower {111} etch rate. However, as pointed out, these authors invoke the possibility of surface reconstruction and/or adsorption of surface species, which, by decreasing the surface free energy, could make faces such as {001} and {110} flat as well but with lower activation energies. The authors also take heart in the fact that CVD experiments often end up showing flat {110}, {100}, {331}, and, strongest of all, {111} planes. Si etching models are improving fast, but especially where the influence of the etchant is concerned, more convincing thermodynamic data are needed to estimate $\Delta\mu$ and $\gamma$.

### *Madou Model for Isotropic versus Anisotropic Etching of Silicon*

In contrast to alkaline etching, with an acidic etchant such as HF, holes are needed for etching Si. An n-type Si electrode immersed in HF in the dark will not etch because of lack of holes. The same electrode in an alkaline medium etches readily. A p-type electrode in an HF solution, where holes are available under the proper bias, will etch even in the dark. For HF etchants, one might assume that the Ghandi and Kern model,[77,181] relying on the injection of holes, applies. In terms of the band model, this must mean that the silicon/electrolyte interface in acidic solutions exhibits different energetics from the alkaline case. It is not directly obvious why the energetics of the silicon/electrolyte interface would be pH-dependent. To the contrary, because the flat-band potential of most oxide semiconductors and most oxide-covered semiconductors (such as Si) and the Fermi level of the $H_2O/OH^-$ redox couple in an aqueous solution are both expected to change by 59 mV for each pH unit change,[186] one would expect the energetics of the interface to be pH independent. Because the electronegativity in a Si-F bond is higher than for a Si-OH bond, one might even expect the back-bond surface states to be raised higher in an HF medium, making an electron injection mechanism even more likely than in alkaline media. To clarify this contradiction, we analyze the band model of a Si electrode in an acidic medium in more detail. The band model shown in Figure 4.61 was constructed based on a set of impedance measurements on an n-type Si electrode in a set of aqueous solutions at different pH values. Mott-Schottky plots were constructed from the impedance measurements to determine the pH dependency of the flat-band potential. From that, the position of the conduction band $E_{cs}$, as well as valence band edges $E_{vs}$, of the Si electrode in an acidic medium at a fixed pH of 2.2 (the point of zero charge) was calculated at 0.74 eV versus SCE (an often-used reference electrode in electrochemistry) for $E_{cs}$, and −0.36 eV versus SCE for $E_{vs}$.[187,188]

We have assumed in Figure 4.61 that the bands are bent upward at open circuit (see below for justification); thus, holes in the valence band are driven to the interface where they can react with Si atoms or with competing reducing agents from the electrolyte. Because we want to etch Si, we are only interested in the reactions where Si itself is consumed. Reactions of holes with reducing agents are of great importance in photoelectrochemical solar cells.[188] For n-type Si, holes can be 1) injected by oxidants from the solution (e.g., by adding nitric acid to the HF solution); 2) supplied at the electrolyte/semiconductor interface by shining light on a properly biased n-type Si wafer; or 3) created by impact ionization, that is, Zener breakdown, of a sufficiently high reverse biased n-type Si electrode.[189] With a p-type Si wafer, a small forward bias supplies all the holes needed for the oxidation of the lattice even without light because the conduction happens via a hole mechanism. An important finding, explaining the different reaction paths in acidic and alkaline media, comes from plotting the flat-band potential as a function of pH. It was found that the band diagram of Si shifts with less than 59 mV per pH unit. Actually, the average shift is only about 30 mV per pH unit.[188] As shown in Figure 4.62, with increasing pH, the energy levels of the solution increase faster than the energy levels in the semiconductor. As a consequence, it is more likely that electron injection takes place in alkaline media as the filled levels associated with the $OH^-$ are closer to the conduction band, whereas in acidic media the filled levels of the redox system overlap better with the valence band, favoring a hole reaction. A lower position of the redox couple with respect to the conduction band edge, $E_{cs}$, in acidic media explains the upward bending of the bands as drawn in Figure 4.62. With isotropic etching in acidic media, the reaction starts with a hole in the valence band, equivalent to a broken Si-Si bond. In this case, the relative position of backbond-related surface states in different crystal orientations is of no consequence, and all planes etch at the same rate. A study of the interfacial energetics helps to understand why isotropic etching occurs in acidic media and anisotropic etching in alkaline media.

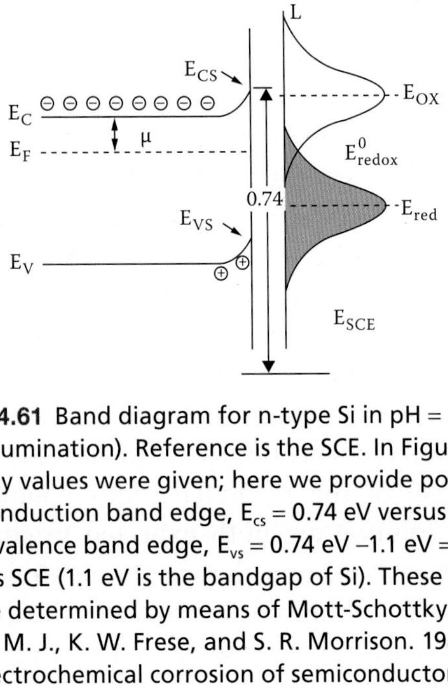


**FIGURE 4.61** Band diagram for n-type Si in pH = 2.2 (no bias or illumination). Reference is the SCE. In Figure 4.40 no energy values were given; here we provide positions of the conduction band edge, $E_{cs}$ = 0.74 eV versus SCE, and the valence band edge, $E_{vs}$ = 0.74 eV −1.1 eV = −0.36 eV versus SCE (1.1 eV is the bandgap of Si). These values were determined by means of Mott-Schottky plots. (Madou, M. J., K. W. Frese, and S. R. Morrison. 1980. Photoelectrochemical corrosion of semiconductors for solar cells. *SPIE* 248:88–95.[187]) The separation between the Fermi energy and the bottom of the conduction band is indicated by μ.

A few words of caution regarding our explanation for the reactivity difference between acidic and alkaline media are in order. Little is known about the width of the bell-shaped curves describing the redox levels in solution.[186] Not knowing the surface concentrations of the reactive redox species involved in the etching reactions further hinders a better understanding of the surface energetics because the bell-shaped curves for oxidant and reductant will only be the same height (as drawn in Figures 4.61 and 4.62) if the concentrations of oxidant and reductant are the same. Clearly, the above picture is oversimplified;

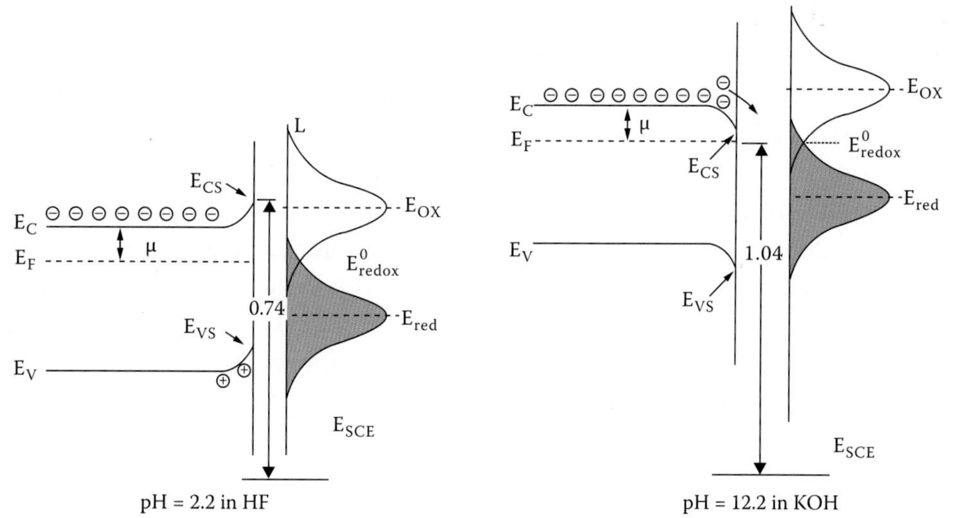


**FIGURE 4.62** Band model comparison of the Si/electrolyte interface at low and high pH. Increasing the pH by 10 units shifts the redox levels up by 600 mV, whereas the Si bands only move up by 300 mV. This leads to a different band-bending and a different reaction mechanism, that is, electron injection in the alkaline media (anisotropic) and hole injection in acidic media (isotropic).

several authors have found that the dissolution of Si in HF might involve both the conduction and valence band, a claim confirmed by photocurrent multiplication experiments.[102,190] These photocurrent multiplication experiments showed that one or two holes generated by light in the Si valence band were sufficient to dislodge one Si unit, meaning that the rest of the charges were injected into the conduction band.[191] Our contention here is only that the low pH dependence of the flat-band potential of a silicon electrode makes a conduction band mechanism more favorable with alkaline-type etchants and a hole mechanism more favorable in acidic media.

Continued attempts at modeling the etch rates of all Si planes are underway. For example, Hesketh et al.[192] attempted to model the etch rates of the different planes developing on silicon spheres in etching experiments with KOH and CsOH by calculating the surface free energy. The density of surface bonds on a Si plane is indicative of the surface free energy, which can be estimated by counting the bond density and multiplying by the bond energy. Using the unit cell dimension *a* of Si of 5.431 Å and a silicon-to-silicon bond energy of 42.2 kcal/mol, the surface free energy, $\Delta G$, can be related to $N_B$, the bond density, by the following expression:

$$\Delta G = \frac{N_B}{2} \times 2.94\ 10^{-19} \frac{J}{m^2} \tag{4.11}$$

Although Hesketh et al. could not explain the etching differences observed between CsOH and KOH (these authors identified a cation effect on the etch rate!), a plot of the calculated surface free energy versus orientation yielded minima for all low index planes such as {100}, {110}, and {111}, as well as for the high index {522} planes. Fewer bonds per unit area on the low index planes produce a lower surface energy and lower etching rate. When Hesketh et al. added the in-plane bond density to the surface bond density, producing a total bond density, a correlation with the hierarchy of etch rates in CsOH and KOH was found, that is, {311}, {522} > {100} > {111}. The surfaces with the higher bond density etched faster, suggesting that the etch rate might be a function of the number of electrons available at the surface. Hesketh et al. imply that their result falls in line with the Seidel et al.'s model, although it is unclear how the total bond density relates to surface state energies of backbonds. Moreover, Hesketh's model does not take into account the angles of the bonds, and in Elwenspoek's view, the surface free energy does not determine the anisotropy.

More research should focus on the modeling of Si etch rates. The semiconductor electrochemistry of corroding Si electrodes will be a major tool in further developments. Interested readers may consult Sundaram et al.[193] on Si etching in hydrazine and Palik et al.[128] on the etch-stop mechanism in heavily doped silicon; both explain in some detail the silicon/electrolyte energetics. A more generic treatise on semiconductor electrochemistry can be found in Morrison.[194] A free etch simulator from the University of Illinois can be found at http://mass.micro.uiuc.edu/research/completed/aces/pages/home.html. For more modeling software of Si etching, consult Volume III, Chapter 6.

## Etching with Bias and/or Illumination of the Semiconductor

### *Introduction*

The isotropic and anisotropic etching of Si we discussed thus far requires neither a bias nor illumination of the semiconductor. The etching in such cases proceeds at open circuit, and the semiconductor is shielded from light. In a cyclic voltammogram, as shown in Figure 4.63, the operational potential in this case is the rest potential, $V_r$, where anodic and cathodic currents are equal in magnitude and opposite in sign, resulting in the absence of flow of current in an external circuit. This does not mean that macroscopic changes do not occur at the electrode surface because the anodic and cathodic currents may be part of different chemical reactions just as in the corrosion of a metal exposed to, for example, seawater. Earlier, in the context of Si etching, we called this type of open circuit etching "electroless" etching (see page 266). As an example of "electroless" etching, consider again the isotropic etching of Si in an $HF/HNO_3$ etchant at open circuit where the local anodic reaction is associated with corrosion of the semiconductor:

$$\text{Local anode: } Si + 2H_2O + nh^+ \rightarrow SiO_2 + 4H^+ + (4\text{-}n)e^- \qquad \text{Reaction 4.19}$$

$$SiO_2 + 6HF \rightarrow H_2SiF_6 + 2H_2O \qquad \text{Reaction 4.20}$$

(the involvement of holes $h^+$ in the acidic reaction was discussed in the preceding sections), whereas the local cathodic reaction could be associated with reduction of $HNO_3$:

$$\text{Local cathode: } NO_3^- + 4H^+ + 3e^- \rightarrow NO + 2H_2O \qquad \text{Reaction 4.21}$$

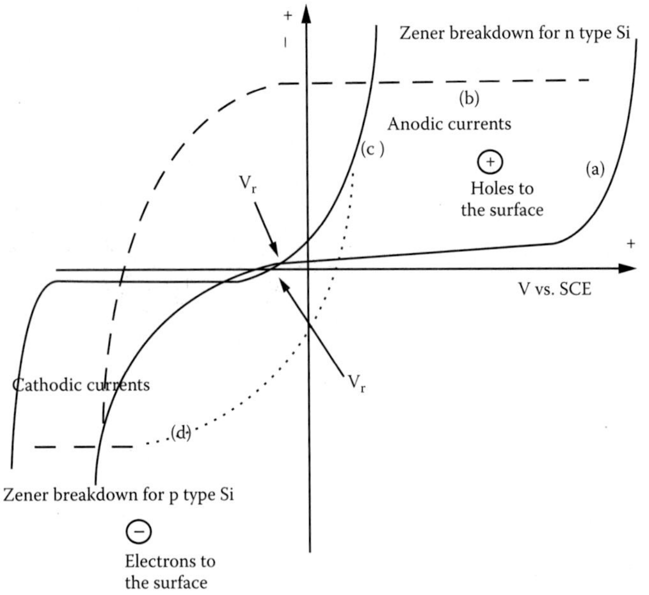


**FIGURE 4.63** Basic cyclic voltammograms (*I* vs. *V*) for n- and p-type Si in an HF solution in the absence of a hole-injecting oxidant: n-type Si without illumination (a); n-type Si under illumination (b); p-type Si without illumination (c); p-type Si under illumination (d). If the reactions on the dark Si electrode determining the rest potential $V_r$ for n- and p-type are the same, then $V_r$ is expected to be the same as well. For clarity of the figure we have chosen $V_r$ differently here; in practice, $V_r$ for n- and p-type Si in HF are found to be identical.

We now explore Si etching while illuminating and/or applying a bias to the silicon sample. To simplify the situation, we first consider the case of an oxidant-free solution so that all the holes must come from within the semiconductor. Anodic dissolution of n-type Si in an HF-containing solution requires a supply of holes to the surface. For an n-type wafer under reverse bias, very few holes will show at the surface unless the high reverse (anodic) bias is sufficient to induce impact ionization or Zener breakdown (see Figure 4.63a). Alternatively, the interface can be illuminated, creating holes in the space charge region (SCR) that the field pushes toward the semiconductor/electrolyte interface (Figure 4.63b). In the forward direction, electrons from the Si conduction band (majority carriers) reduce oxidizing species in the solution (e.g., reduction of protons to hydrogen). A p-type Si sample exhibits high anodic currents even without illumination at small anodic (forward) bias (Figure 4.63c). Here, the current is carried by holes. A p-type electrode illuminated under reverse bias gives rise to a cathodic photocurrent (Figure 4.63d). At relatively low light intensities, the photocurrent plateaus for both n- and p-types (Figure 4.63b and d respectively) depend linearly on light intensity. The photocurrent is cathodic for p-type Si (species are reduced by photoproduced electrons at the surface, e.g., hydrogen formation) and anodic for n-type Si (species are oxidized by photoproduced holes at the surface; either the lattice itself is consumed or reducing compounds in solution are). In Figure 4.64, we show the cyclic voltammograms of n-type and p-type Si in the presence of a hole-injecting oxidant. The most obvious effect is on the dark p-type silicon electrode. The injection of holes in the valence band increases the cathodic dark current dramatically. The current level measured in this manner for varying oxidant concentration or different oxidants could be used to estimate the efficiency of different isotropic Si etchants, a pointer to the fact that semiconductor electrochemistry has been underused as a tool to study Si etching. When n-type Si is consumed under illumination, we experience photocorrosion (see Figure 4.63b). This photocorrosion phenomenon has been a major barrier to the long-term viability of photoelectrochemical cells.[187]

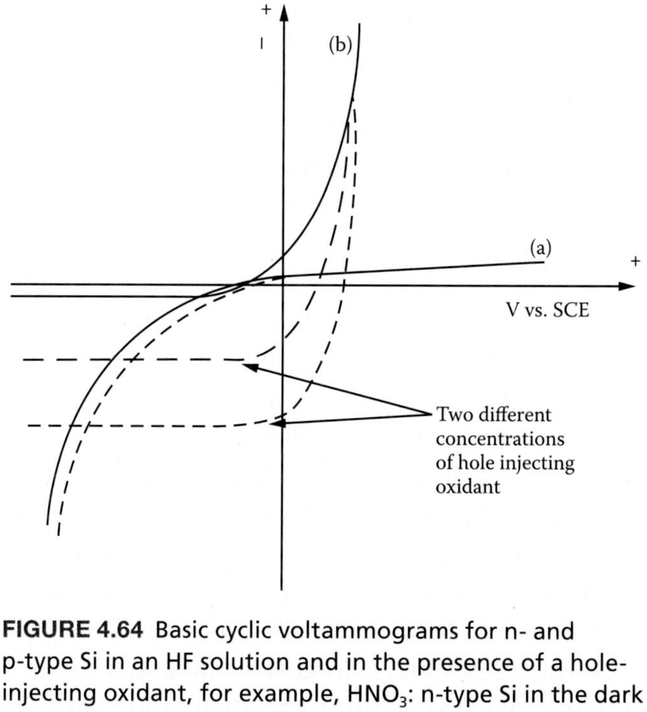


**FIGURE 4.64** Basic cyclic voltammograms for n- and p-type Si in an HF solution and in the presence of a hole-injecting oxidant, for example, $HNO_3$: n-type Si in the dark (a); p-type Si in the dark (b). An increase of the cathodic dark current on the p-type electrodes is most obvious. The current level is proportional to the oxidant concentration.

In what follows, photocorrosion is put to good use for electropolishing and formation of microporous and macroporous layers.[189]

### *Electropolishing and Porous Silicon*

*Electropolishing* In Figure 4.65 we show cyclic voltammograms for p- and n-type Si in an HF solution in the dark and under illumination. The anodic current in the dark for a p-type Si electrode presents two peaks, characterized by $i_{CRIT}$ and $v_p$ (first peak) and $i_{MAX}$ and $v_{MAX}$ (second peak). This was also illustrated, in somewhat less detail, in Figure 4.37, where we only marked the first peak with $i_p$ and $v_p$ ($i_{CRIT}$ was called $i_p$ in that graph) and did not show the oscillations at potentials positive of $v_{MAX}$. As can be deduced from this figure, at high light intensities, in so-called photoelectrochemical etching (PEC), the anodic curves for n-type and p-type Si are similar. The two differences are 1) a potential shift of a few hundred millivolts, and 2) the illuminated n-type Si in the anodic range only shows the first peak and then transitions into a photoplateau. At very high illuminations, a second anodic peak also appears in this case. Because of this equivalence, most of the processes described below apply for both forward-biased p-type and for n-type Si under strong illumination. PEC etching involves photocorrosion in

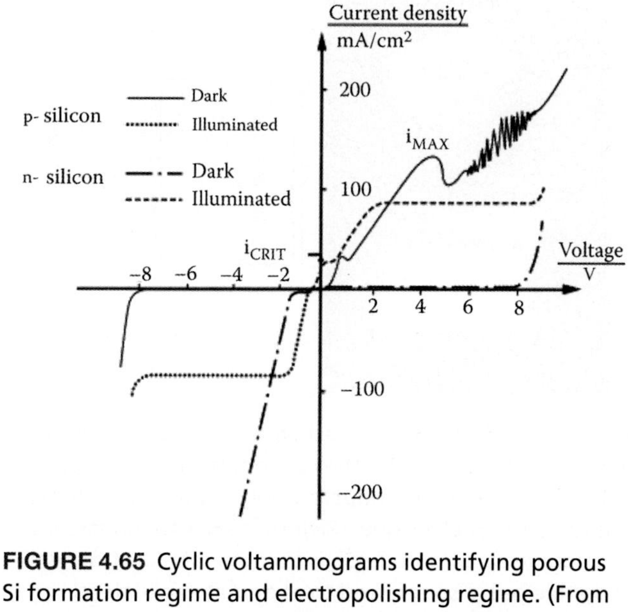


**FIGURE 4.65** Cyclic voltammograms identifying porous Si formation regime and electropolishing regime. (From Levy-Clement, C., A. Lagoubi, R. Tenne, and M. Neumann-Spallart. 1992. Photoelectrochemical etching of silicon. *Electrochim Acta* 37:877–88. With permission.[189])

an electrolyte in which the semiconductor is generally chemically stable in the dark; that is, no hole-injecting oxidants are present, as in the case of an HF solution (see also experiments described in Figure 4.63). For carrying out PEC, a setup as shown in Figure 4.66a may be used with a provision to illuminate the semiconductor electrode. The Al strip shown here is used to make ohmic contact to the back side of the Si electrode. For fabricating porous Si and electropolishing, one often uses a so-called electrochemical double cell, also referred to as double tank lateral cell, as shown in Figure 4.66b. Such a cell has two platinum working electrodes, one in each part of the cell, with the Si wafer in the middle separating the two parts. When current is passed through the wafer, its anodic side will be etched. The main advantages compared with the cell in Figure 4.66a are that no back-side metallization is required and that the uniformity of the process is typically better.

The potential range to the left (more positive) of the first current peak ($v_p$) in Figure 4.65 was clarified earlier as the electropolishing range. Here, the current density exceeds the so-called critical current density, $i_{CRIT}$ (see also Equation 4.1), marked by the first peak in the *I-V* curve. The silicon etches in this domain, leaving behind a smooth (electropolished)

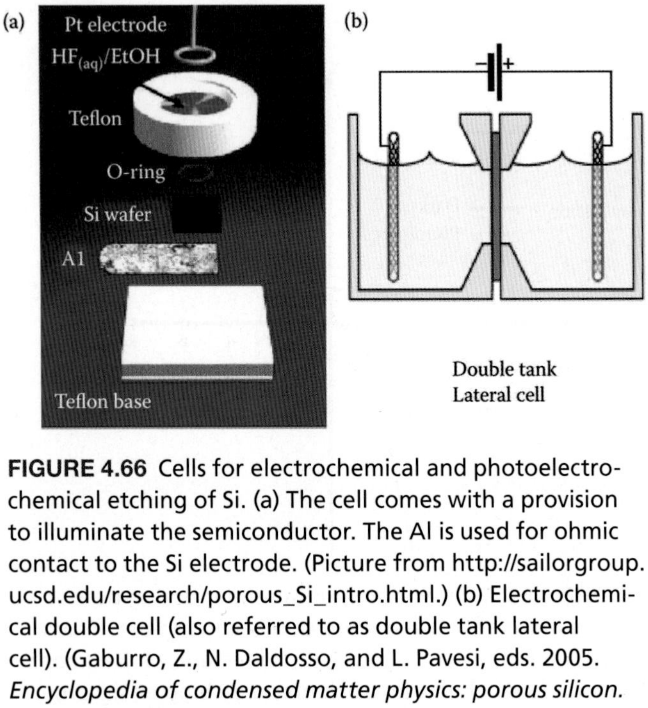


**FIGURE 4.66** Cells for electrochemical and photoelectrochemical etching of Si. (a) The cell comes with a provision to illuminate the semiconductor. The Al is used for ohmic contact to the Si electrode. (Picture from http://sailorgroup.ucsd.edu/research/porous_Si_intro.html.) (b) Electrochemical double cell (also referred to as double tank lateral cell). (Gaburro, Z., N. Daldosso, and L. Pavesi, eds. 2005. *Encyclopedia of condensed matter physics: porous silicon.* Amsterdam: Elsevier Ltd.[195])

surface. It is generally assumed that the electropolishing of silicon takes place in two steps. In the first step an oxide layer forms, and this oxide layer subsequently dissolves in the HF (see Reactions 4.8 and 4.9). Hills on the surface dissolve faster than depressions because the current density is higher on high spots. As a result, the surface becomes smoother, that is, electropolishing takes place.[91] Electropolishing in this regime can be used to smooth silicon surfaces or to thin epitaxially grown silicon layers. The second peak and the oscillations in Figure 4.65 are explained as follows: at current densities exceeding $i_{MAX}$ another type of oxide grows on top of the silicon, leading to a decrease of the anodic photocurrent (explaining the $i_{MAX}$ peak), until a steady state is reached in which dissolution of the oxide by HF through formation of a fluoride complex in solution ($SiF_6^{2-}$) equals again the oxide growth rate. The oscillations sometimes observed in the anodic curve in Figure 4.65 can be explained by a nonlinear correlation between formation and dissolution of the oxide.[189]

Electropolishing is also possible under a layer of porous Si already formed at the porous Si/bulk silicon interface, and the porous layer may be underetched this way. The latter is the enabling technology in the Bosch advanced porous Si membrane process (APSM) (see page 293). The APSM technology makes use of porous Si to form a cavity underneath the surface of monocrystalline silicon and turns the traditional bulk micromachining process into a surface micromachining technology.

### *Porous Silicon*

*Historical Note* Uhlir first discovered porous Si (PS) in 1956 at the Bell Laboratories.[7] He found that, with the appropriate current and solution composition, silicon does not dissolve uniformly, and pores are produced instead. These pores, he found, propagate primarily in the <100> direction of the Si wafer. This did not provide the smooth polish for Si that he sought, and after his curious results were reported in a Bell Labs technical note, the work was pretty much forgotten. Renewed interest in porous Si came about in the 1970s and 1980s: its high surface area was found to be useful as a model in the spectroscopic studies of chemical adsorption/

**FIGURE 4.67** Bright photoluminescence of porous Si at room temperature. (Science and art from Mikael Östling, KTH.)

desorption kinetics on crystalline Si surfaces, and the material also found application as a precursor to generate thick oxide layers on Si and as a dielectric layer in capacitance-based chemical sensors (see above and further below under "Porous Si Applications").

In the late 1980s, Leigh Canham, from the Defense Research Agency in the United Kingdom, argued that the thin Si filaments generated when the pores in porous Si become large and numerous enough might display quantum confinement effects. His intuition turned out to be correct, and by 1990 Canham found that electrochemically etched PS material was fluorescent at room temperature with a bright red-orange color (see Figure 4.67).[196,197] Shortly thereafter, in 1991, Richter demonstrated electroluminescence on porous Si in the 560–480-nm range.[198]

The observation that porous Si features a large photoluminescence (PL) effect in the visible range (at room temperature!) was a very surprising result because the PL efficiency of bulk Si (Si) is very low as a result of its indirect energy bandgap (typical room temperature light emission efficiencies for silicon are much less than 0.001%; in contrast, GaAs LEDs routinely have quantum efficiencies of 1–10%, and approaching 30% for specialized LEDs). As expected from the quantum confinement-induced blue shift, the emitted color is at an energy that is significantly larger than the bandgap energy for bulk Si (which occurs in the infrared region of the spectrum $E_g = 1.1$ eV) (see Figure 4.68 and Volume I, Chapter 3). The underlying mechanism for this blue shift is the result of the quantum-confined structures formed in the porous Si pore walls. High-porosity porous Si material emits green/blue light, whereas low porosity emits red light.

The discovery of efficient visible light emission from porous Si initiated an explosion of work focused on creating Si-based optoelectronic switches, displays, lasers, and so on. However, problems with the material's chemical and mechanical stability and the aging of its optoelectronic properties (see Figure 4.73) led to a waningof the interest in porous Si as an optoelectronic material by the mid-1990s (http://chem-faculty.ucsd.edu/sailor/research/porous_Si_intro.html). By the late 1990s, optoelectronic properties research of porous Si got a new lease on life, however, because of the potential of porous Si for photonic crystals (see below and Volume I, Chapter 5).[197] Porous Si might represent a simple way of making photonic crystals as explained in the section on Macroporus Silicon. Other unique features of porous Si, such as its large surface area, its controllable pore size, its convenient surface chemistry, and its compatibility with conventional silicon microfabrication technologies, also inspired research into applications far outside the optoelectronics field. Additional applications include biocompatible interfaces, biosensing, and drug delivery. An overview of porous

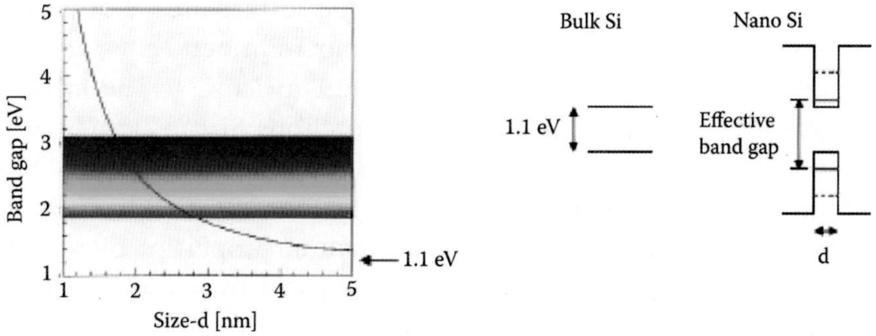


**FIGURE 4.68** Bandgap widening of silicon in porous silicon.

TABLE 4.11 Porous Si Application Fields and Examples[195]

| Application Area | Role of PS | Key Property |
|---|---|---|
| Optoelectronics | LED<br>Waveguide<br>Field emitter<br>Optical memory | Efficient electroluminescence<br>Tunability of refractive index<br>Hot carrier emission<br>Nonlinear properties |
| Microoptics | Fabry-Perot filters<br>Photonic band gap structures<br>All optical switching | Refractive index modulation<br>Regular macropore array |
| Energy conversion | Antireflection coatings<br>Photoelectrochemical cells<br>Solar cells | Low refractive index<br>Photocorrosion cells |
| Environmental monitoring | Gas sensing | Ambient sensitive properties |
| Microelectronics | Microcapacitor<br>Insulator layer<br>Low-k material | High specific surface area<br>High resistance<br>Electrical properties |
| Wafer technology | Buffer layer in heteroepitaxy<br>SOI wafers; see also Eltran process | Variable lattice parameter<br>High etch selectivity |
| Micromachining | Thick sacrificial layer | Highly controllable etching patameters |
| Biotechnology | Tissue bonding<br>Biosensors | Tunable chemical reactivity<br>Enzyme immobilization |

Si applications is listed in Table 4.11. Today, porous Si, together with carbon nanotubes, is often used to illustrate nanotechnology products/applications that are relatively easy to make/implement.

*Porosity in Si* Porosity *P* of a material is the volumetric fraction of air in the material and is defined as:

$$P = \frac{V_{voids}}{V_{material}} = \frac{m_{original} - m_{anodized}}{m_{original} - m_{dissolved}} \tag{4.12}$$

For example, porous Si with $P = 10\%$ means that there is 10% air and 90% silicon in the material. The tunable pore size range in porous Si can be divided into (International Union of Pure and Applied Chemistry definition)[195]:

1. Microporous: $d < 2$ nm
2. Mesoporous: 2 nm $< d <$ 50 nm
3. Macroporous: $d >$ 50 nm

The geometry of porous Si varies from a sponge-like layer consisting of nanoscale branch-shaped fractal-like pores to straight pore arrays. These geometries depend on various conditions, such as the doping level and doping type of the Si substrate, the current density, HF concentration, and crystal orientation. The porous Si porosity can vary between 5% and 85%, and pore size can differ by three orders of magnitude, 20 Å to 10 μm, and the exact underlying formation mechanism, we shall see, is still under debate. The word *nanoporous* is sometimes used for the smallest-pore regime.

Accurate determination of pore size distribution in porous Si is usually performed by the analysis of adsorption isotherms of gases at low temperature in the so-called Barret-Joyner-Halenda, or BJH, method.[199] The physical adsorption of gases by a porous surface is increased relative to a nonporous one because of capillary condensation in the pores. This increase in adsorption starts when the gas pressure is high enough to fill the smallest pores. The specific surface area *A* of porous Si, determined by Brunauer-Emmett-Teller, or BET, ranges from $A_{BET}$ = 1–1000 $m^2/cm^3$.[200] The structure of the pores in nanoporous Si can best be observed by transmission electron microscopy (TEM), and its thickness can be monitored with an infrared (IR) microscope. Structural properties of porous Si have also been investigated by a wide variety of other techniques such as electron microscopy, scanning probe microscopy, x-ray scattering techniques, x-ray absorption techniques, and Raman spectroscopy. For example, Raman measurements were used to demonstrate the crystallinity of the porous Si skeleton and infer information on the size of the Si nanocrystals.

The very reactive porous Si material etches or oxidizes very rapidly. Heat treatment in an oxidizing atmosphere leads to oxidized porous Si (OPS). The oxidation occurs throughout the whole porous volume, and $SiO_2$ layers several micrometers thick can be obtained in times that correspond to the growth of a few hundred nanometers on regular Si surfaces (1100°C in oxygen for 30 minutes is sufficient to make a 4-μm-thick film).[201] Porous Si is low-density and remains single-crystalline, providing a suitable substrate for epitaxial Si film growth. These properties have been used to obtain dielectric isolation (fully isolated by porous oxidized silicon, or FIPOS) and to make silicon-on-insulator (SOI) wafers (see porous Si applications in Table 4.11); in dry etching (see Chapter 3), as prepared, porous Si etches six to seven times faster compared with single-crystal Si.[202] A typical etch rate of porous Si in a reactive ion etching (RIE) station, with $SF_6$ as the gas, is 6.8 μm/min (whereas for nonporous Si it is 1.5 μm/min), and in a high-density plasma station the etch rate is 66 μm/min (Si, 10 μm/min). The dry etching rate of porous Si depends on the porous Si porosity, aging of the layer, and thermal treatment. The etch rate of thermally treated porous Si layers is significantly smaller than that of freshly etched layers (0.33 μm/min in a high-density plasma).[203] By treatment of the porous Si with $NH_3$ at high temperatures, it is possible to make thick $Si_3N_4$ films. Even at 13 μm, these films show little evidence of stress in contrast to stoichiometric LPCVD nitride films.[204]

### *Porous Silicon Fabrication*

*Microporous Silicon* Whether one is in the regime of electropolishing or porous Si formation depends on both the anodic current density and the HF concentration. The surface morphology produced by the Si dissolution process critically depends on whether mass transport or hole supply is the rate-limiting step. Porous Si formation is favored for high HF concentrations and low currents (low anodic bias for p-type and weak light intensities for n-type Si), where the positive charge supply (holes) is limiting, whereas smooth etching is favored for low HF concentration and high currents (higher anodic bias for p-type Si and strong light intensity for n-type Si), where mass transport is limiting.

For current densities less than $i_{CRIT}$ (Figure 4.65), holes are depleted at the surface, and HF accumulates at the electrode/electrolyte interface. As a result, a dense network of fine holes forms.[202,205] Thus, porous Si typically forms when using low current densities in a highly concentrated HF/water or an HF/ethanol solution—in other words, by limiting the oxidation of silicon resulting from hole and $OH^-$ deficiency. In view of the Si etching models introduced earlier, one observes that aqueous HF is less suitable for the porous etching process because the silicon surface is hydrophobic in water. The porous layer can be made more structurally uniform if an ethanolic solution (e.g., 1 part of 48% HF and 1 part ethanol) is used—this increases the wettability of the silicon and allows better surface wetting by the acid. Ethanolic etch solutions also reduce the formation of hydrogen gas bubbles because ethanol acts as a surfactant and prevents bubbles sticking to the silicon surface. The potential range in Figure 4.65 where porous Si forms is at potentials below (toward negative potentials) the first current peak ($v_p$; see also Figure 4.37). During the formation of porous Si, hydrogen gas evolves, and the morphology of the porous Si depends, among other factors, on the doping density of the wafer, orientation, the HF concentration, and the current density (and thus the light intensity for n-type Si and the voltage for p-type Si). As pointed out earlier, porous Si was shown to be monocrystalline,[92] and the suggested reaction mechanism is based on the assumption that fluoride ions close to the surface cause the Si-Si backbonds to be polarized, thus facilitating the breaking of these bonds.[206] The Si backbonds are broken, leaving an H-terminated surface and a $SiH_xF_y$ compound that reacts with water and HF to form $SiF_6^{2-}$ and $H_2$. The overall process, as introduced earlier, is given by Reaction 4.7.

As sketched in Figure 4.69, microporous silicon is easy to obtain: a few minutes of anodization in an HF solution suffices (a typical anodization rate is 0.3 μm/min).

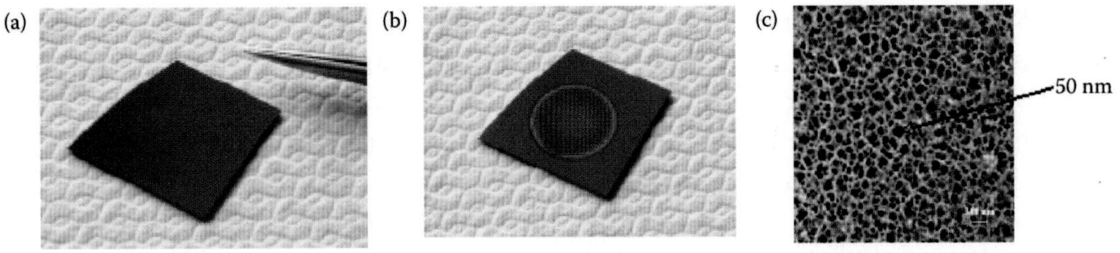


**FIGURE 4.69** (a) Si piece. (b) Porous Si in circular area. (c) SEM of porous Si nanosponge.

Porous Si can also be formed chemically through Reactions 4.19 and 4.21. This is a much less common technique to form porous Si. The purely chemical process is called stain etching, or chemical etching (with no current flow), and is performed in an HF-$HNO_3$ solution. In this case, the difference between chemical polishing and porous Si formation conditions is more subtle. In the chemical polishing case, all reacting surface sites switch constantly from being local anode (Reaction 4.19) to being local cathode (Reaction 4.21), resulting in nonpreferential etching. If surface sites do not switch fast between being local anode and cathode, charges have time to migrate over the surface. In this case, the original local cathode site remains a cathode for a longer time, and the corresponding local anode site, somewhere else on the surface, also remains an anode to keep the overall reaction neutral. A preferential etching results at the localized anode sites, making the surface rough and causing porous Si to form.[207] Any inhomogeneity, for example, some oxide or a kink site at the surface, might increase such preferential etching.[207] Unlike porous Si fabricated by electrochemical means, chemically etched porous Si film thickness is self-limiting.

Depending on porosity and/or pore size, porous Si has the general properties summarized in Table 4.12 (single-crystal Si values are listed as reference). Remember that these properties can be tailored by the process conditions.

The electrical resistivity of porous Si is up to five orders of magnitude higher than that of intrinsic nonporous Si. The reason is that porous Si is depleted of free carriers. This depletion occurs either because 1) energy gap widening caused by quantum confinement, which reduces the thermal generation of free carriers, or 2) trapping of free carriers. Trapping of free carriers takes place during the preparation of the porous Si, and again there are two possible explanations: 1) the binding energy of the dopant impurities is increased, or 2) the formation of more surface states occurs. It has been shown that the dopants are still present in the porous Si in a concentration unchanged from the original Si, but that the dopants are in a neutral state. The electrical transport through porous Si is very much affected by the disordered structure of the porous Si skeleton that restricts the conduction paths to a highly constrained geometry, which for certain types of porosities forms a percolated or fractal geometry. As a consequence, conductivity is thermally activated, strongly frequency-dependent, and highly dispersive. Several models have been proposed to explain the electrical transport properties in porous Si. They differ on the

**Table 4.12** General Properties of Porous Si

| Property | Porous Si Value | Single-Crystal Si Value |
|---|---|---|
| Young's Modules E | 83 GPa (P = 20%)–0.87 GPa (P = 90%) | 160 GPa |
| Thermal conductivity | 1.2 W/mK (nano)–80 W/mK (meso) | 157 W/mK |
| Electrical resistivity | $10^{10}$–$10^{12}$ Ω cm nanoporous | |
| Optical bandgap | 1.4 eV (P = 70%)–2.0 eV (P = 90%) | 1.12 eV |
| Refractive index | 1.2–2.8 | 3.94 |
| Oxidation 1100°C for 30 min in oxygen | 4 μm FASTER | <0.5 μm SLOWER |
| Dry etching ($SF_6$ gas) | 66 μm/min FASTER | 10 μm/min SLOWER |

*Source:* Gaburro, Z., N. Daldosso, and L. Pavesi, eds. 2005. *Encyclopedia of condensed matter physics: porous silicon.* Amsterdam: Elsevier Ltd.[195]

transport paths and mechanisms assumed. The proposed transport paths range from transport in the Si nanocrystals (with diffusion or tunneling between the Si nanocrystals or at their surface) to transport in the amorphous and disordered matrix surrounding the nanocrystals, or through both.

The nanoporous Si structures may have dimensions in the low nanometer range. If the structure size reaches a value less than, e.g., 3 nm, quantum effects can occur; therefore, nanoporous samples can exhibit strong visible photoluminescence and electroluminescence. Raman spectroscopy gives indirect information about the nanostructure of porous Si and has shown that the nanocrystals alter the selection rules relating to the interaction of optical phonons with incident photons (see Volume I, Chapter 5). This results in the efficient photoluminescence and electroluminescence of porous Si compared with nonporous Si. The first and most favored explanation for the visible light emission in porous Si is the quantum confinement of excitons in nanometer-sized Si nanostructures. The nature of the energy gap is still indirect even though a quasidirect bandgap can be formed in ultrasmall Si nanocrystals. The other properties of porous Si listed in Table 4.12 will be highlighted in the text below.

*Macroporous Silicon* In addition to micropores, well-defined macropores ($d > 50$ nm) can also be made in Si by photo and/or bias etching in HF solutions. The macroporous material exhibits very low values of porosity ($P < 10\%$) and high levels of roughness. Macropores have sizes as large as 10 µm, visible with a scanning electron microscope (SEM) rather than the transmission electron microscope (TEM) typically used for nanoporous material. The two types of pores often coexist, with micropores covering the walls of macropores. This is not a matter of a broad fractal-type distribution of pores because the formation mechanism for the large pores is different.

Conditions of lower HF concentration and higher current density tend to offer larger diameter pores. Electrochemical etching of macropores or macroholes has been reported, for example, with n-type silicon in 2.5–5% HF under high voltage (>10 V), low current density (10 $mAcm^{-2}$), under illumination,

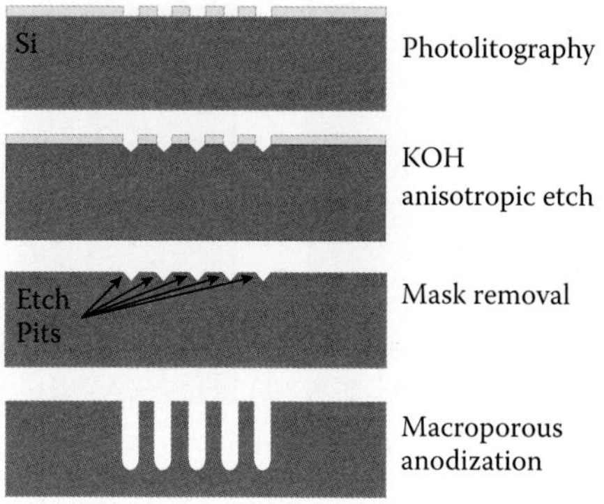


**FIGURE 4.70** Schematic lithographic procedure resulting in the definition of pore-starting etch pits. (Gaburro, Z., N. Daldosso, and L. Pavesi, eds. 2005. *Encyclopedia of condensed matter physics: porous silicon.* Amsterdam: Elsevier Ltd.[195])

and in the dark.[189] In the latter case, Zener breakdown in silicon (electric field strength in excess of $3 \times 10^5$ V/cm) causes the hole formation (see also Volume I, Chapter 4 on Zener diodes and Volume I, Figure 4.43). In the case of lightly doped n-type silicon, the macropores are formed at much higher anodic potentials than those used for micropore formation.[208] More details on models explaining porous Si formation follow below. By using a pore initiation pattern, the macropores can be localized at any desired location. This option is limited to macroporous porous Si because present lithographic resolution lowest limits are typically in the range of 65 nm (Figure 4.70).

The dramatic effect of the use of a pore initiation pattern is illustrated in the comparison of Figures 4.71 and 4.72.[91,209] In Figure 4.71 we show representative porous Si layers without the use of a pore initiation pattern. Figure 4.71a illustrates nanoscale pores (mesoporous Si),[203] and Figure 4.71b shows microscale pores (macroporous Si).[91] The 45° bevel cross-section in Figure 4.71b is of an n-type sample ($10^{15}/cm^3$ phosphorus-doped) and shows a random pattern of macropores. In contrast, the macroscale pores in Figure 4.72 all start at the micropit initiation points. Pores orthogonal to the surface with depths up to a whole wafer thickness can be made, and aspect ratios as large as 250 are possible.

The periodic arrays of macropores shown in Figure 4.72 can form a 2D photonic crystal. Photonic crystals, we recall, are materials with a periodically varying index of refraction, and in porous

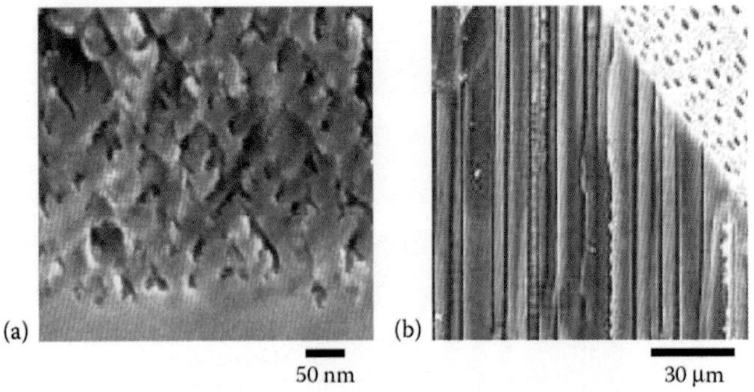


**FIGURE 4.71** Representative porous Si layers. (a) Nanoscale pores (mesoporous Si). (Tserepi, A., et al., 2003. Dry etching of porous silicon in high density plasmas. *Physica Status Solidi (A)* 197:163–67.[203]) (b) Microscale pores (macroporous Si). (Lehmann, V. 1993. The physics of macropore formation in low doped n-type silicon. *J Electrochem Soc* 140:2836–43.[91]) Random: surface, cross-section, and a 45° bevel of an n-type sample ($10^{15}/cm^3$ phosphorus-doped) showing a random pattern of macropores. Pore initiation was enhanced by applying 10-V bias in the first minute of anodization followed by 149 minutes at 3 V. The current density was kept constant at 10 $mA/cm^2$ by adjusting the back-side illumination. A 6% aqueous solution of HF was used as an electrolyte.

Si, air alternates with Si (see Volume I, Chapter 5). Grüning et al.[210] realized such a photonic crystal with a photonic bandgap in the near IR using macropores in n-type silicon etched under back-side illumination.

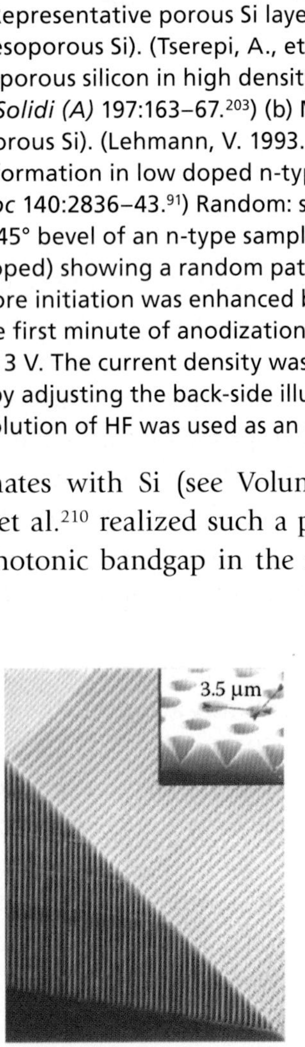

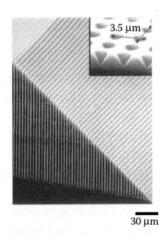


**FIGURE 4.72** SEM image of well-ordered localized or arrayed straight macropores. Inset, SEM image is an artificial micropit array formed in advance of Si electrochemical etching. Localized: surface, cross-section, and a 45° bevel of an n-type sample ($10^{15}/cm^3$ phosphorus-doped) showing a predetermined pattern of macropores (3 V, 350 min, 2.5% HF). Pore growth was induced by a regular pattern of pits produced by standard lithography and subsequent alkaline etching (inset top right). (From Lehmann, V. 1993. The physics of macropore formation in low doped n-type silicon. *J Electrochem Soc* 140:2836–43. With permission.[91])

Summarizing, we have learned that single-crystal silicon can be turned into a porous material with porosity and pore sizes determined by the current density, type, and concentration of the dopant, as well as the hydrofluoric acid concentration. Also, a transition from pore formation to electropolishing is reached by increasing the current density and/or by decreasing the hydrofluoric acid concentration. Electropolishing is even possible under a layer of porous silicon (PS) already formed. The latter is the enabling technology in the Bosch APSM process (see page 293), which makes use of PS to form a cavity underneath the surface of monocrystalline silicon.

In Chapter 7 we learn that PS can also be formed under similar conditions from LPCVD polysilicon, with pores roughly following the grain boundaries of the polysilicon. We will see how, by changing the conditions from pore formation to electropolishing and back to PS, enclosed cavities may be formed with porous polysilicon walls (plugs) and "floor and ceiling" of silicon nitride layers.

*Porous Silicon Drying and Aging* When drying a porous Si film in air, a water meniscus forms in the pores, and this results in stress exerted on the pore walls. Depending on the size of the structure (porosity) and the surface tension of the liquid, cracking of the porous Si layer is always a possibility. This limits the maximum porosity obtainable and also suggests different procedures for drying, for example, by working with very low surface tension liquids in the pores when drying. This may be accomplished by rinsing the freshly made porous Si with pure ethanol that has a lower surface tension than water (22 $mJ/m^2$ compared with 72 $mJ/m^2$); an alternative is to use pentane (with a surface tension of 14 $mJ/m^2$). The best results, however, have been obtained by supercritical drying in $CO_2$ (>95% porosity), where drying is performed above the supercritical point of a liquid, usually $CO_2$.[211] Other methods include freeze drying and using very slow evaporation rates. Pentane drying is the easiest to implement. Pentane has a very low surface tension and shows no chemical interaction with porous Si (unlike ethanol). Using pentane as drying liquid strongly reduces the capillary tension, but because water and pentane are nonmiscible, ethanol or methanol has to be used as intermediate rinsing liquid. Using this drying technique, PS layers with porosity values up to 90% and thickness up to 5 μm exhibit no cracking after drying. Supercritical drying requires a specific apparatus but is more effective in preventing cracking of the film; thus, porous layers of porosity values up to 95% were demonstrated.[211]

In the electrolyte and immediately after drying, the pore walls are mostly H-terminated. The hydrogen termination is replaced by oxygen to form native oxide rapidly in air.[211] This changes the properties of the porous Si over time, making it one of the major problems preventing the implementation of porous Si in practical applications. An example of the nefarious effect of porous Si aging is its effect on photo- and electroluminescence of porous Si. The so-called luminescence fatigue shown in Figure 4.73 is explained by photochemical reactions occurring on the surface of the Si nanocrystals. As the time passes, a shift to the blue region in the peak position and both decreases and increases in the PL intensity have been reported. The blue shift has been attributed to the decrease in dimension of porous Si pillars resulting from the oxidization of the surface with time, which supports the quantum confinement effect.

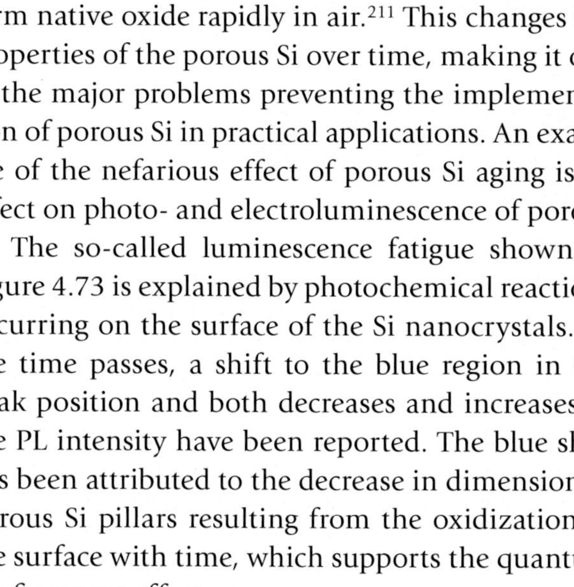

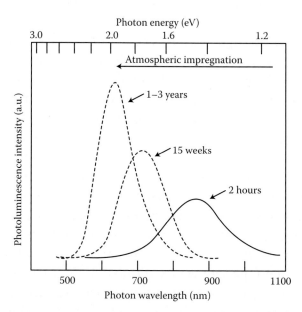


**FIGURE 4.73** Aging of the luminescence of porous Si. (Thönissen, M., and M. G. Burger, eds. 1997. *Properties of porous silicon*. Washington, DC: IEEE.[211])

Because of the large surface area of the porous Si, the silicon/oxygen ratio may be large, resulting in a significant impact on the properties of porous Si. Several methods for surface passivation have been reported in the literature, and they seem to lead to more stable luminescence.[211] To name a few, controlled oxidation by anodic or chemical oxidation, rapid thermal oxidation, capping of the porous Si layer by a dielectric or metal, thermal nitridation, and thermal carbonization.[211]

*Porous Si Models* The critical question about any kind of porous Si formation, nano- or macroporous, is why there is a difference in dissolution rate between pore tip and pore walls? Unagami[205] proposed that insoluble silicates passivate the pore walls, resulting in the formation of a porous Si layer, but this does not explain why such selective passivation takes place in the first case. Parkhutik et al.[212] compared porous Si formation with that of porous alumina formation (see Volume III, Chapter 5). In this model a passive layer (e.g., an oxide) exists at the bottom of each pore and results in the electric field enhancement at the pore tips. Poreformation in this case is a consequence of the dissolution of the passive layer enhanced by the local electrical field, which is stronger at the sharp tips. The competition between growth and dissolution of the passive layer gives rise to a time-evolving electrolyte/passive film interface defining the surface of the pore. This model predicts the correct dependence of pore size on HF concentration. In 1970, Theunissen et al. proposed that porous Si forms in n-type Si as a result of Zener breakdown because of electric field enhancement at the pore tip.[87,163] However, Theunissen's model cannot explain other kinds of porous Si formation without this breakdown condition (e.g., what about porous p-type Si). In 1985, Beale et al.[211] expanded Theunissen's model by introducing the involvement of a space charge region (SCR), which forms at the surface region of the Si in contact with the electrolyte and acts as a Shottky barrier, as illustrated for p-type Si in Figure 4.74. The SCR is depleted of holes and prevents charge transfer at the interface between the electrolyte and the Si surface. When a pore wall becomes sufficiently thin (small d in Figure 4.74), space charge regions from opposite sides of the pore

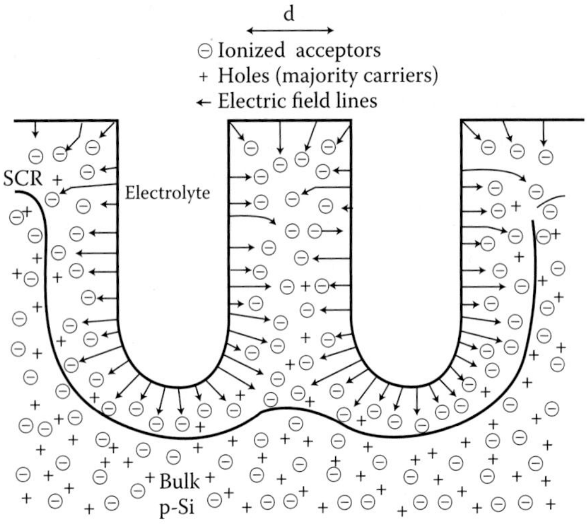


**FIGURE 4.74** Space charge regions (SCRs) at the semiconductor/electrolyte interface form an insulating layer that prevents charge transfer. As two space charge regions in a wall pore wall of thickness *d* overlap that wall will passivate, and current is directed to the bottom of the pore. The example shown here involves p-type Si. (Lehmann, V., et al. 1999. *J Electrochem Soc* 146:2968–75.)

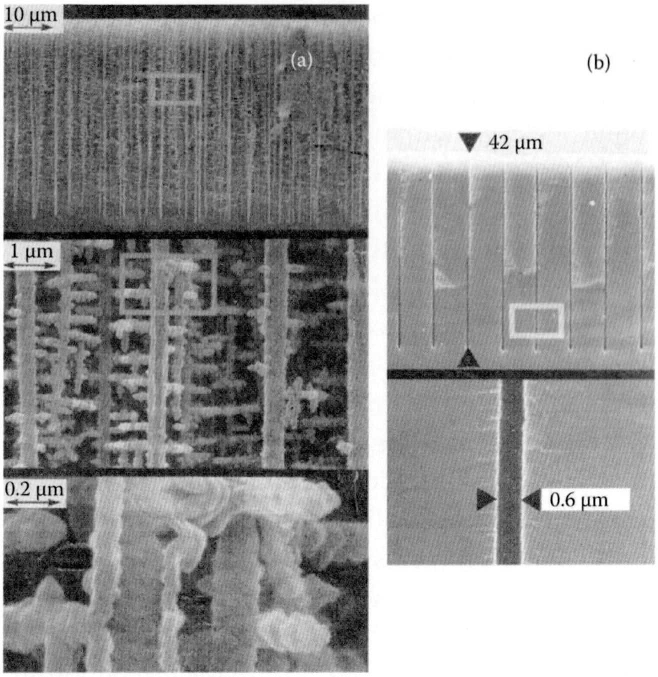


**FIGURE 4.75** Comparison of macropores made with breakdown holes (a) and macropores made with light-created holes (b). (a) An oxide replica of pores etched under weak back-side illumination visualizes the branching of pores produced by generation of charge carriers resulting from electrical breakdown (5 V, 3% HF, room temperature, $10^{15}/cm^3$ phosphorus-doped). (From Lehmann, V. 1993. The physics of macropore formation in low doped n-type silicon. *J Electrochem Soc* 140:2836–43. With permission.[91]) (b) Single pore associated with KOH pit. (From Lehmann, V., and H. Foll. 1990. Formation mechanism and properties of electrochemically etched trenches in n-type silicon. *J Electrochem Soc* 137:653–59. With permission.[213])

wall overlap, thus decreasing charge transfer at the interface even further. Consequently, current is selectively directed to the pore tips' region, and pore walls stop thinning when $d < 1/2$ SCR. Pore formation in this model is a self-adjusting electrochemical mechanism. The limitation is a result of the depletion of holes in the pore walls, causing them to stop etching, whereas holes continue to be collected at the pore tip, where they continue to promote Si dissolution. No passivating layer is involved to protect the pore wall in this model. The only decisive differences between pore tips and pore walls are their geometry and their location. Holes generated by light or Zener breakdown are collected at pore tips. Every depression or pit in the surface initiates pore growth because the electrical field at a curved pore bottom is much larger than that of a flat surface as a result of the effect of the radius of curvature. The latter leads to higher current and enhances local etching.[208] Zener breakdown and illumination of n-type Si lead to different types of pore geometry.[91,163,213] Branched pores with sharp tips form if holes are generated by breakdown (see Figure 4.75a).[91] Unbranched pores with larger tip radii result from holes created by illumination (see Figure 4.75b).[213] The latter difference can be understood as follows: the electric field strength is a function of bias, doping density, and geometry. High doping level density or sharp pore tips will lower the required bias for breakdown; therefore, macropores will tend to follow pores with the sharpest tips. Because every tip causes a new breakdown and hole generation, the position of the original pore tip becomes independent of the other pores, branching of the pores is possible, and fir tree-type fractal-like pores can be observed. With illumination the pore radius may be larger because the breakdown field strength is not necessary to generate charge carriers; thus, the pores remain unbranched.

To obtain the necessary depletion layers in both n- and p-type Si under anodic bias, Beale et al. had to assume Fermi level pinning at midgap (for example, see the energy diagram for n-Si in HF in Figure 4.62).

Yet another model, the diffusion-limited model developed by Smith et al.,[214] describes the formation of porous Si in terms of a stochastic random walk. During pore formation, a hole diffuses to the silicon/electrolyte interface and reacts with a Si surface atom. The nature of the random walk presents the pore tip as the most likely contact site for a hole diffusing from the bulk Si region and provides a similar selective dissolution mechanism of surface irregularities as the Beale model. The rate of pore growth, in this model, is limited by the diffusion of holes to the growing pore tip. A characteristic diffusion length, function of dopant concentration, current density, and so on control the different pore morphologies. Unlike the Beale model, this model applies very generally to all electrochemical phenomena. Smith et al. confirmed that the porous Si geometry simulated by random walk and diffusion-limited model is similar to the geometry of formed porous Si.

In 1990, Lehmann et al.[197] proposed a model for nanoporous Si formation based on the quantum confinement effect. As described above, it is well known that porous Si is luminescent in the visible region. Lehmann et al. expanded the quantum confinement effect to explain porous Si formation. The increase in bandgap for quantum wires and quantum dots in the pore walls decreases the mobile carrier concentration, i.e., produces the barrier to charge transfer.[197] As a result, the current is again limited to the pore tips, resulting in porous layer formation. Obviously this model can only account for nanoporous Si formation.

As for the direction of the pores in porous Si, there will be a natural tendency to follow the direction of the strongest field with the highest supply of holes, i.e., in the <001> direction on a (001) substrate. The <100> and <113> directions have been found to be the preferred growth directions of the macropores in both p- and n-type silicon.[215] The growth direction depends on the surface orientation. When using a {111} surface, three equivalent <113> macropores start to grow into the substrate. The macropore walls consist of (111) facets that can be observed by TEM. These pores can intersect and form a 3D pore network in the silicon, making for a possible 3D photonic crystal. The resulting structure is slightly different from the Yablonovite photonic crystal introduced in Volume I, Chapter 5; the pores are tilted about 29° off the vertical axes, whereas in the Yablonovite they have an angle of about 35° (see Volume I, Chapter 5). Large pore formation for (100) and (111) Si is illustrated in Figure 4.76.

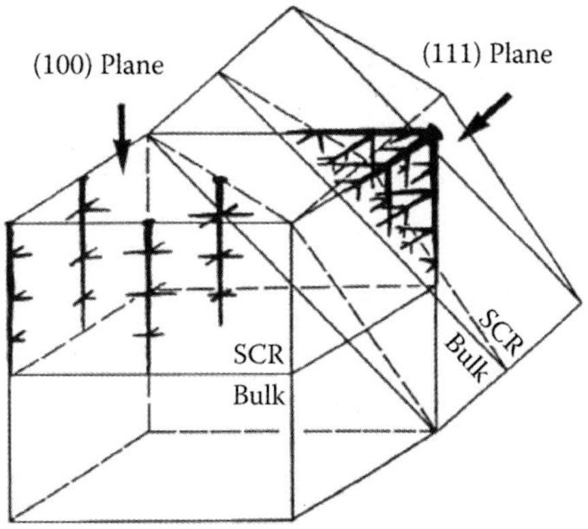


**FIGURE 4.76** Large pore-formation for (100) and (111) Si; SCR, space charge region.

Lehmann has shown that square pore cross-sections and pore arrays of high porosity can be produced by subsequent alkaline etching of macroporous Si.[216] With the proper choice of etchants, the rounded pore walls can be etched to reveal (100) or (110) faces.

It was also determined that the macrohole formation in porous Si depends on the wavelength of the light used. It was found that no pore formation occurs below ~800 nm. Depending on the wavelength, the shape of the pores can be manipulated as well.[213] For wavelengths above 867 nm, the depth profile of the pores changed from conical to cylindrical. The latter was interpreted in terms of the influence of the local minority carrier generation rate. Carriers generated deep in the bulk would promote the pore growth at the tips, whereas near-surface generation would lead to lateral growth.

In 1993, Cahill et al.[217] reported the creation of 1–5-μm-diameter pores with pore spacings (center to center) from 200–1000 μm. Until this finding, the pores typically formed were spaced in the range of 4–30 μm center to center and 0.6–10 μm in diameter. Making highly isolated pores presents another challenge. In the previous work, the relatively close spacing of the pores allowed the authors to conclude that the regions between the pores were almost totally depleted and that practically all carriers were collected by the pore tips. In such a case, neither the pore sidewalls nor the wafer surface etched as all holes was swept to the pore tips. Because the surface was not attacked by pore-forming holes, the quality of the pore initiation mask lost its relevance. In Cahill et al.'s case, on the other hand, a long-lived mask (>20 h) is needed to help prevent pore formation everywhere except at initiation pits. The mask used was a SiC layer sandwiched between two layers of Si nitride. The silicon nitride (insulator) directly atop the Si (semiconductor) serves to insulate the silicon carbide (semiconductor) from the underlying silicon substrate. Because the silicon carbide proves very resistant to HF, loss of thickness does not show during the procedure. The top nitride serves to protect the carbide during anisotropic pit formation. By lowering the bias applied to the Si substrate to less than 2 V with respect to SCE, side branching is avoided.

The Si anisotropy common with alkaline etchants surprisingly shows up here with acidic isotropic etchants. For example, with breakdown-supplied holes, <100>-directed macroholes with <110> branches form (see Figure 4.76),[91] and with <111>-directed wafers, <113> branches form (see Figure 4.76), leading to a complex network beneath the silicon surface. Pyramidal pore tips were also observed when the current density was limited by the bias ($i < i_p$).[91] Isotropic pore pits form when the current is larger than the critical current density, that is, isotropy in HF etching can be changed into anisotropy when the supply of holes is limited. We refer here to the Elwenspoek et al. model (see above), which predicts that in confined spaces etching will tend to be more anisotropic, even when using a normally isotropic etchant such as HF.

The enhanced photoluminescence efficiency of porous Si—compared with Si—has motivated active research toward other porous semiconductor materials. For example, highly porous InP, SiC, GaP, $Si_{1-x}Ge_x$, and Ge structures have all been investigated. Among the first examples of strong anisotropy in porous etching of semiconductors other than Si was that reported by Takizawa et al.,[203] who reported pores in <111> A-oriented InP in HCl with an aspect ratio larger than 100. Some other preliminary evidence was collected on GaAs[218] and GaP.[210] Pore formation in $TiO_2$ anodic and alumina films is described in more detail in Volume III, Chapter 5.

*Porous Silicon Applications* Uhlir's discovery has led to all types of interesting new devices from quantum structures, chemical sensors and biosensors, permeable membranes, and photoluminescent and electroluminescent devices to a basis for making thick $SiO_2$ and $Si_3N_4$ films.[201,204] Some of these applications are listed in Table 4.11 introduced earlier. Besides its use for dielectric isolation and the fabrication of SOI wafers, porous Si has been introduced in a wide variety of other applications: Luggin capillaries for electrochemical reference electrodes, high surface area gas sensors, humidity sensors, sacrificial layers in silicon micromachining, and so on.

We already highlighted the fact that porous Si exhibits photoluminescent and electroluminescent behavior; light-emitting porous Si (LEPOS) was demonstrated. If a LEPOS device could be integrated monolithically with other structures on silicon, a big step in micro-optics, photon data transmission, and processing would be achieved. To explain the blue shift of the absorption edge of LEPOS of about 0.5 eV compared with bulk silicon[197] and room temperature photoluminescence,[196] Searson et al.[219] proposed an energy-level diagram for porous Si where the valence band is lowered with respect to bulk silicon to give a bandgap of about 1.8 eV (see Figure 4.68).

Porous Si can easily be etched in relatively harmless solutions like diluted aqueous KOH solutions, making it an attractive sacrificial material. The surface and bulk properties of porous Si are easily modified; for example, oxidization of the material makes it useful as an insulating material in gas sensors; metallization enables its use in microelectronics; and nitridation of the material is possible as well.

The large specific surface area (e.g., 200 $m^2/cm^3$) and the very small diameters of the pores make porous Si an interesting material for several analytical MEMS and NEMS devices. Given the large specific area per unit volume, the electrical transport is strongly influenced by external factors such as residual electrolyte and ambient atmosphere. The latter property is very interesting for sensor applications. Certain gases—for example, NO—have the capability of modifying the free carrier population. Changes in the electrical conductivity in the presence of subparts per million concentrations of such gases can be detected at room temperature operation. Other gaseous species (for example, polar liquid vapors) also affect electrical transport via electric field interactions with charge carriers.

An optical biosensor based on porous Si also has been demonstrated. PS samples were prepared in such a way that the porous Si films displayed Fabry-Perot fringes in their white-light reflection spectrum. When biological molecules are then chemically attached as recognition elements to the porous Si surface and exposed to the appropriate complementary binding pair, binding occurs and is observed as a shift in the Fabry-Perot fringes.[220,221]

Lammel et al.[222] fabricated tunable optical filters based on porous Si—filters that can be used in reflective or transmission mode. The ratio of voids to total volume of porous Si determines its refractive index, and because porosity can be adjusted by varying the current density, these authors were able to micromachine a multilayer stack of porous Si of different indices by modulating the current density during porosification (in principle, indices between 1.6 and 2.1 can be obtained this way). The process was applied to an area of single-crystal Si delineated by a Si nitride window. The porosified Si plate was released from the substrate by subsurface electropolishing. During chemical release, two suspension arms lift the plate out of the plane automatically by internal stress release. The suspension arms are also provided with Cr/Ni heater wires to actuate the filters. By tilting the freestanding plate of porous Si with the thermal bimorph suspension arms, a wavelength scan is possible. Using this process, Lammel et al. achieved a 20-nm wavelength resolution in the visible part of the spectrum. Figure 4.77 is an SEM of the described porous Si plate optical filter.

Another more recent interest in porous Si stems from the ability to culture mammalian cells directly onto porous Si. Initial research findings from DeMonfort University suggest that the material may be biocompatible (http://www.findarticles.com/m0WVI/ 1999_April_6/54338205/p1/article.jhtml and L.T. Canham[223]). In this vein porous Si can also be used for sieving to filter very small particles. For more on biocompatibility of single-crystal Si, polycrystalline Si, and porous Si, see also Volume III, Chapter 5.

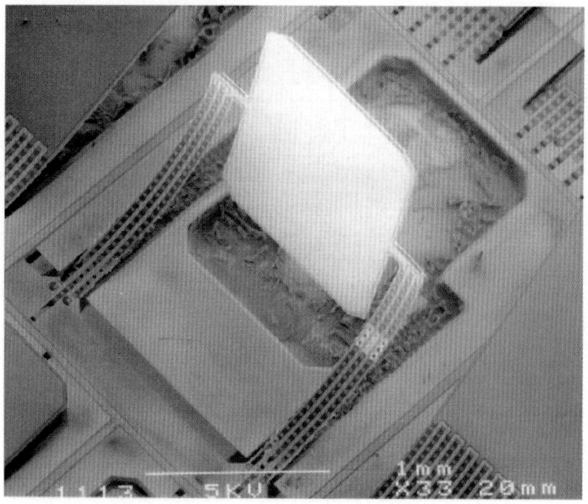


**FIGURE 4.77** Optical filter of porous Si. SEM of a freestanding porous Si microplate containing a multilayered optical interference filter. The wavelength can be tuned by tilting the microplate using the integrated thermal bimorph microactuator. The tilt angle is a function of the actuator DC voltage. This filter element can be used to build a microspectrometer. (Courtesy of Dr. Gerhard Lammel, EPFL, Lausanne, Switzerland.[222])

Porous Si also enabled an elegant new micromachining technology at Bosch with their advanced porous silicon membrane (APSM) process. This technology makes use of porous Si to form a cavity underneath the surface of a single-crystalline silicon membrane, which helps to make the wafer manufacturing process fully CMOS-compatible and turns the traditional bulk micromachining process into a surface micromachining technology. Armbruster et al.[224] reported this new micromachining technology for the fabrication of monocrystalline silicon membranes covering a vacuum cavity for applications such as piezoresistive pressure sensors. In Figure 4.78 we compare a silicon-based absolute pressure sensor made with traditional bulk micromachining with the same device made with the new porous Si-based APSM process. The new process evidently produces not only cavities inside a device in a shorter amount of time, uses no wafer-bonding steps, and constitutes an all-silicon process, but it is also is CMOS-compatible. The main process steps, illustrated in Figure 4.79, are (a) local anodic etching of a layered porous Si (LPCVD $Si_3N_4$ is used as a mask). (b) One of the following processes is used: 1) conversion of part of the porous material to oxidized porous Si and etching it away, 2) direct electropolishing of part of the porous Si, or 3) thermal rearrangement of

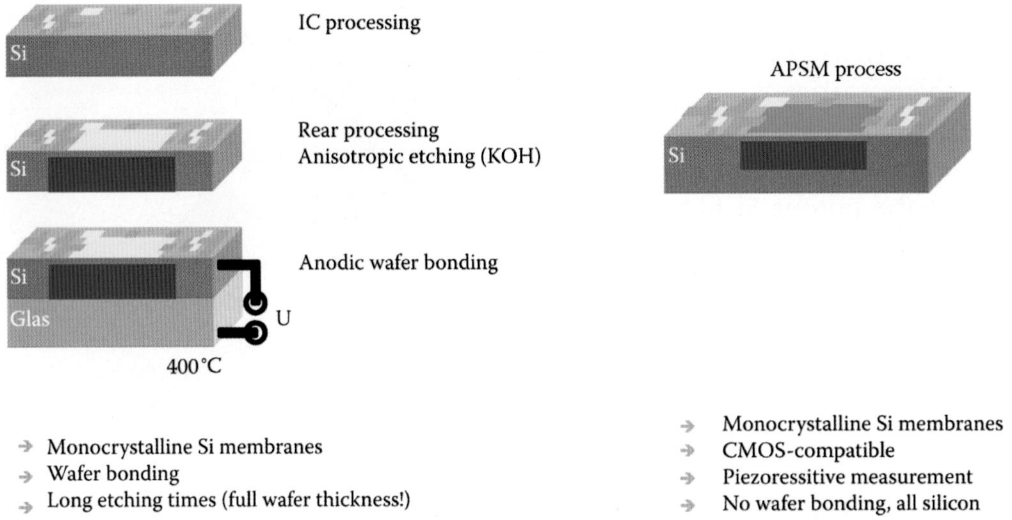


**FIGURE 4.78** Comparison of a bulk Si micromachined absolute pressure sensor (left) with such a sensor made with the new Bosch porous Si-based APSM process (right). (From a Bosch presentation.)

part of the porous layer. These three different techniques all lead to a cavity underneath a suspended porous membrane, and (c) epitaxial growth of a silicon membrane layer over the remaining porous Si membrane. As we saw earlier, porous Si remains single-crystalline, providing a suitable substrate for epitaxial Si growth. The layer thicknesses of the porous layer over the cavity and the cavity layer may be 1–10 μm or more, depending on the application.

The degree of porosity of the porous Si layer in the APSM process deliberately varies in the lateral and vertical direction. The lateral doping of the semiconductor material for the edge region and central region of the porous diaphragm is selected in such a way that mesopores (pores on the order of 5–50 nm) are formed in the edge region and macropores (pores > 50 nm and up to a few micrometers) are formed in the center region. This may be accomplished, for example, with an n-type doping to produce macropores in the center region, whereas the edge region of the diaphragm is provided with $p^+$ doping to obtain mesoporous Si. This makes it possible for the gases produced in etching to escape more easily through regions of the porous layer, which have a comparatively higher porosity, when producing the cavity underneath the porous layer. In the edge region of the diaphragm with the lower porosity, a good layer quality is possible, for example, for the subsequent epitaxial process, whereas in the central region of the diaphragm the epitaxial quality is comparatively lower as a result of higher-porosity sections. However, in the pressure sensor application, this is unimportant because the properties of the pressure sensor are not impaired. The semiconductor substrate is also doped differently in the region of the porous diaphragm layer and in the region of what will become the cavity in such a way that mesopores are produced in the semiconductor material of the diaphragm layer and nanopores (pores from 2–5 nm) are produced as a "precursor structure" in the cavity region. The production of a small-pore precursor structure prevents the formation of comparatively large air bubbles, which makes it possible to adequately remove the gas bubbles through the porous diaphragm layer. In this procedure, use is also made of the fact that the "nanostructure" has a considerably higher internal surface area than the "mesostructure" of the diaphragm region, which is usable in a subsequent process step for a shorter oxidation time. We learned above that the very reactive porous Si etches or oxidizes very rapidly. This makes it possible to subsequently produce a completely oxidized porous Si, which is then selectively removed, e.g., in a vapor etching process. The cavity may also be produced by rearranging the nanoporous layer in a high-temperature process step. Different porosities subjected to heat treatment at high temperatures in a reducing atmosphere induce a structural rearrangement, forming in this way a membrane with a thickness ranging from few tenths of micrometers to 1 μm, as well as the cavity under the membrane.

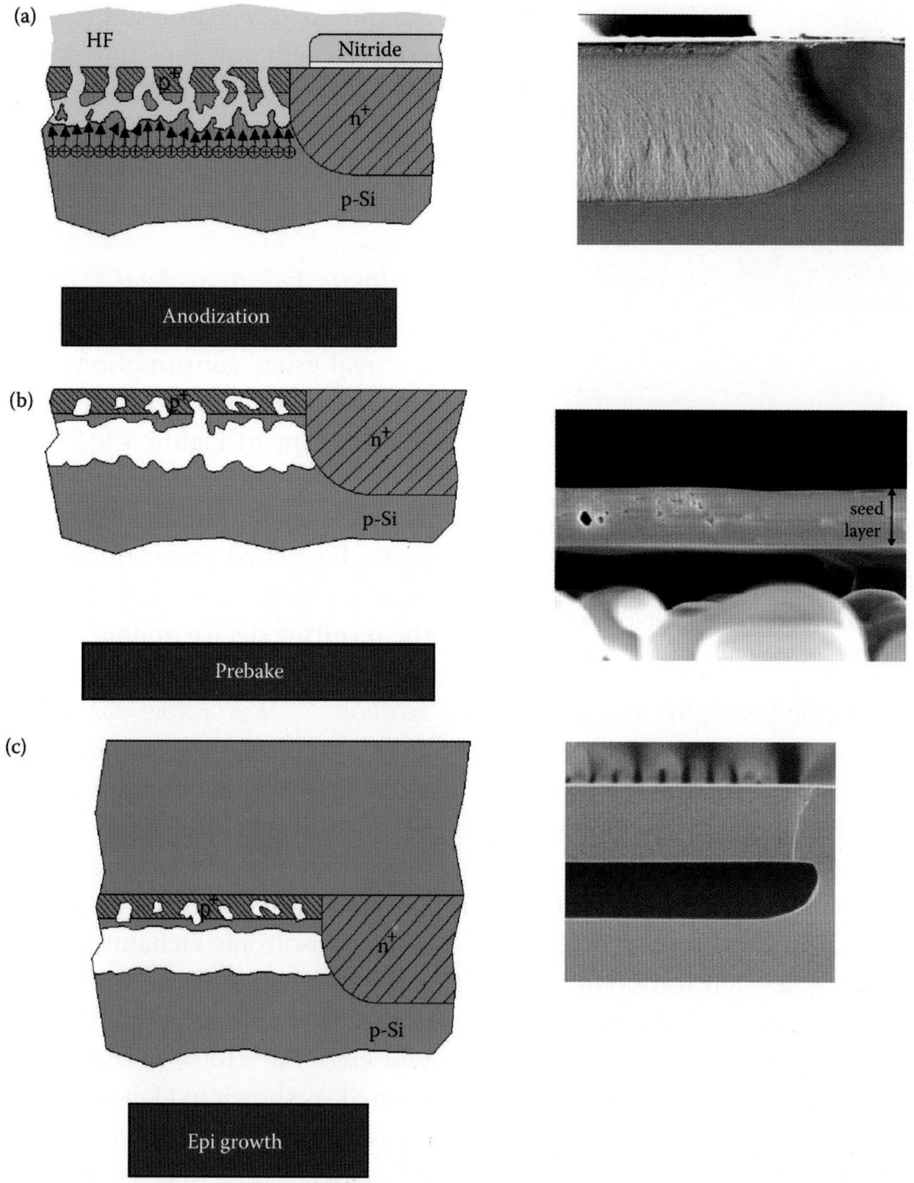


**FIGURE 4.79** The Bosch porous Si-based APSM process. (From a Bosch presentation.) (a) Local anodic etching of layered porous Si with different porosities. (b) Thermal rearrangement of the porous Si. (c) Epitaxial growth of the Si membrane layer. (From a Bosch presentation.)

A third alternative procedure in forming the cavity structure involves directly dissolving the semiconductor material underneath the porous layer using electrolytic polishing, for example, at comparatively higher current densities and lower HF concentrations (see also page 287).

In contrast to conventional bulk micromachining, the new technology has the benefit of more design freedom because the shape/size of the monocrystalline silicon membrane only depends on the porous Si region. Except for the anodic etching, all process steps are part of a standard mixed signal IC production line. We may also compare the APSM process to a surface micromachined membrane pressure sensor. In surface micromachining, we learn in Chapter 7, to make a membrane sensor, a sacrificial layer is normally used, which is applied to the front of a carrier substrate before depositing the sensor diaphragm. The sacrificial layer is later removed from the front of the sensor through "dissolution openings" in the diaphragm, whereby an unsupported structure is obtained. The surface micromachining method is relatively complicated because of the need for separate sacrificial layers. Using APMS,

**FIGURE 4.80** The Bosch SMD 500. The world's smallest digital pressure sensor! With a size of only 5 × 5 × 1.6 mm, the SMD 500 is the smallest pressure sensor for consumer electronics available. Manufactured using the new APSM technology (advanced porous Si membrane), it is much smaller compared with conventional bulk micromachined sensors. (From a Bosch presentation.)

it is possible to considerably simplify the manufacture of such component because no sacrificial layer needs to be additionally applied, and the sacrificial layer and the porous diaphragm are produced from the semiconductor material itself.

A piezoresistive pressure sensor (the SMD 500, the world's smallest digital pressure sensor) with integrated ASIC based on the new fabrication method is shown in Figure 4.80. The sensor that is shown here can measure pressure ranges from ±2.5 kPa (relative) up to 3.3 MPa (absolute). The sensor accuracy is 1.0% over the sensor lifetime, and its high robustness makes it ideal for applications like barometer, altimeter, weather forecast, health care, and fluid level measurements.

A good update on the chemistry of porous Si can be found at Sailor's web site: http://chem-faculty.ucsd.edu/sailor/research/porous_Si_intro.html.

## Problems Associated with Wet Bulk Micromachining of Single-Crystal Silicon

### *Introduction*

Despite the introduction of more controllable etch-stop techniques, bulk micromachining of Si remains a difficult industrial process to control. It is also not an applicable submicrometer technology because wet chemistry is not able to etch reliably on that scale. For submicrometer structure definition, dry etching is required (dry etching is also more environmentally safe). We will now look into some of the other problems associated with Si bulk micromachining, such as the extensive real estate consumption and difficulties in etching at convex corners, and detail current solutions to avoid, control, or alleviate them.

### *Problem: Extensive Real Estate Consumption*

*Introduction* Bulk micromachining involves extensive real estate consumption. This quickly becomes a problem when making arrays of devices. Consider the diagram in Figure 4.81, illustrating two membranes created by etching through a <100> wafer from the back until an etch stop, e.g., a $Si_3N_4$ membrane, is reached. In creating two of these small membranes, a large amount of Si real estate is wasted, and the resulting device might become fragile.

*Solution 1: Real Estate Gain by Etching from the Front* One solution to limit the amount of Si real estate loss is to use thinner wafers, but this solution becomes impractical below 200 μm because such wafers often break during handling. A more elegant solution is to etch from the front rather than from the back. Anisotropic etchants will undercut a masking material by an amount dependent on the orientation of the wafer with respect to the mask. Such an etchant will etch any <100> silicon until a pyramidal pit is formed, as shown in Figure 4.82. These pits have sidewalls with a characteristic 54.7° angle with respect to the surface of a <100> silicon wafer because the delineated planes are {111} planes. This etch property makes it possible to form cantilever structures by etching from the front because the cantilevers will be undercut and eventually will be suspended over a pyramidal pit in the silicon. Once this pyramidal pit is completed, the etch rate of the {111} planes exposed is extremely slow and practically stops. Therefore, process sequences,

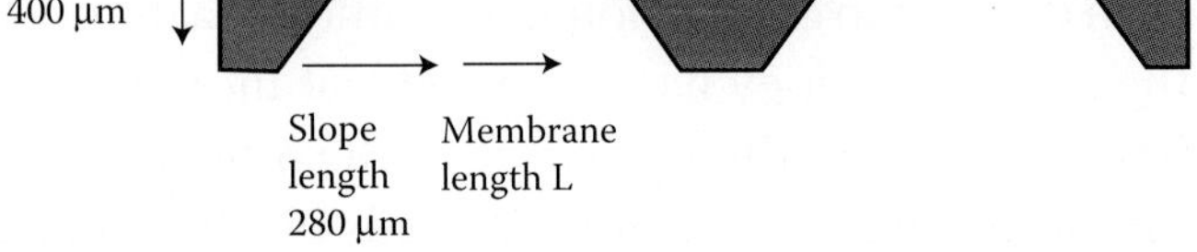

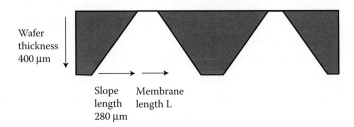


**FIGURE 4.81** Two membranes formed in a <100>-oriented Si wafer.

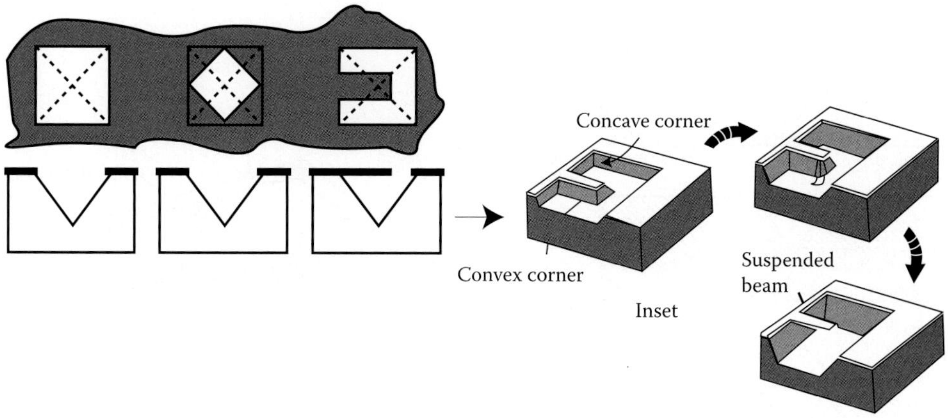


**FIGURE 4.82** Three anisotropically etched pits etched from the front in a <100>-oriented silicon wafer. Inset, illustration of etching at convex corners for the formation of suspended beams.

which depend on achieving this type of a final structure, are very uniform across a wafer and very controllable. The top drawings in Figure 4.82 represent patterned holes in a masking material: a square, a diamond, and a square with a protruding tab. The drawings immediately below represent the etched pit in the silicon produced by anisotropic etching. Note in the first drawing that a square mask aligned with the <110> direction produces a four-sided pyramidal pit. In the second drawing, a similar mask shape oriented at 45° produces an etched pit that is oriented parallel to the pit etched in the first drawing independent of the mask orientation. In the second drawing, the corners of the diamond are undercut by the etchant as it produces the final etch pit. The third drawing illustrates that any protruding member is eventually undercut by the anisotropic etchant, leaving a cantilever structure suspended over an identical etch pit. The inset detailing the third drawing shows how undercutting at the convex corners of the cantilever, in pure KOH, is determined by {411} planes.[225]

*Solution 2: Real Estate Gain by Using Silicon Fusion-Bonded Wafers* Using silicon fusion-bonded (SFB) wafers rather than conventional wafers also makes it possible to fabricate much smaller microsensors and gain real estate. The process, introduced in the 1980s, is clarified in Figure 4.83 for the fabrication of a pressure sensor.[225] The bottom handle wafer has a standard thickness of 525 μm and is anisotropically etched with a square cavity pattern. Next, the etched handle wafer is fusion bonded to a top sensing wafer (the SFB process itself is detailed in Chapter 7, dealing with surface micromachining, and is also touched on in Volume III, Chapter 4, on packaging). The sensor wafer consists of a p-type substrate with an n-type epilayer corresponding to the required thickness of the pressure-transducing membrane. The sensing wafer is thinned all the way to the epilayer by electrochemical etching, and resistors are ion implanted. The handle wafer is ground and polished to the desired thickness. For gauge measurement, the anisotropically etched cavity is truncated by the polishing operation, exposing the back of the diaphragm (Figure 4.83a). For an absolute pressure sensor, the cavity is left enclosed (Figure 4.83b). With the same diaphragm dimensions and the same overall thickness of the chip, an SFB device, because of the inward sloping {111} walls, is almost 50% smaller than a conventional machined device (see Figure 4.84). For example, Lucas NovaSensor (Fremont, CA; now GE Novasensor, http://www.gesensing.com/novasensorproducts) manufactures a sensor that is 400 μm wide, 800 μm long, and 150 μm thick and fits inside the tip of a catheter.

*Solution 3: Real Estate Gain by Using the Bosch Advanced Porous Silicon Membrane Process* As we saw on page 293, the Bosch APSM process makes use of porous Si to form a cavity underneath the surface of a single-crystalline silicon membrane, which turns the traditional bulk micromachining process into a surface micromachining technology (see Figures 4.78–4.80). A piezoresistive pressure sensor made this way, the SMD 500, is the world's smallest digital pressure sensor (Figure 4.80).

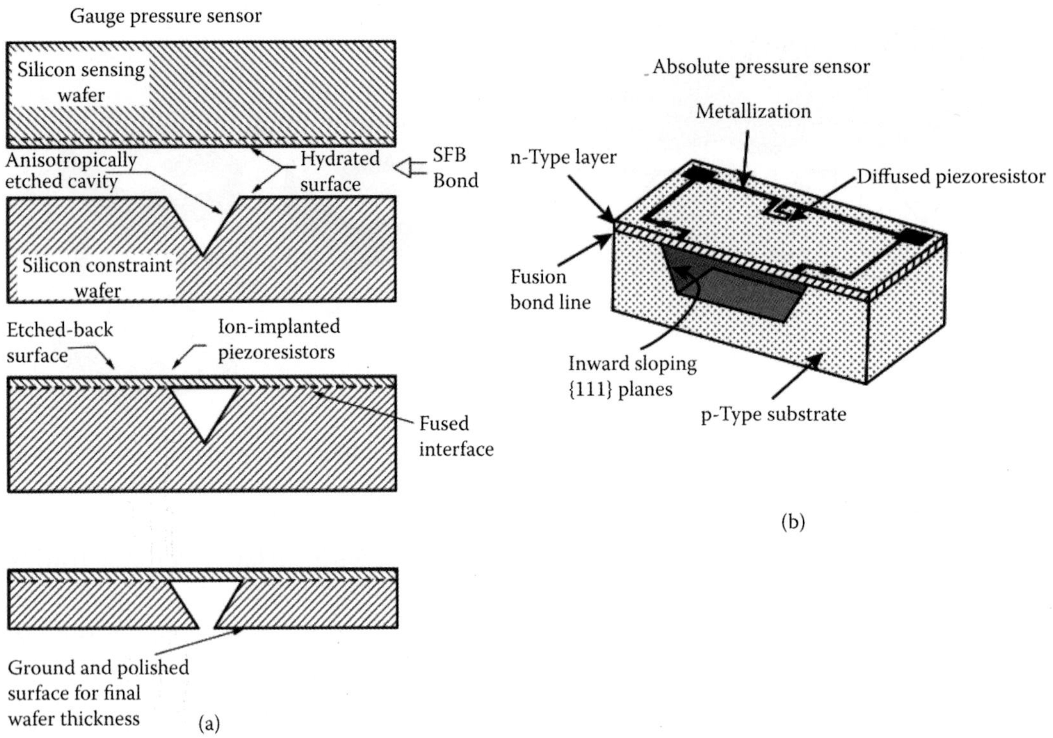


**FIGURE 4.83** Fabrication process of SFB-bonded pressure sensors. (a) For gauge measurement, the anisotropically etched cavity is truncated by the polishing operation, exposing the back of the diaphragm. (b) For an absolute pressure sensor, the cavity is left enclosed. (From Bryzek, J., K. Petersen, J. R. Mallon, L. Christel, and F. Pourahmadi. 1990. *Silicon sensors and microstructures.* Fremont, CA: Novasensor. With permission.[226])

### *Problem: Underetching and Undercutting*

*Mask Underetching* Underetching of a mask that contains no convex corners (that is, corners turning outside in) in principle stems from mask misalignment and/or from a finite etching of the {111} planes. Peeters measured the widening of {111}-walled V-grooves in a (100) Si wafer after etching in 7 M KOH at 80 ± 1° over 24 hours as 9 ± 0.5 μm.[126] The sidewall slopes of the V-groove are a well-defined 54.74°, and the etch rate $R_{111}$ is related to the rate of V-groove widening $R_v$ through:

$$R_{111}\frac{1}{2}\sin(54.74°)\,R_v \text{ or } R_{111} = 0.40R_v \qquad (4.13)$$

with $R_{111}$ the etch rate of the (111) plane in nanometers per minute and $R_v$ the groove widening, also in nanometers per minute. The V-groove widening experiment then results in an $R_{111}$ of 2.55 ± 0.15 nm/min. In practice, this etch rate implies a mask underetching of only 0.9 μm for an etch depth of 360 μm. For a 1-mm-long V-groove and a 1° misalignment angle, a total underetching of 18 μm is theoretically expected, with 95% resulting from misalignment and only 5% caused by etching of the {111} sidewalls.[126] The total underetching will almost always be determined by misalignment rather than by etching of {111} walls.

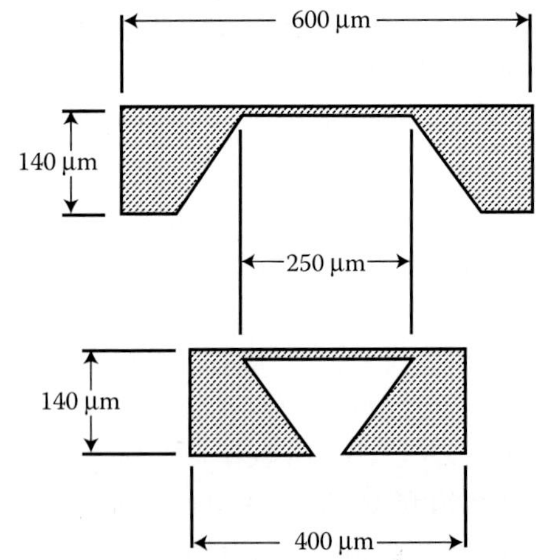


**FIGURE 4.84** Comparison of conventional and SFB processes. The SFB process results in a chip that is at least 50% smaller than the conventional chip. (From Bryzek, J., K. Petersen, J. R. Mallon, L. Christel, and F. Pourahmadi. 1990. *Silicon sensors and microstructures.* Fremont, CA: Novasensor. With permission.[226])

Mask underetching with masks that include convex corners is much larger than the underetching just described because the etchant tends to circumscribe the mask opening with {111} walled cavities. This is usually called *undercutting* rather than underetching.

*Undercutting—Convex Corner Attack* It is advisable to avoid mask layouts with convex corners. However, mesa-type structures often are essential, and in that case there are two possible ways to reduce the undercutting. One is by chemical additives, reducing the undercut at the expense of a reduced ansiotropy ratio, and the other is by a special mask compensating the undercut at the expense of more lost real estate. When etching rectangular convex corners, deformation of the edges occurs because of undercutting. This is often an unwanted effect, especially in the fabrication of, e.g., acceleration sensors, where total symmetry and perfect 90° convex corners on the proof mass are mandatory for good device prediction and specification. The undercutting is a function of etch time and thus directly related to the desired etch depth. An undercut ratio is defined as the ratio of undercut to etch depth ($\delta/H$).

*Solutions: Additives and Corner Compensation Structures* Saturating KOH solutions with isopropyl alcohol (IPA) reduces the convex corner undercutting; unfortunately, this happens at the cost of the anisotropy of the etchant. This additive also often causes the formation of pyramidical or cone-shaped hillocks.[126,141] Peeters claims that these hillocks are the result of carbonate contamination of the etchant, and he advises etching under inert atmosphere and stockpiling all etchant ingredients under an inert nitrogen atmosphere.[126]

Undercutting can also be reduced or even prevented by "corner compensation structures," which are added to the corners on the mask layout to be undercut during etching. Depending on the etching solution, different corner compensation schemes are used, among them square corner compensation (EDP or KOH) and rotated rectangle corner compensation (KOH) as illustrated in Figure 4.85.

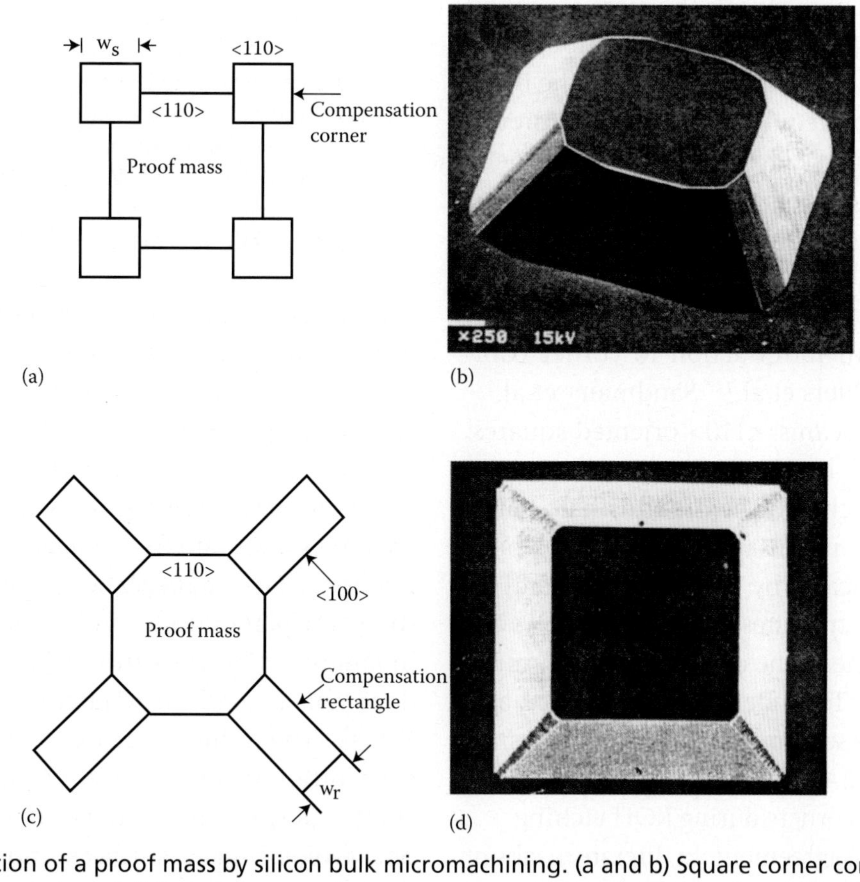


**FIGURE 4.85** Formation of a proof mass by silicon bulk micromachining. (a and b) Square corner compensation method, using EDP as the etchant. (c and d) Rotated rectangle corner compensation method, using KOH as the etchant.

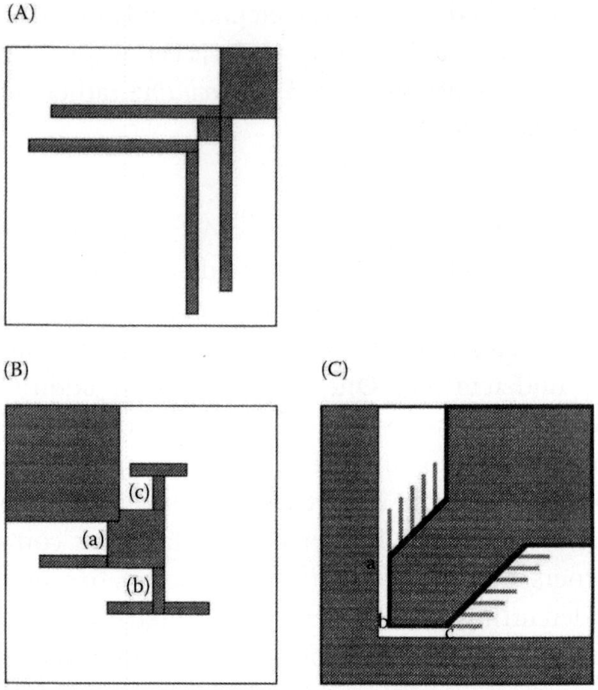


**FIGURE 4.86** Mask layout for various convex corner compensation structures. (A) Splitting of the compensation beam creates concave corners, and by arranging two such double beams, a more symmetrical final structure is achieved. (B) Three convex corners of the compensation square are protected from undercutting by the added <110> beams. The side beam on corner b is made about 30% longer than the other two side beams. In this case, the corner is formed by the etch fronts starting at the corners a and b. (C) <010> bands that are added to convex corners in the <100> direction. (From Sandmaier, H., H. L. Offereins, K. Kuhl, and W. Lang. 1991. *Sixth international conference on solid-state sensors and actuators (Transducers '91).* San Francisco. With permission.[228])

In square corner compensation, the square of $SiO_2$ in the mask, outlining the square proof mass feature for an accelerometer, is enhanced by adding an extra $SiO_2$ square to each corner (Figure 4.85a). Both the proof mass and the compensation squares are aligned with their sides parallel to the <110> direction. In this way, two concave corners are created at the convex corner to be protected. Thus, direct undercutting is prevented. The three "sacrificial" convex corners at the protective square are undercut laterally by the fast etching planes during the etch process. The dimension of the compensation square, $w_s$, depends on the depth of the required cavity; for example, for a 300-μm-deep cavity, a square with a side length of 300 μm is used. The resulting mesa structure after EDP or KOH etching is shown in Figure 4.85b. In rotated rectangle corner compensation, shown in Figure 4.85c, a properly scaled rectangle ($w_r$ should be twice the depth of etching) is added to each of the mask corners. The four sides of the mesa square are still aligned along the <110> direction, but the compensation rectangles are rotated (45°) with their longer sides along the <100> directions. Using KOH as an etchant reveals the mesa shown in Figure 4.85d. A proof mass is frequently dislodged by simultaneously etching from the front and the back. Corner compensation requires a significant amount of real estate in the mask layout around the corners, making the design less compact, and the method is often only applicable for simple geometries.

Different groups using different corner compensation schemes all claim to have optimized spatial requirements. For an introduction to corner compensation, refer to Puers et al.[227] Sandmaier et al.[228] use <110>-oriented beams, <110>-oriented squares, and <010>-oriented bands for corner compensation during KOH etching. They found that spatial requirements for compensation structures could be reduced dramatically by combining several of these compensation structures. Figure 4.86 shows the mask layouts for some of the different compensation schemes they used. To understand the choice and dimensioning of these compensation structures, as well as those in Figure 4.85, we will first look at emerging planes at convex corners during KOH etching.

There is some disagreement in literature about the exact nature of planes that emerge at convex corners as etching progresses. Long et al.[229] in their work assumed {310} to be the dominant etch plane. Mayer et al.[225] found that the undercutting of convex corners in pure KOH etch is determined exclusively by {411} planes. However, the {411} planes of the convex underetching corner, as shown in Figure 4.87, are not entirely laid free; rugged surfaces, where only fractions of the main planes can be detected, overlap the {411} planes under a diagonal line shown as AB in Figure 4.87b. The ratio of {411} to {100} etching does not depend on temperature between 60 and 100°C. The value decreases with increasing KOH concentration from about 1.6 at 15% KOH to 1.3 above 40%, where the curve flattens out.[225] Ideally one avoids rugged surfaces and searches for well-defined planes bounding the convex corner. Figure

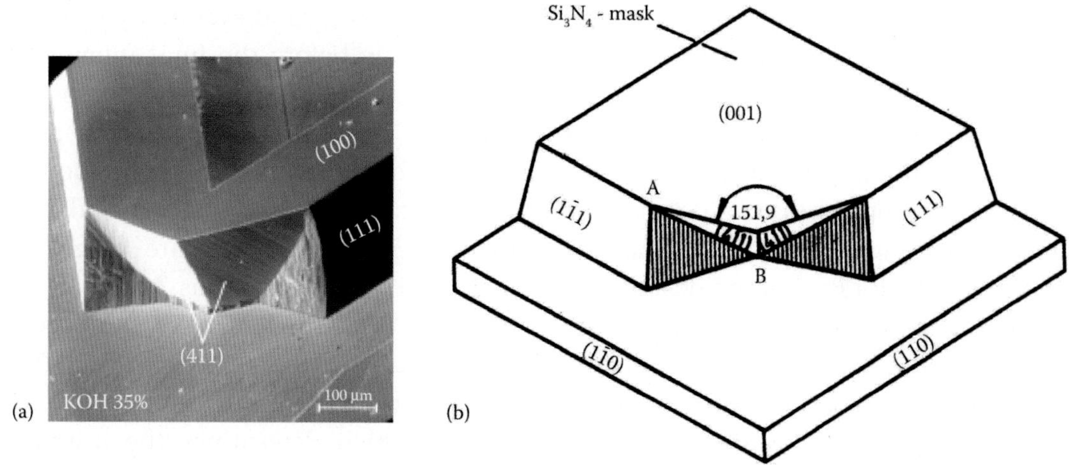


**FIGURE 4.87** Planes occurring at convex corners during KOH etching of Si. (a) SEM picture. (b) Schematic. (From Mayer, G. K., H. L. Offereins, H. Sandmeier, and K. Kuhl. 1990. Fabrication of non-underetched convex corners in ansisotropic etching of (100)-silicon in aqueous KOH with respect to novel micromechanic elements. *J Electrochem Soc* 137:3947–51. With permission.[225])

4.88 shows how a <110> beam is added to the convex corner to be etched. Fast-etching {411} planes, starting at the two convex corners, laterally underetch a <110>-oriented beam (dashed lines in Figure 4.88). The longer this <110>-oriented beam is, the longer the convex corner will be protected from undercutting. It is essential that the beam disappear by the end of the etch to maintain a minimum of rugged surface at the convex edge. On the other hand, as is obvious from Figure 4.88, a complete disappearance of the beam leads to a beveling at the face of the convex corner. The dimensioning of the compensating <110> beam then works as follows: the length of the compensating beam is calculated primarily from the required etch depth ($H$) and the

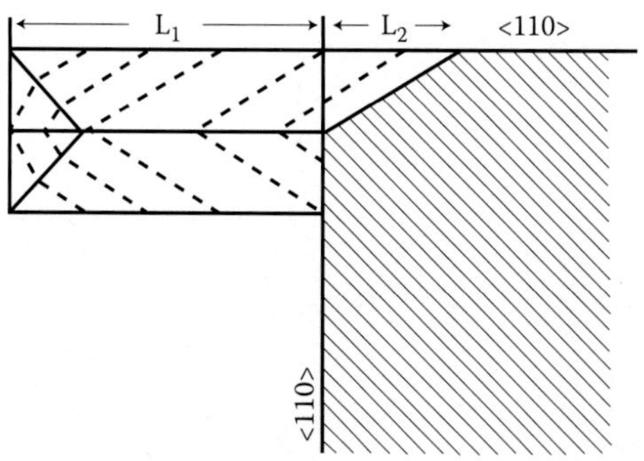


**FIGURE 4.88** Dimensioning of the corner compensation structure with a <110>-oriented beam. (From Sandmaier, H., H. L. Offereins, K. Kuhl, and W. Lang. 1991. *Sixth international conference on solid-state sensors and actuators (Transducers '91)*. San Francisco. With permission.[228])

etch rate ratio $R\{411\}/R\{100\}$ ($\approx\delta/H$) at the concentration of the KOH solution used:

$$L = L_1 - L_2 = 2H\frac{R\{411\}}{R\{100\}} - \frac{B_{<110>}}{2\tan(30.9^\circ)} \qquad (4.14)$$

with $H$ the etch depth, $B_{<110>}$ the width of the <110>-oriented beam, and tan (30.9°) the geometry factor.[225] The factor 2 in the first term of this equation results because the etch rate of the {411} plane is determined normal to the plane and has to be converted to the <110> direction. The second term in Equation 4.15 takes into account that the <110> beam needs to disappear completely by the time the convex corner is reached. How long does etching go in {110} direction when it goes down in {100} direction by a certain depth?

The resulting beveling in Figure 4.88 can be further reduced by altering the etch compensation structures. This is done by decelerating the etch front, which largely determines the corner undercutting. One way to accomplish this is by creating more concave shapes right before the convex corner is reached. In Figure 4.86a, splitting of the compensation beam creates such concave corners, and by arranging two such double beams, a more symmetrical final structure is achieved. By using these split beams, the beveling at the corner is reduced by a factor of 1.4–2 and leads to bevel angles less than 45°.

Corner compensation with <110>-oriented squares (as shown in Figure 4.85a) features considerably higher spatial requirements than the <110>-oriented beams. Because these squares are again undercut by {411} planes that are linked to the rugged surfaces described above, the squares do not easily lead to sharp {111}-defined corners. Dimensioning of the compensation square is done by using Equation 4.15 again, where $L_1$ is half the side length of the square; for $B_{<110>}$, the side length is used. All fast-etching planes have to reach the convex corner at the same time. As before, the spatial requirements of this compensation structure can be reduced if it is combined with <110>-oriented beams. Such a combination is shown in Figure 4.86b. The three convex corners of the compensation square are protected from undercutting by the added <110> beams. During the first etch step, the <110>-oriented side beams are undercut by the etchant. Only after the added beams have been etched does the square itself compensate the convex corner etching. The dimensioning of this combination structure is carried out in two steps. First, the <110>-oriented square is selected with a size that is permitted by the geometry of the device to be etched. From these dimensions, the etch depth corresponding to this size is calculated from Equation 4.15. For the remaining etch depth, the <110>-oriented beams are dimensioned like any other <110>-oriented beam. If the side beam on corner b is made about 30% longer than the other two side beams, the quality of the convex corner can be further improved. In this case, the corner is formed by the etch fronts starting at the corners a and b (Figure 4.86b).

A drawback to all the above-proposed compensation schemes is the impossibility of obtaining a sharp corner in both the top and the bottom of a convex edge as a result of the rugged surfaces associated with {411} planes. Buser et al.[230] introduced a compensation scenario where a convex corner was formed by two {111} planes, which were well defined all the way from the mask to the etch bottom. No rugged, undefined planes show in this case. The mask layout to create such an ideal convex corner uses <010> bands that are added to convex corners in the <100> direction (see Figure 4.86c). These bands will be underetched by vertical {100} planes from both sides. With suitable dimensioning of such a band, a vertically oriented membrane results, thinning and eventually freeing the convex edge shortly before the final intended etch depth is reached. In contrast to compensation structures undercut by {411} planes, posing problems with undefined rugged surfaces (see above), this compensation structure is mainly undercut by {100} planes. For the temperature range of 50–100°C and KOH concentrations ranging from 25–50 wt%, no undefined surfaces could be detected in the case of structures undercut by {100} planes.[228] The width of these <010>-oriented compensation beams, which determines the minimum dimension of the structures to etch, has to be twice the etching depth. These beams can either connect two opposite corners and protect both from undercutting simultaneously or be added to the individual convex corners (open beam). With an open-beam approach (see Figure 4.89), it is extremely important that the {100} planes reach the corner faster than the {411} planes. For that purpose the beams have to be wide enough to avoid complete underetching by {411} planes moving in from the front before they are completely underetched by {100} planes moving in from the side. For example, in a 33% KOH etchant a ratio between beam length and width of at least 1.6 is required. To make these compensation structures smaller while at the same time maintaining {100} undercutting to define the final convex corner, Sandmaier et al.[228] remarked that the shaping {100} planes do not need to be present at

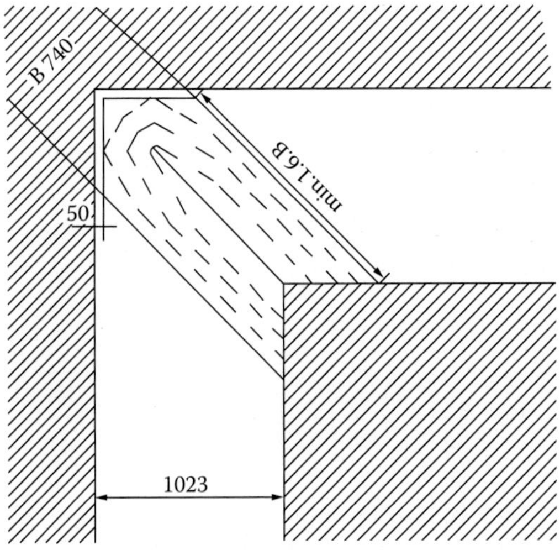


**FIGURE 4.89** Beam structure open on one side. The beam is oriented in the <010> direction. The dimensions are in microns, and *B* is the width of the beam.

the beginning of the etching process. These authors implement delaying techniques by adding fan-like <110>-oriented side beams to a main <100>-oriented beam (see Figure 4.86c). As described above, these narrow beams are underetched by {411} planes and the rugged surfaces they entail until reaching the <100>-oriented beam. Then the {411} planes are decelerated in the concave corners between the side beams by the vertical {100} planes with slower etching characteristics. The length of the <110>-oriented side beams is calculated from:

$$L_{<110>} = \left(H - \frac{B_{<010>}}{2}\right)\frac{R\{411\}}{R\{100\}} \quad (4.15)$$

with $H$ being the etching depth at the deepest position of the device.

The width of the side beams does not influence the calculation of their required length. To avoid the rugged surfaces at the convex corner, the width of the side beams and the spaces between them should be kept as small as possible. For an etching depth of 500 μm, a beam width of 20 μm and a space width of 2 μm are optimal.[228] The biggest drawback of this compensation strategy is that the mask layout is rather complicated, and it takes a lot of time to generate the pattern to fabricate the masks themselves.

Modifications of the compensation scheme that uses <010> bands have been proposed to simplify the mask design. Zhang et al.[231] introduced a modified method that uses a <100> bar with a width greater than twice the etching depth, so that lateral etching would stop at <410> sidewalls, and improved results were obtained. Another novel structure was described by Enoksson[232] and involves a layout that consists of a <010> diagonal band connected to a concave corner on the opposite side of the groove (similar to the situation in Figure 4.89). The band has a slit in the middle of the concave corner, and according to Enoksson, this gives a near-perfect square corner.

Depending on the etchant, different planes are responsible for undercutting. From the above we learned that in pure KOH solutions, undercutting, according to Mayer et al.[225] and Sandmaier et al.,[228] mainly proceeds through {411} planes or {100} planes. That the {411} planes are the fastest undercutting planes was confirmed by Seidel.[233] At the wafer surface, the sectional line of a (411) and a (111) plane point in the <410> direction, forming an angle of 30.96° with the <110> direction, and it was in this direction that he found a maximum in the etch rate. In KOH and EDP etchants, Bean[131] identified the fast undercutting planes as {331} planes. For alkali/alcohol/water, Puers et al.[227] identified the fast underetching planes as {331} planes as well. Mayer et al.,[225] working with pure KOH, could not confirm the occurrence of such planes. Lee[112] indicated that in hydrazine/water, the fastest underetching planes are {211} planes. Abu-Zeid[234] reported that the main beveling planes are {212} planes in ethylenediamine-water solution (no added pyrocatechol). Wu et al.[235] found the main beveling planes at undercut corners to be {212} planes whether using KOH, hydrazine, or EPW solutions. In view of our earlier remarks on the sensitivity of etching rates of higher index rates to a wide variety of parameters (e.g., temperature, concentration, etching size, stirring, cation effect, alcohol addition, and complexing agent), these contradictory results are not too surprising. Along the same line, Wu et al.[236] and Puers et al.[227] have suggested triangles to compensate for underetching, but Mayer et al.[225] found them to lead to rugged surfaces at the convex corner. Combining a chemical etchant with more limited undercutting (IPA in KOH) with Sandmaier's reduced compensation structure schemes could further decrease the required size of the compensation features while retaining an acceptable anisotropy.

Corner compensation for <110>-oriented Si was explored by Ciarlo,[51] who comments that both corner compensation and corner rounding can be minimized by etching from both surfaces so as to minimize the etch time required to achieve the desired features. This requires accurate front-to-back alignment and double-sided polished wafers.

Using corner compensation offers access to completely new applications such as rectangular solids, orbiting V-grooves, truncated pyramids with low cross-sections on the wafer surface, bellow structures for decoupling mechanical stresses between micromechanical devices and their packaging, and so on.[228]

Some additional important references on corner compensation are Ted Hubbard's PhD thesis

(MEMS Design, Caltech 1994)[237] and references by Nikpour et al.[238] and Kim et al.[239]

## Comparing Wet and Dry Etching of Si

For the construction of Si micromechanical devices, etching is a commonly used technique. Silicon can be etched in four principal ways: wet anisotropic, wet isotropic, dry anisotropic, and dry isotropic (see Figure 4.90). For wet anisotropic etching, aqueous KOH solutions are mostly used. Slow-etching (111) planes determine the shape of the etched structure. For dry etching, plasma etching machines are common tools. By varying the conditions, the shape of the channels can be varied from perfectly isotropic to exactly vertical. Positive or negative tapering is also possible.

In bulk micromachining, which requires more extreme topologies (more *z*-axis) than classical IC devices, wet etching of crystalline Si used to dominate the state of the art. The development of the newest fast etching, highly anisotropic DRIE equipment is changing this picture rapidly (see Chapter 3). Wet anisotropic etching of Si results in atomically smooth planes and atomically sharp edges, properties hard to achieve with dry etching. Wet etching also offers the advantage of low-cost batch fabrication—usually 20–25 wafers at a time. Some dry etching advantages include the elimination of handling of dangerous acids and solvents and the use of small amounts of chemicals that can provide both isotropic or anisotropic etch profiles, i.e., directional etching without relying on the crystal orientation of Si. Disadvantages of RIE etching are the fact that expensive equipment must be used, and the etching recipes are machine-dependent and also depend on the etching history of the machine. The walls of the etched structures are also rough compared with wet (electro)chemically etched structures.

Dry etching already is predominantly used in IC manufacture, and with higher-aspect-ratio devices and better materials selectivity being attained, the same is happening in micromachining. Besides selectivity and aspect ratio, problems with and concerns about the slow dry etch rate, mask erosion, and high sensitivity to operating parameters are being addressed. Etch rates as high as 6 μm/min have become possible, and the operation is much more automated. However, the equipment is still very specialized and expensive and requires very clean lines to bring the various gases into the vacuum chamber.

Environmental issues dictated a switch from wet to dry etching, but a recent concern in this context is the search for alternative chemistries because chlorofluorocarbon (CFC) production was banned. According to the Montreal Protocol, all ozone-depleting CFCs had to be eliminated by the year 2000. This included etchants such as $CFCl_3$, $CCl_2F_2$, $CF_3Cl$, $C_2F_5Cl$, and $CF_3Br$.[240] However, the most decisive reason for a switch to dry etching was the need for better CD control.

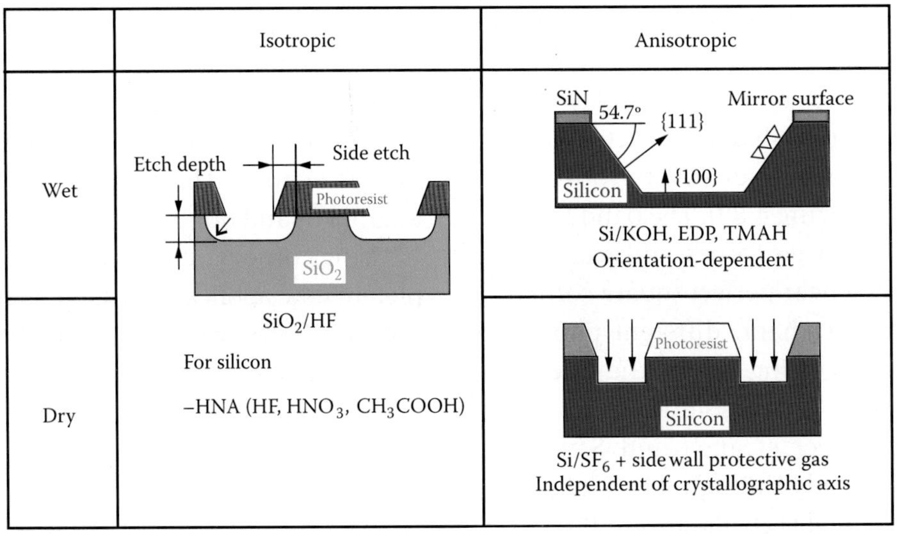


**FIGURE 4.90** Different means for etching Si.

TABLE 4.13 Comparison of Wet versus Dry Etching Techniques

| Parameter | Dry Etching | Wet Etching |
|---|---|---|
| Directionality | Can be highly directional with most materials (aspect ratio of 25 and above) | Only directional with single crystal materials (aspect ratio of 100 and above) |
| Production-line automation | Good ease of automation (e.g., cassette loading) | Poor |
| Environmental impact | Low but some gases are toxic and corrosive | High |
| Masking film adherence | Not as critical | Very critical |
| Cost chemicals | Low | High |
| Selectivity | Poor | Can be very good |
| Materials that can be etched | Only certain materials can be etched (e.g., not Fe, Ni, Co) | All |
| Radiation damage | Can be severe | None |
| Process scale-up | Difficult | Easy |
| Cleanliness | Good under the right operational conditions. Redeposition of nonvolatile compounds might occur. | Good to very good |
| CD control | Very good (<0.1 μm) Faithfully transfers lithographically defined photoresist patterns into underlying layers. High resolution. Less undercutting | Poor |
| Equipment cost | Need for specialized (expensive) equipment | Inexpensive |
| Submicron features | Applicable | Not applicable |
| Typical etch rate | Slow (0.1 μm/min) | Fast (1 μm/min, anis.) |
| Theory | Very complex, not well understood | Better understood (Volume I, Chapter 4) |
| Operating parameters | Many | Few |
| Control of etch rate | Good because of slow etching | Difficult |

Surface micromachining (see Chapter 7) relies on both wet and dry etchants, with processes similar to the ones used in the IC industry. For example, wet etchants are used to etch away sacrificial layers (e.g., HF to etch CVD phosphosilicate glass), whereas dry etching helps to define polysilicon structural elements (e.g., in an $NF_3$-$O_2$ plasma). A combination of wet and dry techniques presently makes up the majority of Si micromachining work. Table 4.13 compares the characteristics of dry versus wet etching techniques.

**Example 4.1: Dissolved Wafer Process**

Figure 4.91 illustrates the dissolved wafer.[241] This process, used by Draper Laboratory for the fabrication of low-cost inertial sensors, involves a sandwich of a silicon sensor anodically bonded to a glass substrate.[242,243] The preparation of the silicon part requires only two masks and three processing steps. A recess is KOH-etched into a p-type (100) silicon wafer (step 1 with mask 1), followed by a high-temperature boron diffusion (step 2 with no mask). In step 1, reactive ion etching (RIE) may be used as well. Cho claims that by maintaining a high-temperature uniformity in the KOH etching bath (±0.1°C), the accuracy and absolute variation of the etch across the wafer, wafer-to-wafer and lot-to-lot, can be maintained to <0.1 μm using premixed, 45 wt% KOH.[241] Cho also uses low-defect oxidation techniques (e.g., nitrogen annealing and dry oxidation) to form

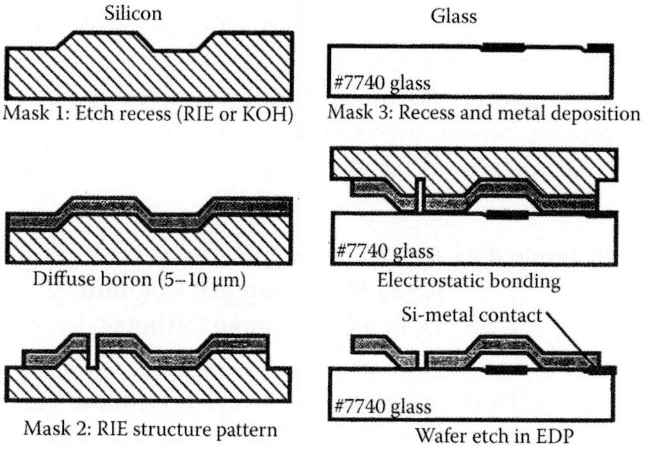


**FIGURE 4.91** Dissolved wafer process. (From Greiff, P. 1995. *Micromachining and microfabrication process technology. Proceedings of the SPIE*. Austin, TX: SPIE. With permission.[242])

defect-free silicon surfaces. In the boron diffusion, the key is optimizing the oxygen content. In general, the optimal flow of oxygen is on the order of 3–5% of the nitrogen flow, in which case the doping uniformity is on the order of ±0.2 μm. Varying the KOH etch depth and the shallow boron diffusion time, a wide variety of operating ranges and sensitivities for sensors can be obtained. Next, the silicon is patterned for a RIE etch (step 3 with mask 2). Aspect ratios greater than 10 are accessible. Using some of the newest dry etching techniques, depths in excess of 500 μm at rates greater than 4 μm/min (with an $SF_6$ chemistry) are possible.[244] The glass substrate (#7740 Corning glass) preparation involves etching a recess, depositing, and, in a one-mask step, patterning a multimetal system of Ti/Pt/Au. The electrostatic bonding of glass to silicon takes place at 335°C with a potential of 1000 V applied between the two parts (electrostatic or anodic bonding is explained in detail in Volume III, Chapter 4). Commercial bonders have alignment accuracies on the order of <1 μm. The lightly doped silicon is dissolved in an EDP solution at 95°C. The keys to uniform EDP etching are temperature uniformity and suppression of etchant depletion through chemical aging or restricted flow (e.g., through bubbles). These effects can be minimized by techniques that optimize temperature control and reduce bubbling (e.g., proper wafer spacing, lower temperature, large bath). The structures are finally rinsed in deionized water and a hot methanol bath.

Draper Laboratory, although obtaining excellent device results with the dissolved wafer process, is now exploring an SOI process as an alternative. The latter yields an all-silicon device while preserving many of the dissolved wafer process advantages (see also Chapter 7 under "SOI Surface Micromachining").

**Example 4.2: An Electrochemical Sensor Array Measuring pH, $CO_2$, and $O_2$ in a Dual Lumen Catheter**

An electrochemical sensor array developed by the author is shown in Figure 4.92, packaged and ready for in vivo monitoring of blood pH, $CO_2$, and $O_2$. The linear electrochemical array fits inside a 20-gauge catheter (750 μm in diameter) without taking up so much space as to distort the pressure signal monitored by a pressure sensor outside the catheter. A classical (macro) reference electrode, making ionic contact with the blood through the saline drip, was used for the pH signal, whereas the $CO_2$ and $O_2$ had their own internal reference electrodes. The high impedance of the small electrochemical probes makes a close integration of the electronics mandatory; otherwise, the high impedance connector leads, in a typical hospital setting, act as antennas for the surrounding electronic noise. As can be seen from the computer-aided design (CAD) picture in Figure 4.93, the thickness of the sensor comes from two silicon pieces: the top piece containing electrochemical cells and the bottom piece containing the active electronics. Each wafer is 250 μm thick. The individual electrochemical cells are etched anisotropically into the top

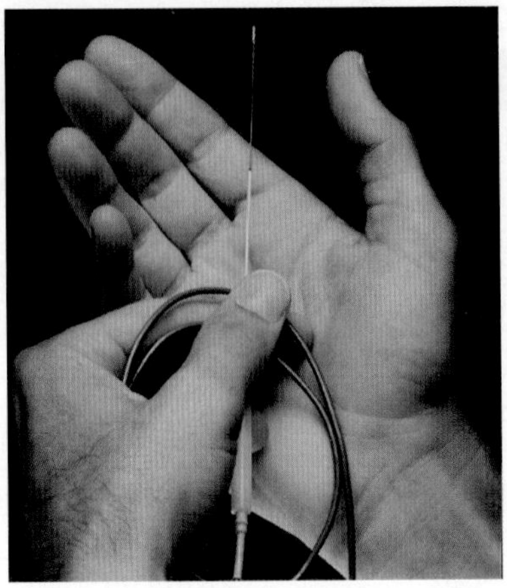

**FIGURE 4.92** An in vivo pH, $CO_2$, and $O_2$ sensor based on a linear array of electrochemical cells.

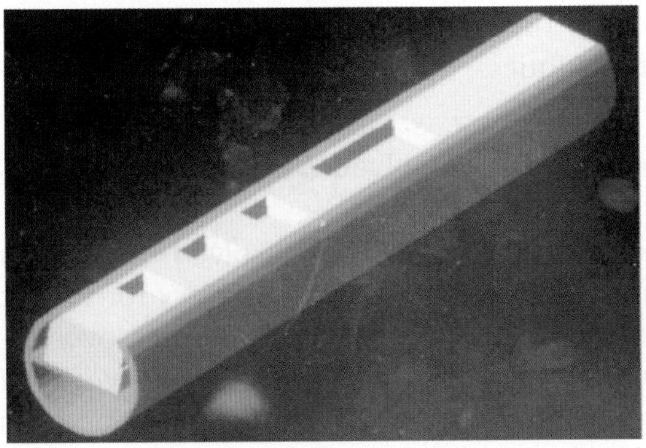

**FIGURE 4.93** CAD of the electrochemical sensor array showing two pieces of silicon (each 250-μm thick) on top of each other, mounted in a dual-lumen catheter. The bottom part of the catheter is left open so pressure can be monitored and blood samples can be taken.

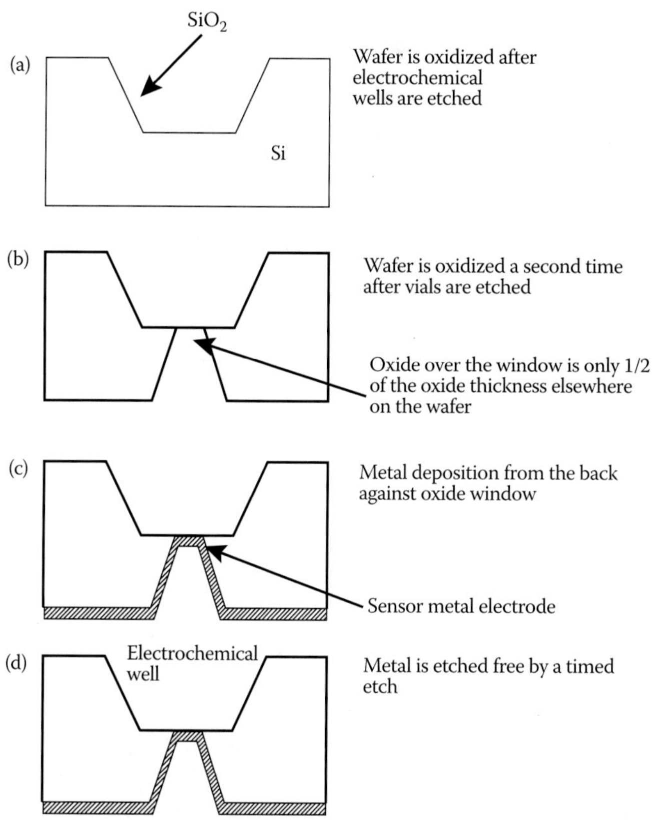


**FIGURE 4.94** (a–d) Fabrication sequence for a generic electrochemical cell in Si. Depth of the electrochemical well, number of electrodes, and electrode materials can be varied.

**FIGURE 4.95** SEM micrograph illustrating process step (d) from Figure 4.94. A 30 × 30 μm Pt electrode is shown at the bottom of an electrochemical well. This Pt electrode is further contacted to the electronics from the back.

silicon wafer. The bottom piece is fabricated in a custom IC housing using standard IC processes. The process sequence to build a generic electrochemical cell in a 250-μm-thick silicon wafer with one or more electrodes at the bottom of each well is illustrated in Figure 4.94. Wells are etched from the front of the wafer, and after an oxidation step access cavities for the metal electrodes are etched from the back. The etching of the vias stops at the oxide-covered bottoms of the electrochemical wells (Figure 4.94a). Next, the wafers are oxidized a second time with the oxide thickness doubling everywhere except in the suspended window areas where no Si can feed further growth (Figure 4.94b). The desired electrode metal is subsequently deposited from the back of the wafer into the access cavities and against the oxide window (Figure 4.94c). Finally, a timed oxide etch removes the sacrificial oxide window from above the underlying metal, while preserving the thicker oxide layer in the other areas on the chip (Figure 4.94d).[245–247] An SEM micrograph demonstrating Figure 4.94d is shown in Figure 4.95. The electrodes in the electrochemical cells of the top wafer are further connected to the bottom wafer electronics by solder balls in the access vias of the top silicon wafer (see Figure 4.96). Separating the chemistry from the electronics in this way provides extra protection for the electronics from the electrolyte, as well as from the electronics, and chemical sensor manufacture can proceed independently. Depending on the type of sensor element, one or more electrodes are fabricated at the bottom of the electrochemical cells. For example, shown in Figure 4.97 is an almost completed (Severinghaus) $CO_2$ sensor with an Ag/AgCl electrode as the reference electrode (left in the figure) and an IrOx pH-sensitive electrode, both at the bottom of one electrochemical well. The metal electrodes are electrically isolated from each other by the $SiO_2$ passivation layer over the surface of the silicon wafer. To complete the $CO_2$ sensor, we silk-screen a hydrogel containing an electrolytic medium into the silicon sensor cavity and dip-coat the sensor into a silicone-polycarbonate rubber solution to form the gas-permeable membrane. For hydrogel inside the micromachined well, we use poly(2-hydroxyethyl)methacrylate or polyvinylalcohol.

The concept of putting the sensor chemistries and the electronics on opposite sides of a substrate is a very important design feature we decided on several years ago in view of the overwhelming problems encountered in building chemical sensors based on ion-sensitive field effect transistors or extended gate field effect transistors.[187]

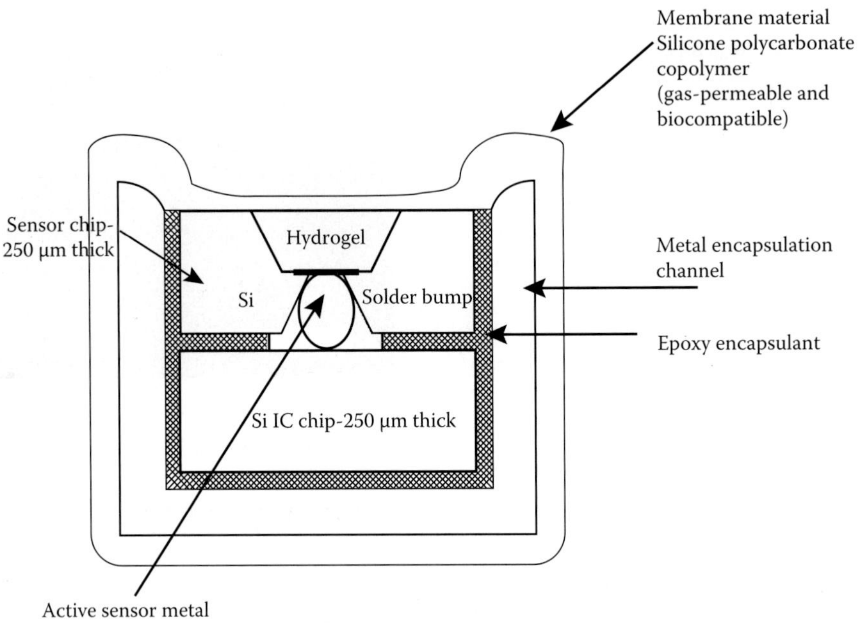


**FIGURE 4.96** Schematic of the bonding scheme between sensor wafer and IC. The schematic is a cut-through of the catheter.

### Example 4.3: Disposable Electrochemical Valves

One of the most difficult aspects of developing a microfluidic system is the miniaturization of valves. Most current MEMS valve technologies are still too complicated, too large, too power hungry, and too expensive for deployment in such applications as disposables (prohibitive cost) or implants (required power and size are too large). A more appropriate technology for a valve in a disposable such as a diagnostic panel for blood electrolytes or for an implantable drug delivery system may be a "sacrificial" or "use-once" valve. Such a valve could ensure device sterility, isolate reagents until they are required, and even entrap a vacuum to draw sample into the device (like a vacutainer blood-collection tube). Most current MEMS valves have moving parts, such as

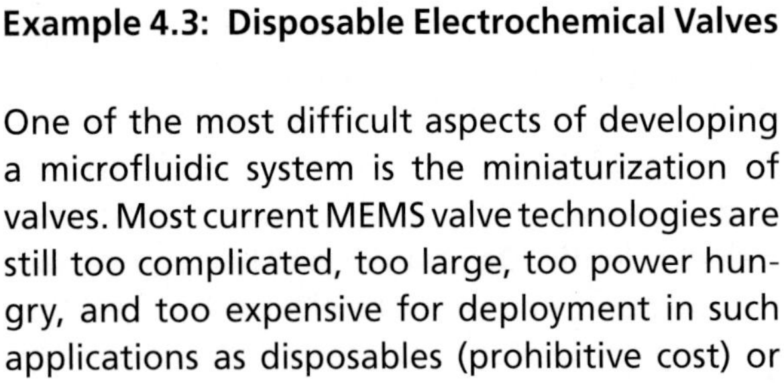

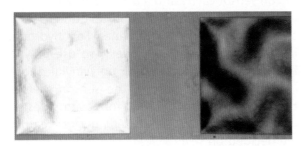

**FIGURE 4.97** SEM micrograph of an Ag/AgCl (left) and $IrO_x$ (right) electrode at the bottom of an anisotropically etched well in silicon. This electrochemical cell forms the basis for a Severinghaus $CO_2$ sensor.

diaphragm valves that are micromachined in silicon, and are prone to malfunction through clogging. Valves without moving parts are obviously preferred, and some such valves have been incorporated in the electrokinetic and centrifugal fluidic platforms covered in Volume I, Chapter 6. These "valves" form a barrier for liquids but not for vapors. There is an even simpler valve, also without moving parts, which forms a barrier for both liquids and vapors. Creating a vapor barrier is essential if liquids are to be stored in small micromachined chambers for extended periods, e.g., for a diagnostic test incorporating different on-chip reagents. In the absence of such vapor barriers, liquid would be distributed over time driven by the gradients in vapor pressure above each individual chemical reservoir.

In researching the structures described in Example 4.2 we discovered that a small current, applied between a counter electrode and a thin suspended metal electrode (see Figure 4.98), causes anodic dissolution of the metal or local electrolysis of water, depending on the nature of the valve metal. In both cases, the applied potential leads to bursting of the thin metal barriers. Thus, thin suspended metal membranes, such as silver or Au, can be burst open by passing a small current from the metal via a contacting electrolyte solution to a counter electrode. A Ag valve, for example, can be opened with an applied bias as low as 1–1.5 V. Although the "use-once"

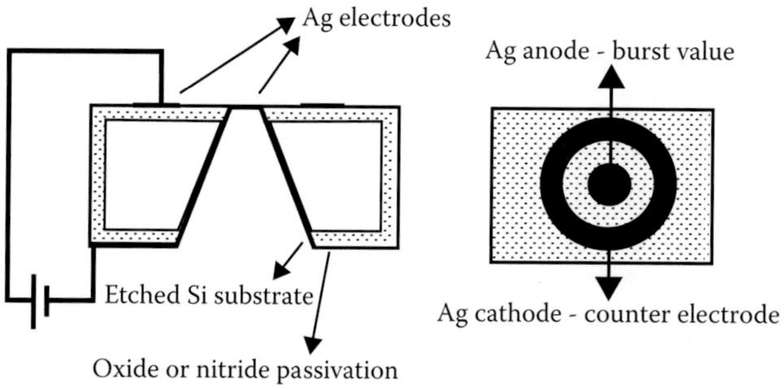


**FIGURE 4.98** Disposable electrochemical valve: principle of operation.

microvalve in Figure 4.98 involves a Ag membrane suspended over an orifice, this patented electrochemical valve technology is generic, and a wide variety of metals may be used.[248] Small amounts of drugs can be kept in the Si micromachined chambers, and the chamber may be closed off using a polymeric laminate. The structure shown in Figure 4.98 has also been fabricated using a dry photoresist replacing the Si as structural material. Arrays of the element shown in Figure 4.98 can be made, and the individually addressable metal covers can then be opened electronically, releasing the drugs stored in the Si or polymer chambers. Different drugs can be released at different times, and by opening more "holes," the rate of drug delivery can be set. This is illustrated in Figure 4.99.[249]

One application for this type of valve is responsive drug delivery in a smart pill as sketched in Figure 4.100 (pharmacy-on-a-chip). In this type of implant, openings in a drug reservoir are regulated by a biological stimulus (e.g., the concentration of glucose sensed by a glucose biosensor); therefore, a patient receives only the amount of drug his/her body requires (e.g., the correct amount of insulin). A doctor may intervene via a telemetric link. This technology is currently being pursued by iGlyko (http://www.iglyko.com/iGlyko/Welcome_to_iGlyko.html).

### Example 4.4: Self-Aligned Vertical Mirrors and V-Grooves for a Magnetic Micro-optical Matrix Switch

Hélin et al.[250] fabricated an elegant, monolithic magnetomechanical optical switch for dynamically reconfigurable dense wavelength division multiplexing networks. Arrays of optical switches allow for the rapid reconfiguration of optical

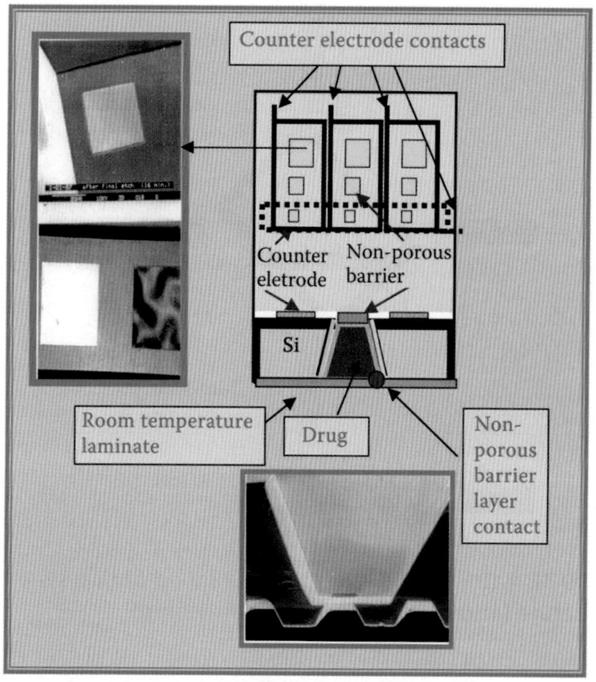


**FIGURE 4.99** Metal electrodes are "blasted" open electrochemically. By making electrode arrays, the number of openings is selectable. The metal valve material may be Ag, Pt, Au, and so on. In the SEM picture on the left, we show a platinum electrode (top) and a set of an Ag/AgCl electrode and an $IrO_x$ electrode (bottom). In the latter case the pH of the drug inside the reservoir can be monitored by measuring the voltage between the Ag/AgCl and $IrO_x$ electrode.

networks by altering the light path in a system of intersecting fibers. A wavelength division multiplexer adds or deletes extra optical channels. A switch routing light from a single fiber into any of $N$ output fibers is a $1 \times N$ switch. An $M \times N$ switch routes any one of $M$ inputs to any one of $N$ outputs. The $M \times N$ designation is the order of the optical switch. Commercially available optical switch arrays from companies such as JDS Uniphase Corp. (Milpitas, CA; http://www.jdsu.

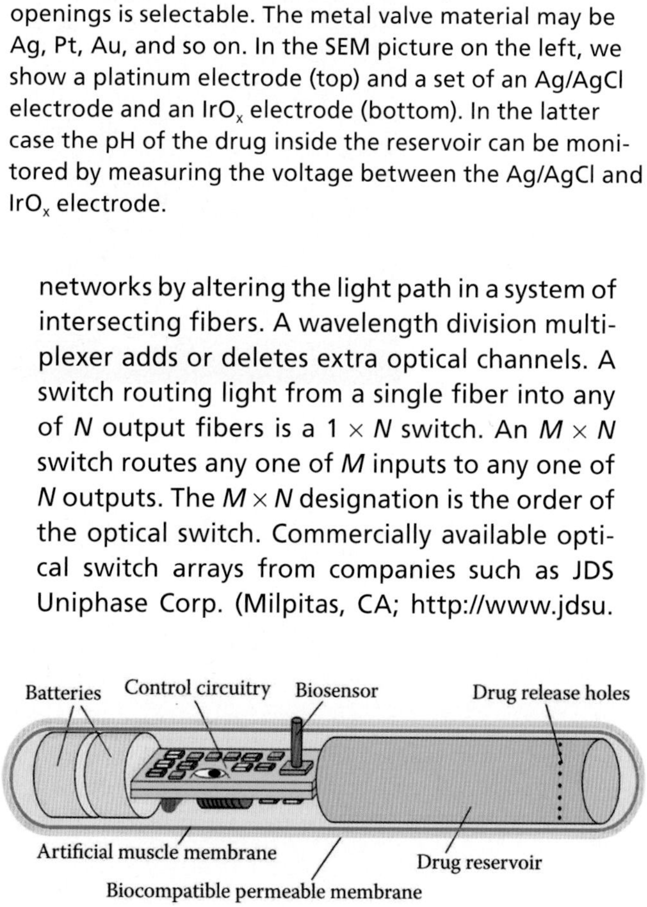


**FIGURE 4.100** Model of responsive drug-delivery system (pharmacy-on-a-chip). Responsive drug-delivery pill-type configuration. A video on the operation of the device can be found at http://www.biomems.net/.

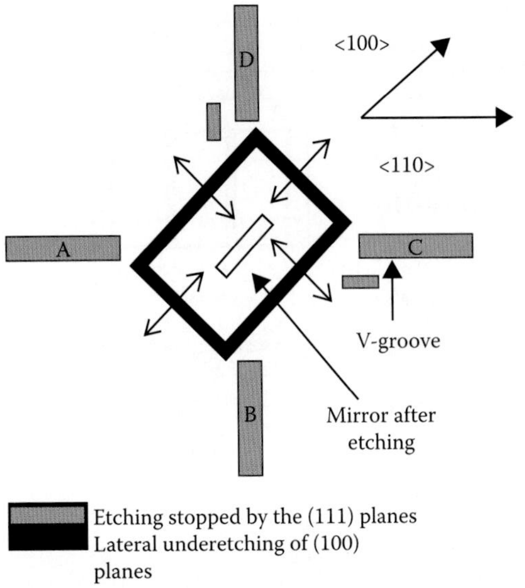


**FIGURE 4.101** Principle of the self-aligned optical switch structure (arrows indicate the underetching).

com) and DiCon Fiberoptics, Inc. (Berkeley, CA; http://www.diconfiberoptics.com) are limited to $1 \times 2$ or $2 \times 2$.[251] Using MEMS, the prospect is to deliver $64 \times 64$ or even larger arrays in the case of Xros (bought by Nortel and then dissolved) (see also Volume III, Chapter 10 on IT applications of MEMS).

To make a low-cost, batch-machined switch, Hélin et al. use a simple, one-level mask process on a (100) silicon wafer, which allows for the simultaneous fabrication of (100) sidewalls for high quality mirrors and (111) V-grooves for self-alignment between V-grooves and mirrors. The etching principle, a nice illustration of the wet etching processes described in this chapter, is shown in Figure 4.101. The 45° angle between the <100> and <110> directions is used to self-align the vertical mirrors and V-grooves. As the

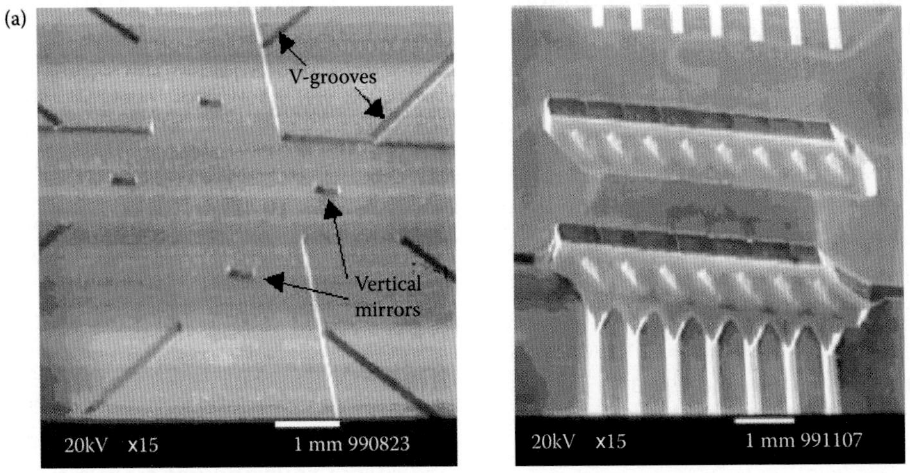


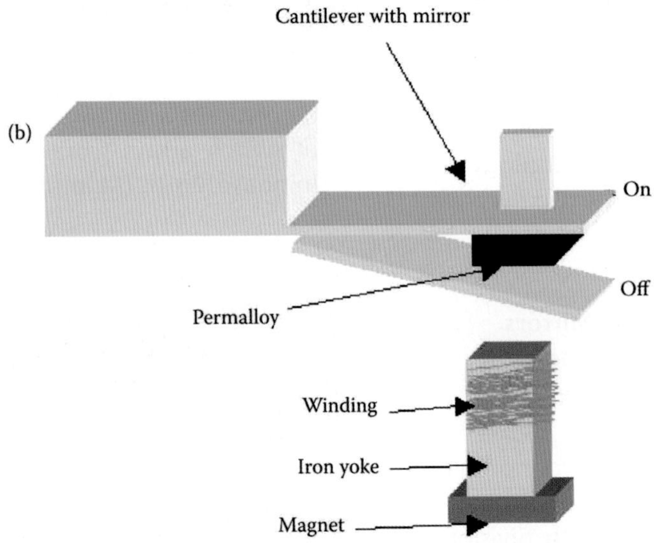


**FIGURE 4.102** (a) Examples of optical switching structures for a self-aligned matrix switch; $2 \times 2$ bidirectional switch (left) and $1 \times 8$ (right). (b) Principle of switch operation with a self-latching system. (Courtesy of Dr. H. Fujita, University of Tokyo.)

bottom and sidewall planes are all from the same {100} family, the lateral underetch rate is equal to the vertical etch rate (see also Figure 4.102). Underetching in the <100> direction creates the upstanding mirror while at the same time V-grooves (A, B, C, and D) are formed in the <110> directions for optical fiber alignment. Contrary to DRIE-based processes, in which ripples etched in the vertical walls are inevitable, wet anisotropic etching fulfills the requirements of high surface quality mirrors, reducing optical losses. Because intersecting (111) planes stop the etching, the width of the V-groove in the mask layout fixes the position of the optical axis. The thickness of the mirror is determined by timing the etch. Cr/Au layers are deposited by vacuum evaporation to finish the mirror manufacture. The etching principle outlined in Figure 4.101 can be expanded to an $M \times N$ matrix switch as shown in Figure 4.102a for a $2 \times 2$ and $1 \times 8$ switch.

The front-side wafer etching process is preceded by a back-side etching of a cantilever beam, whose role is to support the mirror and allow it to move. The thickness of the cantilever support plate is determined by a timed etch, and the cantilever with mirror is actuated using electromagnetic actuation. The actuation principle is demonstrated in Figure 4.102b.

## Acknowledgments

Special thanks to Mr. Chengwu Deng, graduate student UCI.

## Questions

4.1: True or False. Give an explanation for your answer.
  (a) Sputtering technique yields a poor quality of $SiO_2$.
  (b) Etch selectivity of <110> over <111> for KOH is much higher than EDP.
  (c) EDP does not etch $Si_3N_4$.
  (d) KOH does not etch $SiO_2$.
  (e) Doping Si with high levels of boron provides a good etch stop for both EDP and KOH.
  (f) The lift-off technique is used mostly to pattern thick $Si_3N_4$ films.

4.2: A 0.6 μm film of silicon dioxide is to be etched with a buffered oxide etchant (BOE) with an etch rate of 750 Å/min. Process data show that the thickness may vary up to 10% and the etch rate may vary up to 15%.
  (a) Specify a time for the etch process.
  (b) Predict how much undercut will occur at the top of the film.

4.3: Using the isotropic etching diagram for silicon from H. Robbins and B. Schwartz, *J. Electrochem. Soc.*, 107, 108–11, 1960:
  (a) If HF: $HNO_3$: $HC_2H_3O_2$ = 70:10:20, what is the etch rate?
  (b) At this point, which component controls the etch rate?
  (c) If we use water (10%) instead of $HC_2H_3O_2$ (10%) as our diluent, the etch rate will be more sensitive to which components?
  (d) How can we significantly reduce the etch rate?

4.4: You are asked to build a 600 μm square, 25-μm thick diaphragm in an oxidized (100) silicon wafer. Given that the wafer thickness is 300 μm and that you will be using an anisotropic etching solution of KOH/water which attacks the (111) crystal plane 100 times slower than the (100) crystal direction, the etch rate of (100) silicon in this specific KOH/water solution is 10 μm/hr.
  (a) Find the etching time.
  (b) Find the dimensions of the mask opening you would use.

4.5: You are asked to make grooves 50 μm deep in an oxidized (100) silicon wafer. Use first a BOE etch to make an opening in the oxide layer and then an anisotropic etchant like KOH/water. What are the dimensions of the mask opening you would use to make this dimension insensitive to the amount of over-etch?

4.6: Write one paragraph on each of the Si etch stop techniques including electrochemical etch stop, p-n junction etch stop, and dopant-dependent etch stop, etc.

4.8: Explain in detail the various corner compensation schemes to reduce undercutting while etching anisotropically.

4.9: Which are the three orientations of silicon wafers most commonly employed in micromachining? Arrange them in order of most common usage and briefly describe their benefits and drawbacks.

4.10: Derive the relation that connects the size of an etched cavity with the size of the mask opening in the case of anisotropic etching of [100] silicon.

4.11: In a 4″ Si (100) wafer, with a thickness of 525 μm, 2 different etch features must be created. What should the mask opening be to create:

(a) A square orifice of width 200 μm that extends through the wafer?

(b) A V-groove that extends all the way through the wafer?

4.12: Compare the etching characteristics of (110)- and (100)-oriented silicon wafers. Which would you choose for the following applications and why?

(a) A diaphragm-based pressure sensor

(b) Closely spaced packed high-aspect-ratio comb actuators

4.13: Draw a mask with a convex and a concave corner and describe what happens when anisotropically etching with such a mask. What is the effect etch stop layer has on undercutting of a mask?

4.14: We are building a MEMS device using a <100>-oriented silicon substrate. The 4-inch wafer is p-type and 500 μm thick. There is a 200-nm-thick PECVD $SiO_2$ thin film deposited onto the surface. We use photolithography to pattern a 750-μm square onto the wafer backside, and while the photoresist is still in place, employ a 33% KOH bath at 70°C to etch the device (see also the figure below*).

(a) What is the final size of the internal square projected onto the $SiO_2$ on the opposite side of the wafer?

(b) What would have been the minimum photolithography square feature that will etch all the way through the wafer?

(c) How long will it take to etch through the $SiO_2$ layer (see also the figure below)?

(d) Once the $SiO_2$ layer is removed, how long will it take to etch through the Si layer?

(e) The etch rate of $SiO_2$ in buffered hydrofluoric acid (BHF) is 100 nm/min. How long will it take to etch the remaining oxide layers away from the device?

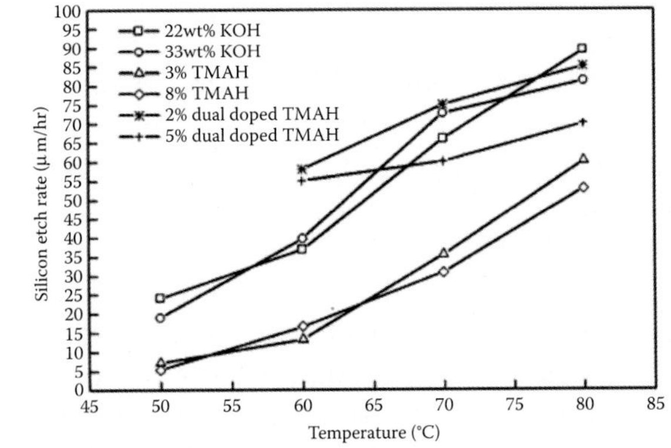


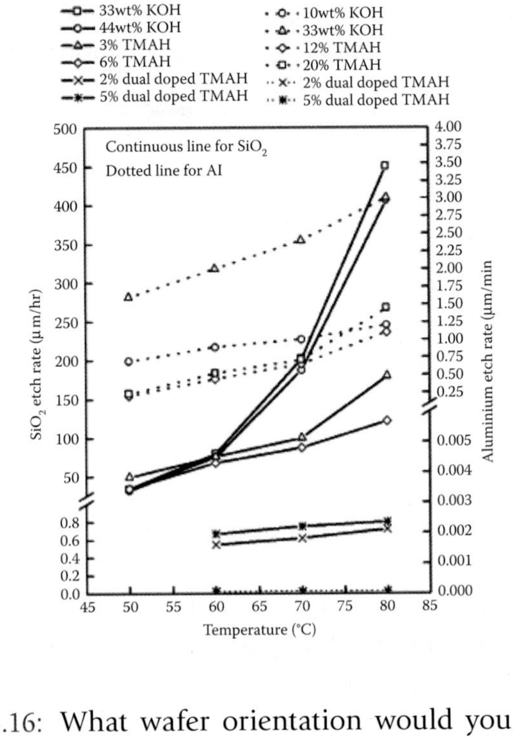


4.16: What wafer orientation would you select if you had to etch under a micro-bridge with a shallow V-groove crossing it at a 90° angle?

4.17: Compare wet and dry etching of Si.

4.18: Prepare a table with the most relevant design parameters for making a very sensitive

* Figures appearing under Question 4.14 reprinted from *Microelectronics Journal*, Volume 37, pp. 519–525, 2006. With kind permission from Elsevier.

piezoresistive Si cantilever to measure bending induced by differential surface stress.

4.19: With what you have learned so far, how would you make a very thin Si membrane in a Si frame and pattern a square Au electrode on that membrane inside the cavity? Where does your process differ from typical IC manufacture?

4.20: How would you make, using lift-off technology, an array of Pt electrodes at the bottom of a deep cavity and contact the Pt electrodes with leads extending over the cavity's edge?

4.21: Mark the wrong statements:

- The factors that affect the etching process are temperature, transport of reactant, reaction rate, diffusion rate, and so on.
- Acetic acid is more polar than water, which helps in achieving proper wetting of slightly hydrophobic Si wafers.
- Photon-pumped etching uses photogenerated electron-hole airs to supply oxidation sites for the etching process.
- Boron doped regions can be used as an etch stop layer.
- KOH etching is used for MOS and CMOS processing.

4.22: In the context of using piezoresistivity to sense strain, what are the advantages/disadvantages of using silicon vs. silicon nitride for pressure sensitive diaphragms? How does residual stress affect the sensitivity?

## References


1. Harris, T. W. 1976. *Chemical milling.* Oxford: Clarendon Press.
2. Durant, W. 1957. *The reformation: a history of European civilization from Wyclif to Calvin, 1300–1564.* New York: MJF Books.
3. Robbins, H., and B. Schwartz. 1959. Chemical etching of silicon-I: the system, HF, $HNO_3$, and $H_2O$. *J Electrochem Soc* 106:505–08.
4. Robbins, H., and B. Schwartz. 1960. Chemical etching of silicon-II: the system HF, $HNO_3$, $H_2O$, and $HC_2C_3O_2$. *J Electrochem Soc* 107:108–11.
5. Schwartz, B., and H. Robbins. 1961. Chemical etching of silicon-III: a temperature study in the acid system. *J Electrochem Soc* 108:365–72.
6. Schwartz, B., and H. Robbins. 1976. Chemical etching of silicon-IV: etching technology. *J Electrochem Soc* 123:1903–9.
7. Uhlir, A. 1956. Electrolytic shaping of germanium and silicon. *Bell Syst Tech J* 35:333–47.
8. Hallas, C. E. 1971. Electropolishing silicon. *Solid State Technol* 14:30–32.
9. Turner, D. R. 1958. Electropolishing silicon in hydrofluoric acid solutions. *J Electrochem Soc* 105:402–08.
10. Stoller, A. I., and N. E. Wolff. 1966. *Proceedings: second international symposium on microelectronics.* Munich, Germany.
11. Stoller, A. I. 1970. The etching of deep vertical-walled patterns in silicon. *RCA Rev* 31: 271–75.
12. Forster, J. H., and J. B. Singleton. 1966. Beam-lead sealed junction integrated circuits. *Bell Lab Rec* 44:313–17.
13. Kenney, D. M. 1967. Methods of isolating chips of a wafer of semiconductor material. US Patent 3,332,137.
14. Lepselter, M. P. 1966. Beam lead technology. *Bell Sys Tech J* 45:233–54.
15. Lepselter, M. P. 1967. Integrated circuit device and method. 1967. US Patent 3,335,338.
16. Waggener, H. A. 1970. Electrochemically controlled thinning of silicon. *Bell Sys Tech J* 49:473–75.
17. Kragness, R. C., and H. A. Waggener. 1973. Precision etching of semiconductors. US Patent 3,765,969.
18. Waggener, H. A., R. C. Kragness, and A. L. Tyler. 1967. *Technical digest: IEEE international electron devices meeting 68.* Washington, DC: IEEE.
19. Waggener, H. A., R. C. Krageness, and A. L. Tyler. 1967. Two-way etch. *Electronics* 40:274.
20. Bean, K. E., and W. R. Runyan. 1977. Dielectric isolation: comprehensive, current and future. *J Electrochem Soc* 124:5C–12C.
21. Rodgers, T. J., W. R. Hiltpold, B. Frederick, J. J. Barnes, F. B. Jenné, and J. D. Trotter. 1977. VMOS memory technology. *IEEE J Solid-State Circuits* SC-12:515–23.
22. Rodgers, T. J., W. R. Hiltpold, J. W. Zimmer, G. Marr, and J. D. Trotter. 1976. VMOS ROM. *IEEE J Solid-State Circuits* SC-11:614–22.
23. Ammar, E. S., and T. J. Rodgers. 1980. UMOS transistors on (110) silicon. *IEEE Trans Electron Devices* ED-27:907–14.
24. Smith, C. S. 1954. Piezoresistance effect in germanium and silicon. *Phys Rev* 94:42–49.
25. Pfann, W. G. 1961. Improvement of semiconducting devices by elastic strain. *Solid State Electron* 3:261–67.
26. McGeough, J. A. 1974. *Principles of electrochemical machining.* New York: Wiley.
27. Romankiw, L. T., and T. A. Palumbo. 1987. *Proceedings: symposium on electrodeposition technology, theory and practice.* Eds. L. T. Romankiw and D. R. Turner. Pennington, NJ: The Electrochemical Society, Inc.
28. Stevens, G. W. W. 1968. *Microphotography: photography and photofabrication at extreme resolution.* New York: Wiley.
29. Humphrey, E. F., and D. H. Tarumoto, eds. 1965. *Fluidics.* Boston: Fluidic Amplifier Associates.
30. Allen, D. M. 1986. *The principles and practice of photochemical machining and photoetching.* Bristol and Boston: Adam Hilger.
31. Trotter, D. M. 1991. Photochromic and photosensitive glass. *Sci Am* 124:124–29.
32. Stookey, S. D. 2000. *Explorations in glass: an autobiography.* Hoboken, NJ: Wiley.
33. Vollmer, J., H. Hein, W. Menz, and F. Walter. 1994. Bistable fluidic elements in liga technique for flow control in fluidic microactuators. *Sensors Actuators A* A43:330–34.
34. Schomburg, W. K., J. Vollmer, B. Bustgens, J. Fahrenberg, H. Hein, and W. Menz. 1994. Microfluidic components in LIGA technique. *J Micromech Microeng* 4:186–91.
35. Ehrfeld W. 1994. Notes from handouts. Banff, Canada.

36. Phillips, R. E. 1985. Electrochemical grinding: What is it? How does it work? *SME Technical Paper* MR-85:383.
37. Boothroyd, G., and W. A. Knight. 1989. *Fundamentals of machining and machine tools.* New York: Marcel Dekker.
38. Sommer, C. 2000. *Non-traditional machining handbook.* Houston: Advance Publishing.
39. Datta, M. 1995. Interface. *The Electrochemical Society.* 4:32.
40. Datta, M., L. T. Romankiw, D. R. Vigliotti, and R. J. von Gutfeld. 1989. Jet and laser-jet electrochemical micromachining of nickel and steel. *J Electrochem Soc* 136:2251.
41. Bard, A. J. 1994. *Integrated chemical systems.* New York: Wiley.
42. Amemiya, S., A. J. Bard, F.-R. F. Fan, M. V. Mirkin, and P. R. Unwin. 2008. Scanning electrochemical microscopy. *Annu. Rev. Anal. Chem.* 1:95–131.
43. Landolt, D. 1994. *Proceedings of symposium on electrochemical microfabrication.* Eds. M. Datta,K. Sheppard, and J. O. Dukovic. Pennington, NJ: The Electrochemical Society.
44. Datta, M. J. 1995. Fabrication of an array of precision nozzles by through-mask electrochemical micromachining. *Electrochem Soc* 142:3801–05.
45. Madore, C., and D. Landolt. 1997. Electrochemical micromachining of controlled topographies on titanium for biological applications. *J Micromech Microeng* 7:270–75.
46. den Braber, E. T., J. E. den Ruijte, L. A. Ginsel, A. F. von Recum, and J. A. Jansen. 1996. Quantitative analysis of fibroblast morphology on microgrooved surfaces with various groove and ridge dimensions. *Biomaterials* 17:2037–44.
47. Gonsalvez, M. 2003. *Atomistic modelling of anisotropic etching of crystalline silicon.* PhD diss., Helsinki University of Technology.
48. Kendall, D. L. 1979. Vertical etching of silicon at very high aspect ratios. *Ann Rev Mater Sci* 9:373–403.
49. Tuckerman, D. B., and R. F. W. Pease. 1981. High-performance heat sinking for VLSI. *IEEE Electron Device Lett* EDL-2:126–29.
50. Barth, P. W., P. J. Shlichta, and J. B. Angel. 1985. *Third international conference on solid-state sensors and actuators.* Philadelphia.
51. Ciarlo, D. R. 1987. *Proceedings: IEEE micro robots and teleoperators workshop.* Hyannis, MA: IEEE.
52. Seidel, H., L. Csepregi, A. Heuberger, and H. Baumgartel. 1990. Anisotropic etching of crystalline silicon in alkaline solutions–part I: orientation dependence and behavior of passivation layers. *J Electrochem Soc* 137:3612–26.
53. Elwenspoek, M., H. Gardeniers, M. de Boer, and A. Prak. 1994. Micromechanics. Report No. 122830. Twente, the Netherlands: University of Twente.
54. Herr, E., and H. Baltes. 1991. *Sixth international conference on solid-state sensors and actuators (Transducers '91).* San Francisco.
55. Elwenspoek, M. 1993. On the mechanism of anisotropic etching of silicon. *J Electrochem Soc* 140:2075–80.
56. Elwenspoek, M., U. Lindberg, H. Kok, and L. Smith. 1994. *IEEE international workshop on micro electro mechanical systems (MEMS '94).* Oiso, Japan: IEEE.
57. Kendall, D. L. 1975. On etching very narrow grooves in silicon. *Appl Phys Lett* 26:195–98.
58. Schumacher, A., H.-J. Wagner, and M. Alavi. 1994. Mit Laser und Kalilauge. *Technische Rundschau* 86:20–23.
59. Seidel, H., and L. Csepregi. 1988. Advanced methods for the micromachining of sensors. *Technical digest of the 7th sensor symposium.* Tokyo, Japan. Reprinted in MBB Jahresbericht 1989, Forschung und Entwicklung - Technisch-wissenschaftliche Veröffentlichungen 1989, S. 41.
60. Alavi, M., S. Buttgenbach, A. Schumacher, and H. J. Wagner. 1991. *Sixth international conference on solid-state sensors and actuators (Transducers '91).* San Francisco.
61. Alavi, M., S. Buttgenbach, A. Schumacher, and H. J. Wagner. 1992. Fabrication of microchannels by laser machining and anisotropic etching of silicon. *Sensors Actuators A* A32:299–302.
62. Kaminsky, G. 1985. Micromachining of silicon mechanical structures. *J Vac Sci Technol* B3:1015–24.
63. Stoller, A. I., R. F. Speers, and S. Opresko. 1970. A new technique for etch thinning silicon wafers. *RCA Rev* 31:265–70.
64. Kern, W., and C. A. Deckert. 1978. *Thin film processes.* Eds. J. L. Vossen, and W. Kern. Orlando, FL: Academic Press.
65. Yang, K. H. 1984. An etch for delineation of defects in silicon. *J Electrochem Soc* 131: 1140–45.
66. Chu, T. L., and J. R. Gavaler. 1965. Dissolution of silicon and junction delineation in silicon by the $CrO_3$-HF-$H_2O$ system. *Electrochim Acta* 10:1141–48.
67. Archer, V. D. 1982. Methods for defect evaluation of thin <100> oriented silicon in epitaxial layers using a wet chemical etch. *J Electrochem Soc* 129:2074–76.
68. Schimmel, D. G., and M. J. Elkind. 1973. An examination of the chemical staining of silicon. *J Electrochem Soc* 125:152–55.
69. Secco d'Aragona, F. 1972. Dislocation etch for (100) planes in silicon. *J Electrochem Soc* 119:948–51.
70. Tuck, B. 1975. Review: the chemical polishing of semiconductors. *J Mater Sci* 10:321–39.
71. Kooij, E. S., K. Butter, and J. J. Kelly. 1999. Silicon etching in HF/HNO solution: charge 3 balance for the oxidation reaction. *Electrochem Solid-State Lett* 2:178–80.
72. Kovacs, G. T. A., N. I. Maluf, and K. E. Petersen. 1998. *Proceedings of the IEEE.* Washington, DC: IEEE.
73. Wise, K. D., M. G. Robinson, and W. J. Hillegas. 1981. Solid state processes to produce hemispherical components for inertial fusion targets. *J Vac Sci Technol* 18:1179–82.
74. Tjerkstra, R. W. 1999. *Isotropic etching of silicon in fluoride containing solutions as a tool for micromachining.* PhD diss., University of Twente, Twente, the Netherlands.
75. Petersen, K. E. 1982. Silicon as a mechanical material. *Proceedings of the IEEE.* 70:420–57.
76. Kuiken, H. K., J. J. Kelly, and P. H. L. Notten. 1986. Etching profiles at resist edges. *J Electrochem Soc* 133:1217–32.
77. Kern, W. 1978. Chemical etching of silicon, germanium, gallium arsenide, and gallium phosphide. *RCA Rev* 39:278–308.
78. Hu, S. M., and D. R. Kerr. Observation of etching of n-type silicon in aqueous HF solutions. *J Electrochem Soc* 114:414.
79. Hoffmeister, W. 1969. Determination of the etch rate of silicon in buffered HF using a $^{31}$Si tracer method. *Int J Appl Radiation Isotopes* 2:139.
80. Muraoka, H., T. Ohashi, and T. Sumitomo. 1973. Controlled preferential etching technology. *Semiconductor Silicon.* Pennington, NJ: The Electrochemical Society, pp. 327–38.
81. Seidel, H. 1989. Mikromechanik. Ed. A. Heuberger. Heidelberg, Germany: Springer Verlag.
82. Bogenschutz, A. F., K.-H. Locherer, W. Mussinger, and W. Krusemark. 1967. Chemical etching of semiconductors in $HNO_3$-HF-$CH_3COOH$. *J Electrochem Soc* 114:970–73.
83. Dash, W. C. 1956. *J Appl Phys* 27:1193.
84. Sirtl, E., and A. Adler. 1961. *Z Metallkd* 52:529.
85. Jenkins, M. W. 1977. *J Electrochem Soc* 124:757.

86. Schimmel, D. G. 1979. Defect etch for <100> silicon evaluation. *J Electrochem Soc* 126:479–83.
87. Theunissen, M. J., J. A. Apples, and W. H. C. G. Verkuylen. 1970. Applications of preferential electrochemical etching of silicon to semiconductor device technology. *J Electrochem Soc* 117:959–65.
88. Meek, R. L. 1971. Electrochemically thinned N/N+ epitaxial silicon: method and applications. *J Electrochem Soc* 118:1240–46.
89. van Dijk, H. J. A., and J. de Jonge. 1970. Preparation of thin silicon crystals by electrochemical thinning of epitaxially grown structures. *J Electrochem Soc* 117:553–54.
90. Shengliang, Z., Z. Zongmin, and L. Enke. 1987. *Fourth international conference on solid-state sensors and actuators (Transducers '87)*. Tokyo, Japan.
91. Lehmann, V. 1993. The physics of macropore formation in low doped n-type silicon. *J Electrochem Soc* 140:2836–43.
92. Barla, K., G. Bomchil, R. Herino, and J. C. Pfister. 1984. X-ray topographic characterization of porous silicon layers. *J Cryst Growth* 68:721–26.
93. Arita, Y. 1978. Fomation and oxidation of porous silicon by anodic reaction. *J Cryst Growth* 45:383–92.
94. Higashi, G. S., Y. J. Chabal, G. W. Trucks, and K. Raghavachari. 1990. Ideal hydrogen termination of the Si (111) surface. *Appl Phys Lett* 56:656–68.
95. Allongue, P., V. Kieling, and G. Gerischer. 1995. Eching mechanism and atomic structure of H-Si (111) surfaces prepared in NH4F. *Electrochim Acta* 40:1353–60.
96. Gerischer, H., and M. Lubke. 1987. The electrochemical behaviour of n-type silicon (111)-surfaces in fluoride containing aqueous electrolytes. *Ber Bunsenges Phys Chem* 91:394–98.
97. Kooij, E. S., and D. Vanmaekelbergh. 1997. Catalysis and pore initiation in the anodic dissolution of silicon in HF. *J Electrochem Soc* 144:1296–301.
98. da Fonseca, C., F. Ozanam, and J.-N. Chazalviel. 1996. In situ infrared characterization of the interfacial oxide during the anodic dissolution of a silicon electrode in fluoride electrolytes. *Surface Sci* 365:1–14.
99. Matsumura, M., and S. R. Morrison. 1983. Anodic properties of n-Si and n-Ge electrodes in HF solution under illumination and in the dark. *J Electroanal Chem* 147:157–66.
100. Gerischer, H., and M. Lubke. 1988. Electrolytic growth and dissolution of oxide layers on silicon in aqueous solutions of fluorides. *Ber Bunsenges Phys Chem* 92:573–77.
101. Serre, C., S. Barret, and R. Herino. 1994. Characterization of the electropolishing layer during anodic etching of p-type silicon in aqueous HF solutions. *J Electrochem Soc* 141:2049–53.
102. Matsumura, M., and S. R. Morrison. 1983. Photoanodic properties of an n-type silicon electrode in aqueous solution containing fluorides. *J Electroanal Chem* 144:113–20.
103. Eddowes, M. J. 1990. Anodic dissolution of p- and n-type silicon: kinetic study of the chemical reaction. *J Electroanal Chem* 280:297–311.
104. Allongue, P., V. Costa-Kieling, and H. Gerischer. 1993. Etching of silicon in NaOH solutions: parts I and II. *J Electrochem Soc* 140:1009–18 (Part I); 1018–26 (Part II).
105. Palik, E. D., O. J. Glembocki, and J. I. Heard. 1987. Study of bias-dependent etching of Si in aqueous KOH. *J Electrochem Soc* 134:404–09.
106. Linde, H., and L. Austin. 1992. Wet silicon etching with aqueous amine gallates. *J Electrochem Soc* 139:1170–74.
107. Price, J. B. 1973. Anisotropic etching of silicon with KOH-$H_2O$- isopropyl alcohol. Eds. H. R. Huff, and. R. R. Burgess, *Semiconductor Silicon*. Princeton, NJ: The Electrochemical Society, pp. 339–53.
108. Finne, R. M., and D. L. Klein. 1967, A water-amine-complexing agent system for etching silicon. *J Electrochem Soc* 114:965–70.
109. Reisman, A., M. Berkenbilt, S. A. Chan, F. B. Kaufman, and D. C. Green. 1979. The controlled etching of silicon in catalyzed ethylene-diamine-pyrocathechol-water solutions. *Electrochem Soc* 126:1406–14.
110. Seidel, H., L. Csepregi, A. Heuberger, and H. Baumgartel. 1990. Anisotropic etching of crystalline silicon in alkaline solutions—part II: influence of dopants. *J Electrochem Soc* 137:3626–32.
111. Kendall, D. L., and G. R. de Guel. 1985. *Micromachining and micropackaging of transducers*. Ed. C. D. Fung. New York: Elsevier.
112. Lee, D. B. 1969. Anisotropic etching of silicon. *J Appl Phys* 40:4569–74.
113. Noworolski, J. M., E. Klaassen, J. Logan, K. Petersen, and N. Maluf. 1995. *Eighth international conference on solid-state sensors and actuators (Transducers '95)*. Stockholm, Sweden.
114. Waggener, H. A., and J. V. Dalton. 1972. Control of silicon etch rates in hot alkaline solutions by externally applied potentials. *J Electrochem Soc* 119:236C.
115. Weirauch, D. F. 1975. Correlation of the anisotropic etching of single-crystal silicon spheres and wafers. *J Appl Phys* 46:1478–83.
116. Clemens, D. P. 1973. Anisotropic etching of silicon on sapphire. *J Electrochem Soc* 407, 120:C240–C240.
117. Bean, K. E., R. L. Yeakley, and T. K. Powell. 1974. Orientation dependent etching and deposition of silicon. *J Electrochem Soc* 121:87C.
118. Declercq, M. J., J. P. DeMoor, and J. P. Lambert. 1975. A comparative study of three anisotropic etchants for silicon. *Electrochem Soc Abstracts* 75-2:446.
119. Wu, X. P., Q. H. Wu, and W. H. Ko. 1986. A study on deep etching of silicon using ethylene-diamine-pyrocathechol-water. *Sensors Actuators* 9:333–43.
120. Declercq, M. J., L. Gerzberg, and J. D. Meindl. 1975. Optimization of the hydrazine-water solution for anisostropic etching of silicon in integrated circuit technology. *J Electrochem Soc* 122:545–52.
121. Mehregany, M., and S. D. Senturia. 1988. Anisotropic etching of silicon in hydrazine. *Sensors Actuators* 13:375–90.
122. Asano, M., T. Cho, and H. Muraoko. 1976. Applications of choline in semiconductor technology. *Electrochem Soc Extend Abstr* 354:911–13.
123. Tabata, O., R. Asahi, and S. Sugiyama. 1990. Anisotropic etching with quaternary ammonium hydroxide solutions. *Technical Digest: Ninth Sensor Symposium*. Tokyo, Japan.
124. Pugacz-Muraszkiewicz, I. J., and B. R. Hammond. 1977. Application of silicates to the detection of flaws in glassy passivation films deposited on silicon substrates. *J Vac Sci Technol* 14:49–53.
125. Williams, K. R., and R. S. Muller. 1996. Etch rates for micromachining processes. *J Electrochem Soc* 137:3612–32.
126. Peeters, E. 1994. Process development for 3D silicon microstructures, with application to mechanical sensor design. Ph.D. diss., Catholic University of Louvain, Belgium.
127. Lambrechts, M., and W. Sansen. 1992. *Biosensors: microelectrochemical devices.* Philadelphia: Institute of Physics Publishing.

128. Palik, E. D., V. M. Bermudez, and O. J. Glembocki. 1985. Ellipsometric study of the etch-stop mechanism in heavily doped silicon. *J Electrochem Soc* 132:135–41.
129. Clark, L. D., and D. J. Edell. 19897. *Proceedings: IEEE micro robots and teleoperators workshop*. Hyannis, MA.
130. Schnakenberg, U., W. Benecke, and B. Lochel. 1990. $NH_4OH$-based etchants for silicon micromachining. *Sensors Actuators A* A23:1031–35.
131. Bean, K. 1978. Anisotropic etching of silicon. *IEEE Trans Electron Devices* ED–25:1185–93.
132. Barth, P. 1984. Si in biomedical applications. Microelectronics—photonics, materials, sensors and technology.
133. Linde, H., and L. Austin. 1992. Wet silicon etching with aqueous amine gallates. *J Electrochem Soc* 139:1170–74.
134. Tabata, O., R. Asahi, H. Funabashi, K. Shimaoka, and S. Sugiyama. 1992. Anisotropic etching of silicon in TMAH solutions. *Sensors Actuators A* A34:51–57.
135. Tabata, O. 1995. *Eighth international conference on solid-state sensors and actuators (Transducers '95)*. Stockholm, Sweden.
136. Reay, R. J., E. H. Klaassen, and G. T. A. Kovacs. 1994. Thermally and electrically isolated single-crystal silicon structures in CMOS technology. *IEEE Electron Device Lett* 15:309–401.
137. Steinsland, E., M. Nese, A. Hanneborg, R. W. Bernstein, H. Sandmo, and G. Kittilsland. 1995. *Eighth international conference on solid-state sensors and actuators (Transducers '95)*. Stockholm, Sweden.
138. Wise, K. D. 1985. *Micromachining and micropackaging of transducers*. Eds. C. D. Fung, P. W. Cheung, W. H. Ko, and D. G. Fleming. New York: Elsevier.
139. Seidel, H. 1987. The mechanism of anisotropic silicon etching and its relevance for micromachining. *Fourth international conference on solid-state sensors and actuators (Transducers '87)*. Tokyo, Japan.
140. Ternez, L. 1988. Ph.D. diss., Uppsala University.
141. Gravesen, P. 1986. *Silicon sensors. Status report for the industrial engineering thesis.* Lyngby, Denmark: DTH.
142. Ohwada, K., Y. Negoro, U. Konaka, and T. Oguchi. 1995. *Proceedings: IEEE micro electro mechanical systems conference.* Amsterdam, the Netherlands.
143. Baum, T., and D. J. Schiffrin. 1997. AFM study of surface finish improvement by ultrasound in the anisotropic etching of Si <100> in KOH for micromachining applications. *J Micromech Microeng* 7:338–42.
144. Bressers, P. M. M. C., J. J. Kelly, J. G. E. Gardeniers, and M. Elwenspoek. 1996. Surface morphology of p-type (100) silicon etched in aqueous alkaline solution. *J Electrochem Soc* 143:1744–50.
145. Klaassen, E. H., R. J. Reay, C. Storment, J. Audy, P. Henry, P. A. P. Brokaw, and G. T. A. Kovacs. 1996. *Proceedings: solid-state sensors and actuators workshop.* Hilton Head Island, SC.
146. Puers, R. 1991. *Proceedings: themadag sensoren.* Rotterdam, the Netherlands.
147. Buttgenbach, S. 1991. *Mikromechanik.* Stuttgart, Germany: Teubner Studienbucher.
148. Samaun, S., K. D. Wise, and J. B. Angell. 1973. An IC piezoresistive pressure sensor for biomedical instrumentation. *IEEE Trans Biomed Engr* 20:101–09.
149. Nunn, T., and J. Angell. 1975. *Workshop on indwelling pressure transducers and systems.* Cleveland, OH.
150. Greenwood, J. C. 1969. Ethylene diamine-cathechol-water mixture shows preferential etching of p-n junction. *J Electrochem Soc* 116:1325–26.
151. Bogh, A. 1971. Ethylene diamine-pyrocatechol-water mixture shows etching anomaly in boron-doped silicon. *J Electrochem Soc* 118:401–2.
152. Brodie, I., and J. J. Muray. 1982. *The physics of microfabrication.* New York: Plenum Press.
153. Heuberger, A. 1989. *Mikromechanik.* Heidelberg, Germany: Springer Verlag.
154. Rehrig, D. L. 1990. In search of precise epi thickness measurements. *Semicond Int* 13:90–95.
155. Palik, E. D., J. W. Faust, H. F. Gray, and R. F. Green. 1982. Study of the etch-stop mechanism in silicon. *J Electrochem Soc* 129:2051–59.
156. Jackson, T. N., M. A. Tischler, and K. D. Wise. 1981. An electrochemical P-N junction etch-stop for the formation of silicon microstructures. *IEEE Electron Device Lett* EDL–2:44–45.
157. Faust, J. W., and E. D. Palik. 1983. Study of the orientation dependent etching and initial anodization of Si in aqueous KOH. *J Electrochem Soc* 130:1413–20.
158. Kim, S. C., and K. Wise. 1983. Temperature sensitivity in silicon piezoresistive pressure transducers. *IEEE Trans Electron Devices* ED–30:802–10.
159. Kloeck, B., S. D. Collins, N. F. de Rooij, and R. L. Smith. 1989. Study of electrochemical etch-stop for high-precision thickness control of silicon membranes. *IEEE Trans Electron Devices* 36:663–69.
160. McNeil, V. M., S. S. Wang, K.-Y. Ng, and M. A. Schmidt. 1990. *Technical digest: 1990 solid state sensor and actuator workshop.* Hilton Head Island, SC.
161. Palik, E. D., O. J. Glembocki, and R. E. Stahlbush. 1988. Fabrication and characterization of Si membranes. *J Electrochem Soc* 135:3126–34.
162. Hirata, M., K. Suzuki, and H. Tanigawa. 1988. Silicon diaphragm pressure sensors fabricated by anodic oxidation etch-stop. *Sensors Actuators* 13:63–70.
163. Theunissen, M. J. 1972. Etch channel formation during anodic dissolution of n-type silicon in aqueous hydrofluoric acid. *J Electrochem Soc* 119:351–60.
164. Wen, C. P., and K. P. Weller. 1972. Preferential electrochemical etching of p+ silicon in an aqueous $HF–H_2SO_4$ electrolyte. *J Electrochem Soc* 119:547–48.
165. van Dijk, H. J. A. 1972. Method of manufacturing a semiconductor device and semiconductor device manufactured by said method. US Patent 3,640,807.
166. Saro, P. M., and A. W. van Herwaarden. 1986. Silicon cantilever beams fabricated by electrochemically controlled etching for sensor applications. *J Electrochem Soc* 133:1722–29.
167. Mlcak, R., H. L. Tuller, P. Greiff, and J. Sohn. *Proceedings: IEEE micro electro mechanical systems (MEMS '93).* Fort Lauderdale, FL: IEEE.
168. Yoshida, T., T. Kudo, and K. Ikeda. 1992. *Proceedings: IEEE micro electro mechanical systems (MEMS '92).* Travemunde, Germany: IEEE.
169. Shockley, W. 1963. Method of making thin slices of semiconductive material. US Patent 3,096,262.
170. Peeters, E., D. Lapadatu, W. Sansen, and B. Puers. 1993. *Seventh international conference on solid-state sensors and actuators (Transducers '93).* Yokohama, Japan.
171. Peeters, E., D. Lapadatu, R. Puers, and W. Sansen. 1994. Phet, an electrodeless photovoltaic electrochemical etch-stop technique. *J. Microelectromech. Syst.* 3(3):113–23.
172. Kelly, J. J., and H.G.G.Philipsen. 2005. Anisotropy in the wet-etching of semiconductors. *Curr Opin Solid State Mat Sci* 9:84–90.

173. Baum, T., and D. J. Schiffrin. 1998. Mechanistic aspects of anisotropic dissolution of materials: etching of single-cystal in alkaline solutions. *J Chem Soc Far Trans* 94:691–94.
174. Hines, M. A., Y. J. Chabal, T. D. Harris, and A. L. Harris. 1994. Measuring the structure of etched silicon surfaces with Raman spectroscopy. *Chem Phys* 101:8055.
175. Palik, E. D., H. F. Gray, and P. B. Klein. 1983. A Raman study of etching silicon in aqueous KOH. *J Electrochem Soc* 130:956–59.
176. Raley, N. F., F. Sugiyama, and T. Van Duzer. 1984. (100) Silicon etch-rate dependence on boron concentration in ethylenediamine-pyrocatechol-water solutions. *J Electrochem Soc* 131:161–71.
177. Wu, X. P., Q. H. Wu, and W. H. Ko. 1985. *Third international conference on solid-state sensors and actuators (Transducers '85)*. Philadelphia.
178. Abu-Zeid, M. M., D. L. Kendall, G. R. de Guel, and R. Galeazzi. 1985. Abstract 275. *JECS* 400.
179. Seidel, H. 1990. *Technical digest: IEEE solid-state sensor and actuator workshop.* Hilton Head Island, SC.
180. Steinsland, E., M. Nese, A. Hanneborg, R. W. Bernstein, H. Sandmo, and G. Kittilsland. 1995. *Eighth international conference on solid-state sensors and actuators (Transducers '95)*. Stockholm, Sweden.
181. Ghandi, S. K. 1968. *The theory and practice of micro-electronics.* New York: Wiley.
182. Glembocki, O. J., R. E. Stahlbush, and M. Tomkiewicz. 1985. Bias-dependent etching of silicon in aqueous KOH. *J Electrochem Soc* 132:145–51.
183. Elwenspoek, M., and J. P. van der Weerden. 1987. Kinetic roughening and step free energy in the solid-on-solid model and on naphtalene crystals. *J Phys A Math Gen* 20:669–78.
184. Bennema, P. 1984. Spiral growth and surface roughening: developments since Burton, Cabrera, and Frank. *J Cryst Growth* 69: 182–97).
185. Jackson, K. A. 1966. Crystal growth. *International conference for crystal growth.* Boston, June 20–24, 1966.
186. Madou, M. J., and S. R. Morrison. 1989. *Chemical sensing with solid state devices.* New York: Academic Press.
187. Madou, M. J., K. W. Frese, and S. R. Morrison. 1980. Photoelectrochemical corrosion of semiconductors for solar cells. *SPIE* 248:88–95.
188. Madou, M. J., B. H. Loo, K. W. Frese, and S. R. Morrison. Bulk and surface characterization of the silicon electrode. *Surf Sci* 108:135–52.
189. Levy-Clement, C., A. Lagoubi, R. Tenne, and M. Neumann-Spallart. 1992. Photoelectrochemical etching of silicon. *Electrochim Acta* 37:877–88.
190. Lewerenz, H. J., J. Stumper, and L. M. Peter. 1988. Deconvolution of charge injection steps in quantum yield multiplication on silicon. *Phys Rev Lett* 61:1989–92.
191. Brantley, W. A. 1973. Calculated elastic constants for stress problems associated with semiconductor devices. *J Appl Phys* 44:534–35.
192. Hesketh, P. J., C. Ju, S. Gowda, E. Zanoria, and S. Danyluk. 1993. Surface free energy model of silicon anisotropic etching. *J Electrochem Soc* 140:1080–85.
193. Sundaram, K. B., and H.-W. Chang. 1993. Electrochemical etching of silicon by hydrazine. *J Electrochem Soc* 140:1592–97.
194. Morrison, S. R. 1980. *Electrochemistry on semiconductors and oxidized metal electrodes.* New York: Plenum Press.
195. Gaburro, Z., N. Daldosso, and L. Pavesi, eds. 2005. *Encyclopedia of condensed matter physics: porous silicon.* Amsterdam: Elsevier Ltd.
196. Canham, L. T. 1990. Silicon quantum wire array fabrication by electrochemical and chemical dissolution of wafers. *Appl Phys Lett* 57:1046–48.
197. Lehmann, V., and U. Gosele. 1991. Porous silicon formation: a quantum wire effect. *Appl Phys Lett* 58:856–58.
198. Richter, A., P. Steiner, F. Kozlowski, and W. Lang. 1991. Current-induced light emission from a porous silicon device. *IEEE Electron Device Lett* 12:691–92.
199. Barrett, E. P., L. G. Joyner, and P. P. Halenda. 1951. The determination of pore volume and area distributions in pure substances. *J Am Chem Soc* 73:373.
200. Brunauer, S., P. H. Emmett, and E. Teller. 1938. Adsorption of gases in multimolecular layers. *J Am Chem Soc* 60: 309–19.
201. Bomchil, G., R. Herino, and K. Barla. 1986. *Energy beam-solid interactions and transient thermal processes.* Eds. V. T. Nguyen, and A. Cullis. Les Ulis, France: Les Editions de Physique.
202. Watanabe, Y., Y. Arita, T. Yokoyama, and Y. Igarashi. 1975. Formation and properties of porous silicon and its applications. *J Electrochem Soc* 122:1351–55.
203. Tserepi, A., et al. 2003. Dry etching of porous silicon in high density plasmas. *Physica Status Solidi (A)* 197:163–67.
204. Smith, R. L., and S. D. Collins. 1990. Thick films of silicon nitride. *Sensors Actuators A* A23:830–34.
205. Unagami, T. 1980. Formation mechanism of porous silicon layer by anodization in HF solution.*J Electrochem Soc* 127:476–83.
206. Tjerkstra, R. W. 1999. Isotropic etching of silicon in fluoride containing solutions as a tool for micromachining. PhD diss., University of Twente, Twente, the Netherlands.
207. Jung, K. H., S. Shih, and D. L. Kwong. 1993. Developments in luminescent porous Si. *J Electrochem Soc* 140:3046–64.
208. Zhang, X. G. 1991. Mechanism of pore formation on n-type silicon. *J Electrochem Soc* 138:3750–56.
209. Lehmann, V. 1996. *Ninth annual international workshop on micro electro mechanical systems (MEMS '96)*. San Diego.
210. Grüning, U., V. Lehmann, S. Ottow, and K. Busch. 1996. Macroporous silicon with a complete two-dimensional photonic band gap centered at 5 mm. *Appl Phys Lett* 68:747–49.
211. Thönissen, M., and M. G. Burger, eds. 1997. *Properties of porous silicon.* Washington, DC: IEEE.
212. Parkhutik, V. P., L. K. Glinenko, and V. A. Labunov. 1983. Kinetics and mechanism of porous layer growth during n-type silicon anodization in HF solution. *Surf Technol* 20:265–77.
213. Lehmann, V., and H. Foll. 1990. Formation mechanism and properties of electrochemically etched trenches in n-type silicon. *J* Electrochem *Soc* 137:653–59.
214. Smith, L., S. F. Chuang, and S. D. Collins. 1988. *J Electron Mater* 17:533.
215. Christophersen, M., J. Carstensen, A. Feuerhake, and H. Föll. 2000. Crystal orientation and electrolyte dependence for macropore nucleation and stable growth on p-type-silicon. *Mater Sci Eng B* 69-70:194 .
216. Lehmann, V. 2006. *Porous semiconductors-science and technology 5th international conference.* Sitges, Spain.
217. Cahill, S. S., W. Chu, and K. Ikeda. 1993. *Seventh international conference on solid-state sensors and actuators (Transducers '93)*. Yokohama, Japan.
218. Föll, H., S. Langa, J. Carstensen, M. Christophersen, and I. M. Tiginyanu. 2003. Pores in III-V semiconductors. *Adv Mater* 15:183–98.

219. Searson, P. C., S. M. Prokes, and O. J. Glembocki. 1993. Luminescence at the porous silicon/electrolyte interface. *J Electrochem Soc* 140:3327–31.
220. Doan, V. V., and M. J. Sailor. 1992. Luminescent color image generation on porous Si. *Science* 256:1791–92.
221. Lin, V. S., K. Motesharei, K. S. Dancil, M. J. Sailor, and M. R. Ghadiri. 1997. A porous silicon-based optical interferometric biosensor. *Science* 278:840–43.
222. Lammel, G., and P. Renaud. 2000. Two mask tunable optical filter of porous silicon as microspectrometer. *Eurosensors XIV.* Eds. R. de Reus, and S. Bouwstra. Copenhagen: MIC-Mikroelektronik Center.
223. Canham, L. T. 1995. Bioactive silicon structure fabrication through nanoetching techniques. *Adv Mater* 7:1033.
224. Armbruster, S., F. Schafer, G. Lammel, H. Artmann, C. Schelling, H. Benzel, S. Finkbeiner, F. Larmer, R. Ruther, and O. Paul. 2003 A novel micromachining process for the fabrication of monocrystalline Si-membranes using porous silicon. *The 12th International Conference on Transducers, Solid-State Sensors, Actuators and Microsystems.* 1:246–49.
225. Mayer, G. K., H. L. Offereins, H. Sandmeier, and K. Kuhl. 1990. Fabrication of non-underetched convex corners in ansisotropic etching of (100)-silicon in aqueous KOH with respect to novel micromechanic elements. *J Electrochem Soc* 137:3947–51.
226. Bryzek, J., K. Petersen, J. R. Mallon, L. Christel, and F. Pourahmadi. 1990. *Silicon sensors and microstructures.* Fremont, CA: Novasensor.
227. Puers, B., and W. Sansen. 1990. Compensation structures for convex corner micromachining in silicon. *Sensors Actuators A* A23:1036–41.
228. Sandmaier, H., H. L. Offereins, K. Kuhl, and W. Lang. 1991. *Sixth international conference on solid-state sensors and actuators (Transducers '91).* San Francisco.
229. Long, M. K., J. W. Burdick, and E. K. Antonsson. 1999. *MSM '99: modeling and simulation of microsystems, semiconductors, sensors and actuators.* San Juan, Puerto Rico.
230. Buser, R. A., and N. F. de Rooij. 1988. Monolithishes Kraftsensorfeld. *VDI-Berichte Nr.* 677.
231. Zhang, Q., L. L. Liu, and Z. Li. 1996. A new approach to convex corner compensation for anisotropic etching of (100) Si in KOH. *Sensors Actuators A* 56:251.
232. Enoksson, P. 1997. New structure for corner compensation in anisotropic KOH etching. *J Micromech Microeng* 7:141.
233. Seidel, H. 1986. PhD thesis. FU Berlin, Germany.
234. Abu-Zeid, M. M. 1984. Corner undercutting in anisotropically etched isolation contours. *J Electrochem Soc* 131:2138–42.
235. Wu, X.-P., and W. H. Ko. 1989. Compensating corner undercutting in anisotropic etching of (100) silicon. *Sensors Actuators* 18:207–15.
236. Wu, X., and W. H. Ko. 1987. *Fourth international conference on solid-state sensors and actuators (Transducers '87).* Tokyo, Japan.
237. Hubbard, Ted J. 1994. *MEMS Design*. PhD thesis. Caltech.
238. Nikpour, B., L. M. Landsberger, T. J. Hubbard, M. Kahrizi, and A. Iftimie. 1998. Concave corner compensation between vertical (010)-(001) planes anisotropically etched in Si(100). *Sensors Actuators A* 66:299.
239. Kim, B., and D. C. Dong-II. 1998. Aqueous KOH etching of silicon (110): etch characteristics and compensation methods for convex corners. *J Electrochem Soc* 145:2499.
240. Mocella, M. T. 1991. The CFC-ozone issue in dry etch process development. *Solid State Technol* April:63–64.
241. Cho, S. T. 1995. Micromachining and microfabrication process technology II. *Proceedings of the SPIE.* Austin, TX: SPIE.
242. Greiff, P. 1995. Micromachining and microfabrication process technology. *Proceedings of the SPIE.* Austin, TX: SPIE.
243. Weinberg, M., J. Bernstein, J. Borenstein, J. Campbell, J. Cousens, B. Cunningham, R. Fields, P. Greiff, B. Hugh, L. Niles, and J. Sohn. 1996. *Micromachining and microfabrication process technology II.* Austin, TX: SPIE.
244. Craven, D., K. Yu, and T. Pandhumsoporn. 1995. *Micromachining and microfabrication process technology. Proceedings of the SPIE.* Austin, TX: SPIE.
245. Joseph, J., M. Madou, T. Otagawa, P. Hesketh, and A. Saaman. 1989. *Catheter-based sensing and imaging technology. Proceedings of the SPIE.* Los Angeles: SPIE.
246. Madou, M. J., and T. Otagawa. 1989. Microelectrochemical sensor and sensor array. US Patent 4,874,500.
247. Holland, C. E., E. R. Westerberg, M. J. Madou, and T. Otagawa. 1988. Etching method for producing an electrochemical cell in a crystalline substrate. US Patent 4,764,864.
248. Madou, M. J., and M. Tierney. 1994. Micro-electrochemical valves and methods. US Patent 5,368,704.
249. Seetharaman, S., H. Ke-Qin, and M. Madou. 2000. Microactuators toward microvalves for responsive controlled drug delivery. *Sensors Actuators* B67:149–60.
250. Hélin, P., T. Bourouina, H. Fujita, H. Maekoba, O. Cugat, and G. Reyne. 2000. *Nano et micro technologies.* Ed. D. Hauden. Paris: Hermes Science Publications.
251. Maluf, N. 2000. *An introduction to microelectromechanical systems engineering.* Boston: Artech House.

# 5

# Thermal Energy-Based Removing

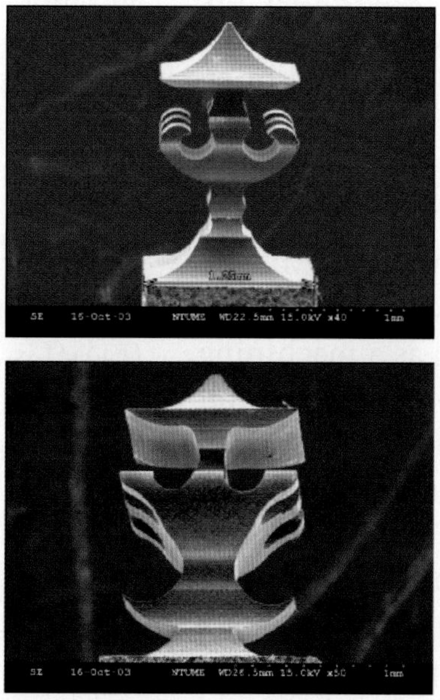

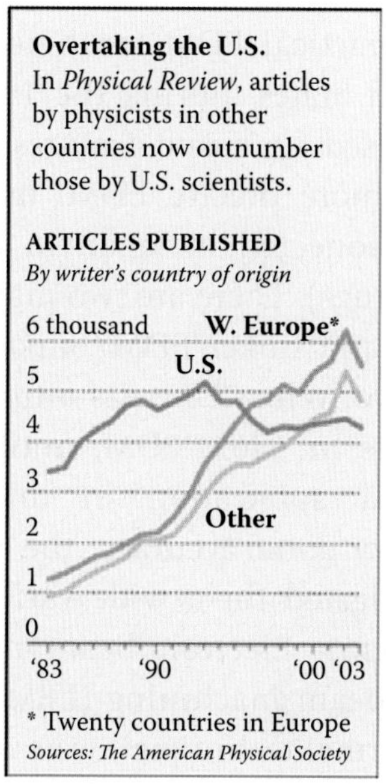


Pagoda (1.25 mm × 1.75 mm; material is Al-6061) produced by microelectrical discharge machining wire cutting (μ-EDM-WC) at the National Taiwan University (Prof. Y. S. Liao); see also Figure 5.10.

## Outline



## Introduction

In thermal removing processes, thermal energy, provided by a heat source, melts and/or vaporizes the volume of the material to be removed. Among thermal removal methods, electrical discharge machining (EDM) is the oldest and most widely used. Electron beam machining (EBM) and laser beam machining (LBM) are newer thermal techniques also widely accepted in industry today. Plasma arc cutting using a plasma arc torch is mostly used for cutting relatively thick materials in the range of 3–75 mm and is less pertinent to most miniaturization science applications. In thermal removal processes, a heat-affected zone (HAZ), sometimes called a *recast layer*, is always left on the work piece. In EBM, LBM, and arc machining, deposition and removal methods are available.

Additive thermal techniques involving EBM, LBM, and arc machining are covered in Chapter 9.

## Historical Note

Among thermal methods, electrical discharge machining (EDM), using electrical sparks for cutting, is one of the oldest. In an "uncontrolled" fashion, EDM has existed ever since the first thunderstorm. The first widespread "controlled" application for shaping some of the toughest metals was developed during World War II. Although the EDM principle was discovered during the early 1940s, it was not until the early 1950s that the first practical EDM machine was designed. In the United States, during the 1960s, EDM machining was used to create features in the 75-μm range; even more precise EDM, using rotating spindles, was pioneered in Japan in the late 1980s (5-μm size range). There are two major types of EDM: wire EDM and sinker EDM. Sinker EDM is the older of the two. Wire EDM was introduced as a machining process in 1969. EDM, until 10 years ago, found limited application in micromachining[1]; given its very good accuracy, the method is now being investigated for a wide variety of miniaturization problems. Electron beam machining (EBM) and laser beam machining (LBM), two other examples of thermal techniques, were developed in the 1960s and 1980s, respectively. In EBM, high velocity electrons are used instead of sparks, and in LBM intense photon pulses are used. Some authors group EBM, LBM, and ion-beam machining together as high-energy beam machining. Ion-beam machining is a mechanical removing technique and as such is covered in Chapter 6 on mechanical energy-based removing techniques. We only touch briefly on plasma arc cutting because the method has only limited utility in miniaturization science.

## Electrical Discharge Machining

### Introduction

Electrical discharge machining, more commonly known as EDM or spark machining, has traditionally been used to machine unusual designs in hard, brittle metals. Metal is removed by high-frequency electrical sparks, which are generated by pulsing a high voltage between the cathode tool and a work piece anode. The work piece and the tool are both submerged in a dielectric fluid as shown in Figure 5.1. Electrodes are usually made from graphite, copper, or silver. We cover two major types of EDM: wire EDM and sinker EDM.

Silicon-based machining often does not achieve the demands of applications like microsurgery, biotechnology, fluidics, or high temperature environments. As a consequence, EDM methods (including μ-EDM) are providing more and more solutions for application, such as fuel-injection nozzles for automobiles (regulations on environmental laws are forcing manufacturers to improve designs and making them smaller and more compact) and miniaturization of medical tools.

### Sinker Electrical Discharge Machining

Sinker or cavity-type electrical discharge machining (EDM) is a thermal mass-reducing process that uses

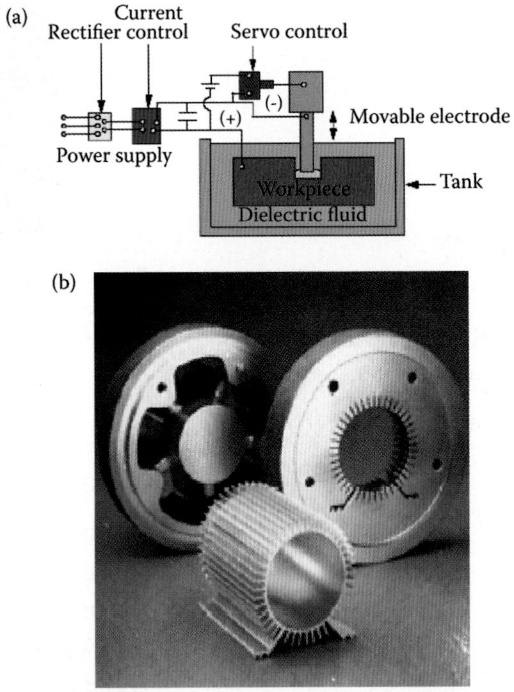


**FIGURE 5.1** (a) Discharge machining. Schematic illustration of the electrical discharge machining (EDM) process. Process is based on erosion of metals by spark discharge. The cavity is formed by the shape of the electrode. (b) Examples of cavities produced by the EDM process using shaped electrodes. Two round parts (rear) are the set of dies for extruding the aluminum piece shown in front.

**Table 5.1** Electrical Discharge Machining Characteristics

| Typical Use | Hard, Machining of Brittle Metals, Tool Making |
|---|---|
| Tool | Carbon, zinc, brass, copper, silver-tungsten, or copper-tungsten |
| Dielectric medium | Distilled water (DI), petroleum oils, silicones, triethylene, glycol water mixtures |
| Aspect ratio of holes | As high as 100:1 |
| Surface finish | 1–3 μm but even 0.25 μm has been reported |
| Gap size/voltage | 25 μm/80 V |
| Removal rate | 0.001–0.1 $cm^3/h$ |
| Work piece | Conductor |

a shaped conductive tool (the cathode) to remove electrically conductive material (anode). It does this by means of rapid, controlled, repetitive spark discharges. A dielectric fluid is used to flush the removed particles, to regulate the discharge, and to keep the tool and work piece cool. With a gap between tool and work piece of about 25 μm and a voltage of, for example, 80 V, intense sparking occurs across the dielectric filled gap, melting and vaporizing metal from both pieces (ideally much more from the work piece than from the tool).

Resolidified small hollow spheres are washed away by the recirculating dielectric, for example, oil or deionized water (DI water). A servomechanism adjusts the work piece–tool gap, and DC current pulses produce sparks at a rate of up to 500,000/s. Each spark generates a localized high temperature on the order of 12,000°C in its immediate vicinity (the power concentration is $10^5$–$10^7$ $W/mm^2$). The tool wear ratio, a measure of work piece material removal to tool material removal, varies from about 3:1 for metallic electrodes to about 70:1 for the better carbon electrodes. The tool wear can be reduced to 1% or less with an adequate selection of tool and work piece materials and appropriate generator settings.[2] Volumetric metal removal rates, ranging from 0.001–0.1 $cm^3/h$, are determined by thermal characteristics, such as conductivity and melting temperature of tool and work piece rather than hardness, and are small compared with more conventional machining methods. Metal removal rates increase as spark intensity (current) increases, but so does the surface roughness. Increasing the spark frequency, while keeping the other parameters constant, results in a decrease in surface roughness because the energy available is shared between more sparks, and smaller-sized surface craters are produced in the work piece. The frequency range of EDM machines is from 180 Hz, for rough cuts, to several hundred kilohertz, for fine finishing. Although sparks are effective for material removal, arcs result in bad surface quality. Arcs are created when the standby time is not long enough for the deionization of the dielectric to take place. Both short circuit and open circuit result in a stoppage of material removal. Thus, the goal of pulse control is to always create sparks and avoid arcs by optimizing the pulse width (on) and standby time (off). With carefully controlled conditions, it is possible to achieve tolerances as fine as 0.005 mm, but 0.02 mm is more typical. The accuracy of the process is closely related to the spark-gap width: the smaller the gap, the higher the accuracy, but a smaller gap results in a lower working voltage and a lower removal rate. A typical surface finish is in the 1–3-μm range. Two layers are of concern in EDM products. One is a layer of melted and resolidified material usually 0.0025–0.05-mm thick, known as *recast*, which is left on the surface produced by EDM. This layer tends to be very hard and brittle and may have to be removed if high levels of fatigue resistance are required (e.g., with shot peening*). A second layer is the so-called *heat-affected zone* (HAZ) that can be removed through electrochemical abrasion. In Table 5.1 some key EDM features are summarized.

With EDM, holes as small as 0.3 mm in materials 20 mm or more in thickness can be readily achieved. With efficient flushing, holes with an aspect ratio as high as 100:1 have been produced.[3]

* Shot peening is a process used as a decorative finish and/or to modify the mechanical properties of metals. It entails impacting a surface with shot (round metallic particles) with a force sufficient to create dimples and with enough shot so that those dimples overlap.

In general, EDM is one of the most widely used nonmechanical, noncontact machining processes in the industry today. Applications of this process include extrusion dies, fuel injector nozzles, aircraft engine turbine blades, and machining of difficult-to-machine materials like tool steel, tungsten carbide, and cemented carbides. EDM provides very good accuracy and repeatability; however, the tradeoff is a very low material removal rate (0.001–0.1 $cm^3/h$) for conductive and semiconductive materials only accompanied by thermal damage of the machined surface as a result of high heat generated during the discharge pulses (recast layer of 20–50 μm). Finally, the severe electrode wear affects machining accuracy, and the process poses some fume and skin hazards to operators.

An excellent starting point on the Internet for studying EDM in more detail is EDM Talk (http://www.edmmachining.com), the original EDM industry bulletin board and online forum. EDM equipment includes Fanuc's Robocut (http://www.fanuc.co.jp/en/product/robocut/index.html), Panasonic's MG-ED82W(http://www.panasonicfa.com/?id=MG-ED82W), and Agie's Agie® Mondo Star™ 50 CNC EDM (http://www.agie.com). Also check SmallTec International (http://www.smalltec.com) for sinker EDM and wire-cutting EDM (EDM-WC). A typical EDM machine may cost $350,000, and a typical service supplier is National Jet Company (Lavelle, MD), performing state-of-the-art micro-EDM (http://www.najet.com). The *EDM Handbook* by E. Bud Guitrau constitutes excellent further reading.[3]

A recent variant, EDM-WC, promises to make yet more impact on the future of microelectromechanical systems and is covered next.

## Electrode Discharge Machining: Wire Cutting

One of the most interesting adaptations of EDM is the development of electrodischarge machining wire cutting (EDM-WC), also called *wire electrode discharge grinding* (WEDG), which can be used for cutting complex two- (2D) and three-dimensional (3D) shapes from electrically conducting materials.[4] EDM-WC, like EDM, is a thermal mass-reducing process that uses a continuously moving wire to remove material by means of rapid controlled repetitive spark discharges. In EDM-WC, the cutting electrode is a continuously moving thin copper or brass wire, 0.15–0.3 mm in diameter, or, for high-precision cutting, stronger molybdenum-steel wire (0.15–0.03 mm). The wire is pulled from a supply reel past two fixed sapphire or diamond guides, one located above and one below the work piece, as shown in Figure 5.2, and collected on a take-up reel. This continuously delivers fresh wire to the work area. As much as 50 hours of machining can be performed with one reel of wire, which is then discarded. Deionized water is used as dielectric and as a flushing fluid. Flushing is essential for achieving high accuracy and preventing short-circuiting between the wire and part. The main advantage of EDM-WC is that it can mill through metals four or five times faster than EDM. Moreover, no expensive, shaped EDM electrodes are required. Linear cutting rates are still relatively slow, especially for thick substrates: from 38–115 mm/h in 25-mm-thick steel.

Numerically controlled EDM-WC has revolutionized die making, particularly for plastic molders. EDM-WC is now common in tool-and-die shops. Shape accuracy in EDM-WC in a working environment with temperature variations of about 3°C is about 4 μm. If temperature control is within ±1°C,

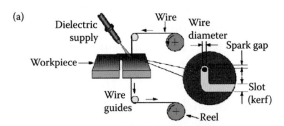

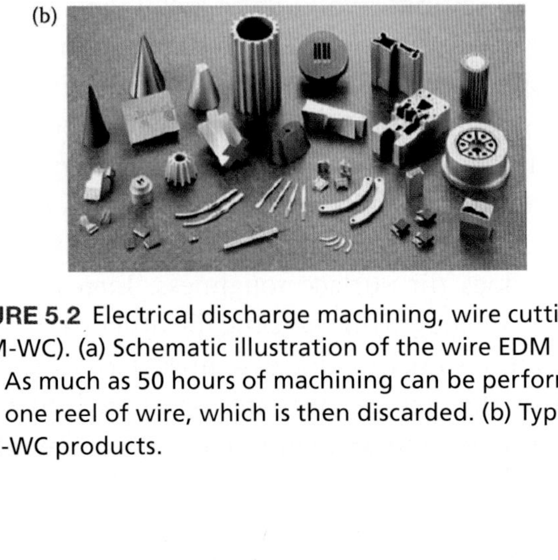


**FIGURE 5.2** Electrical discharge machining, wire cutting (EDM-WC). (a) Schematic illustration of the wire EDM process. As much as 50 hours of machining can be performed with one reel of wire, which is then discarded. (b) Typical EDM-WC products.

the obtainable accuracy is closer to 1 μm.[5] No burrs are generated, and because no cutting forces are present, EDM-WC is ideal for delicate parts. No tooling is required, so delivery times are short. Pieces more than 16-in. thick can be machined. Tools and parts are machined after heat treatment; thus, dimensional accuracy is held and not affected by heat treat distortion.

Service and equipment providers in the EDM-WC arena include Small Tec International (http://www.smalltec.com), Reliable, EDM Inc. (http://www.reliableedm.com), and XACT (http://www.xactedm.com/index.html). Oki is a provider of brass wire for EDM-WC (OB wire) (http://www.okidensen.co.jp/english/e_home.htm). Equipment may be bought from companies such AGIE (http://www.agie.com), Fanuc (http://www.fanuc.co.jp/en/product/index.htm), and Panasonic (http://www.panasonicfa.com/?id=Micro%20EDM). For further reading also consult the *Wire EDM Handbook*.[6]

## Microelectrical Discharge Machining

When referring to microelectrical discharge machining (μ-EDM) one refers either to working with a small EDM machine (see Figure 5.3 for a hand-held EDM at Panasonic) or to working with smaller than usual electrodes (in sinker EDM) or with thinner wires (in EDM-WC).

In die-sinking μ-EDM, an electrode with microfeatures is used to produce its mirror image in the work piece. The Matsushita Research Institute Tokyo, Inc. developed such a microsinker EDM machine. This machine incorporates submicron movement resolution and uses a microtool (microelectrode).

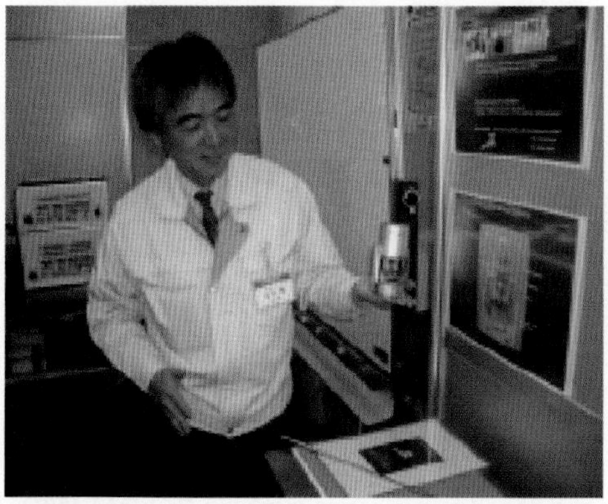

**FIGURE 5.3** Handheld μ-EDM at Panasonic.

The smallest microtools may be produced by the wire electrode discharge grinding method (see above) (http://www.panasonicfa.com/?id=Micro%20EDM). μ-EDM machining can produce a hole of 5–300 μm in diameter, with a depth three to five times the diameter, and at a precision of ± 0.5 μm on its circularity. Under the right conditions, the machine may result in a surface roughness of as little as 0.1 μm $R_{max}$. The latter is achieved by minimizing the electrical discharge energy to an extremely small amount. These microholes can be used as nozzles for ink-jet printers, flow control orifices, and pinholes for x-ray measurements. There are also more applications other than simply drilling microholes, such as microshafts, gears, springs, molds, dies, precision optics, and magnetic heads for digital video cameras.

In μ-EDM drilling, a microelectrode is used to "drill" microholes; a circular electrode is rotated at high speed (to help flushing of the liquid), and holes of 5 μm in diameter are reported. In μ-EDM milling, a simple-shaped electrode, rode or tube, is used to produce 3D cavities by adopting a contour motion.

The μ-EDM-WC technique can be used to make circular rods by rotating the work piece against the cutting wire as shown in Figure 5.4a, resulting in diameters as small as 5 μm and a surface roughness of 0.1 μm. The rods can have cross-sections such as triangular, rectangular, slit-type, and polygonal as well. The way to make a triangular shape is illustrated in Figure 5.4b.

Kuo et al.[7] used μ-EDM-WC to make micropipes with an inner diameter of 23 μm, an outer diameter of 186 μm, and a length of 3 mm. These micropipes came with a variety of micronozzle exit shapes and different sizes at the ends; slit openings of 100 × 10 μm, triangular openings with a base of 50 μm, and circular openings with a radius of 6 μm were demonstrated.[8] The manufacturing process of those micropipes is illustrated in Figure 5.5 (steps a–d). μ-EDM-WC is first used to machine a stainless steel rod (a); then electroforming is used to deposit a Ni coat (b); the Ni deposit is subsequently sized by μ-EDM-WC (c); and finally the stainless steel core is removed (d). To separate the core from the deposit

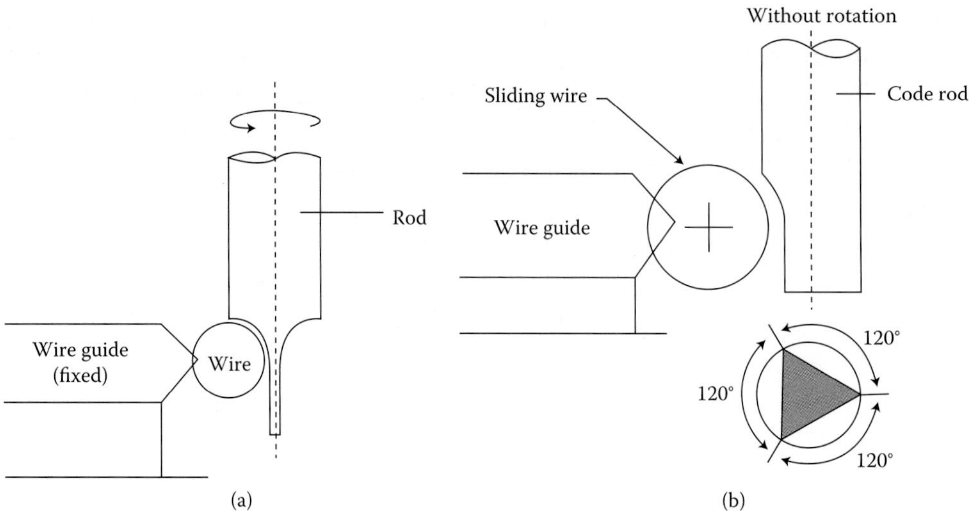


**FIGURE 5.4** Microelectrical discharge machining, wire cutting (μ-EDM-WC) for a round rod profile (a) and for a triangular rod profile (b). (From Kuo, C.-L., T. Masuzawa, and M. Fujino. 1991. *Proceedings: IEEE micro electro mechanical systems (MEMS '91)*. Nara, Japan; and Kuo, C.-L., T. Masuzawa, and M. Fujino. 1992. *Proceedings: IEEE micro electro mechanical systems (MEMS '92)*. Travemunde, Germany. With permission.[7,8])

after plating, it is necessary to reduce the deposit tension stress as much as possible. A sulfamate nickel-plating solution with a plating temperature of 50°C and a pH of 4 was selected for this experiment because it can be operated with high current density and results in low residual stress. A schematic side view of a typical micropipe with a circular exit is shown in Figure 5.6. None of the lithography techniques discussed in the preceding chapters can produce this type of intricate pipe as easily as μ-EDM-WC. It is of great value for a micromachinist to recognize at a glance which shapes are better produced by a lithography-based method and which are better produced by a truly 3D lathe-based method. Microtools produced by EDM-WC may be used as sinker EDM electrodes to enable μ-EDM.

The use of microelectrode arrays enables one to use μ-EDM in batch mode as pioneered by Takahata.[9] Takahata used the LIGA process (Chapter 10) to make the microelectrode array shown in Figure 5.6. Using the setup shown here, 400 holes through 50-μm-thick stainless steel were made in

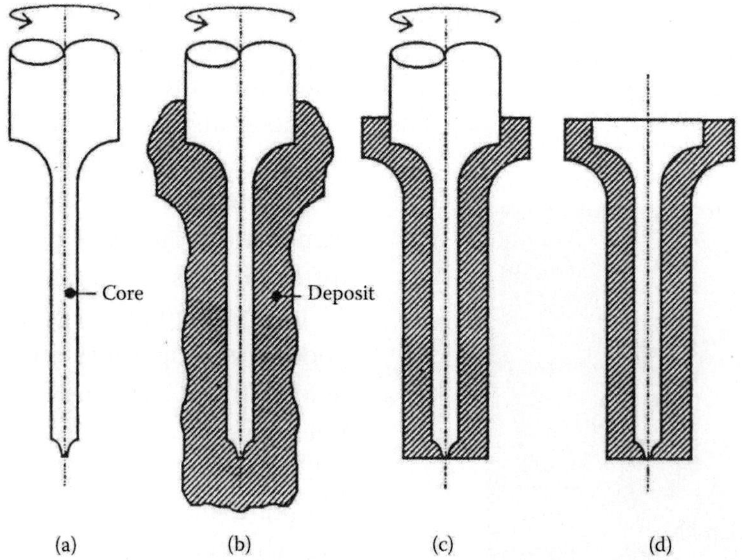


**FIGURE 5.5** Micronozzle fabrication process with microelectrical discharge machining, wire cutting (μ-EDM-WC). For fabrication steps (a–d), see text.

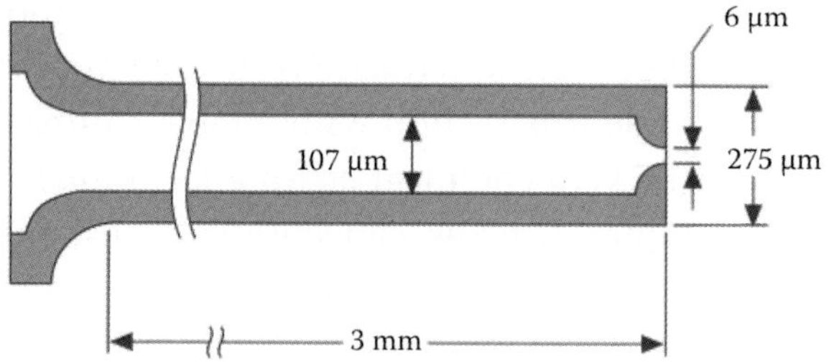


**FIGURE 5.6** Schematic side view of micronozzle (see also Figure 5.5d). Schematic side view of a circular pipe with 6-mm exit hole. (From Kuo, C.-L., T. Masuzawa, and M. Fujino. 1992. *Proceedings: IEEE micro electro mechanical systems (MEMS '92).* Travemunde, Germany. With permission.[8])

5 minutes (20–30 times faster than serial machining with a single electrode). Other structures made with this hybrid LIGA-EDM method are also shown in Figure 5.7.

A desk-top high-precision four-axes computer numerical controlled (CNC) machine for making microparts was developed by Liao et al.[10] at the National Taiwan University. This multifunction high-precision machine (accuracy of 1 μm) does on-machine assembly and on-machine metrology (Figure 5.8). Parts in the size range of 0.1–2 mm can be produced, and all manufacturing processes are completed on the same machine without moving the work piece.

This hybrid desk-top machine is capable of performing the following four functions. 1) Milling with a low-speed (0–3,000 rpm), a middle-speed (1,000–10,000 rpm), and a high-speed (2,000–80,000 rpm) spindle. 2) μ-EDM with microelectrodes with a diameter as small as 8 μm and very high high-aspect ratios. These microelectrodes are made using the on-board μ-EDM-WC with a 0.02 mm ϕ brass wire (see next). 3) μ-EDM-WC that features microwire tension control and wire vibration suppression (Figure 5.9) and a novel mechanism that permits vertical, horizontal, and slanted wire cutting. 4) On-line measuring of the flatness, circularity, concentricity, dimensions, and geometric tolerances of the work piece.

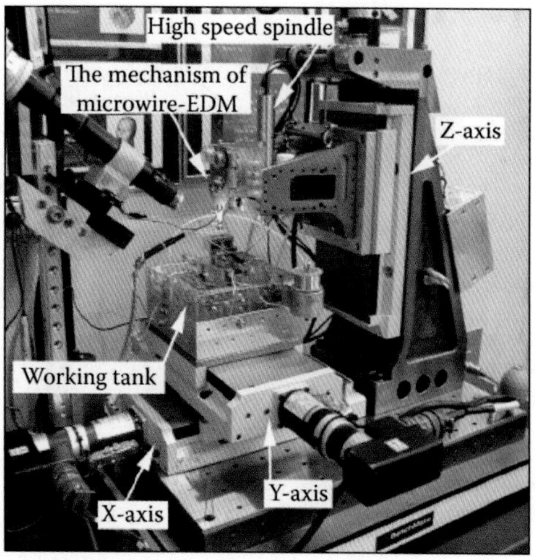


**FIGURE 5.8** High precision four-axes multipurpose CNC machine for making microparts developed by Y. S. Liao et al. at the National Taiwan University. (Ehmann, K. F., D. Bourell, M. L. Culpepper, T. J. Hodgson, T. R. Kurfess, M. Madou, and K. Rajurkar. 2007. *Micromanufacturing: international research and development.* New York: Springer.[10])

The vertical, horizontal, and slanted cutting with the μ-EDM-WC tool has successfully fabricated complex features and parts. An example is the impressive Chinese pagoda (1.25 mm × 1.75 mm) shown in the

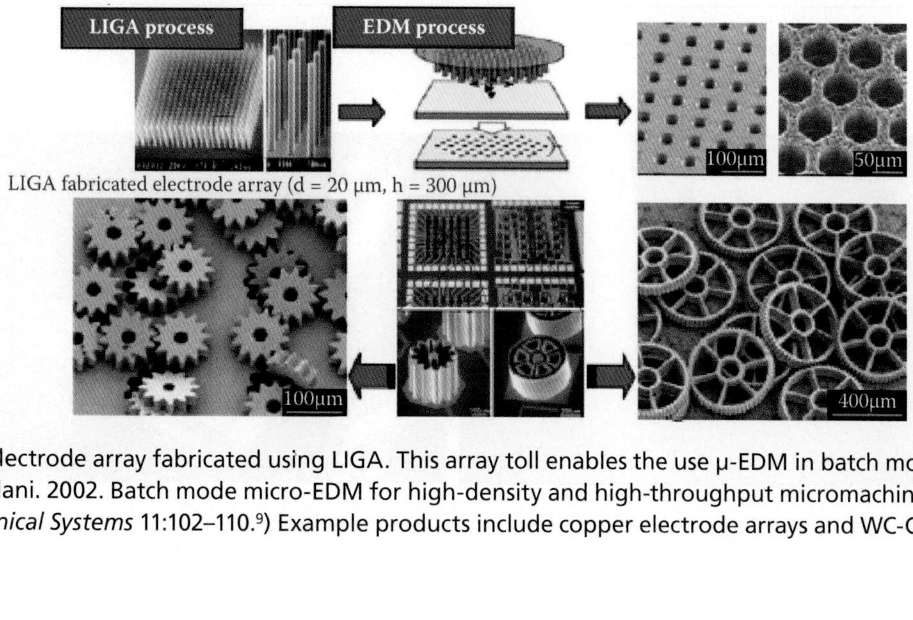


**FIGURE 5.7** Microelectrode array fabricated using LIGA. This array toll enables the use μ-EDM in batch mode. (Takahata, K., and Y. B. Gianchandani. 2002. Batch mode micro-EDM for high-density and high-throughput micromachining. *IEEE/ASME J Microelectromechanical Systems* 11:102–110.[9]) Example products include copper electrode arrays and WC-Cu super hard gears.

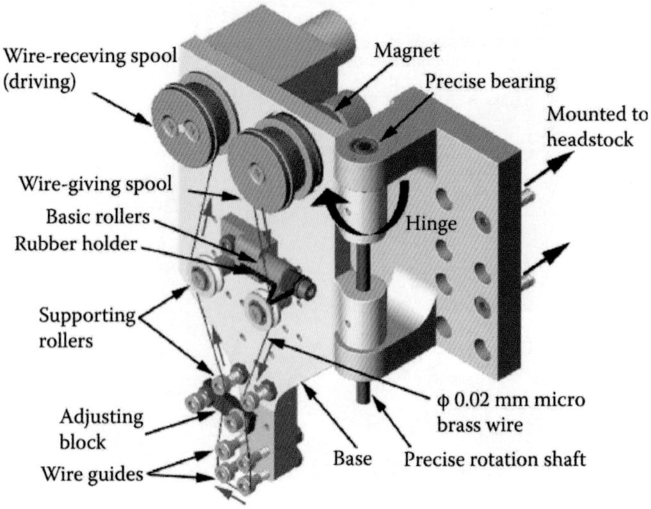


**FIGURE 5.9** The μ-EDM-WC tool features microwire tension control and wire vibration suppression. The tool allows vertical, horizontal, and slanted wire cutting. μ-EDM-WC at the National Taiwan University (Prof. Y. S. Liao).

heading of this chapter. The vertical and horizontal μ-EDM-WC cuts to make this pagoda are shown in Figure 5.10.

High-speed micromilling with a single crystal diamond milling cutter is demonstrated in Figure 5.11. The depth of the cut is 0.03 mm; the rotational speed is 80,000 rpm; and the feed rate is 20 mm/min.

We conclude this section with a table comparing EDM with ECM, a chemical etching technique covered in Chapter 4, using conventional mechanical machining (CNC machining; see Chapter 6) as a reference point. In Table 5.2 we list metal removal rates (MRRs), tolerance, surface finish and damage depth, and required power.

The metal removal rate by ECM is much higher than that of the EDM, with a metal removal rate 0.3 that of CNC, whereas EDM is only a small fraction of the CNC material removal rate. Power requirements for ECM are comparatively high. The tolerance obtained by EDM and ECM is within the range of CNC machining, which means satisfactory dimensional accuracy can be maintained. All the processes obtain satisfactory surface finishes. Depth of surface damage is very small for ECM, whereas it is very high in the case of EDM. For this reason, ECM can be used for making dies and punches. Capital cost for ECM is very high when compared with conventional CNC machining, and

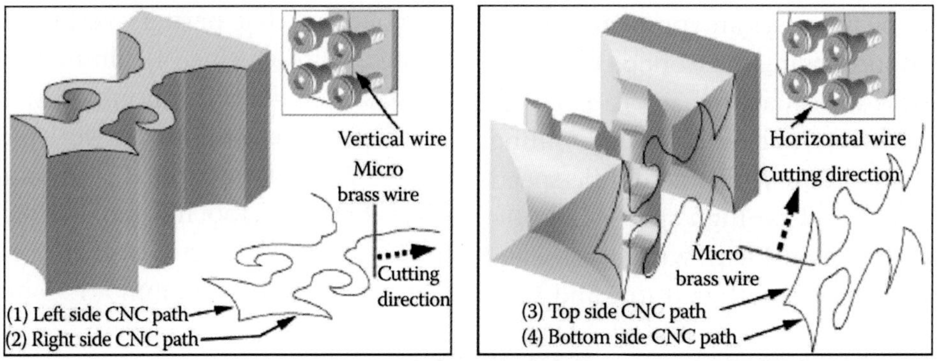


**FIGURE 5.10** Vertical and horizontal μ-EDM-WC to make the pagoda in Al-6061. μ-EDM-WC at the National Taiwan University (Prof. Y. S. Liao).

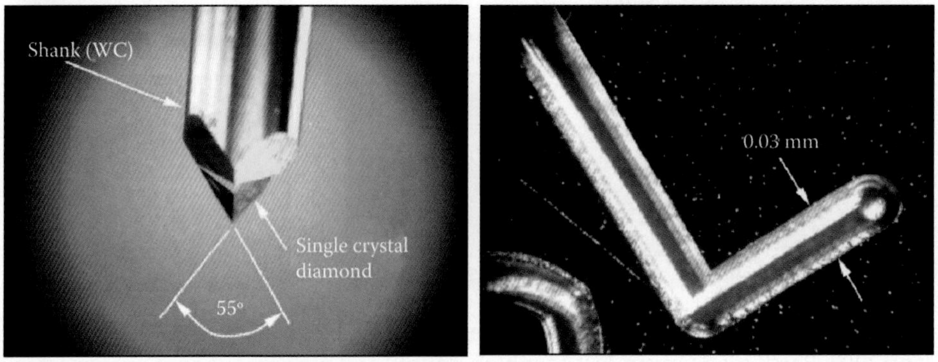


**FIGURE 5.11** High-speed micromilling with a single diamond crystal on a WC shank.

**Table 5.2** Machining Characteristics of EDM and ECM

| Process | MRR ($mm^3$/min) | Tolerance (μm) | Surface Finish (μm) | Damage Depth (μm) | Power (W) |
|---|---|---|---|---|---|
| ECM | 15,000 | 50 | 0.1–2.5 | 5 | 100,000 |
| EDM | 800 | 15 | 0.2–1.2 | 125 | 2,700 |
| CNC | 50,000 | 50 | 0.5–5 | 25 | 3,000 |

MRR, metal removal rate; tolerance, tolerance maintained; surface finish, surface finish required; damage depth, depth of surface damage; ECM, electrochemical machining; EDM, electrical discharge machining; CNC, computer numerical control machining.

EDM has also a higher tooling cost than the other machining processes.

## Laser Beam Machining

### Introduction

The basic operating principle of the laser was put forth by Charles Townes and Arthur Schalow from Bell Telephone Laboratories in 1958, and the first laser, based on a pink ruby crystal, was demonstrated in 1960 by Theodore Maiman at Hughes Research Laboratories. Lasers are defined as narrow beams (a divergence angle might be about 1 mrad, i.e., 1 m spread over 1 km of travel length) of monochromatic and coherent light in the visible, infrared (IR), or ultraviolet (UV) ranges of the spectrum. The key components of a laser unit are the active medium (a gas, liquid, or solid in which the beam is generated), the power supply (which delivers energy to the laser medium in the form needed to stimulate its emission), and the optical cavity (where the beam characteristics are defined). Lasers can be run in continuous mode or in pulsed mode to enhance the irradiance. The word *laser,* we saw in Volume I, Chapter 5, stands for Light Amplification by Stimulated Emission of Radiation. In Volume I, Chapter 5 we explained the lasing action; here we detail laser use in subtractive manufacturing. In 1960, the laser was said to be a "solution waiting for a problem," but today lasers are used in a wide range of applications, and some laser types have even become commodity products (e.g., laser pointers). In Table 5.3 we show the operating mode, wavelength, power, and example applications of some popular lasers (the list of potential applications is far from complete).

The $CO_2$ laser is an example of a gas laser based on a mixture of helium (83%) and nitrogen (16%) with carbon dioxide (6%) as the lasing material. The $CO_2$ laser produces a collimated coherent beam in the IR region (10.6 μm) characteristic of the active material ($CO_2$). The wavelength is strongly absorbed by glasses or polymers; thus, either mirrors or ZnSe lenses with excellent IR transparency are used to handle the beam.

The neodymium-doped yttrium aluminum garnet (Nd:YAG) laser is a solid-state laser producing a collimated coherent beam in the near-IR region of wavelength 1.06 μm, and can be pulsed or continuous. In pulsed mode, peak power pulses of up to 20 kW are now available. The shorter wavelength allows the use of optical glasses to control the beam path.

Excimer (EXCIted diMER) lasers, invented in 1975, are a family of pulsed lasers with a fast electrical discharge in a high-pressure mixture of a rare gas (krypton, argon, and xenon) and a halogen gas (fluorine and hydrogen chloride) as the source of emission. The particular combination of rare gas and halogen determines the output wavelength. Excimer lasers cover a range of wavelengths in the UV range from 157 nm ($F_2$) to 353 nm (XeF) with 193 nm (ArF), 248 nm (KrF), and 308 nm (XeCl) being particularly useful intermediate wavelengths.* Peak powers typically range from 3 MW ($F_2$) to 50 MW (KrF or XeCl) in pulses lasting a few tens of nanoseconds.

Machining with laser beams, first introduced in the early 1970s, is now used routinely in many industries. Laser machining, with long or continuous wave (CW), short, and ultrashort pulses, includes the following applications:

- Heat treatment
- Welding (see Volume III, Chapter 4 on packaging)
- Ablation or cutting of plastics, glasses, ceramics, semiconductors, and metals (this chapter)

* Recently, the 308-nm xenon chloride laser has gained much attention as a very effective treatment modality in dermatological disorders.

**Table 5.3** Laser Types with Characteristics and Applications

| | Wavelength | Maximumm Power Output (W) | Operation Mode | Applications |
|---|---|---|---|---|
| **Semiconductor lasers, also injection or junction laser** | | | | |
| Single diodes | Infrared to visible (0.33–40 μm) | 0.6 W (CW) 100 W (P) | CW and pulsed | Compact discs, laser printers, bar code scanners, optical communications |
| Diode laser bars | Infrared to visible | Up to 100 W | CW and pulsed | Pumping light source for solid state lasers |
| **Solid-state lasers** | | | | |
| Ti-sapphire | 690–960 | | Tunable | |
| Nd:YAG | 1.064 μm | 1000 W (CW) 2 $10^8$ W (P) | CW and pulsed | Welding, hole piercing, cutting |
| Ruby | 694.3 nm | Several MW | Pulsed | Measurement instrumentation, pulsed holography |
| **Gas/liquid lasers** | | | | |
| $CO_2$ (gas) | 9.6 μm, 10.6 μm | 1 W–40 kW (CW) 100 MW(P) | CW and pulsed | Heat treatment, scribing, welding, cutting |
| Excimer laser (gas) | 193 nm, 248 nm, 308 nm (and others) | 1 kW–100 MW | Pulse mode, pulse length 10–100 ns | Micromachining such as flex circuit via drilling and wire stripping, laser chemistry, medicine |
| HeNe (gas) | 632.8 nm (most prominent), 1.15 μm, 3.39 μm | 0.0005–0.05 (CW) | CW | Measurement instrumentation, holography, line-of-sight communications |
| Argon ion laser (gas) | 515–488 nm (several) | 0.005–20 (CW) | CW | Printing technology, pumping laser for dye laser stimulation, surgery |
| Dye laser (liquid) | Continuous between infrared and UV (different dyes): 0.38–1.0 μm | 0.01 (CW) and 1 $10^6$ (P) | CW and pulsed | Measurement instrumentation, spectroscopy, medicine, pollution detection |

*Source:* Based on http://www.ilt.fraunhofer.de/eng/100048.html.

- Material deposition (see Chapter 9)
- Etching with chemical assist, i.e., laser-assisted chemical etching (LACE; this chapter)
- Laser-enhanced jet plating and etching (see Chapter 9 and this chapter, respectively)
- Lithography (see Chapters 1 and 2)
- Surgery (see Volume III, Chapter 10)
- Photopolymerization (e.g., μ-stereolithography described in Chapter 9)

From this list, it is clear that lasers can be used for removal and for additive processes. We only consider ablative or subtractive laser processes here, as illustrated in Figure 5.12; additive laser processes are reviewed in Chapter 9. Interestingly, the laser may also serve as its own process monitor. This multiuse, multirole, and in situ capability is not offered with any of the other advanced material processing techniques such as molecular beam epitaxy, chemical vapor deposition, deep reactive ion etching, and so on.[11]

Earlier uses of laser beam machining (LBM) were mostly scientific in nature, and industrial applications involved macroprocesses, like sheet metal cutting and welding for the automobile industry (Figure 5.13). Today, industrial lasers are also used in a wide variety of microfabrication tasks. R. Srinivasan and V. Mayne-Banton of IBM Research first reported micromachining with laser ablation in 1982.[12] Initial studies by Srinivasan and coworkers showed that pulsed UV laser radiation at 193 nm (an ArF excimer laser) can be used to etch organic polymers several micrometers in depth. To illustrate how fast laser machining has improved and how little yet is known of the fundamental underlying processes in this field, consider that it was only recently recognized that femtosecond laser pulses interact with matter in a manner that is totally different from traditional lasers with longer pulses. During machining with femtosecond lasers, much less heat dissipates in the substrate, and, as

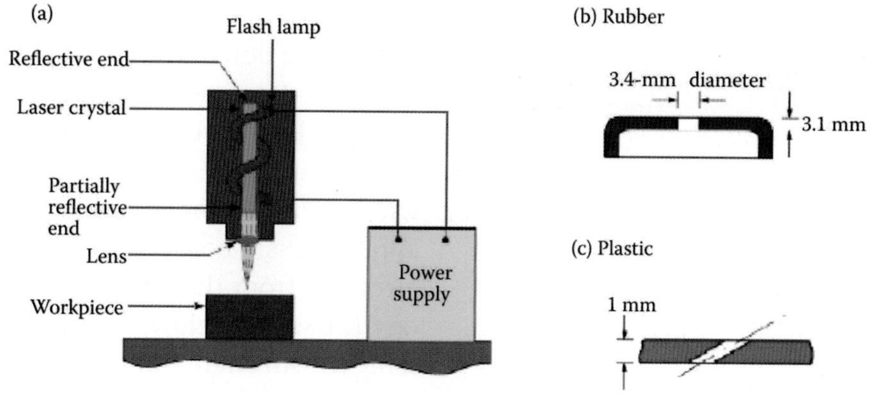


**FIGURE 5.12** (a) Schematic illustration of the laser-beam machining process. (b and c) Examples of holes produced in nonmetallic parts by LBM.

a consequence, ultrafast lasers present the potential for micromachined products of superior quality (http://www.clark-mxr.com/industrial/handbook/index.htm).

Although the number of lasers delivered for strict micromachining applications has been low compared with others so far, it is predicted that the next 10 years will see a phenomenal market growth for this application.

## Laser Parameters for Laser Material Removing—Laser Ablation

### *Introduction*

Laser ablation is the process of removal of matter from a solid by means of an energy-induced transient disequilibrium in the lattice. The characteristics of the released atoms, molecules, clusters, and fragments (the dry aerosol) depend on the efficiency of the energy coupling to the sample structure, i.e., the material-specific absorbance of a certain wavelength, the velocity of energy delivery (laser pulse width), and the laser characteristics (beam energy profile, energy density or fluency, and the wavelength). Table 5.4 summarizes all the different laser parameters that can be controlled and their effect on materials processing.[13] More specifically for micromachining purposes, the wavelength, spot size [i.e., the minimum diameter of the focused laser beam, $d_0$ (Equation 5.9)], average laser beam intensity, depth of focus, laser pulse length, and shot-to-shot repeatability (stability and reliability in Table 5.4) are the six most important parameters to control. Additional parameters we will cover in the section below, not listed in Table 5.4, concern laser machining in a jet of water and laser-assisted chemical etching (LACE).

Like other thermal removing techniques, laser machining leaves a heat-affected zone (HAZ) on the work piece where molten material resolidified in situ or where material was sufficiently heated and cooled rapidly enough to result in embrittlement. This change in material properties can alter subsequent laser ablation and affect material performance.

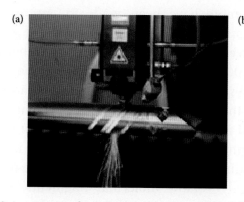

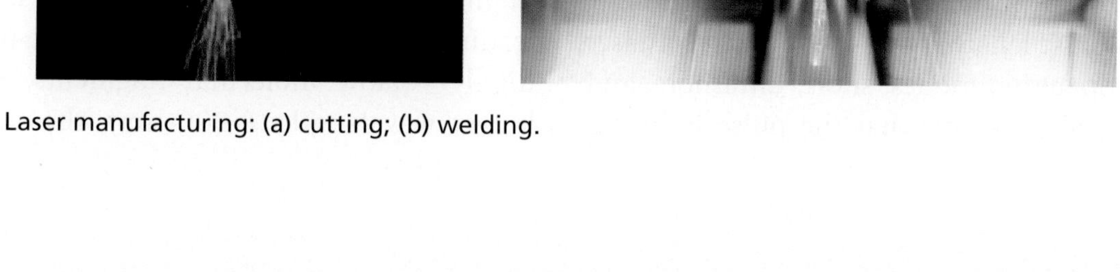

**FIGURE 5.13** Laser manufacturing: (a) cutting; (b) welding.

**TABLE 5.4** Laser Parameter and Related Processing Parameters

| Laser Parameter | Influence on Material Processing |
|---|---|
| Power (average) | Temperature (steady state)<br>Process throughput |
| Wavelength (μm) | Optical absorption, reflection, transmission, resolution, and photochemical effects |
| Spectral line width (nm) | Temporal coherence<br>Chromatic aberration |
| Beam size (mm) | Focal spot size<br>Depth of focus<br>Intensity |
| Lasing modes | Intensity distribution<br>Spatial uniformity<br>Speckle<br>Spatial coherence<br>Modulation transfer function |
| Peak power (W) | Peak temperature<br>Damage/induced stress<br>Nonlinear effects |
| Pulse width (s) | Interaction time<br>Transient processes |
| Stability (%) | Process latitude |
| Efficiency (%) | Cost |
| Reliability | Cost |

*Source:* Liu, Y. S. 1989. *Laser microfabrication: thin film processes and lithography.* Eds. J. Y. Tsao, and D. J. Ehrlich. New York: Academic Press. With permission.[13]

The size of the HAZ is a function of the laser pulse duration and the thermal conductivity and specific heat of the work piece. The HAZ depends on the distance the heat is conducted within the material and varies with material and laser wavelength. The better the material conduction (thermal diffusivity), the greater the extent of the HAZ is. Associated with the HAZ are a recast layer and the formation of burrs. During the ablation process, the expulsion of material in the plasma jet creates a compressive force on the molten pool of material under the laser spot. This will cause a portion of the liquefied material to be forced out of the ablation zone, and it will deposit onto the surrounding region. Debris is also a related problem of the ablation process. Particulates redeposit onto the surface as solid fallout of higher boiling point second phases from the plume and as condensation of the primary material. Taken together, all the preceding effects reduce the quality of the ablated region.

Based on pulse length, there are three major laser machining operation modes: short, ultrashort, and long or CW. *Short* means that the pulse is longer than 10 ps. *Ultrafast* or *ultrashort* means that the laser pulse has a duration somewhat less than 10 ps, usually some fraction of a picosecond. *Long* or *continuous wave* lasers are lasers that are operated in a CW (pulse duration > 0.25 s) rather than a pulsed mode. Each operation mode has its own micromachining applications. Short pulses of coherent light replace electrons from the electron beam vacuum technology (EBM; see below) as the cutting tool by vaporizing the substrate (subtractive process). No vacuum is required, but the removal rate is much slower than in EBM. For "hard to drill" materials and when holes are smaller than 150 μm, mechanical drilling becomes less cost effective, and laser removing techniques become an attractive alternative. With ultrashort pulses, so much energy is deposited in the material at such a fast rate that the material is forced into a plasma state; the material goes from a solid to a gas phase without first going through a melt phase. This plasma then expands away from the material as a highly energetic gas, taking almost all the heat away with it. This newest operation mode avoids a lot of the problems associated with the HAZ and promises to open up a new universe of very accurate micromachined parts unavailable with long laser pulsing. Long-pulse or CW lasers can be used to melt the metal for welding (additive process).

### Laser Wavelength

Laser wavelength has been identified as an important parameter influencing the efficiency of the energy transfer to the body of the irradiated sample (ablation characteristics). The selection of a wavelength with a minimum absorption depth helps ensure a high-energy deposition in a small volume for rapid and complete ablation. Several attempts have been made to describe the laser ablation phenomenon theoretically. It appears to be agreed now that, for very short wavelength radiation (~200 nm), direct photochemical bond breaking plays an important role in the ablation process. At that wavelength, photons have enough energy, and a one- or two-photon absorption by materials might cause direct bond cleavage in the solid, leading to rapid decomposition of the material into highly volatile molecular fragments. However, at longer wavelengths (>300 nm), the single-photon energy is not sufficient for bond breaking; two-photon or multiphoton avalanche effect processes must be considered (see below). As mentioned above, a shorter absorption depth in the material to be machined means a higher laser energy density in metal.

The decay of the intensity of an electromagnetic wave with E along the *x*-axis and propagating in the *z*-direction in an absorbing material we derived in Volume I, Chapter 5 as:

$$I(z) = I_m e^{-\alpha z} \quad \text{(Volume I, Equation 5.110)}$$

where $\alpha = \frac{4\pi n''}{\lambda}$ (with *n*″ the imaginary part of refractive index). Thus, the penetration depth *d* is given as [when d = z, then $I(z)/I_m = e$]:

$$d = 1/\alpha = \lambda / 4\pi n'' \quad (5.1)$$

The surface condition also has some effect on the laser energy coupling. Rougher surface helps absorbing more, and special surface treatments can enhance the absorption.

**Example 5.1**

For copper the complex refractive index at 5893 Å is given as n = 0.62 + i2.57 and its reflectivity R = 0.73. For a λ = 100 Å one obtains a d = 6.2 Å and at a λ = 10μm, d = 62 Å.

### *Laser Spot Size*

The laser beam spot size is another relevant factor influencing the efficiency and spatial resolution of laser ablation. Assuming cylindrical drilling, the ablated mass *M*, is given by:

$$M = \rho\pi L r^2 \quad (5.2)$$

where ρ = the mass density
L = the ablation depth
r = the crater radius

This means that a variation of the spot size *r*, at constant ablation efficiency, leads to a quadratic variation of the signal intensity as a direct consequence of the ablated mass involved.

If the laser beam diameter is not of a controlled diameter, the ablation region may be larger than desired and result in excessive slope of the sidewalls.

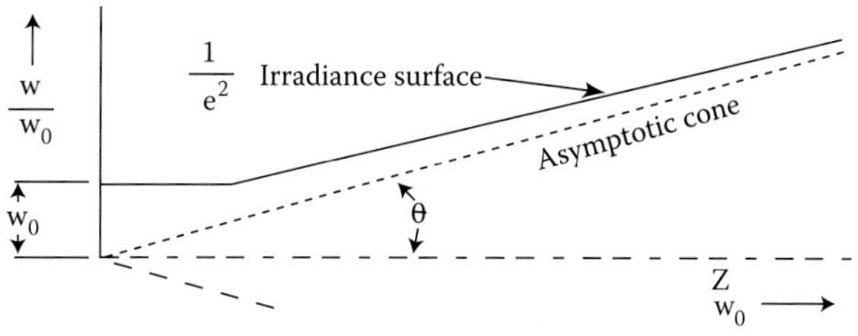


**FIGURE 5.14** Growth in beam diameter as a function of distance from the beam waist for a Gaussian laser beam profile.

Diffraction causes light waves to spread transversely as they propagate; therefore, it is impossible to have a perfectly collimated beam. The spreading of a laser beam is in accord with the predictions of diffraction theory. If the objective is to create the smallest possible feature, it is important to be able to focus the laser beam to the smallest spot size. The latter depends on several factors, but crudely one can say that the smallest spot that one can create is about half the wavelength of the light used; if the wavelength of light is 1 μm, then the smallest spot one can create is about 0.5 μm (see Volume I, Equation 5.1 with NA = 1).

In a more rigorous description, we start from the irradiance distribution of a Gaussian $TEM_{00}$* laser beam (Figure 5.14):

$$I(r) = I_0 e^{-\frac{2r^2}{w(z)^2}} = \frac{2P}{\pi w(z)^2} e^{\frac{2r^2}{w(z)^2}} \quad (5.3)$$

where r = the radial distance across the beam
P = total power in the beam in watts
w(z) = the radius or one-half the diameter of the $1/e^2$ irradiance contour after the wave has propagated a distance z

Gaussian beams propagate as Gaussians, and the expression for the spot size *w*(*z*)—the radius of the $1/e^2$ contour after the wave has propagated a distance *z*—is derived as:

$$w(z) = w_0 \sqrt{1 + \left(\frac{\lambda z}{\pi w_0^2}\right)^2} = w_0 \sqrt{1 + \left(\frac{z}{z_R}\right)^2} \quad (5.4)$$

where λ = wavelength of light
$w_0$ = the waist radius of the $1/e^2$ irradiance contour at the plane where the wavefront is

* For the significance of the $TEM_{00}$ laser beam designation, see Figure 5.19.

flat with all rays there moving in precisely parallel directions

$z_R$ = *Rayleigh range* or *confocal parameter*, i.e., the distance over which the beam remains about the same size (or "collimated")

In Figure 5.14 we show this growth in beam diameter as a function of distance from the beam waist for a Gaussian laser beam profile. The point at z = 0 marks the location of a beam waist, or a place where the wavefront is flat.

The Raleigh range ($z_R$), also called the confocal parameter, defined as the distance over which the beam radius spreads by a factor of $\sqrt{2}$, is given by (see Figure 5.15):

$$z_R = \frac{\pi w_0^2}{\lambda} \tag{5.5}$$

Far from the waist ($z >> z_R$), the laser beam asymptotically approaches the value:

$$w(z) = \frac{\lambda z}{\pi w_0} \tag{5.6}$$

From Figure 5.14 we see that a Gaussian beam does not diverge linearly; near the laser, the divergence θ angle is extremely small; far from the laser, half the divergence angle is given as θ ≈ tan (θ); thus, the $1/e^2$ irradiance contour asymptotically approaches a cone of angular radius:

$$\theta \approx \tan(\theta) = \frac{w(z)}{z} = \frac{\lambda}{\pi w_0} = 0.637\frac{\lambda}{w_0} = 1.27\frac{\lambda}{d_o} \tag{5.7}$$

This laser beam divergence is illustrated in Table 5.5, where we show the waist radius size $w_0$ of the beam and the collimation distances for two lasers, one at λ = 10.6 μm and one at λ = 0.633 μm.

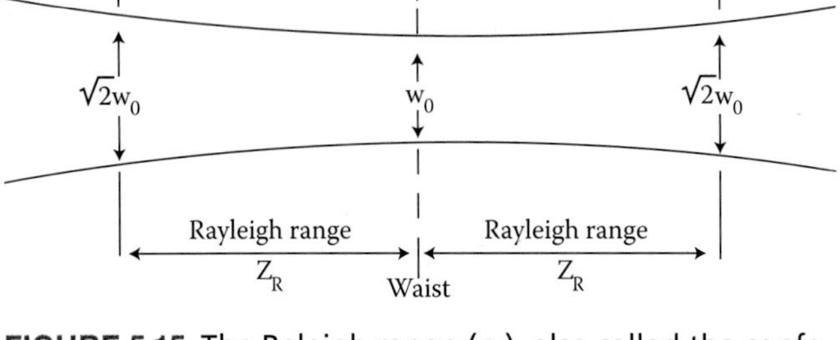


**FIGURE 5.15** The Raleigh range ($z_R$), also called the confocal parameter.

**TABLE 5.5** Waist Spot Size $w_0$ and Collimation Distances for Two Lasers at λ = 10.6 μm and λ = 0.633 μm

| Waist Spot Size $w_0$ | Collimation Distance λ = 10.6 μm | Collimation Distance λ = 0.633 μm |
|---|---|---|
| 0.225 cm | 0.003 km | 0.045 km |
| 2.25 cm | 0.3 km | 5 km |
| 22.5 cm | 30 km | 500 km |

We learn that tightly focused laser beams expand quickly and weakly focused beams expand less quickly, but still expand. As a result, it is very difficult to shoot down a missile with a laser. We also recognize that longer wavelengths expand faster than shorter ones.

Now let us focus this Gaussian laser beam using a lens (Figure 5.16) and calculate what its electric field will be one focal length *f* behind the lens. Gaussian beams focus as Gaussians.

If a "perfect" lens (no spherical aberration) is used to focus a collimated laser beam, the minimum spot size radius or the focused waist ($w_0$) is limited by diffraction only and is given by:

$$w_0 = \frac{\lambda f}{\pi w_{lens}} \tag{5.8}$$

where f = the focal length of the lens.

This is the same result as for a diffracting beam with z = f (Equation 5.6). Light being focused behaves just like diverging light, only reversed, and the divergence angle θ, given by Equation 5.7, is equally valid. With $d_o = 1/e^2$ the diameter of the focus (= $2w_0$) and with the diameter of the lens $D_{lens} = 2w_{lens}$ (or the diameter of the laser beam at the lens—whatever is the smallest), we obtain:

$$d_o = \frac{4\lambda f}{\pi D_{lens}} = \frac{1.27\lambda f}{D_{lens}} \tag{5.9}$$

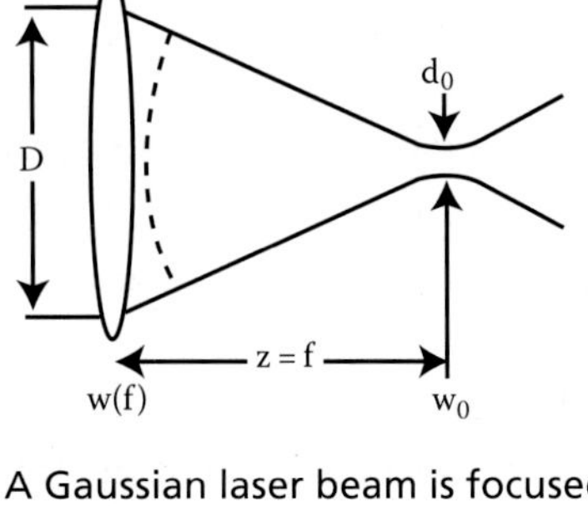


**FIGURE 5.16** A Gaussian laser beam is focused using a lens.

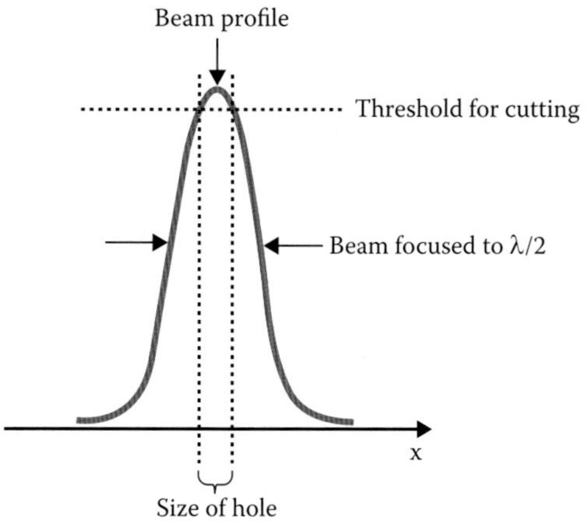


**FIGURE 5.17** Size of hole cut with a laser can be substantially smaller than the beam size.

Thus, the principal way of increasing the resolution in laser machining, as in photolithography, is by reducing the wavelength, and the smallest focal spot will be achieved with a large-diameter beam entering a lens with a short focal length.

It is worth noting that ultrafast laser pulsing can create substantially smaller features than that of the central wavelength of the laser pulse itself. Because high-order nonlinear-optical processes are responsible for cutting, there is effectively a threshold for cutting (see Figure 5.17). This means that careful control of the intensity can yield a very small hole. By focusing the ultrafast laser pulse into a small spot with a Gaussian intensity profile, one can adjust the intensity of the laser spot on the surface of the material so that just the peak of the beam is above the ablation threshold; as a consequence, only material in that very limited area of the laser spot will be removed. That limited area can be as small as one-tenth the size of the spot itself; therefore, the unique combination of multiphoton absorption and saturated avalanche ionization provided by ultrafast laser pulses makes it possible to machine materials on dimensions much smaller than 1 μm: holes as small as 100 nm have been made.

### *Depth of Focus or Depth of Field*

Twice the Raleigh range or 2 $z_R$ is called the "depth of focus" because this is the total distance over which the beam remains relatively parallel, or "in focus" (see Figure 5.18). Or also, the depth of focus or depth of field (DOF) is the distance between the

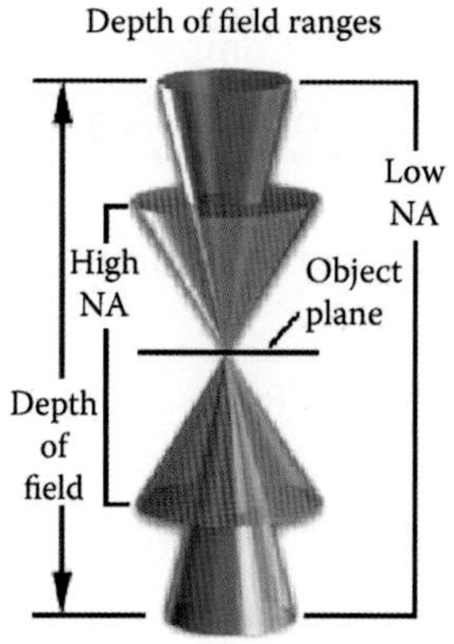


**FIGURE 5.18** Depth of field (DOF) ranges.

values where the beam is √2 times larger than it is at the beam waist (see again Figure 5.18). Based on Equations 5.5, 5.8, and 5.9, we then obtain:

$$\mathrm{DOF} = 2z_R = \frac{2\pi w_0^2}{\lambda} = \left(\frac{8\lambda}{\pi}\right)\frac{f^2}{D_{lens}^2} \tag{5.10}$$

The numerical aperture (NA) of a lens is given as $n\sin\theta$, and for small angles, one can derive that NA = D/2f with *D* the diameter of the lens and *n* the refractive index which in air = 1 (see Volume I, Chapter 5). Thus, we can recalculate the DOF also as:

$$\mathrm{DOF} = 1.27\lambda/\mathrm{NA}^2 \tag{5.11}$$

Material processing with a very short depth of focus requires a very flat surface. If the surface has a corrugated topology, a servo loop connected with an interferometric auto-ranging device must be used.[11]

We want to contrast the results for resolution and depth of focus of laser beams obtained here with the resolution and depth of focus we calculated for other light sources in Volume I, Chapter 5 and in this volume, Chapter 1. In Volume I, Chapter 5 we saw that the Rayleigh criterion defines two incoherent light sources, separated by a distance *d*, as resolved when:

$$d = 1.22\frac{f\lambda}{D} \qquad \text{(Volume I, Equation 5.12)}$$

With an ideal lens system of focal length *f*, the Rayleigh criterion further yields a minimum spatial resolution as:

$$d = \frac{1.22\lambda}{2\mathrm{NA}} \qquad \text{(Volume I, Equation 5.13)}$$

**Table 5.6** Comparing Standard Lens Parameters for Coherent and Incoherent Light

| | Incoherent | Coherent (Gaussian Laser Beam) |
|---|---|---|
| Lens diameter | D | $D_{lens} = 2w_{lens}$ |
| Focal spot radius size | $d = 1.22\frac{f\lambda}{D}$ | $d_0 = \frac{4\lambda f}{\pi D_{lens}} = \frac{1.27\lambda f}{D_{lens}}$ |
| Depth of focus | $DOF = 1.22\lambda\left(\frac{2f}{D}\right)^2 = 1.22\lambda/NA^2$ | $DOF = 1.27\lambda\left(\frac{2f}{D_{lens}}\right)^2 = 1.27\lambda/NA^2$ |

In this expression *d* is the smallest object that the lens can resolve and also the radius of the smallest spot to which a collimated beam of light can be focused. From Equation 1.17 in Chapter 1 the DOF for an incoherent light source can be derived as:

$$DOF = 1.22\lambda\left(\frac{2f}{D}\right)^2 = 1.22\lambda/NA^2 \quad (1.17)$$

We contrast these earlier results with those obtained for laser beams in Table 5.6. Whereas resolution is slightly worse for coherent light, its DOF is slightly better.

Laser beams can have any pattern, not just a Gaussian. The beam shape can even change with distance. Some beam shapes do not change with distance. These laser beam shapes are referred to as transverse electromagnetic (TEM) modes. Lasers can operate in various transverse modes as shown in Figure 5.19. Laser specifications will usually refer to the $TEM_{00}$ mode. The 00 mode is the Gaussian beam. Higher-order modes involve multiplication of a Gaussian by a "Hermite polynomial." The $TEM_{00}$ results in a single beam. The long narrow bore of a typical laser forces this mode of oscillation, and Equation 5.9 is only valid for this mode.

*Laser Power*

The laser power is also a critical parameter because it determines the damage impact on the irradiated solid. However, it must be considered in combination with the wavelength (see above). To clarify this, a distinction must be made between the optical penetration depth (Equation 5.1, wavelength-dependent) and the damage penetration depth (influenced by the power density as well). As we saw in Equation 5.1, the optical penetration depth is defined as the distance below the irradiated surface over which the laser energy is deposited. The fraction of this distance over which the irradiation exceeds the ablation threshold is defined as the damage penetration depth.

The laser intensity ($W/cm^2$) is given by the peak power (W) divided by the focal spot area ($cm^2$), and the peak power (W) is given as the pulse energy (J) divided by pulse duration (s). The intensity levels of a highly collimated, monochromatic, and coherent light beam at the work piece, which can be held in open air, may reach $10^6$ $W/mm^2$, enough to melt any metal. The laser fluency or energy density ($J/cm^2$) equals the laser pulse energy (J) divided by focal spot area ($cm^2$). A typical 50-μJ output of a KrF laser

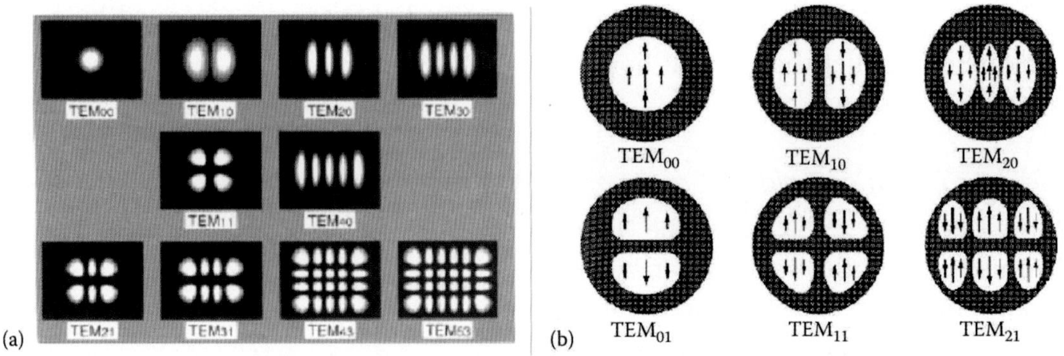


**FIGURE 5.19** Some transverse electromagnetic (TEM) modes: (a) irradiance; (b) electric field.

focused by an optics system with NA of 0.5 will produce a focal spot of 15–20-μm diameter and a corresponding fluence of 10–20 J/cm$^2$.

The average on-axis laser beam intensity is related to the total laser power, *P* in watts, by:

$$I(r) = \frac{2P}{\pi w_0^2} \quad (5.12)$$

where P = the total power in watts

$w_0$ = the Gaussian beam radius—the radius at which the intensity has decreased to $1/e^2$ or 0.135 of its value on the axis

The average beam energy is of no use if it cannot be properly and efficiently delivered to the ablation region; thus, beam quality is not only measured by the brightness (energy) but also by its focusability and homogeneity.

Now we will have a look at a fifth important laser parameter influencing micromachining quality, that is, laser pulse length.

### *Laser Pulse Length*

*Long Pulses* A further parameter with pulsed laser systems is the laser repetition rate, which expresses the number of pulses that are triggered per unit time. The laser pulse width determines the kinetics of the photon-induced disequilibrium on irradiated surfaces (for pulses longer than 100 ns the energy can be delocalized in thermal modes). It influences the role of the laser microplasma, too. The most fundamental feature of laser/material interaction in the long pulse regime (e.g., pulse duration 8 ns, energy 0.5 mJ) is that the heat deposited by the laser in the material diffuses away during the pulse duration; that is, the laser pulse duration is longer than the heat diffusion time. This may be desirable for laser welding, but for most micromachining jobs, heat diffusion into the surrounding material is undesirable and detrimental to the quality of the machining (http://www.clark-mxr.com). Here are a couple of reasons why one should avoid heat diffusion for precise micromachining:

- Heat diffusion reduces the efficiency of the micromachining process because it takes energy away from the work spot—energy that would otherwise go into removing work piece material. The higher the heat conductivity of the material, the more the machining efficiency is reduced.
- Heat diffusion reduces the temperature at the machining spot, pinning the working temperature to not much above the melting point of the material, which is removed by melting. Boiling of the substrate ejects globs of molten material away from the work zone. Those globs fall back onto the surface, contaminating the sample. The droplets can be rather large and retain a fair amount of residual heat so that they bind strongly to the sample surface. Removal of these contaminants is often difficult or impossible without damaging the work piece. An air jet is sometimes used to physically project the melt phase away from the work zone. This results in a cleaner-looking cut but contaminates the sample downstream.
- Heat diffusion reduces the accuracy of the micromachining operation. Heat diffuses away from the focal spot and melts an area that is much larger than the laser spot size. Therefore, it is difficult to do very fine machining; whereas the minimum laser spot size might be in the range of 1 μm or less, in many materials it is not possible to create features with dimensions much smaller than 10-μm diameter.
- Heat diffusion affects a large zone around the machining spot, a zone referred to as the *heat-affected zone* or HAZ. The heating (and subsequent cooling) waves propagating through the HAZ cause mechanical stress and may create microcracks (or in some cases, macrocracks) in the surrounding material. These defects are "frozen" in the structure when the material cools, and in subsequent routine use these cracks may propagate deep into the bulk of the material and cause premature device failure. A closely associated phenomenon is the formation of a recast layer of material around the machined feature. This resolidified material often has a physical and/or chemical structure that is very different from the unmelted material. This recast layer may be mechanically weaker and must often be removed.

- Heat diffusion is sometimes associated with the formation of surface shock waves. These shock waves can damage nearby device structures or delaminate multilayer materials. Whereas the amplitude of the shock waves varies with the material being processed, it is generally true that the more energy deposited in the micromachining process, the stronger the associated shock waves.

The various undesirable effects associated with long laser pulse etching are illustrated in Figure 5.20a and b.

*Short Pulses* We are particularly interested in femtosecond lasers with pulses more than 1 million times shorter than the several nanosecond-duration pulses used in traditional industrial laser micromachining systems (see above). Femtosecond lasers deliver an incredible amount of peak power, routinely up to 5–10 GW (this is more than the average power delivered by a large nuclear plant) (http://www.clark-mxr.com/industrial/handbook/introduction.htm). The laser intensity easily reaches the hundreds of terawatts per square centimeter at the work spot itself. No material can withstand the ablation forces at work at these power densities. This means that, with ultrafast laser pulses, very hard materials, such as diamond, as well as materials with extremely high melting points, such as molybdenum and rhenium, can be machined. The most fundamental feature of laser-matter interaction in the very fast pulse regime is that the heat deposited by the laser into the material does not have time to move away from the work spot during the time of the laser pulse. The duration of the laser pulse is shorter than the heat diffusion time. This regime has numerous advantages as listed below (http://www.clark-mxr.com/industrial/handbook/introduction.htm):

- Because the energy does not have the time to diffuse away, the efficiency of the machining process is high. Laser energy piles up at the level of the working spot, whose temperature increases instantly past the melting point of the material and keeps on climbing into what is called the *plasma* regime.
- After the ultrafast laser pulse creates the plasma at the surface of the work piece, the pressures created by the forces within it cause the material to expand outward from the surface in a highly energetic plume or gas. The internal forces that previously held the material together are vastly insufficient to contain this expansion of highly ionized atoms and electrons from the surface. Consequently, there are no droplets that condense onto the surrounding material. Additionally, because there is no melt phase, there is no splattering of material onto the surrounding surface.

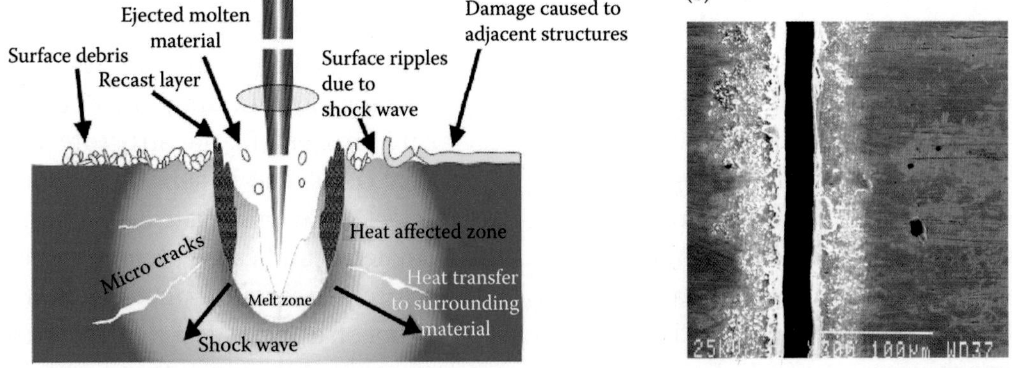

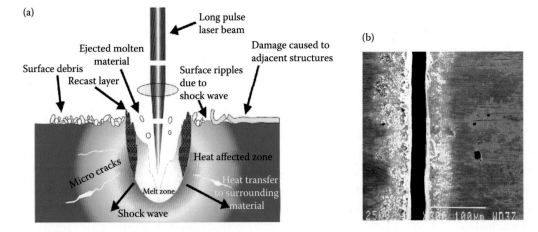


**FIGURE 5.20** (a) Various undesirable effects associated with laser pulses that are too long, for example, pulse duration 8 ns, energy 0.5 mJ (http://www.clark-mxr.com). (b) Example of a 25-μm (1 mil) channel machined in 1-mm (40 mils) thick INVAR with a nanosecond laser. INVAR is often called for in the design of machinery that must be extremely stable. This sample was machined using a "long" pulse laser. The laser pulse parameters are pulse duration 8 ns, energy 0.5 mJ. A recast layer can be clearly seen near the edges of the channel. Large debris is also seen in the vicinity of the cut.

- Heating of the surrounding area is significantly reduced, and, consequently, all the negatives associated with a HAZ are no longer present—no melt zone, no microcracks, no shock wave that can delaminate multilayer materials, no stress that can damage adjacent structures, and no recast layer.

The many advantages associated with working in the femtosecond pulse range are illustrated in Figure 5.21a and b.

With ultrashort laser pulsing, it is possible to accurately control the depth of the cut by counting the number of pulses. Deep cuts (hundreds of microns) have been made this way, mostly by using an excimer laser, although it is now known that clean ablative etching, once the hallmark of excimer lasers, can also be achieved with pulsed laser sources at other wavelengths, including $CO_2$, copper vapor, and Nd:YAG and its harmonics. The latter lasers can be as effective for high-quality ablative micromachining provided laser photons are strongly absorbed in submicron depths at the surface of the material on time scales that are less than the time it takes for heat to diffuse away from the irradiated region. Lasers providing ultrafast pulse durations of less than 10 ps have, in recent years, demonstrated remarkably clean micromachining of many materials.

Ultrashort laser pulses have opened up many new possibilities in laser-matter interaction and materials processing. The extremely short pulse width makes it easy to achieve very high peak laser intensity with low pulse energies. The laser intensity can reach $10^{14}$–$10^{15}$ W/cm$^2$ with a pulse $<1$ mJ when a subpicosecond pulse is focused to a spot size of a few tens of micrometers. The extremely high intensity and short laser/matter interaction times lead to a highly nonequilibrium plasma state in the material under irradiation, where electrons are driven to much higher temperatures than ions. The short duration means that the hydrodynamic motion of the matter under laser irradiation can be ignored, and there is essentially no fluid dynamics to consider during the laser/matter interaction. The catch with ultrafast laser ablation is that throughput is low because the rate of removal of material is dependent on average power. The technology that makes these ultrafast laser pulses does not produce high average power. Additionally, the technology is very expensive (a system specifically tailored to an application can cost in the six- to seven-figure range), and the covered areas are very small.

In the past two decades lasers have become a precise, fast, wearless, flexible tool used for cutting, drilling, marking, and welding. Many of today's production processes are unimaginable without the laser.

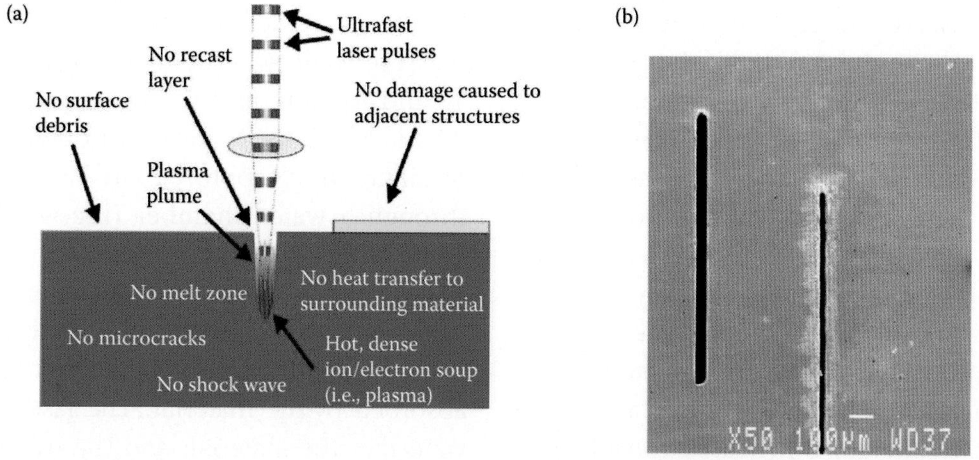


**FIGURE 5.21** (a) Advantages associated with working with very short laser pulses (<10 ps). (b) A channel machined in 1-mm-thick INVAR, nickel/iron alloy under the same experimental conditions as the long pulse channel in Figure 5.20b, but with ultrafast pulses. The channel on the left was machined with 200-fs pulses, 0.5 mJ energy per pulse. It is obvious that the channel machined with femtosecond pulses is cleaner than the sample machined with nanosecond pulses (channel on the right). Note also the absence of a recast layer on the left channel. It is also clear that the machining process was more efficient—the channel is larger. The edges are straighter. Overall the quality of the micromachining is much higher.

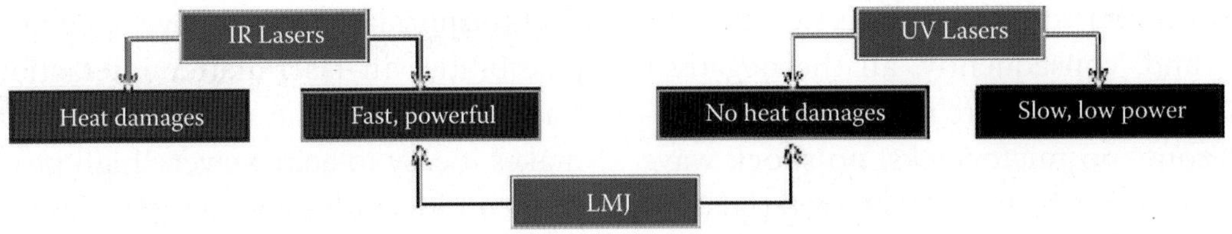


**FIGURE 5.22** The laser dilemma. According to Synova, S.A., their water jet-guided laser, the Laser-Microjet (LMJ) (see also Figure 5.23), promises a solution.

Nevertheless, in some applications, as for example in the electronic or medical field, the laser is still too imprecise, too slow, or unacceptable because of material damages. High-power IR lasers, widely used in industrial material processing, are too "hot," creating heat damage and debris; meanwhile, shorter wavelengths found in UV lasers create less heat and debris but lack power for efficient material processing. This dilemma is summarized in Figure 5.22. According to some, water jet-guided laser machining, discussed below, might provide the solution for this dilemma.

### *Shot-to-Shot Repeatability*

The process of laser ablation requires threshold fluency, and, as we have seen above, fluency is the laser pulse energy per unit area, and it can be changed either by changing the laser pulse energy or by changing the ablation area. At fluencies below the threshold, only miniscule etching (<0.05 μm/laser pulse) is observed, if any at all. However, when laser fluency is above the threshold, the etch depth per pulse increases rapidly with increasing laser fluency. The threshold fluency depends mainly on the target material structure and target absorption coefficients at the laser wavelengths. Absorption of laser radiation leads to a temperature increase in the target material. When laser pulse fluency is below the ablation threshold, the temperature increase is directly proportional to the laser pulse fluency, whereas when laser pulse fluency rises above the threshold, the temperature increases considerably more slowly. The absence of free electrons in a given substrate prevents the avalanche process from significantly starting and ultimately hinders the material ablation. In metals, there are always plenty of free electrons, and the avalanche process starts immediately, leading to reproducible machining (as we have seen, there may be other problems associated with heat diffusion). In semiconductors or insulators, there are naturally very few free electrons. The avalanche process may start right away or may not, depending on the presence or absence of a free electron in the beam path. This variability, which is inherent to the physical process, leads to unstable machining rates. The laser may be perfectly stable, and the beam spot size and amount of energy in the pulse may be precisely the same from shot to shot, yet the material ablation can vary significantly. Using lasers working in the UV range, or using ultrafast lasers, the electromagnetic field is so high that "bound" electrons are knocked free so that a large quantity of free electrons can start the avalanche ionization process immediately, reliably, and reproducibly. The latter ensures an excellent shot-to-shot reliability independent of the material.

### *Water Jet-Guided Laser Machining*

In water jet-guided laser machining, a thin jet of high-pressure water (the diameter of the jet is between 40 and 100 μm, and the water pressure is between 20 and 500 bars) is forced through a nozzle (made of diamond or sapphire). The laser beam is focused through a water chamber (the water is deionized and filtered) into a nozzle as shown in Figure 5.23.

Leaving the nozzle, the laser beam is guided inside the water jet by total reflection at the water-air interface and is directed onto the work piece where it is absorbed by the material. The laser beam melts or vaporizes the material, and the water jet expels the molten debris. At the end of each laser pulse, the plasma disappears, and only the water jet remains in contact with the surface. Before the next pulse is emitted, the water jet removes any heat induced in the material during the laser ablation. Thus, the pulsed mode of the laser ensures continuous alternation of

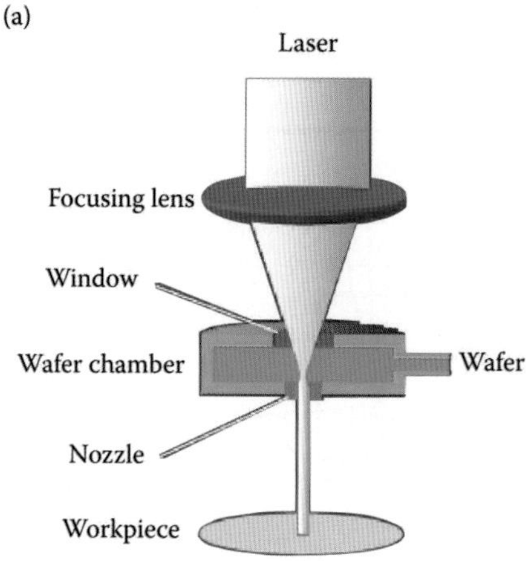

**FIGURE 5.23** (a) Schematic water jet-guided laser machining setup. See text. (b) Water jet-guided setup from Synova, S.A., the Laser-Microjet®.

heating (during laser pulse) and cooling (between laser pulses). The water pressure value (20–500 bars) depends on the application, but the resulting force acting on the work piece is always very small, smaller than the force as a result of the cutting gas of conventional lasers. Thus, the water jet is obviously not abrasive in this case (as it is in mechanical water jet machining described in Chapter 6); the material is solely ablated by the laser (thermally). The high speed of the water, up to 300 m/s at 500 bars, ensures fast removal of debris. Water jet-guided laser etching prevents the problem of heat damage (microcracks, structural changes) inherent in any laser ablation process. The result is a heat damage-free material ablation.

The laser dilemma of high-power IR lasers that are too "hot," creating heat damage and debris, and the shorter wavelengths UV lasers that create less heat and debris, but lack power for efficient material processing, was summarized in Figure 5.22. Water jet-guided laser machining discussed here might provide a solution for this dilemma because it transforms a "hot" IR laser into a "cool" laser, avoiding heat damage. Water jet-guided machining combines high ablation efficiency, speed, and precision while avoiding material damage as demonstrated in Figure 5.24, where we compare laser ablation in Si using a conventional laser (Figure 5.24a) with jet-guided laser machining using the Laser-Microjet from Synova (Figure 5.24b). The Laser-Microjet was conceived in 1993 by scientists at the Institute of Applied Optics of the Swiss Federal Institute of Technology in Lausanne, Switzerland. Ten years later, this technology has proven its efficiency and reliability in many applications, especially of the semiconductor and microelectronics industries. The water jet-guided laser has indisputable advantages over the classical methods. The conventional laser kerv in Si exhibits typical heat damage with burrs and deposits, whereas the Laser-Microjet result exhibits neither burrs nor deposits.

### *Laser Etching with Chemical Assist*

We just saw how water jet-guided laser machining helps keep a work piece clean; an air jet also may help to physically project the melt phase away from the work zone. This again results in a cleaner-looking cut but may contaminate the sample downstream. A third way to keep the work piece clean is the use of laser-assisted chemical etching (LACE).

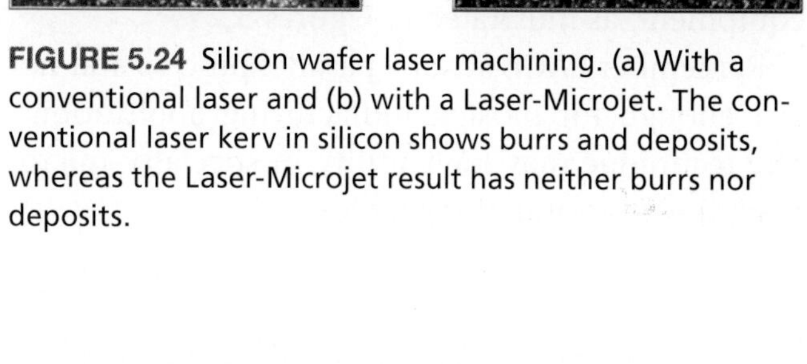

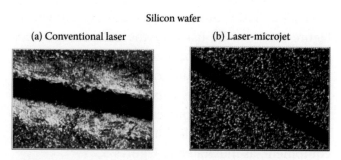


**FIGURE 5.24** Silicon wafer laser machining. (a) With a conventional laser and (b) with a Laser-Microjet. The conventional laser kerv in silicon shows burrs and deposits, whereas the Laser-Microjet result has neither burrs nor deposits.

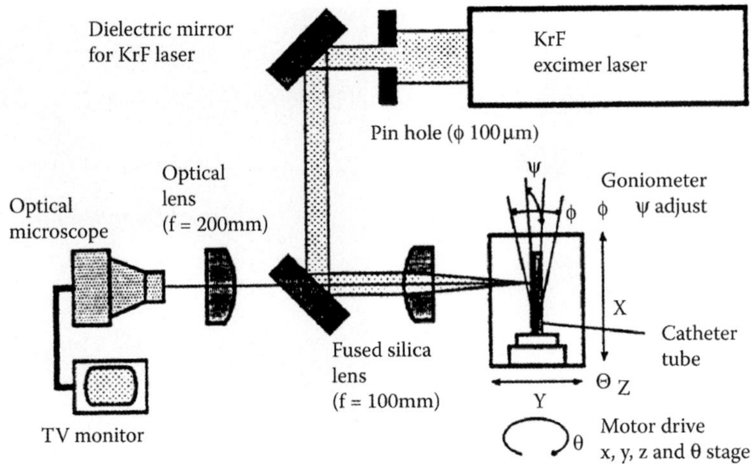


**FIGURE 5.25** Laser microetching experimental setup: laser-assisted chemical etching (LACE). Etching/ablation of a catheter tube.

The laser cutting action can be assisted by dry or wet chemical etchants for building of higher-quality complex 3D micromachines. The process is also called *laser-induced chemical etching*, in which, for example, chlorine is used as the etchant and is photodissociated to react with the substrate.[14,15] If laser etching is performed in chlorine gas instead of open air, the silicon vapor and droplets leaving the work area combine with the chlorine to form $SiCl_4$ gas and are carried away from the work area. Silicon, when heated to its melting point, reacts nearly at the gas transport-limited rate with chlorine, enabling removal rates exceeding $10^5$ $\mu m^3/s$.[16] The resulting hole is cleaner and more sharply defined, and debris does not splatter on the surface of the substrate; thus, an excellent spatial resolution results (1 μm *x,y*-resolution at a removal rate of $2.10^4$ $\mu m^3/s$).[16,17] The latter removal rate is several orders of magnitude faster than with electrical discharge machining (EDM) methods.

Flexible laser machining setups as shown in Figure 5.25 can be used for the machining of 3D parts by using highly localized laser microchemical reactions and an accurate computer-controlled *x*, *y*, *z* and θ translation table. This same type of setup can accommodate additive processes (see Chapter 8), but it is more convenient to build separate, dedicated equipment, as illustrated in Figures 5.25.[18]

Remember LACE is not a parallel process and not fast enough for most manufacturing applications. The technique may have utility in specialty micromachining or for making molds.

At the Massachusetts Institute of Technology Lincoln Laboratories, scanning of the laser beam at a rate of 7500 μm/s is used to remove Si layers in a $Cl_2$ etchant. After a plane of 1 μm is removed, the focusing objective is lowered 1 μm, and a new pattern is etched. With this technique, it is very easy to produce multilevel-type structures. Some example devices made this way are shown in Figure 5.26. The Lincoln group also demonstrated that structures as shown can be replicated by using simple hot embossing molding techniques.[16]

In addition to $Cl_2$, reagents such as $Br_2$, HCl, $XeF_2$, KOH, and so on can also be used as the etchant in the laser photochemical etching of silicon. Table 5.7 summarizes several kinds of substrates etched by laser processes with different etchants and laser wavelengths.[11,20]

## Laser Machining Applications

### *Overview*

Advantages of laser cutting over mechanical cutting vary according to the situation, but important factors are lack of physical contact (because there is no cutting edge that can become contaminated by the material or contaminate the material) and, to some extent, precision (because there is no wear on the laser). There is also a reduced chance of warping the material that is being cut because laser systems have a small heat-affected zone (HAZ). Some materials are also very difficult or impossible to cut by more traditional means. The laser has other distinct

(a) (b)

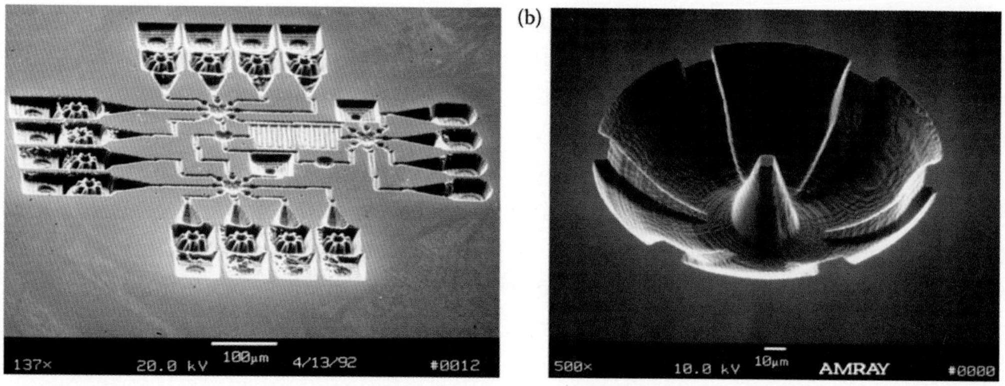


**FIGURE 5.26** Some examples of laser-machined shapes. [Bloomstein, T. M., and D. J. Ehrlich. 1991. *Sixth international conference on solid-state sensors and actuators (Transducers '91).* San Francisco.[19]] The devices are built using chlorine-assisted laser etching (100 Torr $Cl_2$). The beam spot used is 1 μm, and the objective is lowered in 1-μm increments after each plane is scanned at 7500 mm/s. Etch rate 100,000 $mm^3/s$. (a) SEM micrograph of a microfluidic device layer cut into silicon. The white bar is a 100-μm marker. (b) SEM micrograph of a cone inside a dish cut into silicon. The dish has an upper radius of 80 μm and a lower radius of 35 μm with a radius curvature of 82 μm and spacer layers extending 8 μm inward. The clipped cone has an upper radius of 4 μm and lower radius of 16 μm. The laser was swept in a circular scan in each plane. The white bar is a 10-μm marker. (Courtesy of Dr. D. Ehrlich, Revise, Inc.)

advantages: aspect ratios for hole diameter to depth on the order of 1:50 are possible, and holes can be drilled in hard-to-reach areas and at difficult angles (e.g., 10° to the surface). A laser-based scribe can make narrower trenches than a mechanical diamond stylus. Disadvantages of laser cutting may include the high energy required. Also, when hundreds of holes all at the same angle are to be drilled, the job is better handled by an electron beam machine (EBM). However, if holes have to be drilled at many

**Table 5.7** Examples of Etchants and Substrates Etched by the Laser Photochemical Process

| Substrate | Etchant(s) | Etch Rate | Laser/Intensity |
|---|---|---|---|
| **Semiconductors** | | | |
| Si | $Cl_2$,HCl | 7 μm/s | $Ar^+$/>5 $MW/cm^2$ |
| Si | KOH | 15 μm/s | $Ar^+$/~$10^7$ $W/cm^2$ |
| Si | NaOH | 2 μm/min | $CO_2$/NA |
| Ge | $Br_2$ | 36 μm/s | $Ar^+$/0.1 $kW/cm^2$ |
| GaAs | $HNO_3$ | 2 μm/s | $Ar^+$/60 $MW/cm^2$ |
| GaAs | $CH_3Br$ (750 Torr) | 60 Å/s | $Ar^+$ (257 nm)/1 $KW/cm^2$) |
| GaP | KOH | 600 Å/s | $Ar^+$ (351 nm)/3.5 $KW/cm^2$ |
| InP | $CH_3Br$ (750 Torr) | 9.4 Å/s | $Ar^+$ (257 nm)/100 $W/cm^2$ |
| **Insulators** | | | |
| $SiO_2$ | $SiH_4$ | 40 Å/s | KrF/0.3 $J/cm^2$ |
| $SiO_2$ | $NF_3$:$H_2$ | 0.12 nm/s | ArF/7.7 $mJ/cm^2$ |
| Polyimide | KOH | 0.3 μm/s | $Ar^+$/0.03 $MW/cm^2$ |
| Polyimide | Air | 0.15 μm/pulse | KrF/0.3 $J/cm^2$ |
| PMMA | Air | 0.3 μm/pulse | ArF/0.25 $J/cm^2$ |
| Diamond | $Cl_2$, $O_2$,$NO_2$ | 1,400 Å/pulse | ArF/30 $J/cm^2$ |
| **Metals** | | | |
| Ti | $NF_3$ | 0.29 Å/pulse | 0.115 $J/cm^2$ |
| W | $Cl_2$ (0.1 Torr) | 240 Å/min | 0.12 $J/cm^2$ |
| Al | $Cl_2$ (0.1 Torr) | 1.4 μm/s | 1.0 $J/cm^2$ |
| Ag | $Cl_2$ (0.1 Torr) | 500 Å/min | $N_2$ (337 nm)/0.12 $J/cm^2$ |

*Source:* Helvajian, H. 1995. *Microengineering technology for space systems.* Ed. H. Helvajian. El Segundo, CA: Aerospace Corporation; and Eden, J. G. 1986. Photochemical processing of semiconductors: new applications for visible and ultraviolet lasers. *IEEE Circuits Devices Mag* 2:18–24.[11,20]

angles, the laser may be the most cost-effective method. Lasers are most cost-effective for drilling holes between 0.01 and 1.5 mm. For larger holes, the cost of the large laser oscillator needed to produce the energy for vaporization becomes prohibitive, and because of the large amounts of material being vaporized, the hole produced suffers in terms of quality and controllable tolerance. With the very small holes, the high-power focusing lenses required to withstand the energies necessary for drilling holes smaller than 0.01 mm are limiting factors. The requirement for even smaller features necessitates the use of EBM (a thermal technique, see below), focused ion beam (FIB, a mechanical technique; see Chapter 6), or ultrafast laser pulsing (see above).

Both solid-state and gas lasers are commercially available and are touted for traditional machining operations in electronics such as postassembly processing, where site-specific action is necessary, for example, for drilling via holes, repair of embedded interconnects, and thick/thin film resistor and capacitor trimming. The most popular lasers for cutting materials are $CO_2$ and Nd:YAG, although semiconductor lasers are gaining prominence because of their higher efficiency. Because laser micromachining or direct ablation needs high laser intensity or fluence, usually only excimer lasers are used, especially for hard materials.

In recent years, excimer lasers, providing ultrafast durations of $< 10$ ps ($10^{-11}$ s) based on diode-pumped solid-state gain media, have demonstrated remarkably clean micromachining of many materials. Despite initially gaining a poor reputation for reliability, during the past 25 years, excimer lasers have developed into mature industrial tools. For laser ablation of polymers, argon ion lasers, a kind of continuous laser, can also be used, and they can be modulated into different pulse widths, frequencies, and wavelengths. However, the excimer laser has several distinct advantages, even for the dry etching of polymers, over the $CO_2$ and Nd:YAG lasers. For example, the wavelengths are more compatible with the chemical bond energies in organic compounds and tend to produce less thermal damage. Excimer lasers, as we ascertained in Chapter 1, are also used in the newest lithography processes. A visual illustration is obtained comparing the hole-drilling capabilities of mechanical twist drill bits and excimer lasers in Figure 5.27. Here we compare 100-μm-diameter holes drilled in 75-μm-thick high-density polyethylene with a drill bit and a KrF excimer laser. Although mechanically the hole is close to the minimum size that can be drilled, the improved quality of this and much smaller holes drilled by excimer lasers is obvious.

Laser machining with chemical assist takes care of some of the problems associated with long laser pulse operation, but, as we have already seen, ultrashort laser pulsing takes care of most of them. Laser cutting machines have been used extensively in industry for cutting complex profiles in sheet and plate materials. Hard materials such as diamond, tungsten, and titanium can be cut without problems of tool wear because there is no contact between tool and work piece. For many applications, laser micromachining offers the advantage of more rapid prototyping as compared to, e.g., FIB machining or lithography techniques.

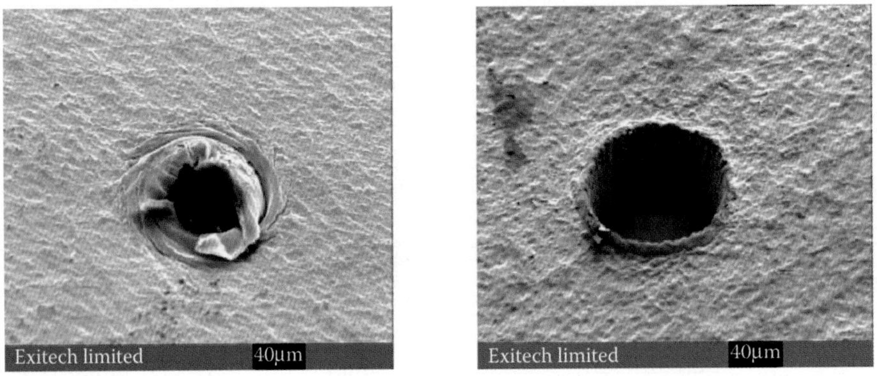


(a) (b)

**FIGURE 5.27** Holes drilled with twist drill bit (a) and KrF laser (b). (From Exitech.)

For a particular micromachining application the choice of laser type is now judged as much by criteria such as process speed, part throughput, reliability, service intervals, capital and operating costs of the overall machine tool rather than solely by the quality of the processed part.

### *Long Laser Pulse Machining Applications*

A simple example of long laser pulse machining is a hole, laser drilled through glass or Si, that can be used as a high-aspect-ratio, through-the-wafer via.[21,22] This is illustrated for the electrical contact via of a solid-state capacitive pressure sensor in Figure 5.28.[21] This compact pressure sensor incorporates electrically conducting vias drilled through the glass cap by a $CO_2$ laser. In contrast, the through-the-wafer via connection fabricated by anisotropic etching has a large opening angle (54.7°) associated with the slow etching crystallographic planes of the Si substrate. Typically, the closest spacing for vias that are wet etched in a 300-μm-thick wafer is approximately 425 μm, thus limiting this method to a low density of via holes (different means of making vias are discussed in Volume III, Chapter 4).

When holes as shown in Figure 5.28 are laser drilled in a substrate, the expelled material solidifies into dust; therefore, a moving protective tape is used in front of the laser lens to keep it clean. Debris also collects around the perimeter of the openings in, e.g., a silicon substrate, when drilling, scribing, or trenching is done in air or an inert atmosphere.

To avoid the identified problems associated with long laser pulses, two methods can be used. One is to use laser-assisted chemical etching (LACE), as described above, which takes care of some of the mentioned problems; the other approach, taking care of most of the mentioned problems, is to use femtosecond laser pulses. The latter technique, as we have explained above, results in superior micromachining.

### *Short Laser Pulse or Excimer Laser Applications*

*Introduction* Energy-deposition time with ultrafast excimer lasers is short compared with the electron-phonon time (a few picoseconds); thus, thermal conduction and hydrodynamic motion are generally negligible, and ejected material carries away most of the deposited energy. As a result, there is minimal collateral damage to remaining material. In metals, laser energy is deposited over the skin depth rather than spread over a thermal conduction length, yielding very high energy densities, temperatures (10 eV), and pressures (Mbar) during the pulse that are not that different from those in the sun (Figure 5.29). In Figure 5.29 we compare existence domains for various types of plasmas as a function of temperature and density. In dielectrics, electrons that are produced by nonlinear light absorption initiate the breakdown, and full breakdown is achieved during each pulse.

Excimer laser materials processing, by providing solutions to critical problems in manufacturing of integrated circuits, hard disks, displays, interconnects, desktop printers, and telecommunication

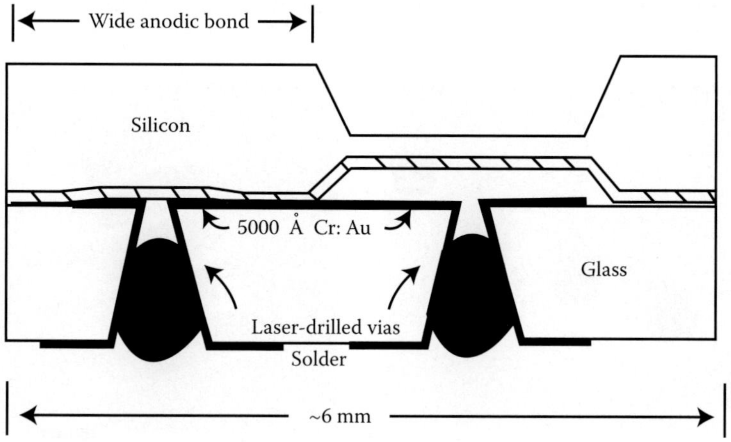


**FIGURE 5.28** Electrical contact in a silicon capacitive pressure sensor using laser-drilled vias. (From Bowman, L., J. M. Schmitt, and J. D. Meindl. 1985. *Micromachining and micropackaging of transducers.* Eds. C. D. Fung, P. W. Cheung, W. H. Ko, and D. G. Fleming. Amsterdam: Elsevier. With permission.[21])

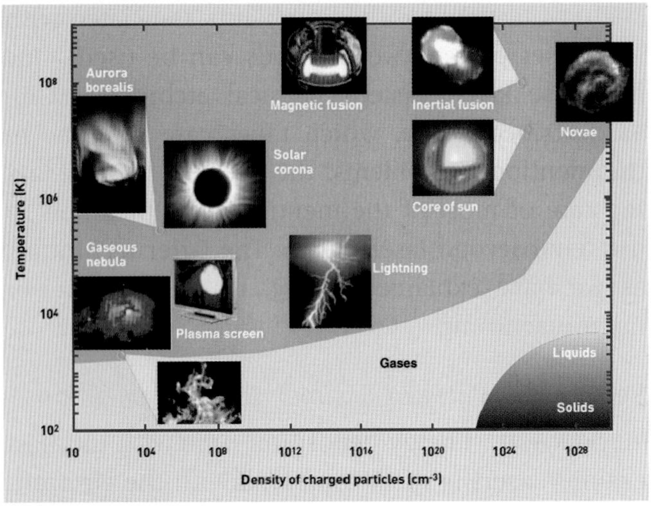


**FIGURE 5.29** Existence domains for various types of plasmas as a function of temperature and density. (Extreme measurements in laser generated plasmas, CLEFS CEA-No. 4, Autumn 2006.[25])

devices, has become a key enabling technology for the current revolution in information technology.

*Ink-Jet Printers, Microvias, Catheters, and DVDs* Excimer lasers on production lines in Asia, Europe, and the United States now drill nozzle-hole arrays for most ink-jet printers. Earlier 300-dpi printers consisted of a 100-nozzle row of 50-μm-diameter holes made by electroforming thin nickel foil. Trying to fabricate more holes with smaller diameters reduced the already low 70–85% production yield even further. Increased printer quality is achieved by simultaneously reducing the nozzle diameter and decreasing the hole pitch. Modern printers like Hewlett Packard's (Palo Alto, CA) DeskJet 800C and 1600C have nozzles with 28-μm diameters, which provide resolution of 600 dpi. Laser drilling of nozzle arrays allows average yields greater than 99%. Excimer laser tools with appropriate CNC programming can readily engineer custom-designed reverse-tapered, $2\frac{1}{2D}$ and 3D structures. Some applications need a rifled, tapered hole to spin the droplet and optimize its accuracy of trajectory. Most of the ink-jet printer heads being currently sold (e.g., by Hewlett Packard and Canon) are excimer laser drilled on production lines in the United States and Asia. Figure 5.30 shows an excimer laser-drilled nozzle array in a modern print head.

Another example application of excimer lasers is drilling of microvia much smaller than the ones depicted in Figure 5.30. Packages on which chips are mounted for connection to other devices have to keep pace with the rapid advances made in ICs, so that speed, power, and area (real estate) do not become compromised. In multilayer sandwiches, blind via holes provide high-speed connections between surface-mounted components on the board and underlying power and signal planes while minimizing valuable real estate occupation. The cost of drilling these vias on such high-density packages can represent 30% of the overall cost of the board. The process was first applied to volume production when the Nixdorf computer plant of Siemens introduced polyimide ablative drilling of 80-μm-diameter vias in multichip modules. Other mainframe computer manufacturers such as IBM rapidly followed suit and installed their own production lines for this application. Trillions of vias have now been drilled with excimer lasers at

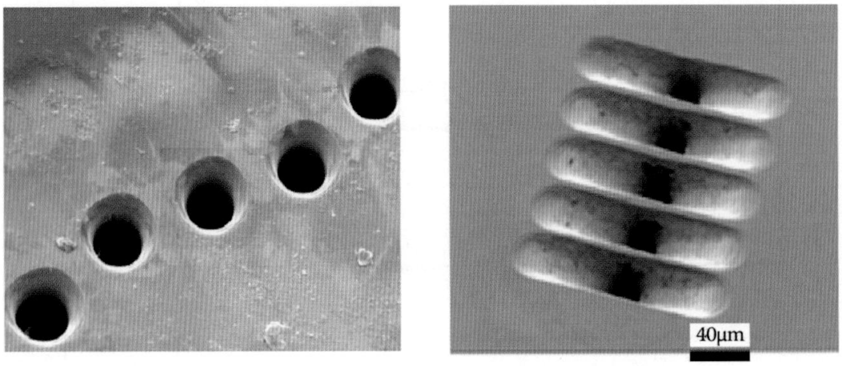


(a) (b)

**FIGURE 5.30** KrF laser-drilled ink-jet printer nozzles in polyimide. (a) 30-μm diameter nozzle array. (b) Nozzles with nonlinear tapers to aid the laminar flow of the droplet through the orifice. (From Excitech.)

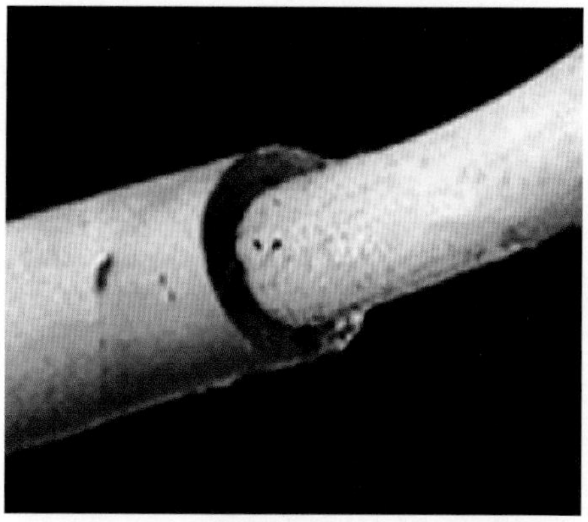

**FIGURE 5.31** Stripping of wires with excimer lasers.

yields greater than 99.99% and mean time between failure at greater than 1000 hours.

Preferential excimer laser etching of plastics against metals is illustrated by the stripping of insulation from insulation sleeving of wires with widespread use in preparing connection wires to computer hard-disk reader heads (see Figure 5.31).

In the biomedical industry, the drive for increased miniaturization with improved device functionality is crucial to rapid progress, for example, in novel catheter-based sensors and actuators. Precision microdrilling with excimer lasers is routine when making delicate catheter-based probes. Here we explore, as a futuristic example, a fiberoptic-based sensor for analyzing arterial blood gases (ABGs) and pH. The use of fiberoptic sensors for ABG analysis provides clinical diagnostics at the patient's bedside without the need for taking blood samples or doing an analysis in the blood gas laboratory (except for a much less frequent calibration of the in vivo sensors). Envision one lumen of a bilumen catheter for blood drawing and one for the sensing optical fiber as shown in Figure 5.32.

The hole at the side of a polyvinyl chloride bilumen sleeving tube through which blood is drawn (to collect blood for the blood gas lab for calibration of the in vivo sensor) is machined using a krypton fluoride (KrF) laser. The clean-cutting capability of the laser provides the catheter with the necessary rigidity that prevents kinking and blockage of the tube when inserted into the artery. The other important components of this in vivo catheter are the oxygen, $CO_2$ sensors, and pH sensors on the side of a fiberoptic cable placed in the other lumen. These consist of ~50 × 15-μm rectangular holes machined in a 100-μm-diameter acrylic [poly(methylmethacrylate) (PMMA)] optical fiber with an argon fluoride (ArF) laser. The holes are filled with reagents whose optical transmission depends on the $O_2$, $CO_2$, and pH levels of the surrounding blood.

For an alternative ABG and pH sensor made with Si bulk micromachining, see Figure 4.92 depicting an electrochemical ABG and pH sensor.

Another typical application of laser micromachining is to increase the density of optical storage media such as DVDs. Currently, the track spacing on a DVD is 400 nm, yielding approximately 6 GB of data storage space. The initial target spacing for the next generation of laser micromachining systems is 100 nm, resulting in 25-GB storage capacity on a DVD.

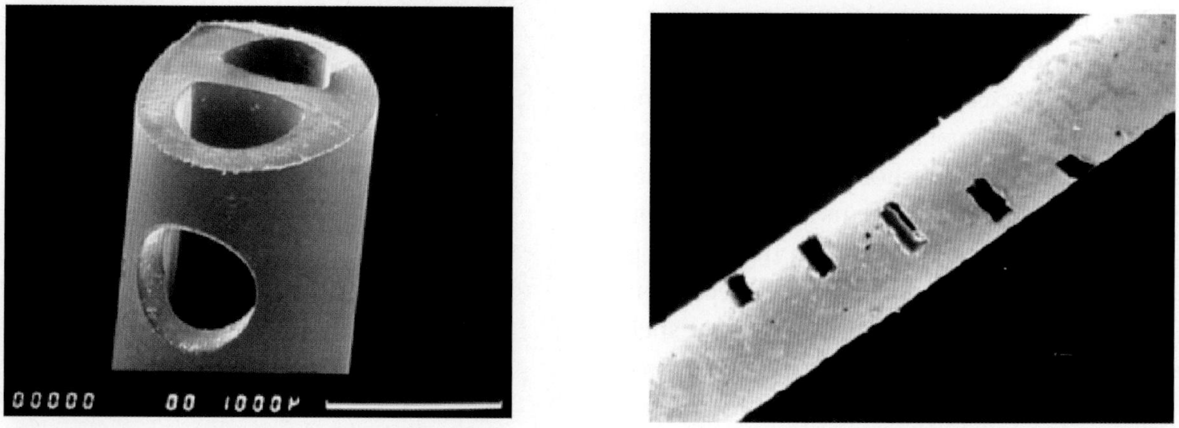

(a) (b)

**FIGURE 5.32** Excimer laser manufacture of blood gas sensor. (a) KrF laser-drilled hole in side of polyvinyl chloride bilumen catheter. (b) ~50 × 15-μm rectangular holes machined in a 100-μm diameter acrylic [poly(methylmethacrylate) (PMMA)] optical fiber with an ArF laser. (From Excitech.)

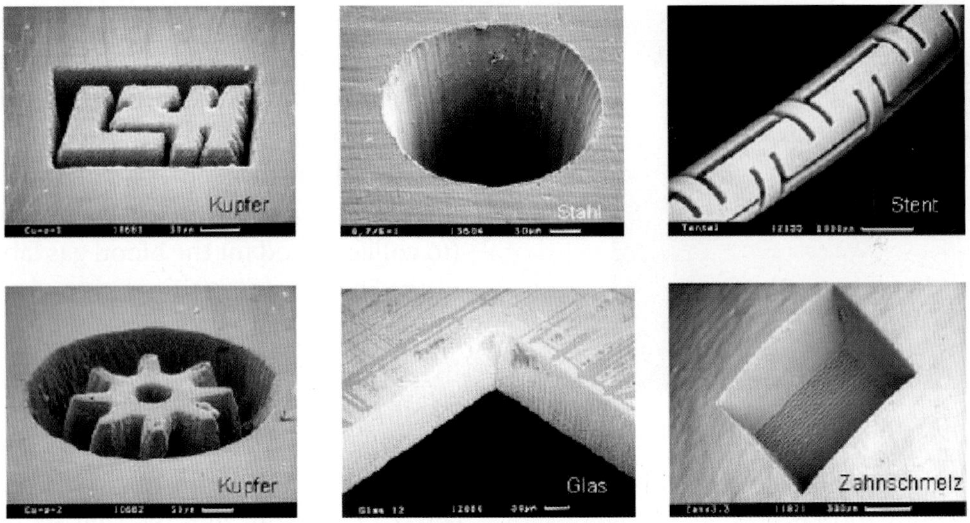


**FIGURE 5.33** From the Laser Centrum at Hannover, Germany (LZH): laser machining in a wide variety of materials.

The excimer laser also has brought more accuracy to corneal surgery and vision correction than ever before. One pulse of the excimer laser removes 0.25 µm of tissue. To put this into perspective, a typical human hair is 70 µm in thickness. Surgeons have never had a device as accurate as the excimer laser in eye surgery.

At the Laser Centrum in Hannover (LZH), Germany, researchers use femtosecond lasers for hole drilling with pulse durations shorter than 100 fs and pulse peak powers in the gigawatt range.[10] Grooves of only 200-nm thickness can be cut. Some demonstration pieces are shown in Figures 5.33 and 5.34.

*Machining inside Transparent Materials* Ultrafast laser pulsing enables machining inside transparent bulk materials. A material such as a glass is transparent and does not absorb visible light—provided the intensity stays below the threshold for multiphoton absorption. One can exceed that threshold by focusing ultrafast laser pulses to a spot inside the material. When the intensity exceeds the threshold for plasma formation, very localized absorption occurs at the focal spot, and, as before, the plasma expands, but this time it is confined by the surrounding material. The effect of the expansion is to create a void within a very dense shell of material—a cavity within the glass itself. The process is not limited to glass. Pits can be created in any material by focusing an ultrafast laser pulse inside the material, whether it is amorphous or crystalline.

Based on the above properties of femtosecond lasers, one can manufacture single-mode waveguides in glasses. Glass is transparent to incident, ultrafast laser pulses at 775 nm, enabling the local melting of the glass via confined multiphoton absorption and avalanche ionization. The glass then resolidifies, changing its physical properties. The result is an index gradient that acts like a waveguide. Only ultrafast lasers are capable of producing this effect in transparent materials; with longer pulse lasers (nanoseconds), the glass incurs damage before the intensity reaches the threshold at which guides are formed. Femtosecond laser pulses can be used to induce localized refractive index increase in a wide variety of glasses, and this way optical waveguides have been produced in silicate, borosilicate, chalcogenide, and fluoride glasses.

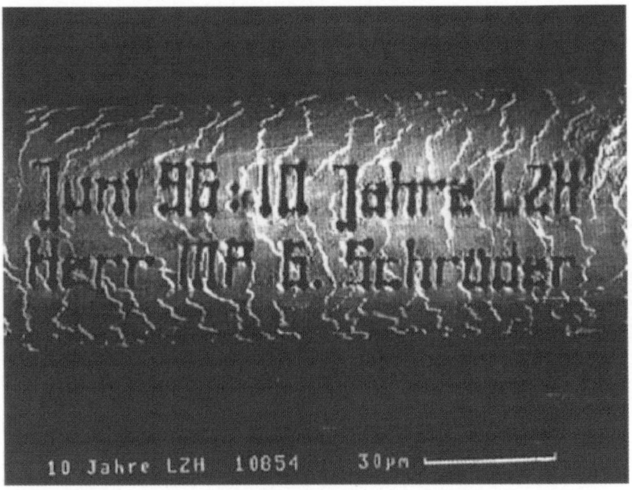


**FIGURE 5.34** From the Laser Centrum at Hannover, Germany: Micromarking (G. Schröder's hair). ArF laser-micromachined 40-µm square holes through a human hair.

**Table 5.8** Laser Applications in Various Industrial Fields

| Heavy Manufacturing | Light Manufacturing | Electronics | Medical | General |
|---|---|---|---|---|
| Profile cutting in sheet and plate metals | Profile cutting in plastics and wood | Via formation insulating material: Tab and MCM | Flow orifices <100 μm diameter | CVD diamond cutting |
| Seam and spot welding | Engraving | High-accuracy wire stripping | Drilling and cutting delicate or thermally sensitive materials | Ceramic and glass micromachining |
| Cladding and drilling | Drilling | Skiving of flexible circuits | Micromachining applications where edge quality and cleanliness are critical | Thin film patterning; microlithography |
| | | IC repair | | |

### *Summary: Laser Machining Applications and Comparisons*

Subtractive and additive laser machining (see Chapter 9) present several application opportunities in micromachining because of their versatility, site-specific operation, and rapid prototyping capability. Ultrafast laser machining, when lasers become less expensive, will make laser micromachining yet more attractive. There are some obvious drawbacks with these direct write techniques: they are serial processes, and the deposition or removal rates limit the speed of the micromanufacture. However, for rapid prototyping, mold fabrication, and site-specific manufacturing, laser machining has a very bright future, and, as we have remarked before, many microsystems might carry a bigger price tag than an IC, making a more expensive manufacturing technology acceptable.

In Table 5.8 we list typical laser applications in different fields, and in Table 5.9 we compare laser micromachining with some other typical competing machining technologies.

A good further introduction to laser machining can be found in *Laser Machining: Theory and Practice* by George Chryssolouris[23] and in an excellent chapter by Henry Helvajian, "Laser Material Processing: A Multifunctional in situ Processing Tool for Microinstrument Development."[11] A very informative laser handbook on the Internet can be found at http://www.clark-mxr.com/industrial/handbook/introduction.htm. An excellent tutorial on the Internet is by Dieter Schuöcker, Chair of Nonconventional Processing, Forming, and Laser Technology at the TU Vienna, Austria, in cooperation with Arnold Braunsteiner, at http://www.argelas.org/files/tutorium/tutorium.htm. Laser tutorials may also befound at http://www.ilt.fraunhofer.de/eng/100048.html by Fraunhofer Institut Lasertechnik, http://www.coherentinc.com/Literature/ by Coherent, and http://www.colemaneye.com/fundamentals/laser.htm by Aldrich Computers & Consultants.

Also Rami Arieli's "The Laser Adventure" at http://stwww.weizmann.ac.il/lasers/laserweb/index.htm is a good laser introduction. A service

**Table 5.9** Comparing Laser Micromachining with EDM, Mechanical Machining, and Chemical Etching

| | Practical Resolution Limit | Attainable Aspect Ratio | Taper | Undesirable Side Effects | Status of Technology Development |
|---|---|---|---|---|---|
| Excimer laser | 5 μm | >100:1 | Yes | Recast layer | Low |
| $CO_2$ laser | 200 μm | 100:1 | Yes | Recast layer, burring, thermal | High |
| Nd:YAG | 50 μm | 100:1 | Yes | Recast layer, burring, thermal | High |
| EDM | 100 μm | 20:1 | No | Surface finish | Moderate |
| Chemical etch | 250 μm | 1:1.5 | Yes | Undercutting | Moderate |
| Mechanical | Ø 100 μm | 10:1 | No | Burring | Moderate |

provider in the laser micromachining area with an excellent educative web site is Resonetics (http://www.resonetics.com/MDmfg.htm). Another web site to bookmark is Sam's Laser FAQ (http://www.repairfaq.org/sam/lasersam.htm). A laser link page can be found at http://www.mysteries-megasite.com/main/bigsearch/laser.html.

## Electron Beam Machining

Electron beam removal of materials is another fast-growing thermal technique. Instead of electrical sparks, this method uses a stream of focused, high-velocity electrons from an electron gun to melt and vaporize the work piece material.[24] In EBM, electrons are accelerated to a velocity of 200,000 km/s or nearly three-fourths that of light. As shown in Figure 5.35, the process is performed in a vacuum chamber to reduce the scattering of electrons by gas molecules in the atmosphere. The stream of electrons is directed against a precisely limited area of the work piece; on impact, the kinetic energy of the electrons is converted into thermal energy that melts and vaporizes the material to be removed, forming holes or cuts. Originally, in the early 1960s, the method was used for welding and annealing, but today it is also used for many IC and micromachining tasks. EBM equipment now is commonly used by the electronics industry to aid in the etching of masks and the fabrication of circuits in microprocessors;

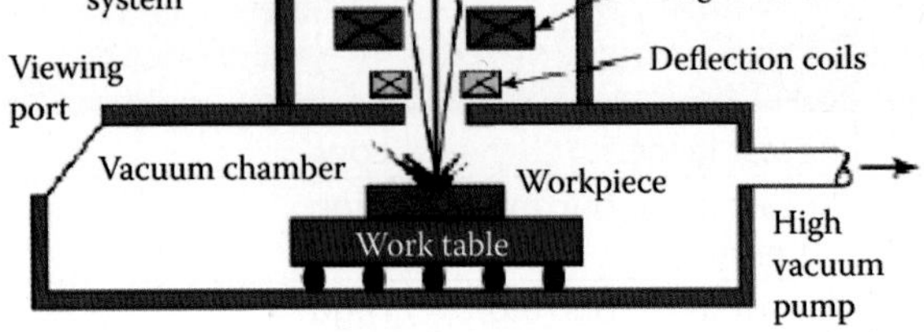

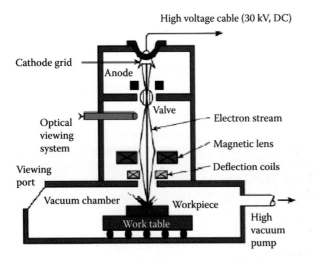


**FIGURE 5.35** Schematic illustration of the electron beam machining process. Unlike LBM, this process requires a vacuum; therefore, work piece size is limited to the size of the vacuum chamber.

today, scanning electron beam lithography remains one of the best-demonstrated methods for microfabrication in the sub-100-nm regime (see Part I). We will not repeat the electron beam lithography aspects of electron beam machines here but rather emphasize other machining applications. An electromagnetic coil or lens focuses the diverging electron beam to a power density of about 6500 $GW/mm^2$, high enough to vaporize the work piece material. A typical cross-sectional diameter of the beam is 10–200 μm at the point of impact on the work piece. A bias electrode switches the electron beam energy from the cathode (electron gun) on and off for pulsed drilling operations. The bias electrode also controls beam intensity. With additional coils, the beam can be deflected in any direction. With mechanical means, holes of a diameter of 0.8 mm are difficult to drill, and those less than 0.1 mm are nearly impossible. In contrast, EBM works well for deep holes of diameters less than 0.1 mm and is generally effective with a diameter-to-depth ratio of 1:10 in hard-to-machine materials. A hole in a sheet 1.25-mm thick up to 125 μm in diameter can be cut almost instantly with a taper of 2–4°. With multiple pulses, diameter-to-depth ratios can reach 1:100. For example, hole diameters of 0.1 mm can be drilled in materials up to 10-mm thick.

Once the vacuum is established ($10^{-4}$ mm Hg), EBM can be extremely fast. The material removal rate is about 0.1 mg/s; hence drilling rates of 2000 holes/s are routine with EBM. Typical tolerances are about 10% of the hole's diameter (e.g., 0.005 mm on a 0.05-mm hole). The most suitable application is for work pieces requiring large numbers of simple small holes to be drilled or for drilling holes in materials that are hard and difficult to machine with other processes. The equipment is very expensive, and the need to operate in vacuum adds considerably to the machining time. The cost of an EBM facility ranges from $75,000 to $1.5 million.

## Plasma Beam Machining

Plasma arc cutting (also plasma arc machining) is mainly used for cutting thick sections of electrically conductive materials (Figure 5.36). A high-temperature plasma stream (up to 60,000°F)

**FIGURE 5.36** Plasma arc cutting.

interacts with the work piece, causing rapid melting. A typical plasma torch is constructed in such a way that the plasma is confined in a narrow column about 1 mm in diameter. The electrically conductive work piece is positively charged, and the electrode is negatively charged. Relatively large cutting speeds can be obtained: for example, 380 mm/min for a stainless steel plate 75-mm thick at an arc current of 800 A. Tolerances of ±0.8 mm can be achieved in materials of thicknesses less than 25 mm, and tolerances of ±3 mm are obtained for greater thicknesses. The HAZ for plasma arc cutting varies between 0.7 and 5 mm in thickness, and the method is used primarily for ferrous and nonferrous metals. The inert gas to form plasma also serves as a shielding gas. The torch and sometimes the work piece require cooling, and the process produces high noise levels. The technique is of marginal use in micromachining, except perhaps to cut the plasma arc-coated sensor substrates discussed in Chapter 9.

For a detailed history of plasma cutting, visit http://www.hypertherm.com/technology/plasma_history.htm. A detailed analysis of the quality of the cut of a plasma arc is presented at http://archive.metalformingmagazine.com/1998/05/plasma/plasma.htm.

## Acknowledgments

Special thanks to Dr. Rodrigo Martinez, UCI Graduate now at ETH in Laussane.

## Questions

*Questions by Mr. Chuan Zhang, UC Irvine*

5.1: In the electrical discharge machining (EDM) process, the spark-gap width is an important parameter in determining the accuracy of the process. Discuss the pros and cons of a very small spark-gap width.

5.2: What are the merits and drawbacks of electrical discharge machining? List some applications of EDM.

5.3: Sketch a way to make a 300-μm-diameter Ni rod with a polygonal cross-section using electro discharge machining wire cutting.

5.4: Draw two shapes that are better produced by lithography, and two shapes that are better made by wire cutting discharge machining (WC-EDM).

5.5: How can we perform μ-EDM in batch mode by relying on LIGA?

5.6: Explain why the metal removal rate in EDM is much slower than that of ECM.

5.7: Explain the energy conversion process in laser beam machining (LBM). How can we thermally remove materials by consuming electrical power?

5.8: Why is the mechanism of laser ablation different for different laser light wavelengths?

5.9: Why is a laser more efficient in machining than visible light? List the common applications of laser machining.

5.10: How can the quality of the ablated region be improved in the laser ablation process? Which factors affect the thickness of the heat-affected zone (HAZ)?

5.11: What are the important factors influencing the efficiency and spatial resolution of laser ablation? Explain why.

5.12: What is the difference between optical penetration depth and damage penetration depth in laser machining?

5.13: How does the laser pulse length affect laser beam machining? Compare long laser pulse etching and short laser pulse etching.

5.14: How can water jet guided laser machining keep a workpiece clean? How does it work?

5.15: What are the merits of laser etching with chemical assist (LACE)? What are the technical barriers for applying this technique in most manufacturing fields?

5.16: Laser machining offers advantages compared to methods such as EDM, ECM, chemical etching, or stamping. A laser can easily cut complex parts with high tolerances. Excessively hard, pressure- and force-sensitive, or very thin materials can be cut with a small focused beam. In many cases, laser cutting is preferred over stamping due to its high flexibility and noncontact characteristics. Lasers also enable low-temperature deposition processes, rapid prototyping, mold fabrication, and site-specific manufacturing. It also has obvious drawbacks. What are they?

## References


1. Kalpajian, S. 1984. *Manufacturing processes for engineering materials.* Reading, MA: Addison-Wesley.
2. Snoeys, R., F. Staelens, and D. F. Dauw. 1986. Adaptive control optimization as basis for intelligent EDM die-sinking machines. *Proceedings of the Winter Annual Meeting of ASME.* Anaheim, CA, 22:63–78.
3. Guitrau, E. B. 1997. *The EDM handbook.* Cincinnati, OH: Hanser Gardner Publications.
4. Rain, C. 1981. Nontraditional methods advance machining industry. *High Technology* November/ December:55–61.
5. Saito, N. 1984. Recent electrical discharge machining (EDM) techniques in Japan. *Bull Jpn Soc Precis Eng* 18:110–16.
6. Sommer, C., and S. Sommer. 2000. *Wire EDM handbook.* Houston: *Advance Publishing.*
7. Kuo, C.-L., T. Masuzawa, and M. Fujino. 1991. *Proceedings: IEEE micro electro mechanical systems (MEMS '91).* Nara, Japan.
8. Kuo, C.-L., T. Masuzawa, and M. Fujino. 1992. *Proceedings: IEEE micro electro mechanical systems (MEMS '92).* Travemunde, Germany.
9. Takahata, K., and Y. B. Gianchandani. 2002. Batch mode micro-EDM for high-density and high-throughput micromachining. *IEEE/ASME J Microelectromechanical Systems* 11:102–110.
10. Ehmann, K. F., D. Bourell, M. L. Culpepper, T. J. Hodgson, T. R. Kurfess, M. Madou, and K. Rajurkar. 2007. *Micromanufacturing: international research and development.* New York: Springer.
11. Helvajian, H. 1995. *Microengineering technology for space systems.* Ed. H. Helvajian. El Segundo, CA: Aerospace Corporation.
12. Srinivasan, R., and V. Mayne-Banton. 1982. Self-developing photoetching of poly(ethylene terephthalate) films by far-ultraviolet excimer laser radiation. *Appl Phys Lett* 41:576–78.
13. Liu, Y. S. 1989. *Laser microfabrication: thin film processes and lithography.* Eds. J. Y. Tsao, and D. J. Ehrlich. New York: Academic Press.
14. Ehrlich, D. J., D. J. Silversmith, R. W. Mountain, and J. Tsao. 1982. Fabrication of through-wafer via conductors in Si by laser photochemical processing. *IEEE Trans Compon Hybrids Manuf Technol* CHMT-5:520–21.
15. von Gutfeld, R. J., and R. T. Hodgson. 1982. Laser enhanced etching in KOH. *Appl Phys Lett* 40:352–54.
16. Bloomstein, T. M., and D. J. Ehrlich. 1992. Laser-chemical three-dimensional writing for microelectromechanics and application to standard-cell microfluidics. *J Vac Sci Technol B* 10:2671–74.
17. Shlichta, P. J. 1988. Laser micromachining in a reactive atmosphere. *NASA Tech Briefs* 12:84.
18. Maeda, S., K. Minami, and M. Esashi. 1994. *IEEE international workshop on micro electro mechanical systems (MEMS '94).* Oiso, Japan: IEEE.
19. Bloomstein, T. M., and D. J. Ehrlich. 1991. *Sixth international conference on solid-state sensors and actuators (Transducers '91).* San Francisco.
20. Eden, J. G. 1986. Photochemical processing of semiconductors: new applications for visible and ultraviolet lasers. *IEEE Circuits Devices Mag* 2:18–24.
21. Bowman, L., J. M. Schmitt, and J. D. Meindl. 1985. *Micromachining and micropackaging of transducers.* Eds. C. D. Fung, P. W. Cheung, W. H. Ko, and D. G. Fleming. Amsterdam: Elsevier.
22. Anthony, T. R., and H. R. Cline. 1976. Deep-diode arrays. *J Appl Phys* 47:2550.
23. Chryssolouris, G. 1991. *Laser machining.* New York: Springer Verlag.
24. Taniguchi, N. 1984. Research on and development of energy beam processing of materials in Japan. *Bull Jpn Soc Precis Eng* 18:117–25.
25. Extreme measurements in laser generated plasmas, CLEFS CEA-No. 4, Autumn 2006.

# 6

# Mechanical Energy-Based Removing

Manufacturing dominates world trade. It is the main wealth creating activity of all industrialized nations and many developing nations. A manufacturing industry based on advanced technologies with the capability of competing in world markets can ensure a higher standard of living for an industrial nation.

**P. McKeown**
*(1996)*

## Outline



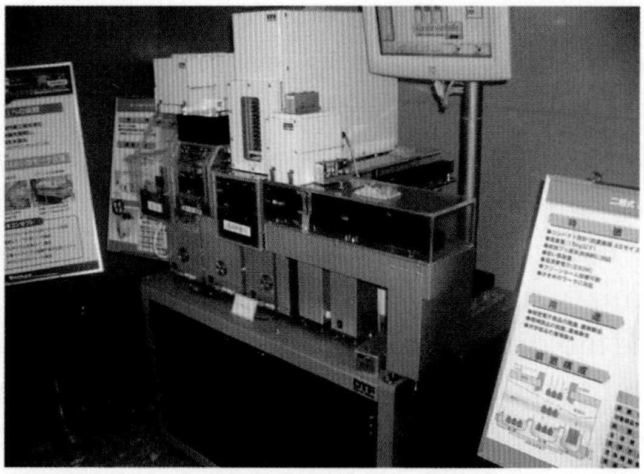

Commercial desktop factories (DTFs) at Sankyo Seiki.

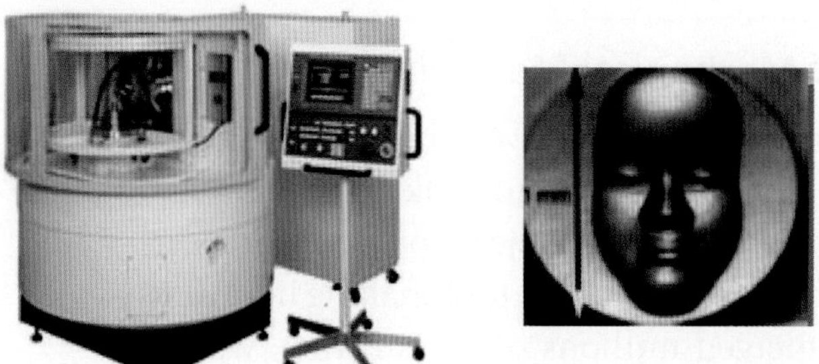

Fanuc's the ROBOnano $U_i$, an ultraprecision micromachining station (cost $1 million) and a Noh mask made with this machine (http://www.fanuc.co.jp/en/product/robonano/index.htm).

## Introduction

In mechanical subtractive machining, physical removal of unwanted material is achieved by mechanical energy applied at the work piece. In the ultraprecision range, the two major categories of mechanical material-removing technologies we detail in this chapter are single-point machining and abrasive machining (basically multipoint machining). Abrasive machining is a process that removes small amounts of material from a work piece in the form of tiny chips through the contact of small, hard, sharp, nonmetallic particles. In abrasive machining we review abrasive wheel machining, in particular electrolytic in-process dressing (ELID), ultrasonic machining (USM), water jet, abrasive water jet (AWJ), and abrasive jet machining (AJM). We also consider briefly two energetic beam mechanical machining methods, i.e., focused ion beam (FIB) and fast atom beam (FAB) machining. We also cover the miniaturization of mechanical machining equipment in a section on "desk-top factories." The fact that it often takes a two-ton machine tool to fabricate microparts, where cutting forces are in the milli- to micro-Newton range, is a clear indication that a complete machine tool redesign is required for the fabrication of micromachines.

Whereas primary manufacturing processes involve casting and molding, secondary manufacturing processes constitute the main mechanical removing techniques involving turning, drilling, and milling. Abrasive processes to superfinish a work piece are called tertiary manufacturing processes.

## History of Mechanical Machining

Tool making for mechanical machining began perhaps when humans learned to walk erectly and had their hands free to grasp objects of wood or stone and to shape them. Although tool-making hominids probably emerged millions of years earlier, we can only trace tool-making history back 2 million years to the oldest surviving and recognizable flint hand axes.[1] Stronger, longer lasting tools resulted when metals such as copper and iron replaced wood and stone. Humans have been cutting metals for thousands of years to make the myriad tools, machines, and other devices that civilization demands.

The birth of more precise mechanical machining dates back to the invention of the lathe, attributed to Anacharis the Scythian and first depicted in an Egyptian low relief about 300 B.C. (Figure 6.1).[1] Hipparchus, in the second century B.C., and Ptolemy, in 150 A.D., used graduated instruments for astronomy and sailing. Ptolemy's astrolabe became the basic tool for observers of the skies for the next thousand years; the Arabs perfected it in the ninth century, and it remained in use for another 700 years (Figure 6.2).[2] Precision machining with a hand-operated lathe got a boost in the mid-1600s from demands for more precision in building better instruments for time measurement.[3] Ruling engines for the control of diffraction gratings have driven state-of-the-art precision machine tool design since the late 1700s. More than a century ago, Rowland at the John Hopkins University built ruling engines with part-per-million accuracy.[4,5] Today, the mechanical technology on which photolithographic alignment systems are based is still grounded in the same ruling engine design. An American astronomer, David Rittenhouse, made the first diffraction grating in 1785, when he fabricated a half-inch-wide grating with 53 apertures.[6] It is difficult to identify another single device that has provided more important experimental information to every field of science than the diffraction grating.[6] Rowland not only constructed sophisticated ruling engines, but he also invented

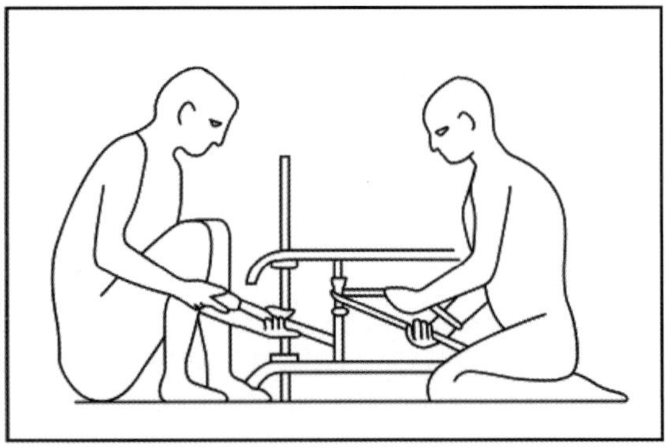

**FIGURE 6.1** First lathe as depicted in an Egyptian bas relief, about 300 B.C. Shown here in a line drawing. The man at left is holding the cutting tool. The man at right is making the work piece rotate back and forth by pulling on a cord or thong.

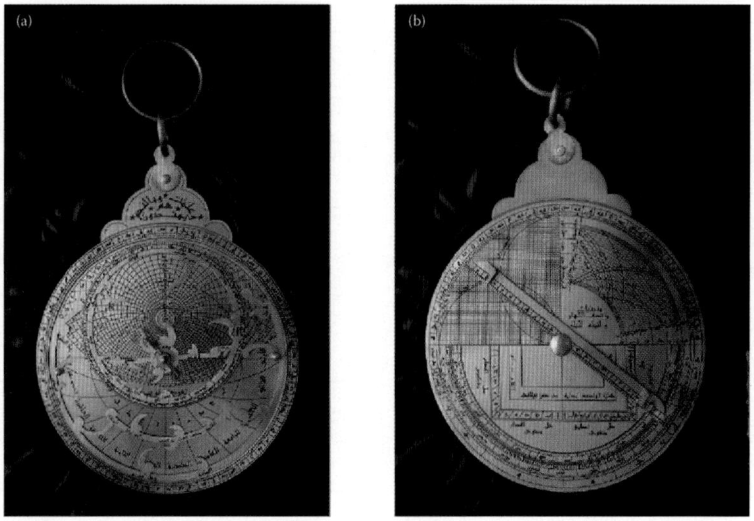

**FIGURE 6.2** Ptolemy's astrolabe; front view (a) and back view (b).

the concave grating, a device of spectacular value to modern spectroscopists.

Tools driven by mechanical power brought on the machine age in the late 1700s.[7] Diamond as a cutting tool was apparently first used in 1779 for making threads in hardened steel. The significance of diamond tools in precision machining was realized in the 1850s. By 1920, the Lord's Prayer had been engraved with a diamond tool into an area of 100 μm × 40 μm. In the 1930s, watchmakers used diamond turning to make watch dial components with excellent surface finishes. In the 1940s, Moore Special Tool Company designed a jig grinder that allowed toolmakers to work with accuracies that were not achievable by any of the existing methods (http://www.mooretool.com).

In modern times, precision engineering was initially pushed by nuclear programs and eventually by semiconductor manufacturing. By the 1960s demands from electronics, defense, and energy applications led to pioneering work on diamond turning from Lawrence Livermore National Laboratory and Oak Ridge National Lab. In the 1970s, the need for large optics in space telescopes and "defense" systems triggered the application of precision machining to complex optical components.

Norio Taniguchi started working on processing of materials to nanoscale precision as early as 1940 and worked on achieving precision finishes of quartz crystals, silicon, and other hard and brittle materials by using ultrasonic machining (USM). Taniguchi coined the term *nanotechnology* and in 1974 used the term to define ultraprecision machining. According to Taniguchi, ultraprecision machining is "the process by which the highest possible dimensional accuracy is achieved at a given point in time." In an ultraprecision manufacturing system, all the material processing methods used in a conventional manufacturing system, like removal, addition, joining, surface treatment, and assembly, are scaled down by orders of magnitude. The machine tool capabilities, metrology, and instrumentation used in these processes range in the orders of micrometers to nanometers.

Figure 6.3 includes the machine accuracy capabilities Norio Taniguchi predicted along with the processes or tools used to achieve it.

In the past 25 years, great progress has been made in building machines that can be operated and controlled automatically.

By 1977, highly precise instruments such as servomotors, feedback devices, and computers were implemented, paving the way for computer numerical control (CNC) machining, which is now standard in many types of machine tools. At the start, the smallest movement these machines could reproducibly make was 0.5 μm.

A crucial development came in the 1980s, when advanced machine tools became equipped with precision metrology and control tools. These machines used laser interferometer and capacitance probe feedback controls, temperature control, and hydrostatic

(a)

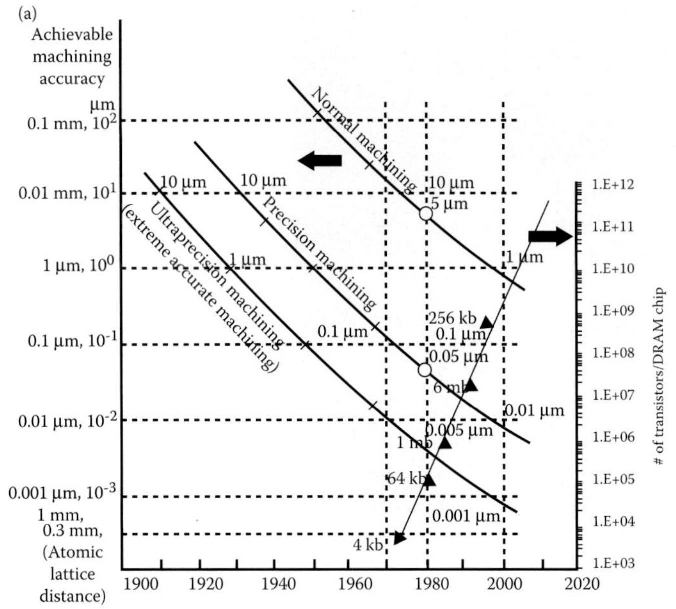


(b)

| Machine Tools (Processing equipment) | Measuring Instruments (Inspection equipment) |
|---|---|
| Lathes, milling machines | Vernier calipers |
| Precision lathes, grinding machines, lapping machines, honing machines | Mechanical comparators, micrometers, dial indicators |
| (Curve generators) Precision jig borers, jig grinding machines, super finishing machines | (Electric or pneumatic micrometers) Optical comparators |
| Precision grinding machines, precision lapping machines (optical lens grinding machines) Precision diamond lathes | Optical or magnetic scales, electrical comparators Electronics comparators (noncontact) |
| (DUV mask aligner) Optical lens finish grinding machines, precision diamond grinding machines, ultraprecision grinding machines | Laser measuring instruments Optical fibers, Talysurfs, Talyronds |
| (Diffraction grating ruling engine) (Electron beam lithographer) Super high-precision grinding machines, super high-precision lapping and polishing machines | High-precision laser measuring instruments (Doppler, multireflection) Talysteps |
| Atom, molecule, or ion beam machining, atom or molecule deposition (Substance synthesizing) | Scanning electron microscopes, transmission electron microscopes, electron diffraction equipment, ion analyzers, x-ray microanalyzers, Auger analyzers, ESCA |
| Atom manipulation with proximal probes | STM and AFM |

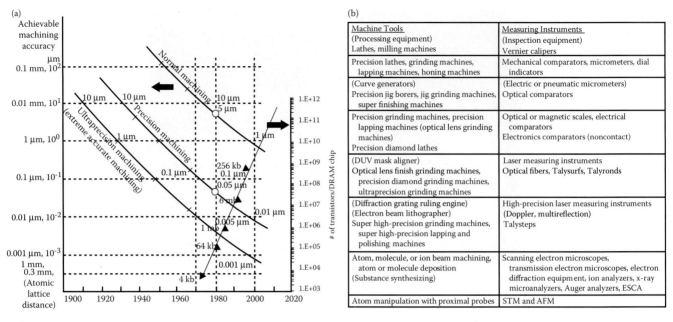

**FIGURE 6.3** Progress of accuracy in machining. The development of achievable machining accuracy during the past 100 years is shown in this figure under the generalized classification of normal machining, precision machining, and ultraprecision machining. (a) Left ordinate, achievable manufacturing accuracy over time based on Taniguchi. (Taniguchi, N. 1983. Current status in, and future trends of, ultraprecision machining and ultrafine materials processing. *Annals CIRP* 32:S573–82.[8]) Right ordinate, increase in transistor density over time according to Moore's Law. (Gibbs, W. W. 1997. *Scientific American* September 22.[9]) (b) Machines, processing equipment, and metrology equipment by which the indicated resolution can be achieved and checked. By "ultraprecision machining," Taniguchi means those processes/machines by which the highest possible dimensional accuracy is or has been achieved at a given point in time.

bearings and featured accuracies better than 0.1 μm. Precision manufacturing methods were extended for industrial use for cutting aluminum, which was used for making components for scanners, photocopying machines, and computer memory disks. Also in the 1980s, cutting with very small diamond tools (e.g., 22-μm diameter) was developed in Japan.

In the 1990s, precision manufacturing became accepted as another advanced tool in manufacturing systems with some machining stations even allowing freeform machining. Large-scale adaptation of CNC machining also occurred during this period as a result of the surge of the electronics and computer industries.

By 1993, 0.05 μm became possible, and today there is equipment available featuring 0.01 μm and even nanometer step resolution (http://www.fanuc.co.jp/eindex.htm).[10]

This evolution closely follows the predictions sketched in Figure 6.3, where the Taniguchi curves show a machining accuracy for ultraprecision machining of subnanometer resolution for 2008. The resolution of the steps a machine can make, of course, is a determining factor for the manufacturing accuracy of the work piece.

Historically important milestones in new machining concepts include the introduction of ultrasonic techniques, focused ion beam, and desktop factories (DTFs).

The roots of ultrasonic technology can be traced back to research on the piezoelectric effect conducted by Pierre Curie around 1880. He found that asymmetrical crystals such as quartz and Rochelle salt (potassium sodium titrate) generate an electric charge when mechanical pressure is applied. Conversely, mechanical vibrations are obtained by applying electrical oscillations to the same crystals. This led to the discovery of SONAR (sound navigating and ranging). In USM, the tool, made of softer material than that of the work piece, is oscillated at a frequency of about 20 kHz with an amplitude of about 25.4 μm.

A liquid metal ion source used in focused ion beam machining (FIBM) was demonstrated between 1978 and 1980. The latter technique, although very slow and expensive, represents the latest and most accurate means of mechanical precision machining.

**FIGURE 6.4** An example of a desktop factory at AIST, Japan.

In this case, a powerful, very thin beam of ions replaces the familiar mechanical tool. A variation of FIBM is fast-atom beam (FAB) machining, pioneered by Hatamura et al.[11] in the early 1990s. In FAB, ionized atoms, accelerated to about 100 km/s, are neutralized and used as the mechanical cutting tool.

Desktop factories (DTFs) constitute a rather interesting new manufacturing philosophy involving flexible and modular table top-sized automated factories that feature minimal human participation in the manufacturing process. An example of such a factory is shown in Figure 6.4. Since the early 1990s progress has been made toward making such DTFs a reality. A DTF as shown here has the potential of becoming the factory of the future: a totally self-contained, robotic, desktop-size machine tool that only requires materials, power, and water as outside inputs, and out come the finished machined products. The first DTFs in research and development incorporated lathes, cleaning, gluing, punching, and drilling stations. The work piece is transported between these different machining functions by a "cart" moving from station to station. A first commercial manufacturing unit by Sankyo Seiki is shown as the inset in the header of this chapter. Sankyo Seiki Mfg. Co. Ltd. believes that DTFs might help manufacturing survive in Japan.[12]

## Absolute and Relative Tolerances in Manufacturing

The application ranges of normal, precision, and ultraprecision mechanical machining, according to Norio Taniguchi, are shown in Figure 6.3. The Taniguchi curves, which dramatically illustrate improvement in mechanical manufacturing accuracies achieved in the second half of the twentieth century (left coordinate) are compared with Moore's Law (right coordinate), which demonstrates the logarithmic progress in density of transistors on a chip during the past few decades. Mechanical machining and integrated circuit (IC) technology-based machining methods, as well as appropriate metrology tools, are listed in the table on the right side of this figure (see also Evans[13]). As illustrated, ultraprecision diamond machining reached a manufacturing accuracy of better than 1 nm in the year 2000. In Chapter 2 we mentioned that the mirrors for an extreme ultraviolet camera are machined flat with an accuracy of 0.25 nm rms or better. At the same time, lithography methods only reached critical dimensions of 0.18 μm. However, this comparison is a bit misleading because the accuracy for mechanical machining, which is truly 3D, pertains to the product surface finish, whereas for lithography-based methods, it is based on the accuracy of carving out very small features (e.g., 0.18 μm) in the *x,y*-plane. Photolithography-based IC methods are limited to 2–2.5D shapes (projected shapes), and although the dimensions along the *z*-axis can also be controlled to the nanometer and subnanometer range, in the *x,y*-plane, the accuracy, which is mostly determined by the alignment accuracy of a large number of photomasks, is considerably less. In terms of attaining the smallest absolute sizes, mechanical machining is considerably inferior to lithography-based methods because the writing tool in mechanical machining is crude compared with writing with light or electrons—the smallest diamond styluses are still about 22 μm in diameter! In mechanical machining (for example, polishing of a mirror), the size of the grinding tool does not influence the resulting accuracy in the *z*-direction.

The discrepancy between the progress in tolerances predicted by the Taniguchi and Moore plots and the tolerances that are being achieved in practice have been increasing, i.e., progress is slowing down (see also Part I on lithography). Progress in mechanical machining is in large part based on continued improvement in positioning stages and computer control. Because positioning stages for lithography and ultraprecision machining are the same, the fact that the progress is slowing for both is not that surprising. Reasons for the slowdown of IC technology are a bit more complicated, however. In lithography, subsequent layers need to be aligned, an operation not needed when diamond milling, e.g., a high-precision lens. The alignment operation in lithography may be checked by laser interferometer-controlled stages to an absolute positioning accuracy of <10 nm, thereby allowing the registration of successive layers to a similar precision. The reasons for the slowdown in electronic device miniaturization are obviously not because of lack of appropriate metrology equipment, but rather hinge on securing more mechanical stability in the positioning stages and even more so on manufacturing cost and the fact that transistors are becoming too leaky at less than 30 nm, where quantum effects start playing a role.

To further extend manufacturing accuracy in electronics, one has to look beyond current manufacturing technologies and consider both top-down and bottom-up manufacturing methods or perhaps hybrid approaches. Bottom-up methods, such as molecular engineering, and top-down methods, such as e-beam nanofabrication of quantum devices, are bound to lead the way beyond the current limitations dictated by the Moore and Taniguchi curves. Micromachining of special manufacturing tools will play a pivotal role for both those new bottom-up and top-down techniques. For example, the scanning tunneling microscope, based on micromachining ultrasharp tips on Si cantilevers, creates machining opportunities with accuracies on the atomic level, well beyond the predictions of the Taniguchi and Moore curves.

For mechanical structures, operation in the micro- and nanodomain imposes barriers other than those described for electronic devices. New design philosophies that are more error-insensitive with simpler mechanisms, smaller numbers of parts, and, wherever possible, flexible members rather than rigid ones must be adopted. Also here bottom-up machining methods will play an important role (see Volume III, Chapter 3).

In Figure 6.5, we show, in a most generalized fashion, human manufactured objects (from an IC fabrication to a house) in terms of absolute size and relative tolerances (feature tolerance/object size). Mechanical machining is typically used to machine the bigger objects, e.g., a lens of an absolute size of 10 cm or a parabolic astronomic mirror with a diameter of several meters.[3] Mechanical engineers define precision machining as machining in which the relative accuracy (tolerance/object size) is $10^{-4}$ or less

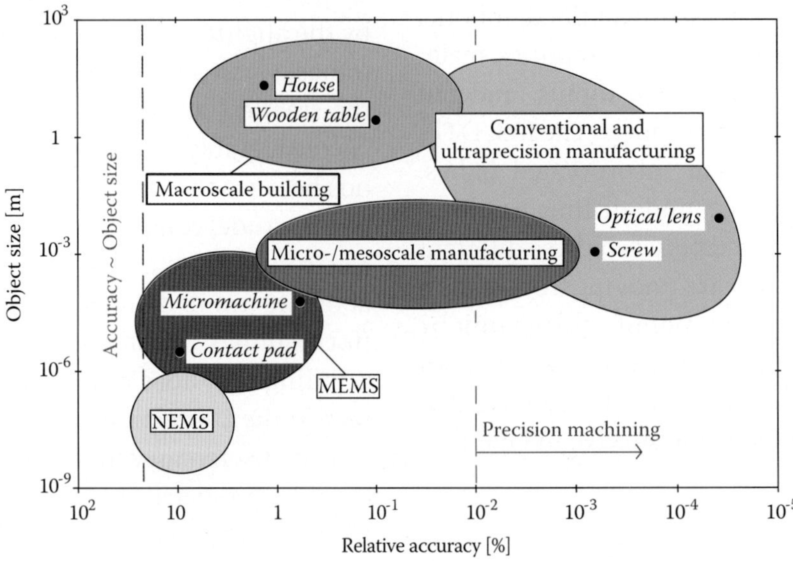


**FIGURE 6.5** Various examples of human-engineered objects as a function of their relative (tolerance/object size) and absolute tolerance (size). The application field for precision machining at relative tolerances of $10^{-4}$ and less is also indicated. Object size versus relative accuracy for MEMS, NEMS, micro-/mesoscale manufacturing, and conventional/ultraprecision engineering.

of a feature/part size (see dividing line in Figure 6.5 at a relative accuracy of $10^{-2}$%).[4,5] For comparison, a relative accuracy of $10^{-3}$ in the construction of a house is considered excellent. It is important to realize that, although IC techniques and silicon micro- and nanomachining (MEMS and NEMS) can achieve excellent absolute tolerances, relative tolerances here are rather poor compared with those achieved by most mechanical machining techniques. The decrease in manufacturing accuracy with decreasing size is rarely mentioned in discussions of Si micromachines; this probably is because Si micromachining originated from electrical engineering practice rather than mechanical engineering. The contrast is striking; in mechanical machining, relative tolerances of $10^{-6}$ (ppm) are becoming standard, whereas in the IC industry, a $10^{-2}$ relative tolerance is considered good. The definition of precision machining, with relative tolerances of $10^{-4}$, excludes micromachining! To describe micromachining with the same terminology, it must be expanded to cover all machining methods where the relative accuracies are at least $10^{-4}$ or where absolute size in one or more dimensions is in the micrometer range.

In Figure 6.5 we also show the domains covered by the various machining options such as MEMS, NEMS, micro-/mesoscale manufacturing, and conventional/ultraprecision engineering. We see that IC, MEMS, and NEMS processes cover smaller absolute sizes, whereas their relative accuracies remain poor compared with those of conventional and ultraprecision manufacturing. IC-based machining in the relatively poor $10^{-1}$ to $10^{-2}$ relative accuracy range is also restricted in the types of engineering materials that can be used (Si, $SiO_2$, $Si_3N_4$, Al, ...). There is an important micro-/mesoscale machining size range for applications with sizes between 10 μm and 1 mm, bridging the gap between silicon-based IC manufacturing and MEMS/NEMS and conventional miniature machining (Figure 6.6). This is a difficult size range to conquer: too large for IC manufacturing techniques and too small for traditional machining. It forms the needed hand-shake technology between the micro- and the macroworld. The bandwidth of sizes that can be attained with ultraprecision machining makes it an attractive candidate for this application range. Ultrahigh-precision techniques cover a wide range of materials, and they are—in contrast to IC methods—truly 3D and may even produce freeform surfaces. A freeform surface is a kind of complex, irregular, asymmetric aspherical surface. Generally, it cannot be described with an equation but usually only by a series of digital points. Unfortunately, whereas microscale technologies are well established in the microelectronics industry, the same cannot be said for mesoscale products involving complex 3D geometry and high relative accuracies in a range of nonsilicon materials. Thus, there is need for a strong focus on mesoscale manufacturing methods because our NEMS and MEMS devices would be rendered worthless if we cannot contact or package them.

"Precision machining," when used by mechanical engineers, has typically been reserved for removal processes only, whereas "micromachining," as used to describe IC, MEMS, and NEMS fabrication technology, covers both removal and forming processes. Precision machining and IC, MEMS, and NEMS are

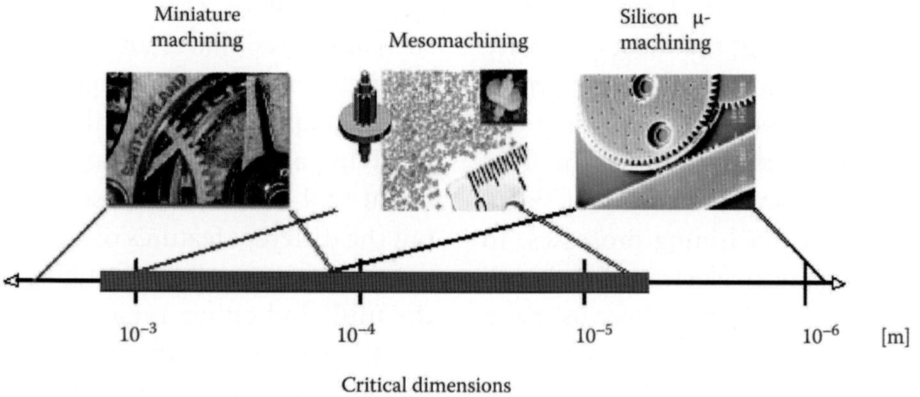


**FIGURE 6.6** Mesoscale machining.

complementary; all strive to improve absolute and relative tolerances, with IC-, MEMS-, and NEMS-based technology better at obtaining smaller features (absolute tolerance) and traditional methods better at obtaining tighter relative tolerances. In this book, precision and ultraprecision machining encompasses removal and forming processes, with MEMS and NEMS as one of its newest disciplines.

The adjectives *nontraditional* and *nonconventional* have been in use by mechanical engineers for at least the past 30 years to describe nonmechanical machining methods such as thermal, electrochemical, and chemical machining of high strength and corrosion- and wear-resistant materials. However, the name *nontraditional* has become obsolete because "nonconventional" machining methods have found a wide range of commercial applications even with more readily machined materials.[14] In this book we refer to all top-down, non-IC manufacturing techniques as *traditional* machining techniques. In our nomenclature, IC-, MEMS-, and NEMS-based micromachining techniques are the nontraditional machining methods of our day. In making a manufacturing decision, it is not so much a question of which manufacturing technique is better or newer (very often, since the 1960s, silicon was the answer, sometimes independent of the question), but more which one is appropriate to use for the problem at hand.

Ultraprecision machining is significantly more expensive than normal machining and precision machining, especially when many abrasive machining steps are involved. Abrasive processes are finishing processes that are used to remove surplus material from the work piece surface. It is usually used on almost any surface that has been previously rough machined and is among the most expensive processes because it is generally slow in removing material. However, as we will learn, they can machine extremely hard surfaces and can be used on materials that are beyond the limits of virtually all other mechanical machining processes. In Table 6.1 available mechanical, electronic, and optical products are listed and categorized according to their tolerance band.

In general, to obtain the best possible accuracies with a given technique, materials such as wood or brick cannot be used; form-stable and workable materials such as aluminum, stainless steel, ceramics, or glasses must be used. In other words, material properties and machining accuracy are intertwined.

## Ultraprecision Mechanical Machining

### Introduction

In mechanical removal processes, stresses induced by a tool overcome the strength of the material. The process can produce complex 3D shapes, with very good dimensional control and good surface finishes. The method tends to be wasteful of material and expensive in terms of labor and capital. How well a part made from a given material holds its shape with time and stress is referred to as the *dimensional stability* of the part and the material. To maximize dimensional stability, the machine design engineer tries to minimize the ratios of applied and residual stress to yield strength of the material. A good rule of thumb is to keep the static stress less than 10–20% of yield. Increased heat at the work piece causes uneven dimensional changes in the part being machined, making it difficult to control its dimensional accuracy and tolerances. Thermal errors are often the dominant type of error in a precision machine, and thermal characteristics such as thermal expansion coefficient and thermal conductivity deserve special attention.[5]

Mechanical removal processes can be broken down into four commonly recognized categories: turning, milling, drilling, and grinding. *Turning* operations involve rotating a part and bringing it into contact with a fixed cutting tool. Mostly, turning makes parts that have a round cross-section such as screws, shafts, and pistons. *Mills* are used on a variety of materials, ranging from wood to metals, to shape a part. A mill's working feature is its revolving armature that can have a number of different tool-bit shapes. These shapes are used to cut out the different features of a part. A "block" of bulk material (e.g., steel) or a rough casting is secured to the mill, and either a manual operator or program (CNC) operates the machine, maintaining the cutting processes until the final milled product is produced. Machine *drilling* is normally done with a drill

**TABLE 6.1** Tolerances of Components of Some Commonly Available Products

| | Tolerance Band (±) | Mechanical | Electronic | Optical |
|---|---|---|---|---|
| Normal machining | 200 μm | Normal domestic appliances and automotive fittings, etc. | General purpose electric parts, e.g., switches | Camera, telescope, and binocular bodies |
| | 50 μm | General purpose mechanical parts for typewriters, engines, etc. | Transistors, diodes, magnetic heads for tape recorders | Camera shutters, lens holders for cameras and microscopes |
| Precision machining | 5 μm | Mechanical watch parts, machine tool bearings, gears, ball screws, rotary compressor parts | Electrical relays, resistors, condensers, silicon wafers TV color masks | Lenses, prisms, optical fiber and connectors (multimode) |
| | 0.5 μm | Ball and roller bearings, precision drawing wire, hydraulic servo-valves, aerostatic bearings, ink jet nozzles, aerodynamic gyro bearings | Magnetic scales, CCD (charged-coupled device), quartz oscillators, magnetic memory bubbles, magnetron, IC line width, thin film pressure transducers, thermal printer heads, thin film head discs | Precision lenses, optical scales, IC exposure masks (photo, x-ray), laser polygon mirrors, x-ray mirrors, monomode optical fiber and connectors |
| Ultraprecision machining | 0.05 μm | Gauge blocks, diamond indenter tip radius, microtome cutter edge radius, ultraprecision *x-y* tables | IC memories, electronic video discs, LSI | Optical flats, precision Fresnel lenses, optical diffraction gratings, optical video discs |
| | 0.005 μm | | VLSI, superlattice thin films | Ultraprecision diffraction gratings |

*Source:* McKeown, P.A. 1986. *Institute of Mechanical Engineers Proceedings* 200:76.[15]

press or a boring machine. Tools that are used to produce holes are called bits. They are held in place in the drill press by clamping them in a chuck or by inserting them into a tapered sleeve. *Grinding* is a finishing process that is used to remove surplus material from the work piece surface. It is usually used on almost any surface that has been previously rough machined and is among the most expensive processes because it is generally slow in removing material. These four processes are characterized by the fact that each process generates chips, and each of these process categories has unique characteristics such as available materials, acceptable tolerance ranges, surface finish, part, and tooling costs.*

A lathe is a machine for shaping a piece of material, such as wood or metal, by rotating it rapidly along its axis while pressing a fixed cutting or abrading tool against it. Lathes are among the more versatile of the machine shop tools. They have the capability of generating true round surfaces, flat surfaces, and also screw threads. The various operations of the lathe include turning, facing, boring, drilling, parting, and threading.

Given that we are mostly interested in micro- and nanomachines, we focus our attention here on ultraprecision single-point machining and abrasive machining for finishing a product. The former is a type of ultraprecision machining where submicrometer form accuracy and nanometer-level surface finish are achieved by using high-precision, high-stiffness machine tools and single-point cutting tools. The latter is a process that removes small amounts of material from a work piece in the form of tiny chips through the contact of small, hard, sharp,

* A useful glossary for terms used in metal removal and manufacturing processes in general can be found at http://instruct1.cit.cornell.edu/courses/orie310/mfgproc/mfgprocsummary.html, and an introduction to machine tools can be found at http://me.mit.edu/2.70/machine/outline.html and at http://engide.bizland.com/mfg-5-machine.htm.

nonmetallic particles. In the abrasive machining section we review abrasive wheel machining, in particular electrolytic in-process dressing (ELID), ultrasonic machining (USM), water jet, abrasive water jet (AWJ), and abrasive jet machining (AJM).

## Single-Point Mechanical Machining

In mechanical single-point machining, various factors such as deformation of the work piece and tool, vibration, thermal deformation, inaccuracies of machine tools, and so on affect the machining accuracy. Tools for single-point machining include the following materials: carbon and medium alloy steels, high-speed steels, cast-cobalt alloys, carbides, coated tools, alumina-based ceramics, cubic boron nitride, silicon-nitride-base ceramics, diamond, and whisker-reinforced materials. Single-point machining falls into these four types of machining: 1) straight turning, 2) cutting off, 3) slab milling, and 4) end milling as illustrated in Figure 6.7.

The types of chip formation in these mechanical single-point machining processes are continuous, discontinuous, built-up edge, and oblique cutting. Most machining operations involve oblique cutting and lead to the familiar rolled-up chips. Built-up edges affect the performance of a cutting tool because of their hardness and the perceived dulling of the tool. Chip breakers are used to prevent discontinuous chips from forming (see Figure 6.8).

As in all metalworking processes where plastic deformation is involved, the energy dissipated in cutting is converted into heat that in turn increases

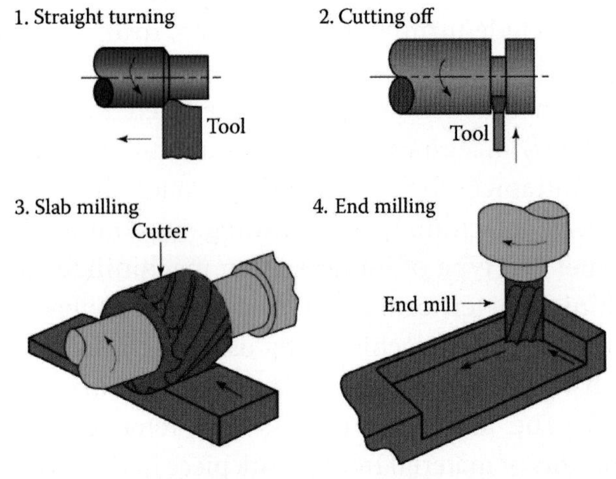


**FIGURE 6.7** Single-point mechanical machining types.

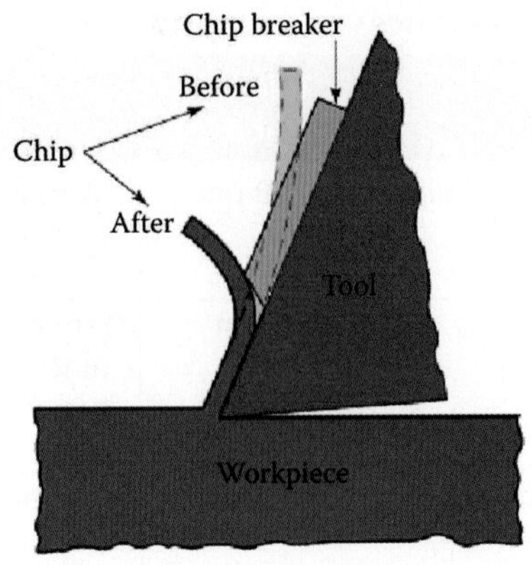


**FIGURE 6.8** Chip formation during diamond milling.

the temperature in the cutting zone. Excessive heat decreases the strength, hardness, stiffness, and wear resistance of the cutting tool; tools also may soften and undergo plastic deformation, and thus the tool shape is altered.

Ultrahigh-precision machines with sharp single-crystal diamond tools have made submicrometer precision machining possible. Single-point diamond turning is less labor intensive than grinding and polishing (see "Abrasive Machining" below). A well-defined single-point tool is more predictable than a multipoint grinding wheel. Diamond-turned optical components have surfaces with a better metallurgical structure compared with the ones made from polishing and lapping. The time required to make optical quality mirrors and spheres using polishing and lapping methods is in the range of weeks, whereas the same or better quality parts can be completely fabricated from blanks using diamond turning in a few days.

A desktop high-precision four-axes CNC machine (four degrees of freedom)* for making microparts using milling and μ-EDM in a desktop machine was introduced in Chapter 5 (Figure 5.8). The single-crystal diamond cutter of this hybrid desktop machine was featured in Figure 5.11. In Figure 6.9a, we show a scanning electron microscope (SEM) photograph of a single-crystal diamond tool bit. The

* The degrees of freedom (DOF) of a machine refers to the number of independent axes along which one can obtain relative motion between the tool and the work piece. One can achieve this either by moving the tool or moving the work piece. The DOF does not include the spindle rotation or tool translation that is responsible for the cutting action. More than three DOF generally allows one to rotate the table or the work piece to achieve the correct angular relationship between the work piece and the cutter. This can help to reduce the number of separate fixturing steps and achieve smoother contours.

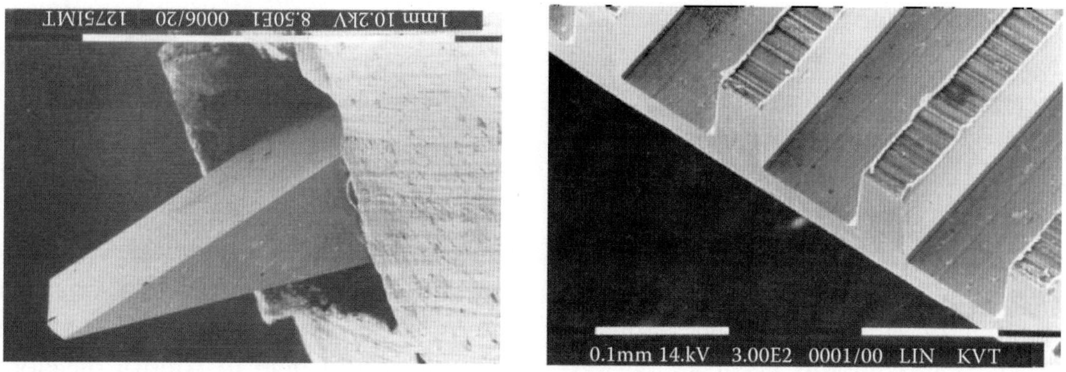


(a) (b)

**FIGURE 6.9** Diamond ultrahigh-precision tools. (a) Single-crystal diamond tool. Angle of tool is 20°. Maximum depth of cut is 500 mm. (b) Microgroove produced by diamond machining a 100-mm-thick Al foil. Width is 85 mm and depth is 70 mm.

radius of the edge on this bit is estimated at less than 0.05 μm; with such a tool, submicrometer grooves equivalent to those produced by silicon micromachining can be generated (Figure 6.9b).[16] The single hard point on the lathe in Figure 6.10 may be made from natural or synthetic diamond. Synthetic diamonds are preferred because natural diamonds have flaws that at times make them unpredictable. Ultraprecision cutting tools need to be hard and sharp, as well as have enhanced thermal properties to maintain their size and shape while cutting. High hot hardness means higher speeds and feed rates (higher production rates and lower costs). Advantages offered by diamond, the hardest material known today, include:

- Crystalline structure with a well-maintained sharp cutting edge
- High thermal conductivity, the highest of any material at room temperature
- Ability to retain high strength at high temperature
- High elastic and shear moduli, which reduce deformation during machining
- High wear resistance
- Low friction
- Produces a very accurate cut and good surface finish
- Most effective in light uninterrupted finishing cuts

However, diamond is chemically attacked by ferrous materials at high temperatures and is generally unsuitable for the machining of steels and nickel alloys. This is because of the very high wear rate of the diamond, resulting in nonviable tool costs for those applications. Diamond bits can be applied to most nonferrous metals, polymers, and crystals. Boron nitride in the forms of single-crystal cubic boron nitride (CBN) and microcrystalline cubic boron nitride (MCBN) under trade names such as Borazon or Borpax is used for precision cutting of hardened ferrous metals and high-temperature alloys.

Diamond machining is used for machining of nonferrous metals such as aluminum and copper, which are difficult materials on which to obtain a mirror surface by grinding, lapping, or polishing. This is because these metals are relatively soft, and the abrasive processes scratch the finished surface and, furthermore, are unable to produce

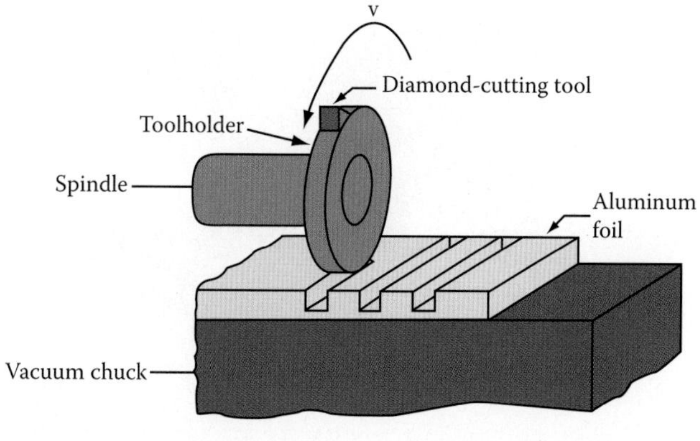


**FIGURE 6.10** Diamond tool on a lathe for precision mechanical machining. Ultrahigh-precision milling of grooves in aluminum foil. The aluminum foil is 100 μm thick, and the grooves are 85 μm wide and 70 mm deep.

high levels of flatness at the edges of the machined surface.

The single-crystal diamond tool refinements are not the only reason for the high precision achieved today; submicrometer precision is also achieved through high-stiffness machine beds, air bearings (air bearings with a rotational precision of 0.01 μm and better are available), and measurement systems such as laser interferometry. Furthermore, highly precise instruments such as servomotors, feedback devices, and computers have been implemented, and many types of machine tools are now equipped for computer numerical control (CNC), further improving precision and reproducibility of the manufactured parts.

In mechanical micromachining processes, the effects of scaling on the process mechanism—including chip formation, cutting forces, vibration, process stability, and the surface integrity of the machined components—are some of the important issues that are not fully understood. The impact of work piece size on factors such as minimum chip thickness, surface quality, burr formation, built-up edge formation, and microstructure effects has been reported as summarized in Ehmann et al.[12] These reports reveal that ploughing and elastic recovery of the work piece plays a more significant role the more closely the chip thickness approaches the cutting edge radius of the cutting tool.

Various modeling techniques—including molecular dynamics (MD) simulation, finite element method (FEM), multiscale simulation modeling, and mechanistic modeling—have been attempted to characterize micromachining processes better. Simulation technologies for mahining processes such as FEM simulation of cutting, kinematical simulation of grinding, computational fluid dynamics (CFD) simulation of abrasive flow machining, and CFD simulation of electrical discharge machining (EDM) have been used for machining at the microscale level. These methods expand the process know-how and help to control development costs and define process limits. Using a diamond tip and numerical control (NC), rice grain-sized cars were machined at Nippondenso. One of these cars was 4.785 mm long, 1.73 mm wide, and 1.736 mm tall as shown in Figure 6.11.[17] In 1995, the Guinness Book of Records awarded a world record certificate to Nippondenso Co., Ltd. (now DENSO Corporation) for the development of this world's smallest motorized car. The DENSO Micro-Car is a miniature version of Toyota's first passenger car, the 1936 Model AA sedan. Its size is astounding: 1/1000th the size of the original car, or about the size of a grain of rice.

The Micro-Car has 24 parts that come in 13 different types, including body, tires, spare tire, wheels, axle, bearings, headlights, rear lights, front bumper, rear bumper, step, number plate, and emblem. The shell of the car was made by the sacrificial mold technique. A piece of aluminum was NC-machined and plated with 30-μm-thick electroless Ni; after cutting off the lower part of the body with EDM, the Al was removed by KOH etching. Finally, the Ni

**FIGURE 6.11** Micro-Car and rice grains (a) and Micro-Car on sandpaper (grain size = 200 mm) (b). The car is 7 mm long, 2.3 mm wide, and 3 mm high. (From Bier, W., G. Linder, D. Seidel, and K. Schubert. 1991. *KfK Nachrichten* 165;[16] and Teshigahara, A., et al. 1995. *J Microelectromech Syst* 4:76. With permission.[17])

car shell was Au coated to protect it from oxidation. The stainless chassis and wheels were also made by NC machining, and the core shaft and coil of the motor were made by NC machining and EDM. Wet etching of aluminum-coated glass was used for the registration plate and the emblem. A 250-μm bearing hole was drilled in the chassis for the zirconia wheel shaft. A stepping motor drives the car electromagnetically. The tube-shaped barium ferrite rotor, formed using grinding, was permanently magnetized with a special tool. The Micro-Car is powered via two 25-μm copper wires. Unfortunately, lubrication has drawbacks for microsystems, including high viscosity and adhesive effects. Therefore, the wear on the bearings is a limiting factor. Despite this, the car can run as fast as 100 mm/s, which is fast at this scale. The electromagnetic step motor is driven by an external magnetic field, and the car's permanent barium ferrite rotor runs at a maximum speed of 100 mm/s, developing a torque of about $10^{-6}$ Nm (at 3 V and 20 mA). To assemble the various microparts into a complete car, a cyanoacrylate adhesive and a mechanical micromanipulator, ordinarily used for handling biological cells, were used.[17]

FANUC's ROBOnano $U_i$ is an ultraprecision multifunction micromachining station capable of making such high-precision parts as the mold for diffraction gratings or aspheric lenses (http://www.fanuc.co.jp/en/product/robonano/index.htm) (see the machine in heading of this chapter). This superprecision micromachining equipment consists of a diamond milling tool rotating at high speed on a superprecision positioning table with 1-nm step resolution and a rotation resolution of 0.00001°. It is a five-axis mill with all linear slides on air bearings. The machining operation is enclosed in an air shower with an environment controlled to within 0.01°C. The single-crystal diamond cutting tool is capable of generating 3D free curved surfaces. The machine is compact with a footprint of 1.27 m × 1.42 m and a height of 1.5 m and comes at a cost of $1 million. It is not available outside Japan at this time. Using this setup, FANUC succeeded in relief engraving a minute, mirror-surfaced Noh mask in copper with a "diamond end mill," which results in a surface smoother than a mirror. The roughness of the forehead ($R_{max}$) of the mask is 60 nm (see heading of this chapter). Thus, FANUC has realized a superprecision micromachining technique that allows mirror finishing in any direction except in a narrow groove. The machine was commercialized after 17 years of development by roughly 30 people. The ROBOnano is used to machine mirrors, filters, gratings, liquid crystal display panels, small lenses, and other small ultraprecise parts.

FIGURE 6.12 5-Axis-HP-Milling Center Gamma High Performance 303 by WISSNER GmbH, Göttingen, Germany.

A similar milling machine from Germany is the 5-Axis-HP-Milling Center Gamma High Performance 303 by WISSNER GmbH (Göttingen) shown in Figure 6.12. In Figure 6.13 a demonstration microstructure made on this machine, by the Fraunhofer Gesellschaft Institute for Production Systems (IPK) in Berlin, is shown.

A common limitation of the single-point diamond turning and grinding machines is their inherent ability to generate only rotationally symmetrical surfaces. Although these geometries cover the majority of requirements, there is an

FIGURE 6.13 Demonstration on the 5-Axis-HP-Milling Center Gamma High Performance 303 by WISSNER GmbH (Göttingen, Germany) shown in Figure 6.12 by IPK (Berlin).

increasing demand for optics to incorporate more random, freeform geometries. Examples of these "freeform" geometries might be advanced laser printer optics or "conformal" windows that blend into the leading edge of an aircraft's wing. A new breed of multiaxis machines has been developed that are able to generate shapes that are no longer limited to rotational symmetry but extend to these freeform geometries. An example fabrication approach, to make, e.g., a freeform optical lens, is called computer-controlled fabrication (CCF). This may include 1) CNC grinding, 2) CNC lapping, 3) profile measuring, 4) computer-controlled corrective lapping, and 5) CNC polishing. After CNC grinding, a basic freeform surface with "peak-valley" structure is obtained. CNC lapping gets rid of the "peak-valley" and microcrack layer and decreases the surface roughness without lowering accuracy. A profile of the freeform lens is then measured with a computer-controlled coordinate measuring machine, and the machining error is obtained through a comparison between the measured data and theoretical profile data. A computer-controlled corrective lapping (CCCL) process then provides the corrective NC code by using the error data. Profile measuring and CCCL are alternated until the required accuracy is reached. Finally, the freeform lens is polished on a CNC polishing machine tool to lower the surface roughness.

Moore Nanotechnology Systems LLC, a US company, developed the Nanotech 220UPL (Figure 6.14) that is considered by many the most accurate diamond-turning machine in the world because of its groove-compensated air-bearing spindle that provides motion accuracy of less than 50 nm through its 1500-rpm speed range. Single-point diamond turning and deterministic microgrinding technologies are used for the production of plano, spherical, aspherical, conformal, and freeform optics as described above. The Nanotech 220UPL two-axis CNC single-point diamond turning lathe from Moore Nanotechnology Systems (Keene, NH) can be used to make many types of medical devices. Some applications include in vivo miniature pumps, injection-molded electroless nickel mold inserts for syringes, endoscopic lenses, or pipettes and implants.

**FIGURE 6.14** Moore Nanotechnology Systems LLC, a US company, developed The Nanotech 220UPL that is considered by many the most accurate diamond-turning machine in the world because of its groove-compensated air-bearing spindle that provides motion accuracy of less than 50 nm of the whole 1500-rpm range.

Another trend in mechanical machining is toward flexible manufacturing. In response to the need for automation and the demands created by frequent design changes over a broad variety of products, the flexible manufacturing system (FMS) was developed. FMS is a combination of several technologies such as computers, CNC workstations, robots, transport bands, computer-aided design, and automatic storage. The technique was developed to produce many varieties of a certain product in smaller quantities rather than many devices of one type. CNC workstations are linked by automatic work piece transfer and handling, with flexible routing and automatic work piece loading and unloading.[18] This philosophy also underlies desktop factory design (see "Desktop Machining" below).

Ultraprecision manufacturing is the commercially preferred technique for the production of computer hard disks, mirrors for x-ray applications, photocopier drums, commercial optics such as polygon mirrors for laser-beam printers, consumer electronics such as mold inserts for the production of compact disc reader heads and camcorder viewfinders, and high-definition television (HDTV) projection lenses and VCR scanning heads.

A tutorial on mechanical drilling and milling can be found at http://www.me.mtu.edu/~microweb.

For more information on estimating manufacturing accuracies, consult http://www.isr.umd.edu/Labs/CIM/vm/ama/node19.html. Custom houses will typically report as best mechanical machining tolerances ±2 μm for cutting and ±0.4 μm for grinding (e.g., Burman tool at http://www.burman.co.uk/engineering/Machining.htm).

## Abrasive Machining

### *Introduction*

Abrasive machining is a process that removes small amounts of material from a work piece in the form of tiny chips through the contact of small, hard, sharp, nonmetallic abrasive particles. These abrasive particles may be attached to a spinning wheel (bonded), embedded in a slurry (loose), or impacted on a surface in a jet (impact). Abrasive methods in each of these categories are summarized in Figure 6.15.

Common types of abrasive machining processes using bonded abrasives include grinding, honing,* and microhoning. Lapping,† buffing, polishing, and ultrasonic machining (USM) use loose abrasive grit. Honing, microhoning, and lapping are usually termed *finishing operations* (also tertiary manufacturing methods). In USM, also called *ultrasonic impact grinding*, high-frequency vibrations delivered to a tool embedded in an abrasive slurry create accurate cavities of virtually any shape. Water jet, abrasive water jet (AWJ), and abrasive jet machining (AJM) are examples of abrasive impact machining.

The abrasive grit or cutter needs to be harder than the material to be cut (see Table 6.2). Natural abrasives include emery, corundum, quartz, garnet, and diamond. However, natural abrasives contain unknown amounts of impurities and possess nonuniform properties, and as a result, aluminum oxide and silicon carbide abrasives are made synthetically. Conventional synthetic abrasives include aluminum oxide ($Al_2O_3$) and silicon carbide (SiC); superabrasives include cubic boron nitride (CBN) and diamond.

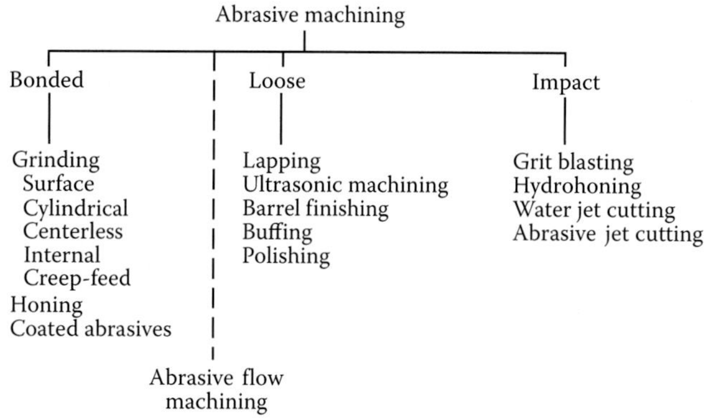


**FIGURE 6.15** Categories of abrasive machining according to the matrix of the abrasive grit.

1. Aluminum oxide ($Al_2O_3$) known as Alundum or Aloxide. Various substances may be added to enhance hardness, toughness, and so on. Plain $Al_2O_3$ is white and used to grind steel, ferrous, and high-strength alloys.
2. Silicon carbide (SiC) known in trade as Carborundum and Crystalon. Harder than $Al_2O_3$ but not as tough. Used to grind aluminum, brass, stainless steel, cast irons, and certain brittle ceramics.
3. Boron nitride in the forms of single-crystal CBN and microcrystalline cubic boron nitride (MCBN) under trade names such as Borazon or Borpax. Used for hard materials such as hardened tool steels and aerospace alloys.
4. Diamond, a pure form of carbon, both natural and artificial. Used on hard materials such as ceramics, cemented carbides, and glass.

The temperature increase during abrasive machining can be very high (3000°F). Chips carry away heat, but a larger fraction of heat is conducted into the work piece. The effect of the temperature increase is more pronounced than in metal cutting, and the excessive temperature increase caused by grinding can temper or soften hardened metals. Abrasive machining is used when a very fine surface finish

---

* In honing, a simultaneous rotating and reciprocating motion between an abrasive stone and the work piece causes the removal of a small amount of material. The resulting work piece surface is one of high precision and uniform finish.

† The process of wearing down the surface of a softer material by rubbing it under pressure against the surface of a harder material that has been formed in the shape opposite to that desired on the softer material. Loose, wet abrasive generally is used between the two parts, but sometimes the abrasive is embedded in the lap in a rigid form (see diamond cutting tool).

TABLE 6.2 Knoop Hardness for Various Materials

| | | | |
|---|---|---|---|
| Common glass | 350–500 | Titanium nitride | 2000 |
| Flint, quartz | 800–1100 | Titanium carbide | 1800–3200 |
| Zirconium oxide | 1000 | Silicon carbide | 2100–3000 |
| Hardened steels | 700–1300 | Boron carbide | 2800 |
| Tungsten carbide | 1800–2400 | Cubic boron nitride | 4000–5000 |
| Aluminum oxide | 2000–3000 | Diamond | 7000–8000 |

and good dimensional accuracy are required and/or when the work piece materials are too hard or too brittle to work by other methods. The method is typically much slower than single-point mechanical machining (see above).

*Grinding*

Here, as a first example of an abrasive process and as an introduction to electrolytic in-process dressing (ELID) (see next section), we survey grinding. In grinding, the chip-removal process uses an abrasive grain or grit as the cutting tool that is bonded to a very fast spinning grinding wheel (see Figure 6.16). The grinding wheel is usually disk shaped and is precisely balanced for high rotational speeds. Grinding wheels incorporate an abrasive from the list in Table 6.2. The properties of a grinding wheel that determine how it acts are the kind and size of abrasive, how closely the grains are packed together, and amount of the bonding material.

The individual abrasive grains have an irregular geometry and are spaced randomly along the periphery of the wheel. The radial position of these cutting grains varies, and the cutting edges are extremely small. The grain size is an important parameter in determining the quality of the surface finish and the material removal rate. Small grit sizes produce better finishes, and larger grit sizes permit larger material removal rates. Also, harder materials need smaller grain sizes to cut effectively, whereas softer materials require larger grit size. The grain size of the abrasives is measured as grit size: coarse grain 8–24, medium 30–60, fine 70–180, and very fine >220. The feeds and depths of the cut in grinding are small, whereas the cutting speed is high (600 ft./min). As mentioned earlier, the temperature increase during abrasive machining can be very high (3000°F). Chips carry away heat, but a larger fraction of heat is conducted into the work piece. The effect of the temperature increase is more pronounced than in metal cutting, and the excessive temperature increase caused by grinding can temper or soften hardened metals. Burning chips are the sparks observed during grinding with no cutting fluid because the chips have heat energy to burn or melt in the atmosphere. Grinding may be classified as nonprecision or precision, according to purpose and procedure. In grinding, the chips are small but are formed by the same basic mechanism of compression and shear as in single-point machining (see above). There are three types of grain action in grinding:

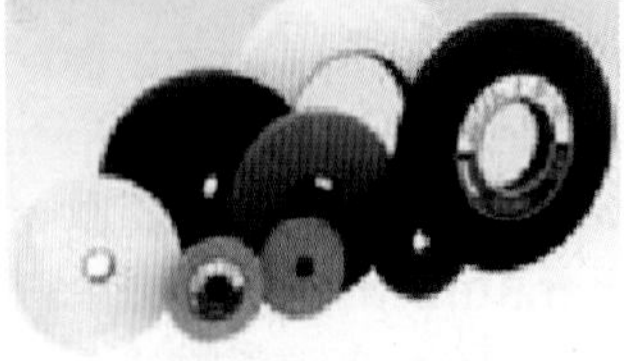

FIGURE 6.16 Common grinding wheels.

1. Cutting: Grit can penetrate to the surface and performs chip removing.
2. Plowing: Grit can penetrate to the surface but cannot perform cutting. The work surface deforms.
3. Rubbing: Grits rub to the surface, and energy is consumed without cutting.

Dressing a grinding wheel is the process of resharpening the tiny cutting edges on a grinding wheel's surface. Thus, dressing a grinding wheel is shaping the grinding face of the wheel; this is often done with computer-controlled shaping features with a diamond-dressing tool kept normal to the surface at point of contact (Figure 6.17). The diamond dresses or turns the grinding wheel.

The machining energy in grinding is typically purely mechanical, but next we review a grinding

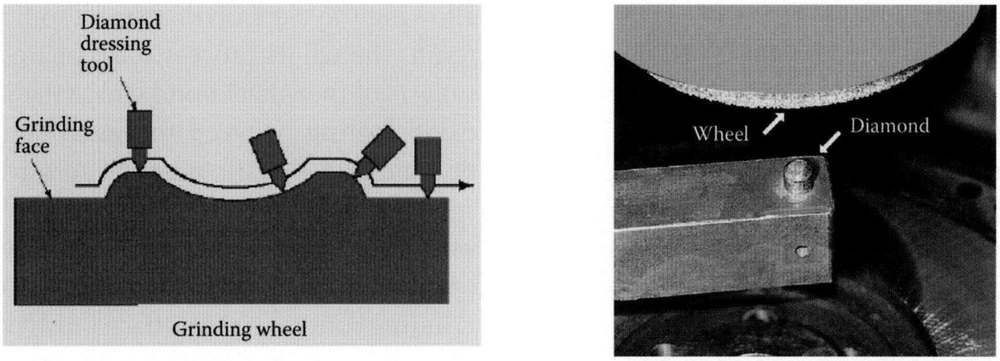


**FIGURE 6.17** Shaping the grinding face of a wheel by dressing it with computer-controlled shaping features. Note that the diamond dressing tool is normal to the surface at point of contact.

process where electrochemistry plays a major role. In ELID, the material removal process (any material in this case) is purely mechanical, but an electrochemical dissolution process of the metal matrix dresses the abrasive wheel, with superabrasives embedded in a metal matrix, continuously.

### *Nanofinishing of Surfaces with Electrolytic In-Process Dressing*

A nanosurface finish of a wide variety of brittle materials is becoming a more important demand in IC fabrication, MEMS, and NEMS. For example, a nanosurface finish is a necessity for printing nanowiring circuits using lithography techniques. The surface finish of Si wafers polished using chemomechanical polishing is found still less than ideal (see Chapter 1). To obtain a nanofinish, chip removal from the surface that is being grinded should be minimized. A *nanosurface* is only possible when the chip size of the material being removed reaches atomic level. Finishing with superabrasive grits [like diamond or cubic boron nitride (CBN)] is the method of choice to achieve these types of surfaces, but superabrasive grits need very good bonding strength to the matrix that holds them for better efficiency. Recent developments of ductile mode machining show that plastically deformed chip removal minimizes the subsurface damage of the work piece and that when chip deformation takes place in the ductile regime, a defect-free nanosurface is possible—completely eliminating the polishing process! Metal-bonded superabrasive grinding wheels in principle offer the solution here: they have the required higher bonding strength of the metal to the superabrasive grit and the required high mechanical stability. The major problems encountered using such superabrasive metal bonded wheels are that during grinding they get loaded and glazed. To avoid these difficulties, the wheel must be dressed periodically (see Figure 6.18), which

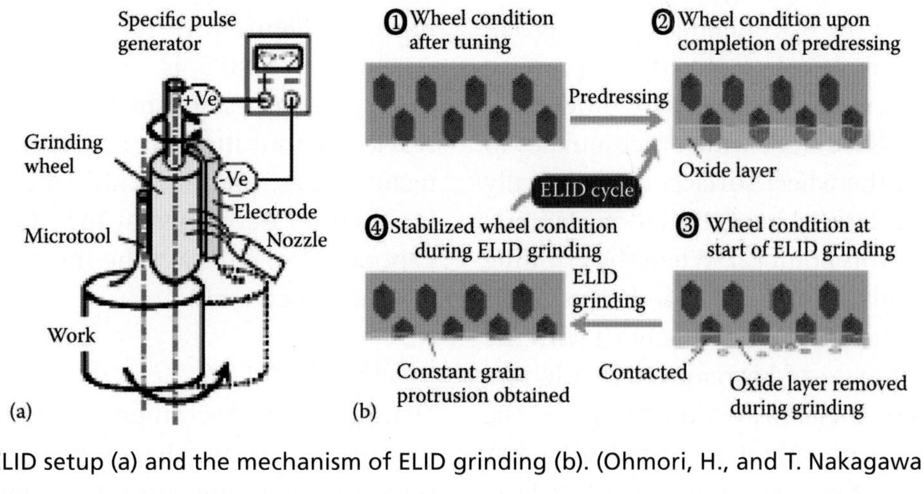


**FIGURE 6.18** The ELID setup (a) and the mechanism of ELID grinding (b). (Ohmori, H., and T. Nakagawa. 1990. *Annals CIRP* 39:329.[19])

makes the grinding process very tedious. It is here that electrolytic in-process dressing (ELID) provides an elegant solution that eliminates both the wheel loading and glazing problems.

ELID is a mechanical grinding operation where the abrasive grit, embedded in a metal matrix, is kept fresh by an electrochemical dissolution process of the metal matrix. The method applies to metal and nonmetal substrates.

The concept of ELID was developed by Ohmori, chief scientist at RIKEN.[19] The system uses a metal bond grinding wheel (tool), a power supply, a grinding fluid (electrolyte), and a counter electrode as sketched in Figure 6.18a. It is important to notice that the work piece is not in the electrical circuit; thus, the grinding action here is purely mechanical, and the electrochemical assist is only there to keep the abrasive grit fresh. The metal bond wheel is made the positive electrode, whereas a negative electrode is positioned about 0.1–0.3 mm below the wheel face. The grinding fluid is supplied between the two. The grinding wheel (tool) used is the metal bond wheel in which bonding material is composed mainly of cast iron and cobalt and is sintered with the abrasive grit. The wheel is dressed using electrolysis during grinding, which helps to maintain larger grain protrusions and grit density. A DC pulse voltage is supplied between the two electrodes to electrolytically remove only the metal bond of the wheel, allowing efficient and automatic dressing of the wheel. This dressing is continued even during the grinding work to prevent reduced wheel sharpness from wear, thereby realizing highly efficient mirror-surface grinding. The mechanism of the ELID grinding method is sketched in Figure 6.18b. First the wheel bond is electrolyzed, which causes the abrasives to protrude appropriately (1). This process at the same time produces a nonconductive oxidized iron, which accumulates to form a coating on the wheel surface, automatically reducing the electrolysis current. Initial or predressing is completed at this point (2). When the grinding work is started in this state, the nonconductive oxide layer on the wheel surface comes in contact with the surface of the work piece and is removed by friction. As this takes places, the abrasives start to grind the work piece, and subsequently start to gradually wear (3). This reduces the insulation of the wheel surface, allowing the electrolytic current to flow again. As a result, the whole process starts again with the electrolysis of the wheel bond where the nonconductive oxide coating between the worn-out abrasives has become thin (4), allowing the abrasives to protrude again. This is called the ELID cycle. The ELID cycle varies according to the type of metal bond of the wheel used, electrolytic conditions, work piece, and grinding conditions. A surface finish of 2 nm may be achieved using ELID (20–50 N force for grinding).

ELID grinding is highly suitable for achieving nanosurface finishes on metals and nonmetals alike. Different types of ELID process are amenable for machining and finishing of all types of microshapes, and a nanosurface finish can be achieved using ductile mode machining on hard and brittle materials. ELID grinding can produce ductile surfaces without any subsurface damage, which eliminates the lapping and polishing processes. Thus, surface finish can be improved with higher accuracy and tolerance because the component is produced using only one process. The method is applied to process material such as silicon, glass, ceramics, extremely hard steel materials, and composite materials. Conventional applications include semiconductor substrates; electronic/optical components, such as lenses, magnetic heads, hard-disk substrates, and optical connectors; and the production of metal molds.

The major remaining difficulties with this technique are the lack of feedback devices to control the in-process dressing, optimization of machining conditions, and modeling of ductile mode grinding.

For up-to-date information on ELID, visit the web site of NEXSYS (http://www.nexsys.co.jp/index_e.htm) see Figure 6.19 for a NEXSYS desktop ELID machine. NEXSYS is an authorized RIKEN venture company established in June 1998 to provide services committed to spreading new fabrication technologies, mainly ELID mirror-surface grinding method developed at the Materials Fabrication Laboratory of RIKEN (The Institute of Physical and Chemical Research).

### *Ultrasonic Machining*

Diamond tool machining is not only limited in the number of feature shapes and sizes that are possible it is also time consuming. Electrical discharge

**FIGURE 6.19** Desktop ELID machine (500 mm × 500 mm × 560 mm) from NEXSYS (http://www.nexsys.co.jp/index_e.htm). ELID grinding is currently the best method to avoid high wheel loading and to generate mirror finish surfaces.

machining [(EDM), see Chapter 5)] is only suitable for use on conductive materials and because, just like laser beam machining [(LBM), see Chapter 5)], it is a thermal process, a heat-affected zone (HAZ) often has a negative impact on the end use of parts, especially in the case of high-reliability applications. In contrast, ultrasonic machining (USM) is a non-thermal, non-chemical and non-electrical machining process that leaves the chemical composition, material microstructure and physical properties of the work piece unchanged. This mechanical process can be used to generate a wide range of intricate features in both conductive and non-metallic materials with hardnesses of greater than 40 HRC (Rockwell Hardness measured in the C scale). The USM process can be used to fabricate precision micro-features, round and odd-shaped holes, blind cavities, and OD/ID features. Multiple features can be drilled in parallel, often reducing the total machining time significantly. Ultrasonic waves are sound waves of frequency higher than 20,000 Hz and can be generated using mechanical, electromagnetic, and thermal energy sources. They can be produced in gases (including air), liquids, and solids. One of the first applications for ultrasound was SONAR (an acronym for sound navigation ranging). It was used on a large scale by the US Navy during World War II to detect enemy submarines. Today's ultrasonic applications include medical imaging (scanning the unborn fetus), testing for cracks in airplane construction, and machining.

In ultrasonic machining (USM), also called *ultrasonic grinding*, high-frequency vibrations delivered to a tool tip, embedded in an abrasive slurry, by a booster or sonotrode, create accurate cavities of virtually any shape; that is, "negatives" of the tool. Almost any hard and brittle material, including aluminum oxides, silicon, silicon carbide, silicon nitride, glass, quartz, sapphire, ferrite, fiberoptics, and so on, can be ultrasonically machined. The tool does not exert any pressure on the work piece (drilling without drills) and is often made from a softer material than the work piece, e.g., from brass, cold-rolled steel, or stainless steel, and wears only slightly.[20] Tool material should be tough and ductile (low carbon steels and stainless steels give good performance). Tools are usually about 25 mm long, with a size equal to the "hole size" minus twice the size of abrasives. The mass of tool should be as small as possible so that it does not absorb the ultrasonic energy. The tool, typically vibrating at a low amplitude of 0.025 mm at a frequency of 20–100 kHz, is gradually fed into the work piece to form a cavity corresponding to the tool shape. The vibration transmits a high velocity to fine abrasive grains between the tool and the surface of the work piece. In the process, material is removed by microchipping or erosion with the abrasive particles. The grains are in a water slurry that also serves to remove debris from the cutting area. The high-frequency power supply for the magnetostrictive or piezoelectric transducer stack that drives the tool is typically rated between 0.1 and 40 kW. Piezoelectric transducers use the inverse piezoelectric effect using natural or synthetic single crystals (such as quartz) or ceramics (such as barium titanate), which have strong piezoelectric behavior. Ceramics have the advantage over crystals in that they are easier to shape by casting, pressing, and extruding.

Because this method is nonthermal, nonelectrical, and nonchemical, it produces virtually stress-free shapes even in hard and brittle work pieces. Ultrasonic drilling is most effective for hard and brittle materials; soft materials absorb too much sound energy and make the process less efficient. In Figure 6.20, important components of a typical USM setup are shown.[21] In the case of hard materials, microchipping is possible because of the high stress produced by particles striking the surface.

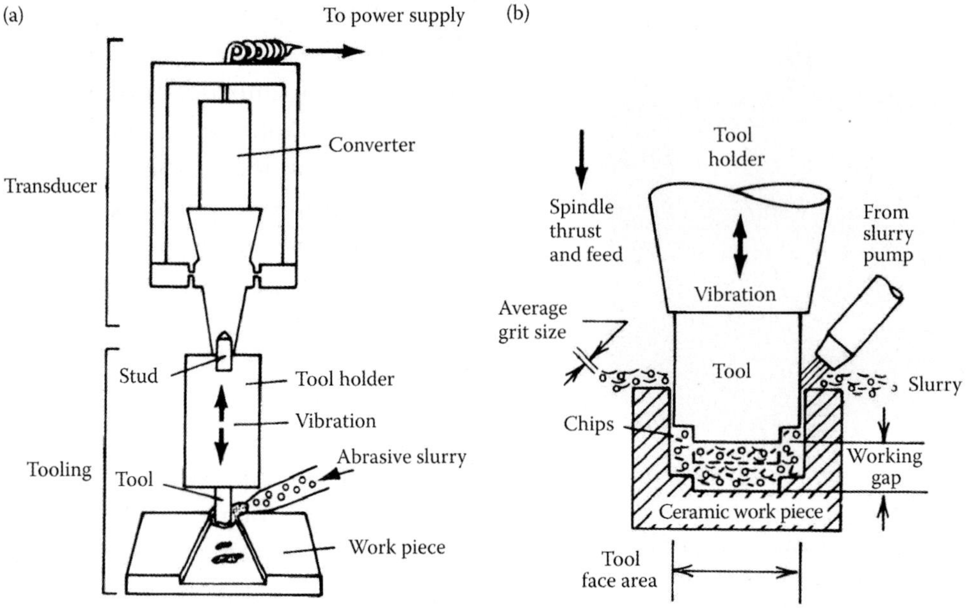


**FIGURE 6.20** Schematic showing key components of a typical ultrasonic machining installation. (a) Transducer assembly coupled to tooling assembly. (b) Close-up view of tooling assembly used to machine ceramics. (From Moreland, M.A. 1992. *Engineered materials handbook*. Ed. S.J. Schneider. Metals Park, OH: ASM International. With permission.[21])

In Figure 6.21, SEM photomicrographs are shown of 640-μm holes drilled in alumina by two different techniques: USM (Figure 6.21a) and laser-beam machining (Figure 6.21b). The edges of the hole are much better defined with the ultrasonic technique. In Figure 6.22a, multiple channels and holes drilled all in parallel in a polycrystalline silicon slab are shown,[21] and in Figure 6.22b, a coin grooved with USM is displayed.

The key to USM, besides the tool itself, is the abrasive: a slurry of water or oil and small abrasive particles (large = 0.5 mm average particle diameter or small = 0.008 mm average particle diameter) of boron carbide, silicon carbide, or aluminum oxide. Boron carbide is the hardest abrasive and lasts the longest. The abrasives are suspended in water or oil at a concentration of 20–50 wt% and recirculated constantly to the work piece to supply fresh abrasive at the cutting site and remove abraded particles. The particle size and the amplitude of the vibrations used are typically made about the same. During one strike, the tool is moved down from its uppermost position with a starting speed of zero; then it speeds up to finally reach a maximum speed at the mean tool position. The tool slows down and eventually reaches zero again. When the grit size is close

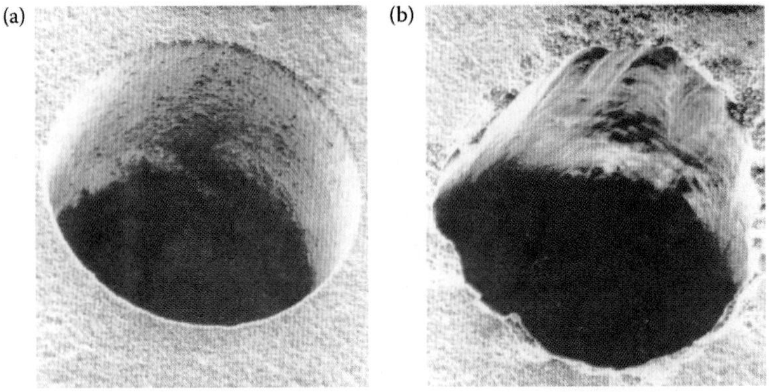


**FIGURE 6.21** Scanning electron microscopy photomicrographs of 640-mm holes drilled into alumina with two different techniques. (a) Ultrasonic machining. (b) Laser-beam machining. (From Moreland, M.A. 1992. *Engineered materials handbook*. Ed. S.J. Schneider. Metals Park, OH: ASM International. With permission.[21])

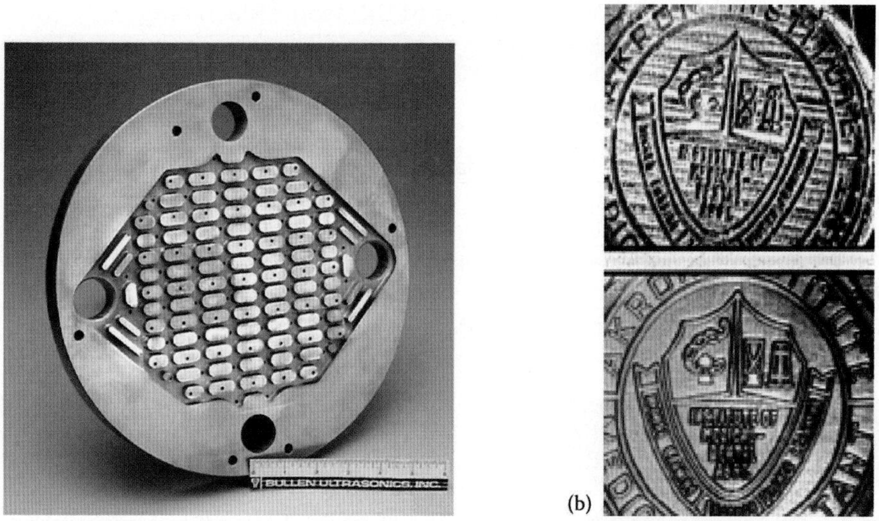


(a) (b)

**FIGURE 6.22** (a) Channels and holes ultrasonically machined in a polycrystalline silicon wafer. (From Moreland, M.A. 1992. *Engineered materials handbook.* Ed. S.J. Schneider. Metals Park, OH: ASM International. With permission.[21]) (b) Coin with grooving carried out with USM.

to the mean position, the tool hits the grit with its full speed. The smaller the grit size, the lesser the momentum it receives from the tool. Therefore, there is an effective speed zone for the tool, and correspondingly there is an effective size range for the grits. The size of the abrasive particles determines the roughness or surface finish and the speed of the cut. Material removal rates are low, usually less than 50 $mm^3$/min. The mechanical properties and fracture behavior of the work piece materials also play a large role in both roughness and cutting speed. For a given grit size of the abrasive, the resulting surface roughness depends on the ratio of the hardness ($H$) to the modulus of elasticity ($E$). As this ratio increases, the surface roughness increases. For example, under identical ultrasonic cutting conditions, carbide with an $H/E$ ratio of 1.26 has a resulting surface roughness of 0.5 μm, and glass with an $H/E$ of 8.1 ends up with a surface finish of 1.55 μm.[21]

Higher $H/E$ ratios also lead to higher removal rates: 4 $mm^3$/min for carbide and 11 $mm^3$/min for glass. Thus, a low fracture toughness will result in a higher material removal rate and an increase in work piece roughness values. For a given material, a larger particle size results in a rougher surface. With a very fine particle size of 0.008 mm and a low $H/E$ ratio (e.g., alumina with $H/E$ 2.36), a 0.5-μm surface roughness can be achieved.[22] Holes as small as 0.076 mm have been drilled, but in production applications 0.25 mm is more the norm. The upper limit on cavity size is approximately 75 mm. With a 250-μm-diameter hole, the aspect ratio is about 2.5; with increasing hole size, the aspect ratio increases. It is safe to assume that a ±10-μm tolerance can be achieved with the finest abrasive powders.

Rotary ultrasonic machining (RUM) enhances material removal rates, finish capabilities, and overall drilling efficiency. A 1/4-in. hole can be drilled 5-in. deep into glass in about 2 minutes with a rotating ultrasonic tool. To make such high-aspect-ratio holes and deep cavities, it is often required that the work piece be removed periodically to ensure uninterrupted abrasive slurry flow to the cutting zone. Conventional USM is not capable of drilling microholes much smaller than 100 μm for lack of corresponding microtools. Using microelectrode discharge machining wire cutting (μ-EDM-WC; see Chapter 5) to make microtools (similar to the pipe structures shown in Figures 5.4 and 5.5), Sun et al. made holes as small as 15 μm in diameter.[23] Surface roughness was typically about 0.2 μm, and ultrasonic drilling speed ranged from 2.0–6.0 μm/min.

Summarizing, with USM, operating costs are reasonably low and required operator skill level is modest (operator must wear eye goggles to protect against abrasive particles and microchips). The method is slow (usually less than 50 $mm^3$/min), but under ideal conditions, penetration rates of 5 mm/min can be obtained. The required power is usually between 0.1 and 40 kW. Normal hole

TABLE 6.3 Advantages and Disadvantages of Ultrasonic Machining

| Advantages | Disadvantages |
|---|---|
| Machining of any material regardless of conductivity | Low material removal rate |
| Precision machining of brittle hard materials | Tool wears fast |
| Does not produce electric, thermal, or chemical defects at the surface | Machining area and depth are restricted |
| Can drill circular or noncircular holes in very hard materials | |
| Less stress because of its nonthermal nature | |

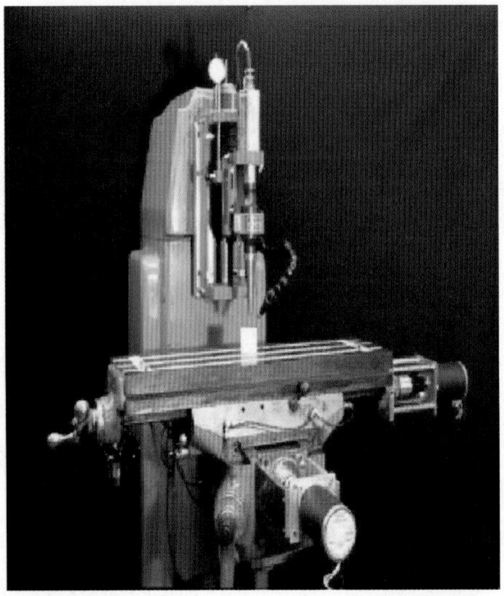

FIGURE 6.23 900-W Sonic Mill ultrasonic mill.

tolerances are 0.007 mm, and a surface finish of 0.02–0.7 μm can be achieved. Computer-controlled standard USM equipment is available with a position resolution on all axes of 0.25 μm (see Figure 6.23).

Machines cost up to $20,000, and production rates of about 2500 parts per machine per day are typical. If the machined part is a complex element (e.g., a fluidic element) of a size >1 cm$^2$ and the best material to be used is an inert, hard ceramic, this machining method might well be the most appropriate. A typical ultrasonic service provider is Bullen Ultrasonics (Eaton, OH; http://www.bullen-ultrasonics.com).

Advantages and disadvantages of USM are summarized in Table 6.3.

### *Water Jet Machining*

In water jet machining, in industrial use since 1970, material is removed through mechanical impact; that is, erosion effects of a high-velocity (600–900 m/s with pressures up to 400 MPa), small-diameter (in the range of 0.07–0.5 mm) jet of water.[24] It is a nonthermal process that eliminates warping and avoids a heat-affected zone (HAZ). Water is most commonly used, but additives such as alcohols, oils, and glycerol are used when they can be dissolved in water. The exit orifice is often made of sapphire and ranges from 0.05–0.020 in. There are some advantages one can list for water jet machining:

- No heat is produced.
- Cut can be started anywhere without the need for predrilled holes.
- Burr produced is minimum.
- Environmentally safe and friendly manufacturing.

The technique is mainly used to cut soft, nonmetallic materials such as Plexiglas®, cardboard, fabrics, foam rubber, wood, food, and so on. With a water jet nozzle of 0.13 mm and a pressure of 379 MPa, a 3-mm-thick, Plexiglas plate can be cut at speeds of 0.9 m/min. Tolerances are a function of the material type and thickness, and they are usually within ± 0.1–0.2 mm.[25] Edges in water jet machining have a sandblasted appearance. Harder materials exhibit a better edge finish. Finishes range from 63 rms* to very coarse depending on the material. In Figure 6.24, we show a schematic illustration of water jet machining and examples of various nonmetallic parts cut by a water jet machine. When dealing with the choice of using laser machining (a thermal technique covered in Chapter 5) or water jet machining, one may use the following decision tree:

* Rms refers to the root mean square which is an average of peaks and valleys of a materials' surface profile (see Equation 6.5). It is given in micrometers or microinches. Rms is also called $R_q$.

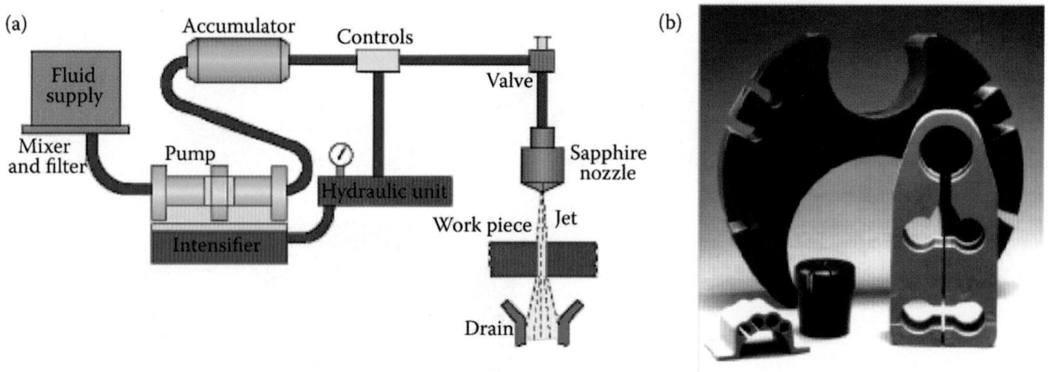


**FIGURE 6.24** (a) Schematic illustration of water jet machining. (b) Examples of various nonmetallic parts cut by a water jet machine. (Courtesy of Possis Corporation.)

1. Is the material reflective or heat conductive? Yes → WATERJET
   No ↓
2. Is material thicker than 3/8″? Yes → WATERJET
   No ↓
3. Is a heat affected zone (HAZ) a concern Yes → WATERJET
   No ↓

LASER WATERJET

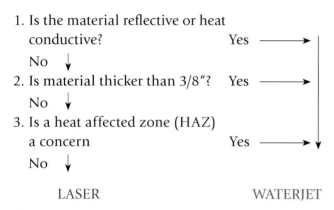

In Table 6.4 we compare advantages and disadvantages of water jet and laser machining. In water jet-guided laser machining (Chapter 5, Figure 5.23) the two techniques are combined. A thin jet of high-pressure water (between 40 and 100 μm) and water at a pressure between 20 and 500 bars is forced through a nozzle (made of diamond or sapphire). The water jet is not abrasive in this case; the material is solely ablated thermally by the laser. The high speed of the water, up to 300 m/s at 500 bars, ensures fast removal

**Table 6.4** Advantages and Disadvantages of Water Jet and Laser Machining

| Waterjet | | Laser | |
|---|---|---|---|
| **Advantages** | **Disadvantages** | **Advantages** | **Disadvantages** |
| Water jet diameter 0.001–0.014 in. | | Spot size as small as 0.0015 in. | |
| Higher material thicknesses can be cut | Equipment cost | Narrow cut (0.020 in.) | Equipment cost |
| No heat-affected zone | Pump maintenance (every 1000 h) | Low maintenance | Requires material that absorbs laser light; material limitations (e.g., laser reflective materials such as Cu and Al) |
| Does not produce electric, thermal, or chemical defects at the surface | Noise (80 db or more) | More compact nesting | Heat-affected zone (HAZ) |
| No part distortion | Need for filters | Faster cutting rates | |
| Not dependent on material property; affected only by resistance to abrasion | Slow cutting rates | Holes as small as 0.003–0.015 in. | |
| | User education in high-pressure plumbing | Tolerance of ±0.0003 in. through thin material | |
| | Finishes will range from 63 rms to very coarse | Edge finish 125 rms | |
| | Water must be highly purified | | |

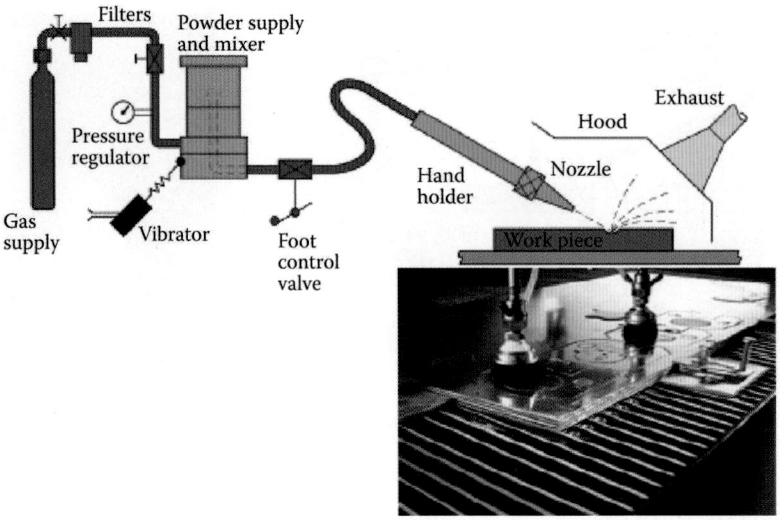


**FIGURE 6.25** Abrasive water jet (AWJ) machining setup.

of debris. Water jet-guided laser etching prevents the problem of heat damage (microcracks, structural changes) inherent in any laser ablation process. The result is a heat damage-free material ablation.

### *Abrasive Water Jet Machining*

By adding abrasives to the water jet, a much wider range of materials, including many more metals and ceramics, can be machined. Abrasive water jet (AWJ) has been in industrial use since 1982. The cutting action is a grinding process where the forces and motions are provided by water rather than a solid grinding wheel. The method is principally used for edge finishing, deburring, and polishing and has been increasingly used for treating hard-to-machine and multilayered materials and as an alternative tool for milling, turning, drilling, and polishing. In some applications AWJ is starting to replace electrical discharge machining (EDM).

An abrasive jet uses water that is pressurized up to 400 MPa and then forced through a small sapphire orifice at 1000–2500 ft./s, or up to about 2.5 times the speed of sound. Some abrasive is then pulled into this high-speed stream of water and mixed with the water in a ceramic mixer as shown in Figure 6.25. This jet of water and abrasive is then directed at the material to be machined. The jet drags the abrasive through the material in a curved path, and the resulting centrifugal forces on the particles press them against the work piece. AWJs can machine a wide range of thicknesses and materials, including metals, plastics, glass, and ceramics. Materials cut by the abrasive jet have a smooth, satin-like finish, similar to a fine sandblasted finish. Abrasive jets abrade material at room temperature. As a result, there are no heat-affected zone (HAZ) or structural changes in the work piece. Abrasive jets can also machine hardened metals and materials with low melting points. Abrasive jets typically use garnet as an abrasive. Garnet is a reddish natural crystal, with a Mohs hardness* of 6.5–7.5. No noxious gases or liquids are used in AWJ, nor are there any oils used in the machining process, so this is again a "green" machining approach. A wide range of conventional processes can be performed with this single tool such as drilling, sawing, broaching, profile milling, and so on. No heavy burrs are produced during the AWJ process. Parts can often be used directly without further deburring.

Common applications of AWJ machining include:

- Fast and precise cutting of fabrics
- Vinyl, foam coverings of car dashboard panels
- Plastic and composite body panels used in the interior of cars
- Cutting glass and ceramic tiles

*Principles of Abrasive Water Jet Machining* provides good further study.[26] For a tutorial on AWJ and

* Mohs hardness is defined by how well a substance will resist scratching by another substance. The scale consists of 10 minerals ranging from 1 for the softest (talc) to 10 for the hardest (diamond).

AJM on the Internet, visit http://www.waterjets.org/about_abrasivejets.html. Common water jet terminology is explained at http://www.omax.com/glossary.php. For a typical equipment supplier, visit Omax at http://www.omax.com.

### *Abrasive Jet Machining—Powder Blasting*

Sandblasting in a pressurized air stream is used to remove paint, clean cathedrals, and decorate glass. On a finer level, powder blasting, using small particles (usually $Al_2O_3$ powder, particle size <100 μm) is used for device demarking in the electronics industry and in surface preparation before plating. On this level, sandblasting is called *abrasive gas jet machining* (AJM) or *microabrasive blasting*. In this type of machining, the work piece material (hard, brittle materials such as glass, silicon, tungsten, and ceramics) is removed by the mechanical impact of a high-velocity ($v_{particle}$ = 80–300 m/s) stream of abrasive-laden inert gas or air. The impact causes crack generation on the target surface.[27,28] Using masks, the method is also put to use in rapid prototyping[29] and etching of shallow intricate holes in electronic components such as resistor paths in insulators and patterns in semiconductors. The process is fast and directional, and removal rates of up to 25 μm/min have been achieved. Another application is the creation of thousands of holes at once at low cost, high speed, and high accuracy for flat panel displays.[14,28]

Recent efforts are in fine-tuning the technique for mesoscale MEMS applications. The required equipment, as schematically illustrated in Figure 6.26, is relatively inexpensive.

The setup consists of the powder blaster that regulates the flow of abrasive powder accelerated to the nozzle. An air dryer is connected to the powder reservoir to avoid moisture-induced particle agglomeration. To obtain an homogeneous scan of the work piece surface, the sample is placed on an *x,y*-translation stage. A dust collector recovers the used alumina powder. The applied pressure can typically be regulated between 1 and 5 bars.

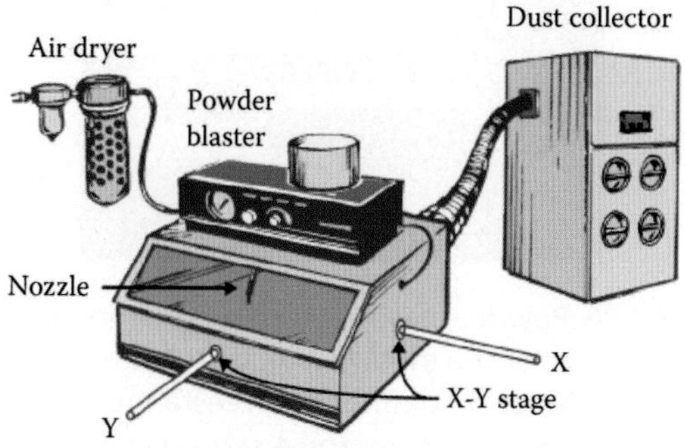


**FIGURE 6.26** Schematic of powder blasting machine. (From Yamahata, C. 2005. EPF de Lausanne.[30])

A schematic diagram showing the principle of powder blasting microerosion process is presented in Figure 6.27. The different elements shown are elucidated next. A nozzle (e.g., from tungsten) provides the focusing and acceleration function for the abrasive stream generated by the blaster. A protective mask with an erosion rate much smaller than that of the substrate is also shown. In combination with suitable masks, for example, an electroplated photolithographically patterned Cu layer, minimum feature sizes of less than 50 μm, an aspect ratio of 2.5, and a depth of >1 mm in glass have been demonstrated. Polymer negative resist foils (e.g., ORDYL BF 400, specially developed for powder blasting) can also be used as mask, but in that case feature sizes are limited to 75 μm and an aspect ratio of 1.5 (http://www.el.utwente.nl/tdm/mmd/education/ d-opdrachten/powderblasting.html). Wensink et al. made a comparison study of various mask materials in powder blasting, and the same group is striving to decrease the minimum feature size to less than 10 μm (http://www.el.utwente.nl/tdm/mmd/education/d-opdrachten/powderblasting.html). The substrate in this type of machining may be glass, silicon, tungsten, or a ceramic. Finally, there is the abrasive powder.

Depending on the application aimed for, different types of abrasive powders with various shapes and hardness can be used. The most common powders are:

- Silicon carbide (SiC), because of its hardness, is the most efficient medium used for microabrasive blasting. It has a hardness of more than 9 on the Mohs scale, just below diamond.
- Crushed glass is a mild abrasive medium (Mohs hardness = 5–6), which is used when only a slight degree of abrasion is desired.
- The spherical shape of glass beads keeps them from cutting into the surface of the part. It is

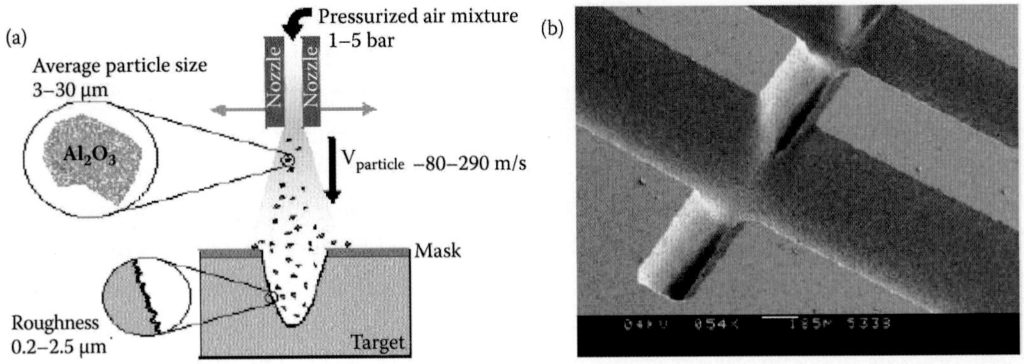


FIGURE 6.27 (a) Schematic diagram demonstrating the principle of the powder blasting microerosion process. (b) 400-μm deep-blasted glass structure. (From Yamahata, C. 2005. EPF de Lausanne[30] and http://www.anteryon.com.)

typically used to perform light deburring or to apply a satin-like finish on a work piece.

- Aluminum oxide is the most commonly used cutting abrasive because of the sharp shape of the particles and good hardness (Mohs hardness = 9).

The rate of material removal in powder blasting is commonly evaluated from the erosion rate $E_{rate}$, which is defined as:

$$E_{rate} = \frac{\text{weight of removed material}}{\text{weight of impacting particles}} \tag{6.1}$$

Slikkerveer[31] studied the mechanics of powder blasting in detail. In his analysis, he introduced the erosion efficiency, $E_{eff}$, as:

$$E_{eff} = \frac{\text{weight of removed material}}{E_k} \tag{6.2}$$

where $E_k$ is the kinetic energy of the incoming particles. Combining Equations 6.1 and 6.2, one obtains:

$$E_{rate} = \frac{1}{2} E_{eff} V_{particle}^2 \tag{6.3}$$

Notice that the erosion efficiency is a function of the kinetic energy only, whereas the erosion rate also depends on the particle velocity. $E_{eff}$ was found to be a more useful parameter for comparison of powder blasting processes because the erosion caused by an individual particle impact directly depends on the kinetic energy of the particle. Slikkerveer et al.[32] in particular studied the erosion of brittle materials by hard angular particles impact and proposed a model that was derived from the indentation fracture theory. From their model, they concluded that the main material characteristics influencing the erosion efficiency are:

- Density (ρ) of the particles, which figures in the evaluation of the kinetic energy $E_k$
- Young's modulus of elasticity ($E$) of the substrate
- Fracture toughness ($K$) of the substrate, which is a measure of the ability of a material to resist the growth of an existing crack or flaw
- Indentation hardness ($H$) of the substrate, which characterizes a material's local resistance to a permanent deformation (e.g., scratching, abrasion, or cutting)

Powder blasting has been shown to be an appropriate micromachining technique for hard, brittle materials. It is a fast, inexpensive, and accurate directional etch technique for brittle materials such as glass, silicon, and ceramics. The time to etch

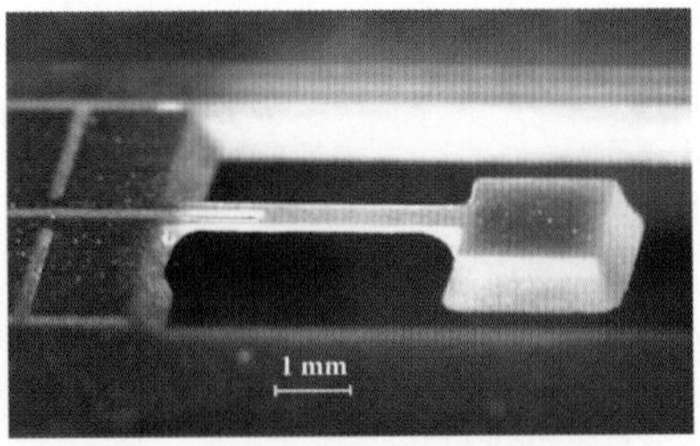


FIGURE 6.28 Powder-blasted accelerometer in glass. A first exposure of the metalized glass substrate to the eroding powder beam defines the freestanding structure and electrical contact geometry. A second exposure from the back is used to thin the cantilever beam. (Belloy, E., A. Sayah, and M. A. M. Gijs. 2000. *Sensors Actuators A* 86:231.[36] Courtesy of Dr. M. Gijs, EPFL.)

through a 500-mm-thick Pyrex wafer with one nozzle is approximately 20 minutes. The erosion process is caused by the impact of accelerated particles on brittle materials, a process that creates cracks. Powder blasting is mostly used with a powder beam at normal incidence angle to the substrate ($\theta = 90°$). Belloy et al.[33] described how, by exploiting oblique powder blasting ($\theta$ = 80°–40°) and underetching phenomena of the mask on a glass substrate, a new family of 3D structures can be realized. Using this technique, a glass cantilever accelerometer beam with inertial mass was structured (Figure 6.28). Using the same method, Solignac et al.[34] fabricated capillaries for electrokinetic experiments in soda lime microscope glass slides, and Bornand et al.[35] fabricated magnetic cores from ferrite wafers for millimeter-size transformers for low power applications.

## Surface Finish

The desired work piece surface determines the type of finishing operation. To rate the final surface finish of a work piece, one uses one of two criteria: the $R_a$ or the $R_q$ (formerly rms) value. Both are generally given in micrometers or microinches. $R_a$ is known as the arithmetic mean value: with reference to Figure 6.29,

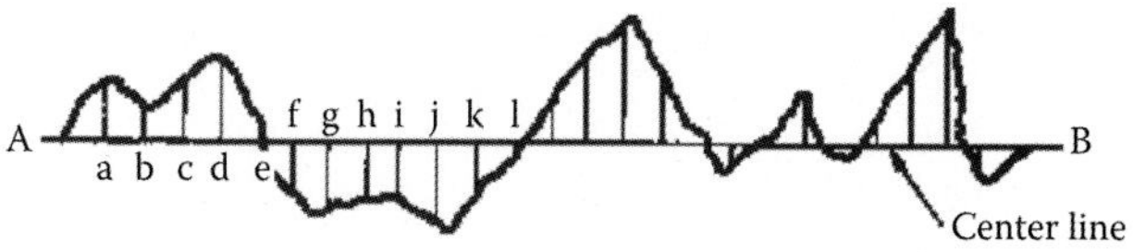


**FIGURE 6.29** Surface roughness and center line.

it takes the average of the absolute values of the deviations for the centerline of the surface:

$$R_a = \left( \frac{|a| + |b| + |c| + \cdots}{n} \right) \tag{6.4}$$

The $R_a$ in an older convention is also called the arithmetic average. The $R_q$ value, which was formerly called the rms, calculates the root mean square of the values of the deviations from the centerline:

$$R_q = \sqrt{\frac{a^2 + b^2 + c^2 + \cdots}{n}} \tag{6.5}$$

Other measurements of surface roughness exist, but these two are the most common. In particular, rms or $R_q$ is the most prevalent. In general, $R_q$ values are 1.1–1.4 times the $R_a$ values, but, of course, it depends on the exact surface topology.

In Figure 6.30 we represent a chart of the typical $R_a$ readings associated with different machining

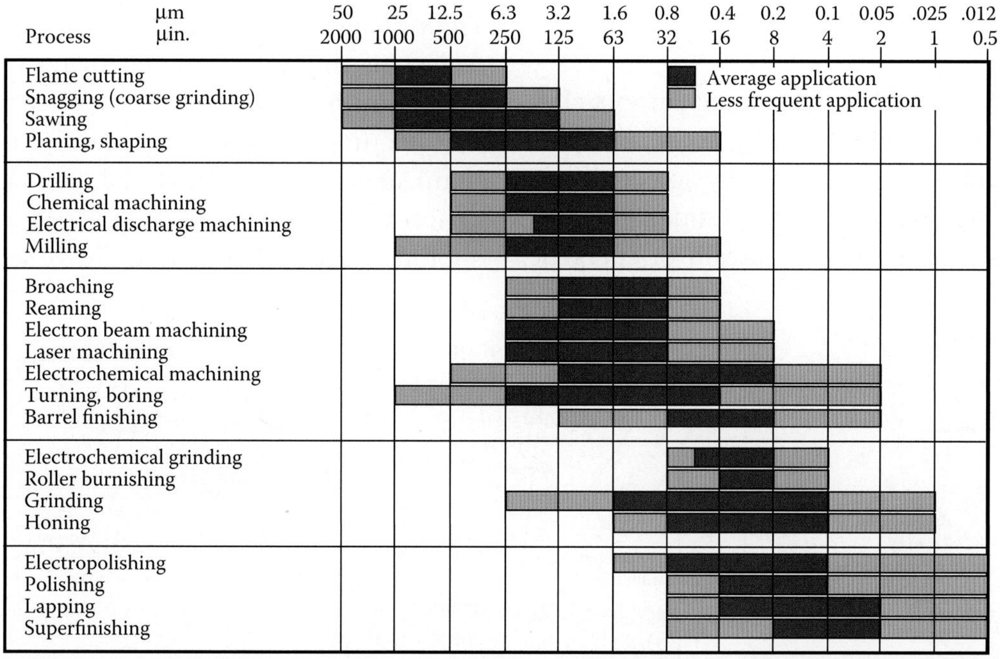


**FIGURE 6.30** Typical arithmetic average roughness, $R_a$, for various machining methods. (Kalpakjian, S. 1984. *Manufacturing processes for engineering materials.* Reading, MA: Addison-Wesley.[37])

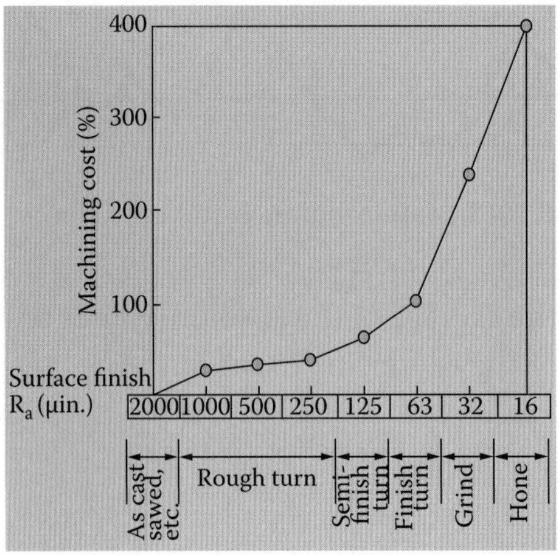


**FIGURE 6.31** Increase in the cost of machining and finishing a part as a function of the surface finish required.

processes. Notice the bottom few processes—lapping and superfinishing—that are typical abrasive machining processes. Electropolishing comes close in surface finish to these abrasive methods but is generally not as good a finish as the abrasives. Also notice the very small numbers with which we are dealing. Remember a human hair is about 80 μm. In some cases we are machining to the hundredth of a micrometer! The ELID process, surveyed above, is even better. In Figure 6.31 we illustrate the increase in the cost of a machining process as a function of the surface finish required.

### *Focused Ion Beam Machining and Fast Atom Beam Machining*

*Focused Ion Beam Machining* Some authors classify focused ion beam (FIB) milling as a thermomechanical technique. It is more widely accepted now as a pure mechanical machining technique in which the drill bit is replaced with a stream of energetic ions—call it ion blasting. A liquid metal ion source (e.g., gallium) is used rather than the wide argon ion beam in dry etching (see Chapter 3), and the ion beam is focused down to a submicrometer diameter. The liquid metal source uses a capillary feed (or needle) system and an acceleration voltage of 5–50 keV. Beam spots smaller than 50 nm are possible. FIB has been used for maskless implantation, metal patterning, and IC repair (see Figure 6.32).

More recently, it has been shown that FIB also constitutes a production technique of micrometer-sized objects in an arbitrary material without the requirements of anisotropy or lithography. Vasile et al., for example, described a 360° rotation mounting device used analogously to a lathe with a 20-keV Ga ion beam as the cutting tool for in-vacuum micromachining.[38,39] The ion beam they used is 0.3 μm in diameter, and they crafted tungsten needles, hooks, tuning forks, and specialized scanning probe tips. Some of these tungsten microstructures are shown in Figure 6.33. The first microstructure (Figure 6.33a and b) is a parasol-shaped scanning probe, which can be used to probe underneath resist lips and other concave features. The disc structure sits on a 1.5-μm square shank and is 4 μm in diameter and about 0.7 μm thick. The shank segment just before the disc is deliberately thinned to 0.4 μm at the thinnest point. The microshovel shown in Figure 6.33c is another illustration of the truly 3D aspect of this new manufacturing technique. The resolution limit to the fabrication process of these various tungsten structures mainly comes from the grain structure of the material. Start-to-finish fabrication time for the objects shown was ~2 hours. It should be noted that this expensive serial technique

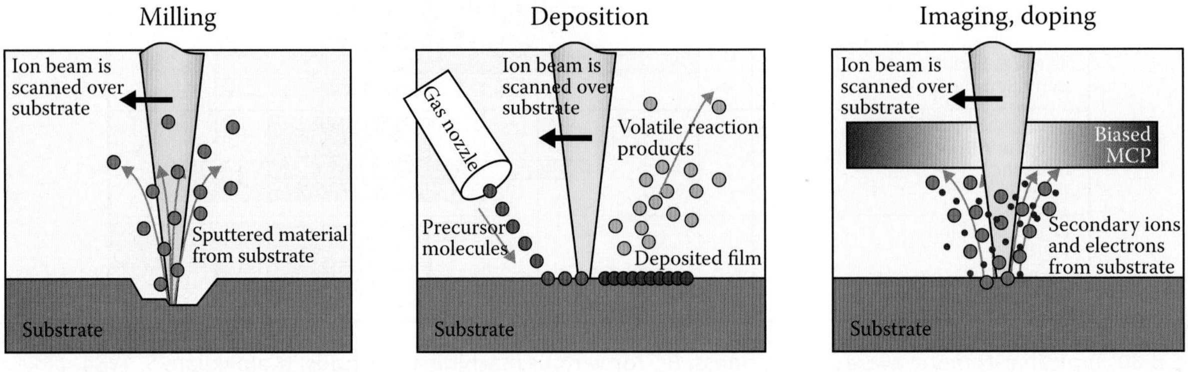


**FIGURE 6.32** Uses of focused ion beam (FIB).

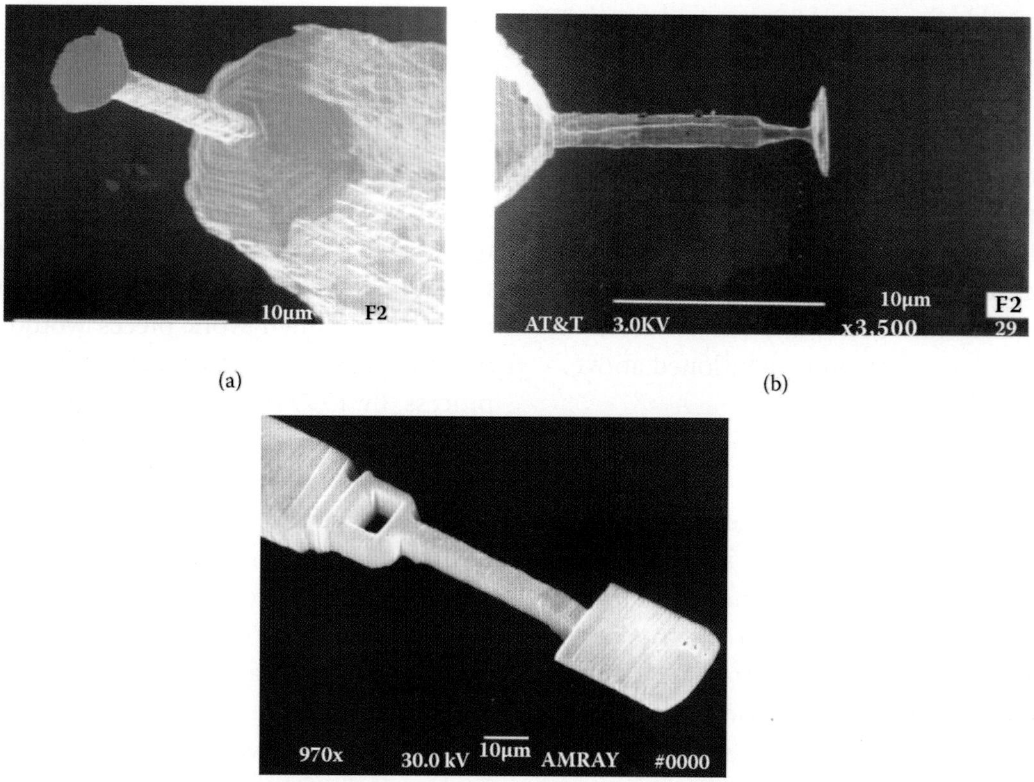


(a) (b) (c)

**FIGURE 6.33** FIB-manufactured tungsten structures. (a) Specialized scanning tip probe parasol. (b) Side view of a parasol structure. The shank is 1.5 $mm^2$ and thinned down locally to 0.4 mm. The disk is 4 mm in diameter. (c) Microshovel. (From Vasile, M. J., C. Biddick, and S. A. Schwalm. 1994. *J Vac Sci Technol* B12:2388. With permission.[39] Courtesy of Dr. M. J. Vasile, Louisiana Tech University.)

might make sense for the fabrication of an atomic force microscope tool set, but to make a microshovel as shown in Figure 6.33c is of course merely meant as a fancy demonstration of the technique.

In ion-beam sputtering, a broad surface area is bombarded with energetic ions from inert gases such as argon, and momentum transfer kicks off atoms from the target. The technique was discussed in detail in Chapter 3. In reactive ion-beam etching, also reviewed in Chapter 3, a flux of reactive particles, such as $CF_4^+$ and $F^+$, is directed at the work piece in a narrow beam. In the latter process, both momentum transfer and chemical reactions are involved in the etching process, and the process cannot be solely characterized as mechanical. In reactive ion etching, the etching is almost exclusively by chemical reaction with the reactive species formed in the plasma; the method must be regarded as a dry chemical etching technique rather than a mechanical technique. The latter techniques are almost exclusively used in semiconductor settings. It is possible with these ion-machining techniques, in combination with a mask, to etch almost any material in a very controlled fashion (dimensions <100 nm).

*Fast Atom Beam Machining* A variation on FIB machining is fast atom beam (FAB) machining. In FAB, ionized atoms, accelerated to about 100 km/s, are neutralized and used as the mechanical cutting tool. For example, the etching rate of GaAs with chlorine atoms at 60°C is 0.15 µm/min, and vertical walls are obtained. The work piece is mounted on a lathe as in FIB, and by rotating it with respect to a Ni mask, the different faces of the GaAs work piece are etched, and 3D structures are created. Hatamura et al.[11] produced a multifaced microstructure, a micro-gojyunoto (named after an old Japanese temple tower) of 280 × 100 × 100 µm this way. A hole etched in a polyimide sheet enabled the erection of the temple with the help of the rotational robot holding the temple. More recently the same group has extended the FAB capabilities into the nanodomain by producing ultrafine stairs, 30 nm wide and 30 nm high. Their results show that these FAB methods are effective in producing 3D microstructures and nanostructures. Combinations of these methods will make it possible to produce functional nanostructures on 3D microstructures. As we will detail in Volume III, Chapter 4 on packaging, assembly, and self-assembly, Hatamura et al. came up with several intriguing microassembly ideas, just as in the erection of the microtemple mentioned above.

## Desktop Factory

Despite all the progress in mechanical micromachining reported thus far, the fact that it often takes a two-ton machine tool to fabricate microparts where cutting forces are in the milli- to micro-Newton range is a clear indication that a complete machine tool redesign is required for the fabrication of micro- and nanomachines.[40] Even with the advancement in the precision and tolerances achieved in mechanical machining, ultraprecision machining is still mostly carried out on large equipment. This equipment consumes significant amounts of floor space and energy. Figure 6.34 illustrates the volumetric utilization [V = volume of machine tool (M)/volume of work piece (W)] of conventional and mesoscale machine tools.[41] The ideal equipment should be of the same scale as the final product to constitute an optimized manufacturing system (V = 1).

The size of the equipment producing precision parts has been reduced, but the optimal size has not yet been attained. Along this line, in Japan, the concept for desktop flexible manufacturing systems, or more simply "desktop factory" (DTF), for building micromachines was proposed in the early 1990s.[42] This effort was supported by Japan's MITI (Agency of Industrial Science and Technology). In Korea, the government started a new DTF national effort at the end of 2004, and similar projects are now underway in the EEC.[12] The proposed manufacturing units would be tabletop size and include universal chuck modules to which work pieces would be continuously clamped through most of the manufacturing process. By 1999 the world's first prototype of a DTF (50 cm × 70 cm) combining miniature machine tools and manipulators was developed at AIST in Japan as shown in Figure 6.4. The factory combined a lathe, milling machine, press machine, a transfer arm, and a two-fingered hand. A next portable factory made at AIST, requiring a single 100-V AC outlet only, was packaged in a suitcase (625 mm long, 490 mm wide, and 380 mm high with a weight of about 34 kg). DTFs like these have the potential of becoming the factory of the future: totally self-contained, robotic, desktop-size machine tools that only require materials, power, and water as outside inputs. A first commercial desktop manufacturing unit by Sankyo Seiki is shown as the inset in the header of this chapter. Sankyo Seiki's management believes that DTFs might help manufacturing survive in Japan.[12]

As an example of a thermal micromachine tool, consider the Panasonic hand-held μ-EDM discussed in Chapter 5 (Figure 5.3). Another DTF reviewed in Chapter 5, combining thermal and mechanical machining, is the high-precision four-axes CNC machine developed by Liao et al. at the National Taiwan University. Parts in the size range

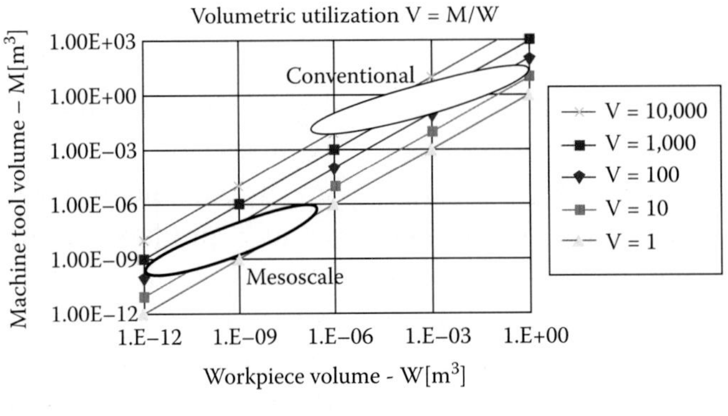


**FIGURE 6.34** Volumetric utilization (V = volume of machine tool/volume work piece) of conventional and mesoscale machines. (Vogler, M.P., et al. 2002. *Transactions of NAMRI/SME* 30:653.[41])

of 0.1–2 mm can be produced on this machine tool, and all manufacturing processes are completed on the same machine without moving the work piece. This multifunction high-precision machine (accuracy of 1 μm) is capable of performing four different functions: milling, μ-EDM, μ-EDM-WC, and on-line measuring of the flatness, circularity, concentricity, dimensions, and geometric tolerances of the work piece (Figure 5.8).

In one scenario, DTFs become very automated and require very few human operators, so that exporting these machines—and thus transferring manufacturing from rich to poor countries—does not make sense anymore because no labor costs are saved. In the capitalistic/nationalistic approach, these micromanufacturing stations are owned by big centralized national entities. The other scenario, promoted by, for example, MIT's Neil Gershenfeld, sees desktop manufacturing becoming as popular as desktop publishing and the manufacturing equipment would be owned by as many people as possible.[43] In the latter, more socially responsible approach, manufacturing becomes much less centralized, and many more people get involved. In either case, the impact on society would be dramatic. The dilemma is similar to that of solar energy versus oil; solar energy is widely spread out and difficult to own/control whereas the Bush family and cohorts try to own oil that, because it is more centralized, is easier to control.

Besides high-accuracy cutting, detailed above, there is also significant progress to report in die punching. The miniature die press from Aoki and Takahashi was one of the first examples of progress in the direction of microdie punching machine tools.[44] Aoki et al. pushed the art of die forming with a press for 3D, metallic medical microcomponents to new limits.[44] Medical forceps, for insertion into an endoscope used to remove diseased tissue, have been fabricated as an example of the new capabilities. The forceps have a diameter of 0.6–1 mm and a length of 10 mm, with assembly holes measuring 0.3 mm, and 0.2-mm long teeth in the cutting tool. The final manufacturing time, including assembly, is about 3 min/piece. Some of the other successes in making DTFs are detailed in Table 6.5.

## Precision Mechanical Machining versus IC-Based Micromachining

In comparing IC-based micro- and nanomachining with precision mechanically machined structures, Slocum makes some important observations.[4] He notes that, although IC-derived micro- and nanomachines are impressive for their small size, their relative accuracy is two orders of magnitude worse than is typically achieved in mechanical precision machining, which, moreover, makes for much more complex designs (see also Figure 6.3). For more than 100 years, precise mechanical machined devices have been designed and built with mechanical removal techniques achieving part-per-million (ppm) relative tolerances, and recently ultraprecision mechanical machining has entered the realm of 0.1 ppm relative tolerances. Typical IC-derived micro- and nanomachines, on the other hand, Slocum points out, are comparable with the macromachines of the early 1700s with respect to complexity and relative tolerance. Although the surface roughness of IC-derived micro- and nanomachines appears high compared with the specular finishes of precision-machined bearing surfaces, in reality, the absolute roughness is about the same. Another key issue Slocum brings up is that of metrology systems; verification of the fabricated geometry and tolerances is much more difficult for IC-derived micromachines because measurement of a displacement of 1 part in 10,000 will typically get one down to the nanometer level.

Higher-accuracy machining is needed to provide computer memory disks and optical mirrors and lenses with accuracies to a fraction of the wavelength of light. The micro- and nanofabrication techniques that are currently used in IC-derived micro- and nanomanufacturing have serious limitations in terms of the materials that can be used and the creation of complicated shapes with the required surface quality.

A better understanding of the limits of ultraprecision mechanical machining and nontraditional IC-based machining methods would help the mechanical engineer make the best choice in machining tools. At the same time, as IC-based micro- and nanomachine complexity increases, it seems appropriate

TABLE 6.5 Desktop Factories

| Developed by | Equipment Category | Dimension of the Equipment | Features of the Equipment |
|---|---|---|---|
| Kitahara et al.[45] | Microlathe | 32 mm length | *x*-*y* driving unit driven by laminated piezoactuators, main shaft device driven by a micromotor |
| Lu and Yoneyama[46] | Microlathe | 200 mm length | Cutting mechanism directly observed by optical microscope |
| Nakao and Hatamura[47] | Nanomanufacturing factory | 1.6 m × 0.9 m × 1.6 m | Fast atom beam (FAB) etching technology, concentrated motion manipulator, multidirectional SEM and an image-driven operating system |
| Vogler et al.[41] | Mesoscale machine tool (mMT) systems | 25 mm × 25 mm × 25 mm | Air turbine spindle, piezoelectric friction drive, voice coil drives |
| McKeown[48] | Three-axis ultraprecision aspheric generator | | High-stiffness servo drives and hydrostatic bearings, servo resolution of 1.25 nm in the *x*- and z-axes |
| Hatamura et al.[11] | Integrated manufacturing system for 3D microstructures | 1.6 m × 0.9 m × 1.6 m | Fast atom beam etching, 3D assembly using a concentric manipulator, multiview scanning electron microscope (SEM) |
| Ruiz-Huerta[49] | CNC micromachining tool | 130 mm × 160 mm × 85 mm | PC-based control |
| Mishima et al.[50] | Microfactory | Floor area: 50 × 70 cm | A microlathe, a micropress machine, a micromilling machine, a microtransfer arm, and a micro two-fingered hand |
| NIST (National Institute for Science and Technology) | Molecular measuring machine (M3) | | Images and measures to nanometer accuracy the positions of features located anywhere within the area of 50 mm × 50 mm and up to 100 μm high |
| Takeuchi et al.[51] | 5-axis control ultraprecision machining center | 20-mm stroke for vertical movement, 1-mm minimum feature size | Air-borne linear guide and air slide for the translational axes |

for the designers of these types of micro- and nanostructures, who are typically electrical engineers, to study ultraprecision machine design.

## Acknowledgments

Special thanks to Dr. Rodrigo Martinez and Dr. Chunlei Wang.

## Questions

*Questions by Dr. Madou assisted by Mr. Chuan Zhang, UC Irvine*

6.1: What are some commonly recognized categories of mechanical removal processes?

6.2: What is the meaning of relative and absolute tolerance? Is it sufficient to use only absolute tolerances in describing the precision of a micromachining process?

6.3: What are the merits of single crystal diamond cutting tools?

6.4: Illustrate the importance of material properties in determining machining accuracy.

6.5: List some important trends and their consequences in the development of automated mechanical machining equipment (consider only single-point mechanical removal) for the manufacturing of the future.

6.6: What are the requirements for a material to be used as an abrasive grit? List some conventional abrasives.

6.7: How does the grit size affect the quality of surface finish and the material removal rates? How should the grit size be chosen when dealing with different materials?

6.8: Why must the grinding wheel be dressed periodically during the grinding process? How does electrolytic in-process dressing (ELID) solve this problem?

6.9: Why are soft materials unsuitable to be machined with ultrasonic drilling? How do hardness and modulus of elasticity impact the removal rates?

6.10: What are the advantages and disadvantages of water jet machining? If we have a piece of copper plate 1 in. thick and the heat-affected zone (HAZ) is of great concern, should we choose laser machining, water jet machining, or abrasive water jet machining? Why?

6.11: How can a wider range of materials be machined by adding abrasives to a water jet? List some applications of abrasive water jet machining.

6.12: Show how the erosion efficiency depends on the kinetic energy in powder blasting. What other factors influence the erosion efficiency?

6.13: What are the major applications of focused ion beam (FIB)? How do we operate FIB in different modes for different applications?

6.14: What are the merits of ultrasonic machining? How does a rotary setup enhance its performance?

6.15: Explain the working principle of ultrasonic machining. Why does it produce stress-free shapes?

6.16: What are the differences and similarities between focused ion beam (FIB) machining and fast atom beam (FAB) machining?

6.17: Why are micromanufacturing researchers developing desk top factories (DTFs)? Why should the ideal equipment be of the same size scale as the final product?

6.18: What are the merits and drawbacks of precision mechanical machining compared with IC-based micromachining? List two examples that are better machined with precision mechanical machining and two with IC-based micromachining.

6.19: What are the advantages of electrodischarge wire cutting (EDWC) compared with sinker EDM?

6.20: Fill in the blanks:

Precision machining is defined at a relative tolerance of or less of a feature/part size and covers both and processes.

6.21: Give a history of mankind's manufacturing methods, including major events, rough dates, and examples.

6.22: What are the underlying reasons that both Taniguchi and Moore curves have started showing signs of a slowdown in progress of manufacturing accuracy and of transistor density over time, respectively?

## References


1. Hodges, H. 1970. *Technology in the ancient world*. New York: Barnes & Noble.
2. Burke, J. 1978, *Connections*. Boston: Back Bay Books.
3. Zuurveen, F. 1994. *Precisie-Technologie-Jaarboek 1994*. Eindhoven, the Netherlands: NVFT.
4. Slocum, A. H. 1992. *Proceedings: IEEE micro electro mechanical systems (MEMS '92)*. Travemunde, Germany.
5. Slocum, A. H. 1992. *Precision machine design*. Englewood Cliffs, NJ: Prentice Hall.
6. Palmer, C. 2000. *Diffraction grating handbook*, 4th ed. Rochester, NY: Richardson Grating Laboratory.
7. *Compton's interactive encyclopedia, 1994*. Elmhurst, IL: Interactive Multimedia.
8. Taniguchi, N. 1983. Current status in, and future trends of, ultraprecision machining and ultrafine materials processing. *Annals CIRP* 32:S573–82.
9. Gibbs, W. W. 1997. Gordon E. Moore interview. *Scientific American* September 22.
10. Szepesi, D. 1993. *Sensoren en Actuatoren in de Werktuigbouw/Machinebouw, Centrum voor Micro-Electronica*. The Hague, the Netherlands.
11. Hatamura, Y., M. Nakao, T. Sato, K. Koyano, K. Ichiki, H. Sangu, M. Hatakeyama, T. Kobata, and K. Nagai. 1994. Construction of 3D micro structure by multi-face FAB, co-focus rotational robot and various mechanical tools. *IEEE international workshop on micro electro mechanical systems (MEMS '94)*. Oiso, Japan.
12. Ehmann, K. F., D. Bourell, M. L. Culpepper, T. J. Hodgson, T. R. Kurfess, M. Madou, K. Rajurkar, and R. DeVor. 2007. *Micromanufacturing: international research and development*. New York: Springer.
13. Evans, C. 1989. *Precision engineering: an evolutionary view*. Bedford, UK: Cranfield Press.
14. Snoeys, R. 1986. *Advances in non-traditional machining*. Anaheim, CA.
15. McKeown, P. A. 1986. High precision manufacturing and the British economy. *Institute of Mechanical Engineers Proceedings* 200:76.
16. Bier, W., G. Linder, D. Seidel, and K. Schubert. 1991. Kernforschungszentrum Karlsruhe. *KfK Nachrichten* 23:2–3, 165–73.
17. Teshigahara, A., M. Watanabe, N. Kawahara, Y. Ohtsuka, and T. Hattori. 1995. Performance of a 7-mm microfabricated car. *J Microelectromech Syst* 4:76–80.

18. Boothroyd, G., and W. A. Knight. 1989. *Fundamentals of machining and machine tools*. New York: Marcel Dekker.
19. Ohmori, H., and T. Nakagawa. 1990. Mirror surface grinding of silicon wafer with electrolytic in-process dressing. *Annals CIRP* 39:329–32.
20. Bellows, G., and J. B. Kohls. 1982. Drilling without drills. *Am Mach Special Rep* 743:173–88.
21. Moreland, M. A. 1992. *Engineered materials handbook*. Ed. S. J. Schneider. Metals Park, OH: ASM International.
22. Humphrey, E. F., and D. H. Tarumoto, eds. *Fluidics*. Boston: Fluidic Amplifier Associates.
23. Sun, X.-Q., T. Masuzawa, and M. Fujino. *Ninth annual international workshop on micro electro mechanical systems (MEMS '96)*. San Diego.
24. Grass, P., W. Koenig, F. U. Meis, C. Wulf, and H. Willerscheid. 1984. Konturbearbeitung faserverstaerkter Kunststoffe. Wasserstrahlschneiden, laserstrahlschneiden und umrissfraesen im wettbewerb. *VDI-Zeitschrift*. 126(21): S.785ff.
25. Benedict, G. 1987. *Nontraditional manufacturing processes*. New York: Marcel Dekker.
26. Momber, A. W., and Kovacevic, R. 1998. *Principles of abrasive water jet machining*. Berlin: Springer.
27. Dombrowski, T. J. 1983. The how and why of abrasive jet machining. *Modern Machine Shop* 76.
28. Ligthart, H. J., P. J. Slikkerveer, F. H. in't Veld, P. H. W. Swinkels, and M. H. Zonneveld. 1996. Glass and glass machining in ZEUS panels. *Philips J Res* 50:475–99.
29. Kruusing, A., S. Leppävuori, A. Uusimäki, and M. Uusimäki. 1999. Rapid prototyping of silicon structures by aid of laser and abrasive-jet machining. *SPIE Proceedings* 3680:870–78.
30. Yamahata, C. 2005. Magnetically actuated micropumps. Thesis. EPF de Lausanne, Lausanne, Switzerland.
31. Slikkerveer, P. J. 1999. Mechanical etching of glass by powder blasting. PhD diss., Technische Universiteit Eindhoven.
32. Slikkerveer, P. J., P. C. P. Bouten, F. H. in't Veld, and H. Scholten. 1998. Erosion and damage by sharp particles. *Wear* 217:237–50.
33. Belloy, E., P. Q. Pham, A. Sayah, and M. A. M. Gijs. 2000. *Eurosensors XIV*. Eds. R. de Reus and S. Bouwstra. Copenhagen: MIC-Mikroelektronik Centret.
34. Solignac, D., A. Sayah, S. Constantin, R. Freitag, and M. A. M. Gijs. 2000. Powder blasting as a novel technique for the realisation of capillary electrophoresis chips. *Eurosensors XIV*. Eds. R. de Reus and S. Bouwstra. Copenhagen: MIC-Mikroelektronik Centret.
35. Bornand, E., F. Amalou, and M. A. M. Gijs. 2000. *Eurosensors XIV*. Eds. R. de Reus and S. Bouwstra. Copenhagen: MIC-Mikroelektronik Centret.
36. Belloy, E., A. Sayah, and M. A. M. Gijs. 2000. Powder blasting for three dimensional microstructuring of glass. *Sensors Actuators A* 86:231–37.
37. Kalpakjian, S. 1984. *Manufacturing processes for engineering materials*. Reading, MA: Addison-Wesley.
38. Vasile, M. J., C. Biddick, and S. A. Schwalm. 1993. *Micromechanical systems*. Eds A. P. Pisano, J. Jara-Almonte, and W. Trimmer. New York: ASME Press, p. 81.
39. Vasile, M. J., C. Biddick, and S. A. Schwalm. 1994. Microfabrication by ion milling: the lathe technique. *J Vac Sci Technol* B12:2388.
40. Ehmann, K. E., D. Bourell, M. L. Culpepper, T. J. Hodgson, T. R. Kurfess, M. Madou, K. Rajurkar, and D. DeVor. 2007. *Micromanufacturing: international assessment of research and development*. Dordrecht, the Netherlands: Springer. www.wtec.org/micromfg/report/micro-report.pdf
41. Vogler, M. P., X. Liu, S. G. Kapoor, R. E. DeVor, and K. F. Ehmann. 2002. Development of meso-scale machine tool (mMT) systems. *Transactions of NAMRI/SME* 30:653–61.
42. Higuchi, T., and Y. Yamagata. 1993. *Proceedings: IEEE micro electro mechanical systems (MEMS '93)*. Fort Lauderdale, FL.
43. Gershenfeld, N. 2005. *FAB: the coming revolution on your desktop—from personal computers to personal fabrication*. New York: Basic Books.
44. Aoki, I., and T. Takahashi. 1996. *SPIE: micromachining and microfabrication process technology II*. Austin, TX.
45. Kitahara, T., Y. Ishikawa, K. Terada, N. Nakajima, and K. Furuta. 1996. Development of microlathe. *J Mechanical Engineering Lab* 50:5, 117–23.
46. Lu, Z., and T. Yoneyama. 1999. Micro cutting in the micro lathe turning system. *Int J Machine Tools Manufact* 39:1171–83.
47. Nakao, M., Y. Hatamura, and S. Tomomasa. 1996. *Microrobotics: components and applications*. Ed. A. Sulzmann. Boston: SPIE.
48. McKeown, P. 1996. From micro- to nano-machining–towards the nanometre era. *Sensor Review* 16:4–10.
49. Ruiz-Huerta, L., A. Caballero Ruiz, and E. Kussul. 2002. Guidelines for design low cost micromechanics. *Proceedings ASPE*. St. Louis, MO, October 20–25, 2002.
50. Mishima, N., K. Ashida, T. Tanikawa, H. Maekawa, K. Kaneko, and M. Tanaka. 2000. Micro-factory and a design evaluation method for miniature machine tools. *Proc. of Int. Workshop on Microfactories*, pp. 155–58.
51. Takeuchi, Y., H. Yonekura, and K. Sawada. 2003. Creation of 3D tiny statue by 5-axis control ultraprecision machining. *Computer Aided Design* 35:403–09.

# Part III

# Pattern Transfer with Additive Techniques

Building structures with additive technologies. *The Tower of Babel, c.1563 by Pieter Bruegel the Elder.*

As an adolescent I aspired to lasting fame, I craved factual certainty, and I thirsted for a meaningful vision of human life—so I became a scientist. This is like becoming an archbishop so you can meet girls.

**Matt Cartmill**

If any student comes to me and says he wants to be useful to mankind and go into research to alleviate human suffering, I advise him to go into charity instead. Research wants real egotists who seek their own pleasure and satisfaction, but find it in solving the puzzles of nature.

**Albert Szent-Györgi**

## Introduction to Part III



## Introduction to Part III

In Part III we cover additive (forming) processes, where materials are added, usually in a selective manner, to a work piece or a device under construction. In manufacturing "forming" creates an original shape from a molten mass, a solution, the gaseous state, plasma, or solid particles, and during such processes cohesion is created among individual particles. Forming techniques are often accompanied or followed by thermal processing to obtain desired material properties and substrate adhesion. Subtractive manufacturing techniques, covered in Part II, are categorized according to the energy used at the work piece as *mechanical*, *chemical*, or *thermal*. This classification—straightforward in the case of subtractive processes—breaks down quickly when applied to additive processes. Take physical vapor deposition (PVD), covered in Chapter 7, a method classified as a mechanical additive technique because material is deposited onto the work piece by direct line-of-sight impingement deposition. However, in additive techniques the depositing material itself might be generated through the application of thermal energy (e.g., thermal evaporation) or mechanical energy (e.g., sputter deposition) to a source material. In such case, the rule is to categorize the technique with the name of the process that brings the depositing material to the substrate, not for the process that generates the depositing material. For example, laser ablation deposition, also called *laser sputter deposition* (Chapter 7), uses thermal energy to ablate the material to be deposited but must be categorized as a mechanical method because the depositing material reaches the surface in a line-of-sight manner (mechanical). Most complicated to categorize are forming methods where chemical and thermal effects are intertwined in the deposition process on the substrate itself. Along this line, laser-assisted chemical vapor deposition (LCVD) (Chapter 9) is called a *thermal technique* because the heat of the laser beam induces the local material deposition, but because it involves a chemical reaction at the surface it could also have been termed a *chemical additive process*. (Remember that subtractive laser machining is a purely thermal technique; see Chapter 5.) In such cases one could refer to the technique as *thermochemical*, as in the case of the thermal oxidation of Si in a stream of humidified oxygen at 900°C, but in practice one classifies such methods as either *chemical* or *thermal*, depending on the primacy of the chemical reaction or the heat application, with the lower-temperature processes typically classified as *chemical* (e.g., sol-gel, screen printing of organics, and ink-jetting) and the higher-temperature processes as *thermal* (e.g., LCVD, screen printing of inorganics, and spray pyrolysis).

Although deposition methods in MEMS and NEMS, especially in the thin-film arena, are generally the same as in integrated circuits (ICs), additive processes in miniaturization science span a much wider and diverse range from inorganic to organic materials. Besides the typical microelectronic elements (Si, Al, Au, Ti, W, Cu, Cr, O, N, and Ni-Fe alloys), miniaturization science involves deposition of several nontypical elements such as Zr, Ta, Ir, C, Pt, Pd, Ag, Zn, In, Nb, and Sn. Moreover, a plethora of exotic compounds ranging from enzymes to shape memory alloys and from hydrogels to piezoelectrics is used in miniaturization science. In comparison, the number of materials and compounds involved in IC fabrication is very limited. In miniaturization science, particularly in chemical sensors and biomedical devices, modular, thick-film deposition technologies are more prevalent than in the IC industry.

In Table III.1 we list IC and MEMS/NEMS materials, typical deposition techniques, and an example

TABLE III.1 IC and MEMS/NEMS Materials, Deposition Method, and Typical Application

| Material | Deposition Technique | Typical Application |
|---|---|---|
| **Organic thin films** | | |
| Hydrogel | Silk screen | Internal electrolyte in chem. sensors |
| Photoresist (0.1–2 μm) | Spin-on | Masking, planarization |
| Thick photoresist (10–1000 μm) | Casting | Mold for metal plating |
| DNA | Ink-jetting | DNA array for molecular diagnostics |
| Polyimide | Spin-on | Electrical isolation, planarization, microstructures |
| **Metal oxides** | | |
| Aluminum oxide | CVD, sputtering, anodization | Electrical isolation |
| Indium oxide | Sputtering | Semiconductor |
| Tantalum oxide | CVD, sputtering, anodization | Electrical isolation |
| Tin oxide ($SnO_2$) | Sputtering | Semiconductor in gas sensors |
| Zinc oxide | Sputtering | Electrical isolation, piezoelectric |
| $ZrO_2$ | Thermal spray deposition from plasmas or flames | Coatings for aircraft engine parts and $ZrO_2$ sensors |
| **Compound semiconductors** | | |
| Cadmium sulphide (II–VI compound) | Spray pyrolysis (CVD) | Solar cell |
| Gallium arsenide (III–V compound) | Metalorganic CVD (MOCVD) | Optoelectronic devices |
| Gallium-aluminum arsenide | Molecular beam epitaxy (MBE) | A semiconductor alloy used as the light confinement layer in both single- and double-heterostructure diode lasers |
| **Noncrystalline silicon compounds** | | |
| α-Si-H | CVD, sputtering, plasma CVD | Semiconductors |
| Polysilicon | CVD, sputtering, plasma CVD | Conductor, microstructures |
| Silicides | CVD, sputtering, plasma CVD, alloying of metal and silicon | Conductors |
| **Metals (thin films) with ρ in μΩ cm in parentheses** | | |
| Silver (Ag) (1.58) | Evaporation, sputtering | Electrochemistry electrode |
| Aluminum (Al) (2.7) | Evaporation, sputtering, plasma CVD | Electrical interconnects, limited to operations <300°C |
| Chromium (Cr) (12.9) | Evaporation, sputtering, electroplating | Electrical conduction, adhesion layer (10–100 nm) |
| Gold (Au) (2.4) | Evaporation, sputtering, electroplating | Electrical interconnects for higher temperatures than Al, optical reflection in the IR |
| Iridium (Ir) (5.1) | Sputtering | Electrochemistry electrode, biopotential measurements |
| Molybdenum | Sputtering | Electrical conduction |
| Platinum (Pt) (10.6) | Sputtering | Electrochemistry electrode, biopotential measurements |
| Palladium (Pd) (10.8) | Sputtering | Electrical conduction, adhesion layer, electrochemistry electrode, solder wetting layer |
| Tungsten (W) (5.5) | Sputtering | Electrical interconnects at higher temperatures |
| Titanium (Ti) (42) | Sputtering | Adhesion layer |
| Copper (Cu) (1.7) | Sputtering | Low resistivity interconnects |
| **Alloys with ρ in μΩ cm in parentheses** | | |
| Al-Si-Cu | Evaporation, sputtering | Electrical conduction |
| Nichrome (NiCr) (200–500) | Evaporation, sputtering | Thin film laser-trimmed resistor |
| Permalloy™ ($Ni_xFe_y$) | Sputtering | Magnetoresistor, thermistor |
| TiNi (80) | Sputtering | Shape memory alloy |

(*Continued*)

TABLE III.1 IC and MEMS/NEMS Materials, Deposition Method, and Typical Application (*Continued*)

| Material | Deposition Technique | Typical Application |
|---|---|---|
| **Chemically/physically modified silicon** | | |
| n/p-type silicon | Implantation, diffusion, incorporation in the melt | Conduction modulation, etch stop |
| Porous silicon | Anodization | Electrical isolation, light-emitting structures, porous junctions |
| Silicon dioxide | Thermal oxidation, sputtering, anodization, implantation, CVD | Electrical and thermal isolation, masking, encapsulation |
| Silicon nitride | Plasma-enhanced CVD (PECVD) | Electrical and thermal isolation, masking, encapsulation |

application for the deposit. As in the case of subtractive processes (Part II), only the underlying physical and chemical principles of the various additive processes dealt with in Part III are explored.

Physical vapor deposition (PVD) processes and chemical vapor deposition (CVD) are topics of Chapter 7. In the same chapter we survey thin-film properties and detail surface micromachining, an important application of thin-film deposition techniques. In surface micromachining, features are built up, layer by layer, on the surface of a substrate (e.g., a single-crystal silicon wafer). Dry etching defines the surface features in the *x,y*-plane, and wet etching releases them from the plane by undercutting. In surface micromachining, shapes in the *x,y*-plane are unrestricted by the crystallography of the substrate.

In Chapter 8 we deal with additive chemical, photochemical, and electrochemical processes.* In chemical forming, chemical reactions create features by forming new compounds. The energy at work in these additive processes is chemical. Some of the best-known examples, with the chemical reaction dominant over the heat application, are sol-gel, reaction injection molding (RIM), colloidal processes, and self-assembled monolayers (SAMs). With the growth of the bio-MEMS field, techniques for depositing organic materials for chemical and biological sensors, often arranged in some type of an array configuration, are gaining importance. The chemical-forming methods covered in Chapter 8 are geared toward bio-MEMS applications. In photochemical forming, photoenergy solidifies a material into a 3D shape, as illustrated in rapid prototyping and microphotoforming (stereolithography). In the case of electrochemical forming, the energy at work is electrochemical. Electrochemical and electroless metal deposition are gaining renewed interest because of their emerging importance in making mold inserts for replication methodologies in both MEMS and NEMS (see also Chapter 10).

Chapter 9 covers thermal forming techniques where thermal energy provided by a heat source "transforms" a material as in crystallization, sintering, alloying, annealing, decomposition, and pyrolizing. Chemical and thermal forming techniques are often intertwined because reactions are induced by combining different chemicals and are aided by heat and catalysts. Along this line, screen printing is treated as a thermal process when dealing with inorganic material deposition (high temperature) and as a chemical process when depositing organics (low temperature). Similarly, the high-temperature ceramic application of tape casting (also called *doctor blading* or *knife coating* when dealing with paper, polymers, and paints) is treated as a thermal process where it is called *green tape technology*.

Replication methods based on plastic molding and LIGA—a sophisticated type of replication technique combining x-ray lithography with electroplating and plastic molding—are discussed in Chapter 10. Plastic molding techniques include RIM, thermoplastic injection molding (IM), and compression molding or hot embossing. Of the three, only RIM is a chemical-forming technique; the other two are thermal methods. Plastic

* Under subtractive chemical processes (Part II), we grouped purely chemical processes and electrochemical and photochemical machining together; in Part III we treat additive chemical, electrochemical, and photochemical as clearly distinct processes.

molding technology often is the least expensive manufacturing technique available. One only has to look around the home or office to recognize the pervasive nature of highly intricate, very low-cost plastic parts surrounding us. This method has the capability of making truly 3D parts with a wide range of feature sizes. Master molds can be made with both nonlithography- (e.g., CNC-machined master mold) and lithography-based (e.g., LIGA mold) techniques.

# 7

# Physical and Chemical Vapor Deposition—Thin Film Properties and Surface Micromachining

## Outline



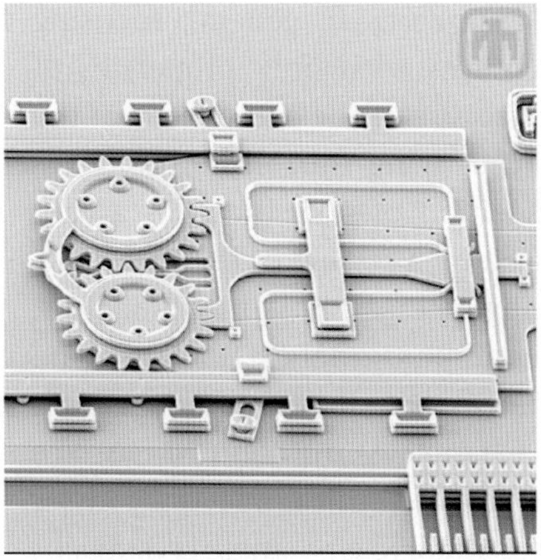

Surface micromachining. Clutch mechanism made in polysilicon. Gears are 50 μm across. Courtesy Sandia National Lab.

## Introduction

Chapter 7 introduces physical vapor deposition (PVD) and chemical vapor deposition (CVD). PVD constitutes a group of direct line-of-sight impingement deposition techniques, and CVD represents a class of diffusive-convective mass transfer deposition techniques. Evaporation, sputtering, laser ablation deposition, ion plating, cluster deposition, molecular beam epitaxy (MBE), and aerosol deposition (AD) represent the PVD techniques discussed. CVD techniques covered are plasma-enhanced CVD (PECVD), high-density plasma CVD (HDPCVD), atmospheric pressure CVD (APCVD), low-pressure CVD (LPCVD), very low-pressure CVD (VLPCVD), electron cyclotron resonance CVD (ECRCVD), metalorganic CVD (MOCVD), and atomic layer deposition (ALD).

Epitaxial techniques arrange atoms in single-crystal fashion on a crystalline substrate that acts as a seed crystal so that the lattice of the newly grown film duplicates that of the substrate. If the film is of the same material as the substrate, the process is called *homoepitaxy, epitaxy,* or simply *epi.* Epi deposition represents one of the cornerstone techniques for building micromachines, and a separate section reviews progress in epitaxial methods. Epitaxial techniques are usually chemical (CVD) in nature but may also be physical (e.g., MBE is a PVD technique).

The mechanical properties of thin CVD and PVD films are discussed next, and their applications in surface micromachining are detailed. We review thin film material properties in general, focusing on significant differences with bulk properties of the same material. Because of the complexity of the many parameters influencing thin film properties, we present the deposition of polysilicon as a case study. This is followed by a review of the main surface micromachining processes. Bulk micromachining means that three-dimensional (3D) features are etched into the bulk of crystalline and noncrystalline materials (see Chapter 4). In contrast, surface micromachined features are built up, layer by layer, on the surface of a substrate (e.g., a single crystal silicon wafer). Dry etching defines the surface features in the *x,y*-plane, and wet etching releases them from the plane by undercutting. In surface micromachining, shapes in the *x,y*-plane are unrestricted by the crystallography of the substrate. For illustration, in Figure 7.1 we compare typical bulk micromachining etching (both dry and wet) with polysilicon surface micromachining. Not reflected in this figure is the fact that the surface micromachined devices typically end up a bit smaller than their bulk micromachined counterparts.

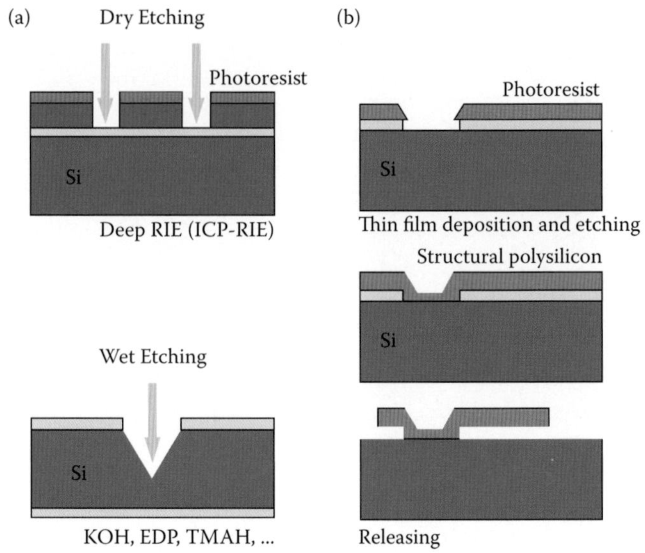


**FIGURE 7.1** Typical bulk micromachining (dry and wet) (a) versus polysilicon surface micromachining process steps (b). (Figure by Dr. Han Xu.) Acronyms such as RIE, ICP-RIE, EDP and TMAH are explained in this chapter.

## Physical Vapor Deposition

### Introduction

Many different kinds of thin films in integrated circuits (ICs) and micro- and nanomachines are deposited by evaporation and sputtering, both of which are examples of physical vapor deposition (PVD). The technique represents a very versatile coating technology applicable to an almost unlimited combination of coating substances and substrate materials. PVD reactors may use solid, liquid, or vapor as raw source material in a variety of configurations and are used to apply a wide variety of metals, alloys, ceramics, and other inorganic compounds, and even certain polymers. However, in practice, the technique is mostly used to deposit metals. Possible substrates include semiconductors, metals, glass, and plastics. At the low pressures used in a PVD reactor, the vaporized material encounters few intermolecular collisions while traveling to the substrate. Consequently, modeling of deposition rates is a relatively straightforward exercise in geometry.

Most PVD processes consist of the following elementary steps: 1) the synthesis of the coating vapor from the source material, 2) transport of the vapor to the substrate, and 3) condensation of vapors onto the surface of the substrate. These steps are generally

TABLE 7.1 Principal Types of PVD Techniques

| PVD Process | Features and Coating Materials |
|---|---|
| Thermal evaporation | Features: Equipment is relatively low cost and simple; deposition of compounds is difficult; coating adhesion not as good as for other PVD processes<br>Typical coating materials: Ag, Al, Au, Cr, Cu, Mo, W |
| Sputtering | Features: Better throwing power and coating adhesion than vacuum evaporation; can coat compounds; slower deposition rates and more difficult process control than vacuum evaporation<br>Typical coating materials: $Al_2O_2$, Au, Cr, Mo, $SiO_2$, $Si_3N_4$, TiC, TiN |
| Ion plating and cluster deposition | Features: Best coverage and coating adhesion of PVD processes; most complex process control; higher deposition rates than sputtering<br>Typical coating materials: Ag, Au, Cr, Mo, $Si_3N_4$, TiC, TiN |
| Laser sputter deposition or laser ablation deposition | Features: Best for complex compound deposition<br>Typical coating materials: high-temperature superconductor $YBa_2Cu_3O_{7-x}$ and biocompatible calcium hydroxylapatite, or $Ca_{10}(PO_4)_6(OH)_2$ |
| Aerosol deposition | Features: Lowest temperature deposition<br>Typical coating materials: ceramic coatings on all types of substrates |

carried out in a vacuum chamber; therefore, evacuation of the chamber precedes the PVD processes. The substrates are held upside down to minimize particulate contamination. The PVD techniques may be grouped into five principal types: 1) thermal evaporation, 2) sputtering, 3) ion plating and cluster deposition, 4) laser sputter deposition or laser ablation deposition, and 5) aerosol deposition as listed in Table 7.1. Ion plating and cluster deposition are based on a combination of evaporation and plasma ionization and offer some of the advantages inherent to both techniques. In aerosol deposition, solid particles are impacted at high speed on a substrate. This promising new process, we will learn, deviates significantly from all other PVD processes.

PVD is applied in the fabrication of electronic devices, microelectromechanical systems (MEMS), and nanoelectromechanical systems (NEMS), principally for depositing metal to form electrical connections. However, applications also include thin decorative coatings on plastic and metal parts such as trophies, toys, pens and pencils, watchcases, and interior trims in automobiles. The coatings in this case are thin films of aluminum (around 150 nm) coated with clear lacquer to give a high-gloss silver or chrome appearance. Another use of PVD is to apply antireflection coatings of magnesium fluoride ($MgF_2$) onto optical lenses. Ion plating is widely used to coat titanium nitride (TiN) onto cutting tools and plastic injection molds for wear resistance, and pulsed laser deposition (PLD) is particularly useful when dealing with complex compounds, as in the case of the deposition of high-temperature superconductor films (HTSC), for example, $YBa_2Cu_3O_{7-x}$. Another example application of PLD is the deposition of the biocompatible calcium-phosphate-based ceramic, calcium hydroxylapatite, or $Ca_{10}(PO_4)_6(OH)_2$. Aerosol impact deposition excels at making ceramic films at low temperature on all types of substrates.

## Thermal Evaporation

Certain materials (mostly pure metals) can be deposited onto a substrate by transforming them first from the solid source to the vapor state in a vacuum and then letting them condense on the substrate surface. Thermal evaporation represents one of the oldest thin film deposition techniques. Evaporation is based on the boiling off (or sublimating) of a heated material onto a substrate in a vacuum. From thermodynamic considerations, the number of molecules leaving a unit area of evaporant per second, i.e., flux *F*, is given by:

$$F = N_0 e^{\frac{\Phi_e}{KT}} \tag{7.1}$$

where $N_0$ is a slowly varying function of temperature ($T$) and $\Phi_e$ is the activation energy (in eV) required to evaporate one molecule of the material. The activation energy for evaporation is related to the enthalpy of formation of the evaporant, *H*, as $H = \Phi_e \times e \times N$ (Avogadro's number) J/mol.

Table 7.2 suggests the need for a good vacuum during evaporation; even at a pressure of $10^{-5}$ Torr,

TABLE 7.2 Kinetic Data for Air as a Function of Pressure

| Pressure (Torr) | Mean Free Path (cm) | Number Impingement Rate ($s^{-1}$. $cm^{-2}$) | Monolayer Impingement Rate ($s^{-1}$) |
|---|---|---|---|
| $10^1$ | 0.5 | $3.8 \times 10^{18}$ | 4400 |
| $10^{-4}$ | 51 | $3.8 \times 10^{16}$ | 44 |
| $10^{-5}$ | 510 | $3.8 \times 10^{15}$ | 4.4 |
| $10^{-7}$ | $5.1 \times 10^4$ | $3.8 \times 10^{13}$ | $4.4 \times 10^{-2}$ |
| $10^{-9}$ | $5.1 \times 10^6$ | $3.8 \times 10^{11}$ | $4.4 \times 10^{-4}$ |

*Source:* Brodie, I., and J. J. Muray. 1982. *The physics of microfabrication.* New York: Plenum Press. With permission.[1]

4.4 contaminating monolayers per second will impinge on the substrate.[1] For reference purposes, the number of atoms per unit area corresponding to a monolayer for a metal is about $10^{15}$ atoms/$cm^2$. Moreover, to avoid reactions at the source (e.g., oxide impurities being formed), the oxygen partial pressure needs to be less than $10^{-8}$ Torr.

In a laboratory setting, metals are often evaporated by passing a high current through a highly refractory metal. A refractory metal (for example, W or Mo) is formed into a suitable container [e.g., a tungsten crucible (boat) or filament] to hold the source material. Current is applied to heat the container, which then heats the material in contact with it. One problem with this heating method is possible alloying between the holder and its contents so that the deposited film becomes contaminated with the metal of the container. This method is called *resistive heating* (Figure 7.2a). Resistive evaporation is simple but easily spreads contaminants present in the filament, and the small size of the filament limits the thickness of the deposited film. Because heating is accomplished in a vacuum, the temperature required for vaporization is significantly below the corresponding temperature required at atmospheric pressure. Also, the absence of air in the chamber prevents oxidation of the source material at the heating temperatures.

In industrial applications, resistive heating has been surpassed by electron beam (e-beam) and radiofrequency (RF) induction evaporation. In the e-beam mode of operation, a high-intensity e-beam gun (3–20 keV) is focused on the target material that is placed in a recess in a water-cooled copper hearth. As shown in Figure 7.2b, the e-beam is magnetically directed onto the evaporant, which melts locally. In this manner, the metal forms its own crucible, and the contact with the hearth is too cool for chemical reactions, resulting in fewer source-contamination problems than in the case of resistive heating. E-beam evaporation not only results in higher quality films, but it also provides a higher deposition rate

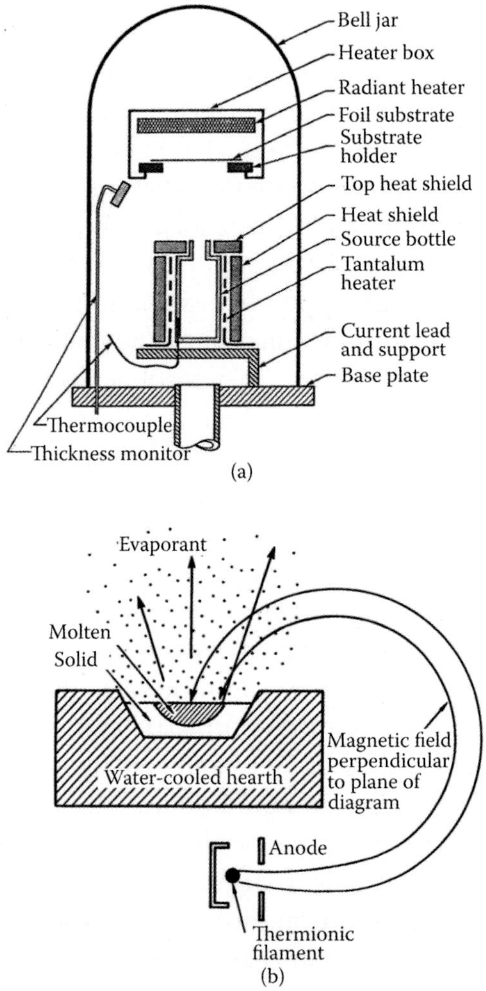


**FIGURE 7.2** Thin film deposition by evaporation. (a) Typical evaporation setup. (b) Diagram of a magnetized deflection electron-beam evaporation system. (From Brodie, I., and J. J. Muray. 1982. *The physics of microfabrication.* New York: Plenum Press. With permission.[1])

(50–500 nm/min). Two disadvantages of e-beam evaporation are that the process might induce x-ray damage and possibly even some ion damage to the substrate (at voltages >10 kV, the incident e-beam causes x-ray emission) and that the deposition equipment is more complex. X-ray damage may be avoided by using a focused, high-power laser beam instead of an e-beam. However, this technique is expensive and has not yet penetrated commercial applications.

In RF induction heating, a water-cooled RF coupling coil surrounds a crucible with the material to be evaporated. Because about two-thirds of the RF energy is absorbed within one skin depth of the surface, the frequency of the RF supply must decrease as the size of the evaporant charge increases. With a charge of a few grams of evaporant, frequencies of several hundred kilohertz are sufficient.

In Table 7.3, we compare the different heat sources available for thermal evaporation.

The substances used most frequently for thin-film formation by evaporation are elements or simple compounds whose vapor pressures range from 1–$10^{-2}$ Torr in the temperature interval from 600–1200°C. Refractory metals such as platinum, molybdenum, tantalum, and tungsten do not easily heat to the temperatures required to reach that vapor pressure range. To obtain a deposition rate high enough for practical applications, the vapor pressure at the source must be above the background pressure and reach at least $10^{-2}$ Torr. An evaporation source operating at a vapor pressure of $10^{-1}$ Torr, for example, can deliver rates of about 1000 atomic layers per second. The rate of atoms or molecules lost from the source as a result of evaporation, flux *F*, in molecules/unit area/unit time (Equation 7.1), can also be expressed in the following relationship derived from kinetic considerations on how the vapor pressure of the evaporant relates to the evaporation rate:

$$F = \frac{P_v(T)}{(2\pi MKT)^{\frac{1}{2}}} = 3.513\, 10^{22} \frac{P_v(T)}{(MT)^{\frac{1}{2}}} \tag{7.2}$$

where M = molecular weight of the evaporant
T = source temperature in Kelvins
$P_V(T)$ = vapor pressure of the evaporant in Torr

Metal deposition rates of 0.5 μm/min are typical when operating in the 10-mTorr range. Whatever the vaporization technique, evaporated atoms leave the source and follow straight-line paths until they collide with other gas molecules or strike the substrate to be coated. If a high vacuum is established, most atoms/molecules will deposit on the substrate without suffering intervening collisions with other gas molecules. The fraction of particles scattered by collisions ($N/N_0$) with atoms of residual gas is proportional to:

$$\frac{N}{N_0} \propto 1 - e^{\frac{d}{\lambda}} \tag{7.3}$$

where d = distance between source and substrate (see Figure 7.3a)
λ = mean free path of the particles

At $10^{-5}$ Torr, the mean free path in air hovers about 5 m, and at about 0.5 m at $10^{-4}$ Torr (see Table 7.2). The source-to-wafer distance must be smaller than the mean free path of the residual gas. Typically, this distance is 25–70 cm. With $\lambda > d$, Equation 7.3 approaches 1, i.e., no particles are scattered ($N = N_0$).

On contact with the relatively cool substrate surface, the energy of the impinging atoms is suddenly reduced to the point where they cannot remain in the vapor state; they condense and become attached to the solid surface, forming a deposited thin film. Reaching the substrate, atoms, depending on their energy, still may diffuse laterally over the surface according to:

$$D = D_0 e^{-\frac{E_D}{KT}} \tag{7.4}$$

**Table 7.3** Comparison of Heat Sources for Evaporation

| Heat Sources | Advantages | Disadvantages |
|---|---|---|
| Resistance | No radiation | Contamination |
| Electron beam | Low contamination | Radiation |
| RF | No radiation | Contamination |
| Laser | No radiation, low contamination | Expensive |

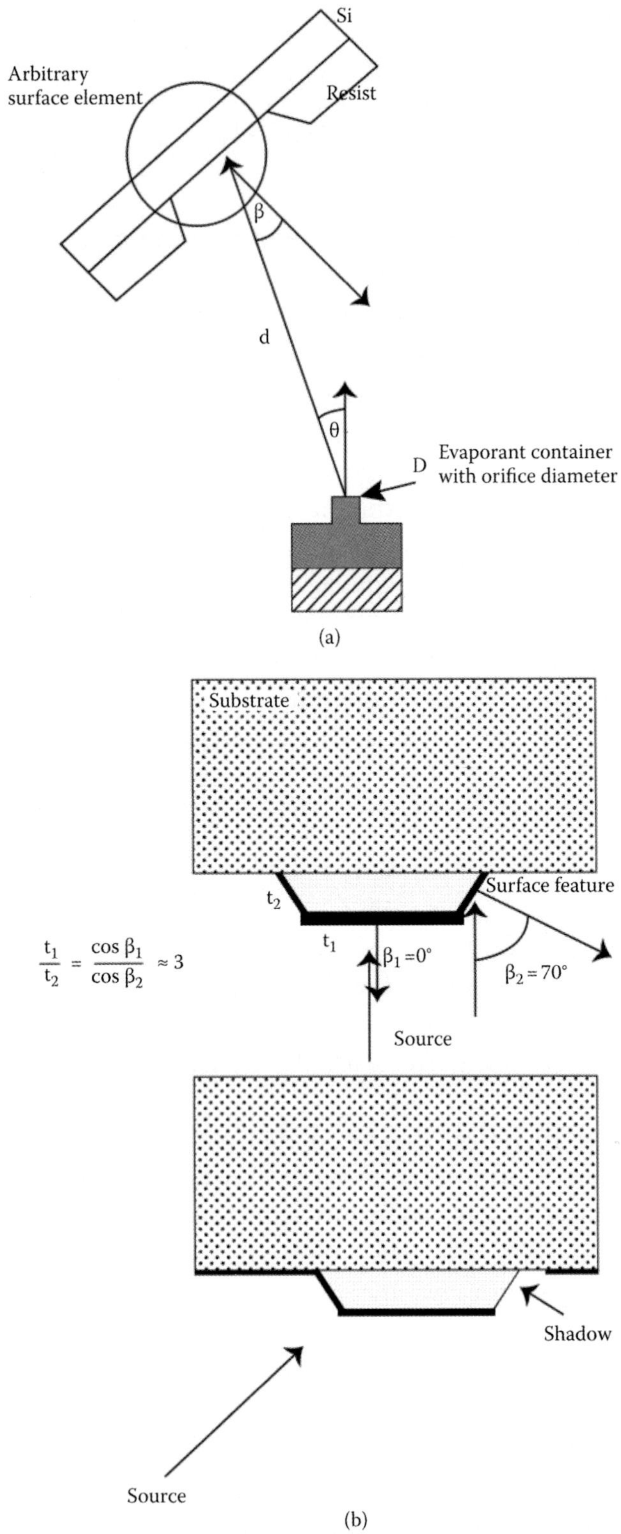


**FIGURE 7.3** Geometric considerations in evaporation. (a) Geometric considerations of arrival rate A at an arbitrary surface element in an evaporation experiment. (b) Nonuniform thickness of deposits over varying topography with θ = 0° (top) and θ ≠ 0°C (bottom).

where $E_D$ is the activation energy for diffusion (~2–3 eV) and $kT$ is the energy of the diffusing species. For atoms, after arrival on the surface, to find the best energetic site one must either use energetic atoms or heat the substrate. In a low energy deposition system, such as molecular beam epitaxy (MBE) (see Ion Cluster Beam Deposition—Nanoparticle Deposition section), arriving atoms (~0.1 eV) may form islands unless one picks the right substrate or the substrate is heated to high temperatures. On the other hand, in a high energy deposition system, such as in sputtering (~1 eV), one gets smoother films at lower substrate temperatures, but intermixing may occur.

In a configuration where the evaporant is held in a container with a small hole (i.e., in an *effusion* or *Knudsen cell* as shown in Figures 7.2a and 7.3a) rather than in an open containment structure, the flow through the exit orifice of the evaporant source can range from free molecular to viscous. If the Knudsen number (*Kn*, ratio of the mean free path, λ, to the orifice diameter, *D*) is greater than 1:

$$\mathrm{Kn} = \frac{\lambda}{\mathrm{D}} > 1 \tag{7.5}$$

an atom or molecule passes the orifice in a single, straight track and the flow is free molecular. If, on the other hand, the Knudsen number is <0.01, the flow is viscous, that is, the atom or molecule, before emerging from the orifice, bounces several times into the orifice's sidewalls. A transition region exists between the two.

By kinetic molecular theory the mean free path relates to the total pressure as:

$$\lambda = \left(\frac{\pi RT}{2M}\right)^{\frac{1}{2}} \frac{\eta}{P_T} \tag{7.6}$$

where $P_T$ = pressure at the point of interest
R = ideal gas constant
η = gas viscosity

Low pressure, low molecular weight, and high viscosity translate into a large mean free path. A high vacuum prevents oxidation of, e.g., aluminum and also ensures a purer material deposit with very little gas inclusions. The arrival rate, $A$, at distance $d$ from a small evaporation source follows the cosine law of deposition:

$$A \sim \frac{\cos\beta \cos\theta}{d^2} \tag{7.7}$$

where $\beta$ = angle between the normal to the substrate and the radial vector joining the source to the arrival point being considered

$\theta$ = angle between the normal to the evaporation source and the same radial vector (see Figure 7.3a)

The deposition, according to Equation 7.7, is not spherical; maximum deposition occurs in directions normal to the evaporation source, where $\cos\theta$ is maximum ($\theta = 0$). With the substrate positioned at a distance $d$ from the aperture and directly in line with the aperture, the arrival rate, $A$, or total number of molecules per second striking a unit area of substrate, is:

$$A = 3.513\ 10^{22} \frac{P_v(T)}{d^2(MT)^{\frac{1}{2}}} \tag{7.8}$$

(See also Equation 7.2 for the evaporation rate $F$.) The straight line or directional deposition of evaporant leads to difficulties in obtaining a continuous coating over topographical steps on a wafer, a problem known as *shadowing*. This is of particular concern in micromachining, where high-aspect-ratio features are common. Figure 7.3b illustrates a typical case. Because the thickness of the deposited film, $t$, is proportional to the $\cos\beta$, the ratio of the film thickness shown in Figure 7.3b, at $\theta = 0°$, is given as:

$$\frac{t_1}{t_2} = \frac{\cos\beta_1}{\cos\beta_2} \tag{7.9}$$

In case of a steep wall, $\beta_2 \sim 70°$ and $\beta_1 \sim 0°$, the ratio of $t_1/t_2$ is approximately 3. The thinner section, $t_2$, is susceptible to cracking at the extreme ends of the interval. Two methods can help overcome this problem. The first requires heating of the substrate during deposition to 300–400°C to increase the surface mobility of the metal atoms. The second relies on rotating the wafers in planetary wafer holders so that the angle $\beta$ varies during deposition (see Figure 7.4).[2] Shadowing prevents the formation of a conformal coat over a step. Mounting the substrates on rotating planetaries aids in making the coat more conformal. A conformal coat is uniform in thickness independent of the underlying topography (see Figure 7.5).

Evaporation is fast (e.g., 0.5 μm/min for Al) and comparatively simple; it registers a low-energy impact on the substrate (~0.1 eV); that is, no surface damage results except when using e-beam evaporation. Under proper experimental conditions, evaporation can provide films of extreme purity and known structure. In cases where the purity of the deposited film is of prime importance, evaporation is the preferred technique. For example, for electrochemical sensors, where the electrocatalytic activity of the top monolayers of the sensing electrode determines proper operation, material purity

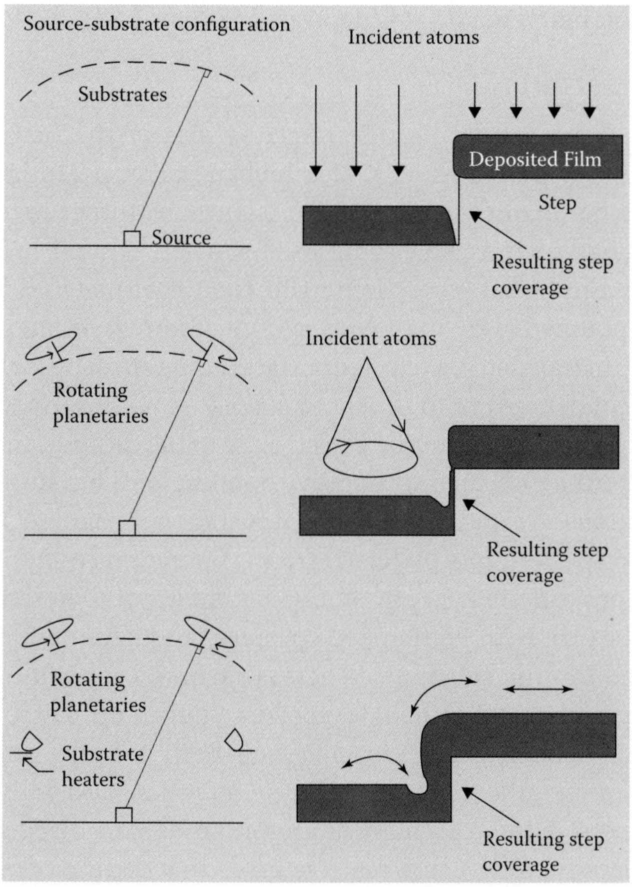


**FIGURE 7.4** Shadowing effect may be overcome by using rotating planetaries and heating of the substrate.

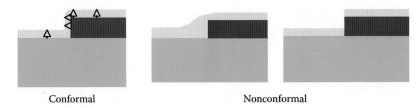


**FIGURE 7.5** Illustration of conformal and nonconformal step coverage.

is a priority consideration. In chemical sensors, the surface purity of films is more important than the bulk resistivity (a criterion often used to evaluate IC films). As we just learned, evaporators emit material from a point source (e.g., a small tungsten filament), resulting in shadowing and sometimes causing problems with deposition, especially on high-aspect structures. Difficulties also arise for large areas where highly homogeneous films are required, unless special setups are chosen. A different problem to overcome arises from the source materials decomposing at high evaporation temperatures. Although this risk does not exist when evaporating pure elements, it becomes a problem in the evaporation of compounds and substance mixtures. The e-beam heating systems have an advantage here because only a small part of the metal source evaporates. The initial evaporant stream is richer in the higher-vapor-pressure component; however, the melt depletes that constituent locally, and eventually an equilibrium rate is established. The best way to deposit complex metal alloys is then to e-beam evaporate the metallic elements from different sources because modern quartz-crystal deposition rate monitors enable excellent control in alloy composition. To form oxides of the deposited metals, evaporation is performed in a low-pressure oxygen atmosphere. This process is known as *reactive evaporation*. The oxygen supply comes from a jet directed at the substrate during deposition. To obtain the correct stoichiometry, the deposition needs to take place on a heated substrate.

Evaporated thin films are usually under tensile stress, and the higher the material's melting point, the higher the stress. Tungsten and nickel, for example, can have stress in excess of 500 MPa, which may lead to curling or peeling.[3] Increasing the temperature of the substrate tends to lower the internal stresses in the thin film.

Theoretical film thickness profiles for deposition into an arbitrary-shaped cavity or trench can be calculated using a computer program named SAMPLE, which gives the thin film profile for particular input parameters (http://www.eecs.berkeley.edu/~neureuth).[4,5] In Figure 7.6a we show a SAMPLE result for the line-of-sight deposition in the case of reentrant side-wall angles. The program correctly predicts a discontinuity in the thin film lines in Figure 7.6b. In Figure 7.6c it is shown how, for the case of wafers rotating on a planetary system, step coverage is significantly improved. However, the solution is not perfect because a keyhole effect might result if too much material is deposited (Figure 7.6d).

## Sputtering

During sputtering, the target (a disc of the material to be deposited), at a high negative potential, is bombarded with positive argon ions (other inert gases such as Xe can be used as well) created in a plasma (also glow discharge). The target material is sputtered away mainly as neutral atoms by momentum transfer, and ejected surface atoms are deposited (condensed) onto the substrate placed on the anode. The substrate must be placed close to the cathode and is usually heated to improve bonding of the coating atoms. A typical arrangement is shown in Figure 7.7.

Sputtering is preferred over evaporation in many applications because of a wider choice of materials to work with, better step coverage, and better adhesion to the substrate. Whereas vacuum evaporation is generally limited to metals, sputtering can be applied to nearly any material, both metallic and nonmetallic elements, alloys, ceramics, and polymers. Films of chemical compounds can also be deposited by using reactive gases that form oxides, carbides, or nitrides with the sputtered metal. Films of alloys and compounds can be sputtered without changing their chemical compositions. Sputtering

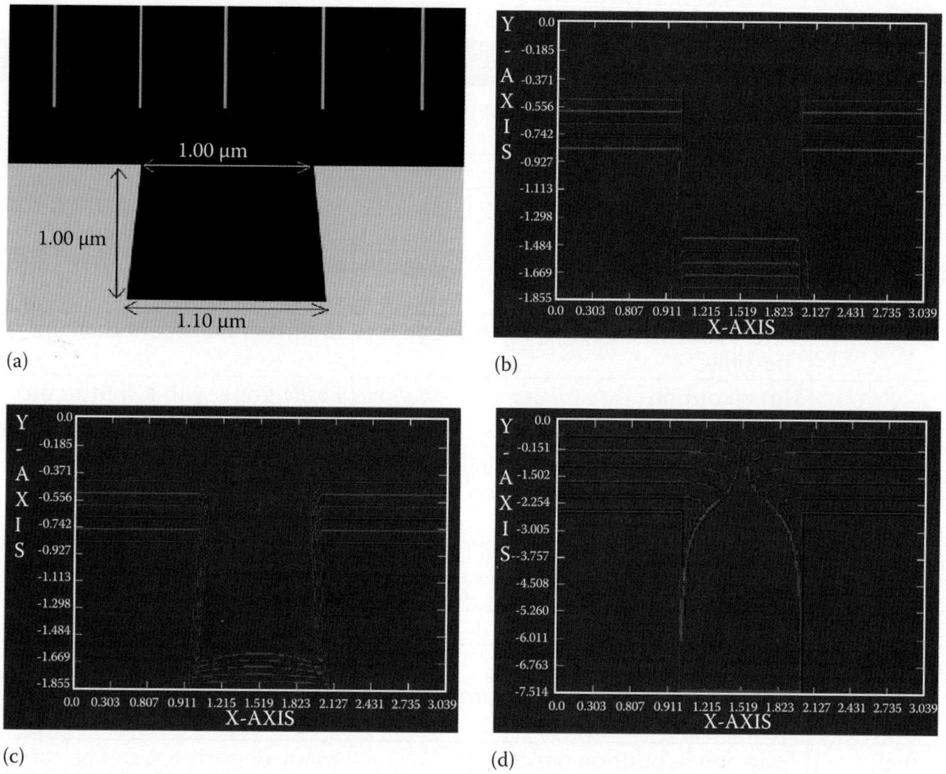


**FIGURE 7.6** The SAMPLE computer program gives the thin film profile for particular input parameters (http://www.eecs.berkeley.edu/~neureuth). (a) dimensions of feature to be coated. (b) predicted line-of-sight deposition profile in the case of re-entrant side-wall angles. Notice the discontinuity in the thin film lines. (c) for wafers rotating on a planetary system, step coverage is improved. (d) the solution is not perfect, however, because a keyhole effect might result.

is used in laboratories and production settings, whereas evaporation mainly remains a laboratory technique. Other reasons to choose sputtering over evaporation can be concluded from a comparison of the two techniques in Table 7.4.

Because most aspects pertaining to the physics and chemistries of DC and RF plasmas were discussed in Chapter 3, we will only amplify these topics here as they apply to material deposition. During ion bombardment, the source is not purposely

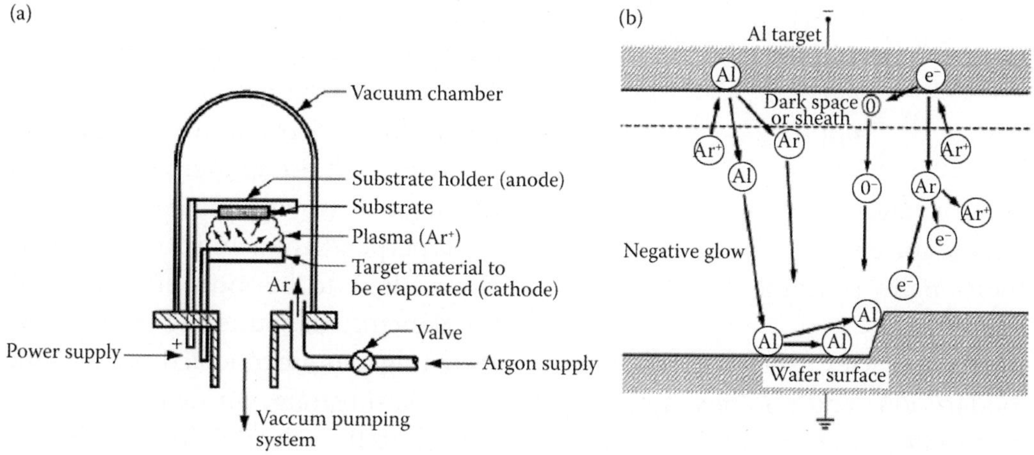


**FIGURE 7.7** (a) One possible setup for sputtering, a form of physical vapor deposition. (b) Gas ions are accelerated by a high voltage, producing a glow discharge or plasma. A source (the *cathode*, also called the *target*) is bombarded in high vacuum by gas ions. Atoms from the target (Al in this case) are ejected by momentum transfer and move across the vacuum chamber. Atoms are deposited on the substrate to be coated and form a thin film.

TABLE 7.4 Comparison of Evaporation and Sputtering Technology

| | Evaporation | Sputtering |
|---|---|---|
| Rate | Thousand atomic layers per second (e.g., 0.5 μm/min for Al) | One atomic layer per second |
| Choice of materials | Limited | Almost unlimited |
| Purity | Better (no gas inclusions, very high vacuum) | Possibility of incorporating impurities (low-medium vacuum range) |
| Substrate heating | Very low | Unless magnetron is used substrate heating can be substantial |
| Surface damage | Very low, with e-beam x-ray damage is possible | Ionic bombardment damage |
| In situ cleaning | Not an option | Easily done with a sputter etch |
| Alloy compositions, stoichiometry | Little or no control | Alloy composition can be tightly controlled |
| X-ray damage | Only with e-beam evaporation | Radiation and particle damage is possible |
| Changes in source material | Easy | Expensive |
| Decomposition of material | High | Low |
| Scaling up | Difficult | Good |
| Uniformity | Difficult | Easy over large areas |
| Capital equipment | Low cost | More expensive |
| Vacuum path | High: few collisions, line-of-sight deposition, little gas in film | Low: many collisions, less line-of-sight, gas inclusions |
| Number of depositions | Only one deposition per charge | Many depositions can be carried out per target |
| Thickness control | Not easy to control | Several controls possible |
| Adhesion | Often poor | Excellent |
| Shadowing effect | Large | Small |
| Film properties (e.g., grain size and step coverage) | Difficult to control, larger grain size, fewer grain orientations | Control by bias, pressure, substrate heat, smaller grain size, many grain orientations |

heated to high temperature, and the vapor pressure of the source is not a consideration as it is in vacuum-evaporation. The amount of material, *W*, sputtered from the cathode is inversely proportional to the total gas pressure, $P_T$, and the anode-cathode distance, *d*[1]:

$$W = \frac{kVi}{P_T d} \tag{7.10}$$

where V = working voltage
i = discharge current
k = proportionality constant

Other energetic particles, such as secondary electrons, secondary ions and photons, and x-rays are created at the target and can be incorporated in the growing film and/or influence its properties through heating, radiation, or chemical reactions. Deposition rate is roughly proportional to sputter yield, *S*, for a given plasma energy. The sputter yield *S*, we saw in Chapter 3, stands for the average number of atoms removed per incident ion (S = number ejected/number incident) and is a function of the bombarding species (for example, with a heavier sputtering gas, *S* increases), the ion energy of the bombarding species, the target material, the incident angle of the bombarding species, and its electronic charge. Target atoms come off with a nonuniform distribution as shown in Figure 3.16, where we plot the sputter yield *S* as a function of θ, the angle at which ions are directed at the surface (0° is perpendicular and 90° is glancing). Sputtering is seen here to be most efficient at about 20–30° off glancing. In sputtering at energies <50 keV, the incoming ions are modeled as hard spheres colliding with the substrate, with 95% of the incident energy going into the substrate (that is why in sputter deposition cooling of the target is required!) and only 5% of the incident energy carried off by substrate/target atoms. Some yield figures for commonly used metals are tabulated in Table 7.5.[6] Sputtering yield is

**TABLE 7.5** Sputter Yields of Several Commonly Used Metals with 500 eV of Argon

| Element | Symbol | Sputter Yield S |
|---|---|---|
| Aluminum | Al | 1.05 |
| Chrome | Cr | 1.18 |
| Gold | Au | 2.4 |
| Nickel | Ni | 1.33 |
| Platinum | Pt | 1.4 |
| Titanium | Ti | 0.51 |

*Source:* Vossen, J. L., and W. Kern, eds. 1978. *Thin film processes.* Orlando, FL: Academic Press. With permission.[6]

low, usually from less than 1% up to 4% depending on the materials and ion energy level. Because of the low sputtering yield, sputter deposition rates are low (slow process). Another drawback is that because the ions bombarding the surface are embedded in a gas, traces of that gas can usually be found in the coated films; the entrapped gases sometimes adversely affect mechanical properties.

A typical sputter yield as a function of ion energy was shown in Figure 3.15.[1] In the low-energy region, the yield increases rapidly from a threshold energy. The sputter threshold, usually in the range of 10–130 eV, is essentially independent of the bombarding ion species used and equals about four times the activation energy for evaporation. This threshold voltage decreases with increasing atomic number within each group of the periodic table. Above a few hundred volts, a changeover region takes hold. Its value depends on the ion species and the target used, but after its occurrence the sputter yield increases more slowly. Ion energies in the range of 0.5–3 kV are typically used for sputter deposition because nuclear collisions are predominant in this range. In this region, the sputter yields typically range from 0.1–20 atoms per ion, and the yield of most metals is about 1 (see Table 7.5). An ion can only penetrate up to a certain depth into the target and still effect a recoil collision that can reach the surface with sufficient energy to eject atoms (e.g., 1 nm/kV penetration for argon ions in copper). Consequently, at ion bombardment energies in the range of 10 keV to 1 MeV, the sputter yield reaches a maximum and then gradually decreases in value as a result of deep ion implantation. Average ejection energies of ions from the target range between 10 and 100 eV. At those energies, the incident ion can penetrate a substrate one to two atomic layers into the surface on which it lands. As a result, the adhesion of sputtered films is superior to films deposited by other methods.[7] Although sputtering is basically a low-temperature process, considerable amounts of energy are dissipated at the target surface. To liberate one atom requires 100–1000 times the activation energy needed for evaporation, limiting useful sputtering rates to about one atomic layer per second (1000 times less than evaporation from a source operating at $10^{-1}$ Torr). The target needs water cooling because most of the energy, estimated at about 95%, dissipates into it, and excessive heating might result. Only 1% of the remaining energy goes into sputtering, and the rest is dissipated by secondary electrons that bombard and heat the substrate.

Figure 7.7 exemplifies a typical sputtering setup. Sputtering systems often have a load-lock connected to the sputtering chamber used for loading and unloading the substrates. With the use of a load-lock system, the sputtering station can be kept under continuous high vacuum; the target and chamber can be kept free of contamination from gaseous atmospheric components. Instead of pumping the large sputtering chamber, only the small load-lock chamber is pumped, resulting in a significant shorter cycle time.

To prevent sputtering of the structural elements of the cathode assembly, a shield of metal at anode potential is placed around all the surfaces to be protected at a distance less than that of the dark space at the cathode in Figure 7.8.[7] No discharge will

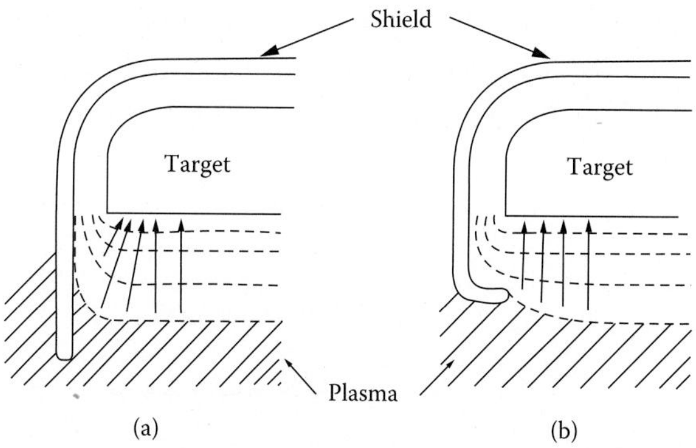


**FIGURE 7.8** Sputtering: use of cathode shield. (a) Reducing rim effect by extending cathode shield. (b) Reducing rim effect by wrapping shield around the cathode. (From Maissel, L. I., and R. Glang, eds. 1970. *Handbook of thin film technology.* New York: McGraw-Hill.[7])

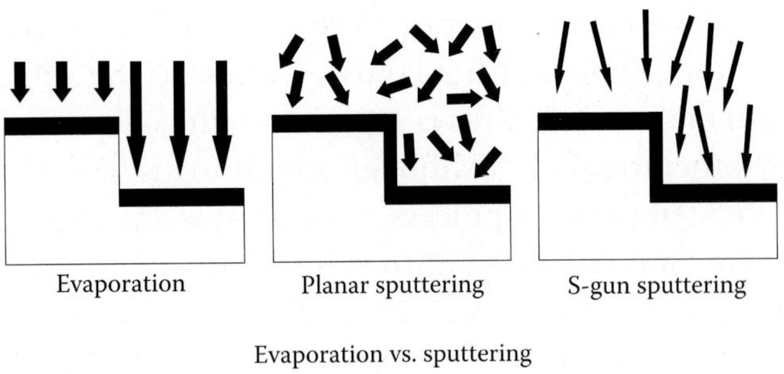


**FIGURE 7.9** Sputtering leads to better step coverage than evaporation. (By Dr. Han Xu.)

take place between two surfaces that have a separation less than the Crookes cathode dark space (see Chapter 3).

Crucial to the formation of sustained DC plasma, as discussed in Chapter 3, is the production of ionizing collisions between secondary electrons released from the cathode and the gas in the sputtering chamber. In effect, this requires a relatively high working pressure of greater than $1 \times 10^{-2}$ Torr. However, if the pressure is too high, significant numbers of the sputtered atoms cannot pass through the sputtering gas and are reflected back to the cathode by collisions. At $10^{-1}$ Torr, the mean free path of the sputtered metal atoms is about 1 mm, which is near the practical limit. Because of the multiple collisions of the metal atoms in the path between the cathode and anode, the metal atoms arrive at the anode at random incident angles. This leads to a better step coverage compared with evaporation (Figure 7.9). Furthermore, because the sputtering target is very broad compared with an evaporation point source, evidence of shadowing decreases drastically. Sputtering at low gas pressure leads to improved film adhesion because the sputtered atoms have a higher energy. Reduced pressure also reduces contamination of the film by trapped gas molecules, resulting in films of higher density and purity. The higher material density causes a slower chemical etch of the deposited film.

For conductors, a DC sputtering setup can be used. Insulating materials require an RF power supply. The two types of setups are schematically shown in Figure 7.10. Before sputter deposition, the substrate may be sputter etched by connecting it to the negative pole of the power supply. Sputter cleaning further promotes adhesion of subsequent metallizations. In situ cleaning can also be done efficiently with an ion gun.

As in dry etching, the plasma can be confined and densified by using electron cyclotron resonance (ECR) in a planar or cylindrical magnetron or an S-gun using a conical magnetron, which produces higher deposition rates than evaporation produced by electron beam (e-beam) or RF induction. Higher ion densities permit higher sputter rates at lower gas pressures (down to 0.5 mTorr). In a planar magnetron sputtering apparatus, the crossed electric and magnetic fields (magnets of 200 Gauss are put behind the target) contain the electrons and force them into long, helical paths, thus increasing the probability of an ionizing collision with an argon atom (see Figure 7.11). Furthermore, secondary electrons emitted by the cathode as a result of ion bombardment are bent by the crossed fields and are collected by ground shields. This eliminates the secondary electron bombardment of the substrates, which is one of the main sources of unplanned substrate heating. The deposition rate reaches hundreds of angstrom per minute. For some materials such as aluminum, the deposition rate can be as high as 1 μm/min. Magnetron sputter enhancement can be

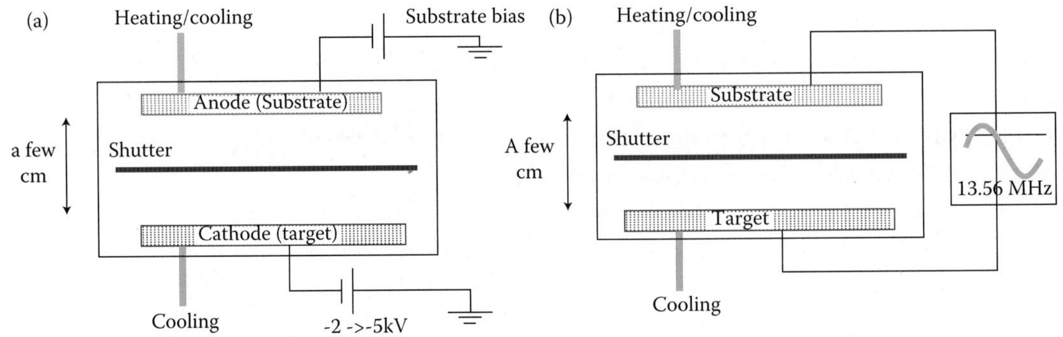


**FIGURE 7.10** The DC (a) and RF (b) sputter deposition setups.

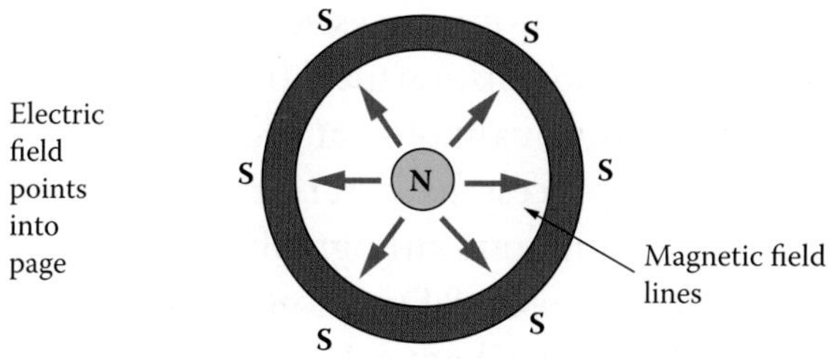


**FIGURE 7.11** Magnets behind the target in a magnetron trap electrons near the cathode, causing more ionization near the cathode (10×) with fewer electrons reaching the substrate (less heating).

used in high-frequency sputtering systems, as well as in DC sputtering systems.

The substrate is also often heated to promote film adhesion and reduce film stress. Stress levels vary from compressive at low pressures (0.1–1 Pa) to tensile at high pressures (1–10 Pa). The transition regime between compressive and tensile is often sharp (more than a few tenths of Pa), making the crossover, an ideal point for zero-stress deposition, difficult to control.[3] In general, it is preferable to have ions strike the substrate while growing a film. The plasma ions add energy to the film and keep the surface atoms mobile enough to fill virtually every void. The influence of plasma on a depositing film is considered in more detail in the section on plasma-enhanced chemical vapor deposition (PECVD) further below.

Sputtering often is carried out in the presence of a reactive gas, ensuring control or modification of the properties of the deposited film in so-called *reactive sputtering*. A sensor material produced this way is $IrO_X$, a low-impedance pH-sensing material arrived at by sputtering from a pure Ir target with argon ions in the presence of oxygen.[8] The exact mechanism for compound formation in reactive sputtering still eludes researchers. It is conjectured that at low pressures the reaction takes place at the substrate as the film is being deposited. At high pressure, the reaction is believed to occur at the cathode with the compound transported to the substrate.

Composite films can also be made by cosputtering or by sputtering from a single composite target. Because the various elements in a target have different sputtering rates and sticking probabilities, the composition of the target might differ somewhat from that on the substrate. However, the

**Table 7.6** Unique Characteristics of Sputter Deposition

| Characteristics |
|---|
| Uniform thickness over large area |
| Simple thickness control (because it is so slow!) |
| The alloy composition maintains stoichiometry with the original target composition |
| Deposition rates do not differ a great deal from one material to another |
| Sputtering cleaning before initiating film deposition is possible; the surface is not again exposed to ambient after such cleaning |
| The lifetime of a sputtering target may be as long as hundreds of runs and is seldom less than 20 |

decomposition is less significant than in evaporation because no thermal equilibrium is established at the target, and atoms are ejected by momentum transfer. High substrate temperatures are needed to obtain the right composition and to promote adhesion.

In Table 7.6 we list some of the salient characteristics of sputter deposition.

Nowadays, sputtering equipment applies films to compact discs, computer disks, large area active-matrix liquid crystal displays, and magneto-optic disks.[9] Bearing gears, saw blades, and so on also can be coated with a number of hard, wear-resistant coatings such as TiN, TiC, TiAlN, NbN, CrN, TiNbN, or CrAlN.[10] Commercial in-line sputtering equipment is available, enabling, in a series of evacuated chambers and load locks, depositions of insulators such as $SiO_2$, $Si_3N_4$, and $Ta_2O_5$, metals, and semiconductors such as indium tin oxide (ITO) on substrate with sizes of 830 × 1500 mm. Some example materials deposited by sputter deposition are listed in Table 7.7.

In one particular setup, one can perform continuous sputter deposition (Figure 7.12) on rolls

**Table 7.7** Example Materials Deposited by Sputtering

| Type of Material: Examples |
|---|
| Metals: Al, Cu, Zn, Au, Ni, Cr, W, Mo, Ti |
| Alloys: Ag-Cu, Pb-Sn, Al-Zn, Ni-Cr |
| Nonmetals: graphite, $MoS_2$, $WS_2$, PTFE |
| Refractory oxides: $Al_2O_3$, $Cr_2O_3$, $Al_2O_3$-$Cr_2O_3$, $SiO_2$, $ZrO_2$-$Y_2O_3$ |
| Refractory carbides: TiC, ZrC, HfC, NbC |
| Refractory nitrides: TiN, $Ti_2N$, ZrN, HfN, TiN-ZrN, TiN-AlN-ZrN |
| Refractory borides: $TiB_2$, $ZrB_2$, $HfB_2$, $CrB_2$, $MoB_2$ |
| Refractory silicides: $MoSi_2$, $WSi_2$, $Cr_3Si_2$ |

(a)

(b)

**FIGURE 7.12** Continuous deposition. (a) Mill Lane offers systems for continuous vacuum web coating (http://www.mill-lane.com). (b) Innovative Sputtering Technology (I.S.T.), Belgium. Southwall Technologies (http://www.southwall.com) uses two-story high sputtering machines to manufacture transparent sputtered films on 2 m wide × 5,000 m long rolls of polyethylene terephthalate (PET) substrate that is 25 μm or 50 μm thick. One roll allows fabrication of up to 10,000 square meters of film.

of substrate, including foils and fabrics. In a high-vacuum chamber, the rolled substrate is unwound and passes over the cathode cylinder where thin layers of target material are deposited. The substrate is continuously rewound on the other side of the deposition site. The sputter target is a long cylinder, and reactive and nonreactive processes are available.[11] Roll-to-roll (R2R) manufacturing is currently used to print, for example, newspapers and labels. One envisions that many other products could be manufactured the same way. For example, organic light-emitting diodes (OLEDs) could be manufactured by patterning all layers with continuous deposition processes (web vs. batch). This would be potentially low cost and be scalable to large areas (see also Volume III, Chapter 1). Innovations like these might help revive the potential of micromachined chemical sensors for which Si batch processes are often too expensive. When manufacturing microfabricated chemical sensors, one often competes with continuous fabrication processes (e.g., continuous printing of glucose sensor paper strips), and classical Si micromachining cannot beat the cost advantages these offer. A process as described here, complemented by continuous lithography, might enable chemical microsensor fabrication at an acceptable cost (see Volume III, Chapter 1). A good reference on continuous, roll-to-roll thin film coating is *Vacuum deposition onto webs, films, and foils* by Charles A. Bishop (William Andrew Publishing, 2007).

The most negative aspects of sputtering are its complexity compared with an evaporation process, the excessive substrate heating resulting from secondary electron bombardment, and, finally, the slow deposition rate. In a regular sputtering process, the rate is one atomic layer per second versus thousand atomic layers per second available from a typical evaporation source at a vapor pressure of $10^{-1}$ Torr. Controls in a sputtering setup include argon pressure, flow rate, substrate temperature, sputter power, bias voltage, and electrode distance.

**Example 7.1: Sputtered Polysilicon**

The quest for low-temperature polysilicon deposition processes is driven in part by the need for compatibility with prefabricated aluminum-metallized complementary metal oxide semiconductor (CMOS) circuitry. This makes the 350°C sputter deposition of polysilicon especially interesting.

Honer et al.[12] introduced sputtered polysilicon to the MEMS community. Sputtered silicon can be used to fabricate polysilicon microstructures (the key mechanical structural element in surface micromachining; see section on Doped Polysilicon) atop standard, aluminum-metallized CMOS at temperatures less than 350°C. A drawback for sputtered boron doped Si films is that to increase their conductivity (to 25 Ω/□) they need to be clad in symmetric, 50-nm-thick layers of titanium-tungsten. Films with stress values less than 100 MPa are routinely obtained.

An important benefit of this low-temperature polysilicon process is that it is not only compatible with conventional oxide sacrificial layers (sacrificial layers, we shall see below, are another key ingredient in surface micromachining) but also with certain organic sacrificial layers (e.g., polyimide). Organic layers can be removed in a dry oxygen plasma etch, thus avoiding the stiction and selectivity problems associated with wet etch releases and at the same time alleviating the concerns over chemical attack on structural elements by HF.

## Ion Plating

In ion plating, evaporation of a material is combined with ionization of the atom flux by an electron filament or plasma. The principle of ion-plating in argon plasma is illustrated in Figure 7.13.[13] The substrate is set up to be the cathode in the upper part of the chamber, and the source material is placed below it. A vacuum is then established in the chamber. Argon gas is admitted, and an electric field is applied to ionize the gas (forming $Ar^+$) and establish plasma. This results in ion bombardment (sputtering) of the substrate so that its surface is atomic clean (interpret this as "very clean"). Next, the source material (Ti in the example case) is heated sufficiently to generate coating vapors. The heating methods used here are the same as those used in vacuum evaporation: resistance heating, e-beam bombardment, and so on. The Ti vapor passes through the plasma and coats the substrate. Sputtering is continued during deposition; thus, the ion bombardment of the substrate consists not only of the original argon ions but also of source material ions (Ti) that have been energized while being subjected to the same energy field as the argon. The effect of these processing conditions is to produce films of uniform thickness and excellent

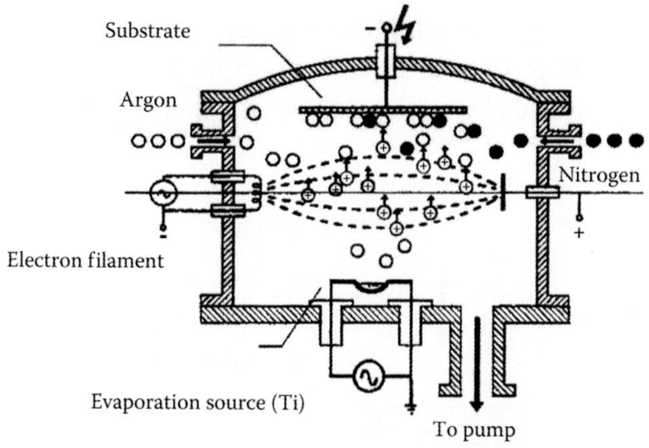


**FIGURE 7.13** Principle of ion plating. (From Menz, W., and P. Bley. 1993. *Mikrosystemtechnik fur Ingenieure.* Weinheim, Germany: VCH Publishers.[13])

adherence to the substrate. Ion plating is applicable to parts with irregular geometries because of the high degree of scattering of the depositing material in the plasma. An example of interest here is TiN coating of high-speed cutting tools (for example, drill bits). As shown in Figure 7.13, the addition of a gas (nitrogen in this case) to the reactor enables one to make new compounds (such as TiN) on the substrate surface, with the gas reacting with the ionized atoms from the evaporation source (Ti). Because of the high kinetic energy of the impacting ions, a very well-adhering, dense TiN film with extraordinarily low friction coefficient and high hardness coefficient (Vickers hardness of 50,000) forms. Because of the thermal nature of the process, very high deposition rates can be achieved. In addition to coating uniformity and good adherence, other advantages of the process include high deposition rates, high film densities, and the capability to coat the inside walls of holes and other hollow shapes.

The NTH-1000 from Nanotec Corp. (http://nanotec-jp.com/www_nanotec/english/nth1000.html) uses a tandem hollow cathode discharge technique to deposit high-quality TiN, CrN, TiCN, and TiCrN thin films on cutting tools, machine parts, and molds.

## Ion Cluster Beam Deposition—Nanoparticle Deposition

Clusters are aggregations of atoms or molecules, intermediate in size between individual atoms and aggregates that are large enough to be called *bulk matter.* Atom clusters are also called *nanoparticles* because their size is on the order of nanometers or tens of nanometers. The ultrasmall size of building blocks leads to dramatically different properties, and it is anticipated that such atomically engineered materials can be tailored to perform as no previous material could. T. Takagi, from Kyoto University, first proposed the idea of ionized cluster beam (ICB) thin film deposition in 1972. Conditions for formation of ion cluster beams, suitable for thin film deposition, were only established following an additional 20 years of effort. When applying cluster beam technology, ionized atom clusters (e.g., 100–1000 atoms) are deposited on a substrate in a high vacuum ($10^{-5}$–$10^{-7}$ mbar). Those atom clusters typically

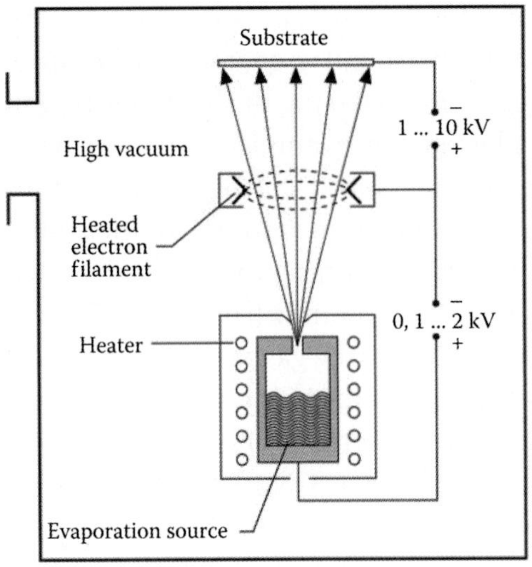


**FIGURE 7.14** Setup for ion-cluster beam deposition. (From Menz, W., and P. Bley. 1993. *Mikrosystemtechnik fur Ingenieure.* Weinheim, Germany: VCH Publishers.[13])

carry one elementary charge per cluster and therefore achieve the same energy in an electrical field as would a single ion.

A special evaporation cell must be used to generate these atom clusters. As shown in Figure 7.14,[13] the heating of the evaporant in an evaporation cell with a small opening causes an adiabatic expansion of more than 100–$10^{-5}$ or $10^{-7}$ mbar on exiting that cell. The expansion causes a sudden cooling, inducing the formation of atom clusters. These clusters are then partially ionized by an electron bombardment from a heated filament. Low-energy neutral clusters (0.1 eV) and somewhat higher energy ionized clusters (a few eV) arrive at the surface where they flatten and form a film (Figure 7.15)[13] of excellent adhesion and purity, with a relatively low number of defects. Cluster beam epitaxy is possible at temperatures as low as 250°C, and no charge buildup occurs when depositing on an insulator. Ion cluster beam technology qualifies as a gas-phase nanoparticle synthesis method. In gas-phase synthesis of nanoparticles supersaturation is achieved by vaporizing material into a background gas. Condensation to form particles is induced by cooling, e.g., by mixing with a cool gas or expansion through a nozzle. In ion cluster beam technology the cooling occurs when the vapor exits the evaporation cell. Other types of gas-phase nanoparticle fabrication techniques are discussed in Chapter 9 on thermal energy-based forming techniques.

The cluster beam technology is also applied to shallow ion implantation, high-yield sputtering, smoothing, surface cleaning, and low-temperature thin film formation. Cluster ion beam processing is now expanding into new industrial fields presently limited by the available atomic and molecular ion beam processes. With the wide variety of selective layers needed in chemical sensors and the frequent need for low-temperature deposition of thick, well-adhering layers, ion plating, laser sputter deposition (see below), and cluster beam technology seem destined to play a more crucial role in chemical sensor development in the future. Also, building devices from atomic constituents (bottom-up approach), with proximal probes for example, is too time consuming when performed one atom at the time. To speed up the process, massive parallelism is required; for example, by using an army of coordinated proximal probes. Cluster beam technology better enables nanomachining because the size of the building blocks lends itself to being built faster.

Epion (http://www.epion.com) is an example of a company selling ion cluster beam equipment, and their units feature surface smoothing to less than 3 Å, surface cleaning, and high-yield sputtering.

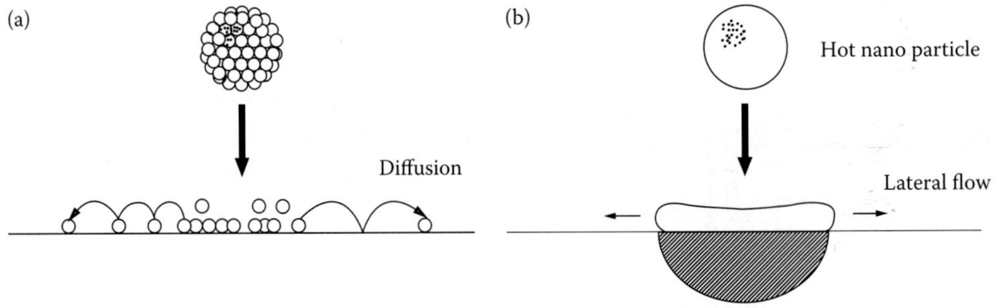


**FIGURE 7.15** Film forming with atom clusters. (a) Old model based on atom diffusion. (b) New model based on particle flow. (From Menz, W., and P. Bley. 1993. *Mikrosystemtechnik fur Ingenieure.* Weinheim, Germany: VCH Publishers.[13])

## Pulsed Laser Deposition, Laser Sputtering, or Laser Ablation Deposition

Pulsed laser deposition (PLD) uses intense laser radiation to erode a target and deposit the eroded material onto a substrate. A high-energy focused laser beam avoids the x-ray damage to the substrate encountered with e-beam evaporation. A high-energy UV excimer laser pulse (10 ns) coming from, for example, a KrF laser at 248 nm with a pulse energy in the focus of 2 J/cm$^2$, is directed onto the material to be deposited, as sketched in Figure 7.16. As shown here, the laser ablates material from the target, forming a high-temperature vapor plume incident on the sample. The target may be just about anything, and one can operate at atmospheric pressures or in a very high vacuum.

The PLD process can be split into the following basic steps:

1. The energy of the very short wavelength radiation is absorbed in the upper surface of the target, i.e., absorption of the laser pulse in the material (for a metal, the absorption depth is ~10 nm, depending on $\lambda$).
2. Energy relaxation to the lattice through electron-phonon interaction (~1 ps).
3. Heat diffusion (microseconds), melting (tens of nanoseconds), and evaporation of a small amount of material.
4. Plasma creation.
5. Absorption of light by plasma and ionization.
6. Interaction of target and ablated species with plasma.
7. Cooling and resolidification between pulses.

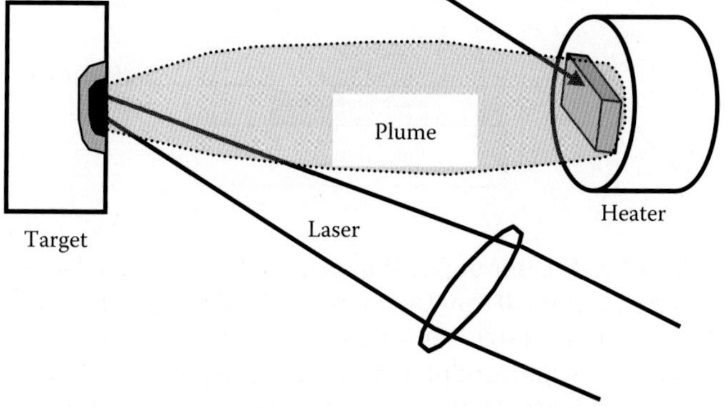


**FIGURE 7.16** Schematic of pulsed laser deposition (PLD).

Laser ablation is a very nonequilibrium process. At the peak of the laser pulse, temperatures on the target can reach >10$^5$ K (>40 eV!) with electric fields >10$^5$ V/cm (accompanied by high magnetic fields) and plasma temperatures of 3000–5000 K. The ablated species come with energies in the 1–100-eV range.

Laser ablation deposition was first mentioned in Chapter 5, under thermal removing techniques, where we learned that laser subtractive machining leaves an undesirable heat-affected zone (HAZ) on a surface that is being machined. The size of the HAZ is a function of the laser pulse duration and the thermal conductivity and specific heat of the work piece. The HAZ constitutes a problem in the case of laser subtractive machining because it degrades the quality of the cut. Moreover, during the ablation process, the expulsion of material in the plasma jet redeposits material onto the surrounding region, further compromising the fidelity of the process. In the case of PLD, we take advantage of this deposition of materials from the plume onto an adjacent substrate surface.

Splashing is the bane of PLD. The laser superheats the subsurface layer before the surface reaches its evaporation point, and, as a consequence, the surface breaks apart into large (micrometer-sized) globular particles when the subsurface expands. This leads to globular deposits. Another problem with PLD is the expansion of the plume, which causes a sudden decrease in the pressure just above the surface. This shock wave pulls droplets of liquid of the surface. The thermal shock also causes irregularities in the surface to break off (exfoliation). Advantages and disadvantages of PLD are summarized in Table 7.8.

As we saw in Chapter 5 with ultrashort laser pulses (<1 ps), so much energy is deposited in the material at such a fast rate that the material is forced directly into a plasma state; the material goes from a solid to a gas phase without first going through a melt phase. The reason behind this is that pulse width is now shorter than the electron lattice-relaxation time; thus, the heat diffusion and melting are significantly reduced. This plasma then expands away from the material as a highly energetic gas, taking almost all the heat away with it. This leads

TABLE 7.8 Advantages and Disadvantages of Pulsed Laser Deposition (PLD)

| Advantages | Disadvantages |
|---|---|
| Flexible, easy to implement | Uneven coverage |
| Growth in any environment | High defect or particulate concentration |
| Exact transfer of complicated materials (YBCO) | Not well suited for large-scale film growth |
| Variable growth rate | Mechanisms and dependence on parameters not well understood |
| Epitaxy at low temperature | Uneven coverage |
| Resonant interactions possible (i.e., plasmons in metals, absorption peaks in dielectrics and semiconductors) | Splashing as a result of subsurface boiling, shock wave recoil, exfoliation |
| Atoms arrive in bunches, allowing for much more controlled deposition | |
| Greater control of growth (e.g., by varying laser parameters) | |

*Source:* Based on *Pulsed laser deposition of thin films.* New York: John Wiley and Sons, 1994.[230]
The table does not take into account pulsed laser deposition with ultrafast pulses improving the technique a lot, for example, smoother films result.

not only to cleaner cutting but also smoother film deposition (fewer particulates, less splashing) and the potential for ultrahigh quality films, circuit writing, isotope enrichment, new materials, and nanoparticle production.

The ablated material, partially ionized in the laser-induced plasma, is deposited onto a substrate almost without decomposition. This technique is particularly useful when dealing with complex compounds, as in the case of the deposition of high-temperature superconductor films (HTSCs), for example, $YBa_2Cu_3O_{7-x}$. PLD faithfully replicates the atomic ratios present in the hot isostatically pressed target disc onto the thin film coating. Achieving complex stoichiometries presents more difficulties with any other deposition technology. Approximately 10,000 pulses (pulse length of 20 ns and a repetition rate of 15 pulses/s) are needed to achieve a film thickness of 0.1 μm on the substrate. Normally, the laser-deposited films are amorphous. The energy necessary to crystallize the film comes from heating the substrate 700–900°C and from the energy transferred from the intense laser beam to the substrate via atomic clusters.

Figure 7.17a shows a schematic setup of a laser deposition system.[14] Here, a pulsed excimer laser is used to deposit superconductor materials such as those based on YbaCuO.[14] Figure 7.17b shows an improved laser deposition setup where substrate heating can be replaced in part or completely by

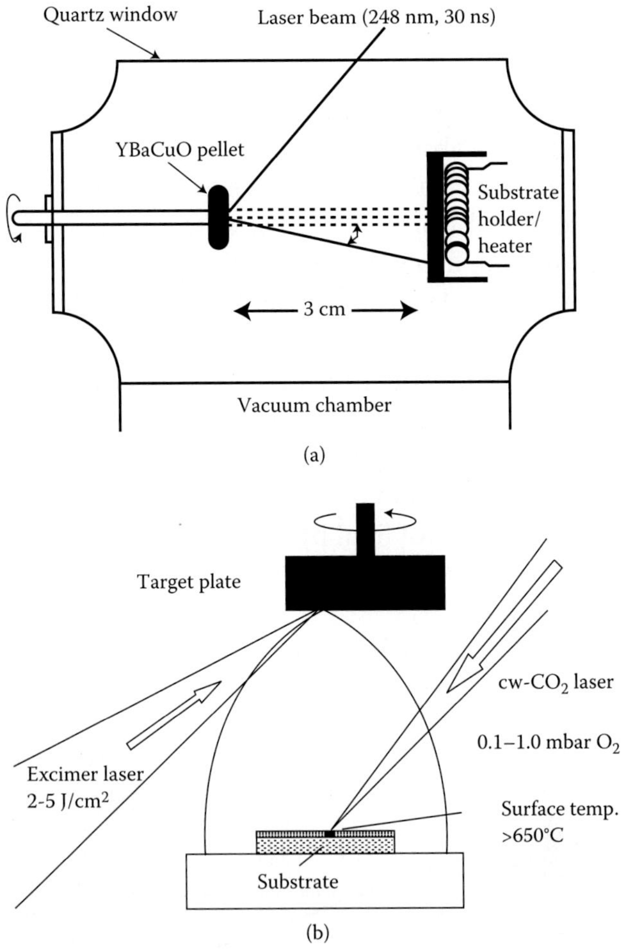


**FIGURE 7.17** Laser deposition system. (a) Traditional laser ablation system. (From Dutta, B., X. D. Wu, A. Inam, and T. Venkatesan. 1989. Pulsed laser deposition: a viable process for superconducting thin films? *Solid State Technol* February:106–10. With permission.[14]) (b) Laser ablation with additional laser for surface heating.

additional laser radiation, in this case, a continuous wave-$CO_2$ (cw-$CO_2$) infrared laser. With this setup, using two crossed laser beams, it is possible to deposit and induce the correct crystalline structure in the growing HTSC film while it is being deposited.[15] Superconductive Components Inc. manufactures superconducting and nonsuperconducting ceramic targets for use in sputtering and laser ablation systems (http://www.superconductivecomp.com/targets2.html).

A sensor-related application of PLD is the protection of components in contact with body fluids using the biocompatible calcium-phosphate-based ceramic calcium hydroxylapatite, or $Ca_{10}(PO_4)_6(OH)_2$. This material exemplifies the most stable calcium phosphate in contact with body fluids. Deposition methods of this material include sputtering, plasma spraying, electrophoretic deposition, and combinations of these techniques. In all cases, a postdeposition treatment is needed to crystallize the partially or wholly amorphous and/or dehydrated films. With the laser pulse technique, in water-vapor-enriched inert gas environments, deposition of the right hydroxyl apatite was observed at temperatures between 400 and 700°C. Adhesion of this material to Si and Ti-6Al-4V was found to be excellent.[16] Laser sputtering also could help prepare complex and stoichiometry-sensitive coatings such as electrochromic devices.

Pulsed laser ablation, like ion cluster beam deposition, can also be used as a nanoparticle gas-phase synthesis method. In this approach the solid precursor is heated by absorption of laser energy. Typically an infrared laser ($CO_2$) is used, and laser energy may be directly absorbed by precursors or by an inert photosensitizer ($SF_6$). The use of a pulsed laser shortens the reaction time and allows for the preparation of the smallest particles. In 1995, at Rice University it was reported that synthesis of carbon nanotubes could be accomplished by laser ablation.[17] A pulsed or continuous laser is used to vaporize a graphite target placed in an oven at 1200°C (Figure 7.18). The oven is filled with argon gas to keep the pressure at 500 Torr. A very hot vapor plume forms, which then expands and cools rapidly. Contact with a cooled copper substrate causes the carbon atoms to be deposited in the form of nanotubes.

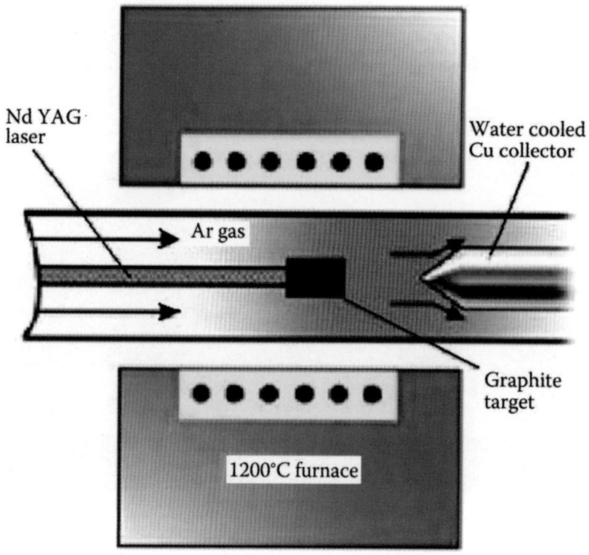


**FIGURE 7.18** Laser ablation system for carbon nanotube production.

Laser vaporization produces a higher yield of single-walled nanotubes (SWNTs) with better properties and with a narrower size distribution than nanotubes produced by arc discharge. Laser ablation also produces nanotubes that are more pure (up to 90% purity) than those produced by arc discharge.

Pulsed laser deposition (PLD) with ultrafast pulses (<1 ps) is opening up all types of new research and application opportunities as evidenced by the Defense Advanced Research Projects Agency's (DARPA's) mesoscopic integrated conformal electronics (MICE) program. The goal of that program is to create electronic circuits and materials on any surface, e.g., to print electrical circuits on the frames of eyeglasses or interwoven with clothing (see Figure 7.19). MICE efforts have demonstrated the ability to print metal lines on curved surfaces, feature sizes as small as 5 μm, and print speeds close to 1 m/s. Other exciting developments include the demonstration of 1) printed zinc-air batteries that have four times more volumetric power density than commercial batteries, 2) direct-write antennas, and 3) direct-write solar cells. A single laser does surface pretreatment, spatially selective material deposition, surface annealing, component trimming, ablative micromachining, dicing, and via drilling.

To sum up, laser ablation is a good technique for preparing thin films of any desired stoichiometry,

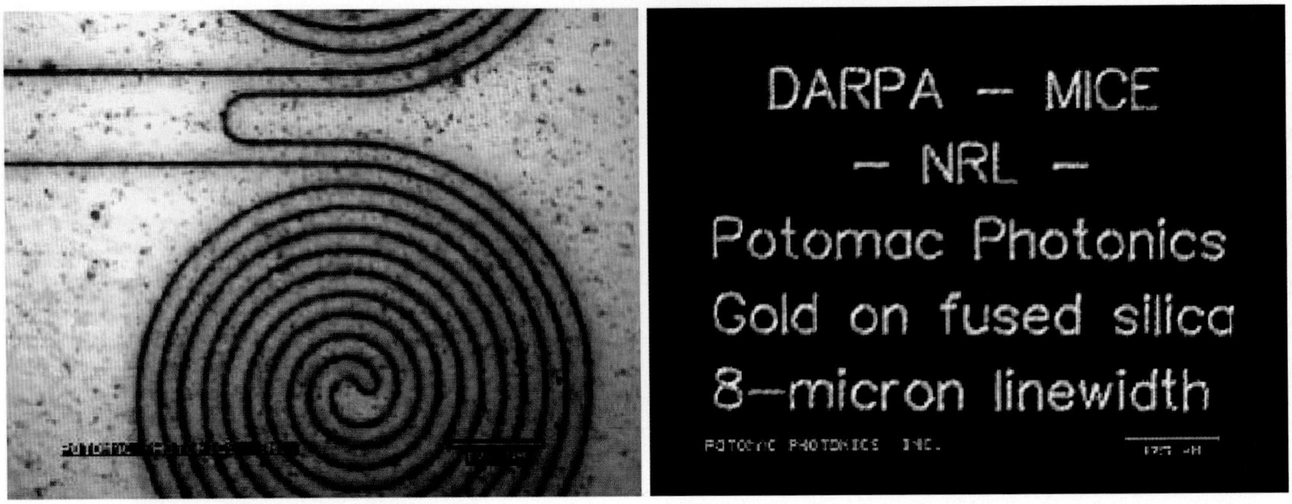


**FIGURE 7.19** Direct writing in air with pulsed laser deposition (PLD).

but, because of the small source size, it is not useful for large-scale coatings. A small source size might make it essential, especially when working with high-aspect-ratio micromachined structures, to rotate the sample at an angle with respect to the material flux so as to obtain coverage of the vertical features in the device. The possibility of using PLD for nanoparticles and carbon nanotubes and the use of PLD with ultrashort laser pulses (<1 ps) has given the technique an added impetus. For further reading on laser sputtering, refer to references 14–16,18,19.

Other good references are found in "Pulsed laser vaporization and deposition," Wilmott and Huber, *Reviews of Modern Physics*, vol. 72, 315 (2000), *Pulsed laser deposition of thin films*, Chrisey and Hubler (Wiley, New York, 1994), and *Laser ablation and desorption*, Miller and Haglund (Academic Press, San Diego, 1998). Epion, for example, sells large-area PLD systems (http://www.pvdproducts.com/products/pld_systems.aspx).

## Molecular Beam Epitaxy

Epitaxial techniques arrange atoms in single-crystal fashion on a crystalline substrate acting as a seed crystal so that the lattice of the newly grown film duplicates that of the substrate. If the film is of the same material as the substrate, the process is called *homoepitaxy*, *epitaxy*, or simply *epi*. Epi deposition represents one of the cornerstone techniques for building micromachines. Of special importance is that Si membranes of a predetermined thickness and doping level can be fashioned. The growth rate of an epilayer depends on the substrate crystal orientation. Si (111) planes have the highest density of atoms on the surface, and the film grows most easily on these planes. Important epi applications are Si on Si substrates and GaAs on GaAs substrates. If the deposit is made on a chemically different substrate, usually of closely matched lattice spacing and thermal expansion, the process is termed *heteroepitaxy*. One important heteroepitaxy application is the deposition of silicon on insulator (SOI), for example, Si on $SiO_2$ or Si on sapphire ($Al_2O_3$). Various sapphire orientations such as <$011\bar{2}$>, <$10\bar{1}\ 2$>, and <$1\ \bar{1}02$> have been used to grow <100>-oriented silicon layers. The lattice mismatch between sapphire and silicon crystals limits the thickness of the Si to about 1 μm. Another example of heteroepitaxy is that of gallium phosphide on gallium arsenide.

Various types of epitaxy techniques exist. Chemical vapor phase and liquid epitaxy are described further below. Here, we briefly discuss PVD epitaxy, that is, molecular beam epitaxy (MBE), invented in late 1960s at Bell Laboratories by J. R. Arthur and A. Y. Cho. In MBE, the heated single-crystal sample (e.g., between 400 and 800°C) is placed in an ultrahigh vacuum ($10^{-11}$ Torr) in the path of streams of atoms from heated cells that contain the materials of interest. These atomic streams impinge, in a line-of-sight fashion, on the surface-creating layers with a structure controlled by the crystal structure of the surface, the thermodynamics of the constituents, and the sample temperature. This technique is the most sophisticated form of PVD. The deposition rate of MBE is very low (i.e., about 1 μm/h or 1 monolayer/s), and considerable attention is devoted to in situ material characterization to obtain high-purity epitaxial layers. Once on the substrate, the atoms or molecules move around until they find an atomic site to chemically bond to. Fast-acting shutters control the deposition. One or two atomic layers of material lie between the shutter action. This becomes important when an ultrasharp profile is called for. The low deposition rate gives the operator better control over the film thickness. Ultrasharp profiles are needed, for example, when making quantum well devices. A quantum well might be 40 Å thick. To uniquely define its energy levels it must be 40 ± 2 Å. Figure 7.20 represents a schematic of a molecular beam deposition setup.[1]

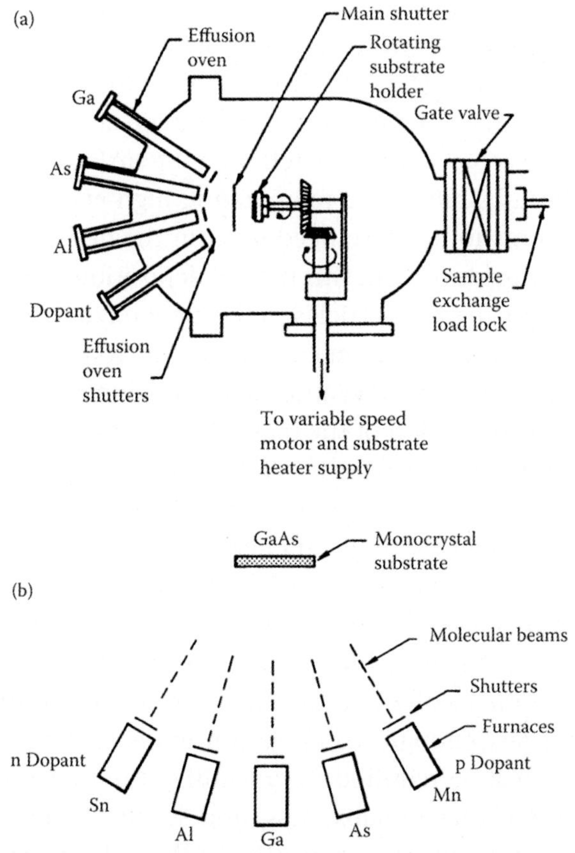


**FIGURE 7.20** (a) Schematic of molecular beam epitaxy growth chamber. (b) Example is the growth of GaAlAs epi on GaAs single crystal. (From Brodie, I., and J. J. Muray. 1982. *The physics of microfabrication.* New York: Plenum Press. With permission.[1])

**TABLE 7.9** Key Points in MBE

| Characteristics |
|---|
| Ultrahigh vacuum |
| High purity layers with good crystal structure |
| No chemical byproducts created at growth surface |
| High uniformity (<1% deviation) |
| Often uses multiples sources to grow alloy films |
| High control of composition |
| Growth rates are very low (0.1–10 μm/h) |
| In situ monitoring and feedback |
| Expensive |
| Mature production technology |

The technique has several potential advantages over CVD epitaxy (see next section): for example, the relatively low growth temperatures reduce diffusion and autodoping effects. Precise control of layer thickness and doping profile on an atomic layer level is possible. Novel structures such as quantum devices, silicon/insulator/metal sandwiches, and superlattices can be made. Today, half of the lasers that are used in CD players, as well as those in pen-sized laser pointers, are grown by MBE. Figure 7.20 shows the deposition of n- and p-type GaAlAs on a GaAs single-crystal surface.[20] With MBE, virtually any device structure can be made. The limitations to consider lie in volume manufacturing and cost. The ultrahigh vacuum requirements make operation very expensive. From the available epi techniques, MBE exemplifies one that is perhaps the most complicated.[21] In Table 7.9 we review the key points concerning MBE, and in Figure 7.21 we show a typical commercial MBE station.

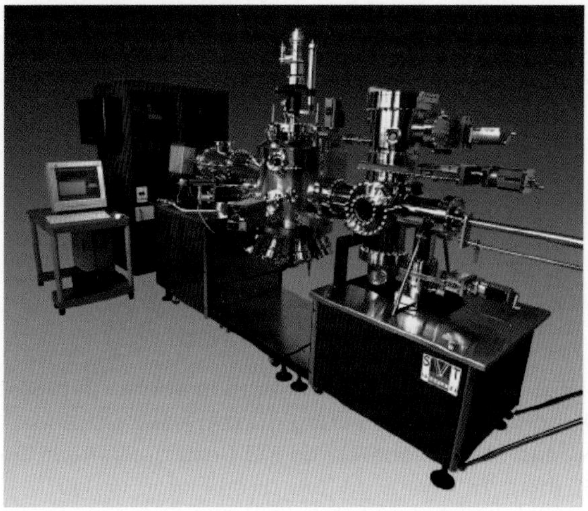

**FIGURE 7.21** Typical MBE system photo.

## Aerosol Deposition—Impact Deposition

Aerosol deposition (AD) is a mechanical film fabrication method invented in 1997 by Dr. Jun Akedo of the National Institute of Advanced Industrial Science and Technology (AIST) in Japan.[22] In Figure 7.22 we illustrate the overall configuration for film fabrication using AD, together with electron micrographs of raw powder and a resulting AD film.

Raw powders with an average powder diameter of 600 nm are mixed and vibrated with a carrier gas to form an aerosol. By using the pressure difference between the aerosol and a vacuum chamber, the aerosol is sprayed from a nozzle and impacted onto a substrate. Using this simple process, AD films with ultrafine crystalline particles of several tens of nanometers can be formed as shown in the cross-sectional image of the AD film in Figure 7.22. The impact of solid-state particles creates a strongly adherent, high-density nanocrystalline film by gas blasting the nanosized particles onto the surface. The deposition rate is 30 times faster (10 μm/min) than traditional deposition rates and can be carried out even at room temperature. The formation mechanism is assumed to be room-temperature shock-compaction. The impacting 600-nm particles are deformed and fractured into nanocrystalline structures of 10–30-nm size. Uniform deposition rates over a 200-mm square area have been achieved. The raw powders may be manufactured by conventional processing methods for most materials.

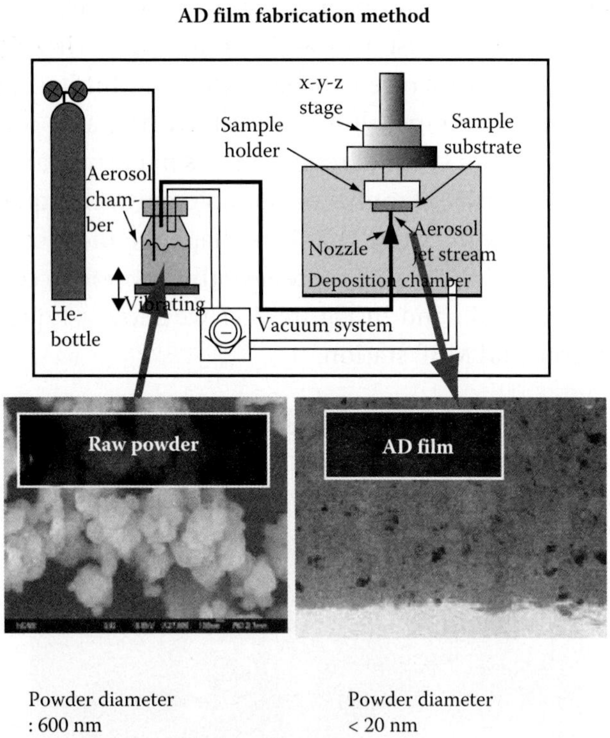


**FIGURE 7.22** Schematic illustration of aerosol deposition (AD), powder, and a resulting AD film [solid solution of PNN (lead nickel niobate) and PZT (lead zirconate titanate)]. (http://www.nec.co.jp/techrep/en/journal/g07/n01/070115.html)

The two most important distinguishing characteristics of the AD process are that 1) the deposited films maintain the crystalline properties of the raw powder even at low deposition temperatures, and 2) relatively thick films can be produced fast. This stands in stark contrast to the thin film-forming methods such as sputtering, molecular beam epitaxy (MBE), and chemical vapor deposition (CVD). When these latter methods are used to deposit a material at low temperatures, for example, a piezoelectric material (typically an oxide with a perovskite crystalline structure) onto a substrate such as glass, an amorphous layer results, and all piezoelectric properties are lost. To maintain the piezoelectric function of this material, it is essential to retain the perovskite crystalline structure. To achieve this, one relies on the surface energy of the substrate and a temperature high enough for the atoms to be rearranged to form the desired perovskite oxide film. In this type of deposition process, restructuring the particles in the crystallization process occurs at the atomic level and far from equilibrium, inducing many lattice defects that result again in the degradation of the piezoelectric effect.

For many piezoelectric actuator applications, a film, several micrometers thick, is essential to output sufficient mechanical energy for a given electrical input. When a thick film of more than several micrometers is formed onto a substrate with the conventional thin film-forming technologies, various problems often occur: the speed of film forming is not fast enough, composition tends to vary, too much heat is required to crystallize the film limiting the choice of substrates, and so on. Because of these concerns, the conventional thin film-forming technologies have been used only for limited applications, such as for thin film sensors. In case thicker films are required, bulk-type piezoelectric ceramics

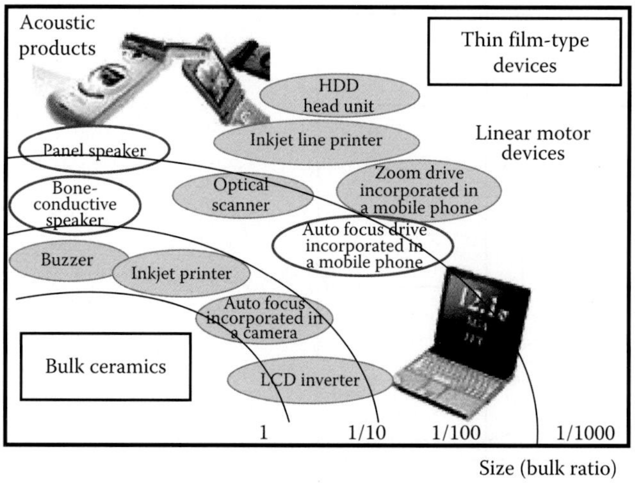


**FIGURE 7.23** Use of piezoelectrics in consumer products. (http://www.nec.co.jp/techrep/en/journal/g07/n01/070115.html.)

made of sintered powders have usually been adopted (see Figure 7.23).

Screen-printing of piezoelectrics is a thick-film technique that, in principle, could cover the meso-range between bulk ceramics and thin films. In screen-printing one uses a mixture of ceramic powder and binder to make an ink. After screen-printing, the ink is fired at 1000°C or more (screen-printing of ceramics is a thermal additive technique covered in Chapter 9). This limits this technology to ceramic substrate materials.

As suggested in Figure 7.23, thick film-formed piezoelectric devices integrated onto a substrate will be essential for many piezoelectric actuators of the future, such as in sound equipment and for products that require a high-precision positioning motion. The AD method is a film fabrication method that can fulfill that need. Piezoelectric AD results in nanocrystalline films that preserve a large dielectric and piezoelectric effect and that are little influenced by the underlying layers and/or substrates. For example, by using ceramic particles that feature chemically unstable surface characteristics, AD technology has been used to form ceramic films at room temperature on resin circuit boards to which ceramic is usually difficult to adhere. The maximum dielectric constant of the newly created ceramic film was 400, i.e. 10 times the dielectric constant of ceramic combined with resin, and is approximately at the same level as that of ceramic film made with the sputtering method that uses heat treating at a high temperature of 600°C. A condenser composed of three layers was also successfully created on an FR4 printed circuit board. The capacitance of the newly created condenser was 300 nF/cm$^2$, a high level that is ideal for practical use (http://www.nec.co.jp/techrep/en/journal/g07/n01/070115.html).

From the above, the AD method is an excellent candidate for the fabrication of piezoelectric thick film actuators for the manufacture of micropumps, ultrasonic mixers in microelectromechanical systems (MEMS) and micrototal analysis systems (μ-TAS), micromanipulators for medical applications, ink-jet printer heads, flapper actuators for high-density hard disk drives, and other applications where large strains and high-speed response are necessary (see Figure 7.23).

An AD of a solid solution of PNN (lead, nickel, and niobate) and PZT (lead zirconate titanate) was attempted on zirconate ceramics and stainless steel substrates. Zirconate ceramics have good high-temperature resistance and stainless steel features, superior corrosion resistance, and processability that lead to low-cost manufacturing of practical piezoelectric devices. The AD film-forming process was carried out at room temperature with a deposition rate of 10 μm/min or higher over an area of 10 × 10 mm$^2$. These process conditions confirm that this process is amenable to mass production. Figure 7.24 shows a comparison of the film properties between the piezoelectric film formed by the AD and the conventional film deposition method. The piezoelectric film fabricated by AD showed a piezoelectric constant (−d31) of 158 mp/V on a stainless substrate, which is almost the same value as that of film fabricated by conventional film-forming technology. On a zirconate substrate, 360 pm/V was achieved, which is similar to that for bulk ceramics. As shown in Figure 7.24, where we plot the piezoelectric constant as a function of film thickness of the actuator, AD covers a broad range of thicknesses—between bulk ceramic material and thin film—with a better piezoelectric constant than screen printing (http://www.nec.co.jp/techrep/en/journal/g07/n01/070115.html).

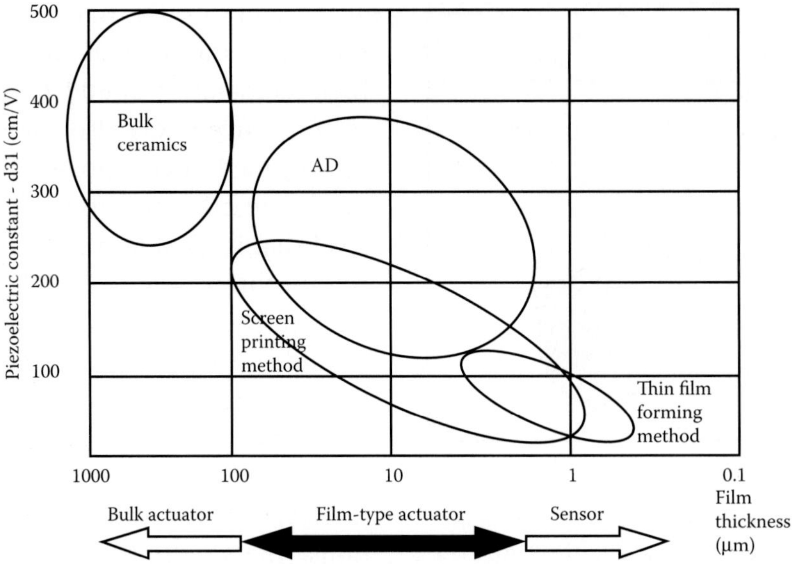


**FIGURE 7.24** Plot of the piezoelectric constant (–d31) as a function of film thickness of the actuator [solid solution of PNN (lead nickel niobate) and PZT (lead zirconate titanate)] and manufacturing means. AD covers the thickness range between bulk and thin film. (Based on http://www.nec.co.jp/techrep/en/journal/g07/n01/070115.html by Dr. Han Xu.)

## Chemical Vapor Deposition

### Introduction

Chemical vapor deposition (CVD) is one of the principal thin film manufacturing options. In Figure 7.25 we show how thin film processes fit in the overall IC fabrication scheme. In the case of thin film of silicon dioxide, if the surface of the wafer is silicon (e.g., at the start of IC fabrication), then thermal oxidation is the appropriate process by which to form a layer of $SiO_2$. If the oxide layer must be grown over materials other than silicon, such as aluminum or silicon nitride, then an alternative technique must be used, such as CVD. During CVD, the constituents of a vapor phase, often diluted with an inert carrier gas, react at a hot surface (typically higher than 300°C) to deposit a solid thin film. In CVD, the diffusive-convective transport to the hot substrate surface involves many intermolecular collisions, and mass and heat transfer modeling of deposition rates is consequently much more complex than in PVD. Chemical vapor deposition of $SiO_2$ is accomplished by reacting a silicon compound such as silane ($SiH_4$) with oxygen onto a heated substrate. The reaction is carried out at around 425°C and can be summarized by:

$$SiH_4 + O_2 \rightarrow SiO_2 + 2H_2 \qquad \text{Reaction 7.1}$$

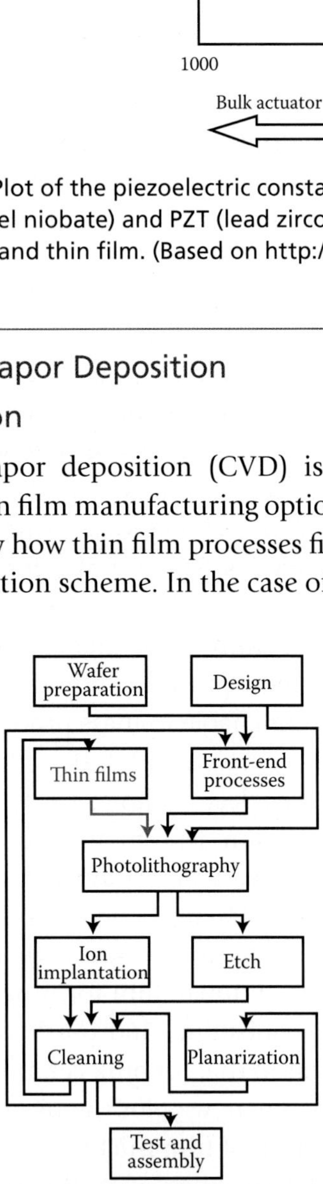


**FIGURE 7.25** Thin films. CVD is a broad class of processes using controlled chemical reactions to create thin films on substrates. A variant of CVD, called *plasma-enhanced chemical vapor deposition* (PECVD), uses gas plasma to lower the temperature required to obtain a chemical reaction and achieve film deposition.

The density of a CVD silicon dioxide film and its bonding to the substrate are generally poorer than that achieved by thermal oxidation. Consequently, CVD is used only when the preferred process is not feasible: when the substrate surface is not silicon or when the high temperature used in thermal oxidation cannot be tolerated.

In the CVD reaction chamber, the reactants are adsorbed on the heated substrate surface, and the adatoms undergo a series of processes resulting in film formation. When a molecule has reached the surface, the required reaction analysis is the same, regardless of deposition method. The molecular phenomena at the surface to be considered include sticking coefficient, surface adsorption, surface diffusion, surface reaction, desorption, and film or crystal growth. Gaseous byproducts are desorbed and removed from the reaction chamber. The reactions forming a solid material do not always occur on or close to the heated substrate (heterogeneous reactions) but can also occur in the gas phase (homogeneous reactions). As homogeneous reactions lead to gas-phase cluster deposition and result in poor adhesion, low density, and high defect films, heterogeneous reactions are preferred. The slowest of any of the CVD steps mentioned, the gas phase or surface process determines the rate of deposition. The sample surface chemistry, its temperature, and thermodynamics determine the compounds deposited. The most favorable end product of the physical and chemical interactions on the substrate surface is a stoichiometric-correct film. Several activation barriers need to be surmounted to arrive at this end product. Some energy source, such as thermal, photon, or ion bombardment, is required to achieve this.

The CVD method is very versatile and works over a wide range of temperatures and pressures. Amorphous, polycrystalline, epitaxial, and uniaxially oriented polycrystalline layers can be deposited with a high degree of purity, control, and economy. CVD is used extensively in the semiconductor industry and has played an important role in transistor miniaturization, for example, by introducing very thin film deposition of silicon. More recently, CVD copper and low dielectric insulators ($\varepsilon < 3$) also have become important CVD applications. CVD embodies the principal building technique in surface micromachining (see section on Doped Polysilicon). Modern interest in CVD is also focused on coating applications, such as coated carbide tools, solar cells, depositing refractory metals on jet engine turbine blades, and other applications where resistance to wear, corrosion, erosion, and thermal shock

**Table 7.10** Example CVD Reactions

| Process Description Reaction |
|---|
| 1. The Mond process includes a CVD process for decomposition of nickel from nickel carbonyl [$Ni(CO)_4$], which is an intermediate compound formed in reducing nickel ore: $Ni(CO)_4 \xrightarrow{400\,°F\,(200\,°C)} Ni + 4CO$ |
| 2. Coating of titanium carbide (TiC) onto a substrate of cemented tungsten carbide (WC-Co) to produce a high-performance cutting tool: $TiCl_4 \xrightarrow{1800\,°F\,(1000\,°C),\ excess\ H_2} TiC + 4HCl$ |
| 3. Coating of titanium carbide (TiC) onto a substrate of cemented tungsten carbide (WC-Co) to produce a high-performance cutting tool: $TiCl_4 + 0.5N_2 + 2H_2 \xrightarrow{1650\,°F\,(900\,°C)} TiN + 4HCl$ |
| 4. Coating of aluminum oxide ($Al_2O_3$) onto a substrate of cemented tungsten carbide (WC-Co) to produce a high-performance cutting tool: $2AlCl_3 + 3CO_2 + 3H_2 \xrightarrow{900\,°F\,(500\,°C)} Al_2O_3 + 3CO + 6HCl$ |
| 5. Coating of silicon nitride ($Si_3N_4$) onto silicon (Si), a process in semiconductor manufacturing: $3SiF_4 + 4NH_3 \xrightarrow{1800\,°F\,(1000\,°C)} Si_3N_4 + 12HF$ |
| 6. Coating of silicon dioxide ($SiO_2$) onto silicon (Si), a process in semiconductor manufacturing: $2SiCl_3 + 3H_2O + 0.5O_2 \xrightarrow{1600\,°F\,(900\,°C)} 2SiO_2 + 6HCl$ |
| 7. Coating of the refractory metal tungsten (W) onto a substrate, such as a jet engine turbine blade: $WF_6 + 3H_2 \xrightarrow{1100\,°F\,(600\,°C)} W + 6HF$ |

Table 7.11 Comparison of a Typical PVD Method (Sputtering) with a CVD Method

| | PVD | CVD |
|---|---|---|
| Example processes | Sputtering | PECVD |
| Advantages | Clean process<br>Easy to control composition<br>Easy mechanism<br>Good adhesion<br>Safe process<br>High deposition rate | Good step coverage |
| Disadvantages | Poor step coverage | Expensive<br>High impurity using toxic gas hardware complexity |
| Example of deposited films | Ti, TiN, Co, W | W, TiN (MOCVD), Ti, TiN, ... $SiO_2$, Dielectrics |

are important. Table 7.10 presents some examples of CVD reactions that result in deposition of a metal or ceramic coating onto a suitable substrate.

A few words about the difference between physical vapor deposition (PVD) and chemical vapor deposition processes follow. PVD constitutes a group of direct line-of-sight impingement deposition techniques, and CVD represents a class of diffusive-convective mass transfer deposition techniques. Whereas CVD includes a wide range of pressures and temperatures and can be applied to a great variety of coating and substrate materials, PVD is mostly limited to metal film deposition. With PVD techniques we saw that sputtering leads to a better conformal coating than evaporation, but CVD is better than sputtering in this respect. Unlike CVD, which operates at elevated temperature, PVD operates at room temperature; consequently, CVD often produces better crystalline films than PVD. Advantages typically cited for CVD include that 1) it is possible to deposit refractory materials at temperatures below their melting or sintering temperatures, 2) control of grain size is possible, 3) the process can be carried out at atmospheric pressure (it does not require vacuum equipment); and 4) there is good bonding of coating to substrate surface.

Disadvantages include that 1) the corrosive and/or toxic nature of chemicals generally necessitates a closed chamber and special pumping and disposal equipment, 2) certain reaction ingredients are relatively expensive, and 3) material utilization is low. Many gases used in CVD systems are toxic (hazardous to humans), corrosive (causes corrosion to stainless steel and other metals), flammable (burns when exposed to an ignition source and an oxygen source), and explosive and/or pyrophoric (spontaneously burn or explode in air, moisture, or when exposed to oxygen). In Table 7.11 we compare a specific PVD technique (sputtering) with a specific CVD technique, i.e., plasma-enhanced chemical vapor deposition (PECVD). With CVD, layer thicknesses range from 1 monolayer to 100 μm.

## Reaction Mechanisms and CVD Reactor

Figure 7.26 illustrates the various transport and reaction processes underlying CVD schematically[23]:

- Mass transport of reactant and diluent gases (if present) in the bulk gas flow region from the reactor inlet to the deposition zone (1)
- Gas-phase reactions (homogeneous) leading to film precursors and byproducts (often unselective and undesirable) (2)
- Mass transport of film precursors and reactants to the growth surface (3)
- Adsorption of film precursors and reactants on the growth surface (4)
- Surface migration of film formers to the growth sites (5)
- Surface reactions (heterogeneous) of adatoms occurring selectively on the heated surface, and incorporation of film constituents into the growing film, that is, nucleation (island formation) (6)
- Desorption of byproducts of the surface reactions (7)
- Mass transport of byproducts in the bulk gas flow region away from the deposition zone toward the reactor exit (8)

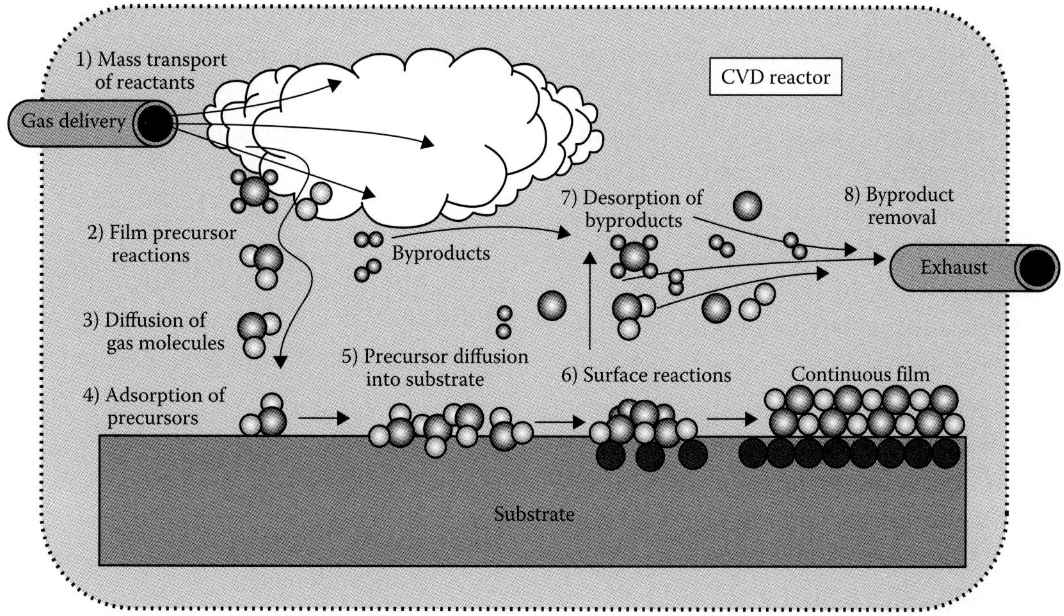


**FIGURE 7.26** Schematic of transport and reaction processes underlying CVD. (Based on Jensen, K. F. 1989. *Microelectronics processing: chemical engineering aspects.* Eds. D. W. Hess, and K. F. Jensen. Washington, DC: American Chemical Society.[23])

CVD processes are carried out in a reactor as shown in Figure 7.27, which consists of 1) reactant supply system, 2) deposition chamber, and 3) recycle/disposal system. The purpose of the reactant supply system is to deliver reactants to the deposition chamber in the proper proportions. The deposition chamber contains the substrates and the chemical reactions that lead to the deposition of reaction products onto the substrate surfaces. Deposition occurs at elevated temperatures, and the substrate may be heated by induction heating, radiant heat, or other means. Deposition temperatures for different CVD reactions can range from around 250–1950°C; therefore, the chamber must be designed to meet these temperature demands. The recycle/disposal system is to render harmless the byproducts of the CVD reaction.

In the case of a thermally driven CVD reaction, a temperature gradient is imposed on the reactor; the gas-phase species (e.g., $SiH_4$) forms in a hot region; and the equilibrium shifts toward the desired solid (e.g., Si) in a slightly colder region. Either the gas phase or the surface processes can determine rate. Thermal energy is the sole driving force in high-temperature CVD reactors; for lower-temperature deposition, an additional energy source is needed. A radiofrequency (RF) plasma, photoradiation, or laser radiation can be used to enhance the process,

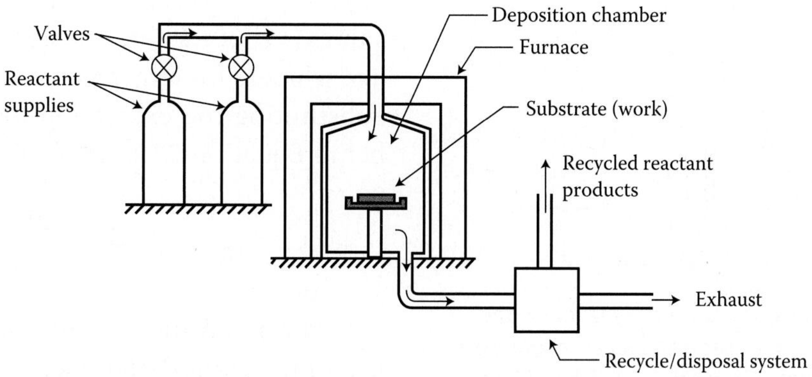


**FIGURE 7.27** Schematic of a CVD reactor.

known as *plasma-enhanced chemical vapor deposition* (PECVD), photon-assisted CVD,[24] and laser-assisted CVD (LCVD),[25] respectively.

With photon- and laser-assisted CVD systems, part of the energy needed for deposition is provided by photons. This method fills the need for lower-temperature deposition processes. With a laser source, it is possible to write a pattern on a surface directly by scanning the micrometer-size light beam over the substrate in the presence of the suitable reactive gases. By adjusting the focal point of the laser continuously, it is even possible to grow 3D microstructures such as fibers and springs in a wide variety of materials such as boron, carbon, tungsten, silicon, SiC, $Si_3N_4$, and so on. Laser CVD is treated as a thermal additive technique in Chapter 9.[25]

In PECVD, which also encompasses high-density plasma CVD (HDPCVD), plasma activation provides the radicals that result in the deposited films, and ion bombardment of the substrate provides the energy required to arrive at the stable desired end products. Operational temperatures are lower because part of the activation energy needed for deposition now comes from the plasma.

Because CVD processes involve carrier gas flow over the substrate's surface, the issue of the "fluid/solid interface" must be dealt with in assessing the effectiveness of the CVD process. Whenever a fluid is flowing over a solid surface, a "boundary layer" is created, as shown in Figure 7.28. The boundary layer acts as a barrier to transfer of heat or transport of medium. Transport in the gas phase takes place through diffusion proportional to the diffusivity of the gas, *D*, and

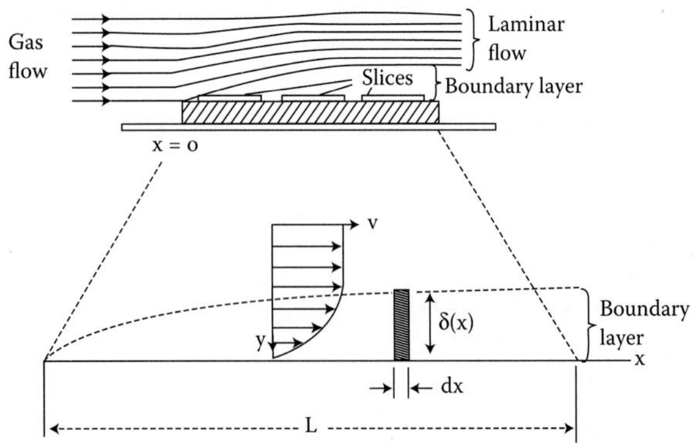


**FIGURE 7.28** Development of a boundary layer in gas flowing over a flat plate. The figure shows an expanded view of the boundary layer. (Granger, R. A. 1995. *Fluid mechanics*. New York: Dover Publications.[26])

the concentration gradient across the boundary layer that separates the bulk flow (source) and substrate surface (sink). The flux of depositing material is given by Volume I, Equation 4.20 (Fick's first law):

$$J = -D\frac{\delta N(x,t)}{\delta x}$$

The boundary layer thickness, $\delta(x)$, as a function of distance along the substrate, *x* (see Figure 7.28), can be calculated from (x = 0 is the leading edge):

$$\delta(x) = \left(\frac{\eta x}{\rho V}\right)^{\frac{1}{2}} \quad (7.11)$$

where $\eta$ = gas viscosity
$\rho$ = gas density
V = gas stream velocity parallel to the substrate

The average boundary layer thickness, in the boundary layer model from Prandtl,[26] can then be calculated over the whole plate as follows:

$$\delta = \frac{1}{L}\int_0^L \delta(x)dx = \frac{2}{3}L\left(\frac{\eta}{\rho VL}\right)^{\frac{1}{2}} \quad (7.12)$$

where *L* stands for the length of the plate receiving the deposit.

The dimensionless Reynolds number, *Re*, is a most important figure in fluid dynamics characterizing the nature of a fluid flow (gas or liquid) (see Volume I, Chapter 6). It is given by:

$$Re = \frac{\rho VL}{\eta} \quad \text{(Volume I, Equation 6.71)}$$

It equals the ratio of the magnitude of inertial effects to viscous effects in fluid motion. For low values (<2000–3000), the flow regime is called *laminar*, whereas for larger values the regime is *turbulent*. Substituting the expression for the Reynolds number in Equation 7.12, we obtain:

$$\delta = \frac{2L}{3\sqrt{Re}} \quad (7.13)$$

Control of $\delta$ and its uniformity over the substrate is an important design factor. In a horizontal CVD reactor (see Figure 7.29a) the substrate is often tilted

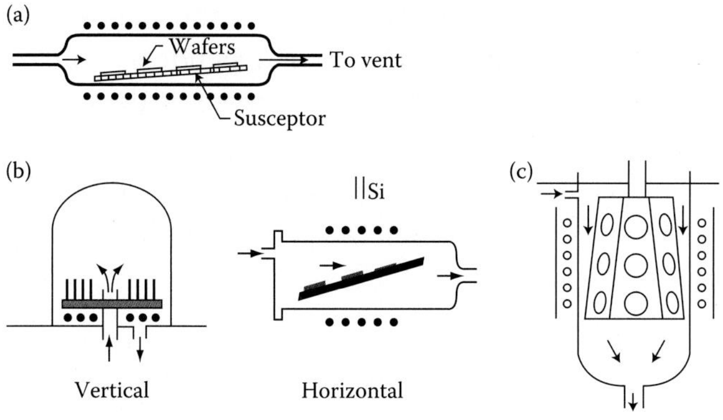


**FIGURE 7.29** CVD reactor configurations: horizontal (a), vertical reactor (b), and barrel CVD reactor (c).

1–3° for Si and 5–15° for GaAs. The susceptor is tilted so that the cross-sectional area of the chamber is decreased, increasing the gas velocity along the susceptor. This compensates for both the boundary layer and depletion effects.

In a vertical reactor (see Figure 7.29b and c) one may use a baffle to deflect the gas from the center of the susceptor in addition to substrate rotation, again leading to a more uniform δ. The stability of the flow in a CVD reactor is crucial in achieving uniform deposition. The criterion for flow stability depends on whether the flow is fully developed (i.e., laminar) before it reaches the susceptor. Gas flows in containment structures (internal flows), e.g., a cylindrical tube CVD reactor, differ from external flows in that there is a boundary layer and a uniform free stream in the entry region of an internal flow that accelerates according to the rate of growth of the boundary layer as shown in Figure 7.30. The velocity is uniform at the entrance, and the boundary layer grows with distance from the entrance until the flow becomes fully developed at $X_L$. From the continuity equation (see Volume I, Chapter 6) it follows that the frictionless core must accelerate, whereas from applying the Bernoulli equation (see Volume I, Chapter 6) to a streamline in the free stream region, we learn that the pressure must decrease. At $X_L$ the velocity varies over the entire channel, and there is no free stream anymore. The laminar development length $X_L$ for the flow to become fully developed can be empirically determined as:

$$X_L = 0.056\,\mathrm{Re}\,D \tag{7.14}$$

where $D$ is the diameter of the tube. The thermal entrance length $X_T$ is the length beyond which the dimensionless temperature profile is constant with $x$ and before which it varies with $x$ (i.e., is developing). For laminar flows in a tube, the thermal entrance length is a function of the Reynolds number and the Prandtl number, and for uniform surface temperature it is given as:

$$X_T = 0.033\,\mathrm{Pr}\,\mathrm{Re}\,D \tag{7.15}$$

where the Prandtl number is defined as $\mathrm{Pr} = \nu/\lambda_d$ (see Volume I, Equation 6.61), i.e., the ratio of kinematic viscosity and thermal diffusivity. The interpretation of this number is the ratio of the thickness of the fluid velocity boundary layer to the fluid temperature boundary layer (for water $Pr$ is 10, and for air it is 0.71). Both velocity and thermal entrance length are considerably longer when the flow is turbulent.

By substituting the value for the average boundary layer thickness from Equation 7.12 in Volume I, Equation 4.20 (i.e. Fick's first law with dx = δ), the following expression for the material flux to the substrate results:

$$J = -D\frac{\Delta N}{2L}3\sqrt{\mathrm{Re}} \tag{7.16}$$

According to Equation 7.16, the film growth rate in the mass flow-controlled regime should depend on the square root of the gas velocity, $V$ (because the Reynolds number is proportional to gas velocity $V$).

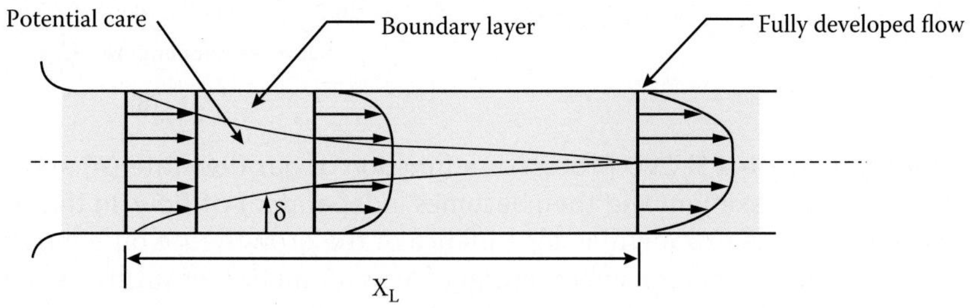


**FIGURE 7.30** Flow in the entry region of a pipe for laminar flow.

**Table 7.12** Mass Transport: Fundamental and Experimental Parameters

| Fundamental Parameters | Experimental Parameters |
|---|---|
| Reactant concentration | Pressure |
| Diffusivity | Gas velocity |
| Boundary layer thickness | Temperature distribution |
| | Reactor geometry |
| | Gas properties (viscosity...) |

Because gas-phase mass transport (see Table 7.12) is not an activated process, it is not strongly temperature-dependent. In Figure 7.31A we show a plot of silicon growth rate (in micrometers per minute) as a function of the gas flow rate (in liters per minute and proportional to the gas velocity in a fixed volume reaction chamber) with a substrate temperature of 1270°C, and the predicted square root dependence is clearly observed over a wide range of flow rates.[27]

At high flow rates, the growth rate reaches a maximum and then becomes independent of flow. In this regime, the reaction rate controls the deposition, evidenced by the exponential dependence of the growth rate on temperature observed at those flow rates (see Figure 7.31Ba).[27]

Surface reactions can be modeled by a thermally activated phenomenon proceeding at a rate, *k*, given by:

$$k = k_0 e^{\frac{E_a}{kT}} \tag{7.17}$$

where $k_0$ = frequency factor
$E_a$ = activation energy in electronvolts

In Figure 7.31Bb, the growth rate (in micrometers per minute) of Si as a function of substrate temperature and a variety of precursor gases is illustrated in an APCVD reactor. From the slope of an Arrhenius plot, as demonstrated here, the activation energy of the rate-determining surface process can be deduced (for most surface reactions, $E_a$ ranges from 1–4 eV).

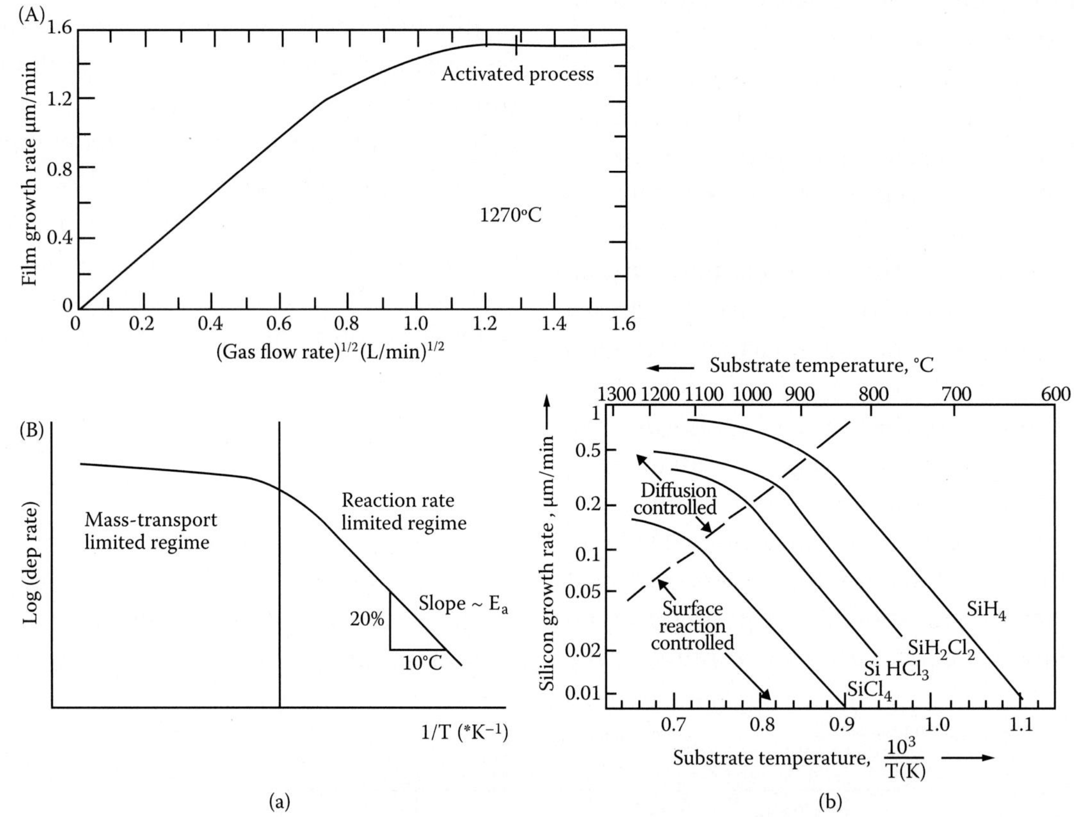


**FIGURE 7.31** Growth rate dependence in a Si CVD process as a function of gas flow rate (A) and temperature (B). At high flow rates, the growth rate reaches a maximum and then becomes independent of flow. In this regime, the reaction rate controls the deposition, evidenced by the exponential dependence of the growth rate on temperature observed at those flow rates (see a). The growth rate (in micrometers per minute) of Si as a function of substrate temperature and a variety of precursor gases is illustrated in an APCVD reactor (see b). (Bb) Various precursor gases are used. (Wolf, S., and R. N. Tauber. 1987. *Silicon processing for the VLSI era.* Sunset Beach, CA: Lattice Press.[27])

We want the deposition reaction to occur at the gas/wafer boundary (heterogenous reaction). However, the higher temperatures (650–800°C for silane) used to enhance surface migration also result in gas-phase decomposition of silane, resulting in particles (homogeneous reaction). This is undesirable because particles tend to fall onto the wafers. Higher-quality (fewer defects) material is obtained at temperatures higher than 800°C. To be able to operate at higher temperatures, instead of silane, chlorosilanes that are stable up to higher temperatures are commonly used. Dichlorosilane ($SiH_2Cl_2$) is the most common and allows growth at ~800–1050°C. For a certain rate-limiting reaction, the temperature may increase high enough for the reaction rate to exceed the rate at which reactant species arrive at the surface. In such a case, the reaction rate cannot proceed any faster than the rate at which the reactant gases are supplied to the substrate by mass transport, no matter how high the temperature is increased (see plateau in Figure 7.31Ba and b). This situation is referred to as a mass transport-limited deposition process. Temperature is less important in this regime than it is in the reaction rate-limited (Arrhenius) one. In the latter case, the arrival rate of reactants is less important because their concentration does not limit the growth rate.

A direct practical application of these two possible rate-limiting processes is the way substrates are stacked in low-pressure CVD (LPCVD) (<10 Pa) versus atmospheric pressure to slightly reduced pressure CVD (APCVD) (±100–10 kPa) reactors. In an LPCVD reactor (e.g., at ~1 Torr), the diffusivity of the gas species is increased by a factor of 1000 over that at atmospheric pressure, resulting in a one order of magnitude increase in the transport of reactants to the substrate. The rate-limiting step becomes the surface reaction. LPCVD reactors enable wafers to be stacked vertically at very close spacings as the rate of arrivals of reactants is less important (Figure 7.32). On the other hand, APCVD, operating in the mass transport-limited regime, must be designed such that all locations on the wafer and all wafers are supplied with an equal flux of reactant species. In this case, the wafers often are placed horizontally or at a slight incline (Figure 7.29a). In APCVD, reactor walls typically are cooled in a so-called *cold wall reactor type* (where only the susceptor holding the substrates is heated) to minimize particulate and impurity problems caused by deposition on them. Horizontal tube, hot wall reactors are the most widely used LPCVD reactors. In this case, not only the wafers but also the reaction chamber walls get coated. Such systems require frequent cleaning to avoid serious particulate contamination. They find wide application because of their economy, throughput, and uniformity.

**FIGURE 7.32** Vertical stacking of substrates in LPCVD reactors.

The Reynolds number in a typical LPCVD reactor is smaller than that for APCVD. Consequently, the thermal entrance length for APCVD is longer than that for LPCVD reactors (see Equation 7.15).

## Step Coverage

Thin deposited films grow more readily on horizontal, exposed surfaces than within trenches or holes, resulting in poor coverage of the surface with the film. The gaps that are formed can lead to shorts, reliability problems, etching failures, and other generally undesirable phenomena. Poor step coverage results from different deposition rates in different parts of a microstructure. The profile of the thin films deposited by any CVD or PVD techniques depends on:

1. Equipment configuration
2. Deposition method (e.g., LPCVD, PECVD, PVD)
3. Reaction chemistry
4. Reactant transport mechanism

TABLE 7.13 Mean Free Path in Various Gases at 20°C

| Gas | $P_T$ (Torr) λ (m) (cm·Torr) at 20°C | |
|---|---|---|
| $O_2$ | $4.9 \times 10^{-3}$ | Atmospheric pressure (760 Torr) ~40 nm (smaller than most chip features) |
| $N_2$ | $4.6 \times 10^{-3}$ | |
| $H_2$ | $9 \times 10^{-3}$ | Low vacuum (0.76 Torr) ~40 μm (larger than most chip features) |
| He | $13.6 \times 10^{-3}$ | |
| Ar | $4.8 \times 10^{-3}$ | Medium vacuum (7.6 Torr) ~4.0 mm (larger than working gap) |
| Ne | $9.3 \times 10^{-3}$ | |
| Xe | $2.7 \times 10^{-3}$ | High vacuum (7.6 μTorr) ~4.0 m (larger than chamber) |
| $CO_2$ | $3.0 \times 10^{-3}$ | |
| $H_2O$ | $3.0 \times 10^{-3}$ | Evaporated material travels straight to wafer (line of sight; thus, is very directional) |
| Cl | $2.3 \times 10^{-3}$ | |
| $NH_3$ | $3.5 \times 10^{-3}$ | |

Here we emphasize step coverage in CVD processes. The mean free path of a molecule, λ, based on a slightly modified Equation 7.6, is given by:

$$\lambda = \frac{kT}{2^{\frac{1}{2}} P_T \pi a^2} \tag{7.18}$$

where $a$ is the molecular diameter.[28] If we assume a = 3 Å and T = 300 K, then $P_T(\text{torr})\lambda(\text{m}) = 7.810^{-5}$. Therefore, if λ = 30 cm, then a pressure of $2.6 \times 10^{-4}$ Torr is required (see also Table 7.13).

Equation 7.18 is of crucial importance in understanding CVD coating of micromachined features. CVD films may electrically isolate underlying features from subsequent layers or protect them against the atmosphere, determined by the degree at which edges and pits of the underlying features can be covered uniformly. As demonstrated in Figure 7.33A, three cases can be distinguished. Ideally, a uniform, dense, and conformal coating should form (Figure 7.33Aa). This can occur in instances where the reactants, after first hitting the solid, have enough energy left for surface migration before a bond is established with the underlying substrate. Coatings in which equal film thickness exists over all substrate topography, regardless of its slope, are called *conformal* (see also Figure 7.5). In a second case, as shown in Figure 7.33Ab, the mean free path of the molecules is large enough to reach the bottom of the trench, but little energy remains for surface migration. Finally, in the third case, shown in Figure 7.33Ac, the mean free path length is too short to reach the bottom, and there is little surface migration (temperature too low). Whereas sputtering is better than evaporation, for the best conformal coating CVD is preferred. Evaporated and sputtered metal films often have trench profiles as shown in the Figure 7.33Ab, whereas CVD (especially LPCVD) deposited films are often more uniform and conformal, as demonstrated in Figure 7.33Aa.

The value of the integral of the material flux in Equation 7.16 (∫Jdθ), and thus the CVD film thickness, is directly proportional to the range of feasible angles of arrival, θ, of the depositing species (in the absence of surface migration). Different

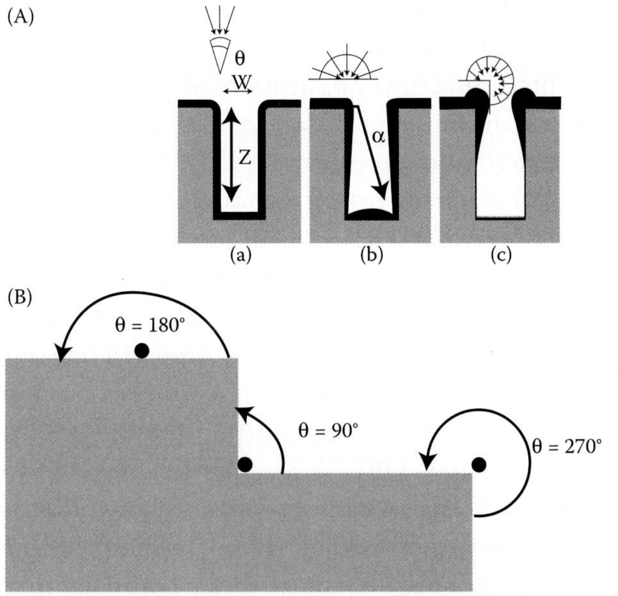


**FIGURE 7.33** Step coverage cases of deposited film. (Aa) Uniform coverage resulting from rapid surface migration. (Ab) Nonconformal step coverage for long mean free path and no surface migration. Distance a is the longest path a molecule travels to reach the corner of the trench. (Ac) Nonconformal step coverage for short mean free path and no surface migration. (B) Different arrival angles in two dimensions.

arrival angles in two dimensions are illustrated in Figure 7.33B. The arrival angle at a planar surface is 180°. At the top of a vertical step, the arrival rate is nonzero over a range of 270°; the resultant film thickness is 270/180, or 1.5 times greater than that for the planar case. At the bottom corner of a trench, the arrival angle is only 90°, and the film thickness is 90/180 or one-half that of the planar case. The CVD profile in Figure 7.33Ac, where the mean free path is short compared with the trench dimensions and there is no surface migration, reflects the 180°, 90°, and 270° arrival angles. The thick cusp at the top of the step and the thin crevice at the bottom combine to give a re-entrant shape that is particularly difficult to cover with evaporated or sputtered metal. Gas depletion effects also are observed along the trench walls. Along the vertical walls, the arrival angle, θ, is determined by the width of the opening, and the distance from the top and can be calculated from:

$$\theta = \arctan \frac{w}{z} \tag{7.19}$$

where $w$ is the width of the opening and $z$ the distance from the bottom of the trench to the top surface (Figure 7.33Aa). This type of step coverage, as shown in Figure 7.33Ab, thins along the vertical walls and may have a crack at the bottom of the step. For uniformity of deposition, in the case of Figure 7.33Ab, where the mean free path is longer than the distance α (the longest path a molecule must travel to reach the corner of the trench), the rate of surface migration of adspecies should exceed the rate of adsorption of adspecies. The condition of $\lambda > \alpha$ can be met by working at low pressures, based on Equation 7.18:

$$\frac{kT}{2^{\frac{1}{2}} P_T \pi a^2} > \alpha \tag{7.20}$$

For a given gas, Equation 7.20 gives us the pressures and temperatures at which $\lambda > \alpha$. The most important variable influencing the requirement for large surface diffusion is the reactor temperature. The mobility of surface species is largest on metallic and semiconducting surfaces, where bonding is not very directional, and rather limited on dielectrics, where highly directional covalent bonds tend to hold a molecule in one place once it is chemisorbed.

There is another factor influencing conformality, namely, the reactive stiction coefficient, $S_c$, which might be even more important than surface diffusion. To appreciate the importance of this parameter we need to reintroduce some simple facts about the mean free path in various deposition systems. Remember that the mean free path for APCVD is 40 nm at 760 Torr (see Table 7.13), smaller than most features on an IC or a micromachine. However, the mean free path is 10 μm in typical working conditions for LPCVD and PECVD, and for sputtering it is a few centimeters; therefore, in this case one can ignore collisions between the gas particles when they are inside a via or a trench-like structure. The only collisions are those with the surface of the wafer. The particular flow characteristics around an IC, MEMS, or NEMS feature on a substrate are given by the value of the Knudsen number ($Kn$), defined in this case as the ratio of the mean free path of the molecules (λ) to the characteristic dimension of the structure to be coated (e.g., the width $D$ of the structure) (see also Equation 7.5). If the Knudsen number is greater than 1, the flow is free molecular. If, on the other hand, the Knudsen number is $<0.01$, the flow is viscous, that is, the atom or molecule bounces several times into the feature's sidewalls. A transition region exists between the two. The flow around micromachined features is typically in this transition regime ($0.1 < Kn = \lambda/w < 10$) or in the free molecular regime ($Kn > 10$).

The transport of reactants to the surface at low pressures is sketched in Figure 7.34. The three processes shown here are 1) direct deposition, 2) re-emission, and 3) surface diffusion. In most applications, it has been found that the surface diffusion (3) is less important than re-emission (2), which is controlled

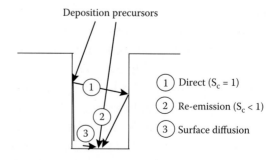

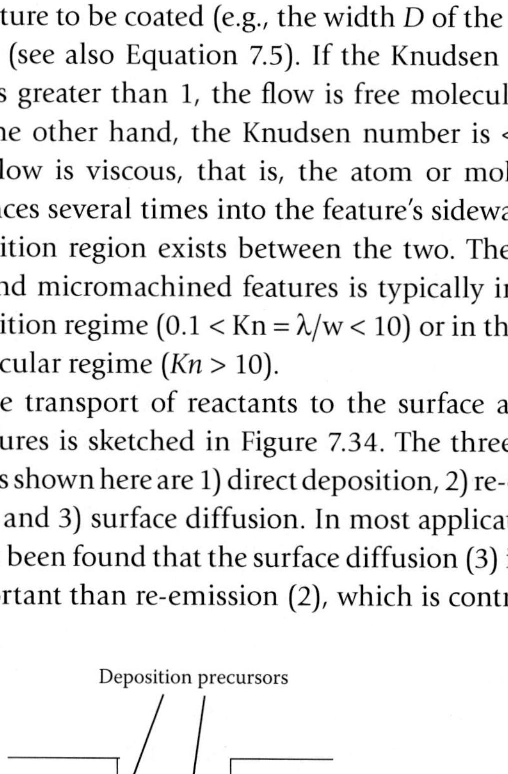


**FIGURE 7.34** The transport of reactants to the surface at low pressures.

by the reactive stiction coefficient, $S_c$. The stiction coefficient is the ratio of incident precursor species minus re-emitted ones, i.e., all the species that remain on the surface react, over the total of incident precursor species:

$$S_c = \frac{\Gamma_{incident} - \Gamma_{re\text{-}emission}}{\Gamma_{incident}} = \frac{\Gamma_{reaction}}{\Gamma_{incident}} \quad (7.21)$$

The reactive sticking coefficient (RSC) $S_c$ depends on the nature of the incoming species (molecule, ion, neutral, radical, size of molecule) and on the surface conditions. A radical or other unstable precursor molecule is likely to react readily with a surface site and stick wherever it strikes ($S_c = 1$). A more stable precursor may not react readily, particularly at higher temperatures or when the surface is already saturated with adsorbents ($S_c < 1$). Particularly in dielectric deposition, it has been shown that an important mechanism for transport of the material into gaps and holes is through multiple adsorption and re-emission from the surface: that is, from precursors having a low RSC. When such "reflections" occur many times, the flux of precursor to the surface inside a hole may differ very little from the flux onto the horizontal surfaces. In other words, the likelihood of reaction is important because successive adsorption and re-emission is one of the key mechanisms determining the effect of topography of a surface on the thickness of a film deposited on that surface, and is one of the key advantages of CVD over evaporation or sputtering of thin films. Therefore, for a direct deposition (1) $S_c = 1$ and with any degree of re-emission (2) $S_c < 1$. Shadowing of the direct flux by the walls reduces the flux at the bottom corners of a via or a trench, resulting in thinner deposition. Lower reaction stiction coefficients increase the number of bounces in the trench or via and hence increase the probability of deposition precursors reaching the bottom. Thus, step coverage is better for lower $S_c$ (see Figure 7.35).

To illustrate the effect of a lower $S_c$, imagine that we have a couple of metal lines (e.g., aluminum) on a substrate that are being used as wires, and that we need to cover them up with an insulating dielectric so that a subsequent layer of metal can be deposited without being shunted out. When leaving voids

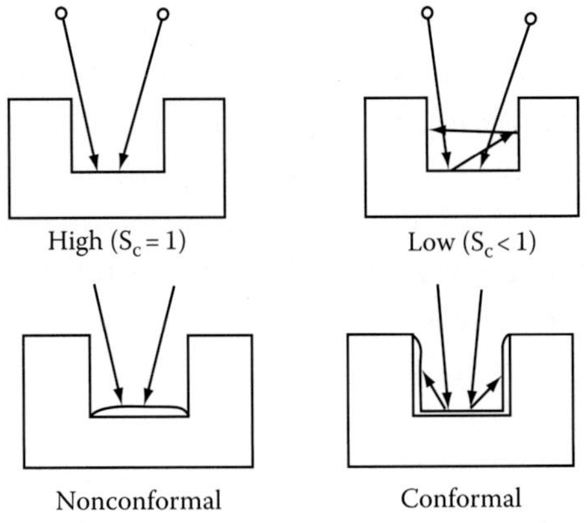


**FIGURE 7.35** A low $S_c$ can produce more conformal coverage because of redeposition (usually more important than surface diffusion)

in a fill they trap chemicals that may lead to cracks (dielectric) or large contact resistance and sheet resistance (metallization). In Figure 7.36 we show how a decreasing $S_c$ fills the gap between the two metal lines better with the dielectric material and that at a value of $S_c = 0.007$ the gap is filled completely. The simulation model used here is SPEEDIE,

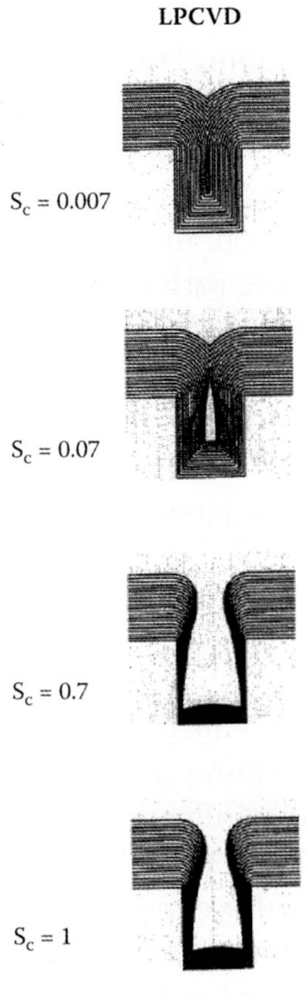


**FIGURE 7.36** Effect of varying $S_c$ on filling a gap between two metal lines. The simulation model used here is SPEEDIE, the Stanford etching and deposition profile simulator.

the Stanford etching and deposition profile simulator. Scaling of interconnects in future ICs will require more metal layers with reduced interconnect pitch. The height of the metals also has to be increased. The larger aspect ratio between the metal lines will increase dramatically the problem of step coverage. For a small gap or opening in a film, the aspect ratio is the ratio of its depth to width. Filling high-aspect-ratio gaps (>3:1) is critical for advanced wafer fabrication. It is anticipated that the thickness of deposited films in the IC industry will reach the order of 10 nm rather than a few micrometers, and thickness uniformity will be harder to maintain.

Although LPCVD reactors have lower deposition rates than APCVD reactors, it is found that this is more than compensated for by the high wafer capacity and coating uniformity. The future trend is toward single-wafer processing systems instead of the batch processes. This is because of the increasing wafer sizes (8 in. or more), and the development in technology to achieve better film uniformity in modern single-wafer reactors. The tendency is toward large single wafers that are transferred from station to station in so-called *cluster tools*.

Besides SPEEDIE, simulation models for nonplanar CVD were presented, for example, by Coronell et al.[29] The latter model again provides a picture of the evolution of the depositing film profile. The parameters investigated include the sticking coefficient, the surface mobility of the adsorbed reactants, the Knudsen number (the ratio of the mean free path to the feature size), the feature aspect ratio, and feature geometry.

## Overview of CVD Process Types

In Table 7.14[23,30] we review some important CVD processes, listing applications and operational pressures and temperatures for the different types.

### *Atmospheric Pressure CVD*

Atmospheric pressure to slightly reduced pressure CVD (±100–10 kPa) is used primarily to grow epitaxial (i.e., single-crystalline) films of Si and to deposit, at high rates, $SiO_2$, for example, from the reaction of $SiH_4$ and oxygen, at low temperatures of 300–450°C [low-temperature oxide (LTO)].[23] The epitaxy processes [also vapor phase epitaxy (VPE)] involve high growth temperatures (>850°C for Si). As we saw above, an APCVD reactor operates in the mass-transport regime so that wafer access, in contrast to the LPCVD process, becomes more important and temperature control less so. APCVD is susceptible to gas-phase reactions, and step coverage is often poor. High gas dilutions help avoid gas-phase nucleation. APCVD, reactor walls typically are cooled in a so-called *cold wall reactor type* (where only the susceptor holding the substrates is heated) to minimize particulate and impurity problems caused by deposition on them. In the early days of semiconductor processing, atmospheric pressure reactors were used mostly to deposit silicon and dielectric films. The advantage of deposition at atmospheric pressure is the simplicity of the reactor design and the high deposition rates (700 Å/min for $SiO_2$). The wafers are put on a hot susceptor, and reactant gases flow over the susceptor. The susceptor is heated by using high-intensity lamps, radiofrequency (RF) induction, or DC electric current (resistive) heating. These atmospheric pressure reactors have low throughput, require excessive wafer handling during wafer loading and unloading, and have poor thickness uniformity (>10%) across the wafer. To overcome some of these disadvantages, continuous throughput atmospheric pressure reactors were developed as shown in Figure 7.37. In these reactors, the wafers are carried through the reactor on a conveyor belt and are heated by convection. Also, there is a more uniform flow of the reactant gases across the wafer surface. These reactors have high throughput, good uniformity, and the ability to handle large-diameter wafers. The main disadvantages of this type of reactor are that they required frequent cleaning and that particles formed on the reactant dispenser head wind up on the wafers, which impaired process yield and device reliability.

**Example 7.2: Epitaxial Si**

As a first example of an APCVD application, consider silicon epitaxy at 1200°C (see also Figure 7.31B):

$$SiCl_4\,(gas) + 2H_2 \rightarrow Si\,(solid) + 4HCl\,(gas)$$

Reaction 7.2

TABLE 7.14 Review of CVD Processes

| Process | Advantages | Disadvantages | Applications | Remark | Pressure/ Temperature |
|---|---|---|---|---|---|
| APCVD | Simple, high deposition rate (700 Å/min for $SiO_2$), low temperature | Poorer step coverage, particle contamination | Doped and undoped, low temperature, thick oxides (mainly dielectrics) | Mass transport-controlled | 100–10 kPa 250–450°C |
| LPCVD | Excellent purity and uniformity, conformable step coverage, large wafer capacity | High temperature and lower deposition rate | Doped and undoped, high-temperature oxides, silicon nitride, polysilicon, W, $WSi_2$ | Surface reaction-controlled | 100 Pa 550–650°C |
| VLPCVD | | | Single-crystalline Si and compound semiconductor superlattices | Surface reaction-controlled | 1.3 Pa |
| MOCVD | Excellent for epi on large surface areas | Safety concerns (highly toxic), very expensive source material | Compound semiconductors for solar cells, laser, photocathodes, LEDs, HEMTs, and quantum wells; also W and Cu | High volume, large surface area production | |
| PECVD | Lower substrate temperatures, fast, good adhesion, good step coverage, low pinhole density | Chemical (e.g., hydrogen) and particulate contamination, plasma damage | Low-temperature insulators over metals, passivation (nitride) | Tends to have more pinholes than LPCVD | 2–5 Torr (pressure higher than in sputter deposition, more gas-phase collisions, less ion bombardment on surface) more collision in gas phase, less ion bombardment on substrate –200–400°C |
| Atomic layer deposition (ALD) | ALD is the deposition method with the greatest potential for producing very thin, conformal films with control of the thickness and composition of the films possible at the atomic level. | A major limitation of ALD is its slowness; usually only a fraction of a monolayer is deposited in one cycle. | A major driving force for the interest in ALD is for scaling down microelectronic devices. | Many technologically important materials such as Si and Ge, several multi-component oxides and certain metals cannot currently be deposited by ALD in a cost-effective way. | RT to 400°C |
| Spray pyrolysis | Inexpensive | Difficult to control, not compatible with IC | Gas sensors, solar cells, ITO, large area | | Atmospheric 100–180°C |

*Source:* Hess, D. W., and K. F. Jensen, eds. 1989. *Microelectronics processing: chemical engineering aspects.* Washington, DC: American Chemical Society; and Sze, S. M., ed. 1988. *VLSI technology.* New York: McGraw-Hill. With permission.[31]

An intermediate species, $SiCl_2$, is necessary for silicon formation, and at less than 1000°C no $SiCl_2$ forms when starting from $SiCl_4$. Lower-temperature epitaxy can be performed by starting with $SiH_2Cl_2$ or $SiHCl_3$, which both decompose more readily to $SiCl_2$. When discussing Figure 7.31Bb, we pointed out that using pure silane (in the lowest temperature range) results in gas-phase decomposition of silane, resulting in particles (homogeneous reaction), and silane also must be avoided. Dichlorosilane ($SiH_2Cl_2$) is the most common and allows growth at ~800–1,050°C.

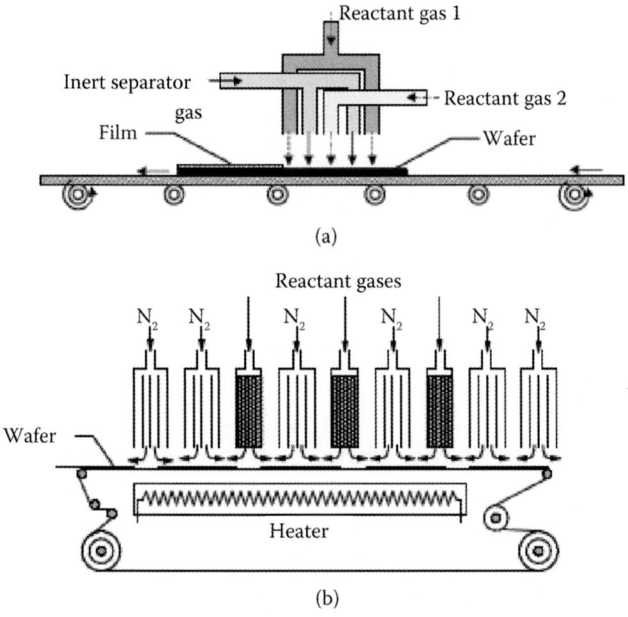


**FIGURE 7.37** Simple continuous-feed atmospheric pressure reactor (APCVD): (a) Gas-injection type; (b) plenum type.

In Figure 7.38, we show the growth rate versus temperature for that reaction; the plot delineates a CVD polysilicon deposition region and an epitaxial monocrystalline silicon deposition region.[27] Nominal growth rates for single-crystal Si varies from 0.2–1.5 μm/min, depending on the source gas and the growth temperature. Impurity atoms can be introduced in the gas stream to grow doped epitaxial layers (Figure 7.39) (e.g., arsine-doped epitaxial silicon).

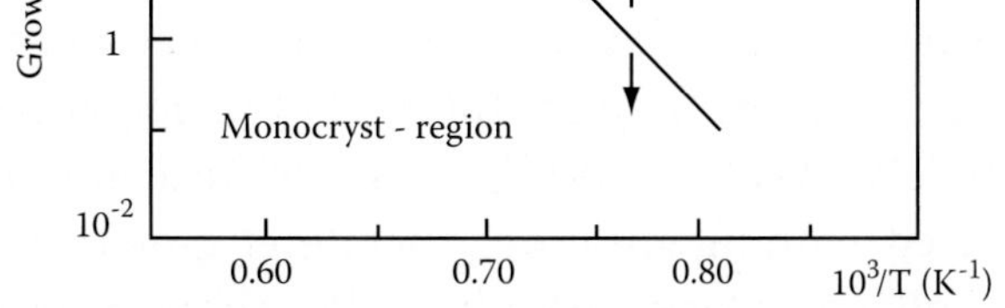

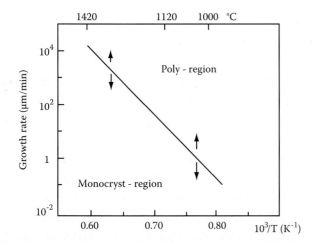


**FIGURE 7.38** Growth rate versus temperature. Epitaxy to CVD transition in CVD polysilicon deposition and epitaxial monocrystalline Si deposition. (From Wolf, S., and R. N. Tauber. 1987. *Silicon processing for the VLSI era*. Sunset Beach, CA: Lattice Press. With permission.[27])

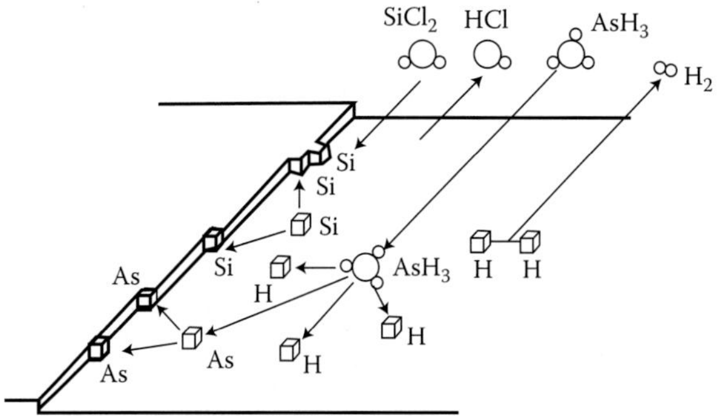


**FIGURE 7.39** Arsine-doped epitaxial silicon.

### Example 7.3: Oxide APCVD with Tetraethylorthosilicate

In low-temperature oxidation (LTO), silicon dioxide is deposited on the wafer out of the vapor phase from the reaction of silane ($SiH_4$) with oxygen at relative low temperatures (400–450°C) and low pressure (200–400 mTorr). Silane spontaneously combusts in the presence of oxygen (in other words, silane is pyrophoric), and to avoid a large depletion of the gases at the entrance of the furnace tube, the gases are introduced in the furnace through injectors along the length of the tube. The silane/oxygen gas mixture has been applied in atmospheric pressure CVD (APCVD) and plasma-enhanced CVD (PECVD) conditions. The main advantage of silane/oxygen is the low deposition temperature; the main disadvantage is the nonconformal step coverage.

Because silane is not safe, now tetraethylorthosilicate (TEOS) with oxygen or ozone is commonly used. TEOS is a liquid at room temperature, with a vapor pressure of about 1.5 Torr. TEOS slowly hydrolyzes into silicon dioxide and ethanol when in contact with ambient moisture, but its flammability and toxicity are similar to that of an alcohol. Ozone is more reactive than oxygen; hence the process can be carried out at low temperature without plasma. In general, TEOS brings about a better starting chemistry than the traditional silane-based CVD technologies. Use of TEOS oxide, to replace $SiH_4$-based oxide in spin-on-glass and photoresist planarization schemes, now has become commonplace for devices with small features. TEOS is especially suited for filling high-aspect-ratio gaps. Trench fill by APCVD of TEOS is illustrated in Figure 7.40.

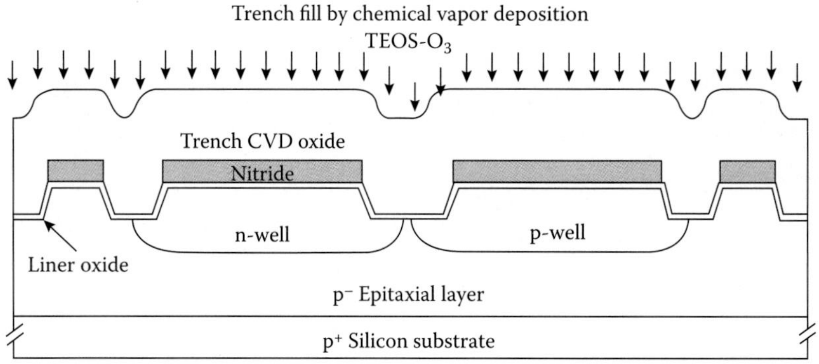


**FIGURE 7.40** Trench fill by atmospheric pressure chemical vapor deposition of TEOS.

In APCVD both undoped glass and borophosphosilicate glass have been deposited from TEOS. The process involves the reaction of ozone with TEOS at a pressure of 600 Torr and at a temperature less than 400°C:

$$Si(C_2H_5O)_4 + 8O_3 \rightarrow SiO_2 + 10H_2O + 8CO_2$$

Reaction 7.3

While offering the same good step coverage over submicrometer gaps, the films from the thermal reaction of ozone and TEOS exhibit relatively neutral stress and have a higher film density compared with low-pressure processes. This increased density gives the oxide greater moisture resistance, lower wet etch rate, and smaller thermal shrinkage. Compared with a lower-pressure process, the film density is increased from 2.09–2.15 g/cm$^3$; the wet etch rate decreased by more than 40%; and the thickness shrinkage changed from 12 to 4% after a 30-min anneal in dry $N_2$ at 1000°C.[33]

**Example 7.4: Carbon Nanotubes Grown by APCVD**

Carbon nanotubes are produced in atmospheric pressure CVD (APCVD) by catalytic disproportionation of CO or $CH_4$ on catalyst particles (e.g., Fe in the form of iron pentacarbonyl). A typical CVD setup consists of a target substrate held in a quartz tube placed inside of a furnace. Typical parameters for carbon nanotube deposition include a pressure of 1 atm and a temperature of 700–900°C. Possible substrates include silicon, mica, quartz, or alumina, and the carbon supply is either $CH_4$ or CO gas. Common catalysts are nickel, iron, or cobalt. In a typical procedure, catalyst is sputtered, layered, or specifically placed onto the substrate. A carbon-containing gas is then passed over the substrate inside the furnace, and growth is initiated on the catalyst particles. Adding 25% hydrogen increases the SWNT yield. Advantages of the APCVD method for making carbon nanotubes include increased length and purity of the carbon nanotubes, large-scale production capability, increased control, and lower temperatures.

In an intriguing development, fast-heating CVD now offers the ability to grow millimeter-long carbon nanotubes while maintaining control over orientation of the carbon nanotubes. The fast heating minimizes the growing tubes' interaction with the substrate; this is called the *kite mechanism*. In this process, a catalyst is applied by photolithography to a silicon wafer. Using this process, Huang et al.[34] grew a square 2D network of SWNTs guided by flow of carrier gas (Fe/Mo catalyst and $CO_2$ gas) (Figure 7.41).

### *Low-Pressure CVD*

Low-pressure CVD (LPCVD) at less than 10 Pa allows large numbers of wafers to be coated simultaneously without detrimental effects to film uniformity. This is the result of the large diffusion coefficient at low pressures leading to a growth limited by the rate of surface reactions rather than by the rate of mass transfer to the substrate. The surface reaction rate is very sensitive to temperature, but temperature is relatively easy to control. Typically, reactants can be used without dilution; therefore, growth rates are only an order of magnitude less than operation at atmospheric pressure allows. LPCVD in some cases can overcome the uniformity, step coverage, and particulate contamination limitations of early APCVD systems. One of the difficulties in the cold

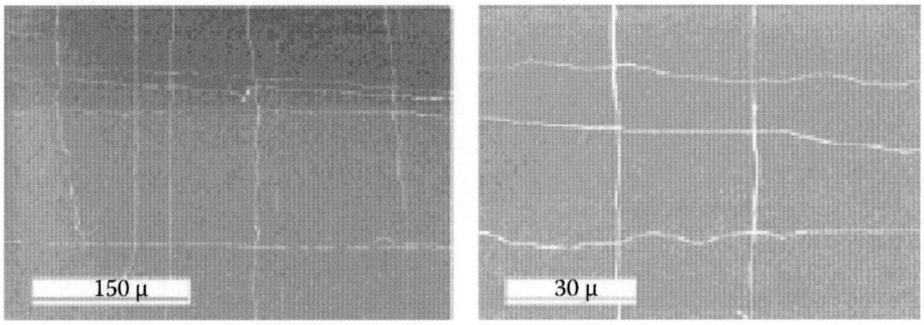


**FIGURE 7.41** Square 2D network of SWNTs guided by flow of carrier gas (Fe/Mo catalyst and $CO_2$ gas). (Huang, S., X. Cai, and J. Liu. 2003. Growth of millimeter-long and horizontally aligned single-walled carbon nanotubes on flat substrates. *J Am Chem Soc* 125:5636–37.[34])

wall systems (where only the susceptor holding the wafer is heated) described earlier is maintaining a uniform temperature across the wafer surface. This can be overcome by putting the entire reaction chamber in a furnace maintained at a uniform temperature, as in the case of standard resistance-heated hot wall tubular furnaces. A large number of wafers can be stacked closely in these furnaces, and good uniformity across the wafer and wafer-to-wafer film thickness can be achieved if they are operated at very low pressures and in the surface reaction-limited regime. Figure 7.42 gives the schematics of a hot-wall, low-pressure CVD reactor used to deposit thin films of materials such as polycrystalline silicon, silicon dioxide, and silicon nitride. The reactor consists of a quartz tube heated by a resistance-heated furnace to maintain a uniform temperature along the reactor. Gases are introduced in one end and pumped out from the other end of the reactor. The operating pressures range from 0.25–2 Torr; temperatures range from 300–900°C; and gas flow rates range from 100–1000 sccm. A large number of wafers (~100) are stacked vertically, perpendicular to the gas flow, in a quartz holder. The inlet gases may undergo homogeneous gas-phase reactions to produce the deposition precursors, which are transported to the wafer surface by gas-phase diffusion. Excellent film uniformities (better than 5%) are obtained in these reactors. Although these reactors have lower deposition rates, it is found that this is more than compensated for by the high wafer capacity. In a hot-wall LPCVD reactor, not only the wafers get coated but also the reaction chamber walls get coated. Such systems require frequent cleaning to avoid serious particulate contamination (in situ cleaning using $NF_3$ or $ClF_3$). They find wide application because of their economy, throughput, and uniformity.

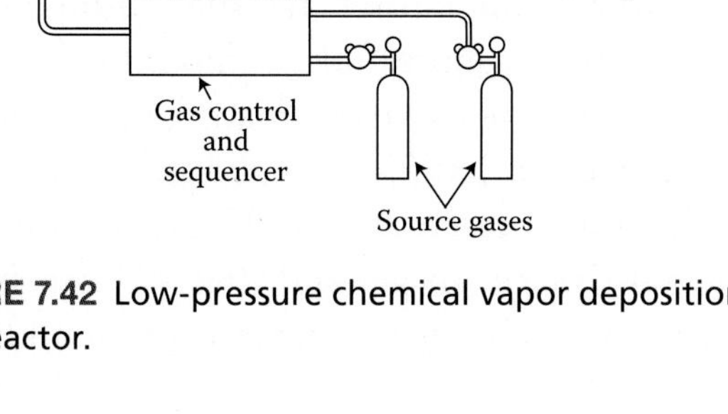

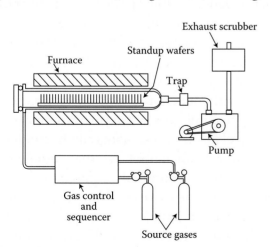


**FIGURE 7.42** Low-pressure chemical vapor deposition hot wall reactor.

LPCVD polysilicon is used for structural layers in surface micromachines, and LPCVD $SiO_2$ and phosphosilicate glass (PSG) are used as sacrificial layers (see below). Two disadvantages of LPCVD are the low deposition rate and the relatively high operating temperatures.

Very low-pressure processes (about 1 Pa) have been used for the growth of single-crystalline Si at relatively low temperatures. Very low-pressure operation is also advantageous for the growth of III–V compound superlattices by reducing flow recirculations and improving interface abruptness.[23]

#### Example 7.5: Polysilicon Surface Micromachining Using LPCVD

Surface micromachining, we will see below, is based mostly on CMOS manufacturing. Alternating structural and sacrificial layers are deposited, patterned, and etched. Sacrificial

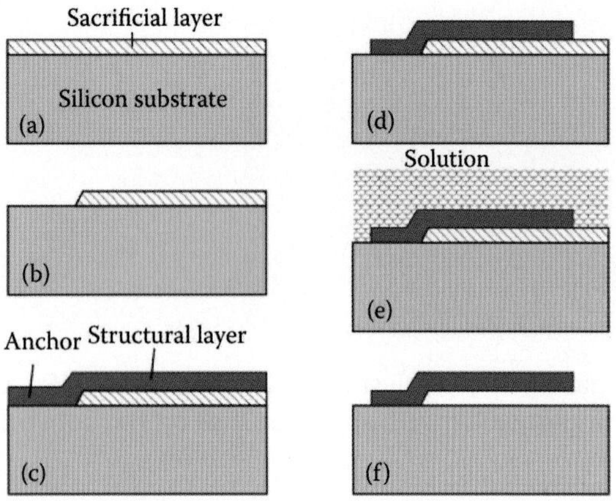


**FIGURE 7.43** Basic surface micromachining process sequence. (a) Spacer layer deposition (the thin dielectric insulator layer is not shown). (b) Base patterning with mask 1. (c) Microstructure layer deposition. (d) Pattern microstructure with mask 2. (e–f) Selective etching of spacer layer.

layers are dissolved away at the end to free the structural layers so that they can be free to move as demonstrated in Figure 7.43. Materials are more or less restricted to CMOS-type materials (polycrystalline silicon, silicon oxide, silicon nitride, boron phosphosilicate glass [BPSG], phosphosilicate glass [PSG]). Structures have low aspect ratios and are sometimes referred to as 2.5D (quite planar). Polycrystalline silicon (called *polysilicon* to distinguish it from silicon having a single-crystal structure) has a number of uses in IC fabrication, including conducting materials for leads, gate electrodes in metal oxide semiconductor devices, and contact material in shallow junction devices. It is also the main structural material in surface micromachines. CVD to coat polysilicon onto a wafer involves reduction of silane at temperatures around 600°C, as expressed by the following reaction:

$$SiH_4 \rightarrow Si + 2H_2 \qquad \text{Reaction 7.4}$$

The surface micromachining process sequence depicted in Figure 7.43 is for the creation of a simple freestanding polysilicon anchor. A sacrificial layer, also called a *spacer layer* or *base*, is deposited on a silicon substrate (Figure 7.43a). CVD can be used to deposit layers of doped $SiO_2$, such as phosphorus-doped silicon dioxide (called *P-glass* or *PSG*). Phosphosilicate glass (PSG) deposited by LPCVD stands out as the best material for the sacrificial layer because it etches even more rapidly in HF than $SiO_2$. To obtain a uniform etch rate, the PSG film must be densified by heating the wafer to 950–1100°C in a furnace or a rapid thermal annealer (RTA). With a first mask, the base is patterned as shown in Figure 7.43b. A window is opened up in the sacrificial layer, and a microstructural thin film, consisting of polysilicon, is conformably deposited over the patterned sacrificial layer (Figure 7.43c). Furnace annealing of the polysilicon film at 1050°C in nitrogen for 1 h reduces stress stemming from thermal expansion coefficient mismatch and nucleation and growth of the film. Rapid thermal annealing has been found effective for reducing stress in polysilicon as well. With a second mask, the microstructure layer is patterned, usually by dry etching in $CF_4$ + $O_2$ or $CF_3Cl$ + $Cl_2$ plasma (Figure 7.43d). Finally, selective wet etching of the sacrificial layer, e.g., in 49% HF, leaves a freestanding micromechanical structure (Figure 7.43e and f). The surface micromachining technique is applicable to combinations of thin films and lateral dimensions where the sacrificial layer can be etched without significant etching or attack of the microstructure, the dielectric, or the substrate. Typically, a surface micromachining stack may contain a total of four to five structural and sacrificial layers, but more are possible; the polysilicon surface machining process at Sandia's SUMMiT, for example, stacks up to five polysilicon and five oxide layers (see page 462).

### *Plasma-Enhanced CVD*

One of the most important limitations of LPCVD is that it requires high deposition temperatures, which may be incompatible with other process steps, e.g., the deposition on an aluminum film. In plasma-enhanced CVD (PECVD), homogenous gas-phase reactions are initiated by collisions of neutrals with nonequilibrium energetic electrons produced by RF plasma instead of by thermal energy. Thus, the deposition precursors can be produced at much lower temperatures by these homogenous reactions. Furthermore, ion bombardment makes the surface more reactive by creating more dangling bonds (active sites) and/or supplying instantaneous energy at the surface for the heterogeneous reaction to take place. In general, PECVD can be carried out at a much lower temperature than LPCVD.

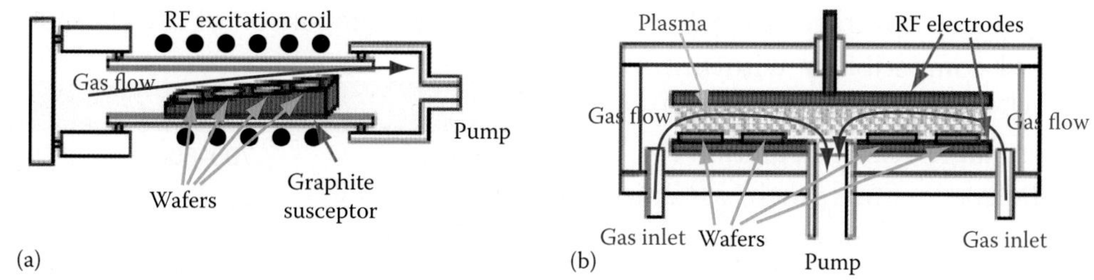


**FIGURE 7.44** (a) Horizontal tube reactor. (b) Schematic of a cold wall, parallel-plate plasma deposition reactor, which is very similar to an etching reactor.

Moreover, the deposition rate is found to be higher in PECVD than in LPCVD. There are a number of different reactor configurations used for PECVD, but the physical principle is the same in all cases. Two basic PECVD setups are illustrated in Figure 7.44. In these setups, RF-induced plasma transfers energy into the reactant gases, allowing the substrate to remain at lower temperatures than in APCVD and LPCVD processes. We will only detail the cold-wall, parallel-plate plasma deposition reactor, which is very similar to an etching reactor.

With a simple parallel-plate reactor (Figure 7.44b), substrates can be placed horizontally or, if the pressure is low enough, vertically. The vertical position is used to increase throughput. No loss of film-thickness uniformity occurs in the latter case because the PECVD method is surface reaction limited. Adequate substrate temperature control ensures uniformity. The chamber consists of two electrodes, one of which is grounded, and the other is powered by an RF source. Wafers are placed on the grounded electrode and are subjected to a less-energetic bombardment than wafers placed on the powered electrode. The wafers are heated from below to the desired temperature by resistive heaters or high-intensity lamps. The reactant gases are introduced at the outer periphery, and they flow radially toward the center from where they are pumped out. Plasma is generated between the electrodes by the RF voltage applied across the electrodes. The electrode spacing is about 5–10 cm, and the operating temperature and pressure are –400°C and 0.1–5 Torr, respectively. Homogenous reaction between the reactant gases is initiated by the electrons produced in the plasma. The ions are accelerated toward the electrodes as a result of the self-bias voltage between the plasma and the electrode. These energetic ions bombard the surface, making it more reactive, leading to higher deposition rates.

In most PECVD systems, the reactor configuration is changed so that the potential of both the powered and the grounded electrode, relative to the plasma, become equal.[27] Compared with sputter deposition, PECVD offers several advantages. The lower power densities, higher pressures, and somewhat higher substrate temperatures (e.g., >200°C) all lead to less-severe radiation damage than in sputter deposition. Moreover, for radiation-sensitive substrates such as compound semiconductors, afterglow or downstream deposition systems can be used where the radicals are formed in the glow discharge and then transported out of the region to a downstream deposition system. Thus, selective activation of reactants becomes possible without damaging the surface of the substrate.[35] PECVD films are not stoichiometric because the deposition reactions vary widely, and particle bombardment during growth of a multicomponent system changes the composition according to the ratios of sputtering yields of the component materials (nonthermal equilibrium). Because of the rapid deposition without the temperatures required for surface migration, film tends to be porous (possess holes like a sponge), leading to lower density than thermally grown films. Film stoichiometry and density can be monitored by a combination of etch rate comparisons to thermally grown films and ellipsometry to determine the index of refraction of the films.

Despite this negative consequence of particle bombardment, in general, the more ion bombardment, the better the film quality. Microstructure, stress, density, and other film properties show marked response, mostly for the better, to ion bombardment during deposition. Good adhesion, low pinhole density, good step coverage, adequate electrical properties, and compatibility with fine line-width pattern transfer processes have led to wide use of PECVD in very large-scale integration (VLSI). PECVD enables dielectric films such as oxides, nitrides, and oxynitrides to be deposited on wafers with small feature sizes and line widths at low temperatures and on devices unable to withstand the high temperatures of a thermally activated reaction. Planarization represents only one of the many applications of this versatile technology. Another application is deposition of amorphous silicon thin films, as used in flat-panel displays, eyeglasses, and photovoltaic panels. The most significant application is probably the deposition of $SiO_2$ or $Si_3N_4$ over metal lines.[35]

Summarizing, in PECVD, low RF excitation frequency, low reactor pressures, and low deposition rate at high temperature contribute to:

1. Lower processing temperature (250–450°C)
2. Excellent gap-fill for high-aspect-ratio gaps (with high-density plasma)
3. Good film adhesion to the wafer
4. High deposition rates
5. High film density as a result of low pinholes and voids
6. Low film stress as a result of lower processing temperature

**Example 7.6: $Si_3N_4$ by PECVD**

Silicon nitride is used as a passivation layer (protecting against sodium diffusion and moisture). $Si_3N_4$ has a low oxidation rate compared with silicon; therefore, a nitride mask can be used to prevent oxidation in coated areas on the silicon surface. Plasma nitride films are almost universally used for final passivation layers in IC fabrication, often in conjunction with a deposited oxide or PSG layer. A conventional CVD process for coating $Si_3N_4$ onto a silicon wafer involves reaction of silane and ammonia ($NH_3$) at around 800°C as follows:

$$3SiH_4 + 4NH_3 \rightarrow Si_3N_4 + 12H_2 \quad \text{Reaction 7.5}$$

Plasma-enhanced CVD is also used for basically the same coating reaction, the advantage being that it can be performed at a much lower temperature, around 300°C, compatible with aluminum metallization. In the latter application, stress control is important to avoid film cracking and degradation of metal reliability. Transparency in the UV must also be ensured in UV-erasable EPROM (erasable programmable read only memory) applications. PECVD silicon nitride is also often used for passivation of GaAs MESFETs [MESFET stands for metal semiconductor field effect transistor. Instead of using a p-n junction for a gate, a Schottky (metal-semiconductor) junction is used] to avoid oxidation of the exposed GaAs surface.

Properties of silicon nitride made by LPCVD versus PECVD are compared in Table 7.15.

Table 7.15 Comparison of LPCVD and PECVD Silicon Nitride

| Property | LPCVD | PECVD |
|---|---|---|
| Deposition temperature (°C) | 700–800 | 300–400 |
| Composition | $Si_3N_4$ | $Si_xN_yH_z$ |
| Step coverage | Conformal | Fair |
| Stress at 23°C on silicon (dyn/cm$^{-2}$) | $1.2–1.8 \times 10^{10}$ (tensile) | $1–8 \times 10^9$ (tensile or compressive) |

**Example 7.7: Polysilicon by PECVD**

We remarked earlier on the quest for low-temperature polysilicon deposition processes, driven by the need for compatibility with prefabricated aluminum-metallized CMOS circuitry. This makes the 320°C PECVD polysilicon interesting (as is the ~350°C sputter deposition method).

PECVD polysilicon films, deposited in a 50-kHz parallel-plate diode reactor, can be doped in situ and crystallized by RTA (1100°C, 100 s). It was shown that small-grained PECVD films annealed by RTA have good electrical properties and gauge factors between 20 and 30, similar to those reported for other alternative types of polycrystalline silicon.[35]

A low-temperature polysilicon PECVD process is not only compatible with conventional oxide sacrificial layers (as used in surface micromachining; see below) but also with certain organic sacrificial layers such as polyimides.

More information on CVD thin films and how they are influenced by the above

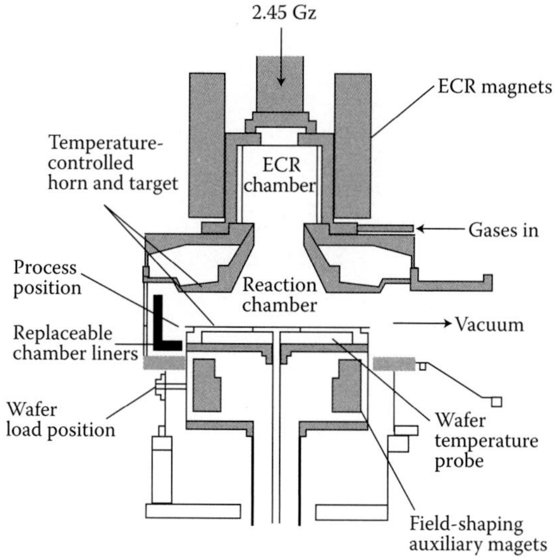


**FIGURE 7.45** In ECRCVD, the plasma can deposit a film while simultaneously sputtering.

deposition parameters, as well as other parameters (e.g., moisture, flow rate, and gas composition), will be given in Volume III, Chapter 5 on selected MEMS materials.

### *High-Density Plasma CVD*

High-density plasma sources such as inductively coupled plasma (ICP) and electron cyclotron resonance (ECR) stations (see Figure 3.36)—machines discussed in Chapter 3—can be used in high-density plasma CVD (HDPCVD) for both enhanced deposition and enhanced etching. Both systems work at low pressures (a few mTorr) and can independently control ion flux and ion energy. In ECRCVD, one uses ECR to generate a high-density plasma (HDP) that can deposit materials at high rates while pressures and temperatures remain low (Figure 7.45). ECRCVD technology provides a way to ensure durable ultrathin films. Alternating deposition and etch back is used for very effective planarization of topography. In submicrometer device fabrication, interlevel dielectric gap filling of high-aspect-ratio topography generally requires the use of multistep deposition/etch or spin-on dielectric processing to produce void-free, filled structures. With HDPCVD processing, high-aspect-ratio (up to 4:1) sub-half-micrometer structures can be filled and locally planarized in a *single* processing step. For gap filling, HDPCVD processing consists of a simultaneous deposition/etching process in which films over topographical features are sputtered off by reactive ions and radicals during deposition while deposition refills the newly created voids (see Example 7.8). The deposition/sputtering rate ratio (*D*/*S*) is an important measure of the gap-filling capability of the processes. For further reading, see Nguyen and references therein.[36] Alternative planarization techniques are reviewed in Chapter 1.

A typical ECRCVD machine is the Tek-Vac DRIE-1100-LL-ECR (http://www.tekvac.com/prd-ecr-1.htm) designed for both high-density plasma CVD and reactive ion etching of submicrometer integrated circuits (ICs), optical devices, and RF microwave electron devices. This type of equipment is used for CVD of diamond, diamond-like carbon, SiC, $Si_3N_4$, BN, and other refractory or high dielectric materials. The equipment provides high-density plasma with low damage and high etching rates and is well suited for etching devices sensitive to high energy electron damage, such as low-noise amplifiers. Various materials include aluminum, copper, gallium nitride, gallium arsenide, and HeCdTe, and semiconductor dielectrics have been etched this way. The method provides high uniformity of deposition and etching across large area substrates.

**Example 7.8: Gap Filling by ECRCVD**

ECRCVD simultaneously deposits and etches films to prevent bread-loaf and key-hole effects typical for PECVD processes, as shown in Figure 7.46. As shown in Figure 7.46b, ion-induced deposition is filling the gap between two metal lines. Argon ions then sputter-etch the excess film at the gap entrance, resulting in a beveled appearance of the film (Figure 7.46c). The etched material is redeposited, and the process is repeated, resulting in an equal "bottom-up" profile (Figure 7.46d).

### *Metalorganic CVD*

Many materials that one wishes to deposit have too low a vapor pressure and thus are difficult to transport via gases. One solution is to chemically attach the material (e.g., Ga, Al, Cu) to an organic compound that has a high vapor pressure. Organic

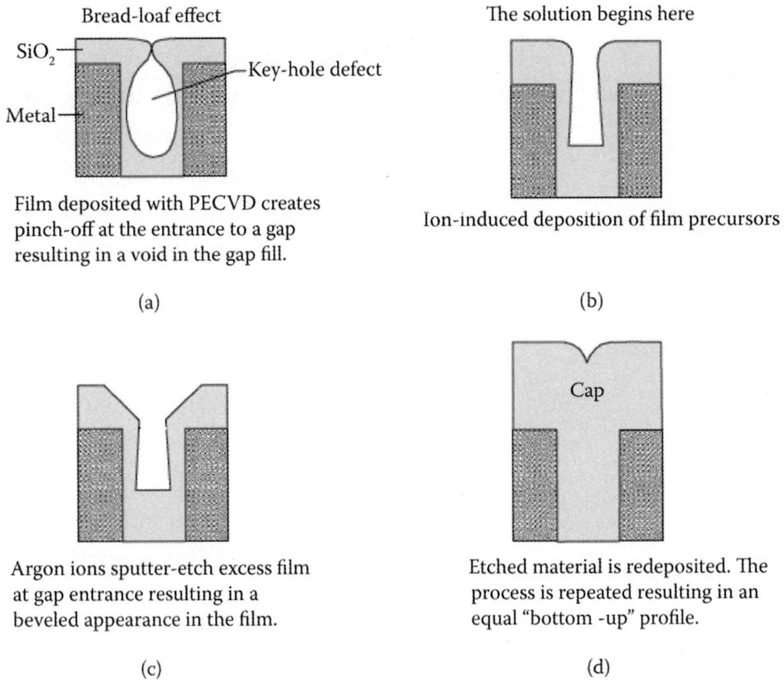


**FIGURE 7.46** ECRCVD planarization.

compounds often have very high vapor pressure. The organic-metal bond is very weak and can be broken via thermal means on a heated wafer, depositing the metal with the high vapor pressure organic being pumped away. Care must be taken to ensure little of the organic byproducts are incorporated. Carbon contamination and unintentional hydrogen incorporation are sometimes a problem. Crystal growth by metalorganic CVD (MOCVD) began in the late 1960s. Metalorganic CVD, sometimes called *organometallic vapor phase epitaxy,* relies on the flow of gases (hydrides such as arsine and phosphine and organometallics such as trimethyl gallium and trimethyl aluminum) past heated samples placed in the gas stream. Typically, the reactive metal alkyls [e.g., $Ga(CH_3)_3$] react with a hydride of the nonmetal component (e.g., $AsH_3$) to produce the pure films (e.g., GaAs) and some byproducts (e.g., $CH_4$). For example:

$$(CH_3)_3Ga\ (g) + AsH_3\ (g) \rightarrow GaAs(s) + 3CH_4\ (g)$$
Reaction 7.6

The biggest advantage of MOCVD is the low growth temperature (cold wall reactor); the biggest disadvantage is the high cost of the very pure chemicals. MOCVD provides thickness control within one atomic layer and has become the preferred epitaxial process, a cost-effective manufacturing process for a variety of compound semiconductor devices. Foremost, MOCVD plays a key role in the manufacture of many optoelectronic devices with III–V compounds for solar cells, lasers, photocathodes, light-emitting diodes (LEDs), high electron mobility transistors (HEMTs), and quantum wells.[37] By using MOCVD, it is easier to scale it up to larger growth areas than it is to scale up liquid-phase epitaxy (LPE) (see below), and the uniformity and surface morphology of MOCVD grown lasers are superior to those attainable with LPE.

### Atomic Layer Deposition

Atomic layer deposition (ALD) is a method of applying thin films to various substrates with atomic scale precision and constitutes one of the most promising "new" nanotechnology fabrication techniques. Dr. Tuomo Suntola and coworkers in Finland introduced ALD in 1974 to improve the quality of ZnS films used in electroluminescent displays.* It turns

* Picosun in Espoo, Finland, develops and manufactures ALD reactors for microelectronics and nanotechnology. With the help of Dr. Tuomo Suntola, Picosun has succeeded in developing inexpensive yet highly efficient ALD tools for research and development.

Table 7.16 List of ALD Applications

| |
|---|
| • High-k dielectrics ($Al_2O_3$, $HfO_2$, $ZrO_2$, $Ta_2O_5$, $La_2O_3$, etc.) for transistor gates and DRAM capacitors in Si, heterostructures, compound semiconductors, MESFETs, III–V semiconductor materials, organic transistors, graphene, graphite, nanotubes, nanowires, and molecular electronics |
| • Conductive gate electrodes (Ir, Pt, Ru, TiN, etc.) |
| • Metal interconnects and liners (Cu, WN, TaN, WNC, Ru, Ir, etc.); metallic diffusion barrier layers for copper interconnects and semiconductor vias for transistor gate and memory cell applications, DRAM capacitors, and passivation layers |
| • Catalytic materials (Pt, Ir, Co, $TiO_2$, $V_2O_5$); coatings inside filters, membranes, catalysts (thin economical Pt for automobile catalytic converters), fuel cells, ion exchange coatings |
| • Nanostructures (all materials); conformal deposition around and inside nanostructures and MEMS |
| • Biomedical coatings: (TiN, ZrN, CrN, TiAlN, AlTiN, etc.); biocompatible materials for in vivo medical devices and instruments |
| • Piezoelectric layers (ZnO, AlN, ZnS, etc.) |
| • Transparent electrical conductors (ZnO:Al, ITO) |
| • UV blocking layers (ZnO, $TiO_2$) |
| • OLED passivation ($Al_2O_3$) |
| • Solid lubricant layers (WS2) |
| • Photonic crystals (ZnO, ZnS:Mn, $TiO_2$, $Ta_2N_5$); coatings inside porous alumina, inverted opals |
| • Antireflection and optical filters ($Al_2O_3$, ZnS, $SnO_2$, $Ta_2O_5$); Fabry-Perot, flip-flop optical filters |
| • Electroluminescent devices (SrS:Cu, ZnS:Mn, ZnS:Tb, SrS:Ce) |
| • Processing layers ($Al_2O_3$, $ZrO_2$); etch barriers, ion diffusion barriers, fill layers for magnetic read heads |
| • Optical applications (AlTiO, $SnO_2$, ZnO); photonics, nanophotonics, solar cells, integrated optics, optical coatings, lasers, variable dielectric constant nanolaminates |
| • Sensors ($SnO_2$, $Ta_2O_5$, etc.); gas sensors, pH sensors |
| • Wear and corrosion inhibiting layers ($Al_2O_3$, $ZrO_2$) |

out that ALD also produces outstanding high-k* dielectrics, metal electrodes, barriers, and spacers (see list of applications in Table 7.16). Objects of almost any size and shape can be coated, even atomic force microscope (AFM) tips. The method is similar in chemistry to CVD, except that the ALD reaction breaks up the CVD reaction into two half-reactions, keeping the precursor reagents separate during their reaction at the surface. As opposed to conventional CVD where one has continuous deposition and concurrent flow of precursors, ALD is based on the sequential deposition of individual monolayers at the surface in a well-controlled manner: the growth surface is alternately exposed to only one of two complementary chemical environments, i.e., individual precursors are supplied to the reactor one at a time. The precursors need to be thermally stable and should chemisorb fast and react aggressively with each other. Between exposure steps, an inert gas purge or a pump-down step is used to remove any residual chemically active source gas or byproducts before another precursor is introduced into the reactor. The overall ALD process then consists of a repetition of a number of growth cycles. Each cycle is made up of a typical sequence: 1) flow of precursor 1, 2) purge, 3) flow of precursor 2 and 4) purge. During a precursor exposure step, precursor molecules react with the surface until all available surface sites are saturated. Precursor chemistries and process conditions are chosen such that no further reaction takes place once the surface is completely saturated. This surface saturation guarantees the self-limiting nature of ALD. Precursors are preferably overdosed so that process results become independent of potential slight variations in the amount of precursor supplied to the surface. Thus, surface chemistry governs film growth rather than a precise control of tool-specific process parameters such as precursor flow and partial pressure. A known and constant thickness is deposited per growth cycle. Typically, deposition rates on the order of 0.1–1.0 Å/cycle are obtained, with cycle times ranging from 1–10 s. Cycle time critically depends on saturation behavior, chamber volume, and reactor design. The film is grown in a layer-by-layer mode, and the total film thickness is given by the number of cycles. Exceptional across-wafer thickness uniformity and excellent step coverage on high-aspect-ratio features can be achieved with ALD, making this technique one of the most exciting nanotechnology methods. In Figure 7.47 we illustrate a typical ALD setup, and in Figure 7.48 the ALD cycle for $Al_2O_3$ deposition is demonstrated.

* Many materials are currently under consideration as potential candidates for gate dielectrics for sub-0.1-μm CMOS technology. Silicon dioxide with a dielectric constant of 3.9 will be replaced by so-called *high-k materials*. Candidates for gate dielectrics include the oxides of yttrium, lanthanum, zirconium, and hafnium, which have much higher dielectric constants than $SiO_2$.

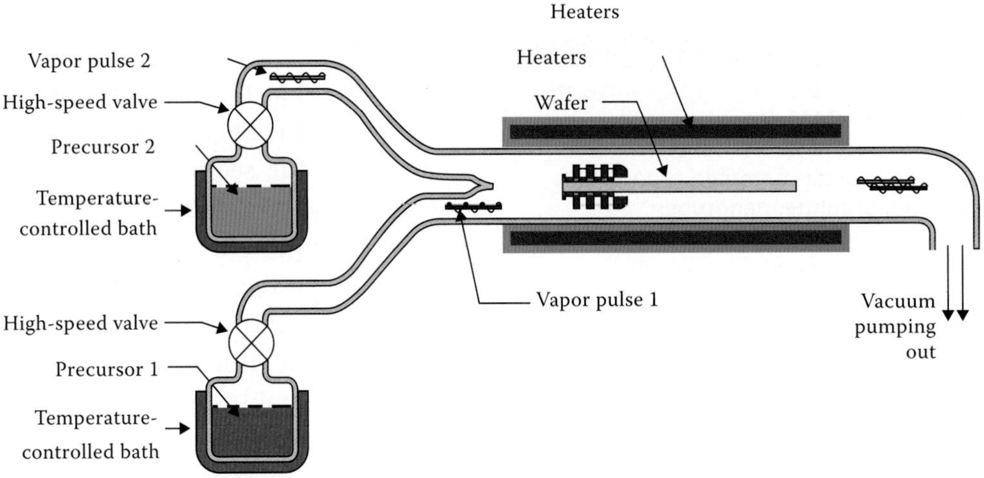


**FIGURE 7.47** Typical atomic layer deposition (ALD) setup. (From *Atomic layer deposition.* Cambridge NanoTech Inc., April 24, 2006. http://www.cambridgenanotech.com.[231])

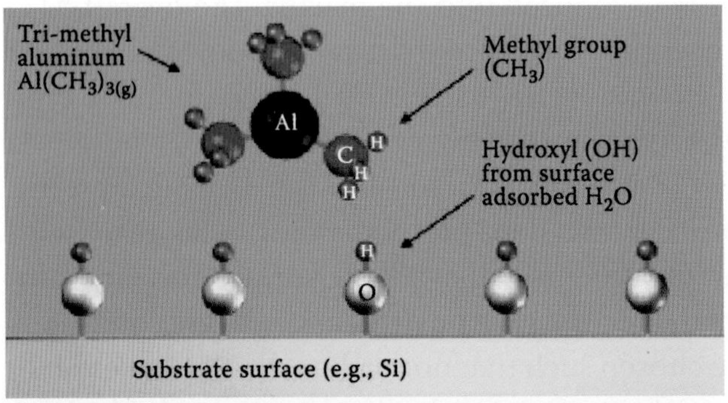


1. Flow of precursor 1: In air, water vapor is adsorbed on most surfaces, forming a hydroxyl group. In the case of Si this forms Si-OH. After placing the substrate in the reactor, trimethyl aluminum (TMA) is pulsed into the reaction chamber.

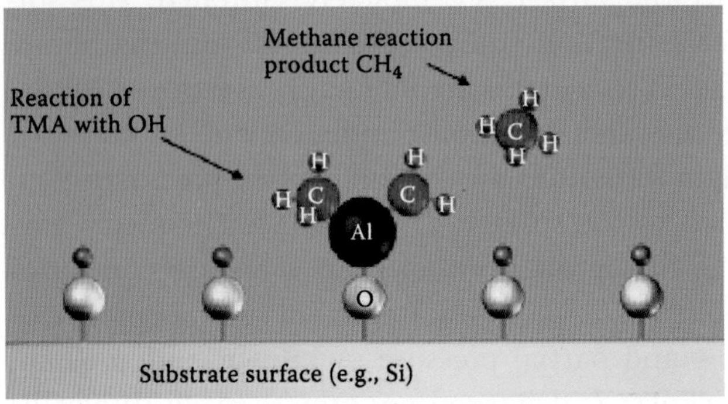


TMA reacts with the adsorbed hydroxyl groups, producing methane as a reaction product.

$$Al(CH_3)_3(g) + :Si-O-H(s) \rightarrow :Si-O-Al(CH_3)_2(s) + CH_4(g) \qquad \text{Reaction 7.7}$$

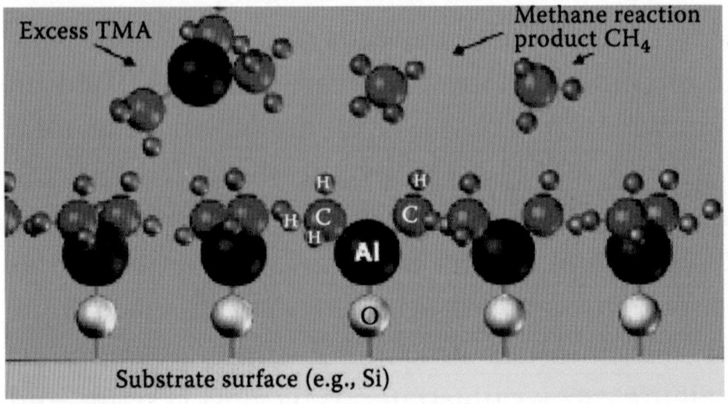


2. Purge cycle: TMA reacts with the adsorbed hydroxyl groups until the surface is passivated. TMA does not react with itself, limiting the reaction to one monolayer. This leads to the perfect uniformity and conformality of the ADL process. In the purge pulse, excess TMA and the methane reaction product are pumped away.

**FIGURE 7.48** The atomic layer deposition (ALD) process illustrated for the $Al_2O_3$ process. (From *Atomic layer deposition.* Cambridge NanoTech Inc., April 24, 2006. http://www.cambridgenanotech.com.[231])

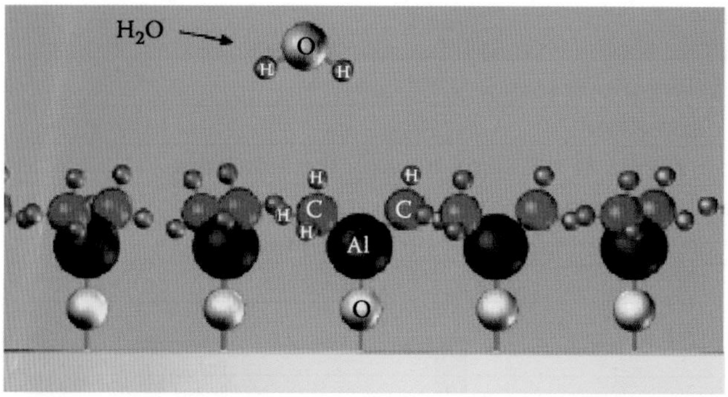


3. Flow of precursor 2: After all the TMA and methane have been pumped away, water is pulsed into the reactor chamber.

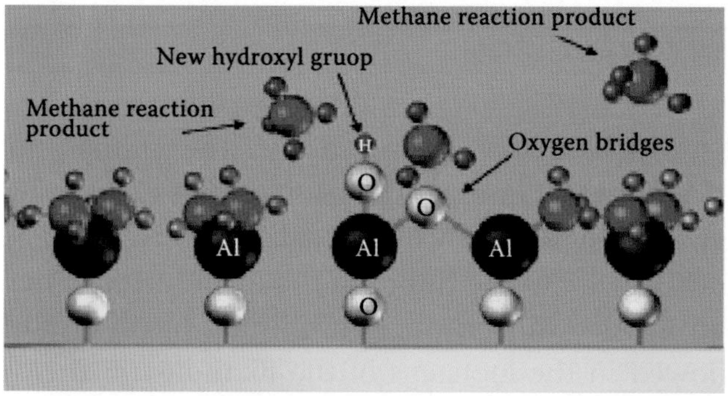


Water reacts with the dangling methyl groups on the new surface, forming aluminum-oxygen (Al-O) bridges and hydroxyl surface groups waiting for a new TMA pulse.

$$2H_2O(g) + :Si{-}O{-}Al(CH_3)_2(s) \rightarrow :Si{-}O{-}Al(OH)_2(s) + 2CH_4(g) \qquad \text{Reaction 7.8}$$

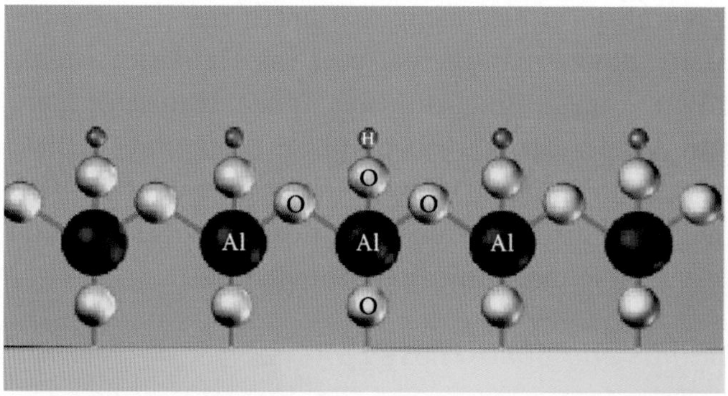


4. New purge cycle: The reaction product methane is pumped away. Excess water does not react with the hydroxyl surface groups, again causing perfect passivation limited to one monolayer.

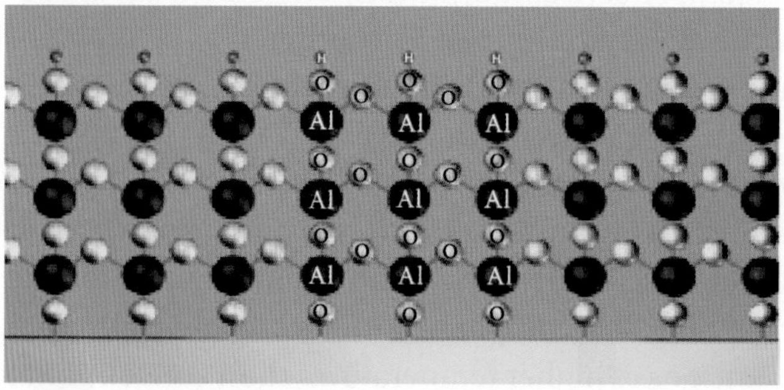


One TMA pulse, one purge, and one water pulse form one cycle. Here three cycles are shown, with approximately 1 Å/cycle. Each cycle, including pulsing and purging, takes about 3 s.

**FIGURE 7.48** *(Continued)*

**Table 7.17** Comparison of ALD with CVD

| ALD | CVD |
|---|---|
| Highly reactive precursors | Less reactive precursors |
| Slower deposition | Faster deposition |
| Precursors react separately on the substrate | Precursors react at the same time on the substrate |
| Low-temperature deposition possible | Higher deposition temperature |
| Precursors must not decompose at process temperature | Precursors can decompose at process temperature |
| Uniformity ensured by the saturation mechanism | Uniformity requires uniform flux of reactant and temperature |
| Stoichiometric films with large area uniformity and 3D conformality | |
| Gentle deposition process for sensitive substrates | |
| Thickness control by counting the number of reaction cycles | Thickness control by precise process control and monitoring |
| Surplus precursor dosing acceptable | Precursor dosing important |

The list of applications for ALD keeps expanding, as evident from Table 7.16. In Table 7.17 we compare ALD with CVD.

## Epitaxy

### *Introduction*

Epitaxial techniques arrange atoms in single-crystal fashion on a crystalline substrate acting as a seed crystal so that the lattice of the newly grown film duplicates that of the substrate. The lattice constant of the epitaxially grown layer needs to be close to the lattice constant of the substrate wafer. Otherwise, the bonds cannot stretch far enough, and dislocations will result.

Si epitaxy is used for high-purity layer growth and can form very thick doped structures (30–100 μm) not possible with implantation or diffusion. Such thick, pure layers are often used in power devices with thinner, 1–5-μm structures, commonly used for CMOS and bipolar technology. In surface micromachining, where polycrystalline silicon functional layers are built up on a substrate rather than etched in the bulk of a single-crystal substrate as in bulk micromachining, epi-grown SOI wafers are a very attractive alternative. Surface micromachines built from layers of epi silicon and isolated from the substrate by a $SiO_2$ layer combine the most attractive features of surface micromachining (i.e., CMOS compatibility and freedom in types of structural shapes) with the superior single-crystal properties of the epilayer. For ICs, SOI is compatible with existing wafer processes, provides 50% faster circuit speed than bulk silicon, allows easier scaling to finer line widths, and can reduce the number of required mask levels for a given design by ~30%. SOI may very well represent the wave of the future in the IC industry rather than GaAs.[38] Silicon micromachining has typically followed in the footsteps of the IC industry; therefore, the prevalence of SOI in IC is a good indicator of coming trends in micromachining. SOI simplifies building micromachines by reducing the number of masks, reducing packaging concerns, and making integration of electronics easier. In the IC industry, many device parameters such as transistor breakdown voltage, junction capacitance, transistor gain, and AC performance depend on the epilayer thickness, thus necessitating precise control of it. This precise control also enables more reproducible and predictable mechanical micromachines.

### *Liquid- and Solid-Phase Epitaxy*

In the section on PVD techniques we discussed molecular beam epitaxy (MBE) as the most advanced. In the CVD section we encountered two types of vapor phase epitaxy (VPE) techniques, APCVD and MOCVD. For silicon processing, VPE has met with the widest acceptance because excellent control of impurity concentration and crystalline perfection can be achieved. In growing an epilayer with VPE, the processing is accomplished under closely controlled conditions at higher temperatures than conventional CVD of Si, using diluted reacting gases to slow the process so that an epitaxial layer can be successfully formed. Here, we briefly discuss liquid- and solid-phase epitaxy. For depositing multilayer structures

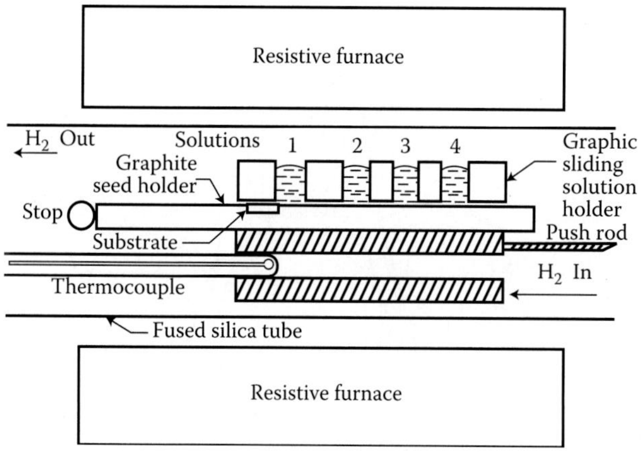


**FIGURE 7.49** Liquid-phase epitaxy setup. (From Brodie, I., and J. J. Muray. 1982. *The physics of microfabrication.* New York: Plenum Press. With permission.[1])

of different materials on the same substrate, liquid phase epitaxy (LPE) is used. LPE films grow from a liquid solution very near the equilibrium state, making the technique reproducible and resulting in films with low concentrations of growth-induced defects. A schematic for a typical LPE setup is presented in Figure 7.49.[1] In operation, a graphite slider plate moves relative to a multiple-well assembly to bring the substrate in a recess in the slider plate in contact with the different solutions. LPE has found its widest application in producing epitaxial layers of III–V compounds (e.g., InP, GaAs) for lasers and photodetectors. Anderson reviews the technique well.[39]

Solid-phase epitaxy describes the crystalline regrowth of amorphous layers that extend continuously to the underlying single-crystal substrate. At temperatures between 500 and 600°C, in the case of silicon, a recrystallization process occurs on the underlying crystalline substrate, and regrowth proceeds toward the surface. Regrowth is faster on (100) than on (111) Si, and impurities such as B, P, and As enhance the regrowth, whereas O, C, N, and Ar retard it.[27]

### *Selective Epitaxy*

The incorporation of selective epi in the micromachining arsenal is making more versatile microstructures possible. Under the correct growth conditions and/or surface treatment, it is possible to initiate Si growth in selected areas.[27,40] Selective epitaxial growth allows the formation of closely spaced silicon features isolated by $SiO_2$ (see Figure 7.50). Besides increased density over other insulation techniques (important mainly for the IC industry), structures as shown in Figure 7.50a and b, might enable a host of interesting mechanical microstructures. For example, imagine that the $SiO_2$ in Figure 7.50b is selectively etched away, resulting in an epi anchor with suspended polybeams. In selective epitaxy of the type shown in Figure 7.50a, silicon atoms possessing high surface mobility migrate to sites on the single crystal where nucleation is favored. In the ideal case, all the epi grows exclusively in the oxide openings. Silicon mobility improves in the presence of halides. The higher the number of chlorine atoms in the silicon source, the better the degree of selectivity (e.g., $SiHCl_3$ is a better source than $SiH_2Cl_2$). The selective deposition shown in Figure 7.50b is accomplished by simultaneous deposition of epitaxial silicon in the oxide openings and polysilicon on the oxide surfaces. The angle between the epi and the poly in Figure 7.50b depends on the crystallographic orientation of the

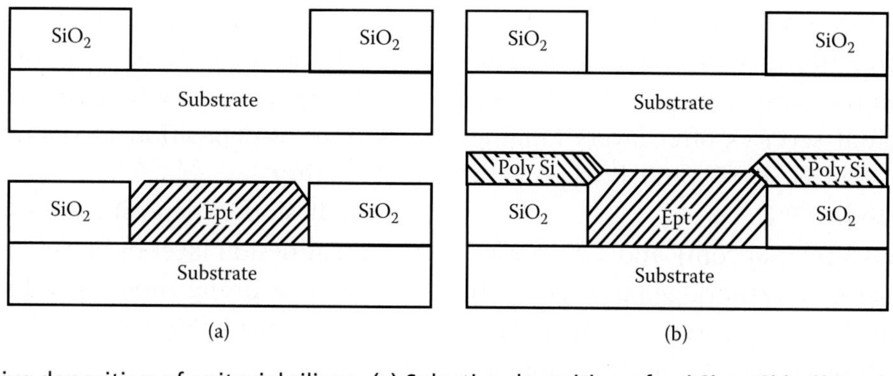


**FIGURE 7.50** Selective deposition of epitaxial silicon. (a) Selective deposition of epi Si on Si in $SiO_2$ windows. (b) Simultaneous deposition of epi Si on Si, and polysilicon on $SiO_2$. (From Wolf, S., and R. N. Tauber. 1987. *Silicon processing for the VLSI era.* Sunset Beach, CA: Lattice Press.[27])

substrate. The angle is 90° for <110> orientations, 72° toward the polysilicon for <100> orientations, or tapered toward the single-crystal silicon at 70° for the <111> orientation.[27] These figures represent interesting angles from which to construct micromechanical structures. This selective deposition process also lends itself to CVD, for example, for selective deposition of W.

### *Epilayer Thickness*

Epitaxial layer thickness is a critical parameter both in IC applications and micromachines and therefore must be accurately measured and controlled. Many devices, from discrete transistors and 16-megabyte dynamic random access memory (DRAM) to membrane-based micromechanical structures, use silicon epitaxial layer wafers as their starting material. The thickness of an epitaxial layer forms an integral part in the design of many micromachined devices. For example, in a piezoresistive pressure sensor, epilayer thickness control ultimately determines the pressure sensitivity control.

Epilayer thickness can be measured from infrared reflectance, angle lap and stain, tapered groove, weighing, capacitance-voltage measurements, and profilometry. The most widely used, nondestructive method of measuring epi thickness is with infrared instruments. Fourier transform infrared offers automated epilayer thickness measurements.[41]

## CVD Equipment

For a detailed historical timeline of vacuum coating and vacuum/plasma technology, visit http://www.svc.org/H/H_HistoricalTime.html, posted on the Internet by the Society of Vacuum Coaters.

Typical sources for CVD equipment are CVD Equipment Corporation (http://www.cvdequipment.com) and IonBond Inc. (http://www.ionbond.com). Commercial epitaxial services offer layers ranging from 0.5–150 μm and N, P, $N^+$, and $P^+$ type with uniformity better than ±5%.

Planar (http://www.planar.com) and Cambridge NanoTech (http://www.cambridgenanotech.com) manufacture atomic layer deposition (ALD) systems for many coating applications. Vacuum equipment to clean, deposit, and etch is combined more and more in so-called *cluster tools* (Figure 7.51).[42]

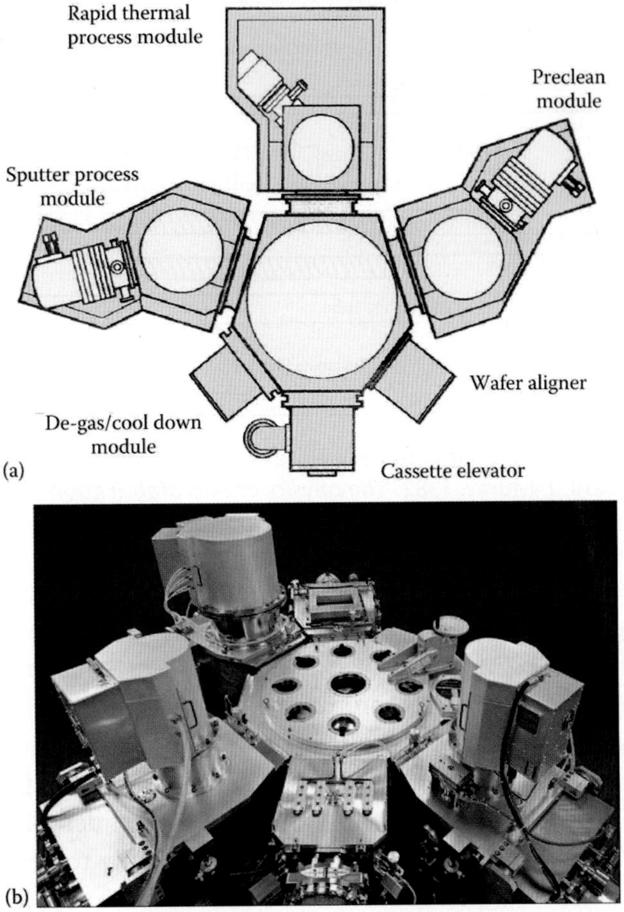


**FIGURE 7.51** Cluster tools. (a) Schematic. (b) Example from Ovonyx (http://www.ovonyx.com/corporate).

Selection criteria for the additive processes reviewed in this chapter depend on a variety of considerations, such as:

1. Limitations imposed by the substrate or the mask material: $T_{max}$ (maximum temperature), surface morphology, substrate structure and geometry, and so on
2. Apparatus requirement and availability
3. Limitations imposed by the material to be deposited: chemistry, purity, thickness, $T_{max}$, morphology, crystal structure, and so on
4. Rate of deposition to obtain the desired film quality
5. Adhesion of deposit to the substrate; necessity of adhesion layer or buffer layer
6. Total running time, including setup time and postcoating processes
7. Cost
8. Ease of automation
9. Safety and ecological considerations

## Mechanical Properties of Thin Films

### Introduction

The study of the mechanical behavior of thin films started in the 1950s. This early work was initiated to better understand the stresses and failures in integrated circuit (IC) structures. The MEMS field, especially surface micromachining, provided an additional impetus in the late 1980s, and the investigations were extended to more general applications of thin films. In this section, advances in the understanding of mechanical properties of thin films are reviewed with special focus on the mechanical behavior of polycrystalline thin films for MEMS and NEMS applications.

Thin films in surface micromachines must satisfy a large set of rigorous chemical, structural, mechanical, and electrical requirements. Excellent adhesion, low residual stress, low pinhole density, good mechanical strength, and chemical resistance all may be required simultaneously. For many microelectronic thin films, the material properties depend strongly on the details of the deposition process and the growth conditions, and some properties may depend on postdeposition thermal processing, referred to as *annealing*. Furthermore, the details of thin-film nucleation and/or growth may depend on the specific substrate or on the specific surface orientation of the substrate. Although the properties of a bulk material might be well characterized, its thin-film form may have properties substantially different from those of the bulk. For example, thin films generally display smaller grain size than bulk materials. An overwhelming reason for the many differences stem from the properties of thin films, which exhibit a higher surface-to-volume ratio than large chunks of material and are strongly influenced by the surface properties.

Examples where thin films are directly "mechanically challenged" include the myriad of structural and moving parts in MEMS devices, such as the membranes in pressure sensors, hinges in micromirror arrays, wear-resistance coatings on cutting tools, and protective coatings for magnetic disks. In many other applications, thin films are selected primarily because of their unique electronic, magnetic, optical, or thermal properties; however, the mechanical characteristics of the materials of choice in such applications are not unimportant because thin films are often subjected to large mechanical stresses during both the manufacturing process and normal operation of the end-use devices.

Some terminology characterizing thin films and their deposition is introduced in Table 7.18.

Because thin films were originally not intended for load-bearing applications, their mechanical properties were ignored for a long time. As we indicated earlier, it was the IC and MEMS world that

**Table 7.18** Thin-Film Terms Used in Characterizing Deposition

| Term | Remark |
|---|---|
| Film | Bond energy <10 kcal/mol |
| Chemisorbed film | Bond energy >20 kcal/mol |
| Nucleation | Adatoms forming stable clusters |
| Condensation | Initial formation of nuclei |
| Island formation | Nuclei grow in three dimensions, especially along the substrate surface |
| Coalescence | Nuclei contact each other and larger, rounded shapes form |
| Secondary nucleation | Areas between islands are filled in by secondary nucleation, resulting in a continuous film |
| Grain size of thin film | Generally smaller than for bulk materials and function of deposition and annealing conditions (higher T, larger grains) |
| Surface roughness | Lower at high temperatures except when crystallization starts; at low temperature the roughness is higher for thicker films, also oblique deposition and contamination increases roughness |
| Epitaxial and amorphous films | Very low surface roughness |
| Density | More porous deposits are less dense; density reveals a lot about the film structure |
| Crystallographic structure | Adatom mobility: amorphous, polycrystalline, single crystal or fiber texture, or preferred orientation |

precipitated a strong appreciation for the need of an understanding of the mechanical properties as essential for improving the reliability and lifetime of thin films, even in nonstructural applications.[43] Generally, mechanical stresses in thin films can be divided into intrinsic or growth stresses that develop during the deposition process and extrinsic stresses that are induced by external physical effects. The extrinsic stresses are induced by various physical effects after the film is grown.

Thin-film materials are fabricated using very different processing methods than those for bulk materials, such as the various CVD and PVD deposition techniques studied in the current chapter. These processing methods often result in development of intrinsic mechanical stresses in the films. We saw earlier that metal films sputtered at room temperature typically consist of very fine grains, whereas evaporated films are often highly textured with larger grain sizes than sputtered films. It was also pointed out that mobility of the atoms of the deposited material significantly affects the microstructure formation in both deposition and postdeposition processes. The microstructure evolution during the growth process typically involves 1) nucleation of crystallite islands from the condensed material atoms at many sites on the substrate surface, 2) growth of individual crystallite islands until they coalescence with other islands, and 3) coarsening of grains during thickening of the film. For bulk materials, the crystal grain growth occurs at the expense of small grains through the motion of grain boundaries, driven by the reduction in the total grain boundary energy. In thin films the situation is a bit more complex because free surfaces and interfaces also play an important role in the microstructure development. Mismatch strains often develop between the film and the substrate, and this leads to higher strain energy in the film. Grain growth and crystallographic texture development in thin films are driven by the reduction of the total energy, including the grain boundary energy, the surface and interface energy, and the strain energy. This is a thermally activated kinetic process, and for films grown at room temperature, grain growth and texture development are often slow because of lack of thermal activation, resulting in a metastable structure with very fine grains (see section on sputtered films). The microstructure of thin films can be further modified through a postdeposition process, such as annealing. During annealing, grains grow and crystallographic textures develop to minimize the total energy at a rate determined by the annealing temperature. When a grain grows to a size on the order of the film's thickness, the grain boundaries start intersecting the film surface, forming grooves at the surface. These grooves suppress further growth of the grain. As a result, a columnar grain structure with grain boundaries traversing the film thickness results. Thus, the grain size in the plane of the film is on the order of the film thickness.

There have been many mechanisms proposed for the development of growth stresses, which depend sensitively on the material systems, deposition technique, and process parameters. Mechanics at small length scales in micro- and nanotechnology is a challenge for those who recognize its existence and want to optimize the mechanical performance of micro- and nanosystems and a threat for those who ignore it. We will see in Volume III, Chapter 6 on metrology and analytical techniques that surface micromachined devices contribute strongly to the understanding of mechanical properties of thin films.

## Strength of Thin Films

Most thin film properties can be simply explained through extrapolation of bulk behavior, but other properties depend on mechanisms that remain unexplained. The strength of thin films is such an example. A substantial body of work has shown that thin metal films can support much higher stresses than the same material in bulk form. This strengthening has generally been attributed to dimensional and microstructural constraints on dislocation activity in these films. Dimensional constraints are imposed by the interfaces and the small dimensions typically encountered in thin films, whereas microstructural constraints arise from the very fine grains that often typify thin films. In bulk materials, microstructural constraints dominate the plastic behavior of the material. However, when material dimensions are comparable to microstructural length scales—as is typically the case for thin films—free surfaces and interfaces play an important role as well. For example, dislocations can exit the material through free surfaces, whereas interfaces may prevent them from doing so. Consequently, strong interfaces lead to a higher cumulative dislocation density in the film and result in a higher flow stress and a greater strain-hardening rate. In addition to dimensional constraints, the microstructure of thin films also affects their mechanical properties. Because thin films are fabricated using different techniques than those for bulk materials, the microstructure of thin films is often very different from that of bulk materials.

Let us remember what controls the strength of a material. Based on their bond strength most materials should be much stronger than they are in practice. For example, the strength for an ionic bond should be about $10^6$ psi, but a more typical strength for an ionic material is 40 $10^3$ psi. The reason, as we saw in Volume I, Chapter 2, is that materials do not usually fail by breaking bonds. Real materials have lots of dislocations; therefore, the strength of the material depends on the force required to make the dislocation move, not the bonding energy. When dislocations move, slip occurs and the direction of the movement is the same as the Burgers vector (= the deformation direction, e.g., for an edge dislocation it is perpendicular to the dislocation line, and for a screw dislocation it is parallel to the line of the dislocation). When dislocations run into each other a traffic jam effect results; thus, more dislocations increase the strength of a material. Dislocations may be pinned by solutes, interstitials, and precipitates. An example of a solute is Zn atoms in a Cu matrix: the bigger Zn atoms make the slip plane "rougher," and thus increase the resistance to dislocation motion. Precipitates (small particles) also can promote strengthening by impeding dislocation motion. Applying a force to the material increases the number of dislocations, and this is called *strain hardening* or *cold work*. Work hardening accounts for the higher strength of cold rolled steel. Grain boundaries act to impede dislocation motion because the slip systems in adjoining grains will usually not line up; increases in yield strength arising from this mechanism are called *boundary strengthening*. In a small grain, a dislocation gets to the boundary and stops. In a large grain, the dislocation can travel farther. Therefore, a small grain size equates to more strength. Hall and Petch, working independently around 1950, found that the yield strength of a polycrystalline material follows the following dependence:

$$\sigma_Y = \sigma_0 + \frac{K}{\sqrt{d}}$$

(Volume I, Equation 2.70)

where $\sigma_0$ is the lattice friction stress needed to move dislocations and $K$ is a constant (e.g., the elastic limit of copper doubles when the grain size decreases from 100 μm to 25 μm). This relation is essentially empirical, but it can be rationalized by viewing the second term as related to the stress needed to activate a new mobile dislocation in an unfavorably oriented grain. The Hall-Petch relation is illustrated in Figure 7.52 for steel at room temperature. The grain size is often given in ASTM grain size number ($n$)—a measure of the size of the grains in a crystalline material obtained by counting the number of grains per square inch at a magnification × 100. Instead of yield strength one might also plot the hardness ($H$) of the material in Figure 7.52. Indentation (hardness) testing is very common for bulk materials where the direct relationship between bulk hardness and yield strength is well known. Pressing a hard, specially shaped point into the surface and observing the indentation can measure hardness. This type of measurement is of little use for measuring thin films less than 5 × $10^4$ Å. Consequently, very little is known

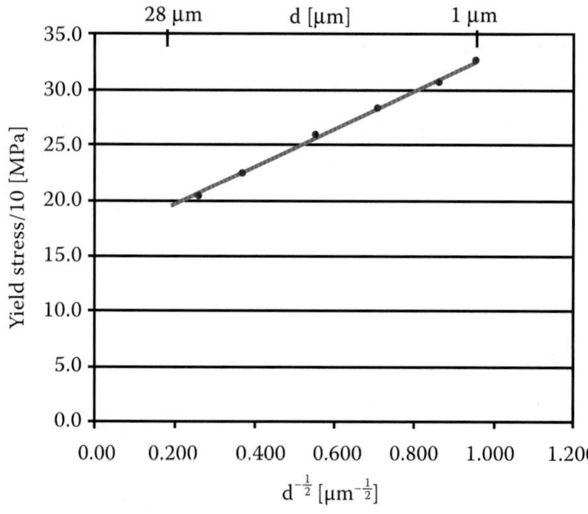


**FIGURE 7.52** The Hall-Petch equation—the relationship between yield strength and grain size—in steel at room temperature.

about the hardness of thin films. Recently, specialized instruments have been constructed (e.g., the nanoindenter) in which load and displacement data are collected while the indentation is introduced in a thin film. This eliminates the errors associated with later measurement of indentation size and provides continuous monitoring of load/displacement data similar to a standard tensile test. Load resolution may be 0.25 μN, displacement resolution 0.2–0.4 nm, and *x*-*y* sample position accuracy 0.5 μm. Empirical relations have correlated hardness with Young's modulus and with uniaxial strength of thin films. Hardness calculations must include both plastic and long-distance elastic deformation. If the indentation is deeper than 10% of the film, corrections for elastic hardness contribution of the substrate must also be included.[43] Mechanical properties such as hardness and modulus of elasticity can also be determined on the micro- to picoscales using an atomic force microscope (AFM).[44] Bushan[45] provides an excellent introduction to this field in the *Handbook of Micro/Nanotribology*.

The Hall-Petch relationship has been well established experimentally for grain sizes in the millimeter through submicrometer regimes. Consequently, it was thought that nanosized grains would produce materials with even greater mechanical integrity. The above relation indicates that simply by decreasing the grain size one can make a material arbitrarily strong! Of course, this is not sensible because a polycrystalline material with grain size of a single atom would just be amorphous or a single crystal. In other words, there must be a limit where Hall-Petch breaks down! It is easily recognized that the Hall-Petch behavior is expected to break down, for example, if the smallest dislocation loop no longer fits in a grain (Figure 7.53). For very small grain sizes one can perform atomistic simulations of the deformation, and these simulations suggest that for very small grain sizes there is a reverse Hall-Petch effect, i.e., the strength decreases with increasing grain size; in other words, an optimal grain size ($d_c$) exists as suggested by the plot in Figure 7.54. The classic Hall-Petch relationship is based on the idea that grain boundaries act as obstacles to dislocation movement and that dislocations require greater amounts of energy to overcome these barriers. Because dislocations are carriers of plastic deformation, this manifests itself macroscopically as an increase in material strength. The deformation mechanism for materials with very small grains (<20 nm) is different, and it has been suggested that plastic deformation in this case is no longer dominated by dislocation motion but by atomic sliding of grain boundaries.[46] For very small grain sizes, this sliding effect might conceivably constitute the dominant effect because of the larger ratio of grain boundary to crystal lattice, and this mode of deformation could then lead to the observed softening of a nanomaterial. Other explanations to rationalize the apparent softening of metals with nanosized grains include poor sample quality and the suppression of dislocation pileups.[47]

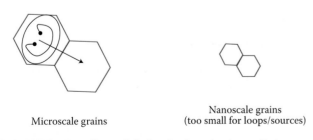


**FIGURE 7.53** Hall-Petch behavior breaks down if the smallest dislocation loop no longer fits in a grain.

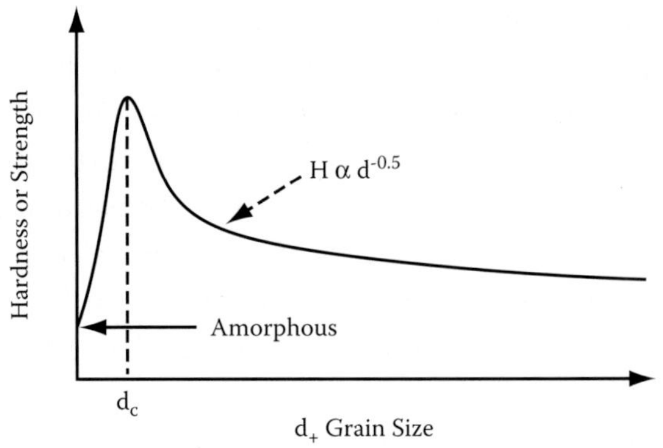


**FIGURE 7.54** Hall-Petch and reverse Hall-Petch.

A reverse Hall-Petch behavior has been observed experimentally in nanocrystalline materials with sufficiently small grain sizes.[48] There are also observations of "kinked" Hall-Petch graphs, i.e., cases where the slope is reduced (but still positive) below a certain grain size. However, today there is still a lot of uncertainty about the exact interpretation of a reverse Hall-Petch effect. It is possible that many of the early measurements of a reverse Hall-Petch effect were likely the result of unrecognized pores in the samples. The presence of voids in nanocrystalline metals would undoubtedly lead to their having weaker mechanical properties. It is not entirely clear what exactly the dependency of yield stress should be on grain sizes less than a diameter *d* of 20 nm. Nanocrystalline materials have been the subject of widespread research in recent decades; since the review by Gleiter[49] in 1989, thousands of papers have been published on the topic.

The yield strength of metallic films is affected by several microstructural dimensions: grain size, film thickness, free-standing or deposited on a substrate, and obstacle spacing (e.g., precipitate concentration). A common approach to the quantification of the role of different parts of the microstructure to the yield stress of a deposited film is to assume that their contributions have a power-law form and can be superimposed:

$$\sigma_Y = \sigma_0 + Kd^{-n} + K't^{-m} \tag{7.22}$$

where the first term represents the bulk yield stress for a large-grained polycrystal, the second represents the contribution from the grain boundaries (with *d* the grain size), and the third represents the contribution from the film surface or interface with *t*, the film thickness. The first two terms together form the well-known Hall-Petch relation, in which, commonly, n = 0.5.[50] For thin films with very fine grains ($d \ll t$) or in multilayers with very thin constituent layers, the deposition of individual dislocation segments becomes more important than the formation of pileups, which makes *n* approach 1. In several studies of thin films deposited on a substrate, an increase of the yield strength with decreasing film thickness has been observed. For example, for thin films of copper, with thickness between 0.1 and 3 µm, vapor deposited on 12.7- or 7.6-µm-thick polyimide Kapton substrates, Yu et al.[50] find that the yield stress depends strongly on the film thickness and is fit by $\sigma_Y = 116 + 355t^{-0.473}$ where *t* is the thickness in meters and $\sigma_Y$ is in MPa. The origin of the hardening induced by the finite size of the film is the deposition of dislocation segments at the film surface if oxidized or otherwise passivated and at the film substrate interface. The blocking of the dislocations at the film substrate interface leads to an inhomogeneous strain distribution, in which the plastic strain approaches zero near the interface.

For free-standing films, the effect of film thickness is reversed, with thinner films being weaker. It has been found that grain-size effects are diminished when Cu test sample size decreases to less than about 1 mm, an observation that has important consequences for micromanufacturing. When the grain size *d* varied between 18 µm and 200 µm, it was found that strength is reduced with an increase in *d*/*t*, not the Hall-Petch grain size *d*.[51] The tensile strength may apparently be described by a rule of mixtures between surface grains with a grain-size independent strength and interior grains that obey the Hall-Petch relation. As shown in Figure 7.55, the tensile strength of Al films decreases with decreasing part thickness in the range of 0.1–0.5 mm.[51]

From the above, size effects may strengthen or weaken the material, and interface strengthening or weakening will play a dominant role. Let us consider Si yield strength as an example. Because of the high activation energy for dislocation motion in silicon (2.2 eV), hardly any plastic flow occurs in single crystalline silicon for temperatures lower than 673°C. Grain boundaries in polysilicon further enhance this blocking of dislocation motion; hence polysilicon films can be treated as an ideal brittle material at room temperature.[52] From the above, we now understand why high yield strengths often are obtained in thin deposited films with values up to 200 times as large as those found in the corresponding bulk material.[53] In this light, the fact that fracture stress of polysilicon is between 2 and 10 times smaller than that of bulk single crystal is surprising. Greek et al.[54] explain this deviation by pointing at the high surface roughness of polysilicon films compared with single-crystal Si. They believe that a reduction in surface roughness would improve the tensile fracture strength considerably.

## Adhesion

The importance of adhesion of various films to one another and to the substrate in overall IC performance and reliability cannot be stressed enough. Because

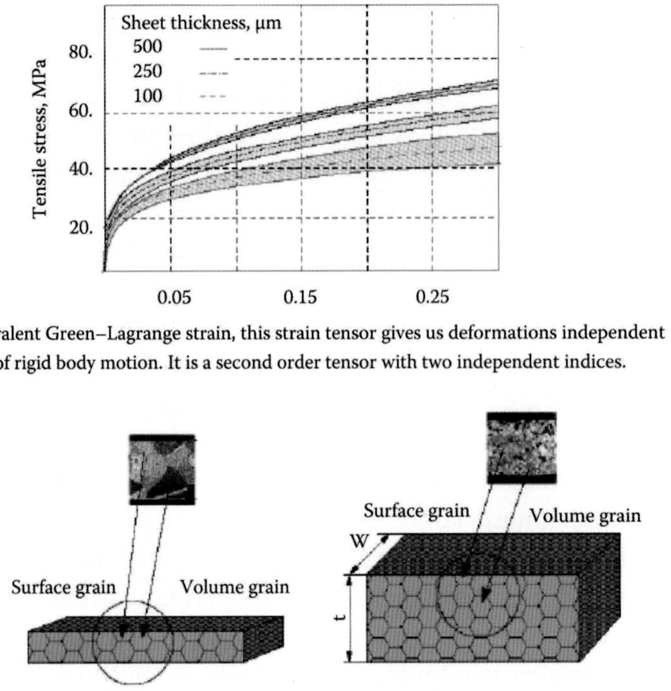


**FIGURE 7.55** Thinner Al films mean weaker.[51]

mechanical pulling forces might be involved, adhesion is even more crucial in micromachining. If films lift from the substrate under a repetitive, applied mechanical force, the device will fail. Classical adhesion tests include the scotch tape test, abrasion method, scratching, deceleration (ultrasonic and ultracentrifuge techniques), bending, pulling, and so on (see also Volume III, Chapter 6 on metrology and analytical techniques).[55] Micromachined structures, because of their sensitivity to thin-film properties, enable some innovative new ways of in situ adhesion measurement. Figure 7.56 illustrates how a suspended membrane may be used for adhesion measurements. In Figure 7.56a, the membrane is suspended but still adherent to the substrate. Figure 7.56b shows the membrane after it has been peeled partially from its substrate by an applied load (gas pressure). Figure 7.56c illustrates the accompanying pressure-volume (*P-V*) cycle, in which the membrane is inflated, peeled, and then deflated. The shaded portion of Figure 7.56c illustrates the *P-V* work creating the new surface, which equals the average work of adhesion for the film-substrate interface times the area peeled during the test.[55]

Cleanliness of a substrate is a *conditio sine qua non* for good film adhesion. Roughness, providing more bonding surface area and mechanical interlocking, further improves it. Adhesion also improves with increasing adsorption energy of the deposit and/or increasing number of nucleation sites in the early growth stage of the film. Sticking energies between film and substrate range from less than 10 kcal/mol in physisorption to more than 20 kcal/mol for chemisorption. The weakest form of

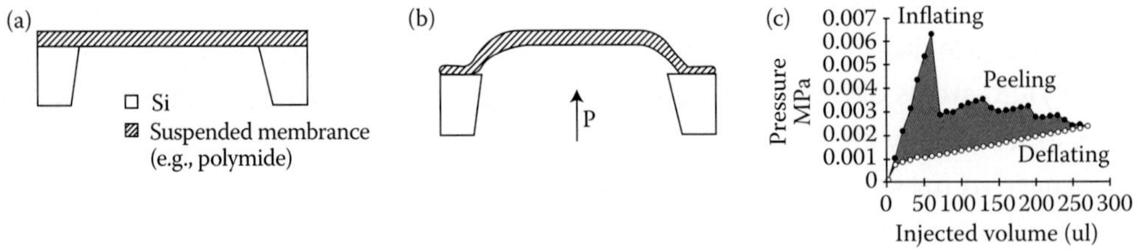


**FIGURE 7.56** Micromachined structure to evaluate adhesion. (a) Suspended membrane. (b) Partially detached membrane-outward peel. (c) Pressure-volume curve during inflate-deflate cycle. (From Senturia, S. 1987. *Proceedings: IEEE micro robots and teleoperators workshop.* Hyannis, MA: IEEE. With permission.[55])

adhesion involves Van der Waals forces only (see also Table 7.17).

It is highly advantageous to include a layer of oxide-forming elements between a metal and a substrate. These adhesion layers, such as Cr, Ti, and Al, provide good anchors for subsequent metallization. Intermediate film formation allows for a continuous transition from one lattice to the other and results in the best adhesion. Adhesion also improves when formation of intermetallic metal alloys takes place.

## Stress in Thin Films

### *Qualitative Description*

Film cracking, delamination, and void formation may all be linked to film stress. Nearly all films foster a state of residual stress as a result of mismatch in the thermal expansion coefficient, nonuniform plastic deformation, lattice mismatch, substitutional or interstitial impurities, and growth processes. Figure 7.57 illustrates stress-causing factors categorized as either intrinsic or extrinsic.[55] The intrinsic stresses (also growth stresses) develop during the film nucleation. Extrinsic stresses are imposed by unintended external factors such as temperature gradients or sensor package-induced stresses. Thermal stresses, the most common type of extrinsic stresses, are well understood and often rather easy to calculate (see below on page 448). They arise either in a structure with inhomogeneous thermal expansion coefficients subjected to a uniform temperature change or in a homogeneous material exposed to a thermal gradient.[56] Intrinsic stresses in thin films often are larger than thermal stresses. They usually are a consequence of the nonequilibrium nature of the thin-film deposition process. For example, in chemical vapor deposition (CVD), depositing atoms (adatoms) may at first occupy positions other than the lowest energy configuration. With too high a deposition rate and/or too low adatom surface mobility, these first adatoms may become pinned by newly arriving adatoms, resulting in the development of intrinsic stress. Other types of intrinsic stresses illustrated in Figure 7.57 include transformation stresses occurring when part of a material undergoes a volume change during a phase transformation; misfit stresses arising in epitaxial

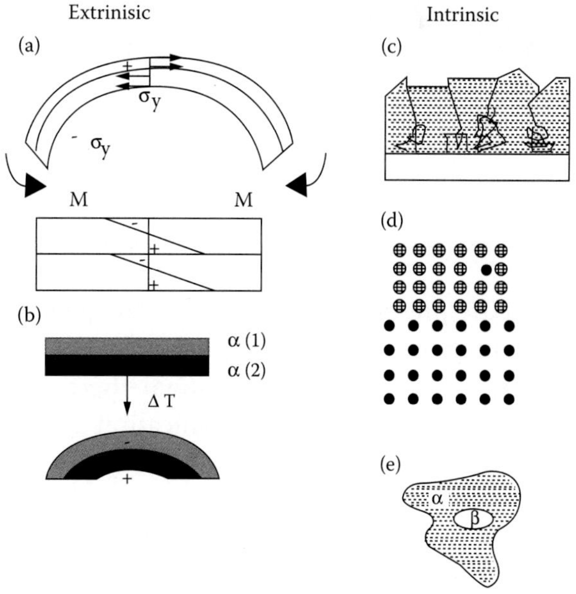


**FIGURE 7.57** Examples of intrinsic and extrinsic residual stresses. (a) Nonuniform plastic deformation results in residual stresses on unloading. M = bonding moment. (b) Thermal expansion mismatch between two materials bonded together. α (1) and α (2) are thermal expansion coefficients. (c) Growth stresses evolve during film deposition. (d) Misfit stresses resulting from mismatches in lattice parameters in an epitaxial film and stresses from substitutional or interstitial impurities. (e) Volume changes accompanying phase transformations cause residual stresses. (From Krulevitch, P. A. 1994. *Micromechanical investigations of silicon and Ni–Ti–Cu thin films.* PhD thesis, University of California, Berkeley.[56])

films as a result of lattice mismatch between film and substrate; and impurities, either interstitial or substitutional, which cause intrinsic residual stresses as a result of the local expansion or contraction associated with point defects. Intrinsic stress in a thin film does usually not suffice to result in delamination unless the film is thick. For example, to overcome a low adsorption energy of 0.2 eV, a relatively high stress of about $5 \times 10^9$ dyn/cm$^2$ ($10^7$ dyn/cm$^2$ = 1 MPa) is required.[53] High stress can result in buckling or cracking of films.

The stress developing in a film during the initial phases of a deposition may be compressive (i.e., the film tends to expand parallel to the surface), causing buckling and blistering or delamination in extreme cases (especially with thick films). Alternatively, thin films may be in tensile stress (i.e., the film tends to contract), which may lead to cracking if forces high enough to exceed the fracture limit of the film

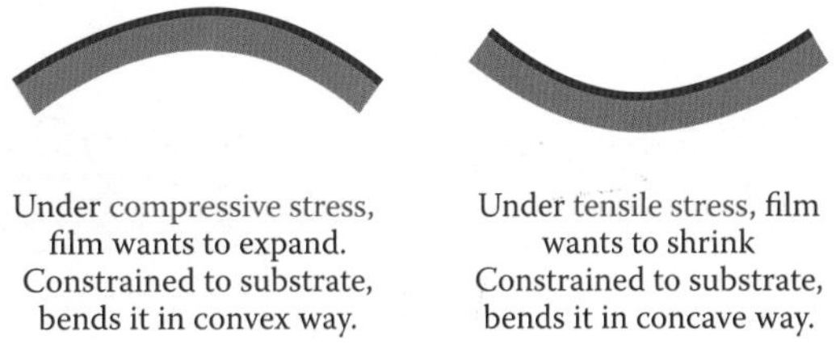

Under compressive stress, film wants to expand. Constrained to substrate, bends it in convex way.
Under tensile stress, film wants to shrink Constrained to substrate, bends it in concave way.


**FIGURE 7.58** Tensile and compressive stress in thin films.

material are present (Figure 7.58). Subsequent rearrangement of the atoms, either during the remainder of the deposition or with additional thermal processing, can lead to further densification or expansion, decreasing remaining tensile or compressive stresses, respectively.

The mechanical response of thin-film structures is affected by the residual stress, even if the structures do not fail. For example, if the residual stress varies in the direction of film growth—in other words, if there is a stress gradient—the resulting built-in bending moment will warp released structures, such as cantilever beams (Figure 7.59). The presence of residual stress also alters the resonant frequency of thin-film, resonant microstructures (see Equation 7.29 below).[57] In addition, residual stress can lead to degradation of electrical characteristics and yield loss through defect generation.[56] For example, it was found that the resistivity of stressed metallic films is higher than that of their annealed counterparts. Residual stress has been used advantageously in a few cases, such as in self-adjusting microstructures,[58] and for altering the shape-set configuration in shape memory alloy films.[56]

In general, the stresses in thin films, by whatever means produced, are in the range of $10^8$ to $5 \times 10^{10}$ dyn/cm$^2$ and can be either tensile or compressive. For normal deposition temperatures (50 to a few hundred °C), the stress in metal films typically

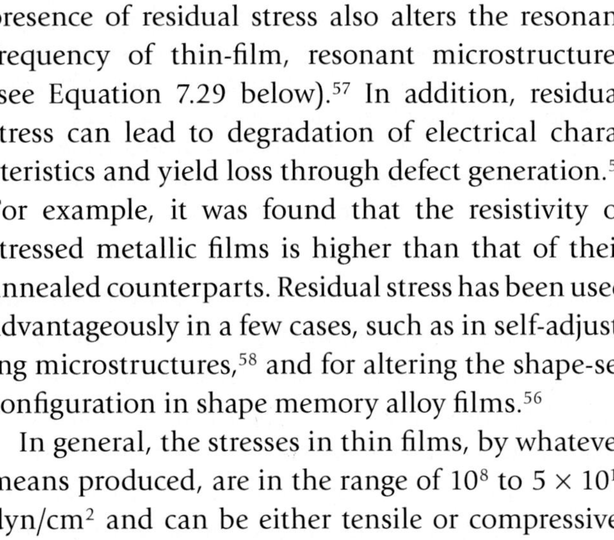

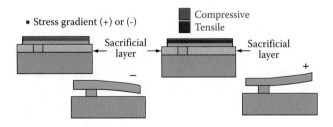

**FIGURE 7.59** Stress gradients cause a bending moment in a cantilever upon release, i.e., removal of the sacrificial material.

ranges from $10^8$–$10^{10}$ dyn/cm$^2$ and is tensile, with the refractory metals (Mo, Ta, Nb, W, and Ta) at the upper end and the soft ones (Cu, Ag, Au, and Al) at the lower end. At low substrate temperatures, metal films tend to exhibit tensile stress. This often decreases in a linear fashion with increasing substrate temperature, finally going through zero or even becoming compressive. The changeover to compressive stress occurs at lower temperatures for lower melting point metals. The mobility of the adatoms is key to understanding the ranking for refractory and soft metals. A metal such as aluminum has a low melting temperature and a corresponding high diffusion rate even at room temperature; thus, usually it is fairly stress free. By comparison, tungsten has a relatively high melting point and a low diffusion rate, and tends to accumulate more stress when sputter deposited. With dielectric films, stresses often are compressive and have slightly lower values than commonly noted in metals.

Tensile films result, for example, when a process byproduct is present during deposition and later driven off as a gas. If the deposited atoms are not sufficiently mobile to fill in the holes left by these departing byproducts, the film will contract and go in tension. Nitrides deposited by plasma CVD are usually compressive because of the presence of hydrogen atoms in the lattice. By annealing, which drives out the hydrogen, the films can become highly tensile. Annealing also has a dramatic effect on most oxides. Oxides often are porous enough to absorb or give off a large amount of water. Full of water, they are compressive; devoid of water, they are tensile. Thermal $SiO_2$ is compressive even when dry. If atoms are jammed in place (such as with sputtering), the film tends to act compressively. The stress in a thin film also varies with depth.

When reviewing dry etching in Chapter 3 we discussed plasma chemistry and physics. The following listing further completes that information. The RF power of a plasma-enhanced CVD (PECVD) deposition influences stress, e.g., a thin film may start out tensile, decrease as the power increases, and finally become compressive with further RF power increase. CVD equipment manufacturers concentrate on building stress-control capabilities into new equipment by controlling plasma frequency.

Therefore, in what follows we pay specific attention to PECVD parameter settings influencing thin-film properties, important in building surface micromachined microstructures, i.e., film stress and density. Surface micromachining uses deposited thin films, such as polysilicon, silicon nitride, and silicon dioxide. One of the challenges of any surface-micromachined process is to control the intrinsic stresses in those deposited films. Plasma settings act as critical controlling parameters of intrinsic stress in CVD films. Recall that the plasmas of interest are low-pressure glow discharges consisting of ions, electrons, and neutral species. The neutral species, molecules, and radicals greatly outnumber the electrons and ions and are not influenced by external electrical fields. Just as etching occurs mainly through neutral species, deposition also almost exclusively involves neutrals. In etching, the ions impart anisotropy to the etching process, whereas in deposition they alter the properties of the deposited films. The different reactor parameter settings have the following influence on the CVD deposited films:

1. *Total reactor pressure*: Because gas density varies with pressure, the mean free path is longer at lower pressures. Consequently, ions accelerated toward the cathode at lower pressure can gain more energy before a collision takes place. Therefore, the effect of ion bombardment is more pronounced at lower pressures, and better quality CVD films ensue, characterized as films with a low wet etch rate (high film density) and low compressive stress. Experimental results show that as the reactor pressure is lowered, film stress goes from tensile to compressive and wet etch rates decrease. Often, stress dynamically changes when the film is exposed to the atmosphere and subsequent heating. In the case of oxide films, for example, tensile films take up water, swell, and become more compressive (the substrate bends down). Compressive oxide films have a greater moisture resistance. A nitride film, on the other hand, absorbs no water and shows no tendency toward compressive stress with time. In the latter case, the amount of Si-H bonds, controlled by annealing at 490°, seems to dominate the stress behavior.[59] Too high pressures promote gas-phase polymerization, increasing defect density in the deposited material. In the other extreme, too low pressures (alternatively, the reactant gas is too diluted) change the process from CVD-like to PVD-like, giving way to a columnar film morphology with more defects.
2. *Frequency of the RF excitation*: At low frequencies, ions experience the full amplitude of the RF voltage, whereas above the ion-transition frequency (>3 MHz), where ions cannot follow anymore, the ion energy is determined by the time average of the RF amplitude. Consequently, lower frequency shifts the ion energy upward. At the lower frequencies, wet etch rates decrease, and compressive film stress results. Again, higher-energy bombardment yields better films. Higher ion bombardment also improves the film quality on sidewalls. Multiple-frequency plasma equipment is emerging rapidly because it allows the user precise control over film properties (particularly stress) for a wide range of process conditions and can also improve step coverage.
3. *RF power effects*: An increase in RF power leads to more intense ion bombardment due to the increase in ion current. With a higher ion current, the film deposition rate goes up. To separate the effect of RF power on growth rate and film quality, the ratio of power density to deposition rate must be evaluated. This ratio represents a rough measure of ion bombardment per deposited molecule. In a plot of wet etch rate versus power density divided by the deposition rate, the maximum ion bombardment, corresponding to highest energy density, leads to the lowest wet etch rate.[60]
4. *Growth temperature*: The growth temperature has a strong influence on the structure of the film. At low temperatures (and high growth rates), the surface diffusion is slow relative to the arrival rate of film precursors. In this situation, the adsorbed precursor molecule is likely to interact with an impinging precursor molecule before it has a chance to diffuse away on the surface, and an amorphous film is formed. At high temperatures (and low growth rates),

the surface diffusion is fast relative to the incoming flux. The adsorbed species can diffuse to step growth sites, forming single crystalline materials.

Summarizing, low RF excitation frequency, low reactor pressures, and low deposition rate at high temperature contribute to improved film quality as evidenced by stress and (wet) etch rate measurements. More information on CVD thin films and how they are influenced by the above deposition parameters, as well as other parameters (e.g., moisture, flow rate, and gas composition) will be given in Volume III, Chapter 5 on selected MEMS and NEMS materials.

### Stress in Thin Films on Thick Substrates—Quantitative Analysis

The total stress in a thin film typically is given by:

$$\sigma_{tot} = \sigma_{th} + \sigma_{int} + \sigma_{ext} \tag{7.23}$$

i.e., the sum of any intentional external applied stress ($\sigma_{ext}$), the thermal stress ($\sigma_{th}$, an unintended external stress), and different intrinsic components ($\sigma_{int}$). With constant stress through the film thickness, the stress components retain the form of:

$$\begin{aligned} \sigma_x &= \sigma_x(x,y) \\ \sigma_y &= \sigma_y(x,y) \\ \tau_{x,y} &= \tau_{x,y}(x,y) \\ \tau_{xz} &= \tau_{yz} = \sigma_z = 0 \end{aligned} \tag{7.24}$$

In other words, the three nonvanishing stress components are functions of *x* and *y* alone. No stress occurs in the direction normal to the substrate (*z*). With *x*,*y* as principal axes, the shear stress $\tau_{xy}$ also vanishes,[61] and Equation 7.24 reduces to the following strain-stress relationships:

$$\begin{aligned} \varepsilon_x &= \frac{\sigma_x}{E} - \frac{\nu\sigma_y}{E} \\ \varepsilon_y &= \frac{\sigma_y}{E} - \frac{\nu\sigma_x}{E} \\ \sigma_z &= 0 \end{aligned} \tag{7.25}$$

In the isotropic case, $\varepsilon = \varepsilon_x = \varepsilon_y$, so that $\sigma_x = \sigma_y = \sigma$, or:

$$\sigma = \left(\frac{E}{1-\nu}\right)\varepsilon \tag{7.26}$$

where the Young's modulus *E* of the film and the Poisson ratio of the film act independently of orientation. The quantity $E/1 - \nu$ often is called the *biaxial modulus*. Uniaxial testing of thin films is difficult, prompting the use of the biaxial modulus rather than Young's uniaxial modulus (see Volume III, Chapter 6 on metrology and analytical techniques). Plane stress, as described here, is only a good approximation when several thicknesses (e.g., three) away from the edge of the film.

### Thermal Stress

The most commonly encountered example of thin film stress is the thermal mismatch stress due to the thermal expansion mismatch between a film and its substrate. The manufacturing process and normal operation of most thin film-based devices often involve large temperature cycles. Therefore, thermal mismatch stresses are inevitable, and they can be very high if the thermal mismatch is large and the elastic modulus of the film is large, such as is the case for Cu films used in integrated circuits (ICs). Thermal stresses develop in thin films when high-temperature deposition or annealing is involved, and they usually are unavoidable because of mismatch of thermal expansion coefficients between film and substrate. The problem of a thin film under residual thermal stress can be modeled by considering a thought experiment involving a stress-free film at high temperature on a thick substrate. Imagine detaching the film from the high-temperature substrate and cooling the system to room temperature. Usually, the substrate dimensions undergo minor shrinkage in the plane, whereas the film's dimensions may reduce significantly. To reapply the film to the substrate with complete coverage, the film needs stretching with a biaxial tensile load to a uniform radial strain, ε, followed by perfect bondage to the rigid substrate and load removal. The film stress is assumed to be the same in the stretched and freestanding film as in the film bonded to the substrate, i.e., no relaxation occurs in the bonding process. To calculate the thermal residual stress from Equation 7.26, the elastic moduli of the film must be known, as well as the volume change associated with the residual stress, i.e., the thermal strain, $\varepsilon_{th}$, resulting from the difference in the coefficients of thermal expansion between the film and the substrate.

Let us now consider whether, qualitatively, the above assumptions apply to the measurement of thin films on Si wafers. Such films typically measure 1 μm thick and are deposited on 4-in. wafers, nominally 550 μm thick. In this case, the substrate measures nearly three orders of magnitude thicker than the film, and, because the bending stiffness is proportional to the thickness cubed, the substrate essentially is rigid relative to the film. The earlier assumptions clearly apply.

Figure 7.60 portrays a quantitative example where a polyimide film, strain-free at the deposition temperature ($T_d$) of 400°C, is cooled to room temperature $T_r$ (25°C) on a Si substrate with a different coefficient of thermal expansion. The resulting strain is given by:

$$\varepsilon_{th} = \int [\alpha_f(T) - \alpha_s(T)]dT \qquad (7.27)$$

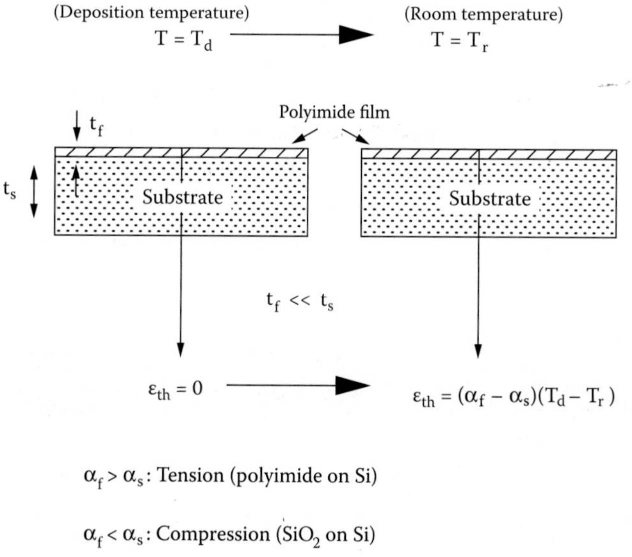


**FIGURE 7.60** Thermal stress. Tension and compression are determined by the relative size of thermal expansion coefficients of film and substrate. Suppose a strain-free film at deposition temperature, $T_d$, is cooled to room temperature, $T_r$, on a substrate with a different coefficient of thermal expansion.

where $\alpha_f$ and $\alpha_s$ represent the coefficients of thermal expansion for the polyimide film and the Si substrate, respectively. The thermal strain can be of either sign; based on the relative values of $\alpha_f$ and $\alpha_s$, positive is tensile, and negative is compressive. Polyimide features a thermal expansion ($\alpha_f = 70 \times 10^{-6}$°C$^{-1}$) larger than the thermal expansion coefficient of Si ($\alpha_s = 2.6 \times 10^{-6}$°C$^{-1}$); hence a tensile stress is expected. With $SiO_2$ grown or deposited on silicon at elevated temperatures, a compressive component [$\alpha_f(0.35 \times 10^{-6}$°C$^{-1}) < a_s(2.6 \times 10^{-6}$°C$^{-1})$] is expected. Assuming that the coefficients of thermal expansion are temperature independent, Equation 7.27 simplifies to:

$$\varepsilon_{th} = [\alpha_f - \alpha_s](T_d - T_r) \qquad (7.28)$$

The calculated thermal strain for polyimide on Si then measures $25 \times 10^{-3}$ at room temperature. The biaxial modulus ($E/1 - \nu$), with E = 3 GPa and ν = 0.4, equals 5 GPa, and the residual stress σ, from Equation 7.26, is 125 MPa and tensile.

### *Intrinsic Stress*

The intrinsic stress, $\sigma_i$, reflects the internal structure of a material and is less clearly understood than the thermal stress, which it often dominates.[62] Several phenomena may contribute to $\sigma_i$, making its analysis very complex. Intrinsic stress depends on thickness, deposition rate (locking in defects), deposition temperature, ambient pressure, method of film preparation, type of substrate used (lattice mismatch), incorporation of impurities during growth, and so on. Some semiquantitative descriptions of various intrinsic stress-causing factors follow:

- Doping ($\sigma_{int} > 0$ or $\sigma_{int} < 0$): When doping Si, the atomic or ionic radius of the dopant and the substitutional site determine the positive or negative intrinsic stress ($\sigma_{int} > 0$ or tensile and $\sigma_{int} <$ 0 or compressive). With boron-doped polysilicon, a small atom compared with Si, the film is expected to be tensile ($\sigma_{int} > 0$); with phosphorous doping, a large atom compared with Si, the film is expected to be compressive ($\sigma_{int} < 0$).
- Atomic peening ($\sigma_{int} < 0$): Ion bombardment by sputtered atoms and working gas densifies thin films, rendering them more compressive.

Magnetron sputtered films at low working pressure (<1 Pa) and low temperature often exhibit compressive stress. This topic was discussed in some detail higher above.

- Microvoids ($\sigma_{int} > 0$): Microvoids may arise when byproducts during deposition escape as gases and the lateral diffusion of atoms evolves too slowly to fill all the gaps, resulting in a tensile film.
- Gas entrapment ($\sigma_{int} < 0$): As an example we can cite the hydrogen trapped in $Si_3N_4$. Annealing removes the hydrogen, and a nitride film, compressive at first, may become tensile if the hydrogen content is sufficiently low.
- Shrinkage of polymers during cure ($\sigma_{int} > 0$): The shrinkage of polymers during curing may lead to severe tensile stress, as becomes clear in the case of polyimides. Special problems are associated with measuring the mechanical properties of polymers because they exhibit a time-dependent mechanical response (viscoelasticity), a potentially significant factor in the design of mechanical structures where polymers are subjected to sustained loads.[63]
- Grain boundaries ($\sigma_{int}$ = ?): Based on intuition it is expected that the interatomic spacing in grain boundaries differs depending on the amount of strain, thus contributing to the intrinsic stress. However, the origin of, for example, the compressive stress in polysilicon and how it relates to the grain structure and interatomic spacing are not yet completely clear (see also below on coarse and fine-grated Si).

For further reading on thin-film stress, refer to Hoffman.[64,65] For a short tutorial, visit http://www.uccs.edu/~tchriste/courses/PHYS549/549lectures/mechchar.html.

### *Case Study: Polysilicon Deposition and Film Microstructure*

*Introduction* Thin-film properties prove not only difficult to measure but also cumbersome to reproduce, given the many influencing parameters. Evaporation, sputtering, and molecular beam techniques can deposit dielectric and polysilicon films. In VLSI and surface micromachining, none of these techniques is as widely used as CVD. The major problems associated with the former methods are defects caused by excessive wafer handling, low throughput, poor step coverage, and nonuniform depositions. From the comparison of CVD techniques for typical IC and MEMS materials in Table 7.19, we can conclude that LPCVD, at medium temperatures, prevails above all others. VLSI devices and integrated surface micromachines require low processing temperatures to prevent movement of shallow junctions, uniform step coverage, few process-induced defects (mainly from particles generated during wafer handling and loading), and high wafer throughput to reduce cost. These requirements are best met by hot-wall, low-pressure depositions.[66] While depositing a material with LPCVD the following process parameters can be varied: deposition temperature, gas pressure, flow rate, and deposition time.

The IC industry applies LPCVD polysilicon in applications ranging from simple resistors, gates

**Table 7.19** Comparison of Different Deposition Techniques

| | Atmospheric CVD (APCVD) | Low T LPCVD | Medium T LPCVD | Plasma-Assisted CVD (PECVD) |
|---|---|---|---|---|
| Temp (°C) | 300–500 | 300–500 | 500–900 | 100–350 |
| Materials | $SiO_2$, P-glass | $SiO_2$, P-glass, BP-glass | Polysilicon, $SiO_2$, P-glass, BP-glass $Si_3N_4$, SiON | SiN, $SiO_2$, $SiO_2$, SiON |
| Uses | Passivation, insulation, spacer | Passivation, insulation, spacer | Passivation, gate metal, structural element, spacer | Passivation, insulation, structural elements |
| Throughput | High | High | High | Low |
| Step coverage | Poor | Poor | Conformal | Poor |
| Particles | Many | Few | Few | Many |
| Film properties | Good | Good | Excellent | Poor |

*Source:* Adapted from Adams, A. C. 1988. *VLSI technology*. Ed. S. M. Sze. New York: McGraw-Hill.[30]

for MOS transistors, thin-film transistors (TFT) based on amorphous hydrogenated silicon (a-Si:H), DRAM cell plates, and trench fills, as well as in emitters in bipolar transistors and conductors for interconnects. For the last application, highly doped polysilicon is especially suited: it is easy to establish ohmic contact; it is light insensitive and corrosion resistant; and its rough surface promotes adhesion of subsequent layers. Doping elements such as arsenic, phosphorous, or boron reduce the resistivity of the polysilicon. Commercial LPCVD equipment is available to deposit polysilicon on wafers up to 8 in. in diameter. A typical batch is 100–200 wafers. The mechanical properties of polycrystalline silicon are very good:

- Stronger than stainless steel: fracture strength of polysilicon ~ 2–3 GPa (steel ~ 0.2 GPa–1 GPa)
- Young's modulus E ~ 140–190 GPa
- Extremely flexible: maximum strain before fracture ~ 0.5%
- Does not fatigue readily
- Compatible with IC fabrication processes: process parameters are well known

These properties explain how polysilicon also has emerged as the central structural/mechanical material in surface micromachining, and a closer look at the influence of deposition methodology on its material characteristics is warranted. In this case study we tackle polysilicon deposition by LPCVD and compare it with PECVD and sputtered polysilicon.

### *Undoped Polysilicon*

The properties of low-pressure CVD (LPCVD) undoped polysilicon films are determined by the nucleation and growth of the silicon grains. In most cases, the Si wafers to deposit polysilicon on are placed vertically and closely spaced together in quartz boats. Stress depends on crystal structure. LPCVD Si films, grown slightly below the crystallization temperature (about 600°C for LPCVD), initially form an amorphous solid that subsequently may crystallize during the deposition process.[56,68] The CVD method results in amorphous films when the deposition temperatures are well below the melting temperature of Si (1410°C). The subsequent transition from amorphous to crystalline depends on atomic surface mobility and deposition rate. At low temperatures, surface mobility is low, and nucleation and growth are limited. Newly deposited atoms become trapped in random positions and, once buried, require a substantial amount of time to crystallize as solid-state diffusion is significantly lower than surface mobility. That is why, for low temperature deposition, amorphous layers only start to crystallize after sufficient time at temperature in the reactor. Working at temperatures between 580 and 591.5°C, Guckel et al.[68] produced mostly amorphous films. However, Krulevitch,[56] working at only slightly higher temperatures (605°C) and probably leaving the films longer in the LPCVD setup, produced crystallized films. On crossing the transition temperature between amorphous and crystalline growth (see Figures 7.61 and 7.38), crystalline growth immediately initiates at the substrate due to the increased surface mobility, which allows adatoms to find low-energy, crystalline positions from the start of the deposition process.

The deposition temperature at which the transition from amorphous to a crystalline structure occurs depends on many parameters, such as deposition rate, partial pressure of hydrogen, total pressure, presence of dopants, and presence of impurities (O, N, or C).[67] In the crystalline regime, numerous nucleation sites form, resulting in a transition zone of a multitude of small grains at the film/substrate interface to columnar crystallites on top, as shown in the schematic of a 620–650°C columnar film in Figure 7.62. In this figure, a transition zone of small, randomly oriented grains is sketched near the $SiO_2$ layer. The rate of crystallization is faster here than the deposition rate. Columnar grains ranging between 0.03 and 0.3 μm in diameter form on top of the small grains.[67] The columnar coarse grain structure arises from a process of growth competition among the small grains, during which those grains preferentially oriented for fast vertical growth survive at the expense of misoriented, slowly growing grains.[69,70] The lower the deposition temperature, the smaller the initial grain size will be. At 700°C, films also are columnar; however, the grains are cylindrical, extending through the thickness of the entire film, and there is no transition zone near the

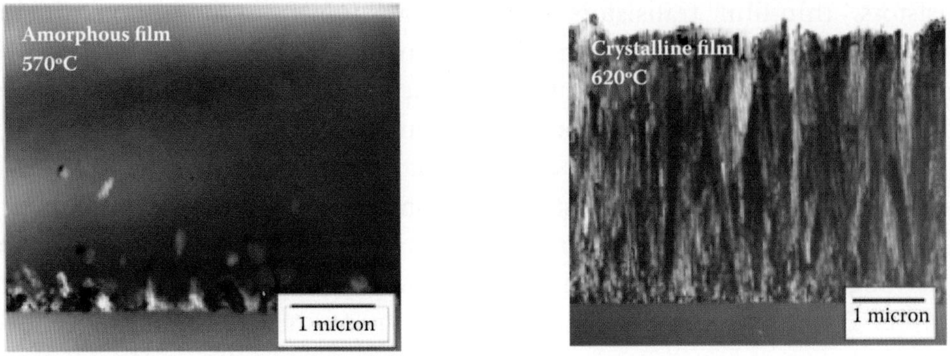


Effect of temperature:

| | |
|---|---|
| Amorphous to crystalline: | 570°C |
| Equi-axed grains: | 600°C |
| Columnar grains: | 625°C |
| (110) Crystal orientation | 625 – 650°C |
| (100) Crystal orientation | 650 – 700°C |

**FIGURE 7.61** From amorphous to polycrystalline Si.

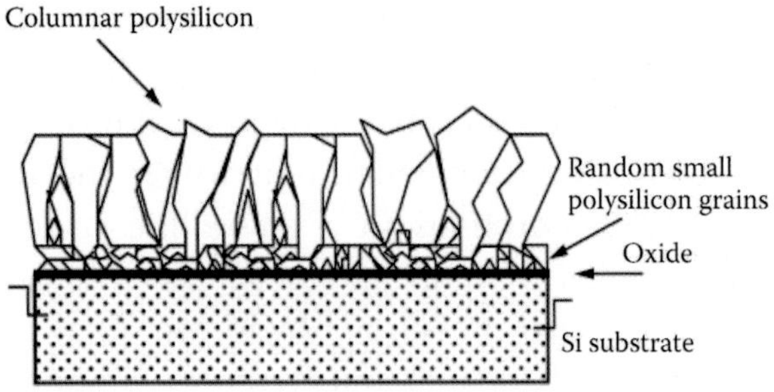


**FIGURE 7.62** Schematic of compressive polysilicon formed at 620–650°C. The columnar coarse-grain structure arises from a process of grain growth competition among the small grains, during which those grains preferentially oriented for fast vertical growth survive at the expense of misoriented, slowly growing grains. (From Krulevitch, P. A. 1994. *Micromechanical investigations of silicon and Ni–Ti–Cu thin films.* PhD thesis, University of California, Berkeley.[56])

$SiO_2$ interface.[56] Temperature has to be very accurately controlled as grains grow with temperature, increasing surface roughness, causing loss of pattern resolution and stresses in MEMS.

Stress in polysilicon films was found to vary significantly with deposition temperature and silane pressure. Guckel et al.[68] found that their mainly amorphous films, deposited at temperatures less than 600°C, proved highly compressive with strain levels as high as −0.67%. At temperatures barely more than 600°C, Krulevitch reports tensile films, whereas for yet higher temperatures (≥620°C), the stress again turns compressive. Whereas films deposited at temperatures more than 630°C all turned out compressive, the magnitude of the compression decreased with increasing temperature. The stress gradient in the polysilicon films explains why compressive undoped and unannealed polysilicon beams tend to curl upward (positive stress gradient) when released from the substrate.[71]

Using high-resolution transmission electron microscopy, Guckel et al.[68] and Krulevitch[56] found a strong correlation between the material's microstructure and the exhibited stress. Guckel et al. found that in their mainly amorphous films, deposited at temperatures less than 600°C, a region near the substrate interface crystallized during growth with grains between 100 and 4000 Å. Krulevitch found that tensile, low-temperature films (605°C) have Si grains dispersed throughout the film thickness. These equiaxed crystals in an isotropic, higher-density film result in a tensile stress with a very small stress gradient. Krulevitch suggests that the compressive stress in the higher temperature compressive films (≥620°C) relates to the competitive growth mechanism of the columnar grains. As crystals grow vertically and in-plane they push on neighbors, resulting in compressive stress and a positive stress gradient. The same author concluded that thermal sources of stress are insignificant.

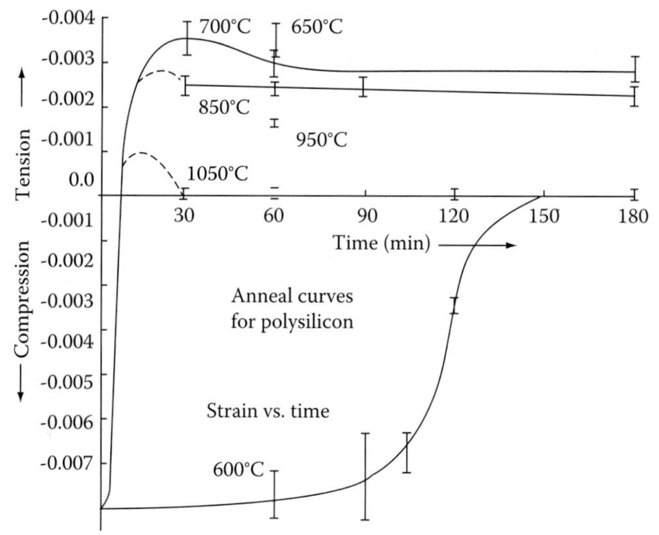


**FIGURE 7.63** Anneal curves for polysilicon. Strain versus anneal time. Upper curves: low-temperature film. Lower curve: high-temperature film. (From Guckel, H., D. W. Burns, C. C. G. Visser, H. A. C. Tilmans, and D. Deroo. 1988. Fine-grained polysilicon films with built-in tensile strain. *IEEE Trans Electron Devices* 35:800–01. With permission.[74])

Importantly, Guckel et al. discovered that annealing in nitrogen or under vacuum converts the low-temperature films with compressive built-in strain (−0.007) to a tensile strain with controllable strain levels between 0 and +0.003 (see Figure 7.63). Annealing can reduce stresses by a factor of 10–100.

During anneal, no grain size increase was noticed (100–4000 Å), but a slight increase in surface roughness was measured. This type of polysilicon is referred to as *fine-grain polysilicon* (also Wisconsin poly-Si). Guckel et al. explain this strain field reversal as follows: as the amorphous region of the film crystallizes, it attempts to contract, but because of the substrate-constrained newly crystallized region, a tensile stress results. Higher temperature films, during an anneal, also become less strained, but the strain remains compressive (see lower curve in Figure 7.63). Moreover, in this case grain size increases, and the surface turns considerably rougher. The latter is called *coarse-grain polysilicon*. Fine-grain polysilicon, with its tensile strain, is preferable; however, it cannot be doped to as low a resistivity as coarse-grain polysilicon. Hence fine-grain polysilicon should be considered a structural material rather than an electronic material.

The above picture is further complicated by the controlling nature of the substrate. For example, depositing amorphous Si at even lower temperatures of 480°C from disilane ($Si_2H_6$) and crystallizing it by subsequent annealing at 600°C demonstrated a large dependency of crystallite size on the underlying $SiO_2$ surface condition. Treating the surface with $HF:H_2O$ or $NH_4OH:H_2O_2:H_2O$ leads to polysilicon films with a large grain size, two or three times as large as without $SiO_2$ treatment, believed to be the consequence of nucleation rate suppression.[72]

Abe and Reed made low-strain polysilicon thin film by DC-magnetron sputtering and postannealing. The films showed very small regional stress and very smooth texture. The deposition rate was 193 Å/min, and the substrate was neither cooled nor heated. The average roughness was found to be comparable with the surface roughness of polished, bare silicon substrates.[73]

Summarizing, stress in polysilicon depends on the material's microstructure, with tension arising from the amorphous to crystalline transformation during deposition and compression from the competitive grain growth mechanism.

Polysilicon deposited at 600–650°C has a {110}-preferred orientation. At higher temperatures, the {100} orientation dominates. Dopants, impurities, and temperature influence this preferred orientation.[67] Drosd and Washburn[75] introduced a model explaining the experimental observation that regrowth of amorphous Si is faster for {100} surfaces, followed by {110} and {111}, which are 2.3 and 20 times slower, respectively. Interestingly, the latter also pinpoints the order of fastest to slowest etching of the crystallographic planes in alkaline etchants. As discussed in Chapter 4, Elwenspoek et al.[76] used this observation of symmetry between etching and growing of Si planes as an important insight to develop a new theory explaining anisotropy in etching. In Table 7.20 we compare the discussed coarse- and fine-grain polysilicon forms.

### Doped Polysilicon

To produce micromachines, doped polysilicon is used far more frequently than undoped polysilicon. Dopants decrease the resistivity to produce conductors and control stress. Polysilicon can be doped by diffusion, implantation, or the addition of dopant gases during deposition (in situ doping).

TABLE 7.20 Comparison of Coarse-Grain and Fine-Grain Polysilicon

| | Coarse-Grain Polysilicon | Fine-Grain Polysilicon |
|---|---|---|
| Temperature of deposition (°C) | 620–650 | 570–591.5 |
| Surface roughness | Rough >50 Å | Smooth <15 Å |
| Grain size | Undoped: 160–320 Å as deposited<br>In situ P-doped: 240–400 Å | Very small grains |
| As deposited strain | −0.002 (compressive) | −0.007 (compressive) |
| Effect of high temperature anneal | Grains size increases<br>Residual strain decreases but remains compressive<br>Reduced bending moment | Grain size increases to 100 Å;[74] others have found 700–900 Å large variation in strain: from compressive to tensile |
| Dry and wet etch rate | Higher for doped material, depends on dopant concentration | Higher for doped material, depends on dopant concentration |
| Texture | <110> as deposited<br><311> in situ P-doped | No texture as deposited, depends on dopant concentration, <111> after 900–1,000°C anneal[77] |

Doping polysilicon films in situ reduces the number of processing steps required for producing doped microdevices by eliminating the need for a subsequent high-temperature step associated with a diffusion or ion-implantation anneal and also provides more uniform doping throughout the film thickness. In situ doping of polysilicon is accomplished by maintaining a constant $PH_3$ to $SiH_4$ gas flow ratio of about 1 vol% in a hot-wall LPCVD setup. At this ratio, the phosphorous content in the film appears above the saturation limit, and the excess dopant segregates at the grain boundaries.[67] Phosphorus diffuses significantly faster in polysilicon than in single-crystalline silicon. The diffusion takes place primarily along the grain boundaries. The diffusivity in thin films of polysilicon (i.e., small equiaxed grains) is about 1 $10^{-12}$ cm$^2$/s. The dopant concentration of in situ doped films is normally very high (~$10^{20}$ cm$^{-3}$). At more than about 1 $10^{21}$ cm$^{-3}$, the film resistivity reaches a plateau of $4 \times 10^{-4}$ Ω·cm because of the low mobility of electrons or holes. The maximum mobility for the highest phosphorus-doped polysilicon is about 30 cm$^2$/Vs.[78]

In situ phosphorus-doped polysilicon undergoes the same amorphous to crystalline growth transformation observed in the undoped film, with the material's microstructure depending on deposition temperature and deposition pressure. The temperature of transformation is lower for the doped films than for the undoped polysilicon and occurs between 580 and 620°C.[79,80] Thus, phosphorus doping enhances crystallization in amorphous silicon,[81] and because of passivation of the polysilicon surface by the phosphine gas, reduces the polysilicon deposition rate.[79] Decreases in deposition rate by as much as a factor of 25 have been reported.[82] Slower deposition rates allow more time for adatoms to find crystalline sites, resulting in crystalline growth at lower temperatures. From Table 7.20, we read that the grain size of phosphorus-doped polysilicon tends to be larger (240–400 Å) than for the undoped material and that {311} planes show up as a texture facet in the doped material. In contrast to in situ phosphine and arsine doping, which both decrease the deposition rate, diborane doping of polysilicon to make it $p^+$ accelerates the deposition rate.[67]

At lower deposition temperatures and higher pressures, the microstructure again consists of amorphous and crystalline regions, whereas at higher temperatures and lower pressures columnar films result and as deposited films exhibit compressive residual stress. The columnar films have a stress gradient that increases toward the film surface, as opposed to the gradient found in undoped columnar polysilicon. This gradient in stress most likely is to the result of nonuniform distribution of phosphorus throughout the film. Annealing at 950°C for 1 h results in the same stress and stress gradient for initially columnar and initially amorphous/crystalline films [i.e., $\sigma_f = -45$ MPa and $\Gamma = +0.2$ mm$^{-1}$ (a linear stress gradient with physical dimensions 1/length), respectively].[56]

As in the case of undoped polysilicon, phosphorus-doped polysilicon films with smooth surfaces (fine grain) are obtained by depositing in situ doped films in the amorphous state and then annealing.[77,83] Phosphorus-doped polysilicon oxidizes faster than undoped polysilicon. The rate of oxidation is determined by the dopant concentration at the polysilicon surface.[67] The addition of oxygen to polysilicon increases the film's resistivity, and the resulting coating, semi-insulating polysilicon (SIPOS) acts as a passivating coating for high-voltage devices in the IC industry. SISPOS has not yet emerged in surface micromachining.

Four drawbacks of in situ phosphorus doping are the complexity of the deposition process, slower deposition rates,[84] reduced film thickness uniformity,[82] and the cleaning of the reactor, which is more demanding for the doped process.

Diffusion is often a more effective method for doping of polysilicon than in situ doping, especially for very heavy dopings (e.g., polysilicon resistivities down to the $10^{-4}$ Ω·cm) of thick (e.g., 2 μm) films. However, diffusion is a high-temperature (e.g., 900–1,000°C) process. If the diffusion step is performed for long durations (a few hours) to achieve uniform doping throughout film thickness a few micrometers thick, it will destroy electronics that are fabricated on the wafer before polysilicon surface micromachining. If performed for too short a time, dopant distribution through the film thickness will not be sufficiently uniform, resulting in difficulties with mechanical property variations through the film thickness.

Ion implantation can also be used for doping polysilicon. The implantation energy is typically adjusted so that the peak of the concentration profile is at the center of the film thickness. However, the resistivity of implanted polysilicon films is not as low as that possible by diffusion.

As with undoped polysilicon, the intrinsic stresses in as-deposited doped polysilicon films are large (>500 MPa) and can result in warping or curling of released micromechanical structures. The values for the fracture stress of boron-, arsenic-, and phosphorus-doped polysilicon are 2.77 ± 0.08 GPa, 2.70 ± 0.09 GPa, and 2.11 ± 0.1 GPa, respectively, compared with 2.84 ± 0.09 GPa for undoped polysilicon. The lower value for phosphorus-doped material has been attributed to high surface roughness and with the large number of defects associated with extensive grain growth in highly phosphorus-doped films.[85] Maluf[86] reports that the operation of polysilicon at elevated temperatures is subject to long-term instabilities, drift, and hysteresis as a result of slow stress-annealing effects.

Several sensors using polysilicon piezoresistive sense elements have been demonstrated. The piezoresistive coefficient is an average over all the orientations in polycrystalline silicon. The gauge factor ranges between 20 and 40, which is about a factor of five smaller than in single-crystal Si, and quickly decreases as the doping exceeds $10^{19}$ $cm^{-3}$. At doping levels of $10^{20}$ $cm^{-3}$, the temperature coefficient of resistance of polysilicon is approximately 0.04% per °C compared with 0.14% per °C for crystalline Si. This is an advantage for the use of polysilicon, even though the gauge factor is low at these high doping levels.

### *PECVD and Sputtered Polysilicon*

The quest for low-temperature polysilicon deposition processes is driven by the need for compatibility with prefabricated aluminum-metallized CMOS circuitry. This makes the 320°C PECVD and ~350°C sputter deposition methods of polysilicon especially interesting, although both methods result in amorphous films.

PECVD films, deposited in a 50-kHz parallel-plate diode reactor, can be doped in situ and crystallized by rapid thermal annealing (RTA; 1100°C, 100 s). It was shown that small-grained PECVD films annealed by RTA have good electrical properties and gauge factors between 20 and 30, similar to those reported for LPCVD polycrystalline silicon (see above).[87]

Honer et al.[88] introduced sputtered polysilicon to the MEMS community. Sputtered silicon can be used to fabricate polysilicon microstructures atop standard, aluminum metallized CMOS at temperatures less than 350°C. Films with stress values less than 100 MPa are routinely obtained.

An important benefit of both these low-temperature processes is that they are not only compatible with conventional oxide sacrificial layers but also with certain organic sacrificial layers (e.g., polyimide). Organic layers can be removed in a dry oxygen plasma etch, thus avoiding the stiction and selectivity problems associated with wet etch releases and at the same time alleviating the concerns over chemical attack on structural elements by HF.

A drawback for the sputtered boron-doped Si films is that to increase their conductivity (to 25 mho/□), they need to be clad in symmetric, 50-nm-thick layers of titanium-tungsten.

Kamins[78] provides a detailed study of polysilicon physical properties. Another good resource on polysilicon morphology and doping can be found at http://www.iue.tuwien.ac.at/22.0.html. Sharpe and Edwards[89] at Johns Hopkins have perhaps provided the most detailed polysilicon mechanical property studies.

## Thin Film Characterization

In Volume III, Chapter 6 on metrology, we detail thin film characterization. For material structure we rely on scanning electron microscopy (SEM), transmission electron microscopy (TEM), scanning probe microscopies (STM, AFM), and x-ray diffraction (XRD). For chemical composition we use auger electron spectroscopy (AES), energy dispersive analysis of x-rays (EDAX or EDS), x-ray photoelectron spectroscopy (XPS), secondary ion mass spectrometry (SIMS), and Rutherford backscattering (RBS). Mechanical properties of thin films are studied using residual stress measurement and micro-/nanoindentation. Electrical properties are derived from resistance (four-point probe) and capacitance measurements. For optical properties we rely on ellipsometry, and for film thickness we rely on stylus profilometer and ellipsometry.

We also introduce a wide variety of surface micromachined devices that are used in characterization of thin films.

## Surface Micromachining

### Introduction

Bulk micromachining means that 3D features are etched into the bulk of crystalline and noncrystalline materials (see Chapter 4). In contrast, surface-micromachined features are built up, layer by layer, on the surface of a substrate (e.g., a single-crystal silicon wafer). Dry etching defines the surface features in the $x$, $y$ plane, and wet etching releases them from the plane by undercutting. In surface micromachining,

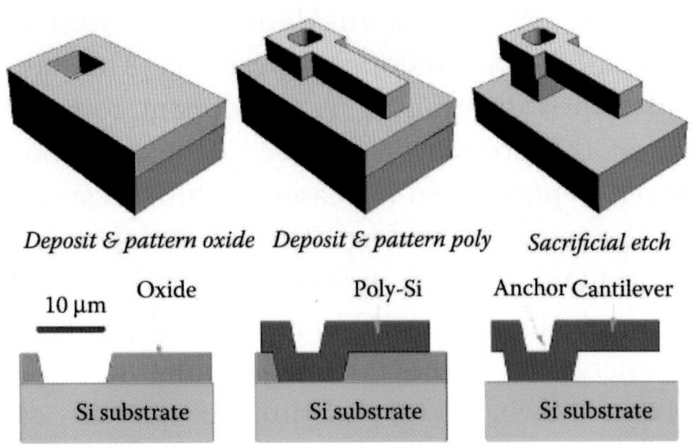


**FIGURE 7.64** Surface micromachining of a polysilicon anchor.

shapes in the $x$, $y$ plane are unrestricted by the crystallography of the substrate (see Figure 7.1). In Figure 7.64 a polysilicon anchor is made using typical surface micromachining techniques.

The nature of the deposition processes involved determines the limited height of surface-micromachined features (Hal Jerman, from EG&G's IC Sensors, called them *2.5D features*[90]). Specifically, low-pressure CVD (LPCVD) polycrystalline silicon (polysilicon) films generally are only a few micrometers high (low $z$), in contrast with wet bulk micromachining where only the wafer thickness limits the feature height. A low $z$ may be a drawback for some mechanical sensors. For example, it would be difficult to fashion a large inertial mass for an accelerometer from thin polysilicon plates (a commercial surface-micromachined accelerometer, the ADLX05, has an inertial mass of only 0.3 μg). Not only do many parameters in the LPCVD polysilicon process need to be controlled very precisely, but subsequent high-temperature annealing (e.g., at temperatures of about 580°C) also is needed to transform the deposited amorphous silicon into polysilicon—the main structural material in surface micromachining. Even with the best possible process control, polysilicon has some material disadvantages over single-crystal Si. For example, it generates a somewhat smaller yield strength (values between 2 and 10 times smaller have been reported)[91,92] and has a lower piezoresistivity.[93] Moreover, because a single polysilicon grain diameter may constitute a significant fraction of the thickness of a mechanical member, the effective Young's modulus may exhibit variability from sensor to sensor.[94] An important positive attribute of polysilicon is that its material properties, although somewhat inferior to single crystal, are far superior compared with those of metal films, and, most of all, because they are isotropic, design is rendered dramatically simpler than with single-crystal material. Dimensional uncertainties may be of greater concern than material issues. Although absolute dimensional tolerances obtained with lithography techniques can be submicrometer, relative tolerances are poor, perhaps 1% on the length of a 100-μm-long feature. The situation becomes critical with yet smaller feature sizes (see Figures 6.3 and 6.4). Although the relatively coarse dimensional control in the microdomain is not specific to surface micromachining, there is no crystallography to rely on for improved dimensional control, as in the case of wet bulk micromachining. Moreover, because the mechanical members in surface micromachining tend to be smaller, more postfabrication adjustment of the features is required to achieve reproducible characteristics. Finally, the wet process for releasing structural elements from a substrate tends to cause sticking of suspended structures to the substrate, or *stiction*, introducing another disadvantage associated with surface micromachining.

Some of the problems associated with surface micromachining mentioned above have been resolved by process modifications and/or alternative designs, and the technique has rapidly gained commercial interest, mainly because it is the most IC-compatible micromachining process developed to date. Moreover, in the past 10–15 years, processes such as silicon on insulator (SOI) (1),[95] hinged polysilicon (2),[96] Keller's molded milliscale polysilicon (3),[97] thick (10 μm and beyond) polysilicon (4),[98] LIGA and LIGA-like processes (5), and porous polysilicon (6) have further enriched the surface micromachining arsenal. Some preliminary remarks on each of these surface micromachining extensions follow.

1. Silicon crystalline features, anywhere between fractions of a micrometer to 100 μm high, can be readily obtained by surface micromachining of the epi silicon or fusion-bonded silicon layer of SOI wafers.[99] Structural elements made from these single-crystalline Si layers result in more reproducible and reliable sensors. SOI or epi-micromachining combines the best features of surface micromachining (i.e., IC compatibility and freedom in $x,y$ shapes) with the best features of bulk micromachining (superior single-crystal Si properties). Moreover, SOI surface micromachining frequently involves fewer process steps and offers better control over the thickness of crucial building blocks. Given the poor reproducibility of mechanical properties and generally poor electronic characteristics of polysilicon films, SOI machining may surpass the polysilicon technology for fabricating high performance devices.
2. The fabrication of polysilicon planar structures for subsequent vertical assembly by mechanical rotation around micromachined hinges, introduced by Pister et al.[96] in 1991, dramatically increases the plethora of designs feasible with polysilicon. Today, erecting these polysilicon structures with the probes of an electrical probe station, or occasionally assembly by chance in an HF etch or DI water rinse,[100] represents a too-complicated or unreliable postrelease assembly method for commercial acceptance, and alternative self-assembly means are an urgent need.
3. While at the University of California, Berkeley, Keller,[97] now CEO of MEMS Precision Instruments (http://www.memspi.com), introduced a combination of surface micromachining and LIGA-like molding processes in the HEXSIL (HEXagonal honeycomb polySILicon) process, a technology enabling the fabrication of tall 3D microstructures without postrelease assembly. Using CVD processes, generally only thin films (~2–5 μm) can be deposited on flat surfaces. If, however, these surfaces are the opposing faces of deep narrow trenches, the growing films will merge to form solid beams. Releasing of such polysilicon structures and the incorporation of electroplating steps expand the surface micromachining bandwidth in terms of choice of materials and accessible feature heights. In this fashion, high-aspect-ratio structures normally associated with LIGA can now also be made of CVD polysilicon.

4. Applying classical LPCVD to obtain polysilicon deposition is a slow process. For example, a layer of 10 μm typically requires a deposition time of 10 h. Consequently, most micromachined structures are based on layer thicknesses in the 2–5-μm range. Based on dichlorosilane ($SiH_2Cl_2$) chemistry, Lange et al.[98] developed a CVD process in a vertical epitaxy batch reactor with deposition rates as high as 0.55 μm/min at 1000°C. The process yields acceptable deposition times for thicknesses in the 10-μm range. The highly columnar polysilicon films are deposited on sacrificial $SiO_2$ layers and exhibit low internal tensile stress, making them suitable for surface micromachining.
5. Thick layers of polyimide and other new UV resists have also received a lot of attention as important new extensions of surface micromachining. Because of their transparency to exposing UV light, they can be transformed into tall surface structures with LIGA-like high-aspect ratios. They may also be electroplated and micromolded in any plastic of choice.
6. In Chapter 4 we discussed the transformation of single-crystal Si into a porous material with porosity and pore sizes determined by the current density, type, and concentration of the dopant, as well as the hydrofluoric acid concentration. A transition from pore formation to electropolishing is reached by increasing the current density and/or by decreasing the hydrofluoric acid concentration.[101,102] Porous silicon can also be formed under similar conditions from LPCVD polysilicon.[103]

In this section, we first review the main surface micromachining process steps and clarify the extensions of the surface micromachining technique listed above. Next we compare surface micromachining with bulk micromachining. In the surface micromachining examples at the end of this section, we first look at a lateral resonator. Resonators have found an important industrial application in accelerometers and gyros introduced by Analog Devices. The second example involves Texas Instruments' (TI's) Digital Micromirror Device™ (DMD), a chip now found in many projectors.

### Historical Note

The first example of a surface micromachine for an electromechanical application consisted of an underetched metal cantilever beam for a resonant gate transistor made by Nathanson in 1967.[104] By 1970, a first suggestion for a magnetically actuated metallic micromotor emerged.[105] Because of fatigue problems, metals are not typically used as mechanical components. The surface micromachining method as we know it today was first demonstrated by Howe and Muller in the early 1980s and relied on polysilicon as the structural material.[106] These pioneers, and the late Guckel,[107] an early contributor to the field, produced free-standing LPCVD polysilicon structures by removing the oxide layers on which the polysilicon features were formed. Howe's first device consisted of a resonator designed to measure the change of mass on adsorption of chemicals from the surrounding air. However, this gas sensor does not necessarily represent a good application of a surface-micromachined electrostatic structure because humidity and dust foul the thin air gap of such a microstructure in a minimal amount of time. Later, mechanical structures, especially hermetically sealed mechanical devices, provided proof that the IC revolution could be extended to electromechanical systems.[104] In these structures, the height ($z$-direction) typically is limited to less than 10 μm, ergo the name *surface micromachining*.

Gabriel et al.[108] in 1989 made the first survey of possible applications of polysilicon surface micromachining. Microscale movable mechanical pin joints, springs, gears, cranks, sliders, sealed cavities, and many other mechanical and optical components have been demonstrated in the laboratory.[109,110] For awhile, in the early 1990s, it seemed that every MEMS group in the United States was trying to make surface-micromachined micromotors of the type shown in Figure 7.65.

Micromotors may lack practical use as of yet, but, just as the ion-sensitive field effect transistor (ISFET) galvanized the chemical sensor community in trying out new chemical sensing approaches, micromotors energized the micromachining research community to fervently explore miniaturization of a wide variety of mechanical sensors and actuators. Micromotors

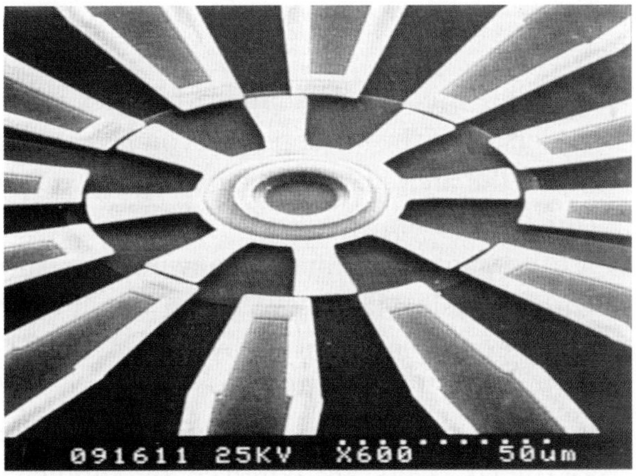


**FIGURE 7.65** Surface micromachined motor. (M. Mehregany, 1990. PhD dissertation, MIT.)

also brought about the christening of the micromachining field into MEMS. In 1991, Analog Devices (Norwood, MA) announced the first commercial product based on surface micromachining, namely, the ADXL-50, a 50-g accelerometer for activating air-bag deployment.[111] The ADXL-50 chip is shown in Figure 7.66a. The SEM inset in this figure depicts the polysilicon micromachined mechanical structure. By 2001, Analog Devices was making 2 million surface-micromachined accelerometers a month (at $4 per device in volume). A second commercial success for surface micromachining was based on Texas Instruments' Digital Mirror Device™ (DMD). This surface-micromachined movable aluminum mirror is a digital light switch that precisely controls a light source for projection displays and hard copy applications.[112] A schematic of the tilting mirrors in a DMD chip is shown in Figure 7.66b. The commercial acceptance of this second application confirmed the staying power of surface micromachining.

Surface micromachining is also an established manufacturing process at Cronos (Research Triangle Park, NC; now a JDS Uniphase Company) and Robert Bosch (Stuttgart, Germany).

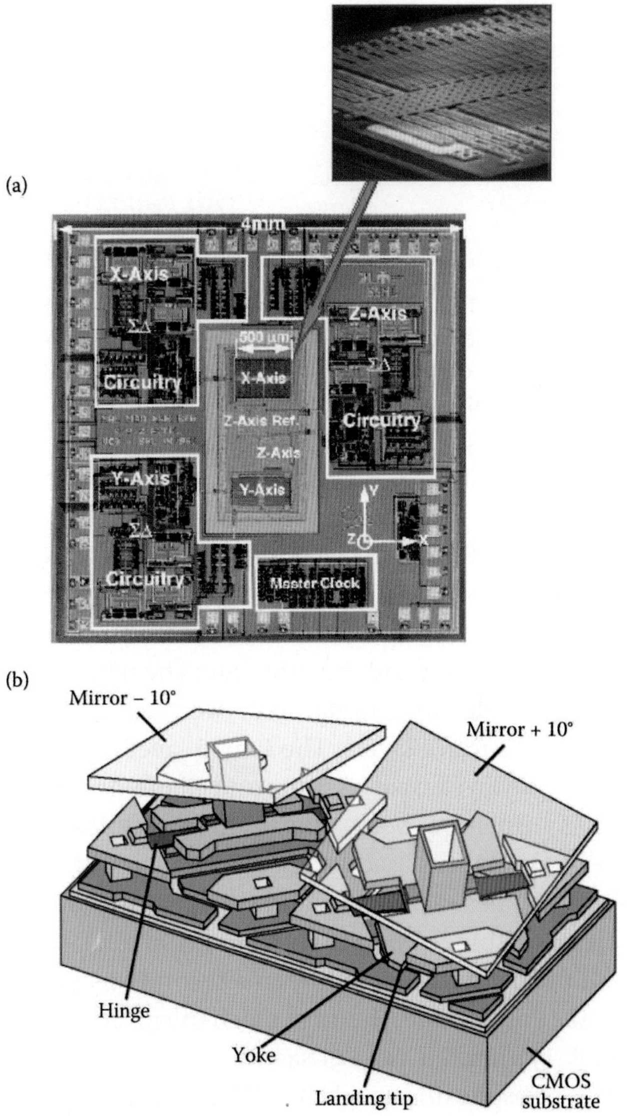


**FIGURE 7.66** (a) The ADXL-50 chip from Analog Devices. The H-shaped structure in this photo is the polysilicon micromachined mechanical structure (http://www.analog.com/library/techArticles/mems/xlbckgdr4.html). (b) Schematic of the tilting mirrors in a DMD chip from TI.

## Surface Micromachining Processes

### *Basic Process Sequence—Overview*

A first simple surface micromachining process sequence for the creation of a freestanding polysilicon anchor was illustrated in Figure 7.1. We will now detail typical surface micromachining fabrication steps using the DARPA-sponsored multiuser MEMS process (MUMPs®). The baseline process of the MUMPs program is a three-layer polysilicon (Poly 0, Poly 1, and Poly 2) surface micromachining process ("polysurf") known as *PolyMUMPs*. The process includes eight lithography levels and seven physical layers. A physical layer represents a layer of material deposited during the fabrication process and is usually represented in mixed-case letters. A lithographic level is used to pattern a physical layer and is always represented in CAPITAL letters. The physical layers illustrated in Figure 7.67 are two mechanical and one electrical layer of polysilicon,

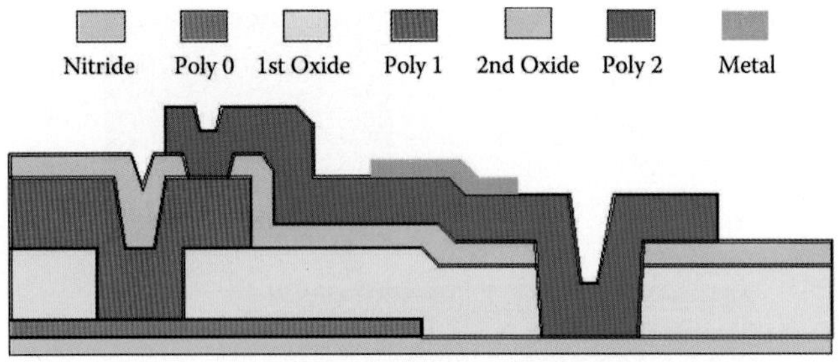


**FIGURE 7.67** The seven physical: two mechanical (Poly 1 and Poly 2) and one electrical layer (Poly 0) of polysilicon, two sacrificial oxide layers, one metal conduction and one electrical isolation nitride layer.

two sacrificial oxide layers, one metal conduction, and one electrical isolation nitride layer (also buffer or isolation layer). A sacrificial oxide layer of fast etching silicon dioxide is used to define the separation between mechanical layers and between mechanical layer and the substrate. The nitride layer is used for electrical isolation between substrate and electrical surface layers. The Poly 0 layer is the electrical polylayer used as a ground plane or for electrode formation below the first mechanical layer (Poly 1). The first sacrificial oxide layer will establish a gap between Poly 1 and substrate/nitride after release etching. The second sacrificial oxide layer will make for the gap between a second mechanical layer (Poly 2) and the first mechanical polysilicon (Poly 1) after release. For sacrificial oxide one often uses phosphosilicate glass (PSG), a phosphorous-containing silicon dioxide layer generated by CVD and used for its fast etching properties. A metal layer provides electrical connection to the package.

The eight lithography levels are:

1. POLY ZERO: Defines the polysilicon zero features (Poly 0)
2. ANCHOR 1: Opens points of contact between first polysilicon and substrate (nitride or Poly 0)
3. DIMPLE: Generates "bumps" in undersurface of Poly 1 to minimize stiction
4. POLY 1: Defines first polysilicon features
5. POLY 1-POLY 2-VIA: Opens points of contact between first and second polysilicon
6. ANCHOR 2: Opens points of contact between second polysilicon and substrate/nitride
7. POLY 2: Defines second polysilicon features
8. METAL: Defines location of metal features

Remarks:

- Dimples are small, shallow features on the underside of the lower polysilicon layer to minimize the area of contact between the polysilicon and the substrate.
- A lithography level may or may not correspond with a physical layer: e.g., Poly 1 = POLY 1, but second oxide is patterned by both ANCHOR 2 and POLY 1-POLY 2-VIA.
- Release is the last step of the process where the sacrificial layers are removed by submersion into HF.
- Stiction is the sticking effect between polysilicon and the substrate that occurs during the removal of the sacrificial oxide. Many attempts to limit stiction, including dimples, special release chemicals, and processes, are tried, some successfully (see below).

Since its inception in December 1992, the multiuser MEMS process (MUMPs) quickly became a well-established, commercial program that provides customers with cost-effective access to surface micromachining for prototyping activities. It caused many researchers to focus on design and testing of new MEMS concepts rather than building yet another clean room. MUMPs was approaching its 40th run in December 2000. Another source for surface micromachining prototyping is at Sandia, with the ultraplanar multilevel MEMS technology (SUMMiT), e.g., SUMMiT-IV and SUMMiT-V (http://www.mems.sandia.gov/tech-info/summit-v.html) introduced in 1998 – (four and five polysilicon level processes). SUMMiT is reviewed further below.

### *Process Steps for Making a Polysilicon Micromotor*

The following is a description of the PolyMUMPs process applied to the fabrication of a surface-micromachined electrostatic micromotor. A blanket n$^+$ diffusion of the silicon substrate (e.g., with $POCl_3$), defining a ground plane, often outlines the very first step in surface micromachining (Step I in Figure 7.68), followed by a passivation step of the substrate, for example, with 0.15-μm-thick LPCVD low stress nitride (sometimes on top of a 0.5-μm

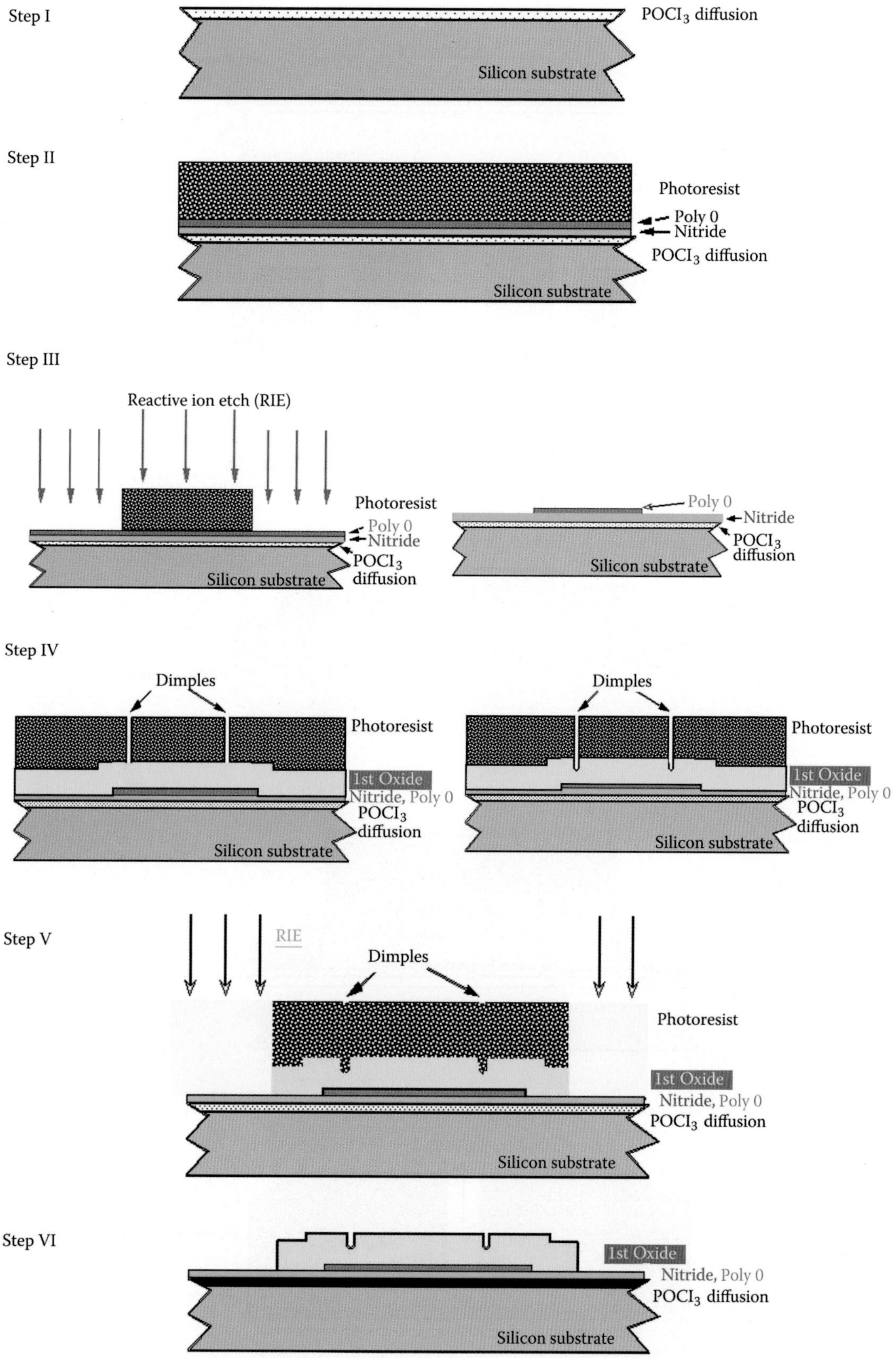


**FIGURE 7.68** The PolyMUMPs process applied to the fabrication of a surface-micromachined micromotor.

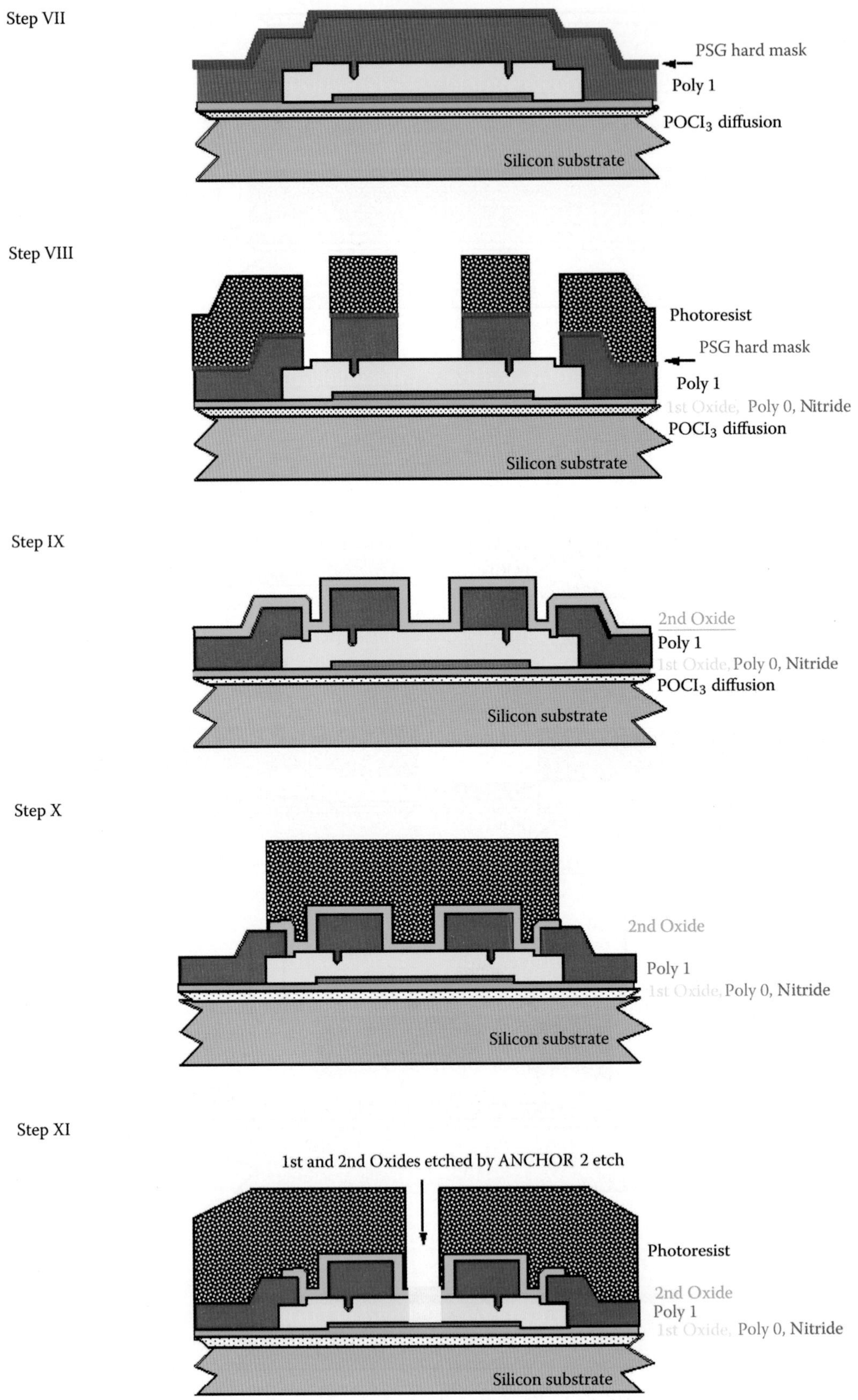


**FIGURE 7.68** *(Continued)*

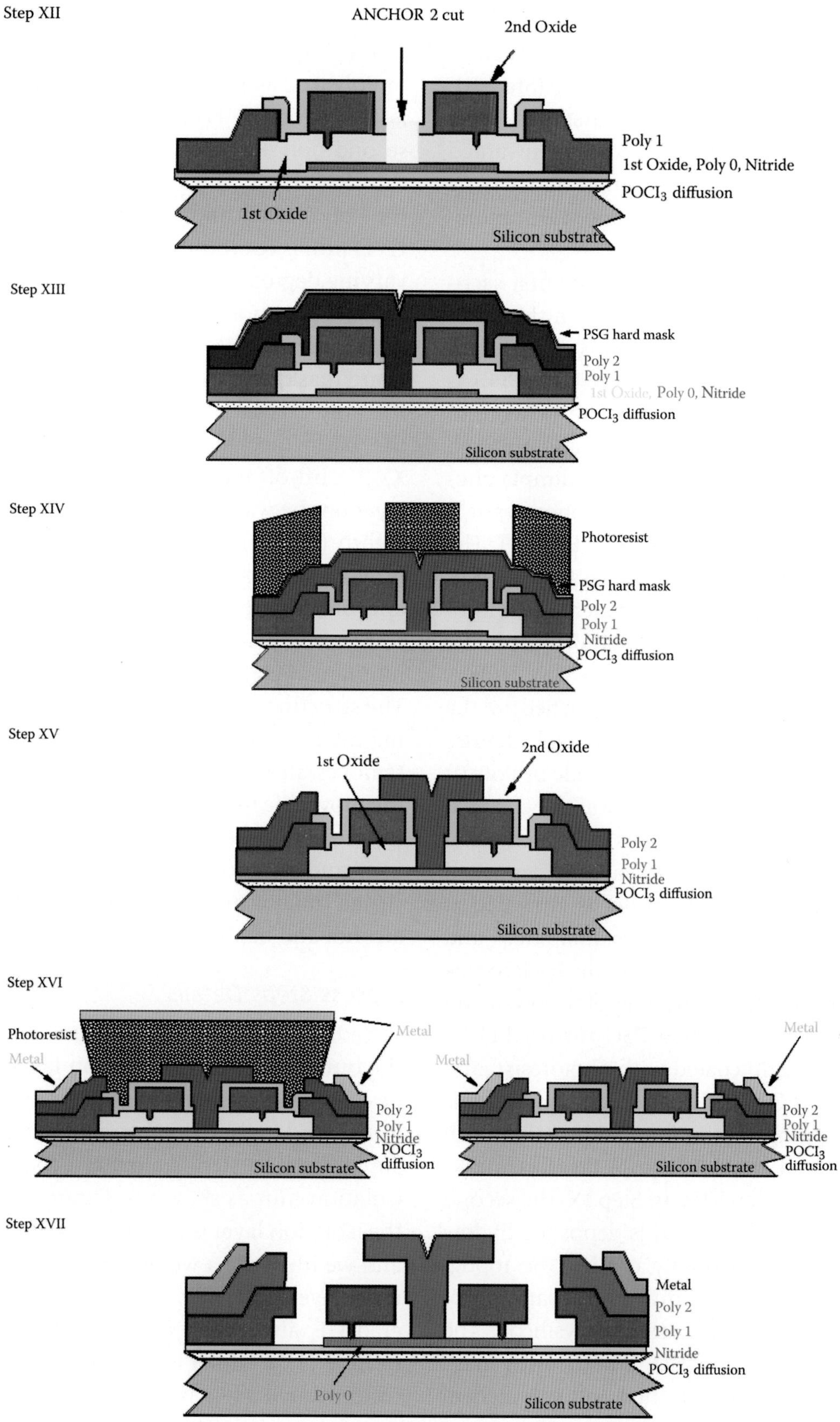


**FIGURE 7.68** *(Continued)*

thermal oxide). This is followed by a blanket deposition of polysilicon (Poly 0), and the wafer is spin-coated with a photoresist for patterning (Step II). The photoresist is exposed using the first mask level (POLY 0), and the image is developed. The unwanted polysilicon is removed by RIE etching, transferring the POLY 0 pattern onto the wafer, and the photoresist is stripped in a solvent (Step III).

Step IV involves the deposition of the first sacrificial oxide (PSG) layer (2.0 μm thick) on the wafer by low temperature CVD. The wafer is coated with photoresist, and a second mask level (DIMPLE) is exposed and developed. The dimples are etched into the first oxide by a combination of RIE and buffered oxide etch (BOE) etching. In Step V, the dimple photoresist is stripped, and a new layer of photoresist is applied for the third mask level (ANCHOR 1). The first oxide is patterned and then processed with RIE to remove the oxide from the anchor area. The photoresist is stripped in a solvent bath (Step VI). The wafer is now ready for Poly 1 processing. The ANCHOR 1 level defines where Poly 1 will be attached to the substrate, and the thickness of the first oxide defines how far above the substrate (either nitride or Poly 0) Poly 1 will sit after release. After the wafer is cleaned, the first polysilicon layer (Poly 1) is deposited using LPCVD. An additional thin PSG layer is deposited on top of the Poly 1, and the wafer is annealed at high temperature to reduce the residual stress and dope the Poly 1 (Step VII). We will come back to this stress reduction matter and doping of a polysilicon layer sandwiched between two PSG further below. The wafer is now again coated with photoresist, and the fourth mask (POLY 1) is patterned. The wafer is RIE etched, stopping on the first oxide, using both the photoresist and thin PSG (top) layer as masks for the Poly 1 layer (Step VII). In Step IX, the second sacrificial oxide layer (0.75 μm) is deposited by low temperature CVD, conformally coating the topography on the wafer and defining the separation of the first polylayer from the second polysilicon. The wafer is coated with photoresist, and the fifth mask (POLY 1-POLY 2-VIA) is patterned and RIE etched. This defines the contact regions between Poly 1 and Poly 2 (Step X). The photoresist is stripped, and the wafer is recoated. The sixth mask (ANCHOR 2) is patterned, and both first oxide and second oxide are etched in one step (Step XI). This defines the region where Poly 2 will contact the substrate (either nitride or Poly 0). The photoresist is stripped, and the wafer is ready for the Poly 2 deposition (Step XII). The second polysilicon layer is deposited followed by a thin PSG layer. The structure is annealed to reduce the Poly 2 stress and to dope the Poly 2. Because the CVD film is conformal, all the holes will be filled, to varying degrees, with Poly 2 (Step XIII). The wafer is coated with photoresist, and the seventh mask (POLY 2) is patterned. Both the photoresist and thin PSG (hard mask) will mask the RIE etch (Step XIV). After the Poly 2 layer is etched and the wafer is stripped, the basic mechanical structure is complete (Step XV). A lift-off template is used to deposit the metal layer on the wafer. Photoresist is patterned using the eighth (and ninth) masks (METAL), and the metal is deposited, adhering to the Poly 2, where exposed, and breaking continuity as it goes over the photoresist step (lift-off profile), and the photoresist and remnant metal are removed by rinsing in solvent. The structure is now completed and ready for releasing (Step XVI). A 2.5-min soak in straight 49% HF removes all the sacrificial oxide layers and releases the moveable mechanical parts. The rotor is now free to spin about the center pin, and the stator poles are fixed and electrically active (Step XVII). The chips are rinsed in DI water, followed by soaking in isopropyl alcohol and baking in a convection oven.

### *Process Steps Details*

*Pattern Transfer to Buffer/Isolation Layer* Suppose the buffer/isolation passivation layer itself needs to be patterned, perhaps to make a metal contact pad to the underlying Si substrate. Then, the appropriate fabrication step is a pattern transfer to the thin isolation film as shown in Figure 7.69. In this case, the isolation layer is a 1-μm-thick thermal $SiO_2$ film, and we illustrate a wet pattern transfer with a 1-μm resist layer. Typically, an isotropic etch such as buffered HF, also called a *buffered oxide etch* [e.g., BHF or BOE (5:1), which is five parts $NH_4F$ and one part concentrated HF], is used (unbuffered HF attacks the photoresist). This solution etches $SiO_2$ at a rate of 100 nm/min, and the creation of the opening to the underlying substrate takes about 10 min. The etch progress may be monitored optically (color

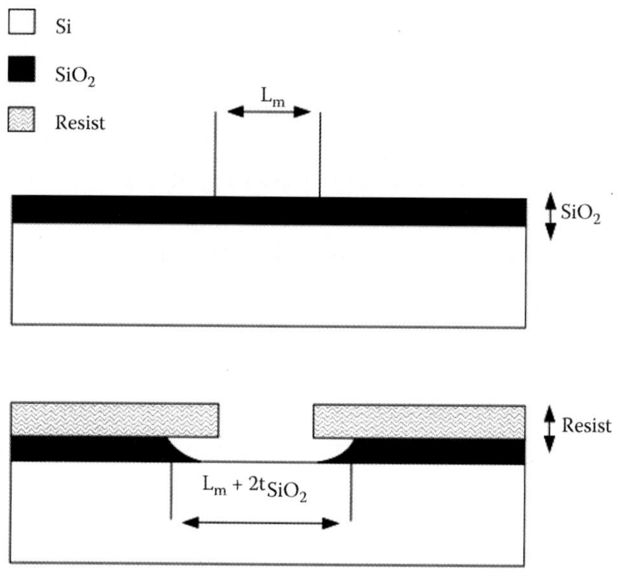


**FIGURE 7.69** Wet etch pattern transfer to a thin thermal $SiO_2$ film for the fabrication of a contact pad to the Si substrate.

change) or by observing the hydrophobic/hydrophilic behavior[113] of the etched layer. With a resist opening, $L_m$, the undercut typically measures the same thickness as the oxide thickness, $t_{SiO2}$. In other words, the contact pad will have a size of $L_m + 2t_{SiO2}$. The undercut worsens with loss of photoresist adhesion during etching. Using an adhesion promoter such as HMDS (hexamethyldisilazane) proves useful in such cases. A new bake after 5 min of etching is a good procedure to maintain the resist integrity. After the isotropic etch, the resist is stripped in a piranha etch bath. This strong oxidizer grows about 3 nm of oxide back in the cleared window. To remove the oxide resulting from the piranha, a dip in diluted BHF suffices. After cleaning and drying, the substrate is ready for contact metal and base material deposition. Applying a dry etch (e.g., $CF_4$-$H_2$ plasma) to open up the window in the oxide would eliminate undercutting of the resist but requires a longer setup time. With an LPCVD low-stress nitride or a nitride on top of a thermal $SiO_2$, an often-used combination for etching the buffer/isolation layer is a dry etch (e.g., $SF_6$ plasma) followed by 5:1 BHF.[114]

*Deposition of Structural Material* For the best step coverage of a structural material over underlying topography, chemical vapor deposition (CVD) is preferred. If a physical deposition method must be used, sputtering is preferred over evaporation (line-of-sight deposition technique), which leads to the poorest step coverage. In the latter case, edge taper could be introduced advantageously.

The most widely used structural material in surface micromachining is polysilicon (poly-Si or simply "poly"). Polysilicon is deposited by low pressure (25–150 Pa) chemical vapor deposition (LPCVD) in a furnace (a polychamber) at about 600°C. The undoped material is usually deposited from pure silane, which thermally decomposes according to Reaction 7.2. Typical process conditions may consist of a temperature of 605°C, a pressure of 550 mTorr (73 Pa), and a silane flow rate of 125 sccm. Under those conditions, a normal deposition rate is 100 Å/min. To deposit a 1-μm film will take about 90 min. Sometimes the silane is diluted by 70–80% nitrogen. The silicon is deposited at temperatures ranging from 570–580°C for fine-grained polysilicon to 620–650°C for coarse-grained polysilicon. The characteristics of these two types of polysilicon materials were compared above (see Table 7.20). Furnace annealing of the polysilicon film at 1050°C in nitrogen for 1 h is commonly used to reduce stress stemming from thermal expansion coefficient mismatch and nucleation and growth of the polysilicon film. Low-temperature deposition of silicon at 320°C in PECVD and at ~350°C in sputter deposition is also possible, although this results in amorphous films.

To make parts of the microstructure conductive, dopants can be introduced in the polysilicon film by adding dopant gases to the silane gas stream, by drive-in from a solid dopant source, or by ion implantation. When doping from the gas phase, the dopant can be readily controlled in the range of $10^{19}$–$10^{21}$/cm$^3$. Polysilicon deposition rates, in the case of gas-phase doping (in situ doping), may be significantly impacted. For example, decreases in polysilicon deposition rate by as much as a factor of 25 have been reported in phosphine and arsine doping. The effect is associated with the poisoning of reaction sites by phosphine and arsine.[115] The lower deposition rate of in situ phosphine doping can be mitigated by reducing the ratio of phosphine to silane flow by one third.[116] With the latter flow regime, deposition at 585°C (for a 2-μm-thick film), followed by 900°C rapid thermal annealing for 7 min, results in a polysilicon with low residual stress, negligible stress gradient, and low resistivity.[116,117] In situ boron doping, in contrast to arsine and phosphine doping, accelerates the polysilicon deposition rate through an enhancement of silane adsorption induced by the boron presence.[118] Film thickness uniformity for doped films typically is less than 1%, and sheet resistance uniformity is less than 2%. Alternatively, polysilicon may be doped from PSG films sandwiching the undoped polysilicon film as we alluded to when discussing Steps VII and XIII in Figure 7.68. By annealing such a PSG sandwich at 1050°C in $N_2$ for 1 h, the polysilicon is symmetrically doped by diffusion of dopant from the top and the bottom layers of PSG. Symmetric doping results in a polysilicon film with a moderate compressive stress. The resulting uniform grain texture avoids gradients in the residual stress, which would cause bending moments, warping microstructures on release. Finally, ion implantation of undoped polysilicon, followed by high-temperature dopant drive-in, also leads to conductive polysilicon. This polysilicon has a moderate tensile stress, with a strain gradient that causes cantilevers to deflect toward the substrate.[119] The polysilicon is now ready for patterning by RIE in, e.g., $CF_4$-$O_2$ plasma.

Although the mechanical properties are less understood than for single-crystal Si, microstructures based on polysilicon as a mechanical member have been commercialized swiftly.[111,119]

Surface micromachining of thin single-crystalline Si layers in epitaxial layers or fusion-bonded and etched back Si, as well as surface micromachining involving resists such as polyimides, is also very common and is discussed separately below. Other structural materials used in surface micromachining include aluminum, $SiO_2$, silicon nitride, silicon oxynitride, diamond, SiC, sputtered silicon, GaAs, tungsten, α-Si:H, Ni, W, Al, etc. More information on the material properties of these materials in thin film form is provided in the section on Surface Micromachining Modifications and in Volume III, Chapter 5 (Selected Materials and Processes for MEMS and NEMS).

*Base Layer (also Spacer or Sacrificial Layer) Deposition* A thin LPCVD phosphosilicate glass (PSG) layer (e.g., 2 μm thick) is a preferred base, spacer, or sacrificial layer material. Adding phosphorous to $SiO_2$ to produce PSG enhances the etch rate in HF.[120,121] Other advantages to using doped $SiO_2$ include its utility as a solid-state diffusion dopant source to make subsequent polysilicon layers electrically conductive and helping to control window taper (planarization). As deposited phosphosilicate displays a nonuniform etch rate in HF and must be densified, typically carried out in a furnace at 950°C for 30 min to 1 h in a wet oxygen ambient. The etch rate in BHF can be used as the measure of the densification quality. The base window etching stops at the buffer isolation layer, often a $Si_3N_4$/$SiO_2$ layer that also forms the permanent passivation of the device. Windows in the base layer are used to make anchors onto the buffer/isolation layer for mechanical structures.

The edges of the etched windows in the base may need to be tapered to minimize coverage problems with subsequent structural layers, especially if these layers are deposited with a line-of-sight deposition technique. An edge taper is introduced through an optimization of the plasma etch conditions (see Chapter 3), through the introduction of a gradient of the etch rate, or by reflow of the etched spacer. Viscous flow at higher temperature smoothes the edge taper. The ability of PSG to undergo viscous deformation at a given temperature primarily is a function of the phosphorous content in the glass; reflow profiles get progressively smoother the higher the phosphorous concentration—reflecting the corresponding enhancement in viscous flow.[122] Doped oxides used as diffusion sources contain 5–15 wt% of the dopant. Doped oxides for passivation or interlevel insulation contain 2–8 wt% phosphorus to prevent the diffusion of ionic impurities to the device. PSG used for the reflow process contains 6–8 wt% phosphorus and reflows at temperatures between 950 and 1100°C. Oxides with lower phosphorus concentrations will not soften and flow, but higher phosphorus concentrations can give rise to deleterious effects because phosphorus can react with atmospheric moisture to form phosphoric acid, which can consequently corrode the aluminum metallization. Addition of boron to PSG further reduces the reflow temperature without exacerbating this corrosion problem [undoped $SiO_2$ is often used above

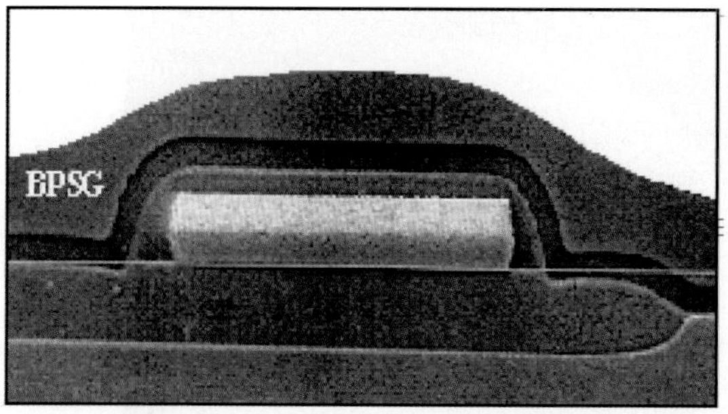


**FIGURE 7.70** Edge taper reflow of BPSG layer.

and below PSG or borophosphosilicate glass (BPSG) to prevent corrosion of Al]. BPSG typically contains 4–6 wt% P and 1–4 wt% B and reflows at 800°C. In Figure 7.70 we depict the reflow of a BPSG spacer after patterning in a dry etch; the SEM shows the BPSG oxide layer after an 800°C reflow step, showing smooth topography over the step.

Ion implantation of PSG creates a rapidly etching, damaged PSG layer.[123–125] The steady-state taper is a function of the etch rate in PSG etchants (e.g., BHF).

*Selective Etching of Spacer Layer*

*Selective Etching* To create movable micromachines, the microstructures must be freed from the spacer layers. The challenge in freeing microstructures by undercutting is evident from Figure 7.71. After patterning the polysilicon by RIE in, e.g., $SF_6$ plasma, it is immersed in an HF solution to remove the underlying sacrificial layer, releasing the structure from the substrate. The release step is the source of several technological woes. Commonly, a layer of sacrificial phosphosilicate glass, between 1 and 2000 μm long and 0.1–5 μm thick, is etched in concentrated, dilute, or buffered HF (BHF = BOE). The spacer etch rate, $R_s$, should be faster than the attack on the microstructural element, $R_m$, and that of the insulator layer, $R_i$. For this type of complete undercutting, only wet etchants can be used. Etching narrow gaps and undercutting wide areas with BHF can take hours. The removal rate is this slow because the sacrificial layer is only a few micrometers thick, and the reaction becomes quickly diffusion limited. As a consequence, the depth of sacrificial layer dissolved under the structure will increase slowly with the etching time as $d_{release} \propto \sqrt{t_{etch}}$ Simply said, releasing a structure twice as wide will take four times more time. However, if the etching lasts too long, the chemical may start attacking the device structural material, too. Detailed studies on the etching mechanism of oxide spacer layers were undertaken by Monk et al.[126,127] They found that the etching reaction shifts from kinetic controlled to diffusion controlled as the etch channel becomes longer. This affects mainly large-area structures because diffusion limitations were observed only after approximately 200 μm of channel etching or 15 min in concentrated HF. Eaton and Smith[128] developed a release etch model that is an extension of the work done by Monk et al.[126,127] and Liu et al.[129] To shorten the etch time, extra apertures in the microstructures are sometimes provided for additional access to the spacer layer (see Figure 7.72).

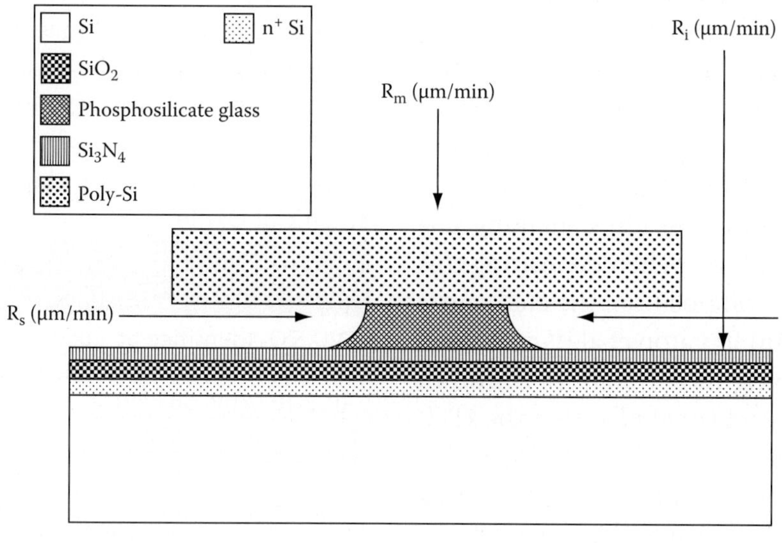


**FIGURE 7.71** Selective etching of spacer layer.

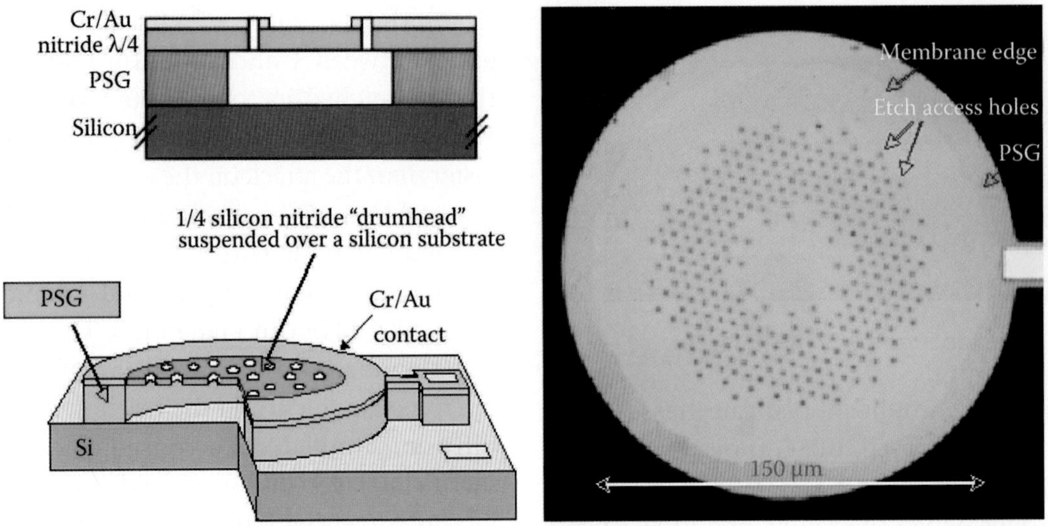


**FIGURE 7.72** Analog optical modulator. The 1/4 silicon nitride "drumhead" suspended over a silicon substrate is equipped with access holes to etch the BSG faster. (Goossen, K., J. Walker, and S. Arney. 1994. Silicon modulator based on mechanically-active antireflection layer with 1 Mbit/sec capability. *IEEE Photonics Tech Lett* 6:1119.[130])

These access holes in the structural layer are typically spaced by roughly 30 μm, allowing for the HF to etch all the oxide beneath in less than 5 min.

The etch rate of PSG, the most common spacer material, increases monotonically with dopant concentration, and thicker sacrificial layers etch faster than thinner layers.[126]

The selectivity ratios for spacer layer, microstructure, and buffer layer are not infinite,[131] and in some instances even silicon substrate attack by BHF is observed under polysilicon/spacer regions.[132,133] It also has been shown that during an HF release step the mechanical properties of polysilicon, including residual stress, Young's modulus, and fracture strain, are affected.[134] Heavily phosphorous-doped polysilicon is especially prone to attack by BHF. In general, the Young's modulus and fracture strain of a thin polysilicon film decrease with increasing exposure time to HF and increasing HF concentration. Silicon nitride deposited by LPCVD etches much more slowly in HF than oxide films, making it a more desirable isolation film. When depositing this film with a silicon-rich composition, the etch rate is even slower (15 nm/min).[114] Eaton et al.[135] compared oxide and nitride etching in a 1:1 HF/$H_2O$ and in a 1:1 HF/HCl solution and concluded that the HCl-based etch yielded both faster oxide etch rates (617 vs. 330 nm/min) and slower nitride etch rates (2 vs. 3.6 nm/min), providing a much greater selectivity of the oxide to silicon nitride (310 vs. 91!). The same authors also studied the optimum composition of a sacrificial oxide for the fastest possible etching in their most selective 1:1 HF/HCl etch. The faster sacrificial layer etch limited the damage to nitride structural elements. Their results are summarized in Table 7.21. A densified CVD $SiO_2$ was used as a control, and a 5%/5% borophosphosilicate glass (BPSG) was found to etch the fastest.

Using low-pressure vapor HF, Watanabe et al.[136] found high etch ratios of PSG and BPSG to thermal oxides of more than 2000, with the BPSG etching slightly faster than the PSG. We will see further that low-pressure vapor HF also leads to less stiction of structural elements to the substrate. In Table 7.22 we present etch rate and etch ratios for $R_i$ and $R_s$ in BHF (7:1) for a few selected materials.

Etching is followed by rinsing and drying. The problems with wet release continue during this drying and

**Table 7.21** Etch Rate in 1:1 HF/HCl of a Variety of Sacrificial Oxides

| Thin Oxide | Lateral Etch Rate (Å/min) |
|---|---|
| CVD $SiO_2$ (densified at 1,050 °C for 30 min) | 6,170 |
| Ion implanted and densified CVD $SiO_2$ (P, 8 $10^{15}/cm^2$, 50 keV) | 8,330 |
| Phosphosilicate (PSG) | 11,330 |
| 5%/5% borophosphosilicate (BPSG) | 41,670 |

*Source:* Adapted from Eaton, W. P., and J. H. Smith. 1995. *Smart structures and materials 1995: smart electronics (Proceedings of the SPIE).* San Diego: SPIE.[135]

**Table 7.22** Etching of Spacer Layer and Buffer Layer in BHF (7:1)

| | Material | | |
|---|---|---|---|
| Property | LPCVD $Si_3N_4$ | LP CVD $SiO_2$ | LP CVD 7% PSG |
| Etch rate | 7–12 Å/min | 700 Å/min | ~10,000 Å/min |
| Selectivity ratio | $R_i = 1$ | $R_s = 60–100$ | $R_s$ ~800–1,200 Å/min |

rinsing step. The meniscus created by the receding liquid/air interface tends to pull the structure against the substrate. This intimate contact gives rise to surface forces like Van der Walls force, which will irremediably pin one's structure to the substrate when the drying is complete, effectively destroying the device. This phenomenon is referred to as stiction (see Stiction below). Extended rinsing causes a native oxide to form on the surface of the polysilicon structure. Such a passivation layer often is desirable and can be formed more easily by a short dip in 30% $H_2O_2$.

*Etchant-Spacer-Microstructure Combinations* A wide variety of etchant, spacer, and structural material combinations has been used; a limited listing is presented in Table 7.23. One interesting case concerns polysilicon as the sacrificial layer. This was used, for example, in the fabrication of a vibration sensor at Nissan Motor Co.[137] In this case, polysilicon is etched in KOH from underneath a nitride/polysilicon/nitride sandwich cantilever. Also, a solution of $HNO_3$ and BHF can be used to etch polysilicon, but it proves difficult to control. Using aqueous solutions of $NR_4OH$, where *R* is an alkyl group, provides a better etching solution for polysilicon, with greater selectivity with respect to silicon dioxide and phosphosilicate glass. The relatively slow etch rate enables better process control,[138] and the etchant does not contain alkali ions, making it more CMOS compatible. With tetramethyl ammonium hydroxide (TMAH), the etch rate of CVD polysilicon, deposited at 600°C from $SiH_4$, follows the rates of the (100) face of single-crystal silicon and is dopant dependent. The selectivity of $Si/SiO_2$ and Si/PSG, at temperatures less than 45°C, is measured to be about 1000. Hence a layer of 500 Å of PSG can be used as the etch mask for 10,000 Å of polysilicon. In addition, porous Si has been used as a sacrificial layer in micromachining. The high surface area of this material makes for rapid etching in KOH.[139]

In Texas Instruments' Digital Micro Device (DMD) (Figure 7.66b) the structural layer is aluminum and the sacrificial layer is a polymer. The polymer is removed with oxygen plasma, and prolonged release time will only slightly affect the metal.

Figure 7.73 illustrates a fabrication sequence for a polyimide corrugated structure. The sacrificial Al in Step 4 may be etched away by a mixture of phosphoric acidic, acetic acid, and nitric acid (see Table 7.23).

### Stiction

*Stiction during Release* The use of sacrificial layers enables the creation of very intricate movable polysilicon surface structures. An important limitation of such polysilicon shapes is that large-area structures tend to deflect through stress gradients or surface tension induced by trapped liquids and attach to the substrate/isolation layer during the final rinsing

**Table 7.23** Etchants—Spacer and Microstructural Layer

| Etchant | Buffer/Isolation | Spacer | Microstructure |
|---|---|---|---|
| Buffered HF (5:1,$NH_4F$/conc. HF) | LPCVD $Si_3N_4$/thermal $SiO_2$ | PSG | Polysilicon[133] |
| RIE using $CHF_3$ BHF (6:1) | LPCVD $Si_3N_4$ | LP CVD $SiO_2$ | CVD Tungsten[140] |
| KOH | LPCVD $Si_3N_4$/thermal $SiO_2$ | Polysilicon, porous Si (at room temperature) | $Si_3N_4$[141,142] SOI[139] |
| Ferric chloride | Thermal $SiO_2$ | Cu | Polyimide[143] |
| HF | LPCVD $Si_3N_4$/thermal $SiO_2$ | PSG | Polyimide[144] |
| Phosphoric/acetic acid/nitric acid (PAN or 5:8:1:1 water/phosphoric/acetic/nitric) | Thermal $SiO_2$ | Al | PE CVD $Si_3N_4$ Nickel[145,146] |
| Ammonium iodide/iodine alcohol | Thermal $SiO_2$ | Au | Ti[147] |
| Ethylene-diamine/pyrocathecol (EDP) | Thermal $SiO_2$ | Polysilicon | $SiO_2$ |

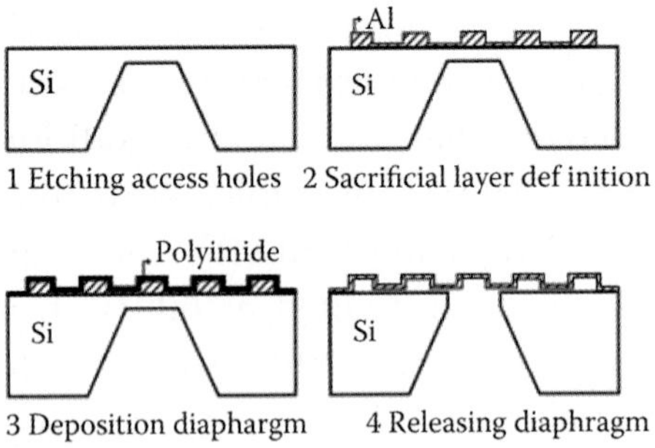


**FIGURE 7.73** Schematic view of the fabrication process of a polyimide corrugated diaphragm. (From van Mullem, C. J., K. J. Gabriel, and H. Fujita. 1991. *Sixth international conference on solid-state sensors and actuators (Transducers '91).* San Francisco: IEEE. With permission.[148])

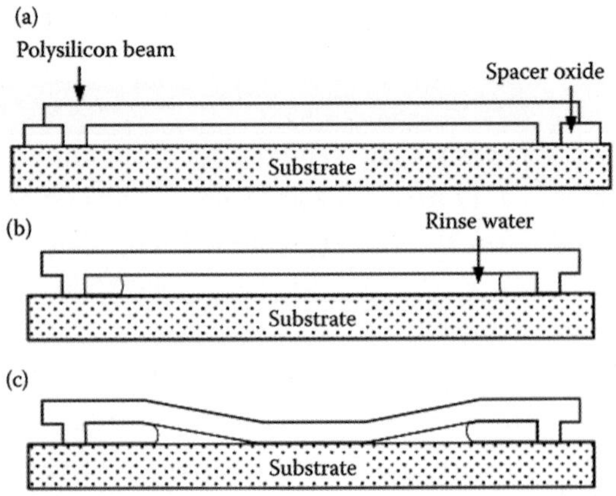


**FIGURE 7.74** Stiction phenomenon in surface micromachining and the effect of surface tension on micromechanical structures. (a) Unreleased beam. (b) Released beam before drying. (c) Released beam pulled to the substrate by capillary forces as the wafer dries.

and drying step, a stiction phenomenon that may be related to hydrogen bonding, residual contamination, and the Casimir force (see also Volume I, Chapter 3). In Volume III, Chapter 7 on scaling laws, we will come to appreciate that surface forces become dominant in the microdomain, and not only stiction but also friction are of great concern to the micromachinist. One of the principal causes of malfunctioning in MEMS is stiction, namely, the collapse of movable elements into nearby surfaces, resulting in their permanent adhesion. This can occur during fabrication, especially as a result of capillary forces, or during operation (in-use stiction, see below).

The sacrificial layer removal with a buffered oxide etch (BOE) followed by a long, thorough rinse in deionized water and drying under an infrared lamp typically represent the last steps in the surface micromachining sequence. As the wafer dries, the surface tension of the rinse water pulls the delicate microstructure to the substrate where a combination of forces—van der Waals forces, hydrogen bonding, and Casimir forces—keeps it firmly attached (see Figure 7.74).[119] Once the structure is attached to the substrate by stiction, the mechanical force needed to dislodge it usually is large enough to damage the micromechanical structure.[125,141] The same phenomena are thought to be involved in room temperature wafer bonding (Volume III, Chapter 7). Van der Waals forces and hydrogen bonding are covered in Volume III, Chapter 3 on nanochemistry: bottom-up manufacturing. Casimir forces are quantum mechanical in nature (see Volume I, Chapter 3) and inversely proportional to the fourth power of the distance between the plates. One can quickly establish that two mirrors of 1 $cm^2$ separated by a distance of 1 μm have an attractive Casimir force of 130 nN (see Volume III, Chapter 8). Although this force appears small, at distances less than 1 μm the Casimir force becomes the dominant force between two neutral objects. At separations of 10 nm, the Casimir effect produces the equivalent of 1 atm! Obviously this is a force to reckon with in MEMS and NEMS devices (see Volume III, Chapter 8 on actuators). We will not further dwell on the mechanics of the stiction process here, but the reader should refer to the theoretical and experimental analysis of the mechanical stability and adhesion of microstructures under capillary forces by Mastrangelo et al.[149,150]

Strategies that are used to overcome stiction have tackled it at the design and fabrication level. An important fabrication idea has been to reduce the contact surface by introducing dimples under the structure. Creating stand-off bumps on the underside of a polysilicon plate[114,151] or adding meniscus-shaping microstructures to the perimeter of the microstructure are mechanical means to help reduce sticking[152] (see Figure 7.75).

Fedder et al.[153] used another mechanical approach to avoid stiction by temporarily stiffening the microstructures with polysilicon links. The very stiff polysilicon structures are not affected by liquid surface tension forces, and the links are severed afterward with a high current pulse once the potentially

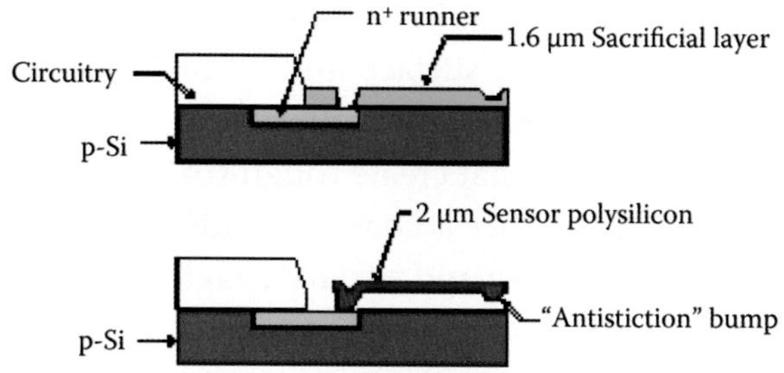


**FIGURE 7.75** Stand-off bump or antistiction bump.

destructive processing is complete. Yet another mechanical approach to avoid stiction involves the use of sacrificial supporting polymer columns. A portion of the sacrificial layer is substituted by polymer spacer material, spun on after partial etch of the oxide glass. After completion of the oxide etch, the polymer spacer prevents stiction during evaporative drying. Finally, an isotropic oxygen plasma etches the polymer to release the structure.[154]

Ideally, to ensure high yields, contact between structural elements and the substrate should be avoided during processing. In a liquid environment, however, this may become impossible because of the large surface tension effects. Consequently, most solutions to the stiction problem involve reducing the surface tension of the final rinse solution by physicochemical means. Lober et al.,[131] for example, tried HF vapor, and Guckel et al.[155,156] used freeze-drying of water/methanol mixtures. Freezing and sublimating the rinse fluid in a low-pressure environment gives improved results by circumventing the liquid phase. Takeshima et al.[157] used t-butyl alcohol freeze-drying. Because the freezing point of this alcohol lies at 25.6°C, it is possible to perform freeze-drying without special cooling equipment. Supercritical drying results in increased microstructure yields.[158] With this technique, the rinse fluid is displaced with a liquid that can be driven into a supercritical phase under high pressure. This supercritical phase does not exhibit surface tension, which allows for the drying of released microstructures without sticking. Typically, $CO_2$ at 35°C and 1100 psi are used (see also Chapter 1).

Kozlowski et al.[159] substituted HF in successive exchange steps by the monomer divinylbenzene to fabricate very thin (500 nm) micromachined polysilicon bridges and cantilevers. The monomer was polymerized under UV light at room temperature and was removed in oxygen plasma. Analog Devices applied a proprietary technique involving only standard IC process technology in the fabrication of a microaccelerometer to eliminate yield losses due to stiction.[118]

Coating the structure with a nonsticking layer [e.g., fluorocarbon, hydrophobic self-assembled mono-layer (SAM)] has also proved successful, and this method, albeit more complex, has the added advantage to provide long-lasting protection (see below under "In-Use Stiction").

The two most popular methods today to avoid stiction are etching a polymer sacrificial layer in an $O_2$ plasma and using xenon difluoride ($XeF_2$) to etch sacrificial silicon (see also Chapter 3). Xenon difluoride has been used successfully to release very compliant or nanosized oxide structures where silicon was used as the sacrificial material. The process does not use plasma, making the chamber rather simple, and several manufacturers like XactiX (http://www.xactix.com) (in cooperation with STS) in the United States and PentaVacuum in Singapore offer tools to exploit the technology.

*In-Use Stiction* Stiction remains a fundamental reliability issue because of contact with adjacent surfaces after release. Stiction-free passivation that can survive the packaging temperature cycle is not known at present.[94] Attempted solutions are summarized below.

Adhesive or stiction energy may be minimized in a variety of ways: for example, by forming bumps on surfaces (see above) or roughening of opposite surface plates.[160] Also, self-assembled monolayer (SAM) coatings have been shown to reduce surface adhesion and to be effective at friction reduction in bearings at the same time.[161] Making the silicon surface very hydrophobic or coating it with diamond-like carbon are other potential solutions for preventing postrelease stiction of polysilicon microstructures. Man et al.[162] eliminate postrelease adhesion in microstructures by using a thin conformal fluorocarbon film. The film eliminates the adhesion of polysilicon beams up to 230 μm long even after direct immersion in water. The film withstands temperatures as high as 400°C, and wear tests show that the film remains effective after $10^8$ contact cycles. Along the same line, ammonium fluoride-treated Si surfaces are thought to be superior to the HF-treated surfaces because of a more complete hydrogen termination, which leads to a cleaner hydrophobic surface.[163] Gogoi et al.[164] introduced electromagnetic pulses for postprocessing release of stuck microstructures.

Wong et al.[165] at the Georgia Institute of Technology are mimicking one of nature's best nonstick surfaces—lotus plant leaves—to help create more reliable MEMS structures unaffected by water and even claim improved biocompatible surfaces able to prevent cells from adhering to implanted medical devices. As we discuss in Volume I, Chapter 6, the lotus plant's ability to repel water and dirt results from an unusual combination of a superhydrophobic (water-repelling) surface and a combination of micrometer-scale hills and valleys and nanometer-scale waxy bumps that create rough surfaces that do not give water or dirt a chance to adhere. With its superhydrophobicity and surface roughness, a lotus surface coating can prevent stiction.

## Control of Film Stress

After reviewing typical surface micromachining process sequences, we are ready to investigate some of the mechanical properties of the fabricated mechanical members. Consider a straight lateral resonator as shown in Figure 7.76a. Electrostatic force is applied by a drive comb to a suspended shuttle. Its motion

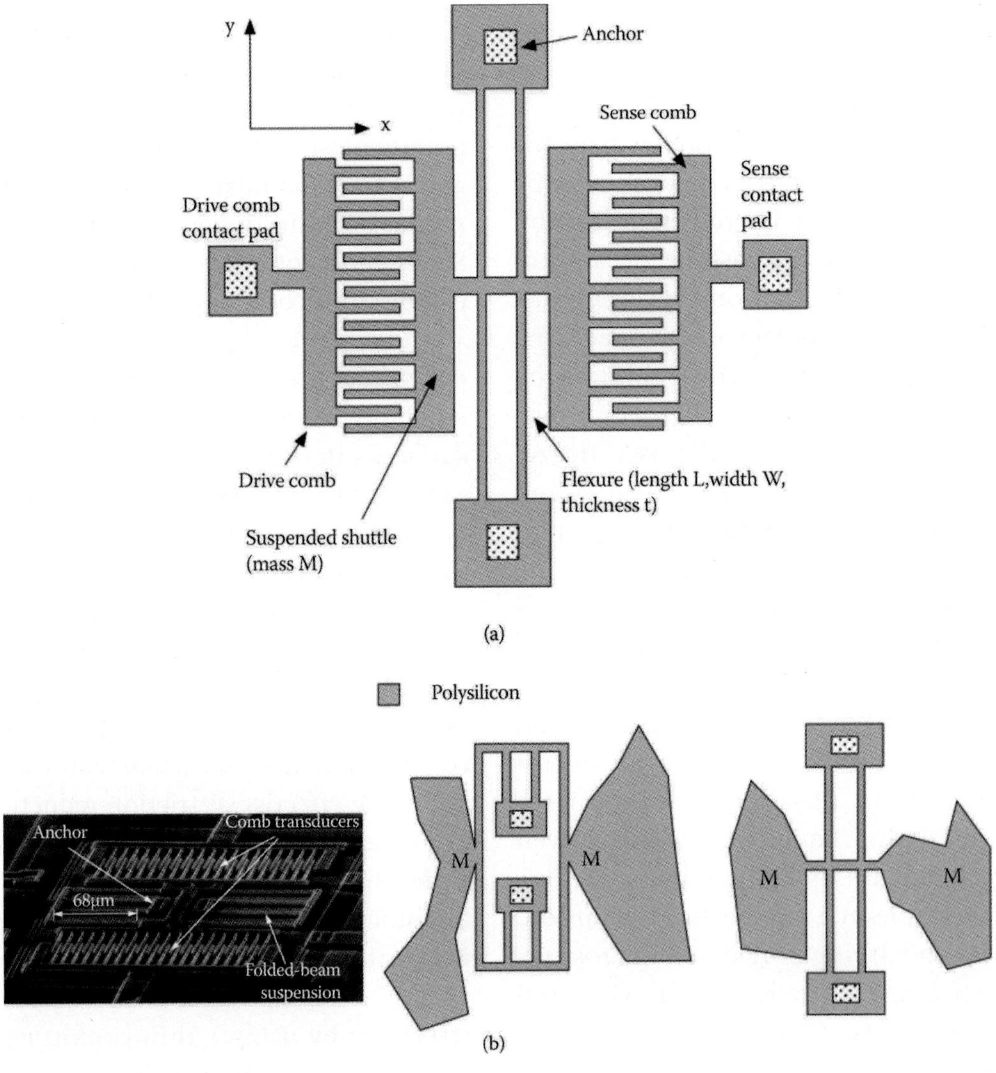


**FIGURE 7.76** (a) Layout of a lateral resonator with straight flexures. (b) Folded flexures (left) to release stress are compared with straight flexures (right). *M* is shuttle mass. Scanning electron microscope picture represents a folded beam lateral resonator. (Nguyen, C. T. C., and R. T. Howe. 1993. *Electron devices meeting, 1993. IEDM '93. Technical Digest, International.* Washington DC: IEEE.[166])

is detected capacitively by a sense comb. For many applications, a tight control over the resonant frequency, $f_0$, is required. An analytical approximation for $f_0$ of this type of resonator can be deduced from Rayleigh's method:

$$f_0 = \frac{1}{2\pi}\sqrt{\frac{2EtW^3}{mL^3} + \frac{24\sigma_r tW}{5\,mL}} \qquad (7.29)$$

where

E = the Young's modulus of polysilicon

L, W, and t = the length, width, and thickness of the flexures, respectively

m = the mass of the suspended shuttle (of the order of $10^{-9}$ kg or less)

For typical values (L = 150 μm and W = t = 2 μm) and a small tensile residual stress, the resonant frequency $f_0$ is between 10 and 100 kHz.[94] For typical values of *L*/*W*, the stress term in Equation 7.29 dominates the bending term. Any residual stress, $\sigma_r$, obviously will affect the resonant frequency. Consequently, stress and stress gradients represent critical stages for microstructural design.

Ignoring the stress term the resonant frequency in Equation 7.29 scales as the square root of the ratio of the stiffness and mass (Figure 7.77) (see also scaling in Volume III, Chapter 7). Thus, frequency scales as $L^{-1}$, and hence micro- and nanoresonators can lead to very high frequencies.

One of the many challenges of any surface-micromachining process is to control the intrinsic stresses in the deposited films. Several techniques can be used to control film stress. Some we detailed before, but we list them again for completeness sake:

- Large-grained polysilicon films, deposited around 625°C, have a columnar structure and are always compressive. Compressive stress can cause buckling in constrained structures. Annealing at high temperatures, between 900 and 1150°C, in nitrogen significantly reduces the compressive stress in as-deposited polysilicon[167,168] and can eliminate stress gradients. No significant structural changes occur when annealing a columnar polysilicon film. The annealing process is not without danger in cases where active electronics are integrated on the same chip. Rapid thermal annealing might provide a solution (see "IC Compatibility" below).
- Undoped polysilicon films are in an amorphous state when deposited at 580°C or less. The stress and the structure of this low-temperature material depend on temperature and partial pressure of the silane. A low-temperature anneal leads to a fine-grained polysilicon with low tensile stress and very smooth surface texture.[155] Tensile rather than compressive films are a necessity if lateral dimensions of clamped structures are not to be restricted by compressive buckling.
- Phosphorous,[169–171] boron,[170,172,173] arsenic,[170] and carbon[174] doping have all been shown to affect the state of residual stress in polysilicon films. In the case of single-crystal Si, to compensate for strain induced by dopants, one can implant with atoms with the opposite atomic radius versus Si. Similar approaches would most likely be effective for polysilicon as well.
- Tang et al.[114,175] developed a technique that sandwiches a polysilicon structural layer between a top and bottom layer of PSG and lets the high-temperature anneal drive in the phosphorous symmetrically, producing low-stress polysilicon with a negligible stress gradient.
- Another stress reduction method is to vary the materials composition, something readily done in CVD processes. An example of this method is the Si enrichment of $Si_3N_4$, which reduces the tensile stress.[176,177]
- During plasma-assisted film deposition processes, stress can be influenced dramatically. In a physical deposition process such as sputtering, stress control involves varying gas pressure and substrate bias. In plasma-enhanced

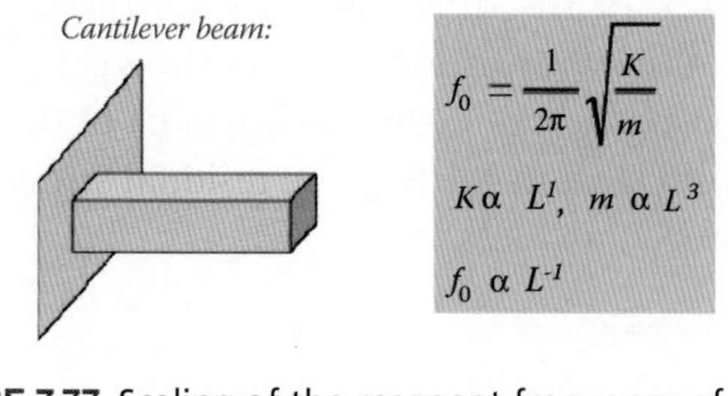


**FIGURE 7.77** Scaling of the resonant frequency of simple cantilever beam.

chemical vapor deposition (PECVD), the RF power, through increased ion bombardment, influences stress. In this way, the stress in a thin film starts out tensile, decreases as the power increases, and finally becomes compressive with further RF power increase. PECVD equipment manufacturers are also working to build stress control capabilities into new equipment by controlling plasma frequency (see above). In CVD, stress control involves all types of temperature treatment programs.

- A clever mechanical design might facilitate structural stress relief.[178] By folding the flexures in the lateral resonator in Figure 7.76b and by the overall structural symmetry, the relaxation of residual polysilicon stress is possible without structural distortion. By folding the flexures, the resonant frequency, $f_0$, becomes independent of $\sigma_r$ (see Equation 7.29). The springs in the resonator structure provide freedom of travel along the direction of the comb-finger motions ($x$) while restraining the structure from moving sideways ($y$), thus preventing the comb fingers from shorting out the drive electrodes. In this design, the spring constant along the $y$-direction must be higher than along the $x$-direction, i.e., $k_y >> k_x$. The suspension should allow for the relief of the built-in stress of the polysilicon film and the axial stress induced by large vibrational amplitudes. The folded-beam suspensions meet both criteria. They enable large deflections in the $x$-direction (perpendicular to the length of the beams) while providing stiffness in the $y$-direction (along the length of the beams). Furthermore, the only anchor points (see Figure 7.76b) for the whole structure reside near the center, thus allowing the parallel beams to expand or contract in the $y$-direction, relieving most of the built-in and induced stress.[114] Tang also modeled and built spiral and serpentine springs supporting torsional resonant plates. An advantage of the torsional resonant structures is that they are anchored only at the center, enabling radial relaxation of the built-in stress in the polysilicon film.[114] For some applications, the design approach with folded flexures is an attractive way to eliminate residual stress. However, a penalty for using flexures is increased susceptibility to out-of-plane warpage from residual stress gradients through the thickness of the polysilicon microstructure.[94]
- Corrugated structural members, introduced in the MEMS field by Jerman for bulk micromachined sensors,[179] also reduce stress effectively. In the case of a single-crystal Si membrane, stress may be reduced by a factor of 1,000–10,000.[180] One of the applications of such corrugated structures is the decoupling of a mechanical sensor from its encapsulation by reducing the influence of temperature changes and packaging stress.[180,181] Thermal stress alone can be reduced by a factor of 120.[182] Besides stress release, corrugated structures enable much larger deflections than do similar planar structures. This type of structural stress release was studied in some detail for single-crystal Si,[183] polyimide,[148] and LPCVD silicon nitride membranes,[184] but the quantitative influence of corrugated polysilicon structures still requires further investigation. A schematic view of the fabrication process of a polyimide corrugated diaphragm was shown in Figure 7.73.

## Dimensional Uncertainties

The often-expressed concerns about run-to-run variability in material properties of polysilicon or other surface-micromachined materials are somewhat misplaced, Howe points out.[114] He contrasts the relatively large dimensional uncertainties inherent to any lithography technique with polysilicon quality factors of up to 100,000 and long-term (>3 years) resonator frequency variation of less than 0.02 Hz. We follow his calculations here to prove the relative importance of the dimensional uncertainties. The shuttle mass $M$ of a resonator as shown in Figure 7.76 is proportional to the thickness ($t$) of the polysilicon film, and neglecting the residual stress term, Equation 7.29 reduces to:

$$f_0 \propto \left(\frac{W}{L}\right)^{\frac{3}{2}} \tag{7.30}$$

In case the residual stress term dominates in Equation 7.29, the resonant frequency is expressed as:

$$f_0 \propto \left(\frac{W}{L}\right)^{\frac{1}{2}} \tag{7.31}$$

The width-to-length ratio is affected by systematic and random variations in the masking and etching of the microstructural polysilicon. For 2-μm-thick structural polysilicon, patterned by a wafer stepper and etched in a reactive-ion etcher, a reasonable estimate for the variation in linear dimension of etched features, Δ is about 0.2 μm (10% relative tolerance).

From Equation 7.30 the variation Δ in lateral dimensions will result in an uncertainty $\delta f_0$ in the lateral frequency of:

$$\frac{\delta f_0}{f_0} \approx \frac{3}{2}\left(\frac{\Delta}{W}\right) \tag{7.32}$$

for a case where the residual stress can be ignored. With a nominal flexure width of W = 2 μm, the resulting uncertainty in resonant frequency is 15%. For the stress-dominated case, Equation 7.31 indicates that the uncertainty is now given as:

$$\frac{\delta f_0}{f_0} \propto \frac{1}{2}\left(\frac{\Delta}{W}\right) \tag{7.33}$$

The same 2-μm-wide flexure would now lead to a 5% uncertainty in resonant frequency. Interestingly, the stress-free case exhibits the most significant variation in the resonant frequency. In either case, resonant frequencies must be set by some postfabrication frequency trimming or other adjustment.

In Chapter 6 we draw further attention to the increasing loss of relative manufacturing tolerance with decreasing structure size (see Figure 6.3).

## Sealing Processes in Surface Micromachining

Sealing cavities to hermetically enclose sensor structures is a significant attribute of surface micromachining. Sealing cavities in surface micromachines often embodies an integral part of the overall fabrication process and presents a desirable chip level, batch-packaging technique. The resulting surface packages (microshells) are much smaller than typical bulk micromachined ones. A sealed cavity made with epi-micromachining is shown in Figure 7.82 further below, and more details on packaging using surface micromachining techniques are provided in Volume III, Chapter 4.

## IC Compatibility

Putting detection and signal conditioning circuits right next to the sensing element enhances the performance of the sensing system, especially when dealing with high impedance sensors. A key benefit of surface micromachining, besides small device size and single-sided wafer processing, is its compatibility with CMOS processing. IC compatibility implies simplicity and economy of manufacturing. In the examples at the end of this chapter, we will discuss, for example, how Analog Devices used a mature 4-μm BiCMOS process to integrate electronics with a surface-micromachined accelerometer.

To develop an appreciation of integration issues involved in combining a CMOS line with surface micromachining, we highlight Yun's[185] comparison of CMOS circuitry and surface micromachining processes in Table 7.24A. Surface micromachining processes are similar to IC processes in several aspects. Both processes use similar materials, lithography, and etching techniques. CMOS processes involve at least 10 lithography steps where lateral small feature size plays an important role. Some processing steps, such as gate and contact patterning, are critical to the functionality and performance of the CMOS circuits. Furthermore, each processing step is strongly correlated with other steps. Change in any one of the processing steps will lead to modifications in a number of other steps in the process. In contrast, surface micromachining appears relatively simple. It usually consists of two to six masks, and the feature sizes are much larger. The critical processing steps, such as structural polysilicon, often are self-aligned, which eliminates lithographic alignment. The CMOS process is mature, generic, and fine-tuned, whereas surface micromachining strongly depends on the application and still needs maturing.

Table 7.24B presents the critical temperatures associated with the LPCVD deposition of a

Table 7.24 Surface Micromachining and CMOS

| A. Comparison of CMOS and Surface Micromachining | | |
|---|---|---|
| | CMOS | Surface Micromachining |
| Common features | Silicon-based processes Same materials Same etching principles | |
| Process flow | Standard | Application specific |
| Vertical dimension | ~1 μm | ~1–5 μm |
| Lateral dimension | <1 μm | 2–10 μm |
| Complexity (number of masks) | >10 | 2–6 |

| B. Critical Process Temperatures for Microstructures | | |
|---|---|---|
| | Temperature (°C) | Material |
| LP CVD deposition | 450 | Low-temperature oxide (LTO)/PSG |
| LP CVD deposition | 610 | Low-stress polysilicon |
| LP CVD deposition | 650 | Doped polysilicon |
| LP CVD deposition | 800 | Nitride |
| Annealing | 950<br>1,050 | PSG densification<br>Polysilicon stress annealing |

*Source:* Yun, W. 1992. *A surface micromachined accelerometer with integrated CMOS detection circuitry*. PhD thesis. University of California, Berkeley.[185]

variety of frequently used materials in surface micromachining. Polysilicon is used for structural layers, and thermal $SiO_2$, LPCVD $SiO_2$, and PSG are used as sacrificial layers; silicon nitride is used for passivation. The highest temperature process in Table 7.24B is 1050°C and is associated with the annealing step to release stress in the polysilicon layers. Doped polysilicon films deposited by LPCVD under conventional IC conditions usually are in a state of compression that can cause mechanically constrained structures such as bridges and diaphragms to buckle. The annealing step above at about 1000°C promotes crystallite growth and reduces the strain. If a polysilicon microstructure is built after the CMOS active electronics have been implemented (a so-called *post-CMOS procedure*), temperatures greater than 950°C must be avoided because junction migration will take place at those temperatures. This is especially true with devices incorporating shallow junctions where migration might be a problem at temperatures as low as 800°C. The degradation of the aluminum metallization presents a bigger problem. Aluminum typically is used as the interconnect material in the conventional CMOS process. At temperatures of 400–450°C, the aluminum metallization will start suffering. Anneal temperatures (densification of the PSG and stress anneal of the polysilicon) only account for some of the concerns; in general, several compatibility issues must be considered: 1) deposition and anneal temperatures, 2) passivation during micromachining etching steps, and 3) surface topography.

Yun[185] compared three possible approaches to build integrated microdynamic systems: pre-, mixed, and post-CMOS microstructural processes. He concluded that building up the microstructures after implementation of the active electronics offers the best results.

To avoid problems with microstructure topography, which commonly includes step heights of 2–3 μm, the CMOS module is fabricated before the microstructure module in a post-CMOS process. Although this solves topography problems, it introduces constraints on the CMOS. In a post-CMOS process, the electronic circuitry is passivated to protect it from the subsequent micromachining processes. The standard IC processing may be performed at a regular IC foundry, whereas the surface micromachining occurs as an add-on in a specialized sensor fabrication facility. LPCVD silicon nitride (deposited at 800°C; see Table 7.24B) is stable in HF solutions and is the preferred passivation layer for the IC during the long-release etching step. PECVD nitride can be deposited at around 320°C, but it displays relatively poor step coverage, whereas pinholes in the film allow HF to diffuse through and react with the oxide underneath. LPCVD nitride is conformably deposited. Although it shows fewer pinholes, circuitry needs to be able to survive the 800°C deposition temperature.

Aluminum metallization must be replaced by another interconnect scheme to increase the post-CMOS temperature ceiling higher than 450°C. Tungsten, which is refractory, shows low resistivity, and has a thermal expansion coefficient matching that of silicon, is an obvious choice. One problem with tungsten metallization is that tungsten reacts with silicon at about 600°C to form $WSi_2$, implying the need for a diffusion barrier. The process sequence for the tungsten metallization developed at the University of California, Berkeley is shown in Figure 7.78. A diffusion barrier consisting of $TiSi_2$ and TiN is used. The TiN film forms during a 30-s sintering step to 600°C in $N_2$. Rapid thermal annealing with its reduced time at high temperatures (10 s to 2 min) and high ramp rates (~150°C/s) allows very precise process control and a dramatic reduction of thermal

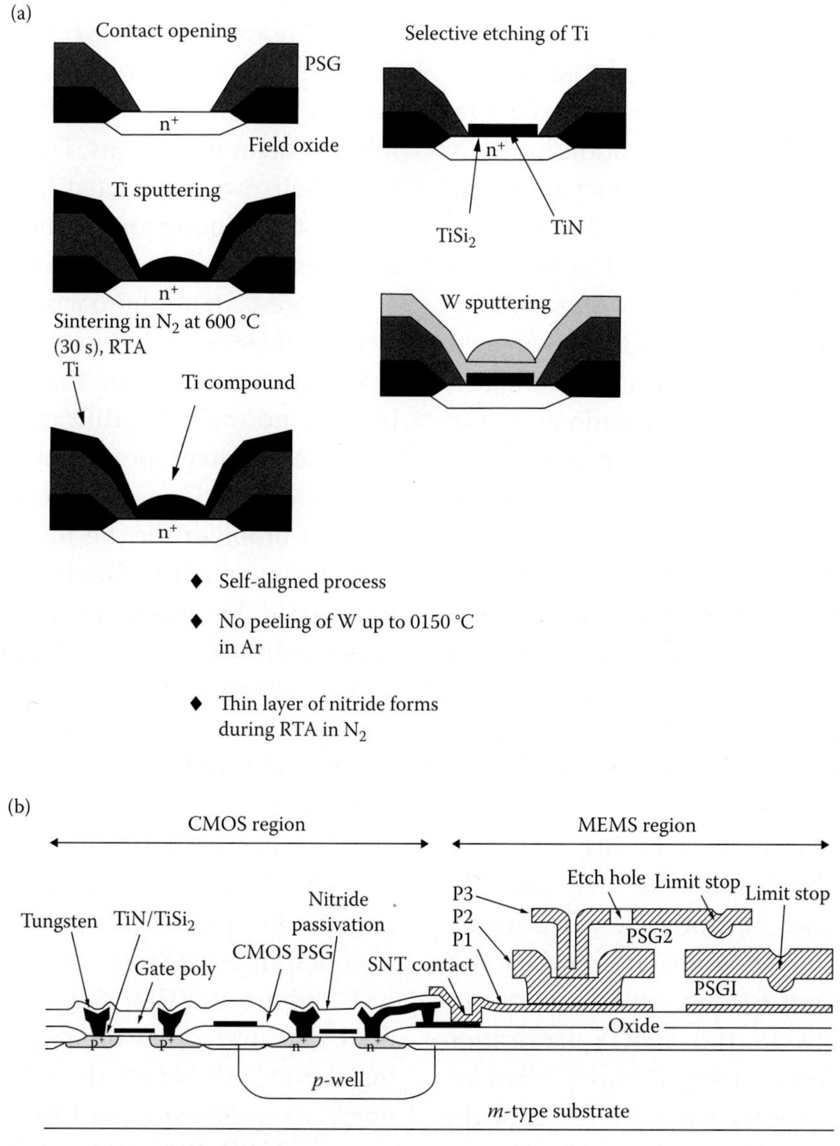


**FIGURE 7.78** (a) Tungsten metallization process in a modified CMOS process. (Adapted from Yun, W. 1992. *A surface micromachined accelerometer with integrated CMOS detection circuitry.* PhD thesis. University of California, Berkeley. With permission.[185]) (b) CMOS electronics combined with postprocessed MEMS device.

budgets, reducing duress for the active on-chip electronics. Titanium silicide is formed at the interface of titanium and silicon, whereas titanium nitride forms simultaneously at the exposed surface of the titanium film. The $TiSi_2$/TiN forms a good diffusion barrier against the formation of $WSi_2$ and at the same time provides an adhesion and contact layer for the W metallization.

To avoid the junction migration in a post-CMOS process, rapid thermal annealing is used for both the PSG densification and polysilicon stress anneal: 950°C for 30 s for the PSG densification and 1000°C for 60 s for the stress anneal of the polysilicon. Alternatively, fine-grained polysilicon can be used because it yields a controlled tensile strain with low-temperature annealing.[186]

Despite some advantages, the post-CMOS process with tungsten metallization is not the preferred implementation. Hillock formation in the W lines during annealing and high contact resistance remains problematic.[116] The heavily doped structural and sacrificial layers may also affect the finely tuned CMOS fabrication sequence. Most importantly, the use of tungsten for circuit interconnects is not consistent with mainstream IC technologies, where aluminum interconnects predominate. Given that IC manufacturers have already invested enormous resources into the development of multilevel aluminum interconnect technologies, and further given the inferior resistivity of tungsten versus aluminum, the described tungsten-based post-CMOS process, although useful as a demonstration tool, is not likely to be adapted in industry.[187] A post-CMOS process for high-aspect-ratio integrated single-crystal silicon microstructures was described in Example 3.3.

The mixed CMOS/micromachining approach implements a processing sequence that puts the processes in a sequence to minimize performance degradation for both electronic and mechanical components. According to Yun, this requires significant modifications to the CMOS fabrication sequence. Nevertheless, Analog Devices relied on such an interleaved process sequence to build the first commercially available integrated microaccelerometer (see Example 7.1). The modifications required on a standard BiCMOS line are minimal: to facilitate integration of the IC and to surface micromachine the thickness of deposited microstructural films, the line is limited to 1–4 μm. Relatively deep junctions permit thermal processing for the sensor polysilicon anneal, and interconnections to the sensor are made only via $n^+$ underpasses. Therefore, no metallization is present in the sensor area. This industrial solution remains truer to the traditional IC process experience than the post-CMOS procedure. Howe[116] recently detailed another example of such a mixed process In Howe's scenario, the micromachining sequence is inserted after the completion of the electronic structures but before contact etching or aluminum metallization. By limiting the polysilicon annealing to 7 min at 900°C, only minor dopant redistribution is expected. Contact and metallization lithography and etching are more complex now because of the severe topography of the polysilicon microstructural elements. These processes, however, have their own associated limitations: mixed processes often require longer, more expensive development periods for new product lines.

The pre-CMOS approach is to fabricate microstructural elements before any CMOS process steps. At first glance, this seems like an attractive approach because no major modifications would be needed for process integration. However, because of the vertical dimensions of microstructures, step coverage is a problem for the interconnection between the sensor and the circuitry (the latest approach introduced by Howe faces the same dilemma). Passivation of the microstructure during the CMOS process can also become problematic. Furthermore, the heavily doped structural layers can affect the fine-tuned CMOS fabrication sequences, such as gate oxidation. Consequently, this approach is only used for special applications.[185] Precircuit processes may place limitations on foundry-based fabrication schemes because circuit foundries may be sensitive to contamination from MEMS foundries.

A unique pre-CMOS process (micromechanics first) that overcomes the planarity issues of building the MEMS before the CMOS has been developed at Sandia National Labs (http://mems.sandia.gov/tech-info/summit-v.html).[188] The SUMMiT-V™ fabrication process is a batch fabrication process using conventional IC processing tools. Using this

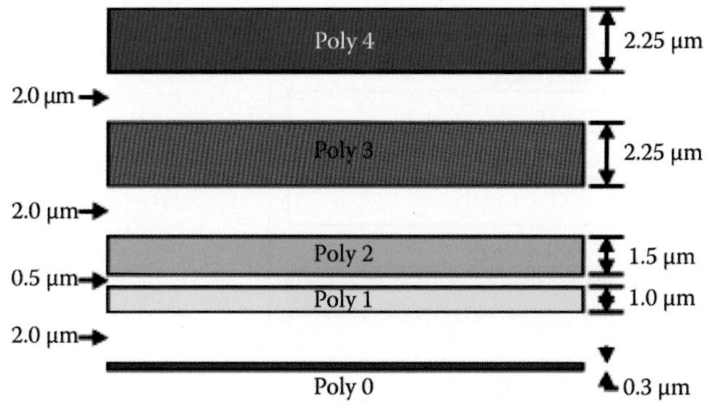


**FIGURE 7.79** Sandia's Ultraplanar Multilevel MEMS Technology (SUMMiT-V™) (http://mems.sandia.gov/tech-info/summit-v.html).

technology, high-volume, low-cost production can be achieved. Topography issues are mitigated using chemical-mechanical polishing (CMP) to achieve planarization (see Chapter 1 on lithography), and stress is maintained at low levels using a proprietary process. The process uses 14 individual masks, approximately the same quantity as in many CMOS IC processes (Figure 7.79).

In this approach, micromechanical devices are fabricated in a trench etched in a Si epilayer. After the mechanical components are complete, the trench is filled with oxide, planarized using CMP, and sealed with a nitride membrane. The flat wafer with the embedded micromechanical devices is then processed by means of conventional CMOS processing. Additional steps are added at the end of the CMOS process to expose and release the embedded micromechanical devices and to hook them up with the CMOS. A cross-section of a structure made with this technology is shown in Figure 7.80.

The SPIE "Smart Structures and Materials 1996" meeting in San Diego, CA, had two complete sessions dedicated to the crucial issue of integrating electronics with polysilicon surface micromachining.[189] Research aimed at achieving truly modular merged circuits + microstructures technology is still ongoing.

## Surface Micromachining Modifications

### *Silicon on Insulator—SOI Surface Micromachining*

*Introduction* SOI is an exciting new approach to both IC chip making and MEMS. In SOI, bulk silicon wafers are replaced with wafers that have three layers: a thin surface layer of silicon (from a few hundred angstroms to several micrometers thick), an underlying layer of insulating material, and a support or "handle" silicon wafer. The insulating layer, usually made of silicon dioxide, is called the *buried oxide* or *BOX* and is typically a few thousand angstroms thick. As we will see below, this thin buried oxide layer, sandwiched between two layers of Si, can be achieved in several different ways.

When transistors are built within the thin top silicon layer, they switch signals faster (up to 10 GHz), run at lower voltages, and are much less vulnerable to signal noise from background cosmic ray particles. Furthermore, on an SOI wafer each transistor is isolated from its neighbor by a complete layer of silicon dioxide, which makes them immune to "latch-up" problems, and they can be spaced closer together than transistors built on bulk silicon wafers. Thus, building circuits on SOI allows for more compact

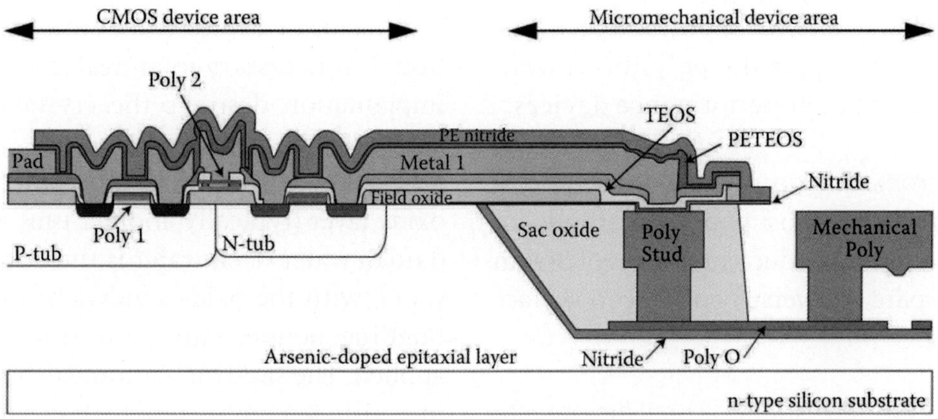


**FIGURE 7.80** Schematic cross-section of the embedded micromechanics approach to Sandia's CMOS/MEMS integration.

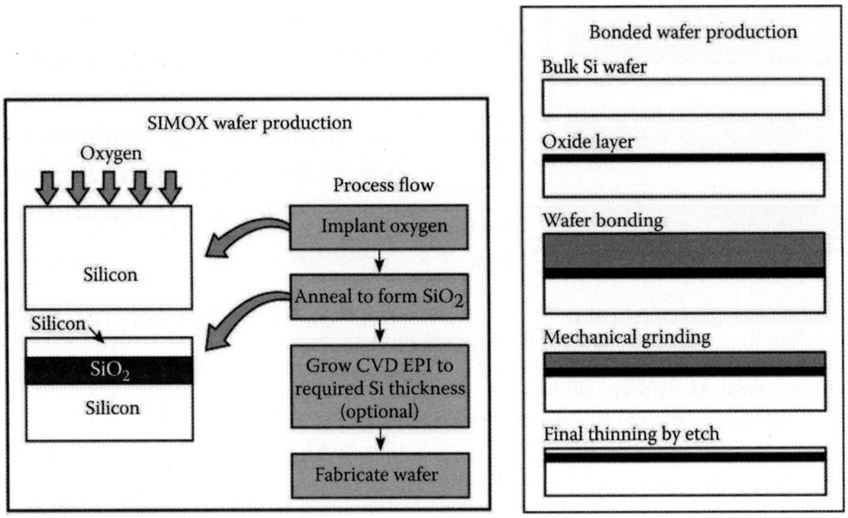


**FIGURE 7.81** How SOI wafers are made.

chip designs, resulting in smaller IC devices (with higher production yield) and more chips per wafer (increasing fabrication productivity).

In MEMS, silicon crystalline features, anywhere between fractions of a micrometer to 100 μm high, can be readily obtained by surface micromachining of the epi silicon or fusion-bonded silicon layer of SOI wafers.[99] Structural elements made from these single-crystalline Si layers result in more reproducible and reliable sensors than those made from polysilicon. SOI or epi-micromachining combines the best features of surface micromachining (i.e., IC compatibility and freedom in $x,y$ shapes) with the best features of bulk micromachining (superior single-crystal Si properties). Moreover, SOI surface micromachining frequently involves fewer process steps and offers better control over the thickness of crucial building blocks. Given the poor reproducibility of mechanical properties and generally poor electronic characteristics of polysilicon films, SOI machining may surpass the polysilicon technology for fabricating high-performance devices.

*SOI Surface Micromachining* Below, we present a short review of the three most popular methods for producing SOI wafers, introduce some uses of SOI in MEMS, and compare SOI versus polysilicon surface micromachining.

*SOI Wafer Fabrication Techniques* Three major techniques currently are applied to produce SOI wafers (Figure 7.81) (see also under "Epitaxy" above): SIMOX (Separated by IMplanted OXygen), the Si fusion bonded (SFB) wafer technique, and zone-melt recrystallized (ZMR) polysilicon. With SIMOX, standard Si wafers are implanted with oxygen ions and then annealed at high temperatures (1300°C). The oxygen and silicon combine to form a silicon oxide layer beneath the silicon surface. The oxide layer's thickness and depth are controlled by varying the energy and dose of the implant and the anneal temperature. In some cases, a CVD process deposits additional epitaxial silicon on the top silicon layer. Attempts have also been made to implant nitrogen in Si to create abrupt etch stops [in the SIMON process (Separation by IMplantation Of Nitrogen]. At high-enough energies, the implanted nitrogen is buried 0.5–1 μm deep. At a high-enough dose, the etching in that region stops. It is not necessary to implant a stoichiometric amount of nitrogen; a dose lower by a factor of two to three suffices. After implantation, it is necessary to anneal the wafer because the implantation destroys the crystal structure at the surface of the wafer.

The Si fusion bonded wafer process starts with an oxide layer (typically about 1 μm) grown on a standard Si wafer. That wafer is then bonded to another wafer, with the oxide sandwiched between. For the bonding, no mechanical pressure or other forces are applied. The sandwich is annealed at 1100°C for 2 h in a nitrogen ambient, which creates a strong bond between the two wafers. One of the wafers is then ground to a thickness of a few micrometers using mechanical and CMP.

A third process for making SOI structures is to recrystallize polysilicon (e.g., with a laser, an electron beam, or a narrow strip heater), which has been deposited on an oxidized silicon wafer. This process is called *zone melting recrystallization* (ZMR) and is used primarily for local recrystallization but has not yet been explored much for use in micromachining applications.

The crystalline perfection of conventional silicon wafers in SFB and ZMR is completely maintained in the SOI layer because the wafers do not suffer from implant-induced defects. By using plasma etching, wafers with a top Si layer thickness of as little as 1000–3000 Å with total thickness variations of less than 200 Å can be made.[190] SOI top Si layers of 2 μm thick are more standard.[191] A tremendous amount of effort is spent in the IC industry on controlling the SOI thin Si layer thickness, which benefits any narrow tolerance IC and micromachine.

Etched-back fusion-bonded Si wafers and SIMOX are used extensively to build both ICs and micromachines, and both are commercially available. In 1989 TI introduced its commercial 64K static random access memory (SRAM) based on SOI, and AMD's 90 nm is also based on SOI technology (introduced in late 2004). One vendor is SOITEC (http://www.soitec.com/en/products).

### SOI Use in MEMS

*Overview* Kanda[192] reviews different types of SOI wafers in terms of their micromachining and IC applications. Working with SOI wafers, he points out, offers several advantages over bulk Si wafers: fewer process steps are needed for feature isolation; parasitic capacitance is reduced; and power consumption is lowered. In the IC industry, SOI wafers are principally used for high-speed CMOS ICs (<10 GHz), smart power ICs (voltages up to 100 V), 3D ICs, and radiation-hardened devices.[193]

An important use of silicon on insulator wafers is in the production of high-temperature sensors. Compared with p-n junction isolation, which is limited to about 125°C, much higher temperature (up to 300°C) devices are possible based on the dielectric insulation of SOI. A wide variety of SOI surface-micromachined structures have been explored, including pressure sensors, accelerometers, torsional micromirrors, light sources, and optical choppers.[95,99] Often in those micromachining applications, SOI wafers are used to produce an etch stop. The silicon fusion bonded (SFB) method offers the more versatile MEMS approach because of the associated potential for thicker single-crystal layers and the option of incorporating buried cavities, facilitating micromachine packaging. Sensors manufactured by means of SFB now are commercially available.[194] The SIMOX approach is less labor intensive and holds better membrane thickness control. An important expansion of the SOI technique is selective epitaxy. The latter enables a wide range of new mechanical structures and enables novel etch-stop methods,[195] as well as electrical and/or thermal separation and independent optimization of active sensor and readout electronics.[196] In most cases, SOI machining involves dry anisotropic etching to etch a pattern into the Si layer on top of the insulator. These structures then are released by etching the sacrificial buried $SiO_2$ insulator layer, which displays a thickness with very high reproducibility (400 ± 5 nm) and uniformity (<±5 nm), especially in the case of SIMOX. Etched free cantilevers and membranes consist of single-crystalline silicon with thicknesses ranging from micrometers and submicrometers (SIMOX) up to hundreds of micrometers (SFB).

The increasing use of SOI technology for the fabrication of MEMS devices has brought with it its own set of unique problems that must be overcome for applications where CD control is important. For example, the etching of deep trenches in SOI wafers necessitates a fast etch process, followed by a slow etch process as soon as the buried dioxide layer is reached.[197] The main problem in plasma processing of SOI is the notching at the oxide interface, which results from the deflection of the incident ions by charging up the insulating oxide surface. The notch width increases with time, and it is here that etch uniformity begins to play an important role because overetching times to clear the wafer can become significant. For example, a 400-μm-deep etch with +5% uniformity requires a >10% overetch to ensure that all features etch to the required depth. Thus, some areas of the wafer will be subject to a 40-μm overetch, and precise control of the profile during overetching becomes essential.

*Creating Buried Cavities in Silicon by Silicon Fusion Bonded Micromachining* Silicon fusion bonding enables the formation of thick single-crystal layers with cavities built in. An example is shown in Figure 7.82.[99] The device pictured involves two 4-in. <100> wafers: a handle wafer and a wafer used for the SOI surface. The p-type (3–7 Ωcm) handle wafer is thermally oxidized at 1100°C to obtain a 1-μm-thick oxide. Thermal oxidation enables thicker oxides than the ones formed in SIMOX by ion implantation and avoids the potential implantation damage in the working material. To make a buried cavity, the oxide is patterned and etched. To produce yet deeper cavities, the Si handle wafer may be etched as well (as in Figure 7.82). In the case shown, the top wafer consists of the same p-type substrate material as the handle wafer with a 2–30-μm-thick n-type epitaxial layer. The epitaxial layer determines the thickness of the final mechanical/structural element. The epitaxial layer is fusion bonded to the cavity side of the handle wafer (2 h at 1100°C). The top wafer is then partially thinned by grinding and polishing (Figure 7.82a). An insulator is deposited and patterned on the back side of the handle wafer to etch access holes to the insulator. After the insulator at the bottom of the etch hole is removed by a buffered oxide etch (BOE), aluminum is sputtered and sintered to make contact to the n-type epilayer for the electrochemical etch back of the remaining p-type material (Figure 7.82b). The final single-crystal silicon thickness is uniform to within ±0.05 μm (standard deviation) and does not require a costly, high-accuracy polish step.

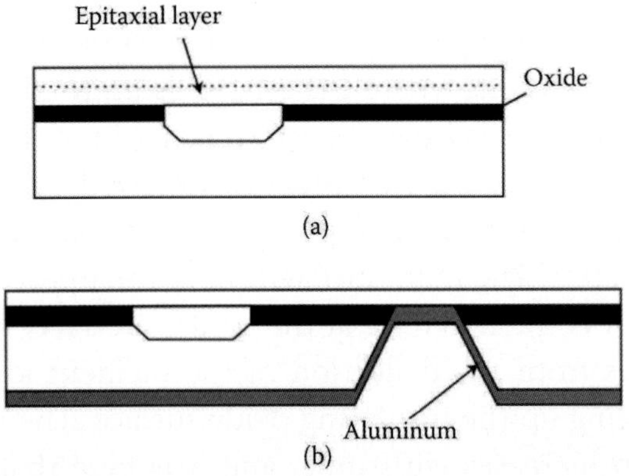


**FIGURE 7.82** (a) Wafer sandwich after grind-and-polish step. (b) Wafer after electrochemical etch-back in KOH, buried oxide removal, and aluminum deposition. [From Noworolski, J. M., E. Klaassen, J. Logan, K. Petersen, and N. Maluf. 1995. *Eighth international conference on solid-state sensors and actuators (Transducers '95).* Stockholm, Sweden: IEEE. With permission.[99]]

Draper Laboratory is using SOI processes in the development of inertial sensors, gyros, and accelerometers as an alternative to devices fabricated by the dissolved wafer process (see Example 4.1). The main advantage is that the former consists of an all-Si process rather than a Si/Pyrex sandwich; an all Si device will exhibit less thermal stress.[198]

*Fabricating Pressure Sensors in SIMOX Surface Micromachining* Both capacitive and piezoresistive pressure sensors were microfabricated from SIMOX wafers.[95] Figure 7.83 illustrates the process sequence by Diem et al.[95] for fabricating an absolute capacitive pressure sensor. The 0.2-μm silicon surface layer of the SIMOX wafer is thickened with doped epi-silicon to 4 μm. An access hole is RIE etched in the Si layer, and vacuum cavity and electrode gap are obtained by etching the $SiO_2$ buried layer. Because the buried thick oxide layer exhibits a very high reproducibility and homogeneity over the whole wafer (0.4 μm ± 5 nm), the resulting vacuum cavity and electrode gap after etching also are very well controlled. The small gap results in relatively high capacitance values between the free membrane and bulk substrate (20 pf/mm²). Diaphragm diameter, controlled by the $SiO_2$ etching, is up to several hundred micrometers (±2 μm). The etching hole is hermetically sealed under vacuum by plasma CVD deposition of nonstressed dielectric layer plugs.

With the above scheme, Diem et al. realized an absolute pressure sensor with a size of less than 1.5 mm². The temperature dependence of a capacitive sensor is mainly a result of the temperature coefficient of the offset capacitance. Therefore, temperature compensation is needed for high accuracy sensors. A drastic reduction of the temperature dependence is obtained by a differential measurement, especially if the reference capacitor resembles the sensing capacitor. A reference capacitor is designed with the membrane blocked by several plugs for pressure

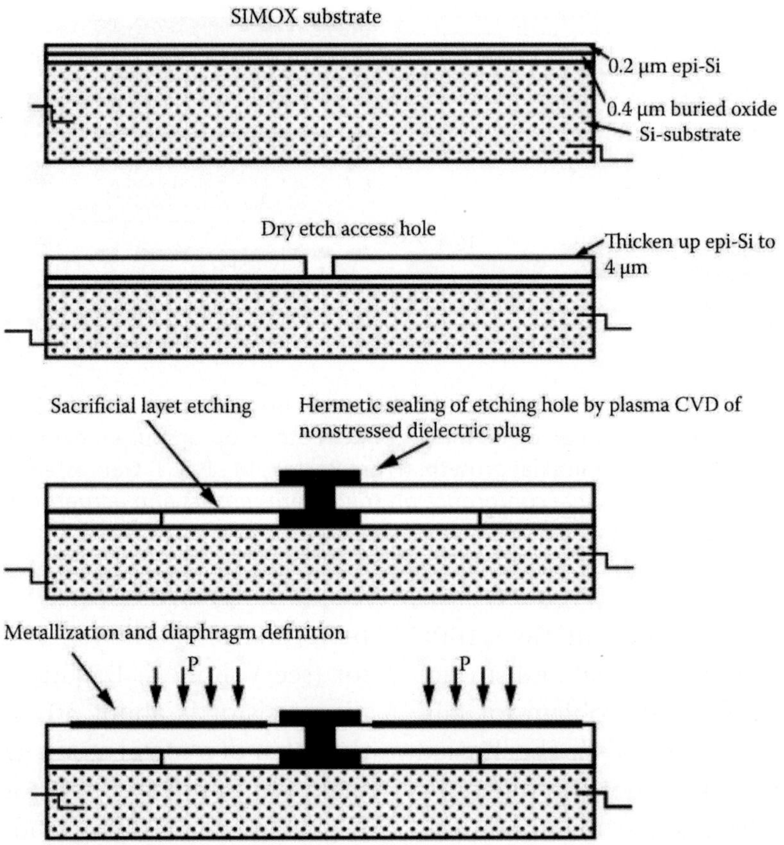


**FIGURE 7.83** Process sequence of a SIMOX absolute capacitive pressure sensor by Diem et al. [From Diem, B., M. T. Delaye, F. Michel, S. Renard, and G. Delapierre. 1993. *Seventh international conference on solid-state sensors and actuators (Transducers '93).* Yokohama, Japan: IEEE. With permission.[95]]

insensitivity. The localization and the number of plugs are modeled by finite element analysis (FEA; ANSYS software was used) to get a deformation lower than 1% of the active sensor's deformation. Even without temperature calibration, the high output of the differential signal results in an overall output error better than ±2% over the whole temperature range (−40°C to +125°C) compared with 10% for nondifferential measurements. The temperature coefficient of the sensitivity is about 100 ppm/°C, which agrees with the theoretical variation of the Young's modulus of silicon. A piezoresistive sensor can be achieved by implanting strain gauges in the membrane. Although SIMOX wafers are more expensive than regular wafers, they come with several process steps embedded, and they make packaging easier.

*Selective Epitaxy Surface Micromachining* In the discussion on epitaxy, we drew attention to the potential of selective epitaxy for creating novel microstructures. The example in Figure 7.50 illustrates the selective deposition of epi-silicon on a Si substrate through a $SiO_2$ window. The same figure also demonstrates the simultaneous deposition of polysilicon on $SiO_2$ and crystalline epi-silicon on silicon, creating the basis for a structure featuring an epi-silicon anchor with polysilicon side arms.

Neudeck et al.[199,200] at Purdue and Gennissen et al.[195,196] in Twente proved that selective epitaxy can also be applied for automatic etch stop on buried oxide islands. Figure 7.84a demonstrates how epitaxial lateral overgrowth (ELO) can bury oxide islands. After removal of the native oxides from the seed windows, epi is grown for 20 min at 950°C and at 60 Torr using a $Si_2H_2Cl_2$-HCl-$H_2$ gas system. The epi growth front moves parallel to the wafer surface while growing in the lateral direction, leaving a smooth planar surface. During epi growth, the HCl prevents polynucleation on the nonsilicon areas. The epi quality is strongly dependent on the orientation of the seed holes

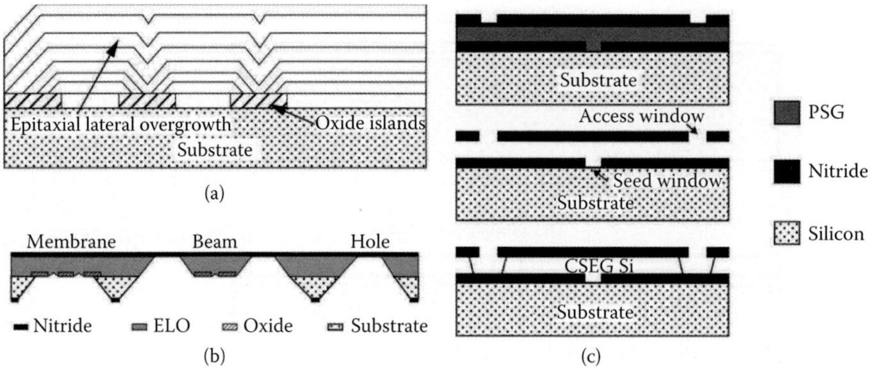


**FIGURE 7.84** Micromachining with epi-silicon. (a) Lateral overgrowth process of epi-silicon (ELO, epitaxial lateral overgrowth) out of <100>-oriented holes in an oxide mask. (b) KOH etch stop on buried oxide islands or front side nitride. (c) Principle of confined selective epitaxial growth. (From Bartek, M., P. T. J. Gennissen, P. J. French, and R. F. Wolffenbuttel. 1995. *Eighth international conference on solid-state sensors and actuators (Transducers '95).* Stockholm, Sweden: IEEE. With permission.[196])

in the oxide. Seed holes oriented in the <100> direction lead to the best epi material and surface quality. Selective epi's other big problem for fabrication remains sidewall defects.[201] The buried oxide islands stop the KOH etch of the substrate, enabling formation of beams and membranes as shown in Figure 7.84b. This technique might form the basis of many high-performance microstructures. The Purdue and Twente groups also work on confined selective epitaxial growth (CSEG), a process pioneered by Neudeck et al.[200] In this process a micromachined cavity is formed above a silicon substrate with a seed contact window to the silicon substrate and access windows for epi-silicon (Figure 7.84c).[196] Low-stress, silicon-rich nitride layers act as structural layers to confine epitaxial growth; PSG is used as sacrificial material. This confined selective epitaxial growth technique allows electrical and/or thermal isolation separation, as well as independent optimization of active sensor and readout electronic areas.

*SOI versus Polysilicon Surface Micromachining*
The power of polysilicon surface micromachining mainly lays in its CMOS compatibility. When deposited on an insulator, both polysilicon and single crystal layers enable higher operating temperatures (>200°C) than bulk micromachined sensors featuring p-n junction isolation only (130°C max).[202] An additional benefit for SOI-based micromachining is IC compatibility combined with single-crystal Si performance excellence. The maximum gauge factor (see Volume I, Equation 4.93) of a polysilicon piezoresistor is about 30, roughly 15 times larger than that of a metal strain gauge but only one third that of an indiffused resistor in single-crystal Si.[203] Higher piezoresistivity and fracture stress would seem to favor SOI for sensor manufacture. However, there is an important counterargument: the piezoresistivity and fracture stress in polysilicon are isotropic, a major design simplification. Moreover, by laser recrystallization, the gauge factor of polysilicon might increase to greater than 50,[204] and by appropriate boron doping the temperature coefficient of resistance (TCR) can reach 0 versus a TCR of, e.g., $1.7 \times 10^{-3}\ K^{-1}$ for single-crystal p-type silicon. Neither technical nor cost issues will be the deciding factor in determining which technology will become dominant in the next few years. Micromachining is very much a hostage to trends in the IC industry; promising technologies such as GaAs and micromachining do not necessarily take off, in no small part because of the invested capital in some limited sets of standard silicon technologies. On this basis, SOI surface micromachining is the favored candidate; SOI extends silicon's technological relevance and experiences, increasing investment from the IC industry and thus benefiting SOI micromachining.[192]

Based on the above, we believe that SOI micromachining not only introduces an improved method of making many simple micromachines, but it also will probably become the favored approach of the IC industry. A summary of SOI advantages is listed below:

- IC industry use in all type of applications such as MOS, bipolar digital, bipolar linear, power devices, BiCMOS, CCDs, and heterojunction bipolar[205]
- Batch packaging through embedded cavities
- CMOS compatibility
- Substrate industrially available at increasingly lower costs (about $200 today)
- Excellent mechanical properties of the single-crystalline surface layer
- Freedom of shapes in the *x*-*y* dimensions and continually improving dry etching techniques, resulting in larger aspect ratios and higher features
- Freedom to choose a very well-controlled range of thicknesses of epi surface layers
- $SiO_2$ buried layer as sacrificial and insulating layer and excellent etch stop
- Dramatic reduction of process steps as the SOI wafer comes with several "embedded" process steps
- High-temperature operation

A tutorial on SOI can be found at http://www.freescale.com/webapp/sps/site/overview.jsp?code=TM_SOI.

### *Hinged Polysilicon*

One way to achieve high vertical structures with surface micromachining is to build large, flat structures horizontally and then rotate them on a hinge to an upright position. Pister et al.[194–203] developed the polysilicon hinges shown in Figure 7.85A; on these hinges, long structural polysilicon features (1 mm and beyond) can be rotated out of the plane of a substrate. To make these hinged structures, a 2-μm-thick PSG oxide layer (PSG-1) is deposited on the Si substrate as the sacrificial material, followed by the deposition of the first polysilicon layer (2-μm-thick Poly 1). Photolithography and dry etching to form the desired structural elements, including hinge pins to rotate them, pattern the polysilicon structural layer. Following the deposition and patterning of Poly 1, another layer of sacrificial material (PSG-2) of 0.5-μm thickness is deposited (Figure 7.85Aa). Contacts are made through both PSG layers to the Si substrate, and a second layer of polysilicon is deposited and patterned (Poly 2), forming a staple to hold

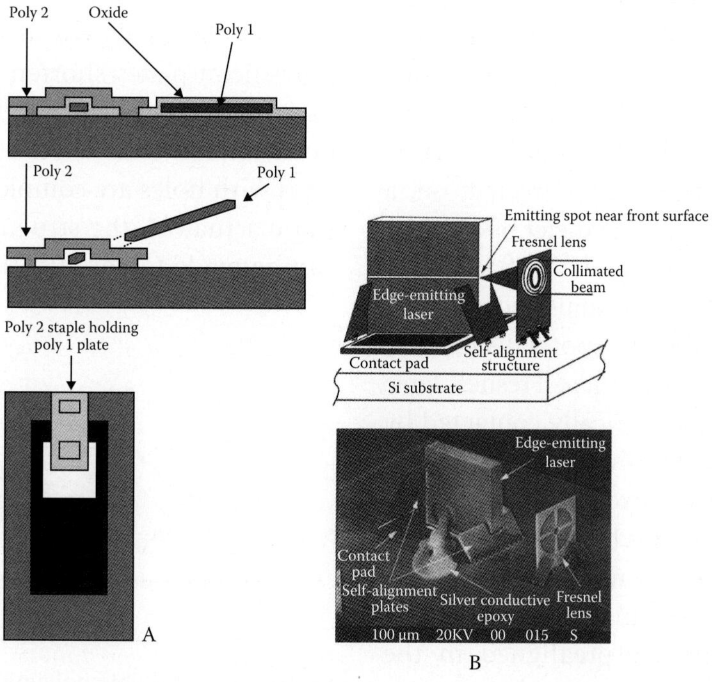


FIGURE 7.85 Microfabricated hinges. (A) Cross-section, side view, and top view of a single-hinged plate before and after the sacrificial etch. SEM of a hinge (MCNC). (B) Schematic (top) and SEM micrograph of the self-aligned hybrid integration of an edge-emitting laser with a micro-Fresnel lens. (From Lin, L. Y., S. S. Lee, M. C. Wu, and K. S. J. Pister. 1995. *Proceedings: IEEE micro electro mechanical systems (MEMS '95)*. Amsterdam, the Netherlands: IEEE. With permission.[206])

the first polysilicon layer hinge to the surface (Figure 7.85Ab). The first and second layers of poly are separated everywhere by PSG-2 for the first polysilicon layer to freely rotate off the wafer surface when the PSG is removed in a sacrificial etch. After the sacrificial etch, the structures are rotated in their respective positions (Figure 7.85Ac). This is accomplished in an electrical probe station by skillfully manipulating the movable parts with the probe needles. Once the components are in position, high friction in the hinges tends to keep them in the same position. To obtain more precise and stable control of position, additional hinges and supports are incorporated. To provide electrical contact to the vertical polysilicon structures, one can rely on the mechanical contact in the hinges, or polysilicon beams (cables) can be attached from the vertical structure to the substrate.

Pister's research team made a wide variety of hinged microstructures, including hot wire anemometers, a box dynamometer to measure forces exerted by embryonic tissue, a parallel-plate gripper,[207] a microwindmill,[208] a microoptical bench (MOB) for free-space integrated optics,[206] and a standard CMOS single piezoresistive sensor to quantify the single heart cell contractile forces of a rat.[209] One example from this group's efforts is illustrated in Figure 7.85B, showing an SEM photograph of an edge-emitting laser diode shining light onto a collimating micro-Fresnel lens.[206] The micro-Fresnel lens in the SEM photo is surface micromachined in the plane and erected on a polysilicon hinge. The lens has a diameter of 280 μm. Alignment plates at the front and the back sides of the laser are used for height adjustment of the laser spot so that the emitting spot falls exactly onto the optical axis of the micro-Fresnel lens. After assembly, the laser is electrically contacted by silver epoxy. Although this hardly outlines standard IC manufacturing practices, excellent collimating ability for the Fresnel lenses has been achieved. The eventual goal of this work is a microoptical bench (MOB), in which microlenses, mirrors, gratings, and other optical components are prealigned in the mask layout stage using computer-aided design. Additional fine adjustment would be achieved by on-chip microactuators and micropositioners such as rotational and translational stages. Erecting these polysilicon structures with the probes of an electrical probe station and, occasionally, assembly by chance in the HF etch or deionized water rinse represent too complicated or too unreliable postrelease assembly methods for commercial acceptance. At the University of Illinois (Urbana), the Micro Actuator, Sensors, and Systems Group headed by Dr. Chang Liu uses magnetic actuation to position a large array of hinged microstructures in parallel. A magnetic material, Permalloy, is electroplated and integrated into the hinged microstructures. The result may be improved manufacturability and stability of these delicate structures.

Friction in polysilicon joints is high because friction is proportional to the surface area ($s^2$) and becomes dominant over inertial forces ($s^3$) in the microdomain (see Volume III, Chapter 7). Such joints are not suitable for microrobotic applications. Although attempts have been made to incorporate polysilicon hinges in such applications,[210] plastically deformable hinges make more sense for microrobot machinery involving rotation of rigid components. Noting that the external skeleton of insects incorporates hard cuticles connected by elastic hinges, Suzuki et al.[146] fabricated rigid polysilicon plates (E = 140 GPa) connected by elastic polyimide hinges (E = 3 GPa) as shown in Figure 7.86. Holes in the polysilicon plates shorten the PSG etch time compared with plates without holes. The plates without holes remain attached to the substrate, whereas the ones with holes are completely freed. Using electrostatic actuators, the structure shown in Figure 7.85 can be made to flap like the wing of a butterfly. By

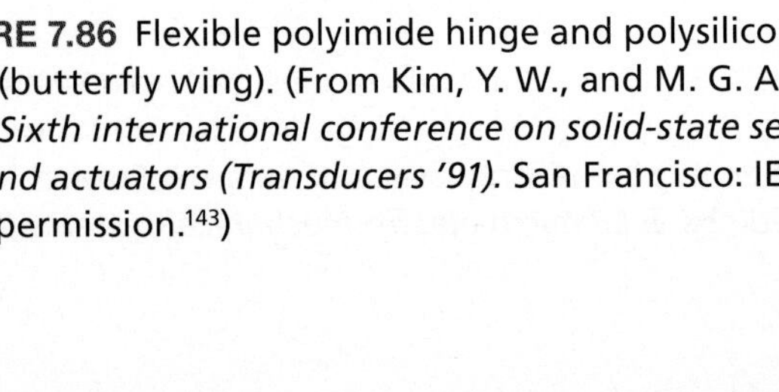

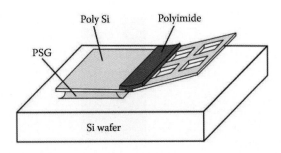


**FIGURE 7.86** Flexible polyimide hinge and polysilicon plate (butterfly wing). (From Kim, Y. W., and M. G. Allen. 1991. *Sixth international conference on solid-state sensors and actuators (Transducers '91).* San Francisco: IEEE. With permission.[143])

applying an AC voltage of 10 kHz, resonant vibration of such a flapping wing was observed.[144]

Hoffman et al.[211] demonstrated aluminum plastically deformable hinges on oxide movable thin plates. Oxide plates and Al hinges were etched free from a Si substrate by using $XeF_2$, a vapor phase etchant exhibiting excellent selectivity of Si over Al and oxide. According to the authors, this process, because of its excellent CMOS compatibility, might open the way to designing and fabricating sophisticated integrated CMOS-based sensors with rapid turnaround time (see also Chapter 3).

### *Milliscale Molded Polysilicon Structures*

The manual assembly of fabricated microparts complicates assembly of tall 3D features in the hinged polysilicon approach we just described, which is similar to building a miniature boat in a bottle. Keller, while at the University of California, Berkeley, came up with an elegant alternative for building tall, high-aspect-ratio microstructures in a process that does not require postrelease assembly steps.[97] The technique involves deep dry etching of trenches in a Si substrate, deposition of sacrificial and structural materials in those trenches, and demolding of the deposited structural materials by etching away the sacrificial materials. CVD processes can typically only deposit thin films (~1–2 μm) on flat surfaces. If, however, these surfaces are the opposing faces of deep narrow trenches, the growing films will merge to form solid beams. In this fashion, high-aspect-ratio structures that would normally be associated with LIGA now also can be made of CVD polysilicon. The procedure is illustrated in Figure 7.87.[97,212] The first step is to etch deep trenches into a silicon wafer. The depth of the trenches equals the height of the desired beams and is limited to about 100 μm with aspect

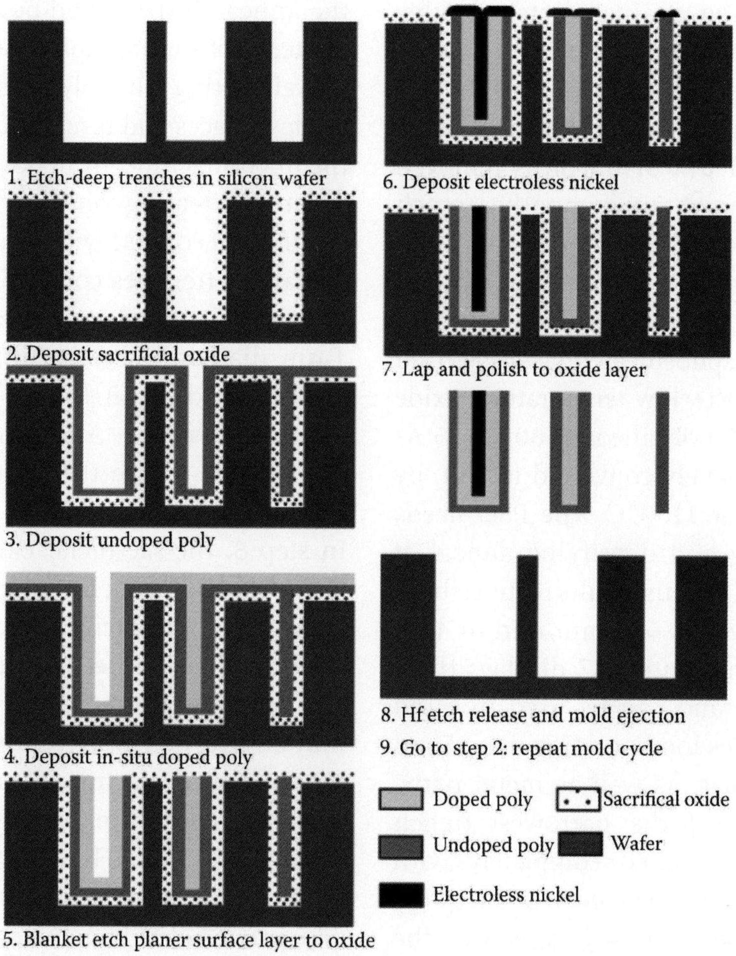


**FIGURE 7.87** Schematic illustration of HEXSIL process. The mold wafer may be part of an infinite loop. (Courtesy of Dr. C. Keller, MEMS Precision Instruments.)

ratios of about 10 (e.g., a 10-μm-diameter hole with a depth of 100 μm). For trench etching, Keller uses a $Cl_2$ plasma etch with the following approximate etching conditions: flow rates of 200 sccm for He and 180 sccm for $Cl_2$, a working pressure of 425 mTorr, a power setting of 400 W, and an electrode gap of 0.8 cm. The etch rate for Si in this mode equals 1 μm/min. Thermal oxide and CVD oxide act as masks with 1 μm of oxide needed for each 20 μm of etch depth. Before the $Cl_2$ etch, a short 7-s $SF_6$ pre-etch removes any remaining native oxide in the mask openings. During the chlorine etch, a white sidewall passivating layer must be controlled to maintain perfect vertical sidewalls. After every 30 min of plasma etching, the wafers are submerged in a silicon isotropic etch long enough to remove the residue.[213] Beyond 100 μm, severe undercutting occurs, and the trench cross-section becomes sufficiently ellipsoidal to prevent molded parts from being pulled out. Advances in dry cryogenic etching are continually improving attainable etch depths, trench profiles, and minimum trench diameter. We can expect continuous improvements in the tolerances of this novel technique. After plasma etching, an additional 1 μm of silicon is removed by an isotropic wet etch to obtain a smoother trench wall surface. Alternatively, to smooth sidewalls and bottom of the trenches, a thermal wet oxide is grown and etched away. The sacrificial oxide in step 2 is made by CVD phosphosilicate glass (PSG) (at 450°C, 140 Å/min), CVD low-temperature oxide (LTO at 450°C), or CVD polysilicon (580°C, 65 Å/min). The latter is completely converted to $SiO_2$ by wet thermal oxidation at 1100°C. The PSG needs an additional reflowing and densifying anneal at 1000°C in nitrogen for 1 h. This results in an etching rate of the sacrificial layer of ~20 mm/min in 49% HF. The mold shown in Figure 7.87 displays three different trench widths and can be used to build integrated micromachines incorporating doped and undoped polysilicon parts, as well as metal parts. The remaining volume of the narrowest trench after oxide deposition is filled completely with the first deposition of undoped polysilicon (Poly 1) in step 3. The undoped poly will constitute the insulating regions in the micromachine. Undoped CVD polysilicon was formed in this case at 580°C, with a 100-sccm silane flow rate, and a 300-mTorr reactor pressure, resulting in a deposition rate of 0.39 μm/h. The deposited film under these conditions is amorphous or very fine grained. Generally speaking, CVD polysilicon films conform well to the underlying topography on the wafer and show good step coverage. With trenches of an aspect ratio in excess of 10, some thinning of the film on the sidewalls occurs. Because the narrowest trenches are completely filled in by the first deposition, they cannot accept material from later depositions. The trenches of intermediate width are lined with the first material and then completely filled in by the second deposition. In the case illustrated, the second deposition (step 4) consists of in situ doped polysilicon and forms the resistive region in the micromachine under construction. To prevent diffusion of *P* from the doped poly deposited on top of the narrow undoped beams, a blanket etch in step 5 is used to remove the doped surface layer before the anneal of the doped poly. The third deposition, in step 6 of the example case, consists of electroless nickel plating on polysilicon surfaces but not on oxide surfaces and results in the conducting parts of the micromachine. By depositing structural layers in order of increasing conductivity, as done here, regions of different conductivity can be separated by regions of narrow trenches containing only nonconducting material. Lapping and polishing in step 7 with a 1-μm diamond abrasive in oil planarizes the top surface, readying it for HF etch release and mold ejection in step 8. Annealing of the polysilicon is required to relieve the stress before removing the parts from the wafer so they remain straight and flat. In step 8, the sacrificial oxide is dissolved in 49% HF. A surfactant such as Triton X100 is added to the etch solution to facilitate part ejection by reducing surface adhesion between the part and the mold. The parts are removed from the wafer, and the wafer may be returned to step 2 for another mold cycle. An example micromachine, resulting from the described process, is the thermally actuated tweezers shown in Figure 7.88. These HEXSIL tweezers measure 4 mm long, 2 mm wide, and 80 μm tall. The thermal expansion beam to actuate the tweezers consists of the in situ doped polysilicon; the insulating parts are made from the undoped polysilicon material. Nickel-

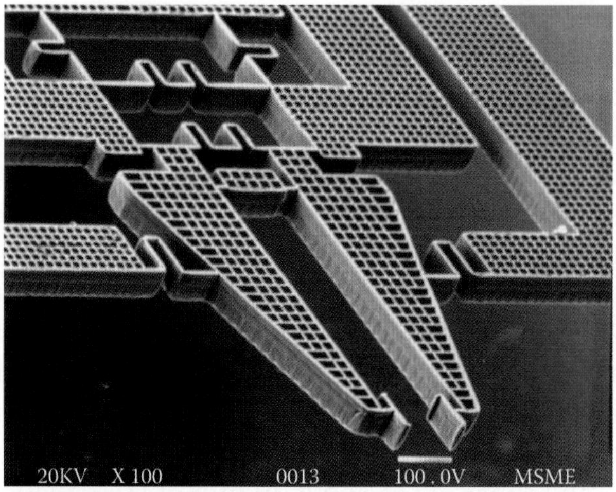


**FIGURE 7.88** SEM micrograph of HEXSIL tweezers: 4 mm long, 2 mm wide, and 80 µm tall. Lead wires for current supply are made from nickel-filled polysilicon beams; in situ phosphorus-doped polysilicon provides the resistor part for actuation. The width of the beam is 8 mm: 2 mm polysilicon, 4 mm nickel, and 2 mm polysilicon. (Courtesy of Dr. C. Keller, MEMS Precision Instruments.)

filled polysilicon beams are used for the current supply leads. It is possible to combine the HEXSIL process with classical polysilicon micromachining, as illustrated in Figure 7.89a, where HEXSIL forms a stiffening rib for a membrane filter fashioned by surface micromachining of a surface polysilicon layer. The surface polysilicon is deposited after HEXSIL. A critical need in HEXSIL technology is controlled mold ejection. Keller et al.[213] have experimented with HEXSIL-produced bimorphs, making the structure spring up after release. Keller at MEMS Precision Instruments is exploiting HEXSIL technology commercially (http://www.memspi.com). Keller did his early HEXSIL work at the University of California, Berkeley and later incorporated the Bosch deep RIE process (see Chapter 3) to create tweezers structures similar to the one shown in Figure 7.88, which can now be fabricated in single-crystal silicon (see Figure 7.89b). MEMS Precision Instruments today makes microtweezers with one of three processes: HEXSIL, deep RIE Bosch process, or Sandia's SUMMiT-V process.

### *Thick Polysilicon*

Applying classical LPCVD to obtain polysilicon deposition is a slow process. For example, a layer of 10 µm typically requires a deposition time of 10 h.

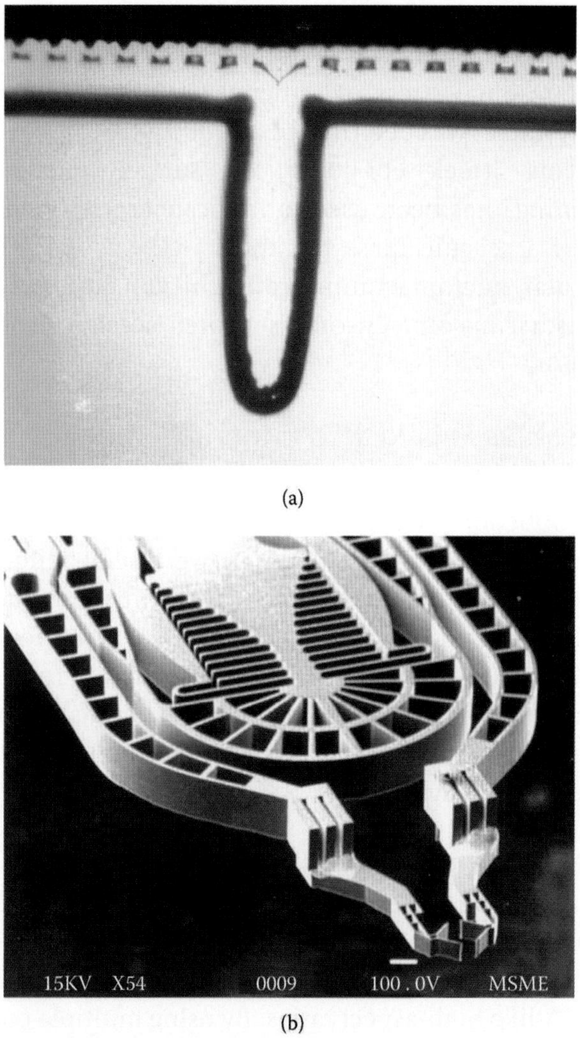


**FIGURE 7.89** (a) SEM micrograph of surface-micromachined membrane filter with a stiffening rib (50 mm high). Original magnification 1000×. (b) Single-crystal microtweezers made by the Bosch deep RIE process. (Courtesy of Dr. C. Keller, MEMS Precision Instruments.)

Consequently, most micromachined structures are based on layer thicknesses in the 2–5-µm range. Basing their process on dichlorosilane ($SiH_2Cl_2$) chemistry, Lange et al.[98] developed a CVD process in a vertical epitaxy batch reactor with deposition rates as high as 0.55 µm/min at 1000°C. The process yields acceptable deposition times for thicknesses in the 10-µm range (20 min). The highly columnar polysilicon films are deposited on sacrificial $SiO_2$ layers and exhibit low internal tensile stress, making them suitable for surface micromachining. The surface roughness comprises about 3% of the thickness, which might preclude some applications.

Kahn et al.[214] made mechanical property test structures from thick undoped and in situ B-doped polysilicon films. The elastic modulus of the B-doped polysilicon films was determined as 150 ± 30 GPa. The residual stress of as-deposited undoped thick polysilicon was determined as 200 ± 10 MPa.

This "thick epi-polysilicon surface micromachining" has been used to fabricate accelerometers with a seismic mass thickness well beyond what regular micromachining can achieve (http://www.sensorscan.com/sensoren/sensoren/accelerometer_1.htm).

*Resists as Structural Elements and Molds in Surface Micromachining*

*Introduction* Deep UV photoresists were covered in Chapter 1. In this section, we briefly reiterate some of the material covered there in the context of surface micromachining. Deep UV photoresists enable the molding of high-aspect-ratio microstructures in a wide variety of moldable materials, or they are used directly as structural elements in the case of permanent photoresists.

*UV Depth Lithography*

*Polyimide Surface Structures* Because of their transparency to exposing UV light, polyimide surface structures can be made very high and exhibit LIGA-like high-aspect ratios. By using multiple coats of spun-on polyimide, thick suspended plates are possible. Moreover, composite polyimide plates can be made, depositing and patterning a metal film between polyimide coats. Polyimide surface microstructures are typically released from the substrate by selectively etching an aluminum sacrificial layer, although Cu and PSG (e.g., in the polyimide flapping wing in Figure 7.86) have been used as well (see also Table 7.23).

Polyimides, because they are easily deformed, usually do not qualify as mechanical members but have been used, for example, in plastically deformable hinges (see Figure 7.86).[144,211] An early result in polyimide surface micromachining was obtained at SRI International, where polyimide pillars (spacers) about 100 μm in height were used to separate a Si wafer, which was equipped with a field emitter array, from the display glass plate of a field emission flat panel display.[215] The concept of a field emission display is illustrated in Figure 7.90a. The flat panel display and an SEM picture of the pillars are shown in Figure 7.90b. The field-emitter tip arrays are positioned between these supporting polyimide posts. The Probimide 348 FC formulation of Ciba-Geigy was used. This viscous precursor formulation (48% by weight of a polyamic ester, a surfactant for wetting, and a sensitizer) with a 3500 cs viscosity was applied to the Si substrate and formed into a film of a 125-μm thickness by spinning. A 30- to 40-min prebake at about 100°C removed the organic solvents from the precursor. The mask with the pillar pattern was then aligned to the wafer coated with the precursor and subjected to about 20 min of UV radiation. After driving off moisture by another baking operation, the coating, still warm, was spray developed (QZ 3301 from Ciba-Geigy), revealing the desired spacer matrix. By baking the polyimide at 100°C in a high vacuum ($10^{-9}$ Torr), the pillars shrunk to about 100 μm, whereas the polyimide became more dense and exhibited greater structural integrity.

Frazier et al.[216] obtained a height-to-width aspect ratio of about 7 with polyimide structures. Ultraviolet was used to produce structures with heights in the range of 30–50 μm. At greater heights, the verticality of the sidewalls became relatively poor. Spun-on thickness in excess of 60 μm in a single coat was obtained for both Ciba-Geigy and Dupont commercial UV-exposable, negative-tone polyimides. Using a G-line mask aligner, exposure energy of 350 mJ/cm$^2$ was sufficient to develop a pattern with the Ciba-Geigy QZ 3301 developer. Allen and his team combined polyimide insert molds with electrodeposition to make a wide variety of metal structures.[217]

*Other Resists Used for UV Deep Lithography Surface Micromachining* Besides polyimides, research on Novolak-type resists also is leading to higher 3D features. Lochel et al.[206] use Novolak, positive tone resists of high viscosity (e.g., AZ 4000 series, Hoechst). They deposit, in multiple-coating process, layers up to 200 μm thick in a specially designed spin coater that incorporates a corotating cover. The subsequent deep UV lithography yields patterns with aspect ratios up to 10, steep edges (more than 88°), and a minimum feature size down to 3 μm. By combining this resist technology with sacrificial

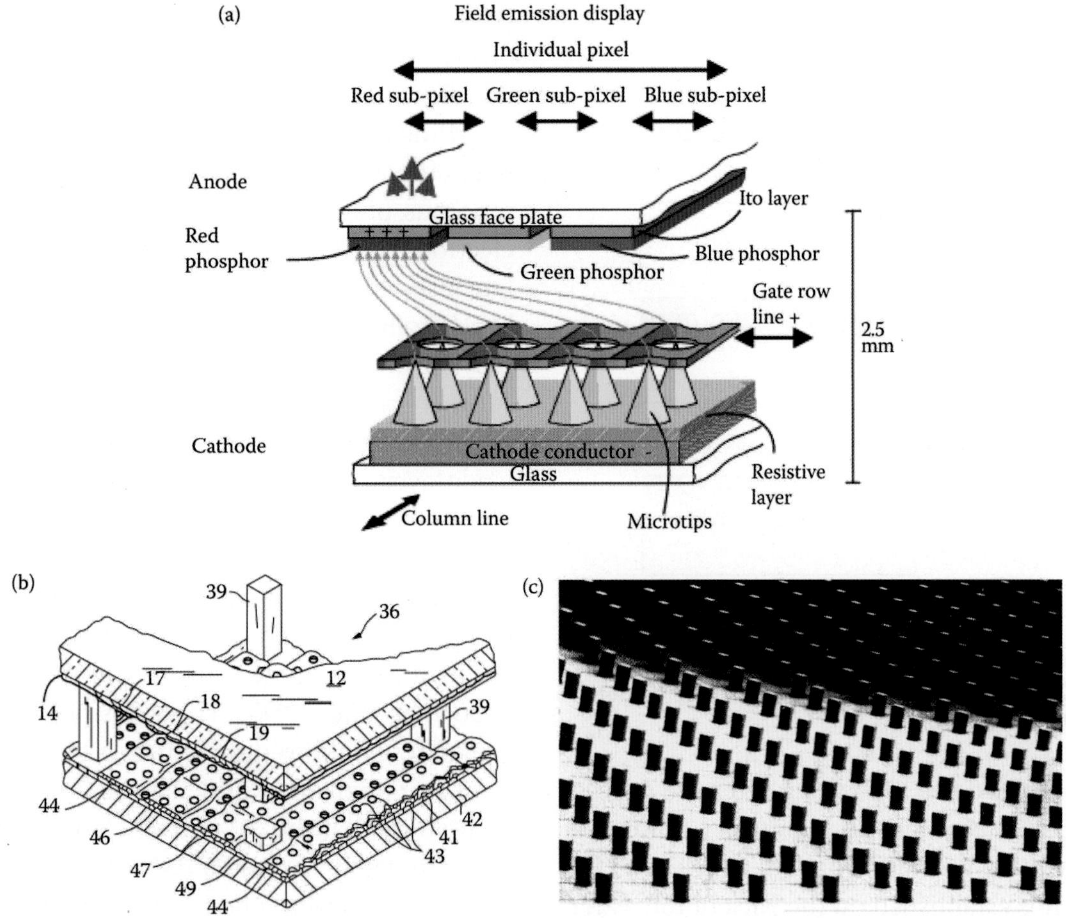


**FIGURE 7.90** Polyimide structural elements. (a) Principle of a field emission display. (b) Micro-machined flat panel display. Number 39 represents one of the spacer pillars in the matrix of polyimide pillars (100 mm high). The spacer array separates the emitter plate from the front display plate. (c) SEM of the spacer matrix. (Courtesy of Dr. I. Brodie.) The height of the pillars is similar to what can be accomplished with LIGA. Because only a simple UV exposure was used, this polyimide is referred to as poor man's LIGA process, or pseudo-LIGA. (From I. Brodie et al., US Patent 4,923,421, 1990; and Brodie, I., H. R. Gurnick, C. E. Holland, and H. A. Moessner. 1990. *Method for providing polyimide spacers in a field emission panel display.*[215])

layers and electroplating, a wide variety of 3D microstructures result.

Along the same line, researchers at IBM experimented with Epon SU-8 (Shell Chemical), an epoxy-based, onium-sensitized, UV transparent negative photoresist used to produce high-aspect-ratio (>10:1) features and straight sidewalled images in thick film (>200 μm) using standard lithography.[218,219] SU-8 imaged films were used as stencils to plate Permalloy for magnetic motors.[218]

Patterns generated with these thick resist technologies should be compared with LIGA-generated patterns, not only in terms of aspect ratio but also in terms of sidewall roughness and sidewall run-out. Such a comparison will determine which surface machining technique to use for the job.

### *Porous Polysilicon*

In Chapter 4, we discussed the transformation of single-crystal Si into a porous material with porosity and pore sizes determined by the current density, type, and concentration of the dopant, and the hydrofluoric acid concentration. A transition from pore formation to electropolishing is reached by increasing the current density and/or by decreasing the hydrofluoric acid concentration.[101,102] Porous silicon can also be formed under similar conditions from LPCVD polysilicon.[103] In this case, pores roughly follow the grain boundaries of the polysilicon. Figure 7.91 illustrates the masking and process sequence to prepare a wafer to make thin layers of porous Si between two insulating layers of low

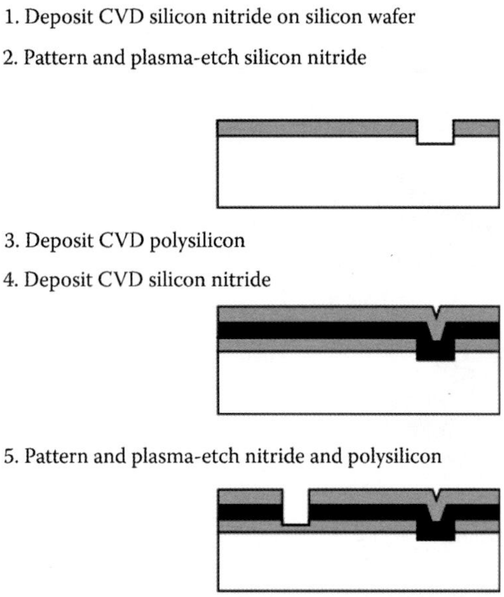

1. Deposit CVD silicon nitride on silicon wafer
2. Pattern and plasma-etch silicon nitride
3. Deposit CVD polysilicon
4. Deposit CVD silicon nitride
5. Pattern and plasma-etch nitride and polysilicon


**FIGURE 7.91** Masking and process sequence preparing a wafer for laterally grown porous polysilicon. (From Field, L. 1987. EECS/ERL Res. summary. University of California, Berkeley. With permission.[220])

stress silicon nitride.[220] The wafer, after the process steps outlined in Figure 7.91, is put in a Teflon test fixture, protecting the back from HF attack. An electrical contact is established on the back of the wafer, and a potential is applied with respect to a platinum-wire counter electrode immersed in the same HF solution. Electrolytes consisting of 5–49 wt% HF and current densities from 0.1–50 A cm$^{-2}$ are used. The advance of a pore-etching front, growing parallel to the wafer surface, may be monitored using a line-width measurement tool. The highest observed rate of porous-silicon formation is 15 μm min$^{-1}$ (in 25 wt% HF). In the electropolishing regime, at the highest currents, the etch rate is diffusion limited, but the reaction is controlled by surface reaction kinetics in the porous Si growth regime.

By changing the conditions from pore formation to electropolishing and back to porous Si, an enclosed chamber may be formed with porous polysilicon walls (plugs) and "floor and ceiling" silicon nitride layers. Sealing of the cavities by clogging the microporous polysilicon was attempted by room temperature oxidation in air and in an $H_2O_2$ solution. Leakage through the porous plug persisted after those room temperature oxidation treatments, but with a Ag deposition from a 400-mM $AgNO_3$ solution and subsequent atmospheric tarnishing (48 h, $Ag_2S$) the chambers appeared to be sealed, as determined from the lack of penetration by methanol. This technology might open up possibilities for filling cavities with liquids and gases under low-temperature conditions. The chamber provided with a porous plug might also make a suitable on-chip electrochemical reference electrode with an electrolyte reservoir.

Hydrofluoric acid can penetrate thin layers of polysilicon either at foreign particle inclusion sites or at other critical film defects, such as grain boundaries (see above). This way, the HF can etch underlying oxide layers, creating, for example, circular regions of freestanding polysilicon, so-called *blisters*. The polysilicon permeability associated with blistering of polysilicon films has been applied successfully by Judy et al.[221,222] to produce thin-shelled hollow beam electrostatic resonators from thin polysilicon films deposited onto PSG. The possible advantage of using these hollowed structures is to obtain a higher resonator quality factor Q. The devices were made in such a way that the 0.3-μm-thick undoped polysilicon completely encased a PSG core. After annealing, the structures were placed in HF, which penetrated the polysilicon shell and dissolved away the PSG, eliminating the need for etch windows. It was not possible to discern the pathways through the polysilicon using TEM.

Lebouitz et al.[223] applied permeable polysilicon etch-access windows to increase the speed of creating microshells for packaging surface-micromachined components. After etching the PSG through the many permeable Si windows, the shell is sealed with 0.8 μm of low-stress LPCVD nitride.

Porous Si can be used very effectively as a sacrificial layer both in polysilicon surface micromachining and in SOI micromachining (see Table 7.23).[139] The high surface area of the porous Si results in rapid etching in KOH at room temperature. The porous layer can be formed directly under a deposited layer or under the epilayer in SOI surface micromachining.

Silicon is not the only material that can be converted into a porous sponge-like material. Most semiconductor materials can be modified this way, and in Volume III, Chapter 5 on selected MEMS and NEMS materials we discuss micromachining of high-aspect-ratio structures in porous anodic aluminum oxide. The latter enables a class of MEMS applications such as high-temperature gas microsensors, vacuum microelectronics, RF-MEMS, nozzles, filters, membranes, and so on.[224]

## Comparison of Bulk Micromachining with Surface Micromachining

Surface and bulk micromachining have many processes in common. Both techniques rely heavily on photolithography; oxidation; diffusion and ion implantation; LPCVD and PECVD for oxide, nitride, and oxynitride; plasma etching; use of polysilicon; and metallizations with sputtered, evaporated, and plated Al, Au, Ti, Pt, Cr, and Ni. Where the techniques differ is in the use of anisotropic etchants, anodic and fusion bonding, (100) versus (110) starting material, $p^+$ etch stops, double-sided processing and electrochemical etching in bulk micromachining, and the use of dry etching in patterning and isotropic etchants in release steps for surface micromachines. Combinations of substrate and surface micromachining also frequently appear. The use of polysilicon avoids many challenging processing difficulties associated with bulk micromachining and offers new degrees of freedom for the design of integrated sensors and actuators. Design freedom includes many more possible shapes in the *x,y* plane and the ease of integration of several sensors on one die (e.g., a two-axis accelerometer). The technology combined with sacrificial layers also allows the nearly indispensable further advantage of in situ assembly of the tiny mechanical structures because the structures are preassembled as a consequence of the fabrication sequence. Another advantage focuses on thermal and electrical isolation of polysilicon elements. Polycrystalline piezoresistors can be deposited and patterned on membranes of other materials, for example, on a $SiO_2$ dielectric. This configuration is particularly useful for high-temperature applications. The p-n junctions act as the only electrical insulation in the single-crystal sensors, resulting in high leakage currents at high temperatures, whereas current leakage for the polysilicon/$SiO_2$ structure virtually does not exist. The limits of surface micromachining are striking. CVD silicon usually caps at layers no thicker than 1–2 μm because of residual stress in the films and the slow deposition process (thick polysilicon needs further investigation).* A combination of a large variety of layers may produce complicated structures, but each layer is still limited in thickness. Also, the wet chemistry needed to remove the interleaved layers may require many hours of etching (except when using the porous Si option discussed above), and even then stiction often results.

The structures made from polycrystal silicon exhibit inferior electronic and slightly inferior mechanical properties compared with single-crystal silicon. For example, polysilicon has a lower piezoresistive coefficient (resulting in a gauge factor of 30 vs. 90 for single-crystal Si), and it has a somewhat lower mechanical fracture strength. Polysilicon also warps because of the difference of thermal expansion coefficient between polysilicon and single-crystal silicon. Its mechanical properties strongly depend on processing procedures and parameters.

Table 7.25 introduces a comparison of surface micromachining with wet bulk micromachining. The status depicted reflects the mid-1990s and only includes polysilicon surface micromachining. As discussed, SOI micromachining, thick polysilicon, hinged polysilicon, polyimide, and millimeter-molded polysilicon structures have dramatically expanded the application bandwidth of surface micromachining. In Chapter 10 on replication techniques-LIGA, we will see how x-ray lithography further expands the *z*-direction for new surface-micromachined devices with unprecedented aspect ratios and extremely low surface roughness. In Table 7.26 we compare physical properties of single-crystal silicon with those of polysilicon.

Although polysilicon can be an excellent mechanical material, it remains a poor electronic material. Reproducible mechanical characteristics are difficult and complex to consistently realize. Fortunately, SOI surface micromachining and other newly emerging surface micromachining techniques can alleviate many of the problems.[229]

* As a direct consequence, the inertial mass in a surface-micromachined accelerometer, such as the ADXL05 from Analog Devices, is only 0.3 μg, and the corresponding noise, dominated by Brownian noise, is 500 μg/√Hz. By contrast, the mass for a bulk-micromachined accelerometer can easily be made 100 μg.

**Table 7.25** Comparison of Bulk Micromachining with Surface Micromachining

| Bulk Micromachining | Surface Micromachining |
|---|---|
| Large features with substantial mass and thickness | Small features with low thickness and mass |
| Uses both sides of the wafer | Multiple deposition and etching required to build up structures |
| Vertical dimensions: one or more wafer thicknesses | Vertical dimensions are limited to the thickness of the deposited layers (~2 μm) leading to compliant suspended structures with the tendency to stick to the support |
| Generally involves laminating Si wafer to Si or glass | Surface micromachined device has its built-in support and is more cost effective |
| Piezoresistive or capacitive sensing | Capacitive and resonant sensing mechanisms |
| Wafers may be fragile near the end of the production | Cleanliness critical near end of process |
| Sawing, packaging, testing is difficult | Sawing, packaging, testing is difficult |
| Some mature products and producers | No mature products or producers |
| Not very compatible with IC technology | Natural but complicated integration with circuitry; integration is often required due to the tiny capacitive signals |

*Source:* Adapted from Jerman, H. 1994. Hard copies of viewgraphs presented in Banff, Canada.[90]

**Table 7.26** Comparison of Materials Properties of Si Single Crystal with Crystalline Polysilicon

| Material Property | Single-Crystal Si (SCS) | Polysilicon |
|---|---|---|
| Thermal conductivity (W/cm°K) | 1.57 | Strong function of the grain structure of the film; 0.30–0.35 (for fine grains and double that for larger grains) |
| Thermal expansion ($10^{-6}$/°K) | 2.33 | 2–2.8 |
| Specific heat (cal/g°K) | 0.169 | 0.169 |
| Piezoresistive coefficients | n-Si ($\pi_{11}$ = −102.2)<br>p-Si ($\pi_{44}$ = +138.1)<br>e.g., gauge factor of 90 | e.g., gauge factor of 30<br>(>50 with laser recrystallization) |
| Refractive index | | 4.1 at 600 nm |
| Mobility ($cm^2$/V/s) | Holes: 600<br>Electrons: 1500 | Maximum for electrons: 30 |
| Density ($cm^{-3}$) | 2.32 | 2.32 |
| Fracture strength (GPa) | 6 | 0.8–2.84 (Undoped polysilicon) |
| Dielectric constant | 11.9 | Sharp maxima of 4.2 and 3.4 eV at 295 and 365 nm, respectively |
| Residual stress | None | Depends on structure as deposited films are compressive |
| Temperature resistivity coefficient (°$C^{-1}$)TCR | 0.0017 (p-type) | 0.0012 nonlinear, + or – through selective doping, increases with decreasing doping level can be made 0! |
| Poisson ratio | 0.262 max for (111) | 0.23 |
| Young's modulus ($10^{11}$ N/$m^2$) | 1.90 (111) | 1.61 |
| Resistivity at room temperature (ohm.cm) | Depends on doping | Strong function of the grain structure of the film. Plateaus at 4 $10^{-4}$ above 1 $10^{21}$ $cm^{-3}$ P (always higher than for SCS) |

*Source:* Based on Lin, L. 1993. *Selective encapsulations of MEMS: micro channels, needles, resonators, and electromechanical filters.* PhD thesis. University of California, Berkeley;[225] Adams, A. C. 1988. *VLSI technology.* Ed. S. M. Sze. New York: McGraw-Hill;[30] Kamins, T. 1988. *Polycrystalline silicon for integrated circuits.* Boston: Kluwer;[78] and Heuberger, A. 1989. *Mikromechanik.* Heidelberg: Springer Verlag.[227] (See also Volume I, Table 4.15.)

## Examples

### Example 7.9: Analog Devices Accelerometer

Both Robert Bosch GmbH (Stuttgart, Germany) and Analog Devices (Norwood, MA) offer surface-micromachined accelerometers based on lateral resonators. We will only review the Analog Devices ADXL accelerometer product family here. The ADXL-50 constituted the first commercially available surface-micromachined MEMS structures. Today, surface-micromachined accelerometers are incorporated in Ford and General Motors cars, among others, as well as inside joysticks for computer games, robots, watches, and shoes.

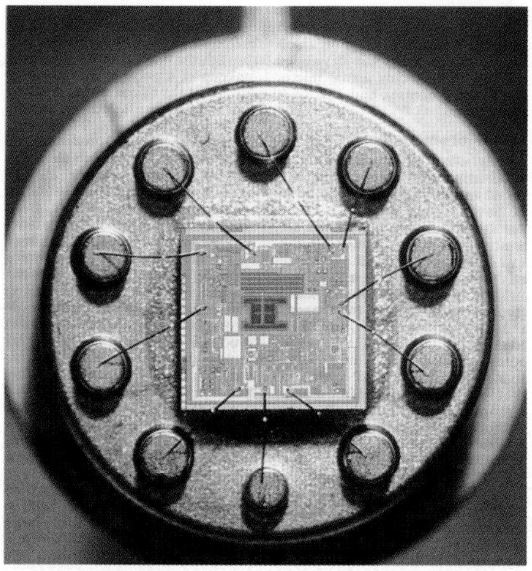

**FIGURE 7.92** Analog Devices' ADXL-50 accelerometer with a surface-micromachined capacitive sensor (center), on-chip excitation, self-test, and signal-conditioning circuitry. (From Core, T. A., W. K. Tsang, and S. J. Sherman. 1993. Fabrication technology for an integrated surface-micromachined sensor. *Solid State Technol* 36:39–47. With permission.[119])

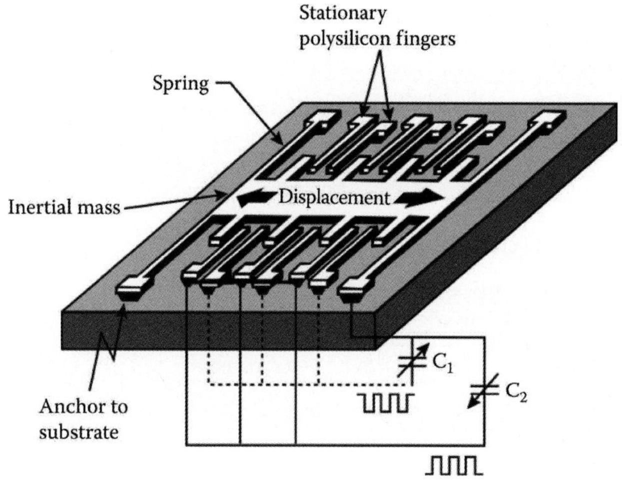


**FIGURE 7.93** Illustration of the basic mechanical structure of Analog Devices' ADXL family of surface-micromachined accelerometers. A comb-like plate suspended from springs forms the inertial mass. Displacements of the mass are measured capacitively with respect to two sets of stationary finger-like electrodes. (Based on Maluf, N. 2000. *An introduction to microelectromechanical systems engineering.* Boston: Artech House.[3])

To facilitate integration of their surface-micromachined accelerometers with on-board electronics, Analog Devices opted for a mature 4-μm BiCMOS process.[119] BiCMOS is a manufacturing process for semiconductor devices that combines bipolar and CMOS to give the best balance between available output current and power consumption. Figure 7.92 presents a photograph of the finished accelerometer with on-chip excitation, self-test, and signal conditioning circuitry.

The suspended comb-like H-shaped structure in the center of the die is the sensitive element, and its primary axis of sensitivity lies in the plane of the die (*x-y* plane). In bulk micromachining, the sense axis is more often orthogonal to the plane of the die (*z*-axis). The polysilicon-sensing element of the ADLAX-50 only occupies 5% of the total die area and consists of three sets of 2-μm-thick polysilicon finger-like electrodes (Figure 7.93).

Two sets are anchored to the substrate, and a third set is suspended about 1 μm above the surface by means of two folded polysilicon beams acting as suspension springs. The fingers of the movable shuttle mass are interlaced with the fingers of the two fixed sets. The whole chip measures 500 × 625 μm and operates as an automotive airbag deployment sensor. The measurement accuracy is 5% over the ±50-g range. Deceleration in the axis of sensitivity exerts a force on the central mass that, in turn, displaces the interleaved capacitor plates, causing a fractional change in capacitance. The overall capacitance is small, typically on the order of 100 fF, and for the ADXL-05 (rated at ±5 g), with an inertial mass of 0.3 μg only, the capacitance change is as small as 100 aF.[197] These small capacitance changes necessitate on-chip integrated electronics to reduce the impact of parasitic sources. In operation, the ADXL family has a force-balance electronic control loop to prevent the mass from macroscopic movements, greatly improving output linearity because the center element never moves by more than a few nanometers. Applying a large-amplitude, low-frequency voltage, below the natural frequency of the sensor, allows one to compensate for accelerometer plate movement by external acceleration. At the same time, the sensing excitation frequency (1 MHz) is much higher than the resonant frequency, so that it does not produce an actuation force on the capacitor plates. As long as sense and actuation signals do not interfere, sense and actuation plates may be the same. Straight flexures were used for the layout of the lateral resonator shuttle mass (see also Figure 7.76a).

In the sensor design, $n^+$ underpasses connect the sensor area to the electronic circuitry, replacing the usual heat sensitive aluminum connect lines. Most of the sensor processing is

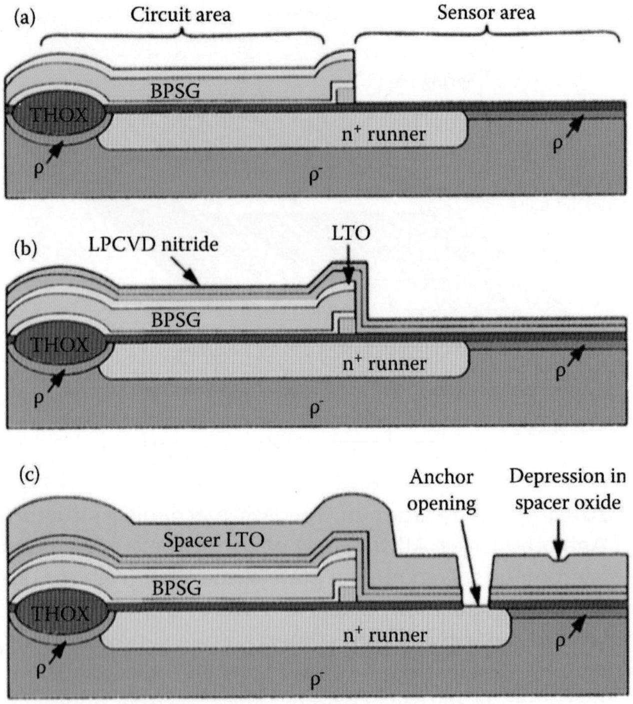


**FIGURE 7.94** Preparation of IC chip for polysilicon. (a) Sensor area post-BPSG planarization and moat mask. (b) Blanket deposition of thin oxide and thin nitride layer. (c) Bumps and anchors made in LTO spacer layer. (From Core, T. A., W. K. Tsang, and S. J. Sherman. 1993. Fabrication technology for an integrated surface-micromachined sensor. *Solid State Technol* 36:39–47. With permission.[119])

inserted into the BiCMOS process right after the borophosphosilicate glass (BPSG) planarization. After planarization, a designated sensor region or moat is cleared in the center of the die (Figure 7.94a). A thin oxide is then deposited to passivate the n+ underpass connects, followed by a thin, low-pressure, vapor-deposited nitride to act as an etch stop (buffer layer) for the final polysilicon release etch (Figure 7.94b). The spacer or sacrificial oxide used is a 1.6-μm densified low-temperature oxide (LTO) deposited over the whole die (Figure 7.94c). In a first timed etch, small depressions that will form bumps or dimples on the underside of the polysilicon sensor are created in the LTO layer. These will limit stiction in case the sensor comes in contact with the substrate. A subsequent etch cuts anchors into the spacer layer to provide regions of electrical and mechanical contact (Figure 7.94c). The 2-μm-thick sensor polysilicon is then deposited, implanted, annealed, and patterned (Figure 7.95a). The relatively deep junctions of the BiCMOS process permit the polysilicon thermal anneal and brief dielectric densifications without resulting in degradation of the electronic functions. Next is the IC metallization, which starts with the removal of the sacrificial spacer oxide from the circuit area along with the LPCVD nitride and LTO layer. A low-temperature oxide is deposited on the polysilicon sensor part, and contact openings appear in the IC part of the die where platinum is deposited to form a platinum silicide (Figure 7.95b). The trimmable thin-film material, TiW barrier metal, and Al/Cu interconnect metal are sputtered on and patterned in the IC area. The circuit area is then passivated in two separate deposition steps. First, plasma oxide is deposited and patterned (Figure 7.95c), followed by a plasma nitride (Figure 7.96a) to form a seal with the earlier deposited LPCVD nitride. The nitride acts as an HF barrier in the subsequent long etch release. The plasma oxide left on the sensor acts as an etch stop for the removal of the plasma nitride. Subsequently, the sensor area is prepared for the final release etch.

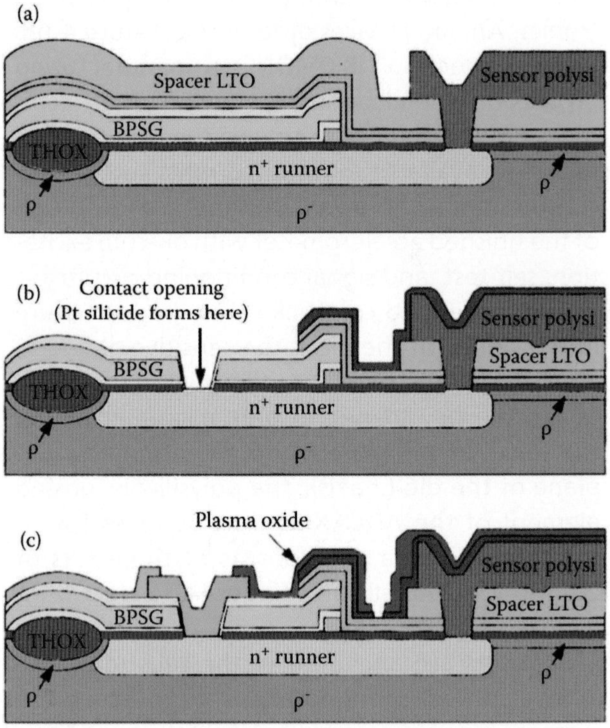


**FIGURE 7.95** Polysilicon deposition and IC metallization. (a) Cross-sectional view after polysilicon deposition, implant, anneal, and patterning. (b) Sensor area after removal of dielectrics from circuit area, contact mask, and Pt silicide. (c) Metallization scheme and plasma oxide passivation and patterning. (From Core, T. A., W. K. Tsang, and S. J. Sherman. 1993. Fabrication technology for an integrated surface-micromachined sensor. *Solid State Technol* 36:39–47. With permission.[119])

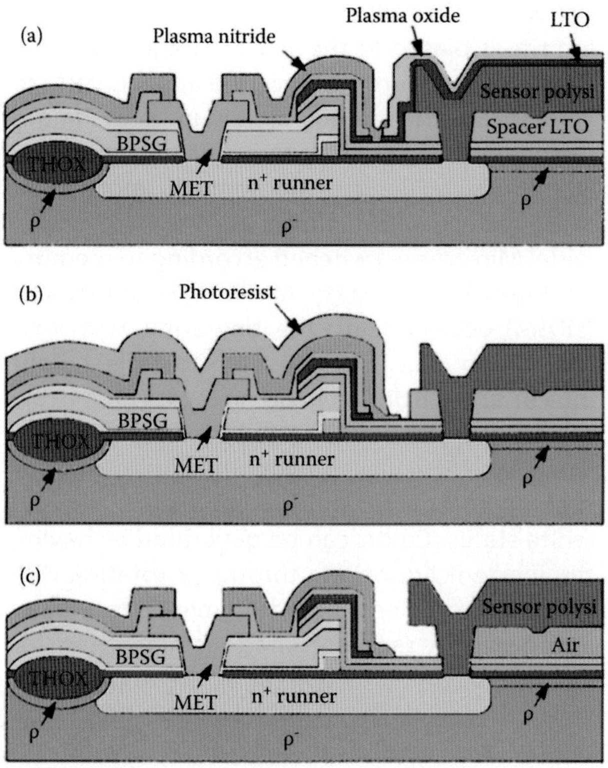


**FIGURE 7.96** Prerelease preparation and release. (a) Postplasma nitride passivation and patterning. (b) Photoresist protection of the IC. (c) Freestanding, released polysilicon beam. (From Core, T. A., W. K. Tsang, and S. J. Sherman. 1993. Fabrication technology for an integrated surface-micromachined sensor. *Solid State Technol* 36:39–47. With permission.[119])

The undensified dielectrics are removed from the sensor, and the final protective resist mask is applied. The photoresist protects the circuit area from the long-term buffered oxide etch (Figure 7.96b). The final device cross-section is shown in Figure 7.96c.

### Example 7.10: TI Micromirrors

In 1987 the first digital micromirror device (DMD™) was developed at TI by Dr. Larry Hornbeck [US Patent 4,615,595 (October 7, 1986)]. A typical DMD consists of a 2D array of optical switching elements (pixels) on a silicon substrate, as shown in Figure 7.97.

Two pixels are schematically illustrated in Figure 7.98 together with the underlying Si chip and circuitry. Each pixel is made up of a reflective aluminum micromirror supported from a central post. The central mirror post at the back of the mirror is mounted on a lower aluminum metal platform—the yoke. The yoke is suspended above the silicon substrate by thin compliant L-shaped hinges (made from a proprietary Al alloy) anchored to the underlying substrate by two stationary posts. The different components of an individual micromirror are illustrated in Figure 7.99. Two bias electrodes tilt the mirror either +10° or −10° by applying 24 V between one electrode or the other and the yoke. Off-axis illumination of the Al mirror reflects into the projection lens only when the micromirror is in its +10° state, producing a bright appearance or ON state. In the flat position and in the −10° state the pixel appears dark. In the fully deflected position, the yoke touches a landing

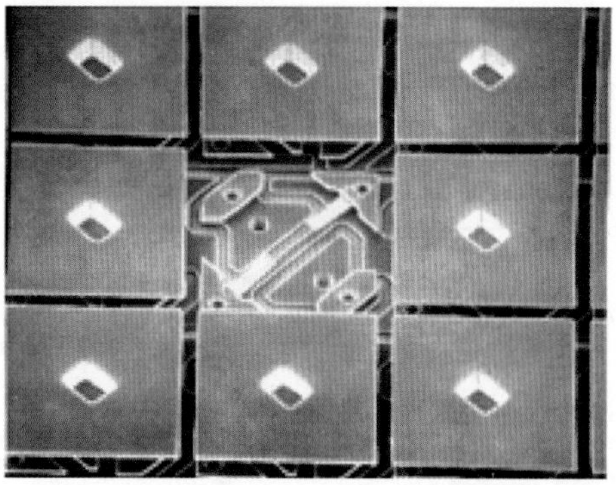

**FIGURE 7.97** Texas Instruments DMD pixel array. One pixel has been removed to show the silicon chip below the reflective aluminum mirrors.

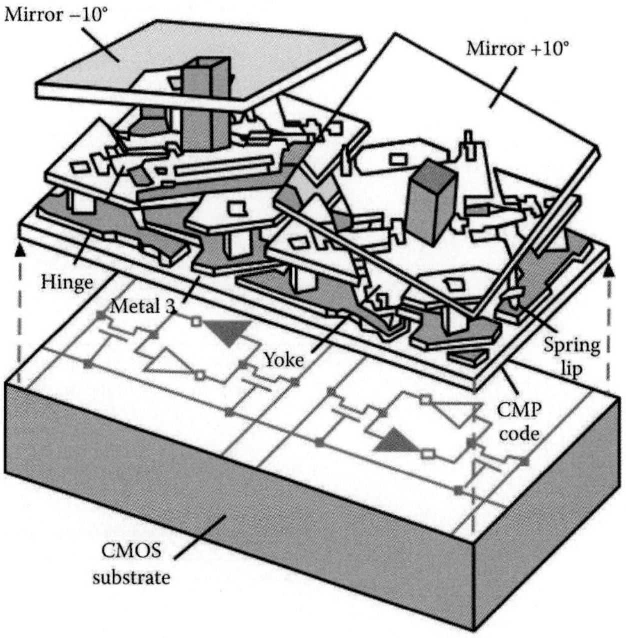


**FIGURE 7.98** Schematic of two pixels in a Texas Instruments of a DMD. The mirrors are made transparent for clarity of the drawing. The +10° mirror is in the ON position, and the −10° is in the OFF position.

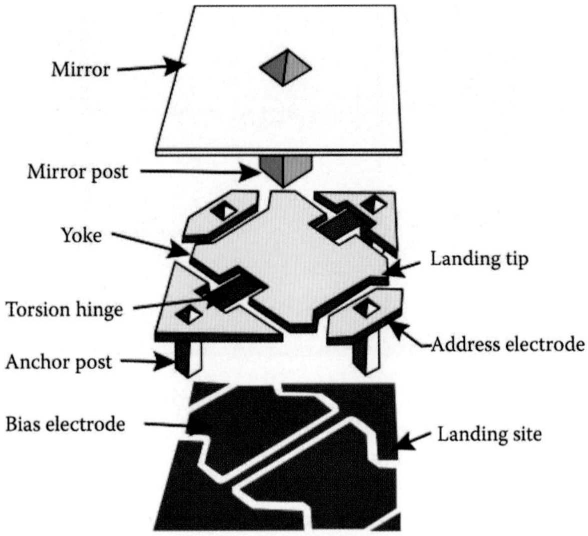


**FIGURE 7.99** Illustration of the various components of a single DMD pixel. The basic structure consists of a bottom aluminum layer containing two electrodes, a middle aluminum layer containing a yoke suspended by two torsional hinges, and a top reflective aluminum mirror. An applied electrostatic voltage on a bias electrode deflects the yoke and the mirror toward that electrode. A pixel measures approximately 17 μm on a side. (Adapted from Maluf, N. 2000. *An introduction to microelectromechanical systems engineering.* Boston: Artech House.[3])

site with its landing tips, and because the landing site is biased at the same voltage, electrical shorting is prevented. Once the applied voltage is removed, the springy aluminum alloy hinges restore the micromirror to its initial position. A standard DMD microchip contains more than 442,000 switchable mirrors on a surface 5/8 in. wide. Mirrors are switched according to memory impulses stored in static-random-access memory (SRAM) cells beneath the tiny array, with mirrors tilting plus or minus 10° to reflect light into or away from an imaging lens. The mirrors can be independently cycled at 100,000 flips/s. Grays can be achieved by multiple mirrors or by flipping the mirror quickly between black and white states. Colors can be generated by having the incident light shine through a rotating disk divided into three colored segments: red, blue, and green. A separate image is generated for each color, timed to appear as the appropriate segment covers the light source. Printing and display technology can now enjoy the advantage of digital fidelity and digital stability.

In Figure 7.99, one single pixel is illustrated with its various components.[197]

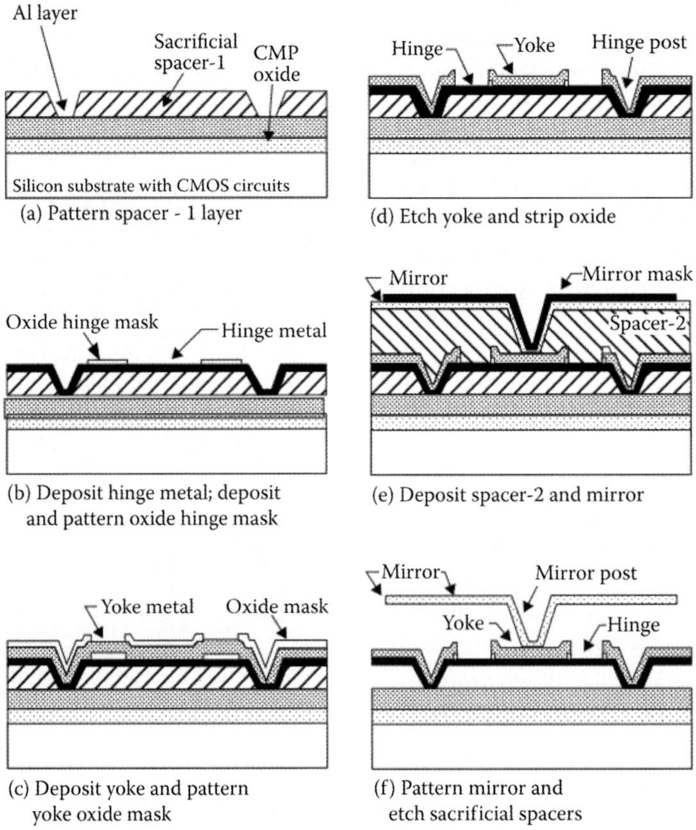


**FIGURE 7.100** Fabrication steps of the Texas Instruments DMD. Steps (a) through (f) are explained in the text. (Based on Maluf, N. 2000. *An introduction to microelectromechanical systems engineering.* Boston: Artech House.[3])

The surface micromachining process to fabricate DMDs on wafers incorporating CMOS electronic address and control circuitry is illustrated in Figure 7.100. Because of the underlying active circuitry and the presence of Al metal for connectors and MEMS structures, all micromachining process steps are carried out at temperatures less than 400°C. After completing the CMOS circuitry, a thick oxide is deposited over the whole Si wafer and is chemomechanically polished to provide a flat surface to start building the mirror array. A sputter-deposited Al layer is patterned to provide bias and address electrodes, landing pads, and electrical interconnects to the underlying electronics. Hardened photoresist is used as the sacrificial material (Figure 7.100a). A proprietary Al alloy is sputter deposited to form the hinges for the mirror. It is the nature of this aluminum alloy that secures the mechanical integrity of the mirror actuation. Subsequently, the torsion hinge regions are protected by a patterned thin PECVD-deposited silicon dioxide (Figure 7.100b). In the next step, a thicker coat of another proprietary aluminum alloy is deposited to form the yoke structure; this new coat of Al buries the thin oxide hinge mask. A second PECVD oxide mask is deposited over this second level of Al metal and patterned in the shape of the yoke and anchor posts (Figure 7.100c). In a dry etch step, the exposed aluminum areas are removed down to the organic sacrificial resist except where the oxide hinge mask remains. In those regions, only the thick yoke metal is removed, stopping on the $SiO_2$ mask and so preserving the underlying hinge structure (Figure 7.100d). Both thin layers of PECVD oxide mask are stripped before a second layer of sacrificial resists is deposited, UV-hardened, and patterned. A third aluminum alloy is sputter deposited and defines the mirror and the central mirror post. Again a thin layer of PECVD silicon dioxide is deposited and patterned to define the mirror (Figure 7.100e). An oxygen plasma etch removes both sacrificial layers and releases the micromirrors (Figure 7.100f). Finally after release a special passivation step deposits a thin antistiction layer to prevent adhesion between the yoke tips and the landing pads. Because the weight of the micromirrors is insignificant, the DMD micromirrors can withstand 1500-G mechanical shocks. Optimization of the hinge metal alloy and fabrication processes have resulted in a mean time between failure of more than 100,000 h. Invented by Larry Hornbeck at TI, DMD is the key component in more than 17 projector brands and brought the first digital cinema to *Star Wars* and *Toy Story II*. A competing technology, the actuated mirror array (AMA; now Daewoo's TMA) invented by Gregory Um, is also a MEMS technology. Unlike DMD, TMA is built with piezo materials, and for high resolution its large array size and consequent cost still pose "sticky issues," but it achieves brightness 15% higher than any other projector. DMDs are also used in place of masks to map the human genome. At the University of Wisconsin, arrays for "gene expression analysis," usually produced with lithographic masks like displays, are being replaced with the same DMD used for projection display to make "virtual masks" containing nearly half a million features.

For his amazing piece of engineering, TI's Larry J. Hornbeck received the first Emmy Award ever bestowed for a projection display technology.

## Questions

7.1: What is the mean free path (MFP)? How can you increase the MFP in a vacuum chamber? For metal deposition in an evaporation system, compare the distance between target and evaporation source with the working MFP. Which one has the smaller dimension? 1 atmosphere pressure = … mm Hg = … torr. What are the physical dimensions of impingement rate?

7.2: If we want to deposit a metal film on a substrate by resistance heated evaporation (as opposed to E-beam), what kind of metals are preferred? How is the thickness of a deposited thin film measured during evaporation?

7.3: Why is sputter deposition so much slower than evaporation deposition? Make a detailed comparison of the two deposition methods.

7.4: Develop the principal equation for the material flux to a substrate in a CVD process, and indicate how one moves from a mass transport limited to reaction-rate limited regime.

Explain why, in one case, wafers can be stacked close and vertically, while in the other a horizontal stacking is preferred.

7.5: Describe step coverage in CVD processes. Explain how gas pressure and surface temperature may influence the different coverage profiles.

7.6: Compare sputter deposition with evaporation for a simple metal such as Ag. Compare the two techniques for as many different parameters as you remember. Give examples where you would use one technique over the other.

7.7: CVD: (a) Show different types of step coverage and explain the most important parameters influencing each. (b) Describe the difference between PECVD and LPCVD.

7.8: What are the characteristics of electron beam heated evaporation compared to resistance heated evaporation? Mark correct answers with an X.

( ) It is more complex.
( ) It is very versatile.
( ) It only works under lower temperatures.
( ) Everything a resistance heated evaporator can deposit can also be accommodated by electron beam heated evaporation.
( ) A magnetic field can be used to increase the temperature.
( ) It is not possible to provide a large evaporant surface area in electron beam heated evaporation.
( ) The adhesion between an evaporant and a substrate is accomplished by local reactions (sticking). Titanium and chromium are often used as a "glue" to improve the adhesion between an evaporant and a substrate.
( ) When the partial pressure of an evaporant vapor exceeds its equilibrium vapor pressure, it will condense.
( ) When a system is under equilibrium vapor pressure, the net transfer rate of material from one state to the other state is equal to 1.
( ) The cosine law is the underlying principle that a Kundsen cell can deposit a perfectly uniform coating inside a spherical glass jar.

7.9: Design at least three micromachines enabling the testing of thin film mechanical properties.

7.10: Detail several processes to create vacuum shells in a solid.

7.11: Consider a MEMS-based condenser acoustic transducer as shown in the figure below. The diaphragm is made of silicon nitride, and the back plate is perforated. A stress gauge was used to measure the stress in the silicon nitride and was found to be $1.5 \times 10^8 N/m^2$.

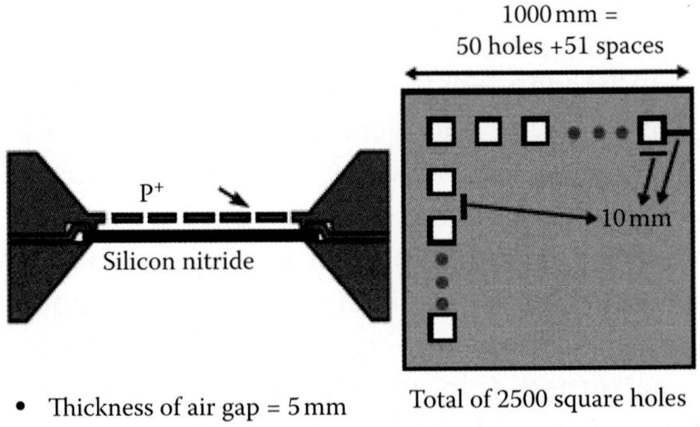


- Thickness of air gap = 5 mm
- Thickness of diaphragm = 5 mm
- Density of $Si_3N_4 = 3 \times 10^3 kg/m^3$

(a) Write the differential equation governing the dynamic behavior of the movable diaphragm.
(b) Solve the differential equation under the 3 cases of overdamped, underdamped, and critically damped conditions.
(c) Calculate the mass of the diaphragm, the small signal capacitance, and the air-streaming resistance as per the layout shown if the device is to be operated in air and water.
(d) Calculate the cut-off frequency for the overdamped case in air and in water.

7.12: Discuss the advantages/disadvantages of using surface vs. bulk micromachining.

7.13: Define the following terms in 3–4 lines with examples where applicable:
LPCVD
PECVD
DRIE
Difference between RIE and plasma etching
LIGA process
SCREAM process

Silicon fusion bonding
Dissolved wafer process

7.14: Describe 5 methods you could implement to prevent stiction between surface micromachined components.

7.15: Design a comb drive for the MUMPs process.
(a) Show the mask layouts (top view) of the relevant layers.
(b) Calculate the maximum force, maximum displacement, and resonance frequency that you expect to achieve with this design.

Make sure the following constraints are satisfied:

- The total number of comb fingers should be less than 80.
- The device fits into a $200 \times 200\ \mu m^2$ square area.
- The maximum applied voltage is 30V.
- The device satisfies all design rules for MUMPs processing.

7.16: For pressure sensing diaphragms of the order of 100 μm square, why is it impractical to use capacitance to measure deflection?

7.17: List five methods you could use to release stress in a poly-Si comb resonator and describe the relative merits of each approach.

7.18: Take a look at the Sandia website at http://mems.sandia.gov/scripts/images.asp. Many of the surface micromachined gears have tiny holes in their surfaces. Why are these there?

## References


1. Brodie, I., and J. J. Muray. 1982. *The physics of microfabrication.* New York: Plenum Press.
2. Murarka, S. P. 1988. *VLSI technology.* Ed. S. M. Sze. New York: McGraw-Hill.
3. Maluf, N. 2000. *An introduction to microelectromechanical systems engineering.* Boston: Artech House.
4. ERL. 1985. *SAMPLE version 1.6a user's guide.* Berkeley, CA: Electronics Research Laboratory (ERL), University of California, Berkeley.
5. Fichtner, W. 1988. *VLSI technology.* Ed. S. M. Sze. New York: McGraw-Hill.
6. Vossen, J. L., and W. Kern, eds. 1978. *Thin film processes.* Orlando, FL: Academic Press.
7. Maissel, L. I., and R. Glang, eds. 1970. *Handbook of thin film technology.* New York: McGraw-Hill.
8. Kinoshita, K., and M. J. Madou. 1984. Electrochemical measurement on Pt, Ir, and Ti oxides as pH probes. *J Electrochem Soc* 131:1089–94.
9. Studt, T. 1992. Better living through sputtering. *Res Dev* 54.
10. Comello, V. 1992. Tough coatings are a cinch with new PVD method. *Res Dev* January:59–60.
11. Editorial. 1994. *Brochure B-9870.* Zulte, Belgium: Innovative Sputtering Technology (IST).
12. Honer, K. A., and G. T. A. Kovacs. 2000. *Solid-state sensor and actuator workshop.* Hilton Head Island, SC.
13. Menz, W., and P. Bley. 1993. *Mikrosystemtechnik fur Ingenieure.* Weinheim, Germany: VCH Publishers.
14. Dutta, B., X. D. Wu, A. Inam, and T. Venkatesan. 1989. Pulsed laser deposition: a viable process for superconducting thin films? *Solid State Technol* February:106–10.
15. Inam, A. 1991. Pulsed laser takes the heat off HTS materials deposition. *Res Dev* February:90–92.
16. Cotell, C. M. 1993. Pulsed laser deposition and processing of biocompatible hydroxylapatite thin films. *App Surf Sci* 69:140–48.
17. Guo, T., P. Nikolaev, A. Thess, D. T. Colbert, and R. E. Smalley. 1995. Catalytic growth of single-walled nanotubes by laser vaporization. *Chem Phys Lett* 243:49–54.
18. Cotell, C. M., and K. S. Grabowski. 1992. Novel materials applications of pulsed laser deposition. *MRS Bull* 17:44–53.
19. Cotell, C. M., D. B. Chrisey, K. S. Grabowski, J. A. Sprague, and C. R. Gossett. 1992. Pulsed laser deposition of hydroxylapatite thin films on Ti-6Al-4V. *J Appl Biomater* 3:87–93.
20. Parker, E. H. C. 1985. *The technology and physics of molecular beam epitaxy.* New York: Plenum Press.
21. Comello, V. 1992. Silicon MBE research races toward production. *Res Dev* April:87–88.
22. Akedo, J., and M. Lebedev. 1999. Microstructure and electrical properties of lead zirconate titanate ($Pb(Zr_{52}/Ti_{48})O_3$) thick films deposited by aerosol deposition method. *Jpn J Appl Phys* 38:5397.
23. Jensen, K. F. 1989. *Microelectronics processing: chemical engineering aspects.* Eds. D. W. Hess, and K. F. Jensen. Washington, DC: American Chemical Society.
24. Ehrlich, D. J., R. M. J. Osgood, and J. Deutsch. 1982. Photodeposition of metal films with ultraviolet laser light. *J Vac Sci Technol* 21:23–32.
25. Wallenberger, F. T., and P. C. Nordine. 1993. Inorganic fibers and microstructures by laser assisted chemical vapor deposition. *Mater Technol* 8:198–202.
26. Granger, R. A. 1995. *Fluid mechanics.* New York: Dover Publications.
27. Wolf, S., and R. N. Tauber. 1987. *Silicon processing for the VLSI era.* Sunset Beach, CA: Lattice Press.
28. Lee, H. H. 1990. *Fundamentals of microelectronics processing.* New York: McGraw-Hill.
29. Coronell, D. G., and K. F. Jensen. 1994. Simulation of rarified gas transport and profile evolution in nonplanar substrate chemical vapor deposition. *J Electrochem Soc* 141:2545–51.
30. Adams, A. C. 1988. *VLSI technology.* Ed. S. M. Sze. New York: McGraw-Hill.
31. Hess, D. W., and K. F. Jensen, eds. 1989. *Microelectronics processing: chemical engineering aspects.* Washington, DC: American Chemical Society.
32. Sze, S. M., ed. 1988. *VLSI technology.* New York: McGraw-Hill.
33. Lee, J. G., S. H. Choi, T. C. Ahn, C. G. Hong, P. Lee, K. Law, M. Galiano, P. Keswick, and B. Shin. 1992. SA CVD: a new approach for 16 Mb dielectrics. *Semicond Int* May:115–20.
34. Huang, S., X. Cai, and J. Liu. 2003. Growth of millimeter-long and horizontally aligned single-walled carbon nanotubes on flat substrates. *J Am Chem Soc* 125:5636–37.

35. Compton, R. D. 1992. PECVD: a versatile technology. *Semicond Int* July:60–65.
36. Nguyen, S. V. 1999. High-density plasma chemical vapor deposition of silicon-based dielectric films for integrated circuits. *IBM J Res Dev* 43:109–26.
37. Burggraaf, P. 1993. The status of MOCVD technology. *Semicond Int* 16:80–83.
38. Peters, L. 1993. SOI takes over where silicon leaves off. *Semicond Int* March:48–51.
39. Anderson, T. J. 1989. *Microelectronics processing: chemical engineering aspects.* Eds. Hess, D. W., and K. F. Jensen. Washington, DC: American Chemical Society.
40. Borland, J., R. Wise, Y. Oka, M. Gangani, S. Fong, and Y. Matsumoto. 1988. Silicon epitaxial growth for advanced device structures. *Solid State Technol* January:111–19.
41. Rehrig, D. L. 1990. In search of precise epi thickness measurements. *Semicond Int* 13:90–95.
42. Singer, P. 1992. CVC builds cluster tool through partnering. *Semicond Int* 15:46.
43. Vinci, R. P., and J. C. Braveman. 1991. Mechanical testing of thin films. *Sixth international conference on solid-state sensors and actuators (Transducers '91).* San Francisco.
44. Bushan, B. 1996. *Proceedings: IEEE ninth annual international workshop on micro electro mechanical systems.* San Diego, CA.
45. Bushan, B., ed. 1995. *Handbook of micro/nanotribology.* Boca Raton, FL: CRC Press.
46. Schiotz, J., and K. W. Jacobsen. 2003. A maximum in the strength of nanocrystalline copper. *Science* 301:1357–59.
47. Schiotz, J., F. D. D. Tolla, and K. W. Jacobsen. 1998. Softening of nanocrystalline metals at very small grains. *Nature* 391:561.
48. Chokshi, A. H., A. Rosen, J. Karch, and H. Gleiter. 1989. On the validity of the hall-petch relationship in nanocrystalline materials. *Scripta Metallurgica* 23:1679–83.
49. Gleiter, H. 1989. Nanocrystalline materials. *Prog Mater Sci* 33:223–315.
50. Yu, D. Y. W., and F. Spaepen. 2004. The yield strength of thin copper films on Kapton. *J Appl Phys* 95:2991–97.
51. Ehmann, K. F., D. Bourell, M. L. Culpepper, T. J. Hodgson, T. R. Kurfess, M. Madou, K. Rajukar, and R. DeVor. 2007. *Micromanufacturing: international research and development.* New York: Springer.
52. Biebl, M., G. Brandl, and R. T. Howe. 1995. Young's modulus of in-situ phosphorus-doped polysilicon. *Eighth international conference on solid-state sensors and actuators (Transducers '95).* Stockholm, Sweden.
53. Campbell, D. S. 1970. *Handbook of thin film technology.* Eds. L. I. Maissel, and R. Glang. New York: McGraw-Hill.
54. Greek, S., F. Ericson, S. Johansson, and J.-Å. Schweitz. 1995. In situ tensile strength measurement of thick-film and thin-film micromachined structures. *Eighth international conference on solid-state sensors and actuators (Transducers '95).* Stockholm, Sweden.
55. Senturia, S. 1987. *Proceedings: IEEE micro robots and teleoperators workshop.* Hyannis, MA.
56. Krulevitch, P. A. 1994. *Micromechanical investigations of silicon and Ni–Ti–Cu thin films.* PhD thesis, University of California, Berkeley.
57. Pratt, R. I., G. C. Johnson, R. T. Howe, and D. J. J. Nikkel. Characterization of thin films using micromechanical structures. *Mat Res Soc Symp Proc* 276:197–202.
58. Judy, M. W., Y. H. Cho, R. T. Howe, and A. P. Pisano. 1991. *Proceedings: IEEE micro electro mechanical systems (MEMS '91).* Nara, Japan.
59. Wu, T. H. T., and R. S. Rosler. 1992. Stress in PSG and nitride films as related to film properties and annealing. *Solid State Technol* May:65–71.
60. Hey, H. P. W., B. G. Sluijk, and D. G. Hemmes. 1990. Ion bombardment factor in plasma CVD. *Solid State Technol* April:139–44.
61. Chou, P. C., and N. J. Pagano. 1967. *Elasticity: tensor, dyadic, and engineering approaches.* New York: Dover Publications.
62. Guckel, H., D. W. Burns, H. A. C. Tilmans, C. C. G. Visser, D. W. DeRoo, T. R. Christenson, P. J. Klomberg, J. J. Sniegowski, and D. H. Jones. 1988. Processing conditions for polysilicon films with tensile strain for large aspect ratio microstructures. *Technical digest: 1988 solid state sensor and actuator workshop.* Hilton Head Island, SC.
63. Maseeh, F., and S. D. Senturia. 1990. *Technical digest: 1990 solid state sensor and actuator workshop.* Hilton Head Island, SC.
64. Hoffman, R. W. 1976. *Physics of nonmetallic thin films (NATO advanced study institutes series: series B, physics.* Eds. C. H. S. Dupuy, and A. A. Cachard. New York: Plenum Press.
65. Hoffman, R. W. 1975. Stresses in thin films: the relevance of grain boundaries and impurities. *Thin Solid Films* 34:185–90.
66. Iscoff, R. 1991. Hotwall LP CVD reactors: considering the choices. *Semicond Int* June:60–64.
67. Adams, A. C. 1988. *VLSI technology.* Ed. S. M. Sze. New York: McGraw-Hill.
68. Guckel, H., J. J. Sniegowski, T. R. Christenson, and F. Raissi. 1990. The application of fine-grained, tensile polysilicon to mechanically resonant transducers. *Sensors Actuators A* A21:346–51.
69. van der Drift, A. 1967. Evolutionary selection, a principle governing growth orientation in vapour-deposited layers. *Philips Res Rep* 22:267–88.
70. Matson, E. A., and S. A. Polysakov. 1977. On the evolutionary selection principle in relation to the growth of polycrystalline silicon films. *Phys Sta Sol (A)* 41:K93–K95.
71. Lober, T. A., J. Huang, M. A. Schmidt, and S. D. Senturia. 1988. Characterization of the mechanisms producing bending moments in polysilicon micro-cantilever beams by interferometric deflection measurements. *Technical digest: 1988 solid state sensor and actuator workshop.* Hilton Head Island, SC.
72. Shimizu, T., and S. Ishihara. 1995. Effect of $SiO_2$ surface treatment on the solid-phase crystallization of amorphous silicon films. *J Electrochem Soc* 142:298–302.
73. Abe, T., and M. L. Reed. 1996. *Ninth annual international workshop on micro electro mechanical systems.* San Diego, CA.
74. Guckel, H., D. W. Burns, C. C. G. Visser, H. A. C. Tilmans, and D. Deroo. 1988. Fine-grained polysilicon films with built-in tensile strain. *IEEE Trans Electron Devices* 35: 800–01.
75. Drosd, R., and J. Washburn. 1982. Some observation on the amorphous to crystalline transformation in silicon. *J Appl Phys* 53:397–403.
76. Elwenspoek, M., U. Lindberg, H. Kok, and L. Smith. 1994. *IEEE international workshop on micro electro mechanical systems (MEMS '94).* Oiso, Japan.
77. Harbeke, G., L. Krausbauer, E. F. Steigmeier, and A. E. Widmer. 1983. LP CVD polycrystalline silicon: growth and physical properties of in-situ phosphorous doped and undoped films. *RCA Rev* 44:287–313.
78. Kamins, T. 1988. *Polycrystalline silicon for integrated circuits.* Boston: Kluwer.

79. Mulder, J. G. M., P. Eppenga, M. Hendriks, and J. E. Tong. 1990. An industrial LP CVD process for in situ phosphorus-doped polysilicon. *J Electrochem Soc* 137:273–79.
80. Kinsbron, E., M. Sternheim, and R. Knoell. 1983. Crystallization of amorphous silicon films during low pressure chemical vapor deposition. *Appl Phys Lett* 42: 835–37.
81. Lietoila, A., A. Wakita, T. W. Sigmon, and J. F. Gibbons. 1982. Epitaxial regrowth of intrinsic, $^{31}$P-doped and compensated ($^{31P+11}$B-Doped) amorphous Si. *J Appl Phys* 53:4399–405.
82. Meyerson, B. S., and W. Olbricht. 1984. Phosphorous-doped polycrystalline silicon via LPCVD I. process characterization. *J Electrochem Soc* 131:2361–65.
83. Hendriks, M., and C. Mavero. 1991. Phosphorous doped polysilicon for double poly structures. Part I. morphology and microstructure. *J Electrochem Soc* 138:1466–70.
84. Kurokawa, H. 1982. P-doped polysilicon film growth technology. *J Electrochem Soc* 129:2620–24.
85. Biebl, M., and H. von Philipsborn. 1993. Fracture strength of doped and undoped panical filters. PhD thesis. University of California, Berkeley.
86. Maluf, N. 2000. An introduction to microelectromechanical systems engineering. Boston: Artech House.
87. Compton, R., D. 1992. PECVD: a versatile technology. *Semicond Int* July:60–65.
88. Honer, K. A., and G. T. A. Kovacs. 2000. *Solid-state sensor and actuator workshop.* Hilton Head Island, SC: Transducers Research Foundation.
89. Sharpe, W. N., B. Yuan, R. Vaidyanathan, and R. L. Edwards. 1997. *Tenth IEEE international workshop on microelectromechanical systems.* Nagoya, Japan: IEEE.
90. Jerman, H. 1994. Hard copies of viewgraphs presented in Banff, Canada.
91. Biebl, M., and H. von Philipsborn. 1993. *Fracture strength of doped and undoped panical filters.* PhD thesis, University of California, Berkeley.
92. Greek, S., F. Ericson, S. Johansson, and J.-Å. Schweitz. 1995. In situ tensile strength measurement of thick-film and thin-film micromachined structures. *Eighth international conference on solid-state sensors and actuators (Transducers '95).* Stockholm, Sweden.
93. Le Berre, M., P. Kleinmann, B. Semmache, D. Barbier, and P. Pinard. 1996. *Smart electronics and MEMS: proceedings of the smart structures and materials 1996 meeting.* Eds. V. K. Varadan, and P. J. McWhorter. San Diego: SPIE.
94. Howe, R. T. 1995. Recent advances in surface micromachining. *Technical digest: 13th sensor symposium.* Tokyo, Japan.
95. Diem, B., M. T. Delaye, F. Michel, S. Renard, and G. Delapierre. 1993. SOI(SIMOX) as a substrate for surface micromachining of single crystalline silicon sensors and actuators. *Seventh international conference on solid-state sensors and actuators (Transducers '93).* Yokohama, Japan: IEEE.
96. Pister, K. S. J. 1992. Hinged polysilicon structures with integrated CMOS TFTS. *Technical digest: 1992 solid state sensor and actuator workshop.* Hilton Head Island, SC.
97. Keller, C., and M. Ferrari. 1994. Milli-scale polysilicon structures. *Technical digest: 1994 solid state sensor and actuator workshop.* Hilton Head Island, SC.
98. Lange, P., M. Kirsten, W. Riethmuller, B. Wenk, G. Zwicker, J. R. Morante, F. Ericson, and J.-Å. Schweitz. 1995. Thick polycrystalline silicon for surface micromechanical applications: deposition, structuring and mechanical. *Eighth international conference on solid-state sensors and actuators (Transducers '95).* Stockholm, Sweden.
99. Noworolski, J. M., E. Klaassen, J. Logan, K. Petersen, and N. Maluf. 1995. *Eighth international conference on solid-state sensors and actuators (Transducers '95).* Stockholm, Sweden: IEEE.
100. Chu, P. B., P. R. Nelson, M. L. Tachiki, and K. S. J. Pister. 1995. Dynamics of polysilicon parallel-plate electrostatic actuators. *Eighth international conference on solid-state sensors and actuators (Transducers '95).* Stockholm, Sweden.
101. Memming, R., and G. Schwandt. 1966. Anodic dissolution of silicon in hydrofluoric acid solutions. *Surf Sci* 4:109–24.
102. Zhang, X. G., S. D. Collins, and R. L. Smith. 1989. Porous silicon formation and electropolishing of silicon by anodic polarization in HF solution. *J Electrochem Soc* 136:1561–65.
103. Anderson, R. C., R. S. Muller, and C. W. Tobias. 1994. Porous polycrystalline silicon: a new material for MEMS. *J Microelectromech Syst* 3:10–18.
104. Nathanson, H. C., W. E. Newell, R. A. Wickstrom, and J. R. Davis. 1967. The resonant gate transistor. *IEEE Trans Electron Devices* ED–14:117–33.
105. Dutta, B., P. Dev, P. Dewilde, B. Sharma, and R. Newcomb. 1970. Integrated micromotors concepts. *International conference on microelectronic circuits and systems theory.* Sydney, Australia.
106. Howe, R. T., and R. S. Muller. 1982. Polycrystalline silicon micromechanical beams. *Spring meeting of the electrochemical society.* Montreal, Canada.
107. Guckel, H., and D. W. Burns. 1985. A technology for integrated transducers. *International conference on solid-state sensors and actuators.* Philadelphia.
108. Gabriel, K., J. Jarvis, and W. Trimmer. 1988. *Small machines, large opportunities: a report on the emerging field of microdynamics.* Murray Hill, NJ: National Science Foundation, published by AT&T Bell Laboratories.
109. Muller, R. S. 1987. *Proceedings: IEEE micro robots and teleoperators workshop.* Hyannis, MA: IEEE.
110. Fan, L.-S., Y. C. Tai, and R. S. Muller. 1987. Pin joints, gears, springs, cranks, and other novel micromechanical structures. *Fourth international conference on solid-state sensors and actuators (Transducers '87).* Tokyo, Japan, June 2–5, 1987, pp. 849–52.
111. Editor. 1991. Analog devices combine micromachining with BICMOS. *Semicond Int*14:17.
112. Hornbeck, L. J. 1995. *Micromachining and microfabrication process technology.* Austin, TX.
113. Hermansson, K., U. Lindberg, B. Hok, and G. Palmskog. 1991. *Sixth international conference on solid-state sensors and actuators (Transducers '91).* San Francisco.
114. Tang, W. C. K. 1990. *Electrostatic comb drive for resonant sensor and actuator applications.* PhD thesis, University of California, Berkeley.
115. Adams, A. C. 1988. *VLSI technology.* Ed. S. M. Sze. New York: McGraw-Hill.
116. Howe, R. T. 1995. *Eighth international conference on solid-state sensors and actuators (Transducers '95).* Stockholm, Sweden.
117. Biebl, M., G. T. Mulhern, and R. T. Howe. 1995. In situ phosphorus doped polysilicon for integrated MEMS. *Eighth international conference on solid-state sensors and actuators (Transducers '95).* Stockholm, Sweden.
118. Fresquet, G., C. Azzaro, and J.-P. Couderc. 1995. Analysis and modeling of in situ boron-doped polysilicon deposition by LP CVD. *J Electrochem Soc* 142:538–47.

119. Core, T. A., W. K. Tsang, and S. J. Sherman. 1993. Fabrication technology for an integrated surface-micromachined sensor. *Solid State Technol* 36:39–47.
120. Monk, D. J., D. S. Soane, and R. T. Howe. 1992. LPCVD silicon dioxide sacrificial layer etching for surface micromachining. *Symposium on "smart" materials fabrication/materials for micro-electro-mechanical systems.* Materials Research Society Spring Meeting, San Francisco, April 27–May 1, 1992, MRS Symposium Proceedings, pp. 303–10.
121. Tenney, A. S., and M. Ghezzo. 1973. Etch rates of doped oxides in solutions of buffered HF. *J Electrochem Soc* 120:1091–95.
122. Levy, R. A., and K. Nassau. 1986. Viscous behavior of phosphosilicate and borophosphosilicate glasses in VLSI processing. *Solid State Technol* October:123–30.
123. North, J. C., T. E. McGahan, D. W. Rice, and A. C. Adams. 1978. Tapered windows in phosphorous-doped silicon dioxide by ion implantation. *IEEE Trans Electron Devices* ED–25:809–12.
124. Goetzlich, J., and H. Ryssel. 1981. Tapered windows in silicon dioxide, silicon nitride, and polysilicon layers by ion implantation. *J Electrochem Soc* 128:617–19.
125. White, L. K. 1980. Bilayer taper etching of field oxides and passivation layers. *J Electrochem Soc* 127:2687–93.
126. Monk, D. J., D. S. Soane, and R. T. Howe. 1994. Hydrofluoric acid etching of silicon dioxide sacrificial layers-part I: experimental observations. *J Electrochem Soc* 141:264–69.
127. Monk, D. J., D. S. Soane, and R. T. Howe. 1994. Hydrofluoric acid etching of silicon dioxide sacrificial layers-part II: modeling. *J Electrochem Soc* 141:270–74.
128. Eaton, W. P., and J. H. Smith. 1996. *SPIE proceedings: micromachining and microfabrication process technology.* Austin, TX: SPIE.
129. Liu, J., Y.-C. Tai, J. Lee, K.-C. Pong, Y. Zohar, and C.-H. Ho. 1993. In situ monitoring and universal modelling of sacrificial PSG etching using hydrofluoric acid. *Proceedings: micro electro mechanical systems (MEMS '93).* Fort Lauderdale, FL.
130. Goossen, K., J. Walker, and S. Arney. 1994. Silicon modulator based on mechanically-active antireflection layer with 1 Mbit/sec capability. *IEEE Photonics Tech Lett* 6:1119.
131. Lober, T. A., and R. T. Howe. 1988. Surface-micromachining processes for electrostatic microactuator fabrication. *Technical digest: 1988 solid state sensor and actuator workshop.* Hilton Head Island, SC.
132. Fan, L.-S., Y. C. Tai, and R. S. Muller. 1988. Integrated movable micromechanical structures for sensors and actuators. *IEEE Trans Electron Devices* 35:724–30.
133. Mehregany, M., K. J. Gabriel, and W. S. N. Trimmer. 1988. Integrated fabrication of polysilicon mechanisms. *IEEE Trans Electron Devices* 35:719–23.
134. Walker, J. A., K. J. Gabriel, and M. Mehregany. 1991. Mechanical integrity of polysilicon films exposed to hydrofluoric acid solutions. *J Electron Mater* 20:665–70.
135. Eaton, W. P., and J. H. Smith. 1995. *Smart structures and materials 1995: smart electronics (Proceedings of the SPIE).* San Diego, CA.
136. Watanabe, H., S. Ohnishi, I. Honma, H. Kitajima, H. Ono, R. J. Wilhelm, and A. J. L. Sophie. 1995. Selective etching of phosphosilicate glass with low pressure vapor HF. *J Electrochem Soc* 142:237–43.
137. Nakamura, M., S. Hoshino, and H. Muro. 1985. Monolithic sensor device for detecting mechanical vibration. *Densi Tokyo* (*IEEE Tokyo* Section) 24:87–88.
138. Bassous, E., and C.-Y. Liu. 1978. Polycrystalline silicon etching with tetramethylammonium hydroxide.US Patent 4113551.
139. Gennissen, P. T. J., P. J. French, D. P. A. D. Munter, T. E. Bell, H. Kaneko, and P. M. Sarro. 1995. Porous silicon micromachining techniques for acceleration fabrication. *Proceedings of the European Solid-State Device Research Conference. ESSDERC '95.* The Hague, the Netherlands.
140. Chen, L.-Y., and N. C. MacDonald. 1991. Selective tungsten process. *Sixth international conference on solid-state sensors and actuators (Transducers '91).* San Francisco.
141. Sugiyama, S., T. Suzuki, K. Kawahata, K. Shimaoka, M. Takigawa, and I. Igarashi. 1986. Micro-diaphragm pressure sensor. *Technical digest: IEEE international electron devices meeting (IEDM '86).*
142. Sugiyama, S., K. Kawakata, M. Abe, H. Funabashi, and I. Igarashi. 1987. *Fourth international conference on solid-state sensors and actuators (Transducers '87).* Tokyo, Japan.
143. Kim, Y. W., and M. G. Allen. 1991. Surface micromachined platforms using electroplated sacrificial layers. *Sixth international conference on solid-state sensors and actuators (Transducers '91).* San Francisco: IEEE.
144. Suzuki, K., I. Shimoyama, and H. Miura. 1994. Insect–model based microrobot with elastic hinges. *J Microelectromech Syst* 3, 4–9.
145. Chang, S., W. Eaton, J. Fulmer, C. Gonzalez, B. Underwood, J. Wong, and R. L. Smith. 1991. Micromechanical structures in amorphous silicon. *Sixth international conference on solid-state sensors and actuators (Transducers '91).* San Francisco.
146. Scheeper, P. R., W. Olthuis, and P. Bergveld. 1991. Fabrication of a subminiature silicon condenser microphone using the sacrificial layer technique. *Sixth international conference on solid-state sensors and actuators (Transducers '91).* San Francisco.
147. Yamada, K., and T. Kuriyama. 1991. *Sixth international conference on solid-state sensors and actuators (Transducers '91).* San Francisco.
148. van Mullem, C. J., K. J. Gabriel, and H. Fujita. 1991. Large deflection performance of surface micromachined corrugated diaphragms. *Sixth international conference on solid-state sensors and actuators (Transducers '91).* San Francisco: IEEE.
149. Mastrangelo, C. H., and C. H. Hsu. 1993. Mechanical stability and adhesion of microstructures under capillary forces—part I: basic theory. *J Microelectromech Syst* 2:33–43.
150. Mastrangelo, C. H., and C. H. Hsu. 1993. Mechanical stability and adhesion of microstructures under capillary forces—part II: experiments. *J Microelectromech Syst* 2:44–55.
151. Fan, L.-S. 1989. *Integrated micromachinery: moving structures on silicon chips.* PhD thesis. University of California, Berkeley.
152. Abe, T., W. C. Messner, and M. L. Reed. 1995. *Proceedings: IEEE micro electro mechanical systems (MEMS '95).* Amsterdam, the Netherlands.
153. Fedder, G. K., J. C. Chang, and R. T. Howe. 1992. Thermal assembly of polysilicon microstructures with narrow-gap electrostatic comb drive. *Technical digest: 1992 solid state sensor and actuator workshop.* Hilton Head Island, SC.
154. Mastrangelo, C. H., and G. S. Saloka. 1993. *Proceedings: IEEE micro electro mechanical systems (MEMS '93).* Fort Lauderdale, FL: IEEE.
155. Guckel, H., J. J. Sniegowski, T. R. Christenson, S. Mohney, and T. F. Kelly. 1989. Fabrication of micromechanical devices from polysilicon films with smooth surfaces. *Sensors Actuators* 20:117–21.

156. Guckel, H., J. J. Sniegowski, T. R. Christenson, and F. Raissi. 1990. The application of fine-grained, tensile polysilicon to mechanically resonant transducers. *Sensors Actuators A* A21:346–51.
157. Takeshima, N., K. J. Gabriel, M. Ozaki, J. Takahashi, H. Horiguchi, and H. Fujita. 1991. Electrostatic parallelogram actuators. *Sixth international conference on solid-state sensors and actuators (Transducers '91)*. San Francisco.
158. Mulhern, G. T., D. S. Soane, and R. T. Howe. 1993. Supercritical carbon dioxide drying of microstructures. *Seventh international conference on solid-state sensors and actuators (Transducers '93)*. Yokohama, Japan.
159. Kozlowski, F., N. Lindmair, T. Scheiter, C. Hierold, and W. Lang. 1995. A novel method to avoid sticking of surface micromachine structures. *Eighth international conference on solid-state sensors and actuators (Transducers '95)*. Stockholm, Sweden.
160. Alley, R. L., P. Mai, K. Komvopoulos, and R. T. Howe. 1993. Surface roughness modification of interfacial contacts in polysilicon microstructures. *Seventh international conference on solid-state sensors and actuators (Transducers '93)*. Yokohama, Japan.
161. Alley, R. L., R. T. Howe, and K. Komvopoulos. 1992. The effect of release etch processing on the surface microstructure stiction. *Technical digest: 1992 solid state sensor and actuator workshop*. Hilton Head Island, SC.
162. Man, P. F., B. P. Gogoi, and C. H. Mastrangelo. 1996. Elimination of post-release adhesion in microstructures using thin conformal fluorocarbon films. *Ninth annual international workshop on micro electro mechanical systems (MEMS '96)*. San Diego, CA..
163. Houston, M. R., R. Maboudian, and R. Howe. 1995. Ammonium fluoride antistiction treatments for polysilicon microstructures. *Eighth international conference on solid-state sensors and actuators (Transducers '95)*. Stockholm, Sweden.
164. Gogoi, B. P., and C. H. Mastrangelo. 1995. Post-processing release of microstructures by electromagnetic pulses. *Eighth international conference on solid-state sensors and actuators (Transducers '95)*. Stockholm, Sweden.
165. Li, J., J. Xu, and C. P. Wong. 2004. Lotus effect coating and its application for microelectromechanical systems stiction prevention. *Proceedings of the 54th Electronic Components and Technology Conference (ECTC)*. Las Vegas, NV.
166. Nguyen, C. T. C., and R. T. Howe. 1993. *Electron devices meeting, 1993. IEDM '93. Technical Digest, International.* Washington DC: IEEE.
167. Guckel, H., T. Randazzo, and D. W. Burns. 1985. A simple technique for the determination of mechanical strain in thin films with applications to polysilicon. *J Appl Phys* 57:1671–75.
168. Howe, R. T., and R. S. Muller. 1983. Stress in polycrystalline and amorphous silicon thin films. *J Appl Phys* 54:4674–75.
169. Murarka, S. P., and T. F. J. Retajczyk. 1983. Effect of phosphorous doping on stress in silicon and polycrystalline silicon. *J Appl Phys* 54:2069–72.
170. Orpana, M., and A. O. Korhonen. 1991. Control of residual stress of polysilicon thin films by heavy doping in surface micromachining. *Sixth international conference on solid-state sensors and actuators (Transducers '91)*. San Francisco.
171. Lin, L., R. T. Howe, and A. P. Pisano. 1993. *Proceedings: IEEE micro electro mechanical systems (MEMS '93)*. Fort Lauderdale, FL: IEEE.
172. Choi, M. S., and E. W. Hearn. 1984. Stress effects in boron-implanted polysilicon films. *J Electrochem Soc* 131:2443–46.
173. Ding, X., and W. Ko. 1991. Buckling behavior of boron-doped P+ silicon diaphragms. *Sixth international conference on solid-state sensors and actuators (Transducers '91)*. San Francisco.
174. Hendriks, M., R. Delhez, and S. Radelaar. 1983. *Studies in inorganic chemistry.* Amsterdam: Elsevier.
175. Tang, W. C., T.-C. H. Nguyen, and R. T. Howe. 1989. Proceedings: IEEE micro electro mechanical systems (MEMS '89). Salt Lake City, UT.
176. Sekimoto, M., H. Yoshihara, and T. Ohkubo. 1982. Silicon nitride single-layer x-ray mask. *J Vac Sci Technol* 21:1017–21.
177. Guckel, H., D. K. Showers, D. W. Burns, C. R. Rutigliano, and C. G. Nesler. 1986. Deposition techniques and properties of strain compensated LP CVD silicon nitride. *Technical digest: 1986 solid state sensor and actuator workshop*. Hilton Head Island, SC.
178. Tang, W. C., T. H. Nguyen, and R. T. Howe. 1989. Laterally driven polysilicon resonant microstructures. *Sensors Actuators* 20:25–32.
179. Jerman, J. H. 1990. The fabrication and use of micromachined corrugated silicon diaphragms. *Sensors Actuators A* A23:988–92.
180. Spiering, V. L., S. Bouwstra, R. M. E. J. Spiering, and M. Elwenspoek. 1991. On-chip decoupling zone for package-stress reduction. *Sixth international conference on solid-state sensors and actuators (Transducers '91)*. San Francisco.
181. Offereins, H. L., H. Sandmaier, B. Folkmer, U. Steger, and W. Lang. Stress-free assembly technique for a silicon-based pressure sensor. *Sixth international conference on solid-state sensors and actuators (Transducers '91)*. San Francisco.
182. Spiering, V. L., S. Bouwstra, J. Burger, and M. Elwenspoek. 1993. Membranes fabricated with a deep single corrugation for package stress reduction and residual stress relief. *Fourth European workshop on micromechanics (MME '93)*. Neuchatel, Switzerland.
183. Zhang, Y., and K. D. Wise. 1994. Performance of non-planar silicon diaphragms under large deflections. *J Microelectromech Syst* 3:59–68.
184. Scheeper, P. R., W. Olthuis, and P. Bergveld. 1994. The design, fabrication, and testing of corrugated silicon nitride diaphragms. *J Microelectromech Syst* 3:36–42.
185. Yun, W. 1992. *A surface micromachined accelerometer with integrated CMOS detection circuitry.* PhD thesis. University of California, Berkeley.
186. Guckel, H., D. W. Burns, H. A. C. Tilmans, D. W. DeRoo, and C. R. Rutigliano. 1988. Mechanical properties of fine grained polysilicon—the repeatability issue. *Technical digest: 1988 solid state sensor and actuator workshop*. Hilton Head Island, SC.
187. Nguyen, C. T.-C. 1998. *Aerospace conference.* Snowmass, CO: IEEE.
188. Smith, J., S. Montague, J. Sniegowski, and P. McWhorter. 1995. *Technical digest: IEEE international electron devices meeting (IEDM '95)*. Washington, DC.
189. Varadan, V. K., and P. J. McWhorter, eds. 1996. *Smart electronics and MEMS: proceedings of the smart structures and materials 1996 meeting.* San Diego: SPIE.
190. Dunn, P. N. 1993. SOI: ready to meet CMOS challenge. *Solid State Technol* October32–35.
191. Abe, T., and J. H. Matlock. 1990. Wafer bonding technique for silicon-on-insulator technology. *Solid State Technol* November:39–40.

192. Kanda, Y. 1991. What kind of SOI wafers are suitable for what type of machining purposes? *Sixth international conference on solid-state sensors and actuators (Transducers '91)*. San Francisco.
193. Kuhn, G. L., and C. J. Rhee. 1973. Thin silicon film on insulating substrate. *J Electrochem Soc* 120:1563–66.
194. Pourahmadi, F., L. Christel, and K. Petersen. 1992. Silicon accelerometer with new thermal self-test mechanism. *Technical digest: 1992 solid state sensor and actuator workshop*. Hilton Head Island, SC.
195. Gennissen, P. T. J., M. Bartke, P. J. French, P. M. Sarro, and R. F. Wolffenbuttel. 1995. Automatic etch stop on buried oxide using epitaxial lateral overgrowth. *Eighth international conference on solid-state sensors and actuators (Transducers '95)*. Stockholm, Sweden.
196. Bartek, M., P. T. J. Gennissen, P. J. French, and R. F. Wolffenbuttel. 1995. Confined selective epitaxial growth: potential for smart silicon sensor fabrication *Eighth international conference on solid-state sensors and actuators (Transducers '95)*. Stockholm, Sweden: IEEE.
197. Maluf, N. 2000. *An introduction to microelectromechanical systems engineering*. Boston: Artech House.
198. Greiff, P. 1995. *Micromachining and microfabrication process technology. Proceedings of the SPIE*. Austin, TX: SPIE.
199. Neudeck, G. W., P. J. Schuberts, J. L. Glenn, Jr., J. A. Friedrich, W. A. Klaasan, R. P. Zingg, and J. P. Denton. 1990. Three dimensional devices fabricated by silicon epitaxial lateral overgrowth. *J Electron Mater* 19:1111–17.
200. Schubert, P. J., and G. W. Neudeck. 1990. Confined lateral selective epitaxial growth of silicon for device fabrication. *IEEE Electron Device Lett* 11:181–83.
201. Bashir, R., G. W. Neudeck, H. Yen, E. P. Kvam, and J. P. Denton. 1995. Characterization of sidewall defects in selective epitaxial growth of silicon. *J Vac Sci Technol* 11:923–28.
202. Luder, E. 1986. Polycrystalline silicon-based sensors. *Sensors Actuators* 10:9–23.
203. Obermeier, E., and P. Kopystynski. 1992. Polysilicon as a material for microsensor applications. *Sensors Actuators A* A30:149–55.
204. Voronin, V. A., A. A. Druzhinin, I. I. Marjamora, V. G. Kostur, and J. M. Pankov. 1992. Laser-recrystallized polysilicon layers in sensors. *Sensors Actuators A* A30:143–47.
205. Burggraaf, P. 1991. Epi's leading edge. *Semicond Int* June:67–71.
206. Lin, L. Y., S. S. Lee, M. C. Wu, and K. S. J. Pister. 1995. *Proceedings: IEEE micro electro mechanical systems (MEMS '95)*. Amsterdam, the Netherlands.
207. Burgett, S. R., K. S. Pister, and R. S. Fearing. 1992. *ASME 1992, micromechanical sensors, actuators, and systems*. Anaheim, CA: ASME.
208. Ross, M., and K. Pister. 1994. Micro-windmill for optical scanning and flow measurement. *Sensors Actuators A* 46–47:576–79.
209. Lin, G., K. S. J. Pister, and K. P. Roos. 1996. Standard CMOS piezoresistive sensor to quantify heart cell contractile forces. *Ninth annual international workshop on micro electro mechanical systems (MEMS '96)*. San Diego, CA.
210. Yeh, R., E. J. Kruglick, and K. S. J. Pister. 1994. *ASME 1994, micromechanical sensors, actuators, and systems*. Chicago: ASME.
211. Hoffman, E., B. Warneke, E. Kruglick, J. Weigold, and K. S. J. Pister. 1995. *Proceedings: IEEE micro electro mechanical systems (MEMS '95)*. Amsterdam, the Netherlands: IEEE.
212. Keller, C. G., and R. T. Howe. 1995. *Eighth international conference on solid-state sensors and actuators (Transducers '95)*, 376–79. Stockholm, Sweden.
213. Keller, C. G., and R. T. Howe. 1995. *Eighth international conference on solid-state sensors and actuators (Transducers '95)*, 99–102. Stockholm, Sweden.
214. Kahn, H., S. Stemmer, K. Nandakumar, A. H. Heuer, R. L. Mullen, R. Ballarini, and M. A. Huff. 1996. Mechanical properties of thick, surface micromachined polysilicon films. *Ninth annual international workshop on micro electro mechanical systems (MEMS '96)*. San Diego, CA.
215. Brodie, I., H. R. Gurnick, C. E. Holland, and H. A. Moessner. 1990. *Method for providing polyimide spacers in a field emission panel display. US Patent 4923421*.
216. Frazier, A. B., and M. G. Allen. 1993. Metallic microstructures fabricated using photosensitive polyimide electroplating molds. *J Microelectromech Syst* 2:87–94.
217. Ahn, C. H., Y. J. Kim, and M. G. Allen. 1993. *Proceedings: IEEE micro electro mechanical systems (MEMS '93)*. Fort Lauderdale, FL: IEEE.
218. Acosta, R. E., C. Ahn, I. V. Babich, E. I. Cooper, J. M. Cotte, W. J. Horkans, C. Jahnes, S. Krongelb, K. T. Kwietniak, N. C. Labianca, et al. 1995. Integrated variable reluctance magnetic mini–motor. *J Electrochem Soc* 95-2:494–95.
219. LaBianca, N. C., J. D. Gelorme, E. Cooper, E. O'Sullivan, and J. Shaw. 1995. *JECS 188th Meeting*. Chicago: JECS.
220. Field, L. 1987. Electrical Engineering and Computer Sciences/Electronics Research Lab; Research summary. University of California, Berkeley.
221. Judy, M. W., and R. T. Howe. 1993. *Proceedings: IEEE micro electro mechanical systems (MEMS '93)*. Fort Lauderdale, FL.
222. Judy, M. W., and R. T. Howe. 1993. Highly compliant lateral suspensions using sidewall beams. *Seventh international conference on solid-state sensors and actuators (Transducers '93)*. Yokohama, Japan.
223. Lebouitz, K. S., R. T. Howe, and A. P. Pisano. 1995. Permeable polysilicon etch-access windows for microshell fabrication. *Eighth international conference on solid-state sensors and actuators (Transducers '95)*. Stockholm, Sweden.
224. Mardilovich, P., D. Routkevitch, and A. Govyadinov. 2000. Eds. P. J. Hesketh, S. S. Ang, W. E. Bailey, J. L. Davidson, H. G. Hughes, and D. Misra. *Micro-fabricated systems and MEMS*. Electrochemical Society: Pennington, Vol. 2000–19, p. 33.
225. Lin, L. 1993. *Selective encapsulations of MEMS: micro channels, needles, resonators, and electromechanical filters*. PhD thesis. University of California, Berkeley.
226. Kamins, T. 1988. *Polycrystalline silicon for integrated circuits*. Boston: Kluwer.
227. Heuberger, A. 1989. *Mikromechanik*. Heidelberg: Springer Verlag.
228. Sharpe, W. N., B. Yuan, R. Vaidyanathan, and R. L. Edwards. 1997. *Tenth IEEE international workshop on microelectromechanical systems*. Nagoya, Japan: IEEE.
229. Petersen, K., D. Gee, F. Pourahmadi, R. Craddock, J. Brown, and L. Christel. 1991. Resonant beam pressure sensor fabricated with silicon fusion bonding. *Sixth international conference on solid-state sensors and actuators (Transducers '91)*. San Francisco.
230. Chrisey, D. B., and G. K. Hubler. (Eds.) 1994. *Pulsed laser deposition of thin films*. New York: John Wiley and Sons.
231. 2006. *Atomic layer deposition*. Cambridge NanoTech Inc., April 24. http://www.cambridgenanotech.com

# 8

# Chemical, Photochemical, and Electrochemical Forming Techniques

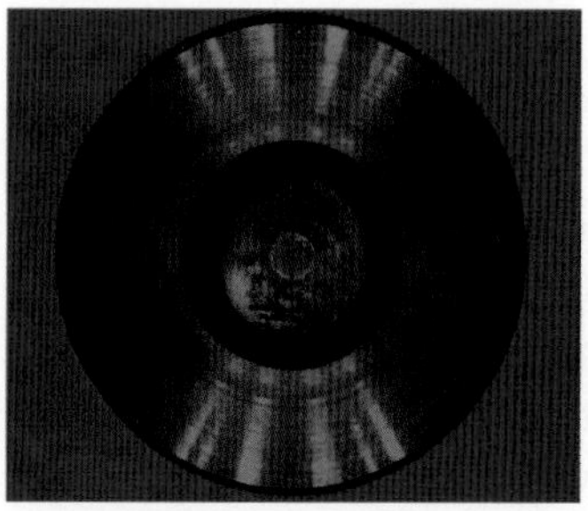

"The Sounds of Earth," a gold-plated copper record placed on Voyager. Two hours of sound and a movie plus some pictures and a message from President Jimmy Carter (From http://voyager.jpl.nasa.gov/spacecraft/goldenrec.html.)

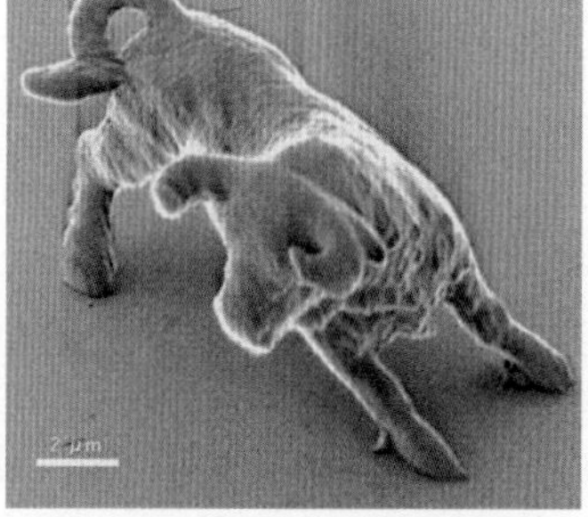

Photochemical forming: Light causes a polymer to solidify. Structural details of 120 nm (as a result of two-photon photopolymerization). (From Kawata, S., H.-B. Sun, T. Tanaka, and K. Takada, 2001. Finer features for functional microdevices. *Nature* 412:697–698.[86])

## Outline

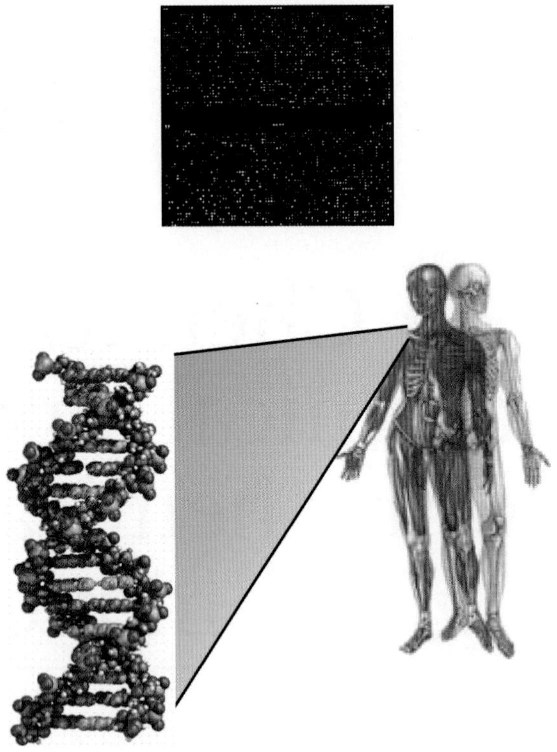

DNA array and the human genome.

> Elderly: In physics, mathematics and astronautics it means over thirty; in other disciplines, senile decay is sometimes postponed to the forties. There are of course, glorious exceptions; but as every researcher just out of college knows, scientists of over fifty are good for nothing but board meetings, and should at all costs be kept out of the laboratory.
>
> **Arthur C. Clarke**
> *Profiles of the Future*

## Introduction

In this chapter we review chemical, photochemical, and electrochemical forming.* In chemical forming, chemical reactions create features by forming new compounds. The energy source at the work piece in this additive process is chemical. The method is extremely broad and covers fabrication with metals, ceramics, and polymers. Reactions are induced by combining different reagents and are often aided by heat and/or catalysts and also may be accompanied by a shape-forming step using, for example, a spraying nozzle, a mold, or a template. Almost all manufacturing processes include one or more chemical forming steps and can be part of lithography- or nonlithography-based schemes. Chemical and thermal forming processes are difficult to distinguish because they are almost always intertwined. Some chemical forming processes involve an exothermic reaction, and such processes we classify as purely chemical (they do not require the application of external heat). This is the case, for example, in reaction injection molding (RIM; see Chapter 10) and for self-assembled monolayers (SAMs; see page 523 below). Other chemical forming processes are endothermic and require a significant increase in temperature for the reaction to initiate such as in the oxidation of Si (see Volume I, Chapter 4) or chemical vapor deposition of polycrystalline Si (CVD; see Chapter 7). The distinction between chemical and thermal forming is somewhat arbitrary, and in this chapter we cover as additional chemical forming methods those low temperature material deposition techniques that are used in chemical and biological sensor manufacture, often arranged in some type of an array configuration. The additional chemical forming methods covered are sol-gel deposition from alkoxide-based precursors, organic film spin coating, polymer dry film lamination, polymer dip coating, polymer spraying, polymer casting, organic film doctor's blade (also called *knife coating*), glow discharge polymerization, low temperature silk screening of organics, Langmuir-Blodgett deposition, and SAMs. Some of the chemical forming processes discussed here involve self-assembly such as in the case of SAMs, in which case one refers to bottom-up manufacturing. Whereas top-down nanofabrication is principally based on lithography and traditional mechanical machining, etching, and grinding, bottom-up manufacturing, which we also call nanochemistry in this book, is based in chemical synthesis, self-assembly, and positional assembly (see Volume III, Chapter 3). Arraying of low temperature materials (often organics) for sensor arrays may be accomplished by lithography, digital mirror deposition (a type of lithography as well), ink-jetting, mechanical microspotting, and

* In Part II on subtractive manufacturing techniques, we did make the same distinction between chemical, photochemical, and electrochemical etching. All three subtractive manufacturing techniques were covered in Chapter 4 on wet chemical etching and wet bulk micromachining-pools as tools.

microcontact printing. Some of the same additive methods discussed in this section will reappear in Chapter 9 on thermal forming where they involve the manufacture of ceramics in which heat application is more essential.

In photochemical forming, photoenergy solidifies a material into a three-dimensional (3D) shape as we illustrate with our description of rapid prototyping and microphotoforming (stereolithography) with, as newest embodiments, parallel stereolithography and two-photon polymerization.

In the case of electrochemical forming, the energy at the work piece is electrochemical. Electrochemical and electroless metal deposition are gaining renewed interest because of their emerging importance in making mold inserts for replication meth-odologies in both microelectromechanical sys-tems (MEMS) and nanoelectromechanical systems (NEMS) (see also Chapter 10). The foundations underlying electrochemical deposition were covered in detail in Volume I, Chapter 7; here we contrast electroless metal deposition with electrodeposition and introduce some additional electrochemical techniques. These new electrodeposition methods sort out into through-mask plating, such as in the fabrication of thin-film read-write heads, the damascene process, and in anodization; instant masking in EFAB (Electrochemical FABrication); and maskless metal deposition methods such as microjet plating, laser-enhanced jet plating, local electrochemical deposition using scanning electrochemical microscopes (SECMs), and techniques to electrodeposit slanted and curved surfaces.

## Chemical Forming

### Introduction

It should come as no surprise that chemical forming techniques are gaining importance because today bottom-up manufacturing or nanochemistry, a type of chemical forming, constitutes one of the most intensely researched topics. Bottom-up fabrication or nanochemistry constitutes a special class of chemical forming incorporating chemical synthesis of small particles, self-assembly, and positional assembly. In nanochemistry device structures are created via systematic assembly of atoms, molecules, or other basic units of matter. Instead of manipulating Si wafers or other solid substrates, we manipulate small amounts of chemicals, even molecules and atoms. Molecular engineers, manipulating and modifying natural polymers such as DNA and proteins, led the way into nanochemistry and practiced this type of nanotechnology long before electrical and mechanical engineers got involved. Now we are designing and manipulating at the molecular level, whereas before it was either evolution that did it for us (e.g., a virus as a nanomachine) or results happened that we never really understood and so could not optimize (e.g., stained glass window—small metal nanoparticle comparable in size with the wavelength of light).

Although there is some overlap, not all chemical forming techniques involve minute amounts of reacting chemicals, nor do they all entail self-assembly or positional assembly. The chemical forming methods covered in this chapter are sol-gel, organic film spin coating, polymer dry film lamination, polymer dip coating, polymer spraying, polymer casting, organic film doctor's blade (or knife coating), glow discharge polymerization, low temperature silk screening of organics, Langmuir-Blodgett deposition, and self-assembled monolayers (SAMs).

Arraying of organics for sensor arrays may be accomplished by lithography, digital mirror deposition, ink-jetting, mechanical microspotting, and microcontact printing.

## Sol-Gel Technique

### *Introduction*

Colloids, we saw in Volume I, Chapter 7, are mixtures in which one material is dispersed in another, with the dispersed material divided into particles ranging in size from 1 nm to 10 μm. They have many properties distinct from those of solutions of ordinary-sized molecules and from mixtures of macroscopic phases such as slurries of visible particles. The dispersed material may either be dissolved (i.e., in the case of solutions of macromolecules), giving a lyophilic colloid, or they may constitute a separate phase, giving a lyophobic colloid. We only deal with lyophobic colloids here,

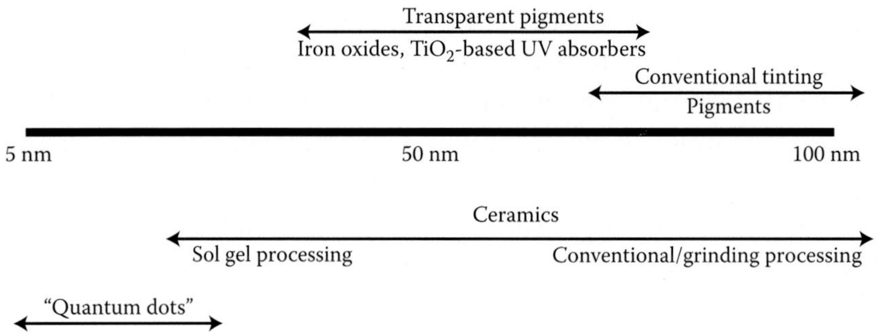


**FIGURE 8.1** Nanoparticle products and their manufacture.

which may be further classified according to the types of phases that compose them, as shown in Volume I, Chapter 7. Some important definitions pertaining to the sol-gel field follow and might help in the understanding of the sol-gel process.

- Colloid: A colloid is a suspension in which the dispersed phase is so small (1 nm–10 μm) that gravitational force is negligible, and interactions are dominated by short-range forces, such as van der Waals attraction and surface charges. The suspended particles usually exhibit Brownian motion.
- Sol: A sol is a colloidal suspension of solid particle or clusters in a liquid.
- Gel: Semirigid mass of a lyophilic sol in which all the dispersion medium is penetrated into the sol particles.
- Emulsion: An emulsion is a suspension of liquid droplets in another liquid.
- Aerosol: An aerosol is a collection of very small particles suspended in a gas. The particles can be liquid (mist) or solid (dust or fume). The term *aerosol* is also commonly used for a pressurized container (aerosol can) designed to release a fine spray of a material such as paint.
- Precursors: The starting compound of the sol-gel process.
- Alkoxide: A basic salt derived from an alcohol by the replacement of the hydroxyl hydrogen with a metal. They are formed by the reaction of an alcohol and an alkali metal. For example, silicon tetraethoxide (TEOS), $Si(OC_2H_5)_4$.
- Hydrolysis: Reaction in which a hydroxyl ion is attached to the metal atom [$Si(OR)_4 + H_2O \rightarrow HO{-}Si(OR)_3 + ROH$, where R is an alkyl].
- Condensation reaction: Reagents combine splitting off water [$HO{-}Si(OR)_3 + HO{-}Si(OR)_3 \rightarrow (OR)_3\,SiO{-}Si(OR)_3 + H_2O$].

The sol-gel process is a wet-chemical additive technique for the fabrication of materials (typically metal oxides) starting from a chemical solution that reacts to produce colloidal particles (a sol). The Derjaguin, Landau, Verwey, and Overbeek theory (the DLVO theory) was introduced in Volume I, Chapter 7 to describe the stability of these colloids.

Sol-gel processing received increasing attention since the 1990s from both the research community and industry as nanoparticles became more important. Colloids form a most important gateway toward bottom-up nanotechnology: the sol-gel technique constitutes a chemical forming process for making very small particles, e.g., 20–40-nm diameter, a particle size that is virtually impossible to make by conventional mechanical ball milling or grinding* (Figure 8.1).

### *Sol-Gel Process*

The first metal alkoxides were prepared in 1846 by Ebelmen using $SiCl_4$. Ebelmen noticed that on exposure to the atmosphere, these chemicals formed a gel. This process was then used in the 1930s by the Schott glass company in Germany to produce metal oxides (http://www.solgelchemist.com/web1.doc). In a sol-gel process, as illustrated in Figure 8.2, chemical precursors, mainly alkoxides (organic epoxides are an alternative for alkoxides) and metal chlorides, are dissolved in an aqueous or alcohol-based solution. Hydrolysis and condensation reactions nucleate

* High-energy ball milling is required to make particles less than 50 nm (see Volume III, Chapter 3).

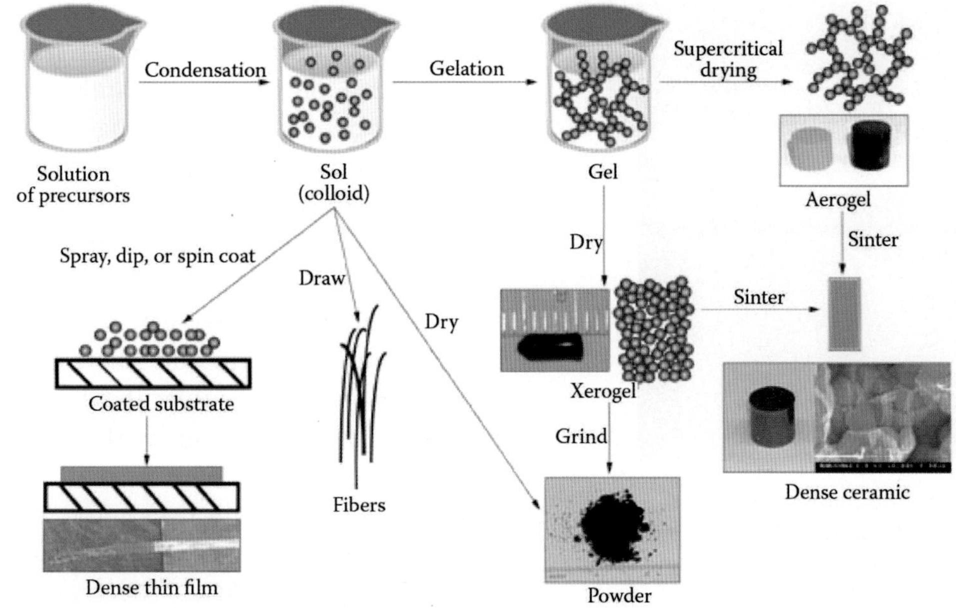


**FIGURE 8.2** Sol-gel processes and sol-gel products. In sol-gel chemistry, molecular precursors are converted to nanometer-sized particles, to form a colloidal suspension, or sol. The gel can be processed by various drying methods (shown by the arrows) to develop materials with distinct properties.

and generate many small nucleation seeds (groups of atoms that build the basic crystal structure). The seeds grow by taking more atoms/molecules from the solution to form clusters. The growth pattern is random, with the atoms or molecules coming from every direction, and fractal-like cluster structures result. The clusters are nanosized, forming a colloid. As we saw in Volume I, Chapter 7, the stability of this colloid depends on the electrokinetic potential $\zeta$ (also the zeta potential), from which all electrokinetic effects can be derived. When the absolute value of the zeta potential is more than 50 mV, the dispersion is very stable, whereas with a $\zeta$ potential close to zero, the coagulation of particles leads to precipitation.

When growth continues, the nanoclusters start to build up a porous network. This is when gelation takes place. The gel point is the point in the phase diagram where the sol abruptly changes from a viscous liquid to a gelatinous, polymerized network and where the dispersion medium is fully penetrated into the sol network (Figure 8.3). Both sol and gel formation are low-temperature processes.

The transformation to a gel can be initiated in several ways, but the most convenient approach is to change the pH of the reaction solution. Formation of a metal oxide involves connecting the metal centers with oxo (M-O-M) or hydroxo (M-OH-M) bridges, therefore generating metal-oxo or metal-hydroxo polymers in solution. The drying process serves to remove the liquid phase from the gel, thus forming a porous material; then a thermal treatment (firing) may be performed to favor further polycondensation and enhance mechanical properties. The method used to remove the solvent from the gel strongly affects the end product's properties. For example, from Figure 8.2 it can be seen that if the solvent in a wet "gel" is removed under a supercritical condition, the gel's original 3D structure is preserved, and a highly porous and extremely low-density material called an *aerogel* is obtained. The supercritical processing condition is realized by exchanging the liquid solvent with $CO_2$. The latter is then removed at slightly elevated temperatures and under high pressure. This way aerogels can be

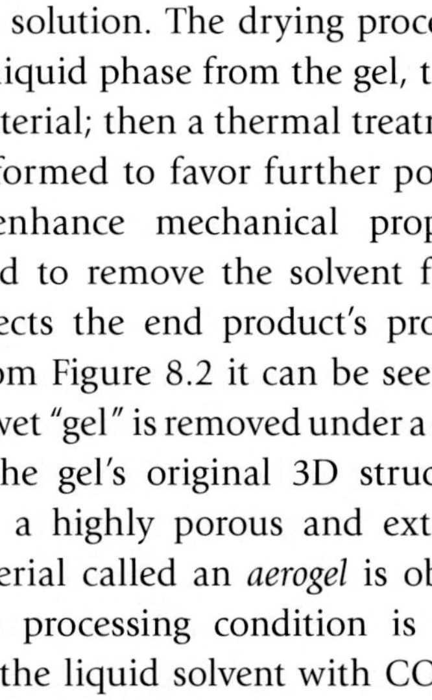

**FIGURE 8.3** Gelation: $TiO_2$ from sol (left) to wet gel (right).

**FIGURE 8.4** Aerogel. This photo illustrates the excellent insulating properties of aerogel. The crayons on top of the aerogel are protected from the flame underneath and are not melting.

produced where 99% of the bulk material consists of pores (Figure 8.4). A Lawrence Livermore group created an aerogel weighing only 1.0 mg/cm$^3$, which is listed in the *Guinness Book of World Records 2005* as the lightest material on Earth.[1]

If, in an alternative process route, the gel is dried slowly in a fluid-evaporation process, the gel's structural network collapses, which creates a high-density material known as a *xerogel*. When a xerogel is further densified by sintering, a dense ceramic material forms.

As the viscosity of a sol is adjusted into a proper viscosity range, ceramic fibers can be drawn; it can be applied as a film onto a substrate by spinning, dipping, or spraying and cast into a suitable container with the desired shape (e.g., to obtain a monolithic ceramics, glasses, fibers, membranes, and aerogels); or finally, ultrafine and uniform ceramic powders may be formed by grinding of the xerogel or by slowly drying the sol.

For a silica sol formation, an appropriate chemical precursor is dissolved in a liquid, for example, tetraethylsiloxane (TEOS) in water. In this case, a silica gel is formed by hydrolysis and condensation using hydrochloric acid as the catalyst. Drying and sintering at temperatures between 200 and 600°C transforms the gel into a glass and then densification into silicon dioxide.

Alkoxides can be very reactive and are commercially available for only a select number of elements. Livermore chemists have found that organic epoxides also initiate the sol-gel reaction. With this approach, precursors that are more widely available can be used, thus increasing the number of potential materials that can be developed. In addition, the starting materials, solvents, and gelling agents used with epoxides are less expensive than those used with alkoxides.

Summarizing, sol-gel is a cheap and low-temperature technique that enables mixing of precursors at the molecular level, leading to finer control and higher purity than most other ceramic synthesis methods (even small quantities of dopants, such as organic dyes and rare earth metals, can be introduced in the sol and end up finely dispersed in the final product). The method is particularly suited for the production of nanosized multicomponent ceramic powders densified through thermal annealing (sintering at relatively low temperature) to an inorganic product like a glass or polycrystal or a dry gel. Applying the sol-gel process, it is possible to fabricate ceramic or glass materials in a wide variety of forms: ultrafine or spherical-shaped powders, thin film coatings, ceramic fibers, microporous inorganic membranes, monolithic ceramics and glasses, or extremely porous aerogel materials. Different material compositions with different porosity may be fabricated depending on the process parameters (such as temperature) and the chemicals used. Its main traditional use is for optical coatings, where the finer particles give better optical clarity, and for the manufacture of fine ceramic fibers.

Sol-gel-derived materials have diverse applications in optics, electronics, energy, space, (bio)sensors, medicine (e.g., controlled drug release), and separation (e.g., chromatography) technology.

Sol-gel methods have been used in MEMS, for example, in the fabrication of piezoelectrics such as lead-zirconate-titanate (see Volume III, Chapter 8). Also, a commercially available, room temperature chemical gas sensor (a CO fire alarm) is fabricated with the sol-gel process (available from Quantum Group Inc.; http://www.qginc.com). The Sol-Gel Gateway (http://www.solgel.com/bookstore/bookstore.htm) is an excellent take-off point for further study of the sol-gel technique on the Internet (in particular, check http://www.solgel.com/Tutorials/tutorials.htm).

## Deposition and Arraying Methods of Organic Layers in Bio-MEMS

### *Introduction*

Today, miniaturization methods are often applied to biotechnology and biomedical problems. With the growth of the "bio-MEMS" field, techniques for depositing organic materials for chemical and biological sensors, often arranged in some type of an array configuration, are gaining importance. Organic gas-permeable membranes, ion selective membranes, hydrogels, and organic monolayers are needed for the manufacture of room temperature gas sensors, ion-selective electrodes (ISEs), enzyme-based sensors, immunosensors, and DNA and protein arrays. Their deposition presents challenges not typically encountered in the integrated circuit (IC) world. Membranes may be based on classical polymers, such as PVC (polyvinylchloride), PVA (polyvinylalcohol), PHEMA (polyhydroxyethylmethacrylate), or silicone rubber, and incorporate pH and temperature-sensitive biological materials such as enzymes, antigens, and antibodies. Single layers of molecules may be deposited by Langmuir-Blodgett (LB) deposition techniques or as self-assembled monolayers (SAMs). Those LB films or SAMs may then be further used as anchor points for proteins or DNA probes, and the latter natural polymer molecules may be synthesized in situ. Not only are the materials dealt with very different, the fact that most of the bio-MEMS sensors are disposable requires new manufacturing approaches as well. Although the number of new tools available for microfabrication has grown dramatically, few methods have been gainfully applied to disposable biosensor construction for clinical applications. The low cost requirements and the tremendous variety and fragmentation of biomedical sensor applications often necessitate a nonsilicon, modular approach. The Si approach is mostly based on thin film techniques with a limited number of materials, too expensive cost, all steps tightly integrated, and too many requirements on process and materials compatibility. Having addressed materials choice and modularity, various competing manufacturing processes must be compared.

Because of the importance of this emerging field, we review the different options available to coat and pattern organic materials on various substrates for bio-MEMS in the following section. For the organic film deposition techniques reviewed earlier in this book, we provide a short summary only.

### *Deposition Methods for Organic Materials*

*Organic Film Spin Coating* As we know from Chapter 1 on lithography, spin coating technology has been optimized for the deposition of thin layers of photoresist, about 1–2 μm thick, on round and nearly ideally flat Si wafers. A typical spin coater schematic is shown in Figure 1.14. Resists are applied by dropping the resist solution, a polymer, a sensitizer (for two-component resists), and a solvent on the wafer. The wafer is then rotated on a spinning wheel at high speed so that centrifugal forces push the excess solution over the edge of the wafer, and a residue on the wafer remains as a result of surface tension. In this way, films down to 0.1 μm can be made. An empirical expression relating film thickness to solution viscosity and rotation speed was given in Chapter 1 (Equation 1.3). However, biosensor substrates rarely are round or flat, and many chemical sensor membranes require a thickness considerably thicker than 1 μm for proper functioning. For example, a typical ion-selective electrode (ISE) membrane is 50 μm thick (see Volume I, Chapter 7). Consequently, spin coating technology does not necessarily fit in with thick chemical membranes on a variety of substrates, and for monolayer deposition the method is not adequate. For building ISEs, dip coating or membrane casting is preferred, and for monolayer deposition one chooses Langmuir-Blodgett or self-assembled monolayer (SAM) methods (see below). For a tutorial on spin coating, visit http://www.mse.arizona.edu/faculty/birnie/Coatings/index.htm.

In Chapter 1, we compared several alternatives to spin coating for depositing photoresist (see Figure 1.15). Obviously all these techniques can be adapted to deposit other organic coatings. This includes spray coating, curtain coating, extrusion coating, roller coating, dip coating, plasma deposited, electrodeposited, silk screening, and meniscus coating. We provide some additional information here complementing the material presented in Chapter 1.

*Polymer Dip Coating* In the dip-coating method a substrate is slowly dipped into and withdrawn from a tank containing the dissolved polymer, with a uniform velocity, to obtain a uniform coating. The dip-coating process may be divided in five stages: immersion, startup, deposition, evaporation, and drainage. Dip coating of a substrate in a solution of a dissolved polymer typifies the simplest method to apply an organic layer to a substrate. It is especially suited for wire-type ISEs and enzyme-based biosensors where the membrane forms a droplet at the end of a wire (see Volume I, Chapter 7). A substrate, for example, a chloridized silver wire, is dipped into a solution containing the polymer and a solvent. After evaporation of the solvent, a thin membrane forms on the surface of the sensor. To obtain pinhole-free membranes, the dipping is repeated several times, interspersed with drying periods. Even though the eventual goal is more typically the production of a planar sensor structure, a Ag wire may be applied in the research phase to quickly evaluate a new membrane composition. The method is difficult to commercialize because of the variability in coating thickness and uniformity. For further reading on dip coating, visit http://www.solgel.com/articles/Nov00/mennig.htm.

Dipping and spinning are also used in the fabrication of sol-gel layers (see Figure 8.2).

*Polymer Dry Film Lamination* The use of dry film lamination, such as the dry photoresist films explored in Chapter 1, is introduced as a better approach for the mass manufacture of chemical and biological sensors than spin coating in Volume III, Example 1.1. Most resists in IC and MEMS fabrication are deposited as liquids, whereas resists used in printed wiring board manufacture are usually dry film resists that come in rolls (ranging from 2–60 in. wide and 125–1000 ft. long) and are laminated onto the substrate instead of being spin-coated on it. A typical dry film lamination setup is shown in Figure 1.33. The heat and pressure of the laminating rollers cause the dry film to soften and adapt to surface topologies. The resist is then exposed to a UV light source. The lamination approach can be extended to other polymer coatings as exemplified in the manufacture of the compact disc (CD)-based fluidic platform shown in Volume III, Chapter 5. In the manufacture of this CD, we use a layering approach, with different sheets of thin machined plastic CDs (polycarbonate-PC) laminated together using pressure-sensitive adhesive layers (PSA-FLEX mount DFM 200 clear V-95). The PSA layers are as thin as 100 μm and are cut using a computer numerical controlled (CNC) cutter-plotter (in Volume III, Figure 5.27, CNC plotter, Graphtec CE-2000-60). The use of PSA layers allows for channels as narrow as 250 μm. When sandwiched between two plastic sheets, these channels in the PSA connect the larger CNC-machined reservoirs in the PC disks.

*Polymer Spraying* Polymer or plastic spray-coating techniques are the most widely used methods for applying organic coatings from the liquid, gas, or the solid phase. Spray coating of a liquid involves pressurization by compressed air or, in an airless method, pushing liquid mechanically through tiny orifices. Vapors are carried in an inert dry vapor carrier. In the case of a solid, a powdered plastic resin is melted and blown through a flame-shrouded nozzle. Such thermal spray coating involves heating a material, in powder or wire form, to a molten or semimolten state. The material is propelled using a stream of gas or compressed air to deposit it, creating a surface structure on a substrate. The coating material may consist of a single element but is often a composite with unique physical properties that are only achievable through the thermal spray process.

There are two main classes of powder coatings: thermosetting and thermoplastic coatings. In a thermosetting film, cross-linking occurs between the molecules in the powder during baking. This cross-linking turns the baked film into a single giant molecule that cannot melt or flow. In a thermoplastic film, thermal energy makes the binder molecules mobile enough to become entangled; thus, a continuous film forms, and this film hardens on cooling. Although a thermoplastic film can still melt or flow, it can do that only at elevated temperatures. In electrostatic spraying, a negatively charged plastic powder is spray gunned onto grounded conductive parts (see Figure 8.5).[2] Electrostatic spray is the primary technique used for thermoset powders. The particles of powder are given their electrical charge in the powder coating gun, and the target part is

**FIGURE 8.5** Electrostatic spray gun from ITW Ransburg. (From http://www.itwransburg.com/products.html.)

attached to a fixture that is grounded. The electrically charged powder particles are attracted to the grounded part and build up on the surface until the part is covered with charged particles. At this point, the oncoming particles are repelled by the part, and the coating process stops. This provides for an even film thickness.

Thermoplastic and thermosetting coatings may also be formed by dipping a heated part into a container of resin particles set in motion by a stream of low-pressure air (fluidized-solid bed). This method is only of use if an inert coating is desired, for example, for biocompatibility, but not for heat-sensitive functionalized coatings. The fluidized bed is the original powder-coating technique. It is still the primary technique used for the application of thermoplastic powders.

Thermal spraying is ideally suited for large structures that otherwise could not be dipped in a polymer suspension. Unlike electrostatic powder coatings, nonconductive components can be coated, and unlike fluidized bed coatings, heat-sensitive materials (aircraft skins) can be sprayed. Polymer coatings can be repaired by heating (for remelting or curing) and by applying additional material to the desired location. Polymer coatings can be applied in high humidity and at temperatures below freezing. Certain polymers have excellent adhesion to metallic surfaces as a result of interfacial bonding. Metals, ceramics, or other polymers can be incorporated into the polymer matrix during spraying to act as a fill.

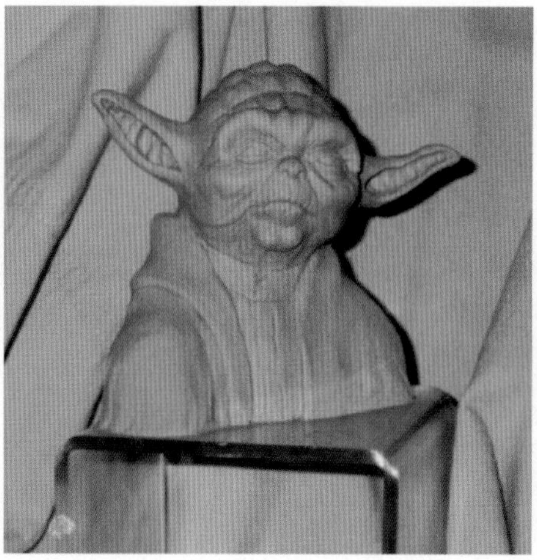

**FIGURE 8.6** Yoda, typical polymer casting application.

*Polymer Casting* Cast polymer operations are operations where a gel coat resin is sprayed to a mold, after which a casting resin is applied without spraying. These are used in the manufacture of pools, bathtubs, marble-look countertops, masks, and toys (Figure 8.6). Casting in bio-MEMS is often based on the application of a given amount of dissolved organic material on the surface of a sensor substrate and letting the solvent evaporate. A rim structure is fashioned around the substrate, providing a "flat beaker" for the solution. This method provides a more uniform and a more reproducible membrane than dip coating. Membranes in planar ion-selective electrodes (ISEs) are often made this way. Casting is also often used to obtain thicker photoresist layers than typically possible with spin coating (see also Chapter 1). Casting is the main technology used in soft lithography (see Chapter 2).

*Polymer Doctor's Blade or Knife Coating* Doctor's blade or knife coating relies on a coating being applied to a substrate that then passes through a slit between a knife and a support roller (Figure 8.7). The "doctor's blade" refers to the scraping knife blade for the removal of excess substance from the moving surface being coated. The doctor's blade is adjustable with a precisely controlled height, limiting the dispensed material to a known thickness. This process can be used for high-viscosity coatings, and there are innumerable variants of this relatively simple process, which is rugged but somewhat inaccurate. It is well known in many industries, including paper,

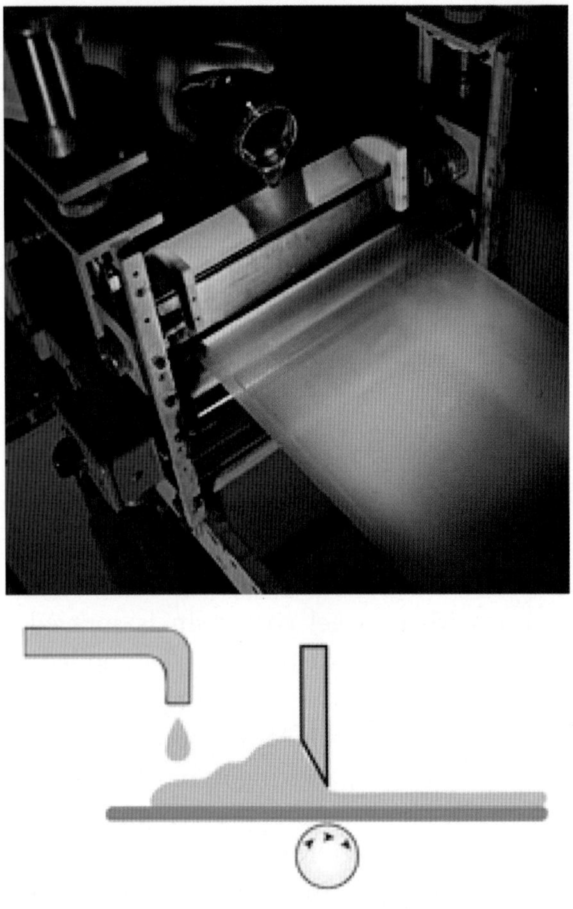

FIGURE 8.7 Doctor blading or knife coating of organic materials on a moving substrate.

plastic, and paint manufacturing. Doctor's blade technology is a continuous fabrication method, which can also be used to apply all types of organic cocktails, for example, to make enzyme-based biosensors or to deposit ceramics, in which case one also refers to tape casting and green tape technology. Howatt was the first to obtain a patent to use the technique for "forming ceramic materials into flat plates, especially useful in the electric and radio fields."[3] This is still the principal application of tape casting today, although its use extends far beyond the field envisioned in 1952 (http://www.drblade.com/history.htm). The ceramic application of doctor blading is a thermal forming process well suited for fabrication of mesoscale devices (layers with a thickness between several micrometers and millimeters) and is covered in Chapter 9 on thermal energy-based forming.

As a specific application of a doctor's blade process in bio-MEMS, consider the mass production of amperometric glucose sensors discussed in Volume III, Chapter 1. In that chapter, as an illustration of an informed choice of manufacturing options, we find out that using a silicon batch approach, it is almost impossible to make a glucose sensor for less than $1 (or any disposable biosensor for that matter!). The current industrial process, and our conclusion, is that for mass-producing glucose sensor strips, the optimal choice is doctor's blade on a continuous moving web, making a 10-cents cost per glucose sensor possible.

From this example of a glucose sensor strip, it appears that one of the major challenges in bio-MEMS is the fabrication of affordable disposable biosensors. In the case of a glucose sensor strip, this may be accomplished using doctor's blade in a continuous process. An example of a generic approach for biosensor manufacture on large plastic sheets and eventually on a moving web is shown in Volume III, Figure 1.19. The process illustrated allows for the fabrication of a sensor array composed of sensors that may otherwise have fabrication incompatibilities. We call this futuristic bio-MEMS approach "beyond batch."

*Glow Discharge (Plasma) Polymerization* Plasma polymerization is a type of plasma-enhanced chemical vapor deposition procedure (PECVD; see Chapter 7), in which gaseous precursor monomers, activated by the plasma, deposit on freely selectable substrates as highly cross-linked polymer layers. A polymerizable gas can, in principle, include any substance that can be brought into the gas phase at reduced pressure. The *sine qua non* for this process to take place is the presence in the gas phase of polymer-producing atoms, such as carbon, silicon, or sulfur. These react with one another and with any surface groups to form a polymer film, the chemical structure of which is dependent on the process conditions used. The monomer molecules in the plasma, for the most part, become fragmented into reactive segments, and remain only partially preserved in the chemical structure of the final product, which results in a highly cross-linked and disordered structure compared with a conventional polymer (Figure 8.8). The degree of structural preservation and cross-linking gradients in the deposit are controlled

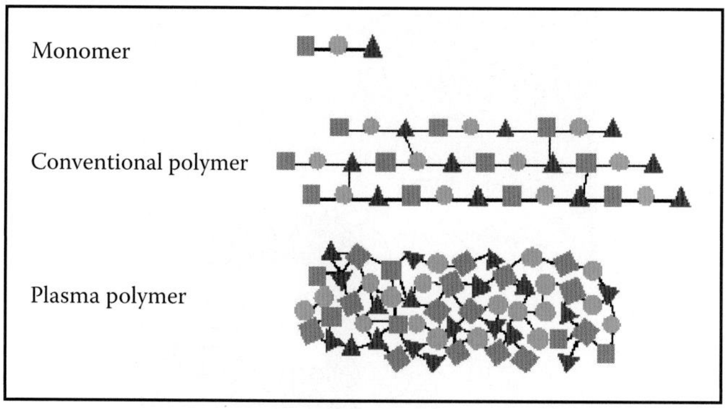


**FIGURE 8.8** Polymers deposited with plasma polymerization. Comparison of the structures of plasma polymers and conventional polymers.

through process parameters, such as gas pressure, working gas flow, and the applied electrical field. For example, it has been found that pulsing the plasma in the millisecond regime constitutes a very simple and unique method to control the film chemistry.[4] Unlike the more conventional continuous wave plasma deposition processes, pulsed plasma deposition enables the polymerization of monomers containing labile groups such as amines, ethers, and anhydrides, to name but a few. By selecting the monomer type and the energy density per monomer, known as the *Yasuda parameter*, the chemical composition and structure of the resulting thin film can be varied over a wide range.[5]

The so-obtained polymer coats can be used to further attach specific molecules by conventional wet chemistry techniques. One can also construct so-called *gradient layers*, i.e., with increasing degree of cross-linking over the film thickness.

Figure 8.9 illustrates the setup we used in our own work on plasma polymerization of doped polymers. The apparatus consisted of a quartz tube reaction chamber; a gas-handling system to introduce the carrier gas, monomer, and dopant into the system, as well as to remove the unreacted material; and a power supply (operating at 27 MHz) to provide the radio-frequency (RF) energy necessary to create and maintain a plasma within the system. Despite the complex chemistry of this process, good conformal coatings often result. Guckel applied this technique to obtain deposition of poly(methylmethacrylate) (PMMA) in layers more than 100 μm thick.[6] In research at SRI International, we synthesized electroactive plasma polymers using $I_2$ or $N_2O$ as dopants from the following monomers: thiophene, furan, aniline, benzaldehyde, benzene, indole, diphenylacetylene, and 1-methylpyrrole.

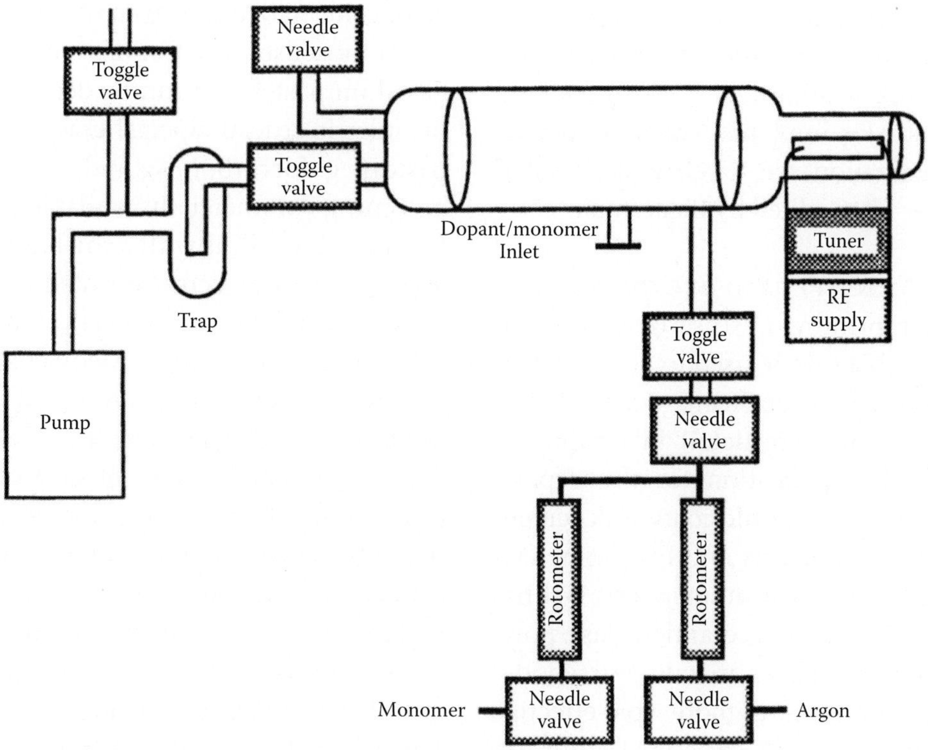


**FIGURE 8.9** Schematic of plasma deposition system.

In general, plasma-polymerized materials offer the following advantages:

- The plasma-polymerized films are uniform, pinhole-free, chemically resistant, and mechanically strong.
- A thin to thick film (200 Å to >100 μm) can be formed in a flawless manner at ambient temperature onto any substrate.
- The organic films deposited by the plasma process adhere strongly to the substrate.
- The choice of monomers is unlimited; almost any organic compound convertible into vapor can be polymerized.
- The plasma process (one-step process from a vapor source) is compatible with conventional CMOS technology.
- Some functional groups can be introduced onto the surface of the organic film by subsequent glow discharge treatment in a reactive gas atmosphere.
- Highly irregular surfaces can be coated and patterned by depositing a light-sensitive polymer by plasma polymerization (e.g., polymerized PMMA).

Applications from plasma polymer coatings include scratch-resistant coatings, corrosion protection, antibonding, antisoiling coatings, barrier layers, and biocompatible coats for implants. For more details on plasma polymerization, see, for example, Yasuda[7] and *Plasma Polymer Films* edited by Biederman.[8]

*Silk Screening or Screen Printing of Organics* Silk screening or screen printing (also serigraphy) is an additive technique that can be used to create sharp-edged organic coatings/images using a stencil. The method is reviewed in more detail in Chapter 9, where we discuss its application for the deposition of higher temperate ceramic coats. Silk screening is popular both in fine arts and in commercial printing, where it is used to print images on T-shirts (Figure 8.10), hats, CDs, DVDs, ceramics, glass, polyethylene, polypropylene, paper, metals, and wood.

**FIGURE 8.10** Silk screening of T-shirts.

Screen printing presents a more cost-effective means of depositing a wide variety of films on planar substrates than does integrated circuit (IC) technology, especially when fabricating devices at relatively low production volumes. The technique constitutes one of several thick film or hybrid technologies used for selective coating of flat surfaces (e.g., a ceramic substrate). The technology was originally developed for the production of miniature, robust, and, above all, cheap electronic circuits. The up-front investment in a thick film facility is low compared with that of IC manufacturing. For disposable chemical sensors, recent industrial experience indicates that screen printing thick films is a viable alternative to Si thin film technologies.

For biosensor applications, thick film technology based on pastes that can be deposited at room temperature is crucial. Special grades of polymer-based pastes (e.g., for carbon, Ag, and Ag/AgCl electrodes) are becoming commercially available for this purpose.[9] Polymer thick films with a thickness anywhere from 5–50 μm can be screen printed on cheap polymer substrates. In the commercial planar electrochemical glucose sensor from MediSense/Abbott sensor, the Precision QID, shown in Volume I, Chapter 7, all films are silk-screened onto a PVC substrate.

In the research phase of new chemical sensors, pastes must be developed from their pure components. Some examples follow. Pace et al.[10] screen printed a PVC/ionophore layer for a pH sensor. Belford et al.[11] investigated pH-sensitive glass mixtures and proceeded to screen print them on a multilayer metal conductor to make planar pH sensors. In Pace et al.,[12] a thick film, multilayered oxygen sensor with screen-printed chemical membranes of PVA and silicone rubber is detailed (see also Karagounis et al.[13]). A thick film glucose sensor is presented by Lewandowski et al.[14] and Lambrechts et al.[15] The latter authors developed an enzyme-based thick film glucose sensor with $RuO_2$ electrodes. All the above thick film sensors were fabricated on $Al_2O_3$ substrates. Weetall et al.[16] present an extremely low-cost, silk-screened immunosensor on cardboard. Cha et al.[17] compare the performance of thick film Au and Pt electrodes with conventional bulk electrodes.

When faced with adapting a biosensor membrane to an IC process versus adaptation to a thick film process, it now seems that, once the specialty inks are available, a thick film approach is easier, less expensive, and more accessible. For in vitro chemical sensors where small size is not as important, we expect more research to result in a switch from the overly ambitious IC approach to the more realistic thick film approaches as sketched above. In Volume III, Chapter 1, we compare the pros and cons of thin and thick film technologies.

Important other driving forces for the development of new low-temperature inks for silk screening come from the exciting market opportunity of organic light-emitting diodes (OLEDs), OLED flexible displays, polymer transistors, and electronic paper. There are two types of OLED displays: the ones based on small molecule OLEDs and the polymer-based ones. The small molecule OLEDs use organic emissive materials that do not contain long polymer chains. "Small" molecules often used include organometallic chelates, flourescent and phosphorescent dyes, and conjugated dendrimers. The production of small molecule devices and dispalys usually involves thin film deposition such as a separation in a vacuum. Small molecule OLEDs are more mature, but the polymer based OLEDs have the potential of being lower cost because of the applicability of screen printing, which enables high throughput of large area devices, such as large area displays. Lifetime improvement and encapsulation are progressing fast; thus, polymer OLEDs now have the potential of being very bright, extremely thin, low power, and low cost. There is the eventual possibility of producing OLED displays using roll to roll (R2R) manufacturing (see Volume III, Chapter 1), which would dramatically further reduce their manufacturing cost. The aim is also to produce organic thin film transistors (TFTs) using printing techniques, but there are a number of issues that still need to be resolved in this case. However, reduction in cost can already be achieved today by using hybrid-processing technologies, i.e., using low-cost printing and dispensing techniques wherever it is possible, whereas other processing steps remain "conventional," such as photolithography or vapor deposition processes.

In Figure 8.11 some typical thin film organic devices are shown. Using thin sheets of plastic—similar to overhead transparencies—as the base, one can print the multiple layers of OLEDs or transistors with silk screening one layer at a time. The squeegee pushes a liquid plastic mixture over a stainless-steel mesh, and after the solvent evaporates, a new plastic feature remains.

With electronic paper, the ink is a liquid that can be printed onto nearly any surface. Within the liquid are suspended tiny microcapsules, each containing white and black particles (Figure 8.12). The relative movement of negatively charged black and positively charged white particles inside their microcapsules is controlled by the direction of an applied voltage. In an electric field, the white particles move to one end of the microcapsule and make the surface of the electronic paper appear white at that spot. An opposite electric field pulls the white particles to the other end of the microcapsules, where they are obscured by the black particles, making the surface appear dark at that spot.[18] To form an electronic ink display, the ink is screen printed onto a sheet of plastic film that is laminated to a layer of circuitry. The circuitry forms a pattern of pixels that can then be controlled by a display driver. The microcapsules are suspended in a liquid "carrier medium," allowing them to be printed using existing screen-printing processes onto virtually any surface, including glass, plastic, fabric, and even paper. Ultimately electronic ink will permit most any surface to become a display, bringing information out of the confines of traditional devices and into the world around us.

Applications of electronic ink include e-book readers capable of displaying digital versions of books and e-paper magazines, electronic pricing labels in retail shops, timetables at bus stations, and

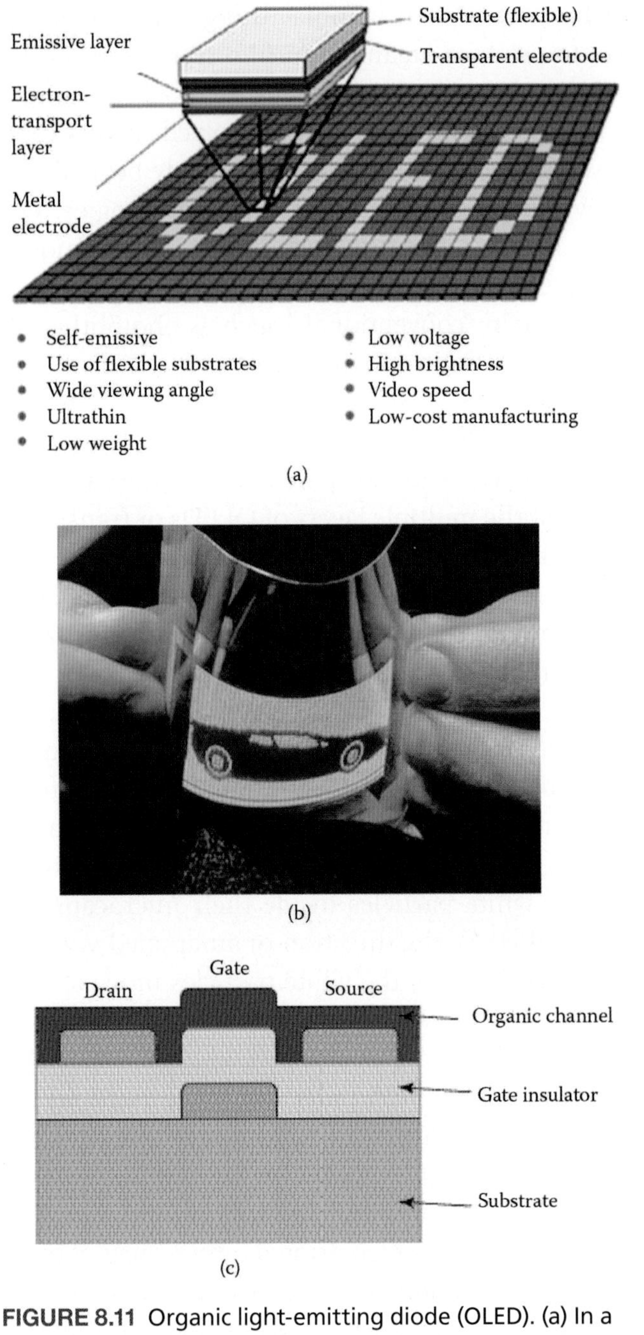


**FIGURE 8.11** Organic light-emitting diode (OLED). (a) In a typical organic light-emitting device, luminescent molecular excited states are generated in the electron transport and luminescent layer near its contact region with the hole transport layer. (b) OLED display; an organic passive matrix display on a substrate of polyethylene terephthalate, a lightweight plastic, will bend around a diameter of less than 1 cm. The 1.8-mm-thick, 5 × 10-cm monochrome display consists of 128 × 64 pixels, each measuring 400 × 500 μm, and is operated at conventional video brightness of 100 cd/m$^2$. It was fabricated by Universal Display Corp. (Ewing, NJ), with a moisture barrier built into the plastic that prevents degradation of the pixels (http://www.universaldisplay.com). (c) Organic thin-film transistors (OTFTs) are constructed of an organic or inorganic gate insulator and an organic semiconducting channel linking the source and drain.

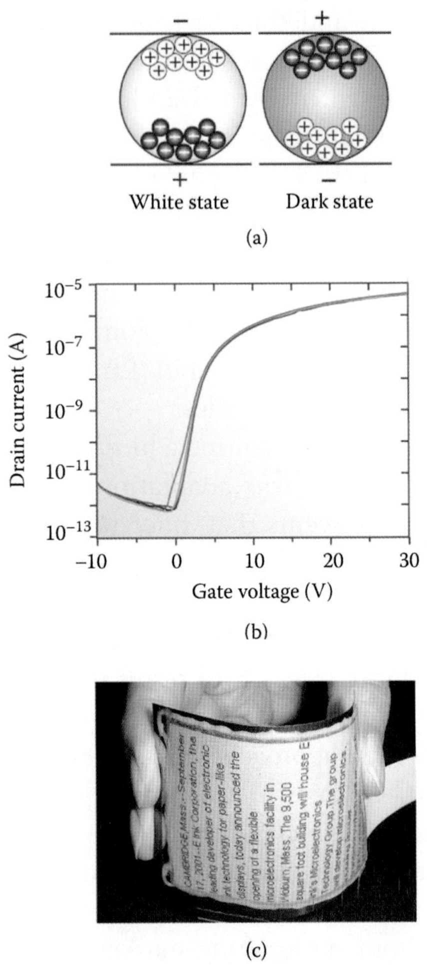

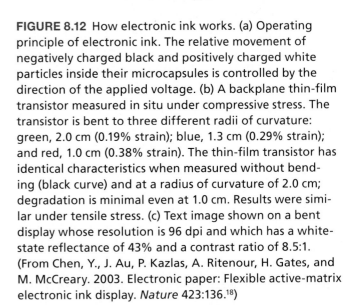

**FIGURE 8.12** How electronic ink works. (a) Operating principle of electronic ink. The relative movement of negatively charged black and positively charged white particles inside their microcapsules is controlled by the direction of the applied voltage. (b) A backplane thin-film transistor measured in situ under compressive stress. The transistor is bent to three different radii of curvature: green, 2.0 cm (0.19% strain); blue, 1.3 cm (0.29% strain); and red, 1.0 cm (0.38% strain). The thin-film transistor has identical characteristics when measured without bending (black curve) and at a radius of curvature of 2.0 cm; degradation is minimal even at 1.0 cm. Results were similar under tensile stress. (c) Text image shown on a bent display whose resolution is 96 dpi and which has a white-state reflectance of 43% and a contrast ratio of 8.5:1. (From Chen, Y., J. Au, P. Kazlas, A. Ritenour, H. Gates, and M. McCreary. 2003. Electronic paper: Flexible active-matrix electronic ink display. *Nature* 423:136.[18])

electronic billboards. In February 2006, the Flemish daily *De Tijd* distributed an electronic version of the paper to select subscribers in a limited marketing study. This was the first recorded application of electronic ink to newspaper publishing.

*Langmuir-Blodgett Deposition and Self-Assembled Monolayers* The Langmuir-Blodgett technique, invented by Irving Langmuir and Katharine Blodgett in 1935, allows the controlled deposition of monomolecular layers on a wide variety of substrates. This ultrathin film deposition technique is limited to materials that consist of amphilic long chain molecules with a hydrophobic molecule at one end and a hydrophilic molecule at the other. In the Langmuir-Blodgett process, a monolayer of film-forming molecules (stearic acid is a model molecule) on an aqueous surface is compressed into a compact floating film and transferred to a solid substrate by passing a substrate through the water surface at a constant speed and film surface tension (Figure 8.13). Thus, layered films can be built up in thickness (up to 100 layers) by consecutive dipping in the Langmuir trough. For example, phthalocyanine thin films sensitive to oxidizing gases such as $NO_2$ and biological materials sensitive to odors resulted via this method. Most of the difficulties with Langmuir-Blodgett films stem from the need to make the material pinhole-free and to overcome the problem of their lack of mechanical, chemical, and thermal stability.

In the early 1980s at the Bell Labs, David L. Allara (now at Pennsylvania State University) and Ralph G. Nuzzo (now at University of Illinois, Urbana-Champaign) discovered the self-assembly of disulfide and, soon thereafter, of alkanethiol monolayers (SAMs) on metal surfaces.[19,20] SAM films, especially on Au, spontaneously assemble into stable and highly organized molecular layers, bonding with the sulfur atoms onto the gold and resulting in a new surface with properties determined by the alkane head group. The preferred crystal face for alkanethiolate SAM preparation on gold substrates is the (111) surface, which is obtained either by using single-crystal substrates or by evaporation of thin

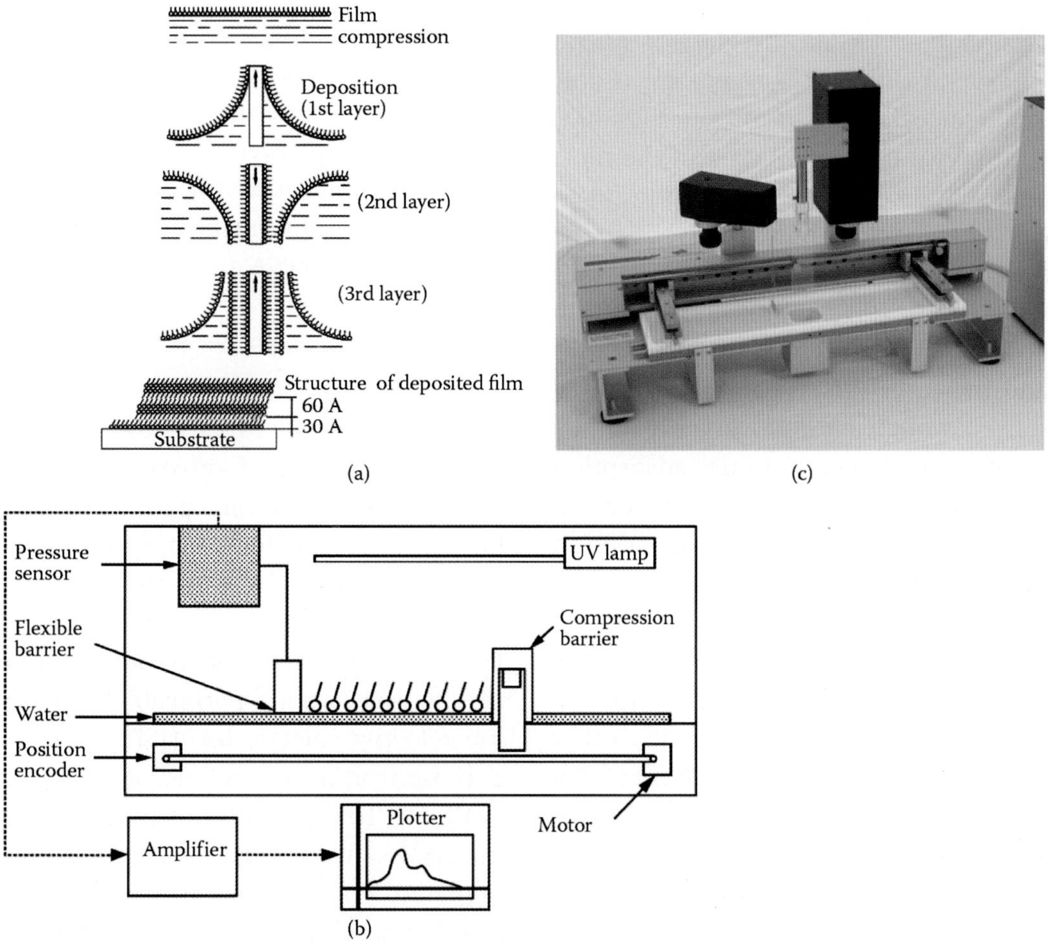


**FIGURE 8.13** Langmuir-Blodgett film deposition: sequence of a deposition (a), schematic of apparatus (b), and photo of apparatus (c).

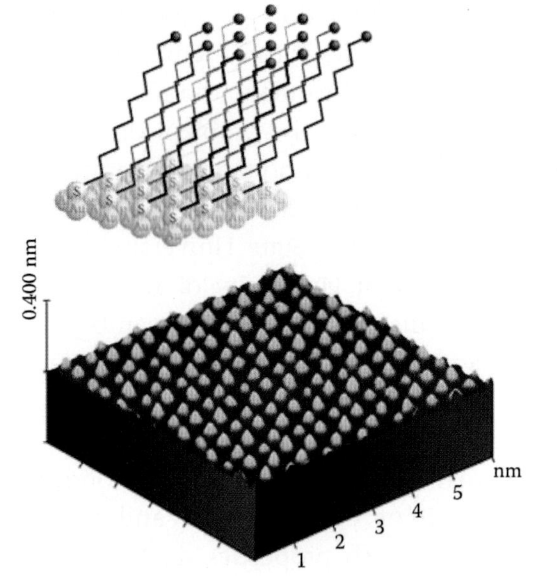


**FIGURE 8.14** Preparation of SAMs. The substrate, Au on Si, is immersed into an ethanol solution of the desired thiol. Initial adsorption is fast (seconds); then an organization phase follows, which should be allowed to continue for >15 h for best results. An STM of a fully assembled SAM is also shown.

Au films on flat supports, typically glass or silicon (Figure 8.14). Several different solvents are usable at the low thiol concentrations (typically 1–2 mM) that are used in preparation of SAMs, but the most commonly used one is ethanol. It is important to minimize the water content in the solvent if the SAMs are to be used in ultrahigh vacuum (UHV) experiments. Drying the solvent limits incorporation of water into the SAM structure, which reduces outgassing and increases the repeatability of UHV experiments. Even though a SAM forms very rapidly on a Au substrate, it is necessary to use adsorption times of 15 h or more to obtain well-ordered, defect-free SAMs. Multilayers do not form, and adsorption times of 2–3 days are optimal in obtaining the highest-quality monolayers.

The tail group that provides the functionality of the SAM can be widely varied. $CH_3$-terminated SAMs are commercially available, and other functional groups can be easily synthesized by any well-equipped chemical laboratory, providing almost infinite possibilities. In addition, chemical modification of the tail group is also possible after formation of the SAM, expanding the available range of functionalities even further. Examples of functionalities are $-CH_3$, $-OH$, $-(C{=}O)OCH_3$, $-O(C{=}O)CH_3$, $-O(C{=}O)CF_3$, $-O(C{=}O)C_6H_5$, $-COOH$, $-OSO_3H$, and so on. SAMs on gold are generally stable and can withstand strong acids and bases; they are not destroyed by solvents and can withstand physiological environments.[21]

This method of building monolayers forms an important alternative to the Langmuir-Blodgett method. The use of monolayer electron beam resists, as discussed in Chapter 2, affords nanometer-scale lithography resolution. The chemical sensor industry exploits monolayers of organics, for example, in immunosensors and to provide anchor points for subsequent organic molecules or membranes.

## Patterning of Organic Materials

### *Introduction*

In the previous sections, we summarized some of the different methods available to deposit organic thin layers. In what follows, we will encounter some of the same techniques and introduce some new ones as we look into the process of depositing small amounts of material in a well-defined, small spot on a substrate, i.e., the patterning of organic films. Patterning of organic thin layers into discrete array elements has recently spawned a wide variety of applications. These range from sensor arrays for electronic noses and tongues to DNA and protein arrays.

### *Patterning through Photolithography*

*Patterning of Hydrogels, Gas-Permeable Membranes and Ion-Selective Electrodes* Spinning, UV exposure, and development of resists are well-known, low-cost, mass-production patterning procedures in photolithography for IC fabrication. Naturally, this approach became one of the first methods to be applied to the patterning of other organic materials. However, photosensitized organic materials such as hydrogels, gas-permeable membranes, and ion-selective electrodes (ISEs), e.g., to fabricate a biosensor array, are usually not commercially available. To cope with this problem, photosensitive materials from high-purity materials must be prepared. For example, to pattern a hydrogel, e.g., a water-soluble polymer like polyvinyl alcohol (PVA), a photosensitizer such as $(NH_4)_2Cr_2O_7$ must be added.[15] After spin coating, the polymer film is

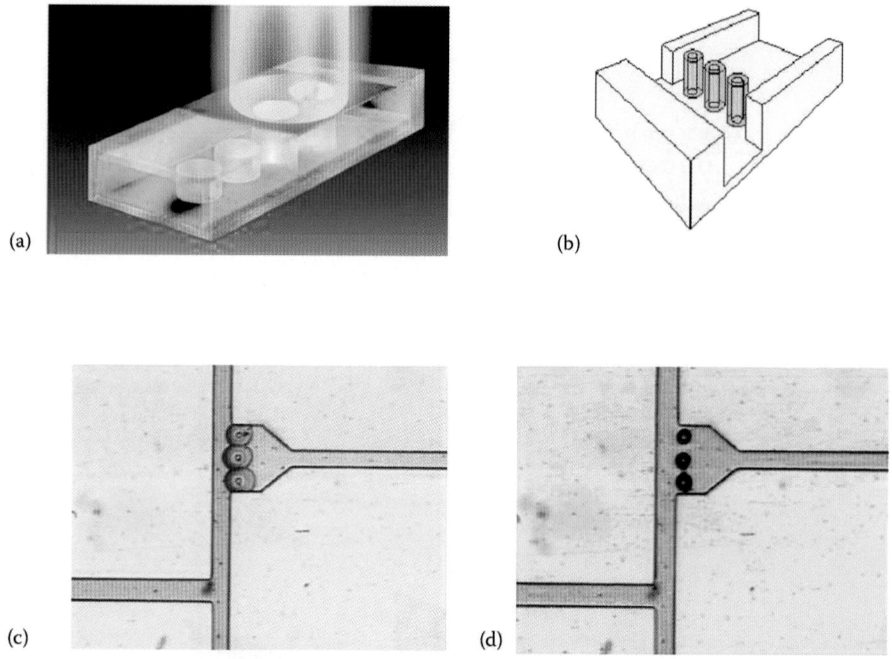


**FIGURE 8.15** Patterning hydrogels. (a) Schematic showing the in-channel photo-polymerziation technique. (b) Schematic of a 2D shut-off microvalve consisting of hydrogel "jackets" (50 μm thick) around three prefabricated SU-8 posts. (c) Micrograph of the hydrogel jackets blocking the regulated channel (pH 7) in their expanded state in a pH 12 solution (dyed). (d) Micrograph showing the contracted hydrogels allowing fluid to flow down the side branch. (Courtesy of Dr. D. Beebe, University of Madison, Wisconsin.[22])

photochemically cross-linked by UV light. In the exposed regions, an insoluble hydrogel materializes. The development or removal of unexposed regions is carried out in warm water. An example of patterning a hydrogel actuator material inside a fluidic network through the use of photolithography is described by Liu et al.[22] The actuator hydrogel, after in situ photopolymerization, responds to changes in local pH by changing its volume and thus provides a means for valving action in the fluidic network.[23] The in situ polymerization of these polymeric valves and their function in a fluidic network are demonstrated in Figure 8.15.[22]

An alternative approach to patterning organic materials without modifying the material through the addition of a photosensitizer is through lift-off. Lift-off to lithographically pattern hard to etch materials (e.g., Pt), or in general to pattern materials that are not photosensitive, was introduced in Chapter 1. At one time, a problem with lift-off for patterning organic materials was the thickness required for the patterning photoresist layer. As discussed in Chapter 1, a typical resist layer used to measure only about 1 μm thick, whereas the thickness needed for the organic layers involved in chemical sensors frequently can reach up to 50–100 μm, demanding resist layers that are thicker. The many new thick resist technologies (e.g., SU-8) now available make this approach feasible today. However, a remaining problem with lift-off is that it can be used only with materials resistant to the solvent necessary for the resist removal; therefore, this technology probably will be limited to specific cases.

Hydrogels, gas-permeable membranes, and ISEs are all relatively thick, from several micrometers to 100 μm. Next we consider patterning of much thinner organic layers, such as monolayers of DNA and proteins.

*Very Large-Scale Immobilized Polymer Patterning and Synthesis* Affymetrix's *very large-scale immobilized polymer synthesis* (VLSIPS) is illustrated in Figure 8.16. The technique patterns and synthesizes biopolymers at the same time.[24] In light-directed oligonucleotide synthesis for DNA arrays (GeneChips), a solid support (e.g., a 1.28 cm × 1.28 cm Si chip) is derivatized with a covalently linked aminosilanated layer and terminated with a photolabile protecting

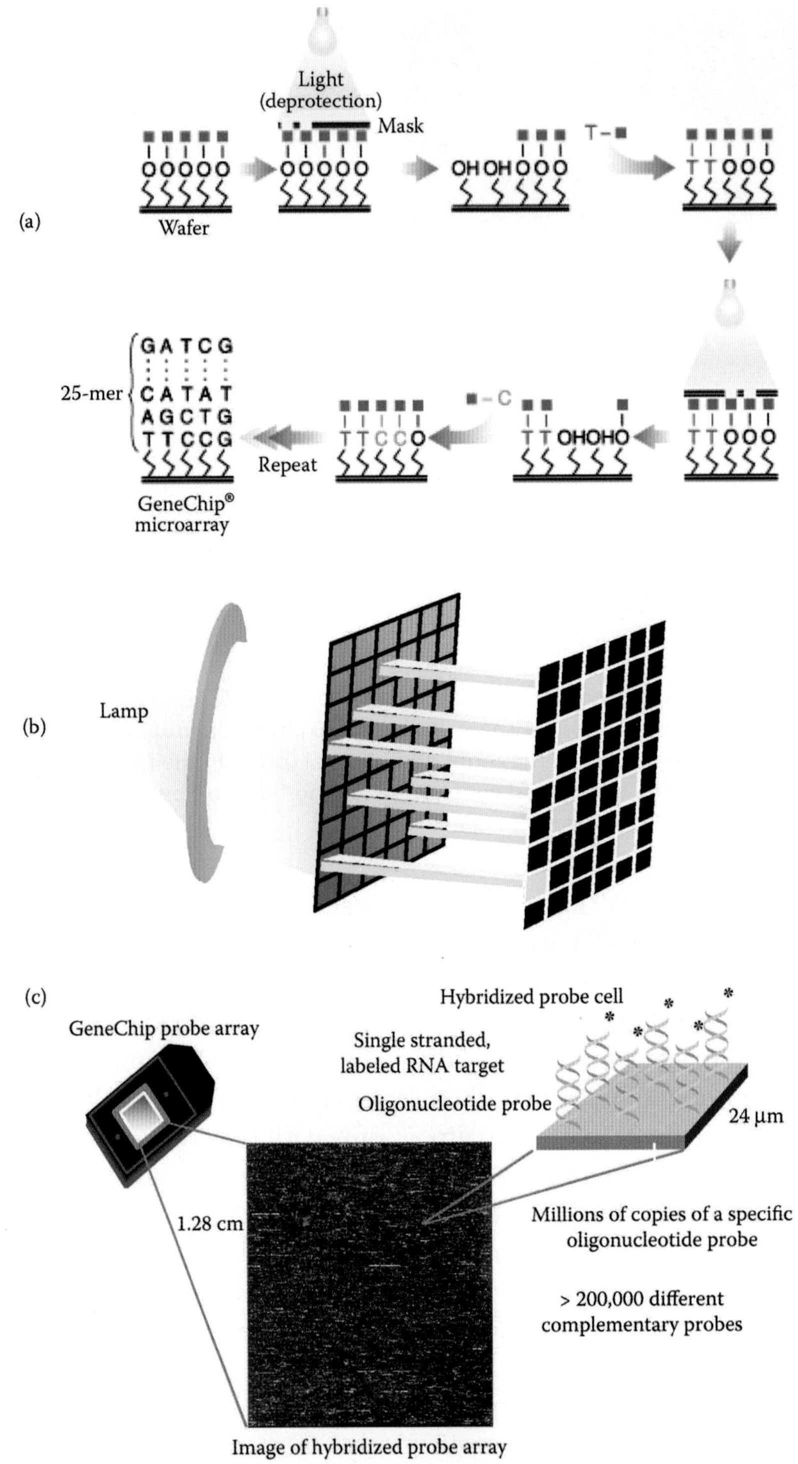


**FIGURE 8.16** Very large-scale immobilized polymer synthesis (VLSIPS). Light-directed oligonucleotide synthesis. (a) A solid support is derivatized with a covalent linker molecule terminated with a photolabile protecting group. Light is directed through a mask to deprotect and activate selected sites, and protected nucleotides couple to the activated sites. The process is repeated, activating different sets of sites and coupling different bases, allowing arbitrary DNA probes to be constructed at each site. (b) Schematic representation of the lamp, mask, and array. (From Fodor, P. A. S., J. L. Read, M. C. Pirrung, L. Sryer, A. T. Lu, and D. Solas. 1991. Light-directed, spatially addressable parallel chemical synthesis. *Science* 251:767–73.[24]) (c) Finished GeneChip probe array with a fluorescent image of a hybridized probe array and details of an individual hybridized probe cell (24 μm).

group. The on-chip combinatorial synthesis proceeds by photolithographically deprotecting all array elements that are to receive a common nucleotide, coupling that nucleotide by exposing the entire array to the appropriate phosphoramidite, and then after the oxidation and washing steps, repeating the procedure for the next nucleotide. The in situ synthesis of oligonucleotides occurs in parallel, resulting in consecutive addition of A, C, G, and T nucleotides to the appropriate gene sequences on the array. To make an array of N-mers requires 4N cycles of deprotection and coupling, one for each of the four bases, times N base positions. This photolithographic method requires 4N masks, which are specific to the pattern of sequences on the array but can be used to make many copies of the array. The masks add considerable expense, and the procedure is best suited for generating large numbers of identical arrays. By going through 32 iterations of the oligonucleotide synthesis, 65,536 oligos containing eight units can be fabricated in about 1 day.

In Figure 8.17 we show how an Affymetrix oligo microarray can be used to establish the amount of messenger RNA (mRNA) in a sample. mRNA is extracted from a cell and is converted to complementary DNA (cDNA) using reverse transcriptase. After transcription, the cRNA is biotin-labeled and fragmented. This single-stranded and labeled RNA target is then put in contact with the GeneChip with millions of copies of different complementary probes. After hybridization, the chip is washed and stained, and the array is scanned and quantitated. Each gene is represented on the GeneChip using 16–20 (preferably nonoverlapping) 25-mers, and each oligonucleotide has a single-base mismatch partner for internal control of hybridization specificity.

The combinatorial approach described here to fabricate DNA arrays can also be used for peptide synthesis to create an assortment of peptides of almost any length. The approach is a generic and powerful method to create large numbers of compounds in a very small area in a reasonable amount of time.

*Proteins Patterning with Lithography* A wide range of proteins has been patterned with features in the micrometer range by making a surface locally hydrophobic or hydrophilic through lithography (see Chapter 2, "Very Thin Resist Layers"). Such patterns may be observed by a number of techniques including fluorescence microscopy, atomic force microscopy (AFM), and the growth response of cultured cells.[25] There are also efforts to produce nanopatterned protein arrays by using proteins that self-assemble into molecular lattices, such as the bacterial S protein. Douglas et al.[26] used metal-decorated crystals of the S proteins of *Sulfolobus acidocaldarius* as a

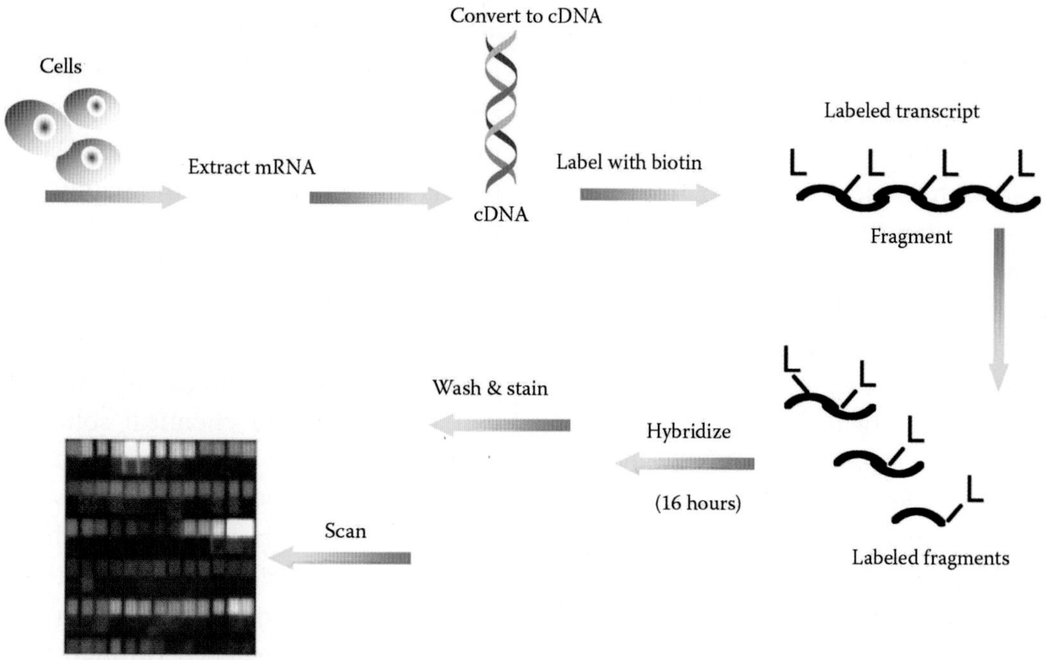


**FIGURE 8.17** How an oligo microarray works.

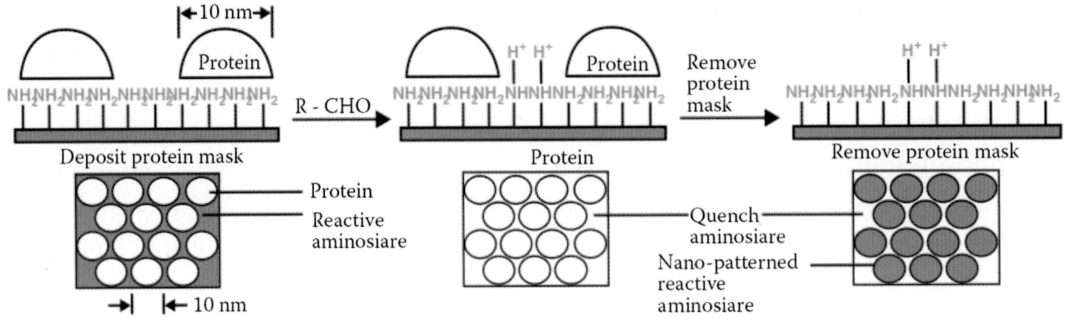


**FIGURE 8.18** Nanopatterning of surface chemistry using a self-assembled protein mask. (From Douglas, K., G. Deavaud, and N. A. Clark. 1992. Transfer of biologically derived nanometer scale patterns to smooth substrates. *Science* 257:642–44.[26])

protein mask to pattern hexagonal arrays of 10-nm-diameter holes onto graphite surfaces. Figure 8.18 demonstrates how the self-assembled protein lattice may be used for nanopatterned protein surfaces.

UV lithography, in some special cases, may also work directly for patterning thin protein layers as demonstrated in Example 8.1. In this example a protein grating was created by shining UV, in the presence of oxygen, through a grating photomask. Besides UV lithography, soft and hard x-rays, e-beam, and ion projection lithography can all be used for direct protein film structuring.

*Digital Mirror Array Patterning* With VLSIPS (see above), a DNA chip of 25-mers may require as many as 100 different masks, leading to a very high cost and a too-long fabrication time.

Singh-Gasson et al.[27] came up with an elegant maskless fabrication alternative for light-directed oligonucleotide microarrays by using a digital micromirror array. Specifically, this group used the digital micromirror device (DMD) from Texas Instruments (http://www.dlp.com/dlp/default.asp), the fabrication of which is described in Chapter 7. This specific DMD consists of a 600 × 800 array of 16-μm-wide micromirrors (the same type used in projection systems). The tiny mirrors are individually addressable and can be used to create any given pattern or image in a broad range of wavelengths. With a 1:1 imaging system, the DMD can be exploited to address 480,000 pixels on a 10 × 14-mm area. This maskless array synthesizer (MAS) or virtual mask array synthesizer is shown in Figure 8.19. An added benefit of this clever approach is that the active surface of the glass substrate can be mounted in a flow cell reaction chamber connected to a DNA synthesizer. Chemical coupling cycles follow light exposure, and these steps are repeated with different virtual masks to grow desired oligonucleotides in any desired pattern. It is obvious that the MAS can also be used for other array patterning/synthesis experiments.

Virtual masks were discussed in Chapter 1 under "Inexpensive Masks and Maskless Optical Projection Lithography for Research and Development." A maskless imaging method was also used by Bertsch et al.[28] at the Swiss Federal Institute of Technology (EPLF) in microstereolithography. This research group uses a computer-controlled liquid crystal display (LCD) as a dynamic pattern generator in microstereolithography (see further below).

### *Ink Jetting and Microspotting*

*Introduction* Drop delivery systems present an alternative mechanical approach to pattern organic materials on a planar substrate. Although resolution is lower than with lithography methods, these mechanical methods are fast and less expensive.

*Ink-Jet Printing* Drop delivery in bio-MEMS is based on the same principles as commercial ink-jet printing. The ink-jet nozzle is connected to a reservoir filled with the chemical solution and placed above a computer-controlled *x*-*y* stage. Depending on the ink expulsion method, even temperature-sensitive enzyme formulations can be delivered. The substrate to be coated is placed on the *x*-*y* stage, and under computer control, liquid drops (e.g., 50 μm in diameter) are expelled through the nozzle onto a well-defined place on the wafer (see Figure 8.20).

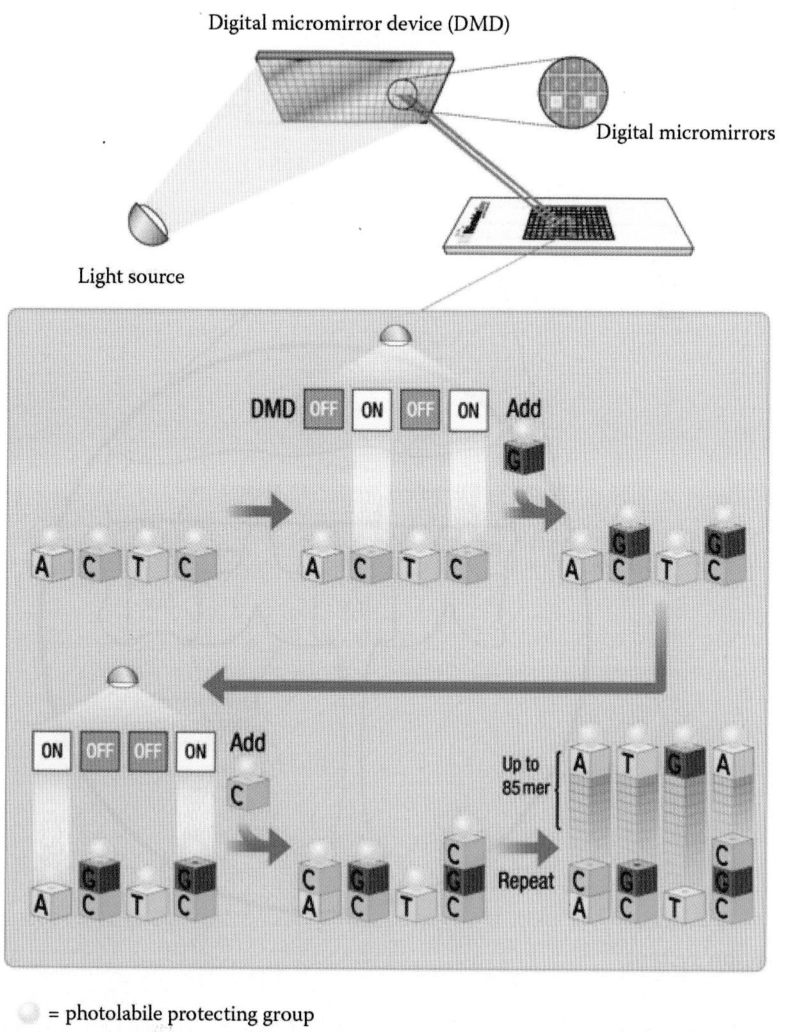


**FIGURE 8.19** Schematic of the maskless array synthesizer (MAS). A UV light source is used to illuminate the digital micromirror array. A set of reflective OFF/ON mirrors and a 1:1 imaging system with a numerical aperture (NA) of 0.08 form an image of the pattern on the digital micromirror array on the active surface of the glass substrate. The glass substrate is enclosed in a flow cell connected to a DNA synthesizer. (Courtesy of Dr. R. Green, NimbleGen Systems.)

Different nozzles may print different spots in parallel. Although these drop delivery systems are serial, they can be very fast, as evidenced by epoxy delivery stations in IC manufacturing lines.

The ink-jet mechanism in Figure 8.20 is thermal and is often called a *bubble jet*, with droplets as small as a few picoliters (2–4 pL).[29] Some ink-jet printers can vary the drop size. Five years ago a typical drop size was 30–50 pL; today drop sizes of 2–4 pL are state of the art. This is comparable to the size difference between a softball and a ping pong ball. Smaller drops produce smoother-looking prints because they make smaller dots on the paper. Hewlett Packard's (HP) thermal ink jet (TIJ) represents the state of the art in this arena. In the TIJ, tiny resistors are used to rapidly heat (1,000,000°C/s) a thin (0.1 μm) layer of liquid ink to about 340°C. A superheated vapor explosion vaporizes a tiny fraction of the ink to form an expanding bubble that ejects a drop of ink (and any trapped air) from the ink cartridge onto the paper or other substrate. It is noteworthy that the ink does not boil. The bubble collapse and break off draws fresh ink over the resistor. The bubble formation, expansion, break off, and refill (all very fast) are illustrated in Figure 8.21a. The HP TIJ has all the active power electronics and orifice addressing integrated on the same Si chip with the drop generator as shown in Figure 8.21b. The orifices may be spaced at 300/in. in a single column for a 300-dpi printhead, or for a 600-dpi printhead,

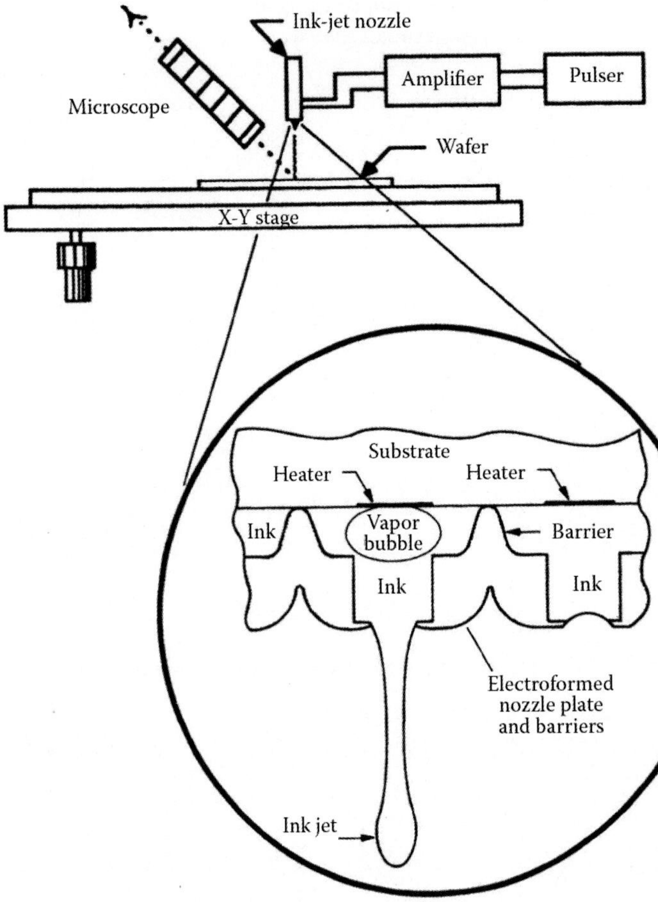


**FIGURE 8.20** Ink-jet deposition system.

two offset columns of 300 orifices may be used. It is even possible to put 600 TIJ orifices/in. in a single column. Current piezo ink jets (see next paragraph) only have 90 orifices/in. The TIJ technology is a remarkable piece of engineering. To put this in perspective, engineers trying to develop a commercial ion-sensitive field effect transistor (ISFET) have struggled for more than 30 years to integrate liquids with electronics with little success or market acceptance until very recently (see Volume I, Chapter 7 and Volume III, Chapter 10). Not only did HP succeed in integrating liquid with electronics, they were also able to find an effective method of repeatedly heating the liquid (ink). This way HP has also carved out a very lucrative business in disposable ink cartridges. Perhaps their efforts will also pay off in other fluidic areas, and it may even be worth reconsidering the ISFET using HP technology.

A piezoelectric ink-jet head is used for another type of ink-jet printing and consists of a small reservoir with an inlet port and a nozzle at the other end. One wall of the reservoir consists of a thin diaphragm with an attached piezoelectric crystal. When voltage is applied to the crystal, it contracts laterally, thus deflecting the diaphragm and ejecting a small drop of fluid from the nozzle. The reservoir then refills via capillary action through the inlet. One, and only one, drop is ejected for each voltage pulse applied to the crystal, thus allowing complete control over when a drop is ejected. Such devices are inexpensive and can deliver drops with volumes of tens of picoliters at rates of thousands of drops per second. In conjunction with a computer-controlled $x$-$y$ stepping stage to position the array with respect to the ink-jet nozzles, it is possible to deliver different reagents to different spots on the array. Arrays of 150,000 spots can be addressed in less than 1 min, with each spot receiving one drop of reagent. As pointed out above, thermal ink jets have a higher orifice density than piezoelectric printing. For more details on both types of ink-jet printing, visit http://www.hp.com/oeminkjet/tij/about.htm.

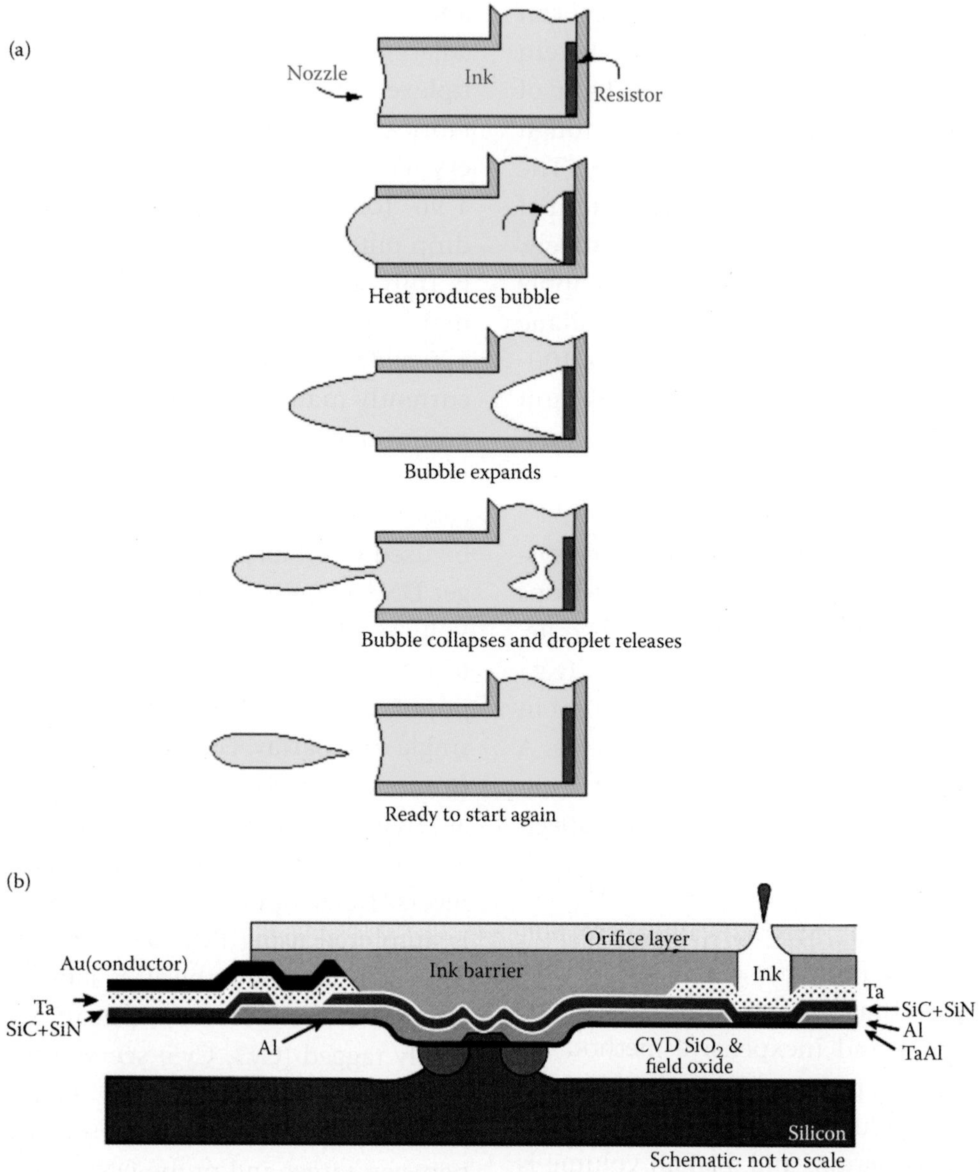


**FIGURE 8.21** HP's thermal ink-jet (TIJ) technology. (a) Bubble formation, expansion, collapse, and refill. (b) Integrated power electronics and orifice addressing.

Because a computer controls the pattern of reagents as an array is being made with ink jetting, it is as easy to make 10 arrays with different sequences as it is to make 10 identical arrays. This flexibility is perhaps the main advantage of the ink-jet approach. Achieving high density with the ink-jet approach requires one more trick. Two drops of liquid applied too closely together on a surface will tend to spread into each other and mix. For 40-pL drops, the minimal center-to-center spacing is about 600 μm. This limits the array density achievable with the ink-jet method. One way around this is to engineer patterns in the surface chemistry of the array to produce spots of a relatively hydrophilic character surrounded by hydrophobic barriers. At the small length scales involved (approximately 100 μm), surface tension is the dominant force on a drop of liquid, and a hydrophobic surface will effectively prevent a drop from spreading out beyond the confines of the hydrophilic surface. There are several ways to engineer such a surface. Modern techniques use fluorinated alkyl silanes, which covalently couple to glass and present an extremely hydrophobic surface. They can be patterned by masking the areas to remain hydrophilic, derivatizing the exposed surface with the appropriate silane and then removing the mask by dissolving it with various organic solvents. The mask itself can be formed by a photolithographic process wherein the array is covered with a thin, uniform layer of photoresist, which is then exposed to light through a photographic negative to define the array. The photoresist is developed, leaving behind a pattern of protective photoresist to act as a mask. Alternatively, a "Whitesides" rubber stamp can be used to apply either the protective mask or the hydrophobic silane itself. Any of these methods will easily produce 100-μm-diameter hydrophilic wells separated by 40-μm hydrophobic barriers, or 5,000 array sites/cm$^2$. Ink-jet technology has been used to prepare microarrays of single cDNA at a density of 10,000 spots/cm$^2$.

On a somewhat larger scale, epoxy-delivery systems used in the IC industry similarly deliver drops in a serial fashion on specific spots on a substrate. A typical commercial drop-dispensing system (e.g., the IVEK Digispense 2000) delivers 0.20–0.50 μl in a drop and has a cycle time per dispense of 1 s. A vision system verifies substrate position and accurate dispense location to within ±25 μm of a specified location.

*Mechanical Microspotting* Mechanically microspotting, as developed by Brown et al. (Stanford University), is simpler than ink-jet printing. In this approach, a popular and inexpensive method for making DNA arrays, a prepared DNA sample (e.g., cDNA or a PCR product) is loaded into a spotting pin by capillary action, and a small volume is transferred to a solid surface by physical contact between the pin and the solid substrate (see Figure 8.22). After a first spotting cycle, the pin is washed, and a second sample is loaded and deposited to an adjacent address. Robotic control systems and multiplexed printheads allow automated microarray fabrication. This very flexible mechanical drop delivery technique takes a small amount of liquid from a 96- (or 384-) well plate and places a tiny (1 nL) drop onto a microscope slide. Because the machine is fully automated, it requires little extra work to make additional slides. One can typically make 120 slides at a time. The microspotted microarrays currently manufactured contain as much as 10,000 groups of cDNA in an area of 3.6 cm$^2$.

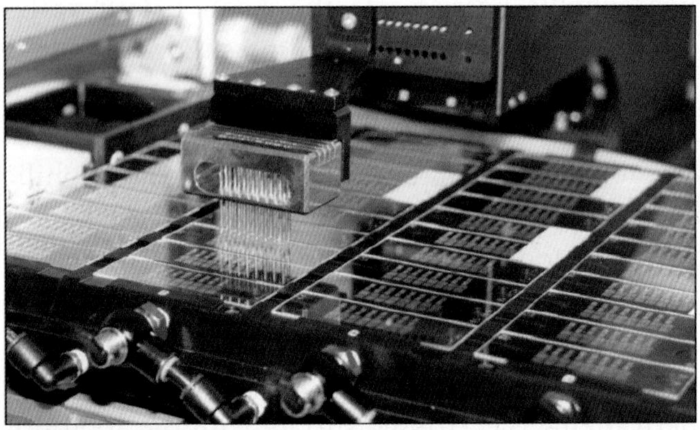

**FIGURE 8.22** Spotted arrays of cDNA on 50 glass slides.

As sketched in Figure 8.23, a cDNA array as fabricated with the equipment shown in Figure 8.22 may be used to establish the relative amount of messenger DNA expression in a biological sample (see also Volume III, Chapter 2). For example, we might want to compare the gene expression levels for two cell populations (e.g., from tumor and normal cells) on a single microarray. For that application, both sample (tumor cell) and control DNA (normal cell) are used as targets. One first extracts mRNA from both sample and control and converts the mRNA to cDNA using reverse transcriptase. In a subsequent step, the cDNA is amplified using PCR. DNA probe molecules are then deposited onto the glass slide substrate. Finally, target DNA from tumor and healthy cells are fluorescently tagged (Cy3, Cy5; see Volume I, Chapter 7), and an equal amount of each is added to the cDNA probe array. Depending whether there is a match between target and probe DNA, hybridization does or does not take place. If no hybridization occurs, no color results. If the sample (red) DNA hybridizes and the control does not, a red color results, and if only control (green) DNA hybridizes, a green color results. If both hybridize equally, a yellow color results. The color intensity measured in a fluorescence reader reflects the abundance of the mRNA in the sample. The assumption in this experiment is that cellular mRNA levels directly reflect gene expression.

In Table 8.1 we compare the procedure for measuring mRNA as depicted in Figure 8.23, where we use a spotted cDNA array, with the one in Figure 8.17, where we use a lithographically synthesized oligo array.

In Figure 8.24, we compare three arraying methods: lithography, microspotting, and ink jetting.

**TABLE 8.1** Spotted cDNA Array versus Synthesized Oligo Array

| | Microarray: Spotted | DNA Chips: Synthesized |
|---|---|---|
| Probes | Probes (0.6–2.4 kb) are PCR-amplified full-length cDNA sequences | Probes are 20–25 deoxyoligonucleotides |
| Manufacture | Spotted by "robo-arms" on nonporous, solid support | Synthesized on glass by solid-phase DNA synthesis coupled with selectively masked light protection and deprotection (photolithography) |
| Number of spots | About 10,000 "spots" on a microscope glass slide. Maximum 24,000 features per array | Commercial GeneChips have about 300,000 probes on $1.28 \times 1.28$ cm surface; experimental versions exceed 1,000,000 probes per array |
| Quality of spots | Variability in spot quality from slide to slide | High quality with little variability between slides |
| Gene representation | Each gene represented by its purified PCR product | Each gene represented multiply using 16–20 (preferably nonoverlapping) 25-mers |
| Internal control | Simultaneous analysis of two samples (treated vs. untreated cells) provides internal control | Each oligonucleotide has single-base mismatch partner for internal control of hybridization specificity |
| Gene expression amount | Relative | Absolute |
| Cost/flexibility | More flexible and cheaper, homemade | More expensive yet less flexible |

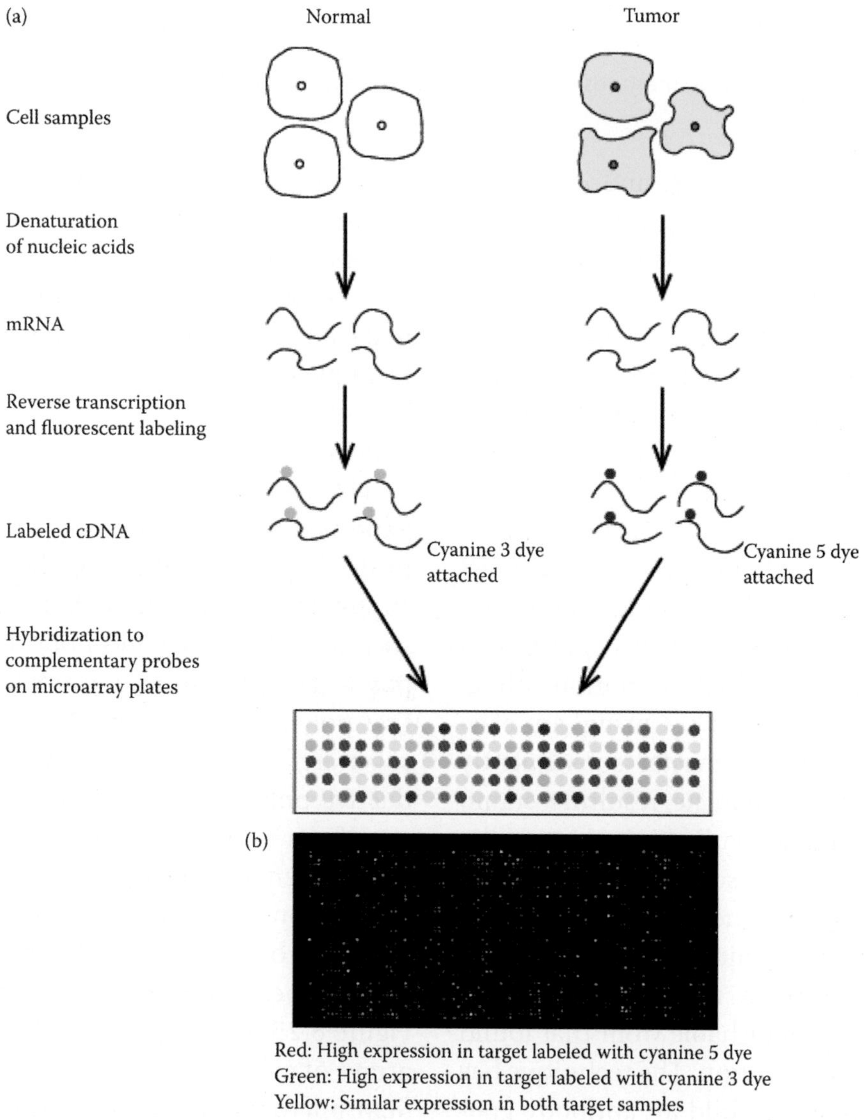


**FIGURE 8.23** Example of spotted cDNA microarray analysis: experimental procedure (a) and microarray image (b).

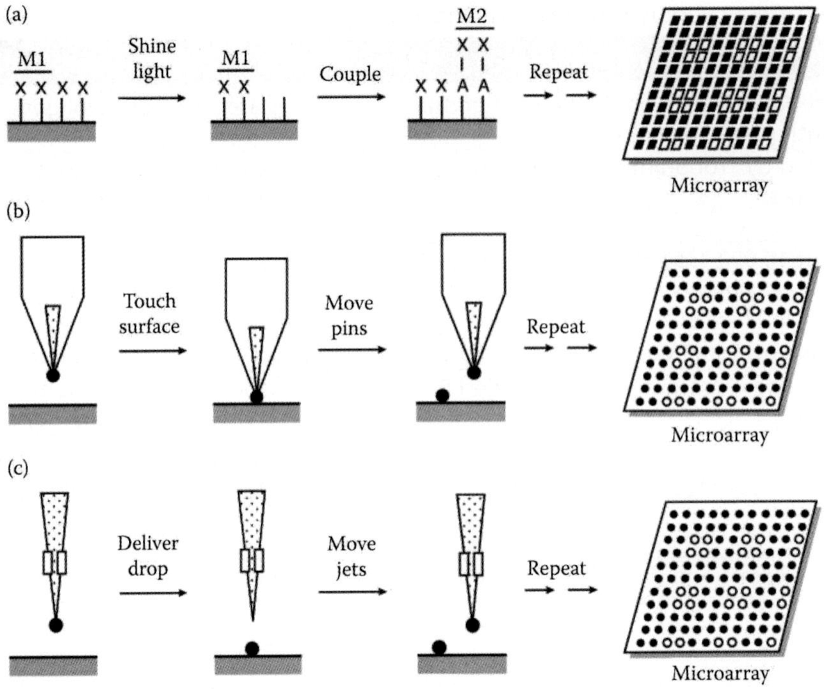


**FIGURE 8.24** The three main arraying methods: lithography (a), microspotting (b), and ink-jetting (c).

One disadvantage of microspotting and ink jetting compared with VLSIPS is that each sample must be synthesized, purified, and stored before microarray fabrication.

Some good web sites for further information on arraying technology are Stanford University's *Brown's Lab Guide to Microarraying* at http://cmgm.stanford.edu/pbrown/mguide and BASF's Leming Shi's http://www.gene-chips.com.

*Microcontact Printing* Micro- or nanocontact printing with a polydimethylsiloxane stamp was reviewed in detail in Chapter 2. In this approach, a Si master is created using a high-resolution lithography technique (e.g., e-beam). Curing an elastomeric prepolymer on this master subsequently forms an elastomeric stamp. The resulting stamp can be inked, for example, with thiols that can be subsequently transferred to a solid support, forming highly localized structured monolayers (see Figure 2.53). Scanning tunneling microscope (STM) studies of the transferred SAMs reveal an achievable monolayer order indistinguishable from that found for SAMs prepared from solution. The technique has been used, for example, to build an antibody grating on a Si wafer by inking the rubber stamp with an antibody solution.[30] The antibody grating alone produces insignificant diffraction, but on immunocapture of cells, the optical phase change produces diffraction. This is an alternative method to make an immunograting than the one described in Example 8.1.

*Conductive Polymer Patterning* Polymers that can be electropolymerized may also be deposited and patterned by first patterning a thin film metal electrode in a desired pattern on the substrate. In the late 1980s, this author patterned conductive polymers such as polypyrrole and polyaniline with lithography techniques. In one example, arrays of conductive polymer posts were formed on a conductive substrate. Increased ion access to individual conductive polymer posts as compared with access to a uniform film of the same polymer leads to higher electrochemical reversibility, which might find applications in faster polymer battery electrodes, electrochromic devices, enzyme-based biosensors, and microelectronic or molecular electronic devices. Figure 8.25 depicts the procedure for fabricating 3D arrays of electronically conductive polymers and an SEM micrograph of the resulting patterned conductive polymer posts.[31,32]

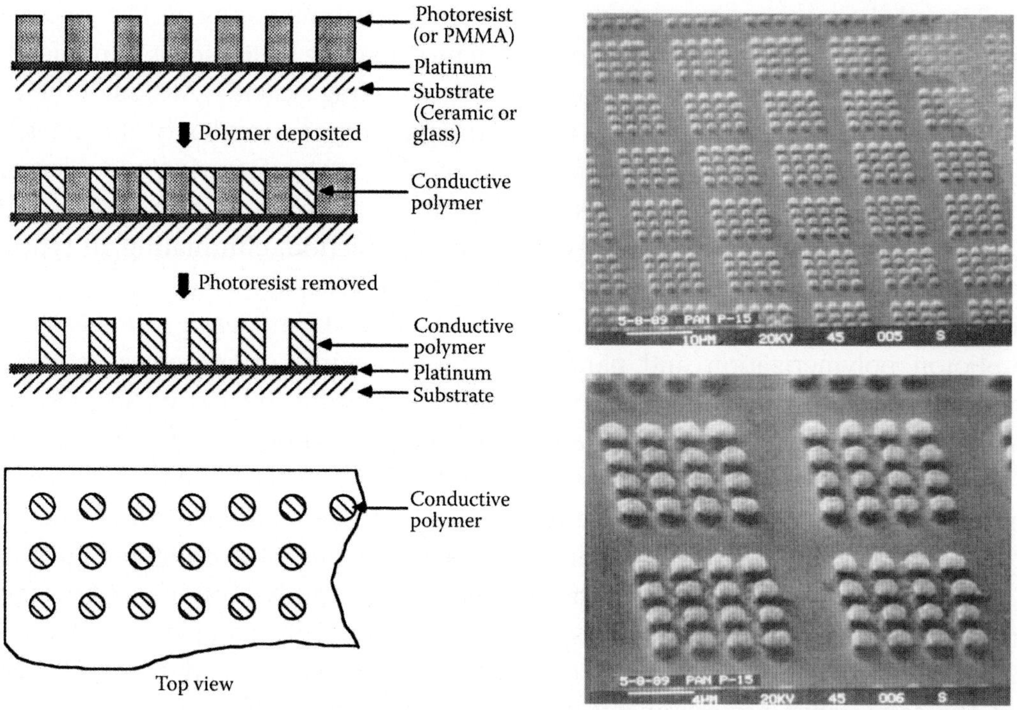


**FIGURE 8.25** Micropatterning of conductive polymers. (a) Process sequence. (b) SEM micrograph of conductive polymer submicrometer electrode array (polyaniline doped with tosylate).

Table 8.2 summarizes some of the most popular membrane deposition and patterning techniques.[15] At first glance, photolithography seems very promising, but the chemistry is complex and few results have been obtained to date, the most important exceptions being the i-STAT Abbott Point of Care blood electrolyte and blood gas sensors (http://www.abbottpointofcare.com/) and the Affymetrix DNA arrays (http://www.affymetrix.com). More array developers today are opting for microspotting, ink-jet printing, and screen printing as the safest and least expensive approach.

## Photochemical Forming

### Introduction

In photochemical forming the energy at the work piece is provided by photons. A prominent example is photolithography, where negative-tone photoresists are cross-linked by photons of the right energy (Chapters 1 and 2). Traditional photolithography techniques result in projected shapes. A photoresist structure can take on any shape in the *x*-*y* plane and possess some limited height (in the *z*-direction), but

**Table 8.2** Deposition and Patterning Techniques for Planar Chemical Membranes

| | Typical Use | Thickness Range (μm) | Cost | Uniformity | Reproducibility | Patterning |
|---|---|---|---|---|---|---|
| Dip coating | Wire ISEs | 0.1–50 | Low | Poor | Poor | No |
| Casting | Planar ISEs | 0.1–>100 | Low | Moderate | Moderate | No |
| Photolithography | Planar sensors (PVA,PHEMA) | 1–10 μm | Moderate | Good | Good | Yes |
| Lift-off | Immunosensors | 0.1–3 μm | Moderate | Moderate | Good | Yes |
| Plasma etching | PVC, Teflon | 1–10 μm | High | Good | Good | Yes |
| Ink-jet printing | Universal | 1–5 μm | Moderate | Poor | Moderate | Yes |
| Screen printing | Universal | 5–50 μm | Low | Moderate | Moderate | Yes |

*Source:* Lambrechts, M., and W. Sansen. 1992. *Biosensors: microelectrical devices.* Philadelphia: Institute of Physics Publishing. With permission.[15]

they remain projected shapes that are not truly 3D because the method does not lend much control over the resist sidewalls along the *z*-axis; such structures are sometimes referred to as 2 1/2D. In miniaturization science, truly 3D lithography is the subject of increased research. Here we review methods of photochemical forming of truly micro- and even nano-3D shapes, including stereolithography or microphotoforming and its newest embodiments such as two-photon polymerization and parallel stereolithography. We start this section with a brief summary of rapid prototyping techniques. In rapid prototyping, additive layering or 3D polymerization techniques have long played an important role for quickly making a nonfunctional mock-up of a real component. Recently improved versions of those methods have also became promising candidates for micro- and nanomanufacturing. It is also important to notice that the term rapid prototyping is less and less the right word to use. Indeed quite often these additive or 3D polymerization manufacturing processes are leading more and more to final functional products rather than nonfunctional prototypes. We classify free-photoforming additive as either chemical or thermal, depending on the primacy of the chemical reaction or the heat application. Thus, low-temperature polymer photoforming is covered in this chapter, whereas the higher-temperature laser-assisted chemical vapor deposition process is covered in Chapter 9.

## Rapid Prototyping

The past 20 years witnessed the emergence of a set of new manufacturing techniques with parts made on a layer-by-layer basis or by 3D polymerization. With these technologies, manufacturing time for even the most complex parts is measured in hours instead of days or weeks. Hence these technologies are called *rapid prototyping* (RP). RP, based on layered manufacturing (LM), and direct 3D polymerization in two-photon lithography where no layering is needed form a set of manufacturing techniques by which a solid physical model of a structure is made directly from 3D computer-aided design (CAD) model data, without any special tooling. The CAD data may be generated from 3D CAD modelers, computer tomography (CT),* and magnetic resonance imaging (MRI) scans or model data created by 3D digitizing systems. Although there is no limit to the prototype's complexity, its size is limited by the size of the manufacturing bed. The same data used for the rapid prototype creation can be used to go into production, eliminating all types of sources of human errors.

Besides 3D modeling, RP needs specific fluids, powders, wires, or laminates, as well as a machine tool that builds the desired 3D prototypes using different physical principles such as lasers, laminators, and 3D printers. This type of fabrication is referred to as additive manufacturing versus the subtractive manufacturing of milling and turning. Some very new RP methods combine additive and subtractive techniques. For example, the laser sintering/milling hybrid machine, the LUMEX 25C, developed by Matsuura Machinery Corporation, successfully combines freeform manufacturing and high-speed milling (see the photo in the heading of Chapter 9 on thermal energy-based forming).

The first commercial RP method appeared on the market in 1987 from 3D Systems (http://www.3dsystems.com) and was based on stereolithography (SL) where liquid polymers are solidified when exposed to UV light. Stereolithography remains the best known RP system and is widely used in automotive and aerospace industries to fabricate parts from basic design for "show and tell" before the parts are produced. As listed in Table 8.3, there are many different RP methods available that offer varying degree of accuracy, speed, product size (see Table 8.3, inset), as well as a large choice of materials. Two-photon lithography, the newest RP technique that does not rely on layering the part under construction, is dealt with separately below and is not listed in Table 8.3. In all RP methods listed in Table 8.3, a designer, using CAD software, exports a model of the intended product to a standardized portable file format for 3D objects, such as the stereolithography

* Computer tomography (CT) is a powerful nondestructive evaluation (NDE) technique for producing 2D and 3D cross-sectional images of an object from flat x-ray images. Characteristics of the internal structure of an object (such as dimensions, shape, internal defects, and density) are readily available from CT images.

TABLE 8.3 Rapid Prototyping Techniques

- SL: Stereolithography (see below for details).
- SLS: Selective laser sintering laser beam melts the layer of metal powder to create solid objects (see Chapter 9).
- 3DP: In 3D printing layers of ceramic (plaster) powder are glued to form objects. 3DP uses ink-jet printer technology (e.g., the Z402c from the Z Corporation) to print fine patterns of glue onto a smooth bed of plaster powder (so smooth it looks like paper). The bed is lowered, and a roller spreads a very fine layer (0.004″) over it to print a new layer.
- LOM: Sheets of material (paper, plastic), either precut or on a roll, are glued (laminated) together and laser cut in the laminated object manufacture process.
- FDM: In fused deposition modeling each cross-section is produced by melting a plastic filament (ABS) that solidifies on cooling. The FDM (e.g., the Stratasys FDM 1650) builds layers by extruding a hot, viscous strand of ABS plastic through a very fine nozzle,like squeezing toothpaste out of a tube. The fine strands (0.012″) stick together as they cool.

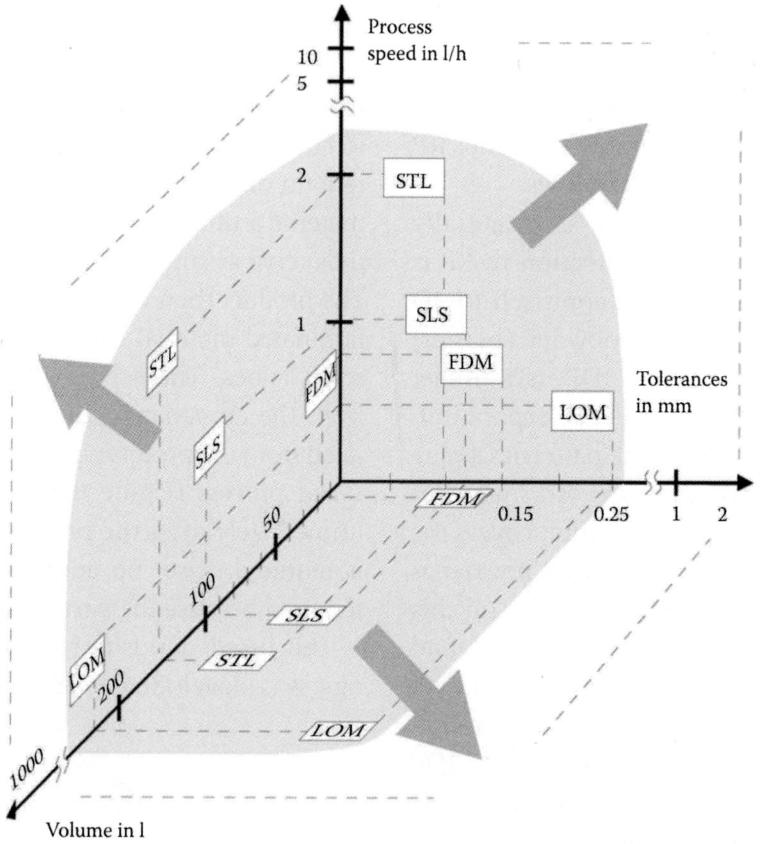

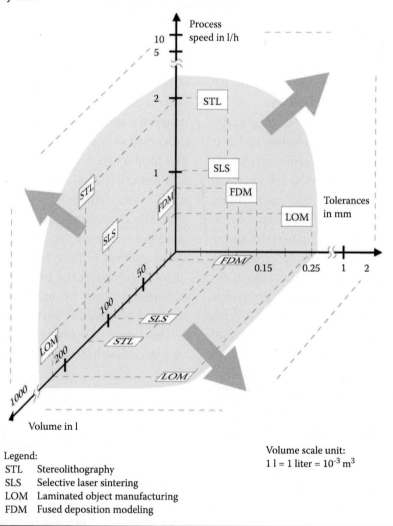

Legend:
STL Stereolithography
SLS Selective laser sintering
LOM Laminated object manufacturing
FDM Fused deposition modeling

Volume scale unit:
1 l = 1 liter = $10^{-3}$ $m^3$

software (http://stl-software.qarchive.org) that interfaces with the RP machines, analyzes the surface vectors (boundary) of the model, and "slices" the part into many thin layers. Stereolithography files describe only the surface geometry of a 3D object without any representation of color, texture, or other common CAD model attributes. RP machines make parts by producing one of these layers at a time and eventually building up a complex, 3D object. As mentioned earlier, in two-photon lithography there is no need for this layering step.

Stereolithography (SL) is a photoforming-based RP process, and its application to MEMS and NEMS, as well as its newest embodiments such as two-photon polymerization and parallel stereolithography, is detailed below. Selective laser sintering (SLS)

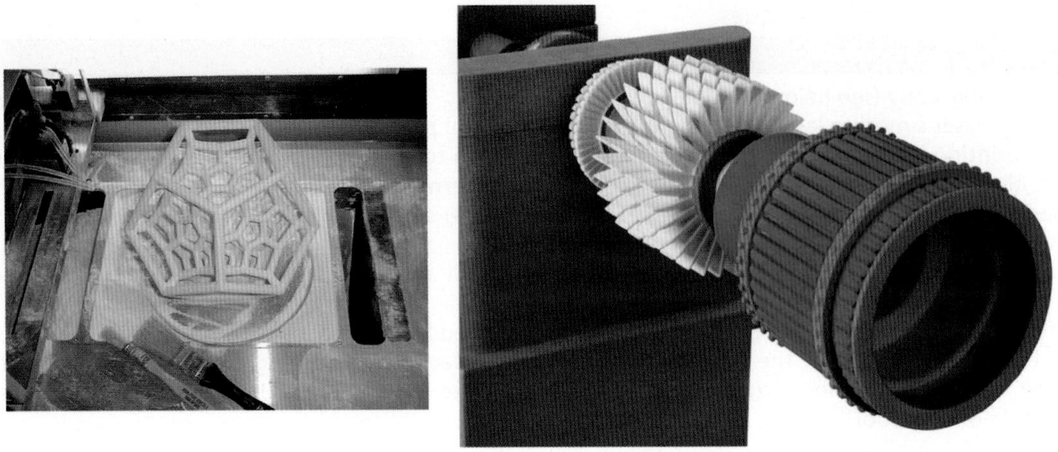

**FIGURE 8.26** A 3D-printer by Z Corporation that creates models from plaster powder (http://www.zcorp.com).

is a thermal additive technique expanded on in Chapter 9. A brief explanation of how the other RP techniques listed in Table 8.3 work follows.

Ink-jet printing, discussed earlier, is essentially a 2D method; an important new direction in RP is the concept of 3D printing. In one approach to 3D printing (3DP), layers of ceramic powder (plaster) are glued/bound to form objects. 3DP uses ink-jet printer technology (e.g., from the Z Corporation; http://www.zcorp.com) to print fine patterns of glue onto a smooth bed of plaster powder (so smooth it looks like paper). In a first step the printer covers the bed with a 1/8th inch (3.18 mm) layer of a powder as support so that the parts when finished rest on this base for easy removal. The bed is then lowered, and a roller spreads a next very fine layer (0.004″) over it to print a first layer (Figure 8.26). Z Corporation first introduced high-resolution, 24-color, 3DP (HD3DP™) in 2005 (600 dpi). The printer accurately and precisely deposits colored binder in the desired areas. The binder solidifies the powder in that cross-section of the model, leaving the rest of the powder dry for recycling. The 3D printers control the printhead movement while positioning the head extremely close to the powder, reducing inaccuracies related to fanning of the binder spray.

Sheets of material (e.g., paper, plastic, ceramic, composite), either precut or on a roll, are glued (laminated) together and laser cut in the laminated object manufacture (LOM) process. The laminated part is constructed on a platform with vertical incremental movement. Above the platform, a heated roller heats and compresses each new sheet to the top of the construct. A laser scans the upper surface of the laminated stack to cut the very top layer. Scrap pieces of the laminate remain on the platform as the part is built. The scrap material is diced by the laser into cross-hatched squares that serve as support for the part under construction. The product then comes out as a rectangular block of laminated material containing the prototype and the scrap cubes. The scrap, support material—resulting from the cross-hatch cut by the laser—is easily separated from the prototype part. LOM is a relatively high-speed process (Figure 8.27), and parts can be used immediately after the process because no postcuring is required. Also, no additional support structure is required because the part is supported on its own.

The fused deposition modeling (FDM) technology was developed by S. Scott Crump in the late

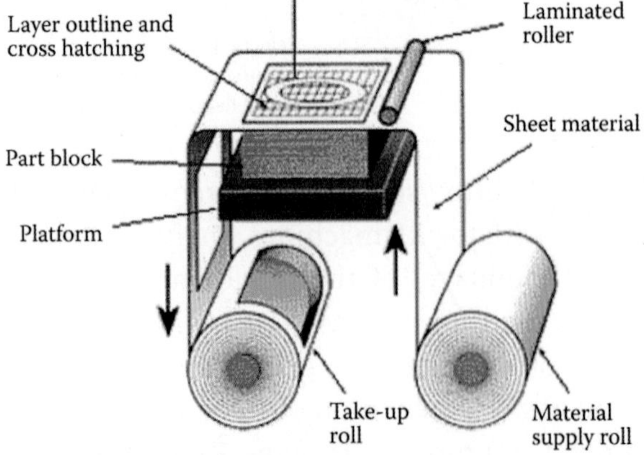

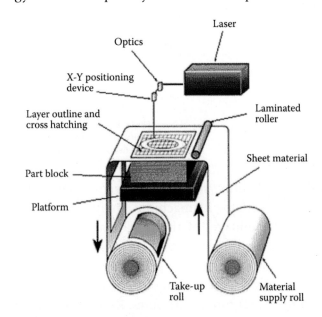


**FIGURE 8.27** LOM process.

1980s and was commercialized in 1990. The FDM technology is marketed commercially by Stratasys Inc. (http://www.stratasys.com/selectapart.html). Stratasys-patented technology builds solid models by melting a continuously supplied plastic wire (see Figure 8.28). The supplied wire typically is acrylonitrile butadiene styrene (ABS, a thermoplastic) with a 0.012-in. diameter and is extruded as a hot, viscous strand through a fine nozzle (like squeezing toothpaste out of a tube). A computer controls the nozzle movement along the *x*- and *y*-axes, and each cross-section of the prototype is produced by melting a plastic filament (ABS) that solidifies on cooling. After one layer is done, the stage is moved downward, and another layer of deposition starts. The newest FDM models feature two nozzles; one carries the construction material and the other a support wax that can easily be removed afterward, allowing construction of more complex parts. Marketed under the name WaterWorks by Stratasys, this soluble support material is dissolved in a heated sodium hydroxide solution with the assistance of ultrasonic agitation.

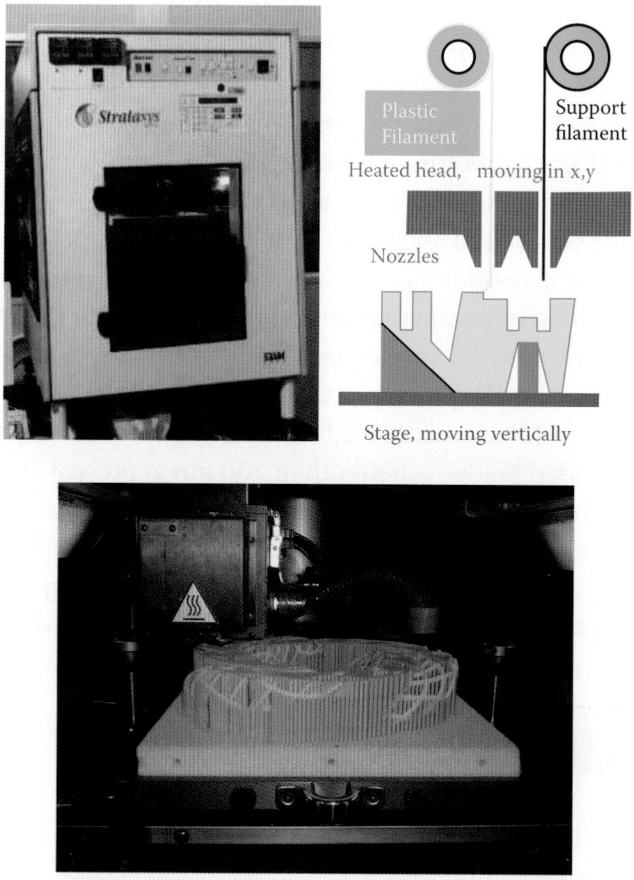


**FIGURE 8.28** FDM machine; models are made from molten ABS plastic (http://www.stratasys.com).

The FDM method is environmentally friendly but has a somewhat restricted accuracy because of the nozzle size, which needs to accommodate a viscous material. Also, few types of materials can be incorporated in FDM.

## Stereolithography/ Microphotoforming Process

### *Introduction*

The application of rapid prototyping (RP) techniques to MEMS and NEMS requires higher accuracy than what is normally achievable with commercial RP equipment. From the RP techniques listed in Table 8.3, laminated object manufacturing (LOM), fused deposition modeling (FDM), and selective laser sintering (SLN) all must be excluded as microfabrication candidates on that basis. Only stereolithography has the potential to achieve the fabrication tolerances required to qualify as a MEMS or NEMS tool. In this section we explore the basic principles underlying stereolithography and detail the latest enhancements that have made it an attractive high-resolution micro- and nanofabrication method.

### *Basic Principles Underlying Stereolithography and Microstereolithography*

In stereolithography, light exposure solidifies a special liquid resin into a desired 3D shape. Besides producing industrial 3D mockups (see above under RP), micromachinists are exploring this same technology for the production of functional micro- and nanomachines. Some basic concepts illustrating stereolithography are shown in Figure 8.29. A liquid resin is kept either in the free surface mode (see Figure 8.29b) or in the fixed surface mode (see Figure 8.29c). The latter has a resin container with a transparent window plate for exposure. The solidification happens at the stable window/resin interface. An elevator is pulled up over the thickness of one additional layer above the window for each new exposure (Figure 8.30). In the case of free surface, solidification occurs at the resin/air interface, and more care needs to be taken to avoid waves or a slant of the liquid surface. Microstereolithography, derived from conventional stereolithography, was

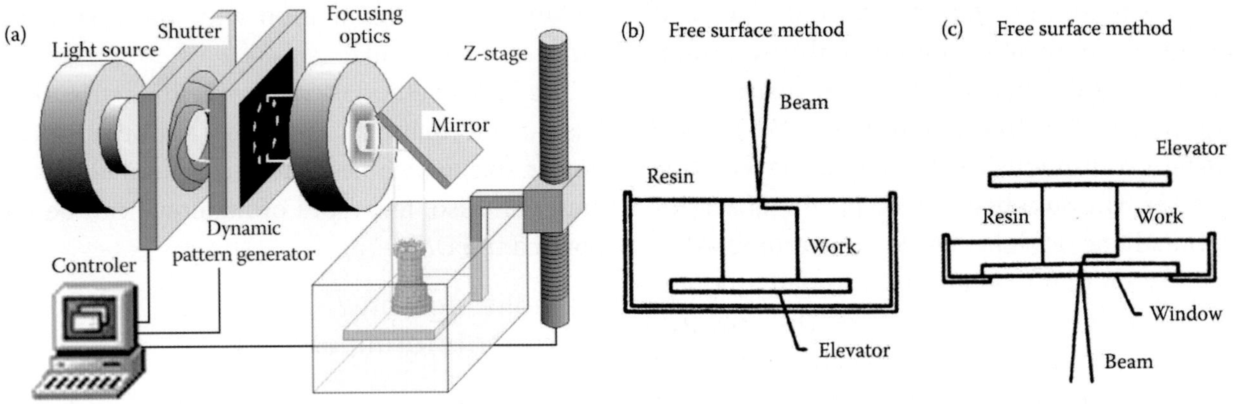


**FIGURE 8.29** (a) Stereolithography or photoforming. Free surface method (b) and fixed surface method (c).

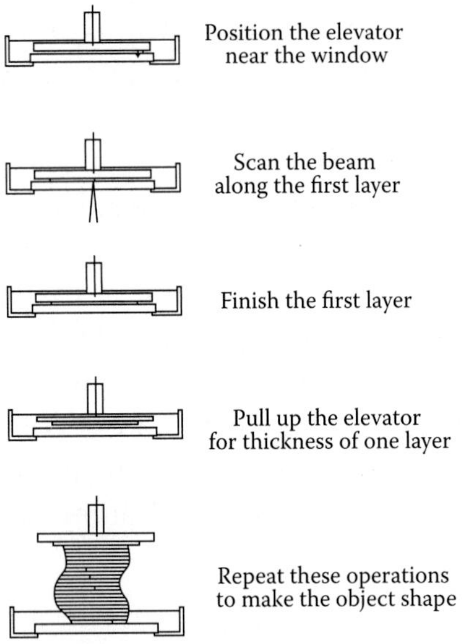


**FIGURE 8.30** Stereolithography forming process with the fixed surface method.

introduced by Ikuta et al.[33] in 1993. Whereas in conventional stereolithography the laser spot size and layer thickness are both in the 100-μm range, in microstereolithography, a UV laser beam is focused to a 1–2-μm spot size to solidify material in a thin layer of 1–10 μm. The monomers used in RP and microstereolithography are both UV-curable systems, but the viscosity in the latter case is much lower (e.g., 6 cPs vs. 2000 cPs). In microstereolithography the viscosity should be as low as possible for optimal flat layer formation because high surface tension hinders both efficient crevice filling and flat surface formation in the microscale. Another difference resides in the fact that in microstereolithography the solidified polymer is light enough so that it does not require a support as is required for the heavier pieces made in RP. In Table 8.4 we summarize this comparison between RP and microstereolithography.

Ikuta's group named their scanning microstereolithography "integrated harden polymer stereolithography" or the "IH process." The original IH process used single-photon polymerization to develop a photo-curable polymer with an UV laser beam. A glass slide was placed above the layer being cured, and raising the glass slide small distances above the previously solidified layer generated an extremely thin layer (fixed surface method mode; see Figure 8.29c and Figure 8.30). The IH process resolution in Ikuta's lab was 5 μm in 1992, and, through the various enhancements discussed below, is better than 100 nm at present.[34]

The two major types of stereolithography are scanning stereolithography and projection stereolithography. The scanning stereolithography parts are constructed in a point-by-point and line-by-line

**TABLE 8.4** Comparison of Rapid Prototyping and Microstereolithography

| | Rapid Prototyping | Microstereolithography |
|---|---|---|
| Need for a support | Yes | No |
| Layer thickness | Hundreds of micrometers | Several micrometers to tens of micrometers |
| Spot size | Hundreds of micrometers | Micrometer or submicrometer |
| Type of monomers | UV curable | UV and visible light curable |
| Viscosity of monomer system | High | Low |

fashion, with the sliced shapes written directly from a computerized design of the cross-sectional shapes by a beam in the liquid (Figure 8.30). Projection stereolithography is a parallel fabrication process that enables sets of truly 3D solid structures made of a UV polymer by exposing the polymer with a set of 2D cross-sectional shapes (masks) of the final structures. These 2D shapes are either a set of real photomasks used to subsequently expose the work (Figure 8.31), or they involve a dynamic mask projection system instead of a physical mask. We will compare two types of dynamic mask projection systems for their merit in microstereolithography: a liquid crystal display (LCD) and a digital micromirror display (TI's DMD chip). As we saw in Chapter 1, the DMD chip is also used in maskless lithography (see Chapter 1).

### *Scanning Microstereolithography, Parallel Microstereolithography and Two-Photon Polymerization*

Most of the microstereolithography practiced today is based on the scanning laser beam approach introduced by Ikuta et al.[33] from Nagoya University (Japan) in 1992. The scanning method has the advantage of point-by-point controllability, thereby avoiding unevenness of solidification, which leads to nonuniform shrinking of the work piece. When applying the scanning technique, a laser beam (e.g., a He-Cd laser) is used to solidify one microscopic

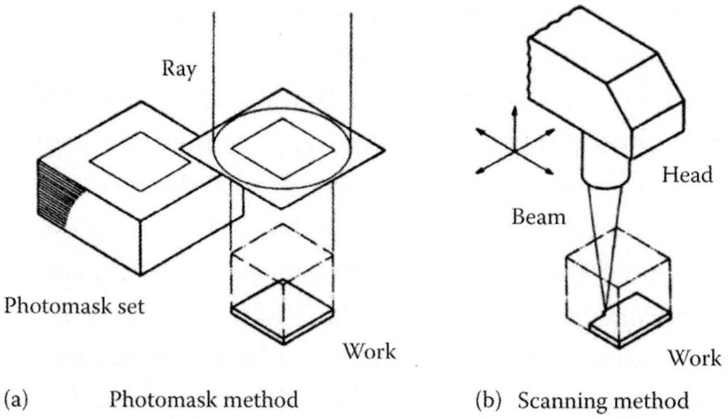


**FIGURE 8.31** Exposure with photomask set (a) and exposure with a scanning beam (b). [From Ikuta, K., and K. Hirowatari. 1993. *Proceedings: IEEE micro electro mechanical systems (MEMS '93).* Fort Lauderdale, FL: IEEE;[33] and Ikuta, K., K. Hirowatari, and T. Ogata. 1994. *IEEE international workshop on micro electro mechanical systems (MEMS '94).* Oiso, Japan: IEEE.[35]]

polymer area at a time to arrive at complicated 3D shapes by stacking thin films of hardened polymer layer upon layer. Process control in this case simply is directed from a CAD system containing the "slice data." The laser beam is focused down to a spot size of a few micrometers into the polymer through a glass window (Figure 8.30).

Ikuta et al. used an *x-y* stage to move the work piece rather than galvanometric mirrors to deflect the laser beam, and this led to a smaller focus spot. They attached the glass window to the *z*-stage for precise layer thickness control. The position accuracy for the laser beam spot (5-μm diameter in the early days) reached 1 μm in the *z*-axis and 0.25 μm in the *x*- and *y*-directions. Takagi et al.[36] obtained an 8-μm resolution with their photoforming setup, and Ikuta et al.[33,35] reported a minimum solidification unit size of 5 × 5 × 3 μm and a maximum size of fabricated structures of 10 × 10 × 10 mm. As we will learn, this technique has undergone several improvements since these early days.

In scanning microstereolithography, no physical contact occurs between tools and works, and a very large number of layers can be achieved (e.g., 1000 layers) with typical machining times ranging from 30 min to 1 h. High-aspect ratio and very complex shapes, including curved surfaces, can be made with this type of desktop microfabrication method. The objects realized this way used to feature a medium-range accuracy of 3–5 μm.

Professor Ikuta's group and other researchers have introduced various IH process enhancements, including a "mass IH process," a "hybrid IH process," a "super IH process," and finally the "two-photon IH process."

Given the serial nature of the IH process, fabrication speed is a concern, and in 1996 Ikuta et al. introduced a mass IH process. In the mass IH process, multiple parts are produced in parallel, using multiple scanning fiberoptic beams.[37] An array of five single-mode optical fibers (4-μm core diameter) was used to demonstrate this concept. The mass IH process needs to be further developed using optical fiber arrays and by improving the resolution.

In a hybrid IH process other non-IH microcomponents are introduced in the IH structure. As an example of a hybrid IH process, consider Takagi et al.'s work[36,38] wherein a combination of IH with

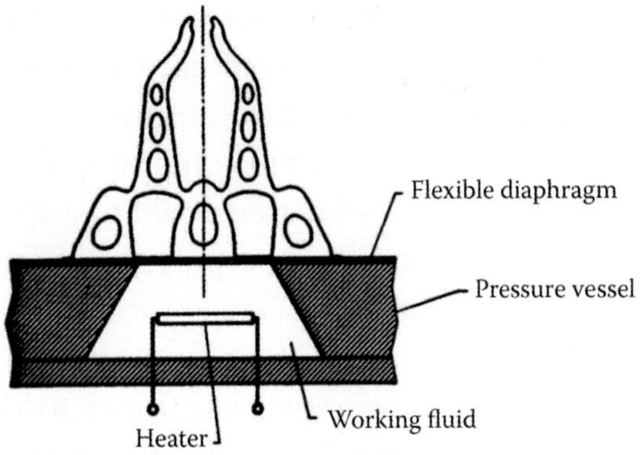


**FIGURE 8.32** Thermally driven microclamping tool. The clamping tool is made by photoforming, and the silicon substrate is by silicon bulk micromachining (see text for dimensions). [From Takagi, T., and N. Nakajima. 1994. *IEEE international workshop on micro electro mechanical systems (MEMS '94).* Oiso, Japan: IEEE. With permission.[38]]

more traditional Si micromachined substrates is introduced. In Figure 8.32, a schematic for a photoformed plastic clamp anchored to a Si substrate is represented. The photoformed plastic clamp measures about 2 mm long and 2 mm high and is 250 μm thick. The Si substrate is bulk micromachined to hold a working fluid reservoir and a heater element to actuate the polymer clamp. On heating the working fluid in the micromachined chamber, the thin Si membrane cover bulges out and opens the clamp. Cooling down closes the clamp.

The original IH process and the mass IH process both involve laser scanning and layer preparation. Two major drawbacks associated with this layer-by-layer approach are the limited depth resolution (limited by the thickness of the stacked layers) and the viscosity of the liquid UV-curable monomers (limited form filling and deformation of solidified structures in the microdomain). In 1997, Ikuta and his team upgraded the IH and mass IH process to the "super IH process." [39] The enabling technology for this new process was "deep site pin-point solidification" of a UV polymer. Unlike the typical rapid prototyping (RP) stereolithography process, the polymer in this super IH process is cured below the polymer/air surface where the UV beam is focused. The entire structure is cured into a single body by moving the laser focus, and the curing occurs not only at the liquid surface but also at arbitrary locations inside the liquid resin. A shallow focus lens is used [low depth of focus (DOF) and large numerical aperture (NA) lens] such that the beam's energy is small enough not to cure polymer outside of its focal point. A 3D microstructure is made by scanning the focused laser beam in 3D space into liquid UV monomers, enabling 3D fabrication without any support or sacrificial layers. Because no layer preparation steps are involved, viscosity and surface tension influences are minimized. The properties of the UV-curable monomer system are precisely tuned to ensure that polymerization only happens in the focal point of the laser beam. The UV monomer system used in the super IH process is a mixture of urethane acrylate oligomers, monomers, and photoinitiators. Microparts with freely moving parts can be made this way—no assembly is necessary—allowing for a more advanced type of MEMS device. The resolution of the super IH process is less than 1 μm.

Summarizing, the super IH process has several advantages compared with the layer-by-layer method, such as 1) higher resolution, 2) movable 3D microstructures can be fabricated without supporting or sacrificial layer, and 3) the influence of viscosity or surface tension can be ignored during fabrication.

In 1997, "nanostereolithography" based on two-photon absorption was implemented.[40,41] As we learned in Chapter 2, in the section on quantum lithography, the use of linear absorption of laser light for cross-linking a monomer limits the resolution roughly to the wavelength of the laser. Moreover, deep penetration of the light into the resin is difficult because of the strong photon absorption; therefore, unless very tight focusing of the laser is used as in super IH (see above), the process must be carried out thin layer by thin layer (see Figure 8.30). Using two-photon excitation, one can penetrate deeper into the bulk of the resin and enhance the resolution even further than with super IH. It has been demonstrated that quantum lithography with entangled N-photon states beats the Rayleigh diffraction limit by a factor of N, and in two-photon lithography the resolution is thus improved by a factor of two (as if one used a classical source with wavelength $\lambda/2$).[42,43] When high-intensity light shines on a material, the probability for two-photon simultaneous absorption is proportional to the square of the field intensity, and thus is greatest at the center of a Gaussian laser spot. Subwavelength structures can be produced because the absorption profile for the two-photon process is narrower than the beam profile. Using a high-NA (e.g., 0.85) lens, the laser spot can be made less than 1 μm in diameter, and significantly narrower structures can be obtained because a certain threshold of intensity must be reached to initiate polymerization. With a femtosecond laser beam, tightly focused into a resin, photo-induced reactions such as polymerization occur only in the close vicinity of the focal point, allowing the fabrication of subwavelength 3D polymer structures by directly writing 3D patterns. Importantly, because the energy required to initiate polymerization is twice that of the incoming radiation, it is possible to excite deep into the monomer as the material is transparent to one-photon excitation, whereas at the focus, where the intensity is high, two-photon excitation occurs. The 3D resolution has reached 140 nm in the two-photon IH process.

In two-photon stereolithography, one often works with epoxy resins and urethane acrylates. Several two-photon photoinitiators with very large two-photon absorption cross-sections to be added to the base polymers have been developed.[44] In these photoinitiator molecules, the photon energy that is absorbed is emitted as photons of a higher energy (up-converted fluorescence). These IH processes can be widely used for making polymeric microdevices. Some examples of 3D two-photon lithography were shown in Figure 2.37 (see also the inset in the heading of this chapter). Two-photon 3D lithography has also been used to make photonic crystals.

Microstereolithography is not limited to polymers: micro- or even nanoceramic and metallic powders can be mixed with the photosensitive resin in microstereolithography of ceramics[45] and metals.[46]

### *Projection Microstereolithography*

In projection microstereolithography, a reduced-size image of complicated patterns is reproduced onto a photoresist. Similar to photolithography, an image is transferred to the liquid photopolymer by irradiating UV light through a patterned mask. The photomask approach, which solidifies a whole layer at the time, is significantly faster than the point-by-point technique but requires a large number of expensive photomasks. In dynamic mask projection microstereolithography, a dynamic mask generator is used instead of a real mask. We compare two types of dynamic mask projection systems: a liquid crystal display (LCD) and a digital micromirror display (DMD).

Bertsch et al.[28] at the Swiss Federal Institute of Technology (EPLF) pioneered an interesting variation on the photomask approach, which, given the many layers involved, is often impractical. This research group uses a computer-controlled LCD as a dynamic pattern generator. A light beam passes through the LCD, and a beam reducer focuses it on the surface of a polymerizable medium. Selective polymerization takes place in the irradiated areas corresponding to the transparent pixels of the LCD. Between the irradiation steps, a shutter blocks the beam, and a new layer of fresh resin (about 5 μm thick) spreads over the object under construction. Complex objects, with a resolution better than 5 μm in the *x*-, *y*-, and *z*-directions, as illustrated in Figure 8.33, have been made this way.[47] The same research group pioneered the combination of planar UV/LIGA lithography (e.g., using SU-8) with microstereolithography—to add nonvertical structures with curved and conical surfaces in a postprocessing step onto planar structures without the need of microassembly.[48] The surfaces of the parts made by microstereolithography are smooth enough to enable conformal electroplating to make a metal mold from the obtained 3D microobject. Polymer replication, of course, remains limited to convex and nonreentrant structures (see Chapter 10).

A comparison of the resolution between conventional stereolithography and small-spot and layer-by-layer microstereolithography was presented by Bertsch et al.[49] in *Rapid Prototyping Journal.* To further improve the lithography—3D photoforming in this case—it is necessary to better understand the shape of a "solidified cell," which depends on both the characteristics of the beam and the resin.

In 2005, Sun et al. introduced another dynamic projection microstereolithography approach using TI's DMD chip as the mask generator (Figure 8.34).[50] The authors point out that the LCD technique from Bertsch et al. has some serious disadvantages compared with their higher-resolution DMD approach.

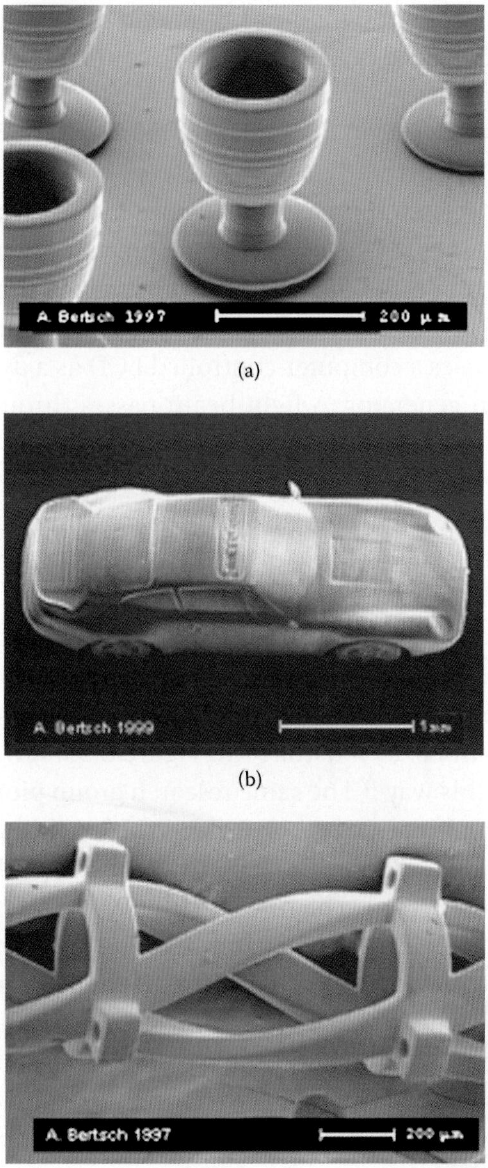


**FIGURE 8.33** Examples of objects made by microstereolithography. SEM photos of microcups (80 layers × 5 μm, external diameter is 200 μm) (a); Porsche (673 layers × 5 μm, the diameter of a wheel is 400 μm) (b); microspring (1000 layers × 5 μm, external diameter is 500 μm) (c).

An LCD display not only features larger pixels and a lower filling factor (pixels vs. space between) than a DMD display, but it also has a slower switching speed (~20 ms), and the low optical density of the refractive elements in the off cycle makes for poor contrast. Moreover, the LCD is inserted between four glass plates that are opaque to UV light, restricting its use to visible light where fewer resins are available. Sun et al. made 3D structures with a smallest feature size of 0.6 μm.

Summarizing, microphotoforming offers several advantages over more classical photolithography-based micromachining processes:

1. The turnaround from CAD to prototype takes only 1 h or less.
2. Photoforming is an additive process accommodating virtually any shape.
3. It is a fully automatic process.
4. It requires a small capital investment (less than $30,000).
5. There is no need for a clean room.

This direct-write lithography technique might well represent an alternative to LIGA in cases where 3D shape versatility outweighs accuracy. As with the LIGA technique, the plastic shapes made by stereolithography may be used as a cast for electroplating metals or for other materials that can be molded into the polymer structures.

## Electrochemical Forming Processes

### Introduction

Both subtractive and additive electrochemical processes are of extraordinary importance in miniaturization science. Subtractive electrochemical techniques were reviewed in Chapter 4; here we review additive/forming electrochemical techniques. Electrochemical forming includes electroless and electrodeposition, and we analyze pros and cons of both. We add information on electrodeposition through masks, working with instant masks in electrochemical fabrication (EFAB) and maskless metal deposition methods. Today, there is a strong push in the IC industry and in micromachining toward dry processing, and deep dry-etching developments are particularly swift. Electrochemical machining, on the other hand, despite its many uses, is often still considered a "black art" and "dirty," with lack of cleanliness and creation of waste particles making it an unacceptable technique. In reality, besides its major technological advantages, electrochemistry offers several intrinsic advantages, such as low capital equipment, high deposition and etch rates, and relatively simple operation. In the sections below, we learn that there are many cases in which only wet

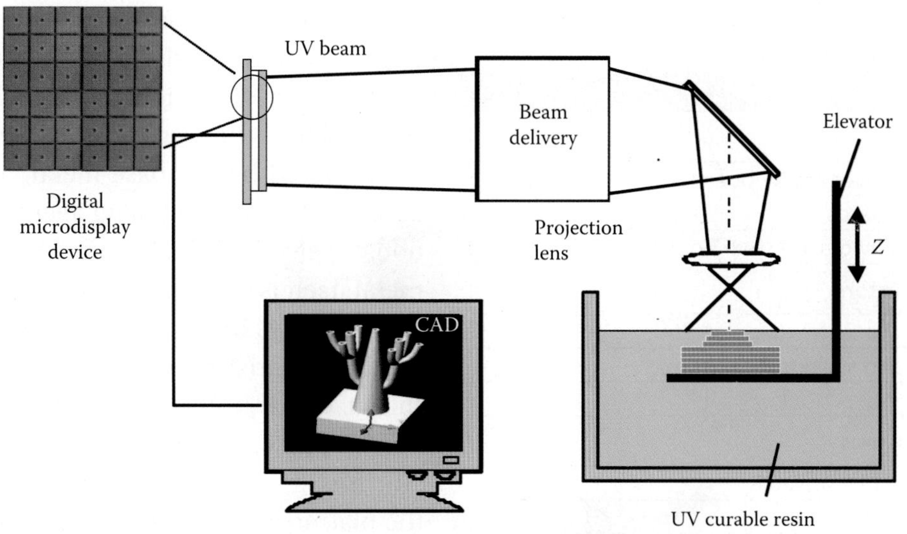


**FIGURE 8.34** A dynamic projection microstereolithography approach using TI's DMD chip as the mask generator. (From Sun, C., N. Fang, D. M. Wu, and X. Zhang. 2005. Projection micro-stereolithography using digital micro-mirror dynamic mask. *Sensors Actuators A* 121:113–20.[50])

chemical and electrochemical processes are feasible; for example, line-of-sight deposition techniques cannot readily be used on curved or irregular surfaces or to plate into via holes or blind vias. CVD often leaves voids when metallizing (e.g., tungsten) in a high-aspect-ratio structure, whereas with electrodeposition such voids can be avoided completely. Multilayer PCBs with through holes connecting the individual metal layers would not be possible without electroless or electrochemical technology. As long as a solution can wet and fill a cavity to be plated, a metal pattern can be generated precisely, replicating the mold down to angstrom dimensions. In Table 8.5, many applications of electrochemistry are reviewed, and

**Table 8.5** Situations for which an Aqueous Electrochemical or Chemical Technique Is the Only Means to Obtain the Desired Result

| |
|---|
| • Uniform plating on irregular surfaces |
| • Leveling of rough surfaces |
| • Plating of via holes and blind vias |
| • Formation of alloys of metastable phases |
| • Compositionally modulated structures |
| • Amorphous metal films |
| • Maskless, high-speed, selective deposition of metals and alloys |
| • Most faithful reproduction of features of polymeric masks |
| • Smallest dimension features with highest height-to-width aspect ratio (highest packing density of conductors) |
| • Metal deposits with incorporated particles (e.g., diamond, WC, TiC, $Al_2O_3$, Teflon, oil) |
| • Wear-resistant surfaces |
| • Corrosion-resistant surfaces |
| • Uniform thickness nonporous anodic films |
| • Anodic films with porous structure |
| • Very uniform, thick, pore-free, polymer films on irregular surfaces |
| • Very uniform thickness, glazed, and devitrified glass surfaces |
| • Shaping of almost any metal or semiconductor and electromachining without introducing stress |
| • Electropolishing of metal surfaces to mirror bright finishes |

*Source:* Based on Romankiw, L. T., and T. A. Palumbo. 1987. *Proceedings: symposium on electrodeposition technology, theory, and practice.* San Diego, CA[51]; and Romankiw, L. T. 1984. Electrochemical technology in electronics today and its future: a review. *Oberflache-Surface* 25:238–47.[52]

**Table 8.6** Electrochemical Applications in the IC Industry and Micromachining

| |
|---|
| **Photocircuit boards** |
| • Single and multilayer epoxy boards |
| • Flexible circuit boards |
| • Electrophoretically glazed steel boards |
| **Contacts and connectors** |
| • Beam leads |
| • Contacts (pins and sockets) |
| • Reed switches, etc. |
| **Cabinets and enclosures** |
| • Corrosion protective surfaces |
| • Electromagnetic shielding |
| • Decorative purposes (anodization, electrophoretic painting) |
| **Auxiliary equipment used in device fabrication** |
| • Paste-screening masks |
| • Metal evaporation masks |
| • X-ray lithography masks |
| • Diamond saws and cutting tools |
| **Active elements** |
| • IC chips |
| • Magnetic recording heads |
| • Recording surfaces |
| • Displays |
| • Wear-resistant surfaces |
| **Chip carriers and packages** |
| • Chip in tape packages |
| • Surface mount boards |
| • Dual in-line packages |
| • Pin-grid array packages |
| • Multilayer ceramic packages |
| • Hybrid packages |

*Source:* Based on Romankiw, L. T., and T. A. Palumbo. 1987. *Proceedings: symposium on electrodeposition technology, theory, and practice.* San Diego, CA[51]; and Romankiw, L. T. 1984. Electrochemical technology in electronics today and its future: a review. *Oberflache-Surface* 25:238–47.[52]

in Table 8.6, we define several situations in which an aqueous chemical or electrochemical technique may be the only means by which a desired property or quality of a deposit can be achieved.

## Electroless Metal Deposition

### *Underlying Principles*

To continuously build thick deposits by chemical means without applying a voltage and without consuming the substrate, i.e., electroless plating, it is essential that a sustainable oxidation reaction be used instead of the dissolution of the substrate itself as in the case of immersion plating. In an immersion plating bath, the electrons for the metal deposition are supplied by the base metal, and effectively the base metal is the reducing agent. As the base metal donates electrons to the metal being plated, the base metal itself is oxidized and goes into solution, which is why this is also called a *replacement reaction*. The key characteristics of the immersion plating bath is that it is self-limiting, which means that once the base metal, e.g., copper, in the case of a printed circuit board (PCB), is covered with the new metal, the plating ceases. This results in a very thin, dense, nonporous coating, which is very economical, from a bath that is extremely stable. In an "electroless" plating bath, on the other hand, the electrons are donated by a reducing agent that is already in the bath. In Figure 8.35, the difference between immersion plating and electroless deposition is illustrated by comparing deposit thickness versus time; immersion plating is self-limiting, but electroless plating is not.[53] In an electroless copper bath, the formaldehyde is the reducing agent. In electroless Ni, it is either hypophosphite or dimethyl amine borane.

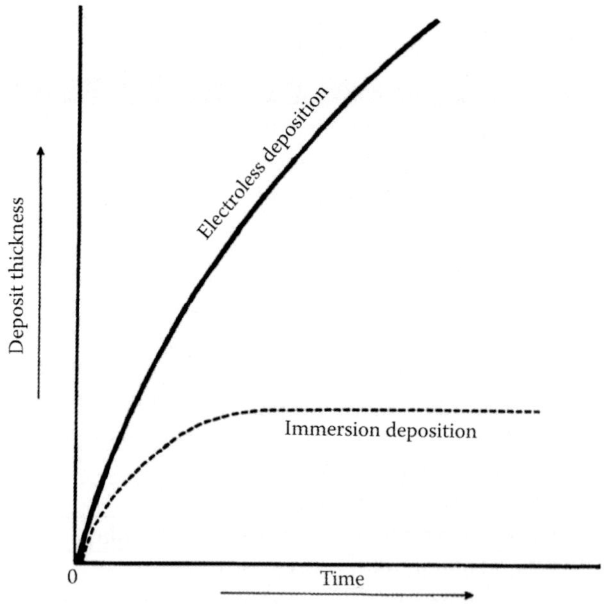


**FIGURE 8.35** Thickness versus time comparison between electroless and immersion deposition. [Based Mallory, G. O., and J. B. Hadju, eds. 1990. *Electroless plating: fundamentals and applications.* Orlando, FL: American Electroplaters and Surface Finishers Society (AESF). With permission.[53]]

The reduction of Ni ions (Reaction 8.1) is fed electrons by an oxidation reaction such as that of hypophosphite (Reaction 8.2), resulting in an overall reaction for Ni deposition given by Reaction 8.3.

Reduction:

$$Ni^{2+} + 2e^- \rightarrow Ni \qquad \text{Reaction 8.1}$$

Oxidation:

$$H_2PO_2^- + H_2O + 2e^- \rightarrow H_2PO_3^- + 2H^+ + 2e^- \qquad \text{Reaction 8.2}$$

Overall reaction:

$$Ni^{2+} + H_2PO_2^- + H_2O \rightarrow Ni + H_2PO_3^- + 2H^+ \qquad \text{Reaction 8.3}$$

Reaction 8.1 continues (in principle) until all the hypophosphite in solution is consumed (Reaction 8.2). In immersion deposition, Reaction 8.1 stops as soon as the whole substrate surface is covered with Ni metal and there is no further dissolution of uncovered substrate possible. In parallel to Reaction 8.1, more or less severe hydrogen reduction goes on:

$$2H^+ + 2e^- \rightarrow H_2 \qquad \text{Reaction 8.4}$$

Copious hydrogen evolution can upset the quality of the depositing metal film and should be avoided. The hydrogen evolution rate is not directly related to that of the metal deposition and mainly originates from the reductant molecules. Stabilizers (i.e., catalytic poisons) are needed in electroless deposition baths because the solutions are thermodynamically unstable; deposition might start spontaneously onto the container walls. Substances that poison hydrogenation* catalysts such as thiourea, $Pb^{2+}$, and mercaptobenzothiazole function as stabilizers in such electroless baths. Besides stabilizers, the metal salt, and a reducing agent, electroless solutions may contain other additives such as complexing agents, buffers, and accelerators. Complexing agents exert a buffering action and prevent the pH from decreasing too fast. They also prevent the precipitation of metal salts and reduce the concentration of free metal ions. Buffers keep the deposition reaction in the desired pH range. Accelerators, also termed *exaltants*, increase the rate of deposition to an acceptable level without causing bath instability. These exaltants are anions, such as $CN^-$, thought to function by making the anodic oxidation process easier. In electroless copper, for example, compounds derived from imidazole, pyrimidine, and pyridine can increase the deposition rate to 40 μm/h. The electroless deposition must occur initially and exclusively on the surface of an active substrate and subsequently continue to deposit on the initial deposit through the catalytic action of the deposit itself. Because the deposit catalyzes the reduction reaction, the term *autocatalytic* is often used to describe the plating process. Electroless plating is an inexpensive technique enabling plating of conductors and nonconductors alike (plastics such as ABS, polypropylene, Teflon, and polycarbonate are plated in huge quantities). A catalyzing procedure is necessary for electroless deposition on nonactive surfaces such as plastics and ceramics. The most common method for sensitizing those surfaces is by dipping into $SnCl_2$/HCl or immersion in $PdCl_2$/HCl.[53] This chemical treatment produces sites that provide a chemical path for the initiation of the plating process.

Metal alloys such as nickel-phosphorus, nickel-boron, cobalt-phosphorus, cobalt-boron, nickel-tungsten, copper-tin-boron, and palladium-nickel can be produced by codeposition. In the case of Ni deposits, with or without phosphorus and boron incorporated, different electroless solutions are used to obtain optimum hardness, effective corrosion protection, or optimum magnetic properties. Recent experimental results show that composite material also can be produced by codeposition. Finely divided, solid particulate material is added and dispersed in the plating bath. Electroless Ni with alumina particles, diamond, silicon carbide, and PTFE have reached the commercial market. Table 8.7 presents a list of electroless plating baths.[53]

Methods to determine the electroless deposition rate can be split into two categories: electrochemical and nonelectrochemical. Electrochemical techniques include Tafel extrapolation, DC polarization, AC impedance, and anodic stripping. Nonelectrochemical methods include weight gain, optical absorption, resistance probe, and acoustic. For more details on each of these techniques, see Ohno[55] and references therein.

* *Hydrogenation* is a class of chemical reactions that result in an addition of hydrogen, usually to unsaturated organic compounds.

**Table 8.7** Typical Electroless Plating Baths

| Component | Concentration (per L) | Application/Remark | pH | Temp (°C) |
|---|---|---|---|---|
| Au | 1.44 g $KAu(CN)_2$, 6.5 g KCN, 8 g NaOH, 10.4 g $KBH_4$ | Plate beam leads on silicon ICs, ohmic contacts on n-GaAs | 13.3 | 70 |
| Co-P | 30 g $CoSO_4 \cdot 7H_2O$, 20 g $NaH_2PO_2 \cdot H_2O$, 80 g $Na_3$citrate$\cdot 2H_2O$, 60 g $NH_4Cl$, 60 g $NH_4OH$ | Magnetic properties | 9.0 | 80 |
| Cu | 10 g $CuSO_4 \cdot 5H_2O$, 50 g Rochelle salt, 10 g NaOH, 25 mL conc. HCHO (37%) | Printed circuit boards | 13.4 | 25 |
| Ni-Co | 3 g $NiSO_4 \cdot 6H_2O$, 30 g $CoSO_4 \cdot 7H_2O$, 30 g $Na_2$malate$\cdot 1/2H_2O$, 180 g $Na_3$citrate$\cdot 2H_2O$, 50 g $NaH_2PO_2 \cdot H_2O$ | | 10 | 30 |
| Ni-P | 30 g $NiCl_2 \cdot 6\ H_2O$, 10 g $NaH_2PO_2 \cdot H_2O$, 30 g glycine | Corrosion and wear resistance on steel | 3.8 | 95 |
| Pd | 5 g $PdCl_2$, 20 g $Na_2EDTA$, 30 g $Na_2CO_3$, 100 mL $NH_4OH$ (28% $NH_3$), 0.0006 g thiourea, 0.3 g hydrazine | Plating rate is 0.26 μm/min | | 80 |
| Pt | 10 g $Na_2Pt(OH)_6$, 5 g NaOH, 10 g ethylamine, 1 g hydrazine hydrate (added now and then to maintain this concentration) | Plating rate is 12.7 μm/h | | 35 |

*Source:* Based on Mallory, G. O., and J. B. Hadju, eds. 1990. *Electroless plating: fundamentals and applications.* Orlando, FL: American Electroplaters and Surface Finishers Society (AESF)[53]; and Romankiw, L. T. 1976. *Etching for pattern definition.* Pennington, NJ: The Electrochemical Society.[54]

### *Electroless Plating in the IC Industry*

The IC industry applies electroless metal deposition for a wide variety of applications, some of which are incorporated in Tables 8.5 and 8.6. In Figure 8.36, we single out the schematic for the electroless Cu deposition of a buried conductor as an illustrative example of an IC application. The fabrication of this buried conductor illustrates many of the processes we have studied so far; lithographic patterning of a dielectric material (Figure 8.36a), pattern transfer by anisotropic dry etching of a groove in the dielectric substrate (Figure 8.36b), evaporative deposition of an Al seed layer (Figure 8.36c), seed layer patterning with lift-off (Figure 8.36d), trench filling with electroless copper (Figure 8.36e), and spin-on glass dielectric layer deposition for planarization (Figure 8.36f).

### *Electroless Plating in Microfabrication*

In micromachining applications, electroless deposition is used for the same purposes as in the IC

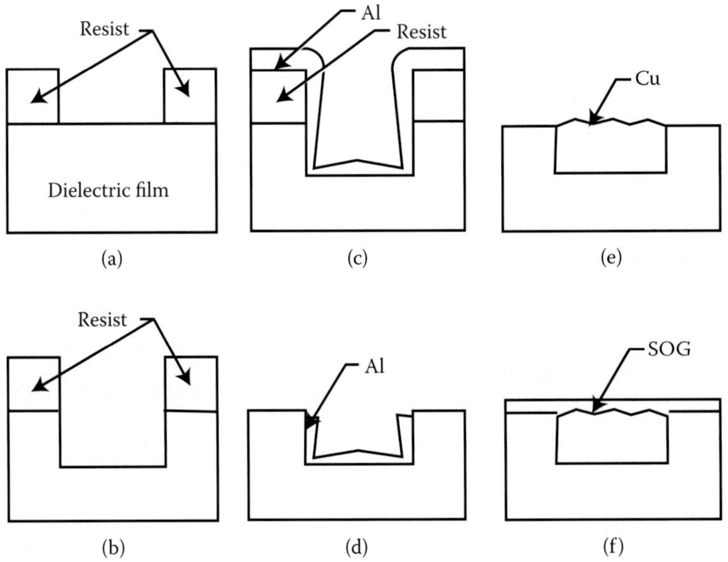


**FIGURE 8.36** Schematic of a buried Cu conductor process. (a) Photoresist pattern over a dielectric. (b) Pattern transfer by anisotropic dry etching. (c) Deposition of a thin aluminum seed layer. (d) Seed metal patterning with lift-off. (e) Trench filling with electroless copper. (f) SOG dielectric layer for planarization.

industry, as well as to make higher-aspect-ratio structural microelements from a wide variety of metals, metal alloys, and even from composite materials. These applications often include polymer molds and introduce new challenges in terms of uniformity of plating rates and removal of the metal form from the master photoresist mold. Electroless plating may be elected above electrodeposition due to the simplicity of the process because no special plating base (electrode) is needed. This represents a major simplification for combining LIGA and pseudo-LIGA structures with active CMOS electronics, where a plating base could short out the active electronics. In addition, electroless Ni exhibits less stress than electrodeposited Ni, a fact of considerable importance in most mechanical structures. The major concern is the temperature of the electroless plating processes, which is often considerably higher than for electrochemical processes. Few studies of electroless plating in LIGA molds have been reported.[56]

We add two interesting MEMS applications of electroless plating: electroless plating of beveled structures and electroless plating using microfluidic structures. Usually, electroless plating is an isotropic process with no means to control the profile of plated features. However, van der Putten et al.[57,58] were able to induce anisotropy in the plating process by taking advantage of the increased poisoning effect a stabilizer has at the edges of any given pattern (see Figure 8.37a). Because the edges of a given pattern experience additional deposition from nonlinear diffusion, mass transport to these edges is enhanced compared with material supply to the bulk of a substrate. As a result, the surface concentration of the adsorbed stabilizer ($Pb^{2+}$) is higher at these edges. By choosing the proper $Pb^{2+}$ concentration, the edges can be selectively poisoned: bevels can be grown with an angle, $\alpha$, which is a function of the stabilizer concentration in the solution (see Figure 8.37b). More important, van der Putten's bevel-plating technique eliminates the commonly observed lateral overgrowth of resist patterns, and it might have broader applications in the area of micromachining. The technique is less controllable than Maciossek's maskless electrodeposition, discussed further below, but does not require a patterned plating base.[59]

Whitesides et al. introduced a means of depositing and patterning metal electrodes inside prefabricated channels by flowing plating and etching solutions as shown in Figure 8.38. Under conditions of laminar flow, reactions, including electroless plating, can take place at the interface between two or more liquids flowing side by side down a channel. In the example in Figure 8.38, the fabrication of a three-electrode system is demonstrated. Two gold electrodes are formed by etching a previously deposited gold strip. The silver reference electrode is formed using a two-phase laminar flow arrangement.[60]

## Electrodeposition

### *Underlying Principles*

Electroplating, we saw in Volume I, Chapter 7, takes place in an electrolytic cell. The reactions involve current flow under an imposed bias. As an example, consider the deposition of Ni from $NiCl_2$ in a KCl solution with a graphite anode (not readily attacked by $Cl_2$) and a Au cathode (inert surface for Ni deposition). With the cathode sufficiently negative and the anode sufficiently positive with respect to the solution, Ni deposits on the cathode and $Cl_2$ evolves at the anode. The process differs from electroless Ni deposition in that the anodic and cathodic processes occur on separate electrodes and that the reduction is affected by the imposed bias rather than a chemical reductant. Important process parameters are pH, current density, temperature, agitation, and solution composition. The amount of hydrogen evolving and competing with metal deposition depends on the pH, the temperature, and the current density. Important causes of defects in electrodeposited metallic microstructures are the appearance of hydrogen bubbles, and these three parameters need very precise control. Besides typical impurities, such as airborne dust or dissolved anode material, the main impurities are metal hydroxide formed at increased pH values in the cathode vicinity and organic decomposition products from the wetting agent. The latter two can be avoided to some degree by monitoring and controlling the pH and by adsorption of the organic decomposition products on activated carbon.

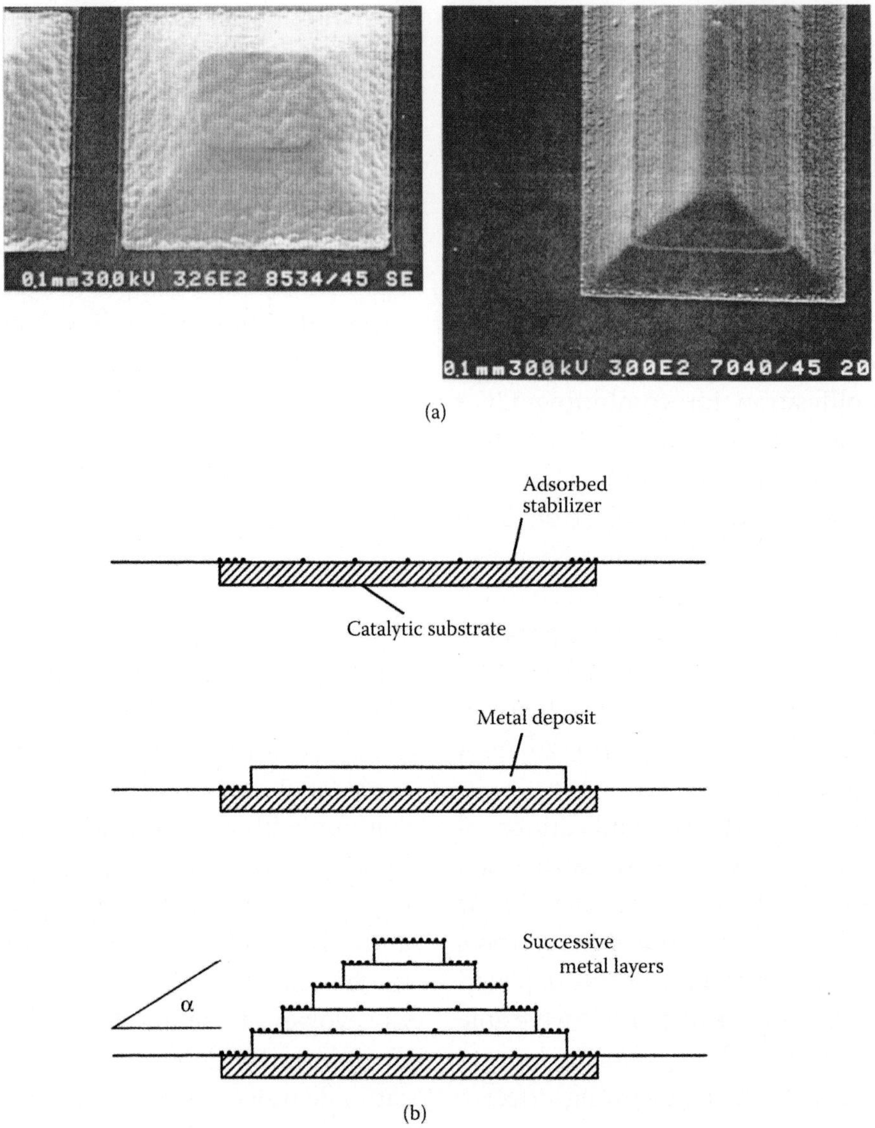


**FIGURE 8.37** Electroless micromachining: bevel plating. (a) Examples of bevel-plated nickel structures. (b) Method to make bevel-plated devices. (From van der Putten, A. M. T., and J. W. G. de Bakker. 1993. Anisotropic deposition of electroless nickel-bevel plating. *J Electrochem Soc* 140:2229–35. With permission.[57])

## *Electrodeposition in ICs, MEMS, and NEMS*

### *Through-Mask Electroplating*

*Overview* In a typical electrodeposition through a polymer mask in IC applications, metal conductors must be deposited on a dielectric substrate. The dielectric is usually made conductive first by sputtering a thin adhesion metal layer (e.g., Ti, Cr) and then a conductive seed layer (e.g., Au, Pt, Cu, Ni, and NiFe). The thickness of the thin refractory metal adhesion layer may be as small as 50–100 Å, whereas the thickness of the conducting seed layer can range from 150–300 Å.[61] The key requirement for the seed layer is that it is electrically continuous and offers low sheet resistance. After forming a pattern in a spin-coated polymer by UV exposure, e-beam, or x-ray radiation and developing away the exposed resist, electrical contact is made to the seed layer, and electrodeposition in the resist mold is carried out. The use of a solvent-containing development agent ensures a substrate surface free of grease and ready for plating. The metal layer growing on the substrate fills the gaps in the resist configuration, thus forming a complementary metal structure. Pollutants may cause hydrogen bubbles to cling to the resist structures, resulting in pores in the metal deposit; thus, the bath must be kept clean, for example, by circulating through a membrane filter with 0.3-μm pore openings.[56]

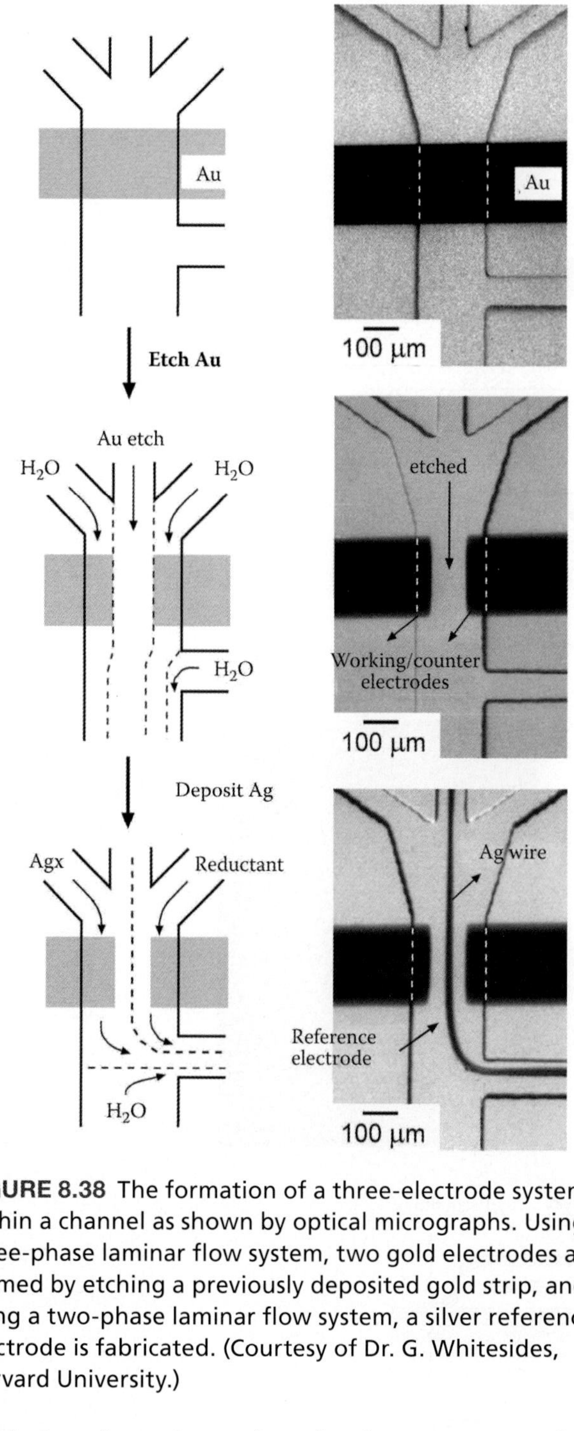


**FIGURE 8.38** The formation of a three-electrode system within a channel as shown by optical micrographs. Using a three-phase laminar flow system, two gold electrodes are formed by etching a previously deposited gold strip, and using a two-phase laminar flow system, a silver reference electrode is fabricated. (Courtesy of Dr. G. Whitesides, Harvard University.)

Plating through-mask technology is successfully used in volume production of beam leads and bumps on IC wafers, fabrication of thin film magnetic heads, fabrication of PCBs, x-ray lithography mask gratings, and diffraction gratings. Electrodeposition in general has been used extensively in the electronics industry in many stages of the manufacturing process, from the device stage, chip carriers, and PCB to corrosion protection and electromagnetic shielding of the electronic enclosures (see Tables 8.5 and 8.6). Processes include electrodeposition and electroless deposition of copper, nickel, tin, tin-lead alloys, and precious metals such as gold, gold alloys, palladium, and palladium alloys, as well as NiFe, CoP, NiCoP, and other magnetic alloys.[51]

Whereas PCBs use electroless or electroplated Cu, contacts and connectors use electroplated Au and Au alloys or alternatively low-cost precious metal substitutes such as Pd, PdNi, PdAg, NiP, NiAs, and NiB with a thin gold overcoat. Separable, low-force, low-voltage contact applications require a contact finish whose resistance is stable for the projected contact life and which has sufficient wear resistance to withstand the projected number of insertions. The surface of one of the two mating parts is usually pure soft gold, whereas the other is a hard gold alloy. In less-critical applications, tin and tin-lead alloys are used as contact materials. Connecting chips with chip carriers or packages also often involves Au contacts. Gold does not adhere directly to silicon dioxide or silicon nitride, and Ti is typically used as an adhesion layer. Because Au and Ti form intermetallics, a Pt barrier is usually deposited between titanium and gold.

*Copper Electroplating for Thin Film Read-Write Heads* In one of many pioneering IBM MEMS efforts, Romankiw used plating through-mask technology to make thin film read-write heads as shown in Figure 8.39.[62] The development of batch-produced, thin film heads is an excellent example of how micromachining has been gainfully practiced in industry—with a definite, practical, well-specified application in mind and using the machining tools best suited for the problem.

Read-write heads, which initially were horseshoe magnets hand wound with insulated copper wire, are of utmost importance in magnetic storage. The fabrication of traditional ferrite heads reached its limit more than 20 years ago, and its further extension was hard to imagine. IBM's objective was to build the next-generation read/write heads using batch fabrication and lithography techniques as much as possible. To develop a multiturn head as shown in Figure 8.39a, it was necessary to develop a technology that would handle dimensions between those of PCBs and semiconductor devices. Plating of copper conductors through thick resist masks (Figure

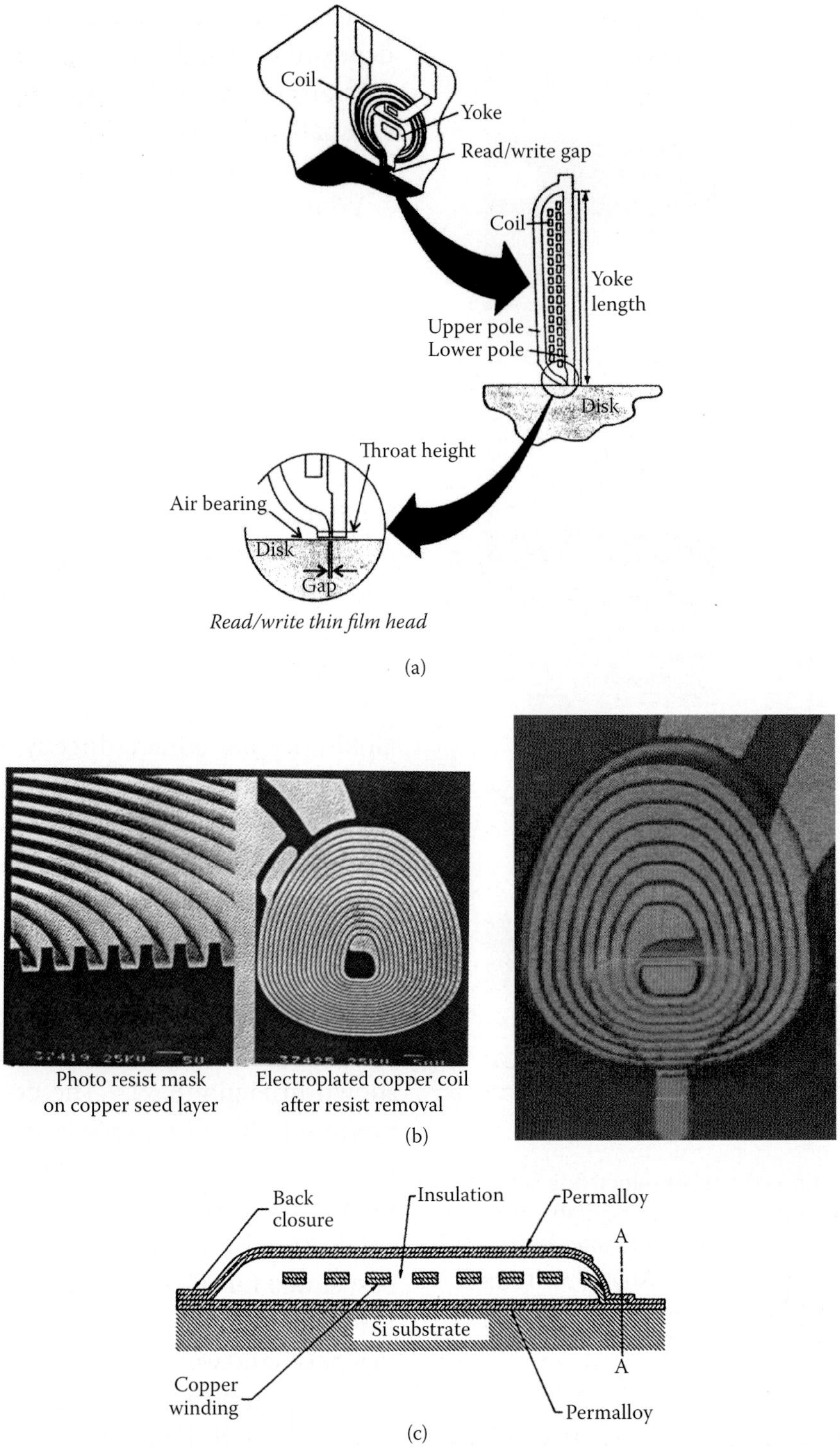


**FIGURE 8.39** Thin film read-write head. (a) Schematic of a multiturn (32) thin film read-write head with inset of pole tip structure on the air-bearing surface. (b) Resist mold and electrodeposited copper coil after resist removal. (c) Schematic cross-section of an eight-turn thin film head. (Romankiw, L. T., I. M. Croll, and M. Hatzakis. 1970. Batch-fabricated thin-film magnetic recording heads. *IEEE Trans Magn* 6:597–601.[62])

8.39b) and of thin films of nickel-iron alloys of 80:20 nominal composition (Permalloy) (Figure 8.39c) made these thin film heads possible. The copper coils in Figure 8.39b carry considerable current and have to be at least 2–3 μm thick and nearly square in cross-section. The more turns in a given length, the higher the writing field at the pole tip gap and the higher the read back signal. The head in Figure 8.39c has eight turns; the 3380-K IBM head has 32 turns (Figure 8.39a). The Permalloy plating in the thin film head is particularly challenging. Because of the very high sensitivity of composition to agitation of the plating solution, a plating cell had to be developed that assured uniform agitation over the entire part. This was achieved by a specially shaped paddle, which moves at a predetermined frequency with a precisely defined separation from the surface being plated. With the right choice of conditions, it was eventually possible to achieve both thickness and composition uniformity better than ±5%.[63] As a dielectric, hard-baked AZ photoresist was used, greatly simplifying the fabrication process. As a substrate, Si was used in the early days; today, $Al_2O_3TiC$ is used.

Without these new micromachined thin film heads, the large increases in the areal densities* of magnetic media in the past few years would not have been possible. According to the data storage market research firm Peripheral Research Inc., market demand for thin film heads was projected to grow nearly 75% from 913 million units in 1998 to more than 1.5 billion units by 2001. This feat makes it probably the most successful micromachined device to date.

As the need for still more memory capacity increased, thin film inductive heads were replaced by newer technologies such as magnetoresistive (MR) heads (introduced in 1991 by IBM) and giant magnetoresistive (GMR) heads (introduced in 1997; see Volume III, Chapter 1). In MR heads, the readback function is now performed by an MR sensor. The inductive portion of the head is used only for the write portion of the data-recording operation.[64]

* Areal density is used as a primary technology growth rate indicator for the memory storage industry. Areal density is defined as the product of the linear bits per inch (BPI) measured along the length of the tracks around a disk, multiplied by the number of tracks per inch (TPI) measured radially on the disk. The results are expressed in units of Mbit per square inch ($Mbit/in.^2$). Current high-end 2.5-in. drives record at areal densities of about 1.5 Gbit per square inch ($Gbit/in.^2$). Prototype drives with densities as high as 10 $Gbit/in.^2$ have been constructed, allowing for capacities of more than 20 GB on a single 2.5-in. platter for notebook drives. Areal density (and, therefore, drive capacity) has been doubling approximately every 2–3 years, and production disk drives are likely to reach areal densities of 10 $Gbit/in.^2$ before the end of 2001. A drive built with this technology will be capable of storing more than 10 GB of data on a single 2.5-in. platter, allowing 20- or 30-GB drives to be constructed that fit in the palm of your hand. The primary challenge in achieving higher densities is manufacturing drive heads and disks to operate at closer tolerances, which clearly is an area in which MEMS could have an impact.

In Volume III, Chapter 10, we learn that it is projected that read-write heads still will represent around 51% of the total MEMS market in 2009. Whereas traditional application of read-write heads in PCs will grow only moderately, the read-write head market is experiencing a renaissance in consumer electronics as hard disks are entering music players (e.g., in every iPod), smart phones (Samsung introduced the first cell phone with HDD in 2004), as well as digital video cameras, set top players, and DVD recorders. Of course, even further down the road we can envision all HDDs may be replaced by flash memory that can support today already up to 8-GB disk space, which is 5600 times more than a 1.44-MB floppy disk (see Volume III, Chapter 10)!

*Copper Electroplating in Printed Circuit Boards* A PCB is made of layers, typically 2–10, that interconnect components via copper pathways. The main PCB in a system is called a *system board* or *motherboard*, whereas smaller ones that plug into the slots in the main board are called *boards* or *cards*. A typical multilayer PCB (which constitutes most of the PCBs used to implement complex circuitry) consists of a sandwich of conducting and insulating layers. Today's PCBs may consist of as many as five to six layers of metals and dielectrics. In Figure 8.40, a typical six-layer PCB is illustrated. By having the signals running on the inside of the board, more components can be more tightly integrated on the board to give a more compact design. PCB fabrication uses either an all electroless copper process or a combination of electroless and electroplating processes. The continuing effort to make active devices smaller and faster, to minimize the length of connecting wire, and to make them narrower and thicker (with smaller spaces between) necessitates more layers per board and smaller diameter holes with a larger ratio of hole length to diameter. A hole length-to-diameter ratio of 10:1 is routine and 20:1 is feasible. Without electrochemical copper plating technology, such a degree of integration would not have been possible.

*Single and Dual Damascene Processes* Copper wiring on IC chips was introduced by IBM in September 1997 with the "damascene process."[65]

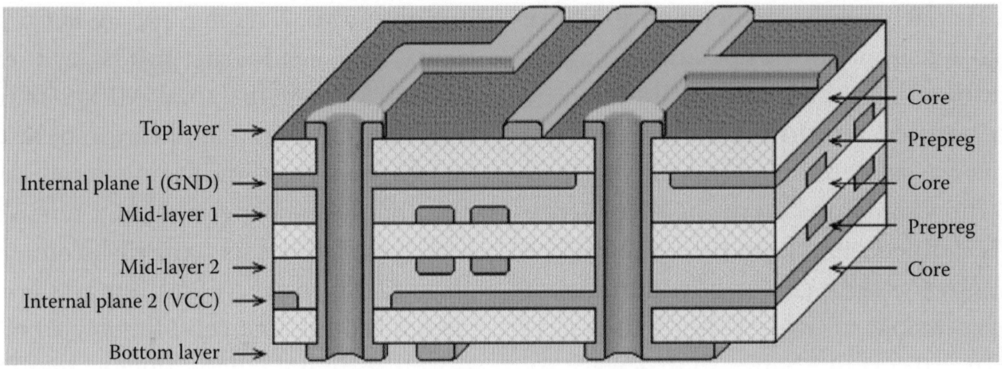


**FIGURE 8.40** A typical six-layer printed circuit board. VCC stands for the positive supply voltage applied to the collectors of a transistor circuit. A "prepreg" is an epoxy-coated glass fabric.

The expertise with copper plating at IBM dates back to the late 1960s at Watson, when Romankiw succeeded in electroplating narrow wires of copper onto thin film read-write heads for memory, using a masking method that deposited the copper only in circuit patterns where it was needed (see above). Copper wires conduct electricity with about 40% less resistance than traditional aluminum wires, leading to faster microprocessors. Copper wires are also less prone to electromigration that ultimately induces wiring failure. Called *damascene* in reference to the metallurgists in Damascus who produced the finest polished swords in medieval times, the technique was initially used to form "vias" linking separate layers of wiring in chips. In the original damascene patterning, the pattern of vias was first formed by etching a silicon dioxide or other insulator layer. The metal was deposited second, and the excess metal was removed by polishing. Electrodeposition was selected as the metal deposition method of choice despite the fact that some people thought the copper patterns would be filled with bubbles, or that the process was too "dirty." It turned out to have a faster rate of deposition, and the evenness of the copper film was better than in the case of electroless deposition. Also, copper CVD, an early favorite option, ran into other severe problems. To polish the copper at an acceptable rate while controlling corrosion, erosion, and other defects in the patterns, IBM pioneered a special chemical-mechanical process (CMP) that proved critical to the copper technology. The latter process is detailed in Chapter 1. The damascene process can also be applied to other good via and interconnect materials such as Al.

In Figure 8.41 we compare through-mask copper plating with the damascene process. Through-mask plating (Figure 8.41a) uses a masking material on top of the seed layer. Electroplating occurs only on those areas of the seed layer that are not covered by the mask. The masking material and the surrounding seed layer are subsequently removed. Damascene plating (Figure 8.41b), in contrast, involves deposition of the seed layer over a patterned material, which, in the case of interconnect structures, is the insulator, a functional part of the device that remains in place. The plated metal covers the entire surface, and excess metal must be removed by CMP.

Damascene electroplating it turned out is ideally suited for the fabrication of complex interconnect structures because it enables inlaying of metal in via holes and overlying lines at the same time in a process called *dual damascene,* as illustrated in Figure 8.42. In a dual-damascene structure as shown here, only a single metal deposition step is used to simultaneously form the main metal lines and the metal in the vias. That is, both trenches and vias are formed in a single dielectric layer. The vias and trenches are defined using two lithography steps. Trenches are typically etched to a depth of 4000–5000 Å, and the vias are typically 5000–7000 Å deep. After the via and trench recesses are etched, the via is filled in the same metal-deposition step that fills the trench. After filling, the excess metal that is deposited outside the trench is removed by a CMP process, and a planar structure with metal inlays is achieved. The process starts with depositing a thin layer (~250 Å) of silicon nitride ($Si_3N_4$) on the silicon surface and between layers of a low

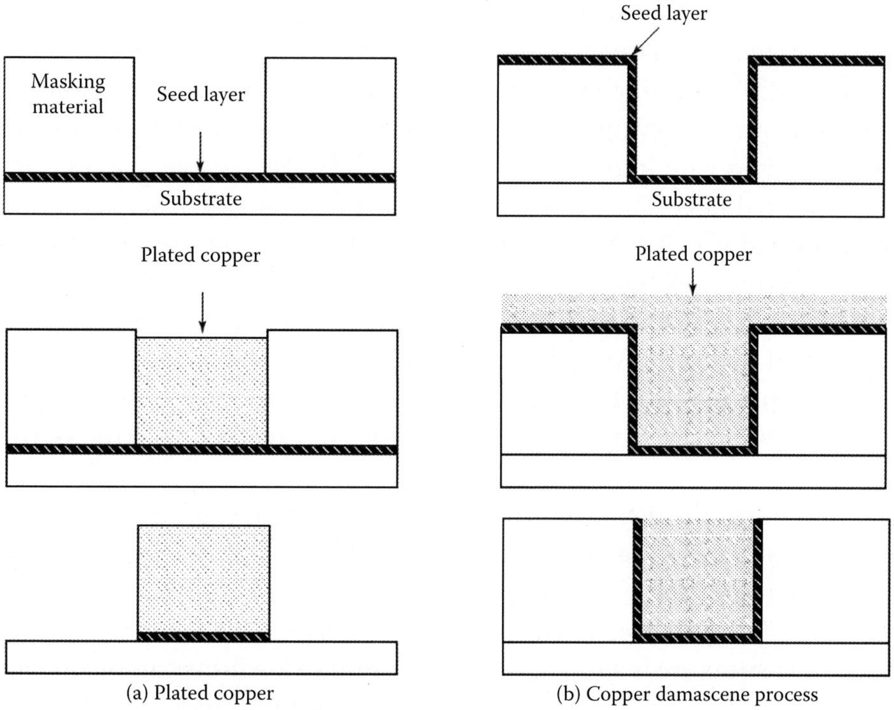


**FIGURE 8.41** (a) Standard copper plating: seed layer deposited, mask layer deposited and patterned, copper plated up, mask layer removed, and seed layer etched away. (b) Damascene process used to obtain highly planar surfaces: dielectric layer (insulator) deposited and patterned, seed layer deposited, copper plated, and surface polished with CMP.

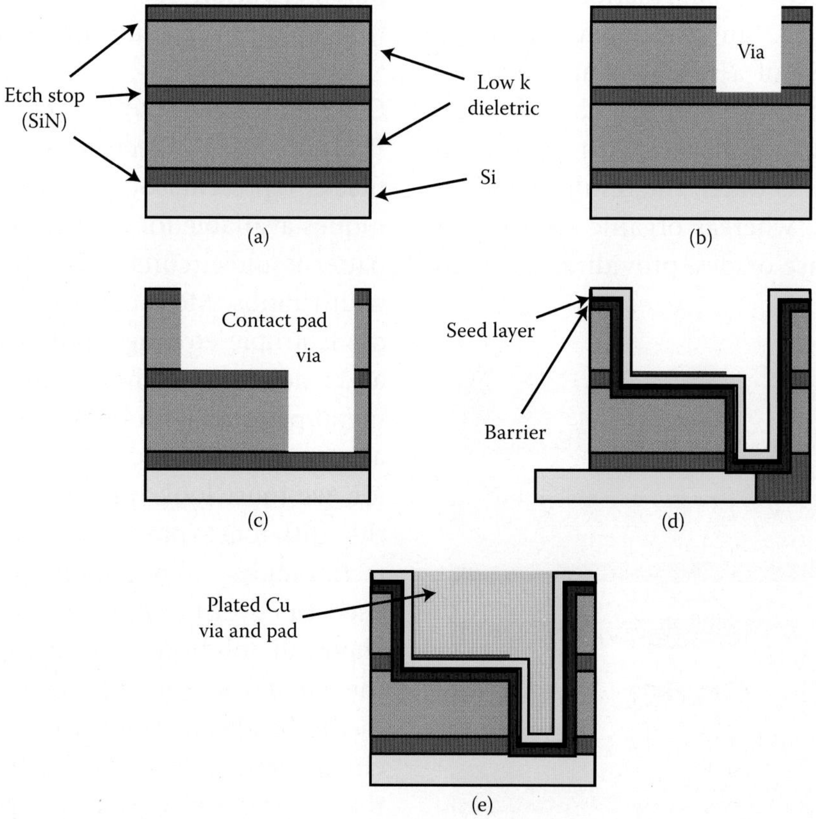


**FIGURE 8.42** Dual-damascene process: process steps for a via and a line. (a) Insulator, low *k* dielectric deposition. (b) Via definition. (c) Line/contact pad definition. (d) Barrier and seed layer deposition. (e) Plating and CMP.

$k$ dielectric, using CVD. The $Si_3N_4$ serves as an etch stop layer during via and trench formation. To overcome copper's tendency to diffuse in silicon, a metal diffusion barrier (e.g., Ta or TaN), preventing atoms from migrating out of a copper wire into surrounding chip material, is deposited next. Then follows a thin layer of copper deposited via CVD as a precursor to the subsequent electrodeposition of copper in both vias and trenches. The term "dual" refers to the formation of second channels, vias, within trenches. Further, it is compatible with the requirement for a barrier layer between the seed layer and the insulator.

As in conventional patterning, the damascene process is repeated many times to form the alternating layers of wires and vias that form the complete wiring system of a silicon chip. In Figure 8.43, an example copper dual-damascene structure is shown.

*Anodization* Anodization is an oxidation process performed in an electrolytic cell. The material to be anodized becomes the anode (+), whereas a noble metal is the cathode (–). Depending on the solubility of the anodic reaction products, an insoluble layer (e.g., an oxide, perhaps $TiO_2$) results, or in the case of a soluble reaction product, the electrode etches. The process can be either global or local through a masked area. If the primary oxidizing agent is water, the resulting oxides generally are porous, whereas organic electrolytes may lead to very dense oxides, providing excellent passivation.[66] Good oxides have been produced this way on W, Al, and Ta. In Volume I, Chapter 4, in the section on thermal oxidation, we learned that anodic oxidation of Si has not led to a commercially acceptable process, mainly because the interface state density at the $SiO_2/Si$ interface is prohibitively high for IC applications. As a sacrificial layer, anodic $SiO_2$ plays a role in micromachining and leads to uniquely structured films that cannot be obtained by chemical means. For example, anodization of Si in a highly concentrated HF solution (excellent etchant for the anodic oxidation product $SiO_2$) leads to porous Si and very high-aspect-ratio pores, with diameters ranging from 20 Å to several micrometers. The growth rate and degree of porosity of the Si can be controlled by the current density (see Chapter 4). Similarly porous aluminum oxide films can be prepared by anodization of aluminum foils.[67]

Through an appropriate choice of electrolyte, films other than oxides can be produced by anodization. For example, sulfide and selenide layers can be produced by anodization of Cd and Zn electrodes, and polymerization of redox polymers can be induced from solutions containing the monomers.

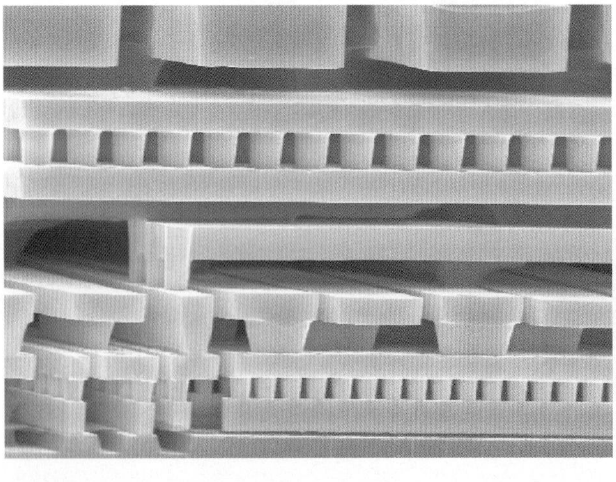

**FIGURE 8.43** Dual damascene copper. Note planarity of structure. (Courtesy of Motorola Inc.)

*Conclusions on Electroless and Electroplating through Polymer Masks* Electrochemistry through polymer masks represents one of the most powerful techniques available for formation of very high-density patterns and circuits with extremely large height-to-width ratios. Most other methods, such as chemical isotropic etching, sputter etching, ion milling, and reactive ion etching, cannot be used to produce metal patterns with better than 1:1 height-to-width aspect ratios. To understand this contrasting behavior, we must look into the fundamental nature of the different approaches. In nonelectrochemical technologies, a polymeric or inorganic mask is used to cover the existing metal, whereas either a chemical solution or an active gas phase removes the metal from the open areas. Hence the limitation in the height-to-width aspect ratio, the maximum circuit density, and the fidelity of reproduction of the features are caused by shadowing, which is the mechanism by which the patterning takes place. In contrast, with electrochemical techniques, one

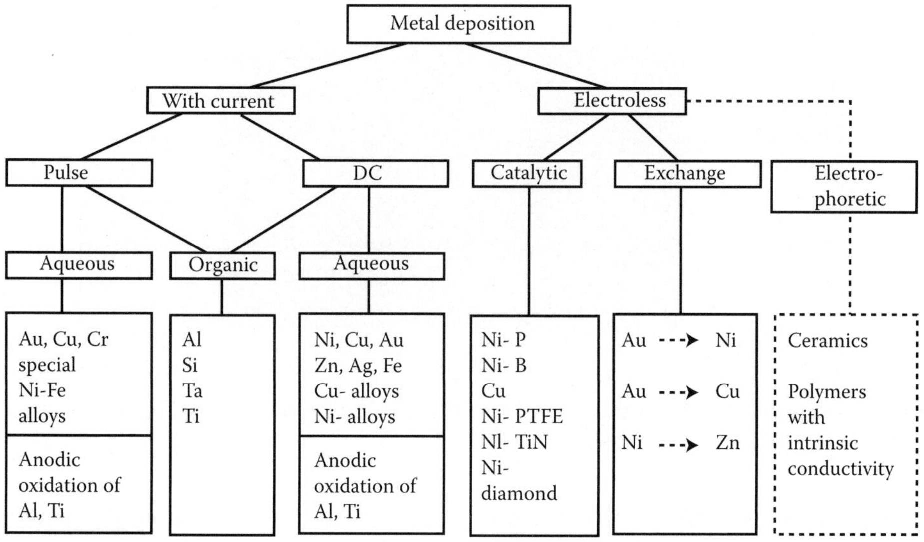


**FIGURE 8.44** Processes and materials for through-mask metal deposition. (From Ehrfeld, W. 1994. Notes from handouts. Banff, Canada. With permission.[68])

obtains the exact replication of the recesses in the mask.[52] Thus, electrodeposition reproduces the finest features of the mask with the greatest fidelity. This is not surprising, considering that, in plating, metal ions discharge from solution present inside the mold. In so doing, the metal displaces the solution from the mold atom by atom, conforming to the smallest features that exist in the mold.[51] A comparison of conventional subtractive etching with electrochemical additive processes was presented as early as 1973 by Romankiw et al.[61] For the fabrication of highly miniaturized magnetic bubble memory devices, this IBM group showed clearly that the additive plating process is superior to dry etching-based techniques.

Figure 8.44 features a list of choices for material deposition using electrochemical and electroless techniques.[68] The first metal LIGA structures (see Chapter 10) consisted of nickel, copper, or gold electrodeposited from suitable electrolytes.[69] Nickel-cobalt and nickel-iron alloys were also experimented with. A nickel-cobalt electrolyte for deposition of the corresponding alloys has been developed especially for the generation of microstructures with increased hardness (400 Vickers at 30% cobalt) and elastic limit.[70] The nickel-iron alloys permit tuning of magnetic and thermal properties of the crafted structures.[70,71] From Figure 8.44 it is obvious that many more materials could be combined with through-mask metal deposition.

*Instant Masking in Electrochemical Fabrication*

Instant masking is a selective electroplating technique similar in some ways to through-mask plating. However, through-mask plating involves up to 10 separate steps, multiple pieces of equipment, multiple liquids, and is time consuming and difficult to automate. Obviously it is not a technique that is very suitable for automatically building complex 3D microdevices consisting of hundreds of layers or more. On the other hand, instant masking operates much more like printing, in which prefabricated plates transfer ink onto a substrate. In instant masking, instead of inks, one deposits materials electrochemically.

Instant masking is the enabling technology of electrochemical fabrication (EFAB), a solid, free-form fabrication technology that creates complex, miniature 3D metal structures impossible or impractical to manufacture using other MEMS technologies, such as electrical discharge machining (EDM), laser machining, or silicon micromachining.[72] The automated EFAB process creates metal structures by electroplating multiple, independently patterned layers. The process is similar in concept to rapid prototyping (RP) techniques, such as stereolithography (SL), in that multiple patterned layers are stacked to build structures. However, unlike stereolithography, EFAB is a batch process suitable for volume production of fully functional devices, not just models and prototypes.

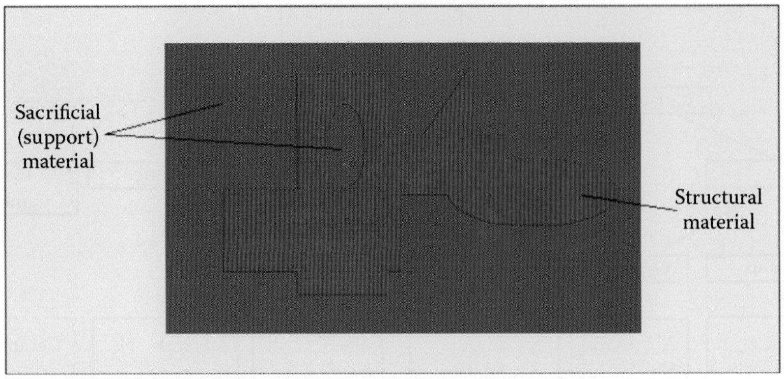


**FIGURE 8.45** EFAB layer example. This 5-μm-thick layer consists of a structural material (e.g., Ni) and a sacrificial (support) material (e.g., Cu). (http://www.isi.edu/efab.)

Furthermore, EFAB provides greater accuracy than stereolithography. The EFAB process was invented at the University of Southern California with funding from the Defense Advanced Research Projects Agency and is commercialized by MEMGen Corp., a company founded in August 1999 (MEMGen Corp. changed its name to Microfabrica in 2003; http://www.microfabrica.com).

The three major fabrication steps in EFAB that are repeated are 1) the instant masking and selective electroplating, 2) blanket deposition, and 3) planarization. The manufacturing process starts with a substrate and grows devices layer by layer. Each thin layer, perhaps 5 μm thick, consists of a structural material (e.g., Ni) and a sacrificial material (e.g., Cu) as shown in the example in Figure 8.45. Microdevices are built by stacking many of these layers (as many as desired). To fabricate a multilayer device, the geometries of the layer cross-sections are automatically determined based on the desired 3D geometry, and a set of instant masks is generated that includes all the unique cross-sections of the device. The device under construction, composed of structural material, is imbedded within the sacrificial material, which, like a scaffold, supports it during fabrication and is later removed. Additional materials can be deposited over entire layers without constraint. Such geometrical freedom makes possible monolithically fabricated assemblies of discrete, interconnected, and rotating parts and eliminates the need for subsequent bonding or assembly steps.

The instant mask consists of an insulator pattern, made, for example, by patterning a resist layer on an anode plate using photolithography. A series of such masks, each representing a thin cross-section through the part being built, is fabricated in a separate process before the formation of the part. Instant masking then patterns a substrate by pressing this insulator pattern on the anode against the cathode substrate (e.g., Ni), and the assembly is immersed in an electrochemical bath (e.g., a Ni bath). Electroplating from the nickel salt solution between anode and cathode only deposits metal (Ni) between the insulator parts on the anode. The result is a layer that has been rapidly deposited and patterned in a single step. After the nickel plating, the mask is removed, and the substrate with its nickel pattern is copper plated, filling in areas left by the insulator. The two materials are then polished flat, providing a smooth layer for the next mask. The process can be repeated indefinitely, which creates the potential for defining arbitrarily complex shapes. Once the final plating has taken place, the sacrificial copper is removed by chemical etching, leaving a 3D nickel structure.

The complete process flow is shown in Figure 8.46. In Figure 8.46a, the first material is patterned onto a substrate, producing a patterned layer. In Figure 8.46b, the second material has been blanket deposited over the first material so that it contacts the substrate in those regions not covered with the first material. Then, as shown in Figure 8.46c, the entire two-material layer is planarized to achieve precise thickness and flatness. After repetition of this process for all layers, the embedded multilayer structure shown in Figure 8.46d is etched to yield the desired device as shown in Figure 8.46e.

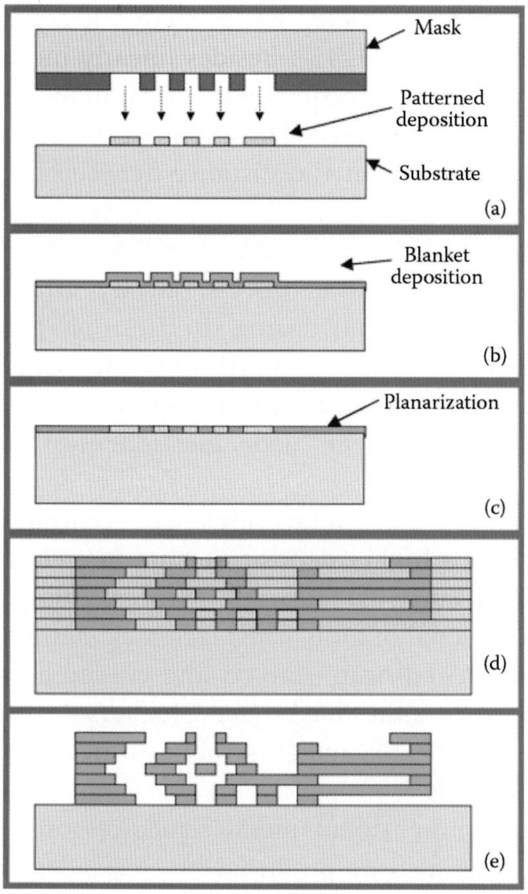


**FIGURE 8.46** The EFAB process. (a) First material is patterned onto a substrate, producing a patterned layer. (b) Second material is blanket deposited over the first material. (c) The entire two-material layer is planarized. (d) Repetition of the process steps (a)–(c). (d) Embedded multilayer structure. (e) Sacrificial material is etched to yield the desired device as shown.

The instant masking process is fast, which makes it possible to fabricate devices with a dozen or more completed layers in 1 day.

The precision EFAB technology allows miniature and microdevices to be generated from 3D computer-aided design (CAD) data. The system uses data from any standard CAD package to determine device cross-sections. These cross-sections are then used to fabricate the instant masks that are then used to rapidly create the devices. EFAB is a precision manufacturing technology that can be used to form structures from any metal or alloy that can be electrodeposited. The only constraint is that the accompanying sacrificial metal can be selectively etched after the layers are formed. EFAB, depending on the lithography used to make

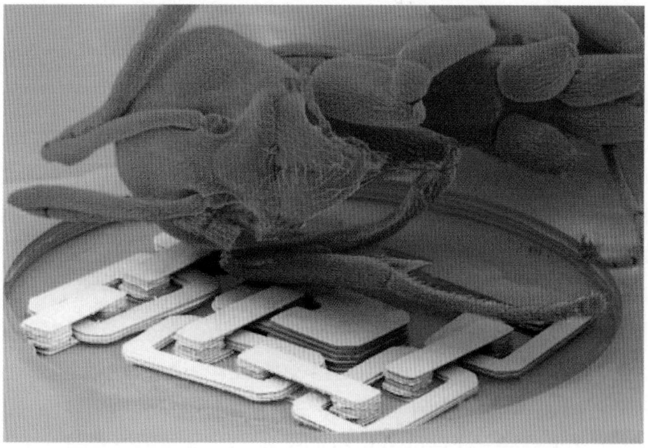

**FIGURE 8.47** A small household ant lies over a 12-layer microchain with independently movable links, fabricated in nickel in a single process without the need for assembly. The chain is about ~100 μm high (about the thickness of a sheet of paper; http://www.isi.edu/efab/Resources).

the instant masks, can produce parts with feature sizes less than 0.001 in. and tolerances better than 0.0001 in. Such precision is difficult to obtain with mechanical precision machining technology. Like a machine tool, the EFAB process is suitable for a wide range of applications (see example microchain in Figure 8.47). Applications include radiofrequency (RF) electronic components such as high-Q inductors and transmission lines, optical components, microfluidic networks, and mold tooling for plastic devices.

*Maskless Plating*

*Microjet Plating and Laser-Enhanced Jet Plating* Through-mask electrochemical deposition, although a very important MEMS technique, is costly and requires extra steps, and several electrochemical deposition techniques have been proposed that obviate the need for masks.

One of the simpler approaches to avoid using masks in electrodeposition uses impinging microjets of electrolyte. This increases the mass transport greatly, and plating selectivity is achieved by virtue of the fact that the jet serves as the current path. The smallest spot size of the area being plated is usually limited to twice the diameter of the jet, which can be operated, if filtering the solution carefully, without too frequent plugging. The nozzle orifice diameter may be anywhere from 1 mm to as little as ~0.01 mm in diameter. The lower limit depends strongly on the degree of filtration of the solution. Micromachining of nozzles will further impact obtainable microjet-plating resolution in the future.

The setup in Figure 4.15 for laser-enhanced jet etching can also be used for faster plating. The local heating of the substrate by the laser (up to 150°C) results in highly increased deposition. Using this technique, plating enhancement of up to 1000 times has been obtained with gold. Plating spots and lines as small as 2 μm wide have been observed. Laser enhancement also works with electroless deposition. Current densities of up to 15 A/cm$^2$ and copper plating rates of 50 μm/s have been obtained. For more background on jet plating and laser-enhanced jet plating, refer to the review by Romankiw et al. and references therein.[51]

Pulse plating, introduced in Chapter 10 in the context of plating in LIGA molds, represents only one of many emerging electrochemical plating techniques considered for microfabrication. For more background on techniques such as laser-enhanced plating, jet plating, laser-enhanced jet plating, and ultrasonically enhanced plating, refer to the review of Romankiw et al.[51] and references therein.

*Localized Electrochemical Deposition* Truly 3D microstructures can be formed using localized electrochemical deposition. The tip of a sharp electrode placed in a plating solution is brought near the substrate where deposition occurs when applying a potential between the tip and the substrate. The electric field is confined to the area beneath the tip, and the spatial resolution is determined by the size of the electrode. Deposition rates of Ni are as high as 6 μm/s, two orders of magnitude higher than in conventional electroplating.[73] The tip of a scanning electrochemical microscope (SECM) can be used for both local etching and deposition with high resolution in the $x$-, $y$-, and $z$-dimensions, basically forming a high-resolution electrochemical machining setup (Figure 4.16).

*Slanted and Curved Plated Metal Shapes* As a final example of maskless electrodeposition, we consider the work by Maciossek, who relies on overplating of a patterned plating base to create a variety of novel slanted and curved metallic microstructures.[59] The angle of the deposited wedge shapes is adjustable in a range from 0–45°. Besides linearly increasing structure profiles, a sinusoidal or parabolic surface can also be generated. The angles of the deposited wedges depend on the distance between and the width of a set of parallel metallic strips on the plating substrate as illustrated in Figure 8.48A. At the start of the plating process, only a central metal strip is biased, and during the isotropic growing process, the outlying metal strips are contacted one at a time by the electrodepositing metal. When the contact has been made, electrodeposition will occur also in this area, whereas further deposition in the original contact area (the central strip) results in a higher deposit on that first electrode. This innovative 3D fabrication process enlarges the MEMS arsenal of tools. The Maciossek method may be used to fabricate slanted and curved absorber patterns for x-ray masks (see Chapters 2 and 10). Two example Maciossek structures are shown in Figure 8.48B.

Three-dimensional metal structures on a substrate may also be produced by electrodeposition through a suitably prepared gelatin layer in a process similar to photolithography. Angus et al.[74] exposed a layer of gelatin on a Cu surface through a gray mask. Cross-linking of the gelatin occurs in the exposed areas, and unexposed material is washed away in a water/isopropanol solution. Thickness of the hardened gelatin layer, and hence the rate of diffusion of depositing metal ions through it, depends on the exposure dose. Metal electrodeposition through the developed gelatin layer film occurs more rapidly at the less exposed sites, thus generating a 3D metal structure on the substrate with a thickness and configuration governed by the exposure time, the mask thickness, and the deposition time.[75] As the mask is permeable to the plating solution, we treat this technique here as a "maskless" technology.

**Example 8.1: Protein Patterning**

Microlithographic techniques can be applied successfully to the field of protein patterning. In this first example, we show how clever use of lithography may help solve the problem

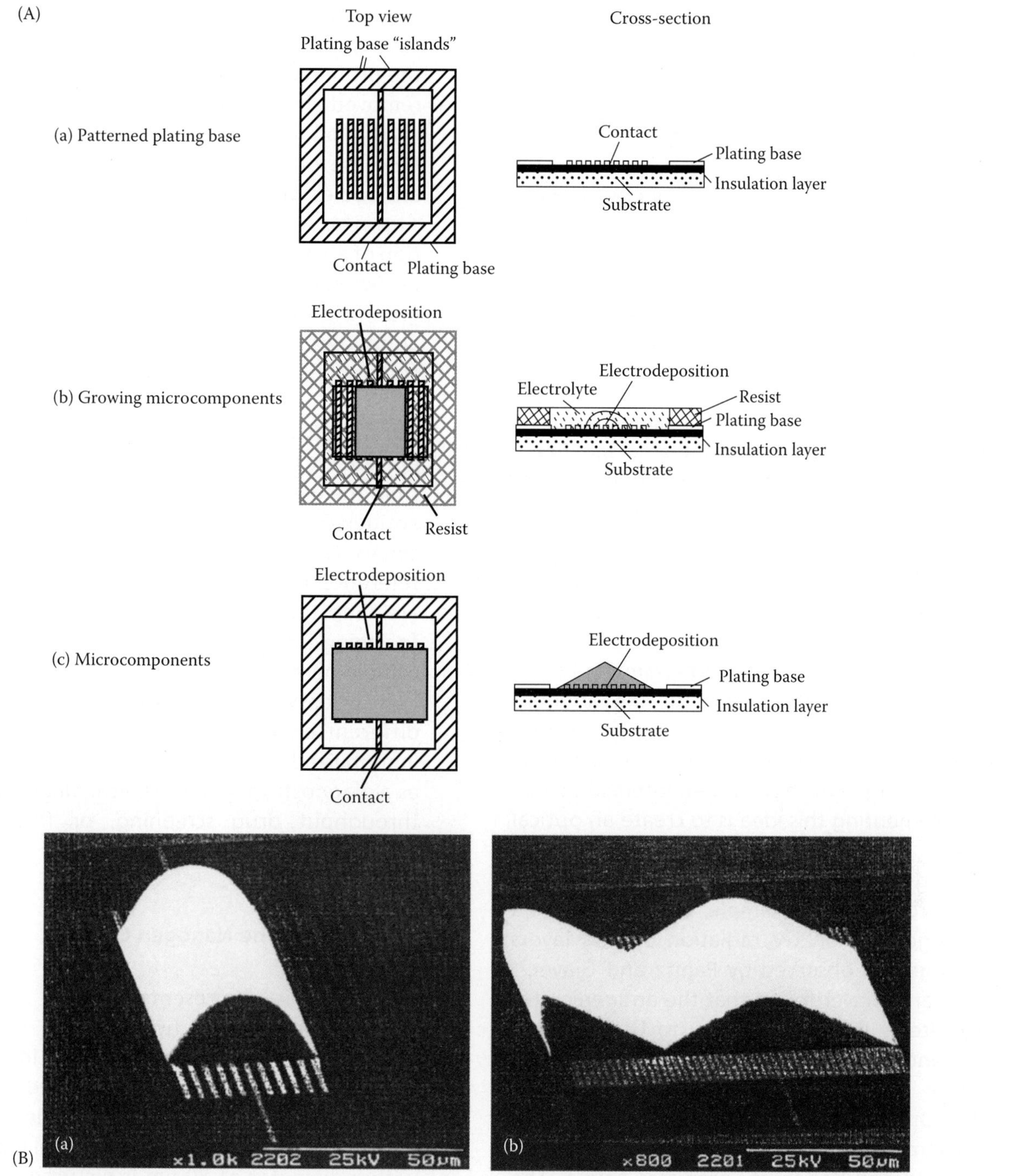


**FIGURE 8.48** (A) Maciossek plating schematic: exposing the patterned plating base (Aa) to a plating solution, the isotropic growing process starts at the contact area and extends until the neighboring plating base islands are contacted. (Ab) Once contacted, electrodeposition also occurs in this area, whereas further deposition in the former contact area results in a higher structure there. On continuing the electroplating process, consecutive outlying "island" electrodes are reached, and electroplating starts later so that wedge-shaped microcomponents are formed (Ac). (B) Electroplated profiles show a very smooth and bright surface. The perpendicular sidewalls are made by a thick surrounding resist. (Ba) Gold microstructure, 18 μm height; (Bb) electroplated gold with two electrical contacts.

of nonspecific protein adsorption competing for detection sites in immunosensors. Proteins bind with considerable avidity to a wide range of surfaces, and a reference surface in an immunosensor should adsorb all the same proteins except for the protein of interest. The best way to avoid nonspecific binding effects in immunosensors is to make the reference surface as similar to the sensing surface as possible, i.e., a reference substrate, subject to all the same

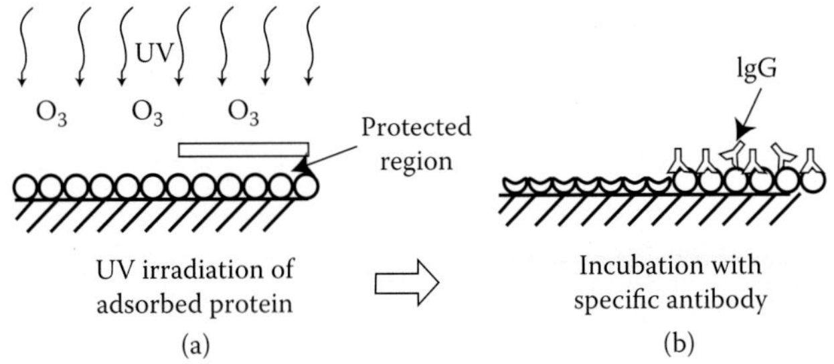


**FIGURE 8.49** UV radiation of antigens leads to bands of "live/dead" protein. (a) Preadsorbed layer of antigen is briefly exposed, in air, to an intense UV source, resulting in the production of ozone ($O_3$). The combination of UV damage and ozone results in partial oxidation of the adsorbed protein. The area of protein shielded from the irradiation is not oxidized. (b) Following this treatment the protein-coated substrate is incubated with antiserum specific to the adsorbed antigen. The partially oxidized protein layer is no longer antigenic, and IgG molecules are able to bind only to the previously shielded portion of the antigen layer. (Clementi, E., G. Corongiu, M. H. Sarma, and R. H. Sarma, eds. 1985. *Structure and motion: membranes, nucleic acids, and proteins.* Guilderland, NY: Adenine Press.[76])

nonspecific protein binding phenomena as the sensor surface itself, except for the antigen-antibody of interest. This way, measuring differentially enables the best possible correction for nonspecific binding. An interesting way of implementing this idea is to create an optical grating pattern, with the antigen-antibody coupling providing the grating structure. This can be realized, for example, based on the loss of antigenicity on UV radiation of thin layers of antigen as observed by Panitz and Giaver.[76] Panitz and Giaver found that the antigenic sites on proteins display sensitivity to UV radiation in air and that a small dose of UV radiation can destroy the antigenicity of antigens (Figure 8.49). One presumes that the UV light breaks up chemical bonds in the adsorbed protein layer and simultaneously produces ozone. The broken bonds are targets for the highly reactive ozone, resulting in the progressive oxidation of the protein layer. The protein can be completely removed from the surface by too long an irradiation (e.g., 10 min). With a short radiation time the optical density does not change, but antibodies will no longer bind to the antigenetic sites. This scheme was put to use in the fabrication of an elegant biograting for immunosensing as shown in Figure 8.50.[77] By using a photomask grating and inactivating alternating bands of antibodies, a biological diffraction grating is created. In this device, the antigen-antibody coupling itself constitutes the grating structure, causing diffraction only if the target antigen is present. A CD-type He-Ne laser beam diffracts from the grating with an intensity related to the antigen concentration. The above grating device comes close to embodying an ideal relative reference because the radiated protein bands are almost identical to the active surface except for being incapable of reacting with the target complementary protein.[78]

It is easy to imagine arrays with several different enzymes, antibodies, or DNA probes immobilized precisely onto a transducer surface as a diagnostic panel for clinical diagnosis, high-throughput drug screening, or for environmental monitoring.[79]

### Example 8.2: The Nanogen Chip

A schematic representation of an open electrophoretic DNA hybridization chip from Nanogen is shown in Figure 8.51a. In this chip DNA, which is negatively charged at most pHs, is collected at biased electrodes through electrophoresis. An array of 25 metal electrodes with a diameter of 80 μm is pictured. Arrays of

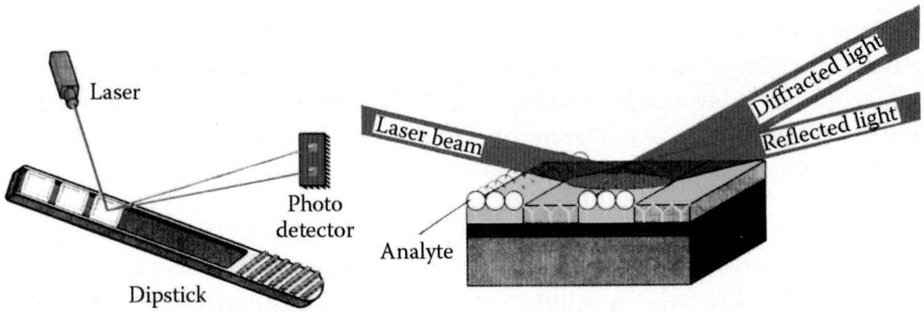


**FIGURE 8.50** Idetek, Inc. immunosensor based on biological grating (Editorial. 1993. Biograting. *Res Dev* 51; Courtesy of Mark Platshon, Idetek, Inc., Sunnyvale, CA.[77])

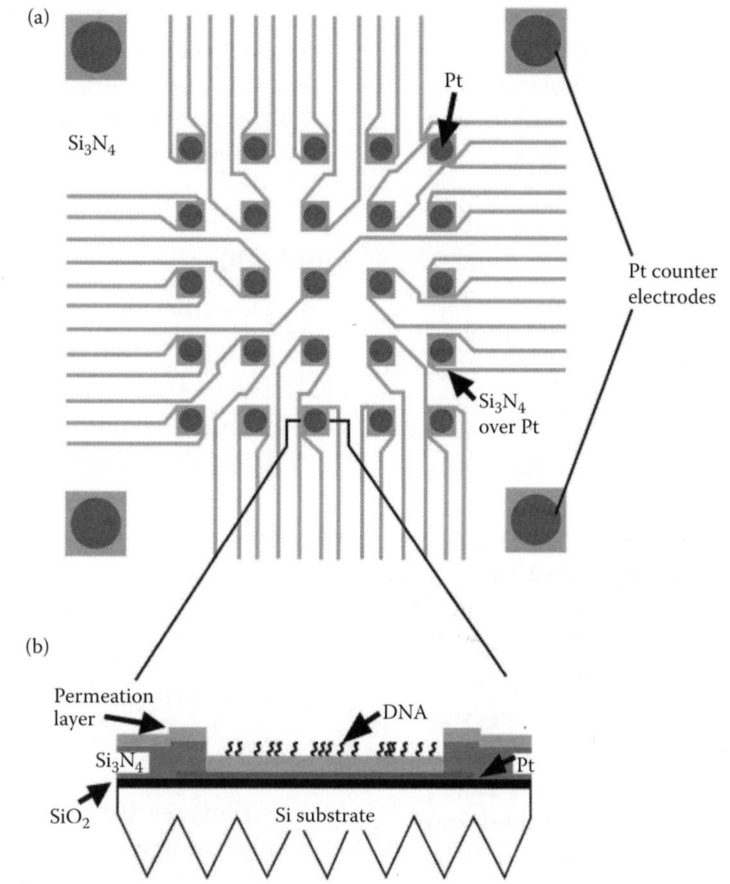


**FIGURE 8.51** Nanogen DNA chip. (a) Electrode array region of the chip. The central 1 × 1-mm test site array region consists of four large (160-μm diameter) corner electrodes and 25 central 80-μm electrodes. Pt, exposed Pt test sites; $Si_3N_4$, dielectric; $Si_3N_4$ over Pt, and Pt insulated by $Si_3N_4$. (b) Cross-section of an electrode test site. Location of section is indicated by lines extending from (a). Pt, $Si_3N_4$, Pt insulated by $Si_3N_4$; $SiO_2$, dielectric layer; Si substrate, wafer material; permeation layer, agarose layer containing streptavidin; DNA, biotinylated oligonucleotides bound to streptavidin.

100 (80-μm diameter), 400 (50-μm diameter), and 10,000 (30-μm diameter) elements have been made as well.[80] Each of the metal electrode sites (microlocations) is a discrete, insulated, and addressable working electrode forming part of an electrochemical cell. An outer group of counter electrodes provides encompassing electric fields for concentrating DNA from the bulk sample solution to specific working electrodes. In Figure 8.51a, there are four such outer electrodes with a diameter of 160 μm. The microlocations are all electrically connected via the overlying electrolyte solution and protected with a permeation layer hydrogel (e.g., agarose or polyacrylamide) (Figure 8.51b). The permeation layer is a 1–10-μm hydrogel layer deposited by spin coating and permits water and ion flow, provides a matrix to attach molecules such as DNA, RNA, or proteins, and separates those sensitive biomacromolecules from the potentially damaging electrochemical reactions that occur on the active Pt electrodes (current levels >100 nA and voltages >1.2 V are possible with such hydrogel layers without damaging the biomacromolecules). Attachment of molecules to the permeation layer such as DNA probes can be achieved, for example, by impregnating the permeation layer with affinity binding substances such as avidin or streptavidin for subsequent attachment of biotinylated DNA or RNA probes.

Net negatively charged molecules such as DNA and RNA are moved to positively biased microlocations, whereas they are repelled from negatively biased ones by electrophoresis. For the electric field to be high enough throughout the solution for electrophoresis to take place, the solution must be desalted. Only low conductivity buffers may be used. The positive microlocations may concentrate target DNA sequences in a very short time (because they reach those sites through electrophoresis rather than by diffusion), and if this location has complementary DNA capture probes, hybridization of the target DNA can occur. The concentrating effect on the positive microlocations facilitates the hybridization as a result of the law of mass action (hybridization in seconds rather than hours). Details of DC current and voltage level, solution conductivity, and buffer species for hybridization can be found in Edman et al.[81] This directed electrophoretic transport and addressing process can be carried out simultaneously at test sites that have different capture sites. Detection of successful hybridization can be accomplished using fluorescent probes and a color, charge-coupled detector (CCD). By reversing the polarity at a microlocation after hybridization, nonspecific sample DNA and unhybridized probes may be selectively removed. The electric field can be adjusted just so to affect selective dehybridization of the DNA sequences from the attached complementary probe. This is called *electronic stringency control.* Discrimination between single base pair mismatches in DNA has been demonstrated this way;[82] for example, Gilles et al. developed a rapid assay for single-nucleotide polymorphisms (SNPs) in a gene implicated in increased susceptibility to infection in pediatric patients.[83]

The power of this open-chip approach was further demonstrated by separating *Escherichia coli* from a mixture containing blood cells by means of dielectrophoresis, lyzing the isolated bacteria by a series of high-voltage pulses, and proteolytic digestion with proteinase K—all on the same chip. The microlocations on the Nanogen chip can be powered with DC and AC.[84] Even more exciting is the demonstration of amplification and detection of multiple targets in this same open format [multiplex strand displacement amplification (SDA)]. SDA is an isothermal (60°C) DNA amplification method. In a regular multiplex amplification in solution, nonspecific interactions between different primer sets reduce the amplification efficiency. Electronic anchoring of sets of amplification primers in distinct areas reduces their interaction, and sets of distinct zones of amplification that only share reagents and enzymes are generated, increasing the efficiency of the multiplex amplification reactions. Isothermal SDA, which uses the combined effects of a restriction endonuclease and DNA polymerase, asynchronously amplifies DNA exponentially and was shown to be ideal for use on the open electrophoresis chip.[85]

## Questions

8.1: Describe two methods to make DNA arrays using a virtual mask approach.

8.2: Briefly sketch the sol-gel technique. Why is this technique so important for nanotechnology?

8.3: Compare Langmuir-Blodgett deposition with self-assembled monolayers (SAMs).

8.4: Sketch the dual damascene process. Where is it used?

8.5: Outline the EFAB process. Where is it used?

8.6: Why is DNA hybridization so much faster in an electronic DNA (Nanogen) than in a passive hybrydization chip?

8.7: Describe how you could make beveled metal structures using both electroless plating and electroplating.

8.8: Describe briefly different types of rapid prototyping (RP) and discuss how they can be scaled down. Which RP technique is of most interest to micromachinists?

8.9: Compare three major arraying methods for making DNA and protein arrays.

8.10: Detail the difference between scanning microstereolithography, parallel microstereolithography, and two-photon stereolithography.

## References


1. Rath, K. 2005. Novel materials from solgel chemistry. *Science Technol Rev*, Lawrence Livermore National Laboratory, May 13, 2005.
2. Licari, J. J. 1970. *Plastic coatings for electronics.* New York: McGraw-Hill.
3. Howatt, G. N. 1952. *Method of producing high-dielectric high-insulation ceramic plates.* US Patent 2582993.
4. Jenkins, A. T. A., J. Hu, Y. Z. Wang, S. Schiller, R. Förch, and W. Knoll. 2000. Pulsed plasma deposited maleic anhydride thin films as supports for lipid bilayers. *Langmuir* 16:6381–84.
5. Bauer, M., T. Schwarz-Selinger, H. Kang, and A. v. Keudell. 2005. Control of the plasma chemistry of a pulsed inductively coupled methane plasma. *Plasma Sources Science Technol* 14:543–48.

6. Guckel, H., J. Uglow, M. Lin, D. Denton, J. Tobin, K. Euch, and M. Juda. 1988. Plasma polymerization of methylmethacrylate: a photoresist for 3-D applications. *Technical digest: 1988 solid state sensor and actuator workshop.* Hilton Head Island, SC.
7. Vossen, J. L., and W. Kern, eds. 1978. *Thin film processes.* Orlando, FL: Academic Press.
8. Biederman, H., ed. 2004. *Plasma polymer films.* London: Imperial College Press.
9. Acheson. 1991. Brochure. Scheemda, the Netherlands: Acheson Coloiden BV.
10. Pace, S. J., and M. A. Jensen. 1986. *Second international meeting on chemical sensors.* Bordeaux, France.
11. Belford, R. E., A. E. Owen, and R. G. Kelly. 1987. Thick-film hybrid pH sensors. *Sensors Actuators* 11:387–98.
12. Pace, S. J., P. P. Zarzycki, R. T. McKeever, and L. Pelosi. 1985. A thick-film multi-layered oxygen sensor. *Third international conference on solid-state sensors and actuators (Transducers '85).* Philadelphia.
13. Karagounis, V., L. Lun, and C. Liu. 1986. A thick-film multiple component cathode three-electrode oxygen sensor. *IEEE Trans Biomed Eng* BME-33:108–12.
14. Lewandowski, J. J., P. S. Malchesky, M. Zborowski, and Y. Nose. 1987. *Proceedings: ninth annual conference of the IEEE engineering in medicine and biology society.* Boston.
15. Lambrechts, M., and W. Sansen. 1992. *Biosensors: microelectrical devices.* Philadelphia: Institute of Physics Publishing.
16. Weetall, H. H., and T. Hotaling. 1987. A simple, inexpensive, disposable electrochemical sensor for clinical and immuno-assay. *Biosensors* 3:57–63.
17. Cha, C. S., M. J. Shao, and C. C. Liu. 1990. Electrochemical behavior of microfabricated thick-film electrodes. *Sensors Actuators B* B2:277–81.
18. Chen, Y., J. Au, P. Kazlas, A. Ritenour, H. Gates, and M. McCreary. 2003. Electronic paper: flexible active-matrix electronic ink display. *Nature* 423:136.
19. Allara, D. L., and R. G. Nuzzo. 1985. Spontaneously organized molecular assemblies; II. Quantitative infrared spectroscopic determination of equilibrium structures of solution adsorbed n-alkanoic acids on an oxidized aluminum surface. *Langmuir* 1:52–66.
20. Allara, D. L., and R. G. Nuzzo. 1985. Spontaneously organized molecular assemblies; I. formation, dynamics, and physical properties of n-alkanoic acids adsorbed from solution on an oxidized aluminum surface. *Langmuir* 1:45–52.
21. Delamarche, E., B. Michel, H. A. Biebuyck, and C. Gerber. 1996. Golden interfaces: the surface of self-assembled monolayers. *Adv Mater* 8:719–29.
22. Liu, R. H., Q. Yu, J. M. Bauer, B.-H. Jo, J. S. Moore, and D. J. Beebe. 2000. *Micro total analysis systems 2000: proceedings, μTAS 2000 symposium.* Eds. A. van den Berg, W. Olthuis, and P. Bergveld. Enschede, the Netherlands: Kluwer.
23. Zubritsky, E. 2000. Microscale muscles. *Anal Chem* August:517–18.
24. Fodor, P. A. S., J. L. Read, M. C. Pirrung, L. Sryer, A. T. Lu, and D. Solas. 1991. Light-directed, spatially addressable parallel chemical synthesis. *Science* 251:767–73.
25. Britland, S. T., G. R. Moores, P. Clark, and P. Connolly. 1990. Patterning and cell adhesion and movement on artificial substrate: a simple method. *J Anat* 170:235–36.
26. Douglas, K., G. Deavaud, and N. A. Clark. 1992. Transfer of biologically derived nanometer scale patterns to smooth substrates. *Science* 257:642–44.
27. Singh-Gasson, S., R. D. Green, Y. Yue, C. Nelson, F. Blattner, M. R. Sussman, and F. Cerrina. 1999. Maskless fabrication of light-directed oligonucleotide micromirrors using a digital micromirror array. *Nat Biotechnol* 17:974–78.
28. Bertsch, A., J. Y. Jézéquel, and J. C. André. 1997. Study of the spatial resolution of a new 3D microfabrication process: the microstereophotolithography using a dynamic mask-generator technique. *Photochem Photobiol A: Chem* 107:275–81.
29. Okamoto, T., T. Suzuki, and N. Yamamoto. 2000. Microarray fabrication with covalent attachment of DNA using bubble jet technology. *Nat Biotechnol* 18:438–41.
30. St. John, P. M., R. Davis, N. Cady, J. Czajka, C. A. Batt, and H. G. Craighead. 1998. Diffraction-based cell detection using a microcontact printed antibody grating. *Anal Chem* 70:1108–11.
31. Madou, M., and T. Otagawa. 1992. *Tetrasulfonated metal phthalocyanine doped electrically conducting electrochromic poly(dithiophene) polymers.* US Patent 5,151,224.
32. Otagawa, T., and M. Madou. 1991. Permanently doped polyaniline and method thereof. US Patent 5,002,700, Osaka Gas Company.
33. Ikuta, K., and K. Hirowatari. 1993. *Proceedings: IEEE micro electro mechanical systems (MEMS '93).* Fort Lauderdale, FL.
34. Ehmann, K. F., D. Bourell, M. L. Culpepper, T. J. Hodgson, T. R. Kurfess, M. Madou, K. Rajurkar, and R. DeVor, eds. 2007. *Micromanufacturing: international research and development.* New York: Springer.
35. Ikuta, K., K. Hirowatari, and T. Ogata. 1994. *IEEE international workshop on micro electro mechanical systems (MEMS '94).* Oiso, Japan: IEEE.
36. Takagi, T., and N. Nakajima. 1993. *Proceedings: IEEE micro electro mechanical systems (MEMS '93).* Fort Lauderdale, FL: IEEE.
37. Ikuta, K., T. Ogata, M. Tsubio, and S. Kojima. 1996. Development of mass productive micro stereo lithography. *Proceedings of IEEE International workshop on micro electro mechanical systems.* MEMS '96. San Diego, CA.
38. Takagi, T., and N. Nakajima. 1994. *IEEE international workshop on micro electro mechanical systems (MEMS '94).* Oiso, Japan.
39. Ikuta, K., S. Maruo, and S. Kojima. 1998. New micro stereo lithography for freely movable 3-D micro structure – Super IH process with submicron resolution. *Proceedings of the 11th IEEE Workshop on micro electro mechanical systems (MEMS'98).* Heidelberg, Germany.
40. Maruo, S., and S. Kawata. 1997. Two-photo-absorbed photopolymerization for three-dimensional microfabrication. *Proceedings of the IEEE micro electro mechanical systems workshop (MEMS '97).* Nagoya, Japan.
41. Maruo, S., and S. Kawata. 1998. Two-photon-absorbed near-infrared photopolymerization for three-dimensional microfabrication. *JMEMS* 7:411–15.
42. Angelo, M. D., M. V. Chekhova, and Y. H. Shih. 2001. Two-photon diffraction and quantum lithography. *Phys Rev Lett* 87:013602.
43. Boto, A. N., P. Kok, D. S. Abrams, S. L. Braunstein, C. P. Williams, and J. P. Dowling. 2000. Quantum interferometric optical lithography: exploiting entanglement to beat the diffraction limit. *Phys Rev Lett* 85:2733–36.
44. Prasad, P. N. 2004. *Nanophotonics.* Hoboken, NJ: Wiley-Interscience.
45. Zhang, X., X. N. Jiang, and C. Sun. 1999. Micro-stereo lithography of polymeric and ceramic microstructures. *Sensors Actuators A* A77:149–56.

46. Taylor, C. S., P. Cherkas, H. Hampton, J. J. Frantzen, B. O. Shah, W. B. Tiffany, L. Nanis, P. Booker, A. Salahieh, and R. Hansen. 1994. A spatial forming—a three dimensional printing process. *Proceedings of IEEE MEMS '94*. Oiso, Japan.
47. Beluze, L., A. Bertsch, and P. Renaud. 1999. *Symposium on design, test, and microfabrication of MEMS and MOEMS*. Paris.
48. Bertsch, A., H. Lorenz, and P. Renaud. 1999. 3D microfabrication by combining microstereolithography and thick resist UV lithography. *Sensors Actuators A* 73:14–23.
49. Bertsch, A., P. Bernhard, C. Vogt, and P. Renaud. 2001. Rapid prototyping of small size objects. *Rapid Prototyping J* 6:259–66.
50. Sun, C., N. Fang, D. M. Wu, and X. Zhang. 2005. Projection micro-stereolithography using digital micro-mirror dynamic mask. *Sensors Actuators A* 121:113–20.
51. Romankiw, L. T., and T. A. Palumbo. 1987. Electrodeposition in the electronic industry. *Proceedings: symposium on electrodeposition technology, theory, and practice.* San Diego, CA.
52. Romankiw, L. T. 1984. Electrochemical technology in electronics today and its future: a review. *Oberflache-Surface* 25:238–47.
53. Mallory, G. O., and J. B. Hadju, eds. 1990. *Electroless plating: fundamentals and applications.* Orlando, FL: American Electroplaters and Surface Finishers Society (AESF).
54. Romankiw, L. T. 1976. Pattern generation in metal films using wet chemical techniques: a review. *Proceedings of the Symposium on Etching for Pattern Definition*. Ed. H. G. Hughes, and M. Rand. Pennington, NJ: The Electrochemical Society.
55. Ohno, I. 1988. Electroless deposition of metals and alloys. 172nd Meeting of The Electrochemical Society, Honolulu, HI. *ECS Proceedings, Vol. 88-12*. Eds. M. Paunovic, and I. Ohno. Pennington, NJ: ECS.
56. Harsch, S., W. Ehrfeld, and A. Maner. 1988. *Untersuchungen zur herstellung von mikrostrukturen grosser strukturhöhe durch galvanoformung im nickelsilfamatelektrolyten. KfK, Report No. 4455.* Karlsruhe, Germany. Kernforschungszentrum Karlsruhe.
57. van der Putten, A. M. T., and J. W. G. de Bakker. 1993. Anisotropic deposition of electroless nickel-bevel plating. *J Electrochem Soc* 140:2229–35.
58. van der Putten, A. M. T., and J. W. G. de Bakker. 1993. Geometrical effects in the electroless metallization of fine metal patterns. *J Electrochem Soc* 140:2221–28.
59. Maciossek, A. 1996. *SPIE: micromachining and microfabrication process technology II.* Austin, TX: SPIE.
60. Kenis, P. J. A., R. F. Ismagilov, and G. M. Whitesides. 1999. Microfabrication inside capillaries using multiphase laminar flow patterning. *Science* 285:83–85.
61. Romankiw, L. T., S. Krongelb, E. E. Castellani, J. Powers, A. Pfeiffer, and B. Stoeber. 1973. Additive electroplating technique for fabrication of magnetic devices. *Proceedings of the international conference on magnetics, ICM-73.*, Vol. 6, pp. 104–111.
62. Romankiw, L. T., I. M. Croll, and M. Hatzakis. 1970. Batch-fabricated thin-film magnetic recording heads. *IEEE Trans Magn* 6:597–601.
63. Romankiw, L. T. 1989. *Proceedings: symposium on magnetic materials, processes, and devices.* Hollywood, FL.
64. Andricacos, P. C., and N. Robertson. 1998. Future directions in electroplated materials for thin-film recording heads. *IBM J Res Dev* 42:671.
65. Editorial. 1997. *Back to the future: copper comes of age. IBM Res* 35.
66. Madou, M. J., W. P. Gomes, F. Fransen, and F. Cardon. 1982. Anodic oxidation of p-type silicon in methanol as compared to glycol. *J Electrochem Soc* 129:2749–52.
67. Miller, C. J., and M. Majda. 1986. Microporous aluminium oxide films at electrodes. Part II: studies of electron transport in the $Al_2O_3$ matrix derivated by adsorption of poly (4-vinylpyridine). *J Electroanal Chem* 207:49–72.
68. Ehrfeld, W. 1994. LIGA at IMM. Notes from handouts. Banff, Canada.
69. Maner, A., S. Harsch, and W. Ehrfeldg. 1988. Mass production of microdevices with extreme aspect ratios by electroforming. *Plating Surface Finishing* March: 60–65.
70. Harsch, S., D. Munchmeyer, and H. Reinecke. 1991. *Proceedings: 78th AESF annual technical conference (SUR/FIN '91)*. Toronto.
71. Thomes, A., W. Stark, H. Goller, and H. Liebscher. 1992. Erste ergebnisse zur galvanoformung von LIGA mikrostrukturen aus eisen-nickel legierungel. *Symp. Mikroelektrochemie Friedrichsroda*. Germany.
72. Cohen, A., G. Zhang, F. Tseng, U. Frodis, F. Mansfeld, and P. Will. 1999. EFAB: rapid, low-cost desktop micromachining of high aspect ratio true 3-D MEMS. *Proceedings of the 12th IEEE micro electro mechanical systems workshop*. Orlando, FL, January 1999, p. 244.
73. Madden, J. D., and I. W. Hunter. 1996. Three-dimensional microfabrication by localized electrochemical deposition. *J Microelectromechan Syst* 5:24–32.
74. Angus, J. C., U. Landau, S. H. Liao, and M. C. Wang. 1986. Controlled electroplating through gelatin films. *J Electrochem Soc* 133:1152–60.
75. Bard, A. J. 1994. *Integrated chemical systems.* New York: Wiley.
76. Clementi, E., G. Corongiu, M. H. Sarma, and R. H. Sarma, eds. 1985. *Structure and motion: membranes, nucleic acids, and proteins.* Guilderland, NY: Adenine Press.
77. Editorial. 1993. Biograting. *Res Dev* 51.
78. Madou, M. J., and J. Joseph. 1993. Immunosensors with commercial potential. *Immunomethods* 3:134–52.
79. Connolly, P., G. R. Moores, W. Monaghan, J. Shen, S. Britland, and P. Clark. 1992. Microelectronic and nanoelectronic interfacing techniques for biological systems. *Sensors Actuators B* B6:113–21.
80. Heller, M. J., A. H. Forster, and E. Tu. 2000. Active microelectronic chip devices which utilize controlled electrophoretic fields for multiplex DNA hybridization and other genomic applications. *Electrophoresis* 21:157–64.
81. Edman, C. F., D. E. Raymond, D. J. Wu, E. Tu, R. G. Sosnowski, W. F. Butler, M. Nerenberg, and M. J. Heller. Electric field directed nucleic acid hybridization on microchips. *Nucleic Acid Res* 25:4907–14.
82. Heller, M. J. 1996. An active microelectronic device for multiplex DNA analysis. *IEEE Eng Med Biol* 15: 100–04.
83. Gilles, P. N., D. J. Wu, C. B. Foster, P. J. Dillon, and S. J. Chanock. 1999. Single nucleotide polymorphic discrimination by an electronic dot blot assay on semiconductor microchips. *Nat Biotechnol* 17:365–70.
84. Cheng, J., E. L. Sheldon, L. Wu, A. Uribe, L. O. Gerrue, J. Carrino, M. J. Heller, and J. P. O'Connell. 1998. Preparation and hybridization analysis of DNA/RNA from E. coli on microfabricated bioelectronic chips. *Nat Biotechnol* 16:501–46.
85. Westin, L., X. Xu, C. Miller, L. Wang, C. F. Edman, and M. Nerenberg. 2000. Anchored multiplex amplification on a microelectronic chip array. *Nat Biotechnol* 18:190–204.
86. Kawata, S., H.-B. Sun, T. Tanaka, and K. Takada, 2001. Finer features for functional microdevices. *Nature* 412: 697–698.

# 9

# Thermal Energy-Based Forming Techniques—Thermoforming

A metal laser sintering/milling hybrid machine, the LUMEX 25C, developed by Matsuura Machinery Corporation in Japan, successfully combines freeform manufacturing and high-speed milling. The integration of laser sintering of metallic powder and high-speed cutting eliminates the finish machining operations (http://www.matsuura.co.jp).

A four-jet lathe burner for glass blowing. This type of burner is usually mounted on a movable carriage attached to a glassblower's lathe.

## Introduction

Thermal forming (or thermoforming) is a method where thermal energy provided by a heat source "transforms" a material's structure and/or shape. Structural material changes occur, for example, during crystallization, sintering, alloying, annealing, decomposition, and pyrolizing. Shaping in thermoforming may be free form, such as in glass blowing and in selective laser sintering (SLS) (see inset in the heading of this chapter). The latter thermoforming method is a rapid prototyping (RP) technique for solid parts, made by solidifying powder-like materials layer by layer through exposure of the surface of a powder bed to a laser beam. However, thermoforming may also involve a step that combines heat and pressure to conform to a final shape in a mold, such as in vacuum forming, a method for thermally shaping a flat plastic sheet into a three-dimensional (3D) shape by heating the sheet and withdrawing the air between the sheet and the mold.

Thermal and chemical forming techniques are often intertwined because heat (and catalysts) usually aids a chemical reaction. The relative contribution of thermal to chemical in a given process may range from 100% thermal to 100% chemical but falls often somewhere in between. Crystallization (see Volume I, Chapter 4), alloying, and sintering, for example, do not involve a chemical reaction and could be viewed as purely thermal forming techniques, but annealing, decomposition, and pyrolizing all typically involve both heat and a chemical reaction. The predominantly thermal techniques detailed in this chapter are spray pyrolysis, silk screening of ceramics, tape casting, plasma spraying, laser and electron beam deposition, and selective laser sintering (SLS). Spray pyrolysis is a simple form of chemical vapor deposition (CVD) and, together with sol-gel, a chemical forming technique that was covered in Chapter 8, a contender for nanoparticle manufacturing. Silk-screening and tape casting technologyconstitutetwoimportantthermalforming processes often referred to as *hybrid* manufacturing methodologies. Tape casting is also known as *doctor blading* or *knife coating*, and the process is well known under these names in many industries, including paper, plastic, and paint manufacturing. Laser and electron beam deposition can be subtractive or additive, and these two options are typically available in the same laser or electron beam reactor. SLS is a rapid prototyping technique (RP) (see also Chapter 8, Table 8.3) method for solid parts, made by solidifying powder-like materials layer by layer through exposure of the surface of a powder bed to a laser beam. The method might also integrate high-speed computer numerical controlled cutting, effectively combining an additive and a subtractive manufacturing technique, in which case RP and real manufacturing are both options.

## Spray Pyrolysis

### Introduction

In spray pyrolysis, a precursor reagent dissolved in a carrier liquid is sprayed through a nozzle or set of nozzles to form an aerosol, and the droplets in the aerosol experience thermal decomposition, i.e., pyrolysis in a hot zone. In the case of film formation, the hot zone is simply a hot substrate surface, and in the case of particle formation, the hot zone may be a furnace. Spray pyrolysis is also known as *aerosol decomposition synthesis*, in which a droplet-to-solid particle conversion takes place.

The production of droplets and their dispersion into a gas environment is called *atomization*. There are a number of different kinds of atomizers in use, as listed in Table 9.1. The atomization variables listed here are droplet size, which determines the size of the final solid particle, the atomization rate, which affects the scalability of the process, and the droplet velocity, which affects the residence time of the droplets in the hot zone. One other important parameter, not listed, is the droplet size dispersion, which relates to the homogeneity of the product.

Spray pyrolysis is a gas-phase synthesis method and constitutes a form of chemical vapor deposition (CVD) without the expensive vacuum reactors. During CVD, as we saw in Chapter Seven, the constituents of a vapor phase, often diluted with an inert carrier gas, react at a hot surface (typically higher than 300°C) to deposit a solid thin film. The CVD

Table 9.1 Characteristics of Atomizers Typically Used in Spray Pyrolysis

| Atomizer | Droplet Size (μm) | Atomization Rate ($cm^3$/min) | Droplet Velocity (m/s) |
|---|---|---|---|
| Pressure | 10–100 | 3–no limit | 5–20 |
| Nebulizer | 0.1–2 | 0.5–5 | 0.2–0.4 |
| Ultrasonic | 1–100 | <2 | 0.2–0.4 |
| Electrostatic | 0.1–10 | | |

*Source:* Messing, G. L., S.-C. Zhang, and G. V. Jayanthi. 1993. Ceramic powder synthesis by spray pyrolysis. *J Am Cer Soc* 76:2707–26.[1]

technologies mostly evolved around the integrated circuit (IC) industry. However, spray pyrolysis, the simplest form of CVD, has never been a contender in that arena. Rather, it is a viable technology for large-area films in solar cells and window panes and, more recently, for the production of nanopowders. Nanoparticle synthesis is a burgeoning field with a broad range of applications. The "holy grail" of research in this field is the ability to generate particles with controlled and narrow size distribution and adequate throughput; thus, the particle size can be tailored as required by the application. In spray pyrolysis when used for nanoparticle synthesis rather than film formation, an organic precursor (liquid or gas) is forced through an orifice at high pressure and burned in a furnace. The resulting ash is air classified to recover the nanoparticles (typically metal oxides).

## How Does It Work?

As mentioned above, spray pyrolysis produces films or powders. A simple spray pyrolysis setup for film deposition is shown in Figure 9.1.[2] Metallic salt solutions are sprayed onto a hot substrate where droplets experience thermal decomposition. Spray pyrolysis tends to work best for ceramics [inorganic, nonmetallic materials that are typically crystalline in nature and are compounds formed between metallic and nonmetallic elements, such as aluminum and oxygen (alumina-$Al_2O_3$)]. All spraying processes display the same significant variables: the substrate temperature, ambient temperature, chemical composition of the carrier gas and/or environment, carrier gas flow rate, nozzle-to-substrate distance, droplet radius, solution concentration, solution flow rate, and—for continuous processes—substrate motion.[2] Because individual droplets

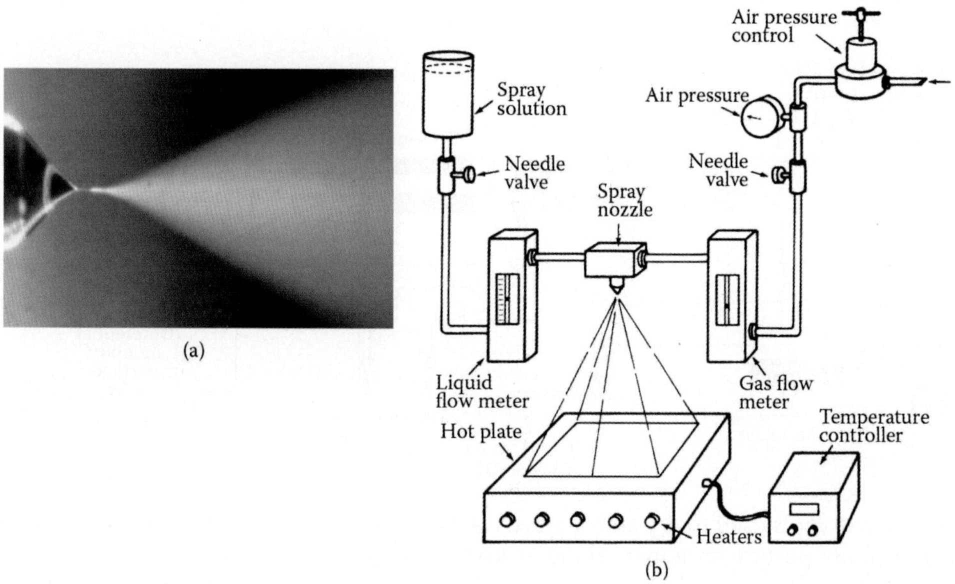


**FIGURE 9.1** (a) Atomizer and aerosol. (b) Spray pyrolysis setup. (Mooney, J. B., and S. B. Radding. 1982. Spray pyrolysis processing. *Ann Rev Mater Sci* 12:81–101. Courtesy of Jack Mooney, SRI International, Menlo Park, CA.[2])

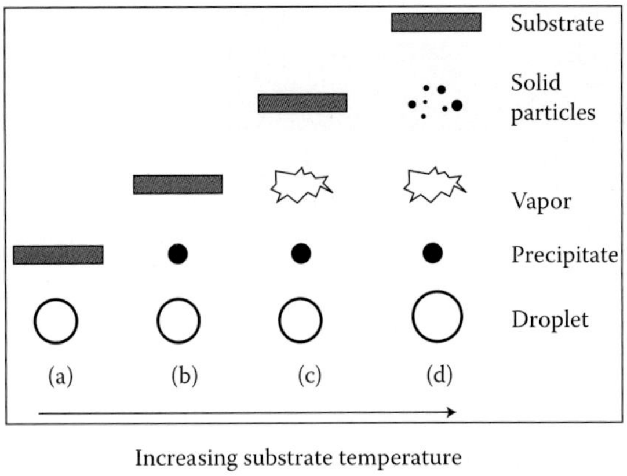


**FIGURE 9.2** Steps involved in the decomposition of the dissolved precursors as the temperature of the substrate is ramped up. Steps a to d are explained in the text.

evaporate and react quickly, grain sizes can be very small (e.g., as small as 0.1 μm). The small grains pose a disadvantage for most semiconductor applications but not necessarily for sensor applications; for example, in $SnO_2$-based gas sensors (Taguchi sensor; see Example 9.1) where sensitivity resides in the intergranular contacts (necks).

In Figure 9.2 we detail the possible scenarios of how droplets laden with solid-particle precursors arrive at a hot substrate as the temperature of the substrate is ramped up. If the temperature is too low, the droplets splash onto substrate and decompose there (Figure 9.2a). At somewhat higher temperature, the solvent evaporates and dry precipitate hits the substrate and decomposes (Figure 9.2b). At higher substrate temperatures, the solvent evaporates and the precipitate melts and vaporizes without decomposition and vapor diffuses to the substrate (Figure 9.2c). At yet higher temperatures, the precursor decomposes to make solid particles before they reach the substrate (Figure 9.2d).

At a fixed substrate temperature, smaller droplets may vaporize before hitting the substrate, and because droplets decrease in size as they approach the substrate, the concentration gradient within the approaching droplet keeps on changing. If the surface concentration of the solute exceeds the solubility limit, precipitation occurs at the surface of the particles, and hollow spheres result, increasing the film roughness.

Spray pyrolysis is a simple, inexpensive, high-throughput technique used to produce several compound semiconductors with utility in various devices, such as solar cells, gas sensors, antireflection coatings, and ion selective electrodes.[3] The technique does not require high-quality targets or substrates and does not need a UHV system. The process produces relatively thick films, is difficult to control, and is not compatible with IC processing. It also tends to form hollow particles (see above) and consumes lots of solvent.

In case spray pyrolysis is used to make a powder instead of a film, a setup as shown in Figure 9.3 may be used. In this apparatus, the precursor solution is atomized into a diffusion dryer, then a thermolysis,

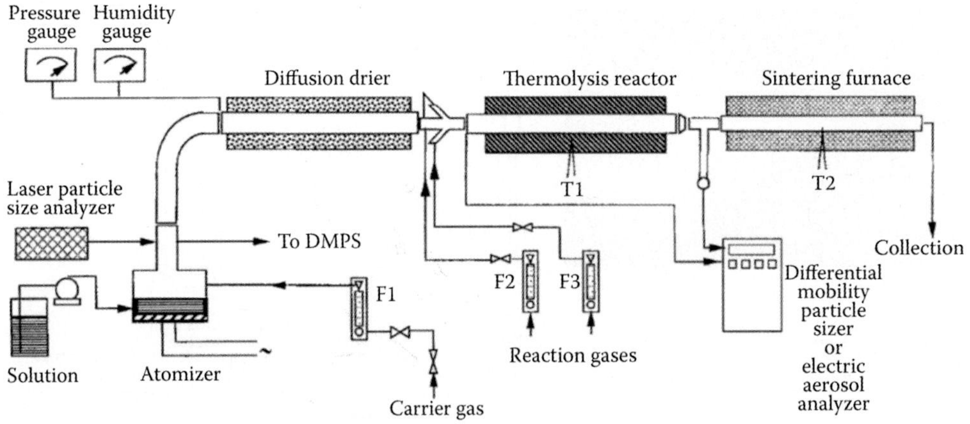


**FIGURE 9.3** The diffusion dryer is used to dry and remove water vapor from the aerosol generated by the atomizer. Thermolysis of the precipitate particle at higher temperature to form a microporous particle occurs in the thermolysis furnace and sintering of the microporous particle to form a dense particle in the sintering furnace. The setup is also equipped with a differential mobility particle sizer (DMPS). (From Messing, G. L., S.-C. Zhang, and G. V. Jayanthi. 1993. Ceramic powder synthesis by spray pyrolysis. *J Am Cer Soc* 76:2707–26.[1])

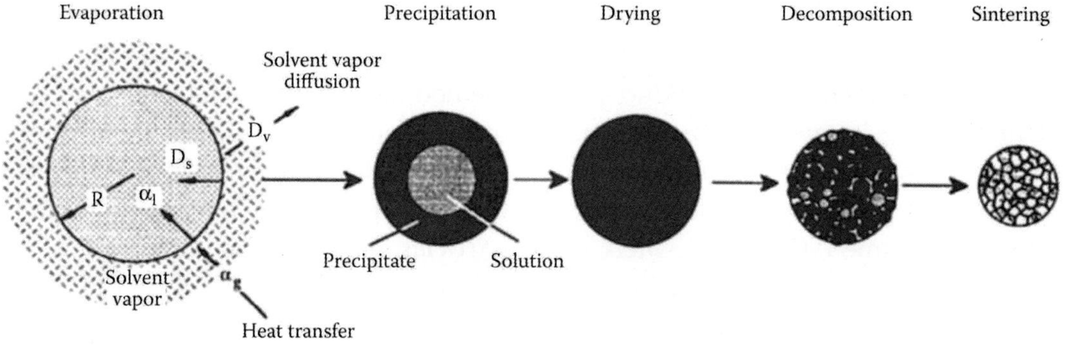


**FIGURE 9.4** Droplet evolution in spray pyrolysis. Thermal decomposition or pyrolysis—forms a nanoporous structure and finally sintering—involves the adhesion/solidification of the crystallites. (From Messing, G. L., S.-C. Zhang, and G. V. Jayanthi. 1993. Ceramic powder synthesis by spray pyrolysis. *J Am Cer Soc* 76:2707–26.[1])

and finally a sintering furnace. In this series of reactors, the aerosol droplets undergo evaporation and solute condensation within the droplet, by drying, thermolysis of the precipitate particle at higher temperature to form a microporous particle, and sintering of the microporous particle to form a dense particle. The setup is also equipped with a differential mobility particle sizer (DMPS). Depending on the droplet size and solution concentration, particles sized from some submicrometers to some hundreds of micrometers are available by spray pyrolysis.

The stages the aerosol droplets go through are sketched in Figure 9.4.

During evaporation of solvent from the droplet surface, there is diffusion of solvent vapor away from the droplet ($D_v$), change in droplet temperature, diffusion of solute toward the center of the droplet ($D_S$), and change in droplet size ($R$). Precipitation/drying involves volume precipitation or surface precipitation of the solute, followed by the evaporation of the solvent through the nanoporous crust. If the solute concentration at the center of the drop is less than the equilibrium saturation of the solute at the droplet temperature, then precipitation occurs only in that part of the drop where the concentration is higher than the equilibrium saturation, i.e., surface precipitation (see also Figure 9.5).

It is one of the unique characteristics of spray pyrolysis that chemical reactions occur within micrometer to submicrometer-sized liquid droplets, microcapsule chemical reactors so to speak. Besides controlling the precipitation in these microreactors

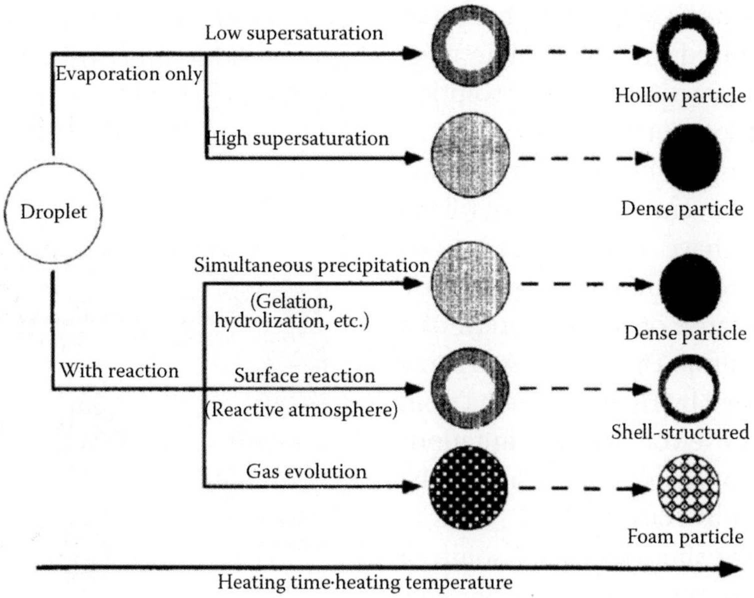


**FIGURE 9.5** Droplet pyrolysis with and without chemical reaction.

by simple evaporation, precipitation can also be induced by a chemical reaction, as illustrated in Figure 9.5. Reaction schemes relied on to induce precipitation might involve gelation, hydrolyzation, gas evolution, reacting with a reactive atmosphere, and so on. We encountered another example of microcapsule reactions in Volume I, Chapter 6 where we describe droplet manipulation fluidic platform. In this case, the droplet microreactors are injected in a nonmixing fluid.

## Salt-Assisted Spray Pyrolysis

Spray pyrolysis, like sol-gel (see Chapter 8), is a contender for making nanoparticles, used for chemical-mechanical polishing (CMP), magnetic recording tapes, sunscreens, and automotive catalysts. Spray pyrolysis is an attractive method for preparing uniform size and composition of nanoparticles because of its high production rate, continuous unit operation, and simple apparatus. However, with conventional spray pyrolysis the production of submicrometer particles faces some challenges. The concept derives from the principle of one particle from one droplet; thus, very fine droplets are needed to produce nanoparticles. The atomizers most commonly used to generate droplets tend to produce average droplet diameters in the range of several micrometers. For a droplet to dry from a typical initial diameter of 5 μm into a solid particle with a diameter of 0.1 μm, the initial volume fraction of dissolved involatile solute must be less than 0.0008%. In practice, these low solution concentrations lead to a low particle generation rate and make it difficult to maintain a high purity of the final particle product. Conventional spray pyrolysis produces particles that have multiple nanosized crystallites, but these crystallite aggregates are virtually inseparable because of the formation of a 3D network within each drop. The salt-assisted spray pyrolysis (SASP) developed by Dr. K. Okuyama from the Hiroshima University overcomes these limitations associated with conventional spray pyrolysis.[4] In the SASP method, salts such as $NaNO_3$, $KNO_3$, $LiNO_3$, and KCl are added to the spray precursor solution where the salt promotes nucleation and prevents nanoparticles from agglomerating. As a result, uniform-sized nanoparticles are produced. The diameters of nanoparticles produced by the SASP method range from several nanometers to several tens of nanometers. These sizes are about one-thirtieth to one-eightieth those of fine particles produced by traditional spray pyrolysis. Size distribution of SASP particles appears to be independent of their parent solution droplets; moreover, the SASP process is less energy consuming because it runs at lower temperatures.

In Figure 9.6 SASP is compared with traditional spray pyrolysis (SP).

## Flame Spray Pyrolysis

Flame spray synthesis has been established as a state-of-the-art method for the production of high-purity mixed metal oxides for a vast range of applications in catalysis, optics, and electronics (Figure 9.7). It is primarily useful for making oxides. In flame spray pyrolysis, one directly sprays a liquid precursor into a flame, and particle synthesis occurs within the flame by a combustion reaction. This allows for the use of low-vapor-pressure precursors. Products are affected by the flame configuration, process temperature, oxidant composition, and additives. With metalorganic precursors, flame spray pyrolysis constitutes a commercial approach for the manufacture of silica and titania powders. Using heavy residual oils as precursor, carbon black is produced (millions of metric tons per year). The production of $TiO_2$ powder from the flame spray pyrolysis of $TiCl_4$ (the chloride process) accounts for ~60% of $TiO_2$ production (Dupont is the world's largest producer with about ~22% of global production). The $TiO_2$

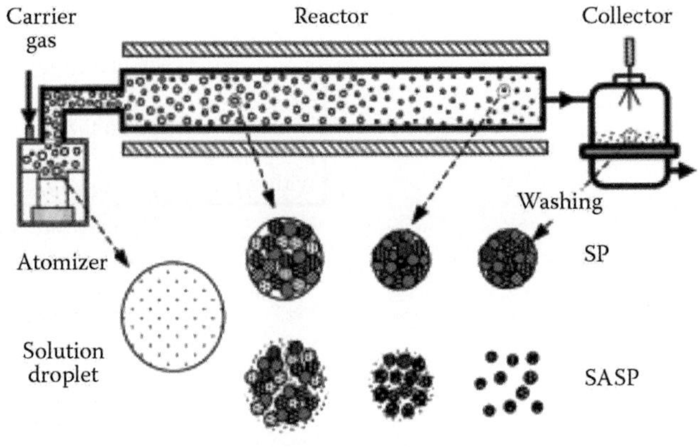


**FIGURE 9.6** Comparison of SASP with traditional SP.

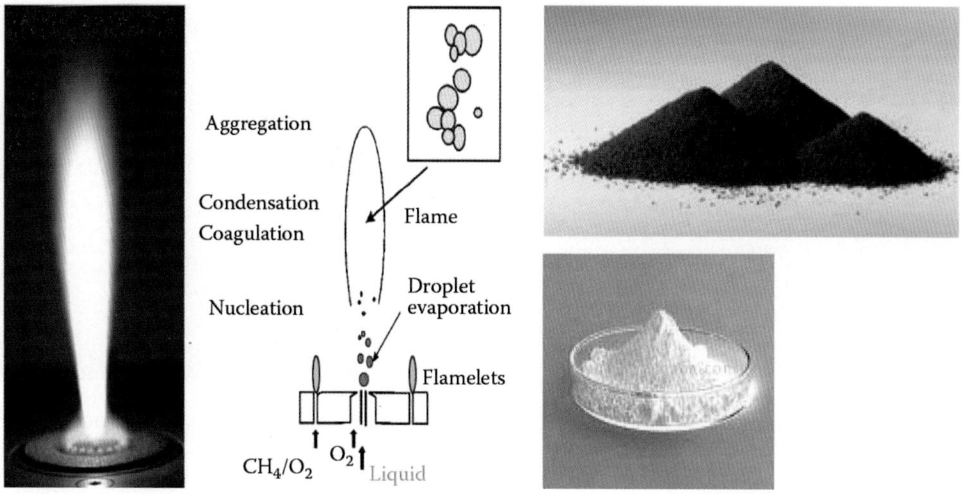


**FIGURE 9.7** Flame spray pyrolysis. Using heavy residual oils as precursor, carbon black is produced using flame spray pyrolysis (millions of metric tons per year). The white powder is $TiO_2$.

powder made this way is the least expensive submicrometer powder available (see bottom right in Figure 9.7). Because of its high index of refraction, it is an ideal opaque dye for paper, paint, pigment, food, and cosmetics. In the chloride process, an electric field can be used to control particle size in the flame.[5] The flame is held between two plates, with one plate connected to a DC source and the other grounded as sketched in Figure 9.8. A flame filled with newly formed $TiO_2$ particles creates both positive and negative ions. As a result, these charged particles that otherwise would have collided with each other and grown to form larger particles are prevented to do so by the presence of the electrical field. The strength of the E field determines the velocity of the ions and essentially the size of the product particles. This method allows very precise control of the size (within 1 nm) and aggregate structure and crystallinity.

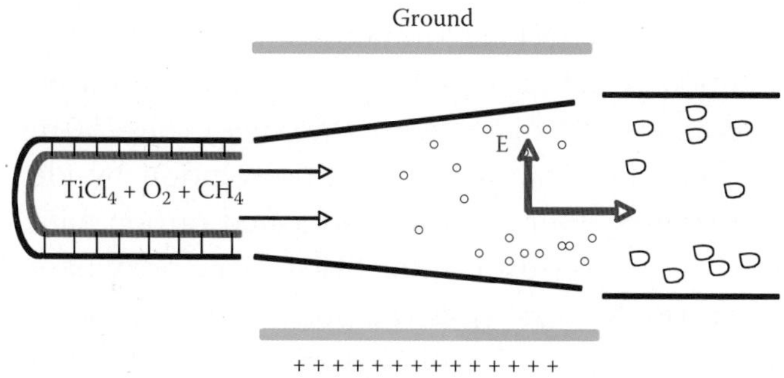


**FIGURE 9.8** Electrical field control of particle size in flame spray pyrolysis of $TiO_2$ nanoparticles.

Pros of this method are its large capacity, proven technique, and high purity of products. Cons include the fact that, unless electrical fields are applied, agglomerates form, and it features a wide particle size distribution.

## Electrospray Pyrolysis

Electrospraying is capable of atomizing a liquid into ultrafine droplets and offers yet another option for converting conventional spray pyrolysis into a nanoparticle production means. In this method, the meniscus of a conducting solution supported at the end of a capillary tube becomes conical when charged to a high-enough voltage (several kV) with respect to a counter electrode. When the electrostatic force becomes stronger than the surface tension of the liquid, the solution exits the capillary as a highly charged jet of fluid. As the solvent evaporates from the droplets, particles form and are deposited onto a heated collector. The droplets are stably formed by the continuous breakup of a jet extending from this liquid cone (Figure 9.9).

Electrospray (ES) pyrolysis can produce monodisperse droplets/particles over a wide size range, from a few nanometers to hundreds of micrometers, depending on liquid flow rate, applied voltage, and liquid electric conductivity. Especially in the small size range, the capability of producing monodisperse particles with relative ease is unmatched by any other aerosol generation scheme. Zinc sulfide

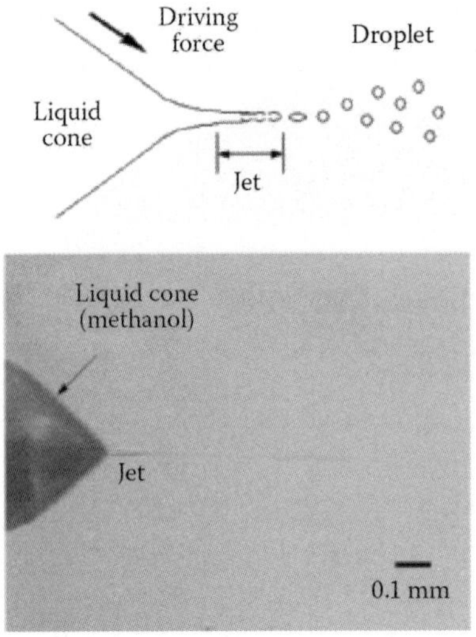


**FIGURE 9.9** Droplet formation in the electrospray technique.

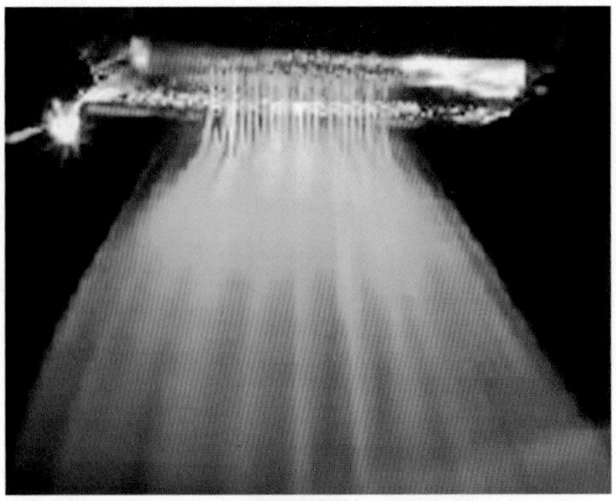

**FIGURE 9.10** Electrospray from an array of nozzles etched in silicon, with a density of 250 sources/cm². (From Deng, W., J. F. Klemic, X. Li, M. A. Reed, and A. Gomez. 2006. Increase of electrospray throughput using multiplexed microfabricated sources for the scalable generation of monodisperse droplets. *Aerosol Sci* 37:696–714.[7])

particles 20–40 nm in diameter were prepared by electrically driven spray pyrolysis from solutions of ethyl alcohol with zinc nitrate [$Zn(NO_3)_2$] and thiourea [$SC(NH_2)_2$] at concentrations from 0.0025–0.2 M and electrical conductivities between $10^{-4}$ and $10^{-1}$ $Sm^{-1}$.[6]

A crucial drawback that has hampered wider ES pyrolysis applications is the low flow rate at which the cone-jet mode is established. Multiplexing the spray source, that is, operating several ESs in parallel is a possible way to address this problem. Deng et al.[7] demonstrated the microfabrication and testing of a compact multiplexed system of ESs as an array of nozzles etched in silicon, with a density of 250 sources/cm² (see Figure 9.10).

It is projected that this arrayed ES technology, once optimized, will be applied in various fields, including the synthesis of biological nanoparticles and biomedical coatings for controlled/targeted drug delivery, the synthesis of nanoparticle ceramics, superconductors, quantum dots, and thin films.

The ES method has also had a dramatic technological impact in the field of analytical chemistry. Electrospray ionization (ESI) is a technique used in mass spectrometry to produce ions. It is especially useful in producing ions from macromolecules because it overcomes the propensity of these molecules to fragment when ionized. The development of ESI for the analysis of biological macromolecules was rewarded with the attribution of the Nobel Prize in Chemistry to John Bennett Fenn in 2002. In the same electro-spraying set-up of Figure 9.9 and 9.10, higher molecular weight solutions enable one to produce continuous fibers instead of beads. At low viscosity, the columbic force dominates the elastic force, leading to a rapid breakup of the polymer jet. However, increased viscoelasticity engendered by chain entanglements does not allow the breakup of the jet, resulting in continuous fibers in the so-called electrospinning process. The electrospinning process is indeed a simple, versatile, and widely used method of producing all types of polymer nanofibers.[8–14] As in the case of particle spraying, pyrolysis may further convert the polymer nanofiber precursor to a final desirable product (e.g., carbon nano fibers through the carbonization process, i.e., the high temperature heating in the absence of oxygen; see also C-MEMS in Volume III, Chapter 5).

Among the precursors used in the synthesis of carbon nanofibers by electrospinning, polyacrylonite (PAN)/N,N-dimethylformamide (DMF) solution is the most commonly employed. Carbon nanofibers with diameters in the range of 80–750 nm are typically obtained from this solution by electrospinning. Our team has fabricated carbon nanofibers by electrospinning SU-8 (an epoxy-based negative photoresist) as the polymer precursor and subsequent pyrolysis of that precursor.[15] We have shown that the electrospun material retains its photoresist properties and that it can be directly patterned in a wide variety of patterns by lithography before pyrolysis. The pyrolysis of SU-8 nanofibers to produce carbon nanofibers will enhance the C-MEMS toolbox considerably and allow for the facile integration of high-surface-area fibers with other photoresist derived carbon structures. This not only will allow for the making of interesting photopatternable carbon composites but also will enable the fabrication of multiscale hierarchal assemblies such as fractal-like electrodes where a high external surface area is coupled with a minimized internal resistance, which is a feature of great utility, in batteries and fuel cells.[16] The biocompatibility of SU-8-derived carbon[17] provides further motivation for the development of multiscale hierarchal scaffolds for tissue engineering and in vivo biosensors.

**Example 9.1: Spray Pyrolysis Application: Taguchi Gas Sensor**

The use of a ceramic cylindrical tube in a classical Taguchi sensor (Figure 9.11) maximizes the utilization of heater power; thus, most of the power is used to heat the gas-sensitive tin oxide ($SnO_2$). A tin oxide paste is applied by dip coating the outside ceramic body to cover the thick film resistance measuring electrodes and is then sintered at high temperature. The resistance of tin oxide at high temperature gives a measure for the amount of reducing gases in the contacting atmosphere. Sintering stabilizes the intergranular contacts (necks) where the sensitivity of the gas sensor resides.[18] The design shown in Figure 9.11 is not suited for mass production because it involves excessive hand labor. The problems with the thick film structure mainly reside in the areas of reproducibility:

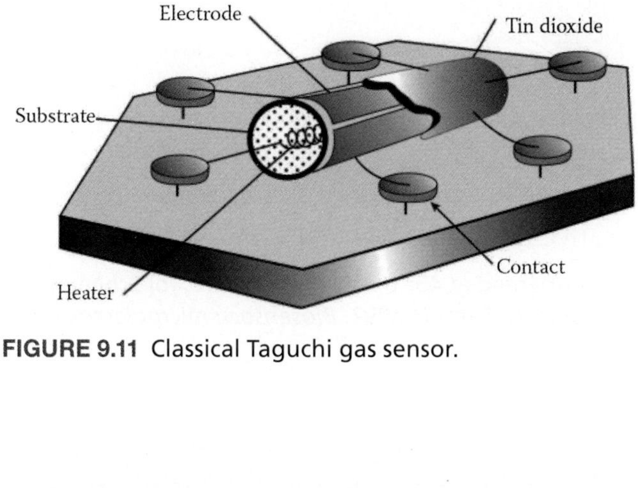


**FIGURE 9.11** Classical Taguchi gas sensor.

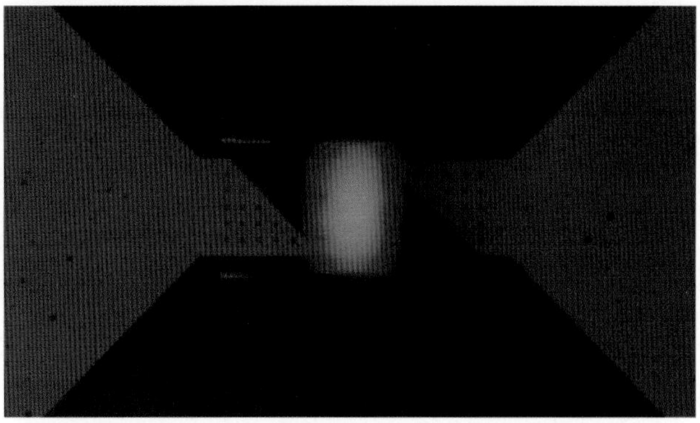

**FIGURE 9.12** Heater element.

compressing and sintering a powder, the deposition of the catalyst, and the use of binders and other ceramics (e.g., for filtering). Application of IC techniques could improve the state of the art dramatically if a thin semiconductor oxide film (e.g., tin oxide) with the same sensitivity as the traditional thick sintered film could be made. Micromachined heater elements, as shown in Figure 9.12, improve the reproducibility and the absolute power budget needed to bring the sensor to the required temperature. Thermal efficiency and low power budget are obtained by thermally isolating the heater element on a thin, suspended membrane. Whereas low power budget sensors have proven feasible, it has been more difficult to make a compatible thin film as sensitive as the ceramic-type, thick film devices. When making thin films with physical vapor deposition (PVD) methods, the films have a lower surface area and buried intergranular contacts (see Figure 9.13). The buried intergranular contacts cannot be reached by the contacting gas as easily as in the powder case, making not only for lower sensitivity but also slower responding gas sensors. The slower response of the thin film sensors is because oxygen diffusion in the grain boundaries between compacted grains involves long time constants. When the grain boundaries are blocked on purpose, the sensor reacts faster but displays even less sensitivity.[18]

Using silk screening of thick films (see following section) on a thin membrane would seem to provide the solution—a low-energy heater combined with a fast and sensitive thick tin oxide film. Unfortunately, during our attempts to deposit thick films this way, usually the thin Si membrane broke as a result of the pushing action of the squeegee (see Figure 9.14).[19] Spray pyrolysis

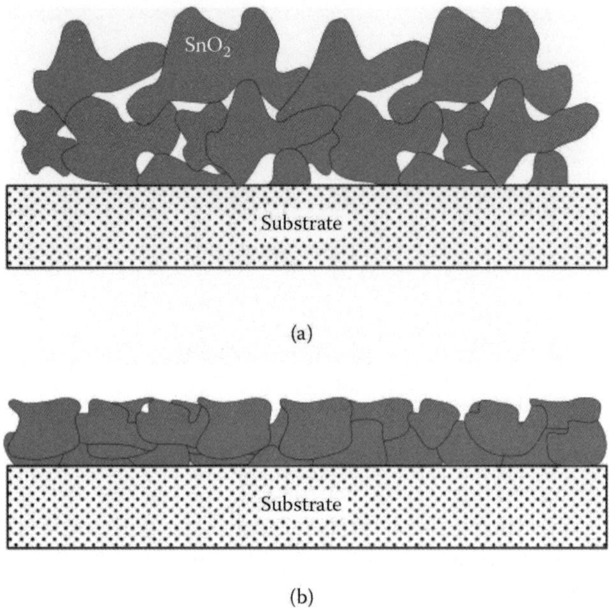


**FIGURE 9.13** Grain structure in a sintered powder, thick $SnO_2$ film (a) and a PVD, $SnO_2$ thin film (b).

provides the possible answer here because its deposition does not involve mechanical pressure on the silicon membrane. Fast-responding, planar Taguchi $SnO_2$ gas sensors were made by spraying, with oxygen as the carrier gas, an organic solution of $(CH_3COO)_2SnCl_2$ in ethyl acetate onto quartz, glass, and $Al_2O_3$ ceramic substrates, heated to 300°C.[3] This result carries significance; other micromachining attempts based on thin PVD $SnO_2$ film fail to reach the sensitivity attained with the classical thick films and sintered $SnO_2$ films. Spray pyrolysis, with its very small grain-size deposits, holds the promise of reaching the objective of providing films with freely accessible polycrystalline grains. As a planar technology, it is also suited for large substrates and considered a viable alternative to the dip-coating process.

## Screen Printing/Silk Screening

### Introduction

Screen printing presents a more cost-effective means of depositing a wide variety of films on planar substrates than does IC technology, especially when fabricating devices at relatively low production volumes. The technique constitutes one of several thick film or hybrid technologies used for selective coating of flat surfaces (e.g., a ceramic substrate). The technology was originally developed for the production of miniature, robust, and (above all) cheap electronic circuits. The up-front investment in a thick film facility is low compared with that of IC manufacturing. For disposable chemical sensors, recent industrial experience indicates that screen printing thick films is a viable alternative to Si thin film technologies. Materials based on low-temperature organic polymers, to be cured at low temperatures or using UV radiation, as well as high-temperature enamel coatings with ceramic paints are coated in screen printing; the latter need sintering at high temperatures. Typical film thicknesses are in the range from several 10 to several 100 μm. Therefore, the coefficients of thermal expansion of the coat have to be matched well to that of the substrate.

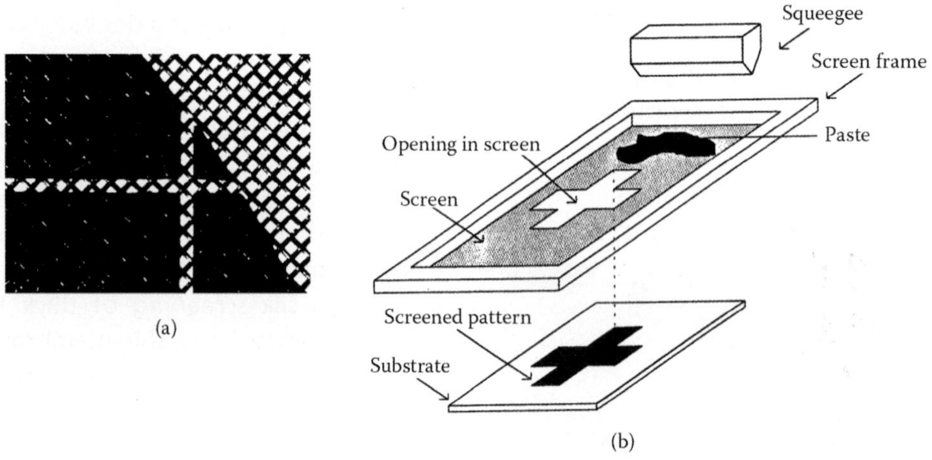


**FIGURE 9.14** Screen-printing. (a) Screen with a 0.002-in. line opening oriented at 45° to the mesh weave. (b) Schematic representation of the screen-printing process. (From Lambrechts, M., and W. Sansen. 1992. *Biosensors: microelectrical devices.* Philadelphia: Institute of Physics Publishing. With permission.[20])

## How Does It Work?

The elements of screen printing are the screen with openings defining the pattern to be printed, the screen frame, a squeegee, the ink material, and the substrate. A paste or ink is pressed onto the substrate through the openings in the photoemulsion fashioned on the stainless steel screen (see Figure 9.14a).[20] The mesh openings are created using photolithography on the photoemulsion in the screen frame. The paste consists of a mixture of the material of interest, an organic binder, and a solvent. The organic vehicle determines the flow properties (rheology) of the paste. The bonding agent provides adhesion of particles to one another and to the substrate. The choice of active particles makes the ink a conductor, a resistor, or an insulator. The lithographic pattern in the screen emulsion is transferred onto a substrate by forcing the paste through the mask openings with a squeegee (Figure 9.14b). This process depends on the mesh opening size and the viscosity of the ink. The squeegee further keeps the mesh in contact with the substrate, a process that depends on the screen tension and the off-contact distance. Squeegees are made from a stiff material like polyurethane to avoid deformation during printing. Relatively hard squeegees are required to withstand the high pressure needed to print at high speed, for example, high-viscosity materials on glass plates. However, the harder the material, the lower the ability to print an even layer of ink on an uneven surface. The squeegee edge must be sharp. The angle of the squeegee to the mesh defines the amount of material to be printed. Pulling the squeegee at 15–20° off vertical ensures an even deposit of material. The setting of the exact angle depends in part on the squeegee's profile and its hardness. The possible speed of the squeegee depends on the hardness of the squeegee and the viscosity of the printed material.

In a first step, paste is put down on the screen (Figure 9.15Aa); then the squeegee lowers and pushes the screen onto the substrate, forcing the paste through openings in the screen during its horizontal motion (Figure 9.15Ab).[20] During the last step, the screen snaps back; the thick film paste that adheres between the screen and the substrate shears; and the printed pattern is formed on the substrate (Figure 9.15Ac). Heavier frames cause the screen to be less deflected and minimize distortions and achieve a better resolution of the printed pattern. Throughput and resolution depend on the dimensions of the mesh openings (holes in the screen) and the nature of the pastes/inks. This leads to compromises in selecting the mesh size: a coarse mesh enables fast processing and high throughput but at low resolution, whereas a fine mesh enables better resolution but causes problems when pressing highly viscous materials through the screen. With a 325-mesh screen (i.e., 325 wires/in. or 40-μm

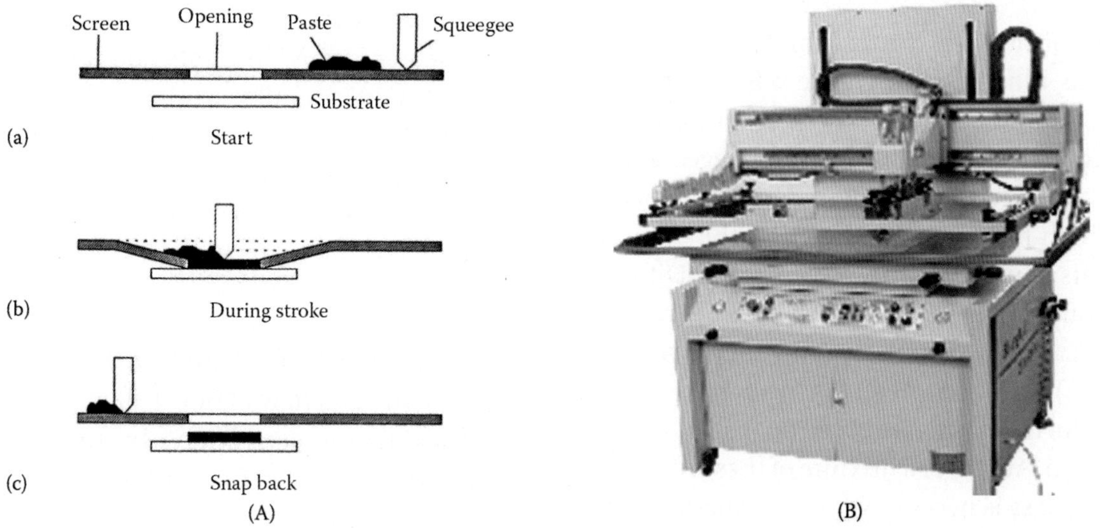


**FIGURE 9.15** (A) The three different steps of the silk screening process. (From Lambrechts, M., and W. Sansen. 1992. *Biosensors: microelectrical devices.* Philadelphia: Institute of Physics Publishing. With permission.[20]) (B) Shanghai Shuoxing Screen Printing Equipment Co., Ltd.

holes) and a typical paste, a lateral resolution of 100 μm can be obtained. For difficult-to-print pastes, a stencil mask may be used, such as a thin metal foil with cut openings. However, the resolution of this method is inferior (>500 μm). Lower-cost and high-throughput printing techniques, such as screen printing, have a great promise in the fabrication of organic light-emitting devices (OLEDs). Jabbour et al. demonstrated, for the first time, the use of screen printing to deposit ultrathin layers in OLEDs of less than 15 nm with root mean square surface roughness of less than 1.5 nm.[21] After printing, the wet films are allowed to settle for 15 min to flatten the surface while drying. This removes the solvents from the paste. Subsequent firing burns off the organic binder; metallic particles are reduced or oxidized; and glass particles are sintered. Typical temperatures range from 500–1000°C. After firing, the thickness of the film ranges from 10–50 μm.

A typical silk screening setup is the one from Universal Instruments Corporation (http://www4.uic.com), the DEK 4265 Horizon Screen (see also The Shanghai Shuoxing screen printing machine in Figure 9.15B). Dupont provides technical tips for screen printing at http://www2.dupont.com/MCM/en_US/techinfo/techtips.html, and the Screen Printing Magazine can be found at http://www.screenweb.com/home.php.

## Types of Inks/Pastes and Applications

Inks are formulated to exhibit pseudoplastic* behavior, as illustrated in Figure 9.16[22,23] (see also Volume I, Chapter 6 on fluidics), to prevent flowing through the screen until the squeegee applies sufficient pressure. Almost all materials compatible with the high firing temperature and the other ink constituents can be used to screen print. Different pastes—conductive (e.g., Au, Pt, Ag/Pd), resistive (e.g., $RuO_2$, $IrO_2$), overglaze, and dielectric (e.g., $Al_2O_3$, $ZrO_2$)—are commercially available.

The conductive pastes are based on metal particles, such as Ag, Pd, Au, Pt, or a mixture of these combined with glass. Glass is necessary for the adhesion of the metal conductor to the ceramic ($Al_2O_3$) substrate. The following conductive pastes can be distinguished by the bonding mechanism used.

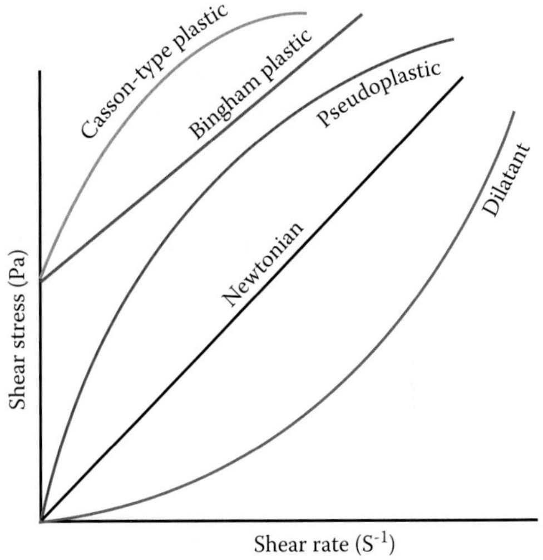


**FIGURE 9.16** Pseudoplastic behavior of inks for silk screening (see also Volume I, Chapter 6).

- Glass-bonded or fritted pastes are inks where adhesion of the metal is achieved with the addition of a glass mixture (30%) (a typical glass composition is 65% PbO, 25% $SiO_2$, and 10% $Bi_2O_3$). A fritted ink contains powdered glass that binds the ink to the substrate material when fired at a temperature of 850°C. Thus, a fritted ink will generally consist of the glass frit, a material defining the desired ink property, and an organic vehicle that renders the ink printable. Vehicles typically consist of solvents mixed with slightly more viscous materials, such as resins, in a ratio designed to give the optimum overall viscosity to the ink.
- Oxides or fritless-bonded inks adhere to the metal via the addition of copper oxide (3%).
- Mixed-bonded pastes are inks for which adhesion is achieved by use of both glass and copper oxide.

Resistive pastes are based on $RuO_2$ or $Bi_2Ru_2O_7$ mixed with glass (65% PbO, 25% $SiO_2$, 10% $Bi_2O_3$). The resistivity is determined by the mixing ratio. Overglaze and dielectric pastes are based on glass mixtures. According to composition, different melting temperatures can be achieved.

Thick film technology with the above type of traditional inks has application in the construction of

* In pseudo-plastic fluids viscosity diminishes as the applied velocity gradient increases (see also Volume I, Chapter 6).

a wide variety of hybrid sensors, such as sensors for radiant signals, pressure sensors, strain gauges, displacement sensors, humidity sensors, thermocouples, capacitive thick-film temperature sensors, and pH sensors (see Middlehoek et al.[23] and references therein). Also, with silicon-based sensors (e.g., pressure sensors and accelerometers) die-mounted on a ceramic substrate, thick film resistors are used for calibration by trimming resistors on the ceramic substrate.

New inks, specifically developed for chemical and biological sensor applications, are available or under development. For example, $SnO_2$ pastes incorporating Pt, Pd, and Sb dopants have been developed for the construction of high-temperature (>300°C) semiconductor gas sensors for reducing gases (so-called *Taguchi sensors*) (see Example 9.1).[18] Thick metal phthalocyanine films have been deposited on alumina to form the active material in relatively low-temperature (<180°C) gas sensors.[24]

In Chapter 8 we saw that for biosensor applications, thick film technology based on pastes (polymer-based inks) that can be deposited at room temperature have been developed.

## Doctor's Blade or Tape Casting

Silk screening and tape casting technology constitute two important thermal forming processes, often referred to as *hybrid* manufacturing methodologies. Here we detail tape casting as yet another example of thermal forming. Tape casting is also known as *doctor blading* or *knife coating*, and the process is well known under these names in many industries, including paper, plastic, and paint manufacturing. The "doctor's blade" is a scraping blade for the removal of excess substance from a moving surface being coated. Howatt was the first to obtain a patent to use the technique for "forming ceramic materials into flat plates, especially useful in the electric and radio fields."[25] This is still the principal application of tape casting today, although its use extends far beyond the field envisioned in 1952 (http://www.drblade.com). Since then, tape casting has been used to manufacture a variety of multilayer ceramic package products, such as low-temperature cofired ceramics, solid oxide fuel cells, and piezoelectric smart devices.

The ceramic application of doctor blading, also referred to as *green tape* technology, is a thermal forming process well suited for fabrication of mesoscale devices (layers with a thickness between several micrometers and millimeters). In green tape technology, a ceramic slurry, either aqueous or organic solvent-based, is continuously dispensed onto a moving substrate from a hopper or other appropriate plumbing. An important requirement of the slurry is that its viscosity is in the 100–130-Poise ($10^{-13}$ Ns/m$^2$) range. The substrate and dispensed product then move under the doctor's blade, an adjustable gate with a precisely controlled height on a smooth casting/coating surface. The purpose of the doctor's blade is to allow a known thickness of material to pass under the blade. In today's technology, very thin ceramic layers are defined in micrometers, with tapes as thin as 5 μm reported by equipment manufacturers (http://www.drblade.com). Standard tape casting machines come with thickness ranges from 12 μm to more than 3 mm. Precise, smooth control of the carrier or web speed is also essential to ensure uniform films. Next, the substrate and bladed ceramic slurry are moved through a precision oven for partial drying. The result is a flexible green ceramic tape that can be easily machined before final drying and sintering. Ceramic in green tape form may be cut, for example, with lasers or razor blades into any desired shape. Tape stacks are produced by lamination of several single green tapes under heat and pressure. The optimum dried thickness range for tape casting is generally accepted as being from 0.025–1.27 mm. The green tape process flow is summarized in Figure 9.17. In Figure 9.18 we show a photo of a typical tape caster.

Tape casting is the best method for forming large-area, thin, flat ceramic parts and even for certain thin metal foils. Such thin parts are virtually impossible to press and are very difficult to extrude. The difficulties are compounded in dry pressing when the plate is to be pierced with numerous holes because of the increased possibility of nonuniform die fill. On the other hand, punching via holes and slots of various sizes and shapes into green tape is relatively easy and essential to the many types of multilayered ceramic packages designed and manufactured today. Tape casting has been

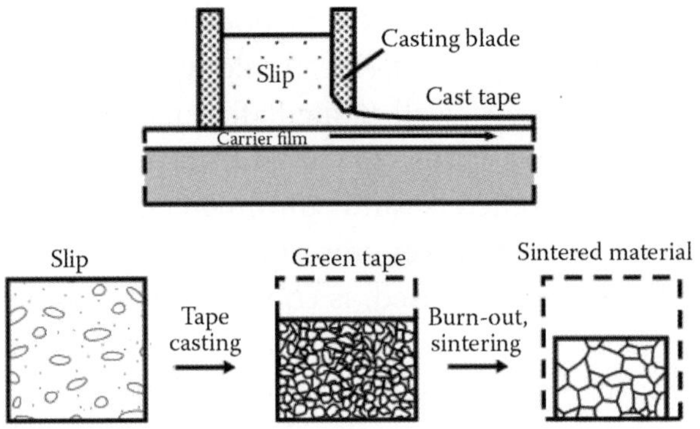


**FIGURE 9.17** Tape casting. The top schematic drawing illustrates tape casting. The bottom sketch shows the different stages during the processing: the slip consisting of water, ceramic particles and binder; the cast, dried green sheet; and finally, the microstructure of the sintered material.

expanded to nonceramic, polymer, and metal applications, such as polyvinylidene fluoride (PVDF) battery separator films and tape automated bonding (TAB) semiconductor adhesive-metals. Tape casting is often the preferred choice in manufacturing these products because of its relative simplicity and low implementation cost compared with other available methods, such as dry pressing, slip casting, or injection molding. Tape casting is a straightforward method of forming uniformly thin sheets of film that is inexpensive, scaleable, and may be used with any ceramics, metals, or polymers that are readily mixed in a liquid suspension or slip.

An industrial green tape setup costs between $100,000 and $120,000. Dupont Microcircuit Materials is a typical supplier of green tape materi-

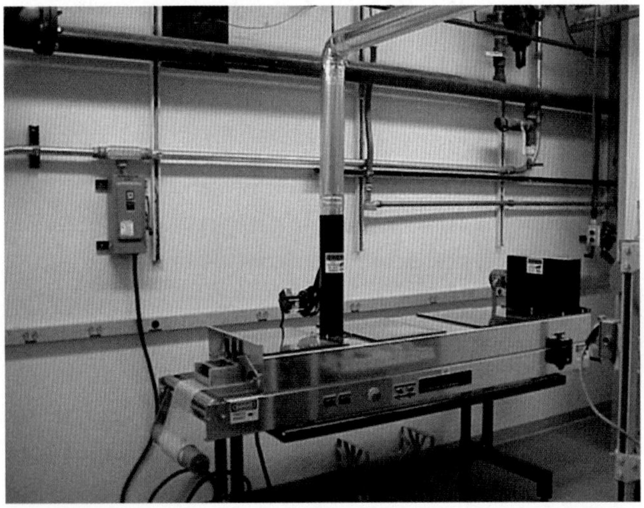

**FIGURE 9.18** A typical tape caster. HED Lab-Cast 7-ft. tape caster.

als for a wide variety of electronic and electrical applications in the wireless and wired telecommunications, automotive, military, medical, instrumentation, industrial, data processing, components, and consumer industry segments (Green Tape™; http://www.mcm.dupont.com).

An excellent recent review of tape casting can be found in *Tape Casting: Theory and Practice* by Richard E. Mistler and Eric R. Twiname.[26] More background information on tape casting is also available at http://www.drblade.com.

**Example 9.2: Green Tape Technology: Multilayer Planar Oxygen Sensor**

Green tape methodology is of special interest in the fabrication of high-temperature gas sensors. As an example, we show Figure 9.19, a multilayer planar oxygen sensor manufactured by NGK Spark Plug Co., Ltd. Also known as a *wide range air-to-fuel ratio* sensor, it incorporates two oxygen-pumping cells and one potentiometric gauge. The operating principle of the NGK sensor is as follows. The potentiometric cell—a lambda sensor—measures EMF between the exhaust in the gap and the atmospheric air. The measured voltage $V_{EMF}$ is compared with $V_o$ (corresponding to EMF of the stoichiometric point). The amplifier applies an appropriate pumping voltage to the pumping cell after comparing $V_{EMF}$ and $V_o$ (if $V_{EMF} > V_o$ then the exhaust is rich; if $V_{EMF} < V_o$ then the exhaust is lean). It is fabricated by laminating and cofiring, at high temperatures, several layers of ceramic green tape, some of which are metal coated and have openings through them.[27] Because both metal and ceramic layers constitute the sensor, the materials should be carefully prepared, and the cofiring process must be tightly controlled to avoid metal diffusion or reaction with the ceramic. Because of the process complexities, this sensor turned out too expensive for common use.

## Plasma Spray Deposition

### Introduction

Because of the need to incorporate more and different materials in a thickness ranging anywhere from monolayers to a few hundred micrometers,

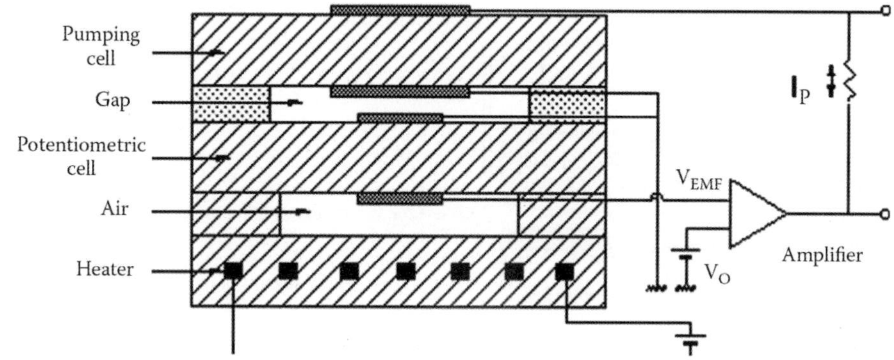


**FIGURE 9.19** Planar Oxygen Sensor from NGK, Spark Plug Co., Ltd.

micromachinists are broadening their horizon beyond IC deposition techniques and increasingly are incorporating hybrid thick film methodology in their tool box. Microstructures can be crafted economically with non-IC equipment and materials, and thick film silk screening, spray pyrolysis, sol-gel, and tape casting are all attractive additive techniques in this regard. Spray pyrolysis (see earlier), drop delivery systems (see Chapter 8), and pick-and-place technology (see Volume III, Chapter 4) are but a few examples. Here we review plasma spraying as another example of a non-IC thermal forming tool. With plasma spraying almost any material can be coated on many types of substrates. Applications include corrosion- and temperature-protective coatings, superconductive materials, and abrasion resistance coatings.[28] Today, turbine blades and other components of aircraft engines are plasma coated with corrosion- and temperature-resistant coatings. In this section, we propose that the technique be expanded to the mass manufacture of ceramic-based sensors.

## How Does It Work?

The CVD and PVD techniques discussed in Chapter 7, except for the cluster deposition method, rely on atomistic deposition, that is, atoms or molecules are individually deposited onto a surface to form a coating. Plasma spraying, like spray pyrolysis, is a typical particle deposition method; particles a few micrometers to 100 μm in diameter are transported from source to substrate. The basic configuration of a plasma-arc torch setup for layer deposition and a plasma spray nozzle are shown in Figures 9.20 and 9.21, respectively.

In plasma spraying, a high-intensity plasma arc is operated between a stick-type cathode and a nozzle-shaped, water-cooled anode as illustrated in Figure 9.21. For atmospheric spraying, one typically works at power levels from 10–100 kW. Plasma gas, pneumatically fed along the cathode, is heated by the arc to plasma temperatures and leaves the anode nozzle as a plasma jet or plasma flame. Argon and mixtures of argon with other noble (He) or molecular gases (e.g., $H_2$, $N_2$, $O_2$) are frequently used for plasma spraying. Fine powder suspended in a carrier gas is injected into the plasma jet where the particles are accelerated and heated. The plasma jet may reach temperatures of 20,000 K and velocities up to 1,000 $ms^{-1}$. The temperature of the particle surface is lower than the plasma temperature, and the dwelling time in the plasma gas is very short. The lower surface temperature and short duration prevent the spray particles from being vaporized in the gas

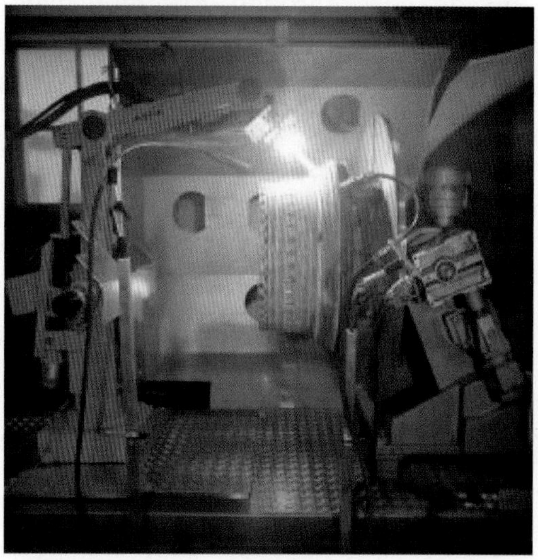

**FIGURE 9.20** Setup for plasma spray. With a robot arm and a sound-proof spraying booth.

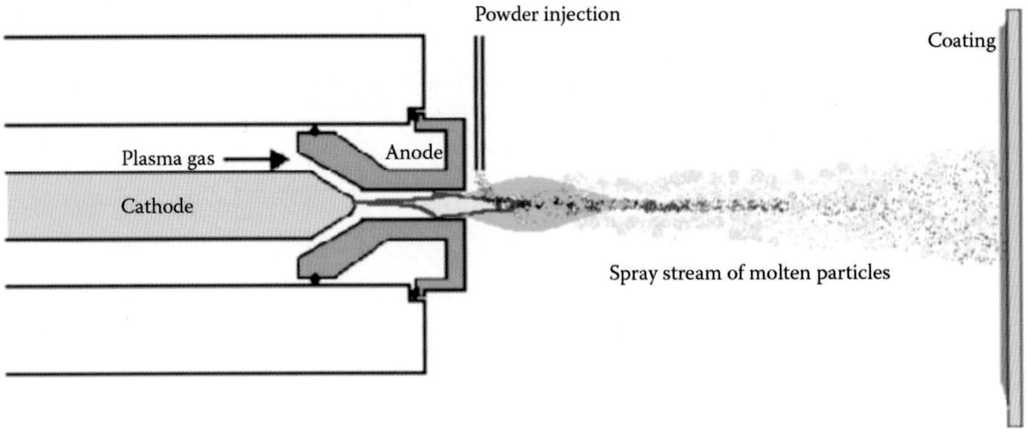


**FIGURE 9.21** Plasma spray nozzle.

plasma. The particles in the plasma assume a negative charge, owing to the different thermal velocities of electrons and ions. As the molten particles splatter with high velocities onto a substrate, they spread, freeze, and form a more or less dense coating, typically forming a good bond with the substrate. The resulting coating is a layered structure (lamellae), as shown in Figure 9.22. Typical coatings for high-temperature applications involve an oxidation-resistant coating and a thermal barrier coating (TBC). The role of TBCs is, as their name suggests, to provide thermal insulation of the blade. A coating of about 1–200 μm can reduce the temperature by up to 200°C. The oxidation-resistant coating is also called *bond coat* because it provides a layer on which the ceramic TBC can adhere.

As shown in Figure 9.23, a particle leaving the spray nozzle goes through different regions of temperature and flow velocity. Ideally, the particles should arrive at the substrate at high velocities in a completely molten state to form the densest coating with little porosity. To produce porous films for the fabrication of oxygen sensors in our own work,[19] we positioned the substrate somewhere between regions 4 and 5. To produce dense films, we positioned the substrate in region 4. With this technique, a minimum thickness is about 25 μm, and very thick coats up to a few millimeters thick are possible. When attempting to deposit gas-sensitive layers, such as $ZrO_2$, on thermally isolated, thin Si membranes to make a power-efficient gas sensor, we found that the kinetic energy of the plasma was too high, and it broke the thin suspended Si membranes. The high temperature and kinetic energy preclude the potential for integrating Si with high-temperature plasma spraying. In Example 9.3 below we demonstrate how plasma spraying of yttria-stabilized zirconia may be used in the batch fabrication of all solid-state oxygen sensors.

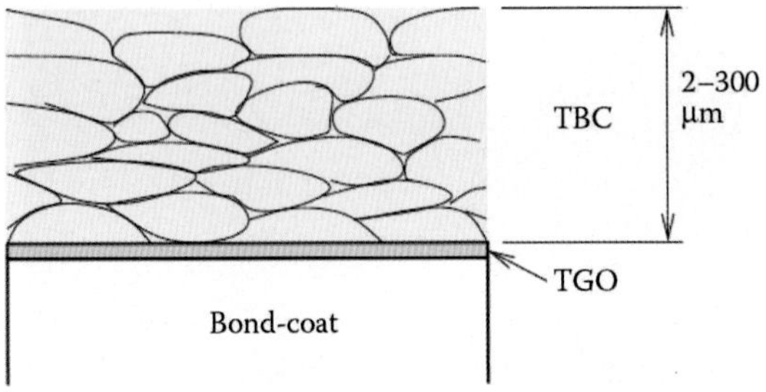


**FIGURE 9.22** Schematic microstructure of a TBC obtained by APS on a TGO.

Plasma spray deposition is perhaps the most flexible of all the thermal spray processes because it can develop sufficient energy to melt any material. Because it uses powder as the coating feedstock, the number of coating materials that can be used in the plasma spray process is almost unlimited. Plasma spraying equipment may be purchased, for example, from Sulzer Metco (http://www.sulzermetco.com/index.html).

#### Example 9.3: Plasma Spraying: Multilayer Planar Oxygen Sensor

Here we present an example of how to use plasma spray technology to batch produce solid-state oxygen sensors based on yttria-stabilized $ZrO_2$ (YSZ) solid electrolyte films.[29] Plasma spraying is a particulate method geared toward

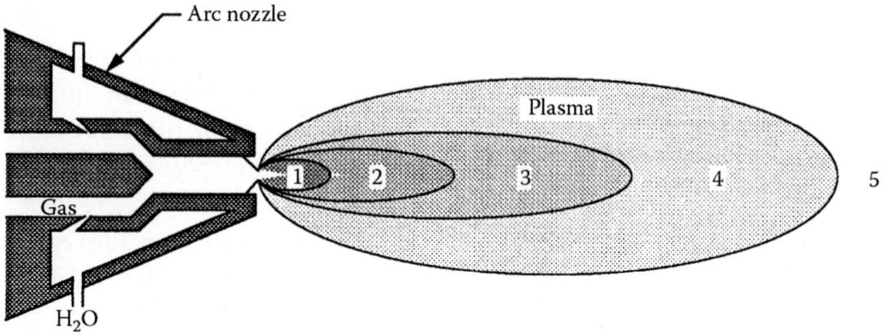


| Region | Distance from nozzle | Temperature range (K) | Linear flow velocity (m/s) |
|---|---|---|---|
| 1 | < 1 cm | $1.5 \times 10^4$ — $1 \times 10^4$ | ~ 400 |
| 2 | 1 < d < 5 cm | $1 \times 10^4$ — $5 \times 10^3$ | ~ 400 — 200 |
| 3 | 5 < d < 10 cm | $5 \times 10^3$ — $2 \times 10^3$ | ~ 200 — 100 |
| 4 | 10 < d < 20 | $2 \times 10^3$ — $1 \times 10^3$ | < 100 |
| 5 | 20 < d | < 500 | |

(a)

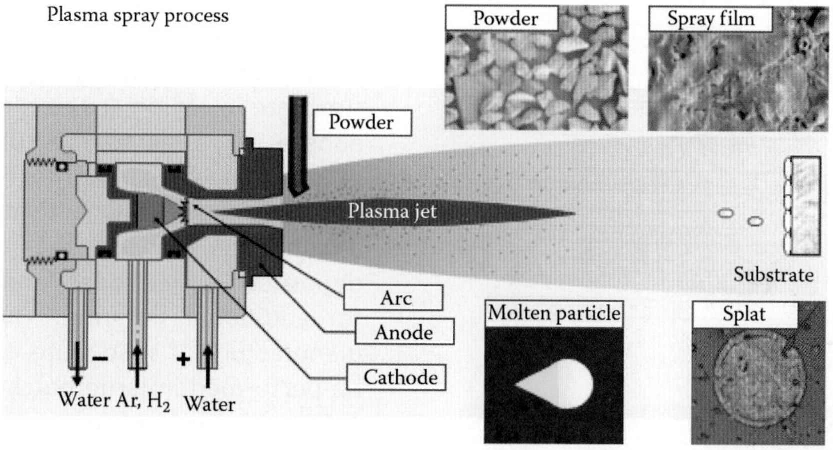


(b)

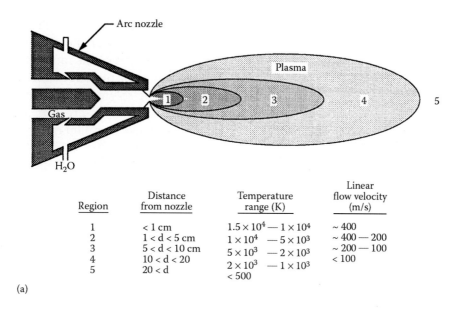

**FIGURE 9.23** (a) Typical ranges of temperature and flow velocity with distance from the nozzle. (b) Plasma spray nozzle.

fast deposition of thicker films (>30 μm), and it might enable the batch fabrication of solid-state oxygen sensors at a fraction of the current cost ($2 vs. $12 and up). Traditional automotive solid-state oxygen sensors for combustion control (so-called *lambda probes*) are nose-shaped, 3D structures fabricated by sintering a molded green tape zirconia body (see Figure 9.24a). A plasma-sprayed, porous, gas-diffusion barrier, typically a spinel structure oxide, protects the resulting dense ceramic YSZ solid electrolyte and the silk-screened oxygen-sensing Pt electrode contacting it. Newer designs for oxygen sensors, called *wide range air-to-fuel ratio sensors*, incorporate two oxygen pumping cells and one potentiometric gauge. As we saw in Example 9.2, these sensors are planar and fabricated by laminating and cofiring at high temperatures several layers of ceramic green tape, some of which are metal coated and have openings through them (see Figures 9.19 and 9.24b).[27] Because both metal and ceramic layers constitute the sensor, the materials should be well prepared, and the cofiring process must be tightly controlled to avoid metal diffusion or reaction with the ceramic. Because of the process complexities, this sensor is too expensive. In Figure 9.24c, an alternative planar oxygen sensor fabricated using plasma-spray deposition is compared with the traditional cone-shaped oxygen sensor.[29,30] In this planar oxygen sensor, the metal electrodes are deposited by sputtering,

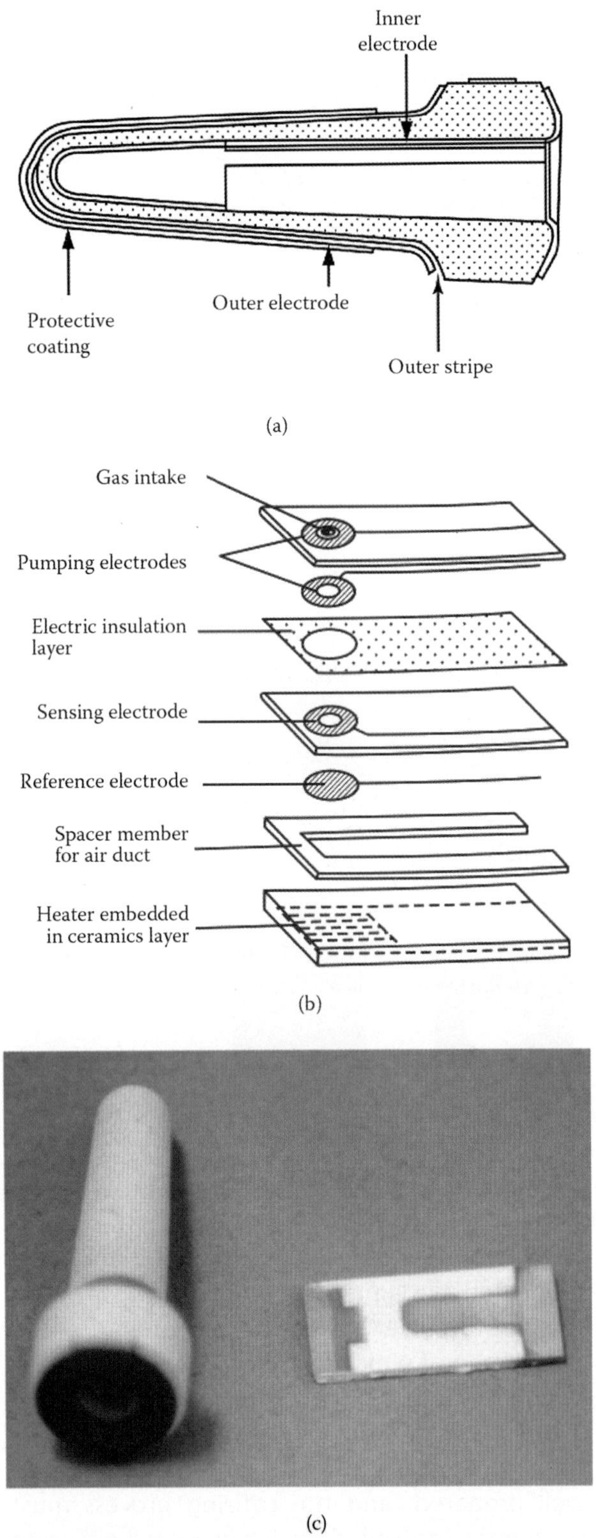


FIGURE 9.24 Solid-state oxygen sensors. (a) Traditional oxygen probe only protective, gas diffusion barrier is applied by plasma spray. (b) Wide-range air-to-fuel ratio sensor. (c) Planar oxygen probe, sensor, and gas diffusion layer made by plasma spray and compared with a classical oxygen probe. [(b) from Suzuki, S., T. Sasayama, M. Miki, M. Ohsuga, S. Tanake, S. Ueno, and N. Ichikawa. 1986. SAE Paper 860408. With permission.[27]]

and the plasma-sprayed YSZ film acts both as a gas-diffusion layer and as an oxygen-conducting electrolyte. The ionic conductivity of the plasma-deposited YSZ films does not reach the same level as YSZ electrolytes sintered at high temperatures because the films are not as dense. However, the relatively thin film geometry of the present sensor allows for using the plasma-sprayed YSZ film as an oxygen-pumping cell for the wide-range, air-to-fuel ratio sensor. The major challenge in the manufacture is the control of the porosity gradient in the gas diffusion barrier and electrolyte material, which is accomplished by controlling the size of the spraying powder and spraying conditions.[30] To change the temperature and flow velocity of the particles, we changed the distance between the nozzle and the substrate. With reference to Figure 9.23, to produce porous YSZ, we positioned the substrate somewhere between regions 4 and 5; to produce dense films, we positioned the substrate in region 4. The plasma-sprayed films adhere very well to the substrates and have exceptionally high integrity. The plasma method produces almost fully activated YSZ films onto a substrate carrying the thin film sputter-deposited Pt electrodes, and there is no need for additional sintering. This straightforward manufacture of solid-state oxygen sensors can be performed in large batches by using simple shadow masks and laser cutting the separate sensor elements.

We believe that this approach—a combination of thin and thick film methods—opens up the potential for planarization of many types of gas sensor devices. In the manufacture of chemical sensors, thick film technology on hybrid substrates is more prevalent than in the IC industry, and plasma deposition of all types of chemical sensor materials is, in the author's opinion, fertile ground for research. Plasma deposition of sensor materials such as $ZrO_2$, $SnO_2$, and ZnO on large inert carrier substrates could provide wafers coated with chemical sensor material very quickly and inexpensively.

## Plasma Arc Deposition

In the plasma reactors discussed above, the energy necessary for evaporation and reaction is provided by an electric arc that forms between a stick-type cathode and a nozzle-shaped, water-cooled anode

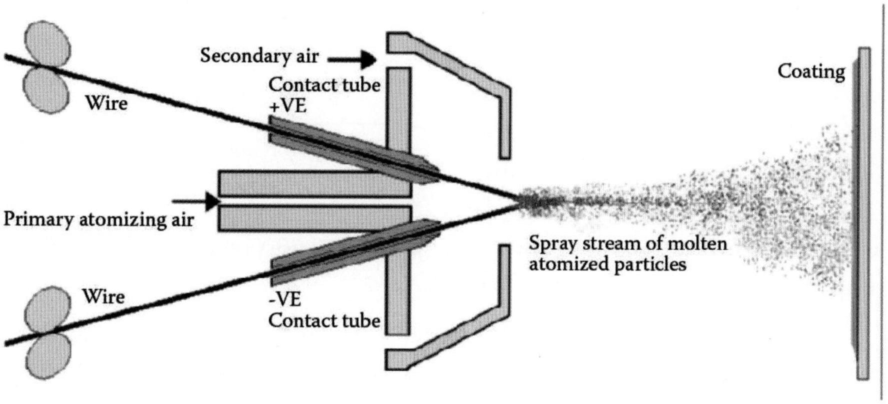


**FIGURE 9.25** Plasma arc spray deposition process.

as illustrated in Figure 9.21. The material to be deposited is a fine powder suspended in a carrier gas and is injected into the plasma jet where the particles are accelerated and heated. The electrodes themselves do not participate in the reaction itself; in contrast, in the arc spray deposition process, illustrated in Figure 9.25, the electrodes are made of the metal to be vaporized. The electrodes are polarized in the presence of an inert background gas until the breakdown voltage is reached, and an arc is formed across the electrodes that vaporizes them. The molten material is propelled by compressed air toward the substrate surface. The impacting molten particles on the substrate rapidly solidify to form a coating. High spray rates and efficiency make plasma arc deposition a good tool for spraying large areas and high production rates. Disadvantages of the electric arc spray process are that only electrically conductive electrodes can be sprayed. The main traditional applications of the arc spray process are anticorrosion coatings of zinc and aluminum.

## Plasmas and Nanoparticles

In the section above on spray pyrolysis, we discovered that this technique can be used for either film formation or for nanopowder generation. The same is true for the plasma equipment discussed above. Plasma temperatures are in the order of 10,000 K; thus, precursors injected in those plasmas are generally decomposed fully into atoms. To form nanoparticles from these, the atoms must then react or condense to form small solid particulates when cooled by mixing with a cool gas or by expansion through a nozzle. We saw an example of that in the ion cluster beam technology discussed in Chapter 7. In ion cluster beam technology the cooling occurs when the vapor exits a special evaporation cell. As shown in Figure 7.14, the heating of the evaporant in an evaporation cell with a small opening causes an adiabatic expansion of more than $100–10^{-5}$ or $10^{-7}$ mbar on exiting that cell. The expansion causes a sudden cooling, inducing the formation of atom clusters.

A very prominent application of arc deposition is that of fullerenes. The first large-scale synthesis of fullerenes was discovered in 1989 by Huffman and Kratschmer; their technique consisted of the arc evaporation of graphite electrodes via resistive heating within an atmosphere of approximately 100 atm of helium (see Volume III, Chapter 3). Spark discharge deposition was also the very technique that led to the discovery of carbon nanotubes in 1991. Carbon nanotubes were found in the carbon soot of graphite electrodes during arc discharge by using a current of 100 amps. In this process, the negative electrode sublimates because of the high temperatures caused by the discharge. Because nanotubes were initially discovered using this technique, it is still the most widely used method for nanotube synthesis. The yield for this method is up to 30% by weight, and it produces both single- and multiwalled nanotubes with lengths of up to 50 μm. The method is also used, for example, to produce fused silica. Silica sand is first vaporized in arc plasma at atmospheric pressure, and the resulting mixture of plasma gas and silica vapor is rapidly cooled by quenching with oxygen, thus ensuring the quality of the fumed silica produced.

In an RF induction plasma, energy coupling to the plasma is accomplished through the electromagnetic field generated by an induction coil. The plasma gas does not come in contact with electrodes, thus eliminating possible sources of contamination and allowing the operation of such plasmas with a wide range of gases, including inert, reducing, oxidizing, and other corrosive atmospheres. The working frequency is typically between 200 kHz and 40 MHz. Because the residence time of the injected feed droplets in the plasma is very short, it is important that the droplet sizes are small enough to obtain complete evaporation. The RF plasma method has been used to synthesize different nanoparticle materials, for example, various ceramic nanoparticles such as oxides and nitrides of Ti and Si.

## Laser and Electron Beam Deposition

### Introduction

Laser and electron beam machining are thermal techniques that are used as either subtractive tools in laser and electron beam etching (see Chapter 5) or as additive tools such as in the applications covered in this section, i.e., laser-assisted CVD (LCVD), SLS (a RP technique), and electron-assisted CVD (ECVD). Lasers also are used in stereolithography to make polymeric microstructures; this additive photochemical technology was reviewed in Chapter 8 and will not be repeated here. Pulsed laser deposition (Chapter 5) is also not covered here; although it uses intense laser radiation (thermal) to erode a target, it deposits the eroded material onto the adjacent substrate mechanically, not thermally.

## Forming Processes with Laser Machining

### *Laser-Assisted Chemical Vapor Deposition*

Laser-assisted chemical vapor deposition (LCVD) is an additive technique that involves the deposition of solids by localized chemical reaction using a laser beam (see Figure 9.26). The method has been used for growing 3D microstructures and thin films in elemental or compound form, and at least 17 elements of the periodic chart have been deposited (e.g., W, Ni, Ti).[31] The reactions are either pyrolytic, i.e., thermally activated, or photolytic, i.e., nonthermally activated. The latter method fills the need for a low-temperature deposition process. Because this chapter is about thermal forming, we will emphasize the former types of reactions.

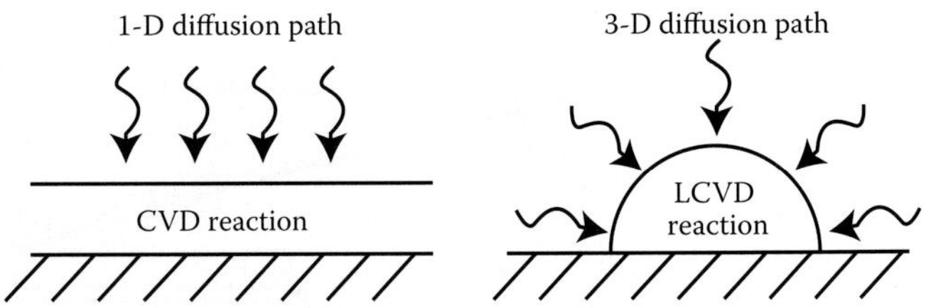


**FIGURE 9.26** Comparison of a CVD and an LCVD reaction.

The setup shown in Figure 5.25 is used to both laser etch and to laser deposit materials. Lasers and electron and ion beams add another important capability to materials processing, namely, the ability to process selected materials at specific sites and at low bulk temperatures. Of the different techniques, the laser is the more versatile. It can be operated in air, is not affected by surface charging, and can more easily be incorporated in a manufacturing assembly line.[32] With a laser source, it is possible to write a pattern on a surface directly by scanning the microsized light beam over the substrate in the presence of the suitable reactive gases. The species to be deposited is usually chemically encapsulated in a precursor gas with carbonyls or alkyl "backbones." Typical laser-driven deposition reactions are:

- Tungsten deposited by:

$$WF_6 + 3H_2 \rightarrow W + 6HF \quad \text{Reaction 9.1}$$

- Nickel deposited by:

$$Ni(CO)_4 \rightarrow Ni + 4CO \quad \text{Reaction 9.2}$$

- Silicon deposited by:

$$SiH_4 \rightarrow Si + 2H_2 \quad \text{Reaction 9.3}$$

Doped polysilicon lines with conductivities as low as $10^{-2}$–$10^{-3}$ Ω-cm may be obtained by adding $PH_3$ to the silane. Typical deposition rates are in the order of micrometers per second. By optimizing the process, very high deposition rates have been obtained in LCVD. For example, in the case

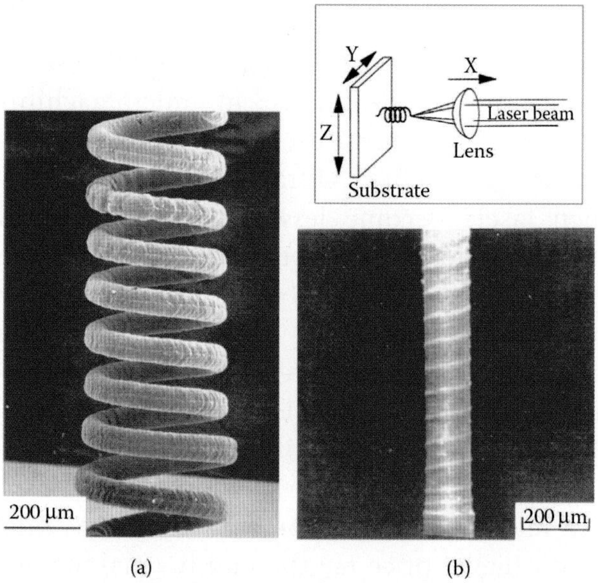


**FIGURE 9.27** Examples of LCVD microfabricated shapes made by adjusting the focal point of a laser (tungsten helix) (a) and rotating goniometer (Si microsolenoid with tungsten helix) (b). [From Boman, M., H. Westberg, S. Johansson, and J.-Å. Schweitz. 1992. *Proceedings: IEEE micro electro mechanical systems (MEMS '92).* Travemunde, Germany: IEEE. With permission.[33]]

of silicon, deposition rates of 500 μm/s have been reported by using 100% $SiH_4$ at 1 atm, and the writing speed of tungsten on silicon substrates can exceed several centimeters per second.[33]

Not only "flat" lines but also more complex 3D structures such as fibers and springs can be grown by LCVD (Figure 9.27). This is accomplished by adjusting the focal point of the laser continuously by moving the substrate using a 3D linear micropositioning system.[33] Microscale Si and boron rods and helical structures, freestanding tungsten coils, and a tungsten helix on a cylindrical silicon substrate (for a microsolenoid) have been demonstrated this way. For deposition of the tungsten helix on the Si rod, a rotating goniometer is used.

Table 9.2 lists examples of localized, electroless, and maskless laser depositions. Besides 2D and 3D writing patterns, there are two other site-specific actions that are possible by the laser irradiation technique: 1) the driving of dopants into semiconductor materials, and 2) the inducing of oxidation at a surface. Examples of the latter two laser machining applications are listed in Table 9.2 as well.

### *Selective Laser Sintering*

Selective laser sintering (SLS®, registered trademark by DTM™ of Austin, TX) is a process that was patented in 1989 by Carl Deckard, a University of Texas graduate student.

SLS is an additive layered rapid manufacturing method that creates solid, 3D objects by using a high-power laser (for example, a carbon dioxide laser) to fuse small particles of plastic, metal, or ceramic powders into a mass representing a desired 3D object. A thin layer of powder material is laid down, and the laser "draws" on the layer, sintering together the particles hit by the laser. The layer is then lowered a small amount, and a new layer of powder is placed on top. This process is repeated one layer at a time until the part is complete. As shown

**Table 9.2** Examples of Localized Laser Deposition, Doping, and Oxidation

| Material Deposition | | |
|---|---|---|
| Deposit | Precursor | Laser |
| Ti | $TiCl_4$ | $Ar^+$ |
| Ti/Al | $TiCl_4$, $Al(CH_3)_3$ | $Ar^+$ |
| Cr | $Cr(CO)_6$ | $Ar^+$ |
| **Laser-induced doping** | | |
| Dopant/substrate or film | Precursor | Laser |
| B in Si | Evaporated B | Ruby |
| P in polysilicon | P-doped glass | $N_2$ dye |
| As in Si | $AsH_3$ | XeCl |
| **Laser oxidation of substrates** | | |
| Substrate | Oxide species | Laser |
| Ti | $TiO_2$ | Nd:YAG |
| Zr | $ZrO_2$ | $CO_2$ |

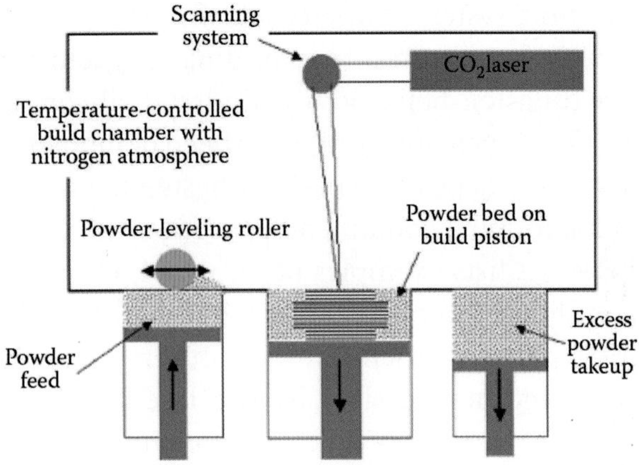


**FIGURE 9.28** Selective laser sintering.

in Figure 9.28, an SLS machine consists of two powder magazines on either side of the work area. A leveling roller moves powder over from one magazine, crossing over the work area to the other magazine. The laser selectively fuses powdered material by scanning cross-sections generated from a 3D digital description of the part (for example, from a CAD file or scan data) on the surface of a powder bed. After each cross-section is scanned, the powder bed is lowered by one layer thickness; a new layer of material is applied on top; and the process is repeated until the part is completed.

The chief advantage of SLS over other RP methods such as stereolithography (SLA) revolves around material properties. Many materials are possible, and these materials can approximate the properties of thermoplastics such as polycarbonate, nylon, or glass-filled nylon.

A new RP method combines additive and subtractive techniques. The laser sintering/milling hybrid machine, the LUMEX 25C, developed by Matsuura Machinery Corporation, successfully combines free-form manufacturing and high-speed milling (see inset in the heading of this chapter). This is the world's first "metal laser modeling hybrid machining unit." The integration of laser sintering of metallic powder and high-speed cutting eliminates the finish machining operations. It layers a powder mixture (90% steel and 10% copper) over the part and then sinters it with a 300-W (500 W maximum) $CO_2$ laser. The edges of the part are milled after sintering five layers, with tolerances in the 25-μm range. For the generated part, a surface roughness of 15 μm can be obtained.

### *Conclusions*

Additive laser machining presents several application opportunities in micromachining because of its versatility, site-specific operation, and RP capability. Ultrafast laser machining, when lasers become less expensive, will make laser micromachining yet more attractive. There are some obvious drawbacks with these direct write techniques; they are serial processes, and the deposition rate limits the speed of the micromanufacture. However, for RP, mold fabrication, and site-specific manufacturing, laser machining has a very bright future, and, as we have remarked before, some microsystems might carry a bigger price tag than an IC, making more expensive manufacturing technology acceptable.

## Electron Beam Forming Processes: Electron Beam-Assisted CVD

Electron beam-induced metal deposition is a slow but flexible and versatile process for the fabrication of microstructures with high lateral accuracies. For example, using a precursor gas at a pressure of $2 \times 10^{-2}$ mbar, W/C needles 0.2 μm in diameter have been deposited. The diameter of the needles is bigger than the electron beam diameter, which is less than 100 nm, because of electron scattering. To increase the growth rate, directing the precursor gas through a small nozzle onto the surface increases the local gas flow. The total gas flow needs to be kept below a maximum value to maintain the base pressure less than $10^{-5}$ mbar. For the deposition of other metals, numerous gases have been reported in the literature as precursors (mainly, for ion beam-induced deposition): $Al(CH_3)_3$ for Al, $C_7H_7O_2F_6Au$ for Au, and (methylcyclopentadienyl)trimethyl platinum for Pt.[34] A schematic of an electron beam fabrication setup is shown in Figure 9.29, along with an SEM photograph of W/C needles produced in this type of setup.[34] The deposition time of the needles shown was 1–2 min. Using the same type of technology, Matsui[35,36] demonstrated tips with a diameter of 15 nm for use in a tunneling microscope.

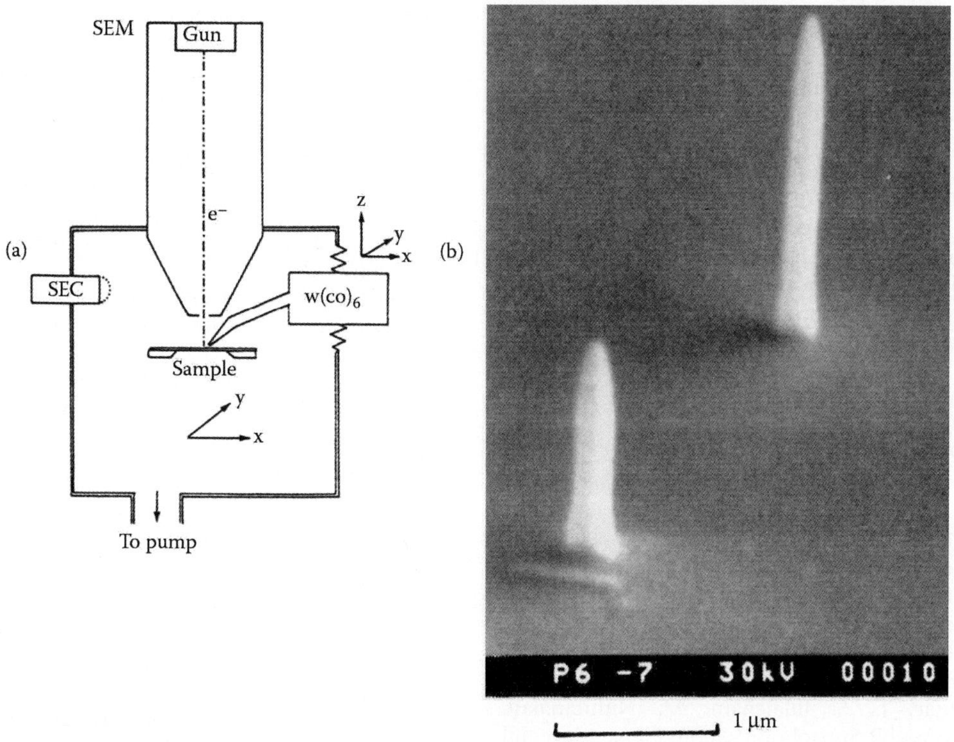


**FIGURE 9.29** Electron beam machining setup (a) and example of an electron beam-formed W/C microstructure (b). [From Brunger, W. H., and K. T. Kohlmann. 1992. *Proceedings: IEEE micro electro mechanical systems (MEMS '92)*. Travemunde, Germany: IEEE. With permission.[34]]

For processes such as controlled submicrometer repair of masks, focused ion beam-induced deposition appears to be the preferred technique, although both lasers[37,38] and electron beams[39] have been used as well.

## Questions

*Questions by Dr. Madou assisted by Mr. Omid Rohani, UC Irvine*

9.1: Explain how salt-assisted spray pyrolysis (SASP) overcomes some of the limitations of traditional spray pyrolysis.

9.2: Explain the flame spray pyrolysis process. What are the advantages and disadvantages of using this method?

9.3: What are typical applications of thick film technology?

9.4: Suppose that you want to form a thin flat ceramic part with a large surface area and various holes and shapes cut into it. Which manufacturing method do you suggest for this purpose? Why?

9.5: Sketch a multilayer ceramic planar oxygen sensor. Explain how this sensor works.

9.6: What might the possible advantages be of plasma depositing (plasma spraying) sensor materials in the manufacture of chemical sensors?

9.7: Explain the plasma spray deposition manufacturing process briefly.

9.8: How do we get nanoparticles from plasma atoms?

9.9: Explain the selective laser sintering (SLS) process.

9.10: How do you convert your chemical vapor deposition (CVD) reactor into a reactor that produces nanoparticles?

9.11: Why does one need a pseudoplastic material to function as an ink in silk screening?

9.12: What is a doctor's blade, and where is it used?

9.13: Compare laser-assisted CVD (LCVD) with E-beam assisted CVD (ECVD).

9.14: Name some of the applications of $TiO_2$ powders.

## References


1. Messing, G. L., S.-C. Zhang, and G. V. Jayanthi. 1993. Ceramic powder synthesis by spray pyrolysis. *J Am Cer Soc* 76:2707–26.
2. Mooney, J. B., and S. B. Radding. 1982. Spray pyrolysis processing. *Ann Rev Mater Sci* 12:81–101.
3. Tomar, M. S., and F. J. Garcia. 1988. Spray pyrolysis in solar cells and gas sensors. *Prog Cryst Growth Charact* 4:221–48.
4. Xia, B., I. W. Lengorro, and K. Okuyama. 2001 Particle formation processes of conventional spray pyrolysis (CSP) and salt-assisted spray pyrolysis (SASP). *Adv Mater* 13:1579.
5. Vemury, S., S. E. Pratsinis, and L. Kibbey. 1977. Electrically-controlled flame synthesis of nanophase $TiO_2$, $SiO_2$, and $SnO_2$ powders. *J Mater Res* 12:1031–42.
6. Lenggoro, I. W., K. Okuyama, J. F. d. l. Mora, and N. Tohge. 2000. Preparation of ZnS nanoparticles by electrospray pyrolysis. *Aerosol Sci* 31:121–36.
7. Deng, W., J. F. Klemic, X. Li, M. A. Reed, and A. Gomez. 2006. Increase of electrospray throughput using multiplexed microfabricated sources for the scalable generation of monodisperse droplets. *Aerosol Sci* 37:696–714.
8. Dzenis, Y. 2004. Spinning continuous fibers for nanotechnology. *Science* 304:1917.
9. Sutasinpromprae, J., S. Jitjaicham, M. Nithitanakul, C. Meechaisue, and P. Supaphol, P. 2006. Preparation and characterization of ultrafine electrospun polyacrylonitrile fibers and their subsequent pyrolysis to carbon fibers. *Polym Int* 55:825.
10. Zussman, E., X. Chen, W. Ding, L. Calabri, D. A. Dikin, J. P. Quintana, and R. S. Ruoff. 2005. Mechanical and structural characterization of PAN-derived carbon nanofibers. *Carbon* 43:2175.
11. Wang, Y., S. Serrano, and J. J. Santiago-Aviles. 2002. Conductivity measurement of electrospun PAN-based carbon nanofiber. *J Mater Sci* 21:1055.
12. Doshi, J. and H. D. Reneker. 1995. Electrospinning process and applications of electrospun fibers. *J Electrost* 35:151.
13. Shin, Y. M., M. M. Hohman, M. P. Brenner, and G. C. Rutledge. 2001. Electrospinning: A whipping fluid jet generates submicron polymer fibers. *Appl Phys Lett* 78:1149.
14. Theron, S. A., E. Zussman, and A. L. Yarin. 2004. Experimental investigation of the governing parameters in the electrospinning of polymer solutions. *Polymer* 45:2017.
15. Sharma, C. S., A. Sharma, and M. Madou. 2010. Multiscale carbon structures fabricated by direct micropatterning of electrospun mats of SU-8 photoresist nanofibers. *Langmuir* 26(4):2218–22.
16. Turon Teixidor, G., B. Park, P. Mukherjee, Q. Kang, and M. Madou. 2009. Modeling fractal electrodes for Li-ion batteries. *Electrochim Acta*, 54(24):5928–36.
17. Turon Teixidor, G., R. Gorkin, P. Tripathi, G. Bisht, M. Kulkarni, T. Maiti, T. Battacharyya, J. Subramaniam, A. Sharma, B. Park, and M. Madou. 2008. Carbon-MEMS as a substratum for cell growth. *J Biomed Mater* 3(3):034116.
18. Madou, M. J., and S. R. Morrison. 1989. *Chemical sensing with solid state devices.* New York: Academic Press.
19. Madou, M. 1994. *NIST special publication 865.* Gaithersburg, MD: NIST.
20. Lambrechts, M., and W. Sansen. 1992. *Biosensors: microelectrical devices.* Philadelphia: Institute of Physics Publishing.
21. Jabbour, G. E., R. Radspinner, and N. Peyghambarian. 2001. Screen printing for the fabrication of organic light-emitting devices. *J IEEE Selected Topics Quantum Electronics* 7:769–73.
22. Riemer, D. E. 1988. Analytical engineering model of the screen printing process: part I. *Solid State Technol* August:107–11.
23. Middlehoek, S., D. J. W. Noorlag, and G. K. Steenvoorden. 1983. Silicon and hybrid micro-electronic sensors. *Electrocomponent Sci Technol* 10:217–29.
24. White, N. M., and A. W. J. Cranny. 1987. Design and fabrication of thick film sensors. *Hybrid Circuits* 12:32–36.
25. Howatt, G. N. 1952. *Method of producing high-dielectric high-insulation ceramic plates.* US Patent 2582993.
26. Mistler, R. E., and E. R. Twiname. 2000. *Tape casting: theory and practice.* Westerville, OH: American Ceramic Society.
27. Suzuki, S., T. Sasayama, M. Miki, M. Ohsuga, S. Tanake, S. Ueno, and N. Ichikawa. 1986. Air-fuel ratio sensor for rich, stoichiometric and lean ranges. SAE Paper 860408.
28. Pfender, E. 1988. Fundamental studies associated with the plasma spray process. *Surf Coat Technol* 34:1–14.
29. Oh, S., and M. Madou. 1992. Planar-type, gas diffusion-controlled oxygen sensor fabricated by the plasma spray method. *Sensors Actuators B* B14:581–82.
30. Oh, S. 1994. A planar-type sensor for detection of oxidizing and reducing gases. *Sensors Actuators B* B20:33–41.
31. Eden, J. G. 1986. Photochemical processing of semiconductors: new applications for visible and ultraviolet lasers. *IEEE Circuits Devices Mag* 2:18–24.
32. Helvajian, H. 1995. *Microengineering technology for space systems.* Ed. H. Helvajian. El Segundo, CA: Aerospace Corporation.
33. Boman, M., H. Westberg, S. Johansson, and J.-Å. Schweitz. 1992. *Proceedings: IEEE micro electro mechanical systems (MEMS '92).* Travemunde, Germany.
34. Brunger, W. H., and K. T. Kohlmann. 1992. *Proceedings: IEEE micro electro mechanical systems (MEMS '92).* Travemunde, Germany: IEEE.
35. Matsui, S. 1987. Direct writing onto Si by electron beam simulated etching. *Appl Phys Lett* 51:1498–99.
36. Matsui, S., and K. Mori. 1986. New selective deposition technology by electron beam induced surface reaction. *J Vac Sci Technol* B4:299–304.
37. Bauerle, D. 1986. *Chemical processing with lasers.* Berlin, Germany.
38. Ehrlich, D. H., and J. Y. Tsao. 1983. A review of laser–microchemical processing. *J Vac Sci Technol* B1:969–84.
39. Kunz, R. R., and T. M. Mayer. 1987. Catalytic growth rate enhancement of electron beam deposited iron films. *Appl Phys Lett* 50:962–64.

# 10

# Micromolding Techniques—LIGA

*The Graduate*, starring Dustin Hoffman, was released by Embassy Pictures in 1967. A memorable poolside scene from the movie:

| | |
|---|---|
| Mr. McGuire: | *Come with me for a minute. I want to talk to you. I just want to say one word to you. Just one word.* |
| Ben: | *Yes, sir.* |
| Mr. McGuire: | *Are you listening?* |
| Ben: | *Yes sir, I am.* |
| Mr. McGuire: | **PLASTICS.** |
| Ben: | *Exactly how do you mean?* |
| Mr. McGuire: | *There is a great future in plastics. Think about it. Will you think about it?* |
| Ben: | *Yes I will.* |

Outline



## Introduction

A wide variety of micro- and nanofabrication techniques such as x-ray lithography (Chapter 2), computer numerical control (CNC) machining (Chapter 6), laser machining (Chapter 5), and many others are either too slow or too expensive for the mass production of inexpensive devices. This is where micromolding/replication technology comes into play because of its capacity for very large volume and inexpensive manufacture.[1]

A key example involving micromolding is LIGA, which, we learned in Chapter 2, is the German acronym for x-ray lithography (*x-ray lithographie*), electrodeposition (*galvanoformung*), and molding (*abformtechnik*). The process involves a thick layer of x-ray resist (from micrometers to centimeters), high-energy x-ray radiation exposure, and development to arrive at a three-dimensional (3D) resist structure. Subsequent metal deposition fills the resist mold with a metal, and, after resist removal, a freestanding metal structure results.[2] The metal shape may be a final product or serve as a mold insert for precision plastic molding. Molded plastic parts may in turn be final products or lost molds (see Figure 2.9). The plastic mold retains the same shape, size, and form as the original resist structure but is produced quickly and inexpensively as part of an infinite loop. The plastic lost mold may generate metal parts in a second electroforming process or generate ceramic parts in a slip casting process.

The bandwidth of possible sizes in all three dimensions renders LIGA useful for manufacture of microstructures (micrometer and submicrometer dimensions) and packages for these microstructures (millimeter and centimeter dimensions), and even for the connectors from those packages to the "macro world" (e.g., electrical, through-vias or physical and gas, in- and outlets). Once LIGA was established in the research community, interest in other micro- and nanomolding/replication methods became more pronounced. Given the cost of the LIGA equipment, various LIGA-like (pseudo-LIGA) processes took center stage. These pseudo-LIGA methods involve micromolding/replication of masters created by alternate means such as deep reactive ion etching (DRIE) and novel deep ultraviolet (UV) thick photoresists (e.g., SU-8). This more generalized lithography and replication procedure is illustrated in Figure 10.1.

Micro- and nanofabrication of molds is reshaping manufacturing approaches for a wide variety of mass-produced small parts. Frequently, integrated circuit (IC)-based batch microfabrication methods are considered together with more traditional, serial machining methods. In this evolution, LIGA and pseudo-LIGA processes constitute "handshake-technologies," bridging IC and classical manufacturing technologies. The capacity of LIGA and pseudo-LIGA for creating a wide variety of shapes from different materials makes these methods akin to classical machining with the added benefit of unprecedented aspect ratios and absolute tolerances rendered possible by lithography or other high-precision mold fabrication techniques.

LIGA and LIGA-like processes can be used directly to fabricate prototype microdevices, but, for commercial-scale manufacture of microdevices, replication by molding from a LIGA or LIGA-like mold is required. If the feature sizes are larger than

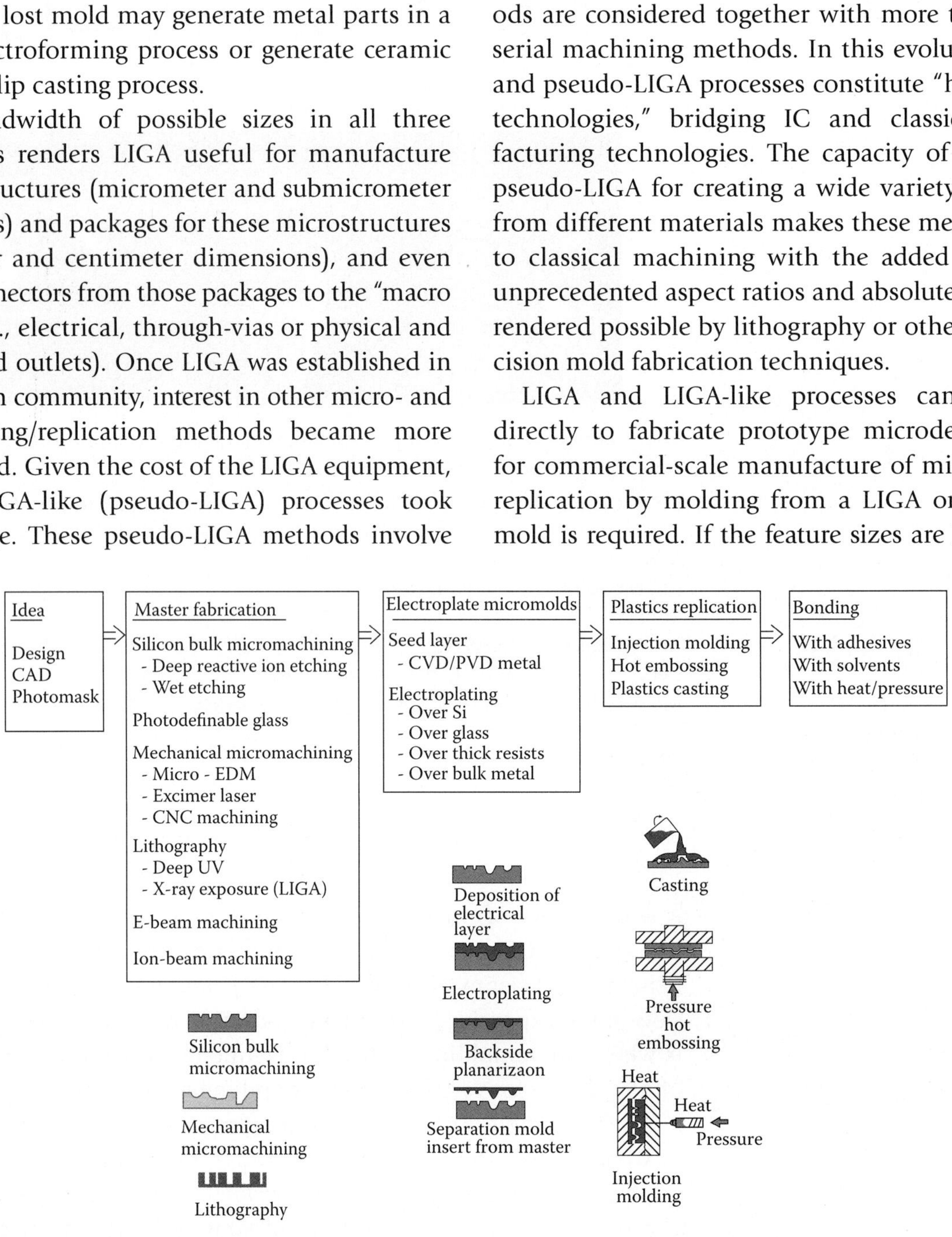


**FIGURE 10.1** Process flow for plastic microfabrication.

50 μm and several levels of feature depths are involved, even CNC machining may be used to manufacture a prototype device. For example, in Figure 10.2, we show a CNC-machined plastic (polycarbonate) fluidic compact disc (CD) platform prototype. Compared with LIGA or LIGA-like processes, CNC machining does not provide as good a surface finish or dimensional control (see Chapter 6). However, it does not have any material limitation; various metals and plastics can be used. For feature sizes smaller than 50 μm (but larger than several micrometers), ultraprecision milling or laser ablation can be used. These prototype processes, just like LIGA and LIGA-like processes, are all intensive, slow, and costly, and for these microfabrication techniques to become economically viable, microstructures must be able to be replicated/micromolded successively, in an "infinite loop," without remaking the master mold. For example, in the plastic molding process, a metal structure produced by metal deposition serves as a mold insert and is used over and over without remaking the metal master. For mass-produced plastic devices, one of the following molding techniques is suitable: liquid resin molding, thermoplastic injection molding (IM), and compression (embossing/imprinting) molding. Our results of IM of a plastic fluidic platform, as shown in Figure 10.2, from a micromachined mold using pseudo-LIGA is shown in Figure 10.6.

In this chapter, we first review the most popular industrially available polymer molding techniques, and then, as examples that push the technology to its extremes, we detail LIGA and LIGA-like processes.

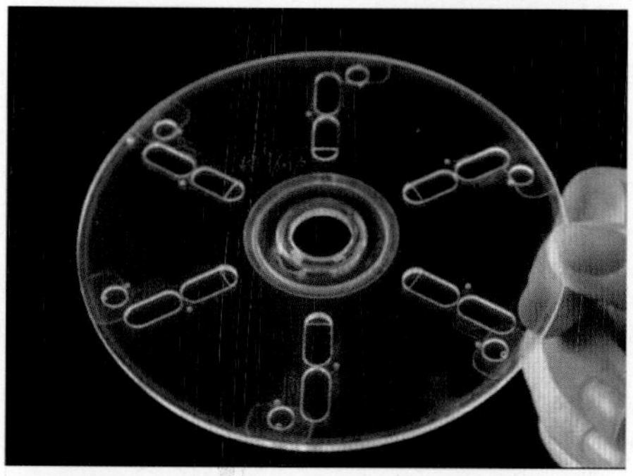

**FIGURE 10.2** CNC-machined plastic lab CD platform: prototype cell disrupter disk, made from machined plastic disks and cut transfer adhesive films. (UCI and RotaPrep, Drs. Madou and Kido.)

The miniature features and the high-aspect ratios of the mold inserts generated from LIGA and LIGA-like methods present new problems compared with the molding and demolding processes of macrocomponents and compared with the production of small but low-aspect-ratio products such as CDs. A variety of the special processes developed to accommodate these new needs are reviewed. We start with a historical introduction to LIGA and analyze the process steps. The different applications and technical characteristics of x-rays from a synchrotron are discussed, and then we present an introduction to the crucial issues involved in making x-ray masks and their alignment for LIGA. This is followed by a treatise on LIGA substrate choice, resist requirements, exposure, development, electroplating, and finally molding. The electroplating and molding sections apply to LIGA and pseudo-LIGA approaches. The section on LIGA is followed by a review of competing mold fabrication techniques and a short treatise on ceramic molding.

Micromolding with an elastomeric mold (e.g., polydimethylsiloxane), or soft lithography, was reviewed in Chapter 2, where we stressed its low cost and convenience in research. Compared with other micromolding techniques reviewed in the current chapter, soft lithography is especially useful for thin microstructure fabrication.

Bonding of plastic molded microparts, including plastic welding, organic solvent bonding, and bonding with thermoset adhesives, is covered in Volume III, Chapter 4 in its discussion of packaging techniques.

## Plastic Molding

### Liquid Resin Molding Techniques

#### *Introduction*

Because mold inserts made by photolithography techniques are typically limited to rather soft metals like nickel or brittle materials such as silicon and glass, micromolding based on low-viscosity liquid resins (instead of high-viscosity polymer melts) is a very attractive approach. During liquid resin molding, the low-viscosity reactive polymer components are mixed shortly before injection into the mold cavity, and polymerization takes place during the molding process. Under liquid resin molding techniques, we distinguish between reaction IM (RIM) and transfer molding.

### *Reaction Injection Molding and Transfer Molding*

Two typical setups for RIM and transfer molding are shown in Figure 10.3a and b, respectively.[3] In the RIM setup shown in Figure 10.3a, two or more highly reactive liquid resins impinge and mix inside a mixer and are injected into the mold. In RIM, the polymerization is usually mixing activated; therefore, it can quickly react at room temperature or slightly above and convert from liquid to solid. The transfer molding setup in Figure 10.3b, for a thermally activated system, is simpler and less expensive. Here, the reactive species have been premixed at room temperature, and the mixture is poured into the "transfer pot." A plunger squeezes the mixture through the sprue into the mold cavity. To cure the mixture, the mold is heated or radiated, depending on the resin system. With the lowest viscosity resins, vacuum-assisted transfer molding is used, which is even less expensive than mechanical transfer molding.

Two complications associated with RIM are 1) an internal mold release agent must be used and 2) polymer shrinkage during polymerization. Release agents are added to the bulk polymer and tend to reduce the thermomechanical properties of the molded polymer. External mold release agents, coating the surfaces of the mold insert, are unsuitable because they are difficult to introduce into all the microfeatures of the mold insert. Because the resin must be fully cured at high temperatures, the cycle time is typically more than 10 min. Hanemann et al.[4] developed a photoinitiated RIM process by using the methylmethacrylate (MMA) resin. They found that the cycle time can be greatly reduced (to 2–3 min), but other disadvantages, such as internal mold release agent and polymerization shrinkage, still exist.

### *Summary: Pros and Cons of Liquid Resin Molding*

The pros and cons of RIM are summarized in Table 10.1. Typical materials are optically clear reactive liquid resins, e.g., epoxies, urethanes, silicone rubbers, cross-linkable acrylics, and hydrogels.

## Injection Molding

### *Introduction*

Injection molding (IM) is based on heating a thermoplastic material until it melts, thermostatting the mold parts, injecting the melt with a controlled injection pressure into the mold cavity, and cooling the manufactured goods. IM is probably the most widely used technique in macroscopic production of polymer parts. Besides conventional IM, one variant of IM—a good candidate for microfabrication—is

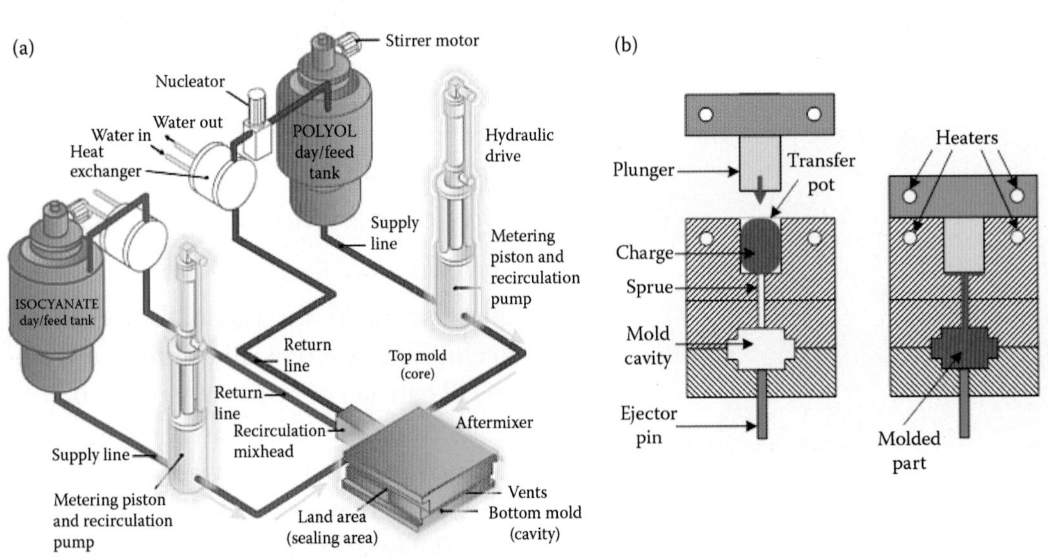


**FIGURE 10.3** Two typical setups for RIM (a) and transfer molding or TM (b).

**TABLE 10.1 Pros and Cons of Liquid Resin Molding**

| Pros | Cons |
|---|---|
| Ease of mold filling (low η) | Long cycle time (minutes, sometimes even hours) |
| Low stress on master (low η) | Polymerization shrinkage |
| High chemical and thermal resistance (because of the cross-linking) | Contamination (resin residue if the reaction is not 100%) |
| Replication extremely good for small, high-aspect ratio, and 3D features | Production cost medium to high |

detailed here[3]: thin-wall IM. A typical injection compression molding setup is shown in Figure 10.4. The mold insert in IM, as in RIM, must have extremely smooth walls to be able to demold the plastic structures. Such smooth walls cannot be produced with a classical mold-making technique such as spark erosion. For some polymers, such as polymethylmethacrylate (PMMA), the polymer pellets need to be dried for 4–6 h at 70°C before use in IM because PMMA absorbs water in air (up to 0.3%), leading to poorer wall quality.

### *Conventional Injection Molding*

In conventional IM of microparts, the mold remains closed, and the walls are kept at a uniform temperature above the glass transition temperature ($T_g$) of the polymer; thus, the injected molten plastic mass does not harden prematurely. To be able to fill very small mold features with aspect ratios higher than 10, the temperature of the tool holding the insertion mold should be higher than what is typically used in IM applications of parts with lower aspect ratios. However, one should stay well below the melting temperature (e.g., 170°C for PMMA with a melting temperature of 240°C) because the microstructure might show temperature-induced defects if one works at a temperature closer to the melting temperature.[5]

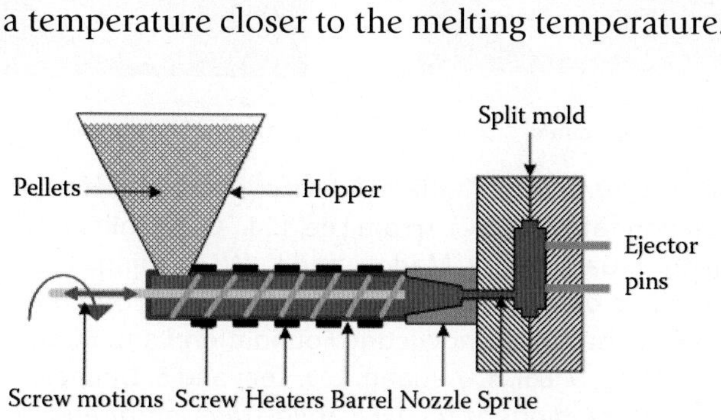


**FIGURE 10.4** Typical injection molding setup.

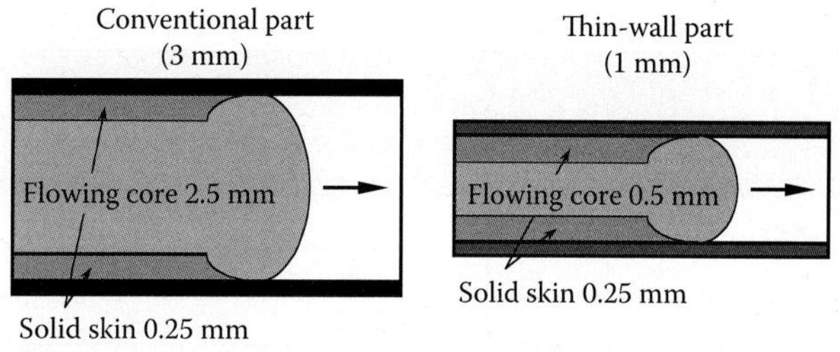


**FIGURE 10.5** Schematics of solid skin buildup in conventional and thin-wall injection molding. (OSU, Drs. Madou and Lee.)

One of the most important factors for replication fidelity in IM is the thickness of the skin, which forms on the mold surface during filling of the cavity (see Figure 10.5).[3]

The injection pressure must penetrate this skin to replicate the shape of the master accurately. The faster the injection, the thinner this skin on the insert mold surfaces. The upper shapes of the master forming the bottom of the parts are usually perfectly formed. It is the bottom of the master and especially inner corners that cause problems. Because in conventional IM the mold is kept closed during the entire mold-filling process, high clamping forces on the mold parts are required, and the method is used for relatively thick substrates with high replication demands and low cost. Parts made this way often end up having high molded-in stresses.

Conventional thermoplastic IM is the technique used, for example, to make compact discs (CDs), involving features about 0.1 μm in depth and minimum lateral dimensions of 0.6 μm; that is, an aspect ratio of 0.16. CDs are made by IM of polycarbonate in a cavity formed between a mirror surface block and a stamper (i.e., mold or master). Conventionally, electroforming from a glass master generates nickel stampers. The principal attributes required of a stamper substrate are toughness, thermal shock resistance, thermal conductivity, and hardness. For IM in particular, toughness, thermal shock resistance, and thermal conductivity are critical. Thermal shock resistance is the product of a material's modulus of rupture or tensile strength and its thermal conductivity, divided by the product of its coefficient of thermal expansion and its modulus of elasticity. In Table 10.2, we compare some materials for their merit as stamper substrates. Ion machining is one of the methods that can be used to make mold

**Table 10.2** Materials Selection for Stamper/Insert Mold Substrates

| Substrate | Knoop Hardness (kg/mm²) | Fracture Toughness (MPa m$^{1/2}$) | Thermal Shock Resistance (W/m) | Thermal Conductivity (W/m K) | Ion Machinability (Surface Quality) |
|---|---|---|---|---|---|
| Nickel | 100 | ~100 | 7,138 | 80 | Poor |
| $Al_2O_3$ | 2,100 | 4 | 3,225 | 30 | Very good |
| SiC | 2,500 | 3 | 19,149 | 90 | Excellent |
| Glassy carbon | 500 | 2 | 135,517 | 120 | Excellent |
| Corning 9647 (glass) | 450 | 1.3 | 546 | 2.5 | Excellent |

*Source:* Bifano, T. G., H. E. Fawcett, and P. A. Bierden. 1997. Precision manufacture of optical disc master stampers. *Precision Eng* 20: 53–62.[6]

inserts (see Figure 10.1 under Master fabrication), and surface quality after ion etching is also listed (ion machinability).

Nickel CD stampers are normally 138 mm in diameter and 0.3 mm thick with an average roughness less than 10 nm. For ceramic stampers, a thickness of 0.9 mm is chosen to avoid fracture in IM. Typical conditions for IM are summarized in Table 10.3.

IM of parts with small features and low-aspect ratios (e.g., CDs) have been widely applied. The main challenge in microelectromechanical system (MEMS) and nanoelectromechanical system (NEMS) replication attempts is to extend this technique to the fabrication of components with the same small feature size but much larger aspect ratios, as needed in many medical and biochemical applications (see later discussions of LIGA and pseudo-LIGA).

### *Thin-Wall Injection Molding*

Thin-wall IM is basically high-speed IM.[3,7] Speeds can be 10 times faster than in conventional IM, reducing the thickness of the polymer skin on the mold (see Figure 10.5). The polymer melt can quickly be injected into the mold to fill all the detail structure of the cavity before solidification occurs.

In our own thin-wall IM experiments for replication of BioMEMS devices, such as the fluidic CD (about 1 mm thick)[3,8] shown in Figure 10.6, we have used a Sumitomo 200-ton high-pressure and high-speed machine.[7] Only at injection speeds of 50 mm/s or higher does the quality of the replication become adequate. A birefringence technique was used to qualitatively examine the stress in the polymer parts as demonstrated in Figure 10.7. Samples showing sharp color contrast have large molded-in stresses. At the lower flow rates the residual stress is higher because, during the flow, the polymer has been stretched and solidified, meaning that there is a lot of flow-induced stress. Increasing the flow

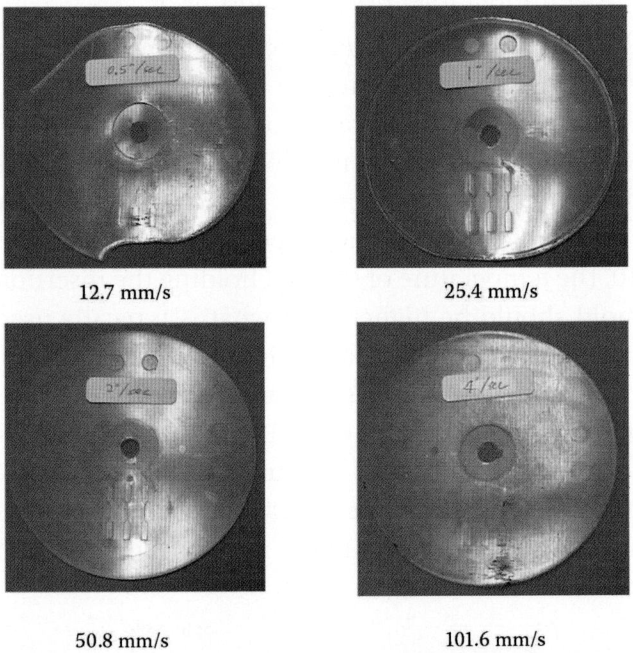


**FIGURE 10.6** Microfluidic CD devices made at different injection rates (OQPC). (From Lee, L. J., C.-H. Shih, Y.-J. Juang, J. Garcia, M. J. Madou, and K. W. Koelling. 1999. *Novel microfabrication options for BioMEMS conference.* San Francisco: The Knowledge Foundation;[3] and Madou, M. J., Y. Lu, S. Lai, Y.-J. Juang, L. J. Lee, and S. Daunert. 2000. *Proceedings of the IEEE: solid-state sensor and actuator workshop.* Hilton Head Island, SC.[8])

**Table 10.3** Typical Conditions for Injection Molding

| | |
|---|---|
| Mold temperature | 85°C |
| Polycarbonate temperature | 330°C |
| Clamping force | 60 tons |
| Injection time | 1 s |
| Cooling time | 2 s |

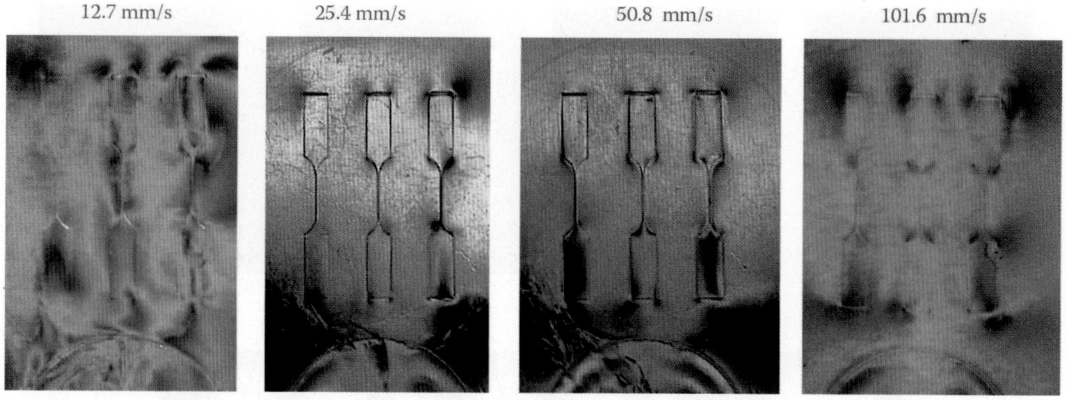


**FIGURE 10.7** Birefringence of microfluidic devices made at different injection rates. (From Madou, M. J., Y. Lu, S. Lai, Y.-J. Juang, L. J. Lee, and S. Daunert. 2000. *Proceedings of the IEEE: solid-state sensor and actuator workshop.* Hilton Head Island, SC;[8] and Lee, L. J., C.-H. Shih, Y.-J. Juang, J. Garcia, M. J. Madou, and K. W. Koelling. 1999. *Novel microfabrication options for BioMEMS conference.* San Francisco: The Knowledge Foundation.[3])

rate decreases the amount of residual stress because the polymer can quickly fill in the mold and relax before solidification occurs. Nevertheless, there is some remaining stress that may cause warpage or less chemical resistance. By comparison, in casting there is essentially no flow: it is stress free.

A comparison of sample birefringence from different replication methods is given in Figure 10.8.[3] One observes that reactive casting with an epoxy resin causes the least amount of residual stress (least sharp color contrasts).

Another example of thin-walled structures made with thin-wall IM is shown in Figure 10.9. High injection speed is needed to completely fill the mold cavities for a thin-wall relay case (0.45 mm thick, flow length is 50.4 mm).

### *Injection Molding Equipment*

For manufacturing MEMS structures, conventional IM machines used for CD fabrication are expanded with special features for the molding of microsized parts or parts with high-aspect ratio. The IM machine periphery then includes a vacuum unit for the evacuation of the split mold cavity and a precise temperature control unit. It is desirable to use one temperature control cycle on each surface of the split mold (see Figure 10.4). The flow temperatures are adapted to the respective half of the tool or to the molding and demolding process cycle to achieve a relatively short overall cycle time and a homogeneous tool temperature. To keep the cycle time as short as possible, the thermal mass of the tool sections to be heated is minimized and

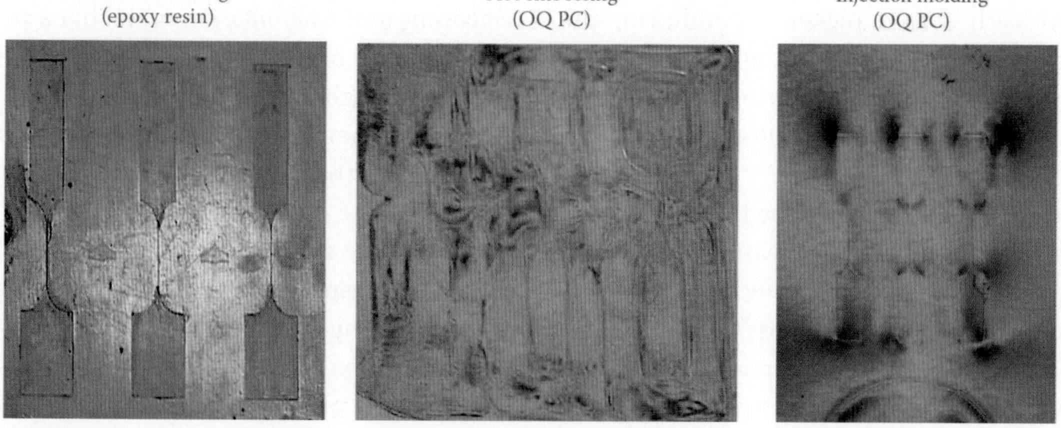


**FIGURE 10.8** Comparison of birefringence patterns of microfluidic platforms made by different replication methods. (From Lee, L. J., C.-H. Shih, Y.-J. Juang, J. Garcia, M. J. Madou, and K. W. Koelling. 1999. *Novel microfabrication options for BioMEMS conference.* San Francisco: The Knowledge Foundation.[3])

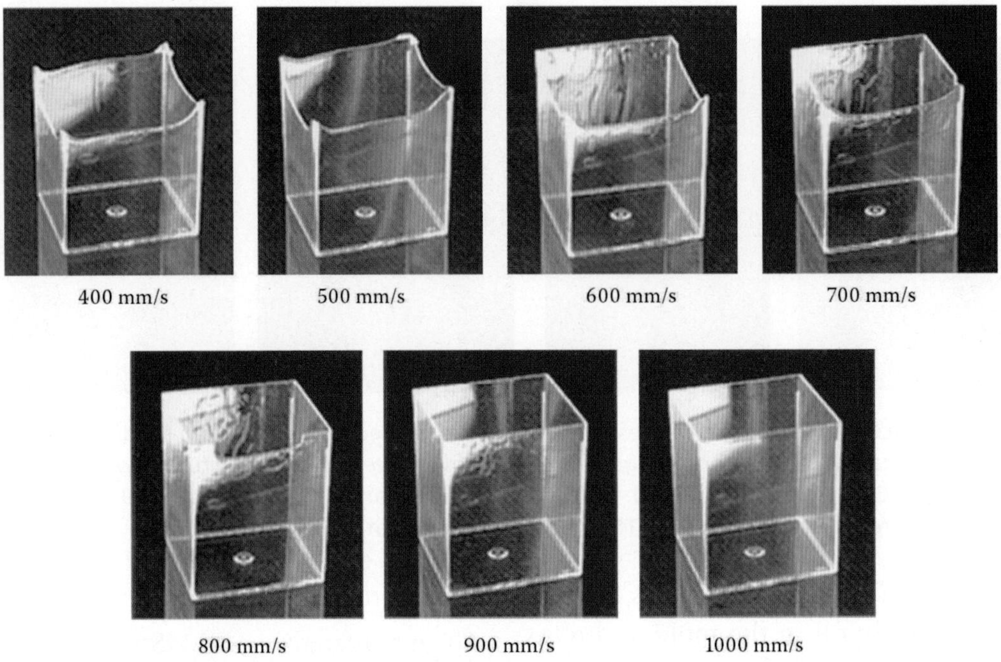


**FIGURE 10.9** Relay case made by thin-wall injection molding at different injection rates (Nissei). (Courtesy Dr. Jim Lee, OSU.)

thermally insulated from the adjacent assemblies. The guiding mechanisms of the tool halves and the ejector system may not be fitted too loosely because small transverse movements might damage the microstructures during demolding. The machines and tools to be evacuated have to comply with tolerances in the micrometer range.[9] Demolding of the microstructures takes place by means of ejector pins, which are located on the sole of the mold tool, the guiding mechanism, or directly at extremities of the microstructure.[9] Under laboratory conditions, the microstructures are removed manually. In small series production, a handling robot is used. It may be equipped with a feed picker. In addition, the machines used for small series and mass production are provided with a granulate drying, supply, and metering system such that fully automatic molding in personnel-free shifts is possible.[10]

Commercial IM machines designed for microsized parts or parts with microsized features are available from Battenfeld (http://www.battenfeld.de), Boy Machines (http://www.dr-boy.de/rightmachine.htm), Engel (http://www.engel.at), Ferromatik Milacron (http://www.ferromatik.com), Murray (http://www.murrayeng.com), and Nissei (http://www.nisseiamerica.com). Figure 10.10 shows a photograph of Battenfeld's Microsystem 50.

### *Summary: Pros and Cons of Injection Molding*

Pros and cons of IM are summarized in Table 10.4. Example materials in IM include polycarbonate (PC), polymethylmethacrylate (PMMA), polystyrene (PS), polyvinylchloride (PVC), polypropylene (PP), polyethylene (PE), and polyacrylnitrilbutadienstyrol (ABS).

## Compression Molding (Also Relief Imprinting or Hot Embossing)

### *Introduction*

The basic principle of compression molding (also *hot embossing* and *relief imprinting*) is that a polymer substrate is first heated above its glass transition temperature, $T_g$ (or softening temperature). A mold (or master) is then pressed against the substrate, fully transferring the pattern onto it (embossing). After a certain time of contact between the mold and the substrate, the system is cooled down below $T_g$, followed by separation of the mold and the substrate (de-embossing). Importantly, the hot embossing process can be achieved in either a cyclic process or a continuous process. Compression molding[11,12] provides several advantages compared with IM, such as relatively low costs for embossing tools, simple process, and high replication accuracy for small features.

(a)

(b)

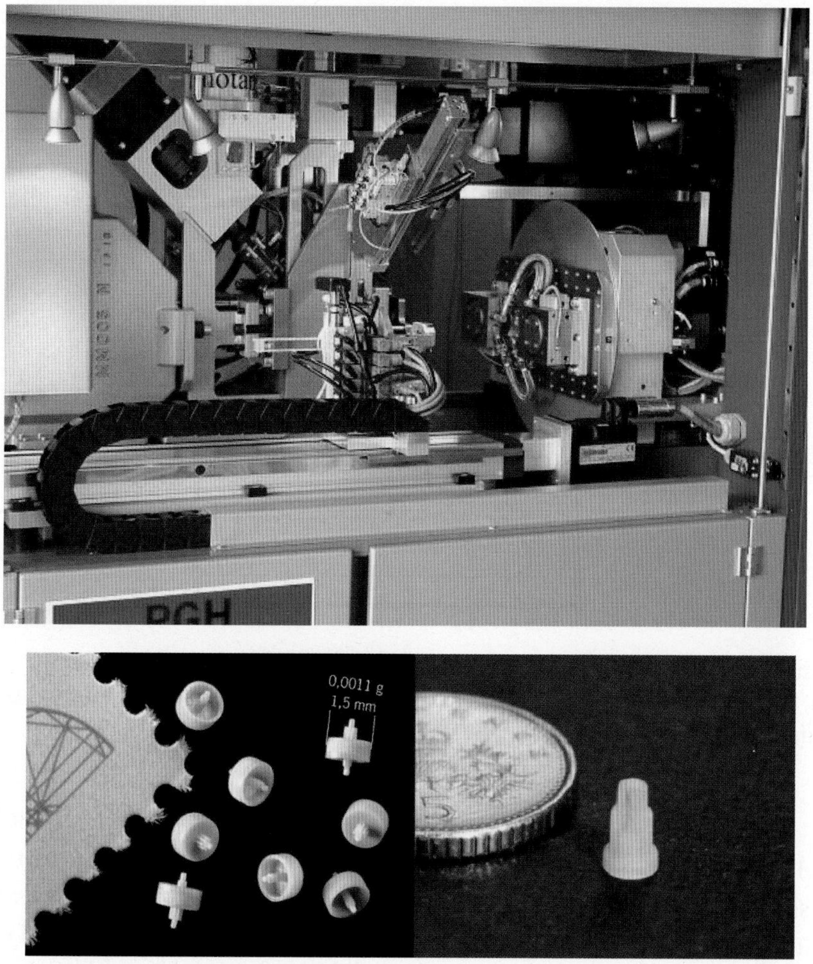


**FIGURE 10.10** (a) Detail of the Battenfeld Microsystem 50 IM machine. (b) Examples of micromolded components with microfeatures.

A typical setup for cyclic and continuous embossing is shown in Figure 10.11a and Figure 10.11b, respectively.[3] In a cyclic process (Figure 10.11a), a metal master is put in a hydraulic press, and applying the appropriate force, thus replicating the structure from the master to the polymer, stamps a heated polymer sheet. This constitutes a low-cost method for making prototypes. For mass production, a continuous process is preferred (Figure 10.11b). Here, a polymer sheet stretches through a temperature chamber, and several masters, mounted on a conveyor belt, continuously produce parts. The process also may incorporate a lamination station to enclose certain features. This continuous-type process is very much in keeping with a vision for continuous manufacture of MEMS and NEMS devices.

**TABLE 10.4** Pros and Cons of Injection Molding

| Pros | Cons |
|---|---|
| Good for small structures with low aspect ratio, e.g., CD and DVD | Only low-molecular-weight polymers (may reduce mechanical and thermal strength) |
| Good for large, high-aspect ratio, and 3D features | More expensive equipment |
| Excellent dimensional control | Cyclic process only |
| Short cycle time (as low as 10 s) | High stress on master |
| High productivity | High residual stresses on molded parts |
| Closed mold process enables packing pressure application | |

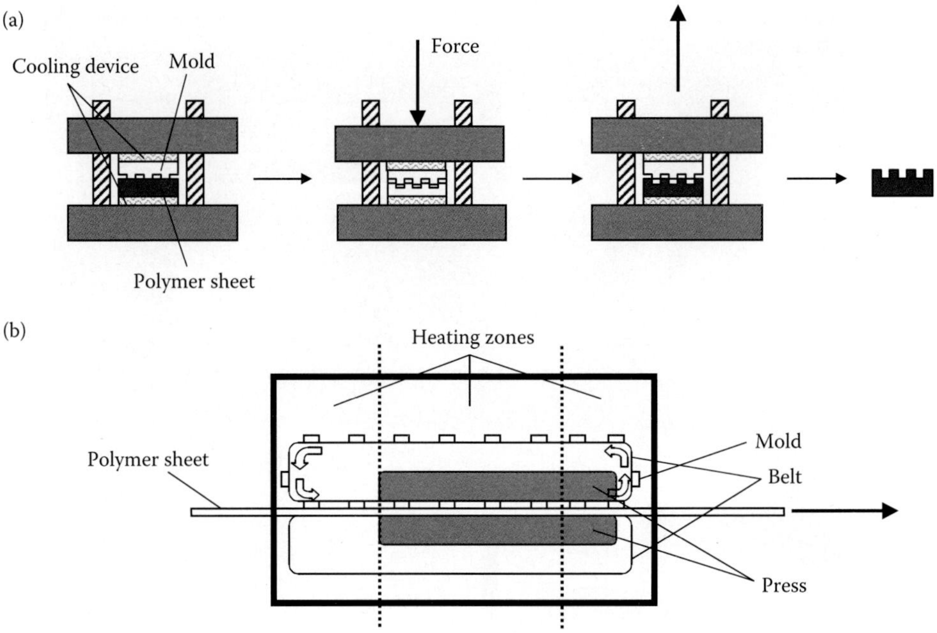


**FIGURE 10.11** Schematic of cyclic (a) and continuous (b) embossing processes. (From Lee, L. J., C.-H. Shih, Y.-J. Juang, J. Garcia, M. J. Madou, and K. W. Koelling. 1999. *Novel microfabrication options for BioMEMS conference.* San Francisco: The Knowledge Foundation.[3])

### *Hot Embossing*

As seen from Figure 10.11a, hot embossing takes place in a machine frame similar to that of a press. The force frame delivers the embossing force—on the order of 5–20 tons. An upper boss holds the molding tool, and the lower boss holds the substrate. Processing parameters include thermal cycle, compression force, and compression speed. The temperature difference between embossing and de-embossing determines the thermal cycle time, typically from 25–40°C. In principle, one could, after hot embossing, cool down the whole device to room temperature before de-embossing, or, at the other extreme, one could de-emboss just below or at the glass-transition temperature. A compromise is needed; the quality of the replication may not be good if one tries to remove the master when the polymer is still soft, whereas cooling all the way down to room temperature takes too long. A small temperature cycle leads to smaller induced thermal stress. Such a smaller temperature cycle also reduces replication errors as a result of different thermal expansion coefficients of tool and substrate. By actively heating and cooling the upper and lower bosses, one gets cycle times of about 5 min. A vacuum for hot embossing ensures longer lifetime for the mold, absorbs any water released during the embossing process from the polymer material, and prevents bubbles from entrapped gases. A Jenoptik hot embossing machine is shown in Figure 10.12.

Replication of micro- and nanosized structures has been successfully achieved using hot embossing.[12–18] Adding an antiadhesive film to reduce the interaction between the mold and the replica during embossing

**FIGURE 10.12** JENOPTIK HEX 04 hot embossing system manufactured by Jenoptik Mikrotechnik GmbH (http://www.jenoptik.com).

TABLE 10.5 Pros and Cons of Compression Molding (Imprinting or Hot Embossing)

| Pros | Cons |
|---|---|
| Low polymer flow | More difficult for structures with high-aspect ratio (near $T_g$ processing) |
| High-molecular-weight polymers (with better mechanical and thermal properties) | Less dimensional control (open mold process) |
| Simple process | Planar features only |
| Continuous or cyclic (see Figure 10.7) | High residual stresses on molded parts |
| Good for small structures | Difficult for large parts and multiple feature depth (too high a pressure and temperature are required) |

has also been studied.[19,20] Instead of the conventional nickel molds, the possibility of using silicon molds because of their excellent surface quality and easy mold release has been demonstrated.[21,22] Also, the use of a plastic mold in the embossing process was recently illustrated.[23]

At the Kernforschungszentrum Karlsruhe, or KfK, hot embossing is mainly used for small series production, whereas IM is applied to mass production.

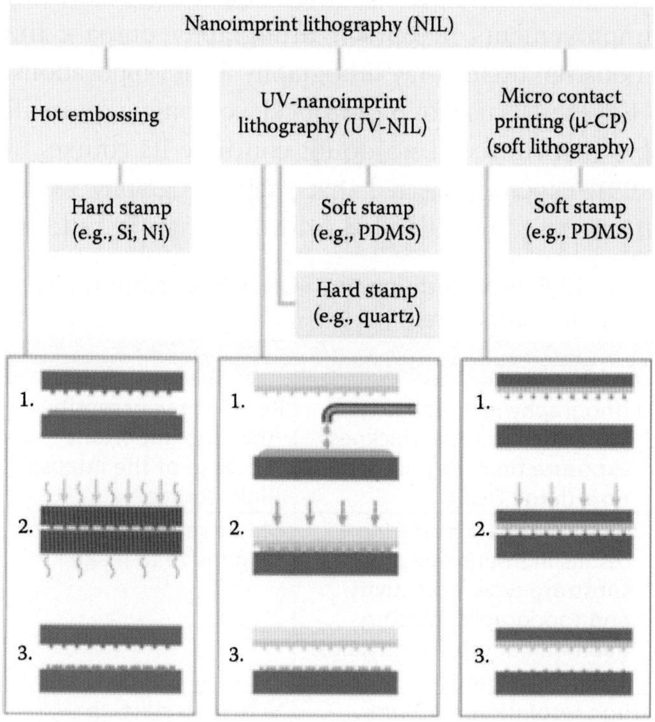


FIGURE 10.13 Nanoimprint lithography (NIL). Comparison of hot embossing with UV-nanoimprint lithography and microcontact printing.

Both techniques are applied to amorphous (PMMA, PC, PSU) and semicrystalline thermoplastics (POM, PA, PVDF, PFA). RIM is used for molding of thermoplastics (PMMA, PA), duroplastics (PMMA, epoxide), or elastomers (silicones).[24]

### *Summary: Pros and Cons of Compression Molding*

Pros and cons of compression molding are summarized in Table 10.5. In Figure 10.13 we compare hot embossing with UV-nanoimprint lithography (see Chapter 2) and microcontact printing (soft lithography, Chapter 2). These techniques can all be grouped under nanoimprint lithography (NIL).

## LIGA—Background

### History

LIGA combines the sacrificial wax molding method, known since the time of the Egyptians, with x-ray lithography and electrodeposition. Combining electrodeposition and x-ray lithography was first carried out by Romankiw and coworkers at IBM as early as 1975.[25] These authors made high-aspect-ratio metal structures by plating gold in x-ray-defined resist patterns of up to 20 μm thick. They had, in other words, already invented "LIG"; that is, LIGA without the *abformung* (molding).[25] This IBM work was an extension of through-mask plating, also pioneered by Romankiw et al. in 1969, and was geared toward the fabrication of thin film magnetic recording heads (see Figure 2.4a).[26] The addition of plastic molding to the lithography and plating process was realized by Ehrfeld et al.[27] at the Kernforschungszentrum Karlsruhe (KfK) in 1982. By adding molding, these pioneers recognized the broader implications of LIGA as a new means of low-cost manufacturing of a wide variety of microparts with unprecedented accuracies from various materials previously impossible to batch fabricate.[27] In Germany, LIGA originally developed almost completely outside of the semiconductor industry. In the United States, it was the late Henry Guckel who, starting in 1988, repositioned the field in light of semiconductor process capabilities and brought it closer to standard manufacturing processes.

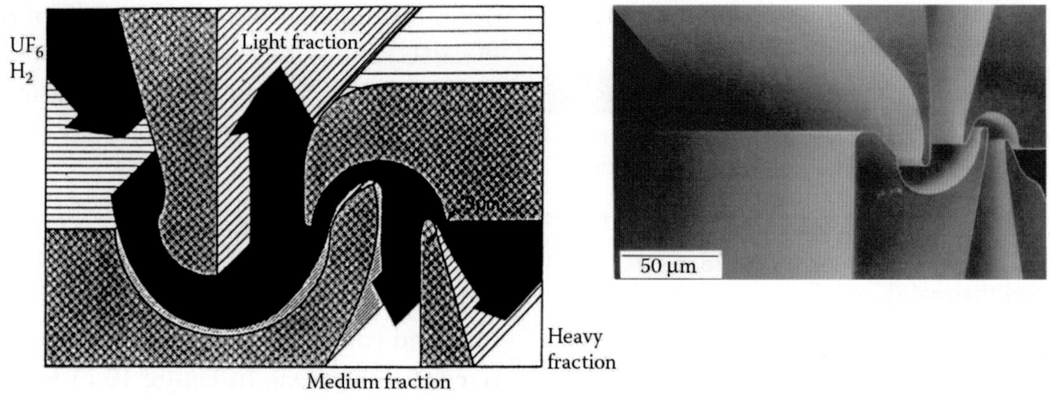


**FIGURE 10.14** Schematic and scanning electron micrograph of a separation nozzle structure produced by electroforming with nickel using a micromolded PMMA template. This nozzle represents the first product ever made by LIGA. (From Hagmann, P., W. Ehrfeld, and H. Vollmer. 1987. *First meeting of the European Polymer Federation, European symposium on polymeric materials.* Lyon, France. http://opac.fzk.de:81/de/oai_frm.html?titlenr=24694&server=//127.0.0.1&port=81; Archiv *KAROLA–OA–Volltextserver_des_Forschungszentrums Karlsruhe* (Germany). With permission.[28])

The development of the LIGA process initiated at KfK was intended for the mass production of micrometer-sized nozzles for uranium-235 enrichment (see Figure 10.14).[27] The German group used synchrotron radiation from a 2.5-GeV storage ring for the exposure of the poly(methylmethacrylate) (PMMA) resist.

Today LIGA and especially LIGA-like processes are researched in many laboratories around the world, and developing the ideal means for fabricating micromolds for the large-scale production of precise micromachines remains an elusive holy grail. In LIGA, mold inserts are made via x-ray lithography, but, depending on the dimensions of the microparts, the accuracy requirements, and the fabrication costs, mold inserts may also be realized by electron beam (e-beam) writing, computer numerically controlled (CNC) machining, wet Si bulk micromachining, deep UV resists, deep reactive ion etching (DRIE), ultrasonic cutting, excimer laser ablation, electrical discharge machining (EDM), and laser cutting (see Figure 10.1).

## Synchrotron Orbital Radiation

### *Introduction*

Lithography based on synchrotron radiation, also called *synchrotron orbital radiation* (SOR), is primarily pursued with the aim of adopting the technology as an industrial tool for the large-scale manufacture of microelectronic circuits with characteristic dimensions in the submicrometer range (see Chapter 2).[29,30] Synchrotron radiation sources "outshine" electron impact and plasmas sources for generating x-rays. They emit a much higher flux of usable collimated x-rays, thereby allowing shorter exposure times and larger throughputs. Pros and cons of x-ray radiation for lithography in IC manufacture are summarized in Table 10.6.

Despite the many promising features of x-ray lithography, the technique today lacks mainstream acceptance in the IC industry, and continued improvements in optical lithography outpace the industrial use of x-ray lithography for IC applications. However, its use for prototype development on small product runs will no doubt continue its course. In 1991, experts projected that x-ray lithography would be in use by 1995 for 64-Mb dynamic random

**Table 10.6** Pros and Cons of SOR X-Ray Lithography for IC Manufacture

| Pros | Cons |
|---|---|
| Lithography process insensitive to resist thickness, exposure time, development time (large DOF) | Resist not very sensitive (not too important because of the intense light source) |
| Absence of backscattering results in insensitivity to substrate type, reflectivity and topography, pattern geometry and proximity, dust and contamination | Masks very difficult and expensive to make |
| High resolution, <0.2 µm | Very high start-up investment |
| Some have suggested high throughput | Not proven as a system<br>Radiation effects on $SiO_2$ |

access memory (DRAM) manufacture with critical dimensions around 0.3–0.4 μm. With more certainty, they projected that the transition to x-rays would occur with the 0.2–0.3-μm critical dimensions of 256-Mb DRAMs, which came online by 1998.[29] Both dates obviously passed without materialization of the wide-scale industrial use of x-rays.

In addition to being an option for the next-generation IC lithography as discussed in more detail in Chapter 2, x-rays are also used in the fabrication of 3D microstructures. In LIGA, synchrotron radiation is used in the lithography step only. However, other micromachining applications for SOR exist. Urisu and his colleagues, for example, explored the use of synchrotron radiation for radiation-excited chemical vapor deposition and etching.[31] Micromachinists are hoping to piggyback x-ray lithography research and development efforts for the fabrication of micromachines onto major IC projects. Today the use of x-ray lithography for fabricating microdevices, other than ICs, does not present a large business opportunity by itself. Not having a major IC product line associated with x-ray lithography makes it extra hard to justify the use of x-ray lithography for micromachining, especially because other, less-expensive micromachining technologies are only very recently starting to open up the type of mass markets one is used to in the IC world. The fact that the x-rays used in LIGA are shorter wavelength than in the IC application (2–10 Å vs. 20–50 Å) also puts micromachinists at a disadvantage. For example, soft x-rays in the IC industry may eventually be generated from a much less-expensive source, such as a transition radiation source.[32] Also, nontraditional IC materials are frequently used in LIGA. The fabrication of x-ray masks poses more difficulties than masks for IC applications. Rotation and slanting of the x-ray masks to craft nonvertical walls further differentiate LIGA exposure stations. All these factors make exploring LIGA a challenge. However, given sufficient research and development money, large markets are likely to emerge in the next 10 years. These markets could be in the manufacture of devices with stringent requirements imposed on resolution, aspect ratio, structural height, and parallelism of structural walls. Optical applications for the information technology field seem particularly attractive early product targets.

Thus far, it is the research community that has primarily benefited from the availability of SOR photon sources. With its continuously tunable radiation across a very wide photon range, highly polarized, and directed into a narrow beam, SOR provides a powerful probe of atomic and molecular resonances. Other types of photon sources prove unsatisfactory for these applications in terms of intensity or energy spread. As can be concluded from Table 10.7, applications of SOR beyond lithography range from structural and chemical analysis to microscopy, angiography, and even to the preparation of new materials.

**Table 10.7** SOR Applications

| Application Area | Instruments / Technologies Needed |
|---|---|
| **Structural analysis** | |
| Atoms<br>Molecules<br>Very large molecules<br>Proteins<br>Cells<br>Crystals<br>Polycrystals | Photoelectron spectrometers<br>Absorption spectrometers<br>Fluorescent spectrometers<br>Diffraction cameras<br>Scanning electron microscope (to view topographical radiographs)<br>Time-resolved x-ray diffractometers |
| **Chemical analysis** | |
| Trace<br>Surface<br>Bulk | Photoelectron spectrometers<br>(Secondary ion) mass spectrometer<br>Absorption/fluorescence spectrometers<br>Vacuum systems |
| **Microscopy** | |
| Photoelectron<br>X-ray | Photoemission microscopes<br>X-ray microscopes SEM (for viewing)<br>Vacuum systems |
| **Micro-/nanofabrication** | |
| X-ray lithography<br>Photochemical deposition of thin films<br>Etching | Steppers, mask making<br>Vacuum systems<br>LIGA process |
| **Medical diagnostics** | |
| Radiography<br>Angiography and tomography | X-ray cameras and equipment<br>Computer-aided display |
| **Photochemical reactions** | |
| Preparation of novel materials | Vacuum systems<br>Gas-handling equipment |

*Source:* Muray, J. J., and I. Brodie. 1991. *Report no. 2019.* Menlo Park, CA: SRI International.[33]

## Synchrotron Radiation: Technical Aspects

Some important concepts associated with synchrotron radiation (such as the bending radius of the synchrotron magnet, magnetic field strength, beam current, critical wavelength, and total radiated power) require introduction. Figure 10.15 presents a schematic of an x-ray exposure station. Electrons are injected into the ring, where they are maintained at energies anywhere from $10^6$–$10^9$ eV. The cone of radiation shown in this figure is the electromagnetic radiation emitted by the circling electrons as a result of the radial acceleration that keeps them in the orbit of the electron synchrotron or storage ring. For high-energy particle studies, this radiation, emitted tangentially to the circular electron path (*Bremsstrahlung**), limits the maximum energy the electrons can attain. Bremsstrahlung is a nuisance for studies of the composition of the atomic nucleus in which high-energy particles are smashed into the nucleus. To minimize the Bremsstrahlung, physicists desire ever bigger synchrotrons.† The energy lost in the emission process is made up in a radiofrequency (RF) cavity, where electrons are accelerated back up to the storage ring energy. Injection of electrons must be repeated a few times a day because the electron current slowly decays as a result of leakage. For x-ray lithography applications, electrical engineers want to maximize the x-ray emission and build small synchrotrons instead (e.g., the radius of curvature for a compact, superconducting synchrotron is 2 m). The operating cost of these magnets is primarily that of the liquid helium refrigeration. Once high-$T_c$ (critical temperature, the temperature at which the resistance falls to zero) superconducting materials can be made in bulk, compact storage rings will become very attractive. The angular opening of the radiation cone in Figure 10.15 is determined by the electron energy, $E$, and is given by:

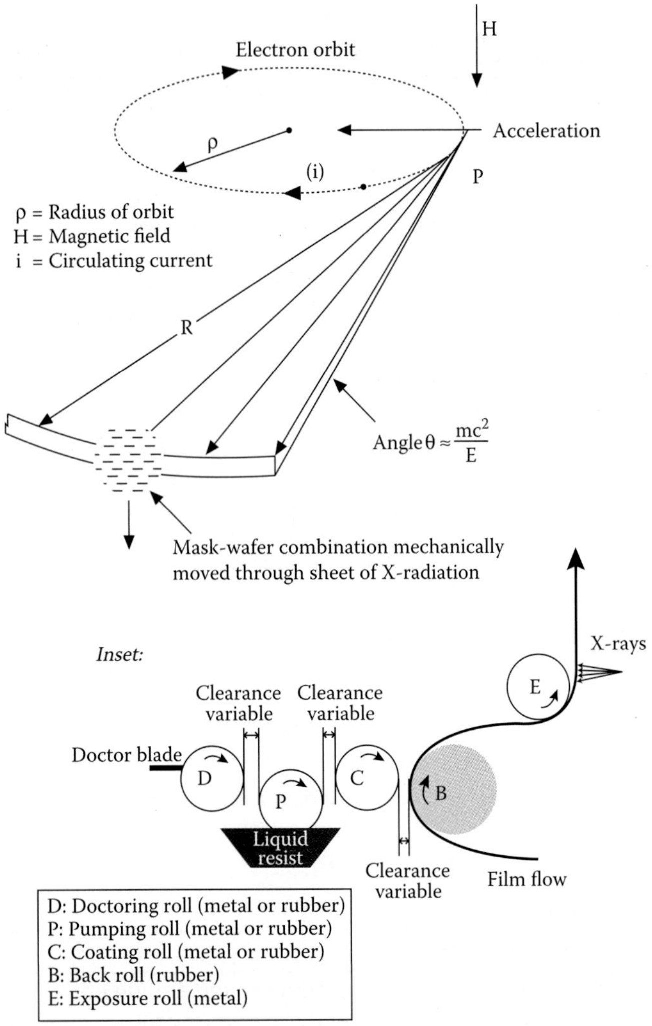


**FIGURE 10.15** Schematic of an x-ray exposure station with a synchrotron radiation source. The x-ray radiation cone (opening θ) is tangential to the electron's path, describing a line on an intersecting substrate. Inset, a vision for the future of x-ray lithography: continuous micromanufacturing.

$$\theta \approx \frac{mc^2}{E} = \frac{0.5}{E(GeV)}(mrad) \tag{10.1}$$

The x-ray light bundle with the cone opening, θ, describes a horizontal line on an intersecting substrate because the x-ray bundle is tangent to the circular electron path. In the vertical direction, the intensity of the beam exhibits a Gaussian distribution, and the vertical exposed height on the intersecting substrate can be calculated knowing θ and $R$, the distance from the radiation point, $P$, to the substrate. With E = 1 GeV, θ = 1 mrad, and R = 10 m, the exposed area in the vertical direction measures about 0.5 cm. To expose a substrate homogeneously over a wider vertical range, the sample is moved vertically through the irradiation band with a precision scanner, for example, at a speed of 10 mm/s over a

* German for "braking radiation."

† The Superconducting Super Collider Laboratory (SSCL or SSC for short), which was never finished, was to be an oval-shaped accelerator, the circumference of which would have been 54 miles. The collider oval was to be located about 200 ft. underground and would have surrounded the city of Waxahachie, Texas.

100-mm scanning distance. Usually, the substrate is stepped up and down repeatedly until the desired x-ray dose is obtained. It is interesting to note that an SOR setup affords continuous lithography, a prospect, this author believes, that will make disposable ICs and MEMS and NEMS a possibility. Rolls of dry x-ray photoresist or x-ray photoresist covered metal foils could pass through the exposure beam, resulting in a continuous lithography process, that is, a "beyond batch" type of approach (see Figure 10.15, inset).

The electron energy $E$ in Equation 10.1 is given by:

$$E(\text{GeV}) = 0.29979\ B(\text{Tesla})\ \rho\ (\text{meters}) \quad (10.2)$$

with $B$ the magnetic field and $\rho$ the radius of the circular path of the electrons in the synchrotron.

The total radiated power can be calculated from the energy loss of the electrons per turn and is given by:

$$P(\text{kW}) = \frac{88.47E^4 i}{\rho} \quad (10.3)$$

with $i$ the beam current.

The emission of the synchrotron electrons is a broad spectrum without characteristic peaks or line enhancements, and its distribution extends from the microwave region through the infrared, visible, ultraviolet (UV), and into the x-ray region. The critical wavelength, $\lambda_c$, is defined so that the total radiated power at lower wavelengths equals the radiated power at higher wavelengths, and is given by:

$$\lambda_c = \frac{5.59\rho(\text{m})}{E^3(\text{GeV})} \quad (10.4)$$

Equation 10.3 shows that the total radiated power increases with the fourth power of the electron energy. From Equation 10.4, we appreciate that the spectrum shifts toward shorter wavelengths with the third power of the electron energy.

The dose variation absorbed in the top versus the bottom of an x-ray resist should be kept small so that the top layer does not deteriorate before the bottom layers are sufficiently exposed. Because the depth of penetration increases with decreasing wavelength, synchrotron radiation of very short wavelengths is needed to pattern thick resist layers. To obtain good aspect ratios in LIGA structures, the critical wavelength ideally should be 2 Å. Bley et al.,[34] at KfK, designed a new synchrotron optimized for LIGA. They proposed a magnetic flux density, $B$, of 1.6285 $T$; a nominal energy, $E$, of 2.3923 GeV; and a bending radius, $\rho$, of 4.9 m. With those parameters, Equation 10.4 results in the desired $\lambda_c$ of 2 Å and an opening angle of radiation, based on Equation 10.1, of 0.2 mrad (in practice this angle will be closer to 0.3 mrad because of e-beam emittance).

The x-rays traveling from the ring to the sample site are held in a high vacuum. The sample itself is either kept in air or in a helium atmosphere. The inert atmosphere prevents corrosion of the exposure chamber, mask, and sample by reactive oxygen species, and removal of heat is much faster than in air (the heat conductivity of He is high compared with air). In He, the x-ray intensity loss is also 500 times less than in air. A Be window separates the high vacuum from the inert atmosphere. For wavelengths shorter than 1 nm, Be is very transparent—that is, an excellent x-ray window. A 25-μm-thick Be window can withstand a 1-atm pressure differential across a small diameter (<1 in.). For large area exposures, windows up to 6-cm diameter have been developed. Be windows age with x-ray exposure and must be replaced periodically.

### *Access to the Technology*

Today, the construction cost for a typical synchrotron totals more than $30 million, restricting the access to LIGA. Obviously, a less-expensive alternative for generating intense x-rays is preferred. Along this line, in Japan, Ishikawajima-Harima Heavy Industries (IHI) is building compact synchrotron x-ray sources (e.g., an 800-MeV synchrotron of about 30 ft./side) (http://www.ihi.co.jp).

By the end of 1993, eight nonprivately owned synchrotrons were in use in the United States. The first privately owned synchrotron was put into service in 1991 at IBM's Advanced Semiconductor Technology Center (East Fishkill, NY). Table 10.8 lists the eight US synchrotron facilities.

Most of the facilities listed in Table 10.8 allow LIGA work to continue. For example, Cronos Integrated Microsystems, Inc., a JDS Uniphase Company and a spinoff from MCNC (Research Triangle Park, NC),

**TABLE 10.8** Access to Synchrotron Radiation Is or Will Soon Be Available at the Following Facilities in the United States

| Facility | Institute | URL |
|---|---|---|
| Advanced Photon Source (APS) | Argonne National Laboratory | http://www.aps.anl.gov |
| Cornell High Energy Synchrotron Source (CHESS) | Cornell University | http://www.tn.cornell.edu |
| National Synchrotron Light Source (NSLS) | Brookhaven National Laboratory | http://www.nsls.bnl.gov |
| Stanford Synchrotron Radiation Laboratory (SSRL) | Stanford University | http://www-ssrl.slac.stanford.edu |
| Synchrotron UV Radiation Facility (SURF) | National Institute of Standards and Technology | http://physics.nist.gov/MajResFac/SURF |
| Synchrotron Radiation Center (SRC) | University of Wisconsin, Madison | http://www.src.wisc.edu |
| Center for Advanced Microstructures and Devices (CAMD) | Louisiana State University | http://www.camd.lsu.edu |
| Advanced Light Source (ALS) | Lawrence Berkeley Laboratory | http://www-als.lbl.gov |

in collaboration with the University of Wisconsin, Madison, announced its first multiuser LIGA process sponsored by Advanced Research Projects Agency in September 1993 (http://www.memsrus.com). CAMD, at Louisiana State University, has three beam lines exclusively dedicated to micromachining work, and the Advanced Light Source at Berkeley has one beam line available for micromachining.

Like Cronos, Forschungszentrum Karlsruhe GmbH offers a multiuser LIGA service (LEMA, or LIGA experiment for multiple applications). The commercial exploitation of LIGA is pursued by at least three German organizations: microParts GmbH STEAG (http://www.microparts.de), IMM (http://www.imm.uni-mainz.de), and Forschungszentrum Karlsruhe, or KfK (http://www.fzk.de). In the United States, Louisiana State University's CAMD (http://www.camd.lsu.edu; Baton Rouge, LA) with its spinoff Mezzo Systems, Inc., founded in 2000, (http://www.mezzosystems.com) is leading the way.

## LIGA and LIGA-Like Process Steps

### X-Ray Masks

#### *Introduction*

With reference to Figure 10.16 we now detail each of the steps in the LIGA process.

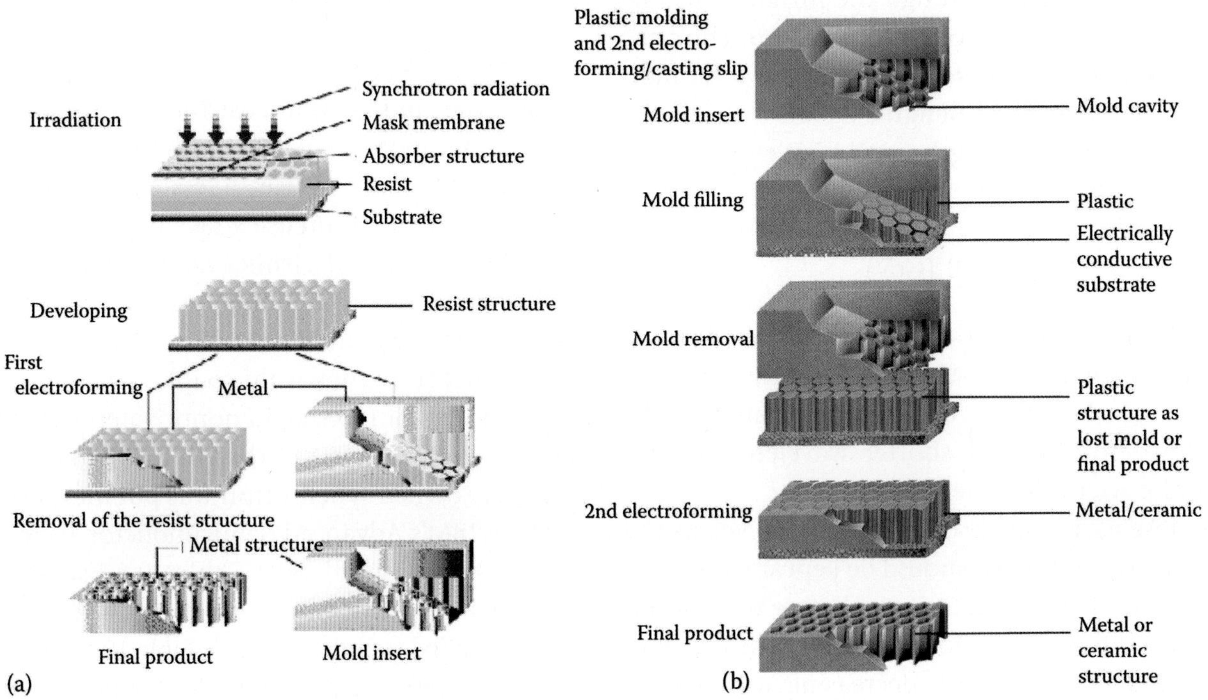


**FIGURE 10.16** (a) Basic LIGA process step: x-ray deep-etch lithography and first electroforming. (b) Plastic molding and second electroforming/slip casting.

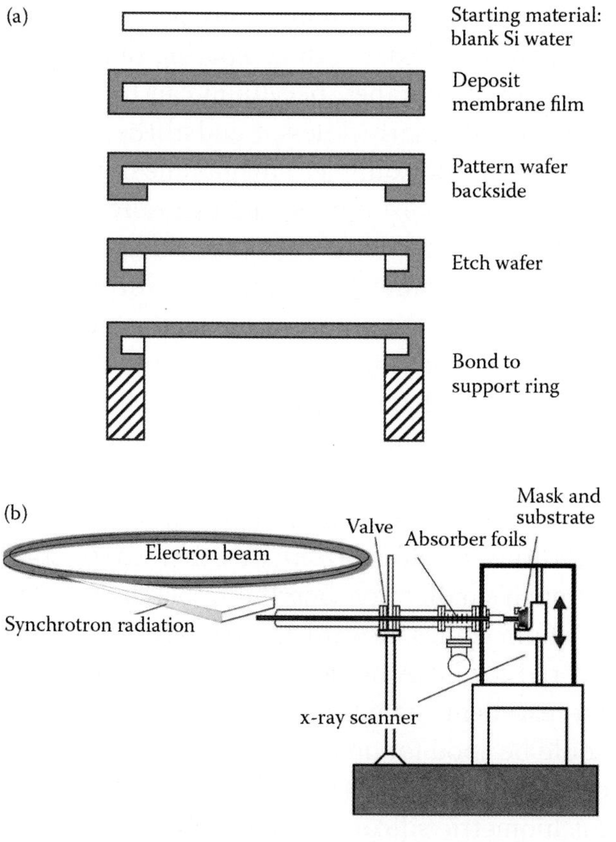


**FIGURE 10.17** Schematic of a typical x-ray mask (a) and mask and substrate assembly in an x-ray scanner (b). [(b) from IMM, http://www.imm-mainz.de/. 1995. Commercial brochure. With permission.[2]]

X-ray mask production is one of the most difficult aspects of x-ray lithography. To be highly transmissive to x-rays, the mask substrate by necessity must be a low $Z$ (atomic number) thin membrane. X-ray masks should withstand many exposures without distortion, be alignable with respect to the sample, and be rugged. A possible x-ray mask architecture and its assembly with a substrate in an x-ray scanner are shown in Figure 10.17. The mask shown here has three major components: an absorber, a membrane or mask blank, and a frame. The absorber contains the information to be imaged onto the resist. It is made up of a material with a high atomic number ($Z$), often Au, patterned onto a membrane material with a low $Z$. The high-$Z$ material absorbs x-rays, whereas the low-$Z$ material transmits x-rays. The frame lends robustness to the membrane/absorber assembly so that the whole can be handled confidently.

The requirements for x-ray masks in LIGA differ substantially from those for the IC industry. A comparison is presented in Table 10.9.[35] The main difference lies in the absorber thickness. To achieve a high contrast (>200), very thick absorbers (>10 μm vs. 1 μm) and highly transparent mask blanks (transparency >80%) must be used because of the low x-ray resist sensitivity and the great depth of the resist. Another difference focuses on the radiation stability of membrane and absorber. For conventional optical lithography, the supporting substrate is a relatively thick, optically flat piece of glass or quartz highly transparent to optical wavelengths. It provides a highly stable (>$10^6$ μm) basis for the thin (0.1 μm) chrome absorber pattern. In contrast, the x-ray mask consists of a very thin membrane (2–4 μm) of low-$Z$ material carrying a high-$Z$ thick absorber pattern.[36] A single exposure in LIGA results in an exposure dose 100 times higher than in the IC case.

We will look into these different mask aspects separately before detailing a process with the potential of obviating the need for a separate x-ray mask altogether, that is, through the use of conformal or transfer masks.

**Table 10.9** Comparison of Masks in LIGA and the IC Industry

| | Semiconductor Lithography | LIGA Process |
|---|---|---|
| Transparency | ≥50% | ≥80% |
| Absorber thickness | ±1 μm | ±10 μm |
| Field size | 50 × 50 mm² | 100 × 100 mm² |
| Radiation resistance | = 1 | = 100 |
| Surface roughness | <0.1 μm | <0.5 μm |
| Waviness | <±1 μm | <±1 μm |
| Dimensional stability | <0.05 μm | <0.1–0.3 μm |
| Residual membrane stress | ~$10^8$ Pa | ~$10^8$ Pa |

*Source:* Ehrfeld, W., W. Glashauer, D. Munchmeyer, and W. Schelb. 1986. Mask making for synchrotron radiation lithography. *Microelectron Eng* 5:463–70.[35]

### *X-Ray Membrane (Mask Blank)*

The low-*Z* membrane material in an x-ray mask must have a transparency for rays with a critical wavelength, $\lambda_c$, from 0.2–0.6 nm of at least 80% and should not induce scattering of those rays. To avoid pattern distortion, the residual stress, $\sigma_r$, in the membrane should be less than $10^8$ dyn/cm$^2$. Mechanical stress in the absorber pattern can cause in-plane distortion of the supporting thin membrane, necessitating a high Young's modulus for the membrane material. Humidity or high deposited doses of x-ray might also distort the membrane directly. During one typical lithography step, the mask may be exposed to 1 MJ/cm$^2$ of x-rays. Because most membranes must be very thin for optimum transparency, a compromise has to be found between transparency, strength, and form stability. Important x-ray membrane materials are listed in Table 10.10. The higher radiation dose in LIGA prevents the use of boron nitride (BN) and compound mask blanks, which incorporate a polyimide layer. Those mask blanks are perfectly appropriate for classical IC lithography work but will not do for LIGA work. Mask blanks of metals such as Ti and Be were specifically developed for LIGA applications because of their radiation hardness.[35,37] In comparing titanium and beryllium membranes, beryllium can have a much greater membrane thickness, *d*, and still be adequately transparent. For example, a membrane transparency of 80%, essential for adequate exposure of a 500-µm-thick PMMA resist layer, is obtained with a thin 2-µm titanium film, whereas with beryllium, a thick 300-µm membrane achieves the same result. The thicker beryllium membrane permits easier processing and handling. In addition, beryllium has a greater Young's modulus *E* than titanium (330 vs. 140 kN/mm$^2$), and because it is the product of $E \times d$ that determines the amount of mask distortion, distortions resulting from absorber stress should be much smaller for beryllium blanks.[37,38] Thus, beryllium comes forward as an excellent membrane material for LIGA because of its high transparency and excellent damage resistance. Such a mask should be good for up to 10,000 exposures and may cost $20,000–30,000 ($10,000–15,000 in quantity). Stoichiometric silicon nitride ($Si_3N_4$) used in x-ray mask membranes may contain numerous oxygen impurities, absorbing x-rays and thus producing

**Table 10.10** Comparison of Membrane Materials for X-Ray Masks

| Material | X-Ray Transparency | Nontoxicity | Dimensional Stability | Remark |
|---|---|---|---|---|
| Si | 0 (50% transmission at 5.5-µm thickness) | ++ | 0 (thermal exp coefficient 2.6°C$^{-1}$ $10^{-6}$) Young's modulus = 1.3 | Single-crystal Si, well developed, rad hard, stacking faults cause scattering, material is brittle |
| $SiN_x$ | (50% transmission at 2.3-µm thickness) | ++ | (thermal exp coefficient 2.7°C$^{-1}$ $10^{-6}$) Young's modulus = 3.36 | Amorphous, well developed, rad hard if free of oxygen, resistant to breakage |
| SiC | (50% transmission at 3.6-µm thickness) | ++ | (thermal exp coefficient 4.7°C$^{-1}$ $10^{-6}$) Young's modulus = 3.8 | Poly and amorphous, rad hard, some resistance to breakage |
| Diamond | +(50% transmission at 4.6-µm thickness) | ++ | ++(thermal exp coefficient 1.0°C$^{-1}$ $10^{-6}$) Young's modulus = 11.2 | Poly, research only, highest stiffness |
| BN | (50% transmission at 3.8-µm thickness) | | (thermal exp coefficient 1.0°C$^{-1}$ $10^{-6}$) Young's modulus = 1.8 | Not rad hard, i.e., not applicable for LIGA |
| Be | ++ | — | ++ | Research, especially suited for LIGA, even at 100 µm the transparency is good, 30 µm typical, difficult to electroplate, toxic material |
| Ti | — | | 0 | Research, used for LIGA, not very transparent, films must not be more than 2–3 µm thick |

heat. This heat often suffices to prevent the use of nitride as a good LIGA mask. Single-crystal silicon masks have been made (1 cm² and 0.4 μm thick and 10 cm² and 2.5 μm thick) by electrochemical etching techniques. Nanostructures, Inc. (http://www.nanostructures.com) is one of the companies that make such thin Si masks. For Si and $Si_3N_4$, the Young's modulus is low compared with CVD-grown diamond, and for SiC films, the Young's modulus is as much as three times higher. These higher stiffness materials are more desirable because the internal stresses of the absorbers, which can distort mask patterns, are less of an issue. Unfortunately, diamond and SiC membranes are also the most difficult to produce.

### *Absorber Materials*

The requirements on the absorber are high attenuation (>10 db), stability under radiation for extended periods, negligible distortion (stress <10⁸ dyn/cm²), ease of patterning, repairability, and low defect density. Typical absorber materials are listed in Table 10.11. Gold is used most commonly, and some research groups are looking at the viability of tungsten and other materials. In the IC industry, an absorber thickness of 0.5 μm might suffice, whereas LIGA deals with thicker layers of resist requiring a thicker absorber to maintain the same resolution.

Figure 10.18 illustrates how x-rays, with a characteristic wavelength of 0.55 nm, are absorbed along their trajectory through a Kapton preabsorber filter, an x-ray mask, and resist.[39] As we saw in Volume I,

**Table 10.11** Comparison of Absorber Materials for X-Ray Masks

| Material | Remark |
|---|---|
| Gold | Not the best stability (grain growth), low stress, electroplating only, defects repairable (thermal exp coefficient 14.2°C⁻¹ 10⁻⁶) (0.7 μm for 10 dB) |
| Tungsten | Refractory and stable, special care is needed for stress control, dry etchable, repairable (thermal exp coefficient 4.5°C⁻¹ 10⁻⁶) (0.8 μm for 10 dB ) |
| Tantalum | Refractory and stable, special care is needed for stress control, dry etchable, repairable |
| Alloys | Easier stress control, greater thickness needed to obtain 10 dB |

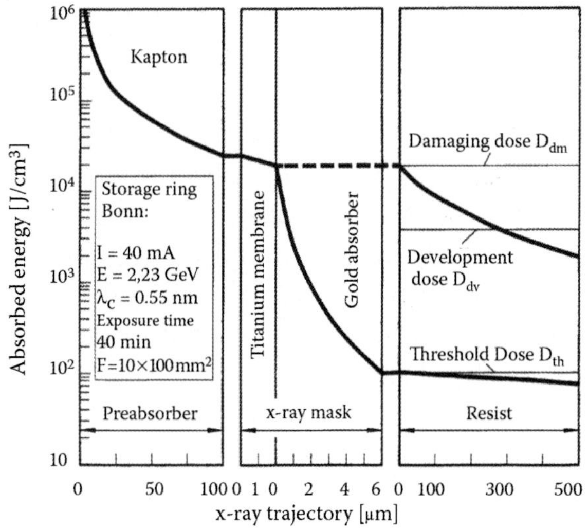


**FIGURE 10.18** Absorbed energy along the x-ray trajectory including a 500-μm-thick PMMA specimen, x-ray mask, and a Kapton preabsorber. (From Bley, P., W. Menz, W. Bacher, K. Feit, M. Harmening, H. Hein, J. Mohr, W. K. Schomburg, and W. Stark. 1991. Proc. 1991 Int. MicroProcess Conf. *J. Appl. Phys.* JJAP Series 5, p. 384, Japan, Tokyo. With permission.[39])

Chapter 5, photon absorption by a solid in the visible region of the spectrum depends on the atomic arrangement of the atoms and their bonding (for example, pure silicon is strongly absorbing, but silicon combined with oxygen is transparent). For the energetic photons in the x-ray regime, photon absorption is much easier to predict and is independent of the details of atomic arrangement. It depends primarily on the electron concentration per unit volume. Because the concentration of atoms per unit volume only differs by factors of two or three from each other, the electron concentration in two materials can be estimated from the atomic number, *Z*. Lead (Z = 82) absorbs x-rays much more efficiently than aluminum (Z = 13) and consequently is used in shielding around x-ray apparatus. X-ray absorption depends on the energy of the x-rays and decreases with increasing x-ray energy, *E*. Absorption decreases nearly proportionally to the cube of the energy (i.e., absorption proportional to $1/E^3$; Equation 10.4). Thus, the low energy portion of the synchrotron radiation is absorbed mainly in the top portion of the resist layer because absorption increases with increasing wavelength. The Kapton preabsorber filters out much of the low energy radiation to prevent overexposure of the top surface of the resist. The x-ray dose at which the resist gets damaged, $D_{dm}$, the dose required for development of the resist, $D_{dv}$, and the "threshold dose" at which the resist starts dissolving in a developer, $D_{th}$, are all indicated in Figure 10.18. In the areas under the absorber pattern of the x-ray mask, the absorbed dose must stay below the threshold dose, $D_{th}$. Otherwise, the structures partly dissolve, resulting in poor feature definition. From Figure 10.18, we can deduce that the height of the gold absorbers must exceed 6 μm to reduce the absorbed radiation dose of the resist under the gold pattern to below the threshold dose, $D_{th}$. In Figure 10.19, the necessary thickness of the gold absorber patterns of an x-ray mask is plotted as a function of the thickness of the resist to be patterned; the Au must be thicker for thicker resist layers and for shorter characteristic wavelengths, $\lambda_c$, of the x-ray radiation. To pattern a 500-mm-high structure with a $\lambda_c$ of 0.225 nm, the gold absorber must be more than 11 μm high.

Exposure of yet more extreme photoresist thicknesses is possible if proper x-ray photon energies are used. At 3,000 eV, the absorption length in PMMA roughly measures 100 μm, which enables the above-mentioned 500-μm exposure depth.[40] Using 20,000-eV photons results in absorption lengths of 1 cm. PMMA structures up to 10 cm (!) thick have been exposed this way.[41] A high-energy mask used by Guckel for these high-energy exposures has an Au absorber 50 μm thick and a blank membrane of 400-μm-thick Si. Guckel obtained an absorption contrast of 400 when exposing a 1000-μm-thick PMMA sheet with this mask. An advantage of using such thick Si blank membranes is that larger resist areas can be exposed because one does not depend on a fragile membrane/absorber combination.[42]

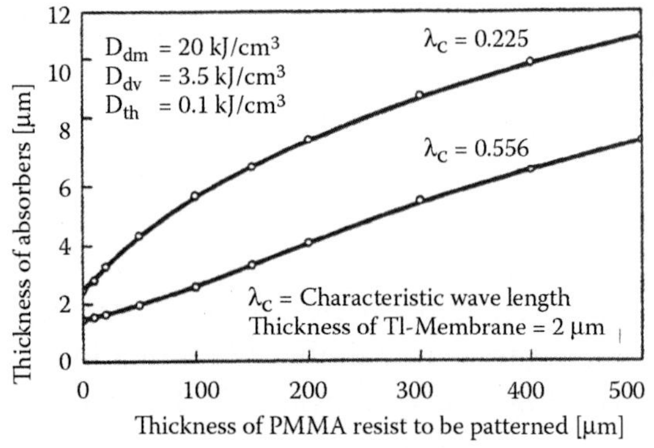


**FIGURE 10.19** Necessary thickness of the gold absorbers of an x-ray mask. (From Bley, P., W. Menz, W. Bacher, K. Feit, M. Harmening, H. Hein, J. Mohr, W. K. Schomburg, and W. Stark. 1991. Proc. 1991 Int. MicroProcess Conf. *J. Appl. Phys*. JJAP Series 5, p. 384, Japan, Tokyo. With permission.[39])

### *Absorber Fabrication*

*Single-Layer Absorbers* To make a mask with gold absorber structures of a height more than 10 μm, one must first succeed in structuring a resist of that thickness. The height of the resist should be a bit higher (e.g., 20%) than the absorber itself to accommodate the electrodeposited metal fully between the resist features. Currently, no means to structure a resist of that height with sufficient accuracy and perfect verticality of the walls exist unless x-rays are used. Different procedures for producing x-ray masks with thicker absorber layers using a two-stage lithography process have been developed.

The KfK solution calls for first making an intermediate mask with photo or e-beam lithography. This intermediate mask starts with a 3-μm-thick resist layer, in which case the needed line-width accuracy and photoresist wall steepness of printed features are achievable. After gold plating between the resist features and stripping the resist, this intermediate mask is used to write a pattern with x-rays in a thicker resist, e.g., 20 μm thick. After electrodepositing and resist stripping, the x-ray mask, that is, the master mask, is obtained.

Because hardly any accuracy is lost in the copying of the intermediate mask with x-rays to obtain the master mask, it is the intermediate mask quality that determines the ultimate quality of the LIGA-produced microstructures. The structuring of the resist in the intermediate mask is handled with optical techniques when the requirements on the LIGA structures are less stringent. The minimal lateral dimensions for optical lithography in a 3-μm-thick resist typically measure about 2.5 μm. Under optimum conditions, a wall angle of 88° is achievable. With e-beam lithography, a minimum lateral dimension of less than 1 μm is feasible. The most accurate pattern transfer is achieved through reactive ion etching of a trilevel resist system. In this approach, a 3–4-μm-thick polyimide resist is first coated onto the titanium or beryllium membrane, followed by a coat of 10–15-nm titanium deposited with magnetron sputtering. The thin layer of titanium is an excellent etch mask for the polyimide; in optimized oxygen plasma, the titanium etches 300 times slower than the polyimide. To structure the thin titanium layer itself, a 0.1-μm-thick optical resist is used. Because this top resist layer is so thin, excellent lateral tolerances result. The thin Ti layer is patterned with optical photolithography and etched in argon plasma. After etching the thin titanium layer, exposing the polyimide locally, oxygen plasma helps to structure the polyimide down to the titanium or beryllium membrane. Lateral dimensions of 0.3 μm can be obtained in this fashion. Patterning the top resist layer with an e-beam increases the accuracy of the three-level resist method even further. Electrodeposition of gold on the titanium or beryllium membrane and stripping of the resist finish the process of making the intermediate LIGA mask. To make a master mask, this intermediate mask is printed by x-ray radiation onto a PMMA resist-coated master mask. The PMMA thickness corresponds to a bit more than the desired absorber thickness. Because the resist layer thickness is in the 10–20-μm range, a synchrotron x-ray wavelength of 10 Å is adequate for making the master mask. A further improvement in LIGA mask making is to fabricate intermediate and master mask on the same substrate, greatly reducing the risk for deviations in dimensions caused, for example, by temperature variations during printing.[43] The ultimate achievement would be to create a one-step process to make the master mask. Along this line, Hein et al.[38] investigated the direct patterning of 10-μm-high resist layers with a 100-kV e-beam.

*Stepped Absorber* In principle, stepped absorber structures may result in stepped LIGA structures by means of x-ray lithography with a single mask. In this manner, variable dose depositions can be achieved at the same resist heights. The variable dose results in different molecular weights and hence in a different developing behavior. This technique unfortunately leads to rounded features and poor step-height control. Further below under "Stepped and Slanted Microstructures" we will learn how to make better resolved stepped features.

*CNC Machined Absorbers* A rather unexpected approach to pattern absorber layers for x-ray masking is pursued by Friedrich et al. from the Michigan Technological University (http://www.me.mtu.edu/~microweb). Friedrich et al. are exploring x-ray mask fabrication by traditional machining methods, such as micromilling, micro-EDM, and lasers.[44] Using micromilling, this group succeeded in making mask features up to 62 μm deep and walls down to 4 μm thick and 10 μm high. The milling is carried out with a 22-μm end mill, fabricated itself by using a 20-keV gallium ion beam. The advantages of this approach are rapid turnaround (less than 1 day/mask), low cost, and flexibility (almost any type of material can be machined) because no intermediate steps interfere. Disadvantages are less dimensional edge acuity and nonsharp interior corners, as well as much lower absolute tolerance.

### *Alignment of X-Ray Mask to Substrate*

The mask and resist-coated substrate must be properly registered to each other before they are put in an x-ray scanner. Alignment of an x-ray mask to the substrate is problematic because no visible light can pass through most x-ray membranes. To solve this problem, Schomburg et al.[37] etched windows in their Ti x-ray membrane. Diamond membranes have a potential advantage here because they are optically transparent and enable easy alignment for multiple irradiations without a need for etched holes.

Figure 10.20 illustrates an alternative, x-ray alignment system involving capacitive pickup between conductive metal fingers on the mask and ridges on a small substrate area; Si, in this case [US Patent 4,607,213 (1986) and 4,654,581 (1987)]. When using

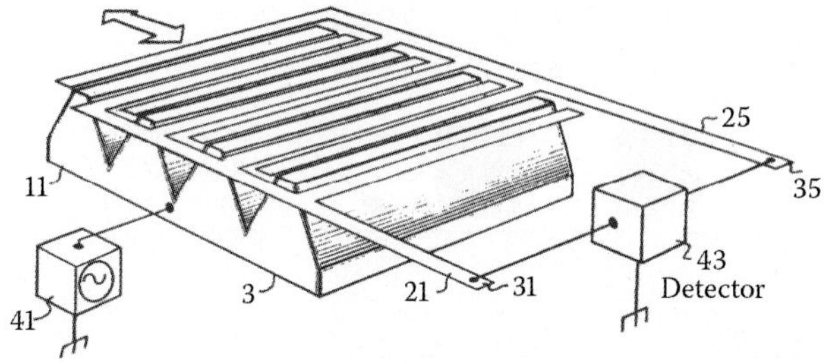


**FIGURE 10.20** Mask alignment system in x-ray lithography. Conductive fingers on the mask and ridges on the Si are used for alignment. [From US Patents 4,654,581 (1987) and 4,607,213 (1986).]

multiple groups of ridges and fingers, two-axis lateral and rotational alignment become possible.

Another alternative may involve liquid nitrogen-cooled Si (Li) x-ray diodes as alignment detectors, eliminating the need for observation with visible light.[45]

### *Conformal, Transfer, or Self-Aligned Mask for High-Aspect-Ratio Microlithography*

Vladimirsky et al.[46,47] developed a procedure to eliminate the need for an x-ray mask membrane. Unlike conventional masks, the so-called *x-ray transfer mask* does not treat a mask as an independent unit. The technique is based on forming an absorber pattern directly on the resist surface forming a conformal, self-aligned, or transfer mask. An example process is shown in Figure 10.21. In this sequence, a transfer mask plating base is first prepared on the PMMA substrate plate by evaporating 70 Å of chromium (as adhesion layer) followed by 500 Å of gold using an e-beam evaporator. A 3-μm-thick layer of standard Novolak-based AZ-type resist S1400-37 (Shipley Co., Marlborough, MA) is then applied over the plating base and exposed in contact mode through an optical mask using an UV exposure station. Three micrometers of electroplated gold on the exposed plating base further completes the transfer mask. A blanket exposure and subsequent development remove the remaining resist. The 500 Å of gold plating base is dissolved by a dip of 20–30 s in a solution of KI (5%) and I (1.25%) in water; the Cr adhesion layer is removed by a standard chromium etch (from KTI, Chemicals Inc., Sunnyvale, CA). Thus, fabrication of the transfer mask can be performed using standard lithography equipment available at almost any lithography shop. Depending on the resolution required, the x-ray transfer mask can be fabricated using known photon, e-beam, or x-ray lithography techniques. The patterning of the PMMA resist with a self-aligned mask is accomplished in multiple steps of exposure and development. An example of a cylindrical resonator made this way is shown in Figure 10.22. Each exposure/development step involves an exposure dose of about 8–12 J/cm$^2$. Subsequent 5-min development steps remove ~30 μm of PMMA. In seven steps, a self-supporting, 1.5-mm-thick PMMA resist is patterned to a depth

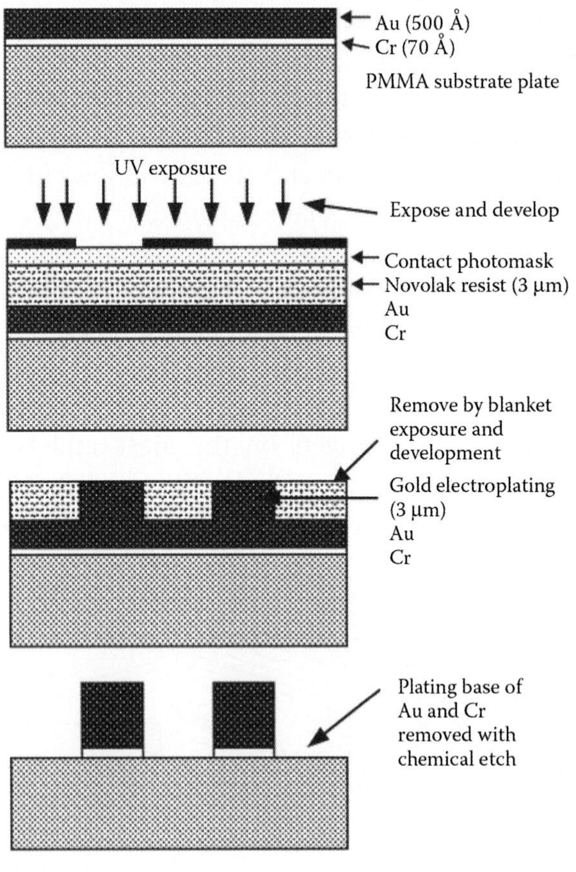


**FIGURE 10.21** Sample transfer mask formation. [From Vladimirsky, Y., O. Vladimirsky, V. Saile, K. Morris, and J. M. Klopf. 1995. *Microlithography '95 (Proceedings of the SPIE).* Santa Clara, CA: SPIE.[46]]

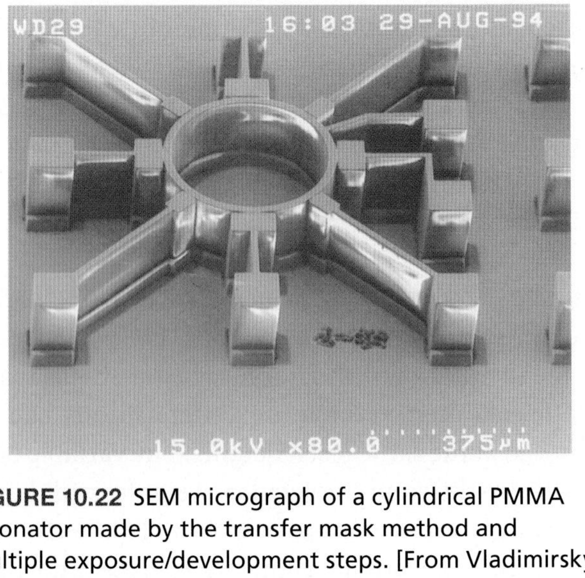


**FIGURE 10.22** SEM micrograph of a cylindrical PMMA resonator made by the transfer mask method and multiple exposure/development steps. [From Vladimirsky, Y., O. Vladimirsky, V. Saile, K. Morris, and J. M. Klopf. 1995. *Microlithography '95 (Proceedings of the SPIE).* Santa Clara, CA: SPIE. With permission.[46]] (Courtesy of Dr. V. Saile.)

of more than 200 μm. The resist pattern shown in Figure 10.22 is 230 μm thick and exhibits a 2-μm gap between the inner cylinder and the pickup electrodes (aspect ratio is 100:1). The resonator pattern was produced using soft x-rays (=10 Å) and a 3-μm-thick Au absorber only.

Vladimirsky et al.[46] suggest that the forming of the transfer mask directly on the sample surface creates several additional new opportunities; besides in situ development, etching, and deposition, these include exposure of samples with curved surfaces and dynamic deformation of a sample surface during the exposure (hemispherical structures for lenses are possible this way). Elegant and cost-saving innovations like these could help mainstream LIGA.

Shih et al.[48] further developed and fine-tuned the conformal mask method for LIGA. In one very attractive embodiment of the technology, a PMMA layer on an Al substrate is coated with a Cu foil (17.5 μm) by cold pressing. The copper foil is further laminated with a dry photoresist foil (48-μm Hitachi H-N650). After exposure and development of the dry resist, gold or tin/lead absorber patterns are electroplated on the exposed Cu foil. After stripping the dry resist in a 3 wt% NaOH solution at around 50°C and the underlying copper foil in a $Cu(NH_3)_4Cl_2$ solution at 45°C, the conformal mask is ready for use. Using this approach, 1000-μm-high structures with an aspect ratio of 6.25 were fabricated with a double-exposure development cycle only. Using yet another transfer mask approach, this same research group made LIGA dies for spinnerets (see Example 10.1).[49] The authors summarize the advantages of the transfer mask method as follows:

- Alleviates the difficulty in fabricating fragile mask membranes
- Avoids alignment requirements during successive exposure steps
- Reduces exposure time and absorber thickness for the same exposure source
- Enhances pattern transfer fidelity because there is almost no proximity gap
- Avoids thermal deformation caused by exposure heat
- Increases photoresist development rate by stepwise elevated exposure dose

## Choice of Primary Substrate

In the LIGA process, the primary substrate, or base plate, must be a conductor or an insulator coated with a conductive top layer. A conductor is required for subsequent electrodeposition. Some examples of primary substrates that have been used successfully are Al,[48] austenite steel plate, Si wafers with a thin Ti or Ag/Cr top layer,[50] and copper plated with gold, titanium, or nickel.[43] Other metal substrates and metal-plated ceramic, plastic, and glass plates have been used.[51] It is important that the plating base provide good adhesion for the resist. For that purpose, before applying the x-ray resist on copper or steel, the surface sometimes is mechanically roughened by microgrit blasting with corundum. Microgrit blasting may lead to an average roughness, $R_a$, of 0.5 μm, resulting in better physical anchoring of the microstructures to the substrate.[52] In the case of a polished metal base, chemical preconditioning may be used to improve adhesion of the resist microstructures. During chemical preconditioning, a titanium layer, sputter-deposited onto the polished metal base plate (e.g., a Cu plate), is oxidized for a few minutes in a solution of 0.5 M NaOH and 0.2 M $H_2O_2$ at 65°C. The oxide produced typically measures 30 nm thick and exhibits a microrough surface instrumental to securing resist to the base plate. The Ti adhesion layer may further be covered with a thin nickel seed layer (~150 Å) for electroless or electroplating of nickel. When using a highly polished Si surface, adhesion promoters need to be added to the resist resin (see "Resist Adhesion" further below). A substrate of special interest is a processed silicon wafer with integrated circuits (ICs). Integrating the LIGA process with IC circuitry on the same wafer will open up additional LIGA applications (see text below and Figure 10.39).

The back of electrodeposited microdevices is attached to the primary substrate but can be removed from the substrate if necessary. In the latter case, the substrate may be treated chemically or electrochemically to induce poor adhesion. Ideally, excellent adhesion exists between substrate and resist, and poor adhesion exists between the electroplated structure and plating base. Achieving these two contradictory demands is one of the main challenges in LIGA.

Thick resist plates can act as plastic substrates themselves. For example, using 20,000-eV rather than the more typical 3,000-eV radiation, Guckel et al.[40,41] exposed plates of PMMA up to 10 cm thick.

## Resist Requirements

An x-ray resist ideally should have high sensitivity to x-rays, high resolution, resistance to dry and wet etching, thermal stability of greater than 140°C, and a matrix or resin absorption of less than 0.35 $\mu m^{-1}$ at the wavelength of interest.[53] These requirements are only those for IC production with x-ray lithography.[54] To produce high-aspect-ratio microstructures with very tight lateral tolerances demands an additional set of requirements. The unexposed resist must be absolutely insoluble during development. This means that a high contrast ($\gamma$) is required. The resist must also exhibit very good adhesion to the substrate and be compatible with the electroforming process. The latter imposes a resist glass transition temperature ($T_g$) greater than the temperature of the electrolyte bath used to electrodeposit metals between the resist features remaining after development (e.g., at 60°C). To avoid mechanical damage to the microstructures induced by stress during development, the resist layers should exhibit low internal stresses.[55] If the resist structure is the end product of the fabrication process, further specifications depend on the application itself, for example, optical transparency and refractive index for optical components or large mechanical yield strength for load-bearing applications. For example, PMMA exhibits good optical properties in the visible and near-infrared range and lends itself to the making of all types of optical components.[56] Because of excellent contrast and good process stability known from e-beam lithography, PMMA is the preferred resist for deep-etch synchrotron radiation lithography. Two major concerns with PMMA as a LIGA resist are a rather low lithographic sensitivity of about 2 J/cm$^2$ at a wavelength $\lambda_c$ of 8.34 Å and susceptibility to stress cracking. For example, even at shorter wavelengths, $\lambda_c$ = 5 Å, more than 90 min of irradiation are required to structure a 500-µm-thick resist layer with an average ring storage current of 40 mA and a power consumption of 2 MW at the 2.3-GeV ELSA synchrotron (Bonn, Germany) (see Figure 10.18).[57] The internal stress arising from the combination of a polymer and a metallic substrate can cause cracking in the microstructures during development, a phenomenon PMMA is especially prone to, as illustrated in the scanning electron microscope (SEM) in Figure 10.23.

To make throughput for deep-etch lithography more acceptable to industry, several avenues to more sensitive x-ray resists have been pursued. For example, copolymers of PMMA were investigated: methylmethacrylate (MMA) combined with methacrylates with longer ester side chains show sensitivity increases of up to 32% (with tertiary butylmethacrylate). Unfortunately, a deterioration in structure quality was observed.[59] Among the other possible approaches for making PMMA more x-ray sensitive, we can count on the incorporation of x-ray-absorbing high-atomic-number atoms or the use of chemically amplified photoresists. X-ray resists explored more recently for LIGA applications are poly(lactides), for example, poly(lactide-co-glycolide) (PLG), polymethacrylimide (PMI), polyoxymethylene (POM), and polyalkensulfone (PAS). PLG is a new positive resist developed by BASF AG, more sensitive to x-rays by a factor of two to three compared with PMMA. From the comparison of different resists for deep x-ray lithography in Table 10.12, PLG emerges as the most promising LIGA

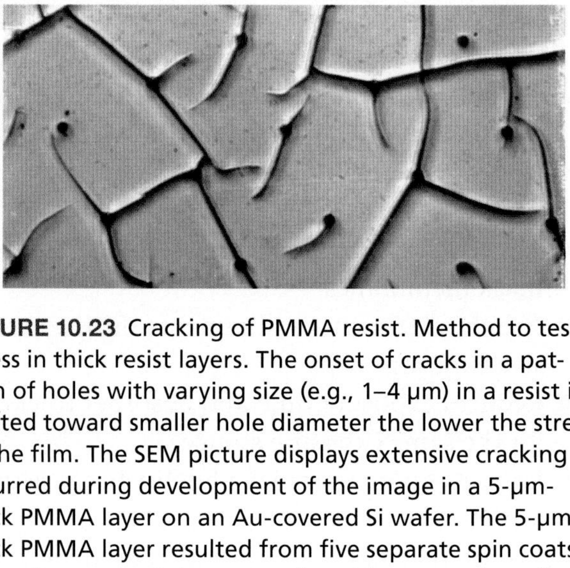

**FIGURE 10.23** Cracking of PMMA resist. Method to test stress in thick resist layers. The onset of cracks in a pattern of holes with varying size (e.g., 1–4 µm) in a resist is shifted toward smaller hole diameter the lower the stress in the film. The SEM picture displays extensive cracking incurred during development of the image in a 5-µm-thick PMMA layer on an Au-covered Si wafer. The 5-µm-thick PMMA layer resulted from five separate spin coats. Annealing pushed the onset of cracking toward smaller holes until the right cycle was reached and no more cracks were visible. (From Madou, M. J., and M. Murphy. 1995. Unpublished results.[58])

**Table 10.12** Properties of Resists for Deep X-Ray Lithography

| | PMMA | POM | PAS | PMI | PLG |
|---|---|---|---|---|---|
| Sensitivity | – | + | ++ | 0 | 0 |
| Resolution | ++ | 0 | –– | + | ++ |
| Sidewall smoothness | ++ | –– | –– | + | ++ |
| Stress corrosion | – | ++ | + | –– | ++ |
| Adhesion on substrate | + | + | + | – | + |

*Source:* Ehrfeld, W. 1994. Notes from handouts. The Commercialization of Microsystems. September 11–16, Banff Conference Center, Banff, Canada.[60]

++, excellent; +, good; 0, reasonable; –, bad; ––, very bad.

**Table 10.13** Few Common Resists for E-Beam and X-Ray Lithography

| Resist | Tone | EBL Sens (μC/cm²) | EBL Contrast | XRL Sens (mJ/cm²) | XRL Contrast |
|---|---|---|---|---|---|
| PMMA | + | 100 | 2.0 | 6500 | 2.0 |
| PBS | + | 1 | 2.0 | 170 | 1.3 |
| EBR-9 | + | 1.2 | 3.0 | | |
| Ray-PF | + | | | 125 | * |
| COP | – | 0.5 | 0.8 | 100 | 1.1 |
| GMCIA | | 7.0 | 1.7 | | |
| DCOPA | – | | | 14 | 1.0 |
| Novakak based | * | 200–500 | 2–3 | 750–2000 | ~2 |

*Source:* Campbell, S. A. 1996. *The science and engineering of microelectronic fabrication.* New York: Oxford University Press.[61]

*Value is process-dependent.

resist. POM, a promising mechanical material, may also be suited for medical applications given its biocompatibility. All the resists shown in Table 10.12 exhibit significantly enhanced sensitivity compared with PMMA, and most exhibit a reduced stress corrosion.[57] Negative x-ray resists have inherently higher sensitivities compared with positive x-ray resists, although their resolution is limited by swelling. Poly(glycidyl methacrylate-co-ethyl acrylate) (PGMA), a negative e-beam resist (not shown in Table 10.12), has also been used in x-ray lithography. In general, resist materials sensitive to e-beam exposure also display sensitivity to x-rays and function in the same fashion; materials positive in tone for e-beam radiation typically are also positive in tone for x-ray radiation. A strong correlation exists between the resist sensitivities observed with these two radiation sources, suggesting that the reaction mechanisms might be similar for both types of irradiation. IMM, in Germany, started developing a negative x-ray resist 20 times more sensitive than PMMA, but the exact chemistry has not yet been disclosed.[2] More common x-ray resists from the IC industry are reviewed in Table 10.13.

## Methods of X-Ray Resist Application

### *Multiple Spin Coats*

Different methods to apply ultrathick layers of x-ray resists such as PMMA have been studied. In the case of multilayer spin coating, high interfacial stresses between the layers can lead to extensive crack propagation on developing the exposed resist. For example, in Figure 10.23, we present an SEM picture of a 5-μm-thick PMMA layer deposited in five sequential spin coatings. Development results in the cracked riverbed mud appearance, with the most intensive cracking propagating from the smallest resist features. The test pattern used to expose the resist consisted of arrays of holes ranging in size from 1–4 μm. Annealing the PMMA films shifted the cracking toward holes with smaller diameter compared with the unannealed film shown in Figure 10.23.[58] CAMD (see Table 10.8) has demonstrated that multiple spin coating of PMMA can be used for up to 15-μm-thick resist layers and that applying the appropriate annealing and developer (see below) eliminates cracking.[62] Below, we will learn that a prerequisite for low stress and small lateral tolerances in PMMA films is a high mean molecular weight. The spin-coated resist films in Figure 10.23 do not have a high-enough molecular weight to lead to good enough selectivity between radiated and nonradiated PMMA during a long development process.

### *Commercial PMMA Sheets*

High-molecular-weight PMMA is commercially available as prefabricated plate (e.g., GS 233, Rohm GmbH, Darmstadt, Germany), and several groups have used freestanding or bonded PMMA resist sheets for producing LIGA structures.[42,52] After overcoming the initial problems encountered when attempting to glue PMMA foils to a metallic base plate with adhesives, this has become the preferred method in several labs.[52] Guckel used commercially available thick PMMA sheets (thickness >3 mm), XY-sized and solvent-bonded them to a substrate, and, after milling the sheet to the desired thickness, exposed the resist plate without cracking problems.[42]

### Casting of PMMA

PMMA also can be purchased in the form of a casting resin, for example, Plexit 60 (PMMA without added cross-linker) and Plexit 74 (PMMA with cross-linker added) from Rohm GmbH. In a typical procedure, PMMA is in situ polymerized from a solution of 35 wt% PMMA of a mean molecular weight of anywhere from 100,000 g/mol up to 106 g/mol in methylmethacrylate (MMA). Polymerization at room temperature takes place with benzoyl peroxide (BPO) catalyst as the hardener (radical builder) and dimethylaniline (DMA) as the starter or initiator.[52,59] The oxygen content in the resin, inhibiting polymerization, and gas bubbles, inducing mechanical defects, are reduced by degassing while mixing the components in a vacuum chamber at room temperature and at a pressure of 100 mbar for 2–3 min.

In a practical application, resin is dispensed on a base plate provided with shims to define pattern and thickness and subsequently covered with a glass plate to avoid oxygen absorption. Because of the hardener, polymerization starts within a few minutes after mixing the components and comes to an end within 5 min. The glass cover plate is coated with an adhesion-preventing layer (e.g., Lusin L39, Firma Lange u. Seidel, Nuremberg, Germany). After polymerization, the antiadhesion material is removed by diamond milling, and a highly polished surface results. In situ polymerization and commercial-cast PMMA sheets top the list of thick resist options in LIGA today.

Plasma polymerization of PMMA in layers (>100 μm) was discussed in Chapters 7 and 8 (see Figure 8.9).

### Resist Adhesion

Adhesion promotion by mechanically or chemically modifying the primary substrate was introduced above, under "Choice of Primary Substrate." Smooth surfaces such as Si wafers with an average roughness, $R_a$, smaller than 20 nm pose additional adhesion challenges often solved by modifying the resist itself. To promote adhesion of resist to polished untreated surfaces, such as metal-coated Si wafers, coupling agents must be used to chemically attach the resist to the substrate. An example of such a coupling agent is methacryloxypropyl trimethoxy silane (MEMO). With 1 wt% of MEMO added to the casting resin, excellent adhesion results. The adherence is brought about by a siloxane bond between the silane and the hydrolyzed oxide layer of the metal. As illustrated in

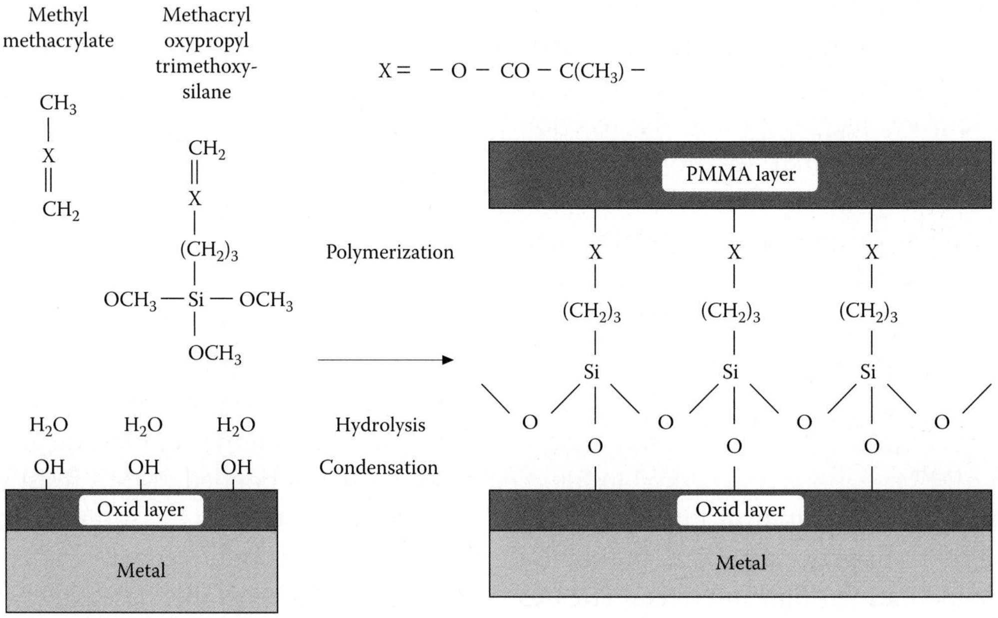


**FIGURE 10.24** Schematic presentation of the adherence mechanism of methacryloxypropyl trimethoxy silane (MEMO). (From Mohr, J., W. Ehrfeld, and D. Munchmeyer. 1988. Requirements on resist layers in deep-etch synchrotron radiation lithography. *J Vac Sci Technol* B6:2264–67. With permission.[59])

Figure 10.24, the integration of this coupling agent in the polymer matrix is achieved via the double bond of the methacryl group of MEMO.[59]

Hydroxyethyl methacrylate (HEMA) can improve PMMA adhesion to smooth surfaces, but higher concentrations are needed to obtain the same adhesion improvement. Silanization of polished surfaces before PMMA casting, instead of adding adhesion promoters to the resin, did not seem to improve the PMMA adhesion. In the case of PMMA sheets, as mentioned before, one option is solvent bonding of the layers to a substrate. In another approach, Galhotra et al.[63] simply mechanically clamped the exposed and developed self-supporting PMMA sheet onto a 1.0-mm-thick Ni sheet for subsequent Ni plating. Rogers et al.[64] have shown that cyanoacrylate can be used to bond PMMA resist sheets to a Ni substrate and that it can be lithographically patterned using the same process sequence used to pattern PMMA. For a 300-μm-thick PMMA sheet on a sputtered Ni coating on a silicon wafer, a 10-μm-thick cyanoacrylate bonding layer was used. Such a thick cyanoacrylate layer caused some problems for subsequent uniform electrodeposition of metal. The dissolution rate of the cyanoacrylate is faster than the PMMA resist, resulting in metal posts with a wide profile at the base.

### *Stress-Induced Cracks in PMMA*

The internal stress arising from the combination of a polymer on a metallic substrate can cause cracking in the microstructures during development. To reduce the number of stress-induced cracks (see Figure 10.23), both the PMMA resist and the development process must be optimized. Detailed measurements of the heat of reaction, the thermomechanical properties, the residual monomer content, and the molecular weight distribution during polymerization and soft baking have shown the necessity to produce resist layers with a high molecular weight and with only a very small residual monomer content.[52,59] Figure 10.25 compares the molecular weight distribution determined by gel permeation chromatography of a polymerized PMMA resist (two hardener concentrations were used) with the molecular weight distribution of the casting resin. The casting resin is unimodal, whereas the polymerized resist layer typically shows a bimodal distribution, with peak molecular weights centered around 90,000 and 300,000 g/mol. The first low-molecular-weight peak belongs to the PMMA oligomer dissolved in the casting resin, and the second molecular-weight peak results from the polymerization of the monomer. The molecular weight distribution is constant across the total resist thickness, except for the boundary layer at the base plate, where the average molecular weight can be significantly higher (~450,000 g/mol).[55]

The amount of the high-molecular-weight portion in the polymerized resist depends on the concentration of the hardener. A low hardener content leads to a high-molecular-weight dominance and vice versa (see Figure 10.25).[59] Because high molecular weight is required for low stress, a hardener concentration of

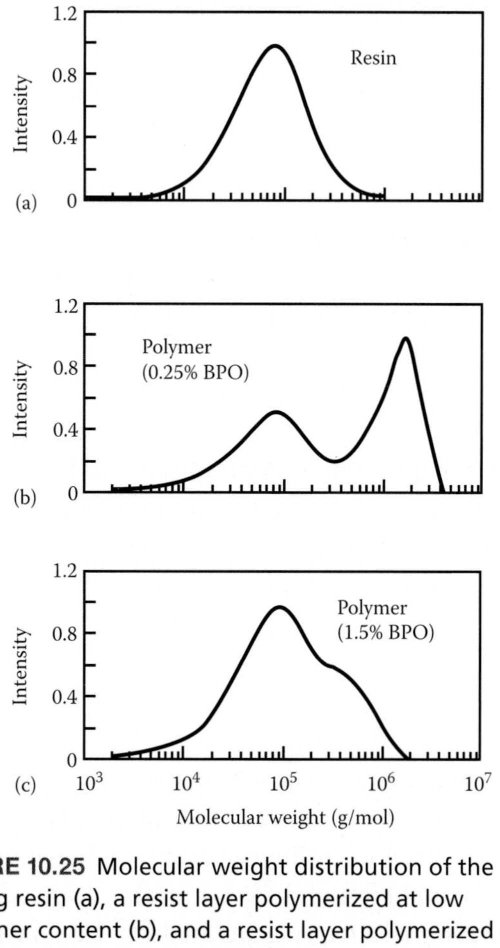


**FIGURE 10.25** Molecular weight distribution of the casting resin (a), a resist layer polymerized at low hardener content (b), and a resist layer polymerized at high hardener content (c) determined by gel permeation chromatography. (From Mohr, J., W. Ehrfeld, and D. Munchmeyer. 1988. Requirements on resist layers in deep-etch synchrotron radiation lithography. *J Vac Sci Technol* B6:2264–67. With permission.[59])

less than 1% BPO must be used. Ideally, for a low stress resist, the residual monomer content should be less than 0.5%. The residual monomer content decreases with increasing hardener content, and >1% BPO is needed to reduce the residual monomer content to less than 0.5%. The problem resulting from these opposite needs can be overcome by the addition of 1% of a cross-linking dimethacrylate (triethylene glycol dimethacrylate) to the resin. In such cross-linked PMMA, a smaller amount of BPO suffices to suppress the residual monomer content; crack-free PMMA can be obtained with 0.8% of BPO.[52]

For solvent removal, and to further minimize the defects caused by stress, the polymerized resin is cured at 110°C for 1 h (soft bake). The measurement of the reaction enthalpy shows that postpolymerization reactions occur at room temperature and during heating to the glass transition temperature.[59] The rate of heating up to that temperature is 20°C/h; after curing, the samples are cooled down from 110°C to room temperature at a very low rate of 5–10°C/h.[52, 55] The soft-bake temperature is slightly below the glass transition temperature measured to be 115°C.

Another important factor that reduces stress in thick PMMA resist layers is the optimization of the developer. Stress-induced cracking can be minimized with solvent mixtures whose dissolution parameters lie near the boundary of the PMMA solubility range; that is, a nonaggressive solvent is preferred. This is discussed in more detail below under "Development." Small amounts of additives such as described above for reducing stress or to promote adhesion do not influence the mechanical stability of the microstructures or the sensitivity of the resist.

## Exposure

### *Optimum Wavelength*

For a given polymer, the lateral dimension variation in a LIGA microstructure could, in principle, result from the combined influence of several mechanisms. These include Fresnel diffraction, the range of high-energy photoelectrons generated by the x-rays, the finite divergence of synchrotron radiation, and the time evolution of the resist profiles during the development process. The theoretical manufacturing precision obtainable by deep x-ray lithography was investigated by means of computer simulation of both the irradiation step and the development step by Becker et al.[65] and Munchmeyer.[66] Results were further tested experimentally and confirmed by Mohr et al.[52] The theoretical results demonstrate that the effect of Fresnel diffraction (edge diffraction), which increases as the wavelength increases, and the effect of secondary electrons in PMMA, which increases as the wavelength decreases, lead to minimal structural deviations when the characteristic wavelength ranges between 0.2 and 0.3 nm (assuming an ideal development process and no x-ray divergence). To fully use the accuracy potential of a 0.2–0.3-nm wavelength, the local divergence of the synchrotron radiation at the sample site should be less than 0.1 mrad. Under these conditions, the variation in critical lateral dimensions likely to occur between the ends of a 500-μm-high structure as a result of diffraction and secondary electrons is estimated to be 0.2 μm. The estimated Fresnel diffraction and secondary electron scattering effects are shown as a function of characteristic wavelength in Figure 10.26.

Using cross-linked PMMA, or linear PMMA with a unimodal and extremely high-molecular-weight distribution (peak molecular weight greater than 1 million g/mol), the experimentally determined lateral tolerances on a test structure, as shown in Figure 7.27, are 0.055 μm/100-μm resist thickness, in good agreement with the 0.2 μm over 500 μm expected

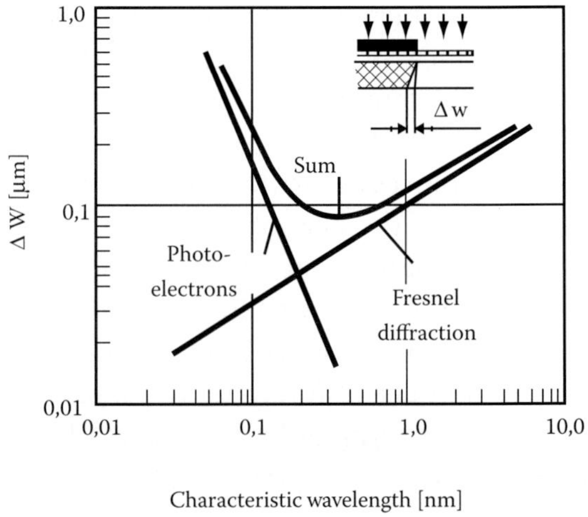


**FIGURE 10.26** Fresnel diffraction and photoelectron generation as a function of characteristic wavelength, $\lambda_c$, and the resulting lateral dimension variation ($\Delta W$). (From Menz, W., and P. Bley. 1993. *Mikrosystemtechnik fur Ingenieure*. Weinheim, Germany: VCH Publishers.[67])

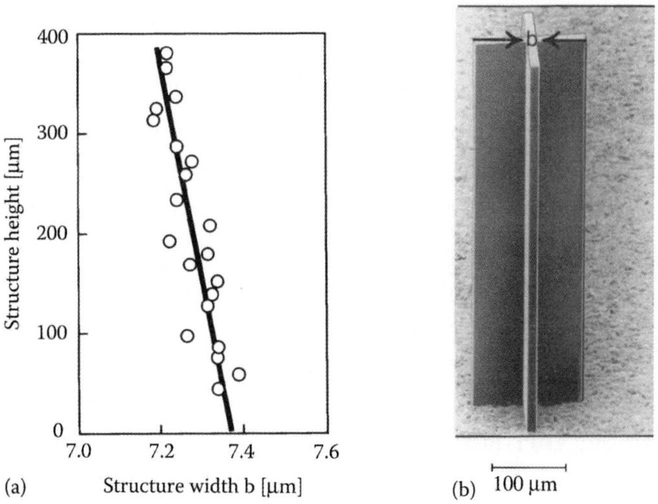


**FIGURE 10.27** Structural tolerances. (a) SEM micrograph of a test structure to determine conical shape. (b) Structural dimensions as a function of structure height. The tolerances of the dimensions are within 0.2 mm over the total structure height of 400 mm. (From Mohr, J., W. Ehrfeld, and D. Munchmeyer. 1988. Requirements on resist layers in deep-etch synchrotron radiation lithography. *J Vac Sci Technol* B6:2264–67.[59]) (Courtesy of the Karlsruhe Nuclear Research Center.)

on a theoretical basis.[52] These results are obtained only when a resist/developer system with a ratio of the dissolution rates in the exposed and unexposed areas of approximately 1000 is used.

The use of resist layers, not cross-linked and displaying a relatively low bimodal molecular-weight distribution, as well as the application of excessively strong solvents such as used to develop thin PMMA resist layers in the IC industry, leads to a more pronounced conical shape in the test structure of Figure 10.23. An illustration of the effect of molecular weight distribution on lateral geometric tolerances is that linear PMMA with a peak molecular weight less than 300,000 g/mol shows structure tolerances of up to 0.15 μm/100 μm.[52] To obtain the best tolerances requires a PMMA with a very high molecular weight, also a prerequisite for low stress in the developed resist. Finally, if the synchrotron beam is not parallel to the absorber wall but at an angle greater than 50 mrad, greater coning angles may also result.[52]

### *Deposited Dose*

As shown in Figure 10.28, depicting the average molecular weight of PMMA as a function of radiation

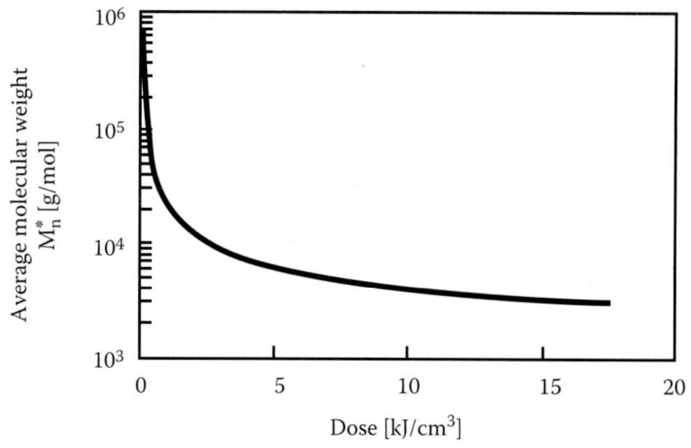


**FIGURE 10.28** Average molecular weight $M_n^*$ versus x-ray radiation dose. (From Menz, W., and P. Bley. 1993. *Mikrosystemtechnik fur Ingenieure.* Weinheim, Germany: VCH Publishers. With permission.[67])

dose, the x-ray irradiation of PMMA reduces the average molecular weight ($M_n'$, see Chapter 1, Equation 1.10).[67] For one-component positive resists, this decrease of the average molecular weight causes the solubility of the resist in the developer to increase dramatically. The average molecular weight making dissolution possible is a sensitive function of the type of developer used and the development temperature. It can be observed from Figure 10.28 that above a certain dose (15–20 kJ/cm$^3$), the average molecular weight does not decrease any further.

The molecular weight distribution, measured after resist exposure, is unimodal with peak molecular weights ranging from 3,000–18,000 g/mol, dependent on the dose deposited during irradiation. The peak molecular weight increases nearly linearly with increasing resist depth; that is, decrease of the absorbed dose.[55] At the bottom of the resist layer, the absorbed dose must be higher than the development dose, $D_{dv}$, whereas at the top of the resist, the absorbed dose must be lower than the damaging dose, $D_{dm}$. In Figure 10.18, where the absorption of x-rays along the path from source to sample was illustrated, the exposure time and the preabsorber were chosen so that the bottom of a 500-μm-thick PMMA layer received the necessary development dose $D_{dv}$, whereas the dose at the top of the layer stayed well below $D_{dm}$. Exposure of PMMA with longer wavelengths results in correspondingly longer exposure times and can lead to an overexposure of the top surface, where the lower energy radiation is mainly absorbed.

Menz and Bley[67] describe the influence of the radiation dose on the quality of the resulting LIGA structures in a slightly different manner. Following their approach, Figure 10.29a illustrates a typical bimodal molecular weight distribution of PMMA before radiation, exhibiting an average molecular weight of 600,000. The gray region in this figure indicates the molecular weight region where PMMA readily dissolves; that is, less than the 20,000 g/mol level for the temperature and developer used. Because the fraction of PMMA with a 20,000 molecular weight is very small in nonirradiated PMMA, the developer hardly attacks the resist. After irradiation with a dose $D_{dv}$ of 4 kJ/cm³, the average molecular weight becomes low enough to dissolve almost all the resist (Figure 10.29b). With a dose $D_{dm}$ of 20 kJ/cm³, all the PMMA dissolves swiftly (Figure 10.29c). At a dose more than $D_{dm}$, the microstructures are destroyed by the formation of bubbles. It follows that to dissolve PMMA completely and to make defect-free microstructures, the radiation dose for the specific type of PMMA used must lie between 4 and 20 kJ/cm³. These two numbers also lock in a maximum value of five for the ratio of the radiation dose at the top and bottom of a PMMA structure. To make this ratio as small as possible, the soft portion of the synchrotron radiation spectrum is usually filtered out by a preabsorber [for example, a 100-µm-thick polyimide foil (Kapton)] to reduce differences in dose deposition in the resist.

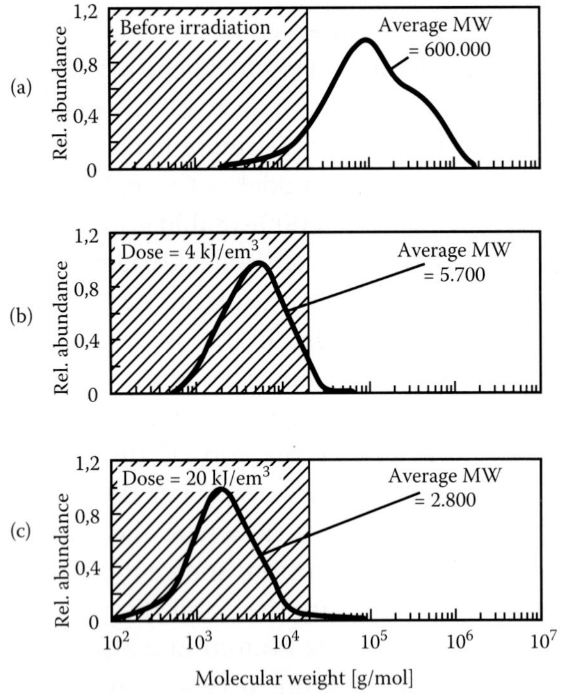


**FIGURE 10.29** Molecular weight distribution of PMMA before (a) and after irradiation with 4 (b) to 20 kJ/cm³ (c). The shaded areas indicate the domain in which PMMA is minimally 50% dissolved (at 38°C in the LIGA developer described in the text). (From Menz, W., and P. Bley. 1993. *Mikrosystemtechnik fur Ingenieure*. Weinheim, Germany: VCH Publishers. With permission.[67])

X-ray exposure equipment developed by KfK has vibration-free bedding; the exposure chamber is under thermostatic control (±0.2°C) and includes a precision scanner for the periodic movement of the sample through the irradiation plane. The polyimide window isolates the vacuum of the accelerator from the helium atmosphere (200 mbar), which serves as coolant for substrate and mask in the irradiation chamber.[43] IMM, in collaboration with Jenoptik GmbH, developed an x-ray scanner for deep lithography, enabling irradiation of up to 1000-µm-thick resists. A mask-to-resist registration within ±0.3 µm is claimed.[68]

### *Stepped and Slanted Microstructures*

For many applications, stepped or inclined resist sidewalls are very useful—consider, for example, the fabrication of multilevel devices or prisms or, more basic yet, angled resist walls to facilitate the release of molded part.* Using stepped absorber layers on a single mask to make stepped multilevel microstructures was discussed above under "Absorber Fabrication" (see also the discussion in Chapter 2 and Figure 2.11). We mentioned that structures made this way are not always very well resolved. To make better-resolved stepped features one can first relief print a PMMA layer, for example, by using a Ni mold insert made from a first x-ray mask. Subsequently, the relief structure may be exposed to synchrotron radiation to further pattern the polymer layer through a precisely adjusted second x-ray mask. To carry out this process, a two-layer resist system needs to be developed consisting of a top PMMA layer, which

* In using LIGA or LIGA-like techniques for plastic micromolding it is important to be able to control the resist wall inclination to facilitate the release of molded devices. Even a 2° angle can make the difference between form locking or easy release.

fulfills the requirement of the relief printing process, and a bottom layer, which fulfills the requirements of the x-ray lithography.[69] The bottom resist layer promotes high molecular weight and adhesion, whereas the top PMMA layer is of lower molecular weight and contains an internal mold-release agent. This process sequence combining plastic impression molding with x-ray lithography is illustrated in Figure 10.30. The two-step resist then facilitates the fabrication of a mold insert by electroforming, which can be used for the molding of two-step plastic structures. Extremely large structural heights can be obtained from the additive nature of the individual microstructure levels.

There are several options to achieve miniaturized features with slanted walls. It is possible to modulate the exposure/development times of the resist, fabricate an inclined absorber, angle the radiation, or, following Tabata et al.,[71] move the mask during exposure in so-called *moving mask deep x-ray lithography* (M$^2$DXL) (Example 10.2).

To make a slanted absorber, a slab of material can be etched into a wedge by pulling it at a linear rate out of an etchant bath. Changing the angle at which synchrotron radiation is incident on the resist, usually 90°, also enables the fabrication of microstructures with inclined sidewalls.[72] This way, slanted microstructures may be produced by a single oblique irradiation or by a swivel irradiation (see Figure 2.11 for an example). One potentially very important application of microstructures incorporating inclined sidewalls is the vertical coupling of

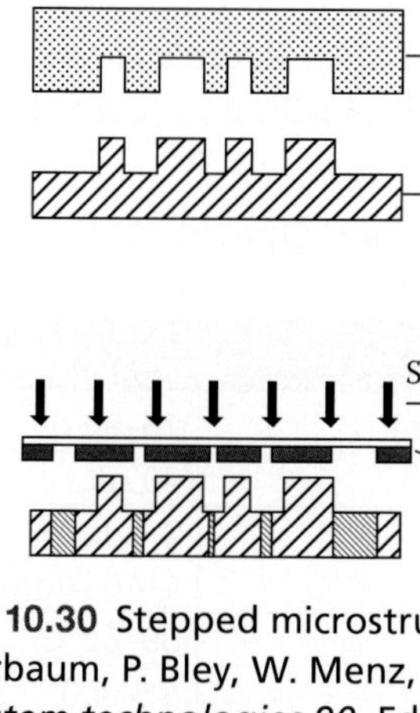


FIGURE 10.30 Stepped microstructures. (From Mohr, J., C. Burbaum, P. Bley, W. Menz, and U. Wallrabe. 1990. *Microsystem technologies 90.* Ed. H. Reichl. Berlin: Springer. With permission.[70])

light into waveguide structures using a 45° prism.[73] Such optical devices must have a wall roughness of less than 50 nm, making LIGA a preferred technique for this application. The sharp decrease of the dose in the resist underneath the edge of the inclined absorber, and the resulting sharp decrease of the dissolution of the resist as a function of the molecular weight in the developer, results in little or no deviation of the inclination of the resist sidewall over the total height of the microstructure.

In Chapter 8, we described additional means to fabricate slanted and curved metal shapes using electrochemical and electroless plating.

## Development

X-ray radiation changes the polymer in the unmasked areas, and chemicals etch away the regions that have been exposed during development. To fully use the accuracy potential of synchrotron radiation lithography, it is essential that the resist/developer system have a ratio of dissolution rate in the exposed and unexposed areas of approximately 1000 (see above). The developer empirically arrived at by KfK consists of a mixture of 20 vol% tetrahydro-1,4-oxazine (an azine), 5 vol% 2-aminoethanol-1 (a primary amine), 60 vol% 2-(2-butoxyethoxy) ethanol (a glycolic ether), and 15 vol% water.[52,55,74] This developer causes an infinitely small dissolution of unexposed, high-molecular-weight, cross-linked PMMA and achieves a sufficient dissolution rate in the exposed area. It also exhibits much less stress-induced cracking than developers conventionally used for thin PMMA resists.

Systematic investigation of different organic solvents and mixtures of the above developer systems shows that solvents with a solubility parameter at the periphery of the solubility range of PMMA dissolve exposed PMMA slowly but selectively and without stress-induced cracking or swelling of unexposed areas. Solvents with a solubility parameter close to those of MMA show a much higher dissolution rate but cause serious problems related to cracking and swelling. As we have already seen, the application of excessively strong solvents also leads to more pronounced conical shapes in the test structures shown in Figure 10.27. An improved developer found in the above systematic investigation is a mixture of tetrahydro-1,4-oxazine and 2-aminoethanol-1. Its sensitivity is 30% higher, but the process latitude is much narrower compared with the developer described above.[55]

KfK has built a machine for the development process, enabling the continuous and homogeneous transport of developing and rinsing agents into deep structure elements and the removal of the dissolved resist from these structures. Several substrates are arranged vertically on a rotor, with each structure surface facing out. During development, the developing agent flows toward the resist surface being developed, circulates, is filtered continuously, and the temperature remains controlled at 35°C. To stop development, less-concentrated developer solutions are applied to prevent the precipitation of already-dissolved resist. Three independent medium circuits are available for immersion and spraying processes.[43]

After development, the microstructures are rinsed with deionized water and dried in a vacuum. Alternatively, drying may be done by spinning and blasting with dry nitrogen. At this stage, the devices can be the final product, for example, as microoptical components, or they can be used for subsequent metal deposition.

## Metal Deposition in LIGA and Pseudo-LIGA

### *Introduction*

Electroplating, which is sometimes also called *electrodeposition*, is a process in which a metal is coated on a conductive surface through electrochemical reactions that are facilitated by an applied electrical potential. In this process, the surface to be coated is immersed into a solution of one or more metal salts. LIGA and pseudo-LIGA depend critically on metal deposition to replicate the master photoresist mold. The additive metal deposition process was detailed in Volume I, Chapter 7, where we discussed the electroplating of metals in high-aspect-ratio crevices, and in Chapter 8, this volume, where we compared electroless with electrodeposition and the crafting of slanted and curved plated metal shapes. In this section we add some modifications to the metal process when applied to high-aspect-ratio resist molds encountered in LIGA and pseudo-LIGA.

### *Nickel Electrodeposition*

#### *Physical Properties of Electrodeposited Nickel*

Nickel, together with copper, is one of the most widely used metals in electroplating LIGA and pseudo-LIGA structures. A nickel sulfamate bath composition, optimized for Ni electrodeposition, is presented in Table 10.14. In addition to nickel sulfamate and boric acid as a buffer, a small quantity of an anion-active wetting agent is usually added. Sulfur-depolarized nickel pellets are used as anode materials (sulfur-depolarized nickel has sulfur cast into it to aid in anode corrosion and keep the anode from going passive) and are held in a titanium basket.[75]

Table 10.14 also lists operational parameters. Nickel sulfamate baths produce low internal stress deposits without the need for additional agents, thus avoiding a cause for more defects. The bath is operated at 50–62°C and at a pH value between 3.5 and 4.0. Metal deposition is carried out at current densities up to 10 A/dm$^2$. Growth rates vary from 12 (at 1 A/dm$^2$) to 120 μm/h (at 10 A/dm$^2$).

The hardness of the Ni deposits can be adjusted from 200–350 Vickers by varying the operating conditions. Hardness decreases with increasing current density. To reach a high hardness of 350 Vickers, the electroforming must proceed at a reduced current of 1 A/dm$^2$. Also, for low compressive stress of 20 N/mm$^2$ or less, a reduced current density must be used. Internal stress in the Ni deposits is not only influenced by current density but also by the layer thickness, pH, temperature, and solution agitation. In the case of pulse plating (see below under Pulse Plating), pulse frequency has a distinct influence as well.[76] From Figure 10.31 we derive that for thin Ni deposits the stress is high and decreases very fast as a function of thickness. For thick Ni deposits (>30 μm),

**Table 10.14** Composition and Operating Conditions of Nickel Sulfamate Bath

| Parameter | Value |
|---|---|
| Nickel metal (as sulfamate) | 76–90 g/L |
| Boric acid | 40 g/L |
| Wetting agent | 2–3 mL/L |
| Current density | 1–10 A/dm$^2$ |
| Temperature | 50–62°C |
| pH | 3.5–4.0 |
| Anodes | Sulfur depolarized |

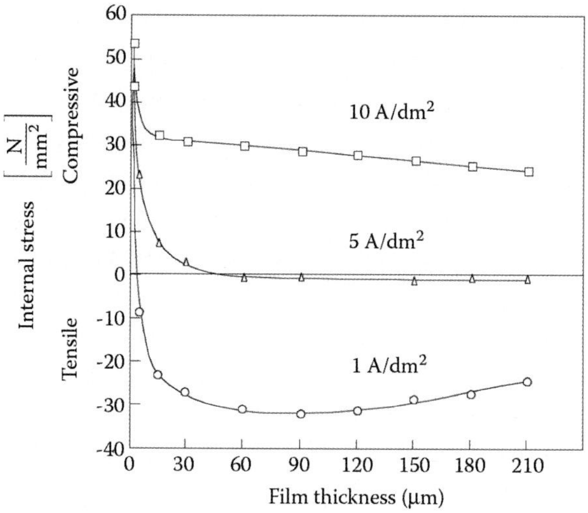


**FIGURE 10.31** Influence of nickel layer thickness on internal stress. The electrolyte used is described in Table 10.14 (pH = 4; bath temperature = 52°C). (From Harsch, S., W. Ehrfeld, and A. Maner. 1988. *KfK report no. 4455.* Karlsruhe, Germany: KfK.[76]) (Courtesy of the Karlsruhe Nuclear Research Center.)

the stress as a function of thickness reaches a plateau. At a current density of 10 A/dm$^2$, these thick Ni films are under compressive stress; at 1 A/dm$^2$, they are under tensile stress; and at 5 A/dm$^2$, the internal stress reduces to practically zero. Stirring of the plating solution reduces stress dramatically, indicating that mass transport to the cathode is an important factor in determining the ultimate internal stress. Consequently, because high-aspect-ratio features do not experience the same agitation of a stirred solution as bigger features, this results in higher stress concentration in the smallest features of the electroplated structure. Because the stress is most severe in the thinnest Ni films, Harsch et al.[76] undertook a separate study of internal stresses in 5-μm-thick Ni films. For three plating temperatures investigated (42°C, 52°C, and 62°C), 5-μm-thick films were found to exhibit minimal or no stress at a current density of 2 A/dm$^2$. At 62°C, the 5-μm-thick films show no stress at 2 A/dm$^2$ and remain at a low compressive stress value of about 20 N/mm$^2$ for the whole current range (1–10 A/dm$^2$). The internal stress at 2 A/dm$^2$ is minimal at a pH value between 3.5 and 4.5 of a sulfamate electrolyte (Table 10.14). Ni concentration (between 76 and 100 g/L) and wetting agent concentration do not seem to influence the internal stress. The higher the frequency of the pulse in pulse plating (see next section), the smaller the internal stress.

The long-term mechanical stability of Ni LIGA structures was investigated by electromagnetic activation of Ni cantilever beams by Mohr et al.[77] The number of stress cycles, *N*, necessary to destroy a mechanical structure depends on the stress amplitude, *S*, and is determined from fatigue curves, or *S-N* curves (applied stress on the *x*-axis and number of cycles necessary to cause breakage on the *y*-axis). Experimental results show that for Ni cantilevers produced by LIGA the long-term stability reaches the range of comparable literature data for bulk annealed and hardened nickel specimens. Usually, stress leads to crack initiation, which often starts at the surface of the structure. Because microstructures have a higher surface-to-volume ratio, one might have expected the *S-N* curves to differ from macroscopic structures, but so far this has not been observed. To the contrary, it seems that the smaller structures are more stable.

### *Pulse Plating*

For the fabrication of microdevices with high deposition rates exceeding 120 μm/h, a reduction of the internal Ni stress can occur only by increasing the temperature of the bath or by using alternative electrodeposition methods. Increasing the bath temperature is not an attractive option, but using alternative electrochemical deposition techniques deserves further exploration. For example, using pulsed (≥500 Hz) galvanic deposition instead of a DC method can influence several important properties of the Ni deposit. Properties such as grain size, purity, and porosity can be manipulated this way without the addition of organic additives.[76] In pulse plating, the current pulse is characterized by three parameters: pulse current density, $i_p$; pulse duration, $t_d$; and pulse pause, $t_p$. These three independent variables determine the average current density, $i_a$, which is the important parameter influencing the deposit quality and is given by:

$$i_a = \frac{t_d}{t_d + t_p} i_p \qquad (10.5)$$

Pulse plating leads to smaller metal grain size and smaller porosity as a result of a higher deposition potential. Because each pulse pause allows some time for metal cation replenishment at the cathode (e.g., $Ni^{2+}$ enrichment) and for diffusion away of undesirable reaction products that might otherwise get entrapped in the metal deposit, a cleaner metal deposit results. The higher the frequency of the pulse, the smaller the internal stress in the resulting metal deposit.[76]

Pulse plating represents only one of many emerging electrochemical plating techniques considered for microfabrication. For more background on techniques such as laser-enhanced plating, jet plating, laser-enhanced jet plating, and ultrasonically enhanced plating, refer to the review of Romankiw et al.[78] and references therein.

### *LIGA and Pseudo-LIGA Plating Issues*

*Electrodeposition in High-Aspect-Ratio Resist Crevices* In Volume I, Chapter 7 we studied how a microelectrode of the same size or less than the diffusion layer thickness $\delta$ could be expected to plate faster than a larger electrode because of the extra current resulting from the nonlinear diffusion (Volume I, Equation 7.131). This turns out to be incorrect: as derived from Volume I, Equation 7.132, describing the current to a metal electrode recessed in a resist layer, the nonlinear diffusion contribution increases with decreasing radius $r$, but it also decreases with increasing resist thickness, $L$. High-aspect-ratio features consequently plate slower than low-aspect-ratio features. Moreover, the hydrogen consumption accompanying cathodic reduction of the depositing metal in the high-aspect-ratio features causes the pH to locally increase. Because agitation is impossible in these deep crevices, an isolating layer might form (e.g., nickel hydroxide, in the case of Ni deposition), preventing further metal deposition. This all contributes to make the metal deposition rate, important for an economical production, much smaller than the rates expected from the linear diffusion model of current density in large, low-aspect-ratio structures.

A major cause of defects in electroplating in high-aspect-ratio features is an incomplete wetting of the microchannels in the resist structure. For example, the contact angle between PMMA and the plating electrolyte at 50°C lies between 70 and 80°. Thus, a wetting agent is indispensable for wetting the surface of the plastic structures for the electrolyte to penetrate into the microchannels. With a wetting agent, the contact angle between PMMA and the plating electrolyte can be reduced from 80 to 5°.[76] In the electroforming of microdevices, a much higher concentration of wetting agent is necessary than in conventional electroplating. A dramatic illustration of this wetting effect can be seen in the case of plating of Ni posts, as shown in Figure 10.32. In the case illustrated in Figure 10.32a, where only 2.5 mL/L of a wetting agent is added to the sulfamate nickel deposition solution (see Table 10.14), nickel posts with a diameter of 50 μm often fail to form. In the same experiment, no posts with a diameter of 5 and 10 μm form. Increasing the wetting agent concentration to 10 mL/L results in perfectly formed nickel posts with a diameter of 5 μm, as shown in Figure 10.32b.

*Plating Compatibility Issues* Two major sources of difficulty associated with plating of any metal structure in photoresist molds, not only the high-aspect ones described above, are chemical and mechanical incompatibility. Chemical incompatibility means that the photoresist mask may be attacked by the plating solution; mechanical incompatibility is film stress in the plated layer, which causes the plated structure to lose adhesion to the substrate.

If the plating solution attacks the photoresist even slightly, considerable damage to the photoresist layer may occur by the time a 200-μm-thick structure has been created. Therefore, the limiting thickness of a plated structure resulting from a chemical interaction is dependent on both the photoresist chemistry and the plating bath itself. In general, the plating bath must have a pH in the range of ~3–9.5 (acidic to mildly alkaline), a fairly wide range that can accommodate many of the commercially available photoresists. A surprising number of plating baths do not fall into that range, however. Baths, either very strongly acidic or more than mildly alkaline, tend to attack and destroy photoresist.

The plated structure must have extremely low stress to avoid cracking or peeling during the plating process. In addition, if the structure contains narrow features, the stress must be tensile because any compressive stress would result in buckling of the structure. The primary concern with regard to

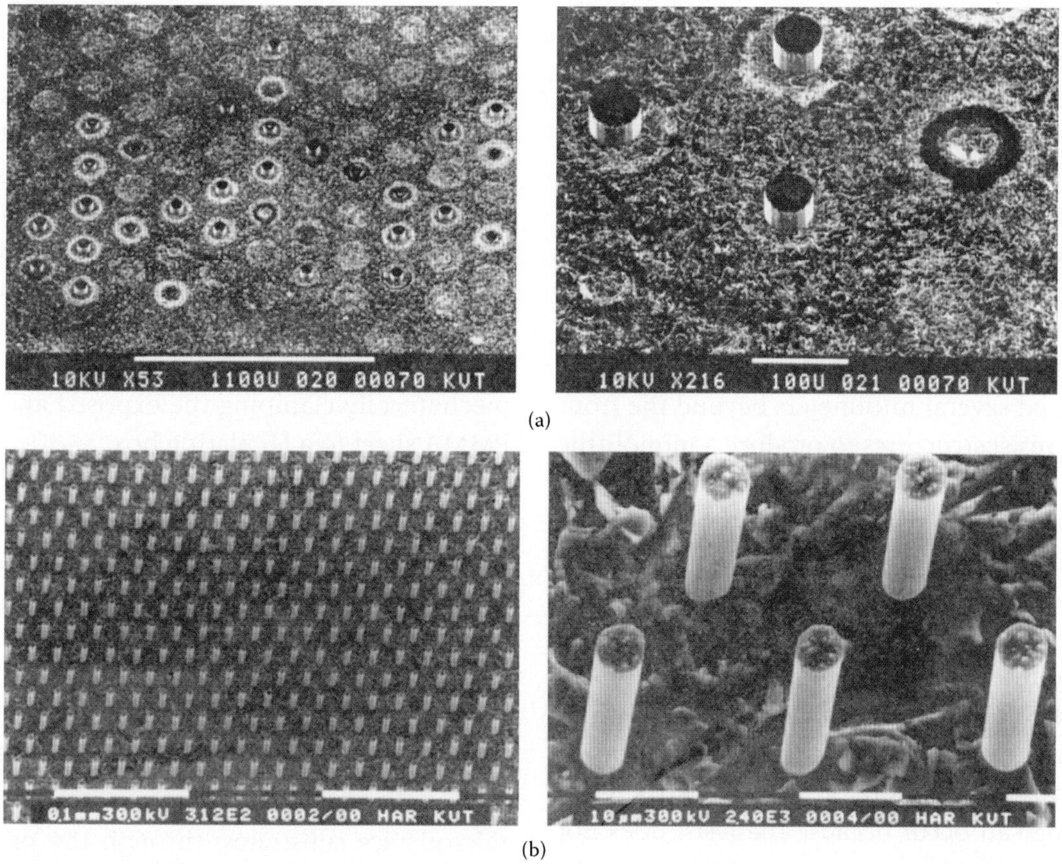


**FIGURE 10.32** Electrodeposition of nickel posts of varying diameter. (a) Nickel posts with diameter of 50 μm. Only 2.5 mL/L wetting agent was added to the nickel sulfamate solution. Many 50-μm posts are missing, and posts with a diameter of 10 or 5 μm do not even form. (b) Nickel posts with diameter of 5 μm; 10 mL/L wetting agent was added, and all posts developed perfectly. (Courtesy of the Karlsruhe Nuclear Research Center.)

stresses is the incorporation of "brightening agents" into the plating solution. These can be selenium, arsenic, thallium, and ammonium ions, as well as others. The brightening materials often exhibit different atomic configurations and sizes compared with most materials that we would like to plate, such as Cu, Au, Ni, FeNi, Pt, Ag, and NiCo. This represents the source of a good deal of the stress intrinsic to the plating process. Ag and FeNi are notorious for their high stresses. The Ni-sulfamate bath, presently used extensively in LIGA, is used without brighteners, leading to Ni deposits with extremely low stress. Electroless plating baths (see Chapter 8) also are a good option: these *autocatalytic* baths coat indiscriminately but cause very low stress. They should be compatible with the LIGA process for certain applications, such as mold insert formation where significant overplating is required.

A third difficulty in finding plating baths compatible with LIGA is that most plating baths are not intended for use in the semiconductor industry. Information (including normal deposition parameters, uniformity, stress, compatibility with various semiconductor processes, and particulate contamination) normally supplied by a vendor is not available. Because the majority of the semiconductor industry does not incorporate these kinds of "thin film" processes, it may be difficult to determine whether a plating bath will be suitable simply from conversations with the vendor. In many instances, trial and error decide.

The treatise of electroplating in micropatterned resists is of importance for both micromachining and the IC industry. Additional reading on this topic is suggested (e.g., Masuko et al.,[79] Dukovic,[80] and Leyendecker et al.[81]).

### *Finishing Metal Parts and Metal Mold Inserts*

Slight differences in final metal layer thickness cannot be avoided in the electroforming process. Finish grinding of the metal samples with diamond paste is used to even out microroughness and slight variations in structural height. After finish grinding an electroplated LIGA or pseudo-LIGA work piece, a primary metal shape results from removing the photoresist by ashing in oxygen plasma or stripping in solvent. In the case of cross-linked PMMA, the resist may be exposed again to synchrotron radiation, guaranteeing sufficient solubility before stripping.

If the metal part needs to function as a mold insert, metal is plated several millimeters beyond the front faces of the resist structures to produce a monolithic micromold (see Figure 10.33). In the latter case, to avoid damage to the mold insert when separating it from the plating base, an intermediate layer sometimes is deposited on the base plate, ensuring adhesion of the resist structures while facilitating the separation of the electroformed mold insert from its plating base. In addition, it helps to prevent burrs from forming at the front face of the mold insert as a result of underplating of the resist structures. Underplating can occur because the resist does not

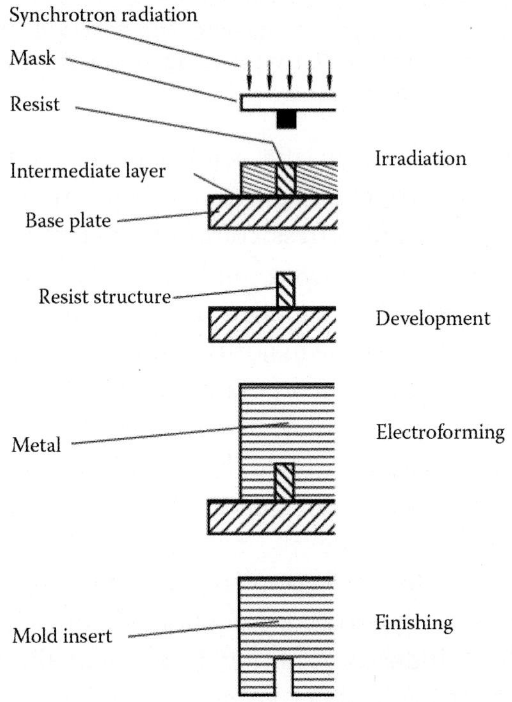


**FIGURE 10.33** Fabrication of a LIGA mold insert. (From Hagmann, P., W. Ehrfeld, and H. Vollmer. 1987. Fabrication of microstructures with extreme structural heights by reaction injection molding. *First meeting of the European Polymer Federation, European symposium on polymeric materials*, pp. 241–51. Lyon, France. With permission.[28])

adhere well to the substrate, allowing electrolyte solution to penetrate between the two, or the plating solution might attack the substrate/resist interface. Finally, microcracks at the interface might contribute to underplating. Burrs are easily eliminated then by dissolving the thin auxiliary metal layer with a selective etchant, removing the mold without the need for a mechanical load. In view of the observed underplating problems, it is surprising that Galhotra et al.[63] obtained good plating results by simply mechanically clamping the exposed and developed PMMA sheet to a Ni plating box.

### *Plating Automation*

An automated galvanoforming facility used by KfK is shown in Figure 10.34.[82] This setup includes provisions for on-line measurement of each electrolyte constituent in flow-through cells and concentration corrections when tolerance limits are exceeded. A computer-controlled transport system moves the individual plating racks, holding the microdevice substrates, through the process stages whereby the substrates are degreased, rinsed, pickled, electroplated, and dried and are then returned to a magazine, which can accommodate up to seven racks. The facility is designed as clean-room equipment because contamination of the microstructures must be prevented.[75] An instrument as described here can be bought, for example, from Reinhard Kissler (the μGLAV 750).[82] Although this type of automation is a must for the eventual commercialization of the LIGA technique, it should be recognized that at this explanatory stage overautomatization may be counterproductive.

## Demolding

### *Introduction*

Even high-aspect-ratio metal structures can be molded and demolded with polymers easily, as long as a polymer with a small adhesive power and rubber-elastic properties, such as silicone rubber, is used. However, rubber-like plastics have low shape stability and would not be adequate. Shape-preserving polymers, on the other hand, after hardening, require a mold with extremely smooth inner surfaces to prevent form-locking between the

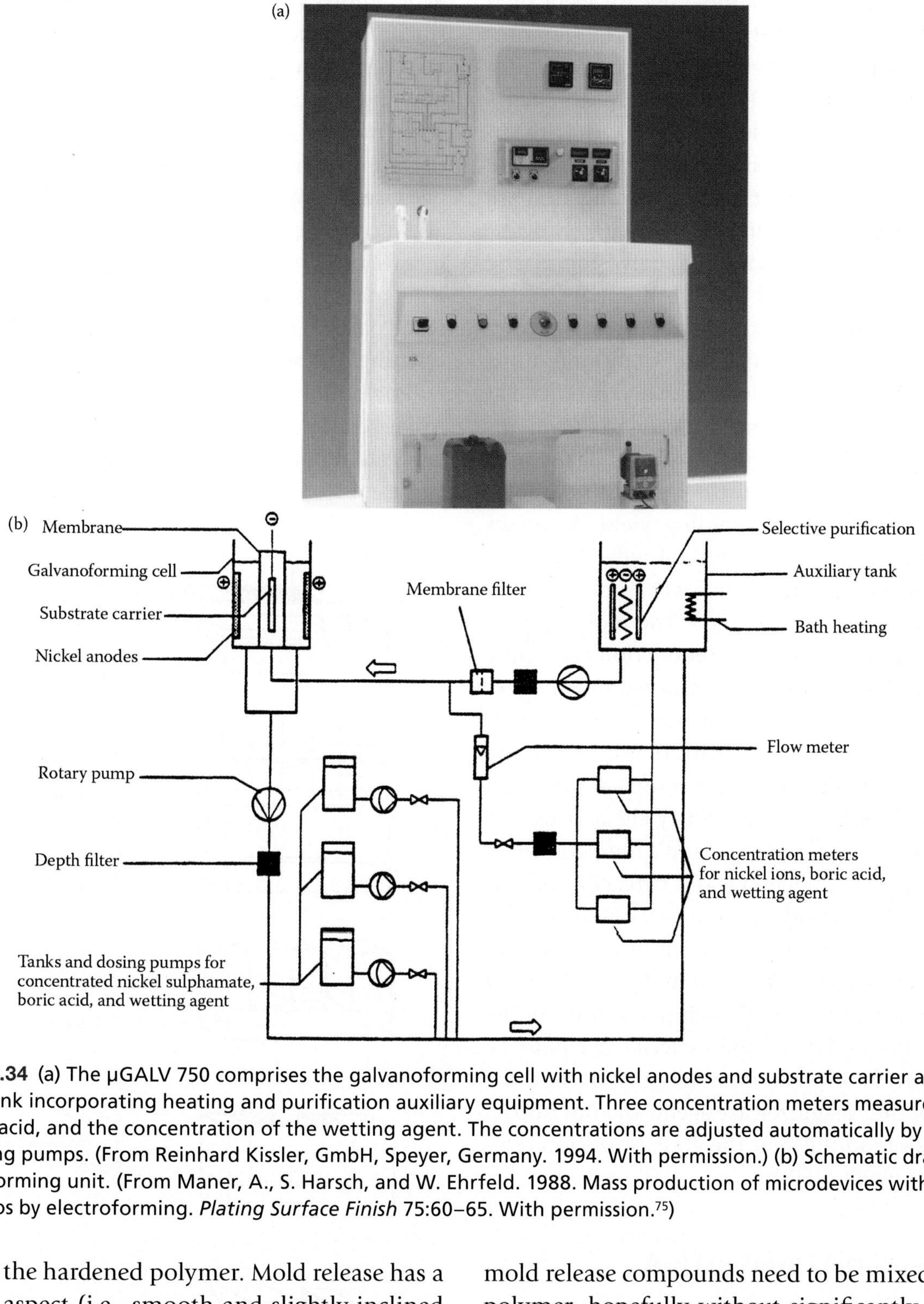


**FIGURE 10.34** (a) The μGALV 750 comprises the galvanoforming cell with nickel anodes and substrate carrier and an auxiliary tank incorporating heating and purification auxiliary equipment. Three concentration meters measure nickel ions, boric acid, and the concentration of the wetting agent. The concentrations are adjusted automatically by adding via-metering pumps. (From Reinhard Kissler, GmbH, Speyer, Germany. 1994. With permission.) (b) Schematic drawing of a galvanoforming unit. (From Maner, A., S. Harsch, and W. Ehrfeld. 1988. Mass production of microdevices with extreme aspect ratios by electroforming. *Plating Surface Finish* 75:60–65. With permission.[75])

mold and the hardened polymer. Mold release has a geometry aspect (i.e., smooth and slightly inclined mold walls) and a chemical aspect (i.e., mold release agents). A mold release agent may be required for demolding. Using external mold release agents is difficult because of the small dimensions of the microstructures to be molded because typically they are sprayed onto the mold. Consequently, internal mold release compounds need to be mixed with the polymer, hopefully without significantly changing the polymer characteristics.

### *Demolding in LIGA and LIGA-Like Applications*

The demolding step in LIGA is often carried out by means of a clamping unit at preset temperatures and rates. Demolding is facilitated by slightly inclined mold walls and an internal mold release agent such as PAT 665 (Wurtz GmbH, Bingen, Germany), normally used with polyester resins.[83] The optimum yield with this agent and Plexit M60 PMMA occurs at 3–6 wt%. The yield decreases very quickly to less than 3 wt% as adhesion between the molded piece and the mold insert becomes stronger. These adhesive forces are estimated by qualitatively determining the forces necessary to remove the plastic structures from the mold insert. An upper limit is 10 wt%, where the MMA does not polymerize anymore. At more than 5 wt%, the Young's modulus and hence the mechanical stability decrease, and at more than 6 wt%, pores start forming in the microstructures. The internal mold release agent also has a marked influence on the optimum demolding temperature. The demolding yield decreases quickly at more than 60°C for a 4 wt% PAT internal mold release agent. With 6 wt%, one can only obtain good yields at 20°C.[28,83] The molding process in LIGA initially led to a production cycle time of 120 min. For a commercial process, much faster cycle times are needed, for example, by optimizing the mold release agent. With that in mind, the KfK group started to work with a special salt of an organic acid, leading to a 100% yield at a release agent content of 0.2 wt% only and a temperature of 40°C (at 0.05 wt%, a 95% yield is still achieved). At 80°C, a 100% demolding yield was obtained, and significantly a cycle time of 11.5 min was reached. During these experiments the mold was filled at 80°C and heated to 110°C within 7.5 min. As the curing occurs at 110°C, the material needs to be cooled down to 80°C for demolding. Moreover, the 0.2 wt% of the "magic release agent" did not impact the Young's modulus and the glass transition temperature of Plexit M60.[83]

## Use of Sacrificial Layers in LIGA

As in surface micromachining, sacrificial layers make it possible to fabricate partially attached and freed metal structures in the primary mold process.[84] The ability to implement these features leads to assembled micromechanisms with submicrometer dimension accuracies, opening up many additional applications for LIGA, especially in the field of actuators. The sacrificial layer may be polyimide, silicon dioxide, polysilicon, or some metal.[85] The sacrificial layer is patterned with photolithography and wet etching before polymerizing the resist layer over it. At KfK, a several-micrometers-thick titanium layer often is used as the sacrificial layer because it provides good adhesion of the polymer, and it can be etched selectively against several other metals used in the process. If before exposure the x-ray mask is adjusted to the sacrificial layer pattern, some parts of the microstructures will lie above the openings in the sacrificial layer, whereas other parts will be built up on it. These latter parts will be able to move after removal of the sacrificial layer.[70] The fabrication of a movable LIGA structure is illustrated in Figure 10.35.

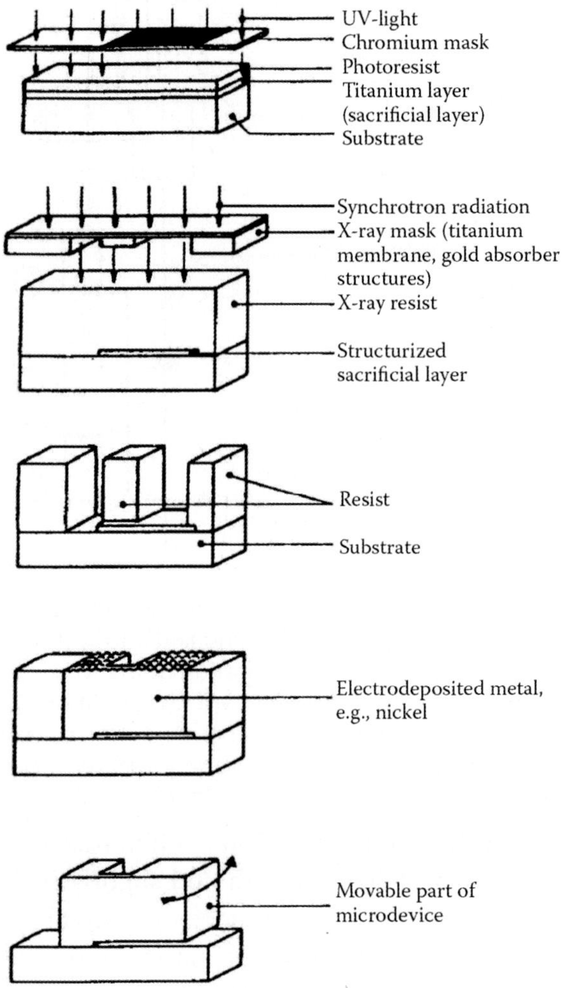


**FIGURE 10.35** Movable LIGA microstructures. (From Mohr, J., C. Burbaum, P. Bley, W. Menz, and U. Wallrabe. 1990. *Microsystem technologies 90.* Ed. H. Reichl. Berlin: Springer. With permission.[70])

## Molding Technology Applied to LIGA

### Application of Reaction Injection Molding to LIGA

A laboratory-sized reaction injection molding (RIM) setup, as shown in Figure 10.36, was used by the KfK group for their LIGA work.[83] The setup consists of a container for mixing the various reactants, a vacuum chamber for evacuation of the mold cavity, the molding tool, and a hydraulic clamping unit to open and close the vacuum chamber and the molding tool. After the vacuum chamber has been closed and evacuated, the tool is closed by the clamping unit. The mixed reagents are degassed in the materials container and, under a gas overpressure of up to 3 MPa, pushed through the opened inlet valve into the tool holding the evacuated insert mold. To compensate for shrinkage caused by polymerization, an overpressure of up to 30 MPa must be applied during hardening of the casting resin. If the holding overpressure is too low, sunken spots appear in the plastic because of shrinkage of the polymer. For PMMA, the volume shrinkage is about 21%. If the mold material is not degassed and the mold cavity is not evacuated, bubbles develop in the molded piece, resulting in defects and possible partial filling of the mold. To harden the mold material and anneal material stress, the RIM machine is operated at temperatures up to 150°C.

To fabricate RIM-based microproducts, a variety of resins were tried, including epoxy resins, silicone resins, and acrylic resins. The most promising results are obtained with resins on an MMA base to which an internal mold release agent is added to reduce adhesion of the molded piece to the walls of the metal mold. The mold insert in the evacuated tool is covered by means of an electrically conductive perforated plate, called the *gate plate* (Figure 10.37a). For filling the mold, injection holes are positioned above large, free spaces in the metal structure, and low-viscosity reactants fill the smaller sections laterally. After hardening of the molded polymer, a form-locking connection between the produced part and the gate plate is established at the injection holes, permitting the demolding of the part from the insert (Figure 10.37b). No damage to the mold insert is observed after up to 100 mold/demold cycles, even at the level of an SEM picture. The secondary plastic structures formed are exact replicas of the original PMMA structures obtained after x-ray irradiation and development. If the final product is plastic, the molding process can be simplified considerably by omitting the gate plate. The mold insert is then cast over with the molding polymer in a vacuum vessel, building a stable gate block over the mold insert after hardening. The polymeric gate block is used to demold the part from the insert.

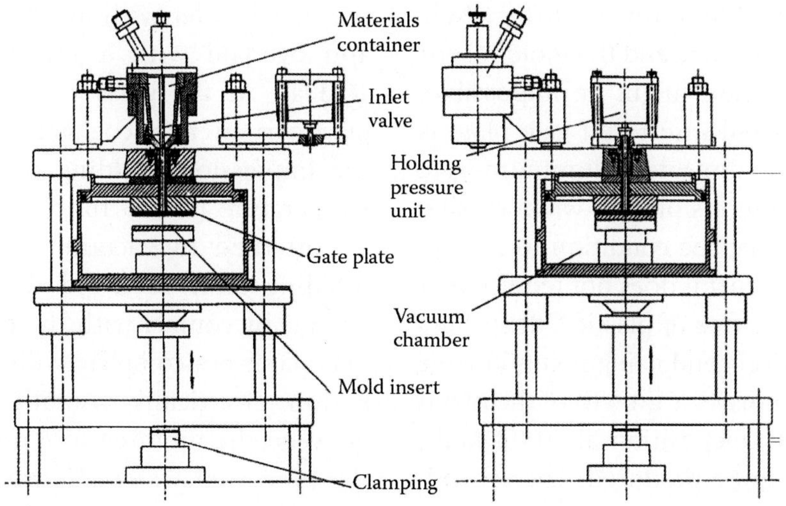


**FIGURE 10.36** Schematic presentation of a vacuum molding setup. With minor changes, this setup can be used for reaction injection molding, thermoplastic injection molding, and compression molding. (From Hagmann, P., and W. Ehrfeld. 1988. Fabrication of microstructures of extreme structural heights by reaction injection molding. *J Polymer Process Soc* 4:188–95. With permission.[83])

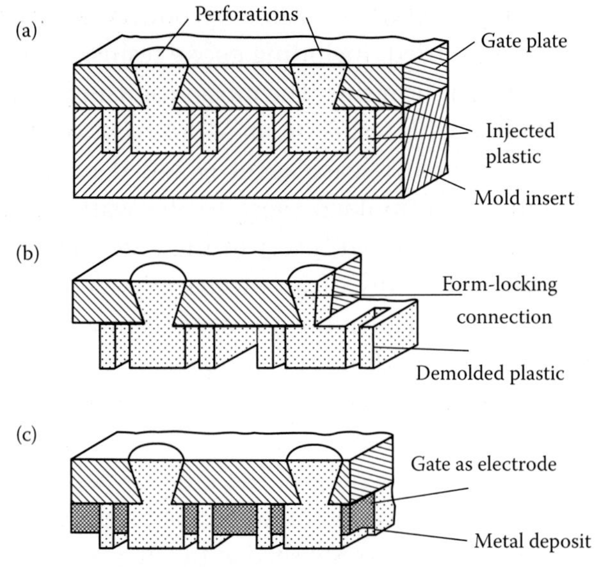


**FIGURE 10.37** Molding process with a perforated gate plate. (a) Filling of the insert mold with plastic material. (b) Demolding. (c) Electrodeposition with the gate plate as plating base. (From Menz, W., and P. Bley. 1993. *Mikrosystemtechnik fur Ingenieure*. Weinheim, Germany: VCH Publishers.[67])

The secondary plastic templates may not be the final product and can be filled with a metal by electrodeposition as with the primary molds. In this case, the secondary plastic templates must be provided with an electrode or plating base. The gate plate can be used directly as the electrode for the deposition of metal in the secondary plastic templates (see Figure 10.37c). It is important that the mold insert is pressed tightly against the gate plate so that no uniform isolating plastic film forms over the whole interface between the gate plate and the mold insert; otherwise, electrodeposition might be impossible. Safe sealing between mold insert and gate plate is achieved by use of soft-annealed aluminum gate plates into which the insert is pressed when closing the tool. It was found that the condition to be met for perfect deposition of metal does not require that the gate plate be entirely free of plastic.[83] It suffices that a number of electrical conducting points emerge from the plastic film. Transverse growth of the metal layer produced in electrodeposition allows a fault-free, continuous layer to be produced. The secondary metal microstructures are replicas of the primary metal structures, and a comparison of, for example, a master separation nozzle mold insert with a secondary separation nozzle structure did not reveal any differences in quality. After the metal forming, the top surface must be polished and the plastic structures must be removed. Depending on the intended use of the secondary metal shape, one can leave it connected to the gate plate, or the gate plate can be selectively dissolved or mechanically removed.

Gate plates can be applied only if the microstructures are interconnected by relatively large openings. The injection holes in the gate plate are produced by mechanical means with openings of about 1 mm in diameter and must be aligned with large openings in the mold insert. Many desirable microstructures cannot be built this way. For example, in Figure 10.38, we show a Ni honeycomb structure with a cell diameter of 80 μm, a wall thickness of 8 μm, and a height of 70 μm.[86] The holes in a gate plate are too large to inject plastic into this structure, and the interconnections in the honeycomb are too small to provide a good transverse movement of the polymerizing resin. In this case an electrically conductive plastic is used instead. The Ni plating occurs on the conductive plastic carrier plate, which is fused with the insulating plastic of the microstructure.

### *Application of Compression Molding to LIGA*

Compression molding or relief printing is the method of choice for incorporating LIGA structures on a processed Si wafer. In a first step, the mold material is applied by polymerization onto the processed wafer with an electrically isolating layer and overlaid with a metal plating base (Figure 10.39a). At the glass transition temperature, the plastic is in a viscoelastic condition suitable for the impression molding step. At more than this temperature (about 160°C for PMMA), the plastic is patterned in vacuum by impression with the molding tool. To avoid damaging the electronic circuits by contact with the molding tool, the mold material is not completely displaced during molding. A thin, electrically insulating residual layer is left between the molding tool and the plating base on the processed wafer (Figure 10.39b). The residual layer is removed by reactive ion etching (RIE) in an oxygen plasma etch (Figure 10.39c), freeing the plating base for subsequent electrodeposition. The oxygen RIE process is as anisotropic as possible so

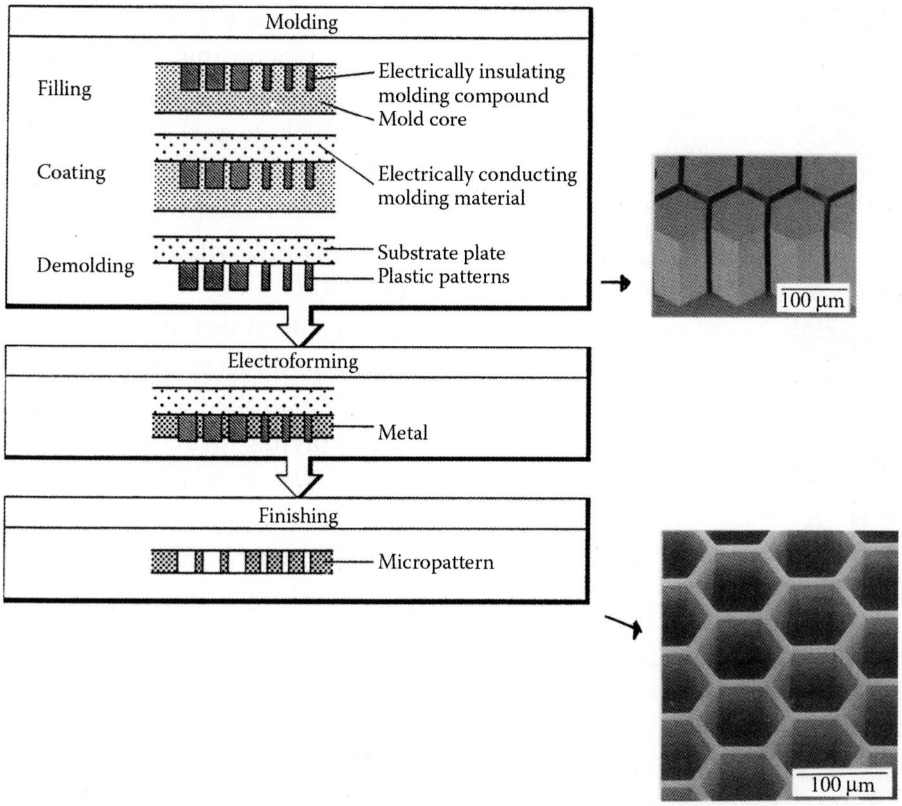


**FIGURE 10.38** Ni honeycomb structure with cell diameter of 80 mm, wall thickness of 8 mm, and wall height of 70 mm. A structure like this could not be produced using a gate plate; a conductive plastic is used instead. [From Harmening, M., W. Bacher, and W. Menz. 1991. Molding plateable micropatterns of electrically insulating and electrically conducting poly(methyl methacrylate)s by the LIGA technique. *Makromol Chem Macromol Symp* 50:277–84. With permission.[86]]

that the sidewalls of the structures do not deteriorate, and the amount of material removed from the top of the plastic microstructures is small compared with the total height. After electrodeposition on the plating base, the metal microstructures are laid bare by dissolution of the mold material (Figure 10.39d). Finally, using argon sputtering or chemical etching, the plating base between the metal microstructures is removed so that the metal parts do not short circuit (Figure 10.39e). In the case of chemical etching, a very fast etch should be used to avoid etching of the plating base from underneath the electrodeposited structures.[5,50]

## Comparison of Master Micromold Fabrication Methods

The high cost of x-ray lithography has caused many miniaturization engineers to search for alternate means for fabricating high-aspect-ratio metal, silicon, or polymer micromasters, or mold inserts.

Polymer micromold inserts are widely used in lost mold micromolding of metals and ceramics where the molds are removed by dissolving or thermal degradation after molding. However, this is not always the case; certain polymer micromold inserts, such as elastomeric mold inserts (e.g., polydimethylsiloxane), can be used repeatedly in soft lithography (see Chapter 2). The polymer mold inserts used for lost micromolding can be fabricated using photolithography, microstereolithography, and so on.

Most of the micromold inserts in micromolding are metal micromold inserts. Nickel is most commonly used because of the well-known nickel electroplating process and its low internal stress. However, the hardness of nickel is relatively low compared with metals such as iron and stainless steel, resulting in a limited lifetime for nickel mold inserts. Other materials such as nickel-iron or tungsten-cobalt alloys with higher hardness are considered for longer-lived metal micromold inserts. Metal micromold inserts (or micromasters) can be

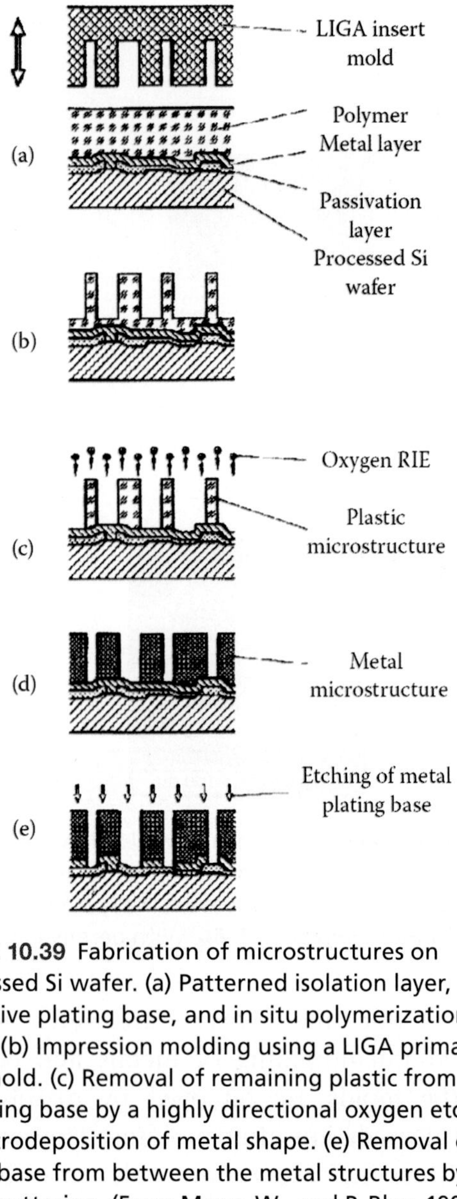


**FIGURE 10.39** Fabrication of microstructures on a processed Si wafer. (a) Patterned isolation layer, conductive plating base, and in situ polymerization of PMMA. (b) Impression molding using a LIGA primary insert mold. (c) Removal of remaining plastic from the plating base by a highly directional oxygen etch. (d) Electrodeposition of metal shape. (e) Removal of plating base from between the metal structures by argon sputtering. (From Menz, W., and P. Bley. 1993. *Mikrosystemtechnik fur Ingenieure*. Weinheim, Germany: VCH Publishers.[67])

fabricated by a variety of alternate techniques, such as CNC machining, e-beam or ion-beam milling, precision EDM or excimer laser ablation of a piece of metal, thick deep UV resist photolithography, resist DRIE, silicon wet bulk micromachining, polymer excimer laser ablation, and polymer e-beam or ion-beam writing, all followed by electroplating (see Figure 10.1).

Both silicon wet and dry etching can be used for silicon mold insert fabrication. Silicon mold inserts have been applied to, for example, the hot embossing of polymer structures and for casting of ceramics. In Table 10.15, LIGA metal molds are compared with metal masters fabricated by other means. For example, comparing metal mold inserts made by spark erosive cutting and x-ray lithography, the latter proves far superior.[83] LIGA PMMA features as small as 0.1 μm are replicated in the metal shape with almost no defects. The electroformed structures have a superior surface quality with a surface roughness, $R_{max}$, of less than 0.02 μm.[28]

DRIE and thick deep UV-sensitive resists (DUV) such as polyimides, AZ-4000, and SU-8 are recent contenders for micromaster mold fabrication. With respect to dry etching, higher and higher aspect-ratio features are being achieved; especially when using highly anisotropic etching conditions as in cryogenic DRIE and in the Bosch process, remarkable results are obtained. Wall roughness, causing form locking, remains a problem with DRIE; the dry etching process was optimized for speed, not for demolding. For small-quantity production, where the lifetime of mold inserts is not crucial, a silicon

**Table 10.15** Comparison of Micromolds

| Parameters | LIGA | DUV | DRIE | LASER | CNC | EDM |
|---|---|---|---|---|---|---|
| Aspect ratio | 100 | 22 | 10–25 | <10 | 14 (hole drilling) | Up to 100 |
| Wall roughness | <20 nm | ~1 μm | ~2 μm | 1 μm–100 nm | Several micrometers | 0.3–1 μm |
| Accuracy | <1 μm | 2–3 μm | <1 μm | A few micrometers | See Figure 6.3 | Some micrometers |
| Mask needed? | Yes | Yes | Yes | No | No | No |
| Maximum height | A few 100 μm up to 1 cm | A few 100 μm | A few 100 μm | A few 100 μm | Unlimited | Micrometers to millimeters |

*Source:* Weber, L., W. Ehrfeld, H. Freimuth, M. Lacher, H. Lehr, and B. Pech. *SPIE: micromachining and microfabrication process technology II*. Austin, TX: SPIE. With permission.[87]

wafer etched by DRIE can be used directly as a mold insert for anywhere from 5–30 molding cycles.[88,89] Figure 10.40a shows such a Si mold for building a two-point calibration fluidic device on a compact disc (CD). Wet etching of Si leads to much smoother surfaces than DRIE (see, for example, the wet anisotropically etched Si nozzle in Figure 4.45b and compare that with the sidewall scalloping caused by Si DRIE in Figure 3.38) and is therefore the preferred method for making master molds out of Si. For much longer-lasting molds, metallizing the Si structure and using the metal as mold is preferred. Photoresist structures on a silicon substrate have also been tested as mold insert in plastic molding because of the simplicity and low cost of the process. Figure 10.40b shows an SU-8 photoresist mold for the same lab CD platform shown in Figure 10.40a. In low-pressure molding processes, such mold inserts work for a limited number of runs (applying a thin metal layer over the top of the resist may further extend the lifetime of the mold), but their applicability in high-pressure processes needs to be further verified. A better approach is to use DUV-photosensitive resists for electroplating to yield a metal tool, usually nickel or nickel-cobalt. Figure 10.40c shows a nickel mold insert made using SU-8 in our lab for the two-point calibration microfluidic platform. Both DUV and DRIE are more accessible than LIGA and will continue to improve, taking more opportunities away from LIGA. Like LIGA, both alternative techniques

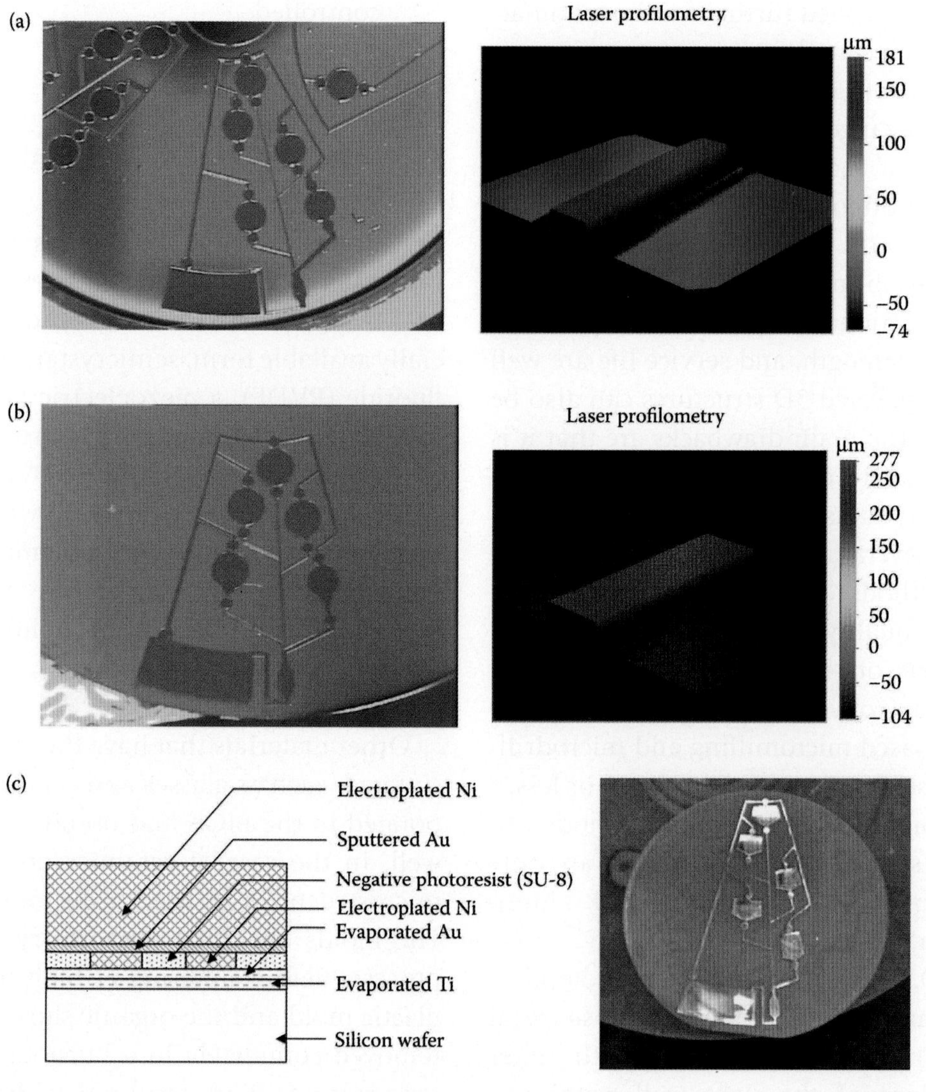


**FIGURE 10.40** (a) Silicon mold insert made by UV photolithography and DRIE at Burstein Technologies/UCLA/OSU. (b) Photoresist (SU-8) mold insert made by UV lithography at Ohio State University by the author. (c) Nickel mold insert made by UV photolithography and electroplating (OSU). The depth of the etched channel is measured by a laser profilometer.

can be coupled with plating, but neither technique can yet achieve the extremely low surface roughness and vertical walls of LIGA.

Other competing technologies for making metal masters are laser ablation methods and ultraprecision CNC machining. The latter two methods are serial processes and rather slow, but because we are considering the production of a master only these technologies are competitive for certain applications.

Laser microablation produces minimum features of about 10 μm width and aspect ratios of 1/10. Challenges include taper and surface finish control. Recast layers around the laser-drilled features cause form locking and infidelity in the replication. Femtosecond pulse lasers promise thinner or even the absence of recast layers and excellent resolution and should be investigated further for the manufacture of micromold inserts.[90]

For large features (>50 μm) with tolerances and repeatability in the range of about 10 μm, traditional CNC machining of materials like tool steel and stainless steel are often accurate enough for making metal mold inserts. The advantage of this technique is that the tool materials used are the same as those in conventional polymer molding; therefore, their design, strength, and service life are well established. Complicated 3D structures can also be machined easily. The main drawbacks are that it is difficult to make sharp corners or right angles, and the surface quality is usually poor (surface roughness is around several micrometers).[8] In contrast, lithographic methods can produce molds with excellent surface quality (surface quality <0.1 μm) and sharp corners or right angles. However, they cannot be used on conventional tool materials like steel. Diamond-based micromilling and microdrilling[91] reduce the surface roughness to 1 μm or less.[92] Although diamond-based machining methods can achieve features smaller than 10 μm, they are only applicable to "soft" metals such as nickel, aluminum, and copper.

From Table 10.15, a significant potential application of LIGA remains the fabrication of those metal molds that cannot be accomplished with other techniques because of the tight wall roughness tolerances,s small size, and high aspect ratios. From the table it is obvious that LIGA micromolds excel both in very low surface roughness and excellent accuracy.

Summarizing, the requirements for an optimal micromold insert fabrication technique are[93]:

- The master has to be removed from the molded structure, so the ease of release through wall inclination control is crucial (undercuts, for example, cannot be tolerated as they cause form locking)
- The most important parameter, including master lifetime and achievable aspect ratios, depends very strongly on the surface quality of the master
- The interface chemistry between master and polymer is a critical factor and must be controlled

## Alternative Molding Materials in LIGA and Pseudo-LIGA

In this last section we discuss alternative molding materials for LIGA and pseudo-LIGA. Besides PMMA and POM (see above), used in a commercially available form, semicrystalline polyvinylidene fluoride (PVDF), a piezoelectric material, has been used to make polymeric microstructures.[69] The optimum molding temperature of PVDF is 180°C, and PVDF structures can be molded without using mold release agents. Fluorinated polymers such as PVDF will enable higher temperature applications than PMMA. PC, well known for molding CDs, is under development for LIGA and LIGA-like processes and should perform well.[51]

Other materials that have the ability to flow or be sintered, such as glasses and ceramics, can be incorporated in the LIGA and pseudo-LIGA processes as well. In the case of ceramics, for example, plastic microstructures are used as disposable or lost molds. The molds are filled with slurry in a slip-casting process. Before sintering at high temperatures, the plastic mold and the organic slurry components are removed completely in a burnout process at lower temperatures. First results have been obtained for zirconium oxide and aluminum oxide. An important application for ceramic microstructures is the fabrication of arrays of piezoceramic columns embedded in a plastic matrix.

An alternative method to make LIGA and pseudo-LIGA ceramic or glass structures is based on sol-gel technology (see Chapter 8). This technique involves relatively low temperatures and may enable LIGA and pseudo-LIGA products such as glass capillaries for gas chromatography, piezoelectric arrays, and even high-$T_C$ superconductor actuators. Sol-gel techniques should work well with LIGA and pseudo-LIGA-type molds if they are filled under a vacuum to eliminate trapped gas bubbles. In the sol-gel technique, a solution is spun on a substrate that is then given an initial firing at around 200°C, driving off the solvent in the film. Subsequently, the substrate is given a high-temperature firing at 800–900°C to drive out the remaining solvents and crystallize the film (see Figure 8.2). A major issue is the large shrinkage of sol-gel films during the initial firing. For example, the maximum thickness of high-$T_C$ superconductor films currently achievable measures approximately 2 μm as a result of the high stress in the film caused during shrinkage. The latter is related to the ceramic yield or the amount of ceramic in the sol-gel compared with the amount of solvent. The sol-gel technique, in general, suits the production of thicker films well, as long as high-ceramic-yield sol-gels are used. Because LIGA-style films tend to be thick, a reduction in the amount of solvents to allow a high ceramic yield after firing would be to make sol-gel compatible with LIGA. This would require some development. Combining sol-gel technology with LIGA and pseudo-LIGA, lead-zirconium-titanate (PZT) piezoelectric devices can be made. In this case it is better to use the metal-plated structure rather than the direct PMMA template for the creation of the device because PZT must be processed in several elevated temperature steps. PZT sol-gel contains no particles that would prevent flow into small channels in a plated mold. The sol-gel contains high-molecular-weight polymer chains that hold the constituent metal salts, which are further processed into the final ceramic film. Thus, the sol-gel could be formed into the mold in a process much like reaction injection molding (RIM), with the mold filled with the chemistry under vacuum. Because the performance of piezoelectric actuator arrays is linked to the height and width ratio of the individual ceramic columns, as well as the distance between the columns in the array, LIGA, with its tremendous capacity for tall, dense, and high-aspect-ratio features, is an ideal fabrication tool.[94,95]

## Examples

### Example 10.1: LIGA Spinneret Nozzles

Profiled capillaries (nozzles) in a spinneret plate for spinning synthetic fibers from a molten or dissolved polymer, as shown in Figure 10.41, normally are produced by micro-EDM (see Chapter 5). This process establishes a practical lower limit of 20–50 μm for the minimal characteristic dimension, and the method cannot satisfy the requirements to produce complexly shaped, high-aspect-ratio spinnerets at a low cost. Smaller, more precise, and high-aspect-ratio nozzles can be produced by LIGA.[96]

Spinneret nozzles lend themselves to LIGA from the technical point of view. Compared with fabrication by micro-EDM, the minimum characteristic dimensions with LIGA can be reduced by an order of magnitude, and a high capillary length with excellent surface finish can easily be obtained. Moreover, the LIGA process makes all the nozzles in parallel, whereas micro-EDM operates as a serial technique. The market might focus on niche applications such as specialty, multilumen catheters. Also, medical use

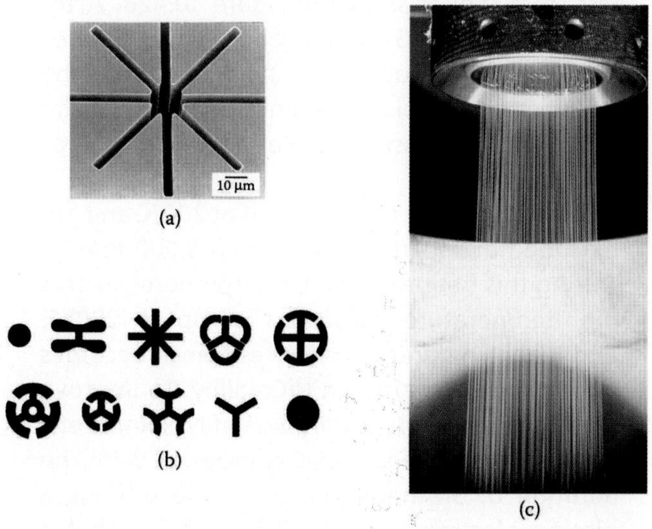


**FIGURE 10.41** LIGA spinneret nozzles. (a) Spinneret plate. (b) Profiled spinneret nozzles. (c) Spinning synthetic fiber. (Courtesy of IMM, Germany.)

of atomizers for dispensing drugs is projected to increase dramatically in the coming years; this might cause a run for a wide variety of precise, inexpensive micronozzles.

Shew et al.[97] used a conformal mask to fabricate a LIGA die with capillaries 2 mm deep and 70 μm wide for the mass production of electroplated spinnerets for the spinning of polyester fibers. The conformal mask eliminates the need of alignment steps accompanying the multiple exposure processes required in the deep x-ray lithography used by these authors. As illustrated in Figure 10.42A, a thin sputtered copper layer covers a thick PMMA layer on an Al substrate (a). A resist layer (JSR 137N, a negative photoresist from Japan Synthesis Rubber Co.) is then patterned on the copper layer by UV lithography, followed by plating gold absorbers (b). After stripping the resist and the Cu layer (c), the conformal mask is ready for multiple x-ray exposure and development steps (d) and die electroforming (e). The die (f) is made by electroforming of a NiCo alloy on the Al substrate from a Ni/Co sulfamate bath at current densities between 1 and 5 A/dm$^2$. The addition of increasing amounts of cobalt sulfamate increases the hardness of the deposit to about 420 Vickers. At a concentration higher than 20 g/L cobalt sulfamate, the hardness saturates. With increasing hardness also comes increased residual stress, which needs to be carefully controlled. Polypropylene (PP) was used to duplicate the NiCo microstructures by injection molding (IM) through a stainless steel tool as shown in Figure 10.42B. After removing sthe LIGA die, the result is a field of polymer microstructures standing up from the stainless steel tool. The stainless steel substrate is subsequently used for plating the spinneret structures.

Polyester fibers typically melt at 260°C and are extruded at speeds of more than 3,000 m/min. During this fiber extrusion, the spinneret suffers high-temperature excursions of up to 280°C, and high-temperature wear resistance becomes a problem when using a NiCo alloy. To improve the high-temperature hardness of Ni spinnerets, Shew et al.[97] plated a Ni/SiC composite. With the addition of SiC nanopowder to the sulfamate plating electrolyte, stable composite hardness of 500 Hv (Vickers hardness) at temperatures as high as 500°C was demonstrated. The authors suggest that these new types of spinnerets will enable the production of a new generation of fibers with ultrafine sizes and new functionalities at a low cost.

**Example 10.2: Inclined LIGA Walls**

Using LIGA, microstructures as high as hundreds or even thousands of micrometers can be achieved with an aspect ratio of 200, lateral

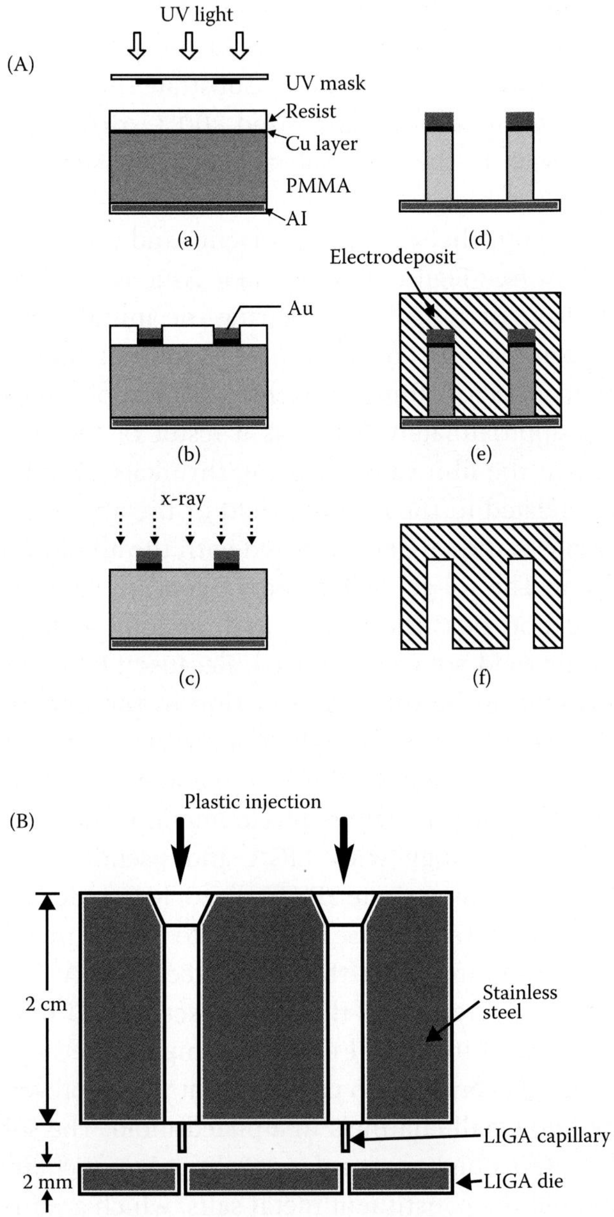


**FIGURE 10.42** (A) Schematic diagram of the modified LIGA process (dimensions in this figure not to scale). (B) Replication process of the LIGA spinneret. (From Cheng, Y., B.-Y. Shew, C.-Y. Lin, D.-H. Wei, and M. K. Chyu. 1999. Ultra-deep LIGA process. *J Micromech Microeng* 9:58–63.[49])

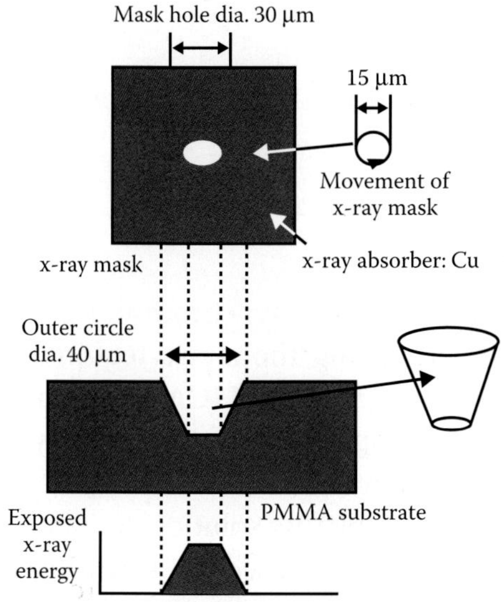


**FIGURE 10.43** Principle of M²DXL. (Based on Tabata, O., H. You, H. Shiraishi, H. Nakanishi, T. Nishimoto, K. Yamamoto, and Y. Baba. 2000. *Micro total analysis systems 2000: proceedings of the μTAS 2000 symposium.* Enschede, the Netherlands: Kluwer Academic.[71])

dimensions smaller than 0.5 μm, accuracy of 0.1 μm, and wall perpendicularity of 90°. For many applications, inclined sidewalls would be a very useful addition to the LIGA tool box; for example, for making lenses or for easier release of molded parts. This second LIGA example concerns both these important applications. When using LIGA for plastic molding, such as through hot embossing, it is important to be able to control the resist wall inclination to facilitate the release of molded devices. To achieve this, one could make an inclined absorber or, following Tabata et al.,[71] one can also move the mask during exposure in M²DXL. The movement of the mask is realized by a precision *x-y* stage with a 10-nm step resolution. The principle of the technique is illustrated in Figure 10.43. Here, the x-ray mask with a circular hole of 30 μm is moved in a circle with a 15-μm diameter. The x-ray dose distributes according to the movement of the mask during exposure. As a result, the energy delivered to the PMMA photoresist shows a high constant value within the inner circle of the trajectory of the hole's movement and decreases with increasing trajectory diameter, becoming zero at the outside region. A truncated conical PMMA structure results with an inclination of the wall controlled by the diameters of the mask hole and its circular movement and the total x-ray dose.

For making inclined microfluidic channels whose depth is 40 μm, a mask with a channel width of 50 μm and mask movement of 10 μm was used. A good inclination for the sidewalls of microfluidic devices for easy release of molded parts is 80°. Tabata et al.[71] demonstrated that their technique may also be used to make microlens arrays. In the latter case, the mask hole diameter was 210 μm, and the mask movement 100 μm.

**Example 10.3: LIGA Fiber-Chip Coupling**

LIGA elements for coupling monomode fibers with integrated optical chips have been developed. These prealignment arrays may use fixed nickel guiding structures in combination with

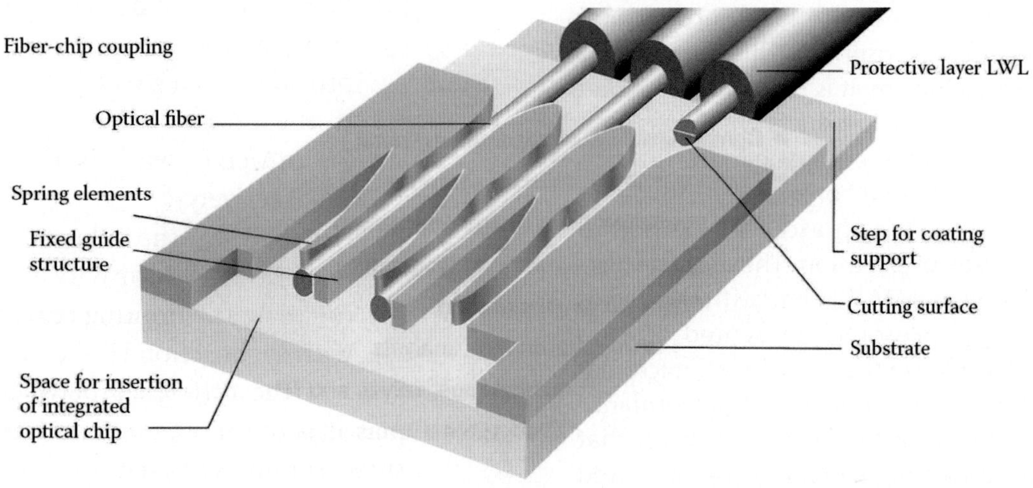


**FIGURE 10.44** Coupling for monomode fibers with nickel spring elements. (Courtesy of IMM, Germany, http://www.imm-mainz.de/.)

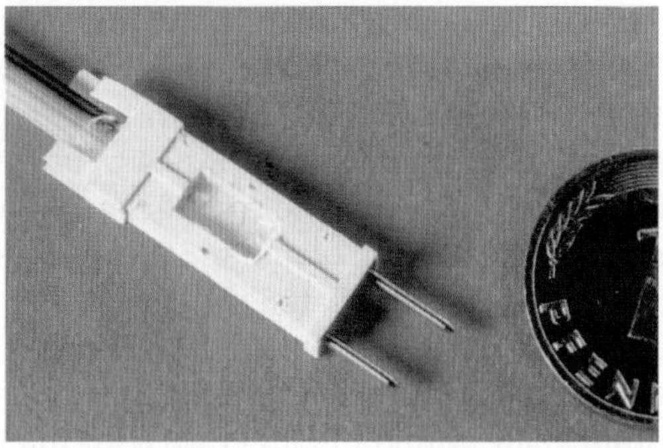

**FIGURE 10.45** Pull-push connector for multimode fiber. (Courtesy of IMM, Germany, http://www.imm-mainz.de/.)

leaf springs to ensure a precise alignment of the optical fibers relative to the optical chip with accuracy in the submicrometer range, as shown in Figure 10.44. The thermal expansion coefficient of the substrate material is matched to the optical chip material—for example, a glass substrate is used for coupling fibers to glass chips—and the use of spring elements simplifies the handling of the fibers.[78] Alternatives involve the use of optical adhesives and silicon V-groove arrays (see Chapter 4). Both alternative technologies are more labor intensive and suffer from thermal expansion problems.

The use of microoptical LIGA components is especially attractive for the coupling of multimode fibers; here, the capability of exact lithographic positioning of mechanical mounting supports is used advantageously to position the multimode fibers very precisely with respect to the microoptical components without need for any additional adjusting operations.[73,98]

A pull-push LIGA connector for single-mode fiber ribbons, shown in Figure 10.45, incorporates a set of precision microsprings coupling up to 12 fibers spaced at 250 μm. It is now on the market and is one of the LIGA products with mass market appeal. To obtain good coupling efficiency, the fibers are positioned horizontally with a precision of 1 μm. LIGA is an excellent method to provide that type of precision. These connectors were developed at IMM.[99] In line with the micromanufacturing philosophy presented in this book (i.e., to optimize the use of each micromachining technique for optimal cost/performance application), those parts of the mold insert that do not require such high accuracy are fabricated using other methods of precision machining such as EDM (see Chapter 5).

## Acknowledgments

Special thanks to Dr. Nahui Kim, formerly at UCI, now at Samsung.

## Questions

10.1: Describe why a synchrotron is the best tool for producing the x-rays required to make the most accurate LIGA parts and make a table comparing x-ray lithography with UV lithography.

10.2: An electroplating solution containing nickel sulphamate is used to electroplate nickel LIGA structures. The formula incorporates 450 g of nickel sulphamate [$Ni(NH_2SO_3)2$] per liter of water (plus some additives). The solution is heated to 50°C. Assuming the plating is occurring on a nickel cathode.

- (a) Calculate the concentration of the $Ni^{++}$ ions in the solution.
- (b) Calculate the equilibrium voltage across the nickel-solution interface at the cathode prior to electroplating.
- (c) If the plating is occurring inside high-aspect-ratio PMMA cavities with a depth of 500 micrometers, and the diffusivity of $Ni^{++}$ ions in water is $5 \times 10^{-6}$ cm$^2$/s, calculate the limiting current density (mA/cm$^2$) assuming the boundary layer thickness is governed by the height of the PMMA.
- (d) Plot the current density versus potential curve at the cathode using the values provided from part (c).
- (e) If the current density is specified at 75 mA/cm$^2$, what is the likely result (qualitatively)?
- (f) If the pH of the solution is 4 and the hydrogen ion diffusivity is $3 \times 10^{-5}$ cm$^2$/s, calculate the limiting current associated with the reaction $H^+ + e \rightarrow 0.5\ H_2$.

10.3: What is the difference between a thermoplastic, a thermoset, and an elastomer? Give three examples of each.

10.4: Assume a polymer is under a tensile load at a variety of temperatures from −20°C to its breakdown temperature at 150°C. Outline the potential stress-strain diagrams that may be found for a thermoplastic polymer under very short loading times.

10.5: What is viscoelastic polymer? Plot the stress-strain curve for a viscoelastic polymer.

10.6: Consider a laboratory setup (Galvanic cell) with a bar of copper (Cu) and a bar of silver (Ag) immersed in an electrolyte solution. The following are the redox half reactions and electrochemical potentials. The Gibbs free energy = (–) number of electron exchanged × electrochemical potential × Faraday constant. For a reaction to spontaneously occur, the Gibbs free energy must be negative.

$$Ag^+ + e^- \rightarrow Ag \qquad E^\circ = 0.80 \text{ V}$$

$$Cu^{++} + 2e^- \rightarrow Cu \qquad E^\circ = 0.34 \text{ V}$$

(a) Will Ag react with Cu++ (i.e., will we be able to plate copper on the silver bar)?

(b) Will Cu react with Ag+ (i.e., will we be able to plate silver on the copper bar)?

(c) If you place a sheet of copper into an $AgNO_3$ solution, what would happen? What is the reaction?

10.7: Electrodeposition:

(a) In an electrochemical cell, we deposit Zn from a $Zn^{2+}$ solution. After the deposition, we monitor the voltage of the Zn deposit with respect to a reference electrode (in this case we are using a standard hydrogen electrode, i.e., SHE). First we measure the potential in a $ZnSO_4$ solution at $10^{-2}$M and then in a ZnSO4 solution of $10^{-3}$M. What will be the change in potential of the Zn electrode between those two solutions? What is the potentsial in the $10^{-2}$M solution? ($E_o$ vs. SHE for Zn is –0.763V.)

$$Zn^{2+} + 2e^- \rightarrow Zn$$

(b) Explain the difference between electroless deposition and electro-deposition.

10.8: What issues do traditional plating and polymer molding companies face when trying to plate and mold LIGA-type structures?

10.9: Compare the pros and cons for injection molding, reaction injection molding, and hot embossing in LIGA. What are the differences in a micro part made with LIGA vs. pseudo-LIGA? How would you find out which technology was used, LIGA or pseudo-LIGA?

10.10: Describe two MEMS applications in which the use of LIGA may be justified.

10.11: Fabricate an x-ray mask detailing all materials and processes required. The mask should be adequate for the fabrication of Au parts with an aspect ratio of 50 (height over width).

10.12: For the following designs, decide which standard MEMS process (CMOS, LIGA, MUMPs) you would use and give a brief justification.

- You want to build a very cheap impact sensor.
- You want to build a rotating disk for a microoptical stroboscope.
- You want to build a strong microgear train.
- You want to build an array of structures with different resonance frequencies to test new theories on viscous damping.

10.13: Describe some techniques that may be employed for the fabrication of MEMS structures with slanted walls.

## References


1. Greener, J., and R. Wimberger-Friedl, eds. 2006. *Precision injection molding-process, materals, and applications*. Munich, Germany: Hanser.
2. IMM. 1995. Commercial brochure. IMM, http://www.imm-mainz.de/.
3. Lee, L. J., C.-H. Shih, Y.-J. Juang, J. Garcia, M. J. Madou, and K. W. Koelling. 1999. *Novel microfabrication options for BioMEMS conference*. San Francisco: The Knowledge Foundation.
4. Hanemann, T., R. Ruprecht, and J. H. Haubelt. 1998. Photomolding in microsystem technology. *Polym Preprints* 39:657–58.
5. Eicher, J., R. P. Peters, and A. Rogner. 1992. VDI-Verlag. Report no. *VDI-Bericht 960*, Dusseldorf, Germany.
6. Bifano, T. G., H. E. Fawcett, and P. A. Bierden. 1997. Precision manufacture of optical disc master stampers. *Precision Eng* 20:53–62.
7. Yu, L., Y.-J. Juang, K. W. Koelling, and L. J. Lee. 2000. *Proceedings: Society of Plastics Engineers ANTEC*. Orlando, FL.
8. Madou, M. J., Y. Lu, S. Lai, Y.-J. Juang, L. J. Lee, and S. Daunert. 2000. A novel design on a CD disc for two-point calibration measurement. *Proceedings of the IEEE: solid-state sensor and actuator workshop*. Hilton Head Island, SC.

9. Dunke, K., H.-D. Bauer, W. Ehrfeld, J. Hobfeld, L. Weber, G. Horcher, and G. Muller. 1998. Injection-molded fiber ribbon connectors for parallel optical links fabricated by the LIGA technique. *J Micromech Microeng* 8:301–6.
10. Fahrenberg, J., W. Bier, D. Mass, W. Menz, R. Ruprecht, and W. K. Schomburg. 1995. A microvalve system fabricated by thermoplasting molding. *J Micromech Microeng* 5:169–71.
11. Ramos, B. L., S. J. Choquette, and F. F. Nicholas. 1996. Embossable grating couplers for planar waveguide optical sensors. *Anal Chem* 68:1245–49.
12. Becker, H., and W. Dietz. 1998. Microfluidic device for μ-TAS applications for fabrication by polymer hot embossing. *Proc SPIE* 3515:177–82.
13. Kopp, M. U., H. J. Crabtree, and A. Manz. 1997. Developments in technology and applications in microsystems. *Curr Opin Chem Biol* 1:410–19.
14. Chou, S. Y., P. R. Krauss, and P. J. Renstrom. 1996. Imprint lithography with 25-nanometer resolution. *Science* 272:85–87.
15. Schift, H., R. W. Jaszewski, C. David, and J. Gobrecht. 1999. Nanostructuring of polymers and fabrication of integrated electrodes by hot embossing lithography. *Microelectron Eng* 46:121–24.
16. Jaszewski, R. W., H. Schift, J. Gobrecht, and P. Smith. 1998. Hot embossing in polymers as a direct way to pattern resist. *Microelectron Eng* 41/42:575–78.
17. Casey, B. G., W. Monaghan, and C. D. W. Wilkinson. 1997. Embossing of nanoscale features and environment. *Microelectron Eng* 35:393–96.
18. Gottschalch, F., T. Hoffman, C. M. Sotomayor Torres, H. Schulz, and H.-C. Scheer. 1999. Polymer issues in nanoimprinting technique. *Solid State Electron* 43:1079–83.
19. Jaszewski, R. W., H. Schift, P. Groning, and G. Margaritondo. 1997. Properties of thin anti-adhesive films used for the replication of microstructures in polymers. *Microelectron Eng* 35:381–84.
20. Jaszewski, R. W., H. Schift, B. Schnyder, A. Schneuwly, and P. Groning. 1999. The deposition of anti-adhesive ultrathin Teflon-like films and their interaction with polymers during hot embossing. *Appl Surface Sci* 143:301–8.
21. Becker, H., and U. Heim. 1999. Silicon as tool material for polymer hot embossing. *12th IEEE International Conference on MEMS '99: Micro Electro Mechanical Systems.* Orlando, FL.
22. Lin, L.-W., C.-J. Chiu, W. Bacher, and M. Heckele. 1996. Microfabrication using silicon mold inserts and hot embossing. *Seventh international symposium on micro machine and human science.* Nagoya, Japan.
23. Casey, B. G., D. R. S. Cumming, I. I. Khandaker, A. G. S. Curtis, and C. D. W. Wilkinson. 1999. Nanoscale embossing of polymers using a thermoplastic die. *Microelectron Eng* 46:125–28.
24. Ruprecht, R., W. Bacher, J. Haubelt, and V. Piotter. 1995. Injection molding of LIGA and LIGA-similar microstructures using filled and unfilled thermoplastics. *Proc SPIE* 2639:146–57.
25. Spiller, E., R. Feder, J. Topalian, E. Castellani, L. Romankiw, and M. Heritage. 1976. X-ray lithography for bubble devices. *Solid State Technol* April:62–68.
26. Romankiw, L. T., I. M. Croll, and M. Hatzakis. 1970. Batch-fabricated thin-film magnetic recording heads. *IEEE Trans Magn* 6:597–601.
27. Becker, E. W., W. Ehrfeld, D. Munchmeyer, H. Betz, A. Heuberger, S. Pongratz, W. Glashauser, H. J. Michel, and V. R. Siemens. 1982. Production of separation nozzle systems for uranium enrichment by a combination of x-ray lithography and galvanoplastics. *Naturwissenschaften* 69:520–23.
28. Hagmann, P., W. Ehrfeld, and H. Vollmer. 1987. Fabrication of microstructures with extreme structural heights by reaction injection molding. *First meeting of the European Polymer Federation, European symposium on polymeric materials.* Lyon, France.
29. Waldo, W. G., and A. W. Yanof. 1991. 0.25 micron imaging by SOR x-ray lithography. *Solid State Technol* 34:29–31.
30. Clemens, J., and R. Hill, ed. 1991. *JTEC Panel Report on X-ray Lithography in Japan*. PB92-100205, Springfield, VA:NTIS.
31. Urisu, T., and H. Kyuragi. 1987. Synchrotron radiation-excited chemical-vapor deposition and etching. *J Vac Sci Technol* B5:1436–40.
32. Goedtkindt, P., J. M. Salome, X. Artru, P. Dhez, N. Maene, F. Poortmans, and L. Wartski. 1991. X-ray lithography with a transition radiation source. *Microelectron Eng* 13:327–30.
33. Muray, J. J., and I. Brodie. 1991. *Report no.* 2019. Menlo Park, CA: SRI International.
34. Bley, P., D. Einfeld, W. Menz, and H. Schweickert. 1992. A dedicated synchrotron light source for micromechanics. *Proceedings of the EPAC 92*. Berlin, Germany.
35. Ehrfeld, W., W. Glashauer, D. Munchmeyer, and W. Schelb. 1986. Mask making for synchrotron radiation lithography. *Microelectron Eng* 5:463–70.
36. Lawes, R. A. 1989. Sub-micron lithography techniques. *Appl Surf Sci* 36:485–99.
37. Schomburg, W. K., H. J. Baving, and P. Bley. 1991. Ti- and Be-x-ray masks with alignment windows for the LIGA process. *Microelectron Eng* 13:323–26.
38. Hein, H., P. Bley, J. Gottert, and U. Klein. 1992. Elektronenstrahllithographie zur Herstellung von Rontgen-masken fur das LIGA-Verfahren. *Feinwtech Messtech* 100:387–89.
39. Bley, P., W. Menz, W. Bacher, K. Feit, M. Harmening, H. Hein, J. Mohr, W. K. Schomburg, and W. Stark. 1991. Application of the LIGA process in fabrication of three-dimensional mechanical microstructures. *Microprocess 91, 1991 International Microprocess Conference.* Kanazawa, Japan.
40. Guckel, H., T. R. Christenson, T. Earles, J. Klein, J. D. Zook, T. Ohnstein, and M. Karnowski. 1994. Laterally driven electromagnetic actuators. *Technical digest: 1994 solid state sensor and actuator workshop*. Hilton Head Island, SC.
41. Siddons, D. P., and E. D. Johnson. 1994. Precision machining using hard x-rays. *Synchrotron Radiation News* 7:16–18.
42. Guckel, H. 1994. Notes from handouts. The Commercialization of Microsystems. September 11–16, Banff Conference Center, Banff, Canada.
43. Becker, E. W., W. Ehrfeld, P. Hagmann, A. Maner, and D. Munchmeyer. 1986. Fabrication of microstructures with high aspect ratios and great structural heights by synchrotron radiation lithography, galvanoforming, and plastic molding (LIGA process). *Microelectron Eng* 4:35–56.
44. Friedrich, C. 1994. Notes from handouts. The Commercialization of Microsystems. September 11–16, Banff Conference Center, Banff, Canada.
45. Henck, R. 1984. Detecteurs au silicium pour electrons et rayons X, principes de fonctionnement, fabrication et performance. *J Microsc Spectrosc Electron* 9:131–33.

46. Vladimirsky, Y., O. Vladimirsky, V. Saile, K. Morris, and J. M. Klopf. 1995. *Microlithography '95 (Proceedings of the SPIE)*. Santa Clara, CA: SPIE.
47. Khan Malek, C., Y. Vladimirsky, O. Vladimirsky, J. Scott, B. Craft, and V. Saile. 1996. X-ray microfabrication activities at the center for advanced microstructures and devices (CAMD). *Rev Sci Instrum* 67:1–6.
48. Shih, W.-P., Y. Cheng, C.-Y. Lin, and G.-J. Hwang. Low-cost x-ray conformal mask using x-ray dry FILM resist. *Microelectron Eng* 40:43–50.
49. Cheng, Y., B.-Y. Shew, C.-Y. Lin, D.-H. Wei, and M. K. Chyu. 1999. Ultra-deep LIGA process. *J Micromech Microeng* 9:58–63.
50. Michel, A., R. Ruprecht, and W. Bacher. 1993. *Abformung von. mikrostrukturen auf prozessierten Wafern. KfK report no. 5171*. Karlsruhe, Germany: Kernforschungszentrum Karlsruhe.
51. Rogner, A., J. Eichner, D. Munchmeyer, R.-P. Peters, and J. Mohr. 1992. The LIGA technique: what are the new opportunities? *J Micromech Microeng* 2:133–40.
52. Mohr, J., W. Ehrfeld, and D. Munchmeyer. 1988. *Analyse der defectursachen und der genauigkeit der structuru-bertragung bei der rontgentiefenlithographie mit synchrotronstrahlung. KfK report no. 4414*. Karlsruhe, Germany: Kernforschungszentrum Karlsruhe.
53. Moreau, W. M. 1988. *Semiconductor lithography*. New York: Plenum Press.
54. Lingnau, J., R. Dammel, and J. Theis. 1989. Recent trends in x-ray resists: part I. *Solid State Technol* 32:105–12.
55. Mohr, J., W. Ehrfeld, D. Munchmeyer, and A. Stutz. 1989. Resist technology for deep-etch synchrotron radiation lithography. *Makromol Chem Macromol Symp* 24:231–51.
56. Gottert, J., J. Mohr, and C. Muller. 1991. Mikrooptische Komponenten aus PMMA, Hergestellt Durch Roentgentiefenlithographie Werkstoffe der Mikrotechnik-Bais fur neue Producte. *VDI Berichte* Düsseldorf, Germany: Deutscher Ingenieur-Verlag, pp. 249–63.
57. Wollersheim, O., H. Zumaque, J. Hormes, J. Langen, P. Hoessel, L. Haussling, and G. Hoffman. 1994. Radiation chemistry of poly(lactides) as new polymer resists for the LIGA process. *J Micromech Microeng* 4:84–93.
58. Madou, M. J., and M. Murphy. 1995. Unpublished results.
59. Mohr, J., W. Ehrfeld, and D. Munchmeyer. 1988. Requirements on resist layers in deep-etch synchrotron radiation lithography. *J Vac Sci Technol* B6:2264–67.
60. Ehrfeld, W. 1994. Notes from handouts. The Commercialization of Microsystems. September 11–16, Banff Conference Center, Banff, Canada.
61. Campbell, S. A. 1996. *The science and engineering of microelectronic fabrication*. New York: Oxford University Press.
62. Vladimirsky, O. 1995. Private communication with the author, LSU.
63. Galhotra, V., C. Marques, Y. Desta, K. Kelly, M. Despa, A. Pendse, and J. Collier. 1996. Fabrication of mold inserts using a modified LIGA procedure. *SPIE: micromachining and microfabrication process technology II*. Austin, TX.
64. Rogers, J., C. Marques, and K. Kelly. 1996. Cyanoacrylate bonding of thick resists for LIGA. *SPIE: microlithography and metrology in micromachining II*. Austin, TX.
65. Becker, E. W., W. Ehrfeld, and D. Munchmeyer. 1984. *Untersuchungen zur abbildungsgenauigkeit der rontgentiefenlithographie mit synchrotronstrahlung. KfK report no. 3732*. Karlsruhe, Germany: KfK. Kernforschungszentrum Karlsruhe.
66. Munchmeyer, D. 1984. PhD thesis. University of Karhsruhe, Germany.
67. Menz, W., and P. Bley. 1993. *Mikrosystemtechnik fur Ingenieure*. Weinheim, Germany: VCH Publishers.
68. Editorial. 1994. Commercial Brochure.
69. Harmening, M., W. Bacher, P. Bley, A. El-Kholi, H. Kalb, B. Kowanz, W. Menz, A. Michel, and J. Mohr. 1992. *Proceedings: IEEE micro electro mechanical systems (MEMS '92)*. Travemunde, Germany: IEEE.
70. Mohr, J., C. Burbaum, P. Bley, W. Menz, and U. Wallrabe. 1990. *Microsystem technologies 90*. Ed. H. Reichl. Berlin: Springer.
71. Tabata, O., H. You, H. Shiraishi, H. Nakanishi, T. Nishimoto, K. Yamamoto, and Y. Baba. 2000. *Micro total analysis systems 2000: proceedings of the μTAS 2000 symposium*. Enschede, the Netherlands: Kluwer Academic.
72. Bley, P., J. Gottert, M. Harmening, M. Himmelhaus, W. Menz, J. Mohr, C. Muller, and U. Wallrabe. 1991. *Microsystem technologies '91*. Eds. R. Krahn and H. Reichl. Berlin: VDE-Verlag.
73. Gottert, J., J. Mohr, and C. Muller. 1992. *Integrierte Optik und Mikrooptik mit Polymeren*. Mainz, Germany: Max-Planck-Institut für Polymerforschung.
74. Ghica, V., and W. Glashauser. 1982. Verfahren fur die Spannungsrissfreie Entwicklung von Bestrahlthen Polymethylmethacrylate-Schichten, in Deutsche Offenlegungsschrift. Patent No. 3039110, Germany.
75. Maner, A., S. Harsch, and W. Ehrfeld. 1988. Mass production of microdevices with extreme aspect ratios by electroforming. *Plating Surface Finish* 75:60–65.
76. Harsch, S., W. Ehrfeld, and A. Maner. 1988. *Untersuchungen zur herstellung von mikrostrukturen grosser strukturhöhe durch galvanoformung im nickelsilfamatelektrolyten. KfK report no.* 4455. Karlsruhe, Germany: Kernforschungszentrum Karlsruhe.
77. Mohr, J., and M. Strohmann. 1992. Examination of long-term stability of metallic LIGA microstructures by electromagnetic activation. *J Micromech Microeng* 2:193–95.
78. Romankiw, L. T., and T. A. Palumbo. 1987. *Proceedings: symposium on electrodeposition technology, theory and practice*. San Diego.
79. Masuko, N., T. Osaka, and Y. Ito, eds. 1996. *Electrochemical technology: innovation and new developments*. Tokyo and Amsterdam: Kodansha and Gordon and Breach Science Publishers, copublishers.
80. Dukovic, J. O. 1993. Feature-scale simulation of resist patterned electrodeposition. *IBM J Res Dev* 37:125–40.
81. Leyendecker, K., W. Bacher, W. Stark, and A. Thommes. 1994. New microelectrodes for the investigation of the electroforming of LIGA microstructures. *Electrochim Acta* 39:1139–43.
82. Sales Brochure, 1994. RK Kissler, Speyer, Germany.
83. Hagmann, P., and W. Ehrfeld. 1988. Fabrication of microstructures of extreme structural heights by reaction injection molding. *J Polymer Process Soc* 4:188–95.
84. Burbaum, C., J. Mohr, P. Bley, and W. Ehrfeld. 1991. Fabrication of capacitive acceleration sensors by the LIGA technique. *Sensors Actuators A* A25:559–63.
85. Guckel, H., K. J. Skrobis, T. R. Christenson, J. Klein, S. Han, B. Choi, and E. G. Lovell. 1991. *Proceedings: IEEE micro electro mechanical systems (MEMS '91)*. Nara, Japan.
86. Harmening, M., W. Bacher, and W. Menz. 1991. Molding plateable micropatterns of electrically insulating and electrically conducting poly(methyl methacrylate)s by the LIGA technique. *Makromol Chem Macromol Symp* 50:277–84.

87. Weber, L., W. Ehrfeld, H. Freimuth, M. Lacher, H. Lehr, and B. Pech. SPIE: *micromachining and microfabrication process technology II*. Austin, TX.
88. Madou, M. 1997. *Fundamentals of microfabrication*. Boca Raton, FL: CRC Press.
89. Wimberger-Friedl, R. 1999. Injection molding of sub-µm grating optical elements. *Proc SPE ANTEC* 57:476–80.
90. Momma, C., S. Nolte, N. Chichkov, B. V. Alvensleben, and F. A. Tunermann. 1997. Precise laser ablation with ultra-short pulses. *Appl Surf Sci* 109/110:15–19.
91. Warrington, R. O. 1999. *Novel microfabrication options for BioMEMS conference (proceedings)*. San Francisco: The Knowledge Foundation.
92. Roberts, M. A., J. S. Rossier, P. Bercier, and H. Girault. 1997. UV laser machined polymer substrates for the development of microdiagnostic systems. *Anal Chem* 69:2035–42.
93. Becker, H., and C. Gärtner. 2000. Polymer microfabrication methods for microfluidic analytical applications. *Electrophoresis* 21:12–26.
94. Preu, G., A. Wolff, D. Cramer, and U. Bast. 1991. *Proceedings: second European ceramic society conference (2nd ECerS '91)*. Augsburg, Germany: European Ceramic Society.
95. Lubitz, K. 1989. *Mikrostrukturierung von Piezokeramik*. Düsseldorf, Germany: VDI-Tagungsbericht.
96. Maner, A., S. Harsch, and W. Ehrfeld. 1987. *Proceedings: 74th AESF annual technical conference (SUR/FIN '87)*. Chicago: AESF.
97. Shew, B.-Y., Y. Cheng, C.-H. Lin, W.-P. Ma, G.-J. Huang, C.-L. Kuo, S.-C. Tseng, D.-S. Lee, and G.-L. Chang. 1999. Manufacturing process for LIGA spinnerets. *Sensors Materials* 11:329–37.
98. Gottert, J., J. Mohr, and C. Muller. 1992. *Proceedings: micro system technologies '92*. Berlin: VDE-Verlag.
99. Editorial. 1994. Commercial brochure. http://www.imm-mainz.de/.

# Index

## D

## E

## K



## L

## N



## O



## P

## Q



## R



## S

## Z

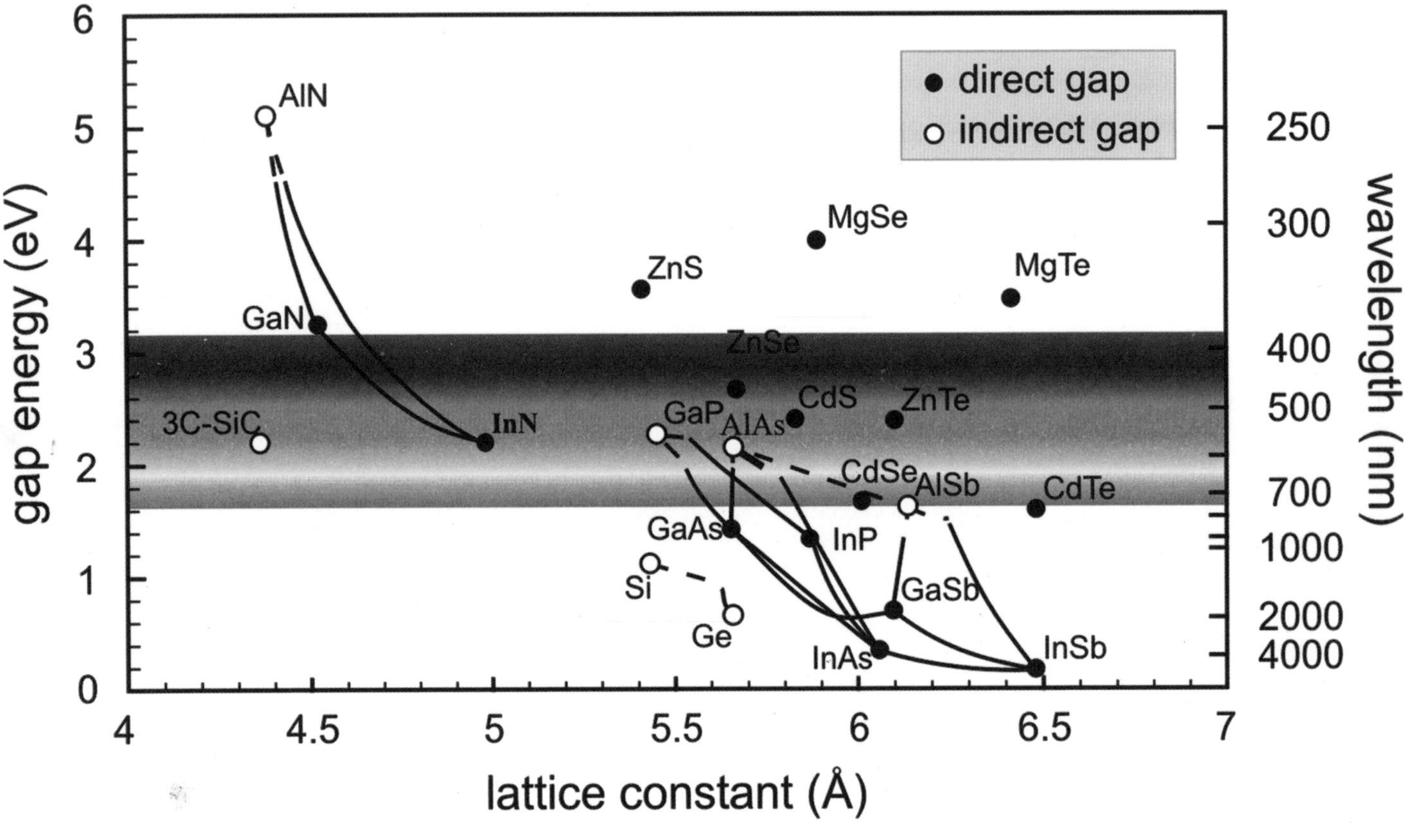
direct gap
indirect gap
gap energy (eV)
wavelength (nm)
lattice constant (Å)
AlN
GaN
InN
3C-SiC
ZnS
MgSe
MgTe
ZnSe
CdS
ZnTe
GaP
AlAs
CdSe
AlSb
CdTe
GaAs
InP
Si
Ge
GaSb
InAs
InSb
0
1
2
3
4
5
6
4
4.5
5
5.5
6
6.5
7
250
300
400
500
700
1000
2000
4000